Some Physical Constants

Quantity	Symbol	Value[a]
Atomic mass unit	u	$1.660\ 538\ 86\ (28) \times 10^{-27}$ kg $931.494\ 043\ (80)$ MeV/c^2
Avogadro's number	N_A	$6.022\ 141\ 5\ (10) \times 10^{23}$ particles/mol
Bohr magneton	$\mu_B = \dfrac{e\hbar}{2m_e}$	$9.274\ 009\ 49\ (80) \times 10^{-24}$ J/T
Bohr radius	$a_0 = \dfrac{\hbar^2}{m_e e^2 k_e}$	$5.291\ 772\ 108\ (18) \times 10^{-11}$ m
Boltzmann's constant	$k_B = \dfrac{R}{N_A}$	$1.380\ 650\ 5\ (24) \times 10^{-23}$ J/K
Compton wavelength	$\lambda_C = \dfrac{h}{m_e c}$	$2.426\ 310\ 238\ (16) \times 10^{-12}$ m
Coulomb constant	$k_e = \dfrac{1}{4\pi\epsilon_0}$	$8.987\ 551\ 788\ldots \times 10^9$ N·m^2/C^2 (exact)
Deuteron mass	m_d	$3.343\ 583\ 35\ (57) \times 10^{-27}$ kg $2.013\ 553\ 212\ 70\ (35)$ u
Electron mass	m_e	$9.109\ 382\ 6\ (16) \times 10^{-31}$ kg $5.485\ 799\ 094\ 5\ (24) \times 10^{-4}$ u $0.510\ 998\ 918\ (44)$ MeV/c^2
Electron volt	eV	$1.602\ 176\ 53\ (14) \times 10^{-19}$ J
Elementary charge	e	$1.602\ 176\ 53\ (14) \times 10^{-19}$ C
Gas constant	R	$8.314\ 472\ (15)$ J/mol·K
Gravitational constant	G	$6.674\ 2\ (10) \times 10^{-11}$ N·m^2/kg^2
Josephson frequency–voltage ratio	$\dfrac{2e}{h}$	$4.835\ 978\ 79\ (41) \times 10^{14}$ Hz/V
Magnetic flux quantum	$\Phi_0 = \dfrac{h}{2e}$	$2.067\ 833\ 72\ (18) \times 10^{-15}$ T·m^2
Neutron mass	m_n	$1.674\ 927\ 28\ (29) \times 10^{-27}$ kg $1.008\ 664\ 915\ 60\ (55)$ u $939.565\ 360\ (81)$ MeV/c^2
Nuclear magneton	$\mu_n = \dfrac{e\hbar}{2m_p}$	$5.050\ 783\ 43\ (43) \times 10^{-27}$ J/T
Permeability of free space	μ_0	$4\pi \times 10^{-7}$ T·m/A (exact)
Permittivity of free space	$\epsilon_0 = \dfrac{1}{\mu_0 c^2}$	$8.854\ 187\ 817\ldots \times 10^{-12}$ C^2/N·m^2 (exact)
Planck's constant	h	$6.626\ 069\ 3\ (11) \times 10^{-34}$ J·s
	$\hbar = \dfrac{h}{2\pi}$	$1.054\ 571\ 68\ (18) \times 10^{-34}$ J·s
Proton mass	m_p	$1.672\ 621\ 71\ (29) \times 10^{-27}$ kg $1.007\ 276\ 466\ 88\ (13)$ u $938.272\ 029\ (80)$ MeV/c^2
Rydberg constant	R_H	$1.097\ 373\ 156\ 852\ 5\ (73) \times 10^7$ m^{-1}
Speed of light in vacuum	c	$2.997\ 924\ 58 \times 10^8$ m/s (exact)

Note: These constants are the values recommended in 2002 by CODATA, based on a least-squares adjustment of data from different measurements. For a more complete list, see P. J. Mohr and B. N. Taylor, "CODATA Recommended Values of the Fundamental Physical Constants: 2002." *Rev. Mod. Phys.* **77**:1, 2005.

[a] The numbers in parentheses for the values represent the uncertainties of the last two digits.

Solar System Data

Body	Mass (kg)	Mean Radius (m)	Period (s)	Distance from the Sun (m)
Mercury	3.18×10^{23}	2.43×10^6	7.60×10^6	5.79×10^{10}
Venus	4.88×10^{24}	6.06×10^6	1.94×10^7	1.08×10^{11}
Earth	5.98×10^{24}	6.37×10^6	3.156×10^7	1.496×10^{11}
Mars	6.42×10^{23}	3.37×10^6	5.94×10^7	2.28×10^{11}
Jupiter	1.90×10^{27}	6.99×10^7	3.74×10^8	7.78×10^{11}
Saturn	5.68×10^{26}	5.85×10^7	9.35×10^8	1.43×10^{12}
Uranus	8.68×10^{25}	2.33×10^7	2.64×10^9	2.87×10^{12}
Neptune	1.03×10^{26}	2.21×10^7	5.22×10^9	4.50×10^{12}
Pluto[a]	$\approx 1.4 \times 10^{22}$	$\approx 1.5 \times 10^6$	7.82×10^9	5.91×10^{12}
Moon	7.36×10^{22}	1.74×10^6	—	—
Sun	1.991×10^{30}	6.96×10^8	—	—

[a] In August 2006, the International Astronomical Union adopted a definition of a planet that separates Pluto from the other eight planets. Pluto is now defined as a "dwarf planet" (like the asteroid Ceres).

Physical Data Often Used

Average Earth–Moon distance	3.84×10^8 m
Average Earth–Sun distance	1.496×10^{11} m
Average radius of the Earth	6.37×10^6 m
Density of air (20°C and 1 atm)	1.20 kg/m^3
Density of water (20°C and 1 atm)	1.00×10^3 kg/m^3
Free-fall acceleration	9.80 m/s^2
Mass of the Earth	5.98×10^{24} kg
Mass of the Moon	7.36×10^{22} kg
Mass of the Sun	1.99×10^{30} kg
Standard atmospheric pressure	1.013×10^5 Pa

Note: These values are the ones used in the text.

Some Prefixes for Powers of Ten

Power	Prefix	Abbreviation	Power	Prefix	Abbreviation
10^{-24}	yocto	y	10^1	deka	da
10^{-21}	zepto	z	10^2	hecto	h
10^{-18}	atto	a	10^3	kilo	k
10^{-15}	femto	f	10^6	mega	M
10^{-12}	pico	p	10^9	giga	G
10^{-9}	nano	n	10^{12}	tera	T
10^{-6}	micro	μ	10^{15}	peta	P
10^{-3}	milli	m	10^{18}	exa	E
10^{-2}	centi	c	10^{21}	zetta	Z
10^{-1}	deci	d	10^{24}	yotta	Y

PHYSICS

for Scientists and Engineers

with Modern Physics

PHYSICS

for Scientists and Engineers
with Modern Physics

Volume 2

Seventh Edition

Raymond A. Serway
Emeritus, James Madison University

John W. Jewett, Jr.
California State Polytechnic University, Pomona

Australia • Brazil • Canada • Mexico • Singapore • Spain • United Kingdom • United States

THOMSON
BROOKS/COLE

Physics for Scientists and Engineers with Modern Physics, Volume 2, Seventh Edition
Raymond A. Serway and John W. Jewett, Jr.

Physics Acquisition Editor: Chris Hall
Publisher: David Harris
Vice President, Editor-in-Chief, Sciences: Michelle Julet
Development Editor: Ed Dodd
Assistant Editor: Brandi Kirksey
Editorial Assistant: Shawn Vasquez
Technology Project Manager: Sam Subity
Marketing Manager: Mark Santee
Marketing Assistant: Melissa Wong
Managing Marketing Communications Manager: Bryan Vann
Project Manager, Editorial Production: Teri Hyde
Creative Director: Rob Hugel
Art Director: Lee Friedman
Print Buyers: Barbara Britton, Karen Hunt

Permissions Editors: Joohee Lee, Bob Kauser
Production Service: Lachina Publishing Services
Text Designer: Patrick Devine Design
Photo Researcher: Jane Sanders Miller
Copy Editor: Kathleen Lafferty
Illustrator: Rolin Graphics, Progressive Information
  Technologies, Lachina Publishing Services
Cover Designer: Patrick Devine Design
Cover Image: Front: © 2005 Tony Dunn; Back: © 2005 Kurt
  Hoffmann, Abra Marketing
Cover Printer: R.R. Donnelley/Willard
Compositor: Lachina Publishing Services
Printer: R.R. Donnelley/Willard

Printed in the United States of America
1 2 3 4 5 6 7 11 10 09 08 07

Library of Congress Control Number: 2006936870

Student Edition:
ISBN-13: 978-0-495-11244-0
ISBN-10: 0-495-11244-5

Thomson Higher Education
10 Davis Drive
Belmont, CA 94002-3098
USA

For more information about our products, contact us at:

Thomson Learning Academic Resource Center
(+1) 1-800-423-0563

For permission to use material from this text or product,
submit a request online at
http://www.thomsonrights.com.

Any additional questions about permissions can be
submitted by e-mail to **thomsonrights@thomson.com.**

We dedicate this book to our wives
Elizabeth and Lisa and all our children
and grandchildren for their loving
understanding when we spent time on
writing instead of being with them.

Part 4 ELECTRICITY AND MAGNETISM 641

23 Electric Fields   642
24 Gauss's Law   673
25 Electric Potential   692
26 Capacitance and Dielectrics   722
27 Current and Resistance   752
28 Direct Current Circuits   775
29 Magnetic Fields   808
30 Sources of the Magnetic Field   837
31 Faraday's Law   867
32 Inductance   897
33 Alternating Current Circuits   923
34 Electromagnetic Waves   952

John W. Jewett, Jr.

© Thomson Learning/Charles D. Winters

Part 5 LIGHT AND OPTICS   977

35 The Nature of Light and the Laws of Geometric Optics   978
36 Image Formation   1008
37 Interference of Light Waves   1051
38 Diffraction Patterns and Polarization   1077

Part 6 MODERN PHYSICS   1111

39 Relativity   1112
40 Introduction to Quantum Physics   1153
41 Quantum Mechanics   1186
42 Atomic Physics   1215
43 Molecules and Solids   1257
44 Nuclear Structure   1293
45 Applications of Nuclear Physics   1329
46 Particle Physics and Cosmology   1357

Appendices   A-1
Answers to Odd-Numbered Problems   A-25
Index   I-1

Courtesy of Henry Leap and Jim Lehman

About the Authors xv

Preface xvii

To the Student xxix

PART 4 ELECTRICITY AND MAGNETISM 641

Chapter 23 Electric Fields 642
23.1 Properties of Electric Charges 642
23.2 Charging Objects by Induction 644
23.3 Coulomb's Law 645
23.4 The Electric Field 651
23.5 Electric Field of a Continuous Charge Distribution 654
23.6 Electric Field Lines 659
23.7 Motion of a Charged Particle in a Uniform Electric Field 661

Chapter 24 Gauss's Law 673
24.1 Electric Flux 673
24.2 Gauss's Law 676
24.3 Application of Gauss's Law to Various Charge Distributions 678
24.4 Conductors in Electrostatic Equilibrium 682

Chapter 25 Electric Potential 692
25.1 Electric Potential and Potential Difference 692
25.2 Potential Difference in a Uniform Electric Field 694
25.3 Electric Potential and Potential Energy Due to Point Charges 697
25.4 Obtaining the Value of the Electric Field from the Electric Potential 701
25.5 Electric Potential Due to Continuous Charge Distributions 703
25.6 Electric Potential Due to a Charged Conductor 707
25.7 The Millikan Oil-Drop Experiment 709
25.8 Applications of Electrostatics 710

Chapter 26 Capacitance and Dielectrics 722
26.1 Definition of Capacitance 722
26.2 Calculating Capacitance 724
26.3 Combinations of Capacitors 727
26.4 Energy Stored in a Charged Capacitor 731
26.5 Capacitors with Dielectrics 735
26.6 Electric Dipole in an Electric Field 738
26.7 An Atomic Description of Dielectrics 740

Chapter 27 Current and Resistance 752
27.1 Electric Current 752
27.2 Resistance 756
27.3 A Model for Electrical Conduction 760
27.4 Resistance and Temperature 762
27.5 Superconductors 762
27.6 Electrical Power 763

Chapter 28 Direct Current Circuits 775
28.1 Electromotive Force 775
28.2 Resistors in Series and Parallel 778
28.3 Kirchhoff's Rules 785
28.4 *RC* Circuits 788
28.5 Electrical Meters 794
28.6 Household Wiring and Electrical Safety 796

Chapter 29 Magnetic Fields 808
29.1 Magnetic Fields and Forces 809
29.2 Motion of a Charged Particle in a Uniform Magnetic Field 813
29.3 Applications Involving Charged Particles Moving in a Magnetic Field 816
29.4 Magnetic Force Acting on a Current-Carrying Conductor 819
29.5 Torque on a Current Loop in a Uniform Magnetic Field 821
29.6 The Hall Effect 825

Chapter 30 Sources of the Magnetic Field 837
30.1 The Biot–Savart Law 837
30.2 The Magnetic Force Between Two Parallel Conductors 842
30.3 Ampère's Law 844
30.4 The Magnetic Field of a Solenoid 848
30.5 Gauss's Law in Magnetism 850
30.6 Magnetism in Matter 852
30.7 The Magnetic Field of the Earth 855

Chapter 31 Faraday's Law 867
31.1 Faraday's Law of Induction 867
31.2 Motional emf 871
31.3 Lenz's Law 876
31.4 Induced emf and Electric Fields 878
31.5 Generators and Motors 880
31.6 Eddy Currents 884

Chapter 32 Inductance 897
32.1 Self-Induction and Inductance 897
32.2 *RL* Circuits 900
32.3 Energy in a Magnetic Field 903
32.4 Mutual Inductance 906
32.5 Oscillations in an *LC* Circuit 907
32.6 The *RLC* Circuit 911

Chapter 33 Alternating Current Circuits 923
33.1 AC Sources 923
33.2 Resistors in an AC Circuit 924
33.3 Inductors in an AC Circuit 927
33.4 Capacitors in an AC Circuit 929

© Thomson Learning/Charles D. Winters

33.5   The *RLC* Series Circuit *932*
33.6   Power in an AC Circuit *935*
33.7   Resonance in a Series *RLC* Circuit *937*
33.8   The Transformer and Power Transmission *939*
33.9   Rectifiers and Filters *942*

Chapter 34   Electromagnetic Waves   952
34.1   Displacement Current and the General Form of Ampère's Law *953*
34.2   Maxwell's Equations and Hertz's Discoveries *955*
34.3   Plane Electromagnetic Waves *957*
34.4   Energy Carried by Electromagnetic Waves *961*
34.5   Momentum and Radiation Pressure *963*
34.6   Production of Electromagnetic Waves by an Antenna *965*
34.7   The Spectrum of Electromagnetic Waves *966*

PART 5   LIGHT AND OPTICS 977

Chapter 35   The Nature of Light and the Laws of Geometric Optics   978
35.1   The Nature of Light *978*
35.2   Measurements of the Speed of Light *979*
35.3   The Ray Approximation in Geometric Optics *981*
35.4   The Wave Under Reflection *981*
35.5   The Wave Under Refraction *985*
35.6   Huygens's Principle *990*
35.7   Dispersion *992*
35.8   Total Internal Reflection *993*

Chapter 36   Image Formation   1008
36.1   Images Formed by Flat Mirrors *1008*
36.2   Images Formed by Spherical Mirrors *1010*
36.3   Images Formed by Refraction *1017*
36.4   Thin Lenses *1021*
36.5   Lens Aberrations *1030*
36.6   The Camera *1031*
36.7   The Eye *1033*
36.8   The Simple Magnifier *1035*
36.9   The Compound Microscope *1037*
36.10   The Telescope *1038*

Chapter 37   Interference of Light Waves   1051
37.1   Conditions for Interference *1051*
37.2   Young's Double-Slit Experiment *1052*
37.3   Light Waves in Interference *1054*
37.4   Intensity Distribution of the Double-Slit Interference Pattern *1056*
37.5   Change of Phase Due to Reflection *1059*
37.6   Interference in Thin Films *1060*
37.7   The Michelson Interferometer *1064*

Chapter 38   Diffraction Patterns and Polarization   1077
38.1   Introduction to Diffraction Patterns *1077*
38.2   Diffraction Patterns from Narrow Slits *1078*
38.3   Resolution of Single-Slit and Circular Apertures *1083*
38.4   The Diffraction Grating *1086*
38.5   Diffraction of X-Rays by Crystals *1091*
38.6   Polarization of Light Waves *1093*

PART 6   MODERN PHYSICS 1111

Chapter 39   Relativity   1112
39.1   The Principle of Galilean Relativity *1113*
39.2   The Michelson–Morley Experiment *1116*
39.3   Einstein's Principle of Relativity *1118*
39.4   Consequences of the Special Theory of Relativity *1119*
39.5   The Lorentz Transformation Equations *1130*
39.6   The Lorentz Velocity Transformation Equations *1131*
39.7   Relativistic Linear Momentum *1134*
39.8   Relativistic Energy *1135*
39.9   Mass and Energy *1139*
39.10   The General Theory of Relativity *1140*

Chapter 40   Introduction to Quantum Physics   1153
40.1   Blackbody Radiation and Planck's Hypothesis *1154*
40.2   The Photoelectric Effect *1160*
40.3   The Compton Effect *1165*
40.4   Photons and Electromagnetic Waves *1167*
40.5   The Wave Properties of Particles *1168*
40.6   The Quantum Particle *1171*
40.7   The Double-Slit Experiment Revisited *1174*
40.8   The Uncertainty Principle *1175*

Chapter 41   Quantum Mechanics   1186
41.1   An Interpretation of Quantum Mechanics *1186*
41.2   The Quantum Particle Under Boundary Conditions *1191*
41.3   The Schrödinger Equation *1196*
41.4   A Particle in a Well of Finite Height *1198*
41.5   Tunneling Through a Potential Energy Barrier *1200*
41.6   Applications of Tunneling *1202*
41.7   The Simple Harmonic Oscillator *1205*

Chapter 42   Atomic Physics   1215
42.1   Atomic Spectra of Gases *1216*
42.2   Early Models of the Atom *1218*

42.3 Bohr's Model of the Hydrogen Atom *1219*

42.4 The Quantum Model of the Hydrogen Atom *1224*

42.5 The Wave Functions for Hydrogen *1227*

42.6 Physical Interpretation of the Quantum Numbers *1230*

42.7 The Exclusion Principle and the Periodic Table *1237*

42.8 More on Atomic Spectra: Visible and X-Ray *1241*

42.9 Spontaneous and Stimulated Transitions *1244*

42.10 Lasers *1245*

Chapter 43 Molecules and Solids   1257

43.1 Molecular Bonds *1258*

43.2 Energy States and Spectra of Molecules *1261*

43.3 Bonding in Solids *1268*

43.4 Free-Electron Theory of Metals *1270*

43.5 Band Theory of Solids *1274*

43.6 Electrical Conduction in Metals, Insulators, and Semiconductors *1276*

43.7 Semiconductor Devices *1279*

43.8 Superconductivity *1283*

Chapter 44 Nuclear Structure   1293

44.1 Some Properties of Nuclei *1294*

44.2 Nuclear Binding Energy *1299*

44.3 Nuclear Models *1300*

44.4 Radioactivity *1304*

44.5 The Decay Processes *1308*

44.6 Natural Radioactivity *1317*

44.7 Nuclear Reactions *1318*

44.8 Nuclear Magnetic Resonance and Magnetic Resonance Imaging *1319*

Chapter 45 Applications of Nuclear Physics   1329

45.1 Interactions Involving Neutrons *1329*

45.2 Nuclear Fission *1330*

45.3 Nuclear Reactors *1332*

45.4 Nuclear Fusion *1335*

45.5 Radiation Damage *1342*

45.6 Radiation Detectors *1344*

45.7 Uses of Radiation *1347*

Chapter 46 Particle Physics and Cosmology   1357

46.1 The Fundamental Forces in Nature *1358*

46.2 Positrons and Other Antiparticles *1358*

46.3 Mesons and the Beginning of Particle Physics *1361*

46.4 Classification of Particles *1363*

46.5 Conservation Laws *1365*

46.6 Strange Particles and Strangeness *1369*

46.7 Finding Patterns in the Particles *1370*

46.8 Quarks *1372*

46.9 Multicolored Quarks *1375*

46.10 The Standard Model *1377*

46.11 The Cosmic Connection *1378*

46.12 Problems and Perspectives *1383*

Appendix A Tables   A-1

Table A.1 Conversion Factors *A-1*

Table A.2 Symbols, Dimensions, and Units of Physical Quantities *A-2*

Appendix B Mathematics Review   A-4

B.1 Scientific Notation *A-4*

B.2 Algebra *A-5*

B.3 Geometry *A-9*

B.4 Trigonometry *A-10*

B.5 Series Expansions *A-12*

B.6 Differential Calculus *A-13*

B.7 Integral Calculus *A-16*

B.8 Propagation of Uncertainty *A-20*

Appendix C Periodic Table of the Elements   A-22

Appendix D SI Units   A-24

D.1 SI Units *A-24*

D.2 Some Derived SI Units *A-24*

Answers to Odd-Numbered Problems   A-25

Index   I-1

Raymond A. Serway received his doctorate at Illinois Institute of Technology and is Professor Emeritus at James Madison University. In 1990, he received the Madison Scholar Award at James Madison University, where he taught for 17 years. Dr. Serway began his teaching career at Clarkson University, where he conducted research and taught from 1967 to 1980. He was the recipient of the Distinguished Teaching Award at Clarkson University in 1977 and of the Alumni Achievement Award from Utica College in 1985. As Guest Scientist at the IBM Research Laboratory in Zurich, Switzerland, he worked with K. Alex Müller, 1987 Nobel Prize recipient. Dr. Serway also was a visiting scientist at Argonne National Laboratory, where he collaborated with his mentor and friend, Sam Marshall. In addition to earlier editions of this textbook, Dr. Serway is the coauthor of *Principles of Physics*, fourth edition; *College Physics*, seventh edition; *Essentials of College Physics;* and *Modern Physics*, third edition. He also is the coauthor of the high school textbook *Physics*, published by Holt, Rinehart, & Winston. In addition, Dr. Serway has published more than 40 research papers in the field of condensed matter physics and has given more than 70 presentations at professional meetings. Dr. Serway and his wife, Elizabeth, enjoy traveling, golf, singing in a church choir, and spending quality time with their four children and eight grandchildren.

John W. Jewett, Jr., earned his doctorate at Ohio State University, specializing in optical and magnetic properties of condensed matter. Dr. Jewett began his academic career at Richard Stockton College of New Jersey, where he taught from 1974 to 1984. He is currently Professor of Physics at California State Polytechnic University, Pomona. Throughout his teaching career, Dr. Jewett has been active in promoting science education. In addition to receiving four National Science Foundation grants, he helped found and direct the Southern California Area Modern Physics Institute. He also directed Science IMPACT (Institute for Modern Pedagogy and Creative Teaching), which works with teachers and schools to develop effective science curricula. Dr. Jewett's honors include the Stockton Merit Award at Richard Stockton College in 1980, the Outstanding Professor Award at California State Polytechnic University for 1991–1992, and the Excellence in Undergraduate Physics Teaching Award from the American Association of Physics Teachers in 1998. He has given more than 80 presentations at professional meetings, including presentations at international conferences in China and Japan. In addition to his work on this textbook, he is coauthor of *Principles of Physics*, fourth edition, with Dr. Serway and author of *The World of Physics . . . Mysteries, Magic, and Myth.* Dr. Jewett enjoys playing keyboard with his all-physicist band, traveling, and collecting antiques that can be used as demonstration apparatus in physics lectures. Most importantly, he relishes spending time with his wife, Lisa, and their children and grandchildren.

In writing this seventh edition of *Physics for Scientists and Engineers,* we continue our ongoing efforts to improve the clarity of presentation and include new pedagogical features that help support the learning and teaching processes. Drawing on positive feedback from users of the sixth edition and reviewers' suggestions, we have refined the text to better meet the needs of students and teachers.

This textbook is intended for a course in introductory physics for students majoring in science or engineering. The entire contents of the book in its extended version could be covered in a three-semester course, but it is possible to use the material in shorter sequences with the omission of selected chapters and sections. The mathematical background of the student taking this course should ideally include one semester of calculus. If that is not possible, the student should be enrolled in a concurrent course in introductory calculus.

Objectives

This introductory physics textbook has two main objectives: to provide the student with a clear and logical presentation of the basic concepts and principles of physics and to strengthen an understanding of the concepts and principles through a broad range of interesting applications to the real world. To meet these objectives, we have placed emphasis on sound physical arguments and problem-solving methodology. At the same time, we have attempted to motivate the student through practical examples that demonstrate the role of physics in other disciplines, including engineering, chemistry, and medicine.

Changes in the Seventh Edition

A large number of changes and improvements have been made in preparing the seventh edition of this text. Some of the new features are based on our experiences and on current trends in science education. Other changes have been incorporated in response to comments and suggestions offered by users of the sixth edition and by reviewers of the manuscript. The features listed here represent the major changes in the seventh edition.

QUESTIONS AND PROBLEMS   A substantial revision to the end-of-chapter questions and problems was made in an effort to improve their variety, interest, and pedagogical value, while maintaining their clarity and quality. Approximately 23% of the questions and problems are new or substantially changed. Several of the questions for each chapter are in objective format. Several problems in each chapter explicitly ask for qualitative reasoning in some parts as well as for quantitative answers in other parts:

> 19. ● Assume a parcel of air in a straight tube moves with a constant acceleration of -4.00 m/s^2 and has a velocity of 13.0 m/s at 10:05:00 a.m. on a certain date. (a) What is its velocity at 10:05:01 a.m.? (b) At 10:05:02 a.m.? (c) At 10:05:02.5 a.m.? (d) At 10:05:04 a.m.? (e) At 10:04:59 a.m.? (f) Describe the shape of a graph of velocity versus time for this parcel of air. (g) Argue for or against the statement, "Knowing the single value of an object's constant acceleration is like knowing a whole list of values for its velocity."

© Thomson Learning/
Charles D. Winters

WORKED EXAMPLES   All in-text worked examples have been recast and are now presented in a two-column format to better reinforce physical concepts. The left column shows textual information that describes the steps for solving the problem. The right column shows the mathematical manipulations and results of taking these steps. This layout facilitates matching the concept with its mathematical execution and helps students organize their work. These reconstituted examples closely follow a General Problem-Solving Strategy introduced in Chapter 2 to reinforce effective problem-solving habits. A sample of a worked example can be found on the next page.

Each solution has been reconstituted to more closely follow the General Problem-Solving Strategy as outlined in Chapter 2, to reinforce good problem-solving habits.

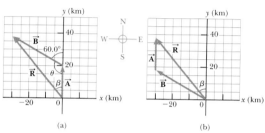

Figure 3.11 (Example 3.2) (a) Graphical method for finding the resultant displacement vector $\vec{R} = \vec{A} + \vec{B}$. (b) Adding the vectors in reverse order $(\vec{B} + \vec{A})$ gives the same result for $\vec{R}$.

EXAMPLE 3.2   **A Vacation Trip**

A car travels 20.0 km due north and then 35.0 km in a direction 60.0° west of north as shown in Figure 3.11a. Find the magnitude and direction of the car's resultant displacement.

SOLUTION

Conceptualize The vectors $\vec{A}$ and $\vec{B}$ drawn in Figure 3.11a help us conceptualize the problem.

Categorize We can categorize this example as a simple analysis problem in vector addition. The displacement $\vec{R}$ is the resultant when the two individual displacements $\vec{A}$ and $\vec{B}$ are added. We can further categorize it as a problem about the analysis of triangles, so we appeal to our expertise in geometry and trigonometry.

Each step of the solution is detailed in a two-column format. The left column provides an explanation for each mathematical step in the right column, to better reinforce the physical concepts.

Analyze In this example, we show two ways to analyze the problem of finding the resultant of two vectors. The first way is to solve the problem geometrically, using graph paper and a protractor to measure the magnitude of $\vec{R}$ and its direction in Figure 3.11a. (In fact, even when you know you are going to be carrying out a calculation, you should sketch the vectors to check your results.) With an ordinary ruler and protractor, a large diagram typically gives answers to two-digit but not to three-digit precision.

The second way to solve the problem is to analyze it algebraically. The magnitude of $\vec{R}$ can be obtained from the law of cosines as applied to the triangle (see Appendix B.4).

Use $R^2 = A^2 + B^2 - 2AB \cos \theta$ from the law of cosines to find R:

$$R = \sqrt{A^2 + B^2 - 2AB \cos \theta}$$

Substitute numerical values, noting that $\theta = 180° - 60° = 120°$:

$$R = \sqrt{(20.0 \text{ km})^2 + (35.0 \text{ km})^2 - 2(20.0 \text{ km})(35.0 \text{ km}) \cos 120°}$$

$$= 48.2 \text{ km}$$

Use the law of sines (Appendix B.4) to find the direction of $\vec{R}$ measured from the northerly direction:

$$\frac{\sin \beta}{B} = \frac{\sin \theta}{R}$$

$$\sin \beta = \frac{B}{R} \sin \theta = \frac{35.0 \text{ km}}{48.2 \text{ km}} \sin 120° = 0.629$$

$$\beta = 38.9°$$

The resultant displacement of the car is 48.2 km in a direction 38.9° west of north.

Finalize Does the angle β that we calculated agree with an estimate made by looking at Figure 3.11a or with an actual angle measured from the diagram using the graphical method? Is it reasonable that the magnitude of $\vec{R}$ is larger than that of both $\vec{A}$ and $\vec{B}$? Are the units of $\vec{R}$ correct?

Although the graphical method of adding vectors works well, it suffers from two disadvantages. First, some people find using the laws of cosines and sines to be awkward. Second, a triangle only results if you are adding two vectors. If you are adding three or more vectors, the resulting geometric shape is usually not a triangle. In Section 3.4, we explore a new method of adding vectors that will address both of these disadvantages.

What If? Suppose the trip were taken with the two vectors in reverse order: 35.0 km at 60.0° west of north first and then 20.0 km due north. How would the magnitude and the direction of the resultant vector change?

Answer They would not change. The commutative law for vector addition tells us that the order of vectors in an addition is irrelevant. Graphically, Figure 3.11b shows that the vectors added in the reverse order give us the same resultant vector.

What If? statements appear in about 1/3 of the worked examples and offer a variation on the situation posed in the text of the example. For instance, this feature might explore the effects of changing the conditions of the situation, determine what happens when a quantity is taken to a particular limiting value, or question whether additional information can be determined about the problem situation. This feature encourages students to think about the results of the example and assists in conceptual understanding of the principles.

All worked examples are also available to be assigned as interactive examples in the Enhanced WebAssign homework management system (visit **www.pse7.com** for more details).

ONLINE HOMEWORK   It is now easier to assign online homework with Serway and Jewett and Enhanced WebAssign. All worked examples, end-of-chapter problems, active figures, quick quizzes, and most questions are available in WebAssign. Most problems include hints and feedback to provide instantaneous reinforcement or direction for that problem. In addition to the text content, we have also added math remediation tools to help students get up to speed in algebra, trigonometry, and calculus.

SUMMARIES   Each chapter contains a summary that reviews the important concepts and equations discussed in that chapter. A marginal note next to each chapter summary directs students to additional quizzes, animations, and interactive exercises for that chapter on the book's companion Web site. The format of the end-of-chapter summary has been completely revised for this edition. The summary is divided into three sections: Definitions, Concepts and Principles, and Analysis Models for Problem-Solving. In each section, flashcard-type boxes focus on each separate definition, concept, principle, or analysis model.

MATH APPENDIX   The math appendix, a valuable tool for students, has been updated to show the math tools in a physics context. This resource is ideal for students who need a quick review on topics such as algebra, trigonometry, and calculus.

CONTENT CHANGES   The content and organization of the textbook are essentially the same as in the sixth edition. Many sections in various chapters have been streamlined, deleted, or combined with other sections to allow for a more balanced presentation. Vectors are now denoted in boldface with an arrow over them (for example, $\vec{v}$), making them easier to recognize. Chapters 7 and 8 have been completely reorganized to prepare students for a unified approach to energy that is used throughout the text. A new section in Chapter 9 teaches students how to analyze deformable systems with the conservation of energy equation and the impulse-momentum theorem. Chapter 34 is longer than in the sixth edition because of the movement into that chapter of the material on displacement current from Chapter 30 and Maxwell's equations from Chapter 31. A more detailed list of content changes can be found on the instructor's companion Web site.

Content

The material in this book covers fundamental topics in classical physics and provides an introduction to modern physics. The book is divided into six parts. Part 1 (Chapters 1 to 14) deals with the fundamentals of Newtonian mechanics and the physics of fluids; Part 2 (Chapters 15 to 18) covers oscillations, mechanical waves, and sound; Part 3 (Chapters 19 to 22) addresses heat and thermodynamics; Part 4 (Chapters 23 to 34) treats electricity and magnetism; Part 5 (Chapters 35 to 38) covers light and optics; and Part 6 (Chapters 39 to 46) deals with relativity and modern physics.

Text Features

Most instructors believe that the textbook selected for a course should be the student's primary guide for understanding and learning the subject matter. Furthermore, the textbook should be easily accessible and should be styled and written to facilitate instruction and learning. With these points in mind, we have included many pedagogical features, listed below, that are intended to enhance its usefulness to both students and instructors.

Problem Solving and Conceptual Understanding

GENERAL PROBLEM-SOLVING STRATEGY   A general strategy outlined at the end of Chapter 2 provides students with a structured process for solving problems. In all remaining chapters, the strategy is employed explicitly in every example so that students learn how it is applied. Students are encouraged to follow this strategy when working end-of-chapter problems.

MODELING Although students are faced with hundreds of problems during their physics courses, instructors realize that a relatively small number of physical situations form the basis of these problems. When faced with a new problem, a physicist forms a *model* of the problem that can be solved in a simple way by identifying the common physical situation that occurs in the problem. For example, many problems involve particles under constant acceleration, isolated systems, or waves under refraction. Because the physicist has studied these situations extensively and understands the associated behavior, he or she can apply this knowledge as a model for solving a new problem. In certain chapters, this edition identifies Analysis Models, which are physical situations (such as the particle under constant acceleration, the isolated system, or the wave under refraction) that occur so often that they can be used as a model for solving an unfamiliar problem. These models are discussed in the chapter text, and the student is reminded of them in the end-of-chapter summary under the heading "Analysis Models for Problem-Solving."

© Thomson Learning/George Semple

PROBLEMS An extensive set of problems is included at the end of each chapter; in all, the text contains approximately three thousand problems. Answers to odd-numbered problems are provided at the end of the book. For the convenience of both the student and the instructor, about two-thirds of the problems are keyed to specific sections of the chapter. The remaining problems, labeled "Additional Problems," are not keyed to specific sections. The problem numbers for straightforward problems are printed in black, intermediate-level problems are in blue, and challenging problems are in magenta.

- **"Not-just-a-number" problems** Each chapter includes several marked problems that require students to think qualitatively in some parts and quantitatively in others. Instructors can assign such problems to guide students to display deeper understanding, practice good problem-solving techniques, and prepare for exams.
- **Problems for developing symbolic reasoning** Each chapter contains problems that ask for solutions in symbolic form as well as many problems asking for numerical answers. To help students develop skill in symbolic reasoning, each chapter contains a pair of otherwise identical problems, one asking for a numerical solution and one asking for a symbolic derivation. In this edition, each chapter also contains a problem giving a numerical value for every datum but one so that the answer displays how the unknown depends on the datum represented symbolically. The answer to such a problem has the form of a function of one variable. Reasoning about the behavior of this function puts emphasis on the *Finalize* step of the General Problem-Solving Strategy. All problems developing symbolic reasoning are identified by a tan background screen:

> 53. ● A light spring has an unstressed length of 15.5 cm. It is described by Hooke's law with spring constant 4.30 N/m. One end of the horizontal spring is held on a fixed vertical axle, and the other end is attached to a puck of mass m that can move without friction over a horizontal surface. The puck is set into motion in a circle with a period of 1.30 s. (a) Find the extension of the spring x as it depends on m. Evaluate x for (b) $m = 0.070\ 0$ kg, (c) $m = 0.140$ kg, (d) $m = 0.180$ kg, and (e) $m = 0.190$ kg. (f) Describe the pattern of variation of x as it depends on m.

- **Review problems** Many chapters include review problems requiring the student to combine concepts covered in the chapter with those discussed in previous chapters. These problems reflect the cohesive nature of the principles in the text and verify that physics is not a scattered set of ideas. When facing a real-world issue such as global warming or nuclear weapons, it may be necessary to call on ideas in physics from several parts of a textbook such as this one.
- **"Fermi problems"** As in previous editions, at least one problem in each chapter asks the student to reason in order-of-magnitude terms.

- **Design problems** Several chapters contain problems that ask the student to determine design parameters for a practical device so that it can function as required.
- **"*Jeopardy!*" problems** Some chapters give students practice in changing between different representations by stating equations and asking for a description of a situation to which they apply as well as for a numerical answer.
- **Calculus-based problems** Every chapter contains at least one problem applying ideas and methods from differential calculus and one problem using integral calculus.

The instructor's Web site, **www.thomsonedu.com/physics/serway,** provides lists of problems using calculus, problems encouraging or requiring computer use, problems with "**What If?**" parts, problems referred to in the chapter text, problems based on experimental data, order-of-magnitude problems, problems about biological applications, design problems, *Jeopardy!* problems, review problems, problems reflecting historical reasoning about confusing ideas, problems developing symbolic reasoning skill, problems with qualitative parts, ranking questions, and other objective questions.

QUESTIONS The questions section at the end of each chapter has been significantly revised. Multiple-choice, ranking, and true–false questions have been added. The instructor may select items to assign as homework or use in the classroom, possibly with "peer instruction" methods and possibly with "clicker" systems. More than eight hundred questions are included in this edition. Answers to selected questions are included in the *Student Solutions Manual/Study Guide,* and answers to all questions are found in the *Instructor's Solutions Manual.*

---

19. **O** **(i)** Rank the gravitational accelerations you would measure for (a) a 2-kg object 5 cm above the floor, (b) a 2-kg object 120 cm above the floor, (c) a 3-kg object 120 cm above the floor, and (d) a 3-kg object 80 cm above the floor. List the one with the largest-magnitude acceleration first. If two are equal, show their equality in your list. **(ii)** Rank the gravitational forces on the same four objects, largest magnitude first. **(iii)** Rank the gravitational potential energies (of the object–Earth system) for the same four objects, largest first, taking $y = 0$ at the floor.

23. **O** An ice cube has been given a push and slides without friction on a level table. Which is correct? (a) It is in stable equilibrium. (b) It is in unstable equilibrium. (c) It is in neutral equilibrium (d) It is not in equilibrium.

WORKED EXAMPLES Two types of worked examples are presented to aid student comprehension. All worked examples in the text may be assigned for homework in WebAssign.

The first example type presents a problem and numerical answer. As discussed earlier, solutions to these examples have been altered in this edition to feature a two-column layout to explain the physical concepts and the mathematical steps side by side. Every example follows the explicit steps of the General Problem-Solving Strategy outlined in Chapter 2.

The second type of example is conceptual in nature. To accommodate increased emphasis on understanding physical concepts, the many conceptual examples are labeled as such, set off in boxes, and designed to focus students on the physical situation in the problem.

WHAT IF? Approximately one-third of the worked examples in the text contain a **What If?** feature. At the completion of the example solution, a **What If?** question offers a variation on the situation posed in the text of the example. For instance, this feature might explore the effects of changing the conditions of the situation, determine what happens when a quantity is taken to a particular limiting value, or question whether additional information can be determined about the situation. This feature encourages students to think about the results of the example, and it also assists in conceptual understanding of the principles. **What If?** questions also prepare students to encounter novel problems that may be included on exams. Some of the end-of-chapter problems also include this feature.

QUICK QUIZZES   Quick Quizzes provide students an opportunity to test their understanding of the physical concepts presented. The questions require students to make decisions on the basis of sound reasoning, and some of the questions have been written to help students overcome common misconceptions. Quick Quizzes have been cast in an objective format, including multiple-choice, true–false, and ranking. Answers to all Quick Quiz questions are found at the end of each chapter. Additional Quick Quizzes that can be used in classroom teaching are available on the instructor's companion Web site. Many instructors choose to use such questions in a "peer instruction" teaching style or with the use of personal response system "clickers," but they can be used in standard quiz format as well. Quick Quizzes are set off from the text by horizontal lines:

Quick Quiz 7.5   A dart is loaded into a spring-loaded toy dart gun by pushing the spring in by a distance x. For the next loading, the spring is compressed a distance $2x$. How much faster does the second dart leave the gun compared with the first? (a) four times as fast   (b) two times as fast   (c) the same   (d) half as fast   (e) one-fourth as fast

---

PITFALL PREVENTION 16.2
Two Kinds of Speed/Velocity

Do not confuse v, the speed of the wave as it propagates along the string, with v_y, the transverse velocity of a point on the string. The speed v is constant for a uniform medium, whereas v_y varies sinusoidally.

PITFALL PREVENTIONS   More than two hundred Pitfall Preventions (such as the one to the left) are provided to help students avoid common mistakes and misunderstandings. These features, which are placed in the margins of the text, address both common student misconceptions and situations in which students often follow unproductive paths.

Helpful Features

STYLE   To facilitate rapid comprehension, we have written the book in a clear, logical, and engaging style. We have chosen a writing style that is somewhat informal and relaxed so that students will find the text appealing and enjoyable to read. New terms are carefully defined, and we have avoided the use of jargon.

IMPORTANT STATEMENTS AND EQUATIONS   Most important statements and definitions are set in **boldface** or are highlighted with a background screen for added emphasis and ease of review. Similarly, important equations are highlighted with a background screen to facilitate location.

MARGINAL NOTES   Comments and notes appearing in the margin with a ▶ icon can be used to locate important statements, equations, and concepts in the text.

PEDAGOGICAL USE OF COLOR   Readers should consult the **pedagogical color chart** (inside the front cover) for a listing of the color-coded symbols used in the text diagrams. This system is followed consistently throughout the text.

MATHEMATICAL LEVEL   We have introduced calculus gradually, keeping in mind that students often take introductory courses in calculus and physics concurrently. Most steps are shown when basic equations are developed, and reference is often made to mathematical appendices near the end of the textbook. Vector products are introduced later in the text, where they are needed in physical applications. The dot product is introduced in Chapter 7, which addresses energy of a system; the cross product is introduced in Chapter 11, which deals with angular momentum.

SIGNIFICANT FIGURES   Significant figures in both worked examples and end-of-chapter problems have been handled with care. Most numerical examples are worked to either two or three significant figures, depending on the precision of the data provided. End-of-chapter problems regularly state data and answers to three-digit precision.

UNITS   The international system of units (SI) is used throughout the text. The U.S. customary system of units is used only to a limited extent in the chapters on mechanics and thermodynamics.

APPENDICES AND ENDPAPERS   Several appendices are provided near the end of the textbook. Most of the appendix material represents a review of mathematical concepts and techniques used in the text, including scientific notation, algebra, geometry, trigonometry, differential calculus, and integral calculus. Reference to these appendices is made throughout the text. Most mathematical review sections in the appendices include worked examples and exercises with answers. In addition to the mathematical reviews, the appendices contain tables of physical data, conversion factors, and the SI units of physical quantities as well as a periodic table of the elements. Other useful information—fundamental constants and physical data, planetary data, a list of standard prefixes, mathematical symbols, the Greek alphabet, and standard abbreviations of units of measure—appears on the endpapers.

Course Solutions That Fit Your Teaching Goals and Your Students' Learning Needs

Recent advances in educational technology have made homework management systems and audience response systems powerful and affordable tools to enhance the way you teach your course. Whether you offer a more traditional text-based course, are interested in using or are currently using an online homework management system such as WebAssign, or are ready to turn your lecture into an interactive learning environment with JoinIn on TurningPoint, you can be confident that the text's proven content provides the foundation for each and every component of our technology and ancillary package.

Homework Management Systems

Enhanced WebAssign   Whether you're an experienced veteran or a beginner, Enhanced WebAssign is the perfect solution to fit your homework management needs. Designed by physicists for physicists, this system is a reliable and user-friendly teaching companion. Enhanced WebAssign is available for *Physics for Scientists and Engineers*, giving you the freedom to assign

- every end-of-chapter Problem and Question, enhanced with hints and feedback
- every worked example, enhanced with hints and feedback, to help strengthen students' problem-solving skills
- every Quick Quiz, giving your students ample opportunity to test their conceptual understanding.

- animated Active Figures, enhanced with hints and feedback, to help students develop their visualization skills
- a math review to help students brush up on key quantitative concepts

Please visit **www.thomsonedu.com/physics/serway** to view a live demonstration of Enhanced WebAssign.

The text also supports the following Homework Management Systems:

LON-CAPA: A Computer-Assisted Personalized Approach
http://www.lon-capa.org/

The University of Texas Homework Service
contact **moore@physics.utexas.edu**

Personal Response Systems

JoinIn on TurningPoint Pose book-specific questions and display students' answers seamlessly within the Microsoft® PowerPoint slides of your own lecture in conjunction with the "clicker" hardware of your choice. JoinIn on TurningPoint works with most infrared or radio frequency keypad systems, including Responsecard, EduCue, H-ITT, and even laptops. Contact your local sales representative to learn more about our personal response software and hardware.

Personal Response System Content Regardless of the response system you are using, we provide the tested content to support it. Our ready-to-go content includes all the questions from the Quick Quizzes, test questions, and a selection of end-of-chapter questions to provide helpful conceptual checkpoints to drop into your lecture. Our series of Active Figure animations have also been enhanced with multiple-choice questions to help test students' observational skills.

We also feature the Assessing to Learn in the Classroom content from the University of Massachusetts at Amherst. This collection of 250 advanced conceptual questions has been tested in the classroom for more than ten years and takes peer learning to a new level.

Visit **www.thomsonedu.com/physics/serway** to download samples of our personal response system content.

Lecture Presentation Resources

The following resources provide support for your presentations in lecture.

MULTIMEDIA MANAGER INSTRUCTOR'S RESOURCE CD An easy-to-use multimedia lecture tool, the Multimedia Manager Instructor's Resource CD allows you to quickly assemble art, animations, digital video, and database files with notes to create fluid lectures. The two-volume set (Volume 1: Chapters 1–22; Volume 2: Chapters 23–46) includes prebuilt PowerPoint lectures, a database of animations, video clips, and digital art from the text as well as editable electronic files of the *Instructor's Solutions Manual* and *Test Bank*.

TRANSPARENCY ACETATES Each volume contains approximately one hundred transparency acetates featuring art from the text. Volume 1 contains Chapters 1 through 22, and Volume 2 contains Chapters 23 through 46.

Assessment and Course Preparation Resources

A number of resources listed below will assist with your assessment and preparation processes.

INSTRUCTOR'S SOLUTIONS MANUAL by Ralph McGrew. This two-volume manual contains complete worked solutions to all end-of-chapter problems in the textbook as well as answers to the even-numbered problems and all the questions. The solutions to problems new to the seventh edition are marked for easy identification. Volume 1 contains

Chapters 1 through 22, and Volume 2 contains Chapters 23 through 46. Electronic files of the Instructor's Solutions are available on the Multimedia Manager CD as well.

PRINTED TEST BANK by Edward Adelson. This two-volume test bank contains approximately 2 200 multiple-choice questions. These questions are also available in electronic format with complete answers and solutions in the ExamView test software and as editable Word® files on the Multimedia Manager CD. Volume 1 contains Chapters 1 through 22, and Volume 2 contains Chapters 23 through 46.

EXAMVIEW   This easy-to-use test generator CD features all of the questions from the printed test bank in an editable format.

WEBCT AND BLACKBOARD CONTENT   For users of either course management system, we provide our test bank questions in the proper format for easy upload into your online course. In addition, you can integrate the ThomsonNOW for Physics student tutorial content into your WebCT or Blackboard course, providing your students a single sign-on to all their Web-based learning resources. Contact your local sales representative to learn more about our WebCT and Blackboard resources.

INSTRUCTOR'S COMPANION WEB SITE   Consult the instructor's site by pointing your browser to **www.thomsonedu.com/physics/serway** for additional Quick Quiz questions, a detailed list of content changes since the sixth edition, a problem correlation guide, images from the text, and sample PowerPoint lectures. Instructors adopting the seventh edition of *Physics for Scientists and Engineers* may download these materials after securing the appropriate password from their local Thomson•Brooks/Cole sales representative.

Student Resources

STUDENT SOLUTIONS MANUAL/STUDY GUIDE by John R. Gordon, Ralph McGrew, Raymond Serway, and John W. Jewett, Jr. This two-volume manual features detailed solutions to 20% of the end-of-chapter problems from the text. The manual also features a list of important equations, concepts, and notes from key sections of the text in addition to answers to selected end-of-chapter questions. Volume 1 contains Chapters 1 through 22, and Volume 2 contains Chapters 23 through 46.

THOMSONNOW PERSONAL STUDY   This assessment-based student tutorial system provides students with a personalized learning plan based on their performance on a series of diagnostic pre-tests. Rich interactive content, including Active Figures, Coached Problems, and Interactive Examples, helps students prepare for tests and exams.

Teaching Options

The topics in this textbook are presented in the following sequence: classical mechanics, oscillations and mechanical waves, and heat and thermodynamics followed by electricity and magnetism, electromagnetic waves, optics, relativity, and modern physics. This presentation represents a traditional sequence, with the subject of mechanical waves being presented before electricity and magnetism. Some instructors may prefer to discuss both mechanical and electromagnetic waves together after completing electricity and magnetism. In this case, Chapters 16 through 18 could be covered along with Chapter 34. The chapter on relativity is placed near the end of the text because this topic often is treated as an introduction to the era of "modern physics." If time permits, instructors may choose to cover Chapter 39 after completing Chapter 13 as a conclusion to the material on Newtonian mechanics.

For those instructors teaching a two-semester sequence, some sections and chapters could be deleted without any loss of continuity. The following sections can be considered optional for this purpose:

2.8	Kinematic Equations Derived from Calculus
4.6	Relative Velocity and Relative Acceleration
6.3	Motion in Accelerated Frames
6.4	Motion in the Presence of Resistive Forces
7.9	Energy Diagrams and Equilibrium of a System
9.8	Rocket Propulsion
11.5	The Motion of Gyroscopes and Tops
14.7	Other Applications of Fluid Dynamics
15.6	Damped Oscillations
15.7	Forced Oscillations
17.5	Digital Sound Recording
17.6	Motion Picture Sound
18.6	Standing Waves in Rods and Membranes
18.8	Nonsinusoidal Wave Patterns
22.8	Entropy on a Microscopic Scale
25.7	The Millikan Oil-Drop Experiment
25.8	Applications of Electrostatics
26.7	An Atomic Description of Dielectrics
27.5	Superconductors
28.5	Electrical Meters
28.6	Household Wiring and Electrical Safety
29.3	Applications Involving Charged Particles Moving in a Magnetic Field
29.6	The Hall Effect
30.6	Magnetism in Matter
30.7	The Magnetic Field of the Earth
31.6	Eddy Currents
33.9	Rectifiers and Filters
34.6	Production of Electromagnetic Waves by an Antenna
36.5	Lens Aberrations
36.6	The Camera
36.7	The Eye
36.8	The Simple Magnifier
36.9	The Compound Microscope
36.10	The Telescope
38.5	Diffraction of X-Rays by Crystals
39.10	The General Theory of Relativity
41.6	Applications of Tunneling
42.9	Spontaneous and Stimulated Transitions
42.10	Lasers
43.7	Semiconductor Devices
43.8	Superconductivity
44.8	Nuclear Magnetic Resonance and Magnetic Resonance Imaging
45.5	Radiation Damage
45.6	Radiation Detectors
45.7	Uses of Radiation

© Thomson Learning/Charles D. Winters

Acknowledgments

This seventh edition of *Physics for Scientists and Engineers* was prepared with the guidance and assistance of many professors who reviewed selections of the manuscript, the prerevision text, or both. We wish to acknowledge the following scholars and express our sincere appreciation for their suggestions, criticisms, and encouragement:

David P. Balogh, *Fresno City College*
Leonard X. Finegold, *Drexel University*
Raymond Hall, *California State University, Fresno*

Bob Jacobsen, *University of California, Berkeley*
Robin Jordan, *Florida Atlantic University*
Rafael Lopez-Mobilia, *University of Texas at San Antonio*
Diana Lininger Markham, *City College of San Francisco*
Steven Morris, *Los Angeles Harbor City College*
Taha Mzoughi, *Kennesaw State University*
Nobel Sanjay Rebello, *Kansas State University*
John Rosendahl, *University of California, Irvine*
Mikolaj Sawicki, *John A. Logan College*
Glenn B. Stracher, *East Georgia College*
Som Tyagi, *Drexel University*
Robert Weidman, *Michigan Technological University*
Edward A. Whittaker, *Stevens Institute of Technology*

This title was carefully checked for accuracy by Zinoviy Akkerman, *City College of New York;* Grant Hart, *Brigham Young University;* Michael Kotlarchyk, *Rochester Institute of Technology;* Andres LaRosa, *Portland State University;* Bruce Mason, *University of Oklahoma at Norman;* Peter Moeck, *Portland State University;* Brian A. Raue, *Florida International University;* James E. Rutledge, *University of California at Irvine;* Bjoern Seipel, *Portland State University;* Z. M. Stadnik, *University of Ottawa;* and Harry W. K. Tom, *University of California at Riverside.* We thank them for their diligent efforts under schedule pressure.

We are grateful to Ralph McGrew for organizing the end-of-chapter problems, writing many new problems, and suggesting improvements in the content of the textbook. Problems and questions new to this edition were written by Duane Deardorff, Thomas Grace, Francisco Izaguirre, John Jewett, Robert Forsythe, Randall Jones, Ralph McGrew, Kurt Vandervoort, and Jerzy Wrobel. Help was very kindly given by Dwight Neuenschwander, Michael Kinney, Amy Smith, Will Mackin, and the Sewer Department of Grand Forks, North Dakota. Daniel Kim, Jennifer Hoffman, Ed Oberhofer, Richard Webb, Wesley Smith, Kevin Kilty, Zinoviy Akkerman, Michael Rudmin, Paul Cox, Robert LaMontagne, Ken Menningen, and Chris Church made corrections to problems taken from previous editions. We are grateful to authors John R. Gordon and Ralph McGrew for preparing the *Student Solutions Manual/Study Guide.* Author Ralph McGrew has prepared an excellent *Instructor's Solutions Manual.* Edward Adelson has carefully edited and improved the test bank. Kurt Vandervoort prepared extra Quick Quiz questions for the instructor's companion Web site.

Special thanks and recognition go to the professional staff at the Brooks/Cole Publishing Company—in particular, Ed Dodd, Brandi Kirksey (who managed the ancillary program and so much more), Shawn Vasquez, Sam Subity, Teri Hyde, Michelle Julet, David Harris, and Chris Hall—for their fine work during the development and production of this textbook. Mark Santee is our seasoned marketing manager, and Bryan Vann coordinates our marketing communications. We recognize the skilled production service and excellent artwork provided by the staff at Lachina Publishing Services, and the dedicated photo research efforts of Jane Sanders Miller.

Finally, we are deeply indebted to our wives, children, and grandchildren for their love, support, and long-term sacrifices.

Raymond A. Serway
St. Petersburg, Florida

John W. Jewett, Jr.
Pomona, California

It is appropriate to offer some words of advice that should be of benefit to you, the student. Before doing so, we assume you have read the Preface, which describes the various features of the text and support materials that will help you through the course.

How to Study

Instructors are often asked, "How should I study physics and prepare for examinations?" There is no simple answer to this question, but we can offer some suggestions based on our own experiences in learning and teaching over the years.

First and foremost, maintain a positive attitude toward the subject matter, keeping in mind that physics is the most fundamental of all natural sciences. Other science courses that follow will use the same physical principles, so it is important that you understand and are able to apply the various concepts and theories discussed in the text.

Concepts and Principles

It is essential that you understand the basic concepts and principles before attempting to solve assigned problems. You can best accomplish this goal by carefully reading the textbook before you attend your lecture on the covered material. When reading the text, you should jot down those points that are not clear to you. Also be sure to make a diligent attempt at answering the questions in the Quick Quizzes as you come to them in your reading. We have worked hard to prepare questions that help you judge for yourself how well you understand the material. Study the **What If?** features that appear in many of the worked examples carefully. They will help you extend your understanding beyond the simple act of arriving at a numerical result. The Pitfall Preventions will also help guide you away from common misunderstandings about physics. During class, take careful notes and ask questions about those ideas that are unclear to you. Keep in mind that few people are able to absorb the full meaning of scientific material after only one reading; several readings of the text and your notes may be necessary. Your lectures and laboratory work supplement the textbook and should clarify some of the more difficult material. You should minimize your memorization of material. Successful memorization of passages from the text, equations, and derivations does not necessarily indicate that you understand the material. Your understanding of the material will be enhanced through a combination of efficient study habits, discussions with other students and with instructors, and your ability to solve the problems presented in the textbook. Ask questions whenever you believe that clarification of a concept is necessary.

© Thomson Learning/Charles D. Winters

Study Schedule

It is important that you set up a regular study schedule, preferably a daily one. Make sure that you read the syllabus for the course and adhere to the schedule set by your instructor. The lectures will make much more sense if you read the corresponding text material *before* attending them. As a general rule, you should devote about two hours of study time for each hour you are in class. If you are having trouble with the course, seek the advice of the instructor or other students who have taken the course. You may find it necessary to seek further instruction from experienced students. Very often, instructors offer review sessions in addition to regular class periods. Avoid the practice of delaying study until a day or two before an exam. More often than not, this approach has disastrous results. Rather than undertake an all-night study session before a test, briefly review the basic concepts and equations, and then get a good night's rest. If you believe that you need additional help in understanding the concepts, in preparing for exams, or in problem solving, we suggest that you acquire a copy of the *Student Solutions Manual/Study Guide* that accompanies this textbook; this manual should be available at your college bookstore or through the publisher.

Use the Features

You should make full use of the various features of the text discussed in the Preface. For example, marginal notes are useful for locating and describing important equations and concepts, and **boldface** indicates important statements and definitions. Many useful tables are contained in the appendices, but most are incorporated in the text where they are most often referenced. Appendix B is a convenient review of mathematical tools used in the text.

Answers to odd-numbered problems are given at the end of the textbook, answers to Quick Quizzes are located at the end of each chapter, and solutions to selected end-of-chapter questions and problems are provided in the *Student Solutions Manual/Study Guide*. The table of contents provides an overview of the entire text, and the index enables you to locate specific material quickly. Footnotes are sometimes used to supplement the text or to cite other references on the subject discussed.

After reading a chapter, you should be able to define any new quantities introduced in that chapter and discuss the principles and assumptions that were used to arrive at certain key relations. The chapter summaries and the review sections of the *Student Solutions Manual/Study Guide* should help you in this regard. In some cases, you may find it necessary to refer to the textbook's index to locate certain topics. You should be able to associate with each physical quantity the correct symbol used to represent that quantity and the unit in which the quantity is specified. Furthermore, you should be able to express each important equation in concise and accurate prose.

Problem Solving

R. P. Feynman, Nobel laureate in physics, once said, "You do not know anything until you have practiced." In keeping with this statement, we strongly advise you to develop the skills necessary to solve a wide range of problems. Your ability to solve problems will be one of the main tests of your knowledge of physics; therefore, you should try to solve as many problems as possible. It is essential that you understand basic concepts and principles before attempting to solve problems. It is good practice to try to find alternate solutions to the same problem. For example, you can solve problems in mechanics using Newton's laws, but very often an alternative method that draws on energy considerations is more direct. You should not deceive yourself into thinking that you understand a problem merely because you have seen it solved in class. You must be able to solve the problem and similar problems on your own.

The approach to solving problems should be carefully planned. A systematic plan is especially important when a problem involves several concepts. First, read the problem several times until you are confident you understand what is being asked. Look for any key words that will help you interpret the problem and perhaps allow you to make certain assumptions. Your ability to interpret a question properly is an integral part of problem solving. Second, you should acquire the habit of writing down the information given in a problem and those quantities that need to be found; for example, you might construct a table listing both the quantities given and the quantities to be found. This procedure is sometimes used in the worked examples of the textbook. Finally, after you have decided on the method you believe is appropriate for a given problem, proceed with your solution. The General Problem-Solving Strategy will guide you through complex problems. If you follow the steps of this procedure (*Conceptualize, Categorize, Analyze, Finalize*), you will find it easier to come up with a solution and gain more from your efforts. This Strategy, located at the end of Chapter 2, is used in all worked examples in the remaining chapters so that you can learn how to apply it. Specific problem-solving strategies for certain types of situations are included in the text and appear with a blue heading. These specific strategies follow the outline of the General Problem-Solving Strategy.

Often, students fail to recognize the limitations of certain equations or physical laws in a particular situation. It is very important that you understand and remember the assumptions that underlie a particular theory or formalism. For example, certain equations in kinematics apply only to a particle moving with constant acceleration. These equations are not valid for describing motion whose acceleration is not constant such as the motion of an object connected to a spring or the motion of an object through a fluid. Study the Analysis Models for Problem-Solving in the chapter summaries carefully so that you know how each model can be applied to a specific situation.

Experiments

Physics is a science based on experimental observations. Therefore, we recommend that you try to supplement the text by performing various types of "hands-on" experiments either at home or in the laboratory. These experiments can be used to test ideas and models discussed in class or in the textbook. For example, the common Slinky toy is excellent for studying traveling waves, a ball swinging on the end of a long string can be used to investigate pendulum motion, various masses attached to the end of a vertical spring or rubber band can be used to determine their elastic nature, an old pair of Polaroid sunglasses and some discarded lenses and a magnifying glass are the components of various experiments in optics, and an approximate measure of the free-fall acceleration can be determined simply by measuring with a stopwatch the time it takes for a ball to drop from a known height. The list of such experiments is endless. When physical models are not available, be imaginative and try to develop models of your own.

New Media

We strongly encourage you to use the **ThomsonNOW** Web-based learning system that accompanies this textbook. It is far easier to understand physics if you see it in action, and these new materials will enable you to become a part of that action. **Thomson-NOW** media described in the Preface and accessed at **www.thomsonedu.com/physics/serway** feature a three-step learning process consisting of a pre-test, a personalized learning plan, and a post-test.

It is our sincere hope that you will find physics an exciting and enjoyable experience and that you will benefit from this experience, regardless of your chosen profession. Welcome to the exciting world of physics!

The scientist does not study nature because it is useful; he studies it because he delights in it, and he delights in it because it is beautiful. If nature were not beautiful, it would not be worth knowing, and if nature were not worth knowing, life would not be worth living.

—**Henri Poincaré**

PART 4

Electricity and Magnetism

We now study the branch of physics concerned with electric and magnetic phenomena. The laws of electricity and magnetism play a central role in the operation of such devices as MP3 players, televisions, electric motors, computers, high-energy accelerators, and other electronic devices. More fundamentally, the interatomic and intermolecular forces responsible for the formation of solids and liquids are electric in origin.

Evidence in Chinese documents suggests magnetism was observed as early as 2000 BC. The ancient Greeks observed electric and magnetic phenomena possibly as early as 700 BC. The Greeks knew about magnetic forces from observations that the naturally occurring stone *magnetite* (Fe_3O_4) is attracted to iron. (The word *electric* comes from *elecktron*, the Greek word for "amber." The word *magnetic* comes from *Magnesia*, the name of the district of Greece where magnetite was first found.)

Not until the early part of the nineteenth century did scientists establish that electricity and magnetism are related phenomena. In 1819, Hans Oersted discovered that a compass needle is deflected when placed near a circuit carrying an electric current. In 1831, Michael Faraday and, almost simultaneously, Joseph Henry showed that when a wire is moved near a magnet (or, equivalently, when a magnet is moved near a wire), an electric current is established in the wire. In 1873, James Clerk Maxwell used these observations and other experimental facts as a basis for formulating the laws of electromagnetism as we know them today. (*Electromagnetism* is a name given to the combined study of electricity and magnetism.)

Maxwell's contributions to the field of electromagnetism were especially significant because the laws he formulated are basic to *all* forms of electromagnetic phenomena. His work is as important as Newton's work on the laws of motion and the theory of gravitation.

Lightning is a dramatic example of electrical phenomena occurring in nature. Although we are most familiar with lightning originating from thunderclouds, it can occur in other situations such as in a volcanic eruption (here, the Sakurajima volcano, Japan). (M. Zhilin/M. Newman/Photo Researchers, Inc.)

Mother and daughter are both enjoying the effects of electrically charging their bodies. Each individual hair on their heads becomes charged and exerts a repulsive force on the other hairs, resulting in the "stand-up" hairdos seen here. (Courtesy of Resonance Research Corporation)

23.1 Properties of Electric Charges

23.2 Charging Objects by Induction

23.3 Coulomb's Law

23.4 The Electric Field

23.5 Electric Field of a Continuous Charge Distribution

23.6 Electric Field Lines

23.7 Motion of a Charged Particle in a Uniform Electric Field

23 Electric Fields

The electromagnetic force between charged particles is one of the fundamental forces of nature. We begin this chapter by describing some basic properties of one manifestation of the electromagnetic force, the electric force. We then discuss Coulomb's law, which is the fundamental law governing the electric force between any two charged particles. Next, we introduce the concept of an electric field associated with a charge distribution and describe its effect on other charged particles. We then show how to use Coulomb's law to calculate the electric field for a given charge distribution. The chapter concludes with a discussion of the motion of a charged particle in a uniform electric field.

23.1 Properties of Electric Charges

A number of simple experiments demonstrate the existence of electric forces. For example, after rubbing a balloon on your hair on a dry day, you will find that the balloon attracts bits of paper. The attractive force is often strong enough to suspend the paper from the balloon.

When materials behave in this way, they are said to be *electrified* or to have become **electrically charged**. You can easily electrify your body by vigorously rubbing your shoes on a wool rug. Evidence of the electric charge on your body can be detected by lightly touching (and startling) a friend. Under the right conditions, you will see a spark when you touch and both of you will feel a slight tingle.

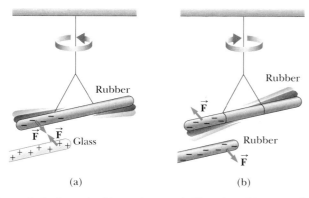

(a)   (b)

Figure 23.1   (a) A negatively charged rubber rod suspended by a thread is attracted to a positively charged glass rod. (b) A negatively charged rubber rod is repelled by another negatively charged rubber rod.

(Experiments such as these work best on a dry day because an excessive amount of moisture in the air can cause any charge you build up to "leak" from your body to the Earth.)

In a series of simple experiments, it was found that there are two kinds of electric charges, which were given the names **positive** and **negative** by Benjamin Franklin (1706–1790). Electrons are identified as having negative charge, while protons are positively charged. To verify that there are two types of charge, suppose a hard rubber rod that has been rubbed on fur is suspended by a sewing thread as shown in Figure 23.1. When a glass rod that has been rubbed on silk is brought near the rubber rod, the two attract each other (Fig. 23.1a). On the other hand, if two charged rubber rods (or two charged glass rods) are brought near each other as shown in Figure 23.1b, the two repel each other. This observation shows that the rubber and glass have two different types of charge on them. On the basis of these observations, we conclude that **charges of the same sign repel one another and charges with opposite signs attract one another**.

Using the convention suggested by Franklin, the electric charge on the glass rod is called positive and that on the rubber rod is called negative. Therefore, any charged object attracted to a charged rubber rod (or repelled by a charged glass rod) must have a positive charge, and any charged object repelled by a charged rubber rod (or attracted to a charged glass rod) must have a negative charge.

Another important aspect of electricity that arises from experimental observations is that **electric charge is always conserved** in an isolated system. That is, when one object is rubbed against another, charge is not created in the process. The electrified state is due to a *transfer* of charge from one object to the other. One object gains some amount of negative charge while the other gains an equal amount of positive charge. For example, when a glass rod is rubbed on silk as in Figure 23.2, the silk obtains a negative charge equal in magnitude to the positive charge on the glass rod. We now know from our understanding of atomic structure that electrons are transferred in the rubbing process from the glass to the silk. Similarly, when rubber is rubbed on fur, electrons are transferred from the fur to the rubber, giving the rubber a net negative charge and the fur a net positive charge. This process is consistent with the fact that neutral, uncharged matter contains as many positive charges (protons within atomic nuclei) as negative charges (electrons).

In 1909, Robert Millikan (1868–1953) discovered that electric charge always occurs as integral multiples of a fundamental amount of charge e (see Section 25.7). In modern terms, the electric charge q is said to be **quantized,** where q is the standard symbol used for charge as a variable. That is, electric charge exists as discrete "packets," and we can write $q = \pm Ne$, where N is some integer. Other experiments in the same period showed that the electron has a charge $-e$ and the proton has a charge of equal magnitude but opposite sign $+e$. Some particles, such as the neutron, have no charge.

◄ Electric charge is conserved

Figure 23.2   When a glass rod is rubbed with silk, electrons are transferred from the glass to the silk. Because of conservation of charge, each electron adds negative charge to the silk and an equal positive charge is left behind on the rod. Also, because the charges are transferred in discrete bundles, the charges on the two objects are $\pm e$, or $\pm 2e$, or $\pm 3e$, and so on.

Quick Quiz 23.1 Three objects are brought close to each other, two at a time. When objects A and B are brought together, they repel. When objects B and C are brought together, they also repel. Which of the following are true? (a) Objects A and C possess charges of the same sign. (b) Objects A and C possess charges of opposite sign. (c) All three objects possess charges of the same sign. (d) One object is neutral. (e) Additional experiments must be performed to determine the signs of the charges.

23.2 Charging Objects by Induction

It is convenient to classify materials in terms of the ability of electrons to move through the material:

> Electrical **conductors** are materials in which some of the electrons are free electrons[1] that are not bound to atoms and can move relatively freely through the material; electrical **insulators** are materials in which all electrons are bound to atoms and cannot move freely through the material.

Materials such as glass, rubber, and dry wood fall into the category of electrical insulators. When such materials are charged by rubbing, only the area rubbed becomes charged and the charged particles are unable to move to other regions of the material.

In contrast, materials such as copper, aluminum, and silver are good electrical conductors. When such materials are charged in some small region, the charge readily distributes itself over the entire surface of the material.

Semiconductors are a third class of materials, and their electrical properties are somewhere between those of insulators and those of conductors. Silicon and germanium are well-known examples of semiconductors commonly used in the fabrication of a variety of electronic chips used in computers, cellular telephones, and stereo systems. The electrical properties of semiconductors can be changed over many orders of magnitude by the addition of controlled amounts of certain atoms to the materials.

To understand how to charge a conductor by a process known as **induction,** consider a neutral (uncharged) conducting sphere insulated from the ground as shown in Figure 23.3a. There are an equal number of electrons and protons in the sphere if the charge on the sphere is exactly zero. When a negatively charged rubber rod is brought near the sphere, electrons in the region nearest the rod experience a repulsive force and migrate to the opposite side of the sphere. This migration leaves the side of the sphere near the rod with an effective positive charge because of the diminished number of electrons as in Figure 23.3b. (The left side of the sphere in Figure 23.3b is positively charged *as if* positive charges moved into this region, but remember that it is only electrons that are free to move.) This process occurs even if the rod never actually touches the sphere. If the same experiment is performed with a conducting wire connected from the sphere to the Earth (Fig. 23.3c), some of the electrons in the conductor are so strongly repelled by the presence of the negative charge in the rod that they move out of the sphere through the wire and into the Earth. The symbol ⏚ at the end of the wire in Figure 23.3c indicates that the wire is connected to **ground,** which means a reservoir, such as the Earth, that can accept or provide electrons freely with negligible effect on its electrical characteristics. If the wire to ground is then removed

Figure 23.3 Charging a metallic object by *induction* (that is, the two objects never touch each other). (a) A neutral metallic sphere, with equal numbers of positive and negative charges. (b) The electrons on the neutral sphere are redistributed when a charged rubber rod is placed near the sphere. (c) When the sphere is grounded, some of its electrons leave through the ground wire. (d) When the ground connection is removed, the sphere has excess positive charge that is nonuniformly distributed. (e) When the rod is removed, the remaining electrons redistribute uniformly and there is a net uniform distribution of positive charge on the sphere.

---

[1] A metal atom contains one or more outer electrons, which are weakly bound to the nucleus. When many atoms combine to form a metal, the *free electrons* are these outer electrons, which are not bound to any one atom. These electrons move about the metal in a manner similar to that of gas molecules moving in a container.

(a)   (b)

Figure 23.4   (a) The charged object on the left induces a charge distribution on the surface of an insulator due to realignment of charges in the molecules. (b) A charged comb attracts bits of paper because charges in molecules in the paper are realigned.

(Fig. 23.3d), the conducting sphere contains an excess of *induced* positive charge because it has fewer electrons than it needs to cancel out the positive charge of the protons. When the rubber rod is removed from the vicinity of the sphere (Fig. 23.3e), this induced positive charge remains on the ungrounded sphere. Notice that the rubber rod loses none of its negative charge during this process.

Charging an object by induction requires no contact with the object inducing the charge. That is in contrast to charging an object by rubbing (that is, by *conduction*), which does require contact between the two objects.

A process similar to induction in conductors takes place in insulators. In most neutral molecules, the center of positive charge coincides with the center of negative charge. In the presence of a charged object, however, these centers inside each molecule in an insulator may shift slightly, resulting in more positive charge on one side of the molecule than on the other. This realignment of charge within individual molecules produces a layer of charge on the surface of the insulator as shown in Figure 23.4a. Your knowledge of induction in insulators should help you explain why a comb that has been drawn through your hair attracts bits of electrically neutral paper as shown in Figure 23.4b.

Quick Quiz 23.2   Three objects are brought close to one another, two at a time. When objects A and B are brought together, they attract. When objects B and C are brought together, they repel. Which of the following are necessarily true? (a) Objects A and C possess charges of the same sign. (b) Objects A and C possess charges of opposite sign. (c) All three objects possess charges of the same sign. (d) One object is neutral. (e) Additional experiments must be performed to determine information about the charges on the objects.

23.3   Coulomb's Law

Charles Coulomb measured the magnitudes of the electric forces between charged objects using the torsion balance, which he invented (Fig. 23.5). The operating principle of the torsion balance is the same as that of the apparatus used by Cavendish to measure the gravitational constant (see Section 13.1), with the electrically neutral spheres replaced by charged ones. The electric force between charged spheres A and B in Figure 23.5 causes the spheres to either attract or repel each other, and the resulting motion causes the suspended fiber to twist. Because the restoring torque of the twisted fiber is proportional to the angle through which the fiber rotates, a measurement of this angle provides a quantitative measure of the electric force of attraction or repulsion. Once the spheres are charged by rubbing, the electric force between them is very large compared with the gravitational attraction, and so the gravitational force can be neglected.

Figure 23.5   Coulomb's torsion balance, used to establish the inverse-square law for the electric force between two charges.

From Coulomb's experiments, we can generalize the properties of the **electric force** between two stationary charged particles. We use the term **point charge** to refer to a charged particle of zero size. The electrical behavior of electrons and protons is very well described by modeling them as point charges. From experimental observations, we find that the magnitude of the electric force (sometimes called the *Coulomb force*) between two point charges is given by **Coulomb's law**:

Coulomb's law ▶

$$F_e = k_e \frac{|q_1||q_2|}{r^2}$$ (23.1)

where k_e is a constant called the **Coulomb constant**. In his experiments, Coulomb was able to show that the value of the exponent of r was 2 to within an uncertainty of a few percent. Modern experiments have shown that the exponent is 2 to within an uncertainty of a few parts in 10^{16}. Experiments also show that the electric force, like the gravitational force, is conservative.

The value of the Coulomb constant depends on the choice of units. The SI unit of charge is the **coulomb** (C). The Coulomb constant k_e in SI units has the value

Coulomb constant ▶

$$k_e = 8.987\ 6 \times 10^9\ \text{N} \cdot \text{m}^2/\text{C}^2$$ (23.2)

This constant is also written in the form

$$k_e = \frac{1}{4\pi\epsilon_0}$$ (23.3)

where the constant ϵ_0 (Greek letter epsilon) is known as the **permittivity of free space** and has the value

$$\epsilon_0 = 8.854\ 2 \times 10^{-12}\ \text{C}^2/\text{N} \cdot \text{m}^2$$ (23.4)

The smallest unit of free charge e known in nature,[2] the charge on an electron $(-e)$ or a proton $(+e)$, has a magnitude

$$e = 1.602\ 18 \times 10^{-19}\ \text{C}$$ (23.5)

Therefore, 1 C of charge is approximately equal to the charge of 6.24×10^{18} electrons or protons. This number is very small when compared with the number of free electrons in 1 cm^3 of copper, which is on the order of 10^{23}. Nevertheless, 1 C is a substantial amount of charge. In typical experiments in which a rubber or glass rod is charged by friction, a net charge on the order of 10^{-6} C is obtained. In other words, only a very small fraction of the total available charge is transferred between the rod and the rubbing material.

The charges and masses of the electron, proton, and neutron are given in Table 23.1.

CHARLES COULOMB
French physicist (1736–1806)
Coulomb's major contributions to science were in the areas of electrostatics and magnetism. During his lifetime, he also investigated the strengths of materials and determined the forces that affect objects on beams, thereby contributing to the field of structural mechanics. In the field of ergonomics, his research provided a fundamental understanding of the ways in which people and animals can best do work.

TABLE 23.1

Charge and Mass of the Electron, Proton, and Neutron

Particle	Charge (C)	Mass (kg)
Electron (e)	$-1.602\ 176\ 5 \times 10^{-19}$	$9.109\ 4 \times 10^{-31}$
Proton (p)	$+1.602\ 176\ 5 \times 10^{-19}$	$1.672\ 62 \times 10^{-27}$
Neutron (n)	0	$1.674\ 93 \times 10^{-27}$

[2] No unit of charge smaller than e has been detected on a free particle; current theories, however, propose the existence of particles called *quarks* having charges $-e/3$ and $2e/3$. Although there is considerable experimental evidence for such particles inside nuclear matter, *free* quarks have never been detected. We discuss other properties of quarks in Chapter 46.

EXAMPLE 23.1   **The Hydrogen Atom**

The electron and proton of a hydrogen atom are separated (on the average) by a distance of approximately 5.3×10^{-11} m. Find the magnitudes of the electric force and the gravitational force between the two particles.

SOLUTION

Conceptualize   Think about the two particles separated by the very small distance given in the problem statement. In Chapter 13, we found the gravitational force between small objects to be weak, so we expect the gravitational force between the electron and proton to be significantly smaller than the electric force.

Categorize   The electric and gravitational forces will be evaluated from universal force laws, so we categorize this example as a substitution problem.

Use Coulomb's law to find the magnitude of the electric force:

$$F_e = k_e \frac{|e||-e|}{r^2} = (8.99 \times 10^9 \text{ N} \cdot \text{m}^2/\text{C}^2) \frac{(1.60 \times 10^{-19} \text{ C})^2}{(5.3 \times 10^{-11} \text{ m})^2}$$

$$= \boxed{8.2 \times 10^{-8} \text{ N}}$$

Use Newton's law of universal gravitation and Table 23.1 (for the particle masses) to find the magnitude of the gravitational force:

$$F_g = G \frac{m_e m_p}{r^2}$$

$$= (6.67 \times 10^{-11} \text{ N} \cdot \text{m}^2/\text{kg}^2) \frac{(9.11 \times 10^{-31} \text{ kg})(1.67 \times 10^{-27} \text{ kg})}{(5.3 \times 10^{-11} \text{ m})^2}$$

$$= \boxed{3.6 \times 10^{-47} \text{ N}}$$

The ratio $F_e/F_g \approx 2 \times 10^{39}$. Therefore, the gravitational force between charged atomic particles is negligible when compared with the electric force. Notice the similar forms of Newton's law of universal gravitation and Coulomb's law of electric forces. Other than magnitude, what is a fundamental difference between the two forces?

When dealing with Coulomb's law, remember that force is a vector quantity and must be treated accordingly. Coulomb's law expressed in vector form for the electric force exerted by a charge q_1 on a second charge q_2, written $\vec{\mathbf{F}}_{12}$, is

$$\vec{\mathbf{F}}_{12} = k_e \frac{q_1 q_2}{r^2} \hat{\mathbf{r}}_{12} \qquad (23.6)$$

◄ Vector form of Coulomb's law

where $\hat{\mathbf{r}}_{12}$ is a unit vector directed from q_1 toward q_2 as shown in Active Figure 23.6a. Because the electric force obeys Newton's third law, the electric force exerted by q_2 on q_1 is equal in magnitude to the force exerted by q_1 on q_2 and in

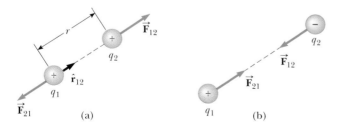

ACTIVE FIGURE 23.6

Two point charges separated by a distance r exert a force on each other that is given by Coulomb's law. The force $\vec{\mathbf{F}}_{21}$ exerted by q_2 on q_1 is equal in magnitude and opposite in direction to the force $\vec{\mathbf{F}}_{12}$ exerted by q_1 on q_2. (a) When the charges are of the same sign, the force is repulsive. (b) When the charges are of opposite signs, the force is attractive.

Sign in at www.thomsonedu.com and go to ThomsonNOW to move the charges to any position in two-dimensional space and observe the electric forces on them.

the opposite direction; that is, $\vec{\mathbf{F}}_{21} = -\vec{\mathbf{F}}_{12}$. Finally, Equation 23.6 shows that if q_1 and q_2 have the same sign as in Active Figure 23.6a, the product $q_1 q_2$ is positive. If q_1 and q_2 are of opposite sign as shown in Active Figure 23.6b, the product $q_1 q_2$ is negative. These signs describe the *relative* direction of the force but not the *absolute* direction. A negative product indicates an attractive force, and each charge experiences a force toward the other. A positive product indicates a repulsive force such that each charge experiences a force away from the other. The *absolute* direction of the force on a charge depends on the location of the other charge. For example, if an x axis lies along the two charges in Active Figure 23.6a, the product $q_1 q_2$ is positive, but $\vec{\mathbf{F}}_{12}$ points in the $+x$ direction and $\vec{\mathbf{F}}_{21}$ points in the $-x$ direction.

When more than two charges are present, the force between any pair of them is given by Equation 23.6. Therefore, the resultant force on any one of them equals the vector sum of the forces exerted by the other individual charges. For example, if four charges are present, the resultant force exerted by particles 2, 3, and 4 on particle 1 is

$$\vec{\mathbf{F}}_1 = \vec{\mathbf{F}}_{21} + \vec{\mathbf{F}}_{31} + \vec{\mathbf{F}}_{41}$$

Quick Quiz 23.3 Object A has a charge of $+2$ μC, and object B has a charge of $+6$ μC. Which statement is true about the electric forces on the objects? (a) $\vec{\mathbf{F}}_{AB} = -3\vec{\mathbf{F}}_{BA}$ (b) $\vec{\mathbf{F}}_{AB} = -\vec{\mathbf{F}}_{BA}$ (c) $3\vec{\mathbf{F}}_{AB} = -\vec{\mathbf{F}}_{BA}$ (d) $\vec{\mathbf{F}}_{AB} = 3\vec{\mathbf{F}}_{BA}$ (e) $\vec{\mathbf{F}}_{AB} = \vec{\mathbf{F}}_{BA}$ (f) $3\vec{\mathbf{F}}_{AB} = \vec{\mathbf{F}}_{BA}$

EXAMPLE 23.2 | Find the Resultant Force

Consider three point charges located at the corners of a right triangle as shown in Figure 23.7, where $q_1 = q_3 = 5.0$ μC, $q_2 = -2.0$ μC, and $a = 0.10$ m. Find the resultant force exerted on q_3.

SOLUTION

Conceptualize Think about the net force on q_3. Because charge q_3 is near two other charges, it will experience two electric forces.

Categorize Because two forces are exerted on charge q_3, we categorize this example as a vector addition problem.

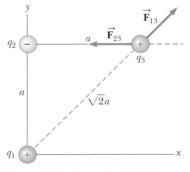

Figure 23.7 (Example 23.2) The force exerted by q_1 on q_3 is $\vec{\mathbf{F}}_{13}$. The force exerted by q_2 on q_3 is $\vec{\mathbf{F}}_{23}$. The resultant force $\vec{\mathbf{F}}_3$ exerted on q_3 is the vector sum $\vec{\mathbf{F}}_{13} + \vec{\mathbf{F}}_{23}$.

Analyze The directions of the individual forces exerted by q_1 and q_2 on q_3 are shown in Figure 23.7. The force $\vec{\mathbf{F}}_{23}$ exerted by q_2 on q_3 is attractive because q_2 and q_3 have opposite signs. In the coordinate system shown in Figure 23.7, the attractive force $\vec{\mathbf{F}}_{23}$ is to the left (in the negative x direction).

The force $\vec{\mathbf{F}}_{13}$ exerted by q_1 on q_3 is repulsive because both charges are positive. The repulsive force $\vec{\mathbf{F}}_{13}$ makes an angle of 45° with the x axis.

Use Equation 23.1 to find the magnitude of $\vec{\mathbf{F}}_{23}$:

$$F_{23} = k_e \frac{|q_2||q_3|}{a^2}$$

$$= (8.99 \times 10^9 \text{ N} \cdot \text{m}^2/\text{C}^2)\frac{(2.0 \times 10^{-6} \text{ C})(5.0 \times 10^{-6} \text{ C})}{(0.10 \text{ m})^2} = 9.0 \text{ N}$$

Find the magnitude of the force $\vec{\mathbf{F}}_{13}$:

$$F_{13} = k_e \frac{|q_1||q_3|}{(\sqrt{2}a)^2}$$

$$= (8.99 \times 10^9 \text{ N} \cdot \text{m}^2/\text{C}^2)\frac{(5.0 \times 10^{-6} \text{ C})(5.0 \times 10^{-6} \text{ C})}{2(0.10 \text{ m})^2} = 11 \text{ N}$$

Find the x and y components of the force $\vec{\mathbf{F}}_{13}$:

$$F_{13x} = F_{13}\cos 45° = 7.9\text{ N}$$

$$F_{13y} = F_{13}\sin 45° = 7.9\text{ N}$$

Find the components of the resultant force acting on q_3:

$$F_{3x} = F_{13x} + F_{23x} = 7.9\text{ N} + (-9.0\text{ N}) = -1.1\text{ N}$$

$$F_{3y} = F_{13y} + F_{23y} = 7.9\text{ N} + 0 = 7.9\text{ N}$$

Express the resultant force acting on q_3 in unit–vector form:

$$\vec{\mathbf{F}}_3 = \boxed{(-1.1\hat{\mathbf{i}} + 7.9\hat{\mathbf{j}})\text{ N}}$$

Finalize　The net force on q_3 is upward and toward the left in Figure 23.7. If q_3 moves in response to the net force, the distances between q_3 and the other charges change, so the net force changes. Therefore, q_3 can be modeled as a particle under a net force as long as it is recognized that the force exerted on q_3 is *not* constant.

What If?　What if the signs of all three charges were changed to the opposite signs? How would that affect the result for $\vec{\mathbf{F}}_3$?

Answer　The charge q_3 would still be attracted toward q_2 and repelled from q_1 with forces of the same magnitude. Therefore, the final result for $\vec{\mathbf{F}}_3$ would be the same.

EXAMPLE 23.3　　Where Is the Net Force Zero?

Three point charges lie along the x axis as shown in Figure 23.8. The positive charge $q_1 = 15.0\ \mu\text{C}$ is at $x = 2.00$ m, the positive charge $q_2 = 6.00\ \mu\text{C}$ is at the origin, and the net force acting on q_3 is zero. What is the x coordinate of q_3?

SOLUTION

Conceptualize　Because q_3 is near two other charges, it experiences two electric forces. Unlike the preceding example, however, the forces lie along the same line in this problem as indicated in Figure 23.8. Because q_3 is negative while q_1 and q_2 are positive, the forces $\vec{\mathbf{F}}_{13}$ and $\vec{\mathbf{F}}_{23}$ are both attractive.

Categorize　Because the net force on q_3 is zero, we model the point charge as a particle in equilibrium.

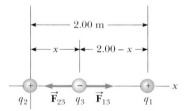

Figure 23.8　(Example 23.3) Three point charges are placed along the x axis. If the resultant force acting on q_3 is zero, the force $\vec{\mathbf{F}}_{13}$ exerted by q_1 on q_3 must be equal in magnitude and opposite in direction to the force $\vec{\mathbf{F}}_{23}$ exerted by q_2 on q_3.

Analyze　Write an expression for the net force on charge q_3 when it is in equilibrium:

$$\vec{\mathbf{F}}_3 = \vec{\mathbf{F}}_{23} + \vec{\mathbf{F}}_{13} = -k_e\frac{|q_2||q_3|}{x^2}\hat{\mathbf{i}} + k_e\frac{|q_1||q_3|}{(2.00-x)^2}\hat{\mathbf{i}} = 0$$

Move the second term to the right side of the equation and set the coefficients of the unit vector $\hat{\mathbf{i}}$ equal:

$$k_e\frac{|q_2||q_3|}{x^2} = k_e\frac{|q_1||q_3|}{(2.00-x)^2}$$

Eliminate k_e and $|q_3|$ and rearrange the equation:

$$(2.00-x)^2|q_2| = x^2|q_1|$$

$$(4.00 - 4.00x + x^2)(6.00\times 10^{-6}\text{ C}) = x^2(15.0\times 10^{-6}\text{ C})$$

Reduce the quadratic equation to a simpler form:

$$3.00x^2 + 8.00x - 8.00 = 0$$

Solve the quadratic equation for the positive root:

$$x = \boxed{0.775\text{ m}}$$

Finalize　The second root to the quadratic equation is $x = -3.44$ m. That is another location where the *magnitudes* of the forces on q_3 are equal, but both forces are in the same direction.

What If? Suppose q_3 is constrained to move only along the x axis. From its initial position at $x = 0.775$ m, it is pulled a small distance along the x axis. When released, does it return to equilibrium, or is it pulled further from equilibrium? That is, is the equilibrium stable or unstable?

Answer   If q_3 is moved to the right, $\vec{\mathbf{F}}_{13}$ becomes larger and $\vec{\mathbf{F}}_{23}$ becomes smaller. The result is a net force to the right, in the same direction as the displacement. Therefore, the charge q_3 would continue to move to the right and the equilibrium is *unstable*. (See Section 7.9 for a review of stable and unstable equilibrium.)

If q_3 is constrained to stay at a *fixed x* coordinate but allowed to move up and down in Figure 23.8, the equilibrium is stable. In this case, if the charge is pulled upward (or downward) and released, it moves back toward the equilibrium position and oscillates about this point.

EXAMPLE 23.4   Find the Charge on the Spheres

Two identical small charged spheres, each having a mass of 3.0×10^{-2} kg, hang in equilibrium as shown in Figure 23.9a. The length of each string is 0.15 m, and the angle θ is 5.0°. Find the magnitude of the charge on each sphere.

SOLUTION

Conceptualize Figure 23.9a helps us conceptualize this example. The two spheres exert repulsive forces on each other. If they are held close to each other and released, they move outward from the center and settle into the configuration in Figure 23.9a after the oscillations have vanished due to air resistance.

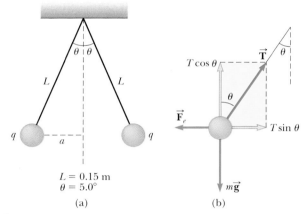

Figure 23.9 (Example 23.4) (a) Two identical spheres, each carrying the same charge q, suspended in equilibrium. (b) The free-body diagram for the sphere on the left of part (a).

Categorize The key phrase "in equilibrium" helps us model each sphere as a particle in equilibrium. This example is similar to the particle in equilibrium problems in Chapter 5 with the added feature that one of the forces on a sphere is an electric force.

Analyze The free-body diagram for the left-hand sphere is shown in Figure 23.9b. The sphere is in equilibrium under the application of the forces $\vec{\mathbf{T}}$ from the string, the electric force $\vec{\mathbf{F}}_e$ from the other sphere, and the gravitational force $m\vec{\mathbf{g}}$.

Write Newton's second law for the left-hand sphere in component form:

(1) $\quad \sum F_x = T \sin \theta - F_e = 0 \quad \rightarrow \quad T \sin \theta = F_e$

(2) $\quad \sum F_y = T \cos \theta - mg = 0 \quad \rightarrow \quad T \cos \theta = mg$

Divide Equation (1) by Equation (2) to find F_e:

$$\tan \theta = \frac{F_e}{mg} \quad \rightarrow \quad F_e = mg \tan \theta$$

Evaluate the electric force numerically:

$$F_e = (3.0 \times 10^{-2} \text{ kg})(9.80 \text{ m/s}^2) \tan (5.0°) = 2.6 \times 10^{-2} \text{ N}$$

Use the geometry of the right triangle in Figure 23.9a to find a relationship between a, L, and θ:

$$\sin \theta = \frac{a}{L} \quad \rightarrow \quad a = L \sin \theta$$

Evaluate a:

$$a = (0.15 \text{ m}) \sin (5.0°) = 0.013 \text{ m}$$

Solve Coulomb's law (Eq. 23.1) for the charge $|q|$ on each sphere:

$$F_e = k_e \frac{|q|^2}{r^2} \quad \rightarrow \quad |q| = \sqrt{\frac{F_e r^2}{k_e}} = \sqrt{\frac{F_e (2a)^2}{k_e}}$$

Substitute numerical values:
$$|q| = \sqrt{\frac{(2.6 \times 10^{-2}\,\text{N})[2(0.013\,\text{m})]^2}{8.99 \times 10^9\,\text{N} \cdot \text{m}^2/\text{C}^2}} = 4.4 \times 10^{-8}\,\text{C}$$

Finalize   We cannot determine the sign of the charge from the information given. In fact, the sign of the charge is not important. The situation is the same whether both spheres are positively charged or negatively charged.

What If?   Suppose your roommate proposes solving this problem without the assumption that the charges are of equal magnitude. She claims the symmetry of the problem is destroyed if the charges are not equal, so the strings would make two different angles with the vertical and the problem would be much more complicated. How would you respond?

Answer   The symmetry is not destroyed and the angles are not different. Newton's third law requires the magnitudes of the electric forces on the two charges to be the same, regardless of the equality or nonequality of the charges. The solution to the example remains the same with one change: the value of $|q|^2$ in the solution is replaced by $|q_1 q_2|$ in the new situation, where q_1 and q_2 are the values of the charges on the two spheres. The symmetry of the problem would be destroyed if the *masses* of the spheres were not the same. In this case, the strings would make different angles with the vertical and the problem would be more complicated.

23.4   The Electric Field

Two field forces—the gravitational force in Chapter 13 and the electric force here—have been introduced into our discussions so far. As pointed out earlier, field forces can act through space, producing an effect even when no physical contact occurs between interacting objects. The gravitational field $\vec{\mathbf{g}}$ at a point in space due to a source particle was defined in Section 13.4 to be equal to the gravitational force $\vec{\mathbf{F}}_g$ acting on a test particle of mass m divided by that mass: $\vec{\mathbf{g}} \equiv \vec{\mathbf{F}}_g/m$. The concept of a field was developed by Michael Faraday (1791–1867) in the context of electric forces and is of such practical value that we shall devote much attention to it in the next several chapters. In this approach, an **electric field** is said to exist in the region of space around a charged object, the **source charge**. When another charged object—the **test charge**—enters this electric field, an electric force acts on it. As an example, consider Figure 23.10, which shows a small positive test charge q_0 placed near a second object carrying a much greater positive charge Q. We define the electric field due to the source charge at the location of the test charge to be the electric force on the test charge *per unit charge*, or, to be more specific, **the electric field vector $\vec{\mathbf{E}}$** at a point in space is defined as the electric force $\vec{\mathbf{F}}_e$ acting on a positive test charge q_0 placed at that point divided by the test charge:[3]

© Johnny Autery

This dramatic photograph captures a lightning bolt striking a tree near some rural homes. Lightning is associated with very strong electric fields in the atmosphere.

$$\vec{\mathbf{E}} \equiv \frac{\vec{\mathbf{F}}_e}{q_0} \tag{23.7}$$

◀ Definition of electric field

The vector $\vec{\mathbf{E}}$ has the SI units of newtons per coulomb (N/C). Note that $\vec{\mathbf{E}}$ is the field produced by some charge or charge distribution *separate from* the test charge; it is not the field produced by the test charge itself. Also note that the existence of an electric field is a property of its source; the presence of the test charge is not necessary for the field to exist. The test charge serves as a *detector* of the electric field.

The direction of $\vec{\mathbf{E}}$ as shown in Figure 23.10 is the direction of the force a positive test charge experiences when placed in the field. **An electric field exists at a point if a test charge at that point experiences an electric force.**

Figure 23.10   A small positive test charge q_0 placed at point P near an object carrying a much larger positive charge Q experiences an electric field $\vec{\mathbf{E}}$ at point P established by the source charge Q.

[3] When using Equation 23.7, we must assume the test charge q_0 is small enough that it does not disturb the charge distribution responsible for the electric field. If the test charge is great enough, the charge on the metallic sphere is redistributed and the electric field it sets up is different from the field it sets up in the presence of the much smaller test charge.

Equation 23.7 can be rearranged as

$$\vec{\mathbf{F}}_e = q\vec{\mathbf{E}} \tag{23.8}$$

This equation gives us the force on a charged particle q placed in an electric field. If q is positive, the force is in the same direction as the field. If q is negative, the force and the field are in opposite directions. Notice the similarity between Equation 23.8 and the corresponding equation for a particle with mass placed in a gravitational field, $\vec{\mathbf{F}}_g = m\vec{\mathbf{g}}$ (Section 5.5). Once the magnitude and direction of the electric field are known at some point, the electric force exerted on *any* charged particle placed at that point can be calculated from Equation 23.8.

To determine the direction of an electric field, consider a point charge q as a source charge. This charge creates an electric field at all points in space surrounding it. A test charge q_0 is placed at point P, a distance r from the source charge, as in Active Figure 23.11a. We imagine using the test charge to determine the direction of the electric force and therefore that of the electric field. According to Coulomb's law, the force exerted by q on the test charge is

$$\vec{\mathbf{F}}_e = k_e \frac{qq_0}{r^2}\hat{\mathbf{r}}$$

where $\hat{\mathbf{r}}$ is a unit vector directed from q toward q_0. This force in Active Figure 23.11a is directed away from the source charge q. Because the electric field at P, the position of the test charge, is defined by $\vec{\mathbf{E}} = \vec{\mathbf{F}}_e/q_0$, the electric field at P created by q is

$$\vec{\mathbf{E}} = k_e \frac{q}{r^2}\hat{\mathbf{r}} \tag{23.9}$$

If the source charge q is positive, Active Figure 23.11b shows the situation with the test charge removed: the source charge sets up an electric field at P, directed away from q. If q is negative, as in Active Figure 23.11c, the force on the test charge is toward the source charge, so the electric field at P is directed toward the source charge as in Active Figure 23.11d.

To calculate the electric field at a point P due to a group of point charges, we first calculate the electric field vectors at P individually using Equation 23.9 and then add them vectorially. In other words, at any point P, the total electric field due to a group of source charges equals the vector sum of the electric fields of all the charges. This superposition principle applied to fields follows directly from the vector addition of electric forces. Therefore, the electric field at point P due to a group of source charges can be expressed as the vector sum

Electric field due to a finite number of point charges ▶

$$\vec{\mathbf{E}} = k_e \sum_i \frac{q_i}{r_i^2}\hat{\mathbf{r}}_i \tag{23.10}$$

ACTIVE FIGURE 23.11

A test charge q_0 at point P is a distance r from a point charge q. (a) If q is positive, the force on the test charge is directed away from q. (b) For the positive source charge, the electric field at P points radially outward from q. (c) If q is negative, the force on the test charge is directed toward q. (d) For the negative source charge, the electric field at P points radially inward toward q.

Sign in at www.thomsonedu.com and go to ThomsonNOW to move point P to any position in two-dimensional space and observe the electric field due to q.

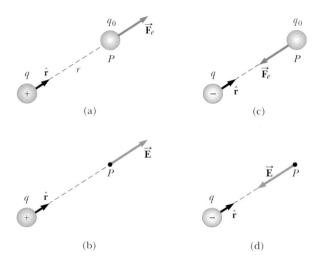

where r_i is the distance from the ith source charge q_i to the point P and $\hat{\mathbf{r}}_i$ is a unit vector directed from q_i toward P.

In Example 23.5, we explore the electric field due to two charges using the superposition principle. Part (B) of the example focuses on an **electric dipole**, which is defined as a positive charge q and a negative charge $-q$ separated by a distance $2a$. The electric dipole is a good model of many molecules, such as hydrochloric acid (HCl). Neutral atoms and molecules behave as dipoles when placed in an external electric field. Furthermore, many molecules, such as HCl, are permanent dipoles. The effect of such dipoles on the behavior of materials subjected to electric fields is discussed in Chapter 26.

Quick Quiz 23.4   A test charge of $+3 \ \mu C$ is at a point P where an external electric field is directed to the right and has a magnitude of 4×10^6 N/C. If the test charge is replaced with another test charge of $-3 \ \mu C$, what happens to the external electric field at P? (a) It is unaffected. (b) It reverses direction. (c) It changes in a way that cannot be determined.

EXAMPLE 23.5	**Electric Field Due to Two Charges**

Charges q_1 and q_2 are located on the x axis, at distances a and b, respectively, from the origin as shown in Figure 23.12.

(A) Find the components of the net electric field at the point P, which is on the y axis.

SOLUTION

Conceptualize   Compare this example to Example 23.2. There, we add vector forces to find the net force on a charged particle. Here, we add electric field vectors to find the net electric field at a point in space.

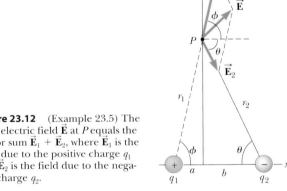

Figure 23.12   (Example 23.5) The total electric field $\vec{\mathbf{E}}$ at P equals the vector sum $\vec{\mathbf{E}}_1 + \vec{\mathbf{E}}_2$, where $\vec{\mathbf{E}}_1$ is the field due to the positive charge q_1 and $\vec{\mathbf{E}}_2$ is the field due to the negative charge q_2.

Categorize   We have two source charges and wish to find the resultant electric field, so we categorize this example as one in which we can use the superposition principle represented by Equation 23.10.

Analyze   Find the magnitude of the electric field at P due to charge q_1:

$$E_1 = k_e \frac{|q_1|}{r_1^{\,2}} = k_e \frac{|q_1|}{(a^2 + y^2)}$$

Find the magnitude of the electric field at P due to charge q_2:

$$E_2 = k_e \frac{|q_2|}{r_2^{\,2}} = k_e \frac{|q_2|}{(b^2 + y^2)}$$

Write the electric field vectors for each charge in unit–vector form:

$$\vec{\mathbf{E}}_1 = k_e \frac{|q_1|}{(a^2 + y^2)} \cos \phi \, \hat{\mathbf{i}} + k_e \frac{|q_1|}{(a^2 + y^2)} \sin \phi \, \hat{\mathbf{j}}$$

$$\vec{\mathbf{E}}_2 = k_e \frac{|q_2|}{(b^2 + y^2)} \cos \theta \, \hat{\mathbf{i}} - k_e \frac{|q_2|}{(b^2 + y^2)} \sin \theta \, \hat{\mathbf{j}}$$

Write the components of the net electric field vector:

$$(1) \quad E_x = E_{1x} + E_{2x} = k_e \frac{|q_1|}{(a^2 + y^2)} \cos \phi + k_e \frac{|q_2|}{(b^2 + y^2)} \cos \theta$$

$$(2) \quad E_y = E_{1y} + E_{2y} = k_e \frac{|q_1|}{(a^2 + y^2)} \sin \phi - k_e \frac{|q_2|}{(b^2 + y^2)} \sin \theta$$

(B) Evaluate the electric field at point P in the special case that $|q_1| = |q_2|$ and $a = b$.

SOLUTION

Conceptualize Figure 23.13 shows the situation in this special case. Notice the symmetry in the situation and that the charge distribution is now an electric dipole.

Categorize Because Figure 23.13 is a special case of the general case shown in Figure 23.12, we can categorize this example as one in which we can take the result of part (A) and substitute the appropriate values of the variables.

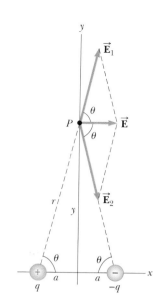

Figure 23.13 (Example 23.5) When the charges in Figure 23.12 are of equal magnitude and equidistant from the origin, the situation becomes symmetric as shown here.

Analyze Based on the symmetry in Figure 23.13, evaluate Equations (1) and (2) from part (A) with $a = b$, $|q_1| = |q_2| = q$, and $\phi = \theta$:

$$(3) \quad E_x = k_e \frac{q}{(a^2 + y^2)} \cos \theta + k_e \frac{q}{(a^2 + y^2)} \cos \theta = 2k_e \frac{q}{(a^2 + y^2)} \cos \theta$$

$$E_y = k_e \frac{q}{(a^2 + y^2)} \sin \theta - k_e \frac{q}{(a^2 + y^2)} \sin \theta = 0$$

From the geometry in Figure 23.13, evaluate $\cos \theta$:

$$(4) \quad \cos \theta = \frac{a}{r} = \frac{a}{(a^2 + y^2)^{1/2}}$$

Substitute Equation (4) into Equation (3):

$$E_x = 2k_e \frac{q}{(a^2 + y^2)} \frac{a}{(a^2 + y^2)^{1/2}} = k_e \frac{2qa}{(a^2 + y^2)^{3/2}}$$

(C) Find the electric field due to the electric dipole when point P is a distance $y \gg a$ from the origin.

SOLUTION

In the solution to part (B), because $y \gg a$, neglect a^2 compared with y^2 and write the expression for E in this case:

$$(5) \quad E \approx k_e \frac{2qa}{y^3}$$

Finalize From Equation (5), we see that at points far from a dipole but along the perpendicular bisector of the line joining the two charges, the magnitude of the electric field created by the dipole varies as $1/r^3$, whereas the more slowly varying field of a point charge varies as $1/r^2$ (see Eq. 23.9). That is because at distant points, the fields of the two charges of equal magnitude and opposite sign almost cancel each other. The $1/r^3$ variation in E for the dipole also is obtained for a distant point along the x axis (see Problem 18) and for any general distant point.

23.5 Electric Field of a Continuous Charge Distribution

Very often, the distances between charges in a group of charges are much smaller than the distance from the group to a point where the electric field is to be calculated. In such situations, the system of charges can be modeled as continuous. That is, the system of closely spaced charges is equivalent to a total charge that is continuously distributed along some line, over some surface, or throughout some volume.

To set up the process for evaluating the electric field created by a continuous charge distribution, let's use the following procedure. First, divide the charge distribution into small elements, each of which contains a small charge Δq as shown in Figure 23.14. Next, use Equation 23.9 to calculate the electric field due to one of these elements at a point P. Finally, evaluate the total electric field at P due to the charge distribution by summing the contributions of all the charge elements (that is, by applying the superposition principle).

The electric field at P due to one charge element carrying charge Δq is

$$\Delta \vec{E} = k_e \frac{\Delta q}{r^2} \hat{r}$$

where r is the distance from the charge element to point P and $\hat{r}$ is a unit vector directed from the element toward P. The total electric field at P due to all elements in the charge distribution is approximately

$$\vec{E} \approx k_e \sum_i \frac{\Delta q_i}{r_i^2} \hat{r}_i$$

where the index i refers to the ith element in the distribution. Because the charge distribution is modeled as continuous, the total field at P in the limit $\Delta q_i \rightarrow 0$ is

$$\vec{E} = k_e \lim_{\Delta q_i \rightarrow 0} \sum_i \frac{\Delta q_i}{r_i^2} \hat{r}_i = k_e \int \frac{dq}{r^2} \hat{r} \qquad (23.11)$$

◀ Electric field due to a continuous charge distribution

where the integration is over the entire charge distribution. The integration in Equation 23.11 is a vector operation and must be treated appropriately.

Let's illustrate this type of calculation with several examples in which the charge is distributed on a line, on a surface, or throughout a volume. When performing such calculations, it is convenient to use the concept of a *charge density* along with the following notations:

■ If a charge Q is uniformly distributed throughout a volume V, the **volume charge density** ρ is defined by

$$\rho \equiv \frac{Q}{V}$$

◀ Volume charge density

where ρ has units of coulombs per cubic meter (C/m³).

■ If a charge Q is uniformly distributed on a surface of area A, the **surface charge density** σ (Greek letter sigma) is defined by

$$\sigma \equiv \frac{Q}{A}$$

◀ Surface charge density

where σ has units of coulombs per square meter (C/m²).

■ If a charge Q is uniformly distributed along a line of length ℓ, the **linear charge density** λ is defined by

$$\lambda \equiv \frac{Q}{\ell}$$

◀ Linear charge density

where λ has units of coulombs per meter (C/m).

■ If the charge is nonuniformly distributed over a volume, surface, or line, the amounts of charge dq in a small volume, surface, or length element are

$$dq = \rho \, dV \qquad dq = \sigma \, dA \qquad dq = \lambda \, d\ell$$

Figure 23.14 The electric field at P due to a continuous charge distribution is the vector sum of the fields $\Delta \vec{E}$ due to all the elements Δq of the charge distribution.

> PROBLEM-SOLVING STRATEGY   **Calculating the Electric Field**
>
> The following procedure is recommended for solving problems that involve the determination of an electric field due to individual charges or a charge distribution:
>
> 1. *Conceptualize.* Establish a mental representation of the problem: think carefully about the individual charges or the charge distribution and imagine what type of electric field they would create. Appeal to any symmetry in the arrangement of charges to help you visualize the electric field.
>
> 2. *Categorize.* Are you analyzing a group of individual charges or a continuous charge distribution? The answer to this question tells you how to proceed in the Analyze step.
>
> 3. *Analyze.*
>
>   (a) *If you are analyzing a group of individual charge,* use the superposition principle: When several point charges are present, the resultant field at a point in space is the *vector sum* of the individual fields due to the individual charges (Eq. 23.10). Be very careful in the manipulation of vector quantities. It may be useful to review the material on vector addition in Chapter 3. Example 23.5 demonstrated this procedure.
>
>   (b) *If you are analyzing a continuous charge distribution,* replace the vector sums for evaluating the total electric field from individual charges by vector integrals. The charge distribution is divided into infinitesimal pieces, and the vector sum is carried out by integrating over the entire charge distribution (Eq. 23.11). Examples 23.6 through 23.8 demonstrate such procedures.
>
>   Consider symmetry when dealing with either a distribution of point charges or a continuous charge distribution. Take advantage of any symmetry in the system you observed in the Conceptualize step to simplify your calculations. The cancellation of field components perpendicular to the axis in Example 23.7 is an example of the application of symmetry.
>
> 4. *Finalize.* Check to see if your electric field expression is consistent with the mental representation and if it reflects any symmetry that you noted previously. Imagine varying parameters such as the distance of the observation point from the charges or the radius of any circular objects to see if the mathematical result changes in a reasonable way.

EXAMPLE 23.6   **The Electric Field Due to a Charged Rod**

A rod of length ℓ has a uniform positive charge per unit length λ and a total charge Q. Calculate the electric field at a point P that is located along the long axis of the rod and a distance a from one end (Fig. 23.15).

SOLUTION

Conceptualize   The field $d\vec{\mathbf{E}}$ at P due to each segment of charge on the rod is in the negative x direction because every segment carries a positive charge.

Categorize   Because the rod is continuous, we are evaluating the field due to a continuous charge distribution rather than a group of individual charges. Because every segment of the rod produces an electric field in the negative x direction, the sum of their contributions can be handled without the need to add vectors.

Analyze   Let's assume the rod is lying along the x axis, dx is the length of one small segment, and dq is the charge on that segment. Because the rod has a charge per unit length λ, the charge dq on the small segment is $dq = \lambda\, dx$.

Figure 23.15   (Example 23.6) The electric field at P due to a uniformly charged rod lying along the x axis. The magnitude of the field at P due to the segment of charge dq is $k_e\, dq/x^2$. The total field at P is the vector sum over all segments of the rod.

Find the magnitude of the electric field at P due to one segment of the rod having a charge dq:

$$dE = k_e \frac{dq}{x^2} = k_e \frac{\lambda\, dx}{x^2}$$

Find the total field at P using[4] Equation 23.11:

$$E = \int_a^{\ell + a} k_e \lambda \frac{dx}{x^2}$$

Noting that k_e and $\lambda = Q/\ell$ are constants and can be removed from the integral, evaluate the integral:

$$E = k_e \lambda \int_a^{\ell + a} \frac{dx}{x^2} = k_e \lambda \left[-\frac{1}{x} \right]_a^{\ell + a}$$

$$(1) \quad E = k_e \frac{Q}{\ell} \left(\frac{1}{a} - \frac{1}{\ell + a} \right) = \boxed{\frac{k_e Q}{a(\ell + a)}}$$

Finalize If ℓ goes to zero, Equation (1) reduces to the electric field due to a point charge as given by Equation 23.9, which is what we expect because the rod has shrunk to zero size.

What If? Suppose point P is very far away from the rod. What is the nature of the electric field at such a point?

Answer If P is far from the rod ($a \gg \ell$), then ℓ in the denominator of Equation (1) can be neglected and $E \approx k_e Q / a^2$. That is exactly the form you would expect for a point charge. Therefore, at large values of a/ℓ, the charge distribution appears to be a point charge of magnitude Q; the point P is so far away from the rod we cannot distinguish that it has a size. The use of the limiting technique ($a/\ell \to \infty$) is often a good method for checking a mathematical expression.

[4]To carry out integrations such as this one, first express the charge element dq in terms of the other variables in the integral. (In this example, there is one variable, x, so we made the change $dq = \lambda\, dx$.) The integral must be over scalar quantities; therefore, express the electric field in terms of components, if necessary. (In this example, the field has only an x component, so this detail is of no concern.) Then, reduce your expression to an integral over a single variable (or to multiple integrals, each over a single variable). In examples that have spherical or cylindrical symmetry, the single variable is a radial coordinate.

EXAMPLE 23.7   The Electric Field of a Uniform Ring of Charge

A ring of radius a carries a uniformly distributed positive total charge Q. Calculate the electric field due to the ring at a point P lying a distance x from its center along the central axis perpendicular to the plane of the ring (Fig. 23.16a).

SOLUTION

Conceptualize Figure 23.16a shows the electric field contribution $d\vec{E}$ at P due to a single segment of charge at the top of the ring. This field vector can be resolved into components dE_x parallel to the axis of the ring and $dE_\perp$ perpendicular to the axis. Figure 23.16b shows the electric field contributions from two segments on opposite sides of the ring. Because of the symmetry of the situation, the perpendicular components of the field cancel. That is true for all pairs of segments around the ring, so we can ignore the perpendicular component of the field and focus solely on the parallel components, which simply add.

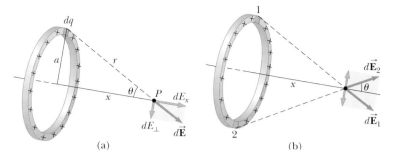

Figure 23.16 (Example 23.7) A uniformly charged ring of radius a. (a) The field at P on the x axis due to an element of charge dq. (b) The total electric field at P is along the x axis. The perpendicular component of the field at P due to segment 1 is canceled by the perpendicular component due to segment 2.

Categorize Because the ring is continuous, we are evaluating the field due to a continuous charge distribution rather than a group of individual charges.

Analyze Evaluate the parallel component of an electric field contribution from a segment of charge dq on the ring:

$$(1) \quad dE_x = k_e \frac{dq}{r^2} \cos\theta = k_e \frac{dq}{(a^2 + x^2)} \cos\theta$$

From the geometry in Figure 23.16a, evaluate $\cos\theta$:

$$(2) \quad \cos\theta = \frac{x}{r} = \frac{x}{(a^2 + x^2)^{1/2}}$$

Substitute Equation (2) into Equation (1):

$$dE_x = k_e \frac{dq}{(a^2 + x^2)} \frac{x}{(a^2 + x^2)^{1/2}} = \frac{k_e x}{(a^2 + x^2)^{3/2}} dq$$

All segments of the ring make the same contribution to the field at P because they are all equidistant from this point. Integrate to obtain the total field at P:

$$E_x = \int \frac{k_e x}{(a^2 + x^2)^{3/2}} dq = \frac{k_e x}{(a^2 + x^2)^{3/2}} \int dq$$

$$(3) \quad E = \frac{k_e x}{(a^2 + x^2)^{3/2}} Q$$

Finalize This result shows that the field is zero at $x = 0$. Is that consistent with the symmetry in the problem? Furthermore, notice that Equation (3) reduces to $k_e Q / x^2$ if $x \gg a$, so the ring acts like a point charge for locations far away from the ring.

What If? Suppose a negative charge is placed at the center of the ring in Figure 23.16 and displaced slightly by a distance $x \ll a$ along the x axis. When the charge is released, what type of motion does it exhibit?

Answer In the expression for the field due to a ring of charge, let $x \ll a$, which results in

$$E_x = \frac{k_e Q}{a^3} x$$

Therefore, from Equation 23.8, the force on a charge $-q$ placed near the center of the ring is

$$F_x = -\frac{k_e q Q}{a^3} x$$

Because this force has the form of Hooke's law (Eq. 15.1), the motion of the negative charge is *simple harmonic!*

EXAMPLE 23.8   **The Electric Field of a Uniformly Charged Disk**

A disk of radius R has a uniform surface charge density σ. Calculate the electric field at a point P that lies along the central perpendicular axis of the disk and a distance x from the center of the disk (Fig. 23.17).

SOLUTION

Conceptualize If the disk is considered to be a set of concentric rings, we can use our result from Example 23.7—which gives the field created by a ring of radius a—and sum the contributions of all rings making up the disk. By symmetry, the field at an axial point must be along the central axis.

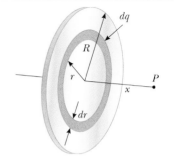

Figure 23.17 (Example 23.8) A uniformly charged disk of radius R. The electric field at an axial point P is directed along the central axis, perpendicular to the plane of the disk.

Categorize Because the disk is continuous, we are evaluating the field due to a continuous charge distribution rather than a group of individual charges.

Analyze Find the amount of charge dq on a ring of radius r and width dr as shown in Figure 23.17:

$$dq = \sigma \, dA = \sigma(2\pi r \, dr) = 2\pi\sigma r \, dr$$

Use this result in the equation given for E_x in Example 23.7 (with a replaced by r and Q replaced by dq) to find the field due to the ring:

$$dE_x = \frac{k_e x}{(r^2 + x^2)^{3/2}}(2\pi\sigma r \, dr)$$

To obtain the total field at P, integrate this expression over the limits $r = 0$ to $r = R$, noting that x is a constant in this situation:

$$E_x = k_e x \pi\sigma \int_0^R \frac{2r \, dr}{(r^2 + x^2)^{3/2}}$$

$$= k_e x \pi\sigma \int_0^R (r^2 + x^2)^{-3/2} \, d(r^2)$$

$$= k_e x \pi\sigma \left[\frac{(r^2 + x^2)^{-1/2}}{-1/2} \right]_0^R = 2\pi k_e \sigma \left[1 - \frac{x}{(R^2 + x^2)^{1/2}} \right]$$

Finalize   This result is valid for all values of $x > 0$. We can calculate the field close to the disk along the axis by assuming that $R \gg x$; therefore, the expression in brackets reduces to unity to give us the near-field approximation

$$E_x = 2\pi k_e \sigma = \frac{\sigma}{2\epsilon_0}$$

where ϵ_0 is the permittivity of free space. In Chapter 24, we obtain the same result for the field created by an infinite plane of charge with uniform surface charge density.

23.6   Electric Field Lines

We have defined the electric field mathematically through Equation 23.7. Let's now explore a means of visualizing the electric field in a pictorial representation. A convenient way of visualizing electric field patterns is to draw lines, called **electric field lines** and first introduced by Faraday, that are related to the electric field in a region of space in the following manner:

- The electric field vector $\vec{E}$ is tangent to the electric field line at each point. The line has a direction, indicated by an arrowhead, that is the same as that of the electric field vector. The direction of the line is that of the force on a positive test charge placed in the field.
- The number of lines per unit area through a surface perpendicular to the lines is proportional to the magnitude of the electric field in that region. Therefore, the field lines are close together where the electric field is strong and far apart where the field is weak.

These properties are illustrated in Figure 23.18. The density of field lines through surface A is greater than the density of lines through surface B. Therefore, the magnitude of the electric field is larger on surface A than on surface B. Furthermore, because the lines at different locations point in different directions, the field is nonuniform.

Is this relationship between strength of the electric field and the density of field lines consistent with Equation 23.9, the expression we obtained for E using Coulomb's law? To answer this question, consider an imaginary spherical surface of radius r concentric with a point charge. From symmetry, we see that the magnitude of the electric field is the same everywhere on the surface of the sphere. The number of lines N that emerge from the charge is equal to the number that penetrate the spherical surface. Hence, the number of lines per unit area on the sphere is $N/4\pi r^2$ (where the surface area of the sphere is $4\pi r^2$). Because E is proportional to the number of lines per unit area, we see that E varies as $1/r^2$; this finding is consistent with Equation 23.9.

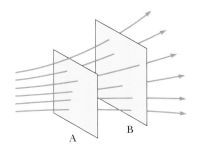

Figure 23.18   Electric field lines penetrating two surfaces. The magnitude of the field is greater on surface A than on surface B.

(a)   (b)   (c)

Douglas C. Johnson, California State Polytechnic University, Pomona

Figure 23.19 The electric field lines for a point charge. (a) For a positive point charge, the lines are directed radially outward. (b) For a negative point charge, the lines are directed radially inward. Notice that the figures show only those field lines that lie in the plane of the page. (c) The dark areas are small particles suspended in oil, which align with the electric field produced by a charged conductor at the center.

Representative electric field lines for the field due to a single positive point charge are shown in Figure 23.19a. This two-dimensional drawing shows only the field lines that lie in the plane containing the point charge. The lines are actually directed radially outward from the charge in all directions; therefore, instead of the flat "wheel" of lines shown, you should picture an entire spherical distribution of lines. Because a positive test charge placed in this field would be repelled by the positive source charge, the lines are directed radially away from the source charge. The electric field lines representing the field due to a single negative point charge are directed toward the charge (Fig. 23.19b). In either case, the lines are along the radial direction and extend all the way to infinity. Notice that the lines become closer together as they approach the charge, indicating that the strength of the field increases as we move toward the source charge.

The rules for drawing electric field lines are as follows:

- The lines must begin on a positive charge and terminate on a negative charge. In the case of an excess of one type of charge, some lines will begin or end infinitely far away.
- The number of lines drawn leaving a positive charge or approaching a negative charge is proportional to the magnitude of the charge.
- No two field lines can cross.

We choose the number of field lines starting from any positively charged object to be Cq and the number of lines ending on any negatively charged object to be $C|q|$, where C is an arbitrary proportionality constant. Once C is chosen, the number of lines is fixed. For example, in a two-charge system, if object 1 has charge Q_1 and object 2 has charge Q_2, the ratio of number of lines in contact with the

(a)   (b)

Douglas C. Johnson, California State Polytechnic University, Pomona

Figure 23.20 (a) The electric field lines for two point charges of equal magnitude and opposite sign (an electric dipole). The number of lines leaving the positive charge equals the number terminating at the negative charge. (b) Small particles suspended in oil align with the electric field.

(a)

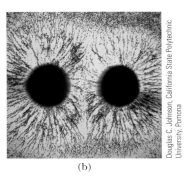

Douglas C. Johnson, California State Polytechnic University, Pomona

(b)

Figure 23.21   (a) The electric field lines for two positive point charges. (The locations A, B, and C are discussed in Quick Quiz 23.5.) (b) Small particles suspended in oil align with the electric field.

charges is $N_2/N_1 = Q_2/Q_1$. The electric field lines for two point charges of equal magnitude but opposite signs (an electric dipole) are shown in Figure 23.20. Because the charges are of equal magnitude, the number of lines that begin at the positive charge must equal the number that terminate at the negative charge. At points very near the charges, the lines are nearly radial. The high density of lines between the charges indicates a region of strong electric field.

Figure 23.21 shows the electric field lines in the vicinity of two equal positive point charges. Again, the lines are nearly radial at points close to either charge, and the same number of lines emerge from each charge because the charges are equal in magnitude. At great distances from the charges, the field is approximately equal to that of a single point charge of magnitude $2q$.

Finally, in Active Figure 23.22, we sketch the electric field lines associated with a positive charge $+2q$ and a negative charge $-q$. In this case, the number of lines leaving $+2q$ is twice the number terminating at $-q$. Hence, only half the lines that leave the positive charge reach the negative charge. The remaining half terminate on a negative charge we assume to be at infinity. At distances much greater than the charge separation, the electric field lines are equivalent to those of a single charge $+q$.

Quick Quiz 23.5   Rank the magnitudes of the electric field at points A, B, and C shown in Figure 23.21a (greatest magnitude first).

23.7   Motion of a Charged Particle in a Uniform Electric Field

When a particle of charge q and mass m is placed in an electric field $\vec{\mathbf{E}}$, the electric force exerted on the charge is $q\vec{\mathbf{E}}$ according to Equation 23.8. If that is the only force exerted on the particle, it must be the net force and it causes the particle to accelerate according to the particle under a net force model. Therefore,

$$\vec{\mathbf{F}}_e = q\vec{\mathbf{E}} = m\vec{\mathbf{a}}$$

and the acceleration of the particle is

$$\vec{\mathbf{a}} = \frac{q\vec{\mathbf{E}}}{m} \qquad (23.12)$$

If $\vec{\mathbf{E}}$ is uniform (that is, constant in magnitude and direction), the electric force on the particle is constant and we can apply the particle under constant acceleration model. If the particle has a positive charge, its acceleration is in the direction of the electric field. If the particle has a negative charge, its acceleration is in the direction opposite the electric field.

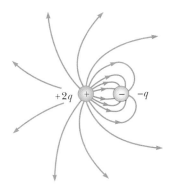

$+2q$ $-q$

ACTIVE FIGURE 23.22

The electric field lines for a point charge $+2q$ and a second point charge $-q$. Notice that two lines leave $+2q$ for every one that terminates on $-q$.

Sign in at www.thomsonedu.com and go to ThomsonNOW to choose the values and signs for the two charges and observe the electric field lines for the configuration you have chosen.

PITFALL PREVENTION 23.4
Just Another Force

Electric forces and fields may seem abstract to you. Once $\vec{\mathbf{F}}_e$ is evaluated, however, it causes a particle to move according to our well-established models of forces and motion from Chapters 2 through 6. Keeping this link with the past in mind should help you solve problems in this chapter.

EXAMPLE 23.9 | An Accelerating Positive Charge

A uniform electric field $\vec{E}$ is directed along the x axis between parallel plates of charge separated by a distance d as shown in Figure 23.23. A positive point charge q of mass m is released from rest at a point Ⓐ next to the positive plate and accelerates to a point Ⓑ next to the negative plate.

(A) Find the speed of the particle at Ⓑ by modeling it as a particle under constant acceleration.

SOLUTION

Conceptualize   When the positive charge is placed at Ⓐ, it experiences an electric force toward the right in Figure 23.23 due to the electric field directed toward the right.

Categorize   Because the electric field is uniform, a constant electric force acts on the charge. Therefore, the example involves a charged particle under constant acceleration.

Figure 23.23   (Example 23.9) A positive point charge q in a uniform electric field $\vec{E}$ undergoes constant acceleration in the direction of the field.

Analyze   Use Equation 2.17 to express the velocity of the particle as a function of position:

$$v_f^2 = v_i^2 + 2a(x_f - x_i) = 0 + 2a(d - 0) = 2ad$$

Solve for v_f and substitute for the magnitude of the acceleration from Equation 23.12:

$$v_f = \sqrt{2ad} = \sqrt{2\left(\frac{qE}{m}\right)d} = \sqrt{\frac{2qEd}{m}}$$

(B) Find the speed of the particle at Ⓑ by modeling it as a nonisolated system.

SOLUTION

Categorize   The problem statement tells us that the charge is a nonisolated system. Energy is transferred to this charge by work done by the electric force exerted on the charge. The initial configuration of the system is when the particle is at Ⓐ, and the final configuration is when it is at Ⓑ.

Analyze   Write the appropriate reduction of the conservation of energy equation, Equation 8.2, for the system of the charged particle:

$$W = \Delta K$$

Replace the work and kinetic energies with values appropriate for this situation:

$$F_e \Delta x = K_Ⓑ - K_Ⓐ = \tfrac{1}{2}mv_f^2 - 0 \quad \rightarrow \quad v_f = \sqrt{\frac{2F_e \Delta x}{m}}$$

Substitute for the electric force F_e and the displacement Δx:

$$v_f = \sqrt{\frac{2(qE)(d)}{m}} = \sqrt{\frac{2qEd}{m}}$$

Finalize   The answer to part (B) is the same as that for part (A), as we expect.

EXAMPLE 23.10   **An Accelerated Electron**

An electron enters the region of a uniform electric field as shown in Active Figure 23.24, with $v_i = 3.00 \times 10^6$ m/s and $E = 200$ N/C. The horizontal length of the plates is $\ell = 0.100$ m.

(A) Find the acceleration of the electron while it is in the electric field.

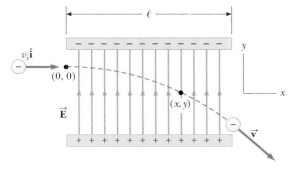

ACTIVE FIGURE 23.24

(Example 23.10) An electron is projected horizontally into a uniform electric field produced by two charged plates. The electron undergoes a downward acceleration (opposite $\vec{E}$), and its motion is parabolic while it is between the plates.

Sign in at www.thomsonedu.com and go to ThomsonNOW to choose the strength of the electric field and the mass and charge of the projected particle.

SOLUTION

Conceptualize This example differs from the preceding one because the velocity of the charged particle is initially perpendicular to the electric field lines. In Example 23.9, the velocity of the charged particle is always parallel to the electric field lines. As a result, the electron in this example follows a curved path as shown in Active Figure 23.24.

Categorize Because the electric field is uniform, a constant electric force is exerted on the electron. To find the acceleration of the electron, we can model it as a particle under a net force.

Analyze The direction of the electron's acceleration is downward in Active Figure 23.24, opposite the direction of the electric field lines.

Combine Newton's second law with the magnitude of the electric force given by Equation 23.8 to find the y component of the acceleration of the electron:

$$\Sigma F_y = ma_y \quad \rightarrow \quad a_y = \frac{\Sigma F_y}{m} = -\frac{eE}{m_e}$$

Substitute numerical values:

$$a_y = -\frac{(1.60 \times 10^{-19} \text{ C})(200 \text{ N/C})}{9.11 \times 10^{-31} \text{ kg}} = \boxed{-3.51 \times 10^{13} \text{ m/s}^2}$$

(B) Assuming the electron enters the field at time $t = 0$, find the time at which it leaves the field.

SOLUTION

Categorize Because the electric force acts only in the vertical direction in Active Figure 23.24, the motion of the particle in the horizontal direction can be analyzed by modeling it as a particle under constant velocity.

Analyze Solve Equation 2.7 for the time at which the electron arrives at the right edges of the plates:

$$x_f = x_i + v_x t \quad \rightarrow \quad t = \frac{x_f - x_i}{v_x}$$

Substitute numerical values:

$$t = \frac{\ell - 0}{v_x} = \frac{0.100 \text{ m}}{3.00 \times 10^6 \text{ m/s}} = \boxed{3.33 \times 10^{-8} \text{ s}}$$

(C) Assuming the vertical position of the electron as it enters the field is $y_i = 0$, what is its vertical position when it leaves the field?

SOLUTION

Categorize Because the electric force is constant in Active Figure 23.24, the motion of the particle in the vertical direction can be analyzed by modeling it as a particle under constant acceleration.

Analyze Use Equation 2.16 to describe the position of the particle at any time t:

$$y_f = y_i + v_{yi} t + \tfrac{1}{2} a_y t^2$$

Substitute numerical values:

$$y_f = 0 + 0 + \tfrac{1}{2}(-3.51 \times 10^{13} \text{ m/s}^2)(3.33 \times 10^{-8} \text{ s})^2$$

$$= -0.019\,5 \text{ m} = \boxed{-1.95 \text{ cm}}$$

Finalize If the electron enters just below the negative plate in Active Figure 23.24 and the separation between the plates is less than the value just calculated, the electron will strike the positive plate.

We have neglected the gravitational force acting on the electron, which represents a good approximation when dealing with atomic particles. For an electric field of 200 N/C, the ratio of the magnitude of the electric force eE to the magnitude of the gravitational force mg is on the order of 10^{12} for an electron and on the order of 10^9 for a proton.

Summary

ThomsonNOW™ Sign in at **www.thomsonedu.com** and go to ThomsonNOW to take a practice test for this chapter.

DEFINITIONS

The **electric field** $\vec{E}$ at some point in space is defined as the electric force $\vec{F}_e$ that acts on a small positive test charge placed at that point divided by the magnitude q_0 of the test charge:

$$\vec{E} \equiv \frac{\vec{F}_e}{q_0} \tag{23.7}$$

CONCEPTS AND PRINCIPLES

Electric charges have the following important properties:

- Charges of opposite sign attract one another, and charges of the same sign repel one another.
- The total charge in an isolated system is conserved.
- Charge is quantized.

Conductors are materials in which electrons move freely. **Insulators** are materials in which electrons do not move freely.

Coulomb's law states that the electric force exerted by a point charge q_1 on a second point charge q_2 is

$$\vec{F}_{12} = k_e \frac{q_1 q_2}{r^2} \hat{\mathbf{r}}_{12} \tag{23.6}$$

where r is the distance between the two charges and $\hat{\mathbf{r}}_{12}$ is a unit vector directed from q_1 toward q_2. The constant k_e, which is called the **Coulomb constant,** has the value $k_e = 8.99 \times 10^9 \text{ N·m}^2/\text{C}^2$.

The electric force on a charge q placed in an electric field $\vec{E}$ is

$$\vec{F}_e = q\vec{E} \tag{23.8}$$

At a distance r from a point charge q, the electric field due to the charge is

$$\vec{E} = k_e \frac{q}{r^2} \hat{\mathbf{r}} \tag{23.9}$$

where $\hat{\mathbf{r}}$ is a unit vector directed from the charge toward the point in question. The electric field is directed radially outward from a positive charge and radially inward toward a negative charge.

The electric field due to a group of point charges can be obtained by using the superposition principle. That is, the total electric field at some point equals the vector sum of the electric fields of all the charges:

$$\vec{E} = k_e \sum_i \frac{q_i}{r_i^2} \hat{\mathbf{r}}_i \tag{23.10}$$

The electric field at some point due to a continuous charge distribution is

$$\vec{E} = k_e \int \frac{dq}{r^2} \hat{\mathbf{r}} \tag{23.11}$$

where dq is the charge on one element of the charge distribution and r is the distance from the element to the point in question.

Questions

☐ denotes answer available in *Student Solutions Manual/Study Guide;* **O** denotes objective question

1. Explain what is meant by the term "a neutral atom." Explain what "a negatively charged atom" means.

2. **O** (i) A metallic coin is given a positive electric charge. Does its mass (a) increase measurably, (b) increase by an amount too small to measure directly, (c) remain unchanged, (d) decrease by an amount too small to measure directly, or (e) decrease measurably? (ii) Now the coin is given a negative electric charge. What happens to its mass? Choose from the same possibilities.

3. A student who grew up in a tropical country and is studying in the United States may have no experience with static electricity sparks and shocks until their first American winter. Explain.

4. Explain the similarities and differences between Newton's law of universal gravitation and Coulomb's law.

5. A balloon is negatively charged by rubbing, and then it clings to a wall. Does that mean the wall is positively charged? Why does the balloon eventually fall?

6. **O** In Figure 23.8, assume the objects with charges q_2 and q_3 are fixed. Notice that there is no sightline from the location of object 2 to the location of object 1. We could say that a bug standing on q_1 is unable to see q_2 because it is behind q_3. How would you calculate the force exerted on the object with charge q_1? (a) Find only the force exerted by q_2 on charge q_1. (b) Find only the force exerted by q_3 on charge q_1. (c) Add the force that q_2 would exert by itself on charge q_1 to the force that q_3 would exert by itself on charge q_1. (d) Add the force that q_3 would exert by itself to a certain fraction of the force that q_2 would exert by itself. (e) There is no definite way to find the force on charge q_1.

7. **O** A charged particle is at the origin of coordinates. The particle produces an electric field of $4\hat{\mathbf{i}}$ kN/C at the point with position vector $36\hat{\mathbf{i}}$ cm. (i) At what location does the field have the value $1\hat{\mathbf{i}}$ kN/C? (a) $9\hat{\mathbf{i}}$ cm (b) $18\hat{\mathbf{i}}$ cm (c) $72\hat{\mathbf{i}}$ cm (d) $144\hat{\mathbf{i}}$ cm (e) nowhere (ii) At what location is the value $16\hat{\mathbf{i}}$ kN/C? Choose from the same possibilities.

8. Is it possible for an electric field to exist in empty space? Explain. Consider point *A* in Figure 23.21a. Does charge exist at this point? Does a force exist at this point? Does a field exist at this point?

9. **O** (i) Rank the magnitude of the forces charged particle A exerts on charged particle B, located at distance *r* away from A, from the largest to the smallest in the following cases. In your ranking, note any cases of equality. (a) $q_A = 20$ nC, $q_B = 20$ nC, $r = 2$ cm (b) $q_A = 30$ nC, $q_B = 10$ nC, $r = 2$ cm (c) $q_A = 10$ nC, $q_B = 30$ nC, $r = 2$ cm (d) $q_A = 30$ nC, $q_B = 20$ nC, $r = 3$ cm (e) $q_A = 45$ nC, $q_B = 20$ nC, $r = 3$ cm (ii) Rank the magnitudes of the electric fields charged particle A creates at the location of charged particle B, a distance *r* away from A, from the largest to the smallest in the same cases. In your ranking, note any cases of equality.

10. **O** Three charged particles are arranged on corners of a square as shown in Figure Q23.10, with charge $-Q$ on both the particle at the upper left corner and the particle at the lower right corner, and charge $+2Q$ on the particle at the lower left corner. (i) What is the direction of the electric field at the upper right corner, which is a point in empty space? (a) It is upward and to the right. (b) It is straight to the right. (c) It is straight downward. (d) It is downward and to the left. (e) It is perpendicular to the plane of the picture and outward. (f) There is no direction; no field exists at that corner because no charge is there. (g) There is no direction; the total field there is zero. (ii) Suppose the $+2Q$ charge at the lower left corner is removed. Then does the magnitude of the field at the upper right corner (a) become larger, (b) become smaller, (c) stay the same, or (d) change unpredictably?

Figure Q23.10

11. **O** Two charged particles A and B are alone in the Universe, 8 cm apart. The charge of A is 40 nC. The net electric field at one certain point 4 cm from A is zero. What can you conclude about the charge of B? Choose every correct answer. (a) It can be 40 nC. (b) It can be 120 nC. (c) It can be 360 nC. (d) It can be -40 nC. (e) It can be -120 nC. (f) It can be -360 nC. (g) It can have any of an infinite number of values. (h) It can have any of several values. (i) It must have one of three values. (j) It must have one of two values. (k) It must have one certain value. (l) No possible value for q_B exists; the situation is impossible.

12. Explain why electric field lines never cross. *Suggestion:* Begin by explaining why the electric field at a particular point must have only one direction.

13. Figures 23.12 and 23.13 show three electric field vectors at the same point. With a little extrapolation, Figure 23.19 would show many electric field lines at the same point. Is it really true that "no two field lines can cross"? Are the diagrams drawn correctly? Explain your answers.

14. **O** A circular ring of charge with radius *b* has total charge *q* uniformly distributed around it. What is the magnitude of the electric field at the center of the ring? (a) 0 (b) $k_e q/b^2$ (c) $k_e q^2/b^2$ (d) $k_e q^2/b$ (c) none of these answers

15. **O** Assume a uniformly charged ring of radius *R* and charge *Q* produces an electric field E_{ring} at a point *P* on its axis, at distance *x* away from the center of the ring. Now the charge *Q* is spread uniformly over the circular area the ring encloses, forming a flat disk of charge with the same radius. How does the field E_{disk} produced by the disk at *P* compare to the field produced by the ring at the same point? (a) $E_{\text{disk}} < E_{\text{ring}}$ (b) $E_{\text{disk}} = E_{\text{ring}}$ (c) $E_{\text{disk}} > E_{\text{ring}}$ (d) impossible to determine

16. **O** A free electron and a free proton are released in identical electric fields. **(i)** How do the magnitudes of the electric force exerted on the two particles compare? (a) It is millions of times greater for the electron. (b) It is thousands of times greater for the electron. (c) They are equal. (d) It is thousands of times smaller for the electron. (e) It is millions of times smaller for the electron. (f) It is zero for the electron. (g) It is zero for the proton. **(ii)** Compare the magnitudes of their accelerations. Choose from the same possibilities.

17. **O** An object with negative charge is placed in a region of space where the electric field is directed vertically upward. What is the direction of the electric force exerted on this charge? (a) It is up. (b) It is down. (c) There is no force. (d) The force can be in any direction.

18. Explain the differences between linear, surface, and volume charge densities and give examples of when each would be used.

19. Would life be different if the electron were positively charged and the proton were negatively charged? Does the choice of signs have any bearing on physical and chemical interactions? Explain.

20. Consider two electric dipoles in empty space. Each dipole has zero net charge. Does an electric force exist between the dipoles; that is, can two objects with zero net charge exert electric forces on each other? If so, is the force one of attraction or of repulsion?

Problems

Section 23.1 Properties of Electric Charges

1. (a) Find to three significant digits the charge and the mass of an ionized hydrogen atom, represented as H^+. *Suggestion:* Begin by looking up the mass of a neutral atom on the periodic table of the elements in Appendix C. (b) Find the charge and the mass of Na^+, a singly ionized sodium atom. (c) Find the charge and the average mass of a chloride ion Cl^- that joins with the Na^+ to make one molecule of table salt. (d) Find the charge and the mass of $Ca^{++} = Ca^{2+}$, a doubly ionized calcium atom. (e) You can model the center of an ammonia molecule as an N^{3-} ion. Find its charge and mass. (f) The plasma in a hot star contains quadruply ionized nitrogen atoms, N^{4+}. Find their charge and mass. (g) Find the charge and the mass of the nucleus of a nitrogen atom. (h) Find the charge and the mass of the molecular ion H_2O^-.

2. (a) Calculate the number of electrons in a small, electrically neutral silver pin that has a mass of 10.0 g. Silver has 47 electrons per atom, and its molar mass is 107.87 g/mol. (b) Imagine adding electrons to the pin until the negative charge has the very large value 1.00 mC. How many electrons are added for every 10^9 electrons already present?

Section 23.2 Charging Objects by Induction

Section 23.3 Coulomb's Law

3. ▲ Nobel laureate Richard Feynman (1918–1988) once said that if two persons stood at arm's length from each other and each person had 1% more electrons than protons, the force of repulsion between them would be enough to lift a "weight" equal to that of the entire Earth. Carry out an order-of-magnitude calculation to substantiate this assertion.

4. A charged particle A exerts a force of 2.62 μN to the right on charged particle B when the particles are 13.7 mm apart. Particle B moves straight away from A to make the distance between them 17.7 mm. What vector force does it then exert on A?

5. ● (a) Two protons in a molecule are 3.80×10^{-10} m apart. Find the electrical force exerted by one proton on the other. (b) State how the magnitude of this force compares with the magnitude of the gravitational force exerted by one proton on the other. (c) **What If?** What must be a particle's charge-to-mass ratio if the magnitude of the gravitational force between two of these particles is equal to the magnitude of electrical force between them?

6. Two small silver spheres, each with a mass of 10.0 g, are separated by 1.00 m. Calculate the fraction of the electrons in one sphere that must be transferred to the other to produce an attractive force of 1.00×10^4 N (about 1 ton) between the spheres. (The number of electrons per atom of silver is 47, and the number of atoms per gram is Avogadro's number divided by the molar mass of silver, 107.87 g/mol.)

7. Three charged particles are located at the corners of an equilateral triangle as shown in Figure P23.7. Calculate the total electric force on the 7.00-μC charge.

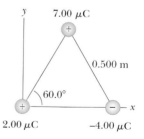

Figure P23.7 Problems 7 and 14.

8. ● Two small beads having positive charges $3q$ and q are fixed at the opposite ends of a horizontal insulating rod, extending from the origin to the point $x = d$. As shown in Figure P23.8, a third small charged bead is free to slide on the rod. At what position is the third bead in equilibrium? Explain whether it can be in stable equilibrium.

+3q   +q

$\longleftarrow d \longrightarrow$

Figure P23.8

9. Two identical conducting small spheres are placed with their centers 0.300 m apart. One is given a charge of 12.0 nC and the other a charge of -18.0 nC. (a) Find the electric force exerted by one sphere on the other. (b) **What If?** The spheres are connected by a conducting wire. Find the electric force each exerts on the other after they have come to equilibrium.

10. **Review problem**. Two identical particles, each having charge $+q$, are fixed in space and separated by a distance d. A third particle with charge $-Q$ is free to move and lies initially at rest on the perpendicular bisector of the two fixed charges a distance x from the midpoint between the two fixed charges (Fig. P23.10). (a) Show that if x is small compared with d, the motion of $-Q$ is simple harmonic along the perpendicular bisector. Determine the period of that motion. (b) How fast will the charge $-Q$ be moving when it is at the midpoint between the two fixed charges if initially it is released at a distance $a \ll d$ from the midpoint?

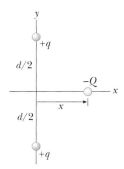

Figure P23.10

11. **Review problem**. In the Bohr theory of the hydrogen atom, an electron moves in a circular orbit about a proton, where the radius of the orbit is 0.529×10^{-10} m. (a) Find the electric force exerted on each particle. (b) If this force causes the centripetal acceleration of the electron, what is the speed of the electron?

Section 23.4 The Electric Field

12. In Figure P23.12, determine the point (other than infinity) at which the electric field is zero.

$\longleftarrow$ 1.00 m $\longrightarrow$

-2.50 μC   6.00 μC

Figure P23.12

13. What are the magnitude and direction of the electric field that balances the weight of (a) an electron and (b) a proton? You may use the data in Table 23.1.

14. Three charged particles are at the corners of an equilateral triangle as shown in Figure P23.7. (a) Calculate the electric field at the position of the 2.00-μC charge due to the 7.00-μC and -4.00-μC charges. (b) Use your answer to part (a) to determine the force on the 2.00-μC charge.

15. ● Two charged particles are located on the x axis. The first is a charge $+Q$ at $x = -a$. The second is an unknown charge located at $x = +3a$. The net electric field these charges produce at the origin has a magnitude of $2k_eQ/a^2$. Explain how many values are possible for the unknown charge and find the possible values.

16. Two 2.00-μC charged particles are located on the x axis. One is at $x = 1.00$ m, and the other is at $x = -1.00$ m. (a) Determine the electric field on the y axis at $y = 0.500$ m. (b) Calculate the electric force on a -3.00-μC charge placed on the y axis at $y = 0.500$ m.

17. Four charged particles are at the corners of a square of side a as shown in Figure P23.17. (a) Determine the magnitude and direction of the electric field at the location of charge q. (b) What is the total electric force exerted on q?

Figure P23.17

18. Consider the electric dipole shown in Figure P23.18. Show that the electric field at a *distant* point on the $+x$ axis is $E_x \approx -4k_eqa/x^3$.

Figure P23.18

19. ● Consider n equal positive charged particles each of magnitude Q/n placed symmetrically around a circle of radius R. (a) Calculate the magnitude of the electric field at a point a distance x from the center of the circle and on the line passing through the center and perpendicular to the plane of the circle. (b) Explain why this result is identical to the result of the calculation done in Example 23.7.

Section 23.5 Electric Field of a Continuous Charge Distribution

20. A continuous line of charge lies along the x axis, extending from $x = +x_0$ to positive infinity. The line carries charge with a uniform linear charge density λ_0. What are

the magnitude and direction of the electric field at the origin?

21. A rod 14.0 cm long is uniformly charged and has a total charge of $-22.0 \mu C$. Determine the magnitude and direction of the electric field along the axis of the rod at a point 36.0 cm from its center.

22. Show that the maximum magnitude E_{max} of the electric field along the axis of a uniformly charged ring occurs at $x = a/\sqrt{2}$ (see Fig. 23.16) and has the value $Q/(6\sqrt{3}\pi\epsilon_0 a^2)$.

23. A uniformly charged ring of radius 10.0 cm has a total charge of $75.0 \mu C$. Find the electric field on the axis of the ring at (a) 1.00 cm, (b) 5.00 cm, (c) 30.0 cm, and (d) 100 cm from the center of the ring.

24. A uniformly charged disk of radius 35.0 cm carries charge with a density of 7.90×10^{-3} C/m². Calculate the electric field on the axis of the disk at (a) 5.00 cm, (b) 10.0 cm, (c) 50.0 cm, and (d) 200 cm from the center of the disk.

25. ● Example 23.8 derives the exact expression for the electric field at a point on the axis of a uniformly charged disk. Consider a disk of radius $R = 3.00$ cm having a uniformly distributed charge of $+5.20 \mu C$. (a) Using the result of Example 23.8, compute the electric field at a point on the axis and 3.00 mm from the center. **What If?** Explain how this answer compares with the field computed from the near-field approximation $E = \sigma/2\epsilon_0$. (b) Using the result of Example 23.8, compute the electric field at a point on the axis and 30.0 cm from the center of the disk. **What If?** Explain how this answer compares with the electric field obtained by treating the disk as a $+5.20$-μC charged particle at a distance of 30.0 cm.

26. The electric field along the axis of a uniformly charged disk of radius R and total charge Q was calculated in Example 23.8. Show that the electric field at distances x that are large compared with R approaches that of a particle with charge $Q = \sigma\pi R^2$. *Suggestion:* First show that $x/(x^2 + R^2)^{1/2} = (1 + R^2/x^2)^{-1/2}$ and use the binomial expansion $(1 + \delta)^n \approx 1 + n\delta$ when $\delta \ll 1$.

27. ▲ A uniformly charged insulating rod of length 14.0 cm is bent into the shape of a semicircle as shown in Figure P23.27. The rod has a total charge of $-7.50 \mu C$. Find the magnitude and direction of the electric field at O, the center of the semicircle.

Figure P23.27

28. (a) Consider a uniformly charged thin-walled right circular cylindrical shell having total charge Q, radius R, and height h. Determine the electric field at a point a distance d from the right side of the cylinder as shown in Figure P23.28. *Suggestion:* Use the result of Example 23.7 and treat the cylinder as a collection of ring charges. (b) **What If?** Consider now a solid cylinder with the same dimensions and carrying the same charge, uniformly distributed

through its volume. Use the result of Example 23.8 to find the field it creates at the same point.

Figure P23.28

29. A thin rod of length ℓ and uniform charge per unit length λ lies along the x axis as shown in Figure P23.29. (a) Show that the electric field at P, a distance y from the rod along its perpendicular bisector, has no x component and is given by $E = 2k_e\lambda \sin \theta_0/y$. (b) **What If?** Using your result to part (a), show that the field of a rod of infinite length is $E = 2k_e\lambda/y$. *Suggestion:* First calculate the field at P due to an element of length dx, which has a charge λdx. Then change variables from x to θ, using the relationships $x = y \tan \theta$ and $dx = y \sec^2 \theta d\theta$, and integrate over θ.

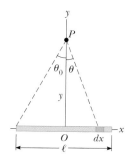

Figure P23.29

30. Three solid plastic cylinders all have radius 2.50 cm and length 6.00 cm. One (a) carries charge with uniform density 15.0 nC/m² everywhere on its surface. Another (b) carries charge with the same uniform density on its curved lateral surface only. The third (c) carries charge with uniform density 500 nC/m³ throughout the plastic. Find the charge of each cylinder.

31. Eight solid plastic cubes, each 3.00 cm on each edge, are glued together to form each one of the objects (i, ii, iii, and iv) shown in Figure P23.31. (a) Assuming each object carries charge with uniform density 400 nC/m³ throughout its volume, find the charge of each object. (b) Assuming each object carries charge with uniform density 15.0 nC/m² everywhere on its exposed surface, find the charge on each object. (c) Assuming charge is placed

Figure P23.31

2 = intermediate;   3 = challenging;   □ = SSM/SG;   ▲ = ThomsonNOW;   = symbolic reasoning;   ● = qualitative reasoning

only on the edges where perpendicular surfaces meet, with uniform density 80.0 pC/m, find the charge of each object.

Section 23.6 Electric Field Lines

32. A positively charged disk has a uniform charge per unit area as described in Example 23.8. Sketch the electric field lines in a plane perpendicular to the plane of the disk passing through its center.

33. A negatively charged rod of finite length carries charge with a uniform charge per unit length. Sketch the electric field lines in a plane containing the rod.

34. Figure P23.34 shows the electric field lines for two charged particles separated by a small distance. (a) Determine the ratio q_1/q_2. (b) What are the signs of q_1 and q_2?

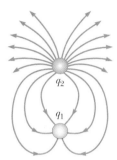

Figure P23.34

35. ▲ Three equal positive charges q are at the corners of an equilateral triangle of side a as shown in Figure P23.35. (a) Assume the three charges together create an electric field. Sketch the field lines in the plane of the charges. Find the location of one point (other than ∞) where the electric field is zero. (b) What are the magnitude and direction of the electric field at P due to the two charges at the base?

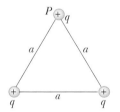

Figure P23.35 Problems 35 and 58.

Section 23.7 Motion of Charged Particles in a Uniform Electric Field

36. A proton is projected in the positive x direction into a region of a uniform electric field $\vec{E} = -6.00 \times 10^5 \hat{i}$ N/C at $t = 0$. The proton travels 7.00 cm as it comes to rest. Determine (a) the acceleration of the proton, (b) its initial speed, and (c) the time interval over which the proton comes to rest.

37. A proton accelerates from rest in a uniform electric field of 640 N/C. At one later moment, its speed is 1.20 Mm/s (nonrelativistic because v is much less than the speed of light). (a) Find the acceleration of the proton. (b) Over what time interval does the proton reach this speed? (c) How far does it move in this time interval? (d) What is its kinetic energy at the end of this interval?

38. ● Two horizontal metal plates, each 100 mm square, are aligned 10.0 mm apart, with one above the other. They are given equal-magnitude charges of opposite sign so that a uniform downward electric field of 2 000 N/C exists in the region between them. A particle of mass 2.00×10^{-16} kg and with a positive charge of 1.00×10^{-6} C leaves the center of the bottom negative plate with an initial speed of 1.00×10^5 m/s at an angle of 37.0° above the horizontal. Describe the trajectory of the particle. Which plate does it strike? Where does it strike, relative to its starting point?

39. ▲ The electrons in a particle beam each have a kinetic energy K. What are the magnitude and direction of the electric field that will stop these electrons in a distance d?

40. Protons are projected with initial speed $v_i = 9.55$ km/s into a region where a uniform electric field $\vec{E} = (-720\hat{j})$ N/C is present as shown in Figure P23.40. The protons are to hit a target that lies at a horizontal distance of 1.27 mm from the point where the protons cross the plane and enter the electric field. Find (a) the two projection angles θ that will result in a hit and (b) the time of flight (the time interval during which the proton is above the plane in Fig. P23.40) for each trajectory.

Figure P23.40

41. A proton moves at 4.50×10^5 m/s in the horizontal direction. It enters a uniform vertical electric field with a magnitude of 9.60×10^3 N/C. Ignoring any gravitational effects, find (a) the time interval required for the proton to travel 5.00 cm horizontally, (b) its vertical displacement during the time interval in which it travels 5.00 cm horizontally, and (c) the horizontal and vertical components of its velocity after it has traveled 5.00 cm horizontally.

Additional Problems

42. ● Two known charges, −12.0 μC and 45.0 μC, and a third unknown charge are located on the x axis. The charge −12.0 μC is at the origin, and the charge 45.0 μC is at $x = 15.0$ cm. The unknown charge is to be placed so that each charge is in equilibrium under the action of the electric forces exerted by the other two charges. Is this situation possible? Is it possible in more than one way? Explain. Find the required location, magnitude, and sign of the unknown charge.

43. A uniform electric field of magnitude 640 N/C exists between two parallel plates that are 4.00 cm apart. A proton is released from the positive plate at the same instant an electron is released from the negative plate. (a) Determine the distance from the positive plate at which the two pass each other. (Ignore the electrical attraction between the proton and electron.) (b) **What If?** Repeat part (a) for a sodium ion (Na$^+$) and a chloride ion (Cl$^-$).

44. Three charged particles are aligned along the x axis as shown in Figure P23.44. Find the electric field at (a) the position (2.00, 0) and (b) the position (0, 2.00).

Figure P23.44

45. ▲ A charged cork ball of mass 1.00 g is suspended on a light string in the presence of a uniform electric field as shown in Figure P23.45. When $\vec{E} = (3.00\hat{i} + 5.00\hat{j}) \times 10^5$ N/C, the ball is in equilibrium at $\theta = 37.0°$. Find (a) the charge on the ball and (b) the tension in the string.

Figure P23.45   Problems 45 and 46.

46. A charged cork ball of mass m is suspended on a light string in the presence of a uniform electric field as shown in Figure P23.45. When $\vec{E} = (A\hat{i} + B\hat{j})$ N/C, where A and B are positive numbers, the ball is in equilibrium at the angle θ. Find (a) the charge on the ball and (b) the tension in the string.

47. Four identical charged particles ($q = +10.0$ μC) are located on the corners of a rectangle as shown in Figure P23.47. The dimensions of the rectangle are $L = 60.0$ cm and $W = 15.0$ cm. Calculate the magnitude and direction of the total electric force exerted on the charge at the lower left corner by the other three charges.

Figure P23.47

48. Inez is putting up decorations for her sister's quinceañera (fifteenth birthday party). She ties three light silk ribbons together to the top of a gateway and hangs a rubber balloon from each ribbon (Fig. P23.48). To include the effects of the gravitational and buoyant forces on it, each balloon can be modeled as a particle of mass 2.00 g, with its center 50.0 cm from the point of support. To show off the colors of the balloons, Inez rubs the whole surface of each balloon with her woolen scarf, making them hang separately with gaps between them. The centers of the hanging balloons form a horizontal equilateral triangle with sides 30.0 cm long. What is the common charge each balloon carries?

Figure P23.48

49. **Review problem**. Two identical metallic blocks resting on a frictionless horizontal surface are connected by a light metallic spring having a spring constant k and an unstretched length L_i as shown in Figure P23.49a. A total charge Q is slowly placed on the system, causing the spring to stretch to an equilibrium length L as shown in Figure P23.49b. Determine the value of Q, assuming all the charge resides on the blocks and modeling the blocks as charged particles.

Figure P23.49

50. Consider a regular polygon with 29 sides. The distance from the center to each vertex is a. Identical charges q are placed at 28 vertices of the polygon. A single charge Q is placed at the center of the polygon. What is the magnitude and direction of the force experienced by the charge Q? *Suggestion:* You may use the result of Problem 60 in Chapter 3.

51. Identical thin rods of length $2a$ carry equal charges $+Q$ uniformly distributed along their lengths. The rods lie along the x axis with their centers separated by a distance $b > 2a$ (Fig. P23.51). Show that the magnitude of the force exerted by the left rod on the right one is

$$F = \left(\frac{k_e Q^2}{4a^2}\right) \ln \left(\frac{b^2}{b^2 - 4a^2}\right)$$

2 = intermediate;   3 = challenging;   □ = SSM/SG;   ▲ = ThomsonNOW;   = symbolic reasoning;   ● = qualitative reasoning

Figure P23.51

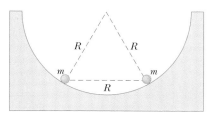

Figure P23.56

52. ⬛ Two small spheres hang in equilibrium at the bottom ends of threads, 40.0 cm long, that have their top ends tied to the same fixed point. One sphere has mass 2.40 g and charge +300 nC. The other sphere has the same mass and a charge of +200 nC. Find the distance between the centers of the spheres. You will need to solve an equation numerically.

53. A line of positive charge is formed into a semicircle of radius $R = 60.0$ cm as shown in Figure P23.53. The charge per unit length along the semicircle is described by the expression $\lambda = \lambda_0 \cos \theta$. The total charge on the semicircle is 12.0 μC. Calculate the total force on a charge of 3.00 μC placed at the center of curvature.

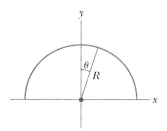

Figure P23.53

54. ● ⬛ Two particles, each with charge 52.0 nC, are located on the y axis at $y = 25.0$ cm and $y = -25.0$ cm. (a) Find the vector electric field at a point on the x axis as a function of x. (b) Find the field at $x = 36.0$ cm. (c) At what location is the field $1.00\hat{\mathbf{i}}$ kN/C? You may need to solve an equation numerically. (d) At what location is the field $16.0\hat{\mathbf{i}}$ kN/C? (e) Compare this problem with Question 7. Describe the similarities and explain the differences.

55. ● Two small spheres of mass m are suspended from strings of length ℓ that are connected at a common point. One sphere has charge Q and the other has charge $2Q$. The strings make angles θ_1 and θ_2 with the vertical. (a) Explain how θ_1 and θ_2 are related. (b) Assume θ_1 and θ_2 are small. Show that the distance r between the spheres is approximately

$$r \approx \left(\frac{4k_eQ^2\ell}{mg} \right)^{1/3}$$

56. Two identical beads each have a mass m and charge q. When placed in a hemispherical bowl of radius R with frictionless, nonconducting walls, the beads move, and at equilibrium they are a distance R apart (Fig. P23.56). Determine the charge on each bead.

57. ● **Review problem**. A 1.00-g cork ball with charge 2.00 μC is suspended vertically on a 0.500-m-long light string in the presence of a uniform, downward-directed electric field of magnitude $E = 1.00 \times 10^5$ N/C. If the ball is displaced slightly from the vertical, it oscillates like a simple pendulum. (a) Determine the period of this oscillation. (b) Should the effect of gravitation be included in the calculation for part (a)? Explain.

58. ● ⬛ Figure P23.35 shows three equal positive charges at the corners of an equilateral triangle of side $a = 3.00$ cm. Add a vertical line through the top charge at P, bisecting the triangle. Along this line label points A, B, C, D, E, and F, with A just below the charge at P; B at the center of the triangle; B, C, D, and E in order and close together with E at the center of the bottom side of the triangle; and F close below E. (a) Identify the direction of the total electric field at A, E, and F. Identify the electric field at B. Identify the direction of the electric field at C. (b) Argue that the answers to part (a) imply that the electric field must be zero at a point close to D. (c) Find the distance from point E on the bottom side of the triangle to the point around D where the electric field is zero. You will need to solve a transcendental equation.

59. Eight charged particles, each of magnitude q, are located on the corners of a cube of edge s as shown in Figure P23.59. (a) Determine the x, y, and z components of the total force exerted by the other charges on the charge located at point A. (b) What are the magnitude and direction of this total force?

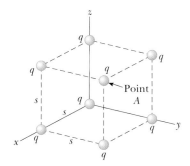

Figure P23.59 Problems 59 and 60.

60. Consider the charge distribution shown in Figure P23.59. (a) Show that the magnitude of the electric field at the center of any face of the cube has a value of $2.18k_eq/s^2$. (b) What is the direction of the electric field at the center of the top face of the cube?

61. **Review problem**. A negatively charged particle $-q$ is placed at the center of a uniformly charged ring, where the ring has a total positive charge Q as shown in Example 23.7. The particle, confined to move along the x axis,

is moved a small distance x along the axis (where $x \ll a$) and released. Show that the particle oscillates in simple harmonic motion with a frequency given by

$$f = \frac{1}{2\pi}\left(\frac{k_e q Q}{ma^3}\right)^{1/2}$$

62. A line of charge with uniform density 35.0 nC/m lies along the line $y = -15.0$ cm between the points with coordinates $x = 0$ and $x = 40.0$ cm. Find the electric field it creates at the origin.

63. **Review problem**. An electric dipole in a uniform electric field is displaced slightly from its equilibrium position as shown in Figure P23.63, where θ is small. The separation of the charges is $2a$, and the moment of inertia of the dipole is I. Assuming the dipole is released from this position, show that its angular orientation exhibits simple harmonic motion with a frequency

$$f = \frac{1}{2\pi}\sqrt{\frac{2qaE}{I}}$$

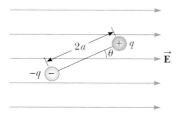

Figure P23.63

64. Consider an infinite number of identical particles, each with charge q, placed along the x axis at distances a, $2a$, $3a$, $4a$, . . . , from the origin. What is the electric field at the origin due to this distribution? *Suggestion:* Use the fact that

$$1 + \frac{1}{2^2} + \frac{1}{3^2} + \frac{1}{4^2} + \cdots = \frac{\pi^2}{6}$$

65. A line of charge starts at $x = +x_0$ and extends to positive infinity. The linear charge density is $\lambda = \lambda_0 x_0 / x$, where λ_0 is a constant. Determine the electric field at the origin.

Answers to Quick Quizzes

23.1 (a), (c), (e). The experiment shows that A and B have charges of the same sign, as do objects B and C. Therefore, all three objects have charges of the same sign. We cannot determine from this information, however, if the charges are positive or negative.

23.2 (e). In the first experiment, objects A and B may have charges with opposite signs or one of the objects may be neutral. The second experiment shows that B and C have charges with the same signs, so B must be charged. We still do not know, however, if A is charged or neutral.

23.3 (b). From Newton's third law, the electric force exerted by object B on object A is equal in magnitude to the force exerted by object A on object B and in the opposite direction.

23.4 (a). There is no effect on the electric field if we assume the source charge producing the field is not disturbed by our actions. Remember that the electric field is created by the source charge(s) (unseen in this case), not the test charge(s).

23.5 A, B, C. The field is greatest at point A because that is where the field lines are closest together. The absence of lines near point C indicates that the electric field there is zero.

In a tabletop plasma ball, the colorful lines emanating from the sphere give evidence of strong electric fields. Using Gauss's law, we show in this chapter that the electric field surrounding a uniformly charged sphere is identical to that of a point charge. (Getty Images)

24.1 Electric Flux

24.2 Gauss's Law

24.3 Application of Gauss's Law to Various Charge Distributions

24.4 Conductors in Electrostatic Equilibrium

24 Gauss's Law

In Chapter 23, we showed how to calculate the electric field due to a given charge distribution. In this chapter, we describe *Gauss's law* and an alternative procedure for calculating electric fields. Gauss's law is based on the inverse-square behavior of the electric force between point charges. Although Gauss's law is a consequence of Coulomb's law, it is more convenient for calculating the electric fields of highly symmetric charge distributions and makes it possible to deal with complicated problems using qualitative reasoning.

24.1 Electric Flux

The concept of electric field lines was described qualitatively in Chapter 23. We now treat electric field lines in a more quantitative way.

Consider an electric field that is uniform in both magnitude and direction as shown in Figure 24.1. The field lines penetrate a rectangular surface of area A, whose plane is oriented perpendicular to the field. Recall from Section 23.6 that the number of lines per unit area (in other words, the *line density*) is proportional to the magnitude of the electric field. Therefore, the total number of lines penetrating the surface is proportional to the product EA. This product of the magnitude of the electric field E and surface area A perpendicular to the field is called the **electric flux** Φ_E (uppercase Greek letter phi):

$$\Phi_E = EA \qquad (24.1)$$

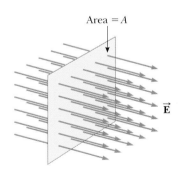

Figure 24.1 Field lines representing a uniform electric field penetrating a plane of area A perpendicular to the field. The electric flux Φ_E through this area is equal to EA.

673

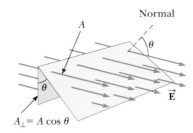

Figure 24.2 Field lines representing a uniform electric field penetrating an area A that is at an angle θ to the field. Because the number of lines that go through the area $A_\perp$ is the same as the number that go through A, the flux through $A_\perp$ is equal to the flux through A and is given by $\Phi_E = EA\cos\theta$.

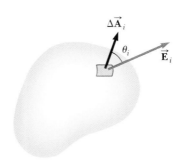

Figure 24.3 A small element of surface area ΔA_i. The electric field makes an angle θ_i with the vector $\Delta\vec{\mathbf{A}}_i$, defined as being normal to the surface element, and the flux through the element is equal to $E_i\,\Delta A_i\cos\theta_i$.

Definition of electric flux ▶

From the SI units of E and A, we see that Φ_E has units of newton meters squared per coulomb $(\text{N}\cdot\text{m}^2/\text{C})$. **Electric flux is proportional to the number of electric field lines penetrating some surface.**

If the surface under consideration is not perpendicular to the field, the flux through it must be less than that given by Equation 24.1. Consider Figure 24.2, where the normal to the surface of area A is at an angle θ to the uniform electric field. Notice that the number of lines that cross this area A is equal to the number of lines that cross the area $A_\perp$, which is a projection of area A onto a plane oriented perpendicular to the field. Figure 24.2 shows that the two areas are related by $A_\perp = A\cos\theta$. Because the flux through A equals the flux through $A_\perp$, the flux through A is

$$\Phi_E = EA_\perp = EA\cos\theta \qquad (24.2)$$

From this result, we see that the flux through a surface of fixed area A has a maximum value EA when the surface is perpendicular to the field (when the normal to the surface is parallel to the field, that is, when $\theta = 0°$ in Fig. 24.2); the flux is zero when the surface is parallel to the field (when the normal to the surface is perpendicular to the field, that is, when $\theta = 90°$).

We assumed a uniform electric field in the preceding discussion. In more general situations, the electric field may vary over a large surface. Therefore, the definition of flux given by Equation 24.2 has meaning only for a small element of area over which the field is approximately constant. Consider a general surface divided into a large number of small elements, each of area ΔA. It is convenient to define a vector $\Delta\vec{\mathbf{A}}_i$ whose magnitude represents the area of the ith element of the large surface and whose direction is defined to be *perpendicular* to the surface element as shown in Figure 24.3. The electric field $\vec{\mathbf{E}}_i$ at the location of this element makes an angle θ_i with the vector $\Delta\vec{\mathbf{A}}_i$. The electric flux $\Delta\Phi_E$ through this element is

$$\Delta\Phi_E = E_i\,\Delta A_i\cos\theta_i = \vec{\mathbf{E}}_i\cdot\Delta\vec{\mathbf{A}}_i$$

where we have used the definition of the scalar product of two vectors $(\vec{\mathbf{A}}\cdot\vec{\mathbf{B}} = AB\cos\theta$; see Chapter 7). Summing the contributions of all elements gives an approximation to the total flux through the surface:

$$\Phi_E \approx \sum \vec{\mathbf{E}}_i\cdot\Delta\vec{\mathbf{A}}_i$$

If the area of each element approaches zero, the number of elements approaches infinity and the sum is replaced by an integral. Therefore, the general definition of electric flux is

$$\Phi_E \equiv \int_{\text{surface}} \vec{\mathbf{E}}\cdot d\vec{\mathbf{A}} \qquad (24.3)$$

Equation 24.3 is a *surface integral*, which means it must be evaluated over the surface in question. In general, the value of Φ_E depends both on the field pattern and on the surface.

We are often interested in evaluating the flux through a *closed surface*, defined as a surface that divides space into an inside and an outside region so that one cannot move from one region to the other without crossing the surface. The surface of a sphere, for example, is a closed surface.

Consider the closed surface in Active Figure 24.4. The vectors $\Delta\vec{\mathbf{A}}_i$ point in different directions for the various surface elements, but at each point they are normal to the surface and, by convention, always point outward. At the element labeled ①, the field lines are crossing the surface from the inside to the outside and $\theta < 90°$; hence, the flux $\Delta\Phi_E = \vec{\mathbf{E}}\cdot\Delta\vec{\mathbf{A}}_1$ through this element is positive. For element ②, the field lines graze the surface (perpendicular to the vector $\Delta\vec{\mathbf{A}}_2$); therefore, $\theta = 90°$ and the flux is zero. For elements such as ③, where the field lines are crossing the surface from outside to inside, $180° > \theta > 90°$ and the flux is negative because $\cos\theta$ is negative. The *net* flux through the surface is proportional to the net number of lines leaving the surface, where the net number

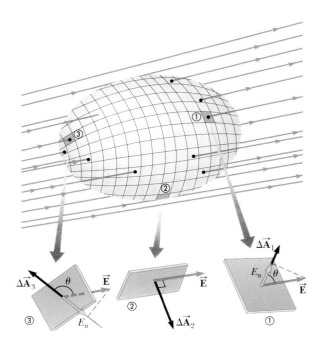

ACTIVE FIGURE 24.4

A closed surface in an electric field. The area vectors are, by convention, normal to the surface and point outward. The flux through an area element can be positive (element ①), zero (element ②), or negative (element ③).

Sign in at www.thomsonedu.com and go to ThomsonNOW to select any segment on the surface and see the relationship between the electric field vector $\vec{\mathbf{E}}$ and the area vector $\Delta\vec{\mathbf{A}}_i$.

means *the number of lines leaving the surface minus the number of lines entering the surface.* If more lines are leaving than entering, the net flux is positive. If more lines are entering than leaving, the net flux is negative. Using the symbol $\oint$ to represent an integral over a closed surface, we can write the net flux Φ_E through a closed surface as

$$\Phi_E = \oint \vec{\mathbf{E}} \cdot d\vec{\mathbf{A}} = \oint E_n \, dA \qquad \text{(24.4)}$$

where E_n represents the component of the electric field normal to the surface.

Quick Quiz 24.1 Suppose a point charge is located at the center of a spherical surface. The electric field at the surface of the sphere and the total flux through the sphere are determined. Now the radius of the sphere is halved. What happens to the flux through the sphere and the magnitude of the electric field at the surface of the sphere? (a) The flux and field both increase. (b) The flux and field both decrease. (c) The flux increases, and the field decreases. (d) The flux decreases, and the field increases. (e) The flux remains the same, and the field increases. (f) The flux decreases, and the field remains the same.

EXAMPLE 24.1   **Flux Through a Cube**

Consider a uniform electric field $\vec{\mathbf{E}}$ oriented in the x direction in empty space. Find the net electric flux through the surface of a cube of edge length ℓ, oriented as shown in Figure 24.5.

SOLUTION

Conceptualize Examine Figure 24.5 carefully. Notice that the electric field lines pass through two faces perpendicularly and are parallel to four other faces of the cube.

Figure 24.5 (Example 24.1) A closed surface in the shape of a cube in a uniform electric field oriented parallel to the x axis. Side ④ is the bottom of the cube, and side ① is opposite side ②.

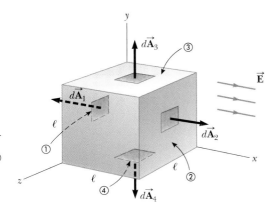

Categorize   We evaluate the flux from its definition, so we categorize this example as a substitution problem.

The flux through four of the faces (③, ④, and the unnumbered ones) is zero because $\vec{\mathbf{E}}$ is parallel to the four faces and therefore perpendicular to $d\vec{\mathbf{A}}$ on these faces.

Write the integrals for the net flux through faces ① and ②:	$$\Phi_E = \int_1 \vec{\mathbf{E}} \cdot d\vec{\mathbf{A}} + \int_2 \vec{\mathbf{E}} \cdot d\vec{\mathbf{A}}$$
For face ①, $\vec{\mathbf{E}}$ is constant and directed inward but $d\vec{\mathbf{A}}_1$ is directed outward ($\theta = 180°$). Find the flux through this face:	$$\int_1 \vec{\mathbf{E}} \cdot d\vec{\mathbf{A}} = \int_1 E(\cos 180°)\, dA = -E \int_1 dA = -EA = -E\ell^2$$
For face ②, $\vec{\mathbf{E}}$ is constant and outward and in the same direction as $d\vec{\mathbf{A}}_2$ ($\theta = 0°$). Find the flux through this face:	$$\int_2 \vec{\mathbf{E}} \cdot d\vec{\mathbf{A}} = \int_2 E(\cos 0°)\, dA = E \int_2 dA = +EA = E\ell^2$$
Find the net flux by adding the flux over all six faces:	$$\Phi_E = -E\ell^2 + E\ell^2 + 0 + 0 + 0 + 0 = \boxed{0}$$

24.2 Gauss's Law

In this section, we describe a general relationship between the net electric flux through a closed surface (often called a *gaussian surface*) and the charge enclosed by the surface. This relationship, known as *Gauss's law*, is of fundamental importance in the study of electric fields.

Consider a positive point charge q located at the center of a sphere of radius r as shown in Figure 24.6. From Equation 23.9, we know that the magnitude of the electric field everywhere on the surface of the sphere is $E = k_e q/r^2$. The field lines are directed radially outward and hence are perpendicular to the surface at every point on the surface. That is, at each surface point, $\vec{\mathbf{E}}$ is parallel to the vector $\Delta\vec{\mathbf{A}}_i$ representing a local element of area ΔA_i surrounding the surface point. Therefore,

$$\vec{\mathbf{E}} \cdot \Delta\vec{\mathbf{A}}_i = E\, \Delta A_i$$

and, from Equation 24.4, we find that the net flux through the gaussian surface is

$$\Phi_E = \oint \vec{\mathbf{E}} \cdot d\vec{\mathbf{A}} = \oint E\, dA = E \oint dA$$

where we have moved E outside of the integral because, by symmetry, E is constant over the surface. The value of E is given by $E = k_e q/r^2$. Furthermore, because the surface is spherical, $\oint dA = A = 4\pi r^2$. Hence, the net flux through the gaussian surface is

$$\Phi_E = k_e \frac{q}{r^2}(4\pi r^2) = 4\pi k_e q$$

Recalling from Section 23.3 that $k_e = 1/4\pi\epsilon_0$, we can write this equation in the form

$$\Phi_E = \frac{q}{\epsilon_0} \qquad\qquad \textbf{(24.5)}$$

Equation 24.5 shows that the net flux through the spherical surface is proportional to the charge inside the surface. The flux is independent of the radius r because the area of the spherical surface is proportional to r^2, whereas the electric field is proportional to $1/r^2$. Therefore, in the product of area and electric field, the dependence on r cancels.

KARL FRIEDRICH GAUSS
German mathematician and astronomer (1777–1855)
Gauss received a doctoral degree in mathematics from the University of Helmstedt in 1799. In addition to his work in electromagnetism, he made contributions to mathematics and science in number theory, statistics, non-Euclidean geometry, and cometary orbital mechanics. He was a founder of the German Magnetic Union, which studies the Earth's magnetic field on a continual basis.

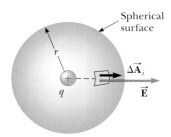

Figure 24.6 A spherical gaussian surface of radius r surrounding a point charge q. When the charge is at the center of the sphere, the electric field is everywhere normal to the surface and constant in magnitude.

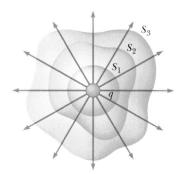

Figure 24.7 Closed surfaces of various shapes surrounding a charge q. The net electric flux is the same through all surfaces.

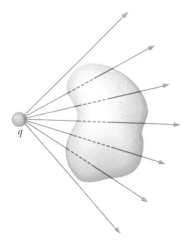

Figure 24.8 A point charge located *outside* a closed surface. The number of lines entering the surface equals the number leaving the surface.

Now consider several closed surfaces surrounding a charge q as shown in Figure 24.7. Surface S_1 is spherical, but surfaces S_2 and S_3 are not. From Equation 24.5, the flux that passes through S_1 has the value q/ϵ_0. As discussed in the preceding section, flux is proportional to the number of electric field lines passing through a surface. The construction shown in Figure 24.7 shows that the number of lines through S_1 is equal to the number of lines through the nonspherical surfaces S_2 and S_3. Therefore, **the net flux through *any* closed surface surrounding a point charge q is given by q/ϵ_0 and is independent of the shape of that surface.**

Now consider a point charge located *outside* a closed surface of arbitrary shape as shown in Figure 24.8. As can be seen from this construction, any electric field line entering the surface leaves the surface at another point. The number of electric field lines entering the surface equals the number leaving the surface. Therefore, **the net electric flux through a closed surface that surrounds no charge is zero.** Applying this result to Example 24.1, we see that the net flux through the cube is zero because there is no charge inside the cube.

Let's extend these arguments to two generalized cases: (1) that of many point charges and (2) that of a continuous distribution of charge. We once again use the superposition principle, which states that **the electric field due to many charges is the vector sum of the electric fields produced by the individual charges.** Therefore, the flux through any closed surface can be expressed as

$$\oint \vec{\mathbf{E}} \cdot d\vec{\mathbf{A}} = \oint (\vec{\mathbf{E}}_1 + \vec{\mathbf{E}}_2 + \cdots) \cdot d\vec{\mathbf{A}}$$

where $\vec{\mathbf{E}}$ is the total electric field at any point on the surface produced by the vector addition of the electric fields at that point due to the individual charges. Consider the system of charges shown in Active Figure 24.9. The surface S surrounds only one charge, q_1; hence, the net flux through S is q_1/ϵ_0. The flux through S due to charges q_2, q_3, and q_4 outside it is zero because each electric field line from these charges that enters S at one point leaves it at another. The surface S' surrounds charges q_2 and q_3; hence, the net flux through it is $(q_2 + q_3)/\epsilon_0$. Finally, the net flux through surface S'' is zero because there is no charge inside this surface. That is, *all* the electric field lines that enter S'' at one point leave at another. Charge q_4 does not contribute to the net flux through any of the surfaces because it is outside all the surfaces.

Gauss's law is a generalization of what we have just described and states that the net flux through *any* closed surface is

$$\Phi_E = \oint \vec{\mathbf{E}} \cdot d\vec{\mathbf{A}} = \frac{q_{\text{in}}}{\epsilon_0} \qquad (24.6) \qquad \blacktriangleleft \text{ Gauss's law}$$

where $\vec{\mathbf{E}}$ represents the electric field at any point on the surface and q_{in} represents the net charge inside the surface.

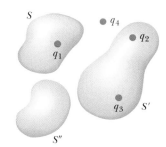

ACTIVE FIGURE 24.9

The net electric flux through any closed surface depends only on the charge *inside* that surface. The net flux through surface S is q_1/ϵ_0, the net flux through surface S' is $(q_2 + q_3)/\epsilon_0$, and the net flux through surface S'' is zero. Charge q_4 does not contribute to the flux through any surface because it is outside all surfaces.

Sign in at www.thomsonedu.com and go to ThomsonNOW to change the size and shape of a closed surface and see the effect on the electric flux of surrounding combinations of charge with that surface.

When using Equation 24.6, you should note that although the charge q_{in} is the net charge inside the gaussian surface, $\vec{E}$ represents the *total electric field,* which includes contributions from charges both inside and outside the surface.

In principle, Gauss's law can be solved for $\vec{E}$ to determine the electric field due to a system of charges or a continuous distribution of charge. In practice, however, this type of solution is applicable only in a limited number of highly symmetric situations. In the next section, we use Gauss's law to evaluate the electric field for charge distributions that have spherical, cylindrical, or planar symmetry. If one chooses the gaussian surface surrounding the charge distribution carefully, the integral in Equation 24.6 can be simplified.

Quick Quiz 24.2 If the net flux through a gaussian surface is *zero,* the following four statements *could be true.* Which of the statements *must be true*? (a) There are no charges inside the surface. (b) The net charge inside the surface is zero. (c) The electric field is zero everywhere on the surface. (d) The number of electric field lines entering the surface equals the number leaving the surface.

CONCEPTUAL EXAMPLE 24.2 | **Flux Due to a Point Charge**

A spherical gaussian surface surrounds a point charge q. Describe what happens to the total flux through the surface if **(A)** the charge is tripled, **(B)** the radius of the sphere is doubled, **(C)** the surface is changed to a cube, and **(D)** the charge is moved to another location inside the surface.

SOLUTION

(A) The flux through the surface is tripled because flux is proportional to the amount of charge inside the surface.

(B) The flux does not change because all electric field lines from the charge pass through the sphere, regardless of its radius.

(C) The flux does not change when the shape of the gaussian surface changes because all electric field lines from the charge pass through the surface, regardless of its shape.

(D) The flux does not change when the charge is moved to another location inside that surface because Gauss's law refers to the total charge enclosed, regardless of where the charge is located inside the surface.

24.3 Application of Gauss's Law to Various Charge Distributions

As mentioned earlier, Gauss's law is useful for determining electric fields when the charge distribution is highly symmetric. The following examples demonstrate ways of choosing the gaussian surface over which the surface integral given by Equation 24.6 can be simplified and the electric field determined. In choosing the surface, always take advantage of the symmetry of the charge distribution so that E can be removed from the integral. The goal in this type of calculation is to determine a surface for which each portion of the surface satisfies one or more of the following conditions:

1. The value of the electric field can be argued by symmetry to be constant over the portion of the surface.
2. The dot product in Equation 24.6 can be expressed as a simple algebraic product $E\, dA$ because $\vec{E}$ and $d\vec{A}$ are parallel.
3. The dot product in Equation 24.6 is zero because $\vec{E}$ and $d\vec{A}$ are perpendicular.
4. The electric field is zero over the portion of the surface.

Different portions of the gaussian surface can satisfy different conditions as long as every portion satisfies at least one condition. All four conditions are used in examples throughout the remainder of this chapter.

| EXAMPLE 24.3 | **A Spherically Symmetric Charge Distribution** |

An insulating solid sphere of radius a has a uniform volume charge density ρ and carries a total positive charge Q (Fig. 24.10).

(A) Calculate the magnitude of the electric field at a point outside the sphere.

SOLUTION

Conceptualize   Note how this problem differs from our previous discussion of Gauss's law. The electric field due to point charges was discussed in Section 24.2. Now we are considering the electric field due to a distribution of charge. We found the field for various distributions of charge in Chapter 23 by integrating over the distribution. In this chapter, we find the electric field using Gauss's law.

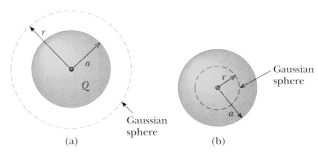

Figure 24.10   (Example 24.3) A uniformly charged insulating sphere of radius a and total charge Q. (a) For points outside the sphere, a large, spherical gaussian surface is drawn concentric with the sphere. In diagrams such as this one, the dotted line represents the intersection of the gaussian surface with the plane of the page. (b) For points inside the sphere, a spherical gaussian surface smaller than the sphere is drawn.

Categorize   Because the charge is distributed uniformly throughout the sphere, the charge distribution has spherical symmetry and we can apply Gauss's law to find the electric field.

Analyze   To reflect the spherical symmetry, let's choose a spherical gaussian surface of radius r, concentric with the sphere, as shown in Figure 24.10a. For this choice, condition (2) is satisfied everywhere on the surface and $\vec{\mathbf{E}} \cdot d\vec{\mathbf{A}} = E\,dA$.

Replace $\vec{\mathbf{E}} \cdot d\vec{\mathbf{A}}$ in Gauss's law with $E\,dA$:

$$\Phi_E = \oint \vec{\mathbf{E}} \cdot d\vec{\mathbf{A}} = \oint E\,dA = \frac{Q}{\epsilon_0}$$

By symmetry, E is constant everywhere on the surface, which satisfies condition (1), so we can remove E from the integral:

$$\oint E\,dA = E \oint dA = E(4\pi r^2) = \frac{Q}{\epsilon_0}$$

Solve for E:

$$(1)\quad E = \frac{Q}{4\pi\epsilon_0 r^2} = k_e \frac{Q}{r^2} \quad \text{(for } r > a\text{)}$$

Finalize   This field is identical to that for a point charge. Therefore, **the electric field due to a uniformly charged sphere in the region external to the sphere is** *equivalent* **to that of a point charge located at the center of the sphere.**

(B) Find the magnitude of the electric field at a point inside the sphere.

SOLUTION

Analyze   In this case, let's choose a spherical gaussian surface having radius $r < a$, concentric with the insulating sphere (Fig. 24.10b). Let V' be the volume of this smaller sphere. To apply Gauss's law in this situation, recognize that the charge q_{in} within the gaussian surface of volume V' is less than Q.

Calculate q_{in} by using $q_{\text{in}} = \rho V'$:

$$q_{\text{in}} = \rho V' = \rho(\tfrac{4}{3}\pi r^3)$$

Notice that conditions (1) and (2) are satisfied everywhere on the gaussian surface in Figure 24.10b. Apply Gauss's law in the region $r < a$:

$$\oint E\,dA = E \oint dA = E(4\pi r^2) = \frac{q_{\text{in}}}{\epsilon_0}$$

Solve for E and substitute for q_{in}:

$$E = \frac{q_{\text{in}}}{4\pi\epsilon_0 r^2} = \frac{\rho(\tfrac{4}{3}\pi r^3)}{4\pi\epsilon_0 r^2} = \frac{\rho}{3\epsilon_0}r$$

Substitute $\rho = Q/\frac{4}{3}\pi a^3$ and $\epsilon_0 = 1/4\pi k_e$:

$$(2) \quad E = \frac{(Q/\frac{4}{3}\pi a^3)}{3(1/4\pi k_e)} \, r = k_e \frac{Q}{a^3} r \quad (\text{for } r < a)$$

Finalize This result for E differs from the one obtained in part (A). It shows that $E \rightarrow 0$ as $r \rightarrow 0$. Therefore, the result eliminates the problem that would exist at $r = 0$ if E varied as $1/r^2$ inside the sphere as it does outside the sphere. That is, if $E \propto 1/r^2$ for $r < a$, the field would be infinite at $r = 0$, which is physically impossible.

What If? Suppose the radial position $r = a$ is approached from inside the sphere and from outside. Do we obtain the same value of the electric field from both directions?

Answer Equation (1) shows that the electric field approaches a value from the outside given by

$$E = \lim_{r \rightarrow a}\left(k_e \frac{Q}{r^2}\right) = k_e \frac{Q}{a^2}$$

From the inside, Equation (2) gives

$$E = \lim_{r \rightarrow a}\left(k_e \frac{Q}{a^3} r\right) = k_e \frac{Q}{a^3} a = k_e \frac{Q}{a^2}$$

Therefore, the value of the field is the same as the surface is approached from both directions. A plot of E versus r is shown in Figure 24.11. Notice that the magnitude of the field is continuous.

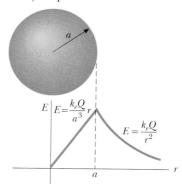

Figure 24.11 (Example 24.3) A plot of E versus r for a uniformly charged insulating sphere. The electric field inside the sphere $(r < a)$ varies linearly with r. The field outside the sphere $(r > a)$ is the same as that of a point charge Q located at $r = 0$.

EXAMPLE 24.4 **A Cylindrically Symmetric Charge Distribution**

Find the electric field a distance r from a line of positive charge of infinite length and constant charge per unit length λ (Fig. 24.12a).

SOLUTION

Conceptualize The line of charge is *infinitely* long. Therefore, the field is the same at all points equidistant from the line, regardless of the vertical position of the point in Figure 24.12a.

Categorize Because the charge is distributed uniformly along the line, the charge distribution has cylindrical symmetry and we can apply Gauss's law to find the electric field.

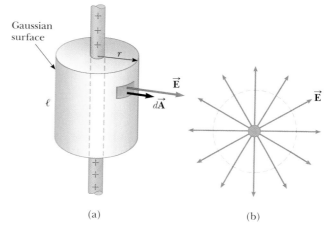

Figure 24.12 (Example 24.4) (a) An infinite line of charge surrounded by a cylindrical gaussian surface concentric with the line. (b) An end view shows that the electric field at the cylindrical surface is constant in magnitude and perpendicular to the surface.

Analyze The symmetry of the charge distribution requires that $\vec{E}$ be perpendicular to the line charge and directed outward as shown in Figures 24.12a and b. To reflect the symmetry of the charge distribution, let's choose a cylindrical gaussian surface of radius r and length ℓ that is coaxial with the line charge. For the curved part of this surface, $\vec{E}$ is constant in magnitude and perpendicular to the surface at each point, satisfying conditions (1) and (2). Furthermore, the flux through the ends of the gaussian cylinder is zero because $\vec{E}$ is parallel to these surfaces. That is the first application we have seen of condition (3).

We must take the surface integral in Gauss's law over the entire gaussian surface. Because $\vec{E} \cdot d\vec{A}$ is zero for the flat ends of the cylinder, however, we restrict our attention to only the curved surface of the cylinder.

Apply Gauss's law and conditions (1) and (2) for the curved surface, noting that the total charge inside our gaussian surface is $\lambda\ell$:

$$\Phi_E = \oint \vec{E} \cdot d\vec{A} = E \oint dA = EA = \frac{q_{in}}{\epsilon_0} = \frac{\lambda\ell}{\epsilon_0}$$

Substitute the area $A = 2\pi r\ell$ of the curved surface:

$$E(2\pi r\ell) = \frac{\lambda\ell}{\epsilon_0}$$

Solve for the magnitude of the electric field:

$$E = \frac{\lambda}{2\pi\epsilon_0 r} = \boxed{2k_e \frac{\lambda}{r}} \qquad (24.7)$$

Finalize   This result shows that the electric field due to a cylindrically symmetric charge distribution varies as $1/r$, whereas the field external to a spherically symmetric charge distribution varies as $1/r^2$. Equation 24.7 can also be derived by direct integration over the charge distribution. (See Problem 29 in Chapter 23.)

What If?   What if the line segment in this example were not infinitely long?

Answer   If the line charge in this example were of finite length, the electric field would not be given by Equation 24.7. A finite line charge does not possess sufficient symmetry to make use of Gauss's law because the magnitude of the electric field is no longer constant over the surface of the gaussian cylinder: the field near the ends of the line would be different from that far from the ends. Therefore, condition (1) would not be satisfied in this situation. Furthermore, $\vec{E}$ is not perpendicular to the cylindrical surface at all points: the field vectors near the ends would have a component parallel to the line. Therefore, condition (2) would not be satisfied. For points close to a finite line charge and far from the ends, Equation 24.7 gives a good approximation of the value of the field.

It is left for you to show (see Problem 27) that the electric field inside a uniformly charged rod of finite radius and infinite length is proportional to r.

EXAMPLE 24.5   A Plane of Charge

Find the electric field due to an infinite plane of positive charge with uniform surface charge density σ.

SOLUTION

Conceptualize   Note that the plane of charge is *infinitely* large. Therefore, the electric field should be the same at all points near the plane.

Categorize   Because the charge is distributed uniformly on the plane, the charge distribution is symmetric; hence, we can use Gauss's law to find the electric field.

Analyze   By symmetry, $\vec{E}$ must be perpendicular to the plane at all points. The direction of $\vec{E}$ is away from positive charges, indicating that the direction of $\vec{E}$ on one side of the plane must be opposite its direction on the other side as shown in Figure 24.13. A gaussian surface that reflects the symmetry is a small cylinder whose axis is perpendicular to the plane and whose ends each have an area A and are equidistant from the plane. Because $\vec{E}$ is parallel to the curved surface—and therefore perpendicular to $d\vec{A}$ everywhere on the surface—condition (3) is satisfied and there is no contribution to the surface integral from this surface. For the flat ends of the cylinder, conditions (1) and (2) are satisfied. The flux through each end of the cylinder is EA; hence, the total flux through the entire gaussian surface is just that through the ends, $\Phi_E = 2EA$.

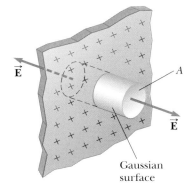

Figure 24.13   (Example 24.5) A cylindrical gaussian surface penetrating an infinite plane of charge. The flux is EA through each end of the gaussian surface and zero through its curved surface.

Write Gauss's law for this surface, noting that the enclosed charge is $q_{in} = \sigma A$:

$$\Phi_E = 2EA = \frac{q_{in}}{\epsilon_0} = \frac{\sigma A}{\epsilon_0}$$

Solve for E:

$$E = \frac{\sigma}{2\epsilon_0} \qquad (24.8)$$

Finalize   Because the distance from each flat end of the cylinder to the plane does not appear in Equation 24.8, we conclude that $E = \sigma/2\epsilon_0$ at *any* distance from the plane. That is, the field is uniform everywhere.

What If?   Suppose two infinite planes of charge are parallel to each other, one positively charged and the other negatively charged. Both planes have the same surface charge density. What does the electric field look like in this situation?

Answer   The electric fields due to the two planes add in the region between the planes, resulting in a uniform field of magnitude σ/ϵ_0, and cancel elsewhere to give a field of zero. This method is a practical way to achieve uniform electric fields.

CONCEPTUAL EXAMPLE 24.6   **Don't Use Gauss's Law Here!**

Explain why Gauss's law cannot be used to calculate the electric field near an electric dipole, a charged disk, or a triangle with a point charge at each corner.

SOLUTION

The charge distributions of all these configurations do not have sufficient symmetry to make the use of Gauss's law practical. We cannot find a closed surface surrounding any of these distributions that satisfies one or more of conditions (1) through (4) listed at the beginning of this section.

24.4 Conductors in Electrostatic Equilibrium

As we learned in Section 23.2, a good electrical conductor contains charges (electrons) that are not bound to any atom and therefore are free to move about within the material. When there is no net motion of charge within a conductor, the conductor is in **electrostatic equilibrium.** A conductor in electrostatic equilibrium has the following properties:

Properties of a conductor ▶
in electrostatic equilibrium

1. The electric field is zero everywhere inside the conductor, whether the conductor is solid or hollow.
2. If an isolated conductor carries a charge, the charge resides on its surface.
3. The electric field just outside a charged conductor is perpendicular to the surface of the conductor and has a magnitude σ/ϵ_0, where σ is the surface charge density at that point.
4. On an irregularly shaped conductor, the surface charge density is greatest at locations where the radius of curvature of the surface is smallest.

We verify the first three properties in the discussion that follows. The fourth property is presented here (but not verified until Chapter 25) to provide a complete list of properties for conductors in electrostatic equilibrium.

We can understand the first property by considering a conducting slab placed in an external field $\vec{\mathbf{E}}$ (Fig. 24.14). The electric field inside the conductor *must* be zero assuming electrostatic equilibrium exists. If the field were not zero, free electrons in the conductor would experience an electric force ($\vec{\mathbf{F}} = q\vec{\mathbf{E}}$) and would accelerate due to this force. This motion of electrons, however, would mean that the conductor is not in electrostatic equilibrium. Therefore, the existence of electrostatic equilibrium is consistent only with a zero field in the conductor.

Let's investigate how this zero field is accomplished. Before the external field is applied, free electrons are uniformly distributed throughout the conductor. When the external field is applied, the free electrons accelerate to the left in Figure 24.14, causing a plane of negative charge to accumulate on the left surface. The movement of electrons to the left results in a plane of positive charge on the right surface. These planes of charge create an additional electric field inside the conductor that opposes the external field. As the electrons move, the surface charge

Figure 24.14   A conducting slab in an external electric field $\vec{\mathbf{E}}$. The charges induced on the two surfaces of the slab produce an electric field that opposes the external field, giving a resultant field of zero inside the slab.

densities on the left and right surfaces increase until the magnitude of the internal field equals that of the external field, resulting in a net field of zero inside the conductor. The time it takes a good conductor to reach equilibrium is on the order of 10^{-16} s, which for most purposes can be considered instantaneous.

If the conductor is hollow, the electric field inside the conductor is also zero, whether we consider points in the conductor or in the cavity within the conductor. The zero value of the electric field in the cavity is easiest to argue with the concept of electric potential, so we will address this issue in Section 25.6.

Gauss's law can be used to verify the second property of a conductor in electrostatic equilibrium. Figure 24.15 shows an arbitrarily shaped conductor. A gaussian surface is drawn inside the conductor and can be very close to the conductor's surface. As we have just shown, the electric field everywhere inside the conductor is zero when it is in electrostatic equilibrium. Therefore, the electric field must be zero at every point on the gaussian surface, in accordance with condition (4) in Section 24.3, and the net flux through this gaussian surface is zero. From this result and Gauss's law, we conclude that the net charge inside the gaussian surface is zero. Because there can be no net charge inside the gaussian surface (which is arbitrarily close to the conductor's surface), **any net charge on the conductor must reside on its surface.** Gauss's law does not indicate how this excess charge is distributed on the conductor's surface, only that it resides exclusively on the surface.

Let's verify the third property. If the field vector $\vec{\mathbf{E}}$ had a component parallel to the conductor's surface, free electrons would experience an electric force and move along the surface; in such a case, the conductor would not be in equilibrium. Therefore, the field vector must be perpendicular to the surface. To determine the magnitude of the electric field, we use Gauss's law and draw a gaussian surface in the shape of a small cylinder whose end faces are parallel to the conductor's surface (Fig. 24.16). Part of the cylinder is just outside the conductor, and part is inside. The field is perpendicular to the conductor's surface from the condition of electrostatic equilibrium. Therefore, condition (3) in Section 24.3 is satisfied for the curved part of the cylindrical gaussian surface: there is no flux through this part of the gaussian surface because $\vec{\mathbf{E}}$ is parallel to the surface. There is no flux through the flat face of the cylinder inside the conductor because here $\vec{\mathbf{E}} = 0$, which satisfies condition (4). Hence, the net flux through the gaussian surface is equal to that through only the flat face outside the conductor, where the field is perpendicular to the gaussian surface. Using conditions (1) and (2) for this face, the flux is EA, where E is the electric field just outside the conductor and A is the area of the cylinder's face. Applying Gauss's law to this surface gives

$$\Phi_E = \oint E\, dA = EA = \frac{q_{\text{in}}}{\epsilon_0} = \frac{\sigma A}{\epsilon_0}$$

where we have used $q_{\text{in}} = \sigma A$. Solving for E gives for the electric field immediately outside a charged conductor

$$E = \frac{\sigma}{\epsilon_0} \tag{24.9}$$

Figure 24.15 A conductor of arbitrary shape. The broken line represents a gaussian surface that can be just inside the conductor's surface.

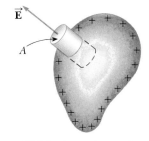

Figure 24.16 A gaussian surface in the shape of a small cylinder is used to calculate the electric field immediately outside a charged conductor. The flux through the gaussian surface is EA. Remember that $\vec{\mathbf{E}}$ is zero inside the conductor.

Quick Quiz 24.3   Your younger brother likes to rub his feet on the carpet and then touch you to give you a shock. While you are trying to escape the shock treatment, you discover a hollow metal cylinder in your basement, large enough to climb inside. In which of the following cases will you *not* be shocked? (a) You climb inside the cylinder, making contact with the inner surface, and your charged brother touches the outer metal surface. (b) Your charged brother is inside touching the inner metal surface and you are outside, touching the outer metal surface. (c) Both of you are outside the cylinder, touching its outer metal surface but not touching each other directly.

EXAMPLE 24.7   **A Sphere Inside a Spherical Shell**

A solid insulating sphere of radius a carries a net positive charge Q uniformly distributed throughout its volume. A conducting spherical shell of inner radius b and outer radius c is concentric with the solid sphere and carries a net charge $-2Q$. Using Gauss's law, find the electric field in the regions labeled ①, ②, ③, and ④ in Active Figure 24.17 and the charge distribution on the shell when the entire system is in electrostatic equilibrium.

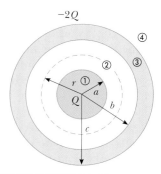

ACTIVE FIGURE 24.17

(Example 24.7) An insulating sphere of radius a and carrying a charge Q surrounded by a conducting spherical shell carrying a charge $-2Q$.

Sign in at www.thomsonedu.com and go to ThomsonNOW to vary the charges on the sphere and shell and see the effect on the electric field.

SOLUTION

Conceptualize   Note how this problem differs from Example 24.3. The charged sphere in Figure 24.10 is now surrounded by a shell carrying a charge of $-2Q$.

Categorize   The charge is distributed uniformly throughout the sphere, and we know that the charge on the conducting shell distributes itself uniformly on the surfaces. Therefore, the system has spherical symmetry and we can apply Gauss's law to find the electric field.

Analyze   In region ②—between the surface of the solid sphere and the inner surface of the shell—we construct a spherical gaussian surface of radius r, where $a < r < b$, noting that the charge inside this surface is $+Q$ (the charge on the solid sphere). Because of the spherical symmetry, the electric field lines must be directed radially outward and be constant in magnitude on the gaussian surface.

The charge on the conducting shell creates zero electric field in the region $r < b$, so the shell has no effect on the field due to the sphere. Therefore, write an expression for the field in region ② as that due to the sphere from part (A) of Example 24.3:

$$E_2 = k_e \frac{Q}{r^2} \quad (\text{for } a < r < b)$$

Because the conducting shell creates zero field inside itself, it also has no effect on the field inside the sphere. Therefore, write an expression for the field in region ① as that due to the sphere from part (B) of Example 24.3:

$$E_1 = k_e \frac{Q}{a^3} r \quad (\text{for } r < a)$$

In region ④, where $r > c$, construct a spherical gaussian surface; this surface surrounds a total charge of $q_{in} = Q + (-2Q) = -Q$. Therefore, model the charge distribution as a sphere with charge $-Q$ and write an expression for the field in region ④ from part (A) of Example 24.3:

$$E_4 = -k_e \frac{Q}{r^2} \quad (\text{for } r > c)$$

In region ③, the electric field must be zero because the spherical shell is a conductor in equilibrium:

$$E_3 = 0 \quad (\text{for } b < r < c)$$

Construct a gaussian surface of radius r, where $b < r < c$, and note that q_{in} must be zero because $E_3 = 0$. Find the amount of charge q_{inner} on the inner surface of the shell:

$$q_{in} = q_{sphere} + q_{inner}$$

$$q_{inner} = q_{in} - q_{sphere} = 0 - Q = -Q$$

Finalize   The charge on the inner surface of the spherical shell must be $-Q$ to cancel the charge $+Q$ on the solid sphere and give zero electric field in the material of the shell. Because the net charge on the shell is $-2Q$, its outer surface must carry a charge $-Q$.

What If? How would the results of this problem differ if the sphere were conducting instead of insulating?

Answer The only change would be in region ①, where $r < a$. Because there can be no charge inside a conductor in electrostatic equilibrium, $q_{in} = 0$ for a gaussian surface of radius $r < a$; therefore, on the basis of Gauss's law and symmetry, $E_1 = 0$. In regions ②, ③, and ④, there would be no way to determine whether the sphere is conducting or insulating.

Summary

ThomsonNOW Sign in at **www.thomsonedu.com** and go to ThomsonNOW to take a practice test for this chapter.

DEFINITION

Electric flux is proportional to the number of electric field lines that penetrate a surface. If the electric field is uniform and makes an angle θ with the normal to a surface of area A, the electric flux through the surface is

$$\Phi_E = EA \cos \theta \tag{24.2}$$

In general, the electric flux through a surface is

$$\Phi_E \equiv \int_{surface} \vec{E} \cdot d\vec{A} \tag{24.3}$$

CONCEPTS AND PRINCIPLES

Gauss's law says that the net electric flux Φ_E through any closed gaussian surface is equal to the *net* charge q_{in} inside the surface divided by ϵ_0:

$$\Phi_E = \oint \vec{E} \cdot d\vec{A} = \frac{q_{in}}{\epsilon_0} \tag{24.6}$$

Using Gauss's law, you can calculate the electric field due to various symmetric charge distributions.

A conductor in **electrostatic equilibrium** has the following properties:

1. The electric field is zero everywhere inside the conductor, whether the conductor is solid or hollow.
2. If an isolated conductor carries a charge, the charge resides on its surface.
3. The electric field just outside a charged conductor is perpendicular to the surface of the conductor and has a magnitude σ/ϵ_0, where σ is the surface charge density at that point.
4. On an irregularly shaped conductor, the surface charge density is greatest at locations where the radius of curvature of the surface is smallest.

Questions

□ denotes answer available in *Student Solutions Manual/Study Guide;* **O** denotes objective question

1. The Sun is lower in the sky during the winter than it is during the summer. How does this change affect the flux of sunlight hitting a given area on the surface of the Earth? How does this change affect the weather?

2. If more electric field lines leave a gaussian surface than enter it, what can you conclude about the net charge enclosed by that surface?

3. A uniform electric field exists in a region of space containing no charges. What can you conclude about the net electric flux through a gaussian surface placed in this region of space?

4. **O** A particle with charge q is located inside a cubical gaussian surface. No other charges are nearby. **(i)** If the particle is at the center of the cube, what is the flux through each one of the faces of the cube? (a) 0 (b) q/ϵ_0 (c) $q/2\epsilon_0$ (d) $q/4\epsilon_0$ (e) $q/6\epsilon_0$ (f) $q/8\epsilon_0$ (g) depends on the size of the cube **(ii)** If the particle can be moved to any point within the cube, what maximum value can the flux through one face approach? Choose from the same possibilities. **(iii)** If the particle can be moved anywhere within the cube or on its surface, what is the minimum possible flux through one face? Choose from the same possibilities.

5. **O** A cubical gaussian surface surrounds a long, straight, charged filament that passes perpendicularly through two opposite faces. No other charges are nearby. **(i)** Over how many of the cube's faces is the electric field zero? (a) 0

(b) 2 (c) 4 (d) 6 **(ii)** Through how many of the cube's faces is the electric flux zero? Choose from the same possibilities.

6. O A cubical gaussian surface is bisected by a large sheet of charge, parallel to its top and bottom faces. No other charges are nearby. **(i)** Over how many of the cube's faces is the electric field zero? (a) 0 (b) 2 (c) 4 (d) 6 **(ii)** Through how many of the cube's faces is the electric flux zero? Choose from the same possibilities.

7. O Two solid spheres, both of radius 5 cm, carry identical total charges of 2 μC. Sphere A is a good conductor. Sphere B is an insulator, and its charge is distributed uniformly throughout its volume. **(i)** How do the magnitudes of the electric fields they separately create at a radial distance of 6 cm compare? (a) $E_A > E_B = 0$ (b) $E_A > E_B > 0$ (c) $E_A = E_B > 0$ (d) $E_A = E_B = 0$ (e) $0 < E_A < E_B$ (f) $0 = E_A < E_B$ **(ii)** How do the magnitudes of the electric fields they separately create at radius 4 cm compare? Choose from the same possibilities.

8. If the total charge inside a closed surface is known but the distribution of the charge is unspecified, can you use Gauss's law to find the electric field? Explain.

9. Explain why the electric flux through a closed surface with a given enclosed charge is independent of the size or shape of the surface.

10. On the basis of the repulsive nature of the force between like charges and the freedom of motion of charge within a conductor, explain why excess charge on an isolated conductor must reside on its surface.

11. O A solid insulating sphere of radius 5 cm carries electric charge uniformly distributed throughout its volume. Concentric with the sphere is a conducting spherical shell with no net charge as shown in Figure Q24.11. The inner radius of the shell is 10 cm, and the outer radius is 15 cm. No other charges are nearby. (a) Rank the magnitude of

Figure Q24.11 Question 11 and Problem 44.

the electric field at points A (at radius 4 cm), B (radius 8 cm), C (radius 12 cm), and D (radius 16 cm) from largest to smallest. Display any cases of equality in your ranking. (b) Similarly rank the electric flux through concentric spherical surfaces through points A, B, C, and D.

12. O A coaxial cable consists of a long, straight filament surrounded by a long, coaxial, cylindrical conducting shell. Assume charge Q is on the filament, zero net charge is on the shell, and the electric field is $E_1\hat{\mathbf{i}}$ at a particular point P midway between the filament and the inner surface of the shell. Next, you place the cable into a uniform external field $-E_1\hat{\mathbf{i}}$. What is the x component of the electric field at P then? (a) 0 (b) between 0 and E_1 (c) E_1 (d) greater than E_1 (e) between 0 and $-E_1$ (f) $-E_1$ (g) less than $-E_1$

13. A person is placed in a large, hollow, metallic sphere that is insulated from ground. If a large charge is placed on the sphere, will the person be harmed upon touching the inside of the sphere? Explain what will happen if the person also has an initial charge whose sign is opposite that of the charge on the sphere.

14. O A large, metallic, spherical shell has no net charge. It is supported on an insulating stand and has a small hole at the top. A small tack with charge Q is lowered on a silk thread through the hole into the interior of the shell. **(i)** What is the charge on the inner surface of the shell now? (a) Q (b) $Q/2$ (c) 0 (d) $-Q/2$ (e) $-Q$ Choose your answers to the following parts from the same possibilities. **(ii)** What is the charge on the outer surface of the shell? **(iii)** The tack is now allowed to touch the interior surface of the shell. After this contact, what is the charge on the tack? **(iv)** What is the charge on the inner surface of the shell now? **(v)** What is the charge on the outer surface of the shell now?

15. A common demonstration involves charging a rubber balloon, which is an insulator, by rubbing it on your hair and then touching the balloon to a ceiling or wall, which is also an insulator. Because of the electrical attraction between the charged balloon and the neutral wall, the balloon sticks to the wall. Imagine now that we have two infinitely large flat sheets of insulating material. One is charged, and the other is neutral. If these sheets are brought into contact, does an attractive force exist between them, as there was for the balloon and the wall?

Problems

WebAssign The Problems from this chapter may be assigned online in WebAssign.

ThomsonNOW Sign in at **www.thomsonedu.com** and go to ThomsonNOW to assess your understanding of this chapter's topics with additional quizzing and conceptual questions.

1, 2, 3 denotes straightforward, intermediate, challenging; ☐ denotes full solution available in *Student Solutions Manual/Study Guide;* ▲ denotes coached solution with hints available at **www.thomsonedu.com;** denotes developing symbolic reasoning; ● denotes asking for qualitative reasoning; ⬚ denotes computer useful in solving problem

Section 24.1 Electric Flux

1. A 40.0-cm-diameter loop is rotated in a uniform electric field until the position of maximum electric flux is found. The flux in this position is 5.20×10^5 N·m²/C. What is the magnitude of the electric field?

2. A vertical electric field of magnitude 2.00×10^4 N/C exists above the Earth's surface on a day when a thunderstorm is brewing. A car with a rectangular size of 6.00 m by 3.00 m is traveling along a dry gravel roadway sloping downward at 10.0°. Determine the electric flux through the bottom of the car.

3. A uniform electric field $a\hat{\mathbf{i}} + b\hat{\mathbf{j}}$ intersects a surface of area A. What is the flux through this area if the surface lies (a) in the yz plane, (b) in the xz plane, and (c) in the xy plane?

4. Consider a closed triangular box resting within a horizontal electric field of magnitude $E = 7.80 \times 10^4$ N/C as shown in Figure P24.4. Calculate the electric flux through (a) the vertical rectangular surface, (b) the slanted surface, and (c) the entire surface of the box.

Figure P24.4

5. A pyramid with horizontal square base, 6.00 m on each side, and a height of 4.00 m is placed in a vertical electric field of 52.0 N/C. Calculate the total electric flux through the pyramid's four slanted surfaces.

Section 24.2 Gauss's Law

6. ● The electric field everywhere on the surface of a thin, spherical shell of radius 0.750 m is measured to be 890 N/C and points radially toward the center of the sphere. (a) What is the net charge within the sphere's surface? (b) What can you conclude about the nature and distribution of the charge inside the spherical shell?

7. The following charges are located inside a submarine: 5.00 μC, −9.00 μC, 27.0 μC, and −84.0 μC. (a) Calculate the net electric flux through the hull of the submarine. (b) Is the number of electric field lines leaving the submarine greater than, equal to, or less than the number entering it?

8. ● (a) A particle with charge q is located a distance d from an infinite plane. Determine the electric flux through the plane due to the charged particle. (b) **What If?** A particle with charge q is located a *very small* distance from the center of a *very large* square on the line perpendicular to the square and going through its center. Determine the approximate electric flux through the square due to the charged particle. (c) Explain why the answers to parts (a) and (b) are identical.

9. Four closed surfaces, S_1 through S_4, together with the charges $-2Q$, Q, and $-Q$ are sketched in Figure P24.9.

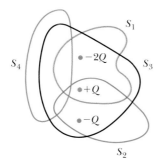

Figure P24.9

(The colored lines are the intersections of the surfaces with the page.) Find the electric flux through each surface.

10. ● A particle with charge of 12.0 μC is placed at the center of a spherical shell of radius 22.0 cm. What is the total electric flux through (a) the surface of the shell and (b) any hemispherical surface of the shell? (c) Do the results depend on the radius? Explain.

11. ▲ A particle with charge Q is located immediately above the center of the flat face of a hemisphere of radius R as shown in Figure P24.11. What is the electric flux (a) through the curved surface and (b) through the flat face?

Figure P24.11

12. In the air over a particular region at an altitude of 500 m above the ground, the electric field is 120 N/C directed downward. At 600 m above the ground, the electric field is 100 N/C downward. What is the average volume charge density in the layer of air between these two elevations? Is it positive or negative?

13. A particle with charge $Q = 5.00$ μC is located at the center of a cube of edge $L = 0.100$ m. In addition, six other identical charged particles having $q = -1.00$ μC are positioned symmetrically around Q as shown in Figure P24.13. Determine the electric flux through one face of the cube.

Figure P24.13   Problems 13 and 14.

14. A particle with charge Q is located at the center of a cube of edge L. In addition, six other identical negative charged particles $-q$ are positioned symmetrically around Q as shown in Figure P24.13. Determine the electric flux through one face of the cube.

15. An infinitely long line charge having a uniform charge per unit length λ lies a distance d from point O as shown in Figure P24.15. Determine the total electric flux

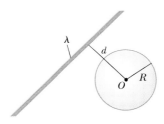

Figure P24.15

2 = intermediate;   3 = challenging;   ☐ = SSM/SG;   ▲ = ThomsonNOW;   ▧ = symbolic reasoning;   ● = qualitative reasoning

through the surface of a sphere of radius R centered at O resulting from this line charge. Consider both cases, where $R < d$ and $R > d$.

16. An uncharged nonconducting hollow sphere of radius 10.0 cm surrounds a 10.0-μC charge located at the origin of a cartesian coordinate system. A drill with a radius of 1.00 mm is aligned along the z axis, and a hole is drilled in the sphere. Calculate the electric flux through the hole.

17. ● A charge of 170 μC is at the center of a cube of edge 80.0 cm. No other charges are nearby. (a) Find the flux through each face of the cube. (b) Find the flux through the whole surface of the cube. (c) **What If?** Would your answers to parts (a) or (b) change if the charge were not at the center? Explain.

Section 24.3 Application of Gauss's Law to Various Charge Distributions

18. A solid sphere of radius 40.0 cm has a total positive charge of 26.0 μC uniformly distributed throughout its volume. Calculate the magnitude of the electric field (a) 0 cm, (b) 10.0 cm, (c) 40.0 cm, and (d) 60.0 cm from the center of the sphere.

19. Determine the magnitude of the electric field at the surface of a lead-208 nucleus, which contains 82 protons and 126 neutrons. Assume the lead nucleus has a volume 208 times that of one proton and consider a proton to be a sphere of radius 1.20×10^{-15} m.

20. The charge per unit length on a long, straight filament is -90.0 μC/m. Find the electric field (a) 10.0 cm, (b) 20.0 cm, and (c) 100 cm from the filament, where distances are measured perpendicular to the length of the filament.

21. A large, flat, horizontal sheet of charge has a charge per unit area of 9.00 μC/m². Find the electric field just above the middle of the sheet.

22. A cylindrical shell of radius 7.00 cm and length 240 cm has its charge uniformly distributed on its curved surface. The magnitude of the electric field at a point 19.0 cm radially outward from its axis (measured from the midpoint of the shell) is 36.0 kN/C. Find (a) the net charge on the shell and (b) the electric field at a point 4.00 cm from the axis, measured radially outward from the midpoint of the shell.

23. A 10.0-g piece of Styrofoam carries a net charge of -0.700 μC and floats above the center of a large, horizontal sheet of plastic that has a uniform charge density on its surface. What is the charge per unit area on the plastic sheet?

24. (a) Write a problem for which the following equation gives the solution. Include the required data in your problem statement and identify the one unknown.

$$2\pi(3\text{ cm})(8\text{ cm})E\cos 0° + 0 + 0$$

$$= \frac{\pi(2\text{ cm})^2(8\text{ cm})(5 \times 10^{-6}\text{ C/m}^3)}{8.85 \times 10^{-12}\text{ C}^2/\text{N}\cdot\text{m}^2}$$

(b) Solve the equation for the unknown.

25. **Review problem**. A particle with a charge of -60.0 nC is placed at the center of a nonconducting spherical shell of inner radius 20.0 cm and outer radius 25.0 cm. The

spherical shell carries charge with a uniform density of -1.33 μC/m³. A proton moves in a circular orbit just outside the spherical shell. Calculate the speed of the proton.

26. ● A nonconducting wall carries charge with a uniform density of 8.60 μC/cm². What is the electric field 7.00 cm in front of the wall? Explain whether your result changes as the distance from the wall is varied.

27. ▲ Consider a long, cylindrical charge distribution of radius R with a uniform charge density ρ. Find the electric field at distance r from the axis, where $r < R$.

28. In nuclear fission, a nucleus of uranium-238, containing 92 protons, can divide into two smaller spheres, each having 46 protons and a radius of 5.90×10^{-15} m. What is the magnitude of the repulsive electric force pushing the two spheres apart?

29. Consider a thin, spherical shell of radius 14.0 cm with a total charge of 32.0 μC distributed uniformly on its surface. Find the electric field (a) 10.0 cm and (b) 20.0 cm from the center of the charge distribution.

30. Fill two rubber balloons with air. Suspend both of them from the same point and let them hang down on strings of equal length. Rub each with wool or on your hair so that the balloons hang apart with a noticeable separation between them. Make order-of-magnitude estimates of (a) the force on each, (b) the charge on each, (c) the field each creates at the center of the other, and (d) the total flux of electric field created by each balloon. In your solution, state the quantities you take as data and the values you measure or estimate for them.

31. A uniformly charged, straight filament 7.00 m in length has a total positive charge of 2.00 μC. An uncharged cardboard cylinder 2.00 cm in length and 10.0 cm in radius surrounds the filament at its center, with the filament as the axis of the cylinder. Using reasonable approximations, find (a) the electric field at the surface of the cylinder and (b) the total electric flux through the cylinder.

Section 24.4 Conductors in Electrostatic Equilibrium

32. A very large, thin, flat plate of aluminum of area A has a total charge Q uniformly distributed over its surfaces. Assuming the same charge is spread uniformly over the *upper* surface of an otherwise identical glass plate, compare the electric fields just above the center of the upper surface of each plate.

33. A long, straight metal rod has a radius of 5.00 cm and a charge per unit length of 30.0 nC/m. Find the electric field (a) 3.00 cm, (b) 10.0 cm, and (c) 100 cm from the axis of the rod, where distances are measured perpendicular to the rod.

34. ● A solid copper sphere of radius 15.0 cm carries a charge of 40.0 nC. Find the electric field (a) 12.0 cm, (b) 17.0 cm, and (c) 75.0 cm from the center of the sphere. (d) **What If?** Explain how your answers would change if the sphere were hollow.

35. A square plate of copper with 50.0-cm sides has no net charge and is placed in a region of uniform electric field of 80.0 kN/C directed perpendicularly to the plate. Find (a) the charge density of each face of the plate and (b) the total charge on each face.

36. In a certain region of space, the electric field is $\vec{E} = 6\,000x^2\hat{i}$ N/C·m². Find the volume density of electric

charge at $x = 0.300$ m. *Suggestion:* Apply Gauss's law to a box between $x = 0.300$ m and $x = 0.300$ m $+ \, dx$.

37. Two identical conducting spheres each having a radius of 0.500 cm are connected by a light 2.00-m-long conducting wire. A charge of 60.0 μC is placed on one of the conductors. Assume the surface distribution of charge on each sphere is uniform. Determine the tension in the wire.

38. ● A solid metallic sphere of radius a carries total charge Q. No other charges are nearby. The electric field just outside its surface is $k_e Q/a^2$ radially outward. Is the electric field here also given by σ/ϵ_0? By $\sigma/2\epsilon_0$? Explain whether you should expect it to be equal to either of these quantities.

39. A long, straight wire is surrounded by a hollow metal cylinder whose axis coincides with that of the wire. The wire has a charge per unit length of λ, and the cylinder has a net charge per unit length of 2λ. From this information, use Gauss's law to find (a) the charge per unit length on the inner and outer surfaces of the cylinder and (b) the electric field outside the cylinder, a distance r from the axis.

40. A positively charged particle is at a distance $R/2$ from the center of an uncharged thin, conducting, spherical shell of radius R. Sketch the electric field lines set up by this arrangement both inside and outside the shell.

41. ▲ A thin, square, conducting plate 50.0 cm on a side lies in the xy plane. A total charge of 4.00×10^{-8} C is placed on the plate. Find (a) the charge density on the plate, (b) the electric field just above the plate, and (c) the electric field just below the plate. You may assume the charge density is uniform.

Additional Problems

42. A nonuniform electric field is given by the expression

$$\vec{\mathbf{E}} = ay\hat{\mathbf{i}} + bz\hat{\mathbf{j}} + cx\hat{\mathbf{k}}$$

where a, b, and c are constants. Determine the electric flux through a rectangular surface in the xy plane, extending from $x = 0$ to $x = w$ and from $y = 0$ to $y = h$.

43. A sphere of radius R surrounds a particle with charge Q, located at its center. (a) Show that the electric flux through a circular cap of half-angle θ (Fig. P24.43) is

$$\Phi_E = \frac{Q}{2\epsilon_0}(1 - \cos\theta)$$

What is the flux for (b) $\theta = 90°$ and (c) $\theta = 180°$?

Figure P24.43

44. A solid insulating sphere of radius 5.00 cm carries a net positive charge of 3.00 μC, uniformly distributed throughout its volume. Concentric with this sphere is a conduct-

ing spherical shell with inner radius 10.0 cm and outer radius 15.0 cm, having net charge -1.00 μC, as shown in Figure Q24.11. (a) Consider a spherical gaussian surface of radius 16.0 cm and find the net charge enclosed by this surface. (b) What is the direction of the electric field at point D, to the right of the shell and at radius 16 cm? (c) Find the magnitude of the electric field at point D. (d) Find the vector electric field at point C, at radius 12.0 cm. (e) Consider a spherical gaussian surface through point C and find the net charge enclosed by this surface. (f) Consider a spherical gaussian surface of radius 8.00 cm and find the net charge enclosed by this surface. (g) Find the vector electric field at point B, at radius 8 cm. (h) Consider a spherical gaussian surface through point A, at radius 4.00 cm, and find the net charge enclosed by this surface. (i) Find the vector electric field at point A. (j) Determine the charge on the inner surface of the conducting shell. (k) Determine the charge on the outer surface of the conducting shell. (l) Sketch a graph of the magnitude of the electric field versus r.

45. ● A hollow, metallic, spherical shell has exterior radius 0.750 m, carries no net charge, and is supported on an insulating stand. The electric field everywhere just outside its surface is 890 N/C radially toward the center of the sphere. (a) Explain what you can conclude about the amount of charge on the exterior surface of the sphere and the distribution of this charge. (b) Explain what you can conclude about the amount of charge on the interior surface of the sphere and its distribution. (c) Explain what you can conclude about the amount of charge inside the shell and its distribution.

46. ● Consider two identical conducting spheres whose surfaces are separated by a small distance. One sphere is given a large net positive charge, and the other is given a small net positive charge. It is found that the force between the spheres is attractive even though they both have net charges of the same sign. Explain how this attraction is possible.

47. ▲ A solid, insulating sphere of radius a has a uniform charge density ρ and a total charge Q. Concentric with this sphere is an uncharged, conducting, hollow sphere whose inner and outer radii are b and c as shown in Figure P24.47. (a) Find the magnitude of the electric field in the regions $r < a$, $a < r < b$, $b < r < c$, and $r > c$. (b) Determine the induced charge per unit area on the inner and outer surfaces of the hollow sphere.

Figure P24.47   Problems 47 and 63.

48. **Review problem.** An early (incorrect) model of the hydrogen atom, suggested by J. J. Thomson, proposed that a positive cloud of charge $+e$ was uniformly distributed throughout the volume of a sphere of radius R, with the electron (an equal-magnitude negatively charged particle $-e$) at the center. (a) Using Gauss's law, show that the

electron would be in equilibrium at the center and, if displaced from the center a distance $r < R$, would experience a restoring force of the form $F = -Kr$, where K is a constant. (b) Show that $K = k_e e^2/R^3$. (c) Find an expression for the frequency f of simple harmonic oscillations that an electron of mass m_e would undergo if displaced a small distance ($< R$) from the center and released. (d) Calculate a numerical value for R that would result in a frequency of 2.47×10^{15} Hz, the frequency of the light radiated in the most intense line in the hydrogen spectrum.

49. A particle of mass m and charge q moves at high speed along the x axis. It is initially near $x = -\infty$, and it ends up near $x = +\infty$. A second charge Q is fixed at the point $x = 0$, $y = -d$. As the moving charge passes the stationary charge, its x component of velocity does not change appreciably, but it acquires a small velocity in the y direction. Determine the angle through which the moving charge is deflected. *Suggestion:* The integral you encounter in determining v_y can be evaluated by applying Gauss's law to a long cylinder of radius d, centered on the stationary charge.

50. Two infinite, nonconducting sheets of charge are parallel to each other as shown in Figure P24.50. The sheet on the left has a uniform surface charge density σ, and the one on the right has a uniform charge density $-\sigma$. Calculate the electric field at points (a) to the left of, (b) in between, and (c) to the right of the two sheets.

Figure P24.50

51. ▲ **What If?** Repeat the calculations for Problem 50 when both sheets have *positive* uniform surface charge densities of value σ.

52. A sphere of radius $2a$ is made of a nonconducting material that has a uniform volume charge density ρ. (Assume the material does not affect the electric field.) A spherical cavity of radius a is now removed from the sphere as shown in Figure P24.52. Show that the electric field within the cavity is uniform and is given by $E_x = 0$ and $E_y = \rho a/3\epsilon_0$. *Suggestion:* The field within the cavity is the superposition of the field due to the original uncut sphere plus the field due to a sphere the size of the cavity with a uniform negative charge density $-\rho$.

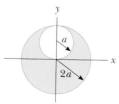

Figure P24.52

53. A uniformly charged spherical shell with surface charge density σ contains a circular hole in its surface. The

radius of the hole is small compared with the radius of the sphere. What is the electric field at the center of the hole? *Suggestion:* This problem, like Problem 52, can be solved by using the idea of superposition.

54. A closed surface with dimensions $a = b = 0.400$ m and $c = 0.600$ m is located as shown in Figure P24.54. The left edge of the closed surface is located at position $x = a$. The electric field throughout the region is nonuniform and is given by $\vec{E} = (3.0 + 2.0x^2)\hat{i}$ N/C, where x is in meters. Calculate the net electric flux leaving the closed surface. What net charge is enclosed by the surface?

Figure P24.54

55. A solid insulating sphere of radius R has a nonuniform charge density that varies with r according to the expression $\rho = Ar^2$, where A is a constant and $r < R$ is measured from the center of the sphere. (a) Show that the magnitude of the electric field outside ($r > R$) the sphere is $E = AR^5/5\epsilon_0 r^2$. (b) Show that the magnitude of the electric field inside ($r < R$) the sphere is $E = Ar^3/5\epsilon_0$. *Suggestion:* The total charge Q on the sphere is equal to the integral of $\rho \, dV$, where r extends from 0 to R; also, the charge q within a radius $r < R$ is less than Q. To evaluate the integrals, note that the volume element dV for a spherical shell of radius r and thickness dr is equal to $4\pi r^2 \, dr$.

56. A particle with charge Q is located on the axis of a disk of radius R at a distance b from the plane of the disk (Fig. P24.56). Show that if one fourth of the electric flux from the charge passes through the disk, then $R = \sqrt{3}b$.

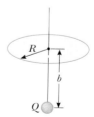

Figure P24.56

57. A spherically symmetric charge distribution has a charge density given by $\rho = a/r$, where a is constant. Find the electric field as a function of r. *Suggestion:* The charge within a sphere of radius R is equal to the integral of $\rho \, dV$, where r extends from 0 to R. To evaluate the integral, note that the volume element dV for a spherical shell of radius r and thickness dr is equal to $4\pi r^2 \, dr$.

58. An infinitely long insulating cylinder of radius R has a volume charge density that varies with the radius as

$$\rho = \rho_0\left(a - \frac{r}{b}\right)$$

2 = intermediate;   3 = challenging;   ☐ = SSM/SG;   ▲ = ThomsonNOW;   = symbolic reasoning;   ● = qualitative reasoning

where ρ_0, a, and b are positive constants and r is the distance from the axis of the cylinder. Use Gauss's law to determine the magnitude of the electric field at radial distances (a) $r < R$ and (b) $r > R$.

59. **Review problem.** A slab of insulating material (infinite in two of its three dimensions) has a uniform positive charge density ρ. An edge view of the slab is shown in Figure P24.59. (a) Show that the magnitude of the electric field a distance x from its center and inside the slab is $E = \rho x/\epsilon_0$. (b) **What If?** Suppose an electron of charge $-e$ and mass m_e can move freely within the slab. It is released from rest at a distance x from the center. Show that the electron exhibits simple harmonic motion with a frequency

$$f = \frac{1}{2\pi}\sqrt{\frac{\rho e}{m_e \epsilon_0}}$$

Figure P24.59 Problems 59 and 60.

60. A slab of insulating material has a nonuniform positive charge density $\rho = Cx^2$, where x is measured from the center of the slab as shown in Figure P24.59 and C is a constant. The slab is infinite in the y and z directions. Derive expressions for the electric field in (a) the exterior regions and (b) the interior region of the slab ($-d/2 < x < d/2$).

61. (a) Using the mathematical similarity between Coulomb's law and Newton's law of universal gravitation, show that Gauss's law for gravitation can be written as

$$\oint \vec{g} \cdot d\vec{A} = -4\pi G m_{in}$$

where m_{in} is the net mass inside the gaussian surface and $\vec{g} = \vec{F}_g / m$ represents the gravitational field at any point on the gaussian surface. (b) Determine the gravitational field at a distance r from the center of the Earth where $r < R_E$, assuming the Earth's mass density is uniform.

62. An insulating solid sphere of radius a has a uniform volume charge density and carries a total positive charge Q. A spherical gaussian surface of radius r, which shares a common center with the insulating sphere, is inflated starting from $r = 0$. (a) Find an expression for the electric flux passing through the surface of the gaussian sphere as a function of r for $r < a$. (b) Find an expression for the electric flux for $r > a$. (c) Plot the flux versus r.

63. For the configuration shown in Figure P24.47, suppose $a = 5.00$ cm, $b = 20.0$ cm, and $c = 25.0$ cm. Furthermore, suppose the electric field at a point 10.0 cm from the center is measured to be 3.60×10^3 N/C radially inward and the electric field at a point 50.0 cm from the center is 2.00×10^2 N/C radially outward. From this information, find (a) the charge on the insulating sphere, (b) the net charge on the hollow conducting sphere, and (c) the charges on the inner and outer surfaces of the hollow conducting sphere.

64. An infinitely long cylindrical insulating shell of inner radius a and outer radius b has a uniform volume charge density ρ. A line of uniform linear charge density λ is placed along the axis of the shell. Determine the electric field everywhere.

65. Consider an electric field that is uniform in direction throughout a certain volume. Can it be uniform in magnitude? Must it be uniform in magnitude? Answer these questions (a) assuming the volume is filled with an insulating material carrying charge described by a volume charge density and (b) assuming the volume is empty space. State reasoning to prove your answers.

Answers to Quick Quizzes

24.1 (e). The same number of field lines pass through a sphere of any size. Because points on the surface of the sphere are closer to the charge, the field is stronger.

24.2 (b) and (d). Statement (a) is not necessarily true because an equal number of positive and negative charges could be present inside the surface. Statement (c) is not necessarily true as can be seen from Figure 24.8: a nonzero electric field exists everywhere on the surface, but the charge is not enclosed within the surface and the net flux is therefore zero.

24.3 (a). Charges added to the metal cylinder by your brother will reside on the outer surface of the conducting cylinder. If you are on the inside, these charges cannot transfer to you from the inner surface. For this same reason, you are safe in a metal automobile during a lightning storm.

Processes occurring during thunderstorms cause large differences in electric potential between a thundercloud and the ground. The result of this potential difference is an electrical discharge that we call lightning, such as this display over Tucson, Arizona. (© Keith Kent/Photo Researchers, Inc.)

25.1 Electric Potential and Potential Difference

25.2 Potential Difference in a Uniform Electric Field

25.3 Electric Potential and Potential Energy Due to Point Charges

25.4 Obtaining the Value of the Electric Field from the Electric Potential

25.5 Electric Potential Due to Continuous Charge Distributions

25.6 Electric Potential Due to a Charged Conductor

25.7 The Millikan Oil-Drop Experiment

25.8 Applications of Electrostatics

25 Electric Potential

The concept of potential energy was introduced in Chapter 7 in connection with such conservative forces as the gravitational force and the elastic force exerted by a spring. By using the law of conservation of energy when solving various problems in mechanics, we were able to avoid working directly with forces. The concept of potential energy is also of great value in the study of electricity. Because the electrostatic force is conservative, electrostatic phenomena can be conveniently described in terms of an electric potential energy. This idea enables us to define a quantity known as *electric potential*. Because the electric potential at any point in an electric field is a scalar quantity, we can use it to describe electrostatic phenomena more simply than if we were to rely only on the electric field and electric forces. The concept of electric potential is of great practical value in the operation of electric circuits and devices that we will study in later chapters.

25.1 Electric Potential and Potential Difference

When a test charge q_0 is placed in an electric field $\vec{E}$ created by some source charge distribution, the electric force acting on the test charge is $q_0\vec{E}$. The force $q_0\vec{E}$ is conservative because the force between charges described by Coulomb's law is conservative. When the test charge is moved in the field by some external agent, the work done by the field on the charge is equal to the negative of the work done by the external agent causing the displacement. This situation is analogous to that of lifting an object with mass in a gravitational field: the work done by the external agent is mgh, and the work done by the gravitational force is $-mgh$.

When analyzing electric and magnetic fields, it is common practice to use the notation $d\vec{s}$ to represent an infinitesimal displacement vector that is oriented tangent to a path through space. This path may be straight or curved, and an integral performed along this path is called either a *path integral* or a *line integral* (the two terms are synonymous).

For an infinitesimal displacement $d\vec{s}$ of a point charge q_0 immersed in an electric field, the work done by the electric field on the charge is $\vec{F} \cdot d\vec{s} = q_0\vec{E} \cdot d\vec{s}$. As this amount of work is done by the field, the potential energy of the charge–field system is changed by an amount $dU = -q_0\vec{E} \cdot d\vec{s}$. For a finite displacement of the charge from point Ⓐ to point Ⓑ, the change in potential energy of the system $\Delta U = U_\text{Ⓑ} - U_\text{Ⓐ}$ is

$$\Delta U = -q_0 \int_\text{Ⓐ}^\text{Ⓑ} \vec{E} \cdot d\vec{s} \qquad \text{(25.1)}$$

◀ Change in electric potential energy of a system

The integration is performed along the path that q_0 follows as it moves from Ⓐ to Ⓑ. Because the force $q_0\vec{E}$ is conservative, **this line integral does not depend on the path taken from Ⓐ to Ⓑ.**

For a given position of the test charge in the field, the charge–field system has a potential energy U relative to the configuration of the system that is defined as $U = 0$. Dividing the potential energy by the test charge gives a physical quantity that depends only on the source charge distribution and has a value at every point in an electric field. This quantity is called the **electric potential** (or simply the **potential**) V:

$$V = \frac{U}{q_0} \qquad \text{(25.2)}$$

Because potential energy is a scalar quantity, electric potential also is a scalar quantity.

As described by Equation 25.1, if the test charge is moved between two positions Ⓐ and Ⓑ in an electric field, the charge–field system experiences a change in potential energy. The **potential difference** $\Delta V = V_\text{Ⓑ} - V_\text{Ⓐ}$ between two points Ⓐ and Ⓑ in an electric field is defined as the change in potential energy of the system when a test charge q_0 is moved between the points divided by the test charge:

$$\Delta V \equiv \frac{\Delta U}{q_0} = -\int_\text{Ⓐ}^\text{Ⓑ} \vec{E} \cdot d\vec{s} \qquad \text{(25.3)}$$

◀ Potential difference between two points

Just as with potential energy, only *differences* in electric potential are meaningful. We often take the value of the electric potential to be zero at some convenient point in an electric field.

Potential difference should not be confused with difference in potential energy. The potential difference between Ⓐ and Ⓑ depends only on the source charge distribution (consider points Ⓐ and Ⓑ *without* the presence of the test charge), whereas the difference in potential energy exists only if a test charge is moved between the points.

If an external agent moves a test charge from Ⓐ to Ⓑ without changing the kinetic energy of the test charge, the agent performs work that changes the potential energy of the system: $W = \Delta U$. Imagine an arbitrary charge q located in an electric field. From Equation 25.3, the work done by an external agent in moving a charge q through an electric field at constant velocity is

$$W = q\Delta V \qquad \text{(25.4)}$$

PITFALL PREVENTION 25.2
Voltage

A variety of phrases are used to describe the potential difference between two points, the most common being **voltage,** arising from the unit for potential. A voltage *applied* to a device, such as a television, or *across* a device is the same as the potential difference across the device.

PITFALL PREVENTION 25.3
The Electron Volt

The electron volt is a unit of *energy,* NOT of potential. The energy of any system may be expressed in eV, but this unit is most convenient for describing the emission and absorption of visible light from atoms. Energies of nuclear processes are often expressed in MeV.

Because electric potential is a measure of potential energy per unit charge, the SI unit of both electric potential and potential difference is joules per coulomb, which is defined as a **volt** (V):

$$1 \text{ V} \equiv 1 \text{ J/C}$$

That is, 1 J of work must be done to move a 1-C charge through a potential difference of 1 V.

Equation 25.3 shows that potential difference also has units of electric field times distance. It follows that the SI unit of electric field (N/C) can also be expressed in volts per meter:

$$1 \text{ N/C} = 1 \text{ V/m}$$

Therefore, **we can interpret the electric field as a measure of the rate of change with position of the electric potential.**

A unit of energy commonly used in atomic and nuclear physics is the **electron volt** (eV), which is defined as **the energy a charge–field system gains or loses when a charge of magnitude e (that is, an electron or a proton) is moved through a potential difference of 1 V.** Because 1 V = 1 J/C and the fundamental charge is 1.60×10^{-19} C, the electron volt is related to the joule as follows:

$$1 \text{ eV} = 1.60 \times 10^{-19} \text{ C} \cdot \text{V} = 1.60 \times 10^{-19} \text{ J} \tag{25.5}$$

For instance, an electron in the beam of a typical television picture tube may have a speed of 3.0×10^7 m/s. This speed corresponds to a kinetic energy equal to 4.1×10^{-16} J, which is equivalent to 2.6×10^3 eV. Such an electron has to be accelerated from rest through a potential difference of 2.6 kV to reach this speed.

Quick Quiz 25.1 In Figure 25.1, two points Ⓐ and Ⓑ are located within a region in which there is an electric field. **(i)** How would you describe the potential difference $\Delta V = V_Ⓑ - V_Ⓐ$? (a) It is positive. (b) It is negative. (c) It is zero. **(ii)** A negative charge is placed at Ⓐ and then moved to Ⓑ. How would you describe the change in potential energy of the charge-field system for this process? Choose from the same possibilities.

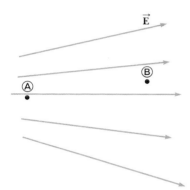

Figure 25.1 (Quick Quiz 25.1) Two points in an electric field.

25.2 Potential Difference in a Uniform Electric Field

Equations 25.1 and 25.3 hold in all electric fields, whether uniform or varying, but they can be simplified for a uniform field. First, consider a uniform electric field directed along the negative y axis as shown in Active Figure 25.2a. Let's calculate the potential difference between two points Ⓐ and Ⓑ separated by a distance $|\vec{\mathbf{s}}| = d$, where $\vec{\mathbf{s}}$ is parallel to the field lines. Equation 25.3 gives

$$V_Ⓑ - V_Ⓐ = \Delta V = -\int_Ⓐ^Ⓑ \vec{\mathbf{E}} \cdot d\vec{\mathbf{s}} = -\int_Ⓐ^Ⓑ (E \cos 0°) ds = -\int_Ⓐ^Ⓑ E \, ds$$

Because E is constant, it can be removed from the integral sign, which gives

▶ Potential difference between two points in a uniform electric field

$$\Delta V = -E \int_Ⓐ^Ⓑ ds = -Ed \tag{25.6}$$

The negative sign indicates that the electric potential at point Ⓑ is lower than at point Ⓐ; that is, $V_Ⓑ < V_Ⓐ$. **Electric field lines always point in the direction of decreasing electric potential** as shown in Active Figure 25.2a.

Now suppose a test charge q_0 moves from Ⓐ to Ⓑ. We can calculate the change in the potential energy of the charge–field system from Equations 25.3 and 25.6:

$$\Delta U = q_0 \Delta V = -q_0 Ed \tag{25.7}$$

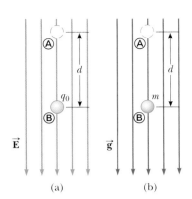

ACTIVE FIGURE 25.2

(a) When the electric field $\vec{\mathbf{E}}$ is directed downward, point Ⓑ is at a lower electric potential than point Ⓐ. When a positive test charge moves from point Ⓐ to point Ⓑ, the electric potential energy of the charge–field system decreases. (b) When an object of mass m moves downward in the direction of the gravitational field $\vec{\mathbf{g}}$, the gravitational potential energy of the object–field system decreases.

Sign in at www.thomsonedu.com and go to ThomsonNOW to observe and compare the motion of the charged object in an electric field and an object with mass in a gravitational field.

This result shows that if q_0 is positive, then U is negative. Therefore, **a system consisting of a positive charge and an electric field loses electric potential energy when the charge moves in the direction of the field.** Equivalently, an electric field does work on a positive charge when the charge moves in the direction of the electric field. (That is analogous to the work done by the gravitational field on a falling object as shown in Active Fig. 25.2b.) If a positive test charge is released from rest in this electric field, it experiences an electric force $q_0\vec{\mathbf{E}}$ in the direction of $\vec{\mathbf{E}}$ (downward in Active Fig. 25.2a). Therefore, it accelerates downward, gaining kinetic energy. **As the charged particle gains kinetic energy, the charge–field system loses an equal amount of potential energy.** This equivalence should not be surprising; it is simply conservation of mechanical energy in an isolated system as introduced in Chapter 8.

If q_0 is negative, then ΔU in Equation 25.7 is positive and the situation is reversed. **A system consisting of a negative charge and an electric field gains electric potential energy when the charge moves in the direction of the field.** If a negative charge is released from rest in an electric field, it accelerates in a direction opposite the direction of the field. For the negative charge to move in the direction of the field, an external agent must apply a force and do positive work on the charge.

Now consider the more general case of a charged particle that moves between Ⓐ and Ⓑ in a uniform electric field such that the vector $\vec{\mathbf{s}}$ is not parallel to the field lines as shown in Figure 25.3. In this case, Equation 25.3 gives

$$\Delta V = -\int_{Ⓐ}^{Ⓑ} \vec{\mathbf{E}} \cdot d\vec{\mathbf{s}} = -\vec{\mathbf{E}} \cdot \int_{Ⓐ}^{Ⓑ} d\vec{\mathbf{s}} = -\vec{\mathbf{E}} \cdot \vec{\mathbf{s}} \tag{25.8}$$

where again $\vec{\mathbf{E}}$ was removed from the integral because it is constant. The change in potential energy of the charge-field system is

$$\Delta U = q_0 \Delta V = -q_0\vec{\mathbf{E}} \cdot \vec{\mathbf{s}} \tag{25.9}$$

Finally, we conclude from Equation 25.8 that all points in a plane perpendicular to a uniform electric field are at the same electric potential. We can see that in Figure 25.3, where the potential difference $V_Ⓑ - V_Ⓐ$ is equal to the potential difference $V_Ⓒ - V_Ⓐ$. (Prove this fact to yourself by working out two dot products for $\vec{\mathbf{E}} \cdot \vec{\mathbf{s}}$: one for $\vec{\mathbf{s}}_{Ⓐ\rightarrowⒷ}$, where the angle θ between $\vec{\mathbf{E}}$ and $\vec{\mathbf{s}}$ is arbitrary as shown in Figure 25.3, and one for $\vec{\mathbf{s}}_{Ⓐ\rightarrowⒸ}$, where $\theta = 0$.) Therefore, $V_Ⓑ = V_Ⓒ$. **The name equipotential surface is given to any surface consisting of a continuous distribution of points having the same electric potential.**

The equipotential surfaces associated with a uniform electric field consist of a family of parallel planes that are all perpendicular to the field. Equipotential surfaces associated with fields having other symmetries are described in later sections.

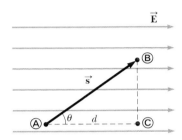

Figure 25.3 A uniform electric field directed along the positive x axis. Point Ⓑ is at a lower electric potential than point Ⓐ. Points Ⓑ and Ⓒ are at the *same* electric potential.

◀ Change in potential energy when a charged particle is moved in a uniform electric field

Quick Quiz 25.2   The labeled points in Figure 25.4 are on a series of equipotential surfaces associated with an electric field. Rank (from greatest to least) the work done by the electric field on a positively charged particle that moves from Ⓐ to Ⓑ, from Ⓑ to Ⓒ, from Ⓒ to Ⓓ, and from Ⓓ to Ⓔ.

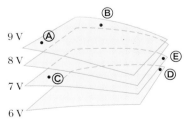

Figure 25.4 (Quick Quiz 25.2) Four equipotential surfaces.

EXAMPLE 25.1 | **The Electric Field Between Two Parallel Plates of Opposite Charge**

A battery has a specified potential difference ΔV between its terminals and establishes that potential difference between conductors attached to the terminals. A 12-V battery is connected between two parallel plates as shown in Figure 25.5. The separation between the plates is $d = 0.30$ cm, and we assume the electric field between the plates to be uniform. (This assumption is reasonable if the plate separation is small relative to the plate dimensions and we do not consider locations near the plate edges.) Find the magnitude of the electric field between the plates.

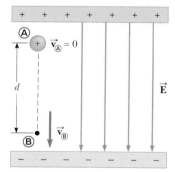

Figure 25.5 (Example 25.1) A 12-V battery connected to two parallel plates. The electric field between the plates has a magnitude given by the potential difference ΔV divided by the plate separation d.

SOLUTION

Conceptualize In earlier chapters, we investigated the uniform electric field between parallel plates. The new feature to this problem is that the electric field is related to the new concept of electric potential.

Categorize The electric field is evaluated from a relationship between field and potential given in this section, so we categorize this example as a substitution problem.

Use Equation 25.6 to evaluate the magnitude of the electric field between the plates:

$$E = \frac{|V_B - V_A|}{d} = \frac{12 \text{ V}}{0.30 \times 10^{-2} \text{ m}} = \boxed{4.0 \times 10^{3} \text{ V/m}}$$

The configuration of plates in Figure 25.5 is called a *parallel-plate capacitor* and is examined in greater detail in Chapter 26.

EXAMPLE 25.2 | **Motion of a Proton in a Uniform Electric Field**

A proton is released from rest at point Ⓐ in a uniform electric field that has a magnitude of 8.0×10^4 V/m (Fig. 25.6). The proton undergoes a displacement of 0.50 m to point Ⓑ in the direction of $\vec{E}$. Find the speed of the proton after completing the 0.50 m displacement.

SOLUTION

Conceptualize Visualize the proton in Figure 25.6 moving downward through the potential difference. The situation is analogous to an object falling through a gravitational field.

Categorize The system of the proton and the two plates in Figure 25.6 does not interact with the environment, so we model it as an isolated system.

Figure 25.6 (Example 25.2) A proton accelerates from Ⓐ to Ⓑ in the direction of the electric field.

Analyze Use Equation 25.6 to find the potential difference between points Ⓐ and Ⓑ:

$$\Delta V = -Ed = -(8.0 \times 10^4 \text{ V/m})(0.50 \text{ m}) = -4.0 \times 10^4 \text{ V}$$

Write the appropriate reduction of Equation 8.2, the conservation of energy equation, for the isolated system of the charge and the electric field:

$$\Delta K + \Delta U = 0$$

Substitute the changes in energy for both terms:

$$\left(\tfrac{1}{2}mv^2 - 0\right) + e\,\Delta V = 0$$

Solve for the final speed of the proton:

$$v = \sqrt{\frac{-2e\,\Delta V}{m}}$$

Substitute numerical values:

$$v = \sqrt{\frac{-2(1.6 \times 10^{-19}\,\text{C})(-4.0 \times 10^{4}\,\text{V})}{1.67 \times 10^{-27}\,\text{kg}}}$$

$$= 2.8 \times 10^{6}\,\text{m/s}$$

Finalize   Because ΔV is negative, ΔU is also negative. The negative value of ΔU means the potential energy of the system decreases as the proton moves in the direction of the electric field. As the proton accelerates in the direction of the field, it gains kinetic energy and the system loses electric potential energy at the same time.

Figure 25.6 is oriented so that the proton falls downward. The proton's motion is analogous to that of an object falling in a gravitational field. Although the gravitational field is always downward at the surface of the Earth, an electric field can be in any direction, depending on the orientation of the plates creating the field. Therefore, Figure 25.6 could be rotated $90°$ or $180°$ and the proton could fall horizontally or upward in the electric field!

25.3   Electric Potential and Potential Energy Due to Point Charges

As discussed in Section 23.4, an isolated positive point charge q produces an electric field directed radially outward from the charge. To find the electric potential at a point located a distance r from the charge, let's begin with the general expression for potential difference,

$$V_{\text{®}} - V_{\text{®}} = -\int_{\text{®}}^{\text{®}} \vec{\mathbf{E}} \cdot d\vec{\mathbf{s}}$$

where ⓐ and ⓑ are the two arbitrary points shown in Figure 25.7. At any point in space, the electric field due to the point charge is $\vec{\mathbf{E}} = (k_e q / r^2)\hat{\mathbf{r}}$ (Eq. 23.9), where $\hat{\mathbf{r}}$ is a unit vector directed from the charge toward the point. The quantity $\vec{\mathbf{E}} \cdot d\vec{\mathbf{s}}$ can be expressed as

$$\vec{\mathbf{E}} \cdot d\vec{\mathbf{s}} = k_e \frac{q}{r^2} \hat{\mathbf{r}} \cdot d\vec{\mathbf{s}}$$

Because the magnitude of $\hat{\mathbf{r}}$ is 1, the dot product $\hat{\mathbf{r}} \cdot d\vec{\mathbf{s}} = ds \cos \theta$, where θ is the angle between $\hat{\mathbf{r}}$ and $d\vec{\mathbf{s}}$. Furthermore, $ds \cos \theta$ is the projection of $d\vec{\mathbf{s}}$ onto $\vec{\mathbf{r}}$; therefore, $ds \cos \theta = dr$. That is, any displacement $d\vec{\mathbf{s}}$ along the path from point ⓐ to point ⓑ produces a change dr in the magnitude of $\vec{\mathbf{r}}$, the position vector of the point relative to the charge creating the field. Making these substitutions, we find that $\vec{\mathbf{E}} \cdot d\vec{\mathbf{s}} = (k_e q/r^2)dr$; hence, the expression for the potential difference becomes

$$V_{\text{®}} - V_{\text{®}} = -k_e q \int_{r_{\text{®}}}^{r_{\text{®}}} \frac{dr}{r^2} = k_e \frac{q}{r} \Big|_{r_{\text{®}}}^{r_{\text{®}}}$$

$$V_{\text{®}} - V_{\text{®}} = k_e q \left[\frac{1}{r_{\text{®}}} - \frac{1}{r_{\text{®}}} \right] \qquad (25.10)$$

Equation 25.10 shows us that the integral of $\vec{\mathbf{E}} \cdot d\vec{\mathbf{s}}$ is *independent* of the path between points ⓐ and ⓑ. Multiplying by a charge q_0 that moves between points ⓐ and ⓑ, we see that the integral of $q_0 \vec{\mathbf{E}} \cdot d\vec{\mathbf{s}}$ is also independent of path. This latter integral, which is the work done by the electric force, shows that the electric force is conservative (see Section 7.7). We define a field that is related to a conservative force as a **conservative field.** Therefore, Equation 25.10 tells us that the electric field of a fixed point charge is conservative. Furthermore, Equation 25.10 expresses the important result that the potential difference between any two

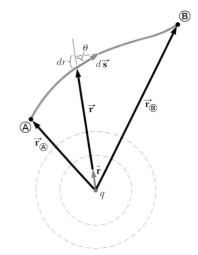

Figure 25.7   The potential difference between points ⓐ and ⓑ due to a point charge q depends *only* on the initial and final radial coordinates $r_{\text{®}}$ and $r_{\text{®}}$. The two dashed circles represent intersections of spherical equipotential surfaces with the page.

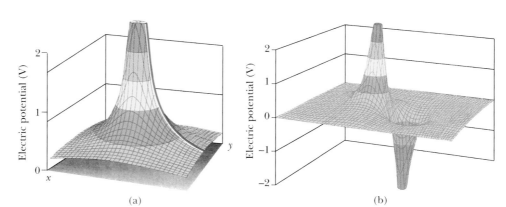

Figure 25.8 (a) The electric potential in the plane around a single positive charge is plotted on the vertical axis. (The electric potential function for a negative charge would look like a hole instead of a hill.) The red line shows the $1/r$ nature of the electric potential as given by Equation 25.11. (b) The electric potential in the plane containing a dipole.

PITFALL PREVENTION 25.4
Similar Equation Warning

Do not confuse Equation 25.11 for the electric potential of a point charge with Equation 23.9 for the electric field of a point charge. Potential is proportional to $1/r$, whereas the field is proportional to $1/r^2$. The effect of a charge on the space surrounding it can be described in two ways. The charge sets up a vector electric field $\vec{\mathbf{E}}$, which is related to the force experienced by a test charge placed in the field. It also sets up a scalar potential V, which is related to the potential energy of the two-charge system when a test charge is placed in the field.

points Ⓐ and Ⓑ in a field created by a point charge depends only on the radial coordinates $r_{Ⓐ}$ and $r_{Ⓑ}$. It is customary to choose the reference of electric potential for a point charge to be $V = 0$ at $r_{Ⓐ} = \infty$. With this reference choice, the electric potential created by a point charge at any distance r from the charge is

$$V = k_e \frac{q}{r} \tag{25.11}$$

Figure 25.8a shows a plot of the electric potential on the vertical axis for a positive charge located in the xy plane. Consider the following analogy to gravitational potential. Imagine trying to roll a marble toward the top of a hill shaped like the surface in Figure 25.8a. Pushing the marble up the hill is analogous to pushing one positively charged object toward another positively charged object. Similarly, the electric potential graph of the region surrounding a negative charge is analogous to a "hole" with respect to any approaching positively charged objects. A charged object must be infinitely distant from another charge before the surface in Figure 25.8a is "flat" and has an electric potential of zero.

We obtain the electric potential resulting from two or more point charges by applying the superposition principle. That is, the total electric potential at some point P due to several point charges is the sum of the potentials due to the individual charges. For a group of point charges, we can write the total electric potential at P as

◀ Electric potential due to several point charges

$$V = k_e \sum_i \frac{q_i}{r_i} \tag{25.12}$$

where the potential is again taken to be zero at infinity and r_i is the distance from the point P to the charge q_i. Notice that the sum in Equation 25.12 is an algebraic sum of scalars rather than a vector sum (which we use to calculate the electric field of a group of charges). Therefore, it is often much easier to evaluate V than $\vec{\mathbf{E}}$. The electric potential around a dipole is illustrated in Figure 25.8b. Notice the steep slope of the potential between the charges, representing a region of strong electric field.

Now consider the potential energy of a system of two charged particles. If V_2 is the electric potential at a point P due to charge q_2, the work an external agent must do to bring a second charge q_1 from infinity to P without acceleration is $q_1 V_2$. This work represents a transfer of energy into the system, and the energy appears in the system as potential energy U when the particles are separated by a distance

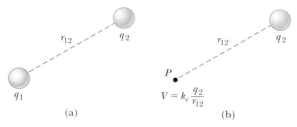

(a)   (b)

ACTIVE FIGURE 25.9

(a) If two point charges are separated by a distance r_{12}, the potential energy of the pair of charges is given by $k_e q_1 q_2 / r_{12}$. (b) If charge q_1 is removed, a potential $k_e q_2 / r_{12}$ exists at point P due to charge q_2.

Sign in at www.thomsonedu.com and go to ThomsonNOW to move charge q_1 or point P and see the result on the electric potential energy of the system for part (a) and the electric potential due to charge q_2 for part (b).

r_{12} (Active Fig. 25.9a). Therefore, the potential energy of the system can be expressed as[1]

$$U = k_e \frac{q_1 q_2}{r_{12}} \qquad (25.13)$$

If the charges are of the same sign, U is positive. Positive work must be done by an external agent on the system to bring the two charges near each other (because charges of the same sign repel). If the charges are of opposite sign, U is negative. Negative work is done by an external agent against the attractive force between the charges of opposite sign as they are brought near each other; a force must be applied opposite the displacement to prevent q_1 from accelerating toward q_2.

In Active Figure 25.9b, we have removed the charge q_1. At the position this charge previously occupied, point P, Equations 25.2 and 25.13 can be used to define a potential due to charge q_2 as $V = U/q_1 = k_e q_2 / r_{12}$. This expression is consistent with Equation 25.11.

If the system consists of more than two charged particles, we can obtain the total potential energy of the system by calculating U for every pair of charges and summing the terms algebraically. For example, the total potential energy of the system of three charges shown in Figure 25.10 is

$$U = k_e \left(\frac{q_1 q_2}{r_{12}} + \frac{q_1 q_3}{r_{13}} + \frac{q_2 q_3}{r_{23}} \right) \qquad (25.14)$$

Physically, this result can be interpreted as follows. Imagine q_1 is fixed at the position shown in Figure 25.10 but q_2 and q_3 are at infinity. The work an external agent must do to bring q_2 from infinity to its position near q_1 is $k_e q_1 q_2 / r_{12}$, which is the first term in Equation 25.14. The last two terms represent the work required to bring q_3 from infinity to its position near q_1 and q_2. (The result is independent of the order in which the charges are transported.)

PITFALL PREVENTION 25.5
Which Work?

There is a difference between work done *by one member of a system on another member* and work done *on a system by an external agent*. In the discussion related to Equation 25.14, we consider the group of charges to be the system; an external agent is doing work on the system to move the charges from an infinite separation to a small separation.

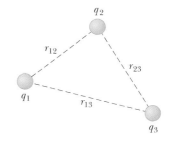

Figure 25.10 Three point charges are fixed at the positions shown. The potential energy of this system of charges is given by Equation 25.14.

Quick Quiz 25.3 In Active Figure 25.9a, take q_1 to be a negative source charge and q_2 to be the test charge. **(i)** If q_2 is initially positive and is changed to a charge of the same magnitude but negative, what happens to the potential at the position of q_2 due to q_1? (a) It increases. (b) It decreases. (c) It remains the same. **(ii)** When q_2 is changed from positive to negative, what happens to the potential energy of the two-charge system? Choose from the same possibilities.

---

[1] The expression for the electric potential energy of a system made up of two point charges, Equation 25.13, is of the *same* form as the equation for the gravitational potential energy of a system made up of two point masses, $-Gm_1 m_2 / r$ (see Chapter 13). The similarity is not surprising considering that both expressions are derived from an inverse-square force law.

EXAMPLE 25.3 **The Electric Potential Due to Two Point Charges**

As shown in Figure 25.11a, a charge $q_1 = 2.00 \mu C$ is located at the origin and a charge $q_2 = -6.00 \mu C$ is located at (0, 3.00) m.

(A) Find the total electric potential due to these charges at the point P, whose coordinates are (4.00, 0) m.

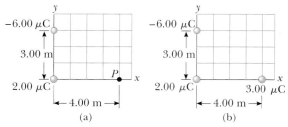

Figure 25.11 (Example 25.3) (a) The electric potential at P due to the two charges q_1 and q_2 is the algebraic sum of the potentials due to the individual charges. (b) A third charge $q_3 = 3.00 \mu C$ is brought from infinity to point P.

SOLUTION

Conceptualize Recognize that the 2.00 μC and $-6.00 \mu C$ charges are source charges and set up an electric field as well as a potential at all points in space, including point P.

Categorize The potential is evaluated using an equation developed in this chapter, so we categorize this example as a substitution problem.

Use Equation 25.12 for the system of two source charges:

$$V_P = k_e \left(\frac{q_1}{r_1} + \frac{q_2}{r_2} \right)$$

Substitute numerical values:

$$V_P = (8.99 \times 10^9 \ N \cdot m^2/C^2) \left(\frac{2.00 \times 10^{-6} \ C}{4.00 \ m} + \frac{-6.00 \times 10^{-6} \ C}{5.00 \ m} \right)$$

$$= \boxed{-6.29 \times 10^3 \ V}$$

(B) Find the change in potential energy of the system of two charges plus a third charge $q_3 = 3.00 \mu C$ as the latter charge moves from infinity to point P (Fig. 25.11b).

SOLUTION

Assign $U_i = 0$ for the system to the configuration in which the charge q_3 is at infinity. Use Equation 25.2 to evaluate the potential energy for the configuration in which the charge is at P:

$$U_f = q_3 V_P$$

Substitute numerical values to evaluate ΔU:

$$\Delta U = U_f - U_i = q_3 V_P - 0 = (3.00 \times 10^{-6} \ C)(-6.29 \times 10^3 \ V)$$

$$= \boxed{-1.89 \times 10^{-2} \ J}$$

Therefore, because the potential energy of the system has decreased, an external agent has to do positive work to remove the charge from point P back to infinity.

What If? You are working through this example with a classmate and she says, "Wait a minute! In part (B), we ignored the potential energy associated with the pair of charges q_1 and q_2!" How would you respond?

Answer Given the statement of the problem, it is not necessary to include this potential energy because part (B) asks for the *change* in potential energy of the system as q_3 is brought in from infinity. Because the configuration of charges q_1 and q_2 does not change in the process, there is no ΔU associated with these charges. Had part (B) asked to find the change in potential energy when *all three* charges start out infinitely far apart and are then brought to the positions in Figure 25.11b, however, you would have to calculate the change using Equation 25.14.

25.4 Obtaining the Value of the Electric Field from the Electric Potential

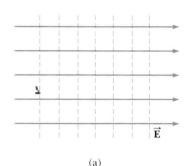

(a)

The electric field $\vec{E}$ and the electric potential V are related as shown in Equation 25.3, which tells us how to find ΔV if the electric field $\vec{E}$ is known. We now show how to calculate the value of the electric field if the electric potential is known in a certain region.

From Equation 25.3, we can express the potential difference dV between two points a distance ds apart as

$$dV = -\vec{E} \cdot d\vec{s} \qquad (25.15)$$

If the electric field has only one component E_x, then $\vec{E} \cdot d\vec{s} = E_x \, dx$. Therefore, Equation 25.15 becomes $dV = -E_x \, dx$, or

$$E_x = -\frac{dV}{dx} \qquad (25.16)$$

That is, the x component of the electric field is equal to the negative of the derivative of the electric potential with respect to x. Similar statements can be made about the y and z components. Equation 25.16 is the mathematical statement of the electric field being a measure of the rate of change with position of the electric potential as mentioned in Section 25.1.

Experimentally, electric potential and position can be measured easily with a voltmeter (see Section 28.5) and a meterstick. Consequently, an electric field can be determined by measuring the electric potential at several positions in the field and making a graph of the results. According to Equation 25.16, the slope of a graph of V versus x at a given point provides the magnitude of the electric field at that point.

When a test charge undergoes a displacement $d\vec{s}$ along an equipotential surface, then $dV = 0$ because the potential is constant along an equipotential surface. From Equation 25.15, we see that $dV = -\vec{E} \cdot d\vec{s} = 0$; therefore, $\vec{E}$ must be perpendicular to the displacement along the equipotential surface. This result shows that the **equipotential surfaces must always be perpendicular to the electric field lines passing through them.**

As mentioned at the end of Section 25.2, the equipotential surfaces associated with a uniform electric field consist of a family of planes perpendicular to the field lines. Figure 25.12a shows some representative equipotential surfaces for this situation.

If the charge distribution creating an electric field has spherical symmetry such that the volume charge density depends only on the radial distance r, the electric field is radial. In this case, $\vec{E} \cdot d\vec{s} = E_r \, dr$, and we can express dV as $dV = -E_r \, dr$. Therefore,

$$E_r = -\frac{dV}{dr} \qquad (25.17)$$

(b)

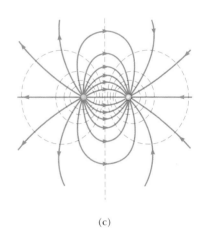

(c)

Figure 25.12 Equipotential surfaces (the dashed blue lines are intersections of these surfaces with the page) and electric field lines for (a) a uniform electric field produced by an infinite sheet of charge, (b) a point charge, and (c) an electric dipole. In all cases, the equipotential surfaces are *perpendicular* to the electric field lines at every point.

For example, the electric potential of a point charge is $V = k_e q/r$. Because V is a function of r only, the potential function has spherical symmetry. Applying Equation 25.17, we find that the electric field due to the point charge is $E_r = k_e q/r^2$, a familiar result. Notice that the potential changes only in the radial direction, not in any direction perpendicular to r. Therefore, V (like E_r) is a function only of r, which is again consistent with the idea that **equipotential surfaces are perpendicular to field lines.** In this case, the equipotential surfaces are a family of spheres concentric with the spherically symmetric charge distribution (Fig. 25.12b). The equipotential surfaces for an electric dipole are sketched in Figure 25.12c.

In general, the electric potential is a function of all three spatial coordinates. If $V(r)$ is given in terms of the Cartesian coordinates, the electric field components E_x, E_y, and E_z can readily be found from $V(x, y, z)$ as the partial derivatives[2]

Finding the electric field ▶
from the potential

$$E_x = -\frac{\partial V}{\partial x} \qquad E_y = -\frac{\partial V}{\partial y} \qquad E_z = -\frac{\partial V}{\partial z} \qquad \textbf{(25.18)}$$

Quick Quiz 25.4 In a certain region of space, the electric potential is zero everywhere along the x axis. From this information, you can conclude that the x component of the electric field in this region is (a) zero, (b) in the $+x$ direction, or (c) in the $-x$ direction.

EXAMPLE 25.4 | **The Electric Potential Due to a Dipole**

An electric dipole consists of two charges of equal magnitude and opposite sign separated by a distance $2a$ as shown in Figure 25.13. The dipole is along the x axis and is centered at the origin.

(A) Calculate the electric potential at point P on the y axis.

SOLUTION

Conceptualize Compare this situation to that in part (B) of Example 23.5. It is the same situation, but here we are seeking the electric potential rather than the electric field.

Figure 25.13 (Example 25.4) An electric dipole located on the x axis.

Categorize Because the dipole consists of only two source charges, the electric potential can be evaluated by summing the potentials due to the individual charges.

Analyze Use Equation 25.12 to find the electric potential at P due to the two charges:

$$V_P = k_e \sum_i \frac{q_i}{r_i} = k_e \left(\frac{q}{\sqrt{a^2 + y^2}} + \frac{-q}{\sqrt{a^2 + y^2}} \right) = 0$$

(B) Calculate the electric potential at point R on the $+x$ axis.

SOLUTION

Use Equation 25.12 to find the electric potential at R due to the two charges:

$$V_R = k_e \sum_i \frac{q_i}{r_i} = k_e \left(\frac{-q}{x - a} + \frac{q}{x + a} \right) = -\frac{2k_e qa}{x^2 - a^2}$$

(C) Calculate V and E_x at a point on the x axis far from the dipole.

SOLUTION

For point R far from the dipole such that $x \gg a$, neglect a^2 in the denominator of the answer to part (B) and write V in this limit:

$$V_R = \lim_{x \gg a} \left(-\frac{2k_e qa}{x^2 - a^2} \right) \approx -\frac{2k_e qa}{x^2} \qquad (x \gg a)$$

---

[2] In vector notation, $\vec{\mathbf{E}}$ is often written in Cartesian coordinate systems as

$$\vec{\mathbf{E}} = -\nabla V = -\left(\hat{\mathbf{i}} \frac{\partial}{\partial x} + \hat{\mathbf{j}} \frac{\partial}{\partial y} + \hat{\mathbf{k}} \frac{\partial}{\partial z} \right) V$$

where ∇ is called the *gradient operator.*

Use Equation 25.16 and this result to calculate the x component of the electric field at a point on the x axis far from the dipole:

$$E_x = -\frac{dV}{dx} = -\frac{d}{dx}\left(-\frac{2k_e qa}{x^2}\right)$$

$$= 2k_e qa \frac{d}{dx}\left(\frac{1}{x^2}\right) = -\frac{4k_e qa}{x^3} \quad (x \gg a)$$

Finalize   The potentials in parts (B) and (C) are negative because points on the $+x$ axis are closer to the negative charge than to the positive charge. For the same reason, the x component of the electric field is negative. Compare the result of part (C) to that of Problem 18 in Chapter 23, in which the electric field on the x axis due to a dipole was calculated directly.

What If?   Suppose you want to find the electric field at a point P on the y axis. In part (A), the electric potential was found to be zero for all values of y. Is the electric field zero at all points on the y axis?

Answer   No. That there is no change in the potential along the y axis tells us only that the y component of the electric field is zero. Look back at Figure 23.13 in Example 23.5. We showed there that the electric field of a dipole on the y axis has only an x component. We could not find the x component in the current example because we do not have an expression for the potential near the y axis as a function of x.

25.5   Electric Potential Due to Continuous Charge Distributions

The electric potential due to a continuous charge distribution can be calculated in two ways. If the charge distribution is known, we consider the potential due to a small charge element dq, treating this element as a point charge (Fig. 25.14). From Equation 25.11, the electric potential dV at some point P due to the charge element dq is

$$dV = k_e \frac{dq}{r} \tag{25.19}$$

where r is the distance from the charge element to point P. To obtain the total potential at point P, we integrate Equation 25.19 to include contributions from all elements of the charge distribution. Because each element is, in general, a different distance from point P and k_e is constant, we can express V as

$$V = k_e \int \frac{dq}{r} \tag{25.20}$$

◀ Electric potential due to a continuous charge distribution

In effect, we have replaced the sum in Equation 25.12 with an integral. In this expression for V, the electric potential is taken to be zero when point P is infinitely far from the charge distribution.

If the electric field is already known from other considerations such as Gauss's law, we can calculate the electric potential due to a continuous charge distribution using Equation 25.3. If the charge distribution has sufficient symmetry, we first evaluate $\vec{\mathbf{E}}$ using Gauss's law and then substitute the value obtained into Equation 25.3 to determine the potential difference ΔV between any two points. We then choose the electric potential V to be zero at some convenient point.

Figure 25.14   The electric potential at point P due to a continuous charge distribution can be calculated by dividing the charge distribution into elements of charge dq and summing the electric potential contributions over all elements.

| PROBLEM-SOLVING STRATEGY | **Calculating Electric Potential** |

The following procedure is recommended for solving problems that involve the determination of an electric potential due to a charge distribution.

1. *Conceptualize.* Think carefully about the individual charges or the charge distribution you have in the problem and imagine what type of potential would be created. Appeal to any symmetry in the arrangement of charges to help you visualize the potential.

2. *Categorize.* Are you analyzing a group of individual charges or a continuous charge distribution? The answer to this question will tell you how to proceed in the *Analyze* step.

3. *Analyze.* When working problems involving electric potential, remember that it is a *scalar quantity*, so there are no components to consider. Therefore, when using the superposition principle to evaluate the electric potential at a point, simply take the algebraic sum of the potentials due to each charge. You must keep track of signs, however.

   As with potential energy in mechanics, only *changes* in electric potential are significant; hence, the point where the potential is set at zero is arbitrary. When dealing with point charges or a finite-sized charge distribution, we usually define $V = 0$ to be at a point infinitely far from the charges. If the charge distribution itself extends to infinity, however, some other nearby point must be selected as the reference point.

   (a) *If you are analyzing a group of individual charges:* Use the superposition principle, which states that when several point charges are present, the resultant potential at a point in space is the *algebraic sum* of the individual potentials due to the individual charges (Eq. 25.12). Example 25.4 demonstrated this procedure.

   (b) *If you are analyzing a continuous charge distribution:* Replace the sums for evaluating the total potential at some point P from individual charges by integrals (Eq. 25.20). The charge distribution is divided into infinitesimal elements of charge dq located at a distance r from the point P. An element is then treated as a point charge, so the potential at P due to the element is $dV = k_e\, dq/r$. The total potential at P is obtained by integrating over the entire charge distribution. For many problems, it is possible in performing the integration to express dq and r in terms of a single variable. To simplify the integration, give careful consideration to the geometry involved in the problem. Examples 25.5 through 25.7 demonstrate such a procedure.

   To obtain the potential from the electric field: Another method used to obtain the potential is to start with the definition of the potential difference given by Equation 25.3. If $\vec{\mathbf{E}}$ is known or can be obtained easily (such as from Gauss's law), the line integral of $\vec{\mathbf{E}} \cdot d\vec{\mathbf{s}}$ can be evaluated.

4. *Finalize.* Check to see if your expression for the potential is consistent with the mental representation and reflects any symmetry you noted previously. Imagine varying parameters such as the distance of the observation point from the charges or the radius of any circular objects to see if the mathematical result changes in a reasonable way.

EXAMPLE 25.5 Electric Potential Due to a Uniformly Charged Ring

(A) Find an expression for the electric potential at a point P located on the perpendicular central axis of a uniformly charged ring of radius a and total charge Q.

SOLUTION

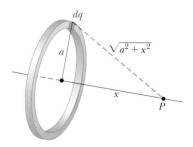

Figure 25.15 (Example 25.5) A uniformly charged ring of radius a lies in a plane perpendicular to the x axis. All elements dq of the ring are the same distance from a point P lying on the x axis.

Conceptualize Study Figure 25.15, in which the ring is oriented so that its plane is perpendicular to the x axis and its center is at the origin.

Categorize Because the ring consists of a continuous distribution of charge rather than a set of discrete charges, we must use the integration technique represented by Equation 25.20 in this example.

Analyze We take point P to be at a distance x from the center of the ring as shown in Figure 25.15. Notice that all charge elements dq are at the same distance $\sqrt{a^2 + x^2}$ from point P.

Use Equation 25.20 to express V in terms of the geometry:

$$V = k_e \int \frac{dq}{r} = k_e \int \frac{dq}{\sqrt{a^2 + x^2}}$$

Noting that a and x are constants, bring $\sqrt{a^2 + x^2}$ in front of the integral sign and integrate over the ring:

$$V = \frac{k_e}{\sqrt{a^2 + x^2}} \int dq = \frac{k_e Q}{\sqrt{a^2 + x^2}} \qquad \text{(25.21)}$$

(B) Find an expression for the magnitude of the electric field at point P.

SOLUTION

From symmetry, notice that along the x axis $\vec{E}$ can have only an x component. Therefore, apply Equation 25.16 to Equation 25.21:

$$E_x = -\frac{dV}{dx} = -k_e Q \frac{d}{dx} (a^2 + x^2)^{-1/2}$$

$$= -k_e Q (-\tfrac{1}{2})(a^2 + x^2)^{-3/2}(2x)$$

$$E_x = \frac{k_e x}{(a^2 + x^2)^{3/2}} Q \qquad \text{(25.22)}$$

Finalize   The only variable in the expressions for V and E_x is x. That is not surprising because our calculation is valid only for points along the x axis, where y and z are both zero. This result for the electric field agrees with that obtained by direct integration (see Example 23.7).

EXAMPLE 25.6   **Electric Potential Due to a Uniformly Charged Disk**

A uniformly charged disk has radius R and surface charge density σ.

(A) Find the electric potential at a point P along the perpendicular central axis of the disk.

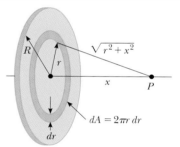

SOLUTION

Conceptualize   If we consider the disk to be a set of concentric rings, we can use our result from Example 25.5—which gives the potential created by a ring of radius a—and sum the contributions of all rings making up the disk.

Figure 25.16   (Example 25.6) A uniformly charged disk of radius R lies in a plane perpendicular to the x axis. The calculation of the electric potential at any point P on the x axis is simplified by dividing the disk into many rings of radius r and width dr, with area $2\pi r\, dr$.

Categorize   Because the disk is continuous, we evaluate the potential due to a continuous charge distribution rather than a group of individual charges.

Analyze   Find the amount of charge dq on a ring of radius r and width dr as shown in Figure 25.16:

$$dq = \sigma\, dA = \sigma (2\pi r\, dr) = 2\pi \sigma r\, dr$$

Use this result in the equation given for V in Example 25.5 (with a replaced by r and Q replaced by dq) to find the potential due to the ring:

$$dV = \frac{k_e\, dq}{\sqrt{r^2 + x^2}} = \frac{k_e 2\pi \sigma r\, dr}{\sqrt{r^2 + x^2}}$$

To obtain the total potential at P, integrate this expression over the limits $r = 0$ to $r = R$, noting that x is a constant:

$$V = \pi k_e \sigma \int_0^R \frac{2r\, dr}{\sqrt{r^2 + x^2}} = \pi k_e \sigma \int_0^R (r^2 + x^2)^{-1/2}\, 2r\, dr$$

This integral is of the common form $\int u^n\, du$ and has the value $u^{n+1}/(n + 1)$, where $n = -\tfrac{1}{2}$ and $u = r^2 + x^2$. Use this result to evaluate the integral:

$$V = 2\pi k_e \sigma [(R^2 + x^2)^{1/2} - x] \qquad \text{(25.23)}$$

(B) Find the x component of the electric field at a point P along the perpendicular central axis of the disk.

SOLUTION

As in Example 25.5, use Equation 25.16 to find the electric field at any axial point:

$$E_x = -\frac{dV}{dx} = 2\pi k_e \sigma \left[1 - \frac{x}{(R^2 + x^2)^{1/2}} \right] \quad \text{(25.24)}$$

Finalize   Compare Equation 25.24 with the result of Example 23.8. The calculation of V and $\vec{E}$ for an arbitrary point off the x axis is more difficult to perform, and we do not treat that situation in this book.

EXAMPLE 25.7   **Electric Potential Due to a Finite Line of Charge**

A rod of length ℓ located along the x axis has a total charge Q and a uniform linear charge density $\lambda = Q/\ell$. Find the electric potential at a point P located on the y axis a distance a from the origin (Fig. 25.17).

SOLUTION

Conceptualize   The potential at P due to every segment of charge on the rod is positive because every segment carries a positive charge.

Categorize   Because the rod is continuous, we evaluate the potential due to a continuous charge distribution rather than a group of individual charges.

Figure 25.17   (Example 25.7) A uniform line charge of length ℓ located along the x axis. To calculate the electric potential at P, the line charge is divided into segments each of length dx and each carrying a charge $dq = \lambda\, dx$.

Analyze   In Figure 25.17, the rod lies along the x axis, dx is the length of one small segment, and dq is the charge on that segment. Because the rod has a charge per unit length λ, the charge dq on the small segment is $dq = \lambda\, dx$.

Find the potential at P due to one segment of the rod:

$$dV = k_e \frac{dq}{r} = k_e \frac{\lambda\, dx}{\sqrt{a^2 + x^2}}$$

Find the total potential at P by integrating this expression over the limits $x = 0$ to $x = \ell$:

$$V = \int_0^\ell k_e \frac{\lambda\, dx}{\sqrt{a^2 + x^2}}$$

Noting that k_e and $\lambda = Q/\ell$ are constants and can be removed from the integral, evaluate the integral with the help of Appendix B:

$$V = k_e \lambda \int_0^\ell \frac{dx}{\sqrt{a^2 + x^2}} = k_e \frac{Q}{\ell} \ln \left(x + \sqrt{a^2 + x^2} \right) \Big|_0^\ell$$

Evaluate the result between the limits: $V = k_e \dfrac{Q}{\ell} [\ln (\ell + \sqrt{a^2 + \ell^2}) - \ln a] = k_e \dfrac{Q}{\ell} \ln \left(\dfrac{\ell + \sqrt{a^2 + \ell^2}}{a} \right)$ **(25.25)**

What If?   What if you were asked to find the electric field at point P? Would that be a simple calculation?

Answer   Calculating the electric field by means of Equation 23.11 would be a little messy. There is no symmetry to appeal to, and the integration over the line of charge would represent a vector addition of electric fields at point P. Using Equation 25.18, you could find E_y by replacing a with y in Equation 25.25 and performing the differentiation with respect to y. Because the charged rod in Figure 25.17 lies entirely to the right of $x = 0$, the electric field at point P would have an x component to the left if the rod is charged positively. You cannot use Equation 25.18 to find the x

component of the field, however, because the potential due to the rod was evaluated at a specific value of x ($x = 0$) rather than a general value of x. You would have to find the potential as a function of both x and y to be able to find the x and y components of the electric field using Equation 25.25.

25.6 Electric Potential Due to a Charged Conductor

In Section 24.4, we found that when a solid conductor in equilibrium carries a net charge, the charge resides on the conductor's outer surface. Furthermore, the electric field just outside the conductor is perpendicular to the surface and the field inside is zero.

We now show that **every point on the surface of a charged conductor in equilibrium is at the same electric potential.** Consider two points Ⓐ and Ⓑ on the surface of a charged conductor as shown in Figure 25.18. Along a surface path connecting these points, $\vec{E}$ is always perpendicular to the displacement $d\vec{s}$; therefore $\vec{E} \cdot d\vec{s} = 0$. Using this result and Equation 25.3, we conclude that the potential difference between Ⓐ and Ⓑ is necessarily zero:

$$V_{\text{Ⓑ}} - V_{\text{Ⓐ}} = -\int_{\text{Ⓐ}}^{\text{Ⓑ}} \vec{E} \cdot d\vec{s} = 0$$

This result applies to any two points on the surface. Therefore, V is constant everywhere on the surface of a charged conductor in equilibrium. That is,

> the surface of any charged conductor in electrostatic equilibrium is an equipotential surface. Furthermore, because the electric field is zero inside the conductor, the electric potential is constant everywhere inside the conductor and equal to its value at the surface.

Because of the constant value of the potential, no work is required to move a test charge from the interior of a charged conductor to its surface.

Consider a solid metal conducting sphere of radius R and total positive charge Q as shown in Figure 25.19a. As determined in part (A) of Example 24.3, the electric field outside the sphere is k_eQ/r^2 and points radially outward. Because the field outside of a spherically symmetric charge distribution is identical to that of a point charge, we expect the potential to also be that of a point charge, k_eQ/r. At the surface of the conducting sphere in Figure 25.19a, the potential must be k_eQ/R. Because the entire sphere must be at the same potential, the potential at any point within the sphere must also be k_eQ/R. Figure 25.19b is a plot of the electric potential as a function of r, and Figure 25.19c shows how the electric field varies with r.

When a net charge is placed on a spherical conductor, the surface charge density is uniform as indicated in Figure 25.19a. If the conductor is nonspherical as in

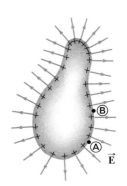

Figure 25.18 An arbitrarily shaped conductor carrying a positive charge. When the conductor is in electrostatic equilibrium, all the charge resides at the surface, $\vec{E} = 0$ inside the conductor, and the direction of $\vec{E}$ immediately outside the conductor is perpendicular to the surface. The electric potential is constant inside the conductor and is equal to the potential at the surface. Notice from the spacing of the positive signs that the surface charge density is nonuniform.

PITFALL PREVENTION 25.6
Potential May Not Be Zero

The electric potential inside the conductor is not necessarily zero in Figure 25.18, even though the electric field is zero. Equation 25.15 shows that a zero value of the field results in no *change* in the potential from one point to another inside the conductor. Therefore, the potential everywhere inside the conductor, including the surface, has the same value, which may or may not be zero, depending on where the zero of potential is defined.

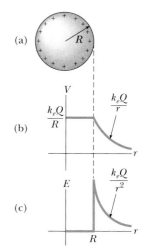

Figure 25.19 (a) The excess charge on a conducting sphere of radius R is uniformly distributed on its surface. (b) Electric potential versus distance r from the center of the charged conducting sphere. (c) Electric field magnitude versus distance r from the center of the charged conducting sphere.

Figure 25.18, however, the surface charge density is high where the radius of curvature is small (as noted in Section 24.4) and low where the radius of curvature is large. Because the electric field immediately outside the conductor is proportional to the surface charge density, **the electric field is large near convex points having small radii of curvature and reaches very high values at sharp points.** In Example 25.8, the relationship between electric field and radius of curvature is explored mathematically.

EXAMPLE 25.8	**Two Connected Charged Spheres**

Two spherical conductors of radii r_1 and r_2 are separated by a distance much greater than the radius of either sphere. The spheres are connected by a conducting wire as shown in Figure 25.20. The charges on the spheres in equilibrium are q_1 and q_2, respectively, and they are uniformly charged. Find the ratio of the magnitudes of the electric fields at the surfaces of the spheres.

SOLUTION

Conceptualize Imagine that the spheres are much farther apart than shown in Figure 25.20. Because they are so far apart, the field of one does not affect the charge distribution on the other. The conducting wire between them ensures that both spheres have the same electric potential.

Categorize Because the spheres are so far apart, we model the charge distribution on them as spherically symmetric, and we can model the field and potential outside the spheres to be that due to point charges.

Figure 25.20 (Example 25.8) Two charged spherical conductors connected by a conducting wire. The spheres are at the *same* electric potential V.

Analyze Set the electric potentials at the surfaces of the spheres equal to each other:

$$V = k_e \frac{q_1}{r_1} = k_e \frac{q_2}{r_2}$$

Solve for the ratio of charges on the spheres:

$$(1) \quad \frac{q_1}{q_2} = \frac{r_1}{r_2}$$

Write expressions for the magnitudes of the electric fields at the surfaces of the spheres:

$$E_1 = k_e \frac{q_1}{r_1^2} \quad \text{and} \quad E_2 = k_e \frac{q_2}{r_2^2}$$

Evaluate the ratio of these two fields:

$$\frac{E_1}{E_2} = \frac{q_1}{q_2} \frac{r_2^2}{r_1^2}$$

Substitute for the ratio of charges from Equation (1):

$$(2) \quad \frac{E_1}{E_2} = \frac{r_1}{r_2} \frac{r_2^2}{r_1^2} = \frac{r_2}{r_1}$$

Finalize The field is stronger in the vicinity of the smaller sphere even though the electric potentials at the surfaces of both spheres are the same.

A Cavity Within a Conductor

Suppose a conductor of arbitrary shape contains a cavity as shown in Figure 25.21. Let's assume no charges are inside the cavity. **In this case, the electric field inside the cavity must be zero** regardless of the charge distribution on the outside surface of the conductor as we mentioned in Section 24.4. Furthermore, the field in the cavity is zero even if an electric field exists outside the conductor.

To prove this point, remember that every point on the conductor is at the same electric potential; therefore, any two points Ⓐ and Ⓑ on the cavity's surface must be at the same potential. Now imagine a field $\vec{\mathbf{E}}$ exists in the cavity and evaluate the potential difference $V_{\circledB} - V_{\circledA}$ defined by Equation 25.3:

$$V_{\circledB} - V_{\circledA} = - \int_{\circledA}^{\circledB} \vec{\mathbf{E}} \cdot d\vec{\mathbf{s}}$$

Because $V_{\circledB} - V_{\circledA} = 0$, the integral of $\vec{\mathbf{E}} \cdot d\vec{\mathbf{s}}$ must be zero for all paths between any two points $\circledA$ and $\circledB$ on the conductor. The only way that can be true for *all* paths is if $\vec{\mathbf{E}}$ is zero *everywhere* in the cavity. Therefore, **a cavity surrounded by conducting walls is a field-free region as long as no charges are inside the cavity.**

Corona Discharge

A phenomenon known as **corona discharge** is often observed near a conductor such as a high-voltage power line. When the electric field in the vicinity of the conductor is sufficiently strong, electrons resulting from random ionizations of air molecules near the conductor accelerate away from their parent molecules. These rapidly moving electrons can ionize additional molecules near the conductor, creating more free electrons. The observed glow (or corona discharge) results from the recombination of these free electrons with the ionized air molecules. If a conductor has an irregular shape, the electric field can be very high near sharp points or edges of the conductor; consequently, the ionization process and corona discharge are most likely to occur around such points.

Corona discharge is used in the electrical transmission industry to locate broken or faulty components. For example, a broken insulator on a transmission tower has sharp edges where corona discharge is likely to occur. Similarly, corona discharge will occur at the sharp end of a broken conductor strand. Observation of these discharges is difficult because the visible radiation emitted is weak and most of the radiation is in the ultraviolet. (We will discuss ultraviolet radiation and other portions of the electromagnetic spectrum in Section 34.7.) Even use of traditional ultraviolet cameras is of little help because the radiation from the corona discharge is overwhelmed by ultraviolet radiation from the Sun. Newly developed dual-spectrum devices combine a narrow-band ultraviolet camera with a visible-light camera to show a daylight view of the corona discharge in the actual location on the transmission tower or cable. The ultraviolet part of the camera is designed to operate in a wavelength range in which radiation from the Sun is very weak.

25.7   The Millikan Oil-Drop Experiment

Robert Millikan performed a brilliant set of experiments from 1909 to 1913 in which he measured e, the magnitude of the elementary charge on an electron, and demonstrated the quantized nature of this charge. His apparatus, diagrammed in Active Figure 25.22, contains two parallel metallic plates. Oil droplets

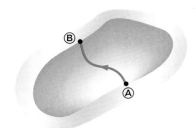

Figure 25.21   A conductor in electrostatic equilibrium containing a cavity. The electric field in the cavity is zero, regardless of the charge on the conductor.

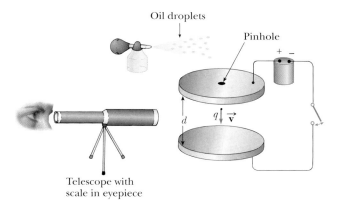

ACTIVE FIGURE 25.22

Schematic drawing of the Millikan oil-drop apparatus.

Sign in at www.thomsonedu.com and go to ThomsonNOW to perform a simplified version of the experiment yourself. You will be able to take data on a number of oil drops and determine the elementary charge from your data.

(a)

(b)

Figure 25.23 The forces acting on a negatively charged oil droplet in the Millikan experiment. (a) With the electric field off, the droplet falls at terminal velocity $\vec{v}$ under the influence of the gravitational and drag forces. (b) When the electric field is turned on, the droplet moves upward at terminal velocity $\vec{v}'$ under the influence of the electric, gravitational, and drag forces.

from an atomizer are allowed to pass through a small hole in the upper plate. Millikan used x-rays to ionize the air in the chamber so that freed electrons would adhere to the oil drops, giving them a negative charge. A horizontally directed light beam is used to illuminate the oil droplets, which are viewed through a telescope whose long axis is perpendicular to the light beam. When viewed in this manner, the droplets appear as shining stars against a dark background and the rate at which individual drops fall can be determined.

Let's assume a single drop having a mass m and carrying a charge q is being viewed and its charge is negative. If no electric field is present between the plates, the two forces acting on the charge are the gravitational force $m\vec{g}$ acting downward[3] and a viscous drag force $\vec{F}_D$ acting upward as indicated in Figure 25.23a. The drag force is proportional to the drop's speed as discussed in Section 6.4. When the drop reaches its terminal speed v, the two forces balance each other ($mg = F_D$).

Now suppose a battery connected to the plates sets up an electric field between the plates such that the upper plate is at the higher electric potential. In this case, a third force $q\vec{E}$ acts on the charged drop. Because q is negative and $\vec{E}$ is directed downward, this electric force is directed upward as shown in Figure 25.23b. If this upward force is strong enough, the drop moves upward and the drag force $\vec{F}'_D$ acts downward. When the upward electric force $q\vec{E}$ balances the sum of the gravitational force and the downward drag force $\vec{F}'_D$, the drop reaches a new terminal speed v' in the upward direction.

With the field turned on, a drop moves slowly upward, typically at rates of hundredths of a centimeter per second. The rate of fall in the absence of a field is comparable. Hence, one can follow a single droplet for hours, alternately rising and falling, by simply turning the electric field on and off.

After recording measurements on thousands of droplets, Millikan and his coworkers found that all droplets, to within about 1% precision, had a charge equal to some integer multiple of the elementary charge e:

$$q = ne \quad n = 0, -1, -2, -3, \ldots$$

where $e = 1.60 \times 10^{-19}$ C. Millikan's experiment yields conclusive evidence that charge is quantized. For this work, he was awarded the Nobel Prize in Physics in 1923.

25.8 Applications of Electrostatics

The practical application of electrostatics is represented by such devices as lightning rods and electrostatic precipitators and by such processes as xerography and the painting of automobiles. Scientific devices based on the principles of electrostatics include electrostatic generators, the field-ion microscope, and ion-drive rocket engines.

The Van de Graaff Generator

Experimental results show that when a charged conductor is placed in contact with the inside of a hollow conductor, all the charge on the charged conductor is transferred to the hollow conductor. In principle, the charge on the hollow conductor and its electric potential can be increased without limit by repetition of the process.

In 1929, Robert J. Van de Graaff (1901–1967) used this principle to design and build an electrostatic generator, and a schematic representation of it is given in Figure 25.24. This type of generator is used extensively in nuclear physics research. Charge is delivered continuously to a high-potential electrode by means of a mov-

---

[3] There is also a buoyant force on the oil drop due to the surrounding air. This force can be incorporated as a correction in the gravitational force $m\vec{g}$ on the drop, so we will not consider it in our analysis.

ing belt of insulating material. The high-voltage electrode is a hollow metal dome mounted on an insulating column. The belt is charged at point Ⓐ by means of a corona discharge between comb-like metallic needles and a grounded grid. The needles are maintained at a positive electric potential of typically 10^4 V. The positive charge on the moving belt is transferred to the dome by a second comb of needles at point Ⓑ. Because the electric field inside the dome is negligible, the positive charge on the belt is easily transferred to the conductor regardless of its potential. In practice, it is possible to increase the electric potential of the dome until electrical discharge occurs through the air. Because the "breakdown" electric field in air is about 3×10^6 V/m, a sphere 1 m in radius can be raised to a maximum potential of 3×10^6 V. The potential can be increased further by increasing the dome's radius and placing the entire system in a container filled with high-pressure gas.

Van de Graaff generators can produce potential differences as large as 20 million volts. Protons accelerated through such large potential differences receive enough energy to initiate nuclear reactions between themselves and various target nuclei. Smaller generators are often seen in science classrooms and museums. If a person insulated from the ground touches the sphere of a Van de Graaff generator, his or her body can be brought to a high electric potential. The person's hair acquires a net positive charge, and each strand is repelled by all the others as in the opening photograph of Chapter 23.

Figure 25.24   Schematic diagram of a Van de Graaff generator. Charge is transferred to the metal dome at the top by means of a moving belt. The charge is deposited on the belt at point Ⓐ and transferred to the hollow conductor at point Ⓑ.

The Electrostatic Precipitator

One important application of electrical discharge in gases is the *electrostatic precipitator*. This device removes particulate matter from combustion gases, thereby reducing air pollution. Precipitators are especially useful in coal-burning power plants and industrial operations that generate large quantities of smoke. Current systems are able to eliminate more than 99% of the ash from smoke.

Figure 25.25a shows a schematic diagram of an electrostatic precipitator. A high potential difference (typically 40 to 100 kV) is maintained between a wire running down the center of a duct and the walls of the duct, which are grounded. The wire is maintained at a negative electric potential with respect to the walls, so the electric field is directed toward the wire. The values of the field near the wire become high enough to cause a corona discharge around the wire; the air near the wire contains positive ions, electrons, and such negative ions as O_2^-. The air to be cleaned enters the duct and moves near the wire. As the electrons and negative

(a)   (b)   (c)

Figure 25.25   (a) Schematic diagram of an electrostatic precipitator. The high negative electric potential maintained on the central coiled wire creates a corona discharge in the vicinity of the wire. Compare the air pollution when the electrostatic precipitator is (b) operating and (c) turned off.

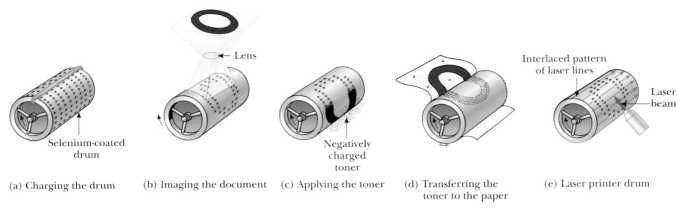

(a) Charging the drum   (b) Imaging the document   (c) Applying the toner   (d) Transferring the toner to the paper   (e) Laser printer drum

Figure 25.26 The xerographic process. (a) The photoconductive surface of the drum is positively charged. (b) Through the use of a light source and lens, an image is created on the surface in the form of positive charges. (c) The surface containing the image is covered with a negatively charged powder, which adheres only to the image area. (d) A piece of paper is placed over the surface and given a positive charge, which transfers the image to the paper as the negatively charged powder particles migrate to the paper. The paper is then thermally treated to "fix" the powder. (e) A laser printer operates similarly except that the image is produced by turning a laser beam on and off as it sweeps across the selenium-coated drum.

ions created by the discharge are accelerated toward the outer wall by the electric field, the dirt particles in the air become charged by collisions and ion capture. Because most of the charged dirt particles are negative, they too are drawn to the duct walls by the electric field. When the duct is periodically shaken, the particles break loose and are collected at the bottom.

In addition to reducing the level of particulate matter in the atmosphere (compare Figs. 25.25b and c), the electrostatic precipitator recovers valuable materials in the form of metal oxides.

Xerography and Laser Printers

The basic idea of xerography[4] was developed by Chester Carlson, who was granted a patent for the xerographic process in 1940. The unique feature of this process is the use of a photoconductive material to form an image. (A *photoconductor* is a material that is a poor electrical conductor in the dark but becomes a good electrical conductor when exposed to light.)

The xerographic process is illustrated in parts (a) through (d) of Figure 25.26. First, the surface of a plate or drum that has been coated with a thin film of photoconductive material (usually selenium or some compound of selenium) is given a positive electrostatic charge in the dark. An image of the page to be copied is then focused by a lens onto the charged surface. The photoconducting surface becomes conducting only in areas where light strikes it. In such areas, the light produces charge carriers in the photoconductor that move the positive charge off the drum. Positive charges, however, remain on those areas of the photoconductor not exposed to light, leaving a latent image of the object in the form of a positive surface charge distribution.

Next, a negatively charged powder called a *toner* is dusted onto the photoconducting surface. The charged powder adheres only to those areas of the surface that contain the positively charged image. The toner (and hence the image) is then transferred to the surface of a sheet of positively charged paper.

Finally, the toner is "fixed" to the surface of the paper as the toner melts while passing through high-temperature rollers. The result is a permanent copy of the original.

A laser printer (Fig. 25.26e) operates by the same principle, with the exception that a computer-directed laser beam is used to illuminate the photoconductor instead of a lens.

---

[4] The prefix *xero-* is from the Greek word meaning "dry." Note that liquid ink is not used in xerography.

Summary

ThomsonNOW Sign in at **www.thomsonedu.com** and go to ThomsonNOW to take a practice test for this chapter.

DEFINITIONS

The **potential difference** ΔV between points Ⓐ and Ⓑ in an electric field $\vec{E}$ is defined as

$$\Delta V \equiv \frac{\Delta U}{q_0} = -\int_{Ⓐ}^{Ⓑ} \vec{E} \cdot d\vec{s} \qquad (25.3)$$

where ΔU is given by Equation 25.1 below. The **electric potential** $V = U/q_0$ is a scalar quantity and has the units of joules per coulomb, where $1 \, \text{J/C} \equiv 1 \, \text{V}$.

An **equipotential surface** is one on which all points are at the same electric potential. Equipotential surfaces are perpendicular to electric field lines.

CONCEPTS AND PRINCIPLES

When a positive test charge q_0 is moved between points Ⓐ and Ⓑ in an electric field $\vec{E}$, the **change in the potential energy of the charge–field system** is

$$\Delta U = -q_0 \int_{Ⓐ}^{Ⓑ} \vec{E} \cdot d\vec{s} \qquad (25.1)$$

The potential difference between two points Ⓐ and Ⓑ separated by a distance d in a uniform electric field $\vec{E}$, where $\vec{s}$ is a vector that points from Ⓐ toward Ⓑ and is parallel to $\vec{E}$, is

$$\Delta V = -E \int_{Ⓐ}^{Ⓑ} ds = -Ed \qquad (25.6)$$

If we define $V = 0$ at $r = \infty$, the electric potential due to a point charge at any distance r from the charge is

$$V = k_e \frac{q}{r} \qquad (25.11)$$

The electric potential associated with a group of point charges is obtained by summing the potentials due to the individual charges.

The **potential energy associated with a pair of point charges** separated by a distance r_{12} is

$$U = k_e \frac{q_1 q_2}{r_{12}} \qquad (25.13)$$

We obtain the potential energy of a distribution of point charges by summing terms like Equation 25.13 over all pairs of particles.

If the electric potential is known as a function of coordinates x, y, and z, we can obtain the components of the electric field by taking the negative derivative of the electric potential with respect to the coordinates. For example, the x component of the electric field is

$$E_x = -\frac{dV}{dx} \qquad (25.16)$$

The **electric potential due to a continuous charge distribution** is

$$V = k_e \int \frac{dq}{r} \qquad (25.20)$$

Every point on the surface of a charged conductor in electrostatic equilibrium is at the same electric potential. The potential is constant everywhere inside the conductor and equal to its value at the surface.

Questions

☐ denotes answer available in *Student Solutions Manual/Study Guide;* **O** denotes objective question

1. Distinguish between electric potential and electric potential energy.

2. **O** In a certain region of space, a uniform electric field is in the x direction. A particle with negative charge is carried from $x = 20$ cm to $x = 60$ cm. **(i)** Does the potential energy of the charge-field system (a) increase, (b) remain constant, (c) decrease, or (d) change unpredictably? **(ii)** Does the particle move to a position where the potential is (a) higher than before, (b) unchanged, (c) lower than before, or (d) unpredictable?

3. **O** Consider the equipotential surfaces shown in Figure 25.4. In this region of space, what is the approximate

direction of the electric field? (a) out of the page (b) into the page (c) toward the right (d) toward the left (e) toward the top of the page (f) toward the bottom of the page (g) the field is zero

4. **O** A particle with charge -40 nC is on the x axis at the point with coordinate $x = 0$. A second particle, with charge -20 nC, is on the x axis at $x = 500$ mm. (i) Is there a point at a finite distance where the electric field is zero? (a) Yes; it is to the left of $x = 0$. (b) Yes; it is between $x = 0$ and $x = 500$ mm. (c) Yes; it is to the right of $x = 500$ mm. (d) No. (ii) Is the electric potential zero at this point? (a) No; it is positive. (b) Yes. (c) No; it is negative. (d) No such point exists. (iii) Is there a point at a finite distance where the electric potential is zero? (a) Yes; it is to the left of $x = 0$. (b) Yes; it is between $x = 0$ and $x = 500$ mm. (c) Yes; it is to the right of $x = 500$ mm. (d) No. (iv) Is the electric field zero at this point? (a) No; it points to the right. (b) Yes. (c) No; it points to the left. (d) No such point exists.

5. The potential energy of a pair of charged particles with the same sign is positive, whereas the potential energy of a pair of charged particles with opposite signs is negative. Give a physical explanation of this statement.

6. Describe the equipotential surfaces for (a) an infinite line of charge and (b) a uniformly charged sphere.

7. **O** In a certain region of space, the electric field is zero. From this fact, what can you conclude about the electric potential in this region? (a) It is zero. (b) It is constant. (c) It is positive. (d) It is negative. (e) None of these answers is necessarily true.

8. **O** A filament running along the x axis from the origin to $x = 80$ cm carries electric charge with uniform density. At the point P with coordinates ($x = 80$ cm, $y = 80$ cm), this filament creates potential 100 V. Now we add another filament along the y axis, running from the origin to $y = 80$ cm, carrying the same amount of charge with the same uniform density. At the same point P, does the pair of filaments create potential (a) greater than 200 V, (b) 200 V, (c) between 141 V and 200 V, (d) 141 V, (e) between 100 V and 141 V, (f) 100 V, (g) between 0 and 100 V, or (h) 0?

9. **O** In different experimental trials, an electron, a proton, or a doubly charged oxygen atom (O^{--}) is fired within a vacuum tube. The particle's trajectory carries it through a point where the electric potential is 40 V and then through a point at a different potential. Rank each of the following cases according to the change in kinetic energy of the particle over this part of its flight, from the largest increase to the largest decrease in kinetic energy. (a) An electron moves from 40 V to 60 V. (b) An electron moves from 40 V to 20 V. (c) A proton moves from 40 V to 20 V. (d) A proton moves from 40 V to 10 V. (e) An O^{--} ion moves from 40 V to 50 V. (f) An O^{--} ion moves from 40 V to 60 V. For comparison, include also in your ranking (g) zero change and (h) $+10$ electron volts of change in kinetic energy. In your ranking, display any cases of equality.

10. What determines the maximum potential to which the dome of a Van de Graaff generator can be raised?

11. **O** (i) A metallic sphere A of radius 1 cm is several centimeters away from a metallic spherical shell B of radius 2 cm. Charge 450 nC is placed on A, with no charge on B or anywhere nearby. Next, the two objects are joined by a long, thin, metallic wire (as shown in Fig. 25.20), and finally the wire is removed. How is the charge shared between A and B? (a) 0 on A, 450 nC on B (b) 50 nC on A and 400 nC on B, with equal volume charge densities (c) 90 nC on A and 360 nC on B, with equal surface charge densities (d) 150 nC on A and 300 nC on B (e) 225 nC on A and 225 nC on B (f) 450 nC on A and 0 on B (g) in some other predictable way (h) in some unpredictable way (ii) A metallic sphere A of radius 1 cm with charge 450 nC hangs on an insulating thread inside an uncharged thin metallic spherical shell B of radius 2 cm. Next, A is made temporarily to touch the inner surface of B. How is the charge then shared between them? Choose from the same possibilities. Arnold Arons, the only physics teacher yet to have his picture on the cover of *Time* magazine, suggested the idea for this question.

12. Study Figure 23.3 and the accompanying text discussion of charging by induction. When the grounding wire is touched to the rightmost point on the sphere in Figure 23.3c, electrons are drained away from the sphere to leave the sphere positively charged. Suppose the grounding wire is touched to the leftmost point on the sphere instead. Will electrons still drain away, moving closer to the negatively charged rod as they do so? What kind of charge, if any, remains on the sphere?

Problems

WebAssign The Problems from this chapter may be assigned online in WebAssign.

ThomsonNOW™ Sign in at **www.thomsonedu.com** and go to ThomsonNOW to assess your understanding of this chapter's topics with additional quizzing and conceptual questions.

1, 2, 3 denotes straightforward, intermediate, challenging; ☐ denotes full solution available in *Student Solutions Manual/Study Guide;* ▲ denotes coached solution with hints available at **www.thomsonedu.com;** denotes developing symbolic reasoning; ● denotes asking for qualitative reasoning; ▪ denotes computer useful in solving problem

Section 25.1 Electric Potential and Potential Difference

1. (a) Calculate the speed of a proton that is accelerated from rest through a potential difference of 120 V. (b) Calculate the speed of an electron that is accelerated through the same potential difference.

2. How much work is done (by a battery, generator, or some other source of potential difference) in moving Avogadro's number of electrons from an initial point where the electric potential is 9.00 V to a point where the poten-

tial is -5.00 V? (The potential in each case is measured relative to a common reference point.)

Section 25.2 Potential Difference in a Uniform Electric Field

3. The difference in potential between the accelerating plates in the electron gun of a television picture tube is about 25 000 V. If the distance between these plates is 1.50 cm, what is the magnitude of the uniform electric field in this region?

4. A uniform electric field of magnitude 325 V/m is directed in the negative y direction in Figure P25.4. The coordinates of point A are $(-0.200, -0.300)$ m and those of point B are $(0.400, 0.500)$ m. Calculate the potential difference $V_B - V_A$, using the blue path.

Figure P25.4

5. ▲ An electron moving parallel to the x axis has an initial speed of 3.70×10^6 m/s at the origin. Its speed is reduced to 1.40×10^5 m/s at the point $x = 2.00$ cm. Calculate the potential difference between the origin and that point. Which point is at the higher potential?

6. ● Starting with the definition of work, prove that at every point on an equipotential surface the surface must be perpendicular to the electric field there.

7. **Review problem.** A block having mass m and charge $+Q$ is connected to an insulating spring having constant k. The block lies on a frictionless, insulating horizontal track, and the system is immersed in a uniform electric field of magnitude E directed as shown in Figure P25.7. If the block is released from rest when the spring is unstretched (at $x = 0$), (a) by what maximum amount does the spring expand? (b) What is the equilibrium position of the block? (c) Show that the block's motion is simple harmonic and determine its period. (d) **What If?** Repeat part (a), assuming the coefficient of kinetic friction between block and surface is μ_k.

Figure P25.7

8. A particle having charge $q = +2.00$ μC and mass $m = 0.010$ 0 kg is connected to a string that is $L = 1.50$ m long and tied to the pivot point P in Figure P25.8. The particle, string, and pivot point all lie on a frictionless, hori-

zontal table. The particle is released from rest when the string makes an angle $\theta = 60.0°$ with a uniform electric field of magnitude $E = 300$ V/m. Determine the speed of the particle when the string is parallel to the electric field (point a in Fig. P25.8).

Figure P25.8

9. ● An insulating rod having linear charge density $\lambda = 40.0$ μC/m and linear mass density $\mu = 0.100$ kg/m is released from rest in a uniform electric field $E = 100$ V/m directed perpendicular to the rod (Fig. P25.9). (a) Determine the speed of the rod after it has traveled 2.00 m. (b) **What If?** How does your answer to part (a) change if the electric field is not perpendicular to the rod? Explain.

Figure P25.9

Section 25.3 Electric Potential and Potential Energy Due to Point Charges

Note: Unless stated otherwise, assume the reference level of potential is $V = 0$ at $r = \infty$.

10. Given two particles with 2.00-μC charges as shown in Figure P25.10 and a particle with charge $q = 1.28 \times 10^{-18}$ C at the origin, (a) what is the net force exerted by the two 2.00-μC charges on the test charge q? (b) What is the electric field at the origin due to the two 2.00-μC particles? (c) What is the electric potential at the origin due to the two 2.00-μC particles?

Figure P25.10

11. (a) Find the potential at a distance of 1.00 cm from a proton. (b) What is the potential difference between two points that are 1.00 cm and 2.00 cm from a proton? (c) **What If?** Repeat parts (a) and (b) for an electron.

12. A particle with charge $+q$ is at the origin. A particle with charge $-2q$ is at $x = 2.00$ m on the x axis. (a) For what finite value(s) of x is the electric field zero? (b) For what finite value(s) of x is the electric potential zero?

2 = intermediate;   3 = challenging;   □ = SSM/SG;   ▲ = ThomsonNOW;   = symbolic reasoning;   ● = qualitative reasoning

13. At a certain distance from a charged particle, the magnitude of the electric field is 500 V/m and the electric potential is −3.00 kV. (a) What is the distance to the particle? (b) What is the magnitude of the charge?

14. ● Two charged particles, $Q_1 = +5.00$ nC and $Q_2 = −3.00$ nC, are separated by 35.0 cm. (a) What is the potential energy of the pair? Explain the significance of the algebraic sign of your answer. (b) What is the electric potential at a point midway between the charged particles?

15. The three charged particles in Figure P25.15 are at the vertices of an isosceles triangle. Calculate the electric potential at the midpoint of the base, taking $q = 7.00$ μC.

Figure P25.15

16. *Compare this problem with Problem 16 in Chapter 23.* Two charged particles each of magnitude 2.00 μC are located on the x axis. One is at $x = 1.00$ m, and the other is at $x = −1.00$ m. (a) Determine the electric potential on the y axis at $y = 0.500$ m. (b) Calculate the change in electric potential energy of the system as a third charged particle of −3.00 μC is brought from infinitely far away to a position on the y axis at $y = 0.500$ m.

17. *Compare this problem with Problem 47 in Chapter 23.* Four identical charged particles ($q = +10.0$ μC) are located on the corners of a rectangle as shown in Figure P23.47. The dimensions of the rectangle are $L = 60.0$ cm and $W = 15.0$ cm. Calculate the change in electric potential energy of the system as the particle at the lower left corner in Figure P23.47 is brought to this position from infinitely far away. Assume the other three particles in Figure P23.47 remain fixed in position.

18. Two charged particles have effects at the origin, described by the expressions

$$8.99 \times 10^9 \text{ N} \cdot \text{m}^2/\text{C}^2 \left[-\frac{7 \times 10^{-9} \text{ C}}{(0.07 \text{ m})^2} \cos 70° \hat{\mathbf{i}} \right.$$

$$\left. -\frac{7 \times 10^{-9} \text{ C}}{(0.07 \text{ m})^2} \sin 70° \hat{\mathbf{j}} + \frac{8 \times 10^{-9} \text{ C}}{(0.03 \text{ m})^2} \hat{\mathbf{j}} \right]$$

and

$$8.99 \times 10^9 \text{ N} \cdot \text{m}^2/\text{C}^2 \left[\frac{7 \times 10^{-9} \text{ C}}{0.07 \text{ m}} - \frac{8 \times 10^{-9} \text{ C}}{0.03 \text{ m}} \right]$$

(a) Identify the locations of the particles and the charges on them. (b) Find the force on a particle with charge −16.0 nC placed at the origin. (c) Find the work required to move this third charged particle to the origin from a very distant point.

19. ▲ Show that the amount of work required to assemble four identical charged particles of magnitude Q at the corners of a square of side s is $5.41 k_e Q^2/s$.

20. *Compare this problem with Problem 19 in Chapter 23.* Five equal negative charged particles $−q$ are placed symmetrically around a circle of radius R. Calculate the electric potential at the center of the circle.

21. *Compare this problem with Problem 35 in Chapter 23.* Three particles with equal positive charges q are at the corners of an equilateral triangle of side a as shown in Figure P23.35. (a) At what point, if any, in the plane of the particles is the electric potential zero? (b) What is the electric potential at the point P due to the two particles at the base of the triangle?

22. Two charged particles of equal magnitude are located along the y axis equal distances above and below the x axis as shown in Figure P25.22. (a) Plot a graph of the potential at points along the x axis over the interval $−3a < x < 3a$. You should plot the potential in units of $k_e Q/a$. (b) Let the charge of the particle located at $y = −a$ be negative. Plot the potential along the y axis over the interval $−4a < y < 4a$.

Figure P25.22

23. ● **Review problem.** Two insulating spheres have radii 0.300 cm and 0.500 cm, masses 0.100 kg and 0.700 kg, and uniformly distributed charges of −2.00 μC and 3.00 μC. They are released from rest when their centers are separated by 1.00 m. (a) How fast will each be moving when they collide? *Suggestion:* Consider conservation of energy and of linear momentum. (b) **What If?** If the spheres were conductors, would the speeds be greater or less than those calculated in part (a)? Explain.

24. ● **Review problem.** Two insulating spheres have radii r_1 and r_2, masses m_1 and m_2, and uniformly distributed charges $−q_1$ and q_2. They are released from rest when their centers are separated by a distance d. (a) How fast is each moving when they collide? *Suggestion:* Consider conservation of energy and conservation of linear momentum. (b) **What If?** If the spheres were conductors, would their speeds be greater or less than those calculated in part (a)? Explain.

25. **Review problem.** A light, unstressed spring has length d. Two identical particles, each with charge q, are connected to the opposite ends of the spring. The particles are held stationary a distance d apart and then released at the same moment. The system then oscillates on a horizontal, frictionless table. The spring has a bit of internal kinetic friction, so the oscillation is damped. The particles eventually stop vibrating when the distance between them is $3d$. Find the increase in internal energy that appears in the spring during the oscillations. Assume the system of the spring and two charged particles is isolated.

26. In 1911, Ernest Rutherford and his assistants Geiger and Marsden conducted an experiment in which they scattered alpha particles from thin sheets of gold. An alpha particle, having charge $+2e$ and mass 6.64×10^{-27} kg, is a product of certain radioactive decays. The results of the experiment led Rutherford to the idea that most of the mass of an atom is in a very small nucleus, with electrons in orbit around it, his planetary model of the atom. Assume an alpha particle, initially very far from a gold nucleus, is fired with a velocity of 2.00×10^7 m/s directly toward the nucleus (charge $+79e$). How close does the alpha particle get to the nucleus before turning around? Assume the gold nucleus remains stationary.

27. Four identical particles each have charge q and mass m. They are released from rest at the vertices of a square of side L. How fast is each particle moving when their distance from the center of the square doubles?

28. How much work is required to assemble eight identical charged particles, each of magnitude q, at the corners of a cube of side s?

Section 25.4 Obtaining the Value of the Electric Field from the Electric Potential

29. The potential in a region between $x = 0$ and $x = 6.00$ m is $V = a + bx$, where $a = 10.0$ V and $b = -7.00$ V/m. Determine (a) the potential at $x = 0$, 3.00 m, and 6.00 m and (b) the magnitude and direction of the electric field at $x = 0$, 3.00 m, and 6.00 m.

30. The electric potential inside a charged spherical conductor of radius R is given by $V = k_e Q/R$, and the potential outside is given by $V = k_e Q/r$. Using $E_r = -dV/dr$, derive the electric field (a) inside and (b) outside this charge distribution.

31. ▲ Over a certain region of space, the electric potential is $V = 5x - 3x^2y + 2yz^2$. Find the expressions for the x, y, and z components of the electric field over this region. What is the magnitude of the field at the point P that has coordinates $(1, 0, -2)$ m?

32. ● Figure P25.32 shows several equipotential lines, each labeled by its potential in volts. The distance between the lines of the square grid represents 1.00 cm. (a) Is the magnitude of the field larger at A or at B? Explain how you can tell. (b) Explain what you can determine about $\vec{\mathbf{E}}$ at B. (c) Represent what the field looks like by drawing at least eight field lines.

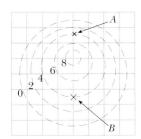

Figure P25.32

33. It is shown in Example 25.7 that the potential at a point P a distance a above one end of a uniformly charged rod of length ℓ lying along the x axis is

$$V = k_e \frac{Q}{\ell} \ln \left(\frac{\ell + \sqrt{a^2 + \ell^2}}{a} \right)$$

Use this result to derive an expression for the y component of the electric field at P. *Suggestion:* Replace a with y.

Section 25.5 Electric Potential Due to Continuous Charge Distributions

34. Consider a ring of radius R with the total charge Q spread uniformly over its perimeter. What is the potential difference between the point at the center of the ring and a point on its axis a distance $2R$ from the center?

35. A rod of length L (Fig. P25.35) lies along the x axis with its left end at the origin. It has a nonuniform charge density $\lambda = \alpha x$, where α is a positive constant. (a) What are the units of α? (b) Calculate the electric potential at A.

Figure P25.35 Problems 35 and 36.

36. For the arrangement described in Problem 35, calculate the electric potential at point B, which lies on the perpendicular bisector of the rod a distance b above the x axis.

37. *Compare this problem with Problem 27 in Chapter 23.* A uniformly charged insulating rod of length 14.0 cm is bent into the shape of a semicircle as shown in Figure P23.27. The rod has a total charge of -7.50 μC. Find the electric potential at O, the center of the semicircle.

38. A wire having a uniform linear charge density λ is bent into the shape shown in Figure P25.38. Find the electric potential at point O.

Figure P25.38

Section 25.6 Electric Potential Due to a Charged Conductor

39. ▲ A spherical conductor has a radius of 14.0 cm and charge of 26.0 μC. Calculate the electric field and the electric potential at (a) $r = 10.0$ cm, (b) $r = 20.0$ cm, and (c) $r = 14.0$ cm from the center.

40. How many electrons should be removed from an initially uncharged spherical conductor of radius 0.300 m to produce a potential of 7.50 kV at the surface?

41. The electric field on the surface of an irregularly shaped conductor varies from 56.0 kN/C to 28.0 kN/C. Calculate the local surface charge density at the point on the surface where the radius of curvature of the surface is (a) greatest and (b) smallest.

2 = intermediate; 3 = challenging; ☐ = SSM/SG; ▲ = ThomsonNOW; = symbolic reasoning; ● = qualitative reasoning

42. Electric charge can accumulate on an airplane in flight. You may have observed needle-shaped metal extensions on the wing tips and tail of an airplane. Their purpose is to allow charge to leak off before much of it accumulates. The electric field around the needle is much larger than the field around the body of the airplane and can become large enough to produce dielectric breakdown of the air, discharging the airplane. To model this process, assume two charged spherical conductors are connected by a long conducting wire and a charge of 1.20 μC is placed on the combination. One sphere, representing the body of the airplane, has a radius of 6.00 cm, and the other, representing the tip of the needle, has a radius of 2.00 cm. (a) What is the electric potential of each sphere? (b) What is the electric field at the surface of each sphere?

Section 25.8 Applications of Electrostatics

43. Lightning can be studied with a Van de Graaff generator, essentially consisting of a spherical dome on which charge is continuously deposited by a moving belt. Charge can be added until the electric field at the surface of the dome becomes equal to the dielectric strength of air. Any more charge leaks off in sparks as shown in Figure P25.43. Assume the dome has a diameter of 30.0 cm and is surrounded by dry air with dielectric strength 3.00×10^6 V/m. (a) What is the maximum potential of the dome? (b) What is the maximum charge on the dome?

Figure P25.43

44. A Geiger-Mueller tube is a radiation detector that consists of a closed, hollow, metal cylinder (the cathode) of inner radius r_a and a coaxial cylindrical wire (the anode) of radius r_b (Fig. P25.44). The charge per unit length on the anode is λ, and the charge per unit length on the cathode is $-\lambda$. A gas fills the space between the electrodes. When a high-energy elementary particle passes through this space, it can ionize an atom of the gas. The strong electric field makes the resulting ion and electron accelerate in opposite directions. They strike other molecules of the gas to ionize them, producing an avalanche of electrical discharge. The pulse of electric current between the wire and the cylinder is counted by an external circuit. (a) Show that the magnitude of the potential difference between the wire and the cylinder is

$$\Delta V = 2k_e \lambda \ln\left(\frac{r_a}{r_b}\right)$$

(b) Show that the magnitude of the electric field in the space between cathode and anode is

$$E = \frac{\Delta V}{\ln\left(r_a/r_b\right)}\left(\frac{1}{r}\right)$$

where r is the distance from the axis of the anode to the point where the field is to be calculated.

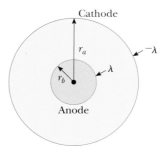

Figure P25.44 Problems 44 and 45.

45. The results of Problem 44 apply also to an electrostatic precipitator (Figs. 25.25 and P25.44). An applied potential difference $\Delta V = V_a - V_b = 50.0$ kV is to produce an electric field of magnitude 5.50 MV/m at the surface of the central wire. Assume the outer cylindrical wall has uniform radius $r_a = 0.850$ m. (a) What should be the radius r_b of the central wire? You will need to solve a transcendental equation. (b) What is the magnitude of the electric field at the outer wall?

Additional Problems

46. ● **Review problem.** From a large distance away, a particle of mass 2.00 g and charge 15.0 μC is fired at $21.0\hat{\mathbf{i}}$ m/s straight toward a second particle, originally stationary but free to move, with mass 5.00 g and charge 8.50 μC. (a) At the instant of closest approach, both particles will be moving at the same velocity. Explain why. (b) Find this velocity. (c) Find the distance of closest approach. (d) Find the velocities of both particles after they separate again.

47. The liquid-drop model of the atomic nucleus suggests high-energy oscillations of certain nuclei can split the nucleus into two unequal fragments plus a few neutrons. The fission products acquire kinetic energy from their mutual Coulomb repulsion. Calculate the electric potential energy (in electron volts) of two spherical fragments from a uranium nucleus having the following charges and radii: $38e$ and 5.50×10^{-15} m, $54e$ and 6.20×10^{-15} m. Assume the charge is distributed uniformly throughout the volume of each spherical fragment and, immediately before separating, each fragment is at rest and their surfaces are in contact. The electrons surrounding the nucleus can be ignored.

48. In fair weather, the electric field in the air at a particular location immediately above the Earth's surface is 120 N/C directed downward. (a) What is the surface charge density on the ground? Is it positive or negative? (b) Imagine the atmosphere is stripped off and the surface charge density is uniform over the planet. What then is the charge of the whole surface of the Earth? (c) What is the Earth's electric potential? (d) What is the difference in potential between the head and the feet of a person 1.75 m tall? (e) Imagine the Moon, with 27.3% of the radius of the Earth, had a charge 27.3% as large, with the same sign. Find the electric force that the Earth would then exert on the Moon. (f) State how the answer to part (e) compares with the gravitational force the Earth exerts on the Moon. (g) A

dust particle of mass 6.00 mg is in the air near the surface of the spherical Earth. What charge must the dust particle carry to be suspended in equilibrium between the electric and gravitational forces exerted on it? Ignore buoyancy. (h) The Earth is not perfectly spherical. It has an equatorial bulge due to its rotation, so the radius of curvature of the ground is slightly larger at the poles than at the equator. Would the dust particle in part (g) require more charge or less charge to be suspended at the equator compared with being suspended at one of the poles? Explain your answer with reference to variations in both the electric force and the gravitational force.

49. The Bohr model of the hydrogen atom states that the single electron can exist only in certain allowed orbits around the proton. The radius of each Bohr orbit is $r = n^2(0.052\ 9\ \text{nm})$, where $n = 1, 2, 3, \ldots$. Calculate the electric potential energy of a hydrogen atom when the electron (a) is in the first allowed orbit, with $n = 1$, (b) is in the second allowed orbit, with $n = 2$, and (c) has escaped from the atom, with $r = \infty$. Express your answers in electron volts.

50. On a dry winter day, you scuff your leather-soled shoes across a carpet and get a shock when you extend the tip of one finger toward a metal doorknob. In a dark room, you see a spark perhaps 5 mm long. Make order-of-magnitude estimates of (a) your electric potential and (b) the charge on your body before you touch the doorknob. Explain your reasoning.

51. The electric potential immediately outside a charged conducting sphere is 200 V, and 10.0 cm farther from the center of the sphere the potential is 150 V. (a) Is this information sufficient to determine the charge on the sphere and its radius? Explain. (b) The electric potential immediately outside another charged conducting sphere is 210 V, and 10.0 cm farther from the center the magnitude of the electric field is 400 V/m. Is this information sufficient to determine the charge on the sphere and its radius? Explain.

52. As shown in Figure P25.52, two large, parallel, vertical conducting plates separated by distance d are charged so that their potentials are $+V_0$ and $-V_0$. A small conducting ball of mass m and radius R (where $R \ll d$) is hung midway between the plates. The thread of length L supporting the ball is a conducting wire connected to ground, so the potential of the ball is fixed at $V = 0$. The ball hangs straight down in stable equilibrium when V_0 is sufficiently small. Show that the equilibrium of the ball is unstable if V_0 exceeds the critical value $k_e d^2 mg/(4RL)$. *Suggestion:* Consider the forces on the ball when it is displaced a distance $x \ll L$.

Figure P25.52

53. The electric potential everywhere on the xy plane is given by

$$V = \frac{36}{\sqrt{(x+1)^2 + y^2}} - \frac{45}{\sqrt{x^2 + (y-2)^2}}$$

where V is in volts and x and y are in meters. Determine the position and charge on each of the particles that create this potential.

54. *Compare this problem with Problem 28 in Chapter 23.* (a) A uniformly charged cylindrical shell has total charge Q, radius R, and height h. Determine the electric potential at a point a distance d from the right end of the cylinder as shown in Figure P25.54. *Suggestion:* Use the result of Example 25.5 by treating the cylinder as a collection of ring charges. (b) **What If?** Use the result of Example 25.6 to solve the same problem for a solid cylinder.

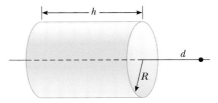

Figure P25.54

55. Calculate the work that must be done to charge a spherical shell of radius R to a total charge Q.

56. ● (a) Use the exact result from Example 25.4 to find the electric potential created by the dipole described at the point $(3a, 0)$. (b) Explain how this answer compares with the result of the approximate expression that is valid when x is much greater than a.

57. ▲ From Gauss's law, the electric field set up by a uniform line of charge is

$$\vec{\mathbf{E}} = \left(\frac{\lambda}{2\pi\epsilon_0 r}\right)\hat{\mathbf{r}}$$

where $\hat{\mathbf{r}}$ is a unit vector pointing radially away from the line and λ is the linear charge density along the line. Derive an expression for the potential difference between $r = r_1$ and $r = r_2$.

58. Four balls, each with mass m, are connected by four non-conducting strings to form a square with side a as shown in Figure P25.58. The assembly is placed on a horizontal, nonconducting, frictionless surface. Balls 1 and 2 each have charge q, and balls 3 and 4 are uncharged. Find the maximum speed of balls 3 and 4 after the string connecting balls 1 and 2 is cut.

Figure P25.58

59. The x axis is the symmetry axis of a stationary, uniformly charged ring of radius R and charge Q (Fig. P25.59). A particle with charge Q and mass M is located initially at

the center of the ring. When it is displaced slightly, the particle accelerates along the x axis to infinity. Show that the ultimate speed of the particle is

$$v = \left(\frac{2k_eQ^2}{MR}\right)^{1/2}$$

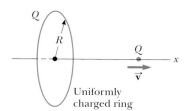

Uniformly charged ring

Figure P25.59

60. The thin, uniformly charged rod shown in Figure P25.60 has a linear charge density λ. Find an expression for the electric potential at P.

Figure P25.60

61. An electric dipole is located along the y axis as shown in Figure P25.61. The magnitude of its electric dipole moment is defined as $p = 2qa$. (a) At a point P, which is far from the dipole $(r \gg a)$, show that the electric potential is

$$V = \frac{k_e p \cos\theta}{r^2}$$

(b) Calculate the radial component E_r and the perpendicular component E_θ of the associated electric field. Note that $E_\theta = -(1/r)(\partial V/\partial\theta)$. Do these results seem reasonable for $\theta = 90°$ and $0°$? For $r = 0$? (c) For the dipole

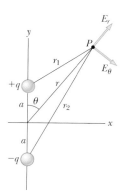

Figure P25.61

arrangement shown, express V in terms of Cartesian coordinates using $r = (x^2 + y^2)^{1/2}$ and

$$\cos\theta = \frac{y}{(x^2 + y^2)^{1/2}}$$

Using these results and again taking $r \gg a$, calculate the field components E_x and E_y.

62. A solid sphere of radius R has a uniform charge density ρ and total charge Q. Derive an expression for its total electric potential energy. *Suggestion:* Imagine the sphere is constructed by adding successive layers of concentric shells of charge $dq = (4\pi r^2 \, dr)\rho$ and use $dU = V \, dq$.

63. A disk of radius R (Fig. P25.63) has a nonuniform surface charge density $\sigma = Cr$, where C is a constant and r is measured from the center of the disk to a point on the surface of the disk. Find (by direct integration) the potential at P.

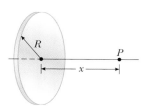

Figure P25.63

64. ● ▪ A uniformly charged filament lies along the x axis between $x = a = 1.00$ m and $x = a + \ell = 3.00$ m as shown in Figure 23.15. The total charge on the filament is 1.60 nC. Calculate successive approximations for the electric potential at the origin by modeling the filament as (a) a single charged particle at $x = 2.00$ m, (b) two 0.800-nC charged particles at $x = 1.5$ m and $x = 2.5$ m, and (c) four 0.400-nC charged particles at $x = 1.25$ m, $x = 1.75$ m, $x = 2.25$ m, and $x = 2.75$ m. Next, write and execute a computer program that will reproduce the results of parts (a), (b), and (c) and extend your calculation to (d) 32 and (e) 64 equally spaced charged particles. (f) Explain how the results compare with the potential given by the exact expression

$$V = \frac{k_e Q}{\ell} \ln\left(\frac{\ell + a}{a}\right)$$

65. Two parallel plates having charges of equal magnitude but opposite sign are separated by 12.0 cm. Each plate has a surface charge density of 36.0 nC/m². A proton is released from rest at the positive plate. Determine (a) the potential difference between the plates, (b) the kinetic energy of the proton when it reaches the negative plate, (c) the speed of the proton just before it strikes the negative plate, (d) the acceleration of the proton, and (e) the force on the proton. (f) From the force, find the magnitude of the electric field and show that it is equal to the electric field found from the charge densities on the plates.

66. A particle with charge q is located at $x = -R$, and a particle with charge $-2q$ is located at the origin. Prove that the equipotential surface that has zero potential is a sphere centered at $(-4R/3, 0, 0)$ and having a radius $r = 2R/3$.

2 = intermediate; 3 = challenging; ☐ = SSM/SG; ▲ = ThomsonNOW; = symbolic reasoning; ● = qualitative reasoning

67. When an uncharged conducting sphere of radius a is placed at the origin of an xyz coordinate system that lies in an initially uniform electric field $\vec{\mathbf{E}} = E_0\hat{\mathbf{k}}$, the resulting electric potential is $V(x, y, z) = V_0$ for points inside the sphere and

$$V(x, y, z) = V_0 - E_0 z + \frac{E_0 a^3 z}{(x^2 + y^2 + z^2)^{3/2}}$$

for points outside the sphere, where V_0 is the (constant) electric potential on the conductor. Use this equation to determine the x, y, and z components of the resulting electric field.

Answers to Quick Quizzes

25.1 **(i),** (b). When moving straight from Ⓐ to Ⓑ, $\vec{\mathbf{E}}$ and $d\vec{\mathbf{s}}$ both point toward the right. Therefore, the dot product $\vec{\mathbf{E}} \cdot d\vec{\mathbf{s}}$ in Equation 25.3 is positive and ΔV is negative. **(ii),** (a). From Equation 25.3, $\Delta U = q_0 \Delta V$, so if a negative test charge is moved through a negative potential difference, the change in potential energy is positive. Work must be done to move the charge in the direction opposite to the electric force on it.

25.2 Ⓑ to Ⓒ, Ⓒ to Ⓓ, Ⓐ to Ⓑ, Ⓓ to Ⓔ. Moving from Ⓑ to Ⓒ decreases the electric potential by 2 V, so the electric field performs 2 J of work on each coulomb of positive charge that moves. Moving from Ⓒ to Ⓓ decreases the electric potential by 1 V, so 1 J of work is done by the field. It takes no work to move the charge from Ⓐ to Ⓑ

because the electric potential does not change. Moving from Ⓓ to Ⓔ increases the electric potential by 1 V; therefore, the field does −1 J of work per unit of positive charge that moves.

25.3 **(i),** (c). The potential is established only by the source charge and is independent of the test charge. **(ii),** (a). The potential energy of the two-charge system is initially negative due to the product of charges of opposite sign in Equation 25.13. When the sign of q_2 is changed, both charges are negative and the potential energy of the system is positive.

25.4 (a). If the potential is constant (zero in this case), its derivative along this direction is zero.

All these devices are capacitors, which store electric charge and energy. A capacitor is one type of circuit element that we can combine with others to make electric circuits. (Paul Silverman/Fundamental Photographs)

26.1 Definition of Capacitance

26.2 Calculating Capacitance

26.3 Combinations of Capacitors

26.4 Energy Stored in a Charged Capacitor

26.5 Capacitors with Dielectrics

26.6 Electric Dipole in an Electric Field

26.7 An Atomic Description of Dielectrics

26 Capacitance and Dielectrics

In this chapter, we introduce the first of three simple *circuit elements* **that can be** connected with wires to form an electric circuit. Electric circuits are the basis for the vast majority of the devices used in our society. Here we shall discuss *capacitors*, devices that store electric charge. This discussion is followed by the study of *resistors* in Chapter 27 and *inductors* in Chapter 32. In later chapters, we will study more sophisticated circuit elements such as *diodes* and *transistors*.

Capacitors are commonly used in a variety of electric circuits. For instance, they are used to tune the frequency of radio receivers, as filters in power supplies, to eliminate sparking in automobile ignition systems, and as energy-storing devices in electronic flash units.

PITFALL PREVENTION 26.1
Capacitance Is a Capacity

To understand capacitance, think of similar notions that use a similar word. The *capacity* of a milk carton is the volume of milk it can store. The *heat capacity* of an object is the amount of energy an object can store per unit of temperature difference. The *capacitance* of a capacitor is the amount of charge the capacitor can store per unit of potential difference.

26.1 Definition of Capacitance

Consider two conductors as shown in Figure 26.1. Such a combination of two conductors is called a **capacitor.** The conductors are called *plates*. If the conductors carry charges of equal magnitude and opposite sign, a potential difference ΔV exists between them.

What determines how much charge is on the plates of a capacitor for a given voltage? Experiments show that the quantity of charge Q on a capacitor[1] is linearly pro-

---

[1] Although the total charge on the capacitor is zero (because there is as much excess positive charge on one conductor as there is excess negative charge on the other), it is common practice to refer to the magnitude of the charge on either conductor as "the charge on the capacitor."

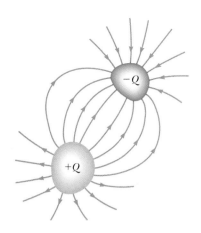

Figure 26.1 A capacitor consists of two conductors. When the capacitor is charged, the conductors carry charges of equal magnitude and opposite sign.

Figure 26.2 A parallel-plate capacitor consists of two parallel conducting plates, each of area A, separated by a distance d. When the capacitor is charged by connecting the plates to the terminals of a battery, the plates carry equal amounts of charge. One plate carries positive charge, and the other carries negative charge.

portional to the potential difference between the conductors; that is, $Q \propto \Delta V$. The proportionality constant depends on the shape and separation of the conductors.[2] This relationship can be written as $Q = C \Delta V$ if we define capacitance as follows:

The **capacitance** C of a capacitor is defined as the ratio of the magnitude of the charge on either conductor to the magnitude of the potential difference between the conductors:

$$C \equiv \frac{Q}{\Delta V} \qquad (26.1)$$

◄   **Definition of capacitance**

By definition *capacitance is always a positive quantity*. Furthermore, the charge Q and the potential difference ΔV are always expressed in Equation 26.1 as positive quantities.

From Equation 26.1, we see that capacitance has SI units of coulombs per volt. Named in honor of Michael Faraday, the SI unit of capacitance is the **farad** (F):

$$1 \text{ F} = 1 \text{ C/V}$$

The farad is a very large unit of capacitance. In practice, typical devices have capacitances ranging from microfarads (10^{-6} F) to picofarads (10^{-12} F). We shall use the symbol μF to represent microfarads. In practice, to avoid the use of Greek letters, physical capacitors are often labeled "mF" for microfarads and "mmF" for micromicrofarads or, equivalently, "pF" for picofarads.

Let's consider a capacitor formed from a pair of parallel plates as shown in Figure 26.2. Each plate is connected to one terminal of a battery, which acts as a source of potential difference. If the capacitor is initially uncharged, the battery establishes an electric field in the connecting wires when the connections are made. Let's focus on the plate connected to the negative terminal of the battery. The electric field in the wire applies a force on electrons in the wire immediately outside this plate; this force causes the electrons to move onto the plate. The movement continues until the plate, the wire, and the terminal are all at the same electric potential. Once this equilibrium situation is attained, a potential difference no longer exists between the terminal and the plate; as a result no electric field is present in the wire and the electrons stop moving. The plate now carries a negative charge. A similar process occurs at the other capacitor plate, where electrons move from the plate to the wire, leaving the plate positively charged. In this final configuration, the potential difference across the capacitor plates is the same as that between the terminals of the battery.

PITFALL PREVENTION 26.2
Potential Difference Is ΔV, Not V

We use the symbol ΔV for the potential difference across a circuit element or a device because this notation is consistent with our definition of potential difference and with the meaning of the delta sign. It is a common but confusing practice to use the symbol V without the delta sign for both a potential and a potential difference! Keep that in mind if you consult other texts.

PITFALL PREVENTION 26.3
Too Many Cs

Do not confuse an italic C for capacitance with a nonitalic C for the unit coulomb.

---

[2] The proportionality between ΔV and Q can be proven from Coulomb's law or by experiment.

Quick Quiz 26.1   A capacitor stores charge Q at a potential difference ΔV. What happens if the voltage applied to the capacitor by a battery is doubled to $2\Delta V$? (a) The capacitance falls to half its initial value, and the charge remains the same. (b) The capacitance and the charge both fall to half their initial values. (c) The capacitance and the charge both double. (d) The capacitance remains the same, and the charge doubles.

26.2   Calculating Capacitance

We can derive an expression for the capacitance of a pair of oppositely charged conductors having a charge of magnitude Q in the following manner. First we calculate the potential difference using the techniques described in Chapter 25. We then use the expression $C = Q/\Delta V$ to evaluate the capacitance. The calculation is relatively easy if the geometry of the capacitor is simple.

Although the most common situation is that of two conductors, a single conductor also has a capacitance. For example, imagine a spherical, charged conductor. The electric field lines around this conductor are exactly the same as if there were a conducting, spherical shell of infinite radius, concentric with the sphere and carrying a charge of the same magnitude but opposite sign. Therefore, we can identify the imaginary shell as the second conductor of a two-conductor capacitor. The electric potential of the sphere of radius a is simply $k_e Q/a$, and setting $V = 0$ for the infinitely large shell gives

Capacitance of an isolated charged sphere ▶

$$C = \frac{Q}{\Delta V} = \frac{Q}{k_e Q/a} = \frac{a}{k_e} = 4\pi\epsilon_0 a \tag{26.2}$$

This expression shows that the capacitance of an isolated, charged sphere is proportional to its radius and is independent of both the charge on the sphere and the potential difference.

The capacitance of a pair of conductors is illustrated below with three familiar geometries, namely, parallel plates, concentric cylinders, and concentric spheres. In these calculations, we assume the charged conductors are separated by a vacuum.

Parallel-Plate Capacitors

Two parallel, metallic plates of equal area A are separated by a distance d as shown in Figure 26.2. One plate carries a charge $+Q$, and the other carries a charge $-Q$. The surface charge density on each plate is $\sigma = Q/A$. If the plates are very close together (in comparison with their length and width), we can assume the electric field is uniform between the plates and zero elsewhere. According to the **What If?** feature of Example 24.5, the value of the electric field between the plates is

$$E = \frac{\sigma}{\epsilon_0} = \frac{Q}{\epsilon_0 A}$$

Because the field between the plates is uniform, the magnitude of the potential difference between the plates equals Ed (see Eq. 25.6); therefore,

$$\Delta V = Ed = \frac{Qd}{\epsilon_0 A}$$

Substituting this result into Equation 26.1, we find that the capacitance is

$$C = \frac{Q}{\Delta V} = \frac{Q}{Qd/\epsilon_0 A}$$

Capacitance of parallel plates ▶

$$C = \frac{\epsilon_0 A}{d} \tag{26.3}$$

That is, **the capacitance of a parallel-plate capacitor is proportional to the area of its plates and inversely proportional to the plate separation.**

Let's consider how the geometry of these conductors influences the capacity of the pair of plates to store charge. As a capacitor is being charged by a battery, electrons flow into the negative plate and out of the positive plate. If the capacitor plates are large, the accumulated charges are able to distribute themselves over a substantial area and the amount of charge that can be stored on a plate for a given potential difference increases as the plate area is increased. Therefore, it is reasonable that the capacitance is proportional to the plate area A as in Equation 26.3.

Now consider the region that separates the plates. Imagine moving the plates closer together. Consider the situation before any charges have had a chance to move in response to this change. Because no charges have moved, the electric field between the plates has the same value but extends over a shorter distance. Therefore, the magnitude of the potential difference between the plates $\Delta V = Ed$ (Eq. 25.6) is smaller. The difference between this new capacitor voltage and the terminal voltage of the battery appears as a potential difference across the wires connecting the battery to the capacitor, resulting in an electric field in the wires that drives more charge onto the plates and increases the potential difference between the plates. When the potential difference between the plates again matches that of the battery, the flow of charge stops. Therefore, moving the plates closer together causes the charge on the capacitor to increase. If d is increased, the charge decreases. As a result, the inverse relationship between C and d in Equation 26.3 is reasonable.

Key

Quick Quiz 26.2   Many computer keyboard buttons are constructed of capacitors as shown in Figure 26.3. When a key is pushed down, the soft insulator between the movable plate and the fixed plate is compressed. When the key is pressed, what happens to the capacitance? (a) It increases. (b) It decreases. (c) It changes in a way you cannot determine because the electric circuit connected to the keyboard button may cause a change in ΔV.

Movable plate

Dielectric

Fixed plate

Figure 26.3   (Quick Quiz 26.2) One type of computer keyboard button.

EXAMPLE 26.1   **The Cylindrical Capacitor**

A solid, cylindrical conductor of radius a and charge Q is coaxial with a cylindrical shell of negligible thickness, radius $b > a$, and charge $-Q$ (Fig. 26.4a). Find the capacitance of this cylindrical capacitor if its length is ℓ.

SOLUTION

Conceptualize   Recall that any pair of conductors qualifies as a capacitor, so the system described in this example therefore qualifies. Figure 26.4b helps visualize the electric field between the conductors.

Categorize   Because of the cylindrical symmetry of the system, we can use results from previous studies of cylindrical systems to find the capacitance.

Analyze   Assuming ℓ is much greater than a and b, we can neglect end effects. In this case, the electric field is perpendicular to the long axis of the cylinders and is confined to the region between them (Fig. 26.4b).

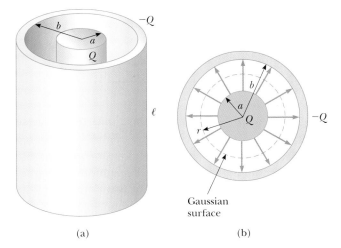

Gaussian surface

(a)   (b)

Figure 26.4   (Example 26.1) (a) A cylindrical capacitor consists of a solid cylindrical conductor of radius a and length ℓ surrounded by a coaxial cylindrical shell of radius b. (b) End view. The electric field lines are radial. The dashed line represents the end of the cylindrical gaussian surface of radius r and length ℓ.

Write an expression for the potential difference between the two cylinders from Equation 25.3:

$$V_b - V_a = -\int_a^b \vec{\mathbf{E}} \cdot d\vec{\mathbf{s}}$$

Apply Equation 24.7 for the electric field outside a cylindrically symmetric charge distribution and notice from Figure 26.4b that $\vec{E}$ is parallel to $d\vec{s}$ along a radial line:

$$V_b - V_a = -\int_a^b E_r \, dr = -2k_e\lambda \int_a^b \frac{dr}{r} = -2k_e\lambda \ln\left(\frac{b}{a}\right)$$

Substitute the absolute value of ΔV into Equation 26.1 and use $\lambda = Q/\ell$:

$$C = \frac{Q}{\Delta V} = \frac{Q}{(2k_e Q/\ell)\ln(b/a)} = \frac{\ell}{2k_e\ln(b/a)} \qquad \textbf{(26.4)}$$

Finalize   The capacitance is proportional to the length of the cylinders. As you might expect, the capacitance also depends on the radii of the two cylindrical conductors. Equation 26.4 shows that the capacitance per unit length of a combination of concentric cylindrical conductors is

$$\frac{C}{\ell} = \frac{1}{2k_e\ln(b/a)} \qquad \textbf{(26.5)}$$

An example of this type of geometric arrangement is a *coaxial cable*, which consists of two concentric cylindrical conductors separated by an insulator. You probably have a coaxial cable attached to your television set or VCR if you are a subscriber to cable television. The coaxial cable is especially useful for shielding electrical signals from any possible external influences.

What If?   Suppose $b = 2.00a$ for the cylindrical capacitor. You would like to increase the capacitance, and you can do so by choosing to increase either ℓ by 10% or a by 10%. Which choice is more effective at increasing the capacitance?

Answer   According to Equation 26.4, C is proportional to ℓ, so increasing ℓ by 10% results in a 10% increase in C. For the result of the change in a, let's use Equation 26.4 to set up a ratio of the capacitance C' for the enlarged cylinder radius a' to the original capacitance:

$$\frac{C'}{C} = \frac{\ell/2k_e\ln(b/a')}{\ell/2k_e\ln(b/a)} = \frac{\ln(b/a)}{\ln(b/a')}$$

We now substitute $b = 2.00a$ and $a' = 1.10a$, representing a 10% increase in a:

$$\frac{C'}{C} = \frac{\ln(2.00a/a)}{\ln(2.00a/1.10a)} = \frac{\ln 2.00}{\ln 1.82} = 1.16$$

which corresponds to a 16% increase in capacitance. Therefore, it is more effective to increase a than to increase ℓ.

Note two more extensions of this problem. First, it is advantageous to increase a only for a range of relationships between a and b. If $b > 2.85a$, increasing ℓ by 10% is more effective than increasing a (see Problem 66). Second, if b decreases, the capacitance increases. Increasing a or decreasing b has the effect of bringing the plates closer together, which increases the capacitance.

EXAMPLE 26.2   **The Spherical Capacitor**

A spherical capacitor consists of a spherical conducting shell of radius b and charge $-Q$ concentric with a smaller conducting sphere of radius a and charge Q (Fig. 26.5). Find the capacitance of this device.

SOLUTION

Conceptualize   As with Example 26.1, this system involves a pair of conductors and qualifies as a capacitor.

Categorize   Because of the spherical symmetry of the system, we can use results from previous studies of spherical systems to find the capacitance.

Analyze   As shown in Chapter 24, the magnitude of the electric field outside a spherically symmetric charge distribution is radial and given by the expression $E = k_e Q/r^2$. In this case, this result applies to the field *between* the spheres ($a < r < b$).

Figure 26.5   (Example 26.2) A spherical capacitor consists of an inner sphere of radius a surrounded by a concentric spherical shell of radius b. The electric field between the spheres is directed radially outward when the inner sphere is positively charged.

Write an expression for the potential difference between the two conductors from Equation 25.3:

$$V_b - V_a = -\int_a^b \vec{\mathbf{E}} \cdot d\vec{\mathbf{s}}$$

Apply the result of Example 24.3 for the electric field outside a spherically symmetric charge distribution and note that $\vec{\mathbf{E}}$ is parallel to $d\vec{\mathbf{s}}$ along a radial line:

$$V_b - V_a = -\int_a^b E_r \, dr = -k_e Q \int_a^b \frac{dr}{r^2} = k_e Q \left[\frac{1}{r}\right]_a^b$$

$$(1) \quad V_b - V_a = k_e Q \left(\frac{1}{b} - \frac{1}{a}\right) = k_e Q \frac{a - b}{ab}$$

Substitute the absolute value of ΔV into Equation 26.1:

$$C = \frac{Q}{\Delta V} = \frac{Q}{|V_b - V_a|} = \frac{ab}{k_e(b - a)} \quad \textbf{(26.6)}$$

Finalize The potential difference between the spheres in Equation (1) is negative because of the choice of signs on the spheres. Therefore, in Equation 26.6, when we take the absolute value, we change $a - b$ to $b - a$. The result is a positive number because $b > a$.

What If? If the radius b of the outer sphere approaches infinity, what does the capacitance become?

Answer In Equation 26.6, we let $b \rightarrow \infty$:

$$C = \lim_{b \to \infty} \frac{ab}{k_e(b - a)} = \frac{ab}{k_e(b)} = \frac{a}{k_e} = 4\pi\epsilon_0 a$$

Notice that this expression is the same as Equation 26.2, the capacitance of an isolated spherical conductor.

26.3 Combinations of Capacitors

Two or more capacitors often are combined in electric circuits. We can calculate the equivalent capacitance of certain combinations using methods described in this section. Throughout this section, we assume the capacitors to be combined are initially uncharged.

In studying electric circuits, we use a simplified pictorial representation called a **circuit diagram.** Such a diagram uses **circuit symbols** to represent various circuit elements. The circuit symbols are connected by straight lines that represent the wires between the circuit elements. The circuit symbols for capacitors, batteries, and switches as well as the color codes used for them in this text are given in Figure 26.6. The symbol for the capacitor reflects the geometry of the most common model for a capacitor, a pair of parallel plates. The positive terminal of the battery is at the higher potential and is represented in the circuit symbol by the longer line.

Parallel Combination

Two capacitors connected as shown in Active Figure 26.7a (page 728) are known as a **parallel combination** of capacitors. Active Figure 26.7b shows a circuit diagram for this combination of capacitors. The left plates of the capacitors are connected to the positive terminal of the battery by a conducting wire and are therefore both at the same electric potential as the positive terminal. Likewise, the right plates are connected to the negative terminal and so are both at the same potential as the negative terminal. Therefore, **the individual potential differences across capacitors connected in parallel are the same and are equal to the potential difference applied across the combination.** That is,

$$\Delta V_1 = \Delta V_2 = \Delta V$$

where ΔV is the battery terminal voltage.

Figure 26.6 Circuit symbols for capacitors, batteries, and switches. Notice that capacitors are in blue and batteries and switches are in red. The closed switch can carry current, whereas the open one cannot.

ACTIVE FIGURE 26.7

(a) A parallel combination of two capacitors in an electric circuit in which the potential difference across the battery terminals is ΔV. (b) The circuit diagram for the parallel combination. (c) The equivalent capacitance is given by Equation 26.8.

Sign in at www.thomsonedu.com and go to ThomsonNOW to adjust the battery voltage and the individual capacitances and see the resulting charges and voltages on the capacitors. You can combine up to four capacitors in parallel.

After the battery is attached to the circuit, the capacitors quickly reach their maximum charge. Let's call the maximum charges on the two capacitors Q_1 and Q_2. The *total charge* Q_{tot} stored by the two capacitors is

$$Q_{tot} = Q_1 + Q_2 \tag{26.7}$$

That is, **the total charge on capacitors connected in parallel is the sum of the charges on the individual capacitors.**

Suppose you wish to replace these two capacitors by one *equivalent capacitor* having a capacitance C_{eq} as in Active Figure 26.7c. The effect this equivalent capacitor has on the circuit must be exactly the same as the effect of the combination of the two individual capacitors. That is, the equivalent capacitor must store charge Q_{tot} when connected to the battery. Active Figure 26.7c shows that the voltage across the equivalent capacitor is ΔV because the equivalent capacitor is connected directly across the battery terminals. Therefore, for the equivalent capacitor,

$$Q_{tot} = C_{eq}\,\Delta V$$

Substituting for the charges in Equation 26.7 gives

$$C_{eq}\,\Delta V = C_1\,\Delta V_1 + C_2\,\Delta V_2$$

$$C_{eq} = C_1 + C_2 \quad \text{(parallel combination)}$$

where we have canceled the voltages because they are all the same. If this treatment is extended to three or more capacitors connected in parallel, the equivalent capacitance is found to be

Capacitors in parallel ▶

$$C_{eq} = C_1 + C_2 + C_3 + \cdots \quad \text{(parallel combination)} \tag{26.8}$$

Therefore, **the equivalent capacitance of a parallel combination of capacitors is (1) the algebraic sum of the individual capacitances and (2) greater than any of the individual capacitances.** Statement (2) makes sense because we are essentially combining the areas of all the capacitor plates when they are connected with conducting wire, and capacitance of parallel plates is proportional to area (Eq. 26.3).

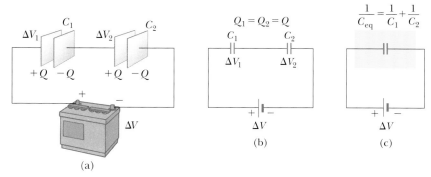

ACTIVE FIGURE 26.8

(a) A series combination of two capacitors. The charges on the two capacitors are the same. (b) The circuit diagram for the series combination. (c) The equivalent capacitance can be calculated from Equation 26.10.

Sign in at www.thomsonedu.com and go to ThomsonNOW to adjust the battery voltage and the individual capacitances and see the resulting charges and voltages on the capacitors. You can combine up to four capacitors in series.

Series Combination

Two capacitors connected as shown in Active Figure 26.8a and the equivalent circuit diagram in Active Figure 26.8b are known as a **series combination** of capacitors. The left plate of capacitor 1 and the right plate of capacitor 2 are connected to the terminals of a battery. The other two plates are connected to each other and to nothing else; hence, they form an isolated system that is initially uncharged and must continue to have zero net charge. To analyze this combination, let's first consider the uncharged capacitors and then follow what happens immediately after a battery is connected to the circuit. When the battery is connected, electrons are transferred out of the left plate of C_1 and into the right plate of C_2. As this negative charge accumulates on the right plate of C_2, an equivalent amount of negative charge is forced off the left plate of C_2, and this left plate therefore has an excess positive charge. The negative charge leaving the left plate of C_2 causes negative charges to accumulate on the right plate of C_1. As a result, all the right plates end up with a charge $-Q$ and all the left plates end up with a charge $+Q$. Therefore, **the charges on capacitors connected in series are the same:**

$$Q_1 = Q_2 = Q$$

where Q is the charge that moved between a wire and the connected outside plate of one of the capacitors.

Active Figure 26.8a shows that the total voltage ΔV_{tot} across the combination is split between the two capacitors:

$$\Delta V_{tot} = \Delta V_1 + \Delta V_2 \tag{26.9}$$

where ΔV_1 and ΔV_2 are the potential differences across capacitors C_1 and C_2, respectively. In general, **the total potential difference across any number of capacitors connected in series is the sum of the potential differences across the individual capacitors.**

Suppose the equivalent single capacitor in Active Figure 26.8c has the same effect on the circuit as the series combination when it is connected to the battery. After it is fully charged, the equivalent capacitor must have a charge of $-Q$ on its right plate and a charge of $+Q$ on its left plate. Applying the definition of capacitance to the circuit in Active Figure 26.8c gives

$$\Delta V_{tot} = \frac{Q}{C_{eq}}$$

Substituting for the voltages in Equation 26.9, we have

$$\frac{Q}{C_{eq}} = \frac{Q_1}{C_1} + \frac{Q_2}{C_2}$$

Canceling the charges because they are all the same gives

$$\frac{1}{C_{eq}} = \frac{1}{C_1} + \frac{1}{C_2} \quad \text{(series combination)}$$

When this analysis is applied to three or more capacitors connected in series, the relationship for the equivalent capacitance is

Capacitors in series ▶

$$\frac{1}{C_{eq}} = \frac{1}{C_1} + \frac{1}{C_2} + \frac{1}{C_3} + \cdots \quad \text{(series combination)} \tag{26.10}$$

This expression shows that **(1) the inverse of the equivalent capacitance is the algebraic sum of the inverses of the individual capacitances and (2) the equivalent capacitance of a series combination is always less than any individual capacitance in the combination.**

Quick Quiz 26.3   Two capacitors are identical. They can be connected in series or in parallel. If you want the *smallest* equivalent capacitance for the combination, how should you connect them? (a) in series   (b) in parallel   (c) either way because both combinations have the same capacitance

---

EXAMPLE 26.3   **Equivalent Capacitance**

Find the equivalent capacitance between a and b for the combination of capacitors shown in Figure 26.9a. All capacitances are in microfarads.

SOLUTION

Conceptualize   Study Figure 26.9a carefully and make sure you understand how the capacitors are connected.

Categorize   Figure 26.9a shows that the circuit contains both series and parallel connections, so we use the rules for series and parallel combinations discussed in this section.

Analyze   Using Equations 26.8 and 26.10, we reduce the combination step by step as indicated in the figure.

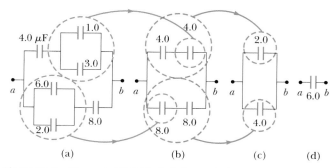

(a)   (b)   (c)   (d)

Figure 26.9   (Example 26.3) To find the equivalent capacitance of the capacitors in (a), we reduce the various combinations in steps as indicated in (b), (c), and (d), using the series and parallel rules described in the text.

The 1.0-μF and 3.0-μF capacitors in Figure 26.9a are in parallel. Find the equivalent capacitance from Equation 26.8:

$$C_{eq} = C_1 + C_2 = 4.0\,\mu\text{F}$$

The 2.0-μF and 6.0-μF capacitors in Figure 26.9a are also in parallel:

$$C_{eq} = C_1 + C_2 = 8.0\,\mu\text{F}$$

The circuit now looks like Figure 26.9b. The two 4.0-μF capacitors in the upper branch are in series. Find the equivalent capacitance from Equation 26.10:

$$\frac{1}{C_{eq}} = \frac{1}{C_1} + \frac{1}{C_2} = \frac{1}{4.0\,\mu\text{F}} + \frac{1}{4.0\,\mu\text{F}} = \frac{1}{2.0\,\mu\text{F}}$$

$$C_{eq} = 2.0\,\mu\text{F}$$

The two 8.0-μF capacitors in the lower branch are also in series. Find the equivalent capacitance from Equation 26.10:

$$\frac{1}{C_{eq}} = \frac{1}{C_1} + \frac{1}{C_2} = \frac{1}{8.0\,\mu\text{F}} + \frac{1}{8.0\,\mu\text{F}} = \frac{1}{4.0\,\mu\text{F}}$$

$$C_{eq} = 4.0\,\mu\text{F}$$

The circuit now looks like Figure 26.9c. The 2.0-μF and 4.0-μF capacitors are in parallel:

$$C_{eq} = C_1 + C_2 = \boxed{6.0 \ \mu F}$$

Finalize   This final value is that of the single equivalent capacitor shown in Figure 26.9d. For further practice in treating circuits with combinations of capacitors, imagine that a battery is connected between points a and b so that a potential difference ΔV is established across the combination. Can you find the voltage across and the charge on each capacitor?

26.4   Energy Stored in a Charged Capacitor

Because positive and negative charges are separated in the system of two conductors in a capacitor, electric potential energy is stored in the system. Many of those who work with electronic equipment have at some time verified that a capacitor can store energy. If the plates of a charged capacitor are connected by a conductor such as a wire, charge moves between each plate and its connecting wire until the capacitor is uncharged. The discharge can often be observed as a visible spark. If you accidentally touch the opposite plates of a charged capacitor, your fingers act as a pathway for discharge and the result is an electric shock. The degree of shock you receive depends on the capacitance and the voltage applied to the capacitor. Such a shock could be fatal if high voltages are present, as in the power supply of a television set. Because the charges can be stored in a capacitor even when the set is turned off, unplugging the television does not make it safe to open the case and touch the components inside.

Active Figure 26.10a shows a battery connected to a single parallel-plate capacitor with a switch in the circuit. Let us identify the circuit as a system. When the switch is closed (Active Fig. 26.10b), the battery establishes an electric field in the wires and charges flow between the wires and the capacitor. As that occurs, there is

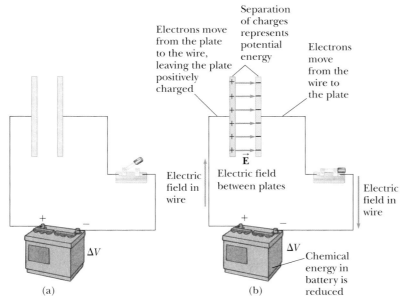

ACTIVE FIGURE 26.10

(a) A circuit consisting of a capacitor, a battery, and a switch. (b) When the switch is closed, the battery establishes an electric field in the wire that causes electrons to move from the left plate into the wire and into the right plate from the wire. As a result, a separation of charge exists on the plates, which represents an increase in electric potential energy of the system of the circuit. This energy in the system has been transformed from chemical energy in the battery.

Sign in at www.thomsonedu.com and go to ThomsonNOW to adjust the battery voltage and see the resulting charge on the plates and the electric field between the plates.

Figure 26.11 A plot of potential difference versus charge for a capacitor is a straight line having slope $1/C$. The work required to move charge dq through the potential difference ΔV existing at the time across the capacitor plates is given approximately by the area of the shaded rectangle. The total work required to charge the capacitor to a final charge Q is the triangular area under the straight line, $W = \frac{1}{2}Q\,\Delta V$. (Don't forget that 1 V = 1 J/C; hence, the unit for the triangular area is the joule.)

a transformation of energy within the system. Before the switch is closed, energy is stored as chemical energy in the battery. This energy is transformed during the chemical reaction that occurs within the battery when it is operating in an electric circuit. When the switch is closed, some of the chemical energy in the battery is converted to electric potential energy associated with the separation of positive and negative charges on the plates.

To calculate the energy stored in the capacitor, we shall assume a charging process that is different from the actual process described in Section 26.1 but that gives the same final result. This assumption is justified because the energy in the final configuration does not depend on the actual charge-transfer process.[3] Imagine that you transfer the charge mechanically through the space between the plates as follows. You grab a small amount of positive charge on the plate connected to the negative terminal and apply a force that causes this positive charge to move over to the plate connected to the positive terminal. Therefore, you do work on the charge as it is transferred from one plate to the other. At first, no work is required to transfer a small amount of charge dq from one plate to the other,[4] but once this charge has been transferred, a small potential difference exists between the plates. Therefore, work must be done to move additional charge through this potential difference. As more and more charge is transferred from one plate to the other, the potential difference increases in proportion and more work is required.

Suppose q is the charge on the capacitor at some instant during the charging process. At the same instant, the potential difference across the capacitor is $\Delta V = q/C$. From Section 25.1, we know that the work necessary to transfer an increment of charge dq from the plate carrying charge $-q$ to the plate carrying charge q (which is at the higher electric potential) is

$$dW = \Delta V\,dq = \frac{q}{C}\,dq$$

This situation is illustrated in Figure 26.11. The total work required to charge the capacitor from $q = 0$ to some final charge $q = Q$ is

$$W = \int_0^Q \frac{q}{C}\,dq = \frac{1}{C}\int_0^Q q\,dq = \frac{Q^2}{2C}$$

The work done in charging the capacitor appears as electric potential energy U stored in the capacitor. Using Equation 26.1, we can express the potential energy stored in a charged capacitor as

Energy stored in a charged capacitor ►

$$U = \frac{Q^2}{2C} = \tfrac{1}{2}Q\,\Delta V = \tfrac{1}{2}C(\Delta V)^2 \qquad (26.11)$$

This result applies to any capacitor, regardless of its geometry. For a given capacitance, the stored energy increases as the charge and the potential difference increase. In practice, there is a limit to the maximum energy (or charge) that can be stored because, at a sufficiently large value of ΔV, discharge ultimately occurs between the plates. For this reason, capacitors are usually labeled with a maximum operating voltage.

We can consider the energy in a capacitor to be stored in the electric field created between the plates as the capacitor is charged. This description is reasonable because the electric field is proportional to the charge on the capacitor. For a parallel-plate capacitor, the potential difference is related to the electric field

---

[3] This discussion is similar to that of state variables in thermodynamics. The change in a state variable such as temperature is independent of the path followed between the initial and final states. The potential energy of a capacitor (or any system) is also a state variable, so it does not depend on the actual process followed to charge the capacitor.

[4] We shall use lowercase q for the time-varying charge on the capacitor while it is charging to distinguish it from uppercase Q, which is the total charge on the capacitor after it is completely charged.

through the relationship $\Delta V = Ed$. Furthermore, its capacitance is $C = \epsilon_0 A/d$ (Eq. 26.3). Substituting these expressions into Equation 26.11 gives

$$U = \tfrac{1}{2}\frac{\epsilon_0 A}{d}(E^2 d^2) = \tfrac{1}{2}(\epsilon_0 Ad)E^2 \qquad (26.12)$$

Because the volume occupied by the electric field is Ad, the *energy per unit volume* $u_E = U/Ad$, known as the *energy density*, is

$$u_E = \tfrac{1}{2}\epsilon_0 E^2 \qquad (26.13)$$

◀ Energy density in an electric field

Although Equation 26.13 was derived for a parallel-plate capacitor, the expression is generally valid regardless of the source of the electric field. That is, **the energy density in any electric field is proportional to the square of the magnitude of the electric field at a given point.**

Quick Quiz 26.4 You have three capacitors and a battery. In which of the following combinations of the three capacitors is the maximum possible energy stored when the combination is attached to the battery? (a) series (b) parallel (c) no difference because both combinations store the same amount of energy

---

PITFALL PREVENTION 26.4
Not a New Kind of Energy

The energy given by Equation 26.13 is not a new kind of energy. The equation describes familiar electric potential energy associated with a system of separated source charges. Equation 26.13 provides a new *interpretation*, or a new way of *modeling* the energy. Furthermore, the equation correctly describes the energy associated with *any* electric field, regardless of the source.

EXAMPLE 26.4   **Rewiring Two Charged Capacitors**

Two capacitors C_1 and C_2 (where $C_1 > C_2$) are charged to the same initial potential difference ΔV_i. The charged capacitors are removed from the battery, and their plates are connected with opposite polarity as in Figure 26.12a. The switches S_1 and S_2 are then closed as in Figure 26.12b.

(A) Find the final potential difference ΔV_f between a and b after the switches are closed.

(a)   (b)

Figure 26.12 (Example 26.4) (a) Two capacitors are charged to the same initial potential difference and connected together with plates of opposite sign to be in contact when the switches are closed. (b) When the switches are closed, the charges redistribute.

SOLUTION

Conceptualize Figure 26.12 helps us understand the initial and final configurations of the system.

Categorize In Figure 26.12b, it might appear as if the capacitors are connected in parallel, but there is no battery in this circuit to apply a voltage across the combination. Therefore, we *cannot* categorize this problem as one in which capacitors are connected in parallel. We *can* categorize it as a problem involving an isolated system for electric charge. The left-hand plates of the capacitors form an isolated system because they are not connected to the right-hand plates by conductors.

Analyze Write an expression for the total charge on the left-hand plates of the system before the switches are closed, noting that a negative sign for Q_{2i} is necessary because the charge on the left plate of capacitor C_2 is negative:

(1) $Q_i = Q_{1i} + Q_{2i} = C_1 \Delta V_i - C_2 \Delta V_i = (C_1 - C_2)\Delta V_i$

After the switches are closed, the charges on the individual capacitors change to new values Q_{1f} and Q_{2f} such that the potential difference is again the same across both capacitors, ΔV_f. Write an expression for the total charge on the left-hand plates of the system after the switches are closed:

(2) $Q_f = Q_{1f} + Q_{2f} = C_1 \Delta V_f + C_2 \Delta V_f = (C_1 + C_2)\Delta V_f$

Because the system is isolated, the initial and final charges on the system must be the same. Use this condition and Equations (1) and (2) to solve for ΔV_f:

$$Q_f = Q_i \;\to\; (C_1 + C_2)\Delta V_f = (C_1 - C_2)\Delta V_i$$

$$(3) \quad \Delta V_f = \left(\frac{C_1 - C_2}{C_1 + C_2}\right)\Delta V_i$$

(B) Find the total energy stored in the capacitors before and after the switches are closed and determine the ratio of the final energy to the initial energy.

SOLUTION

Use Equation 26.11 to find an expression for the total energy stored in the capacitors before the switches are closed:

$$(4) \quad U_i = \tfrac{1}{2}C_1(\Delta V_i)^2 + \tfrac{1}{2}C_2(\Delta V_i)^2 = \tfrac{1}{2}(C_1 + C_2)(\Delta V_i)^2$$

Write an expression for the total energy stored in the capacitors after the switches are closed:

$$U_f = \tfrac{1}{2}C_1(\Delta V_f)^2 + \tfrac{1}{2}C_2(\Delta V_f)^2 = \tfrac{1}{2}(C_1 + C_2)(\Delta V_f)^2$$

Use the results of part (A) to rewrite this expression in terms of ΔV_i:

$$(5) \quad U_f = \tfrac{1}{2}(C_1 + C_2)\left[\left(\frac{C_1 - C_2}{C_1 + C_2}\right)\Delta V_i\right]^2 = \tfrac{1}{2}\frac{(C_1 - C_2)^2(\Delta V_i)^2}{(C_1 + C_2)}$$

Divide Equation (5) by Equation (4) to obtain the ratio of the energies stored in the system:

$$\frac{U_f}{U_i} = \frac{\tfrac{1}{2}(C_1 - C_2)^2(\Delta V_i)^2/(C_1 + C_2)}{\tfrac{1}{2}(C_1 + C_2)(\Delta V_i)^2}$$

$$(6) \quad \frac{U_f}{U_i} = \left(\frac{C_1 - C_2}{C_1 + C_2}\right)^2$$

Finalize   The ratio of energies is *less* than unity, indicating that the final energy is *less* than the initial energy. At first, you might think the law of energy conservation has been violated, but that is not the case. The "missing" energy is transferred out of the system by the mechanism of electromagnetic waves (T_{ER} in Equation 8.2), as we shall see in Chapter 34.

What If?   What if the two capacitors have the same capacitance? What would you expect to happen when the switches are closed?

Answer   Because both capacitors have the same initial potential difference applied to them, the charges on the capacitors have the same magnitude. When the capacitors with opposite polarities are connected together, the equal-magnitude charges should cancel each other, leaving the capacitors uncharged.

Let's test our results to see if that is the case mathematically. In Equation (1), because the capacitances are equal, the initial charge Q_i on the system of left-hand plates is zero. Equation (3) shows that $\Delta V_f = 0$, which is consistent with uncharged capacitors. Finally, Equation (5) shows that $U_f = 0$, which is also consistent with uncharged capacitors.

Figure 26.13 In a hospital or at an emergency scene, you might see a patient being revived with a portable defibrillator. The defibrillator's paddles are applied to the patient's chest, and an electric shock is sent through the chest cavity. The aim of this technique is to restore the heart's normal rhythm pattern.

One device in which capacitors have an important role is the portable *defibrillator* (Fig. 26.13). When cardiac fibrillation (random contractions) occurs, the heart produces a rapid, irregular pattern of beats. A fast discharge of energy through the heart can return the organ to its normal beat pattern. Emergency medical teams use portable defibrillators that contain batteries capable of charging a capacitor to a high voltage. (The circuitry actually permits the capacitor to be charged to a much higher voltage than that of the battery.) Up to 360 J is stored in the electric field of a large capacitor in a defibrillator when it is fully charged. The stored energy is released through the heart by conducting electrodes, called paddles, which are placed on both sides of the victim's chest. The defibrillator can deliver the energy to a patient in about 2 ms (roughly equivalent to 3 000 times the power delivered to a 60-W lightbulb!). The paramedics must wait between applications of the energy because of the time necessary for the capacitors to become fully charged. In this application and others (e.g., camera flash units and lasers used for fusion experiments), capacitors serve as energy reservoirs that can be slowly charged and then quickly discharged to provide large amounts of energy in a short pulse.

26.5 Capacitors with Dielectrics

A **dielectric** is a nonconducting material such as rubber, glass, or waxed paper. We can perform the following experiment to illustrate the effect of a dielectric in a capacitor. Consider a parallel-plate capacitor that without a dielectric has a charge Q_0 and a capacitance C_0. The potential difference across the capacitor is $\Delta V_0 = Q_0/C_0$. Figure 26.14a illustrates this situation. The potential difference is measured by a *voltmeter*, a device discussed in greater detail in Chapter 28. Notice that no battery is shown in the figure; also, we must assume no charge can flow through an ideal voltmeter. Hence, there is no path by which charge can flow and alter the charge on the capacitor. If a dielectric is now inserted between the plates as in Figure 26.14b, the voltmeter indicates that the voltage between the plates decreases to a value ΔV. The voltages with and without the dielectric are related by a factor κ as follows:

$$\Delta V = \frac{\Delta V_0}{\kappa}$$

Because $\Delta V < \Delta V_0$, we see that $\kappa > 1$. The dimensionless factor κ is called the **dielectric constant** of the material. The dielectric constant varies from one material to another. In this section, we analyze this change in capacitance in terms of electrical parameters such as electric charge, electric field, and potential difference; Section 26.7 describes the microscopic origin of these changes.

Because the charge Q_0 on the capacitor does not change, the capacitance must change to the value

$$C = \frac{Q_0}{\Delta V} = \frac{Q_0}{\Delta V_0/\kappa} = \kappa\frac{Q_0}{\Delta V_0}$$

$$C = \kappa C_0 \qquad\qquad\qquad \textbf{(26.14)}$$

◀ Capacitance of a capacitor filled with a material of dielectric constant κ

That is, the capacitance *increases* by the factor κ when the dielectric completely fills the region between the plates.[5] Because $C_0 = \epsilon_0 A/d$ (Eq. 26.3) for a parallel-plate

PITFALL PREVENTION 26.5
Is the Capacitor Connected to a Battery?

For problems in which a capacitor is modified (by insertion of a dielectric, for example), you must note whether modifications to the capacitor are being made while the capacitor is connected to a battery or after it is disconnected. If the capacitor remains connected to the battery, the voltage across the capacitor necessarily remains the same. If you disconnect the capacitor from the battery before making any modifications to the capacitor, the capacitor is an isolated system and its charge remains the same.

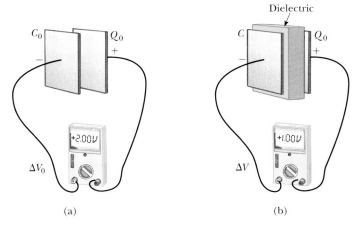

Figure 26.14 A charged capacitor (a) before and (b) after insertion of a dielectric between the plates. The charge on the plates remains unchanged, but the potential difference decreases from ΔV_0 to $\Delta V = \Delta V_0/\kappa$. Therefore, the capacitance increases from C_0 to κC_0.

---

[5] If the dielectric is introduced while the potential difference is held constant by a battery, the charge increases to a value $Q = \kappa Q_0$. The additional charge comes from the wires attached to the capacitor, and the capacitance again increases by the factor κ.

capacitor, we can express the capacitance of a parallel-plate capacitor filled with a dielectric as

$$C = \kappa \frac{\epsilon_0 A}{d}$$ (26.15)

From Equations 26.3 and 26.15, it would appear that the capacitance could be made very large by decreasing d, the distance between the plates. In practice, the lowest value of d is limited by the electric discharge that could occur through the dielectric medium separating the plates. For any given separation d, the maximum voltage that can be applied to a capacitor without causing a discharge depends on the **dielectric strength** (maximum electric field) of the dielectric. If the magnitude of the electric field in the dielectric exceeds the dielectric strength, the insulating properties break down and the dielectric begins to conduct.

Physical capacitors have a specification called by a variety of names, including *working voltage, breakdown voltage,* and *rated voltage.* This parameter represents the largest voltage that can be applied to the capacitor without exceeding the dielectric strength of the dielectric material in the capacitor. Consequently, when selecting a capacitor for a given application, you must consider its capacitance as well as the expected voltage across the capacitor in the circuit, making sure the expected voltage is smaller than the rated voltage of the capacitor. You can see the rated voltage on several of the capacitors in this chapter's opening photograph.

Insulating materials have values of κ greater than unity and dielectric strengths greater than that of air as Table 26.1 indicates. Therefore, a dielectric provides the following advantages:

- An increase in capacitance
- An increase in maximum operating voltage
- Possible mechanical support between the plates, which allows the plates to be close together without touching, thereby decreasing d and increasing C

TABLE 26.1

Approximate Dielectric Constants and Dielectric Strengths of Various Materials at Room Temperature

Material	Dielectric Constant κ	Dielectric Strength[a] (10^6 V/m)
Air (dry)	1.000 59	3
Bakelite	4.9	24
Fused quartz	3.78	8
Mylar	3.2	7
Neoprene rubber	6.7	12
Nylon	3.4	14
Paper	3.7	16
Paraffin-impregnated paper	3.5	11
Polystyrene	2.56	24
Polyvinyl chloride	3.4	40
Porcelain	6	12
Pyrex glass	5.6	14
Silicone oil	2.5	15
Strontium titanate	233	8
Teflon	2.1	60
Vacuum	1.000 00	—
Water	80	—

[a] The dielectric strength equals the maximum electric field that can exist in a dielectric without electrical breakdown. These values depend strongly on the presence of impurities and flaws in the materials.

Metal foil   Plates   Case

Electrolyte

Contacts

Oil

Metallic foil + oxide layer

(a)   Paper   (b)   (c)

Figure 26.15 Three commercial capacitor designs. (a) A tubular capacitor, whose plates are separated by paper and then rolled into a cylinder. (b) A high-voltage capacitor consisting of many parallel plates separated by insulating oil. (c) An electrolytic capacitor.

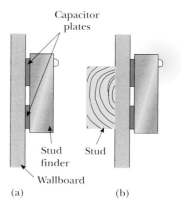

Figure 26.16 A variable capacitor. When one set of metal plates is rotated so as to lie between a fixed set of plates, the capacitance of the device changes.

Types of Capacitors

Commercial capacitors are often made from metallic foil interlaced with thin sheets of either paraffin-impregnated paper or Mylar as the dielectric material. These alternate layers of metallic foil and dielectric are rolled into a cylinder to form a small package (Fig. 26.15a). High-voltage capacitors commonly consist of a number of interwoven metallic plates immersed in silicone oil (Fig. 26.15b). Small capacitors are often constructed from ceramic materials.

Often, an *electrolytic capacitor* is used to store large amounts of charge at relatively low voltages. This device, shown in Figure 26.15c, consists of a metallic foil in contact with an *electrolyte,* a solution that conducts electricity by virtue of the motion of ions contained in the solution. When a voltage is applied between the foil and the electrolyte, a thin layer of metal oxide (an insulator) is formed on the foil, and this layer serves as the dielectric. Very large values of capacitance can be obtained in an electrolytic capacitor because the dielectric layer is very thin and therefore the plate separation is very small.

Electrolytic capacitors are not reversible as are many other capacitors. They have a polarity, which is indicated by positive and negative signs marked on the device. When electrolytic capacitors are used in circuits, the polarity must be correct. If the polarity of the applied voltage is the opposite of what is intended, the oxide layer is removed and the capacitor conducts electricity instead of storing charge.

Variable capacitors (typically 10 to 500 pF) usually consist of two interwoven sets of metallic plates, one fixed and the other movable, and contain air as the dielectric (Fig. 26.16). These types of capacitors are often used in radio tuning circuits.

Capacitor plates

Stud finder   Stud

Wallboard

(a)   (b)

Figure 26.17 (Quick Quiz 26.5) A stud finder. (a) The materials between the plates of the capacitor are the wallboard and air. (b) When the capacitor moves across a stud in the wall, the materials between the plates are the wallboard and the wood. The change in the dielectric constant causes a signal light to illuminate.

Quick Quiz 26.5 If you have ever tried to hang a picture or a mirror, you know it can be difficult to locate a wooden stud in which to anchor your nail or screw. A carpenter's stud finder is a capacitor with its plates arranged side by side instead of facing each other as shown in Figure 26.17. When the device is moved over a stud, does the capacitance (a) increase or (b) decrease?

EXAMPLE 26.5   **Energy Stored Before and After**

A parallel-plate capacitor is charged with a battery to a charge Q_0. The battery is then removed, and a slab of material that has a dielectric constant κ is inserted between the plates. Identify the system as the capacitor and the dielectric. Find the energy stored in the system before and after the dielectric is inserted.

SOLUTION

Conceptualize Think about what happens when the dielectric is inserted between the plates. Because the battery has been removed, the charge on the capacitor must remain the same. We know from our earlier discussion, however, that the capacitance must change. Therefore, we expect a change in the energy of the system.

Categorize Because we expect the energy of the system to change, we model it as a nonisolated system.

Analyze   From Equation 26.11, find the energy stored in the absence of the dielectric:

$$U_0 = \frac{Q_0^2}{2C_0}$$

Find the energy stored in the capacitor after the dielectric is inserted between the plates:

$$U = \frac{Q_0^2}{2C}$$

Use Equation 26.14 to replace the capacitance C:

$$U = \frac{Q_0^2}{2\kappa C_0} = \frac{U_0}{\kappa}$$

Finalize   Because $\kappa > 1$, the final energy is less than the initial energy. We can account for the "missing" energy by noting that the dielectric, when inserted, is pulled into the device. To keep the dielectric from accelerating, an external agent must do negative work (W in Eq. 8.2) on the dielectric, which is simply the difference $U - U_0$.

Figure 26.18   An electric dipole consists of two charges of equal magnitude and opposite sign separated by a distance of $2a$. The electric dipole moment $\vec{\mathbf{p}}$ is directed from $-q$ toward $+q$.

26.6   Electric Dipole in an Electric Field

We have discussed the effect on the capacitance of placing a dielectric between the plates of a capacitor. In Section 26.7, we shall describe the microscopic origin of this effect. Before we can do so, however, let's expand the discussion of the electric dipole introduced in Section 23.4 (see Example 23.5). The electric dipole consists of two charges of equal magnitude and opposite sign separated by a distance $2a$ as shown in Figure 26.18. The **electric dipole moment** of this configuration is defined as the vector $\vec{\mathbf{p}}$ directed from $-q$ toward $+q$ along the line joining the charges and having magnitude $2aq$:

$$p \equiv 2aq \tag{26.16}$$

Now suppose an electric dipole is placed in a uniform electric field $\vec{\mathbf{E}}$ and makes an angle θ with the field as shown in Figure 26.19. We identify $\vec{\mathbf{E}}$ as the field *external* to the dipole, established by some other charge distribution, to distinguish it from the field *due to* the dipole, which we discussed in Section 23.4.

The electric forces acting on the two charges are equal in magnitude ($F = qE$) and opposite in direction as shown in Figure 26.19. Therefore, the net force on the dipole is zero. The two forces produce a net torque on the dipole, however; as a result, the dipole rotates in the direction that brings the dipole moment vector into greater alignment with the field. The torque due to the force on the positive charge about an axis through O in Figure 26.19 has magnitude $Fa \sin \theta$, where $a \sin \theta$ is the moment arm of F about O. This force tends to produce a clockwise rotation. The torque about O on the negative charge is also of magnitude $Fa \sin \theta$; here again, the force tends to produce a clockwise rotation. Therefore, the magnitude of the net torque about O is

$$\tau = 2Fa \sin \theta$$

Figure 26.19   An electric dipole in a uniform external electric field. The dipole moment $\vec{\mathbf{p}}$ is at an angle θ to the field, causing the dipole to experience a torque.

Because $F = qE$ and $p = 2aq$, we can express τ as

$$\tau = 2aqE \sin \theta = pE \sin \theta \tag{26.17}$$

It is convenient to express the torque in vector form as the cross product of the vectors $\vec{\mathbf{p}}$ and $\vec{\mathbf{E}}$:

Torque on an electric ►
dipole in an external
electric field

$$\vec{\boldsymbol{\tau}} = \vec{\mathbf{p}} \times \vec{\mathbf{E}} \tag{26.18}$$

Let's determine the potential energy of the system of an electric dipole in an external electric field as a function of the dipole's orientation with respect to the field. To do so, recognize that work must be done by an external agent to rotate the dipole through an angle so as to cause the dipole moment vector to become less aligned with the field. The work done is then stored as potential energy in the system. The work dW required to rotate the dipole through an angle $d\theta$ is $dW = \tau d\theta$ (Eq. 10.22). Because $\tau = pE \sin \theta$ and the work results in an increase in the potential energy U, we find that for a rotation from θ_i to θ_f, the change in potential energy of the system is

$$U_f - U_i = \int_{\theta_i}^{\theta_f} \tau d\theta = \int_{\theta_i}^{\theta_f} pE \sin \theta \, d\theta = pE \int_{\theta_i}^{\theta_f} \sin \theta \, d\theta$$

$$= pE\left[-\cos \theta\right]_{\theta_i}^{\theta_f} = pE(\cos \theta_i - \cos \theta_f)$$

The term that contains $\cos \theta_i$ is a constant that depends on the initial orientation of the dipole. It is convenient to choose a reference angle of $\theta_i = 90°$ so that $\cos \theta_i = \cos 90° = 0$. Furthermore, let's choose $U_i = 0$ at $\theta_i = 90°$ as our reference of potential energy. Hence, we can express a general value of $U = U_f$ as

$$U = -pE \cos \theta \qquad \textbf{(26.19)}$$

We can write this expression for the potential energy of a dipole in an electric field as the dot product of the vectors $\vec{\mathbf{p}}$ and $\vec{\mathbf{E}}$:

$$U = -\vec{\mathbf{p}} \cdot \vec{\mathbf{E}} \qquad \textbf{(26.20)}$$

◄ Potential energy of the system of an electric dipole in an external electric field

To develop a conceptual understanding of Equation 26.19, compare it with the expression for the potential energy of the system of an object in the Earth's gravitational field, $U = mgy$ (see Chapter 7). This expression includes a parameter associated with the object placed in the gravitational field, its mass m. Likewise, Equation 26.19 includes a parameter of the object in the electric field, its dipole moment p. The gravitational expression includes the magnitude of the gravitational field g. Similarly, Equation 26.19 includes the magnitude of the electric field E. So far, these two contributions to the potential energy expressions appear analogous. The final contribution, however, is somewhat different in the two cases. In the gravitational expression, the potential energy depends on the vertical position of the object, measured by y. In Equation 26.19, the potential energy depends on the angle θ through which the dipole is rotated. In both cases, the configuration of the system is being changed. In the gravitational case, the change involves moving an object in a *translational* sense, whereas in the electrical case, the change involves moving an object in a *rotational* sense. In both cases, however, once the change is made, the system tends to return to the original configuration when the object is released: the object of mass m falls toward the ground, and the dipole begins to rotate back toward the configuration in which it is aligned with the field. Therefore, apart from the type of motion, the expressions for potential energy in these two cases are similar.

Molecules are said to be *polarized* when a separation exists between the average position of the negative charges and the average position of the positive charges in the molecule. In some molecules such as water, this condition is always present; such molecules are called **polar molecules.** Molecules that do not possess a permanent polarization are called **nonpolar molecules.**

We can understand the permanent polarization of water by inspecting the geometry of the water molecule. The oxygen atom in the water molecule is bonded to the hydrogen atoms such that an angle of 105° is formed between the two bonds (Fig. 26.20). The center of the negative charge distribution is near the oxygen atom, and the center of the positive charge distribution lies at a point midway along the line joining the hydrogen atoms (the point labeled × in Fig. 26.20). We can model the water molecule and other polar molecules as dipoles because the average positions of the positive and negative charges act as point charges. As a result, we can apply our discussion of dipoles to the behavior of polar molecules.

Washing with soap and water is a household scenario in which the dipole structure of water is exploited. Grease and oil are made up of nonpolar molecules,

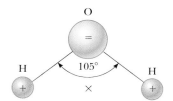

Figure 26.20 The water molecule, H_2O, has a permanent polarization resulting from its nonlinear geometry. The center of the positive charge distribution is at the point ×.

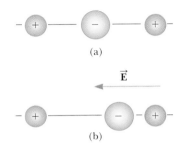

Figure 26.21 (a) A linear symmetric molecule has no permanent polarization. (b) An external electric field induces a polarization in the molecule.

which are generally not attracted to water. Plain water is not very useful for removing this type of grime. Soap contains long molecules called *surfactants*. In a long molecule, the polarity characteristics of one end of the molecule can be different from those at the other end. In a surfactant molecule, one end acts like a nonpolar molecule and the other acts like a polar molecule. The nonpolar end can attach to a grease or oil molecule, and the polar end can attach to a water molecule. Therefore, the soap serves as a chain, linking the dirt and water molecules together. When the water is rinsed away, the grease and oil go with it.

A symmetric molecule (Fig. 26.21a) has no permanent polarization, but polarization can be induced by placing the molecule in an electric field. A field directed to the left as in Figure 26.21b causes the center of the positive charge distribution to shift to the left from its initial position and the center of the negative charge distribution to shift to the right. This *induced polarization* is the effect that predominates in most materials used as dielectrics in capacitors.

EXAMPLE 26.6 | **The H₂O Molecule**

The water (H_2O) molecule has an electric dipole moment of 6.3×10^{-30} C·m. A sample contains 10^{21} water molecules, with the dipole moments all oriented in the direction of an electric field of magnitude 2.5×10^5 N/C. How much work is required to rotate the dipoles from this orientation ($\theta = 0°$) to one in which all the moments are perpendicular to the field ($\theta = 90°$)?

SOLUTION

Conceptualize When all the dipoles are aligned with the electric field, the dipoles–electric field system has the minimum potential energy. This energy has a negative value given by the product of the right side of Equation 26.19, evaluated at $0°$, and the number N of dipoles. Work must be done to rotate all the dipoles of the system by $90°$ because the system's potential energy is raised to a higher value of zero.

Categorize We use Equation 26.19 to evaluate the potential energy, so we categorize this example as a substitution problem.

Write the appropriate reduction of the conservation of energy equation, Equation 8.2, for this situation:

$$(1) \quad \Delta U = W$$

Use Equation 26.19 to evaluate the initial and final potential energies of the system and Equation (1) to calculate the work required to rotate the dipoles:

$$W = U_{90°} - U_{0°} = (-NpE \cos 90°) - (-NpE \cos 0°)$$
$$= NpE = (10^{21})(6.3 \times 10^{-30} \text{ C·m})(2.5 \times 10^5 \text{ N/C})$$
$$= \boxed{1.6 \times 10^{-3} \text{ J}}$$

26.7 An Atomic Description of Dielectrics

In Section 26.5, we found that the potential difference ΔV_0 between the plates of a capacitor is reduced to $\Delta V_0/\kappa$ when a dielectric is introduced. The potential difference is reduced because the magnitude of the electric field decreases between the plates. In particular, if $\vec{E}_0$ is the electric field without the dielectric, the field in the presence of a dielectric is

$$\vec{E} = \frac{\vec{E}_0}{\kappa} \tag{26.21}$$

First consider a dielectric made up of polar molecules placed in the electric field between the plates of a capacitor. The dipoles (that is, the polar molecules making up the dielectric) are randomly oriented in the absence of an electric field

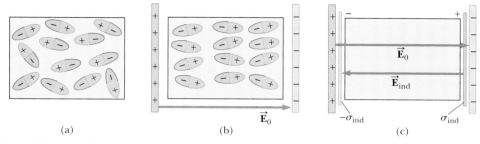

(a)   (b)   (c)

Figure 26.22   (a) Polar molecules are randomly oriented in the absence of an external electric field. (b) When an external electric field is applied, the molecules partially align with the field. (c) The charged edges of the dielectric can be modeled as an additional pair of parallel plates establishing an electric field $\vec{E}_{ind}$ in the direction opposite that of $\vec{E}_0$.

as shown in Figure 26.22a. When an external field $\vec{E}_0$ due to charges on the capacitor plates is applied, a torque is exerted on the dipoles, causing them to partially align with the field as shown in Figure 26.22b. The dielectric is now polarized. The degree of alignment of the molecules with the electric field depends on temperature and the magnitude of the field. In general, the alignment increases with decreasing temperature and with increasing electric field.

If the molecules of the dielectric are nonpolar, the electric field due to the plates produces an induced polarization in the molecule. These induced dipole moments tend to align with the external field, and the dielectric is polarized. Therefore, a dielectric can be polarized by an external field regardless of whether the molecules in the dielectric are polar or nonpolar.

With these ideas in mind, consider a slab of dielectric material placed between the plates of a capacitor so that it is in a uniform electric field $\vec{E}_0$ as shown in Figure 26.22b. The electric field due to the plates is directed to the right and polarizes the dielectric. The net effect on the dielectric is the formation of an *induced* positive surface charge density σ_{ind} on the right face and an equal-magnitude negative surface charge density $-\sigma_{ind}$ on the left face as shown in Figure 26.22c. Because we can model these surface charge distributions as being due to charged parallel plates, the induced surface charges on the dielectric give rise to an induced electric field $\vec{E}_{ind}$ in the direction opposite the external field $\vec{E}_0$. Therefore, the net electric field $\vec{E}$ in the dielectric has a magnitude

$$E = E_0 - E_{ind} \tag{26.22}$$

In the parallel-plate capacitor shown in Figure 26.23, the external field E_0 is related to the charge density σ on the plates through the relationship $E_0 = \sigma/\epsilon_0$. The induced electric field in the dielectric is related to the induced charge density σ_{ind} through the relationship $E_{ind} = \sigma_{ind}/\epsilon_0$. Because $E = E_0/\kappa = \sigma/\kappa\epsilon_0$, substitution into Equation 26.22 gives

$$\frac{\sigma}{\kappa\epsilon_0} = \frac{\sigma}{\epsilon_0} - \frac{\sigma_{ind}}{\epsilon_0}$$

$$\sigma_{ind} = \left(\frac{\kappa - 1}{\kappa}\right)\sigma \tag{26.23}$$

Because $\kappa > 1$, this expression shows that the charge density σ_{ind} induced on the dielectric is less than the charge density σ on the plates. For instance, if $\kappa = 3$, the induced charge density is two-thirds the charge density on the plates. If no dielectric is present, then $\kappa = 1$ and $\sigma_{ind} = 0$ as expected. If the dielectric is replaced by an electrical conductor for which $E = 0$, however, Equation 26.22 indicates that $E_0 = E_{ind}$, which corresponds to $\sigma_{ind} = \sigma$. That is, the surface charge induced on the conductor is equal in magnitude but opposite in sign to that on the plates, resulting in a net electric field of zero in the conductor (see Fig. 24.14).

Figure 26.23   Induced charge on a dielectric placed between the plates of a charged capacitor. Notice that the induced charge density on the dielectric is *less* than the charge density on the plates.

EXAMPLE 26.7 | **Effect of a Metallic Slab**

A parallel-plate capacitor has a plate separation d and plate area A. An uncharged metallic slab of thickness a is inserted midway between the plates.

(A) Find the capacitance of the device.

SOLUTION

Conceptualize Figure 26.24a shows the metallic slab between the plates of the capacitor. Any charge that appears on one plate of the capacitor must induce a charge of equal magnitude and opposite sign on the near side of the slab as shown in Figure 26.24a. Consequently, the net charge on the slab remains zero and the electric field inside the slab is zero.

Categorize The planes of charge on the metallic slab's upper and lower edges are identical to the distribution of charges on the plates of a capacitor. The metal between the slab's edges serves only to make an electri-

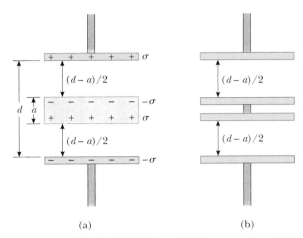

(a)   (b)

Figure 26.24 (Example 26.7) (a) A parallel-plate capacitor of plate separation d partially filled with a metallic slab of thickness a. (b) The equivalent circuit of the device in (a) consists of two capacitors in series, each having a plate separation $(d - a)/2$.

cal connection between the edges. Therefore, we can model the edges of the slab as conducting planes and the bulk of the slab as a wire. As a result, the capacitor in Figure 26.24a is equivalent to two capacitors in series, each having a plate separation $(d - a)/2$ as shown in Figure 26.24b.

Analyze Use Equation 26.3 and the rule for adding two capacitors in series (Eq. 26.10) to find the equivalent capacitance in Figure 26.24b:

$$\frac{1}{C} = \frac{1}{C_1} + \frac{1}{C_2} = \frac{1}{\left[\dfrac{\epsilon_0 A}{(d - a)/2}\right]} + \frac{1}{\left[\dfrac{\epsilon_0 A}{(d - a)/2}\right]}$$

$$C = \frac{\epsilon_0 A}{d - a}$$

(B) Show that the capacitance of the original capacitor is unaffected by the insertion of the metallic slab if the slab is infinitesimally thin.

SOLUTION

In the result for part (A), let $a \to 0$:

$$C = \lim_{a \to 0}\left(\frac{\epsilon_0 A}{d - a}\right) = \frac{\epsilon_0 A}{d}$$

Finalize The result of part (B) is the original capacitance before the slab is inserted, which tells us that we can insert an infinitesimally thin metallic sheet between the plates of a capacitor without affecting the capacitance. We use this fact in the next example.

What If? What if the metallic slab in part (A) is not midway between the plates? How would that affect the capacitance?

Answer Let's imagine moving the slab in Figure 26.24a upward so that the distance between the upper edge of the slab and the upper plate is b. Then, the distance between the lower edge of the slab and the lower plate is $d - b - a$. As in part (A), we find the total capacitance of the series combination:

$$\frac{1}{C} = \frac{1}{C_1} + \frac{1}{C_2} = \frac{1}{\epsilon_0 A/b} + \frac{1}{\epsilon_0 A/(d - b - a)}$$

$$= \frac{b}{\epsilon_0 A} + \frac{d - b - a}{\epsilon_0 A} = \frac{d - a}{\epsilon_0 A} \quad \to \quad C = \frac{\epsilon_0 A}{d - a}$$

which is the same result as found in part (A). The capacitance is independent of the value of b, so it does not matter where the slab is located. In Figure 26.24b, when the central structure is moved up or down, the decrease in plate separation of one capacitor is compensated by the increase in plate separation for the other.

EXAMPLE 26.8 | **A Partially Filled Capacitor**

A parallel-plate capacitor with a plate separation d has a capacitance C_0 in the absence of a dielectric. What is the capacitance when a slab of dielectric material of dielectric constant κ and thickness fd is inserted between the plates (Fig. 26.25a), where f is a fraction between 0 and 1?

SOLUTION

Conceptualize In our previous discussions of dielectrics between the plates of a capacitor, the dielectric filled the volume between the plates. In this example, only part of the volume between the plates contains the dielectric material.

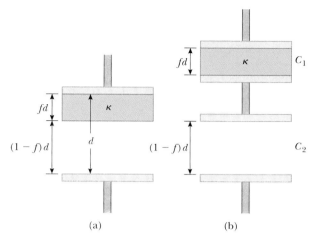

(a)   (b)

Figure 26.25 (Example 26.8) (a) A parallel-plate capacitor of plate separation d partially filled with a dielectric of thickness fd. (b) The equivalent circuit of the capacitor consists of two capacitors connected in series.

Categorize In Example 26.7, we found that an infinitesimally thin metallic sheet inserted between the plates of a capacitor does not affect the capacitance. Imagine sliding an infinitesimally thin metallic slab along the bottom face of the dielectric shown in Figure 26.25a. We can model this system as a series combination of two capacitors as shown in Figure 26.25b. One capacitor has a plate separation fd and is filled with a dielectric; the other has a plate separation $(1 - f)d$ and has air between its plates.

Analyze Evaluate the two capacitances in Figure 26.25b from Equation 26.15:

$$C_1 = \frac{\kappa \epsilon_0 A}{fd} \quad \text{and} \quad C_2 = \frac{\epsilon_0 A}{(1 - f)d}$$

Find the equivalent capacitance C from Equation 26.10 for two capacitors combined in series:

$$\frac{1}{C} = \frac{1}{C_1} + \frac{1}{C_2} = \frac{fd}{\kappa \epsilon_0 A} + \frac{(1 - f)d}{\epsilon_0 A}$$

$$\frac{1}{C} = \frac{fd}{\kappa \epsilon_0 A} + \frac{\kappa(1 - f)d}{\kappa \epsilon_0 A} = \frac{f + \kappa(1 - f)}{\kappa} \frac{d}{\epsilon_0 A}$$

Invert and substitute for the capacitance without the dielectric, $C_0 = \epsilon_0 A/d$:

$$C = \frac{\kappa}{f + \kappa(1 - f)} \frac{\epsilon_0 A}{d} = \frac{\kappa}{f + \kappa(1 - f)} C_0$$

Finalize Let's test this result for some known limits. If $f \to 0$, the dielectric should disappear. In this limit, $C \to C_0$, which is consistent with a capacitor with air between the plates. If $f \to 1$, the dielectric fills the volume between the plates. In this limit, $C \to \kappa C_0$, which is consistent with Equation 26.14.

Summary

ThomsonNOW™ Sign in at **www.thomsonedu.com** and go to ThomsonNOW to take a practice test for this chapter.

DEFINITIONS

A **capacitor** consists of two conductors carrying charges of equal magnitude and opposite sign. The **capacitance** C of any capacitor is the ratio of the charge Q on either conductor to the potential difference ΔV between them:

$$C \equiv \frac{Q}{\Delta V} \qquad (26.1)$$

The capacitance depends only on the geometry of the conductors and not on an external source of charge or potential difference. The SI unit of capacitance is coulombs per volt, or the **farad** (F): $1\ \text{F} = 1\ \text{C/V}$.

The **electric dipole moment** $\vec{\mathbf{p}}$ of an electric dipole has a magnitude

$$p \equiv 2aq \qquad (26.16)$$

where $2a$ is the distance between the charges q and $-q$. The direction of the electric dipole moment vector is from the negative charge toward the positive charge.

CONCEPTS AND PRINCIPLES

If two or more capacitors are connected in parallel, the potential difference is the same across all capacitors. The equivalent capacitance of a **parallel combination** of capacitors is

$$C_{eq} = C_1 + C_2 + C_3 + \cdots \qquad (26.8)$$

If two or more capacitors are connected in series, the charge is the same on all capacitors, and the equivalent capacitance of the **series combination** is given by

$$\frac{1}{C_{eq}} = \frac{1}{C_1} + \frac{1}{C_2} + \frac{1}{C_3} + \cdots \qquad (26.10)$$

These two equations enable you to simplify many electric circuits by replacing multiple capacitors with a single equivalent capacitance.

Energy is stored in a capacitor because the charging process is equivalent to the transfer of charges from one conductor at a lower electric potential to another conductor at a higher potential. The energy stored in a capacitor with charge Q is

$$U = \frac{Q^2}{2C} = \tfrac{1}{2} Q\,\Delta V = \tfrac{1}{2} C(\Delta V)^2 \qquad (26.11)$$

When a dielectric material is inserted between the plates of a capacitor, the capacitance increases by a dimensionless factor κ, called the **dielectric constant:**

$$C = \kappa C_0 \qquad (26.14)$$

where C_0 is the capacitance in the absence of the dielectric.

The torque acting on an electric dipole in a uniform electric field $\vec{\mathbf{E}}$ is

$$\vec{\boldsymbol{\tau}} = \vec{\mathbf{p}} \times \vec{\mathbf{E}} \qquad (26.18)$$

The potential energy of the system of an electric dipole in a uniform external electric field $\vec{\mathbf{E}}$ is

$$U = -\vec{\mathbf{p}} \cdot \vec{\mathbf{E}} \qquad (26.20)$$

Questions

☐ denotes answer available in *Student Solutions Manual/Study Guide;* **O** denotes objective question

1. **O** True or False? (a) From the definition of capacitance $C = Q/\Delta V$, it follows that an uncharged capacitor has a capacitance of zero. (b) As described by the definition of capacitance, the potential difference across an uncharged capacitor is zero.

2. If you are given three different capacitors C_1, C_2, and C_3, how many different combinations of capacitance can you produce?

3. **O** By what factor is the capacitance of a metal sphere multiplied if its volume is tripled? (a) 9 (b) 3 (c) $3^{2/3}$ (d) $3^{1/3}$ (e) 1 (f) $3^{-1/3}$ (g) $3^{-2/3}$ (h) $\tfrac{1}{3}$

4. **O** A capacitor with very large capacitance is in series with another capacitor with very small capacitance. What is the equivalent capacitance of the combination? (a) slightly greater than the capacitance of the large capacitor (b) slightly less than the capacitance of the large capaci-

tor   (c) slightly greater than the capacitance of the small capacitor   (d) slightly less than the capacitance of the small capacitor.

5. O (i) Rank the following six capacitors in order from greatest to smallest capacitance, noting any cases of equality. (a) a 20-μF capacitor with a 4-V potential difference between its plates   (b) a 30-μF capacitor with charges of magnitude 90 μC on each plate   (c) a capacitor with charges of magnitude 80 μC on its plates, differing by 2 V in potential   (d) a 10-μF capacitor storing 125 μJ   (e) a capacitor storing energy 250 μJ with a 10-V potential difference   (f) a capacitor storing charge 120 μC and energy 360 μJ   **(ii)** Rank the same capacitors from largest to smallest according to the potential difference between the plates. **(iii)** Rank the capacitors in the order of the magnitudes of the charges on their plates. **(iv)** Rank the capacitors in the order of the energy they store.

6. The sum of the charges on both plates of a capacitor is zero. What does a capacitor store?

7. O (i) What happens to the magnitude of the charge on each plate of a capacitor if the potential difference between the conductors is doubled? (a) It becomes four times larger. (b) It becomes two times larger. (c) It is unchanged. (d) It becomes one-half as large. (e) It becomes one-fourth as large. **(ii)** If the potential difference across a capacitor is doubled, what happens to the energy stored? Choose from the same possibilities.

8. O A parallel-plate capacitor is charged and then is disconnected from the battery. By what factor does the stored energy change when the plate separation is then doubled? (a) It becomes four times larger. (b) It becomes two times larger. (c) It stays the same. (d) It becomes one-half as large (e) It becomes one-fourth as large.

9. O You charge a parallel-plate capacitor, remove it from the battery, and prevent the wires connected to the plates from touching each other. When you increase the plate separation, does each of the following quantities (a) increase, (b) decrease, or (c) stay the same? **(i)** C **(ii)** Q **(iii)** E between the plates   **(iv)** ΔV **(v)** the energy stored in the capacitor

10. O Repeat Question 9, but this time answer for the situation in which the battery remains connected to the capacitor while you increase the plate separation.

11. Because the charges on the plates of a parallel-plate capacitor are opposite in sign, they attract each other. Hence, it would take positive work to increase the plate separation. What type of energy in the system changes due to the external work done in this process?

12. Explain why the work needed to move a particle with charge Q through a potential difference ΔV is $W = Q \Delta V$, whereas the energy stored in a charged capacitor is $U = \frac{1}{2} Q \Delta V$. Where does the $\frac{1}{2}$ factor come from?

13. O Assume a device is designed to obtain a large potential difference by first charging a bank of capacitors connected in parallel and then activating a switch arrangement that in effect disconnects the capacitors from the charging source and from each other and reconnects them all in a series arrangement. The group of charged capacitors is then discharged in series. What is the maximum potential difference that can be obtained in this manner by using ten capacitors each of 500 μF and a charging source of 800 V? (a) 80 kV   (b) 8 kV   (c) 2.5 kV   (d) 800 V   (e) 80 V   (f) 8 V   (g) 0

14. An air-filled capacitor is charged, then disconnected from the power supply, and finally connected to a voltmeter. Explain how and why the potential difference changes when a dielectric is inserted between the plates of the capacitor.

15. O A fully charged parallel-plate capacitor remains connected to a battery while you slide a dielectric between the plates. Do the following quantities (a) increase, (b) decrease, or (c) stay the same? **(i)** C **(ii)** Q **(iii)** E between the plates   **(iv)** ΔV **(v)** the energy stored in the capacitor

16. Assume you want to increase the maximum operating voltage of a parallel-plate capacitor. Describe how you can do that with a fixed plate separation.

17. If you were asked to design a capacitor in which small size and large capacitance were required, what factors would be important in your design?

Problems

WebAssign The Problems from this chapter may be assigned online in WebAssign.

ThomsonNOW™   Sign in at **www.thomsonedu.com** and go to ThomsonNOW to assess your understanding of this chapter's topics with additional quizzing and conceptual questions.

1, 2, 3 denotes straightforward, intermediate, challenging;   ☐ denotes full solution available in *Student Solutions Manual/Study Guide;*   ▲ denotes coached solution with hints available at **www.thomsonedu.com;**   denotes developing symbolic reasoning;   ● denotes asking for qualitative reasoning;   ▬ denotes computer useful in solving problem

Section 26.1 Definition of Capacitance

1. (a) How much charge is on each plate of a 4.00-μF capacitor when it is connected to a 12.0-V battery? (b) If this same capacitor is connected to a 1.50-V battery, what charge is stored?

2. Two conductors having net charges of +10.0 μC and −10.0 μC have a potential difference of 10.0 V between them. (a) Determine the capacitance of the system. (b) What is the potential difference between the two conductors if the charges on each are increased to +100 μC and −100 μC?

Section 26.2 Calculating Capacitance

3. An isolated, charged conducting sphere of radius 12.0 cm creates an electric field of 4.90×10^4 N/C at a distance 21.0 cm from its center. (a) What is its surface charge density? (b) What is its capacitance?

4. Regarding the Earth and a cloud layer 800 m above the Earth as the "plates" of a capacitor, calculate the capacitance of the Earth-cloud layer system. Assume the cloud layer has an area of 1.00 km² and the air between the cloud and the ground is pure and dry. Assume charge builds up on the cloud and on the ground until a uniform electric field of 3.00×10^6 N/C throughout the space between them makes the air break down and conduct electricity as a lightning bolt. What is the maximum charge the cloud can hold?

5. ▲ An air-filled capacitor consists of two parallel plates, each with an area of 7.60 cm², separated by a distance of 1.80 mm. A 20.0-V potential difference is applied to these plates. Calculate (a) the electric field between the plates, (b) the surface charge density, (c) the capacitance, and (d) the charge on each plate.

6. A variable air capacitor used in a radio tuning circuit is made of N semicircular plates each of radius R and positioned a distance d from its neighbors, to which it is electrically connected. As shown in Figures 26.16 and P26.6, a second identical set of plates is enmeshed with the first set. Each plate in the second set is halfway between two plates of the first set. The second set can rotate as a unit. Determine the capacitance as a function of the angle of rotation θ, where $\theta = 0$ corresponds to the maximum capacitance.

Figure P26.6

7. When a potential difference of 150 V is applied to the plates of a parallel-plate capacitor, the plates carry a surface charge density of 30.0 nC/cm². What is the spacing between the plates?

8. A small object of mass m carries a charge q and is suspended by a thread between the vertical plates of a parallel-plate capacitor. The plate separation is d. If the thread makes an angle θ with the vertical, what is the potential difference between the plates?

9. ▲ A 50.0-m length of coaxial cable has an inner conductor that has a diameter of 2.58 mm and carries a charge of 8.10 μC. The surrounding conductor has an inner diameter of 7.27 mm and a charge of -8.10 μC. (a) What is the capacitance of this cable? (b) What is the potential difference between the two conductors? Assume the region between the conductors is air.

10. ● A 10.0-μF capacitor has plates with vacuum between them. Each plate carries a charge of magnitude 1 000 μC.

A particle with charge -3.00 μC and mass 2.00×10^{-16} kg is fired from the positive plate toward the negative plate with an initial speed of 2.00×10^6 m/s. Does the particle reach the negative plate? Explain how you can tell. If it does, what is its impact speed? If it does not, what fraction of the way across the capacitor does it travel?

11. An air-filled spherical capacitor is constructed with inner and outer shell radii of 7.00 and 14.0 cm, respectively. (a) Calculate the capacitance of the device. (b) What potential difference between the spheres results in a charge of 4.00 μC on the capacitor?

Section 26.3 Combinations of Capacitors

12. Two capacitors, $C_1 = 5.00$ μF and $C_2 = 12.0$ μF, are connected in parallel, and the resulting combination is connected to a 9.00-V battery. Find (a) the equivalent capacitance of the combination, (b) the potential difference across each capacitor, and (c) the charge stored on each capacitor.

13. **What If?** The two capacitors of Problem 12 are now connected in series and to a 9.00-V battery. Find (a) the equivalent capacitance of the combination, (b) the potential difference across each capacitor, and (c) the charge on each capacitor.

14. ● Three capacitors are connected to a battery as shown in Figure P26.14. Their capacitances are $C_1 = 3C$, $C_2 = C$, and $C_3 = 5C$. (a) What is the equivalent capacitance of this set of capacitors? (b) State the ranking of the capacitors according to the charge they store, from largest to smallest. (c) Rank the capacitors according to the potential differences across them, from largest to smallest. (d) **What If?** Assume C_3 is increased. Explain what happens to the charge stored by each of the capacitors.

Figure P26.14

15. Two capacitors give an equivalent capacitance of 9.00 pF when connected in parallel and give an equivalent capacitance of 2.00 pF when connected in series. What is the capacitance of each capacitor?

16. Two capacitors give an equivalent capacitance of C_p when connected in parallel and an equivalent capacitance of C_s when connected in series. What is the capacitance of each capacitor?

17. ▲ Four capacitors are connected as shown in Figure P26.17. (a) Find the equivalent capacitance between

Figure P26.17

points *a* and *b*. (b) Calculate the charge on each capacitor, taking $\Delta V_{ab} = 15.0$ V.

18. According to its design specification, the timer circuit delaying the closing of an elevator door is to have a capacitance of 32.0 μF between two points *A* and *B*. (a) When one circuit is being constructed, the inexpensive but durable capacitor installed between these two points is found to have capacitance 34.8 μF. To meet the specification, one additional capacitor can be placed between the two points. Should it be in series or in parallel with the 34.8-μF capacitor? What should be its capacitance? (b) **What If?** The next circuit comes down the assembly line with capacitance 29.8 μF between *A* and *B*. To meet the specification, what additional capacitor should be installed in series or in parallel in that circuit?

19. Consider the circuit shown in Figure P26.19, where $C_1 = 6.00$ μF, $C_2 = 3.00$ μF, and $\Delta V = 20.0$ V. Capacitor C_1 is first charged by closing switch S_1. Switch S_1 is then opened, and the charged capacitor is connected to the uncharged capacitor by closing S_2. Calculate the initial charge acquired by C_1 and the final charge on each capacitor.

Figure P26.19

20. Consider three capacitors C_1, C_2, C_3, and a battery. If C_1 is connected to the battery, the charge on C_1 is 30.8 μC. Now C_1 is disconnected, discharged, and connected in series with C_2. When the series combination of C_2 and C_1 is connected across the battery, the charge on C_1 is 23.1 μC. The circuit is disconnected and the capacitors discharged. Capacitor C_3, capacitor C_1, and the battery are connected in series, resulting in a charge on C_1 of 25.2 μC. If, after being disconnected and discharged, C_1, C_2, and C_3 are connected in series with one another and with the battery, what is the charge on C_1?

21. A group of identical capacitors is connected first in series and then in parallel. The combined capacitance in parallel is 100 times larger than for the series connection. How many capacitors are in the group?

22. Some physical systems possessing capacitance continuously distributed over space can be modeled as an infinite array of discrete circuit elements. Examples are a microwave waveguide and the axon of a nerve cell. To practice analysis of an infinite array, determine the equivalent capacitance C between terminals X and Y of the infinite set of capacitors represented in Figure P26.22. Each capacitor has capacitance C_0. *Suggestion:* Imagine that the ladder is

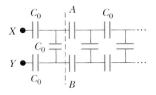

Figure P26.22

cut at the line *AB* and note that the equivalent capacitance of the infinite section to the right of *AB* is also *C*.

23. Find the equivalent capacitance between points *a* and *b* for the group of capacitors connected as shown in Figure P26.23. Take $C_1 = 5.00$ μF, $C_2 = 10.0$ μF, and $C_3 = 2.00$ μF.

Figure P26.23   Problems 23 and 24.

24. For the network described in Problem 23, what charge is stored on C_3 if the potential difference between points *a* and *b* is 60.0 V?

25. Find the equivalent capacitance between points *a* and *b* in the combination of capacitors shown in Figure P26.25.

Figure P26.25

Section 26.4 Energy Stored in a Charged Capacitor

26. The immediate cause of many deaths is ventricular fibrillation, which is an uncoordinated quivering of the heart. An electric shock to the chest can cause momentary paralysis of the heart muscle, after which the heart sometimes resumes its proper beating. One type of *defibrillator* (Fig. 26.13) applies a strong electric shock to the chest over a time interval of a few milliseconds. This device contains a capacitor of several microfarads, charged to several thousand volts. Electrodes called paddles, about 8 cm across and coated with conducting paste, are held against the chest on both sides of the heart. Their handles are insulated to prevent injury to the operator, who calls "Clear!" and pushes a button on one paddle to discharge the capacitor through the patient's chest. Assume an energy of 300 J is to be delivered from a 30.0-μF capacitor. To what potential difference must it be charged?

27. (a) A 3.00-μF capacitor is connected to a 12.0-V battery. How much energy is stored in the capacitor? (b) Had the capacitor been connected to a 6.00-V battery, how much energy would have been stored?

28. Two capacitors, $C_1 = 25.0$ μF and $C_2 = 5.00$ μF, are connected in parallel and charged with a 100-V power supply. (a) Draw a circuit diagram and calculate the total energy stored in the two capacitors. (b) **What If?** What potential difference would be required across the same two capacitors connected in series for the combination to store the same amount of energy as in part (a)? Draw a circuit diagram of this circuit.

29. ▲ A parallel-plate capacitor has a charge Q and plates of area A. What force acts on one plate to attract it toward the other plate? Because the electric field between the plates is $E = Q/A\epsilon_0$, you might think the force is $F = QE = Q^2/A\epsilon_0$. This conclusion is wrong because the field E includes contributions from both plates and the field created by the positive plate cannot exert any force on the positive plate. Show that the force exerted on each plate is actually $F = Q^2/2\epsilon_0 A$. *Suggestion:* Let $C = \epsilon_0 A/x$ for an arbitrary plate separation x and note that the work done in separating the two charged plates is $W = \int F \, dx$.

30. The circuit in Figure P26.30 consists of two identical, parallel metal plates connected by identical metal springs to a 100-V battery. With the switch open, the plates are uncharged, are separated by a distance $d = 8.00$ mm, and have a capacitance $C = 2.00 \ \mu$F. When the switch is closed, the distance between the plates decreases by a factor of 0.500. (a) How much charge collects on each plate? (b) What is the spring constant for each spring? *Suggestion:* Use the result of Problem 29.

Figure P26.30

31. As a person moves about in a dry environment, electric charge accumulates on the person's body. Once it is at high voltage, either positive or negative, the body can discharge via sometimes noticeable sparks and shocks. Consider a human body isolated from ground, with the typical capacitance 150 pF. (a) What charge on the body will produce a potential of 10.0 kV? (b) Sensitive electronic devices can be destroyed by electrostatic discharge from a person. A particular device can be destroyed by a discharge releasing an energy of 250 μJ. To what voltage on the body does this situation correspond?

32. ● Two identical parallel-plate capacitors, each with capacitance C, are charged to potential difference ΔV and connected in parallel. Then the plate separation in one of the capacitors is doubled. (a) Find the total energy of the system of two capacitors *before* the plate separation is doubled. (b) Find the potential difference across each capacitor *after* the plate separation is doubled. (c) Find the total energy of the system *after* the plate separation is doubled. (d) Reconcile the difference in the answers to parts (a) and (c) with the law of conservation of energy.

33. Show that the energy associated with a conducting sphere of radius R and charge Q surrounded by a vacuum is $U = k_e Q^2/2R$.

34. Consider two conducting spheres with radii R_1 and R_2 separated by a distance much greater than either radius. A total charge Q is shared between the spheres, subject to the condition that the electric potential energy of the system has the smallest possible value. The total charge Q is equal to $q_1 + q_2$, where q_1 represents the charge on the first sphere and q_2 the charge on the second. Because the spheres are very far apart, you can assume the charge of each is uniformly distributed over its surface. You may use the result of Problem 33. (a) Determine the values of q_1 and q_2 in terms of Q, R_1, and R_2. (b) Show that the potential difference between the spheres is zero. (We saw in Chapter 25 that two conductors joined by a conducting wire are at the same potential in a static situation. This problem illustrates the general principle that charge on a conductor distributes itself so that the electric potential energy of the system is a minimum.)

35. Review problem. A certain storm cloud has a potential of 1.00×10^8 V relative to a tree. If, during a lightning storm, 50.0 C of charge is transferred through this potential difference and 1.00% of the energy is absorbed by the tree, how much sap in the tree can be boiled away? Model the sap as water initially at 30.0°C. Water has a specific heat of 4 186 J/kg · °C, a boiling point of 100°C, and a latent heat of vaporization of 2.26×10^6 J/kg.

Section 26.5 Capacitors with Dielectrics

36. (a) How much charge can be placed on a capacitor with air between the plates before it breaks down if the area of each of the plates is 5.00 cm²? (b) **What If?** Find the maximum charge if polystyrene is used between the plates instead of air.

37. Determine (a) the capacitance and (b) the maximum potential difference that can be applied to a Teflon-filled parallel-plate capacitor having a plate area of 1.75 cm² and plate separation of 0.040 0 mm.

38. A supermarket sells rolls of aluminum foil, of plastic wrap, and of waxed paper. Describe a capacitor made from such materials. Compute order-of-magnitude estimates for its capacitance and its breakdown voltage.

39. A commercial capacitor is to be constructed as shown in Figure 26.15a. This particular capacitor is made from two strips of aluminum separated by a strip of paraffin-coated paper. Each strip of foil and paper is 7.00 cm wide. The foil is 0.004 00 mm thick, and the paper is 0.025 0 mm thick and has a dielectric constant of 3.70. What length should the strips have if a capacitance of 9.50×10^{-8} F is desired before the capacitor is rolled up? (Adding a second strip of paper and rolling the capacitor effectively doubles its capacitance by allowing charge storage on both sides of each strip of foil.)

40. A parallel-plate capacitor in air has a plate separation of 1.50 cm and a plate area of 25.0 cm². The plates are charged to a potential difference of 250 V and disconnected from the source. The capacitor is then immersed in distilled water. Determine (a) the charge on the plates before and after immersion, (b) the capacitance and potential difference after immersion, and (c) the change in energy of the capacitor. Assume the liquid is an insulator.

41. Each capacitor in the combination shown in Figure P26.41 has a breakdown voltage of 15.0 V. What is the breakdown voltage of the combination?

Figure P26.41

Section 26.6 Electric Dipole in an Electric Field

42. A small, rigid object carries positive and negative 3.50-nC charges. It is oriented so that the positive charge has coordinates $(-1.20$ mm, 1.10 mm) and the negative charge is at the point (1.40 mm, -1.30 mm). (a) Find the electric dipole moment of the object. The object is placed in an electric field $\vec{\mathbf{E}} = (7\,800\,\hat{\mathbf{i}} - 4\,900\,\hat{\mathbf{j}})$ N/C. (b) Find the torque acting on the object. (c) Find the potential energy of the object–field system when the object is in this orientation. (d) Assuming the orientation of the object can change, find the difference between the maximum and minimum potential energies of the system.

43. A small object with electric dipole moment $\vec{\mathbf{p}}$ is placed in a nonuniform electric field $\vec{\mathbf{E}} = E(x)\hat{\mathbf{i}}$. That is, the field is in the x direction and its magnitude depends on the coordinate x. Let θ represent the angle between the dipole moment and the x direction. (a) Prove that the net force on the dipole is

$$F = p\left(\frac{dE}{dx}\right)\cos\theta$$

acting in the direction of increasing field. (b) Consider a spherical balloon centered at the origin, with radius 15.0 cm and carrying charge 2.00 μC. Evaluate dE/dx at the point (16 cm, 0, 0). Assume a water droplet at this point has an induced dipole moment of $6.30\hat{\mathbf{i}}$ nC·m. Find the net force exerted on it.

Section 26.7 An Atomic Description of Dielectrics

44. The general form of Gauss's law describes how a charge creates an electric field in a material, as well as in vacuum:

$$\oint \vec{\mathbf{E}} \cdot d\vec{\mathbf{A}} = \frac{q_{\text{in}}}{\epsilon}$$

where $\epsilon = \kappa\epsilon_0$ is the permittivity of the material. (a) A sheet with charge Q uniformly distributed over its area A is surrounded by a dielectric. Show that the sheet creates a uniform electric field at nearby points, with magnitude $E = Q/2A\epsilon$. (b) Two large sheets of area A, carrying opposite charges of equal magnitude Q, are a small distance d apart. Show that they create uniform electric field in the space between them, with magnitude $E = Q/A\epsilon$. (c) Assume the negative plate is at zero potential. Show that the positive plate is at potential $Qd/A\epsilon$. (d) Show that the capacitance of the pair of plates is $A\epsilon/d = \kappa A\epsilon_0/d$.

45. The inner conductor of a coaxial cable has a radius of 0.800 mm, and the outer conductor's inside radius is 3.00 mm. The space between the conductors is filled with polyethylene, which has a dielectric constant of 2.30 and a dielectric strength of 18.0×10^6 V/m. What is the maximum potential difference that this cable can withstand?

Additional Problems

46. Two large, parallel metal plates each of area A are oriented horizontally and separated by a distance $3d$. A grounded conducting wire joins them, and initially each plate carries no charge. Now a third identical plate carrying charge Q is inserted between the two plates, parallel to them and located a distance d from the upper plate as shown in Figure P26.46. (a) What induced charge appears on each of the two original plates? (b) What potential difference appears between the middle plate and each of the other plates?

Figure P26.46

47. Four parallel metal plates P_1, P_2, P_3, and P_4, each of area 7.50 cm^2, are separated successively by a distance $d = 1.19$ mm as shown in Figure P26.47. P_1 is connected to the negative terminal of a battery, and P_2 is connected to the positive terminal. The battery maintains a potential difference of 12.0 V. (a) If P_3 is connected to the negative terminal, what is the capacitance of the three-plate system $P_1P_2P_3$? (b) What is the charge on P_2? (c) If P_4 is now connected to the positive terminal, what is the capacitance of the four-plate system $P_1P_2P_3P_4$? (d) What is the charge on P_4?

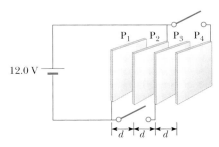

Figure P26.47

48. One conductor of an overhead electric transmission line is a long aluminum wire 2.40 cm in radius. Suppose it carries a charge per length of 1.40 μC/m at a particular moment and its potential is 345 kV. Find the potential 12.0 m below the wire. Ignore the other conductors of the transmission line and assume the electric field is radial everywhere.

49. A 2.00-nF parallel-plate capacitor is charged to an initial potential difference $\Delta V_i = 100$ V and is then isolated. The dielectric material between the plates is mica, with a dielectric constant of 5.00. (a) How much work is required to withdraw the mica sheet? (b) What is the potential difference across the capacitor after the mica is withdrawn?

2 = intermediate; 3 = challenging; ☐ = SSM/SG; ▲ = ThomsonNOW; ▮ = symbolic reasoning; ● = qualitative reasoning

50. (a) Draw a circuit diagram showing four capacitors between two points a and b for which the following expression determines the equivalent capacitance:

$$\frac{1}{\dfrac{1}{30 \ \mu F} + \dfrac{1}{20 \ \mu F + C_1}} + 50 \ \mu F = 70 \ \mu F$$

(b) Find the value of C_1. (c) Assume a 6.00-V battery is connected between a and b. Find the potential difference across each of the individual capacitors and the charge on each.

51. ▲ A parallel-plate capacitor is constructed using a dielectric material whose dielectric constant is 3.00 and whose dielectric strength is 2.00×10^8 V/m. The desired capacitance is 0.250 μF, and the capacitor must withstand a maximum potential difference of 4.00 kV. Find the minimum area of the capacitor plates.

52. ● A horizontal, parallel-plate capacitor with vacuum between its plates has a capacitance of 25.0 μF. A nonconducting liquid with dielectric constant 6.50 is poured into the space between the plates, filling up a fraction f of its volume. (a) Find the new capacitance as a function of f. (b) What should you expect the capacitance to be when $f = 0$? Does your expression from part (a) agree with your answer? (c) What capacitance should you expect when $f = 1$? Does the expression from part (a) agree with your answer? (d) Charges of magnitude 300 μC are placed on the plates of the partially filled capacitor. What can you determine about the induced charge on the free upper surface of the liquid? How does this charge depend on f?

53. (a) Two spheres have radii a and b, and their centers are a distance d apart. Show that the capacitance of this system is

$$C = \frac{4\pi \epsilon_0}{\dfrac{1}{a} + \dfrac{1}{b} - \dfrac{2}{d}}$$

provided d is large compared with a and b. *Suggestion:* Because the spheres are far apart, assume the potential of each equals the sum of the potentials due to each sphere. When calculating those potentials, assume $V = k_e Q/r$ applies. (b) Show that as d approaches infinity, the above result reduces to that of two spherical capacitors in series.

54. A 10.0-μF capacitor is charged to 15.0 V. It is next connected in series with an uncharged 5.00-μF capacitor. The series combination is finally connected across a 50.0-V battery as diagrammed in Figure P26.54. Find the new potential differences across the 5.00-μF and 10.0-μF capacitors.

5.00 μF 10.0 μF

$\Delta V_i = 15.0 \ V$

50.0 V

Figure P26.54

55. ● When considering the energy supply for an automobile, the energy per unit mass (in joules per kilogram) of the energy source is an important parameter. Using the

following data, explain how the energy per unit mass compares among gasoline, lead-acid batteries, and capacitors. (The ampere A will be introduced in Chapter 27 as the SI unit of electric current. 1 A = 1 C/s.)

Gasoline: 126 000 Btu/gal; density = 670 kg/m³.

Lead-acid battery: 12.0 V; 100 A·h; mass = 16.0 kg.

Capacitor: potential difference at full charge = 12.0 V; capacitance = 0.100 F; mass = 0.100 kg.

56. A capacitor is constructed from two square, metallic plates of sides ℓ and separation d. Charges $+Q$ and $-Q$ are placed on the plates, and the power supply is then removed. A material of dielectric constant κ is inserted a distance x into the capacitor as shown in Figure P26.56. Assume d is much smaller than x. (a) Find the equivalent capacitance of the device. (b) Calculate the energy stored in the capacitor. (c) Find the direction and magnitude of the force exerted by the plates on the dielectric. (d) Obtain a numerical value for the force when $x = \ell/2$, assuming $\ell = 5.00$ cm, $d = 2.00$ mm, the dielectric is glass ($\kappa = 4.50$), and the capacitor was charged to 2 000 V before the dielectric was inserted. *Suggestion:* The system can be considered as two capacitors connected in parallel.

Figure P26.56 Problems 56 and 57.

57. ● Two square plates of sides ℓ are placed parallel to each other with separation d as suggested in Figure P26.56. You may assume d is much less than ℓ. The plates carry uniformly distributed static charges $+Q_0$ and $-Q_0$. A block of metal has width ℓ, length ℓ, and thickness slightly less than d. It is inserted a distance x into the space between the plates. The charges on the plates remain uniformly distributed as the block slides in. In a static situation, a metal prevents an electric field from penetrating inside it. The metal can be thought of as a perfect dielectric, with $\kappa \to \infty$. (a) Calculate the stored energy as a function of x. (b) Find the direction and magnitude of the force that acts on the metallic block. (c) The area of the advancing front face of the block is essentially equal to ℓd. Considering the force on the block as acting on this face, find the stress (force per area) on it. (d) Express the energy density in the electric field between the charged plates in terms of Q_0, ℓ, d, and ϵ_0. Explain how the answers to parts (c) and (d) compare with each other.

58. ● To repair a power supply for a stereo amplifier, an electronics technician needs a 100-μF capacitor capable of withstanding a potential difference of 90 V between the plates. The immediately available supply is a box of five 100-μF capacitors, each having a maximum voltage capability of 50 V. Can the technician use one of the capacitors from the box? Can she substitute a combination of these capacitors that has the proper electrical characteristics? Will the technician use all the capacitors in the box? Explain your answers. In a combination of capacitors,

what will be the maximum voltage across each of the capacitors used?

59. An isolated capacitor of unknown capacitance has been charged to a potential difference of 100 V. When the charged capacitor is then connected in parallel to an uncharged 10.0-μF capacitor, the potential difference across the combination is 30.0 V. Calculate the unknown capacitance.

60. A parallel-plate capacitor with plates of area LW and plate separation t has the region between its plates filled with wedges of two dielectric materials as shown in Figure P26.60. Assume t is much less than both L and W. (a) Determine its capacitance. (b) Should the capacitance be the same if the labels κ_1 and κ_2 are interchanged? Demonstrate that your expression does or does not have this property. (c) Show that if κ_1 and κ_2 approach equality to a common value κ, your result becomes the same as the capacitance of a capacitor containing a single dielectric: $C = \kappa \epsilon_0 LW/t$.

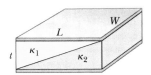

Figure P26.60

61. ● A parallel-plate capacitor of plate separation d is charged to a potential difference ΔV_0. A dielectric slab of thickness d and dielectric constant κ is introduced between the plates while the battery remains connected to the plates. (a) Show that the ratio of energy stored after the dielectric is introduced to the energy stored in the empty capacitor is $U/U_0 = \kappa$. Give a physical explanation for this increase in stored energy. (b) What happens to the charge on the capacitor? (Notice that this situation is not the same as in Example 26.5, in which the battery was removed from the circuit before the dielectric was introduced.)

62. Calculate the equivalent capacitance between points a and b in Figure P26.62. Notice that this system is not a simple series or parallel combination. *Suggestion:* Assume a potential difference ΔV between points a and b. Write expressions for ΔV_{ab} in terms of the charges and capacitances for the various possible pathways from a to b and

require conservation of charge for those capacitor plates that are connected to each other.

Figure P26.62

63. Capacitors $C_1 = 6.00\ \mu$F and $C_2 = 2.00\ \mu$F are charged as a parallel combination across a 250-V battery. The capacitors are disconnected from the battery and from each other. They are then connected positive plate to negative plate and negative plate to positive plate. Calculate the resulting charge on each capacitor.

64. Consider two long, parallel, and oppositely charged wires of radius r with their centers separated by a distance D that is much larger than r. Assuming the charge is distributed uniformly on the surface of each wire, show that the capacitance per unit length of this pair of wires is

$$\frac{C}{\ell} = \frac{\pi \epsilon_0}{\ln (D/r)}$$

65. Determine the equivalent capacitance of the combination shown in Figure P26.65. *Suggestion:* Consider the symmetry involved.

Figure P26.65

66. Example 26.1 explored a cylindrical capacitor of length ℓ with radii a and b for the two conductors. In the **What If?** section of that example, it was claimed that increasing ℓ by 10% is more effective in terms of increasing the capacitance than increasing a by 10% if $b > 2.85a$. Verify this claim mathematically.

Answers to Quick Quizzes

26.1 (d). The capacitance is a property of the physical system and does not vary with applied voltage. According to Equation 26.1, if the voltage is doubled, the charge is doubled.

26.2 (a). When the key is pressed, the plate separation is decreased and the capacitance increases. Capacitance depends only on how a capacitor is constructed and not on the external circuit.

26.3 (a). When connecting capacitors in series, the inverses of the capacitances add, resulting in a smaller overall equivalent capacitance.

26.4 (b). For a given voltage, the energy stored in a capacitor is proportional to C according to $U = C(\Delta V)^2/2$. Therefore, you want to maximize the equivalent capacitance. You do that by connecting the three capacitors in parallel so that the capacitances add.

26.5 (a). The dielectric constant of wood (and of all other insulating materials, for that matter) is greater than 1; therefore, the capacitance increases (Eq. 26.14). This increase is sensed by the stud finder's special circuitry, which causes an indicator on the device to light up.

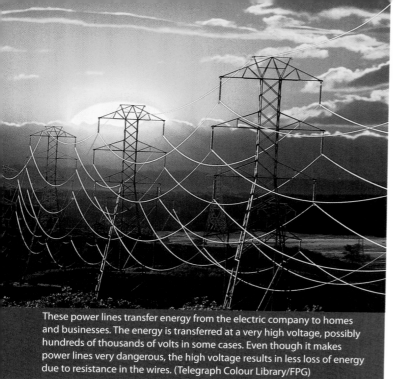

These power lines transfer energy from the electric company to homes and businesses. The energy is transferred at a very high voltage, possibly hundreds of thousands of volts in some cases. Even though it makes power lines very dangerous, the high voltage results in less loss of energy due to resistance in the wires. (Telegraph Colour Library/FPG)

27.1 Electric Current

27.2 Resistance

27.3 A Model for Electrical Conduction

27.4 Resistance and Temperature

27.5 Superconductors

27.6 Electrical Power

27 Current and Resistance

We now consider situations involving electric charges that are in motion through some region of space. We use the term *electric current*, or simply *current*, to describe the rate of flow of charge. Most practical applications of electricity deal with electric currents. For example, the battery in a flashlight produces a current in the filament of the bulb when the switch is turned on. A variety of home appliances operate on alternating current. In these common situations, current exists in a conductor such as a copper wire. Currents can also exist outside a conductor. For instance, a beam of electrons in a television picture tube constitutes a current.

This chapter begins with the definition of current. A microscopic description of current is given, and some factors that contribute to the opposition to the flow of charge in conductors are discussed. A classical model is used to describe electrical conduction in metals, and some limitations of this model are cited. We also define electrical resistance and introduce a new circuit element, the resistor. We conclude by discussing the rate at which energy is transferred to a device in an electric circuit.

27.1 Electric Current

In this section, we study the flow of electric charges through a piece of material. The amount of flow depends on both the material through which the charges are passing and the potential difference across the material. Whenever there is a net flow of charge through some region, an electric **current** is said to exist.

It is instructive to draw an analogy between water flow and current. In many localities, it is common practice to install low-flow showerheads in homes as a water-conservation measure. We quantify the flow of water from these and similar devices by specifying the amount of water that emerges during a given time interval, often measured in liters per minute. On a grander scale, we can characterize a river current by describing the rate at which the water flows past a particular location. For example, the flow over the brink at Niagara Falls is maintained at rates between 1 400 m^3/s and 2 800 m^3/s.

There is also an analogy between thermal conduction and current. In Section 20.7, we discussed the flow of energy by heat through a sample of material. The rate of energy flow is determined by the material as well as the temperature difference across the material as described by Equation 20.15.

To define current more precisely, suppose charges are moving perpendicular to a surface of area A as shown in Figure 27.1. (This area could be the cross-sectional area of a wire, for example.) **The current is the rate at which charge flows through this surface.** If ΔQ is the amount of charge that passes through this surface in a time interval Δt, the **average current** I_{avg} is equal to the charge that passes through A per unit time:

$$I_{avg} = \frac{\Delta Q}{\Delta t} \qquad (27.1)$$

If the rate at which charge flows varies in time, the current varies in time; we define the **instantaneous current** I as the differential limit of average current:

$$I \equiv \frac{dQ}{dt} \qquad (27.2)$$

◄ Electric current

The SI unit of current is the **ampere** (A):

$$1\text{ A} = 1\text{ C/s} \qquad (27.3)$$

That is, 1 A of current is equivalent to 1 C of charge passing through a surface in 1 s.

The charged particles passing through the surface in Figure 27.1 can be positive, negative, or both. **It is conventional to assign to the current the same direction as the flow of positive charge.** In electrical conductors such as copper or aluminum, the current results from the motion of negatively charged electrons. Therefore, in an ordinary conductor, **the direction of the current is opposite the direction of flow of electrons.** For a beam of positively charged protons in an accelerator, however, the current is in the direction of motion of the protons. In some cases—such as those involving gases and electrolytes, for instance—the current is the result of the flow of both positive and negative charges. It is common to refer to a moving charge (positive or negative) as a mobile **charge carrier.**

If the ends of a conducting wire are connected to form a loop, all points on the loop are at the same electric potential; hence, the electric field is zero within and at the surface of the conductor. Because the electric field is zero, there is no net transport of charge through the wire; therefore, there is no current. If the ends of the conducting wire are connected to a battery, however, all points on the loop are not at the same potential. The battery sets up a potential difference between the ends of the loop, creating an electric field within the wire. The electric field exerts forces on the conduction electrons in the wire, causing them to move in the wire and therefore creating a current.

Microscopic Model of Current

We can relate current to the motion of the charge carriers by describing a microscopic model of conduction in a metal. Consider the current in a conductor of

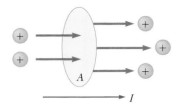

Figure 27.1 Charges in motion through an area A. The time rate at which charge flows through the area is defined as the current I. The direction of the current is the direction in which positive charges flow when free to do so.

PITFALL PREVENTION 27.1
"Current Flow" Is Redundant

The phrase *current flow* is commonly used, although it is technically incorrect because current *is* a flow (of charge). This wording is similar to the phrase *heat transfer*, which is also redundant because heat *is* a transfer (of energy). We will avoid this phrase and speak of *flow of charge* or *charge flow*.

PITFALL PREVENTION 27.2
Batteries Do Not Supply Electrons

A battery does not supply electrons to the circuit. It establishes the electric field that exerts a force on electrons already in the wires and elements of the circuit.

Figure 27.2 A section of a uniform conductor of cross-sectional area A. The mobile charge carriers move with a speed v_d, and the displacement they experience in the x direction in a time interval Δt is $\Delta x = v_d \Delta t$. If we choose Δt to be the time interval during which the charges are displaced, on average, by the length of the cylinder, the number of carriers in the section of length Δx is $nAv_d \Delta t$, where n is the number of carriers per unit volume.

▶ Current in a conductor in terms of microscopic quantities

cross-sectional area A (Fig. 27.2). The volume of a section of the conductor of length Δx (the gray region of the conductor shown in Fig. 27.2) is $A \Delta x$. If n represents the number of mobile charge carriers per unit volume (in other words, the charge carrier density), the number of carriers in the gray section is $nA \Delta x$. Therefore, the total charge ΔQ in this section is

$$\Delta Q = (nA \, \Delta x)q$$

where q is the charge on each carrier. If the carriers move with a speed v_d, the displacement they experience in the x direction in a time interval Δt is $\Delta x = v_d \, \Delta t$. Let Δt be the time interval required for the charge carriers in the cylinder to move through a displacement whose magnitude is equal to the length of the cylinder. This time interval is also the same as that required for all the charge carriers in the cylinder to pass through the circular area at one end. With this choice, we can write ΔQ as

$$\Delta Q = (nAv_d \, \Delta t)q$$

Dividing both sides of this equation by Δt, we find that the average current in the conductor is

$$I_{\text{avg}} = \frac{\Delta Q}{\Delta t} = nqv_d A \qquad (27.4)$$

The speed of the charge carriers v_d is an average speed called the **drift speed.** To understand the meaning of drift speed, consider a conductor in which the charge carriers are free electrons. If the conductor is isolated—that is, the potential difference across it is zero—these electrons undergo random motion that is analogous to the motion of gas molecules. The electrons collide repeatedly with the metal atoms, and their resultant motion is complicated and zigzagged as in Active Figure 27.3a. As discussed earlier, when a potential difference is applied across the conductor (for example, by means of a battery), an electric field is set up in the conductor; this field exerts an electric force on the electrons, producing a current. In addition to the zigzag motion due to the collisions with the metal atoms, the electrons move slowly along the conductor (in a direction opposite that of $\vec{\mathbf{E}}$) at the drift velocity $\vec{\mathbf{v}}_d$ as shown in Active Figure 27.3b.

You can think of the atom–electron collisions in a conductor as an effective internal friction (or drag force) similar to that experienced by a liquid's molecules flowing through a pipe stuffed with steel wool. The energy transferred from the electrons to the metal atoms during collisions causes an increase in the atom's vibrational energy and a corresponding increase in the conductor's temperature.

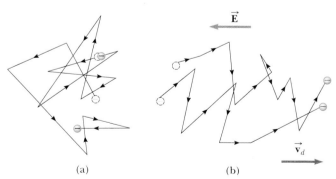

(a) (b)

ACTIVE FIGURE 27.3

(a) A schematic diagram of the random motion of two charge carriers in a conductor in the absence of an electric field. The drift velocity is zero. (b) The motion of the charge carriers in a conductor in the presence of an electric field. Notice that the random motion is modified by the field and the charge carriers have a drift velocity opposite the direction of the electric field. Because of the acceleration of the charge carriers due to the electric force, the paths are actually parabolic. The drift speed, however, is much smaller than the average speed, so the parabolic shape is not visible on this scale.

Sign in at www.thomsonedu.com and go to ThomsonNOW to adjust the electric field and see the resulting effect on the motion of an electron.

Quick Quiz 27.1   Consider positive and negative charges moving horizontally through the four regions shown in Figure 27.4. Rank the current in these four regions from lowest to highest.

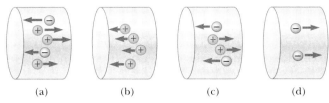

(a)   (b)   (c)   (d)

Figure 27.4   (Quick Quiz 27.1) Charges move through four regions.

| EXAMPLE 27.1 | **Drift Speed in a Copper Wire** |

The 12-gauge copper wire in a typical residential building has a cross-sectional area of 3.31×10^{-6} m^2. It carries a constant current of 10.0 A. What is the drift speed of the electrons in the wire? Assume each copper atom contributes one free electron to the current. The density of copper is 8.92 g/cm^3.

SOLUTION

Conceptualize   Imagine electrons following a zigzag motion such as that in Active Figure 27.3a, with a drift motion parallel to the wire superimposed on the motion as in Active Figure 27.3b. As mentioned earlier, the drift speed is small, and this example helps us quantify the speed.

Categorize   We evaluate the drift speed using Equation 27.4. Because the current is constant, the average current during any time interval is the same as the constant current: $I_{avg} = I$.

Analyze   The periodic table of the elements in Appendix C shows that the molar mass of copper is 63.5 g/mol. Recall that 1 mol of any substance contains Avogadro's number of atoms (6.02×10^{23}).

Use the molar mass and the density of copper to find the volume of 1 mole of copper:

$$V = \frac{m}{\rho} = \frac{63.5 \text{ g}}{8.92 \text{ g/cm}^3} = 7.12 \text{ cm}^3$$

From the assumption that each copper atom contributes one free electron to the current, find the electron density in copper:

$$n = \frac{6.02 \times 10^{23} \text{ electrons}}{7.12 \text{ cm}^3} \left(\frac{1.00 \times 10^6 \text{ cm}^3}{1 \text{ m}^3} \right)$$

$$= 8.46 \times 10^{28} \text{ electrons/m}^3$$

Solve Equation 27.4 for the drift speed:

$$v_d = \frac{I_{avg}}{nqA} = \frac{I}{nqA}$$

Substitute numerical values:

$$v_d = \frac{I}{neA} = \frac{10.0 \text{ A}}{(8.46 \times 10^{28} \text{ m}^{-3})(1.60 \times 10^{-19} \text{ C})(3.31 \times 10^{-6} \text{ m}^2)}$$

$$= 2.23 \times 10^{-4} \text{ m/s}$$

Finalize   This result shows that typical drift speeds are very small. For instance, electrons traveling with a speed of 2.23×10^{-4} m/s would take about 75 min to travel 1 m! You might therefore wonder why a light turns on almost instantaneously when its switch is thrown. In a conductor, changes in the electric field that drives the free electrons travel through the conductor with a speed close to that of light. So, when you flip on a light switch, electrons already in the filament of the lightbulb experience electric forces and begin moving after a time interval on the order of nanoseconds.

27.2 Resistance

In Chapter 24, we found that the electric field inside a conductor is zero. This statement is true, however, *only* if the conductor is in static equilibrium. The purpose of this section is to describe what happens when the charges in the conductor are not in equilibrium, in which case there is an electric field in the conductor.

Consider a conductor of cross-sectional area A carrying a current I. The **current density** J in the conductor is defined as the current per unit area. Because the current $I = nqv_d A$, the current density is

Current density ▶

$$J \equiv \frac{I}{A} = nqv_d \qquad \text{(27.5)}$$

where J has SI units of amperes per meter squared. This expression is valid only if the current density is uniform and only if the surface of cross-sectional area A is perpendicular to the direction of the current.

A current density and an electric field are established in a conductor whenever a potential difference is maintained across the conductor. In some materials, the current density is proportional to the electric field:

$$J = \sigma E \qquad \text{(27.6)}$$

where the constant of proportionality σ is called the **conductivity** of the conductor.[1] Materials that obey Equation 27.6 are said to follow **Ohm's law,** named after Georg Simon Ohm. More specifically, Ohm's law states the following:

> For many materials (including most metals), the ratio of the current density to the electric field is a constant σ that is independent of the electric field producing the current.

Materials that obey Ohm's law and hence demonstrate this simple relationship between E and J are said to be *ohmic.* Experimentally, however, it is found that not all materials have this property. Materials and devices that do not obey Ohm's law are said to be *nonohmic.* Ohm's law is not a fundamental law of nature; rather, it is an empirical relationship valid only for certain materials.

We can obtain an equation useful in practical applications by considering a segment of straight wire of uniform cross-sectional area A and length ℓ as shown in Figure 27.5. A potential difference $\Delta V = V_b - V_a$ is maintained across the wire, creating in the wire an electric field and a current. If the field is assumed to be uniform, the potential difference is related to the field through the relationship[2]

$$\Delta V = E\ell$$

Therefore, we can express the current density in the wire as

$$J = \sigma E = \sigma \frac{\Delta V}{\ell}$$

Because $J = I/A$, the potential difference across the wire is

$$\Delta V = \frac{\ell}{\sigma} J = \left(\frac{\ell}{\sigma A}\right) I = RI$$

GEORG SIMON OHM
German physicist (1789–1854)
Ohm, a high school teacher and later a professor at the University of Munich, formulated the concept of resistance and discovered the proportionalities expressed in Equations 27.6 and 27.7.

Figure 27.5 A uniform conductor of length ℓ and cross-sectional area A. A potential difference $\Delta V = V_b - V_a$ maintained across the conductor sets up an electric field $\vec{E}$, and this field produces a current I that is proportional to the potential difference.

---

[1] Do not confuse conductivity σ with surface charge density, for which the same symbol is used.
[2] This result follows from the definition of potential difference:

$$V_b - V_a = -\int_a^b \vec{E} \cdot d\vec{s} = E \int_a^b dx = E\ell$$

The quantity $R = \ell/\sigma A$ is called the **resistance** of the conductor. We define the resistance as the ratio of the potential difference across a conductor to the current in the conductor:

$$R \equiv \frac{\Delta V}{I} \qquad (27.7)$$

We will use this equation again and again when studying electric circuits. This result shows that resistance has SI units of volts per ampere. One volt per ampere is defined to be one **ohm** (Ω):

$$1\ \Omega \equiv 1\ \text{V/A} \qquad (27.8)$$

This expression shows that if a potential difference of 1 V across a conductor causes a current of 1 A, the resistance of the conductor is 1 Ω. For example, if an electrical appliance connected to a 120-V source of potential difference carries a current of 6 A, its resistance is 20 Ω.

Most electric circuits use circuit elements called **resistors** to control the current in the various parts of the circuit. Two common types are the *composition resistor*, which contains carbon, and the *wire-wound resistor*, which consists of a coil of wire. Values of resistors in ohms are normally indicated by color coding as shown in Figure 27.6 and Table 27.1.

The inverse of conductivity is **resistivity**[3] ρ:

$$\rho = \frac{1}{\sigma} \qquad (27.9)$$

◀ Resistivity is the inverse of conductivity

where ρ has the units ohm meters ($\Omega \cdot$ m). Because $R = \ell/\sigma A$, we can express the resistance of a uniform block of material along the length ℓ as

$$R = \rho \frac{\ell}{A} \qquad (27.10)$$

◀ Resistance of a uniform material along the length ℓ

Every ohmic material has a characteristic resistivity that depends on the properties of the material and on temperature. In addition, as you can see from Equation 27.10, the resistance of a sample depends on geometry as well as on resistivity. Table 27.2 (page 758) gives the resistivities of a variety of materials at 20°C. Notice the enormous range, from very low values for good conductors such as copper and silver to very high values for good insulators such as glass and rubber. An ideal conductor would have zero resistivity, and an ideal insulator would have infinite resistivity.

PITFALL PREVENTION 27.3
Equation 27.7 Is Not Ohm's Law

Many individuals call Equation 27.7 Ohm's law, but that is incorrect. This equation is simply the definition of resistance, and it provides an important relationship between voltage, current, and resistance. Ohm's law is related to a proportionality of J to E (Eq. 27.6) or, equivalently, of I to ΔV, which, from Equation 27.7, indicates that the resistance is constant, independent of the applied voltage.

TABLE 27.1
Color Coding for Resistors

Color	Number	Multiplier	Tolerance
Black	0	1	
Brown	1	10^1	
Red	2	10^2	
Orange	3	10^3	
Yellow	4	10^4	
Green	5	10^5	
Blue	6	10^6	
Violet	7	10^7	
Gray	8	10^8	
White	9	10^9	
Gold		10^{-1}	5%
Silver		10^{-2}	10%
Colorless			20%

Figure 27.6 The colored bands on a resistor represent a code for determining resistance. The first two colors give the first two digits in the resistance value. The third color represents the power of 10 for the multiplier of the resistance value. The last color is the tolerance of the resistance value. As an example, the four colors on the circled resistors are red ($= 2$), black ($= 0$), orange ($= 10^3$), and gold ($= 5\%$), and so the resistance value is $20 \times 10^3\ \Omega = 20\ \text{k}\Omega$ with a tolerance value of $5\% = 1\ \text{k}\Omega$. (The values for the colors are from Table 27.1.)

[3] Do not confuse resistivity ρ with mass density or charge density, for which the same symbol is used.

TABLE 27.2

Resistivities and Temperature Coefficients of Resistivity for Various Materials

Material	Resistivity[a] ($\Omega \cdot m$)	Temperature Coefficient[b] $\alpha[(°C)^{-1}]$
Silver	1.59×10^{-8}	3.8×10^{-3}
Copper	1.7×10^{-8}	3.9×10^{-3}
Gold	2.44×10^{-8}	3.4×10^{-3}
Aluminum	2.82×10^{-8}	3.9×10^{-3}
Tungsten	5.6×10^{-8}	4.5×10^{-3}
Iron	10×10^{-8}	5.0×10^{-3}
Platinum	11×10^{-8}	3.92×10^{-3}
Lead	22×10^{-8}	3.9×10^{-3}
Nichrome[c]	1.50×10^{-6}	0.4×10^{-3}
Carbon	3.5×10^{-5}	-0.5×10^{-3}
Germanium	0.46	-48×10^{-3}
Silicon[d]	2.3×10^3	-75×10^{-3}
Glass	10^{10} to 10^{14}	
Hard rubber	$\sim 10^{13}$	
Sulfur	10^{15}	
Quartz (fused)	75×10^{16}	

[a] All values at 20°C. All elements in this table are assumed to be free of impurities.

[b] See Section 27.4.

[c] A nickel–chromium alloy commonly used in heating elements.

[d] The resistivity of silicon is very sensitive to purity. The value can be changed by several orders of magnitude when it is doped with other atoms.

PITFALL PREVENTION 27.4
Resistance and Resistivity

Resistivity is a property of a *substance*, whereas resistance is a property of an *object*. We have seen similar pairs of variables before. For example, density is a property of a substance, whereas mass is a property of an object. Equation 27.10 relates resistance to resistivity and Equation 1.1 relates mass to density.

Equation 27.10 shows that the resistance of a given cylindrical conductor such as a wire is proportional to its length and inversely proportional to its cross-sectional area. If the length of a wire is doubled, its resistance doubles. If its cross-sectional area is doubled, its resistance decreases by one half. The situation is analogous to the flow of a liquid through a pipe. As the pipe's length is increased, the resistance to flow increases. As the pipe's cross-sectional area is increased, more liquid crosses a given cross section of the pipe per unit time interval. Therefore, more liquid flows for the same pressure differential applied to the pipe, and the resistance to flow decreases.

Ohmic materials and devices have a linear current–potential difference relationship over a broad range of applied potential differences (Fig. 27.7a). The slope of the *I*-versus-ΔV curve in the linear region yields a value for $1/R$. Nonohmic materials have a nonlinear current–potential difference relationship. One common semiconducting device with nonlinear *I*-versus-ΔV characteristics is the *junction diode* (Fig. 27.7b). The resistance of this device is low for currents in one direction (positive ΔV) and high for currents in the reverse direction (negative ΔV). In fact, most modern electronic devices, such as transistors, have nonlin-

Figure 27.7 (a) The current–potential difference curve for an ohmic material. The curve is linear, and the slope is equal to the inverse of the resistance of the conductor. (b) A nonlinear current–potential difference curve for a junction diode. This device does not obey Ohm's law.

ear current–potential difference relationships; their proper operation depends on the particular way they violate Ohm's law.

Quick Quiz 27.2   A cylindrical wire has a radius r and length ℓ. If both r and ℓ are doubled, does the resistance of the wire (a) increase, (b) decrease, or (c) remain the same?

Quick Quiz 27.3   In Figure 27.7b, as the applied voltage increases, does the resistance of the diode (a) increase, (b) decrease, or (c) remain the same?

EXAMPLE 27.2   **The Resistance of Nichrome Wire**

The radius of 22-gauge Nichrome wire is 0.321 mm. **(A)** Calculate the resistance per unit length of this wire.

SOLUTION

Conceptualize   Table 27.2 shows that Nichrome has a resistivity two orders of magnitude larger than the best conductors in the table. Therefore, we expect it to have some special practical applications that the best conductors may not have.

Categorize   We model the wire as a cylinder so that a simple geometric analysis can be applied to find the resistance.

Analyze   Use Equation 27.10 and the resistivity of Nichrome from Table 27.2 to find the resistance per unit length:

$$\frac{R}{\ell} = \frac{\rho}{A} = \frac{\rho}{\pi r^2} = \frac{1.5 \times 10^{-6}\,\Omega \cdot m}{\pi (0.321 \times 10^{-3}\,m)^2} = 4.6\,\Omega/m$$

(B) If a potential difference of 10 V is maintained across a 1.0-m length of the Nichrome wire, what is the current in the wire?

SOLUTION

Analyze   Use Equation 27.7 to find the current:

$$I = \frac{\Delta V}{R} = \frac{\Delta V}{(4.6\,\Omega/m)\ell} = \frac{10\,V}{(4.6\,\Omega/m)(1.0\,m)} = 2.2\,A$$

Finalize   A copper wire of the same radius would have a resistance per unit length of only 0.053 Ω/m. A 1.0-m length of copper wire of the same radius would carry the same current (2.2 A) with an applied potential difference of only 0.12 V.

Because of its high resistivity and resistance to oxidation, Nichrome is often used for heating elements in toasters, irons, and electric heaters.

EXAMPLE 27.3   **The Radial Resistance of a Coaxial Cable**

Coaxial cables are used extensively for cable television and other electronic applications. A coaxial cable consists of two concentric cylindrical conductors. The region between the conductors is completely filled with polyethylene plastic as shown in Figure 27.8a. Current leakage through the plastic, in the *radial* direction, is unwanted. (The cable is designed to conduct current along its length, but that is *not* the current being considered here.) The radius of the inner conductor is $a = 0.500$ cm, the radius of the outer conductor is $b = 1.75$ cm, and the length is $L = 15.0$ cm. The resistivity of the plastic is $1.0 \times 10^{13}\,\Omega \cdot m$. Calculate the resistance of the plastic between the two conductors.

Polyethylene

Inner conductor

Outer conductor

(a)

End view

(b)

Figure 27.8   (Example 27.3) A coaxial cable. (a) Plastic fills the gap between the two conductors. (b) End view, showing current leakage.

SOLUTION

Conceptualize Imagine two currents as suggested in the text of the problem. The desired current is along the cable, carried within the conductors. The undesired current corresponds to charge leakage through the plastic, and its direction is radial.

Categorize Because the resistivity and the geometry of the plastic are known, we categorize this problem as one in which we find the resistance of the plastic from these parameters, using Equation 27.10. Because the area through which the charges pass depends on the radial position, we must use integral calculus to determine the answer.

Analyze We divide the plastic into concentric elements of infinitesimal thickness dr (Fig. 27.8b). Use the differential form of Equation 27.10, replacing ℓ with r for the distance variable: $dR = \rho \, dr/A$, where dR is the resistance of an element of plastic of thickness dr and surface area A. In this example, our representative element is a concentric hollow plastic cylinder of radius r, thickness dr, and length L as in Figure 27.8. Any charge passing from the inner to the outer conductor must move radially through this concentric element. The area through which this charge passes is $A = 2\pi rL$ (the curved surface area—circumference multiplied by length—of our hollow plastic cylinder of thickness dr).

Write an expression for the resistance of our hollow cylinder of plastic:

$$dR = \frac{\rho}{2\pi rL} \, dr$$

Integrate this expression from $r = a$ to $r = b$:

$$(1) \quad R = \int dR = \frac{\rho}{2\pi L} \int_a^b \frac{dr}{r} = \frac{\rho}{2\pi L} \ln\left(\frac{b}{a}\right)$$

Substitute the values given:

$$R = \frac{1.0 \times 10^{13} \, \Omega \cdot m}{2\pi (0.150 \, m)} \ln\left(\frac{1.75 \, cm}{0.500 \, cm}\right) = \boxed{1.33 \times 10^{13} \, \Omega}$$

Finalize Let's compare this resistance to that of the inner copper conductor of the cable along the 15.0-cm length.

Use Equation 27.10 to find the resistance of the copper cylinder:

$$R = \rho \frac{\ell}{A} = (1.7 \times 10^{-8} \, \Omega \cdot m)\left[\frac{0.150 \, m}{\pi (5.00 \times 10^{-3} \, m)^2}\right]$$

$$= 3.2 \times 10^{-5} \, \Omega$$

This resistance is 18 orders of magnitude smaller than the radial resistance. Therefore, almost all the current corresponds to charge moving along the length of the cable, with a very small fraction leaking in the radial direction.

What If? Suppose the coaxial cable is enlarged to twice the overall diameter with two possible choices: (1) the ratio b/a is held fixed or (2) the difference $b - a$ is held fixed. For which choice does the leakage current between the inner and outer conductors increase when the voltage is applied between them?

Answer For the current to increase, the resistance must decrease. For choice (1), in which b/a is held fixed, Equation (1) shows that the resistance is unaffected. For choice (2), we do not have an equation involving the difference $b - a$ to inspect. Looking at Figure 27.8b, however, we see that increasing b and a while holding the voltage constant results in charge flowing through the same thickness of plastic but through a larger area perpendicular to the flow. This larger area results in lower resistance and a higher current.

27.3 A Model for Electrical Conduction

In this section, we describe a classical model of electrical conduction in metals that was first proposed by Paul Drude (1863–1906) in 1900. This model leads to Ohm's law and shows that resistivity can be related to the motion of electrons in metals. Although the Drude model described here has limitations, it introduces concepts that are applied in more elaborate treatments.

Consider a conductor as a regular array of atoms plus a collection of free electrons, which are sometimes called *conduction* electrons. The conduction electrons, although bound to their respective atoms when the atoms are not part of a solid, become free when the atoms condense into a solid. In the absence of an electric field, the conduction electrons move in random directions through the conductor with average speeds on the order of 10^6 m/s (Active Fig. 27.3a). The situation is similar to the motion of gas molecules confined in a vessel. In fact, some scientists refer to conduction electrons in a metal as an *electron gas.*

When an electric field is applied, the free electrons drift slowly in a direction opposite that of the electric field (Active Fig. 27.3b), with an average drift speed v_d that is much smaller (typically 10^{-4} m/s) than their average speed between collisions (typically 10^6 m/s).

In our model, we make the following assumptions:

1. The electron's motion after a collision is independent of its motion before the collision.
2. The excess energy acquired by the electrons in the electric field is lost to the atoms of the conductor when the electrons and atoms collide.

With regard to assumption (2), the energy given up to the atoms increases their vibrational energy, which causes the temperature of the conductor to increase.

We are now in a position to derive an expression for the drift velocity. When a free electron of mass m_e and charge q ($= -e$) is subjected to an electric field $\vec{\mathbf{E}}$, it experiences a force $\vec{\mathbf{F}} = q\vec{\mathbf{E}}$. The electron is a particle under a net force, and its acceleration can be found from Newton's second law, $\Sigma \vec{\mathbf{F}} = m\vec{\mathbf{a}}$:

$$\vec{\mathbf{a}} = \frac{\Sigma \vec{\mathbf{F}}}{m} = \frac{q\vec{\mathbf{E}}}{m_e} \tag{27.11}$$

Because the electric field is uniform, the electron's acceleration is constant, so the electron can be modeled as a particle under constant acceleration. If $\vec{\mathbf{v}}_i$ is the electron's initial velocity the instant after a collision (which occurs at a time defined as $t = 0$), the velocity of the electron at a very short time t later (immediately before the next collision occurs) is, from Equation 4.8,

$$\vec{\mathbf{v}}_f = \vec{\mathbf{v}}_i + \vec{\mathbf{a}}t = \vec{\mathbf{v}}_i + \frac{q\vec{\mathbf{E}}}{m_e}t \tag{27.12}$$

Let's now take the average value of $\vec{\mathbf{v}}_f$ for all the electrons in the wire over all possible collision times t and all possible values of $\vec{\mathbf{v}}_i$. Assuming the initial velocities are randomly distributed over all possible values, the average value of $\vec{\mathbf{v}}_i$ is zero. The average value of the second term of Equation 27.12 is $(q\vec{\mathbf{E}}/m_e)\tau$, where τ is the *average time interval between successive collisions.* Because the average value of $\vec{\mathbf{v}}_f$ is equal to the drift velocity,

$$\vec{\mathbf{v}}_{f,\text{avg}} = \vec{\mathbf{v}}_d = \frac{q\vec{\mathbf{E}}}{m_e}\tau \tag{27.13}$$

◀ Drift velocity in terms of microscopic quantities

The value of τ depends on the size of the metal atoms and the number of electrons per unit volume. We can relate this expression for drift velocity in Equation 27.13 to the current in the conductor. Substituting the magnitude of the velocity from Equation 27.13 into Equation 27.5, the current density becomes

$$J = nqv_d = \frac{nq^2E}{m_e}\tau \tag{27.14}$$

◀ Current density in terms of microscopic quantities

where n is the number of electrons per unit volume. Comparing this expression with Ohm's law, $J = \sigma E$, we obtain the following relationships for conductivity and resistivity of a conductor:

$$\sigma = \frac{nq^2\tau}{m_e} \tag{27.15}$$

◀ Conductivity in terms of microscopic quantities

Resistivity in terms of ▶
microscopic quantities

$$\rho = \frac{1}{\sigma} = \frac{m_e}{nq^2\tau}$$ (27.16)

According to this classical model, conductivity and resistivity do not depend on the strength of the electric field. This feature is characteristic of a conductor obeying Ohm's law.

27.4 Resistance and Temperature

Over a limited temperature range, the resistivity of a conductor varies approximately linearly with temperature according to the expression

Variation of ρ with ▶
temperature

$$\rho = \rho_0[1 + \alpha(T - T_0)]$$ (27.17)

where ρ is the resistivity at some temperature T (in degrees Celsius), ρ_0 is the resistivity at some reference temperature T_0 (usually taken to be 20°C), and α is the **temperature coefficient of resistivity.** From Equation 27.17, the temperature coefficient of resistivity can be expressed as

Temperature coefficient ▶
of resistivity

$$\alpha = \frac{1}{\rho_0}\frac{\Delta\rho}{\Delta T}$$ (27.18)

where $\Delta\rho = \rho - \rho_0$ is the change in resistivity in the temperature interval $\Delta T = T - T_0$.

The temperature coefficients of resistivity for various materials are given in Table 27.2. Notice that the unit for α is degrees Celsius^{-1} [(°C)$^{-1}$]. Because resistance is proportional to resistivity (Eq. 27.10), the variation of resistance of a sample is

$$R = R_0[1 + \alpha(T - T_0)]$$ (27.19)

where R_0 is the resistance at temperature T_0. Use of this property enables precise temperature measurements through careful monitoring of the resistance of a probe made from a particular material.

For some metals such as copper, resistivity is nearly proportional to temperature as shown in Figure 27.9. A nonlinear region always exists at very low temperatures, however, and the resistivity usually reaches some finite value as the temperature approaches absolute zero. This residual resistivity near absolute zero is caused primarily by the collision of electrons with impurities and imperfections in the metal. In contrast, high-temperature resistivity (the linear region) is predominantly characterized by collisions between electrons and metal atoms.

Notice that three of the α values in Table 27.2 are negative, indicating that the resistivity of these materials decreases with increasing temperature. This behavior is indicative of a class of materials called *semiconductors*, first introduced in Section 23.2, and is due to an increase in the density of charge carriers at higher temperatures.

Because the charge carriers in a semiconductor are often associated with impurity atoms, the resistivity of these materials is very sensitive to the type and concentration of such impurities.

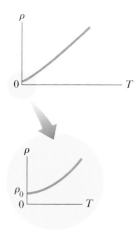

Figure 27.9 Resistivity versus temperature for a metal such as copper. The curve is linear over a wide range of temperatures, and ρ increases with increasing temperature. As T approaches absolute zero (inset), the resistivity approaches a finite value ρ_0.

Quick Quiz 27.4 When does a lightbulb carry more current, (a) immediately after it is turned on and the glow of the metal filament is increasing or (b) after it has been on for a few milliseconds and the glow is steady?

27.5 Superconductors

There is a class of metals and compounds whose resistance decreases to zero when they are below a certain temperature T_c, known as the **critical temperature.** These materials are known as **superconductors.** The resistance–temperature graph for a superconductor follows that of a normal metal at temperatures above T_c (Fig. 27.10).

Figure 27.10 Resistance versus temperature for a sample of mercury (Hg). The graph follows that of a normal metal above the critical temperature T_c. The resistance drops to zero at T_c, which is 4.2 K for mercury.

TABLE 27.3

Critical Temperatures for Various Superconductors

Material	T_c (K)
$HgBa_2Ca_2Cu_3O_8$	134
Tl—Ba—Ca—Cu—O	125
Bi—Sr—Ca—Cu—O	105
$YBa_2Cu_3O_7$	92
Nb_3Ge	23.2
Nb_3Sn	18.05
Nb	9.46
Pb	7.18
Hg	4.15
Sn	3.72
Al	1.19
Zn	0.88

A small permanent magnet levitated above a disk of the superconductor $YBa_2Cu_3O_7$, which is in liquid nitrogen at 77 K.

When the temperature is at or below T_c, the resistivity drops suddenly to zero. This phenomenon was discovered in 1911 by Dutch physicist Heike Kamerlingh-Onnes (1853–1926) as he worked with mercury, which is a superconductor below 4.2 K. Measurements have shown that the resistivities of superconductors below their T_c values are less than $4 \times 10^{-25} \ \Omega \cdot$ m, or approximately 10^{17} times smaller than the resistivity of copper. In practice, these resistivities are considered to be zero.

Today, thousands of superconductors are known, and as Table 27.3 illustrates, the critical temperatures of recently discovered superconductors are substantially higher than initially thought possible. Two kinds of superconductors are recognized. The more recently identified ones are essentially ceramics with high critical temperatures, whereas superconducting materials such as those observed by Kamerlingh-Onnes are metals. If a room-temperature superconductor is ever identified, its effect on technology could be tremendous.

The value of T_c is sensitive to chemical composition, pressure, and molecular structure. Copper, silver, and gold, which are excellent conductors, do not exhibit superconductivity.

One truly remarkable feature of superconductors is that once a current is set up in them, it persists *without any applied potential difference* (because $R = 0$). Steady currents have been observed to persist in superconducting loops for several years with no apparent decay!

An important and useful application of superconductivity is in the development of superconducting magnets, in which the magnitudes of the magnetic field are approximately ten times greater than those produced by the best normal electromagnets. Such superconducting magnets are being considered as a means of storing energy. Superconducting magnets are currently used in medical magnetic resonance imaging, or MRI, units, which produce high-quality images of internal organs without the need for excessive exposure of patients to x-rays or other harmful radiation.

27.6   Electrical Power

In typical electric circuits, energy is transferred from a source such as a battery, to some device such as a lightbulb or a radio receiver. Let's determine an expression that will allow us to calculate the rate of this energy transfer. First, consider the simple circuit in Active Figure 27.11 (page 764), where energy is delivered to a resistor. (Resistors are designated by the circuit symbol ———◇◇◇———.) Because

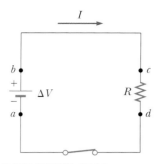

ACTIVE FIGURE 27.11

A circuit consisting of a resistor of resistance R and a battery having a potential difference ΔV across its terminals. Positive charge flows in the clockwise direction.

Sign in at www.thomsonedu.com and go to ThomsonNOW to adjust the battery voltage and the resistance and see the resulting current in the circuit and power delivered to the resistor.

PITFALL PREVENTION 27.5
Charges Do Not Move All the Way Around a Circuit in a Short Time

Because of the very small magnitude of the drift velocity, it might take *hours* for a single electron to make one complete trip around the circuit. In terms of understanding the energy transfer in a circuit, however, it is useful to *imagine* a charge moving all the way around the circuit.

PITFALL PREVENTION 27.6
Misconceptions About Current

Several common misconceptions are associated with current in a circuit like that in Active Figure 27.11. One is that current comes out of one terminal of the battery and is then "used up" as it passes through the resistor, leaving current in only one part of the circuit. The current is actually the same *everywhere* in the circuit. A related misconception has the current coming out of the resistor being smaller than that going in because some of the current is "used up." Yet another misconception has current coming out of both terminals of the battery, in opposite directions, and then "clashing" in the resistor, delivering the energy in this manner. That is not the case; charges flow in the same rotational sense at *all* points in the circuit.

the connecting wires also have resistance, some energy is delivered to the wires and some to the resistor. Unless noted otherwise, we shall assume the resistance of the wires is small compared with the resistance of the circuit element so that the energy delivered to the wires is negligible.

Imagine following a positive quantity of charge Q moving clockwise around the circuit in Active Figure 27.11 from point a through the battery and resistor back to point a. We identify the entire circuit as our system. As the charge moves from a to b through the battery, the electric potential energy of the system *increases* by an amount $Q \Delta V$ while the chemical potential energy in the battery *decreases* by the same amount. (Recall from Eq. 25.3 that $\Delta U = q \Delta V$.) As the charge moves from c to d through the resistor, however, the system *loses* this electric potential energy during collisions of electrons with atoms in the resistor. In this process, the energy is transformed to internal energy corresponding to increased vibrational motion of the atoms in the resistor. Because the resistance of the interconnecting wires is neglected, no energy transformation occurs for paths bc and da. When the charge returns to point a, the net result is that some of the chemical energy in the battery has been delivered to the resistor and resides in the resistor as internal energy associated with molecular vibration.

The resistor is normally in contact with air, so its increased temperature results in a transfer of energy by heat into the air. In addition, the resistor emits thermal radiation, representing another means of escape for the energy. After some time interval has passed, the resistor reaches a constant temperature. At this time, the input of energy from the battery is balanced by the output of energy from the resistor by heat and radiation. Some electrical devices include *heat sinks*[4] connected to parts of the circuit to prevent these parts from reaching dangerously high temperatures. Heat sinks are pieces of metal with many fins. Because the metal's high thermal conductivity provides a rapid transfer of energy by heat away from the hot component and the large number of fins provides a large surface area in contact with the air, energy can transfer by radiation and into the air by heat at a high rate.

Let's now investigate the rate at which the system loses electric potential energy as the charge Q passes through the resistor:

$$\frac{dU}{dt} = \frac{d}{dt}(Q \Delta V) = \frac{dQ}{dt} \Delta V = I \Delta V$$

where I is the current in the circuit. The system regains this potential energy when the charge passes through the battery, at the expense of chemical energy in the battery. The rate at which the system loses potential energy as the charge passes through the resistor is equal to the rate at which the system gains internal energy in the resistor. Therefore, the power $\mathcal{P}$, representing the rate at which energy is delivered to the resistor, is

$$\mathcal{P} = I \Delta V \tag{27.20}$$

We derived this result by considering a battery delivering energy to a resistor. Equation 27.20, however, can be used to calculate the power delivered by a voltage source to *any* device carrying a current I and having a potential difference ΔV between its terminals.

Using Equation 27.20 and $\Delta V = IR$ for a resistor, we can express the power delivered to the resistor in the alternative forms

$$\mathcal{P} = I^2 R = \frac{(\Delta V)^2}{R} \tag{27.21}$$

___
[4] This usage is another misuse of the word *heat* that is ingrained in our common language.

When I is expressed in amperes, ΔV in volts, and R in ohms, the SI unit of power is the watt, as it was in Chapter 8 in our discussion of mechanical power. The process by which power is lost as internal energy in a conductor of resistance R is often called *joule heating*;[5] this transformation is also often referred to as an I^2R loss.

When transporting energy by electricity through power lines such as those shown in the opening photograph for this chapter, you should not assume that the lines have zero resistance. Real power lines do indeed have resistance, and power is delivered to the resistance of these wires. Utility companies seek to minimize the energy transformed to internal energy in the lines and maximize the energy delivered to the consumer. Because $\mathcal{P} = I\,\Delta V$, the same amount of energy can be transported either at high currents and low potential differences or at low currents and high potential differences. Utility companies choose to transport energy at low currents and high potential differences primarily for economic reasons. Copper wire is very expensive, so it is cheaper to use high-resistance wire (that is, wire having a small cross-sectional area; see Eq. 27.10). Therefore, in the expression for the power delivered to a resistor, $\mathcal{P} = I^2R$, the resistance of the wire is fixed at a relatively high value for economic considerations. The I^2R loss can be reduced by keeping the current I as low as possible, which means transferring the energy at a high voltage. In some instances, power is transported at potential differences as great as 765 kV. At the destination of the energy, the potential difference is usually reduced to 4 kV by a device called a *transformer*. Another transformer drops the potential difference to 240 V for use in your home. Of course, each time the potential difference decreases, the current increases by the same factor and the power remains the same. We shall discuss transformers in greater detail in Chapter 33.

Quick Quiz 27.5 For the two lightbulbs shown in Figure 27.12, rank the current values at points a through f from greatest to least.

PITFALL PREVENTION 27.7
Energy Is Not "Dissipated"

In some books, you may see Equation 27.21 described as the power "dissipated in" a resistor, suggesting that energy disappears. Instead, we say energy is "delivered to" a resistor. The notion of *dissipation* arises because a warm resistor expels energy by radiation and heat, so energy delivered by the battery leaves the circuit. (It does not disappear!)

Figure 27.12 (Quick Quiz 27.5) Two lightbulbs connected across the same potential difference.

EXAMPLE 27.4	**Power in an Electric Heater**

An electric heater is constructed by applying a potential difference of 120 V across a Nichrome wire that has a total resistance of 8.00 Ω. Find the current carried by the wire and the power rating of the heater.

SOLUTION

Conceptualize As discussed in Example 27.2, Nichrome wire has high resistivity and is often used for heating elements in toasters, irons, and electric heaters. Therefore, we expect the power delivered to the wire to be relatively high.

Categorize We evaluate the power from Equation 27.21, so we categorize this example as a substitution problem.

Use Equation 27.7 to find the current in the wire:

$$I = \frac{\Delta V}{R} = \frac{120\ \text{V}}{8.00\ \Omega} = \boxed{15.0\ \text{A}}$$

Find the power rating using the expression $\mathcal{P} = I^2R$ from Equation 27.21:

$$\mathcal{P} = I^2R = (15.0\ \text{A})^2(8.00\ \Omega) = 1.80 \times 10^3\ \text{W} = \boxed{1.80\ \text{kW}}$$

What If? What if the heater were accidentally connected to a 240-V supply? (That is difficult to do because the shape and orientation of the metal contacts in 240-V plugs are different from those in 120-V plugs.) How would that affect the current carried by the heater and the power rating of the heater?

---

[5] It is commonly called *joule heating* even though the process of heat does not occur when energy delivered to a resistor appears as internal energy. This is another example of incorrect usage of the word *heat* that has become entrenched in our language.

Answer If the applied potential difference were doubled, Equation 27.7 shows that the current would double. According to Equation 27.21, $\mathcal{P} = (\Delta V)^2/R$, the power would be four times larger.

EXAMPLE 27.5 Linking Electricity and Thermodynamics

An immersion heater must increase the temperature of 1.50 kg of water from 10.0°C to 50.0°C in 10.0 min while operating at 110 V.

(A) What is the required resistance of the heater?

SOLUTION

Conceptualize An immersion heater is a resistor that is inserted into a container of water. As energy is delivered to the immersion heater, raising its temperature, energy leaves the surface of the resistor by heat, going into the water. When the immersion heater reaches a constant temperature, the rate of energy delivered to the resistance by electrical transmission is equal to the rate of energy delivered by heat to the water.

Categorize This example allows us to link our new understanding of power in electricity with our experience with specific heat in thermodynamics (Chapter 20). The water is a nonisolated system. Its internal energy is rising because of energy transferred into the water by heat from the resistor: $\Delta E_{int} = Q$. In our model, we assume the energy that enters the water from the heater remains in the water.

Analyze To simplify the analysis, let's ignore the initial period during which the temperature of the resistor increases and also ignore any variation of resistance with temperature. Therefore, we imagine a constant rate of energy transfer for the entire 10.0 min.

Set the rate of energy delivered to the resistor equal to the rate of energy Q entering the water by heat:

$$\mathcal{P} = \frac{(\Delta V)^2}{R} = \frac{Q}{\Delta t}$$

Use Equation 20.4, $Q = mc\,\Delta T$, to relate the energy input by heat to the resulting temperature change of the water and solve for the resistance:

$$\frac{(\Delta V)^2}{R} = \frac{mc\,\Delta T}{\Delta t} \quad \rightarrow \quad R = \frac{(\Delta V)^2 \Delta t}{mc\,\Delta T}$$

Substitute the values given in the statement of the problem:

$$R = \frac{(110\text{ V})^2(600\text{ s})}{(1.50\text{ kg})(4\,186\text{ J/kg} \cdot {}^\circ\text{C})(50.0°\text{C} - 10.0°\text{C})} = 28.9\ \Omega$$

(B) Estimate the cost of heating the water.

SOLUTION

Multiply the power by the time interval to find the amount of energy transferred:

$$\mathcal{P}\,\Delta t = \frac{(\Delta V)^2}{R}\,\Delta t = \frac{(110\text{ V})^2}{28.9\ \Omega}(10.0\text{ min})\left(\frac{1\text{ h}}{60.0\text{ min}}\right)$$

$$= 69.8\text{ Wh} = 0.069\,8\text{ kWh}$$

Find the cost knowing that energy is purchased at an estimated price of 10¢ per kilowatt-hour:

$$\text{Cost} = (0.069\,8\text{ kWh})(\$0.1/\text{kWh}) = \$0.007 = 0.7¢$$

Finalize The cost to heat the water is very low, less than one cent. In reality, the cost is higher because some energy is transferred from the water into the surroundings by heat and electromagnetic radiation while its temperature is increasing. If you have electrical devices in your home with power ratings on them, use this power rating and an approximate time interval of use to estimate the cost for one use of the device.

Summary

ThomsonNOW Sign in at **www.thomsonedu.com** and go to ThomsonNOW to take a practice test for this chapter.

DEFINITIONS

The electric **current** I in a conductor is defined as

$$I \equiv \frac{dQ}{dt} \tag{27.2}$$

where dQ is the charge that passes through a cross section of the conductor in a time interval dt. The SI unit of current is the **ampere** (A), where 1 A = 1 C/s.

The **current density** J in a conductor is the current per unit area:

$$J \equiv \frac{I}{A} \tag{27.5}$$

The **resistance** R of a conductor is defined as

$$R \equiv \frac{\Delta V}{I} \tag{27.7}$$

where ΔV is the potential difference across it and I is the current it carries. The SI unit of resistance is volts per ampere, which is defined to be 1 **ohm** (Ω); that is, 1 Ω = 1 V/A.

CONCEPTS AND PRINCIPLES

The average current in a conductor is related to the motion of the charge carriers through the relationship

$$I_{\text{avg}} = nqv_d A \tag{27.4}$$

where n is the density of charge carriers, q is the charge on each carrier, v_d is the drift speed, and A is the cross-sectional area of the conductor.

The current density in an ohmic conductor is proportional to the electric field according to the expression

$$J = \sigma E \tag{27.6}$$

The proportionality constant σ is called the **conductivity** of the material of which the conductor is made. The inverse of σ is known as **resistivity** ρ (that is, $\rho = 1/\sigma$). Equation 27.6 is known as **Ohm's law,** and a material is said to obey this law if the ratio of its current density to its applied electric field is a constant that is independent of the applied field.

For a uniform block of material of cross-sectional area A and length ℓ, the resistance over the length ℓ is

$$R = \rho \frac{\ell}{A} \tag{27.10}$$

where ρ is the resistivity of the material.

In a classical model of electrical conduction in metals, the electrons are treated as molecules of a gas. In the absence of an electric field, the average velocity of the electrons is zero. When an electric field is applied, the electrons move (on average) with a **drift velocity** $\vec{\mathbf{v}}_d$ that is opposite the electric field. The drift velocity is given by

$$\vec{\mathbf{v}}_d = \frac{q\vec{\mathbf{E}}}{m_e} \tau \tag{27.13}$$

where q is the electron's charge, m_e is the mass of the electron, and τ is the average time interval between electron–atom collisions. According to this model, the resistivity of the metal is

$$\rho = \frac{m_e}{nq^2\tau} \tag{27.16}$$

where n is the number of free electrons per unit volume.

The resistivity of a conductor varies approximately linearly with temperature according to the expression

$$\rho = \rho_0[1 + \alpha(T - T_0)] \tag{27.17}$$

where ρ_0 is the resistivity at some reference temperature T_0 and α is the **temperature coefficient of resistivity.**

If a potential difference ΔV is maintained across a circuit element, the **power,** or rate at which energy is supplied to the element, is

$$\mathcal{P} = I\,\Delta V \tag{27.20}$$

Because the potential difference across a resistor is given by $\Delta V = IR$, we can express the power delivered to a resistor as

$$\mathcal{P} = I^2 R = \frac{(\Delta V)^2}{R} \tag{27.21}$$

The energy delivered to a resistor by electrical transmission appears in the form of internal energy in the resistor.

Questions

☐ denotes answer available in *Student Solutions Manual/Study Guide;* **O** denotes objective question

1. Newspaper articles often contain a statement such as "10 000 volts of electricity surged through the victim's body." What is wrong with this statement?

2. What factors affect the resistance of a conductor?

3. **O** Two wires A and B with circular cross sections are made of the same metal and have equal lengths, but the resistance of wire A is three times greater than that of wire B. **(i)** What is the ratio of the cross-sectional area of A to that of B? (a) 9 (b) 3 (c) $\sqrt{3}$ (d) 1 (e) $1/\sqrt{3}$ (f) $\frac{1}{3}$ (g) $\frac{1}{9}$ (h) None of these answers is necessarily true. **(ii)** What is the ratio of the radius of A to that of B? Choose from the same possibilities.

4. **O** A metal wire of resistance R is cut into three equal pieces that are then braided together side by side to form a new cable with a length equal to one-third the original length. What is the resistance of this new wire? (a) $R/27$ (b) $R/9$ (c) $R/3$ (d) R (e) $3R$ (f) $9R$ (g) $27R$

5. When the potential difference across a certain conductor is doubled, the current is observed to increase by a factor of three. What can you conclude about the conductor?

6. Use the atomic theory of matter to explain why the resistance of a material should increase as its temperature increases.

7. **O** A current-carrying ohmic metal wire has a cross-sectional area that gradually becomes smaller from one end of the wire to the other. The current has the same value for each section of the wire, so charge does not accumulate at any one point. **(i)** How does the drift speed vary along the wire as the area becomes smaller? (a) It increases. (b) It decreases. (c) It remains constant. **(ii)** How does the resistance per unit length vary along the wire as the area becomes smaller? Choose from the same possibilities.

8. How does the resistance for copper and for silicon change with temperature? Why are the behaviors of these two materials different?

9. Over the time interval after a difference in potential is applied between the ends of a wire, what would happen to the drift velocity of the electrons in a wire and to the current in the wire if the electrons could move freely without resistance through the wire?

10. If charges flow very slowly through a metal, why does it not require several hours for a light to come on when you throw a switch?

11. **O** A cylindrical metal wire at room temperature is carrying electric current between its ends. One end is at potential $V_A = 50$ V, and the other end at potential $V_B = 0$ V. Rank the following actions in terms of the change that each one separately would produce in the current, from the greatest increase to the greatest decrease. In your ranking, note any cases of equality. (a) Make $V_A = 150$ V with $V_B = 0$ V. (b) Make $V_A = 150$ V with $V_B = 100$ V. (c) Adjust V_A to triple the power with which the wire converts electrically transmitted energy into internal energy. (d) Double the radius of the wire. (e) Double the length of the wire. (f) Double the Celsius temperature of the wire. (g) Change the material to an insulator.

12. **O** Two conductors made of the same material are connected across the same potential difference. Conductor A has twice the diameter and twice the length of conductor B. What is the ratio of the power delivered to A to the power delivered to B? (a) 32 (b) 16 (c) 8 (d) 4 (e) 2 (f) 1 (g) $\frac{1}{2}$ (h) $\frac{1}{4}$

13. **O** Two conducting wires A and B of the same length and radius are connected across the same potential difference. Conductor A has twice the resistivity of conductor B. What is the ratio of the power delivered to A to the power delivered to B? (a) 4 (b) 2 (c) $\sqrt{2}$ (d) 1 (e) $1/\sqrt{2}$ (f) $\frac{1}{2}$ (g) $\frac{1}{4}$ (h) None of these answers is necessarily correct.

14. **O** Two lightbulbs both operate from 120 V. One has a power of 25 W and the other 100 W. **(i)** Which lightbulb has higher resistance? (a) The dim 25-W lightbulb does. (b) The bright 100-W lightbulb does. (c) Both are the same. **(ii)** Which lightbulb carries more current? Choose from the same possibilities.

15. **O** Car batteries are often rated in ampere-hours. Does this information designate the amount of (a) current, (b) power, (c) energy, (d) charge, or (e) potential that the battery can supply?

16. If you were to design an electric heater using Nichrome wire as the heating element, what parameters of the wire could you vary to meet a specific power output such as 1 000 W?

Problems

WebAssign The Problems from this chapter may be assigned online in WebAssign.

ThomsonNOW Sign in at **www.thomsonedu.com** and go to ThomsonNOW to assess your understanding of this chapter's topics with additional quizzing and conceptual questions.

1, 2, 3 denotes straightforward, intermediate, challenging; ☐ denotes full solution available in *Student Solutions Manual/Study Guide;* ▲ denotes coached solution with hints available at **www.thomsonedu.com;** denotes developing symbolic reasoning; ● denotes asking for qualitative reasoning; ▪ denotes computer useful in solving problem

2 = intermediate; 3 = challenging; ☐ = SSM/SG; ▲ = ThomsonNOW; = symbolic reasoning; ● = qualitative reasoning

Section 27.1 Electric Current

1. In a particular cathode-ray tube, the measured beam current is 30.0 μA. How many electrons strike the tube screen every 40.0 s?

2. A teapot with a surface area of 700 cm^2 is to be plated with silver. It is attached to the negative electrode of an electrolytic cell containing silver nitrate (Ag$^+$ NO$_3{}^-$). The cell is powered by a 12.0-V battery and has a resistance of 1.80 Ω. Over what time interval does a 0.133-mm layer of silver build up on the teapot? (The density of silver is 10.5×10^3 kg/m^3.)

3. ▲ Suppose the current in a conductor decreases exponentially with time according to the equation $I(t) = I_0 e^{-t/\tau}$, where I_0 is the initial current (at $t = 0$) and τ is a constant having dimensions of time. Consider a fixed observation point within the conductor. (a) How much charge passes this point between $t = 0$ and $t = \tau$? (b) How much charge passes this point between $t = 0$ and $t = 10\tau$? (c) **What If?** How much charge passes this point between $t = 0$ and $t = \infty$?

4. A small sphere that carries a charge q is whirled in a circle at the end of an insulating string. The angular frequency of rotation is ω. What average current does this rotating charge represent?

5. The quantity of charge q (in coulombs) that has passed through a surface of area 2.00 cm^2 varies with time according to the equation $q = 4t^3 + 5t + 6$, where t is in seconds. (a) What is the instantaneous current through the surface at $t = 1.00$ s? (b) What is the value of the current density?

6. An electric current is given by the expression $I(t) = 100 \sin(120\pi t)$, where I is in amperes and t is in seconds. What is the total charge carried by the current from $t = 0$ to $t = \frac{1}{240}$ s?

7. The electron beam emerging from a certain high-energy electron accelerator has a circular cross section of radius 1.00 mm. (a) The beam current is 8.00 μA. Find the current density in the beam assuming it is uniform throughout. (b) The speed of the electrons is so close to the speed of light that their speed can be taken as 300 Mm/s with negligible error. Find the electron density in the beam. (c) Over what time interval does Avogadro's number of electrons emerge from the accelerator?

8. ● Figure P27.8 represents a section of a circular conductor of nonuniform diameter carrying a current of 5.00 A. The radius of cross-section A_1 is 0.400 cm. (a) What is the magnitude of the current density across A_1? (b) The radius at A_2 is larger than the radius at A_1. Is the current at A_2 larger, smaller, or the same? Is the current density larger, smaller, or the same? Assume one of these two quantities is

Figure P27.8

different at A_2 by a factor of 4 from its value at A_1. Specify the current, current density, and radius at A_2.

9. ● A Van de Graaff generator produces a beam of 2.00-MeV deuterons, which are heavy hydrogen nuclei containing a proton and a neutron. (a) If the beam current is 10.0 μA, how far apart are the deuterons? (b) Is the electrical force of repulsion among them a significant factor in beam stability? Explain.

10. An aluminum wire having a cross-sectional area of 4.00×10^{-6} m^2 carries a current of 5.00 A. Find the drift speed of the electrons in the wire. The density of aluminum is 2.70 g/cm^3. Assume each aluminum atom supplies one conduction electron.

Section 27.2 Resistance

11. ▲ A 0.900-V potential difference is maintained across a 1.50-m length of tungsten wire that has a cross-sectional area of 0.600 mm^2. What is the current in the wire?

12. A lightbulb has a resistance of 240 Ω when operating with a potential difference of 120 V across it. What is the current in the lightbulb?

13. Suppose you wish to fabricate a uniform wire from 1.00 g of copper. If the wire is to have a resistance of $R = 0.500 \Omega$ and all the copper is to be used, what must be (a) the length and (b) the diameter of this wire?

14. (a) Make an order-of-magnitude estimate of the resistance between the ends of a rubber band. (b) Make an order-of-magnitude estimate of the resistance between the "heads" and "tails" sides of a penny. In each case, state what quantities you take as data and the values you measure or estimate for them. (c) WARNING: Do not try this part at home! What is the order of magnitude of the current that each would carry if it were connected across a 120-V power supply?

15. A current density of 6.00×10^{-13} A/m^2 exists in the atmosphere at a location where the electric field is 100 V/m. Calculate the electrical conductivity of the Earth's atmosphere in this region.

Section 27.3 A Model for Electrical Conduction

16. ● If the current carried by a conductor is doubled, what happens to (a) the charge carrier density, (b) the current density, (c) the electron drift velocity, and (d) the average time interval between collisions? Explain your answers.

17. ▲ If the magnitude of the drift velocity of free electrons in a copper wire is 7.84×10^{-4} m/s, what is the electric field in the conductor?

Section 27.4 Resistance and Temperature

18. A certain lightbulb has a tungsten filament with a resistance of 19.0 Ω when cold and 140 Ω when hot. Assume the resistivity of tungsten varies linearly with temperature even over the large temperature range involved here. Find the temperature of the hot filament. Assume an initial temperature of 20.0°C.

19. An aluminum wire with a diameter of 0.100 mm has a uniform electric field of 0.200 V/m imposed along its entire length. The temperature of the wire is 50.0°C. Assume one free electron per atom. (a) Use the information in Table 27.2 and determine the resistivity. (b) What is the current density in the wire? (c) What is the total

2 = intermediate;   3 = challenging;   □ = SSM/SG;   ▲ = ThomsonNOW;   = symbolic reasoning;   ● = qualitative reasoning

current in the wire? (d) What is the drift speed of the conduction electrons? (e) What potential difference must exist between the ends of a 2.00-m length of the wire to produce the stated electric field?

20. An engineer needs a resistor with zero overall temperature coefficient of resistance at 20°C. She designs a pair of circular cylinders, one of carbon and one of Nichrome as shown in Figure P27.20. The device must have an overall resistance of $R_1 + R_2 = 10.0\ \Omega$ independent of temperature and a uniform radius of $r = 1.50$ mm. Can she meet the design goal with this method? If so, state what you can determine about the lengths ℓ_1 and ℓ_2 of each segment. You may ignore thermal expansion of the cylinders and assume both are always at the same temperature.

Figure P27.20

21. What is the fractional change in the resistance of an iron filament when its temperature changes from 25.0°C to 50.0°C?

22. **Review problem.** An aluminum rod has a resistance of 1.234 Ω at 20.0°C. Calculate the resistance of the rod at 120°C by accounting for the changes in both the resistivity and the dimensions of the rod.

Section 27.6 Electrical Power

23. A toaster is rated at 600 W when connected to a 120-V source. What current does the toaster carry and what is its resistance?

24. A Van de Graaff generator (see Fig. 25.24) is operating so that the potential difference between the high-potential electrode ⑧ and the charging needles at Ⓐ is 15.0 kV. Calculate the power required to drive the belt against electrical forces at an instant when the effective current delivered to the high-potential electrode is 500 μA.

25. ▲ A well-insulated electric water heater warms 109 kg of water from 20.0°C to 49.0°C in 25.0 min. Find the resistance of its heating element, which is connected across a 220-V potential difference.

26. A 120-V motor has mechanical power output of 2.50 hp. It is 90.0% efficient in converting power that it takes in by electrical transmission into mechanical power. (a) Find the current in the motor. (b) Find the energy delivered to the motor by electrical transmission in 3.00 h of operation. (c) If the electric company charges $0.160/kWh, what does it cost to run the motor for 3.00 h?

27. Suppose a voltage surge produces 140 V for a moment. By what percentage does the power output of a 120-V, 100-W lightbulb increase? Assume its resistance does not change.

28. One rechargeable battery of mass 15.0 g delivers an average current of 18.0 mA to a compact disc player at 1.60 V for 2.40 h before the battery needs to be recharged. The recharger maintains a potential difference of 2.30 V across the battery and delivers a charging current of 13.5 mA for 4.20 h. (a) What is the efficiency of the battery as an energy storage device? (b) How much internal energy is produced in the battery during one charge–discharge cycle? (b) If the battery is surrounded by ideal thermal insulation and has an overall effective specific heat of 975 J/kg · °C, by how much will its temperature increase during the cycle?

29. A 500-W heating coil designed to operate from 110 V is made of Nichrome wire 0.500 mm in diameter. (a) Assuming the resistivity of the Nichrome remains constant at its 20.0°C value, find the length of wire used. (b) **What If?** Now consider the variation of resistivity with temperature. What power is delivered to the coil of part (a) when it is warmed to 1 200°C?

30. A coil of Nichrome wire is 25.0 m long. The wire has a diameter of 0.400 mm and is at 20.0°C. If it carries a current of 0.500 A, what are (a) the magnitude of the electric field in the wire and (b) the power delivered to it? (c) **What If?** If the temperature is increased to 340°C and the potential difference across the wire remains constant, what is the power delivered?

31. Batteries are rated in terms of ampere-hours (A · h). For example, a battery that can produce a current of 2.00 A for 3.00 h is rated at 6.00 A · h. (a) What is the total energy, in kilowatt-hours, stored in a 12.0-V battery rated at 55.0 A · h? (b) At $0.060 0 per kilowatt-hour, what is the value of the electricity produced by this battery?

32. ● Residential building codes typically require the use of 12-gauge copper wire (diameter 0.205 3 cm) for wiring receptacles. Such circuits carry currents as large as 20 A. If a wire of smaller diameter (with a higher gauge number) carried that much current, the wire could rise to a high temperature and cause a fire. (a) Calculate the rate at which internal energy is produced in 1.00 m of 12-gauge copper wire carrying 20.0 A. (b) **What If?** Repeat the calculation for an aluminum wire. Explain whether a 12-gauge aluminum wire would be as safe as a copper wire.

33. An 11.0-W energy-efficient fluorescent lamp is designed to produce the same illumination as a conventional 40.0-W incandescent lightbulb. How much money does the user of the energy-efficient lamp save during 100 h of use? Assume a cost of $0.080 0/kWh for energy from the electric company.

34. We estimate that 270 million plug-in electric clocks are in the United States, approximately one clock for each person. The clocks convert energy at the average rate 2.50 W. To supply this energy, how many metric tons of coal are burned per hour in coal-fired power plants that are, on average, 25.0% efficient? The heat of combustion for coal is 33.0 MJ/kg.

35. Compute the cost per day of operating a lamp that draws a current of 1.70 A from a 110-V line. Assume the cost of energy from the electric company is $0.060 0/kWh.

36. **Review problem.** The heating element of an electric coffee maker operates at 120 V and carries a current of 2.00 A. Assuming the water absorbs all the energy delivered to the resistor, calculate the time interval during which the temperature of 0.500 kg of water rises from room temperature (23.0°C) to the boiling point.

37. A certain toaster has a heating element made of Nichrome wire. When the toaster is first connected to a 120-V source (and the wire is at a temperature of 20.0°C),

the initial current is 1.80 A. The current decreases as the heating element warms up. When the toaster reaches its final operating temperature, the current is 1.53 A. (a) Find the power delivered to the toaster when it is at its operating temperature. (b) What is the final temperature of the heating element?

38. The cost of electricity varies widely through the United States; $0.120/kWh is one typical value. At this unit price, calculate the cost of (a) leaving a 40.0-W porch light on for two weeks while you are on vacation, (b) making a piece of dark toast in 3.00 min with a 970-W toaster, and (c) drying a load of clothes in 40.0 min in a 5 200-W dryer.

39. Make an order-of-magnitude estimate of the cost of one person's routine use of a handheld hair dryer for 1 yr. If you do not use a hair dryer yourself, observe or interview someone who does. State the quantities you estimate and their values.

Additional Problems

40. ● One lightbulb is marked "25 W 120 V," and another is marked "100 W 120 V." These labels mean that each lightbulb has its respective power delivered to it when it is connected to a constant 120-V source. (a) Find the resistance of each lightbulb. (b) During what time interval does 1.00 C pass into the dim lightbulb? Is this charge different upon its exit versus its entry into the lightbulb? Explain. (c) In what time interval does 1.00 J pass into the dim lightbulb? By what mechanisms does this energy enter and exit the lightbulb? Explain. (d) Find the cost of running the dim lightbulb continuously for 30.0 days, assuming the electric company sells its product at $0.070 0 per kWh. What product *does* the electric company sell? What is its price for one SI unit of this quantity?

41. An office worker uses an immersion heater to warm 250 g of water in a light, covered, insulated cup from 20°C to 100°C in 4.00 min. In electrical terms, the heater is a Nichrome resistance wire connected to a 120-V power supply. Specify a diameter and a length that the wire can have. Can it be made from less than 0.5 cm³ of Nichrome? You may assume the wire is at 100°C throughout the time interval.

42. A charge Q is placed on a capacitor of capacitance C. The capacitor is connected into the circuit shown in Figure P27.42, with an open switch, a resistor, and an initially uncharged capacitor of capacitance $3C$. The switch is then closed, and the circuit comes to equilibrium. In terms of Q and C, find (a) the final potential difference between the plates of each capacitor, (b) the charge on each capacitor, and (c) the final energy stored in each capacitor. (d) Find the internal energy appearing in the resistor.

Figure P27.42

43. A more general definition of the temperature coefficient of resistivity is

$$\alpha = \frac{1}{\rho}\frac{d\rho}{dT}$$

where ρ is the resistivity at temperature T. (a) Assuming α is constant, show that

$$\rho = \rho_0 e^{\alpha(T-T_0)}$$

where ρ_0 is the resistivity at temperature T_0. (b) Using the series expansion $e^x \approx 1 + x$ for $x \ll 1$, show that the resistivity is given approximately by the expression $\rho = \rho_0[1 + \alpha(T - T_0)]$ for $\alpha(T - T_0) \ll 1$.

44. A high-voltage transmission line with a diameter of 2.00 cm and a length of 200 km carries a steady current of 1 000 A. If the conductor is copper wire with a free charge density of 8.46×10^{28} electrons/m³, over what time interval does one electron travel the full length of the line?

45. ▲ ● An experiment is conducted to measure the electrical resistivity of Nichrome in the form of wires with different lengths and cross-sectional areas. For one set of measurements, a student uses 30-gauge wire, which has a cross-sectional area of 7.30×10^{-8} m². The student measures the potential difference across the wire and the current in the wire with a voltmeter and an ammeter, respectively. For each set of measurements given in the table taken on wires of three different lengths, calculate the resistance of the wires and the corresponding values of the resistivity. What is the average value of the resistivity? Explain how this value compares with the value given in Table 27.2.

L (m)	ΔV (V)	I (A)	R (Ω)	ρ ($\Omega \cdot$ m)
0.540	5.22	0.500		
1.028	5.82	0.276		
1.543	5.94	0.187		

46. An electric utility company supplies a customer's house from the main power lines (120 V) with two copper wires, each of which is 50.0 m long and has a resistance of 0.108 Ω per 300 m. (a) Find the potential difference at the customer's house for a load current of 110 A. For this load current, find (b) the power delivered to the customer and (c) the rate at which internal energy is produced in the copper wires.

47. A straight, cylindrical wire lying along the x axis has a length of 0.500 m and a diameter of 0.200 mm. It is made of a material described by Ohm's law with a resistivity of $\rho = 4.00 \times 10^{-8}$ $\Omega \cdot$ m. Assume a potential of 4.00 V is maintained at $x = 0$. Also assume $V = 0$ at $x = 0.500$ m. Find (a) the electric field in the wire, (b) the resistance of the wire, (c) the electric current in the wire, and (d) the current density in the wire. State the direction of the electric field and of the current. (e) Show that $E = \rho J$.

48. A straight, cylindrical wire lying along the x axis has a length L and a diameter d. It is made of a material described by Ohm's law with a resistivity ρ. Assume potential V is maintained at $x = 0$. Also assume the potential is zero at $x = L$. In terms of L, d, V, ρ, and physical constants, derive expressions for (a) the electric field in the wire, (b) the resistance of the wire, (c) the electric current in the wire, and (d) the current density in the wire. State the direction of the field and of the current. (e) Prove that $E = \rho J$.

49. An all-electric car (not a hybrid) is designed to run from a bank of 12.0-V batteries with total energy storage of 2.00×10^7 J. (a) If the electric motor draws 8.00 kW, what is the current delivered to the motor? (b) If the electric motor draws 8.00 kW as the car moves at a steady speed of 20.0 m/s, how far can the car travel before it is "out of juice"?

50. ● **Review problem.** When a straight wire is warmed, its resistance is given by $R = R_0 [1 + \alpha (T - T_0)]$ according to Equation 27.19, where α is the temperature coefficient of resistivity. (a) Show that a more precise result, one that includes that the length and area of the wire change when it is warmed, is

$$R = \frac{R_0[1 + \alpha(T - T_0)][1 + \alpha'(T - T_0)]}{[1 + 2\alpha'(T - T_0)]}$$

where α' is the coefficient of linear expansion (see Chapter 19). (b) Explain how these two results compare for a 2.00-m-long copper wire of radius 0.100 mm, first at 20.0°C and then warmed to 100.0°C.

51. The temperature coefficients of resistivity in Table 27.2 were determined at a temperature of 20°C. What would they be at 0°C? Note that the temperature coefficient of resistivity at 20°C satisfies $\rho = \rho_0[1 + \alpha(T - T_0)]$, where ρ_0 is the resistivity of the material at $T_0 = 20°C$. The temperature coefficient of resistivity α' at 0°C must satisfy the expression $\rho = \rho'_0[1 + \alpha' T]$, where ρ'_0 is the resistivity of the material at 0°C.

52. An oceanographer is studying how the ion concentration in seawater depends on depth. She makes a measurement by lowering into the water a pair of concentric metallic cylinders (Fig. P27.52) at the end of a cable and taking data to determine the resistance between these electrodes as a function of depth. The water between the two cylinders forms a cylindrical shell of inner radius r_a, outer radius r_b, and length L much larger than r_b. The scientist applies a potential difference ΔV between the inner and outer surfaces, producing an outward radial current I. Let ρ represent the resistivity of the water. (a) Find the resistance of the water between the cylinders in terms of L, ρ, r_a, and r_b. (b) Express the resistivity of the water in terms of the measured quantities L, r_a, r_b, ΔV, and I.

Figure P27.52

53. ● The strain in a wire can be monitored and computed by measuring the resistance of the wire. Let L_i represent the original length of the wire, A_i its original cross-sectional area, $R_i = \rho L_i / A_i$ the original resistance between

its ends, and $\delta = \Delta L / L_i = (L - L_i)/L_i$ the strain resulting from the application of tension. Assume the resistivity and the volume of the wire do not change as the wire stretches. Show that the resistance between the ends of the wire under strain is given by $R = R_i(1 + 2\delta + \delta^2)$. If the assumptions are precisely true, is this result exact or approximate? Explain your answer.

54. ● In a certain stereo system, each speaker has a resistance of 4.00 Ω. The system is rated at 60.0 W in each channel, and each speaker circuit includes a fuse rated 4.00 A. Is this system adequately protected against overload? Explain your reasoning.

55. ● A close analogy exists between the flow of energy by heat because of a temperature difference (see Section 20.7) and the flow of electric charge because of a potential difference. In a metal, energy dQ and electrical charge dq are both transported by free electrons. Consequently, a good electrical conductor is usually a good thermal conductor as well. Consider a thin conducting slab of thickness dx, area A, and electrical conductivity σ, with a potential difference dV between opposite faces. (a) Show that the current $I = dq/dt$ is given by the equation on the left:

Charge conduction   Thermal conduction

$$\frac{dq}{dt} = \sigma A \left| \frac{dV}{dx} \right| \qquad \frac{dQ}{dt} = kA \left| \frac{dT}{dx} \right| \qquad \textbf{(Eq. 20.15)}$$

In the analogous thermal conduction equation on the right, the rate of energy flow dQ/dt (in SI units of joules per second) is due to a temperature gradient dT/dx, in a material of thermal conductivity k. (b) State analogous rules relating the direction of the electric current to the change in potential and relating the direction of energy flow to the change in temperature.

56. A material of resistivity ρ is formed into the shape of a truncated cone of altitude h as shown in Figure P27.56. The bottom end has radius b, and the top end has radius a. Assume the current is distributed uniformly over any circular cross section of the cone so that the current density does not depend on radial position. (The current density does vary with position along the axis of the cone.) Show that the resistance between the two ends is

$$R = \frac{\rho}{\pi} \left(\frac{h}{ab} \right)$$

Figure P27.56

57. Material with uniform resistivity ρ is formed into a wedge as shown in Figure P27.57. Show that the resistance between face A and face B of this wedge is

$$R = \rho \frac{L}{w(y_2 - y_1)} \ln\left(\frac{y_2}{y_1} \right)$$

Figure P27.57

58. A spherical shell with inner radius r_a and outer radius r_b is formed from a material of resistivity ρ. It carries current radially, with uniform density in all directions. Show that its resistance is

$$R = \frac{\rho}{4\pi}\left(\frac{1}{r_a} - \frac{1}{r_b}\right)$$

59. ● Problems 56, 57, and 58 deal with calculating the resistance between specified surfaces of an oddly shaped resistor. To verify the results experimentally, a potential difference may be applied to the indicated surfaces and the resulting current measured. The resistance can then be calculated from its definition. Describe a method to ensure that the electric potential is uniform over the surface. Explain whether you can then be sure that the current is spread out over the whole surfaces where it enters and exits.

60. The dielectric material between the plates of a parallel-plate capacitor always has some nonzero conductivity σ. Let A represent the area of each plate and d the distance between them. Let κ represent the dielectric constant of the material. (a) Show that the resistance R and the capacitance C of the capacitor are related by

$$RC = \frac{\kappa\epsilon_0}{\sigma}$$

(b) Find the resistance between the plates of a 14.0-nF capacitor with a fused quartz dielectric.

61. **Review problem.** A parallel-plate capacitor consists of square plates of edge length ℓ that are separated by a distance d, where $d \ll \ell$. A potential difference ΔV is maintained between the plates. A material of dielectric constant κ fills half the space between the plates. The

dielectric slab is withdrawn from the capacitor as shown in Figure P27.61. (a) Find the capacitance when the left edge of the dielectric is at a distance x from the center of the capacitor. (b) If the dielectric is removed at a constant speed v, what is the current in the circuit as the dielectric is being withdrawn?

Figure P27.61

62. ☛ The current–voltage characteristic curve for a semiconductor diode as a function of temperature T is given by

$$I = I_0\left(e^{e\,\Delta V/k_B T} - 1\right)$$

Here the first symbol e represents Euler's number, the base of natural logarithms. The second e is the magnitude of the electron charge. The k_B stands for Boltzmann's constant, and T is the absolute temperature. Set up a spreadsheet to calculate I and $R = \Delta V/I$ for $\Delta V = 0.400$ V to 0.600 V in increments of 0.005 V. Assume $I_0 = 1.00$ nA. Plot R versus ΔV for $T = 280$ K, 300 K, and 320 K.

63. Gold is the most ductile of all metals. For example, one gram of gold can be drawn into a wire 2.40 km long. What is the resistance of such a wire at 20°C? You can find the necessary reference information in this textbook.

64. One wire in a high-voltage transmission line carries 1 000 A starting at 700 kV for a distance of 100 mi. If the resistance in the wire is 0.500 Ω/mi, what is the power loss due to the resistance of the wire?

65. The potential difference across the filament of a lamp is maintained at a constant value while equilibrium temperature is being reached. It is observed that the steady-state current in the lamp is only one tenth of the current drawn by the lamp when it is first turned on. If the temperature coefficient of resistivity for the lamp at 20.0°C is 0.004 50 (°C)$^{-1}$ and the resistance increases linearly with increasing temperature, what is the final operating temperature of the filament?

Answers to Quick Quizzes

27.1 (d), (b) = (c), (a). The current in part (d) is equivalent to two positive charges moving to the left. Parts (b) and (c) each represent four positive charges moving in the same direction because negative charges moving to the left are equivalent to positive charges moving to the right. The current in part (a) is equivalent to five positive charges moving to the right.

27.2 (b). The doubling of the radius causes the area A to be four times as large, so Equation 27.10 tells us that the resistance decreases.

27.3 (b). According to Equation 27.7, resistance is the ratio of voltage across a device to current in the device. In Figure 27.7b, a line drawn from the origin to a point on the curve will have a slope equal to $I/\Delta V$, which is the inverse of resistance. As ΔV increases, the slope of the line also increases, so the resistance decreases.

27.4 (a). When the filament is at room temperature, its resistance is low and the current is therefore relatively large. As the filament warms up, its resistance increases and the current decreases. Older lightbulbs often fail just as they are turned on because this large, initial current "spike" produces a rapid temperature increase and mechanical stress on the filament, causing it to break.

27.5 $I_a = I_b > I_c = I_d > I_e = I_f$. The current I_a leaves the positive terminal of the battery and then splits to flow through the two lightbulbs; therefore, $I_a = I_c + I_e$. Looking at Equation 27.21, we see that power rating is inversely related to resistance. Therefore, we know that the current in the 60-W lightbulb is greater than that in the 30-W lightbulb. Because charge does not build up in the lightbulbs, we know that the same amount of charge flowing into a lightbulb from the left must flow out on the right; consequently, $I_c = I_d$ and $I_e = I_f$. The two currents leaving the lightbulbs recombine to form the current back into the battery, $I_f + I_d = I_b$.

In our lives today, we use many items that are generally called "personal electronics," such as MP3 players, cell phones, and digital cameras. These items as well as dozens of others that we use contain electric circuits powered by batteries. In this chapter, we study simple types of circuits and learn how to analyze them. (© Thomson Learning/Charles D. Winters)

28.1 Electromotive Force

28.2 Resistors in Series and Parallel

28.3 Kirchhoff's Rules

28.4 *RC* Circuits

28.5 Electrical Meters

28.6 Household Wiring and Electrical Safety

28 Direct Current Circuits

In this chapter, we analyze simple electric circuits that contain batteries, resistors, and capacitors in various combinations. Some circuits contain resistors that can be combined using simple rules. The analysis of more complicated circuits is simplified using *Kirchhoff's rules*, which follow from the laws of conservation of energy and conservation of electric charge for isolated systems. Most of the circuits analyzed are assumed to be in *steady state*, which means that currents in the circuit are constant in magnitude and direction. A current that is constant in direction is called a *direct current* (DC). We will study *alternating current* (AC), in which the current changes direction periodically, in Chapter 33. Finally, we describe electrical meters for measuring current and potential difference and then discuss electrical circuits in the home.

28.1 Electromotive Force

In Section 27.6, we discussed a circuit in which a battery produces a current. We will generally use a battery as a source of energy for circuits in our discussion. Because the potential difference at the battery terminals is constant in a particular circuit, the current in the circuit is constant in magnitude and direction and is called **direct current.** A battery is called either a *source of electromotive force* or, more commonly, a *source of emf.* (The phrase *electromotive force* is an unfortunate historical term, describing not a force, but rather a potential difference in volts.) **The emf $\mathcal{E}$**

ACTIVE FIGURE 28.1

(a) Circuit diagram of a source of emf $\mathcal{E}$ (in this case, a battery), of internal resistance r, connected to an external resistor of resistance R. (b) Graphical representation showing how the electric potential changes as the circuit in (a) is traversed clockwise.

Sign in at www.thomsonedu.com and go to ThomsonNOW to adjust the emf and resistances r and R and see the effect on the current and on the graph in part (b).

of a battery is the maximum possible voltage the battery can provide between its terminals. You can think of a source of emf as a "charge pump." When an electric potential difference exists between two points, the source moves charges "uphill" from the lower potential to the higher.

We shall generally assume the connecting wires in a circuit have no resistance. The positive terminal of a battery is at a higher potential than the negative terminal. Because a real battery is made of matter, there is resistance to the flow of charge within the battery. This resistance is called **internal resistance** r. For an idealized battery with zero internal resistance, the potential difference across the battery (called its *terminal voltage*) equals its emf. For a real battery, however, the terminal voltage is *not* equal to the emf for a battery in a circuit in which there is a current. To understand why, consider the circuit diagram in Active Figure 28.1a. The battery in this diagram is represented by the dashed rectangle containing an ideal, resistance-free emf $\mathcal{E}$ in series with an internal resistance r. A resistor of resistance R is connected across the terminals of the battery. Now imagine moving through the battery from a to d and measuring the electric potential at various locations. Passing from the negative terminal to the positive terminal, the potential increases by an amount $\mathcal{E}$. As we move through the resistance r, however, the potential *decreases* by an amount Ir, where I is the current in the circuit. Therefore, the terminal voltage of the battery $\Delta V = V_d - V_a$ is

$$\Delta V = \mathcal{E} - Ir \qquad (28.1)$$

From this expression, notice that $\mathcal{E}$ is equivalent to the **open-circuit voltage,** that is, the terminal voltage when the current is zero. The emf is the voltage labeled on a battery; for example, the emf of a D cell is 1.5 V. The actual potential difference between a battery's terminals depends on the current in the battery as described by Equation 28.1.

Active Figure 28.1b is a graphical representation of the changes in electric potential as the circuit is traversed in the clockwise direction. Active Figure 28.1a shows that the terminal voltage ΔV must equal the potential difference across the external resistance R, often called the **load resistance.** The load resistor might be a simple resistive circuit element as in Active Figure 28.1a, or it could be the resistance of some electrical device (such as a toaster, electric heater, or lightbulb) connected to the battery (or, in the case of household devices, to the wall outlet). The resistor represents a *load* on the battery because the battery must supply energy to operate the device containing the resistance. The potential difference across the load resistance is $\Delta V = IR$. Combining this expression with Equation 28.1, we see that

$$\mathcal{E} = IR + Ir \qquad (28.2)$$

Solving for the current gives

$$I = \frac{\mathcal{E}}{R + r} \qquad (28.3)$$

Equation 28.3 shows that the current in this simple circuit depends on both the load resistance R external to the battery and the internal resistance r. If R is much greater than r, as it is in many real-world circuits, we can neglect r.

Multiplying Equation 28.2 by the current I in the circuit gives

$$I\mathcal{E} = I^2R + I^2r \qquad (28.4)$$

Equation 28.4 indicates that because power $\mathcal{P} = I \Delta V$ (see Eq. 27.20), the total power output $I\mathcal{E}$ of the battery is delivered to the external load resistance in the amount I^2R and to the internal resistance in the amount I^2r.

PITFALL PREVENTION 28.1

What Is Constant in a Battery?

It is a common misconception that a battery is a source of constant current. Equation 28.3 shows that is not true. The current in the circuit depends on the resistance R connected to the battery. It is also not true that a battery is a source of constant terminal voltage as shown by Equation 28.1. **A battery is a source of constant emf.**

Quick Quiz 28.1 To maximize the percentage of the power that is delivered from a battery to a device, what should the internal resistance of the battery be? (a) It should be as low as possible. (b) It should be as high as possible. (c) The percentage does not depend on the internal resistance.

| EXAMPLE 28.1 | **Terminal Voltage of a Battery** |

A battery has an emf of 12.0 V and an internal resistance of 0.05 Ω. Its terminals are connected to a load resistance of 3.00 Ω.

(A) Find the current in the circuit and the terminal voltage of the battery.

SOLUTION

Conceptualize   Study Active Figure 28.1a, which shows a circuit consistent with the problem statement. The battery delivers energy to the load resistor.

Categorize   This example involves simple calculations from this section, so we categorize it as a substitution problem.

Use Equation 28.3 to find the current in the circuit:

$$I = \frac{\mathcal{E}}{R + r} = \frac{12.0 \text{ V}}{(3.00 \ \Omega + 0.05 \ \Omega)} = \boxed{3.93 \text{ A}}$$

Use Equation 28.1 to find the terminal voltage:

$$\Delta V = \mathcal{E} - Ir = 12.0 \text{ V} - (3.93 \text{ A})(0.05 \ \Omega) = \boxed{11.8 \text{ V}}$$

To check this result, calculate the voltage across the load resistance R:

$$\Delta V = IR = (3.93 \text{ A})(3.00 \ \Omega) = 11.8 \text{ V}$$

(B) Calculate the power delivered to the load resistor, the power delivered to the internal resistance of the battery, and the power delivered by the battery.

SOLUTION

Use Equation 27.21 to find the power delivered to the load resistor:

$$\mathcal{P}_R = I^2R = (3.93 \text{ A})^2(3.00 \ \Omega) = \boxed{46.3 \text{ W}}$$

Find the power delivered to the internal resistance:

$$\mathcal{P}_r = I^2r = (3.93 \text{ A})^2(0.05 \ \Omega) = \boxed{0.772 \text{ W}}$$

Find the power delivered by the battery by adding these quantities:

$$\mathcal{P} = \mathcal{P}_R + \mathcal{P}_r = 46.3 \text{ W} + 0.772 \text{ W} = \boxed{47.1 \text{ W}}$$

What If?   As a battery ages, its internal resistance increases. Suppose the internal resistance of this battery rises to 2.00 Ω toward the end of its useful life. How does that alter the battery's ability to deliver energy?

Answer   Let's connect the same 3.00-Ω load resistor to the battery.

Find the new current in the battery:

$$I = \frac{\mathcal{E}}{R + r} = \frac{12.0 \text{ V}}{(3.00 \ \Omega + 2.00 \ \Omega)} = 2.40 \text{ A}$$

Find the new terminal voltage:

$$\Delta V = \mathcal{E} - Ir = 12.0 \text{ V} - (2.40 \text{ A})(2.00 \ \Omega) = 7.2 \text{ V}$$

Find the new powers delivered to the load resistor and internal resistance:

$$\mathcal{P}_R = I^2R = (2.40 \text{ A})^2(3.00 \ \Omega) = 17.3 \text{ W}$$

$$\mathcal{P}_r = I^2r = (2.40 \text{ A})^2(2.00 \ \Omega) = 11.5 \text{ W}$$

The terminal voltage is only 60% of the emf. Notice that 40% of the power from the battery is delivered to the internal resistance when r is 2.00 Ω. When r is 0.05 Ω as in part (B), this percentage is only 1.6%. Consequently, even though the emf remains fixed, the increasing internal resistance of the battery significantly reduces the battery's ability to deliver energy.

EXAMPLE 28.2 **Matching the Load**

Find the load resistance R for which the maximum power is delivered to the load resistance in Active Figure 28.1a.

SOLUTION

Conceptualize Think about varying the load resistance in Active Figure 28.1a and the effect on the power delivered to the load resistance. When R is large, there is very little current, so the power I^2R delivered to the load resistor is small. When R is small, the current is large and there is significant loss of power I^2r as energy is delivered to the internal resistance. Therefore, the power delivered to the load resistor is small again. For some intermediate value of the resistance R, the power must maximize.

Categorize The circuit is the same as that in Example 28.1. The load resistance R in this case, however, is a variable.

Figure 28.2 (Example 28.2) Graph of the power $\mathcal{P}$ delivered by a battery to a load resistor of resistance R as a function of R. The power delivered to the resistor is a maximum when the load resistance equals the internal resistance of the battery.

Analyze Find the power delivered to the load resistance using Equation 27.21, with I given by Equation 28.3:

$$(1) \quad \mathcal{P} = I^2R = \frac{\mathcal{E}^2R}{(R+r)^2}$$

Differentiate the power with respect to the load resistance R and set the derivative equal to zero to maximize the power:

$$\frac{d\mathcal{P}}{dR} = \frac{d}{dR}\left[\frac{\mathcal{E}^2R}{(R+r)^2}\right] = \frac{d}{dR}\left[\mathcal{E}^2R(R+r)^{-2}\right] = 0$$

$$\left[\mathcal{E}^2(R+r)^{-2}\right] + \left[\mathcal{E}^2R(-2)(R+r)^{-3}\right] = 0$$

$$\frac{\mathcal{E}^2(R+r)}{(R+r)^3} - \frac{2\mathcal{E}^2R}{(R+r)^3} = \frac{\mathcal{E}^2(r-R)}{(R+r)^3} = 0$$

Solve for R:

$$R = \boxed{r}$$

Finalize To check this result, let's plot $\mathcal{P}$ versus R as in Figure 28.2. The graph shows that $\mathcal{P}$ reaches a maximum value at $R = r$. Equation (1) shows that this maximum value is $\mathcal{P}_{max} = \mathcal{E}^2/4r$.

28.2 Resistors in Series and Parallel

When two or more resistors are connected together as are the lightbulbs in Active Figure 28.3a, they are said to be in a **series combination.** Active Figure 28.3b is the circuit diagram for the lightbulbs, shown as resistors, and the battery. In a series connection, if an amount of charge Q exits resistor R_1, charge Q must also enter the second resistor R_2. Otherwise, charge would accumulate on the wire between the resistors. Therefore, the same amount of charge passes through both resistors in a given time interval and the currents are the same in both resistors:

$$I = I_1 = I_2$$

where I is the current leaving the battery, I_1 is the current in resistor R_1, and I_2 is the current in resistor R_2.

The potential difference applied across the series combination of resistors divides between the resistors. In Active Figure 28.3b, because the voltage drop[1]

---

[1] The term *voltage drop* is synonymous with a decrease in electric potential across a resistor. It is often used by individuals working with electric circuits.

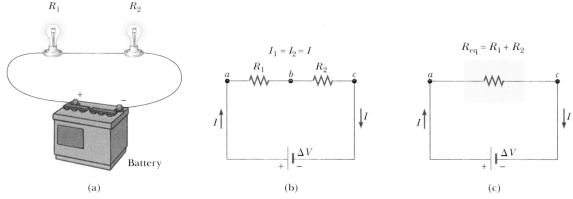

ACTIVE FIGURE 28.3

(a) A series combination of two lightbulbs with resistances R_1 and R_2. (b) Circuit diagram for the two-resistor circuit. The current in R_1 is the same as that in R_2. (c) The resistors replaced with a single resistor having an equivalent resistance $R_{eq} = R_1 + R_2$.

Sign in at www.thomsonedu.com and go to ThomsonNOW to adjust the battery voltage and resistances R_1 and R_2 and see the effect on the currents and voltages in the individual resistors.

from a to b equals $I_1 R_1$ and the voltage drop from b to c equals $I_2 R_2$, the voltage drop from a to c is

$$\Delta V = I_1 R_1 + I_2 R_2$$

The potential difference across the battery is also applied to the **equivalent resistance** R_{eq} in Active Figure 28.3c:

$$\Delta V = I R_{eq}$$

where the equivalent resistance has the same effect on the circuit as the series combination because it results in the same current I in the battery. Combining these equations for ΔV, we see that we can replace the two resistors in series with a single equivalent resistance whose value is the *sum* of the individual resistances:

$$\Delta V = I R_{eq} = I_1 R_1 + I_2 R_2 \quad \rightarrow \quad R_{eq} = R_1 + R_2 \qquad \textbf{(28.5)}$$

where we have canceled the currents I, I_1, and I_2 because they are all the same.

The equivalent resistance of three or more resistors connected in series is

$$R_{eq} = R_1 + R_2 + R_3 + \cdots \qquad \textbf{(28.6)}$$

◄ The equivalent resistance of a series combination of resistors

This relationship indicates that **the equivalent resistance of a series combination of resistors is the numerical sum of the individual resistances and is always greater than any individual resistance.**

Looking back at Equation 28.3, we see that the denominator is the simple algebraic sum of the external and internal resistances. That is consistent with the internal and external resistances being in series in Active Figure 28.1a.

If the filament of one lightbulb in Active Figure 28.3 were to fail, the circuit would no longer be complete (resulting in an open-circuit condition) and the second lightbulb would also go out. This fact is a general feature of a series circuit: if one device in the series creates an open circuit, all devices are inoperative.

PITFALL PREVENTION 28.2
Lightbulbs Don't Burn

We will describe the end of the life of a lightbulb by saying *the filament fails* rather than by saying the lightbulb "burns out." The word *burn* suggests a combustion process, which is not what occurs in a lightbulb. The failure of a lightbulb results from the slow sublimation of tungsten from the very hot filament over the life of the lightbulb. The filament eventually becomes very thin because of this process. The mechanical stress from a sudden temperature increase when the lightbulb is turned on causes the thin filament to break.

Quick Quiz 28.2   With the switch in the circuit of Figure 28.4a (page 780) closed, there is no current in R_2 because the current has an alternate zero-resistance path through the switch. There is current in R_1, and this current is measured with the ammeter (a device for measuring current) at the bottom of the circuit. If the switch is opened (Fig. 28.4b), there is current in R_2. What happens to the reading on the ammeter when the switch is opened? (a) The reading goes up. (b) The reading goes down. (c) The reading does not change.

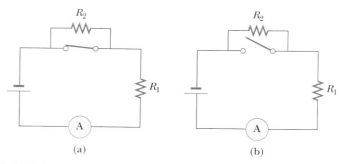

Figure 28.4   (Quick Quiz 28.2) What happens when the switch is opened?

Now consider two resistors in a **parallel combination** as shown in Active Figure 28.5. Notice that both resistors are connected directly across the terminals of the battery. Therefore, the potential differences across the resistors are the same:

$$\Delta V = \Delta V_1 = \Delta V_2$$

where ΔV is the terminal voltage of the battery.

When charges reach point a in Active Figure 28.5b, they split into two parts, with some going toward R_1 and the rest going toward R_2. A **junction** is any such point in a circuit where a current can split. This split results in less current in each individual resistor than the current leaving the battery. Because electric charge is conserved, the current I that enters point a must equal the total current leaving that point:

$$I = I_1 + I_2$$

where I_1 is the current in R_1 and I_2 is the current in R_2.

The current in the **equivalent resistance** R_{eq} in Active Figure 28.5c is

$$I = \frac{\Delta V}{R_{eq}}$$

where the equivalent resistance has the same effect on the circuit as the two resistors in parallel; that is, the equivalent resistance draws the same current I from the

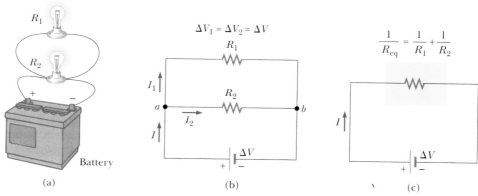

ACTIVE FIGURE 28.5

(a) A parallel combination of two lightbulbs with resistances R_1 and R_2. (b) Circuit diagram for the two-resistor circuit. The potential difference across R_1 is the same as that across R_2. (c) The resistors replaced with a single resistor having an equivalent resistance given by Equation 28.7.

Sign in at www.thomsonedu.com and go to ThomsonNOW to adjust the battery voltage and resistances R_1 and R_2 and see the effect on the currents and voltages in the individual resistors.

battery. Combining these equations for I, we see that the equivalent resistance of two resistors in parallel is given by

$$I = \frac{\Delta V}{R_{eq}} = \frac{\Delta V_1}{R_1} + \frac{\Delta V_2}{R_2} \quad \rightarrow \quad \frac{1}{R_{eq}} = \frac{1}{R_1} + \frac{1}{R_2} \qquad (28.7)$$

where we have canceled ΔV, ΔV_1, and ΔV_2 because they are all the same.

An extension of this analysis to three or more resistors in parallel gives

$$\frac{1}{R_{eq}} = \frac{1}{R_1} + \frac{1}{R_2} + \frac{1}{R_3} + \cdots \qquad (28.8)$$

◀ The equivalent resistance of a parallel combination of resistors

This expression shows that **the inverse of the equivalent resistance of two or more resistors in a parallel combination is equal to the sum of the inverses of the individual resistances. Furthermore, the equivalent resistance is always less than the smallest resistance in the group.**

Household circuits are always wired such that the appliances are connected in parallel. Each device operates independently of the others so that if one is switched off, the others remain on. In addition, in this type of connection, all the devices operate on the same voltage.

Let's consider two examples of practical applications of series and parallel circuits. Figure 28.6 illustrates how a three-way lightbulb is constructed to provide three levels of light intensity.[2] The socket of the lamp is equipped with a three-way switch for selecting different light intensities. The lightbulb contains two filaments. When the lamp is connected to a 120-V source, one filament receives 100 W of power and the other receives 75 W. The three light intensities are made possible by applying the 120 V to one filament alone, to the other filament alone, or to the two filaments in parallel. When switch S_1 is closed and switch S_2 is opened, current exists only in the 75-W filament. When switch S_1 is open and switch S_2 is closed, current exists only in the 100-W filament. When both switches are closed, current exists in both filaments and the total power is 175 W.

If the filaments were connected in series and one of them were to break, no charges could pass through the lightbulb and it would not glow, regardless of the switch position. If, however, the filaments were connected in parallel and one of them (for example, the 75-W filament) were to break, the lightbulb would continue to glow in two of the switch positions because current exists in the other (100-W) filament.

As a second example, consider strings of lights that are used for many ornamental purposes such as decorating Christmas trees. Over the years, both parallel and series connections have been used for strings of lights. Because series-wired lightbulbs operate with less energy per bulb and at a lower temperature, they are safer than parallel-wired lightbulbs for indoor Christmas-tree use. If, however, the filament of a single lightbulb in a series-wired string were to fail (or if the lightbulb were removed from its socket), all the lights on the string would go out. The popularity of series-wired light strings diminished because troubleshooting a failed lightbulb is a tedious, time-consuming chore that involves trial-and-error substitution of a good lightbulb in each socket along the string until the defective one is found.

In a parallel-wired string, each lightbulb operates at 120 V. By design, the lightbulbs are brighter and hotter than those on a series-wired string. As a result, they are inherently more dangerous (more likely to start a fire, for instance), but if one lightbulb in a parallel-wired string fails or is removed, the rest of the lightbulbs continue to glow.

To prevent the failure of one lightbulb from causing the entire string to go out, a new design was developed for so-called miniature lights wired in series. When the filament breaks in one of these miniature lightbulbs, the break in the filament

Figure 28.6   A three-way lightbulb.

---

[2] The three-way lightbulb and other household devices actually operate on alternating current (AC), to be introduced in Chapter 33.

Figure 28.7 (a) Schematic diagram of a modern "miniature" holiday lightbulb, with a jumper connection to provide a current path if the filament breaks. When the filament is intact, charges flow in the filament. (b) A holiday lightbulb with a broken filament. In this case, charges flow in the jumper connection. (c) A Christmas-tree lightbulb.

represents the largest resistance in the series, much larger than that of the intact filaments. As a result, most of the applied 120 V appears across the lightbulb with the broken filament. Inside the lightbulb, a small jumper loop covered by an insulating material is wrapped around the filament leads. When the filament fails and 120 V appears across the lightbulb, an arc burns the insulation on the jumper and connects the filament leads. This connection now completes the circuit through the lightbulb even though its filament is no longer active (Fig. 28.7).

When a lightbulb fails, the resistance across its terminals is reduced to almost zero because of the alternate jumper connection mentioned in the preceding paragraph. All the other lightbulbs not only stay on, but they glow more brightly because the total resistance of the string is reduced and consequently the current in each lightbulb increases. Each lightbulb operates at a slightly higher temperature than before. As more lightbulbs fail, the current keeps rising, the filament of each lightbulb operates at a higher temperature, and the lifetime of the lightbulb is reduced. For this reason, you should check for failed (nonglowing) lightbulbs in such a series-wired string and replace them as soon as possible, thereby maximizing the lifetimes of all the lightbulbs.

Quick Quiz 28.3 With the switch in the circuit of Figure 28.8a open, there is no current in R_2. There is current in R_1, however, and it is measured with the ammeter at the right side of the circuit. If the switch is closed (Fig. 28.8b), there is current in R_2. What happens to the reading on the ammeter when the switch is closed? (a) The reading increases. (b) The reading decreases. (c) The reading does not change.

Quick Quiz 28.4 Consider the following choices: (a) increases, (b) decreases, (c) remains the same. From these choices, choose the best answer for the following situations. **(i)** In Active Figure 28.3, a third resistor is added in series with the first two. What happens to the current in the battery? **(ii)** What happens to the terminal voltage of the battery? **(iii)** In Active Figure 28.5, a third resistor is added in parallel with the first two. What happens to the current in the battery? **(iv)** What happens to the terminal voltage of the battery?

Figure 28.8 (Quick Quiz 28.3) What happens when the switch is closed?

CONCEPTUAL EXAMPLE 28.3 **Landscape Lights**

A homeowner wishes to install low-voltage landscape lighting in his back yard. To save money, he purchases inexpensive 18-gauge cable, which has a relatively high resistance per unit length. This cable consists of two side-by-side wires separated by insulation, like the cord on an appliance. He runs a 200-foot length of this cable from the power supply to the farthest point at which he plans to position a light fixture. He attaches light fixtures across the two wires on the cable at 10-foot intervals so that the light fixtures are in parallel. Because of the cable's resistance, the brightness of the lightbulbs in the fixtures is not as desired. Which of the following problems does the homeowner have? (a) All the lightbulbs glow equally less brightly than they would if lower-resistance cable had been used. (b) The brightness of the lightbulbs decreases as you move farther from the power supply.

Figure 28.9 (Conceptual Example 28.3) The circuit diagram for a set of landscape light fixtures connected in parallel across the two wires of a two-wire cable. The horizontal resistors represent resistance in the wires of the cable. The vertical resistors represent the light fixtures.

SOLUTION

A circuit diagram for the system appears in Figure 28.9. The horizontal resistors with letter subscripts (such as R_A) represent the resistance of the wires in the cable between the light fixtures, and the vertical resistors with number subscripts (such as R_1) represent the resistance of the light fixtures themselves. Part of the terminal voltage of the power supply is dropped across resistors R_A and R_B. Therefore, the voltage across light fixture R_1 is less than the terminal voltage. There is a further voltage drop across resistors R_C and R_D. Consequently, the voltage across light fixture R_2 is smaller than that across R_1. This pattern continues down the line of light fixtures, so the correct choice is (b). Each successive light fixture has a smaller voltage across it and glows less brightly than the one before.

EXAMPLE 28.4   Find the Equivalent Resistance

Four resistors are connected as shown in Figure 28.10a.

(A) Find the equivalent resistance between points a and c.

SOLUTION

Conceptualize Imagine charges flowing into this combination from the left. All charges must pass through the first two resistors, but the charges split into two different paths when encountering the combination of the 6.0-Ω and the 3.0-Ω resistors.

Categorize Because of the simple nature of the combination of resistors in Figure 28.10, we categorize this example as one for which we can use the rules for series and parallel combinations of resistors.

Analyze The combination of resistors can be reduced in steps as shown in Figure 28.10.

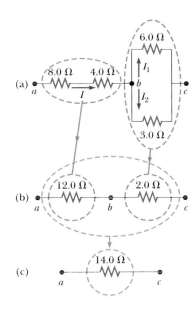

Figure 28.10 (Example 28.4) The original network of resistors is reduced to a single equivalent resistance.

Find the equivalent resistance between a and b of the 8.0-Ω and 4.0-Ω resistors, which are in series:

$$R_{eq} = 8.0 \ \Omega + 4.0 \ \Omega = 12.0 \ \Omega$$

Find the equivalent resistance between b and c of the 6.0-Ω and 3.0-Ω resistors, which are in parallel:

$$\frac{1}{R_{eq}} = \frac{1}{6.0 \ \Omega} + \frac{1}{3.0 \ \Omega} = \frac{3}{6.0 \ \Omega}$$

$$R_{eq} = \frac{6.0 \ \Omega}{3} = 2.0 \ \Omega$$

The circuit of equivalent resistances now looks like Figure 28.10b. Find the equivalent resistance from a to c:

$$R_{eq} = 12.0 \ \Omega + 2.0 \ \Omega = \boxed{14.0 \ \Omega}$$

This resistance is that of the single equivalent resistor in Figure 28.10c.

(B) What is the current in each resistor if a potential difference of 42 V is maintained between a and c?

SOLUTION

The currents in the 8.0-Ω and 4.0-Ω resistors are the same because they are in series. In addition, they carry the same current that would exist in the 14.0-Ω equivalent resistor subject to the 42-V potential difference.

Use Equation 27.7 ($R = \Delta V/I$) and the result from part (A) to find the current in the 8.0-Ω and 4.0-Ω resistors:

$$I = \frac{\Delta V_{ac}}{R_{eq}} = \frac{42\ V}{14.0\ \Omega} = \boxed{3.0\ A}$$

Set the voltages across the resistors in parallel in Figure 28.10a equal to find a relationship between the currents:

$$\Delta V_1 = \Delta V_2 \ \rightarrow\ (6.0\ \Omega)I_1 = (3.0\ \Omega)I_2 \ \rightarrow\ I_2 = 2I_1$$

Use $I_1 + I_2 = 3.0$ A to find I_1:

$$I_1 + I_2 = 3.0\ A \ \rightarrow\ I_1 + 2I_1 = 3.0\ A \ \rightarrow\ I_1 = \boxed{1.0\ A}$$

Find I_2:

$$I_2 = 2I_1 = 2(1.0\ A) = \boxed{2.0\ A}$$

Finalize   As a final check of our results, note that $\Delta V_{bc} = (6.0\ \Omega)I_1 = (3.0\ \Omega)I_2 = 6.0$ V and $\Delta V_{ab} = (12.0\ \Omega)I = 36$ V; therefore, $\Delta V_{ac} = \Delta V_{ab} + \Delta V_{bc} = 42$ V, as it must.

EXAMPLE 28.5   Three Resistors in Parallel

Three resistors are connected in parallel as shown in Figure 28.11a. A potential difference of 18.0 V is maintained between points a and b.

(A) Calculate the equivalent resistance of the circuit.

SOLUTION

Conceptualize   Figure 28.11a shows that we are dealing with a simple parallel combination of three resistors.

Categorize   Because the three resistors are connected in parallel, we can use Equation 28.8 to evaluate the equivalent resistance.

Figure 28.11   (Example 28.5) (a) Three resistors connected in parallel. The voltage across each resistor is 18.0 V. (b) Another circuit with three resistors and a battery. Is it equivalent to the circuit in (a)?

Analyze   Use Equation 28.8 to find R_{eq}:

$$\frac{1}{R_{eq}} = \frac{1}{3.00\ \Omega} + \frac{1}{6.00\ \Omega} + \frac{1}{9.00\ \Omega} = \frac{11.0}{18.0\ \Omega}$$

$$R_{eq} = \frac{18.0\ \Omega}{11.0} = \boxed{1.64\ \Omega}$$

(B) Find the current in each resistor.

SOLUTION

The potential difference across each resistor is 18.0 V. Apply the relationship $\Delta V = IR$ to find the currents:

$$I_1 = \frac{\Delta V}{R_1} = \frac{18.0\ V}{3.00\ \Omega} = 6.00\ A$$

$$I_2 = \frac{\Delta V}{R_2} = \frac{18.0\ V}{6.00\ \Omega} = 3.00\ A$$

$$I_3 = \frac{\Delta V}{R_3} = \frac{18.0\ V}{9.00\ \Omega} = 2.00\ A$$

(C) Calculate the power delivered to each resistor and the total power delivered to the combination of resistors.

SOLUTION

Apply the relationship $\mathcal{P} = I^2R$ to each resistor using the currents calculated in part (B):

3.00-Ω: $\mathcal{P}_1 = I_1{}^2R_1 = (6.00 \text{ A})^2(3.00 \text{ }\Omega) = $ $\boxed{108 \text{ W}}$

6.00-Ω: $\mathcal{P}_2 = I_2{}^2R_2 = (3.00 \text{ A})^2(6.00 \text{ }\Omega) = $ $\boxed{54.0 \text{ W}}$

9.00-Ω: $\mathcal{P}_3 = I_3{}^2R_3 = (2.00 \text{ A})^2(9.00 \text{ }\Omega) = $ $\boxed{36.0 \text{ W}}$

Finalize   Part (C) shows that the smallest resistor receives the most power. Summing the three quantities gives a total power of 198 W. We could have calculated this final result from part (A) by considering the equivalent resistance as follows: $\mathcal{P} = (\Delta V)^2/R_{eq} = (18.0 \text{ V})^2/ 1.64 \text{ }\Omega = 198 \text{ W}$.

What If?   What if the circuit were as shown in Figure 28.11b instead of as in Figure 28.11a? How would that affect the calculation?

Answer   There would be no effect on the calculation. The physical placement of the battery is not important. In Figure 28.11b, the battery still maintains a potential difference of 18.0 V between points *a* and *b*, so the two circuits in the figure are electrically identical.

28.3   Kirchhoff's Rules

As we saw in the preceding section, combinations of resistors can be simplified and analyzed using the expression $\Delta V = IR$ and the rules for series and parallel combinations of resistors. Very often, however, it is not possible to reduce a circuit to a single loop. The procedure for analyzing more complex circuits is made possible by using the two following principles, called **Kirchhoff's rules.**

1. **Junction rule.** At any junction, the sum of the currents must equal zero:

$$\sum_{\text{junction}} I = 0 \qquad\qquad (28.9)$$

2. **Loop rule.** The sum of the potential differences across all elements around any closed circuit loop must be zero:

$$\sum_{\text{closed loop}} \Delta V = 0 \qquad\qquad (28.10)$$

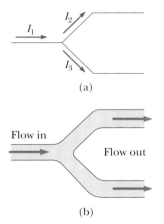

(a)

(b)

Figure 28.12   (a) Kirchhoff's junction rule. Conservation of charge requires that all charges entering a junction must leave that junction. Therefore, $I_1 - I_2 - I_3 = 0$. (b) A mechanical analog of the junction rule. The amount of water flowing out of the branches on the right must equal the amount flowing into the single branch on the left.

Kirchhoff's first rule is a statement of conservation of electric charge. All charges that enter a given point in a circuit must leave that point because charge cannot build up at a point. Currents directed into the junction are entered into the junction rule as $+I$, whereas currents directed out of a junction are entered as $-I$. Applying this rule to the junction in Figure 28.12a gives

$$I_1 - I_2 - I_3 = 0$$

Figure 28.12b represents a mechanical analog of this situation, in which water flows through a branched pipe having no leaks. Because water does not build up anywhere in the pipe, the flow rate into the pipe on the left equals the total flow rate out of the two branches on the right.

Kirchhoff's second rule follows from the law of conservation of energy. Let's imagine moving a charge around a closed loop of a circuit. When the charge returns to the starting point, the charge-circuit system must have the same total energy as it had before the charge was moved. The sum of the increases in energy

(a) $\Delta V = -IR$

(b) $\Delta V = +IR$

(c) $\Delta V = +\mathcal{E}$

(d) $\Delta V = -\mathcal{E}$

Figure 28.13 Rules for determining the potential differences across a resistor and a battery. (The battery is assumed to have no internal resistance.) Each circuit element is traversed from *a* to *b*, left to right.

as the charge passes through some circuit elements must equal the sum of the decreases in energy as it passes through other elements. The potential energy decreases whenever the charge moves through a potential drop $-IR$ across a resistor or whenever it moves in the reverse direction through a source of emf. The potential energy increases whenever the charge passes through a battery from the negative terminal to the positive terminal.

When applying Kirchhoff's second rule, imagine *traveling* around the loop and consider changes in *electric potential* rather than the changes in *potential energy* described in the preceding paragraph. Imagine traveling through the circuit elements in Figure 28.13 toward the right. The following sign conventions apply when using the second rule:

- Charges move from the high-potential end of a resistor toward the low-potential end, so if a resistor is traversed in the direction of the current, the potential difference ΔV across the resistor is $-IR$ (Fig. 28.13a).
- If a resistor is traversed in the direction *opposite* the current, the potential difference ΔV across the resistor is $+IR$ (Fig. 28.13b).
- If a source of emf (assumed to have zero internal resistance) is traversed in the direction of the emf (from negative to positive), the potential difference ΔV is $+\mathcal{E}$ (Fig. 28.13c).
- If a source of emf (assumed to have zero internal resistance) is traversed in the direction opposite the emf (from positive to negative), the potential difference ΔV is $-\mathcal{E}$ (Fig. 28.13d).

There are limits on the numbers of times you can usefully apply Kirchhoff's rules in analyzing a circuit. You can use the junction rule as often as you need as long as you include in it a current that has not been used in a preceding junction-rule equation. In general, the number of times you can use the junction rule is one fewer than the number of junction points in the circuit. You can apply the loop rule as often as needed as long as a new circuit element (resistor or battery) or a new current appears in each new equation. In general, **to solve a particular circuit problem, the number of independent equations you need to obtain from the two rules equals the number of unknown currents.**

Complex networks containing many loops and junctions generate great numbers of independent linear equations and a correspondingly great number of unknowns. Such situations can be handled formally through the use of matrix algebra. Computer software can also be used to solve for the unknowns.

The following examples illustrate how to use Kirchhoff's rules. In all cases, it is assumed the circuits have reached steady-state conditions; in other words, the currents in the various branches are constant. **Any capacitor acts as an open branch in a circuit;** that is, the current in the branch containing the capacitor is zero under steady-state conditions.

GUSTAV KIRCHHOFF
German Physicist (1824–1887)
Kirchhoff, a professor at Heidelberg, and Robert Bunsen invented the spectroscope and founded the science of spectroscopy, which we shall study in Chapter 42. They discovered the elements cesium and rubidium and invented astronomical spectroscopy.

PROBLEM-SOLVING STRATEGY   **Kirchhoff's Rules**

The following procedure is recommended for solving problems that involve circuits that cannot be reduced by the rules for combining resistors in series or parallel.

1. *Conceptualize.* Study the circuit diagram and make sure you recognize all elements in the circuit. Identify the polarity of each battery and try to imagine the directions in which the current would exist in the batteries.

2. *Categorize.* Determine whether the circuit can be reduced by means of combining series and parallel resistors. If so, use the techniques of Section 28.2. If not, apply Kirchhoff's rules according to the *Analyze* step below.

3. *Analyze.* Assign labels to all known quantities and symbols to all unknown quantities. You must assign *directions* to the currents in each part of the circuit.

Although the assignment of current directions is arbitrary, you must adhere *rigorously* to the directions you assign when you apply Kirchhoff's rules.

Apply the junction rule (Kirchhoff's first rule) to all junctions in the circuit except one. Now apply the loop rule (Kirchhoff's second rule) to as many loops in the circuit as are needed to obtain, in combination with the equations from the junction rule, as many equations as there are unknowns. To apply this rule, you must choose a direction in which to travel around the loop (either clockwise or counterclockwise) and correctly identify the change in potential as you cross each element. Be careful with signs!

Solve the equations simultaneously for the unknown quantities.

4. *Finalize.* Check your numerical answers for consistency. Do not be alarmed if any of the resulting currents have a negative value. That only means you have guessed the direction of that current incorrectly, but *its magnitude will be correct.*

EXAMPLE 28.6 | **A Single-Loop Circuit**

A single-loop circuit contains two resistors and two batteries as shown in Figure 28.14. (Neglect the internal resistances of the batteries.) Find the current in the circuit.

SOLUTION

Conceptualize   Figure 28.14 shows the polarities of the batteries and a guess at the direction of the current.

Categorize   We do not need Kirchhoff's rules to analyze this simple circuit, but let's use them anyway simply to see how they are applied. There are no junctions in this single-loop circuit; therefore, the current is the same in all elements.

Figure 28.14   (Example 28.6) A series circuit containing two batteries and two resistors, where the polarities of the batteries are in opposition.

Analyze   Let's assume the current is clockwise as shown in Figure 28.14. Traversing the circuit in the clockwise direction, starting at a, we see that $a \rightarrow b$ represents a potential difference of $+\mathcal{E}_1$, $b \rightarrow c$ represents a potential difference of $-IR_1$, $c \rightarrow d$ represents a potential difference of $-\mathcal{E}_2$, and $d \rightarrow a$ represents a potential difference of $-IR_2$.

Apply Kirchhoff's loop rule to the single loop in the circuit:

$$\sum \Delta V = 0 \quad \rightarrow \quad \mathcal{E}_1 - IR_1 - \mathcal{E}_2 - IR_2 = 0$$

Solve for I and use the values given in Figure 28.14:

$$(1) \quad I = \frac{\mathcal{E}_1 - \mathcal{E}_2}{R_1 + R_2} = \frac{6.0\text{ V} - 12\text{ V}}{8.0\ \Omega + 10\ \Omega} = \boxed{-0.33\text{ A}}$$

Finalize   The negative sign for I indicates that the direction of the current is opposite the assumed direction. The emfs in the numerator subtract because the batteries in Figure 28.14 have opposite polarities. The resistances in the denominator add because the two resistors are in series.

What If?   What if the polarity of the 12.0-V battery were reversed? How would that affect the circuit?

Answer   Although we could repeat the Kirchhoff's rules calculation, let's instead examine Equation (1) and modify it accordingly. Because the polarities of the two batteries are now in the same direction, the signs of $\mathcal{E}_1$ and $\mathcal{E}_2$ are the same and Equation (1) becomes

$$I = \frac{\mathcal{E}_1 + \mathcal{E}_2}{R_1 + R_2} = \frac{6.0\text{ V} + 12\text{ V}}{8.0\ \Omega + 10\ \Omega} = 1.0\text{ A}$$

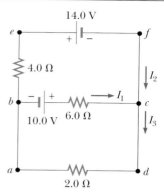

| EXAMPLE 28.7 | **A Multiloop Circuit** |

Find the currents I_1, I_2, and I_3 in the circuit shown in Figure 28.15.

SOLUTION

Conceptualize   We cannot simplify the circuit by the rules associated with combining resistances in series and in parallel. (If the 10.0-V battery were not present, we could reduce the remaining circuit with series and parallel combinations.)

Categorize   Because the circuit is not a simple series and parallel combination of resistances, this problem is one in which we must use Kirchhoff's rules.

Analyze   We arbitrarily choose the directions of the currents as labeled in Figure 28.15.

Figure 28.15   (Example 28.7) A circuit containing different branches.

Apply Kirchhoff's junction rule to junction c:

$$(1) \quad I_1 + I_2 - I_3 = 0$$

We now have one equation with three unknowns: I_1, I_2, and I_3. There are three loops in the circuit: *abcda, befcb,* and *aefda*. We need only two loop equations to determine the unknown currents. (The third loop equation would give no new information.) Let's choose to traverse these loops in the clockwise direction. Apply Kirchhoff's loop rule to loops *abcda* and *befcb*:

abcda: $(2) \quad 10.0\text{ V} - (6.0\ \Omega)I_1 - (2.0\ \Omega)I_3 = 0$

befcb: $\quad -(4.0\ \Omega)I_2 - 14.0\text{ V} + (6.0\ \Omega)I_1 - 10.0\text{ V} = 0$

$$(3) \quad -24.0\text{ V} + (6.0\ \Omega)I_1 - (4.0\ \Omega)I_2 = 0$$

Solve Equation (1) for I_3 and substitute into Equation (2):

$$10.0\text{ V} - (6.0\ \Omega)I_1 - (2.0\ \Omega)(I_1 + I_2) = 0$$

$$(4) \quad 10.0\text{ V} - (8.0\ \Omega)I_1 - (2.0\ \Omega)I_2 = 0$$

Multiply each term in Equation (3) by 4 and each term in Equation (4) by 3:

$$(5) \quad -96.0\text{ V} + (24.0\ \Omega)I_1 - (16.0\ \Omega)I_2 = 0$$

$$(6) \quad 30.0\text{ V} - (24.0\ \Omega)I_1 - (6.0\ \Omega)I_2 = 0$$

Add Equation (6) to Equation (5) to eliminate I_1 and find I_2:

$$-66.0\text{ V} - (22.0\ \Omega)I_2 = 0$$

$$I_2 = \boxed{-3.0\text{ A}}$$

Use this value of I_2 in Equation (3) to find I_1:

$$-24.0\text{ V} + (6.0\ \Omega)I_1 - (4.0\ \Omega)(-3.0\text{ A}) = 0$$

$$-24.0\text{ V} + (6.0\ \Omega)I_1 + 12.0\text{ V} = 0$$

$$I_1 = \boxed{2.0\text{ A}}$$

Use Equation (1) to find I_3:

$$I_3 = I_1 + I_2 = 2.0\text{ A} - 3.0\text{ A} = \boxed{-1.0\text{ A}}$$

Finalize   Because our values for I_2 and I_3 are negative, the directions of these currents are opposite those indicated in Figure 28.15. The numerical values for the currents are correct. Despite the incorrect direction, we *must* continue to use these negative values in subsequent calculations because our equations were established with our original choice of direction. What would have happened had we left the current directions as labeled in Figure 28.15 but traversed the loops in the opposite direction?

28.4   *RC* Circuits

So far, we have analyzed direct current circuits in which the current is constant. In DC circuits containing capacitors, the current is always in the same direction but may vary in time. A circuit containing a series combination of a resistor and a capacitor is called an ***RC* circuit.**

Charging a Capacitor

Active Figure 28.16 shows a simple series *RC* circuit. Let's assume the capacitor in this circuit is initially uncharged. There is no current while the switch is open (Active Fig. 28.16a). If the switch is thrown to position *a* at *t* = 0 (Active Fig. 28.16b), however, charge begins to flow, setting up a current in the circuit, and the capacitor begins to charge.[3] Notice that during charging, charges do not jump across the capacitor plates because the gap between the plates represents an open circuit. Instead, charge is transferred between each plate and its connecting wires due to the electric field established in the wires by the battery until the capacitor is fully charged. As the plates are being charged, the potential difference across the capacitor increases. The value of the maximum charge on the plates depends on the voltage of the battery. Once the maximum charge is reached, the current in the circuit is zero because the potential difference across the capacitor matches that supplied by the battery.

To analyze this circuit quantitatively, let's apply Kirchhoff's loop rule to the circuit after the switch is thrown to position *a*. Traversing the loop in Active Figure 28.16b clockwise gives

$$\mathcal{E} - \frac{q}{C} - IR = 0 \qquad (28.11)$$

where q/C is the potential difference across the capacitor and IR is the potential difference across the resistor. We have used the sign conventions discussed earlier for the signs on $\mathcal{E}$ and IR. The capacitor is traversed in the direction from the positive plate to the negative plate, which represents a decrease in potential. Therefore, we use a negative sign for this potential difference in Equation 28.11. Note that q and I are *instantaneous* values that depend on time (as opposed to steady-state values) as the capacitor is being charged.

We can use Equation 28.11 to find the initial current in the circuit and the maximum charge on the capacitor. At the instant the switch is thrown to position *a* (*t* = 0), the charge on the capacitor is zero. Equation 28.11 shows that the initial current I_i in the circuit is a maximum and is given by

$$I_i = \frac{\mathcal{E}}{R} \quad \text{(current at } t = 0) \qquad (28.12)$$

At this time, the potential difference from the battery terminals appears entirely across the resistor. Later, when the capacitor is charged to its maximum value Q, charges cease to flow, the current in the circuit is zero, and the potential difference from the battery terminals appears entirely across the capacitor. Substituting $I = 0$ into Equation 28.11 gives the maximum charge on the capacitor:

$$Q = C\mathcal{E} \quad \text{(maximum charge)} \qquad (28.13)$$

To determine analytical expressions for the time dependence of the charge and current, we must solve Equation 28.11, a single equation containing two variables q and I. The current in all parts of the series circuit must be the same. Therefore, the current in the resistance R must be the same as the current between each capacitor plate and the wire connected to it. This current is equal to the time rate of change of the charge on the capacitor plates. Therefore, we substitute $I = dq/dt$ into Equation 28.11 and rearrange the equation:

$$\frac{dq}{dt} = \frac{\mathcal{E}}{R} - \frac{q}{RC}$$

ACTIVE FIGURE 28.16

(a) A capacitor in series with a resistor, switch, and battery. (b) When the switch is thrown to position *a*, the capacitor begins to charge up. (c) When the switch is thrown to position *b*, the capacitor discharges.

Sign in at www.thomsonedu.com and go to ThomsonNOW to adjust the values of *R* and *C* and see the effect on the charging and discharging of the capacitor.

---

[3] In previous discussions of capacitors, we assumed a steady-state situation, in which no current was present in any branch of the circuit containing a capacitor. Now we are considering the case *before* the steady-state condition is realized; in this situation, charges are moving and a current exists in the wires connected to the capacitor.

To find an expression for q, we solve this separable differential equation as follows. First combine the terms on the right-hand side:

$$\frac{dq}{dt} = \frac{C\mathcal{E}}{RC} - \frac{q}{RC} = -\frac{q - C\mathcal{E}}{RC}$$

Multiply this equation by dt and divide by $q - C\mathcal{E}$:

$$\frac{dq}{q - C\mathcal{E}} = -\frac{1}{RC}\,dt$$

Integrate this expression, using $q = 0$ at $t = 0$:

$$\int_0^q \frac{dq}{q - C\mathcal{E}} = -\frac{1}{RC}\int_0^t dt$$

$$\ln\left(\frac{q - C\mathcal{E}}{-C\mathcal{E}}\right) = -\frac{t}{RC}$$

From the definition of the natural logarithm, we can write this expression as

$$q(t) = C\mathcal{E}(1 - e^{-t/RC}) = Q(1 - e^{-t/RC}) \tag{28.14}$$

◀ Charge as a function of time for a capacitor being charged

where e is the base of the natural logarithm and we have made the substitution from Equation 28.13.

We can find an expression for the charging current by differentiating Equation 28.14 with respect to time. Using $I = dq/dt$, we find that

◀ Current as a function of time for a capacitor being charged

$$I(t) = \frac{\mathcal{E}}{R}\,e^{-t/RC} \tag{28.15}$$

Plots of capacitor charge and circuit current versus time are shown in Figure 28.17. Notice that the charge is zero at $t = 0$ and approaches the maximum value $C\mathcal{E}$ as $t \to \infty$. The current has its maximum value $I_i = \mathcal{E}/R$ at $t = 0$ and decays exponentially to zero as $t \to \infty$. The quantity RC, which appears in the exponents of Equations 28.14 and 28.15, is called the **time constant** τ of the circuit:

$$\tau = RC \tag{28.16}$$

The time constant represents the time interval during which the current decreases to $1/e$ of its initial value; that is, after a time interval τ, the current decreases to $I = e^{-1}I_i = 0.368I_i$. After a time interval 2τ, the current decreases to $I = e^{-2}I_i = 0.135I_i$, and so forth. Likewise, in a time interval τ, the charge increases from zero to $C\mathcal{E}[1 - e^{-1}] = 0.632C\mathcal{E}$.

The following dimensional analysis shows that τ has units of time:

$$[\tau] = [RC] = \left[\left(\frac{\Delta V}{I}\right)\left(\frac{Q}{\Delta V}\right)\right] = \left[\frac{Q}{Q/\Delta t}\right] = [\Delta t] = \mathrm{T}$$

Because $\tau = RC$ has units of time, the combination t/RC is dimensionless, as it must be to be an exponent of e in Equations 28.14 and 28.15.

The energy output of the battery as the capacitor is fully charged is $Q\mathcal{E} = C\mathcal{E}^2$. After the capacitor is fully charged, the energy stored in the capacitor is $\frac{1}{2}Q\mathcal{E} = \frac{1}{2}C\mathcal{E}^2$, which is only half the energy output of the battery. It is left as a problem (Problem 52) to show that the remaining half of the energy supplied by the battery appears as internal energy in the resistor.

Figure 28.17   (a) Plot of capacitor charge versus time for the circuit shown in Active Figure 28.16. After a time interval equal to one time constant τ has passed, the charge is 63.2% of the maximum value $C\mathcal{E}$. The charge approaches its maximum value as t approaches infinity. (b) Plot of current versus time for the circuit shown in Active Figure 28.16. The current has its maximum value $I_i = \mathcal{E}/R$ at $t = 0$ and decays to zero exponentially as t approaches infinity. After a time interval equal to one time constant τ has passed, the current is 36.8% of its initial value.

Discharging a Capacitor

Imagine that the capacitor in Active Figure 28.16b is completely charged. A potential difference Q/C exists across the capacitor and there is zero potential difference across the resistor because $I = 0$. If the switch is now thrown to position b at $t = 0$ (Active Fig. 28.16c), the capacitor begins to discharge through the resistor.

At some time t during the discharge, the current in the circuit is I and the charge on the capacitor is q. The circuit in Active Figure 28.16c is the same as the circuit in Active Figure 28.16b except for the absence of the battery. Therefore, we eliminate the emf $\mathcal{E}$ from Equation 28.11 to obtain the appropriate loop equation for the circuit in Active Figure 28.16c:

$$-\frac{q}{C} - IR = 0 \tag{28.17}$$

When we substitute $I = dq/dt$ into this expression, it becomes

$$-R\frac{dq}{dt} = \frac{q}{C}$$

$$\frac{dq}{q} = -\frac{1}{RC} dt$$

Integrating this expression using $q = Q$ at $t = 0$ gives

$$\int_Q^q \frac{dq}{q} = -\frac{1}{RC} \int_0^t dt$$

$$\ln\left(\frac{q}{Q}\right) = -\frac{t}{RC}$$

$$q(t) = Qe^{-t/RC} \tag{28.18}$$

◀ Charge as a function of time for a discharging capacitor

Differentiating Equation 28.18 with respect to time gives the instantaneous current as a function of time:

$$I(t) = -\frac{Q}{RC} e^{-t/RC} \tag{28.19}$$

◀ Current as a function of time for a discharging capacitor

where $Q/RC = I_i$ is the initial current. The negative sign indicates that as the capacitor discharges, the current direction is opposite its direction when the capacitor was being charged. (Compare the current directions in Figs. 28.16b and 28.16c.) Both the charge on the capacitor and the current decay exponentially at a rate characterized by the time constant $\tau = RC$.

Quick Quiz 28.5   Consider the circuit in Figure 28.18 and assume the battery has no internal resistance. **(i)** Just after the switch is closed, what is the current in the battery? (a) 0   (b) $\mathcal{E}/2R$ (c) $2\mathcal{E}/R$ (d) $\mathcal{E}/R$ (e) impossible to determine **(ii)** After a very long time, what is the current in the battery? Choose from the same choices.

Figure 28.18   (Quick Quiz 28.5) How does the current vary after the switch is closed?

| CONCEPTUAL EXAMPLE 28.8 | **Intermittent Windshield Wipers** |

Many automobiles are equipped with windshield wipers that can operate intermittently during a light rainfall. How does the operation of such wipers depend on the charging and discharging of a capacitor?

SOLUTION

The wipers are part of an RC circuit whose time constant can be varied by selecting different values of R through a multiposition switch. As the voltage across the capacitor increases, the capacitor reaches a point at which it discharges and triggers the wipers. The circuit then begins another charging cycle. The time interval between the individual sweeps of the wipers is determined by the value of the time constant.

EXAMPLE 28.9 **Charging a Capacitor in an *RC* Circuit**

An uncharged capacitor and a resistor are connected in series to a battery as shown in Active Figure 28.16, where $\mathcal{E} = 12.0$ V, $C = 5.00$ μF, and $R = 8.00 \times 10^5$ Ω. The switch is thrown to position a. Find the time constant of the circuit, the maximum charge on the capacitor, the maximum current in the circuit, and the charge and current as functions of time.

SOLUTION

Conceptualize Study Active Figure 28.16 and imagine throwing the switch to position a as shown in Active Figure 28.16b. Upon doing so, the capacitor begins to charge.

Categorize We evaluate our results using equations developed in this section, so we categorize this example as a substitution problem.

Evaluate the time constant of the circuit from Equation 28.16:

$$\tau = RC = (8.00 \times 10^5 \; \Omega)(5.00 \times 10^{-6} \; \text{F}) = \boxed{4.00 \; \text{s}}$$

Evaluate the maximum charge on the capacitor from Equation 28.13:

$$Q = C\mathcal{E} = (5.00 \; \mu\text{F})(12.0 \; \text{V}) = \boxed{60.0 \; \mu\text{C}}$$

Evaluate the maximum current in the circuit from Equation 28.12:

$$I_i = \frac{\mathcal{E}}{R} = \frac{12.0 \; \text{V}}{8.00 \times 10^5 \; \Omega} = \boxed{15.0 \; \mu\text{A}}$$

Use these values in Equations 28.14 and 28.15 to find the charge and current as functions of time:

$$q(t) = \boxed{(60.0\mu\text{C})(1 - e^{-t/4.00\,\text{s}})}$$

$$I(t) = \boxed{(15.0 \; \mu\text{A})e^{-t/4.00\,\text{s}}}$$

EXAMPLE 28.10 **Discharging a Capacitor in an *RC* Circuit**

Consider a capacitor of capacitance C that is being discharged through a resistor of resistance R as shown in Active Figure 28.16c.

(A) After how many time constants is the charge on the capacitor one-fourth its initial value?

SOLUTION

Conceptualize Study Active Figure 28.16 and imagine throwing the switch to position b as shown in Active Figure 28.16c. Upon doing so, the capacitor begins to discharge.

Categorize We categorize the example as one involving a discharging capacitor and use the appropriate equations.

Analyze Substitute $q(t) = Q/4$ into Equation 28.18:

$$\frac{Q}{4} = Qe^{-t/RC}$$

$$\tfrac{1}{4} = e^{-t/RC}$$

Take the logarithm of both sides of the equation and solve for t:

$$-\ln 4 = -\frac{t}{RC}$$

$$t = RC \ln 4 = 1.39RC = \boxed{1.39\tau}$$

(B) The energy stored in the capacitor decreases with time as the capacitor discharges. After how many time constants is this stored energy one-fourth its initial value?

SOLUTION

Use Equations 26.11 and 28.18 to express the energy stored in the capacitor at any time t:

$$(1) \quad U(t) = \frac{q^2}{2C} = \frac{Q^2}{2C} e^{-2t/RC}$$

Substitute $U(t) = \frac{1}{4}(Q^2/2C)$ into Equation (1):

$$\frac{1}{4}\frac{Q^2}{2C} = \frac{Q^2}{2C} e^{-2t/RC}$$

$$\frac{1}{4} = e^{-2t/RC}$$

Take the logarithm of both sides of the equation and solve for t:

$$-\ln 4 = -\frac{2t}{RC}$$

$$t = \frac{1}{2}RC \ln 4 = 0.693RC = \boxed{0.693\tau}$$

Finalize Notice that because the energy depends on the square of the charge, the energy in the capacitor drops more rapidly than the charge on the capacitor.

What If? What if you want to describe the circuit in terms of the time interval required for the charge to fall to one-half its original value rather than by the time constant τ? That would give a parameter for the circuit called its *half-life* $t_{1/2}$. How is the half-life related to the time constant?

Answer In one half-life, the charge falls from Q to $Q/2$. Therefore, from Equation 28.18,

$$\frac{Q}{2} = Qe^{-t_{1/2}/RC} \quad \rightarrow \quad \frac{1}{2} = e^{-t_{1/2}/RC}$$

which leads to

$$t_{1/2} = 0.693\tau$$

The concept of half-life will be important to us when we study nuclear decay in Chapter 44. The radioactive decay of an unstable sample behaves in a mathematically similar manner to a discharging capacitor in an RC circuit.

EXAMPLE 28.11   **Energy Delivered to a Resistor**

A 5.00-μF capacitor is charged to a potential difference of 800 V and then discharged through a resistor. How much energy is delivered to the resistor in the time interval required to fully discharge the capacitor?

SOLUTION

Conceptualize In Example 28.10, we considered the energy decrease in a discharging capacitor to a value of one-fourth of the initial energy. In this example, the capacitor fully discharges.

Categorize We solve this example using two approaches. The first approach is to model the circuit as an isolated system. Because energy in an isolated system is conserved, the initial electric potential energy U_C stored in the capacitor is transformed into internal energy $E_{int} = E_R$ in the resistor. The second approach is to model the resistor as a nonisolated system. Energy enters the resistor by electrical transmission from the capacitor, causing an increase in the resistor's internal energy.

Analyze We begin with the isolated system approach.

Write the appropriate reduction of the conservation of energy equation, Equation 8.2:

$$\Delta U + \Delta E_{int} = 0$$

Substitute the initial and final values of the energies:

$$(0 - U_C) + (E_{int} - 0) = 0 \quad \rightarrow \quad E_R = U_C$$

Use Equation 26.11 for the electric potential energy in the capacitor:

$$E_R = \tfrac{1}{2}C\mathcal{E}^2$$

Substitute numerical values:

$$E_R = \tfrac{1}{2}(5.00 \times 10^{-6}\ \text{F})(800\ \text{V})^2 = \boxed{1.60\ \text{J}}$$

The second approach, which is more difficult but perhaps more instructive, is to note that as the capacitor discharges through the resistor, the rate at which energy is delivered to the resistor by electrical transmission is I^2R, where I is the instantaneous current given by Equation 28.19.

Evaluate the energy delivered to the resistor by integrating the power over all time because it takes an infinite time interval for the capacitor to completely discharge:

$$\mathcal{P} = \frac{dE}{dt} \quad \rightarrow \quad E_R = \int_0^\infty \mathcal{P}\, dt$$

Substitute for the power delivered to the resistor:

$$E_R = \int_0^\infty I^2 R\, dt$$

Substitute for the current from Equation 28.19:

$$E_R = \int_0^\infty \left(-\frac{Q}{RC}e^{-t/RC}\right)^2 R\, dt = \frac{Q^2}{RC^2}\int_0^\infty e^{-2t/RC}\, dt = \frac{\mathcal{E}^2}{R}\int_0^\infty e^{-2t/RC}\, dt$$

Substitute the value of the integral, which is $RC/2$ (see Problem 30):

$$E_R = \frac{\mathcal{E}^2}{R}\left(\frac{RC}{2}\right) = \tfrac{1}{2}C\mathcal{E}^2$$

Finalize   This result agrees with that obtained using the isolated system approach, as it must. We can use this second approach to find the total energy delivered to the resistor at *any* time after the switch is closed by simply replacing the upper limit in the integral with that specific value of *t*.

28.5   Electrical Meters

In this section, we discuss various electrical meters that are used in the electrical and electronics industries to make electrical measurements.

The Galvanometer

The **galvanometer** is the main component in analog meters for measuring current and voltage. (Many analog meters are still in use, although digital meters, which operate on a different principle, are currently more common.) One type, called the *D'Arsonval galvanometer*, consists of a coil of wire mounted so that it is free to rotate on a pivot in a magnetic field provided by a permanent magnet. The deflection of a needle attached to the coil is proportional to the current in the galvanometer. Once the instrument is properly calibrated, it can be used in conjunction with other circuit elements to measure either currents or potential differences.

Figure 28.19 Current can be measured with an ammeter connected in series with the elements in which the measurement of a current is desired. An ideal ammeter has zero resistance.

The Ammeter

A device that measures current is called an **ammeter.** Because the charges constituting the current to be measured must pass directly through the ammeter, the ammeter must be connected in series with other elements in the circuit as shown in Figure 28.19. When using an ammeter to measure direct currents, you must connect it so that charges enter the instrument at the positive terminal and exit at the negative terminal.

Ideally, an ammeter should have zero resistance so that the current being measured is not altered. In the circuit shown in Figure 28.19, this condition requires that the resistance of the ammeter be much less than $R_1 + R_2$. Because any ammeter always has some internal resistance, the presence of the ammeter in the circuit slightly reduces the current from the value it would have in the meter's absence.

A typical off-the-shelf galvanometer is often not suitable for use as an ammeter primarily because it has a resistance of about 60 Ω. An ammeter resistance this great considerably alters the current in a circuit. Consider the following example. The current in a simple series circuit containing a 3-V battery and a 3-Ω resistor is 1 A. If you insert a 60-Ω galvanometer in this circuit to measure the current, the total resistance becomes 63 Ω and the current is reduced to 0.048 A!

A second factor that limits the use of a galvanometer as an ammeter is that a typical galvanometer gives a full-scale deflection for currents on the order of 1 mA or less. Consequently, such a galvanometer cannot be used directly to measure currents greater than this value. It can, however, be converted to a useful ammeter by placing a shunt resistor R_p in parallel with the galvanometer as shown in Active Figure 28.20. The value of R_p must be much less than the galvanometer resistance so that most of the current to be measured is directed to the shunt resistor.

ACTIVE FIGURE 28.20

A galvanometer is represented here by its internal resistance of 60 Ω. When a galvanometer is to be used as an ammeter, a shunt resistor R_p is connected in parallel with the galvanometer.

Sign in at www.thomsonedu.com and go to ThomsonNOW to predict the value of R_p needed to cause full-scale deflection in the circuit of Figure 28.19 and test your result.

The Voltmeter

A device that measures potential difference is called a **voltmeter.** The potential difference between any two points in a circuit can be measured by attaching the terminals of the voltmeter between these points without breaking the circuit as shown in Figure 28.21. The potential difference across resistor R_2 is measured by connecting the voltmeter in parallel with R_2. Again, it is necessary to observe the instrument's polarity. The voltmeter's positive terminal must be connected to the end of the resistor that is at the higher potential, and its negative terminal must be connected to the end of the resistor at the lower potential.

An ideal voltmeter has infinite resistance so that no current exists in it. In Figure 28.21, this condition requires that the voltmeter have a resistance much greater than R_2. In practice, corrections should be made for the known resistance of the voltmeter if this condition is not met.

A galvanometer can also be used as a voltmeter by adding an external resistor R_s in series with it as shown in Active Figure 28.22. In this case, the external resistor must have a value much greater than the resistance of the galvanometer to ensure that the galvanometer does not significantly alter the voltage being measured.

A digital multimeter is used to measure a voltage across a circuit element.

Figure 28.21 The potential difference across a resistor can be measured with a voltmeter connected in parallel with the resistor. An ideal voltmeter has infinite resistance.

ACTIVE FIGURE 28.22

When the galvanometer is used as a voltmeter, a resistor R_s is connected in series with the galvanometer.

Sign in at www.thomsonedu.com and go to ThomsonNOW to predict the value of R_s needed to cause full-scale deflection in the circuit of Figure 28.21 and test your result.

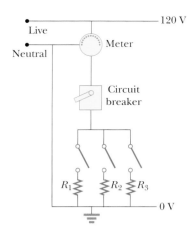

Figure 28.23 Wiring diagram for a household circuit. The resistances represent appliances or other electrical devices that operate with an applied voltage of 120 V.

28.6 Household Wiring and Electrical Safety

Many considerations are important in the design of an electrical system of a home that will provide adequate electrical service for the occupants while maximizing their safety. We discuss some aspects of a home electrical system in this section.

Household Wiring

Household circuits represent a practical application of some of the ideas presented in this chapter. In our world of electrical appliances, it is useful to understand the power requirements and limitations of conventional electrical systems and the safety measures that prevent accidents.

In a conventional installation, the utility company distributes electric power to individual homes by means of a pair of wires, with each home connected in parallel to these wires. One wire is called the *live wire*[4] as illustrated in Figure 28.23, and the other is called the *neutral wire*. The neutral wire is grounded; that is, its electric potential is taken to be zero. The potential difference between the live and neutral wires is approximately 120 V. This voltage alternates in time, and the potential of the live wire oscillates relative to ground. Much of what we have learned so far for the constant-emf situation (direct current) can also be applied to the alternating current that power companies supply to businesses and households. (Alternating voltage and current are discussed in Chapter 33.)

To record a household's energy consumption, a meter is connected in series with the live wire entering the house. After the meter, the wire splits so that there are several separate circuits in parallel distributed throughout the house. Each circuit contains a circuit breaker (or, in older installations, a fuse). The wire and circuit breaker for each circuit are carefully selected to meet the current requirements for that circuit. If a circuit is to carry currents as large as 30 A, a heavy wire and an appropriate circuit breaker must be selected to handle this current. A circuit used to power only lamps and small appliances often requires only 20 A. Each circuit has its own circuit breaker to provide protection for that part of the entire electrical system of the house.

As an example, consider a circuit in which a toaster oven, a microwave oven, and a coffee maker are connected (corresponding to R_1, R_2, and R_3 in Fig. 28.23). We can calculate the current in each appliance by using the expression $\mathcal{P} = I\,\Delta V$. The toaster oven, rated at 1 000 W, draws a current of 1 000 W/120 V = 8.33 A. The microwave oven, rated at 1 300 W, draws 10.8 A, and the coffee maker, rated at 800 W, draws 6.67 A. When the three appliances are operated simultaneously, they draw a total current of 25.8 A. Therefore, the circuit must be wired to handle at least this much current. If the rating of the circuit breaker protecting the circuit is too small—say, 20 A—the breaker will be tripped when the third appliance is turned on, preventing all three appliances from operating. To avoid this situation, the toaster oven and coffee maker can be operated on one 20-A circuit and the microwave oven on a separate 20-A circuit.

Many heavy-duty appliances such as electric ranges and clothes dryers require 240 V for their operation. The power company supplies this voltage by providing a third wire that is 120 V below ground potential (Fig. 28.24). The potential difference between this live wire and the other live wire (which is 120 V above ground potential) is 240 V. An appliance that operates from a 240-V line requires half as much current compared with operating it at 120 V; therefore, smaller wires can be used in the higher-voltage circuit without overheating.

Electrical Safety

When the live wire of an electrical outlet is connected directly to ground, the circuit is completed and a *short-circuit condition* exists. A short circuit occurs when

(a)

(b)

Figure 28.24 (a) An outlet for connection to a 240-V supply. (b) The connections for each of the openings in a 240-V outlet.

---

[4] *Live wire* is a common expression for a conductor whose electric potential is above or below ground potential.

almost zero resistance exists between two points at different potentials, and the result is a very large current. When that happens accidentally, a properly operating circuit breaker opens the circuit and no damage is done. A person in contact with ground, however, can be electrocuted by touching the live wire of a frayed cord or other exposed conductor. An exceptionally effective (and dangerous!) ground contact is made when the person either touches a water pipe (normally at ground potential) or stands on the ground with wet feet. The latter situation represents effective ground contact because normal, nondistilled water is a conductor due to the large number of ions associated with impurities. This situation should be avoided at all cost.

Electric shock can result in fatal burns or can cause the muscles of vital organs such as the heart to malfunction. The degree of damage to the body depends on the magnitude of the current, the length of time it acts, the part of the body touched by the live wire, and the part of the body in which the current exists. Currents of 5 mA or less cause a sensation of shock, but ordinarily do little or no damage. If the current is larger than about 10 mA, the muscles contract and the person may be unable to release the live wire. If the body carries a current of about 100 mA for only a few seconds, the result can be fatal. Such a large current paralyzes the respiratory muscles and prevents breathing. In some cases, currents of approximately 1 A can produce serious (and sometimes fatal) burns. In practice, no contact with live wires is regarded as safe whenever the voltage is greater than 24 V.

Many 120-V outlets are designed to accept a three-pronged power cord. (This feature is required in all new electrical installations.) One of these prongs is the live wire at a nominal potential of 120 V. The second is the neutral wire, nominally at 0 V, which carries current to ground. Figure 28.25a shows a connection to an electric drill with only these two wires. If the live wire accidentally makes contact with the casing of the electric drill (which can occur if the wire insulation wears off), current can be carried to ground by way of the person, resulting in an electric shock. The third wire in a three-pronged power cord, the round prong, is a

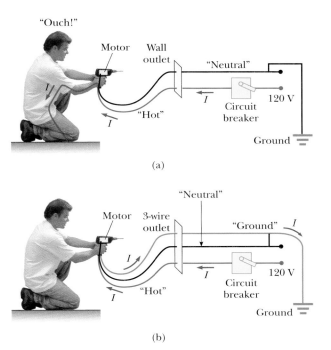

(a)

(b)

Figure 28.25 (a) A diagram of the circuit for an electric drill with only two connecting wires. The normal current path is from the live wire through the motor connections and back to ground through the neutral wire. In the situation shown, the live wire has come into contact with the drill case. As a result, the person holding the drill acts as a current path to ground and receives an electric shock. (b) This shock can be avoided by connecting the drill case to ground through a third ground wire. In this situation, the drill case remains at ground potential and no current exists in the person.

safety ground wire that normally carries no current. It is both grounded and connected directly to the casing of the appliance. If the live wire is accidentally shorted to the casing in this situation, most of the current takes the low-resistance path through the appliance to ground as shown in Figure 28.25b.

Special power outlets called *ground-fault interrupters,* or GFIs, are used in kitchens, bathrooms, basements, exterior outlets, and other hazardous areas of homes. These devices are designed to protect persons from electric shock by sensing small currents (< 5 mA) leaking to ground. (The principle of their operation is described in Chapter 31.) When an excessive leakage current is detected, the current is shut off in less than 1 ms.

Summary

DEFINITIONS

The **emf** of a battery is equal to the voltage across its terminals when the current is zero. That is, the emf is equivalent to the **open-circuit voltage** of the battery.

CONCEPTS AND PRINCIPLES

The **equivalent resistance** of a set of resistors connected in a **series combination** is

$$R_{eq} = R_1 + R_2 + R_3 + \cdots \quad \textbf{(28.6)}$$

The **equivalent resistance** of a set of resistors connected in a **parallel combination** is found from the relationship

$$\frac{1}{R_{eq}} = \frac{1}{R_1} + \frac{1}{R_2} + \frac{1}{R_3} + \cdots \quad \textbf{(28.8)}$$

Circuits involving more than one loop are conveniently analyzed with the use of **Kirchhoff's rules:**

1. **Junction rule.** At any junction, the sum of the currents must equal zero:

$$\sum_{\text{junction}} I = 0 \quad \textbf{(28.9)}$$

2. **Loop rule.** The sum of the potential differences across all elements around any circuit loop must be zero:

$$\sum_{\text{closed loop}} \Delta V = 0 \quad \textbf{(28.10)}$$

When a resistor is traversed in the direction of the current, the potential difference ΔV across the resistor is $-IR$. When a resistor is traversed in the direction opposite the current, $\Delta V = +IR$. When a source of emf is traversed in the direction of the emf (negative terminal to positive terminal), the potential difference is $+\mathcal{E}$. When a source of emf is traversed opposite the emf (positive to negative), the potential difference is $-\mathcal{E}$.

If a capacitor is charged with a battery through a resistor of resistance R, the charge on the capacitor and the current in the circuit vary in time according to the expressions

$$q(t) = Q(1 - e^{-t/RC}) \quad \textbf{(28.14)}$$

$$I(t) = \frac{\mathcal{E}}{R} e^{-t/RC} \quad \textbf{(28.15)}$$

where $Q = C\mathcal{E}$ is the maximum charge on the capacitor. The product RC is called the **time constant** τ of the circuit.

If a charged capacitor is discharged through a resistor of resistance R, the charge and current decrease exponentially in time according to the expressions

$$q(t) = Qe^{-t/RC} \quad \textbf{(28.18)}$$

$$I(t) = -I_i e^{-t/RC} \quad \textbf{(28.19)}$$

where Q is the initial charge on the capacitor and $I_i = Q/RC$ is the initial current in the circuit.

Questions

☐ denotes answer available in *Student Solutions Manual/Study Guide;* **O** denotes objective question

1. Is the direction of current in a battery always from the negative terminal to the positive terminal? Explain.

2. O A certain battery has some internal resistance. **(i)** Can the potential difference across the terminals of a battery be equal to its emf? (a) No. (b) Yes, if the battery is absorbing energy by electrical transmission. (c) Yes, if more than one wire is connected to each terminal. (d) Yes, if the current in the battery is zero. (e) Yes; no special condition is required. **(ii)** Can the terminal voltage exceed the emf? Choose your answer from the same possibilities.

3. Given three lightbulbs and a battery, sketch as many different electric circuits as you can.

4. O When resistors with different resistances are connected in series, which of the following must be the same for each resistor? Choose all correct answers. (a) potential difference (b) current (c) power delivered (d) charge entering (e) none of these answers

5. O When resistors with different resistances are connected in parallel, which of the following must be the same for each resistor? Choose all correct answers. (a) potential difference (b) current (c) power delivered (d) charge entering (e) none of these answers

6. Why is it possible for a bird to sit on a high-voltage wire without being electrocuted?

7. O Are the two headlights of a car wired (a) in series with each other, (b) in parallel, (c) neither in series or in parallel, or (d) is it impossible to tell?

8. A student claims that the second of two lightbulbs in series is less bright than the first, because the first lightbulb uses up some of the current. How would you respond to this statement?

9. O Is a circuit breaker wired (a) in series with the device it is protecting, (b) in parallel, (c) neither in series or in parallel, or (d) is it impossible to tell?

10. O In the circuit shown in Figure Q28.10, each battery is delivering energy to the circuit by electrical transmission. All the resistors have equal resistance. **(i)** Rank the electric potentials at points *a*, *b*, *c*, *d*, *e*, *f*, *g*, and *h* from highest to lowest, noting any cases of equality in the ranking. **(ii)** Rank the magnitudes of the currents at the same points from greatest to least, noting any cases of equality.

Figure Q28.10

11. O A series circuit consists of three identical lamps connected to a battery as shown in Figure Q28.11. The switch S, originally open, is closed. **(i)** What then happens to the brightness of lamp B? (a) It increases. (b) It decreases

somewhat. (c) It does not change. (d) It drops to zero. **(ii)** What happens to the brightness of lamp C? Choose from the same possibilities. **(iii)** What happens to the current in the battery? Choose from the same possibilities. **(iv)** What happens to the potential difference across lamp A? **(v)** What happens to the potential difference across lamp C? **(vi)** What happens to the total power delivered to the lamps by the battery? Choose in each case from the same possibilities (a) through (d).

Figure Q28.11

12. O A circuit consists of three identical lamps connected to a battery having some internal resistance as in Figure Q28.12. The switch S, originally open, is closed. **(i)** What then happens to the brightness of lamp B? (a) It increases. (b) It decreases somewhat. (c) It does not change. (d) It drops to zero. **(ii)** What happens to the brightness of lamp C? Choose from the same possibilities. **(iii)** What happens to the current in the battery? Choose from the same possibilities. **(iv)** What happens to the potential difference across lamp A? **(v)** What happens to the potential difference across lamp C? **(vi)** What happens to the total power delivered to the lamps by the battery? Choose in each case from the same possibilities (a) through (d).

Figure Q28.12

13. A ski resort consists of a few chairlifts and several interconnected downhill runs on the side of a mountain, with a lodge at the bottom. The chairlifts are analogous to batteries, and the runs are analogous to resistors. Describe how two runs can be in series. Describe how three runs can be in parallel. Sketch a junction between one chairlift and two runs. State Kirchhoff's junction rule for ski resorts. One of the skiers happens to be carrying a skydiver's altimeter. She never takes the same set of chairlifts and runs twice, but keeps passing you at the fixed location where you are working. State Kirchhoff's loop rule for ski resorts.

14. Referring to Figure Q28.14, describe what happens to the lightbulb after the switch is closed. Assume the capacitor has a large capacitance and is initially uncharged and assume the light illuminates when connected directly across the battery terminals.

Figure Q28.14

15. So that your grandmother can listen to *A Prairie Home Companion*, you take her bedside radio to the hospital where she is staying. You are required to have a maintenance worker test the radio for electrical safety. Finding that it develops 120 V on one of its knobs, he does not let you take it up to your grandmother's room. She complains that she has had the radio for many years and nobody has ever gotten a shock from it. You end up having to buy a new plastic radio. Is that fair? Will the old radio be safe back in her bedroom?

16. What advantage does 120-V operation offer over 240 V? What disadvantages does it have?

Problems

WebAssign The Problems from this chapter may be assigned online in WebAssign.

ThomsonNOW Sign in at **www.thomsonedu.com** and go to ThomsonNOW to assess your understanding of this chapter's topics with additional quizzing and conceptual questions.

1, 2, 3 denotes straightforward, intermediate, challenging;   ☐ denotes full solution available in *Student Solutions Manual/Study Guide;*   ▲ denotes coached solution with hints available at **www.thomsonedu.com;**   denotes developing symbolic reasoning;   ● denotes asking for qualitative reasoning;   ▪ denotes computer useful in solving problem

Section 28.1 Electromotive Force

1. ▲ A battery has an emf of 15.0 V. The terminal voltage of the battery is 11.6 V when it is delivering 20.0 W of power to an external load resistor R. (a) What is the value of R? (b) What is the internal resistance of the battery?

2. Two 1.50-V batteries—with their positive terminals in the same direction—are inserted in series into the barrel of a flashlight. One battery has an internal resistance of 0.255 Ω, the other an internal resistance of 0.153 Ω. When the switch is closed, a current of 600 mA occurs in the lamp. (a) What is the lamp's resistance? (b) What fraction of the chemical energy transformed appears as internal energy in the batteries?

3. An automobile battery has an emf of 12.6 V and an internal resistance of 0.080 0 Ω. The headlights together have an equivalent resistance of 5.00 Ω (assumed constant). What is the potential difference across the headlight bulbs (a) when they are the only load on the battery and (b) when the starter motor is operated, requiring an additional 35.0 A from the battery?

4. ● As in Example 28.2, consider a power supply with fixed emf $\mathcal{E}$ and internal resistance r causing current in a load resistance R. In this problem, R is fixed and r is a variable. The efficiency is defined as the energy delivered to the load divided by the energy delivered by the emf. (a) When the internal resistance is adjusted for maximum power transfer, what is the efficiency? (b) What should be the internal resistance for maximum possible efficiency? (c) When the electric company sells energy to a customer, does it have a goal of high efficiency or of maximum power transfer? Explain. (d) When a student connects a loudspeaker to an amplifier, does she most want high efficiency or high power transfer? Explain.

Section 28.2 Resistors in Series and Parallel

5. (a) Find the equivalent resistance between points a and b in Figure P28.5. (b) A potential difference of 34.0 V is applied between points a and b. Calculate the current in each resistor.

Figure P28.5

6. ● A lightbulb marked "75 W [at] 120 V" is screwed into a socket at one end of a long extension cord, in which each of the two conductors has resistance 0.800 Ω. The other end of the extension cord is plugged into a 120-V outlet. (a) Explain why the actual power delivered to the lightbulb cannot be 75 W in this situation. (b) What can you reasonably model as constant about the lightbulb? Draw a circuit diagram and find the actual power delivered to the lightbulb in this circuit.

7. ▲ Consider the circuit shown in Figure P28.7. Find (a) the current in the 20.0-Ω resistor and (b) the potential difference between points a and b.

Figure P28.7

8. For the purpose of measuring the electric resistance of shoes through the body of the wearer to a metal ground plate, the American National Standards Institute (ANSI) specifies the circuit shown in Figure P28.8. The potential difference ΔV across the 1.00-MΩ resistor is measured with a high-resistance voltmeter. (a) Show that the resistance of the footwear is

$$R_{\text{shoes}} = 1.00 \text{ M}\Omega \left(\frac{50.0 \text{ V} - \Delta V}{\Delta V} \right)$$

(b) In a medical test, a current through the human body should not exceed 150 μA. Can the current delivered by the ANSI-specified circuit exceed 150 μA? To decide, consider a person standing barefoot on the ground plate.

Figure P28.8

9. Three 100-Ω resistors are connected as shown in Figure P28.9. The maximum power that can safely be delivered to any one resistor is 25.0 W. (a) What is the maximum potential difference that can be applied to the terminals a and b? (b) For the voltage determined in part (a), what is the power delivered to each resistor? What is the total power delivered?

Figure P28.9

10. Using only three resistors—2.00 Ω, 3.00 Ω, and 4.00 Ω—find 17 resistance values that may be obtained by various combinations of one or more resistors. Tabulate the combinations in order of increasing resistance.

11. A 6.00-V battery supplies current to the circuit shown in Figure P28.11. When the double-throw switch S is open as shown in the figure, the current in the battery is 1.00 mA. When the switch is closed in position a, the current in the battery is 1.20 mA. When the switch is closed in position b, the current in the battery is 2.00 mA. Find the resistances R_1, R_2, and R_3.

Figure P28.11

12. Two resistors connected in series have an equivalent resistance of 690 Ω. When they are connected in parallel,

their equivalent resistance is 150 Ω. Find the resistance of each resistor.

13. ● When the switch S in the circuit of Figure P28.13 is closed, will the equivalent resistance between points a and b increase or decrease? State your reasoning. Assume the equivalent resistance changes by a factor of 2. Determine the value of R.

Figure P28.13

14. ● Four resistors are connected to a battery as shown in Figure P28.14. The current in the battery is I; the battery emf is $\mathbf{\mathcal{E}}$; and the resistor values are $R_1 = R$, $R_2 = 2R$, $R_3 = 4R$, and $R_4 = 3R$. (a) Rank the resistors according to the potential difference across them from largest to smallest. Note any cases of equal potential differences. (b) Determine the potential difference across each resistor in terms of $\mathbf{\mathcal{E}}$. (c) Rank the resistors according to the current in them from largest to smallest. Note any cases of equal currents. (d) Determine the current in each resistor in terms of I. (e) **What If?** If R_3 is increased, explain what happens to the current in each of the resistors. (f) In the limit that $R_3 \to \infty$, what are the new values of the current in each resistor in terms of I, the original current in the battery?

Figure P28.14

15. Calculate the power delivered to each resistor in the circuit shown in Figure P28.15.

Figure P28.15

Section 28.3 Kirchhoff's Rules

16. The ammeter shown in Figure P28.16 reads 2.00 A. Find I_1, I_2, and $\mathbf{\mathcal{E}}$.

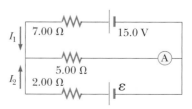

Figure P28.16

2 = intermediate;   3 = challenging;   □ = SSM/SG;   ▲ = ThomsonNOW;   = symbolic reasoning;   ● = qualitative reasoning

17. ▲ Determine the current in each branch of the circuit shown in Figure P28.17.

Figure P28.17 Problems 17, 18, and 19.

18. In Figure P28.17, show how to add only enough ammeters to measure every different current. Show how to add only enough voltmeters to measure the potential difference across each resistor and across each battery.

19. ● The circuit considered in Problem 17 and shown in Figure P28.17 is connected for 2.00 min. (a) Find the energy delivered by each battery. (b) Find the energy delivered to each resistor. (c) Identify the net energy transformation that occurs in the operation of the circuit. Find the total amount of energy transformed.

20. The following equations describe an electric circuit:

$$-I_1(220\ \Omega) + 5.80\ \text{V} - I_2(370\ \Omega) = 0$$

$$I_2(370\ \Omega) + I_3(150\ \Omega) - 3.10\ \text{V} = 0$$

$$I_1 + I_3 - I_2 = 0$$

(a) Draw a diagram of the circuit. (b) Calculate the unknowns and identify the physical meaning of each unknown.

21. Consider the circuit shown in Figure P28.21. What are the expected readings of the ideal ammeter and ideal voltmeter?

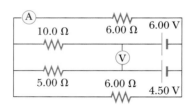

Figure P28.21

22. Taking $R = 1.00$ kΩ and $\mathcal{E} = 250$ V in Figure P28.22, determine the direction and magnitude of the current in the horizontal wire between a and e.

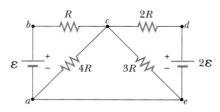

Figure P28.22

23. In the circuit of Figure P28.23, determine the current in each resistor and the potential difference across the 200-Ω resistor.

Figure P28.23

24. A dead battery is charged by connecting it to the live battery of another car with jumper cables (Fig. P28.24). Determine the current in the starter and in the dead battery.

Figure P28.24

25. For the circuit shown in Figure P28.25, calculate (a) the current in the 2.00-Ω resistor and (b) the potential difference between points a and b.

Figure P28.25

26. For the network shown in Figure P28.26, show that the resistance $R_{ab} = \frac{27}{17}\ \Omega$.

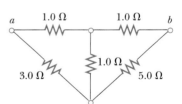

Figure P28.26

Section 28.4 *RC* Circuits

27. ▲ Consider a series *RC* circuit (see Active Fig. 28.16) for which $R = 1.00$ MΩ, $C = 5.00$ μF, and $\mathcal{E} = 30.0$ V. Find (a) the time constant of the circuit and (b) the maximum charge on the capacitor after the switch is thrown to a, connecting the capacitor to the battery. (c) Find the current in the resistor 10.0 s after the switch is thrown to a.

28. A 10.0-μF capacitor is charged by a 10.0-V battery through a resistance R. The capacitor reaches a potential difference of 4.00 V in a time interval of 3.00 s after charging begins. Find R.

29. A 2.00-nF capacitor with an initial charge of 5.10 μC is discharged through a 1.30-kΩ resistor. (a) Calculate the current in the resistor 9.00 μs after the resistor is con-

nected across the terminals of the capacitor. (b) What charge remains on the capacitor after 8.00 μs? (c) What is the maximum current in the resistor?

30. Show that the integral $\int_0^\infty e^{-2t/RC}dt$ in Example 28.11 has the value $RC/2$.

31. The circuit in Figure P28.31 has been connected for a long time. (a) What is the potential difference across the capacitor? (b) If the battery is disconnected from the circuit, over what time interval does the capacitor discharge to one-tenth of its initial voltage?

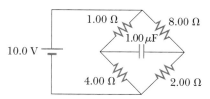

Figure P28.31

32. In the circuit of Figure P28.32, the switch S has been open for a long time. It is then suddenly closed. Determine the time constant (a) before the switch is closed and (b) after the switch is closed. (c) Let the switch be closed at $t = 0$. Determine the current in the switch as a function of time.

Figure P28.32

Section 28.5 Electrical Meters

33. Assume a galvanometer has an internal resistance of 60.0 Ω and requires a current of 0.500 mA to produce full-scale deflection. What resistance must be connected in parallel with the galvanometer if the combination is to serve as an ammeter that has a full-scale deflection for a current of 0.100 A?

34. A particular galvanometer serves as a 2.00-V full-scale voltmeter when a 2 500-Ω resistor is connected in series with it. It serves as a 0.500-A full-scale ammeter when a 0.220-Ω resistor is connected in parallel with it. Determine the internal resistance of the galvanometer and the current required to produce full-scale deflection.

35. A particular galvanometer requires a current of 1.50 mA for full-scale deflection and has a resistance of 75.0 Ω. It can be used to measure voltages by wiring a large resistor in series with the galvanometer as suggested in Active Figure 28.22. The effect is to limit the current in the galvanometer when large voltages are applied. Calculate the value of the resistor that allows the galvanometer to measure an applied voltage of 25.0 V at full-scale deflection.

36. ● *Meter loading.* Work this problem to five-digit precision. Refer to Figure P28.36. (a) When a 180.00-Ω resistor is connected across a battery of emf 6.000 0 V and internal resistance 20.000 Ω, what is the current in the resistor?

What is the potential difference across it? (b) Suppose now an ammeter of resistance 0.500 00 Ω and a voltmeter of resistance 20 000 Ω are added to the circuit as shown in Figure P28.36b. Find the reading of each. (c) **What If?** Now one terminal of one wire is moved as shown in Figure P28.36c. Find the new meter readings. (d) Explain whether one of the voltmeter–ammeter connections is significantly better for taking data to determine the resistance of the resistor.

Figure P28.36

Section 28.6 Household Wiring and Electrical Safety

37. ● ▲ An electric heater is rated at 1 500 W, a toaster at 750 W, and an electric grill at 1 000 W. The three appliances are connected to a common 120-V household circuit. (a) How much current does each draw? (b) Is a circuit with a 25.0-A circuit breaker sufficient in this situation? Explain your answer.

38. Turn on your desk lamp. Pick up the cord, with your thumb and index finger spanning the width of the cord. (a) Compute an order-of-magnitude estimate for the current in your hand. Assume the conductor inside the lamp cord next to your thumb is at potential $\sim 10^2$ V at a typical instant and the conductor next to your index finger is at ground potential (0 V). The resistance of your hand depends strongly on the thickness and the moisture content of the outer layers of your skin. Assume the resistance of your hand between fingertip and thumb tip is $\sim 10^4$ Ω. You may model the cord as having rubber insulation. State the other quantities you measure or estimate and their values. Explain your reasoning. (b) Suppose your body is isolated from any other charges or currents. In order-of-magnitude terms, describe the potential of your thumb where it contacts the cord and the potential of your finger where it touches the cord.

Additional Problems

39. The circuit in Figure P28.39 has been connected for several seconds. Find the current (a) in the 4.00-V battery,

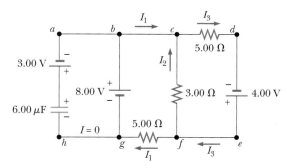

Figure P28.39

2 = intermediate;   3 = challenging;   □ = SSM/SG;   ▲ = ThomsonNOW;   ▧ = symbolic reasoning;   ● = qualitative reasoning

(b) in the 3.00-Ω resistor, (c) in the 8.00-V battery, and (d) in the 3.00-V battery. Find (e) the charge on the capacitor.

40. ● The circuit in Figure P28.40a consists of three resistors and one battery with no internal resistance. (a) Find the current in the 5.00-Ω resistor. (b) Find the power delivered to the 5.00-Ω resistor. (c) In each of the circuits in Figures P28.40b, P28.40c, and P28.40d, an additional 15.0-V battery has been inserted into the circuit. Which diagram or diagrams represent a circuit that requires the use of Kirchhoff's rules to find the currents? Explain why. In which of these three circuits is the smallest amount of power delivered to the 10.0-Ω resistor? You need not calculate the power in each circuit if you explain your answer.

Figure P28.40

41. Four 1.50-V AA batteries in series are used to power a transistor radio. If the batteries can move a charge of 240 C, how long will they last if the radio has a resistance of 200 Ω?

42. ● A battery has an emf of 9.20 V and an internal resistance of 1.20 Ω. (a) What resistance across the battery will extract from it a power of 12.8 W? (b) A power of 21.2 W? Explain your answers.

43. Calculate the potential difference between points a and b in Figure P28.43 and identify which point is at the higher potential.

Figure P28.43

44. Assume you have a battery of emf $\mathcal{E}$ and three identical lightbulbs, each having constant resistance R. What is the total power delivered by the battery if the lightbulbs are connected (a) in series? (b) In parallel? (c) For which connection will the lightbulbs shine the brightest?

45. A rechargeable battery has a constant emf of 13.2 V and an internal resistance of 0.850 Ω. It is charged by a 14.7-V power supply for a time interval of 1.80 h. After charging, the battery returns to its original state as it delivers a constant current to a load resistor over 7.30 h. Find the efficiency of the battery as an energy storage device. (The efficiency here is defined as the energy delivered to the load during discharge divided by the energy delivered by the 14.7-V power supply during the charging process.)

46. A power supply has an open-circuit voltage of 40.0 V and an internal resistance of 2.00 Ω. It is used to charge two storage batteries connected in series, each having an emf of 6.00 V and internal resistance of 0.300 Ω. If the charging current is to be 4.00 A, (a) what additional resistance should be added in series? (b) At what rate does the internal energy increase in the supply, in the batteries, and in the added series resistance? (c) At what rate does the chemical energy increase in the batteries?

47. When two unknown resistors are connected in series with a battery, the battery delivers 225 W and carries a total current of 5.00 A. For the same total current, 50.0 W is delivered when the resistors are connected in parallel. Determine the value of each resistor.

48. When two unknown resistors are connected in series with a battery, the battery delivers total power $\mathcal{P}_s$ and carries a total current of I. For the same total current, a total power $\mathcal{P}_p$ is delivered when the resistors are connected in parallel. Determine the value of each resistor.

49. Two resistors R_1 and R_2 are in parallel with each other. Together they carry total current I. (a) Determine the current in each resistor. (b) Prove that this division of the total current I between the two resistors results in less power delivered to the combination than any other division. It is a general principle that *current in a direct current circuit distributes itself so that the total power delivered to the circuit is a minimum.*

50. ● (a) Determine the equilibrium charge on the capacitor in the circuit of Figure P28.50 as a function of R. (b) Evaluate the charge when $R = 10.0\ \Omega$. (c) Can the charge on the capacitor be zero? If so, for what value of R? (d) What is the maximum possible magnitude of the charge on the capacitor? For what value of R is it achieved? (e) Is it experimentally meaningful to take $R = \infty$? Explain your answer. If so, what charge magnitude does it imply? *Suggestion:* You may do part (b) before part (a), as practice for it.

Figure P28.50

51. The value of a resistor R is to be determined using the ammeter–voltmeter setup shown in Figure P28.51. The ammeter has a resistance of 0.500 Ω, and the voltmeter has a resistance of 20.0 kΩ. Within what range of actual values of R will the measured values be correct to within

5.00% if the measurement is made using the circuit shown in (a) Figure P28.51a and (b) Figure P28.51b?

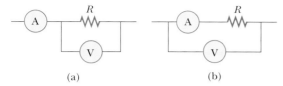

(a)   (b)

Figure P28.51

52. A battery is used to charge a capacitor through a resistor as shown in Active Figure 28.16b. Show that half the energy supplied by the battery appears as internal energy in the resistor and half is stored in the capacitor.

53. The values of the components in a simple series RC circuit containing a switch (Active Fig. 28.16b) are $C = 1.00 \ \mu F$, $R = 2.00 \times 10^6 \ \Omega$, and $\mathcal{E} = 10.0$ V. At the instant 10.0 s after the switch is thrown to a, calculate (a) the charge on the capacitor, (b) the current in the resistor, (c) the rate at which energy is being stored in the capacitor, and (d) the rate at which energy is being delivered by the battery.

54. A young man owns a canister vacuum cleaner marked 535 W at 120 V and a Volkswagen Beetle, which he wishes to clean. He parks the car in his apartment parking lot and uses an inexpensive extension cord 15.0 m long to plug in the vacuum cleaner. You may assume the cleaner has constant resistance. (a) If the resistance of each of the two conductors in the extension cord is 0.900 Ω, what is the actual power delivered to the cleaner? (b) If instead the power is to be at least 525 W, what must be the diameter of each of two identical copper conductors in the cord he buys? (c) Repeat part (b) assuming the power is to be at least 532 W. *Suggestion:* A symbolic solution can simplify the calculations.

55. Three 60.0-W, 120-V lightbulbs are connected across a 120-V power source as shown in Figure P28.55. Find (a) the total power delivered to the three lightbulbs and (b) the potential difference across each. Assume the resistance of each lightbulb is constant (even though in reality the resistance might increase markedly with current).

Figure P28.55

56. Switch S shown in Figure P28.56 has been closed for a long time and the electric circuit carries a constant current. Take $C_1 = 3.00 \ \mu F$, $C_2 = 6.00 \ \mu F$, $R_1 = 4.00$ kΩ, and $R_2 = 7.00$ kΩ. The power delivered to R_2 is 2.40 W. (a) Find the charge on C_1. (b) Now the switch is opened. After many milliseconds, by how much has the charge on C_2 changed?

Figure P28.56

57. ● An ideal voltmeter connected across a certain fresh battery reads 9.30 V, and an ideal ammeter briefly connected across the same battery reads 3.70 A. We say that the battery has an open-circuit voltage of 9.30 V and a short-circuit current of 3.70 A. (a) Model the battery as a source of emf $\mathcal{E}$ in series with an internal resistance r. Determine both $\mathcal{E}$ and r. (b) An irresponsible experimenter connects 20 of these identical batteries together as suggested in Figure P28.57. Do not try this experiment yourself! Find the open-circuit voltage and the short-circuit current of the set of connected batteries. (c) Assume the resistance between the palms of the experimenter's two hands is 120 Ω. Find the current in his body that would result if his palms touched the two exposed terminals of the set of connected batteries. (d) Find the power that would be delivered to his body in this situation. (e) Thinking it is safe to do so, the experimenter threads a copper wire inside his shirt between his hands, like a mitten string. To reduce the current in his body to 5.00 mA when he presses the ends of the wire against the battery poles, what should the resistance of the copper wire be? (f) Find the power delivered to his body in this situation. (g) Find the power delivered to the copper wire. (h) Explain why the sum of the two powers in parts (f) and (g) is much less than the power calculated in part (d). Is it meaningful to ask where the rest of the power is going?

Figure P28.57

58. ● Four resistors are connected in parallel across a 9.20-V battery. They carry currents of 150 mA, 45.0 mA, 14.0 mA, and 4.00 mA. (a) If the resistor with the largest resistance is replaced with one having twice the resistance, what is the ratio of the new current in the battery to the original current? (b) **What If?** If instead the resistor with the smallest resistance is replaced with one having twice the resistance, what is the ratio of the new total current to the original current? (c) On a February night, energy leaves a

house by several energy leaks, including the following: 1 500 W by conduction through the ceiling, 450 W by infiltration (airflow) around the windows, 140 W by conduction through the basement wall above the foundation sill, and 40.0 W by conduction through the plywood door to the attic. To produce the biggest saving in heating bills, which one of these energy transfers should be reduced first? Explain how you decide. Clifford Swartz suggested the idea for this problem.

59. Figure P28.59 shows a circuit model for the transmission of an electrical signal such as cable TV to a large number of subscribers. Each subscriber connects a load resistance R_L between the transmission line and the ground. The ground is assumed to be at zero potential and able to carry any current between any ground connections with negligible resistance. The resistance of the transmission line between the connection points of different subscribers is modeled as the constant resistance R_T. Show that the equivalent resistance across the signal source is

$$R_{eq} = \tfrac{1}{2}[(4R_T R_L + R_T^2)^{1/2} + R_T]$$

Suggestion: Because the number of subscribers is large, the equivalent resistance would not change noticeably if the first subscriber canceled the service. Consequently, the equivalent resistance of the section of the circuit to the right of the first load resistor is nearly equal to R_{eq}.

Figure P28.59

60. A regular tetrahedron is a pyramid with a triangular base. Six 10.0-Ω resistors are placed along its six edges, with junctions at its four vertices. A 12.0-V battery is connected to any two of the vertices. Find (a) the equivalent resistance of the tetrahedron between these vertices and (b) the current in the battery.

61. In Figure P28.61, suppose the switch has been closed for a time interval sufficiently long for the capacitor to become fully charged. Find (a) the steady-state current in each resistor and (b) the charge Q on the capacitor. (c) The switch is now opened at $t = 0$. Write an equation for the current I_{R_2} in R_2 as a function of time and (d) find the time interval required for the charge on the capacitor to fall to one-fifth its initial value.

Figure P28.61

62. The circuit shown in Figure P28.62 is set up in the laboratory to measure an unknown capacitance C with the use of a voltmeter of resistance $R = 10.0$ MΩ and a battery whose emf is 6.19 V. The data given in the table are the measured voltages across the capacitor as a function of time, where $t = 0$ represents the instant at which the switch is opened. (a) Construct a graph of $\ln(\mathcal{E}/\Delta V)$ versus t and perform a linear least-squares fit to the data. (b) From the slope of your graph, obtain a value for the time constant of the circuit and a value for the capacitance.

ΔV (V)	t (s)	$\ln(\mathcal{E}/\Delta V)$
6.19	0	
5.55	4.87	
4.93	11.1	
4.34	19.4	
3.72	30.8	
3.09	46.6	
2.47	67.3	
1.83	102.2	

Figure P28.62

63. The student engineer of a campus radio station wishes to verify the effectiveness of the lightning rod on the antenna mast (Fig. P28.63). The unknown resistance R_x is between points C and E. Point E is a true ground, but it is inaccessible for direct measurement because this stratum is several meters below the Earth's surface. Two identical rods are driven into the ground at A and B, introducing an unknown resistance R_y. The procedure is as follows. Measure resistance R_1 between points A and B, then connect A and B with a heavy conducting wire and measure resistance R_2 between points A and C. (a) Derive an equation for R_x in terms of the observable resistances, R_1 and R_2. (b) A satisfactory ground resistance would be $R_x < 2.00$ Ω. Is the grounding of the station adequate if measurements give $R_1 = 13.0$ Ω and $R_2 = 6.00$ Ω?

Figure P28.63

64. The switch in Figure P28.64a closes when $\Delta V_c > 2\,\Delta V/3$ and opens when $\Delta V_c < \Delta V/3$. The voltmeter reads a

potential difference as plotted in Figure P28.64b. What is the period T of the waveform in terms of R_1, R_2, and C?

(a)

(b)

Figure P28.64

65. An electric teakettle has a multiposition switch and two heating coils. When only one coil is switched on, the well-insulated kettle brings a full pot of water to a boil over the time interval Δt. When only the other coil is switched on, it takes a time interval of $2\Delta t$ to boil the same amount of water. Find the time interval required to boil the same amount of water if both coils are switched on (a) in a parallel connection and (b) in a series connection.

66. In places such as hospital operating rooms or factories for electronic circuit boards, electric sparks must be avoided. A person standing on a grounded floor and touching nothing else can typically have a body capacitance of 150 pF, in parallel with a foot capacitance of 80.0 pF produced by the dielectric soles of his or her shoes. The person acquires static electric charge from interactions with furniture, clothing, equipment, packaging materials, and essentially everything else. The static charge flows to ground through the equivalent resistance of the two shoe soles in parallel with each other. A pair of rubber-soled street shoes can present an equivalent resistance of 5 000 MΩ. A pair of shoes with special static-dissipative soles can have an equivalent resistance of 1.00 MΩ. Consider the person's body and shoes as forming an RC circuit with the ground. (a) How long does it take the rubber-soled shoes to reduce a person's potential from 3 000 V to 100 V? (b) How long does it take the static-dissipative shoes to do the same thing?

Answers to Quick Quizzes

28.1 (a). Power is delivered to the internal resistance of a battery, so decreasing the internal resistance decreases this "lost" power and increase the percentage of the power delivered to the device.

28.2 (b). When the switch is opened, resistors R_1 and R_2 are in series, so the total circuit resistance is larger than when the switch was closed. As a result, the current decreases.

28.3 (a). When the switch is closed, resistors R_1 and R_2 are in parallel, so the total circuit resistance is smaller than when the switch was open. As a result, the current increases.

28.4 **(i)**, (b). Adding another series resistor increases the total resistance of the circuit and therefore reduces the current in the circuit. **(ii)**, (a). The potential difference across the battery terminals increases because the reduced current results in a smaller voltage decrease across the internal resistance. **(iii)**, (a). If a third resistor were connected in parallel, the total resistance of the circuit would decrease and the current in the battery would increase. **(iv)**, (b). The potential difference across the terminals would decrease because the increased current results in a greater voltage drop across the internal resistance.

28.5 **(i)**, (c). Just after the switch is closed, there is no charge on the capacitor. Current exists in both branches of the circuit as the capacitor begins to charge, so the right half of the circuit is equivalent to two resistances R in parallel for an equivalent resistance of $\frac{1}{2}R$. **(ii)**, (d). After a long time, the capacitor is fully charged and the current in the right-hand branch drops to zero. Now, current exists only in a resistance R across the battery.

Magnetic fingerprinting allows fingerprints to be seen on surfaces that otherwise would not allow prints to be lifted. The powder spread on the surface is coated with an organic material that adheres to the greasy residue in a fingerprint. A magnetic "brush" removes the excess powder and makes the fingerprint visible. (James King-Holmes/Photo Researchers, Inc.)

29.1 Magnetic Fields and Forces

29.2 Motion of a Charged Particle in a Uniform Magnetic Field

29.3 Applications Involving Charged Particles Moving in a Magnetic Field

29.4 Magnetic Force Acting on a Current-Carrying Conductor

29.5 Torque on a Current Loop in a Uniform Magnetic Field

29.6 The Hall Effect

29 Magnetic Fields

Many historians of science believe that the compass, which uses a magnetic needle, was used in China as early as the 13th century BC, its invention being of Arabic or Indian origin. The early Greeks knew about magnetism as early as 800 BC. They discovered that the stone magnetite (Fe_3O_4) attracts pieces of iron. Legend ascribes the name *magnetite* to the shepherd Magnes, the nails of whose shoes and the tip of whose staff stuck fast to chunks of magnetite while he pastured his flocks.

In 1269, Pierre de Maricourt of France found that the directions of a needle near a spherical natural magnet formed lines that encircled the sphere and passed through two points diametrically opposite each other, which he called the *poles* of the magnet. Subsequent experiments showed that every magnet, regardless of its shape, has two poles, called *north* (N) and *south* (S) poles, that exert forces on other magnetic poles similar to the way electric charges exert forces on one another. That is, like poles (N–N or S–S) repel each other, and opposite poles (N–S) attract each other.

The poles received their names because of the way a magnet, such as that in a compass, behaves in the presence of the Earth's magnetic field. If a bar magnet is suspended from its midpoint and can swing freely in a horizontal plane, it will rotate until its north pole points to the Earth's geographic North Pole and its south pole points to the Earth's geographic South Pole.[1]

In 1600, William Gilbert (1540–1603) extended de Maricourt's experiments to a variety of materials. He knew that a compass needle orients in preferred directions, so he suggested that the Earth itself is a large, permanent magnet. In 1750, experimenters used a torsion balance to show that magnetic poles exert attractive or repulsive forces on each other and that these forces vary as the inverse square of the distance between interacting poles. Although the force between two magnetic poles is otherwise similar to the force between two electric charges, electric charges can be isolated (witness the electron and proton), whereas **a single magnetic pole has never been isolated.** That is, **magnetic poles are always found in pairs.** All attempts thus far to detect an isolated magnetic pole have been unsuccessful. No matter how many times a permanent magnet is cut in two, each piece always has a north and a south pole.[2]

The relationship between magnetism and electricity was discovered in 1819 when, during a lecture demonstration, Hans Christian Oersted found that an electric current in a wire deflected a nearby compass needle.[3] In the 1820s, further connections between electricity and magnetism were demonstrated independently by Faraday and Joseph Henry (1797–1878). They showed that an electric current can be produced in a circuit either by moving a magnet near the circuit or by changing the current in a nearby circuit. These observations demonstrate that a changing magnetic field creates an electric field. Years later, theoretical work by Maxwell showed that the reverse is also true: a changing electric field creates a magnetic field.

This chapter examines the forces that act on moving charges and on current-carrying wires in the presence of a magnetic field. The source of the magnetic field is described in Chapter 30.

29.1 Magnetic Fields and Forces

In our study of electricity, we described the interactions between charged objects in terms of electric fields. Recall that an electric field surrounds any electric charge. In addition to containing an electric field, the region of space surrounding any *moving* electric charge also contains a magnetic field. A magnetic field also surrounds a magnetic substance making up a permanent magnet.

Historically, the symbol $\vec{\mathbf{B}}$ has been used to represent a magnetic field, and we use this notation in this book. The direction of the magnetic field $\vec{\mathbf{B}}$ at any location is the direction in which a compass needle points at that location. As with the electric field, we can represent the magnetic field by means of drawings with *magnetic field lines.*

Active Figure 29.1 shows how the magnetic field lines of a bar magnet can be traced with the aid of a compass. Notice that the magnetic field lines outside the

---

[1] The Earth's geographic North Pole is magnetically a south pole, whereas the Earth's geographic South Pole is magnetically a north pole. Because *opposite* magnetic poles attract each other, the pole on a magnet that is attracted to the Earth's geographic North Pole is the magnet's *north* pole and the pole attracted to the Earth's geographic South Pole is the magnet's *south* pole.

[2] There is some theoretical basis for speculating that magnetic *monopoles*—isolated north or south poles—may exist in nature, and attempts to detect them are an active experimental field of investigation.

[3] The same discovery was reported in 1802 by an Italian jurist, Gian Domenico Romagnosi, but was overlooked, probably because it was published in an obscure journal.

HANS CHRISTIAN OERSTED
Danish Physicist and Chemist (1777–1851)
Oersted is best known for observing that a compass needle deflects when placed near a wire carrying a current. This important discovery was the first evidence of the connection between electric and magnetic phenomena. Oersted was also the first to prepare pure aluminum.

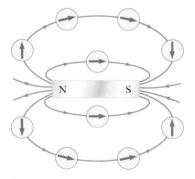

ACTIVE FIGURE 29.1

Compass needles can be used to trace the magnetic field lines in the region outside a bar magnet.

Sign in at www.thomsonedu.com and go to ThomsonNOW to move the compass around and trace the magnetic field lines for yourself.

(a)   (b)   (c)

Figure 29.2   (a) Magnetic field pattern surrounding a bar magnet as displayed with iron filings. (b) Magnetic field pattern between *opposite* poles (N–S) of two bar magnets. (c) Magnetic field pattern between *like* poles (N–N) of two bar magnets.

magnet point away from the north pole and toward the south pole. One can display magnetic field patterns of a bar magnet using small iron filings as shown in Figure 29.2.

We can define a magnetic field $\vec{B}$ at some point in space in terms of the magnetic force $\vec{F}_B$ that the field exerts on a charged particle moving with a velocity $\vec{v}$, which we call the test object. For the time being, let's assume no electric or gravitational fields are present at the location of the test object. Experiments on various charged particles moving in a magnetic field give the following results:

▶ **Properties of the magnetic force on a charge moving in a magnetic field**

- The magnitude F_B of the magnetic force exerted on the particle is proportional to the charge q and to the speed v of the particle.
- When a charged particle moves parallel to the magnetic field vector, the magnetic force acting on the particle is zero.
- When the particle's velocity vector makes any angle $\theta \neq 0$ with the magnetic field, the magnetic force acts in a direction perpendicular to both $\vec{v}$ and $\vec{B}$; that is, $\vec{F}_B$ is perpendicular to the plane formed by $\vec{v}$ and $\vec{B}$ (Fig. 29.3a).
- The magnetic force exerted on a positive charge is in the direction opposite the direction of the magnetic force exerted on a negative charge moving in the same direction (Fig. 29.3b).
- The magnitude of the magnetic force exerted on the moving particle is proportional to $\sin \theta$, where θ is the angle the particle's velocity vector makes with the direction of $\vec{B}$.

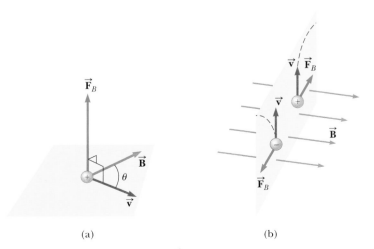

(a)   (b)

Figure 29.3   The direction of the magnetic force $\vec{F}_B$ acting on a charged particle moving with a velocity $\vec{v}$ in the presence of a magnetic field $\vec{B}$. (a) The magnetic force is perpendicular to both $\vec{v}$ and $\vec{B}$. (b) Oppositely directed magnetic forces $\vec{F}_B$ are exerted on two oppositely charged particles moving at the same velocity in a magnetic field. The dashed lines show the paths of the particles, which are investigated in Section 29.2.

Henry Leap and Jim Lehman

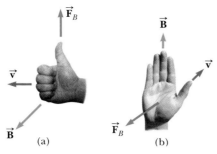

Figure 29.4 Two right-hand rules for determining the direction of the magnetic force $\vec{\mathbf{F}}_B = q\vec{\mathbf{v}} \times \vec{\mathbf{B}}$ acting on a particle with charge q moving with a velocity $\vec{\mathbf{v}}$ in a magnetic field $\vec{\mathbf{B}}$. (a) In this rule, your fingers point in the direction of $\vec{\mathbf{v}}$, with $\vec{\mathbf{B}}$ coming out of your palm, so that you can curl your fingers in the direction of $\vec{\mathbf{B}}$. The direction of $\vec{\mathbf{v}} \times \vec{\mathbf{B}}$, and the force on a positive charge, is the direction in which your thumb points. (b) In this rule, the vector $\vec{\mathbf{v}}$ is in the direction of your thumb and $\vec{\mathbf{B}}$ in the direction of your fingers. The force $\vec{\mathbf{F}}_B$ on a positive charge is in the direction of your palm, as if you are pushing the particle with your hand.

We can summarize these observations by writing the magnetic force in the form

$$\vec{\mathbf{F}}_B = q\vec{\mathbf{v}} \times \vec{\mathbf{B}} \qquad (29.1)$$

◀ Vector expression for the magnetic force on a charged particle moving in a magnetic field

which by definition of the cross product (see Section 11.1) is perpendicular to both $\vec{\mathbf{v}}$ and $\vec{\mathbf{B}}$. We can regard this equation as an operational definition of the magnetic field at some point in space. That is, the magnetic field is defined in terms of the force acting on a moving charged particle.

Figure 29.4 reviews two right-hand rules for determining the direction of the cross product $\vec{\mathbf{v}} \times \vec{\mathbf{B}}$ and determining the direction of $\vec{\mathbf{F}}_B$. The rule in Figure 29.4a depends on our right-hand rule for the cross product in Figure 11.2. Point the four fingers of your right hand along the direction of $\vec{\mathbf{v}}$ with the palm facing $\vec{\mathbf{B}}$ and curl them toward $\vec{\mathbf{B}}$. Your extended thumb, which is at a right angle to your fingers, points in the direction of $\vec{\mathbf{v}} \times \vec{\mathbf{B}}$. Because $\vec{\mathbf{F}}_B = q\vec{\mathbf{v}} \times \vec{\mathbf{B}}$, $\vec{\mathbf{F}}_B$ is in the direction of your thumb if q is positive and is opposite the direction of your thumb if q is negative. (If you need more help understanding the cross product, you should review Section 11.1, including Fig. 11.2.)

An alternative rule is shown in Figure 29.4b. Here the thumb points in the direction of $\vec{\mathbf{v}}$ and the extended fingers in the direction of $\vec{\mathbf{B}}$. Now, the force $\vec{\mathbf{F}}_B$ on a positive charge extends outward from the palm. The advantage of this rule is that the force on the charge is in the direction that you would push on something with your hand: outward from your palm. The force on a negative charge is in the opposite direction. You can use either of these two right-hand rules.

The magnitude of the magnetic force on a charged particle is

$$F_B = |q|vB\sin\theta \qquad (29.2)$$

◀ Magnitude of the magnetic force on a charged particle moving in a magnetic field

where θ is the smaller angle between $\vec{\mathbf{v}}$ and $\vec{\mathbf{B}}$. From this expression, we see that F_B is zero when $\vec{\mathbf{v}}$ is parallel or antiparallel to $\vec{\mathbf{B}}$ ($\theta = 0$ or $180°$) and maximum when $\vec{\mathbf{v}}$ is perpendicular to $\vec{\mathbf{B}}$ ($\theta = 90°$).

Electric and magnetic forces have several important differences:

■ The electric force vector is along the direction of the electric field, whereas the magnetic force vector is perpendicular to the magnetic field.

■ The electric force acts on a charged particle regardless of whether the particle is moving, whereas the magnetic force acts on a charged particle only when the particle is in motion.

■ The electric force does work in displacing a charged particle, whereas the magnetic force associated with a steady magnetic field does no work when a particle is displaced because the force is perpendicular to the displacement.

From the last statement and on the basis of the work–kinetic energy theorem, we conclude that the kinetic energy of a charged particle moving through a magnetic field cannot be altered by the magnetic field alone. The field can alter the

TABLE 29.1

Some Approximate Magnetic Field Magnitudes

Source of Field	Field Magnitude (T)
Strong superconducting laboratory magnet	30
Strong conventional laboratory magnet	2
Medical MRI unit	1.5
Bar magnet	10^{-2}
Surface of the Sun	10^{-2}
Surface of the Earth	0.5×10^{-4}
Inside human brain (due to nerve impulses)	10^{-13}

direction of the velocity vector, but it cannot change the speed or kinetic energy of the particle.

From Equation 29.2, we see that the SI unit of magnetic field is the newton per coulomb-meter per second, which is called the **tesla** (T):

The tesla ▶

$$1 \text{ T} = 1 \frac{\text{N}}{\text{C} \cdot \text{m/s}}$$

Because a coulomb per second is defined to be an ampere,

$$1 \text{ T} = 1 \frac{\text{N}}{\text{A} \cdot \text{m}}$$

A non-SI magnetic-field unit in common use, called the *gauss* (G), is related to the tesla through the conversion $1 \text{ T} = 10^4 \text{ G}$. Table 29.1 shows some typical values of magnetic fields.

Quick Quiz 29.1 An electron moves in the plane of this paper toward the top of the page. A magnetic field is also in the plane of the page and directed toward the right. What is the direction of the magnetic force on the electron? (a) toward the top of the page (b) toward the bottom of the page (c) toward the left edge of the page (d) toward the right edge of the page (e) upward out of the page (f) downward into the page

---

EXAMPLE 29.1 **An Electron Moving in a Magnetic Field**

An electron in a television picture tube moves toward the front of the tube with a speed of 8.0×10^6 m/s along the x axis (Fig. 29.5). Surrounding the neck of the tube are coils of wire that create a magnetic field of magnitude 0.025 T, directed at an angle of $60°$ to the x axis and lying in the xy plane. Calculate the magnetic force on the electron.

SOLUTION

Conceptualize Recall that the magnetic force on a charged particle is perpendicular to the plane formed by the velocity and magnetic field vectors. Use the right-hand rule in Figure 29.4 to convince yourself that the direction of the force on the electron is downward in Figure 29.5.

Categorize We evaluate the magnetic force using an equation developed in this section, so we categorize this example as a substitution problem.

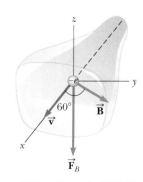

Figure 29.5 (Example 29.1) The magnetic force $\vec{\mathbf{F}}_B$ acting on the electron is in the negative z direction when $\vec{\mathbf{v}}$ and $\vec{\mathbf{B}}$ lie in the xy plane.

Use Equation 29.2 to find the magnitude of the magnetic force:

$$F_B = |q|\,vB\sin\theta$$
$$= (1.6 \times 10^{-19}\,\text{C})(8.0 \times 10^6\,\text{m/s})(0.025\,\text{T})(\sin 60°)$$
$$= \boxed{2.8 \times 10^{-14}\,\text{N}}$$

For practice using the vector product, evaluate this force in vector notation using Equation 29.1.

29.2   Motion of a Charged Particle in a Uniform Magnetic Field

Before we continue our discussion, some explanation of the notation used in this book is in order. To indicate the direction of $\vec{\mathbf{B}}$ in illustrations, we sometimes present perspective views such as those in Figure 29.5. If $\vec{\mathbf{B}}$ lies in the plane of the page or is present in a perspective drawing, we use green vectors or green field lines with arrowheads. In nonperspective illustrations, we depict a magnetic field perpendicular to and directed out of the page with a series of green dots, which represent the tips of arrows coming toward you (see Fig. 29.6a). In this case, the field is labeled $\vec{\mathbf{B}}_{out}$. If $\vec{\mathbf{B}}$ is directed perpendicularly into the page, we use green crosses, which represent the feathered tails of arrows fired away from you, as in Figure 29.6b. In this case, the field is labeled $\vec{\mathbf{B}}_{in}$, where the subscript "in" indicates "into the page." The same notation with crosses and dots is also used for other quantities that might be perpendicular to the page such as forces and current directions.

In Section 29.1, we found that the magnetic force acting on a charged particle moving in a magnetic field is perpendicular to the particle's velocity and consequently the work done by the magnetic force on the particle is zero. Now consider the special case of a positively charged particle moving in a uniform magnetic field with the initial velocity vector of the particle perpendicular to the field. Let's assume that the direction of the magnetic field is into the page as in Active Figure 29.7. As the particle changes the direction of its velocity in response to the magnetic force, the magnetic force remains perpendicular to the velocity. As we found in Section 6.1, if the force is always perpendicular to the velocity, the path of the particle is a circle! Active Figure 29.7 shows the particle moving in a circle in a plane perpendicular to the magnetic field.

The particle moves in a circle because the magnetic force $\vec{\mathbf{F}}_B$ is perpendicular to $\vec{\mathbf{v}}$ and $\vec{\mathbf{B}}$ and has a constant magnitude qvB. As Active Figure 29.7 illustrates, the rotation is counterclockwise for a positive charge in a magnetic field directed into the page. If q were negative, the rotation would be clockwise. We use the particle under a net force model to write Newton's second law for the particle:

$$\sum F = F_B = ma$$

Because the particle moves in a circle, we also model it as a particle in uniform circular motion and we replace the acceleration with centripetal acceleration:

$$F_B = qvB = \frac{mv^2}{r}$$

This expression leads to the following equation for the radius of the circular path:

$$r = \frac{mv}{qB} \tag{29.3}$$

That is, the radius of the path is proportional to the linear momentum mv of the particle and inversely proportional to the magnitude of the charge on the particle and to the magnitude of the magnetic field. The angular speed of the particle (from Eq. 10.10) is

$$\omega = \frac{v}{r} = \frac{qB}{m} \tag{29.4}$$

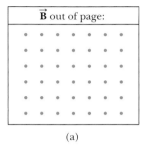

(a)

(b)

Figure 29.6 (a) Magnetic field lines coming out of the paper are indicated by dots, representing the tips of arrows coming outward. (b) Magnetic field lines going into the paper are indicated by crosses, representing the feathers of arrows going inward.

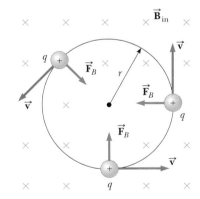

ACTIVE FIGURE 29.7

When the velocity of a charged particle is perpendicular to a uniform magnetic field, the particle moves in a circular path in a plane perpendicular to $\vec{\mathbf{B}}$. The magnetic force $\vec{\mathbf{F}}_B$ acting on the charge is always directed toward the center of the circle.

Sign in at www.thomsonedu.com and go to ThomsonNOW to adjust the mass, speed, and charge of the particle and the magnitude of the magnetic field and observe the resulting circular motion.

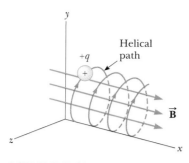

ACTIVE FIGURE 29.8

A charged particle having a velocity vector that has a component parallel to a uniform magnetic field moves in a helical path.

Sign in at www.thomsonedu.com and go to ThomsonNOW to adjust the *x* component of the velocity of the particle and observe the resulting helical motion.

The period of the motion (the time interval the particle requires to complete one revolution) is equal to the circumference of the circle divided by the speed of the particle:

$$T = \frac{2\pi r}{v} = \frac{2\pi}{\omega} = \frac{2\pi m}{qB} \qquad (29.5)$$

These results show that the angular speed of the particle and the period of the circular motion do not depend on the speed of the particle or on the radius of the orbit. The angular speed ω is often referred to as the **cyclotron frequency** because charged particles circulate at this angular frequency in the type of accelerator called a *cyclotron*, which is discussed in Section 29.3.

If a charged particle moves in a uniform magnetic field with its velocity at some arbitrary angle with respect to $\vec{B}$, its path is a helix. For example, if the field is directed in the *x* direction as shown in Active Figure 29.8, there is no component of force in the *x* direction. As a result, $a_x = 0$, and the *x* component of velocity remains constant. The magnetic force $q\vec{v} \times \vec{B}$ causes the components v_y and v_z to change in time, however, and the resulting motion is a helix whose axis is parallel to the magnetic field. The projection of the path onto the *yz* plane (viewed along the *x* axis) is a circle. (The projections of the path onto the *xy* and *xz* planes are sinusoids!) Equations 29.3 to 29.5 still apply provided v is replaced by $v_\perp = \sqrt{v_y^2 + v_z^2}$.

Quick Quiz 29.2 A charged particle is moving perpendicular to a magnetic field in a circle with a radius *r*. **(i)** An identical particle enters the field, with $\vec{v}$ perpendicular to $\vec{B}$, but with a higher speed than the first particle. Compared with the radius of the circle for the first particle, is the radius of the circular path for the second particle (a) smaller, (b) larger, or (c) equal in size? **(ii)** The magnitude of the magnetic field is increased. From the same choices, compare the radius of the new circular path of the first particle with the radius of its initial path.

| EXAMPLE 29.2 | **A Proton Moving Perpendicular to a Uniform Magnetic Field** |

A proton is moving in a circular orbit of radius 14 cm in a uniform 0.35-T magnetic field perpendicular to the velocity of the proton. Find the speed of the proton.

SOLUTION

Conceptualize From our discussion in this section, we know that the proton follows a circular path when moving in a uniform magnetic field.

Categorize We evaluate the speed of the proton using an equation developed in this section, so we categorize this example as a substitution problem.

Solve Equation 29.3 for the speed of the particle:

$$v = \frac{qBr}{m_p}$$

Substitute numerical values:

$$v = \frac{(1.60 \times 10^{-19}\ \text{C})(0.35\ \text{T})(0.14\ \text{m})}{1.67 \times 10^{-27}\ \text{kg}}$$

$$= 4.7 \times 10^6\ \text{m/s}$$

What If? What if an electron, rather than a proton, moves in a direction perpendicular to the same magnetic field with this same speed? Will the radius of its orbit be different?

Answer   An electron has a much smaller mass than a proton, so the magnetic force should be able to change its velocity much more easily than that for the proton. Therefore, we expect the radius to be smaller. Equation 29.3 shows that r is proportional to m with q, B, and v the same for the electron as for the proton. Consequently, the radius will be smaller by the same factor as the ratio of masses m_e/m_p.

EXAMPLE 29.3   **Bending an Electron Beam**

In an experiment designed to measure the magnitude of a uniform magnetic field, electrons are accelerated from rest through a potential difference of 350 V and then enter a uniform magnetic field that is perpendicular to the velocity vector of the electrons. The electrons travel along a curved path because of the magnetic force exerted on them, and the radius of the path is measured to be 7.5 cm. (Such a curved beam of electrons is shown in Fig. 29.9.)

(A) What is the magnitude of the magnetic field?

SOLUTION

Conceptualize   With the help of Figures 29.7 and 29.9, visualize the circular motion of the electrons.

Figure 29.9   (Example 29.3) The bending of an electron beam in a magnetic field.

Categorize   This example involves electrons accelerating from rest due to an electric force and then moving in a circular path due to a magnetic force. Equation 29.3 shows that we need the speed v of the electron to find the magnetic field magnitude, and v is not given. Consequently, we must find the speed of the electron based on the potential difference through which it is accelerated. To do so, we categorize the first part of the problem by modeling an electron and the electric field as an isolated system. Once the electron enters the magnetic field, we categorize the second part of the problem as one similar to those we have studied in this section.

Analyze   Write the appropriate reduction of the conservation of energy equation, Equation 8.2, for the electron–electric field system:

$$\Delta K + \Delta U = 0$$

Substitute the appropriate initial and final energies:

$$(\tfrac{1}{2}m_e v^2 - 0) + (q\Delta V) = 0$$

Solve for the speed of the electron:

$$v = \sqrt{\frac{-2q\,\Delta V}{m_e}}$$

Substitute numerical values:

$$v = \sqrt{\frac{-2(-1.60 \times 10^{-19}\,\text{C})(350\,\text{V})}{9.11 \times 10^{-31}\,\text{kg}}} = 1.11 \times 10^{7}\,\text{m/s}$$

Now imagine the electron entering the magnetic field with this speed. Solve Equation 29.3 for the magnitude of the magnetic field:

$$B = \frac{m_e v}{er}$$

Substitute numerical values:

$$B = \frac{(9.11 \times 10^{-31}\,\text{kg})(1.11 \times 10^{7}\,\text{m/s})}{(1.60 \times 10^{-19}\,\text{C})(0.075\,\text{m})} = 8.4 \times 10^{-4}\,\text{T}$$

(B) What is the angular speed of the electrons?

SOLUTION

Use Equation 10.10:

$$\omega = \frac{v}{r} = \frac{1.11 \times 10^{7}\,\text{m/s}}{0.075\,\text{m}} = 1.5 \times 10^{8}\,\text{rad/s}$$

Finalize The angular speed can be represented as $\omega = (1.5 \times 10^8 \text{ rad/s})(1 \text{ rev}/2\pi \text{ rad}) = 2.4 \times 10^7 \text{ rev/s}$. The electrons travel around the circle 24 million times per second! This answer is consistent with the very high speed found in part (A).

What If? What if a sudden voltage surge causes the accelerating voltage to increase to 400 V? How does that affect the angular speed of the electrons, assuming the magnetic field remains constant?

Answer The increase in accelerating voltage ΔV causes the electrons to enter the magnetic field with a higher speed v. This higher speed causes them to travel in a circle with a larger radius r. The angular speed is the ratio of v to r. Both v and r increase by the same factor, so the effects cancel and the angular speed remains the same. Equation 29.4 is an expression for the cyclotron frequency, which is the same as the angular speed of the electrons. The cyclotron frequency depends only on the charge q, the magnetic field B, and the mass m_e, none of which have changed. Therefore, the voltage surge has no effect on the angular speed. (In reality, however, the voltage surge may also increase the magnetic field if the magnetic field is powered by the same source as the accelerating voltage. In that case, the angular speed increases according to Equation 29.4.)

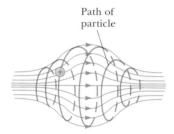

Figure 29.10 A charged particle moving in a nonuniform magnetic field (a magnetic bottle) spirals about the field and oscillates between the endpoints. The magnetic force exerted on the particle near either end of the bottle has a component that causes the particle to spiral back toward the center.

Figure 29.11 The Van Allen belts are made up of charged particles trapped by the Earth's nonuniform magnetic field. The magnetic field lines are in green, and the particle paths are in brown.

When charged particles move in a nonuniform magnetic field, the motion is complex. For example, in a magnetic field that is strong at the ends and weak in the middle such as that shown in Figure 29.10, the particles can oscillate between two positions. A charged particle starting at one end spirals along the field lines until it reaches the other end, where it reverses its path and spirals back. This configuration is known as a *magnetic bottle* because charged particles can be trapped within it. The magnetic bottle has been used to confine a *plasma,* a gas consisting of ions and electrons. Such a plasma-confinement scheme could fulfill a crucial role in the control of nuclear fusion, a process that could supply us in the future with an almost endless source of energy. Unfortunately, the magnetic bottle has its problems. If a large number of particles are trapped, collisions between them cause the particles to eventually leak from the system.

The Van Allen radiation belts consist of charged particles (mostly electrons and protons) surrounding the Earth in doughnut-shaped regions (Fig. 29.11). The particles, trapped by the Earth's nonuniform magnetic field, spiral around the field lines from pole to pole, covering the distance in only a few seconds. These particles originate mainly from the Sun, but some come from stars and other heavenly objects. For this reason, the particles are called *cosmic rays.* Most cosmic rays are deflected by the Earth's magnetic field and never reach the atmosphere. Some of the particles become trapped, however, and it is these particles that make up the Van Allen belts. When the particles are located over the poles, they sometimes collide with atoms in the atmosphere, causing the atoms to emit visible light. Such collisions are the origin of the beautiful Aurora Borealis, or Northern Lights, in the northern hemisphere and the Aurora Australis in the southern hemisphere. Auroras are usually confined to the polar regions because the Van Allen belts are nearest the Earth's surface there. Occasionally, though, solar activity causes larger numbers of charged particles to enter the belts and significantly distort the normal magnetic field lines associated with the Earth. In these situations, an aurora can sometimes be seen at lower latitudes.

29.3 Applications Involving Charged Particles Moving in a Magnetic Field

A charge moving with a velocity $\vec{v}$ in the presence of both an electric field $\vec{E}$ and a magnetic field $\vec{B}$ experiences both an electric force $q\vec{E}$ and a magnetic force $q\vec{v} \times \vec{B}$. The total force (called the Lorentz force) acting on the charge is

◄ Lorentz force

$$\vec{F} = q\vec{E} + q\vec{v} \times \vec{B} \tag{29.6}$$

Velocity Selector

In many experiments involving moving charged particles, it is important that all particles move with essentially the same velocity, which can be achieved by applying a combination of an electric field and a magnetic field oriented as shown in Active Figure 29.12. A uniform electric field is directed to the right (in the plane of the page in Active Fig. 29.12), and a uniform magnetic field is applied in the direction perpendicular to the electric field (into the page in Active Fig. 29.12). If q is positive and the velocity $\vec{v}$ is upward, the magnetic force $q\vec{v} \times \vec{B}$ is to the left and the electric force $q\vec{E}$ is to the right. When the magnitudes of the two fields are chosen so that $qE = qvB$, the charged particle is modeled as a particle in equilibrium and moves in a straight vertical line through the region of the fields. From the expression $qE = qvB$, we find that

$$v = \frac{E}{B} \tag{29.7}$$

Only those particles having this speed pass undeflected through the mutually perpendicular electric and magnetic fields. The magnetic force exerted on particles moving at speeds greater than that is stronger than the electric force, and the particles are deflected to the left. Those moving at slower speeds are deflected to the right.

The Mass Spectrometer

A **mass spectrometer** separates ions according to their mass-to-charge ratio. In one version of this device, known as the *Bainbridge mass spectrometer,* a beam of ions first passes through a velocity selector and then enters a second uniform magnetic field $\vec{B}_0$ that has the same direction as the magnetic field in the selector (Active Fig. 29.13). Upon entering the second magnetic field, the ions move in a semicircle of radius r before striking a detector array at P. If the ions are positively charged, the beam deflects to the left as Active Figure 29.13 shows. If the ions are negatively charged, the beam deflects to the right. From Equation 29.3, we can express the ratio m/q as

$$\frac{m}{q} = \frac{rB_0}{v}$$

Using Equation 29.7 gives

$$\frac{m}{q} = \frac{rB_0B}{E} \tag{29.8}$$

Therefore, we can determine m/q by measuring the radius of curvature and knowing the field magnitudes B, B_0, and E. In practice, one usually measures the masses of various isotopes of a given ion, with the ions all carrying the same charge q. In this way, the mass ratios can be determined even if q is unknown.

A variation of this technique was used by J. J. Thomson (1856–1940) in 1897 to measure the ratio e/m_e for electrons. Figure 29.14a (page 818) shows the basic apparatus he used. Electrons are accelerated from the cathode and pass through two slits. They then drift into a region of perpendicular electric and magnetic fields. The magnitudes of the two fields are first adjusted to produce an undeflected beam. When the magnetic field is turned off, the electric field produces a measurable beam deflection that is recorded on the fluorescent screen. From the size of the deflection and the measured values of E and B, the charge-to-mass ratio can be determined. The results of this crucial experiment represent the discovery of the electron as a fundamental particle of nature.

The Cyclotron

A **cyclotron** is a device that can accelerate charged particles to very high speeds. The energetic particles produced are used to bombard atomic nuclei and thereby

ACTIVE FIGURE 29.12

A velocity selector. When a positively charged particle is moving with velocity $\vec{v}$ in the presence of a magnetic field directed into the page and an electric field directed to the right, it experiences an electric force $q\vec{E}$ to the right and a magnetic force $q\vec{v} \times \vec{B}$ to the left.

Sign in at www.thomsonedu.com and go to ThomsonNOW to adjust the electric and magnetic fields and try to achieve straight-line motion for the charge.

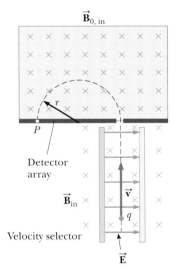

ACTIVE FIGURE 29.13

A mass spectrometer. Positively charged particles are sent first through a velocity selector and then into a region where the magnetic field $\vec{B}_0$ causes the particles to move in a semicircular path and strike a detector array at P.

Sign in at www.thomsonedu.com and go to ThomsonNOW to predict where particles will strike the detector array.

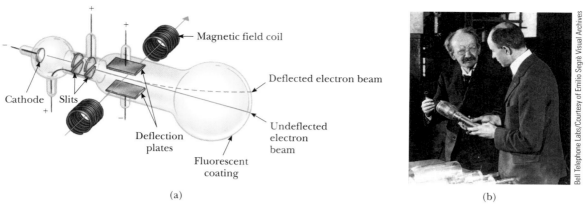

Figure 29.14 (a) Thomson's apparatus for measuring e/m_e. Electrons are accelerated from the cathode, pass through two slits, and are deflected by both an electric field and a magnetic field (directed perpendicular to the electric field). The beam of electrons then strikes a fluorescent screen. (b) J. J. Thomson (*left*) in the Cavendish Laboratory, University of Cambridge. The man on the right, Frank Baldwin Jewett, is a distant relative of John W. Jewett, Jr., coauthor of this text.

PITFALL PREVENTION 29.1

The Cyclotron Is Not State-of-the-Art Technology

The cyclotron is important historically because it was the first particle accelerator to produce particles with very high speeds. Cyclotrons are still in use in medical applications, but most accelerators currently in research use are not cyclotrons. Research accelerators work on a different principle and are generally called *synchrotrons*.

produce nuclear reactions of interest to researchers. A number of hospitals use cyclotron facilities to produce radioactive substances for diagnosis and treatment.

Both electric and magnetic forces play a key role in the operation of a cyclotron, a schematic drawing of which is shown in Figure 29.15a. The charges move inside two semicircular containers D_1 and D_2, referred to as *dees* because of their shape like the letter D. A high-frequency alternating potential difference is applied to the dees, and a uniform magnetic field is directed perpendicular to them. A positive ion released at P near the center of the magnet in one dee moves in a semicircular path (indicated by the dashed brown line in the drawing) and arrives back at the gap in a time interval $T/2$, where T is the time interval needed to make one complete trip around the two dees, given by Equation 29.5. The frequency of the applied potential difference is adjusted so that the polarity of the dees is reversed in the same time interval during which the ion travels around one dee. If the applied potential difference is adjusted such that D_2 is at a lower electric potential than D_1 by an amount ΔV, the ion accelerates across the gap to D_2 and its kinetic energy increases by an amount $q \, \Delta V$. It then moves around D_2 in a semicircular path of greater radius (because its speed has increased). After a time interval $T/2$, it again arrives at the gap between the dees. By this time, the polarity across the dees has again been reversed and the ion is given another "kick" across

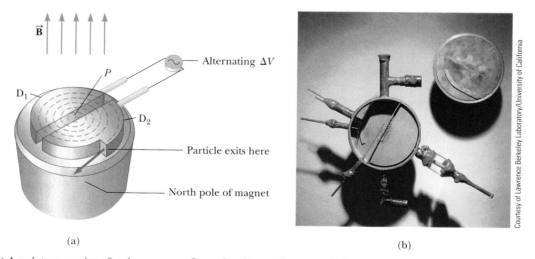

Figure 29.15 (a) A cyclotron consists of an ion source at P, two dees D_1 and D_2 across which an alternating potential difference is applied, and a uniform magnetic field. (The south pole of the magnet is not shown.) The brown, dashed, curved lines represent the path of the particles. (b) The first cyclotron, invented by E. O. Lawrence and M. S. Livingston in 1934.

the gap. The motion continues so that for each half-circle trip around one dee, the ion gains additional kinetic energy equal to $q\,\Delta V$. When the radius of its path is nearly that of the dees, the energetic ion leaves the system through the exit slit. The cyclotron's operation depends on T being independent of the speed of the ion and of the radius of the circular path (Eq. 29.5).

We can obtain an expression for the kinetic energy of the ion when it exits the cyclotron in terms of the radius R of the dees. From Equation 29.3 we know that $v = qBR/m$. Hence, the kinetic energy is

$$K = \tfrac{1}{2}mv^2 = \frac{q^2 B^2 R^2}{2m} \tag{29.9}$$

When the energy of the ions in a cyclotron exceeds about 20 MeV, relativistic effects come into play. (Such effects are discussed in Chapter 39.) Observations show that T increases and the moving ions do not remain in phase with the applied potential difference. Some accelerators overcome this problem by modifying the period of the applied potential difference so that it remains in phase with the moving ions.

29.4 Magnetic Force Acting on a Current-Carrying Conductor

If a magnetic force is exerted on a single charged particle when the particle moves through a magnetic field, it should not surprise you that a current-carrying wire also experiences a force when placed in a magnetic field. The current is a collection of many charged particles in motion; hence, the resultant force exerted by the field on the wire is the vector sum of the individual forces exerted on all the charged particles making up the current. The force exerted on the particles is transmitted to the wire when the particles collide with the atoms making up the wire.

One can demonstrate the magnetic force acting on a current-carrying conductor by hanging a wire between the poles of a magnet as shown in Figure 29.16a. For ease in visualization, part of the horseshoe magnet in part (a) is removed to show the end face of the south pole in parts (b), (c), and (d) of Figure 29.16. The magnetic field is directed into the page and covers the region within the shaded squares. When the current in the wire is zero, the wire remains vertical as in Figure 29.16b. When the wire carries a current directed upward as in Figure 29.16c, however, the wire deflects to the left. If the current is reversed as in Figure 29.16d, the wire deflects to the right.

Let's quantify this discussion by considering a straight segment of wire of length L and cross-sectional area A carrying a current I in a uniform magnetic field $\vec{\mathbf{B}}$ as in

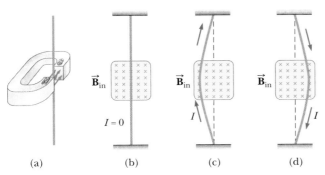

Figure 29.16   (a) A wire suspended vertically between the poles of a magnet. (b) The setup shown in (a) as seen looking at the south pole of the magnet so that the magnetic field (green crosses) is directed into the page. When there is no current in the wire, the wire remains vertical. (c) When the current is upward, the wire deflects to the left. (d) When the current is downward, the wire deflects to the right.

$\vec{\mathbf{F}}_B$

$\vec{\mathbf{B}}_{in}$ A

$\vec{\mathbf{v}}_d$

q +

L

Figure 29.17   A segment of a current-carrying wire in a magnetic field $\vec{\mathbf{B}}$. The magnetic force exerted on each charge making up the current is $q\vec{\mathbf{v}}_d \times \vec{\mathbf{B}}$, and the net force on the segment of length L is $I\vec{\mathbf{L}} \times \vec{\mathbf{B}}$.

Figure 29.18   A wire segment of arbitrary shape carrying a current I in a magnetic field $\vec{\mathbf{B}}$ experiences a magnetic force. The magnetic force on any segment $d\vec{\mathbf{s}}$ is $I\,d\vec{\mathbf{s}} \times \vec{\mathbf{B}}$ and is directed out of the page. You should use the right-hand rule to confirm this force direction.

Figure 29.17. The magnetic force exerted on a charge q moving with a drift velocity $\vec{\mathbf{v}}_d$ is $q\vec{\mathbf{v}}_d \times \vec{\mathbf{B}}$. To find the total force acting on the wire, we multiply the force $q\vec{\mathbf{v}}_d \times \vec{\mathbf{B}}$ exerted on one charge by the number of charges in the segment. Because the volume of the segment is AL, the number of charges in the segment is nAL, where n is the number of charges per unit volume. Hence, the total magnetic force on the wire of length L is

$$\vec{\mathbf{F}}_B = (q\vec{\mathbf{v}}_d \times \vec{\mathbf{B}})\,nAL$$

We can write this expression in a more convenient form by noting that, from Equation 27.4, the current in the wire is $I = nqv_dA$. Therefore,

◄ Force on a segment of current-carrying wire in a uniform magnetic field

$$\vec{\mathbf{F}}_B = I\vec{\mathbf{L}} \times \vec{\mathbf{B}} \tag{29.10}$$

where $\vec{\mathbf{L}}$ is a vector that points in the direction of the current I and has a magnitude equal to the length L of the segment. This expression applies only to a straight segment of wire in a uniform magnetic field.

Now consider an arbitrarily shaped wire segment of uniform cross section in a magnetic field as shown in Figure 29.18. It follows from Equation 29.10 that the magnetic force exerted on a small segment of vector length $d\vec{\mathbf{s}}$ in the presence of a field $\vec{\mathbf{B}}$ is

$$d\vec{\mathbf{F}}_B = I\,d\vec{\mathbf{s}} \times \vec{\mathbf{B}} \tag{29.11}$$

where $d\vec{\mathbf{F}}_B$ is directed out of the page for the directions of $\vec{\mathbf{B}}$ and $d\vec{\mathbf{s}}$ in Figure 29.18. Equation 29.11 can be considered as an alternative definition of $\vec{\mathbf{B}}$. That is, we can define the magnetic field $\vec{\mathbf{B}}$ in terms of a measurable force exerted on a current element, where the force is a maximum when $\vec{\mathbf{B}}$ is perpendicular to the element and zero when $\vec{\mathbf{B}}$ is parallel to the element.

To calculate the total force $\vec{\mathbf{F}}_B$ acting on the wire shown in Figure 29.18, we integrate Equation 29.11 over the length of the wire:

$$\vec{\mathbf{F}}_B = I\int_a^b d\vec{\mathbf{s}} \times \vec{\mathbf{B}} \tag{29.12}$$

where a and b represent the endpoints of the wire. When this integration is carried out, the magnitude of the magnetic field and the direction the field makes with the vector $d\vec{\mathbf{s}}$ may differ at different points.

Quick Quiz 29.3   A wire carries current in the plane of this paper toward the top of the page. The wire experiences a magnetic force toward the right edge of the page. Is the direction of the magnetic field causing this force (a) in the plane of the page and toward the left edge, (b) in the plane of the page and toward the bottom edge, (c) upward out of the page, or (d) downward into the page?

| EXAMPLE 29.4 | **Force on a Semicircular Conductor** |

A wire bent into a semicircle of radius R forms a closed circuit and carries a current I. The wire lies in the xy plane, and a uniform magnetic field is directed along the positive y axis as in Figure 29.19. Find the magnitude and direction of the magnetic force acting on the straight portion of the wire and on the curved portion.

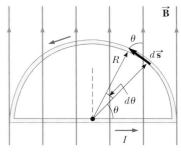

Figure 29.19 (Example 29.4) The magnetic force on the straight portion of the loop is directed out of the page, and the magnetic force on the curved portion is directed into the page.

SOLUTION

Conceptualize Using the right-hand rule for cross products, we see that the force $\vec{\mathbf{F}}_1$ on the straight portion of the wire is out of the page and the force $\vec{\mathbf{F}}_2$ on the curved portion is into the page. Is $\vec{\mathbf{F}}_2$ larger in magnitude than $\vec{\mathbf{F}}_1$ because the length of the curved portion is longer than that of the straight portion?

Categorize Because we are dealing with a current-carrying wire in a magnetic field rather than a single charged particle, we must use Equation 29.12 to find the total force on each portion of the wire.

Analyze Note that $d\vec{\mathbf{s}}$ is perpendicular to $\vec{\mathbf{B}}$ everywhere on the straight portion of the wire. Use Equation 29.12 to find the force on this portion:

$$\vec{\mathbf{F}}_1 = I \int_a^b d\vec{\mathbf{s}} \times \vec{\mathbf{B}} = I \int_0^{2R} B\, ds\, \hat{\mathbf{k}} = \boxed{2IRB\hat{\mathbf{k}}}$$

To find the magnetic force on the curved part, first write an expression for the magnetic force $d\vec{\mathbf{F}}_2$ on the element $d\vec{\mathbf{s}}$ in Figure 29.19:

$$(1) \quad d\vec{\mathbf{F}}_2 = I d\vec{\mathbf{s}} \times \vec{\mathbf{B}} = -IB \sin\theta\, ds\, \hat{\mathbf{k}}$$

From the geometry in Figure 29.19, write an expression for ds:

$$(2) \quad ds = R\, d\theta$$

Substitute Equation (2) into Equation (1) and integrate over the angle θ from 0 to π:

$$\vec{\mathbf{F}}_2 = -\int_0^\pi IRB \sin\theta\, d\theta\, \hat{\mathbf{k}} = -IRB \int_0^\pi \sin\theta\, d\theta\, \hat{\mathbf{k}} = -IRB[-\cos\theta]_0^\pi\, \hat{\mathbf{k}}$$
$$= IRB(\cos\pi - \cos 0)\hat{\mathbf{k}} = IRB(-1 - 1)\hat{\mathbf{k}} = \boxed{-2IRB\hat{\mathbf{k}}}$$

Finalize Two very important general statements follow from this example. First, the force on the curved portion is the same in magnitude as the force on a straight wire between the same two points. In general, **the magnetic force on a curved current-carrying wire in a uniform magnetic field is equal to that on a straight wire connecting the endpoints and carrying the same current.** Furthermore, $\vec{\mathbf{F}}_1 + \vec{\mathbf{F}}_2 = 0$ is also a general result: **the net magnetic force acting on any closed current loop in a uniform magnetic field is zero.**

29.5 Torque on a Current Loop in a Uniform Magnetic Field

In Section 29.4, we showed how a magnetic force is exerted on a current-carrying conductor placed in a magnetic field. With that as a starting point, we now show that a torque is exerted on a current loop placed in a magnetic field.

Consider a rectangular loop carrying a current I in the presence of a uniform magnetic field directed parallel to the plane of the loop as shown in Figure 29.20a (page 822). No magnetic forces act on sides ① and ③ because these wires are parallel to the field; hence, $\vec{\mathbf{L}} \times \vec{\mathbf{B}} = 0$ for these sides. Magnetic forces do, however, act on sides ② and ④ because these sides are oriented perpendicular to the field. The magnitude of these forces is, from Equation 29.10,

$$F_2 = F_4 = IaB$$

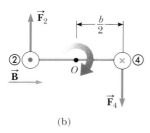

Figure 29.20 (a) Overhead view of a rectangular current loop in a uniform magnetic field. No magnetic forces are acting on sides ① and ③ because these sides are parallel to $\vec{\mathbf{B}}$. Forces are acting on sides ② and ④, however. (b) Edge view of the loop sighting down sides ② and ④ shows that the magnetic forces $\vec{\mathbf{F}}_2$ and $\vec{\mathbf{F}}_4$ exerted on these sides create a torque that tends to twist the loop clockwise. The purple dot in the left circle represents current in wire ② coming toward you; the purple cross in the right circle represents current in wire ④ moving away from you.

The direction of $\vec{\mathbf{F}}_2$, the magnetic force exerted on wire ②, is out of the page in the view shown in Figure 29.20a and that of $\vec{\mathbf{F}}_4$, the magnetic force exerted on wire ④, is into the page in the same view. If we view the loop from side ③ and sight along sides ② and ④, we see the view shown in Figure 29.20b, and the two magnetic forces $\vec{\mathbf{F}}_2$ and $\vec{\mathbf{F}}_4$ are directed as shown. Notice that the two forces point in opposite directions but are *not* directed along the same line of action. If the loop is pivoted so that it can rotate about point O, these two forces produce about O a torque that rotates the loop clockwise. The magnitude of this torque τ_{max} is

$$\tau_{max} = F_2 \frac{b}{2} + F_4 \frac{b}{2} = (IaB) \frac{b}{2} + (IaB) \frac{b}{2} = IabB$$

where the moment arm about O is $b/2$ for each force. Because the area enclosed by the loop is $A = ab$, we can express the maximum torque as

$$\tau_{max} = IAB \tag{29.13}$$

This maximum-torque result is valid only when the magnetic field is parallel to the plane of the loop. The sense of the rotation is clockwise when viewed from side ③ as indicated in Figure 29.20b. If the current direction were reversed, the force directions would also reverse and the rotational tendency would be counterclockwise.

Now suppose the uniform magnetic field makes an angle $\theta < 90°$ with a line perpendicular to the plane of the loop as in Active Figure 29.21. For convenience, let's assume $\vec{\mathbf{B}}$ is perpendicular to sides ② and ④. In this case, the magnetic forces $\vec{\mathbf{F}}_1$ and $\vec{\mathbf{F}}_3$ exerted on sides ① and ③ cancel each other and produce no torque because they pass through a common origin. The magnetic forces $\vec{\mathbf{F}}_2$ and $\vec{\mathbf{F}}_4$ acting on sides ② and ④, however, produce a torque about *any point*. Referring to the end view shown in Active Figure 29.21, we see that the moment arm of $\vec{\mathbf{F}}_2$ about the point O is equal to $(b/2) \sin \theta$. Likewise, the moment arm of $\vec{\mathbf{F}}_4$ about O is also $(b/2) \sin \theta$. Because $F_2 = F_4 = IaB$, the magnitude of the net torque about O is

$$\tau = F_2 \frac{b}{2} \sin \theta + F_4 \frac{b}{2} \sin \theta$$

$$= IaB \left(\frac{b}{2} \sin \theta \right) + IaB \left(\frac{b}{2} \sin \theta \right) = IabB \sin \theta$$

$$= IAB \sin \theta$$

where $A = ab$ is the area of the loop. This result shows that the torque has its maximum value IAB when the field is perpendicular to the normal to the plane of the loop ($\theta = 90°$) as discussed with regard to Figure 29.20 and is zero when the field is parallel to the normal to the plane of the loop ($\theta = 0$).

A convenient expression for the torque exerted on a loop placed in a uniform magnetic field $\vec{\mathbf{B}}$ is

▶ Torque on a current loop in a magnetic field

$$\vec{\boldsymbol{\tau}} = I\vec{\mathbf{A}} \times \vec{\mathbf{B}} \tag{29.14}$$

ACTIVE FIGURE 29.21

An end view of the loop in Figure 29.20b rotated through an angle with respect to the magnetic field. If $\vec{\mathbf{B}}$ is at an angle θ with respect to vector $\vec{\mathbf{A}}$, which is perpendicular to the plane of the loop, the torque is $IAB \sin \theta$ where the magnitude of $\vec{\mathbf{A}}$ is A, the area of the loop.

Sign in at www.thomsonedu.com and go to ThomsonNOW to choose the current in the loop, the magnetic field, and the initial orientation of the loop and observe the subsequent motion.

Figure 29.22 Right-hand rule for determining the direction of the vector $\vec{A}$. The direction of the magnetic moment $\vec{\mu}$ is the same as the direction of $\vec{A}$.

where $\vec{A}$, the vector shown in Active Figure 29.21, is perpendicular to the plane of the loop and has a magnitude equal to the area of the loop. To determine the direction of $\vec{A}$, use the right-hand rule described in Figure 29.22. When you curl the fingers of your right hand in the direction of the current in the loop, your thumb points in the direction of $\vec{A}$. Active Figure 29.21 shows that the loop tends to rotate in the direction of decreasing values of θ (that is, such that the area vector $\vec{A}$ rotates toward the direction of the magnetic field).

The product $I\vec{A}$ is defined to be the **magnetic dipole moment** $\vec{\mu}$ (often simply called the "magnetic moment") of the loop:

$$\vec{\mu} \equiv I\vec{A} \qquad (29.15)$$

◀ Magnetic dipole moment of a current loop

The SI unit of magnetic dipole moment is the ampere-meter^2 (A · m^2). If a coil of wire contains N loops of the same area, the magnetic moment of the coil is

$$\vec{\mu}_{\text{coil}} = NI\vec{A} \qquad (29.16)$$

Using Equation 29.15, we can express the torque exerted on a current-carrying loop in a magnetic field $\vec{B}$ as

$$\vec{\tau} = \vec{\mu} \times \vec{B} \qquad (29.17)$$

◀ Torque on a magnetic moment in a magnetic field

This result is analogous to Equation 26.18, $\vec{\tau} = \vec{p} \times \vec{E}$, for the torque exerted on an electric dipole in the presence of an electric field $\vec{E}$, where $\vec{p}$ is the electric dipole moment.

Although we obtained the torque for a particular orientation of $\vec{B}$ with respect to the loop, the equation $\vec{\tau} = \vec{\mu} \times \vec{B}$ is valid for any orientation. Furthermore, although we derived the torque expression for a rectangular loop, the result is valid for a loop of any shape. The torque on an N-turn coil is given by Equation 29.17 by using Equation 29.16 for the magnetic moment.

In Section 26.6, we found that the potential energy of a system of an electric dipole in an electric field is given by $U = -\vec{p} \cdot \vec{E}$. This energy depends on the orientation of the dipole in the electric field. Likewise, the potential energy of a system of a magnetic dipole in a magnetic field depends on the orientation of the dipole in the magnetic field and is given by

$$U = -\vec{\mu} \cdot \vec{B} \qquad (29.18)$$

◀ Potential energy of a system of a magnetic moment in a magnetic field

This expression shows that the system has its lowest energy $U_{\text{min}} = -\mu B$ when $\vec{\mu}$ points in the same direction as $\vec{B}$. The system has its highest energy $U_{\text{max}} = +\mu B$ when $\vec{\mu}$ points in the direction opposite $\vec{B}$.

The torque on a current loop causes the loop to rotate; this effect is exploited practically in a **motor.** Energy enters the motor by electrical transmission, and the rotating coil can do work on some device external to the motor. For example, the motor in an car's electrical window system does work on the windows, applying a force on them and moving them up or down through some displacement. We will discuss motors in more detail in Section 31.5.

Quick Quiz 29.4 **(i)** Rank the magnitudes of the torques acting on the rectangular loops (a), (b), and (c) shown edge-on in Figure 29.23 from highest to lowest. All loops are identical and carry the same current. **(ii)** Rank the magnitudes of the net forces acting on the rectangular loops shown in Figure 29.23 from highest to lowest.

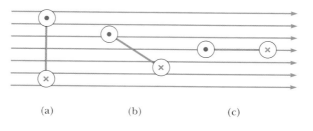

(a)   (b)   (c)

Figure 29.23 (Quick Quiz 29.4) Which current loop (seen edge-on) experiences the greatest torque, (a), (b), or (c)? Which experiences the greatest net force?

EXAMPLE 29.5 The Magnetic Dipole Moment of a Coil

A rectangular coil of dimensions 5.40 cm × 8.50 cm consists of 25 turns of wire and carries a current of 15.0 mA. A 0.350-T magnetic field is applied parallel to the plane of the coil.

(A) Calculate the magnitude of the magnetic dipole moment of the coil.

SOLUTION

Conceptualize The magnetic moment of the coil is independent of any magnetic field in which the loop resides, so it depends only on the geometry of the loop and the current it carries.

Categorize We evaluate quantities based on equations developed in this section, so we categorize this example as a substitution problem.

Use Equation 29.16 to calculate the magnetic moment:

$$\mu_{coil} = NIA = (25)(15.0 \times 10^{-3}\,\text{A})(0.054\,0\,\text{m})(0.085\,0\,\text{m})$$
$$= \boxed{1.72 \times 10^{-3}\,\text{A} \cdot \text{m}^2}$$

(B) What is the magnitude of the torque acting on the loop?

SOLUTION

Use Equation 29.17, noting that $\vec{\mathbf{B}}$ is perpendicular to $\vec{\boldsymbol{\mu}}_{coil}$:

$$\tau = \mu_{coil}B = (1.72 \times 10^{-3}\,\text{A} \cdot \text{m}^2)(0.350\,\text{T})$$
$$= \boxed{6.02 \times 10^{-4}\,\text{N} \cdot \text{m}}$$

EXAMPLE 29.6 Rotating a Coil

Consider the loop of wire in Figure 29.24a. Imagine it is pivoted along side ④, which is parallel to the z axis and fastened so that side ④ remains fixed and the rest of the loop hangs vertically but can rotate around side ④ (Fig. 29.24b). The mass of the loop is 50.0 g, and the sides are of lengths $a = 0.200$ m and $b = 0.100$ m. The loop carries a current of 3.50 A and is immersed in a vertical uniform magnetic field of magnitude 0.010 0 T in the +y direction (Fig. 29.24c). What angle does the plane of the loop make with the vertical?

SOLUTION

Conceptualize   In the side view of Figure 29.24b, notice that the magnetic moment of the loop is to the left. Therefore, when the loop is in the magnetic field, the magnetic torque on the loop causes it to rotate in a clockwise direction around side ④, which we choose as the rotation axis. Imagine the loop making this clockwise rotation so that the plane of the loop is at some angle θ to the vertical as in Figure 29.24c. The gravitational force on the loop exerts a torque that would cause a rotation in the counterclockwise direction if the magnetic field were turned off.

Categorize   At some angle of the loop, the two torques described in the *Conceptualize* step are equal in magnitude and the loop is at rest. We therefore model the loop as a rigid object in equilibrium.

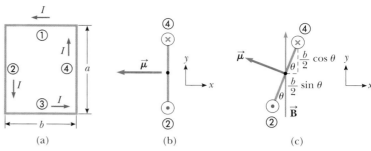

Figure 29.24   (Example 29.6) (a) Overhead view of a rectangular current loop in a uniform magnetic field. This figure is similar to the situations in Figure 29.20 and 29.21. (b) Edge view of the loop sighting down sides ② and ④. The loop hangs vertically and is pivoted so that it can rotate around side ④. (c) An end view of the loop in (b) rotated through an angle with respect to the horizontal when it is placed in a magnetic field. The magnetic torque causes the loop to rotate in a clockwise direction around side ④, whereas the gravitational torque causes a counterclockwise rotation.

Analyze   Evaluate the magnetic torque on the loop from Equation 29.17:

$$\vec{\tau}_B = \vec{\mu} \times \vec{B} = -\mu B \sin(90° - \theta)\hat{\mathbf{k}} = -IAB \cos\theta\,\hat{\mathbf{k}} = -IabB \cos\theta\,\hat{\mathbf{k}}$$

Evaluate the gravitational torque on the loop, noting that the gravitational force can be modeled to act at the center of the loop:

$$\vec{\tau}_g = \vec{\mathbf{r}} \times m\vec{\mathbf{g}} = mg\frac{b}{2}\sin\theta\,\hat{\mathbf{k}}$$

From the rigid body in equilibrium model, add the torques and set the net torque equal to zero:

$$\sum\vec{\tau} = -IabB \cos\theta\,\hat{\mathbf{k}} + mg\frac{b}{2}\sin\theta\,\hat{\mathbf{k}} = 0$$

Solve for $\tan\theta$:

$$IabB \cos\theta = mg\frac{b}{2}\sin\theta \quad\rightarrow\quad \tan\theta = \frac{2IaB}{mg}$$

Substitute numerical values:

$$\theta = \tan^{-1}\left(\frac{2IaB}{mg}\right)$$

$$= \tan^{-1}\left[\frac{2(3.50\text{ A})(0.200\text{ m})(0.010\ 0\text{ T})}{(0.050\ 0\text{ kg})(9.80\text{ m/s}^2)}\right] = \boxed{1.64°}$$

Finalize   The angle is relatively small, so the loop still hangs almost vertically. If the current I or the magnetic field B is increased, however, the angle increases as the magnetic torque becomes stronger.

29.6   The Hall Effect

When a current-carrying conductor is placed in a magnetic field, a potential difference is generated in a direction perpendicular to both the current and the magnetic field. This phenomenon, first observed by Edwin Hall (1855–1938) in 1879, is known as the *Hall effect*. The arrangement for observing the Hall effect consists of a flat conductor carrying a current I in the x direction as shown in Figure 29.25 (page 826). A uniform magnetic field $\vec{\mathbf{B}}$ is applied in the y direction. If the charge

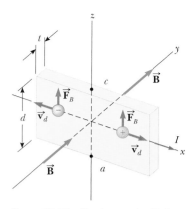

Figure 29.25 To observe the Hall effect, a magnetic field is applied to a current-carrying conductor. When I is in the x direction and $\vec{\mathbf{B}}$ in the y direction, both positive and negative charge carriers are deflected upward in the magnetic field. The Hall voltage is measured between points a and c.

carriers are electrons moving in the negative x direction with a drift velocity $\vec{\mathbf{v}}_d$, they experience an upward magnetic force $\vec{\mathbf{F}}_B = q\vec{\mathbf{v}}_d \times \vec{\mathbf{B}}$, are deflected upward, and accumulate at the upper edge of the flat conductor, leaving an excess of positive charge at the lower edge (Fig. 29.26a). This accumulation of charge at the edges establishes an electric field in the conductor and increases until the electric force on carriers remaining in the bulk of the conductor balances the magnetic force acting on the carriers. When this equilibrium condition is reached, the electrons are no longer deflected upward. A sensitive voltmeter connected across the sample as shown in Figure 29.26 can measure the potential difference, known as the **Hall voltage** ΔV_{H}, generated across the conductor.

If the charge carriers are positive and hence move in the positive x direction (for rightward current) as shown in Figures 29.25 and 29.26b, they also experience an upward magnetic force $q\vec{\mathbf{v}}_d \times \vec{\mathbf{B}}$, which produces a buildup of positive charge on the upper edge and leaves an excess of negative charge on the lower edge. Hence, the sign of the Hall voltage generated in the sample is opposite the sign of the Hall voltage resulting from the deflection of electrons. The sign of the charge carriers can therefore be determined from measuring the polarity of the Hall voltage.

In deriving an expression for the Hall voltage, first note that the magnetic force exerted on the carriers has magnitude qv_dB. In equilibrium, this force is balanced by the electric force qE_{H}, where E_{H} is the magnitude of the electric field due to the charge separation (sometimes referred to as the *Hall field*). Therefore,

$$qv_dB = qE_{\mathrm{H}}$$

$$E_{\mathrm{H}} = v_dB$$

If d is the width of the conductor, the Hall voltage is

$$\Delta V_{\mathrm{H}} = E_{\mathrm{H}}d = v_dBd \qquad (29.19)$$

Therefore, the measured Hall voltage gives a value for the drift speed of the charge carriers if d and B are known.

We can obtain the charge carrier density n by measuring the current in the sample. From Equation 27.4, we can express the drift speed as

$$v_d = \frac{I}{nqA} \qquad (29.20)$$

where A is the cross-sectional area of the conductor. Substituting Equation 29.20 into Equation 29.19 gives

$$\Delta V_{\mathrm{H}} = \frac{IBd}{nqA} \qquad (29.21)$$

(a)

(b)

Figure 29.26 (a) When the charge carriers in a Hall-effect apparatus are negative, the upper edge of the conductor becomes negatively charged and c is at a lower electric potential than a. (b) When the charge carriers are positive, the upper edge becomes positively charged and c is at a higher potential than a. In either case, the charge carriers are no longer deflected when the edges become sufficiently charged that there is a balance on the charge carriers between the electrostatic force qE_{H} and the magnetic deflection force qvB.

Because $A = td$, where t is the thickness of the conductor, we can also express Equation 29.21 as

$$\Delta V_{\text{H}} = \frac{IB}{nqt} = \frac{R_{\text{H}}IB}{t}$$ (29.22)   ◄ The Hall voltage

where $R_{\text{H}} = 1/nq$ is the **Hall coefficient.** This relationship shows that a properly calibrated conductor can be used to measure the magnitude of an unknown magnetic field.

Because all quantities in Equation 29.23 other than nq can be measured, a value for the Hall coefficient is readily obtainable. The sign and magnitude of R_{H} give the sign of the charge carriers and their number density. In most metals, the charge carriers are electrons and the charge-carrier density determined from Hall-effect measurements is in good agreement with calculated values for such metals as lithium (Li), sodium (Na), copper (Cu), and silver (Ag), whose atoms each give up one electron to act as a current carrier. In this case, n is approximately equal to the number of conducting electrons per unit volume. This classical model, however, is not valid for metals such as iron (Fe), bismuth (Bi), and cadmium (Cd) or for semiconductors. These discrepancies can be explained only by using a model based on the quantum nature of solids.

EXAMPLE 29.7   **The Hall Effect for Copper**

A rectangular copper strip 1.5 cm wide and 0.10 cm thick carries a current of 5.0 A. Find the Hall voltage for a 1.2-T magnetic field applied in a direction perpendicular to the strip.

SOLUTION

Conceptualize Study Figures 29.25 and 29.26 carefully and make sure you understand that a Hall voltage is developed between the top and bottom edges of the strip.

Categorize We evaluate the Hall voltage using an equation developed in this section, so we categorize this example as a substitution problem.

Assuming that one electron per atom is available for conduction, we can take the charge carrier density to be 8.46×10^{28} electrons/m^3 (see Example 27.1). Substitute this value and the given data into Equation 29.22:

$$\Delta V_{\text{H}} = \frac{IB}{nqt}$$

$$= \frac{(5.0 \text{ A})(1.2 \text{ T})}{(8.46 \times 10^{28} \text{ m}^{-3})(1.6 \times 10^{-19} \text{ C})(0.001 \text{ 0 m})}$$

$$\Delta V_{\text{H}} = \boxed{0.44 \ \mu\text{V}}$$

Such an extremely small Hall voltage is expected in good conductors. (Notice that the width of the conductor is not needed in this calculation.)

What If? What if the strip has the same dimensions but is made of a semiconductor? Will the Hall voltage be smaller or larger?

Answer In semiconductors, n is much smaller than it is in metals that contribute one electron per atom to the current; hence, the Hall voltage is usually larger because it varies as the inverse of n. Currents on the order of 0.1 mA are generally used for such materials. Consider a piece of silicon that has the same dimensions as the copper strip in this example and whose value for n is 1.0×10^{20} electrons/m^3. Taking $B = 1.2$ T and $I = 0.10$ mA, we find that $\Delta V_{\text{H}} = 7.5$ mV. A potential difference of this magnitude is readily measured.

Summary

ThomsonNOW™ Sign in at **www.thomsonedu.com** and go to ThomsonNOW to take a practice test for this chapter.

DEFINITION

The **magnetic dipole moment** $\vec{\mu}$ of a loop carrying a current I is

$$\vec{\mu} \equiv I\vec{A} \tag{29.5}$$

where the area vector $\vec{A}$ is perpendicular to the plane of the loop and $|\vec{A}|$ is equal to the area of the loop. The SI unit of $\vec{\mu}$ is A · m².

CONCEPTS AND PRINCIPLES

The magnetic force that acts on a charge q moving with a velocity $\vec{v}$ in a magnetic field $\vec{B}$ is

$$\vec{F}_B = q\vec{v} \times \vec{B} \tag{29.1}$$

The direction of this magnetic force is perpendicular both to the velocity of the particle and to the magnetic field. The magnitude of this force is

$$F_B = |q|vB \sin \theta \tag{29.2}$$

where θ is the smaller angle between $\vec{v}$ and $\vec{B}$. The SI unit of $\vec{B}$ is the **tesla** (T), where 1 T = 1 N/A · m.

If a charged particle moves in a uniform magnetic field so that its initial velocity is perpendicular to the field, the particle moves in a circle, the plane of which is perpendicular to the magnetic field. The radius of the circular path is

$$r = \frac{mv}{qB} \tag{29.3}$$

where m is the mass of the particle and q is its charge. The angular speed of the charged particle is

$$\omega = \frac{qB}{m} \tag{29.4}$$

If a straight conductor of length L carries a current I, the force exerted on that conductor when it is placed in a uniform magnetic field $\vec{B}$ is

$$\vec{F}_B = I\vec{L} \times \vec{B} \tag{29.10}$$

where the direction of $\vec{L}$ is in the direction of the current and $|\vec{L}| = L$.

If an arbitrarily shaped wire carrying a current I is placed in a magnetic field, the magnetic force exerted on a very small segment $d\vec{s}$ is

$$d\vec{F}_B = I \, d\vec{s} \times \vec{B} \tag{29.11}$$

To determine the total magnetic force on the wire, one must integrate Equation 29.11 over the wire, keeping in mind that both $\vec{B}$ and $d\vec{s}$ may vary at each point.

The torque $\vec{\tau}$ on a current loop placed in a uniform magnetic field $\vec{B}$ is

$$\vec{\tau} = \vec{\mu} \times \vec{B} \tag{29.17}$$

The potential energy of the system of a magnetic dipole in a magnetic field is

$$U = -\vec{\mu} \cdot \vec{B} \tag{29.18}$$

Questions

☐ denotes answer available in *Student Solutions Manual/Study Guide;* **O** denotes objective question

> Questions 2, 3, and 4 in Chapter 11 can be assigned with this chapter.

1. O Answer each question yes or no. Assume the motions and currents mentioned are along the *x* axis and fields are in the *y* direction. (a) Does an electric field exert a force on a stationary charged object? (b) Does a magnetic field do so? (c) Does an electric field exert a force on a moving charged object? (d) Does a magnetic field do so? (e) Does an electric field exert a force on a straight current-carrying wire? (f) Does a magnetic field do so? (g) Does an electric field exert a force on a beam of electrons? (h) Does a magnetic field do so?

2. O Electron A is fired horizontally with speed 1 Mm/s into a region where a vertical magnetic field exists. Electron B is fired along the same path with speed 2 Mm/s. **(i)** Which electron has a larger magnetic force exerted on it? (a) A does. (b) B does. (c) The forces have the same nonzero magnitude. (d) The forces are both zero. **(ii)** Which electron has a path that curves more sharply? (a) A does. (b) B does. (c) The particles follow the same curved path. (d) The particles continue to go straight.

3. O Classify each of the following as a characteristic (a) of electric forces only, (b) of magnetic forces only, (c) of both electric and magnetic forces, or (d) of neither electric nor magnetic forces. **(i)** The force is proportional to the magnitude of the field exerting it. **(ii)** The force is proportional to the magnitude of the charge of the object on which the force is exerted. **(iii)** The force exerted on a negatively charged object is opposite in direction to the force on a positive charge. **(iv)** The force exerted on a stationary charged object is zero. **(v)** The force exerted on a moving charged object is zero. **(vi)** The force exerted on a charged object is proportional to its speed. **(vii)** The force exerted on a charged object cannot alter the object's speed. **(viii)** The magnitude of the force depends on the charged object's direction of motion.

4. O Rank the magnitudes of the forces exerted on the following particles from the largest to the smallest. In your ranking, display any cases of equality. (a) An electron moving at 1 Mm/s perpendicular to a 1-mT magnetic field. (b) An electron moving at 1 Mm/s parallel to a 1-mT magnetic field. (c) An electron moving at 2 Mm/s perpendicular to a 1-mT magnetic field. (d) An electron moving at 1 Mm/s perpendicular to a 2-mT magnetic field. (e) A proton moving at 1 Mm/s perpendicular to a 1-mT magnetic field. (f) A proton moving at 1 Mm/s at a 45° angle to a 1-mT magnetic field.

5. O At a certain instant, a proton is moving in the positive *x* direction through a magnetic field in the negative *z* direction. What is the direction of the magnetic force exerted on the proton? (a) *x* (b) −*x* (c) *y* (d) −*y* (e) *z* (f) −*z* (g) halfway between the *x* and −*z* axes, at 45° to both (h) The force is zero.

6. O A particle with electric charge is fired into a region of space where the electric field is zero. It moves in a straight line. Can you conclude that the magnetic field in that region is zero? (a) Yes. (b) No; the field might be perpendicular to the particle's velocity. (c) No; the field might be parallel to the particle's velocity. (d) No; the particle might need to have charge of the opposite sign to have a force exerted on it. (e) No; an observation of an object with *electric* charge gives no information about a *magnetic* field.

7. Two charged particles are projected in the same direction into a magnetic field perpendicular to their velocities. If the particles are deflected in opposite directions, what can you say about them?

8. How can the motion of a moving charged particle be used to distinguish between a magnetic field and an electric field? Give a specific example to justify your argument.

9. O In the velocity selector shown in Active Figure 29.12, electrons with speed $v = E/B$ follow a straight path. Electrons moving significantly faster than this speed through the same selector will move along what kind of path? (a) a circle (b) a parabola (c) a straight line (d) a more complicated trajectory

10. Is it possible to orient a current loop in a uniform magnetic field such that the loop does not tend to rotate? Explain.

11. Explain why it is not possible to determine the charge and the mass of a charged particle separately by measuring accelerations produced by electric and magnetic forces on the particle.

12. How can a current loop be used to determine the presence of a magnetic field in a given region of space?

13. Charged particles from outer space, called cosmic rays, strike the Earth more frequently near the poles than near the equator. Why?

14. Can a constant magnetic field set into motion an electron initially at rest? Explain your answer.

Problems

1, 2, 3 denotes straightforward, intermediate, challenging; □ denotes full solution available in *Student Solutions Manual/Study Guide;* ▲ denotes coached solution with hints available at **www.thomsonedu.com;** denotes developing symbolic reasoning; ● denotes asking for qualitative reasoning; ▪ denotes computer useful in solving problem

Section 29.1 Magnetic Fields and Forces

Problems 1, 2, 3, 4, 6, 7, and 10 in Chapter 11 can be assigned with this section.

1. ▲ Determine the initial direction of the deflection of charged particles as they enter the magnetic fields shown in Figure P29.1.

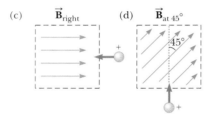

Figure P29.1

2. Consider an electron near the Earth's equator. In which direction does it tend to deflect if its velocity is (a) directed downward? (b) Directed northward? (c) Directed westward? (d) Directed southeastward?

3. A proton moves perpendicular to a uniform magnetic field $\vec{B}$ at a speed of 1.00×10^7 m/s and experiences an acceleration of 2.00×10^{13} m/s² in the $+x$ direction when its velocity is in the $+z$ direction. Determine the magnitude and direction of the field.

4. A proton travels with a speed of 3.00×10^6 m/s at an angle of $37.0°$ with the direction of a magnetic field of 0.300 T in the $+y$ direction. What are (a) the magnitude of the magnetic force on the proton and (b) its acceleration?

5. A proton moving at 4.00×10^6 m/s through a magnetic field of magnitude 1.70 T experiences a magnetic force of magnitude 8.20×10^{-13} N. What is the angle between the proton's velocity and the field?

6. An electron is accelerated through 2 400 V from rest and then enters a uniform 1.70-T magnetic field. What are (a) the maximum and (b) the minimum values of the magnetic force this particle can experience?

7. A proton moves with a velocity of $\vec{v} = (2\hat{i} - 4\hat{j} + \hat{k})$ m/s in a region in which the magnetic field is $\vec{B} = (\hat{i} + 2\hat{j} - 3\hat{k})$ T. What is the magnitude of the magnetic force this particle experiences?

8. ● An electron in a uniform electric and magnetic field has a velocity of 1.20×10^4 m/s (in the positive x direction) and an acceleration of 2.00×10^{12} m/s² (in the positive z direction). If the electric field has a magnitude of 20.0 N/C (in the positive z direction), what can you determine about the magnetic field in the region? What can you not determine?

Section 29.2 Motion of a Charged Particle in a Uniform Magnetic Field

9. The magnetic field of the Earth at a certain location is directed vertically downward and has a magnitude of 50.0 μT. A proton is moving horizontally toward the west in this field with a speed of 6.20×10^6 m/s. (a) What are the direction and magnitude of the magnetic force the field exerts on this particle? (b) What is the radius of the circular arc followed by this proton?

10. ● An accelerating voltage of 2 500 V is applied to an electron gun, producing a beam of electrons originally traveling horizontally north in vacuum toward the center of a viewing screen 35.0 cm away. (a) What are the magnitude and direction of the deflection on the screen caused by the Earth's gravitational field? (b) What are the magnitude and direction of the deflection on the screen caused by the vertical component of the Earth's magnetic field, taken as 20.0 μT down? Does an electron in this vertical magnetic field move as a projectile, with constant vector acceleration perpendicular to a constant northward component of velocity? Is it a good approximation to assume it has this projectile motion? Explain.

11. A proton (charge $+e$, mass m_p), a deuteron (charge $+e$, mass $2m_p$), and an alpha particle (charge $+2e$, mass $4m_p$) are accelerated through a common potential difference ΔV. Each of the particles enters a uniform magnetic field $\vec{B}$, with its velocity in a direction perpendicular to $\vec{B}$. The proton moves in a circular path of radius r_p. Determine the radii of the circular orbits for the deuteron, r_d, and the alpha particle, r_α, in terms of r_p.

12. Review problem. One electron collides elastically with a second electron initially at rest. After the collision, the radii of their trajectories are 1.00 cm and 2.40 cm. The trajectories are perpendicular to a uniform magnetic field of magnitude 0.044 0 T. Determine the energy (in keV) of the incident electron.

13. Review problem. An electron moves in a circular path perpendicular to a constant magnetic field of magnitude 1.00 mT. The angular momentum of the electron about the center of the circle is 4.00×10^{-25} kg·m²/s. Determine (a) the radius of the circular path and (b) the speed of the electron.

14. A singly charged ion of mass m is accelerated from rest by a potential difference ΔV. It is then deflected by a uniform magnetic field (perpendicular to the ion's velocity) into a semicircle of radius R. Now a doubly charged ion of mass m' is accelerated through the same potential difference and deflected by the same magnetic field into a semicircle of radius $R' = 2R$. What is the ratio of the masses of the ions?

15. A cosmic-ray proton in interstellar space has an energy of 10.0 MeV and executes a circular orbit having a radius equal to that of Mercury's orbit around the Sun $(5.80 \times 10^{10}$ m). What is the magnetic field in that region of space?

16. Assume the region to the right of a certain vertical plane contains a vertical magnetic field of magnitude 1.00 mT and the field is zero in the region to the left of the plane. An electron, originally traveling perpendicular to the boundary plane, passes into the region of the field. (a) Determine the time interval required for the electron to leave the "field-filled" region, noting that its path is a semicircle. (b) Find the kinetic energy of the electron, assuming the maximum depth of penetration into the field is 2.00 cm.

Section 29.3 Applications Involving Charged Particles Moving in a Magnetic Field

17. A velocity selector consists of electric and magnetic fields described by the expressions $\vec{E} = E\hat{k}$ and $\vec{B} = B\hat{j}$, with $B = 15.0$ mT. Find the value of E such that a 750-eV electron moving along the positive x axis is undeflected.

18. ● Singly charged uranium-238 ions are accelerated through a potential difference of 2.00 kV and enter a uniform magnetic field of 1.20 T directed perpendicular to their velocities. (a) Determine the radius of their circular path. (b) Repeat for uranium-235 ions. **What If?** How does the ratio of these path radii depend on the accelerating voltage? On the magnitude of the magnetic field?

19. Consider the mass spectrometer shown schematically in Active Figure 29.13. The magnitude of the electric field between the plates of the velocity selector is 2 500 V/m, and the magnetic field in both the velocity selector and the deflection chamber has a magnitude of 0.035 0 T. Calculate the radius of the path for a singly charged ion having a mass $m = 2.18 \times 10^{-26}$ kg.

20. A cyclotron designed to accelerate protons has an outer radius of 0.350 m. The protons are emitted nearly at rest from a source at the center and are accelerated through 600 V each time they cross the gap between the dees. The dees are between the poles of an electromagnet where the field is 0.800 T. (a) Find the cyclotron frequency for the protons in this cyclotron. (b) Find the speed at which protons exit the cyclotron and (c) their maximum kinetic energy. (d) How many revolutions does a proton make in the cyclotron? (e) For what time interval does one proton accelerate?

21. A cyclotron designed to accelerate protons has a magnetic field of magnitude 0.450 T over a region of radius 1.20 m. What are (a) the cyclotron frequency and (b) the maximum speed acquired by the protons?

22. ● A particle in the cyclotron shown in Figure 29.15a gains energy $q \, \Delta V$ from the alternating power supply each time it passes from one dee to the other. The time interval for each full orbit is

$$T = \frac{2\pi}{\omega} = \frac{2\pi m}{qB}$$

so the particle's average rate of increase in energy is

$$\frac{2q \, \Delta V}{T} = \frac{q^2 B \, \Delta V}{\pi m}$$

Note that this power input is constant in time. (a) Show that the rate of increase in the radius r of its path is not constant, but is given by

$$\frac{dr}{dt} = \frac{1}{r} \frac{\Delta V}{\pi B}$$

(b) Describe how the path of the particles in Figure 29.15a could be drawn more realistically. (c) At what rate is the radial position of the protons in Problem 20 increasing immediately before the protons leave the cyclotron? (d) By how much does the radius of the protons' path increase during their last full revolution?

23. ▲ The picture tube in a television uses magnetic deflection coils rather than electric deflection plates. Suppose an electron beam is accelerated through a 50.0-kV potential difference and then through a region of uniform magnetic field 1.00 cm wide. The screen is located 10.0 cm from the center of the coils and is 50.0 cm wide. When the field is turned off, the electron beam hits the center of the screen. What field magnitude is necessary to deflect the beam to the side of the screen? Ignore relativistic corrections.

24. ● In his "discovery of the electron," J. J. Thomson showed that the same beam deflections resulted with tubes having cathodes made of *different* materials and containing *various* gases before evacuation. (a) Are these observations important? Explain your answer. (b) When he applied various potential differences to the deflection plates and turned on the magnetic coils, alone or in combination with the deflection plates, Thomson observed that the fluorescent screen continued to show a *single small* glowing patch. Argue whether his observation is important. (c) Do calculations to show that the charge-to-mass ratio Thomson obtained was huge compared to that of any macroscopic object or of any ionized atom or molecule. How can one make sense of that fact? (d) Could Thomson observe any deflection of the beam due to gravitation? Do a calculation to argue for your answer. (To obtain a visibly glowing patch on the fluorescent screen, the potential difference between the slits and the cathode must be 100 V or more.)

Section 29.4 Magnetic Force Acting on a Current-Carrying Conductor

25. ▲ A wire having a mass per unit length of 0.500 g/cm carries a 2.00-A current horizontally to the south. What are the direction and magnitude of the minimum magnetic field needed to lift this wire vertically upward?

26. A wire carries a steady current of 2.40 A. A straight section of the wire is 0.750 m long and lies along the x axis within a uniform magnetic field, $\vec{B} = 1.60\hat{k}$ T. If the current is in the $+x$ direction, what is the magnetic force on the section of wire?

27. A wire 2.80 m in length carries a current of 5.00 A in a region where a uniform magnetic field has a magnitude of 0.390 T. Calculate the magnitude of the magnetic force on the wire assuming the angle between the magnetic field and the current is (a) 60.0°, (b) 90.0°, and (c) 120°.

28. Imagine a wire with linear mass density 2.40 g/m encircling the Earth at its magnetic equator, where the field is modeled as having the uniform value 28.0 μT horizontally north. What magnitude and direction of the current in the wire will keep the wire levitated immediately above the ground?

29. **Review problem.** A rod of mass 0.720 kg and radius 6.00 cm rests on two parallel rails (Fig. P29.29) that are $d = 12.0$ cm apart and $L = 45.0$ cm long. The rod carries a current of $I = 48.0$ A in the direction shown and rolls along the rails without slipping. A uniform magnetic field of magnitude 0.240 T is directed perpendicular to the rod and the rails. If it starts from rest, what is the speed of the rod as it leaves the rails?

Figure P29.29   Problems 29 and 30.

30. **Review problem.** A rod of mass m and radius R rests on two parallel rails (Fig. P29.29) that are a distance d apart and have a length L. The rod carries a current I in the direction shown and rolls along the rails without slipping. A uniform magnetic field B is directed perpendicular to the rod and the rails. If it starts from rest, what is the speed of the rod as it leaves the rails?

31. ▲ *A nonuniform magnetic field exerts a net force on a magnetic dipole.* A strong magnet is placed under a horizontal conducting ring of radius r that carries current I as shown in Figure P29.31. If the magnetic field $\vec{B}$ makes an angle θ with the vertical at the ring's location, what are the magnitude and direction of the resultant magnetic force on the ring?

Figure P29.31

32. ● In Figure P29.32, the cube is 40.0 cm on each edge. Four straight segments of wire—ab, bc, cd, and da—form a closed loop that carries a current $I = 5.00$ A in the direction shown. A uniform magnetic field of magnitude $B = 0.020\ 0$ T is in the positive y direction. (a) Determine the magnitude and direction of the magnetic force on each segment. (b) Explain how you could find the force exerted on the fourth of these segments from the forces on the other three, without further calculation involving the magnetic field.

Figure P29.32

33. Assume the Earth's magnetic field is 52.0 μT northward at 60.0° below the horizontal in Atlanta, Georgia. A tube in a neon sign—situated between two diagonally opposite corners of a shop window, which lies in a north–south vertical plane—carries current 35.0 mA. The current enters the tube at the bottom south corner of the shop's window. It exits at the opposite corner, which is 1.40 m farther north and 0.850 m higher up. Between these two points, the glowing tube spells out DONUTS. Determine the total vector magnetic force on the tube. You may use the first "important statement" presented in the *Finalize* section of Example 29.4.

Section 29.5 Torque on a Current Loop in a Uniform Magnetic Field

34. A current of 17.0 mA is maintained in a single circular loop of 2.00 m circumference. A magnetic field of 0.800 T is directed parallel to the plane of the loop. (a) Calculate the magnetic moment of the loop. (b) What is the magnitude of the torque exerted by the magnetic field on the loop?

35. ▲ A rectangular coil consists of $N = 100$ closely wrapped turns and has dimensions $a = 0.400$ m and $b = 0.300$ m. The coil is hinged along the y axis, and its plane makes an angle $\theta = 30.0°$ with the x axis (Fig. P29.35). What is the magnitude of the torque exerted on the coil by a uniform magnetic field $B = 0.800$ T directed along the x axis when the current is $I = 1.20$ A in the direction shown? What is the expected direction of rotation of the coil?

Figure P29.35

2 = intermediate;   3 = challenging;   □ = SSM/SG;   ▲ = ThomsonNOW;   ▬ = symbolic reasoning;   ● = qualitative reasoning

36. A current loop with magnetic dipole moment $\vec{\mu}$ is placed in a uniform magnetic field $\vec{B}$, with its moment making angle θ with the field. With the arbitrary choice of $U = 0$ for $\theta = 90°$, prove that the potential energy of the dipole–field system is $U = -\vec{\mu} \cdot \vec{B}$. You may imitate the discussion in Chapter 26 of the potential energy of an electric dipole in an electric field.

37. ● The needle of a magnetic compass has magnetic moment $9.70 \text{ mA} \cdot \text{m}^2$. At its location, the Earth's magnetic field is $55.0 \ \mu\text{T}$ north at $48.0°$ below the horizontal. (a) Identify the orientations of the compass needle that represent minimum potential energy and maximum potential energy of the needle–field system. (b) How much work must be done on the needle to move it from the former to the latter orientation?

38. A wire is formed into a circle having a diameter of 10.0 cm and placed in a uniform magnetic field of 3.00 mT. The wire carries a current of 5.00 A. Find (a) the maximum torque on the wire and (b) the range of potential energies of the wire–field system for different orientations of the circle.

39. ● A wire 1.50 m long carries a current of 30.0 mA when it is connected to a battery. The whole wire can be arranged as a single loop with the shape of a circle, a square, or an equilateral triangle. The whole wire can be made into a flat, compact, circular coil with N turns. Explain how its magnetic moment compares in all these cases. In particular, can its magnetic moment go to infinity? To zero? Does its magnetic moment have a well-defined maximum value? If so, identify it. Does it have minimum value? If so, identify it.

40. The rotor in a certain electric motor is a flat, rectangular coil with 80 turns of wire and dimensions 2.50 cm by 4.00 cm. The rotor rotates in a uniform magnetic field of 0.800 T. When the plane of the rotor is perpendicular to the direction of the magnetic field, it carries a current of 10.0 mA. In this orientation, the magnetic moment of the rotor is directed opposite the magnetic field. The rotor then turns through one-half revolution. This process is repeated to cause the rotor to turn steadily at 3 600 rev/min. (a) Find the maximum torque acting on the rotor. (b) Find the peak power output of the motor. (c) Determine the amount of work performed by the magnetic field on the rotor in every full revolution. (d) What is the average power of the motor?

Section 29.6 The Hall Effect

41. In an experiment designed to measure the Earth's magnetic field using the Hall effect, a copper bar 0.500 cm thick is positioned along an east–west direction. If a current of 8.00 A in the conductor results in a Hall voltage of 5.10×10^{-12} V, what is the magnitude of the Earth's magnetic field? (Assume $n = 8.46 \times 10^{28}$ electrons/m^3 and the plane of the bar is rotated to be perpendicular to the direction of $\vec{B}$.)

42. A Hall-effect probe operates with a 120-mA current. When the probe is placed in a uniform magnetic field of magnitude 0.080 0 T, it produces a Hall voltage of 0.700 μV. (a) When it is used to measure an unknown magnetic field, the Hall voltage is 0.330 μV. What is the magnitude of the unknown field? (b) The thickness of the probe in the direction of $\vec{B}$ is 2.00 mm. Find the den-

sity of the charge carriers, each of which has charge of magnitude e.

Additional Problems

43. ● Heart-lung machines and artificial kidney machines employ blood pumps. A mechanical pump can mangle blood cells. Figure P29.43 represents an electromagnetic pump. The blood is confined to an electrically insulating tube, cylindrical in practice but represented as a rectangle of width w and height h. The simplicity of design makes the pump dependable. The blood is easily kept uncontaminated; the tube is simple to clean or inexpensive to replace. Two electrodes fit into the top and the bottom of the tube. The potential difference between them establishes an electric current through the blood, with current density J over a section of length L. A perpendicular magnetic field exists in the same region. (a) Explain why this arrangement produces on the liquid a force that is directed along the length of the pipe. (b) Show that the section of liquid in the magnetic field experiences a pressure increase JLB. (c) After the blood leaves the pump, is it charged? Is it current carrying? Is it magnetized? The same magnetic pump can be used for any fluid that conducts electricity, such as liquid sodium in a nuclear reactor.

Figure P29.43

44. ● Figure 29.10 shows a charged particle traveling in a nonuniform magnetic field forming a magnetic bottle. (a) Explain why the positively charged particle in the figure must be moving clockwise. The particle travels along a helix whose radius decreases and whose pitch decreases as the particle moves into a stronger magnetic field. If the particle is moving to the right along the x axis, its velocity in this direction will be reduced to zero and it will be reflected from the right-hand side of the bottle, acting as a "magnetic mirror." The particle ends up bouncing back and forth between the ends of the bottle. (b) Explain qualitatively why the axial velocity is reduced to zero as the particle moves into the region of strong magnetic field at the end of the bottle. (c) Explain why the tangential velocity increases as the particle approaches the end of the bottle. (d) Explain why the orbiting particle has a magnetic dipole moment. (e) Sketch the magnetic moment and use the result of Problem 31 to explain again how the nonuniform magnetic field exerts a force on the orbiting particle along the x axis.

45. ● Assume in the plane of the Earth's magnetic equator the planet's field is uniform with the value $25.0 \ \mu\text{T}$ northward perpendicular to this plane, everywhere inside a radius of 100 Mm. Also assume the Earth's field is zero outside this circle. A cosmic-ray proton traveling at one tenth of the speed of light is heading directly toward the center of the Earth in the plane of the magnetic equator. Find the radius of curvature of the path it follows when it enters the region of the planet's assumed field. Explain whether the proton will hit the Earth.

46. A 0.200-kg metal rod carrying a current of 10.0 A glides on two horizontal rails 0.500 m apart. What vertical magnetic field is required to keep the rod moving at a constant speed if the coefficient of kinetic friction between the rod and rails is 0.100?

47. Protons having a kinetic energy of 5.00 MeV are moving in the positive x direction and enter a magnetic field $\vec{B} = 0.050\,0\hat{k}$ T directed out of the plane of the page and extending from $x = 0$ to $x = 1.00$ m as shown in Figure P29.47. (a) Calculate the y component of the protons' momentum as they leave the magnetic field. (b) Find the angle α between the initial velocity vector of the proton beam and the velocity vector after the beam emerges from the field. Ignore relativistic effects and note that 1 eV $= 1.60 \times 10^{-19}$ J.

Figure P29.47

48. ● (a) A proton moving in the $+x$ direction with velocity $\vec{v} = v_i\hat{i}$ experiences a magnetic force $\vec{F} = F_i\hat{j}$ in the $+y$ direction. Explain what you can and cannot infer about $\vec{B}$ from this information. (b) **What If?** In terms of F_i, what would be the force on a proton in the same field moving with velocity $\vec{v} = -v_i\hat{i}$? (c) What would be the force on an electron in the same field moving with velocity $\vec{v} = -v_i\hat{i}$?

49. A particle with positive charge $q = 3.20 \times 10^{-19}$ C moves with a velocity $\vec{v} = (2\hat{i} + 3\hat{j} - \hat{k})$ m/s through a region where both a uniform magnetic field and a uniform electric field exist. (a) Calculate the total force on the moving particle (in unit–vector notation), taking $\vec{B} = (2\hat{i} + 4\hat{j} + \hat{k})$ T and $\vec{E} = (4\hat{i} - \hat{j} - 2\hat{k})$ V/m. (b) What angle does the force vector make with the positive x axis?

50. A proton having an initial velocity of $20.0\hat{i}$ Mm/s enters a uniform magnetic field of magnitude 0.300 T with a direction perpendicular to the proton's velocity. It leaves the field-filled region with velocity $-20.0\hat{j}$ Mm/s. Determine (a) the direction of the magnetic field, (b) the radius of curvature of the proton's path while in the field, (c) the distance the proton traveled in the field, and (d) the time interval for which the proton is in the field.

51. **Review problem.** A wire having a linear mass density of 1.00 g/cm is placed on a horizontal surface that has a coefficient of kinetic friction of 0.200. The wire carries a current of 1.50 A toward the east and slides horizontally to the north. What are the magnitude and direction of the smallest magnetic field that enables the wire to move in this fashion?

52. **Review problem.** A proton is at rest at the plane vertical boundary of a region containing a uniform vertical magnetic field B. An alpha particle moving horizontally makes a head-on elastic collision with the proton. Immediately after the collision, both particles enter the magnetic field, moving perpendicular to the direction of the field. The

radius of the proton's trajectory is R. Find the radius of the alpha particle's trajectory. The mass of the alpha particle is four times that of the proton, and its charge is twice that of the proton.

53. The circuit in Figure P29.53 consists of wires at the top and bottom and identical metal springs in the left and right sides. The upper portion of the circuit is fixed. The wire at the bottom has a mass of 10.0 g and is 5.00 cm long. The springs stretch 0.500 cm under the weight of the wire, and the circuit has a total resistance of 12.0 Ω. When a magnetic field is turned on, directed out of the page, the springs stretch an additional 0.300 cm. What is the magnitude of the magnetic field?

Figure P29.53

54. A handheld electric mixer contains an electric motor. Model the motor as a single flat, compact, circular coil carrying electric current in a region where a magnetic field is produced by an external permanent magnet. You need consider only one instant in the operation of the motor. (We will consider motors again in Chapter 31.) The coil moves because the magnetic field exerts torque on the coil as described in Section 29.5. Make order-of-magnitude estimates of the magnetic field, the torque on the coil, the current in it, its area, and the number of turns in the coil so that they are related according to Equation 29.17. Note that the input power to the motor is electric, given by $\mathcal{P} = I\,\Delta V$, and the useful output power is mechanical, $\mathcal{P} = \tau\omega$.

55. A nonconducting sphere has mass 80.0 g and radius 20.0 cm. A flat, compact coil of wire with five turns is wrapped tightly around it, with each turn concentric with the sphere. As shown in Figure P29.55, the sphere is placed on an inclined plane that slopes downward to the left, making an angle θ with the horizontal so that the coil is parallel to the inclined plane. A uniform magnetic field of 0.350 T vertically upward exists in the region of the sphere. What current in the coil will enable the sphere to rest in equilibrium on the inclined plane? Show that the result does not depend on the value of θ.

Figure P29.55

56. A metal rod having a mass per unit length λ carries a current I. The rod hangs from two vertical wires in a uniform vertical magnetic field as shown in Figure P29.56. The

wires make an angle θ with the vertical when in equilibrium. Determine the magnitude of the magnetic field.

Figure P29.56

57. A cyclotron is sometimes used for carbon dating as will be described in Chapter 44. Carbon-14 and carbon-12 ions are obtained from a sample of the material to be dated and accelerated in the cyclotron. If the cyclotron has a magnetic field of magnitude 2.40 T, what is the difference in cyclotron frequencies for the two ions?

58. A uniform magnetic field of magnitude 0.150 T is directed along the positive x axis. A positron moving at 5.00×10^6 m/s enters the field along a direction that makes an angle of 85.0° with the x axis (Fig. P29.58). The motion of the particle is expected to be a helix as described in Section 29.2. Calculate (a) the pitch p and (b) the radius r of the trajectory.

Figure P29.58

59. Consider an electron orbiting a proton and maintained in a fixed circular path of radius $R = 5.29 \times 10^{-11}$ m by the Coulomb force. Treat the orbiting particle as a current loop and calculate the resulting torque when the system is in a magnetic field of 0.400 T directed perpendicular to the magnetic moment of the electron.

60. A proton moving in the plane of the page has a kinetic energy of 6.00 MeV. A magnetic field of magnitude $B = 1.00$ T is directed into the page. The proton enters the magnetic field with its velocity vector at an angle $\theta = 45.0°$ to the linear boundary of the field as shown in Figure P29.60. (a) Find x, the distance from the point of

Figure P29.60

entry to where the proton will leave the field. (b) Determine θ', the angle between the boundary and the proton's velocity vector as it leaves the field.

61. ● A heart surgeon monitors the flow rate of blood through an artery using an electromagnetic flowmeter (Fig. P29.61). Electrodes A and B make contact with the outer surface of the blood vessel, which has interior diameter 3.00 mm. (a) For a magnetic field magnitude of 0.040 0 T, an emf of 160 μV appears between the electrodes. Calculate the speed of the blood. (b) Verify that electrode A is positive as shown. Does the sign of the emf depend on whether the mobile ions in the blood are predominantly positively or negatively charged? Explain.

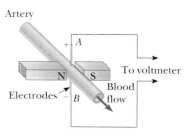

Figure P29.61

62. ▲ The following table shows measurements of a Hall voltage and corresponding magnetic field for a probe used to measure magnetic fields. (a) Plot these data and deduce a relationship between the two variables. (b) If the measurements were taken with a current of 0.200 A and the sample is made from a material having a charge-carrier density of 1.00×10^{26} kg/m^3, what is the thickness of the sample?

ΔV_H (μV)	B (T)
0	0.00
11	0.10
19	0.20
28	0.30
42	0.40
50	0.50
61	0.60
68	0.70
79	0.80
90	0.90
102	1.00

63. ● As shown in Figure P29.63, a particle of mass m having positive charge q is initially traveling with velocity $v\hat{\mathbf{j}}$. At the origin of coordinates it enters a region between $y = 0$ and $y = h$ containing a uniform magnetic field $B\hat{\mathbf{k}}$

Figure P29.63

directed perpendicularly out of the page. (a) What is the critical value of v such that the particle just reaches $y = h$? Describe the path of the particle under this condition and predict its final velocity. (b) Specify the path the particle takes and its final velocity if v is less than the critical value. (c) **What If?** Specify the path the particle takes and its final velocity if v is greater than the critical value.

64. In Niels Bohr's 1913 model of the hydrogen atom, the single electron is in a circular orbit of radius 5.29×10^{-11} m and its speed is 2.19×10^6 m/s. (a) What is the magnitude of the magnetic moment due to the electron's motion? (b) If the electron moves in a horizontal circle, counterclockwise as seen from above, what is the direction of this magnetic moment vector?

65. ● **Review problem**. Review Section 15.5 on torsional pendulums. (a) Show that a magnetic dipole in a uniform magnetic field, displaced from its equilibrium orientation and released, can oscillate as a torsional pendulum in simple harmonic motion. Is this statement true for all angular displacements, for all displacements less than 180°, or only for small angular displacements? Explain. (b) Assume the dipole is a compass needle—a light bar magnet—with a magnetic moment of magnitude μ. It has moment of inertia I about its center, where it is mounted on a frictionless vertical axle, and it is placed in a horizontal magnetic field of magnitude B. Evaluate its frequency of oscillation. (c) Explain how the compass needle can be conveniently used as an indicator of the magnitude of the external magnetic field. If its frequency is 0.680 Hz in the Earth's local field, with a horizontal component of 39.2 μT, what is the magnitude of a field in which its frequency of oscillation is 4.90 Hz?

Answers to Quick Quizzes

29.1 (e). The right-hand rule gives the direction. Be sure to account for the negative charge on the electron.

29.2 **(i),** (b). The magnetic force on the particle increases in proportion to v, but the centripetal acceleration increases according to the square of v. The result is a larger radius, as you can see from Equation 29.3. **(ii),** (a). The magnetic force on the particle increases in proportion to B. The result is a smaller radius as you can see from Equation 29.3.

29.3 (c). Use the right-hand rule to determine the direction of the magnetic field.

29.4 **(i),** (c), (b), (a). Because all loops enclose the same area and carry the same current, the magnitude of $\vec{\mu}$ is the same for all. For (c), $\vec{\mu}$ points upward and is perpendicular to the magnetic field and $\tau = \mu B$, the maximum torque possible. For the loop in (a), $\vec{\mu}$ points along the direction of $\vec{B}$ and the torque is zero. For (b), the torque is intermediate between zero and the maximum value. **(ii),** (a) = (b) = (c). Because the magnetic field is uniform, there is zero net force on all three loops.

A proposed method for launching future payloads into space is the use of *rail guns*, in which projectiles are accelerated by means of magnetic forces. This photo shows the firing of a projectile at a speed of over 3 km/s from an experimental rail gun at Sandia National Research Laboratories, Albuquerque, New Mexico. (Defense Threat Reduction Agency [DTRA])

30.1 The Biot–Savart Law

30.2 The Magnetic Force Between Two Parallel Conductors

30.3 Ampère's Law

30.4 The Magnetic Field of a Solenoid

30.5 Gauss's Law in Magnetism

30.6 Magnetism in Matter

30.7 The Magnetic Field of the Earth

30 Sources of the Magnetic Field

In Chapter 29, we discussed the magnetic force exerted on a charged particle moving in a magnetic field. To complete the description of the magnetic interaction, this chapter explores the origin of the magnetic field, moving charges. We begin by showing how to use the law of Biot and Savart to calculate the magnetic field produced at some point in space by a small current element. This formalism is then used to calculate the total magnetic field due to various current distributions. Next, we show how to determine the force between two current-carrying conductors, leading to the definition of the ampere. We also introduce Ampère's law, which is useful in calculating the magnetic field of a highly symmetric configuration carrying a steady current.

This chapter is also concerned with the complex processes that occur in magnetic materials. All magnetic effects in matter can be explained on the basis of atomic magnetic moments, which arise both from the orbital motion of electrons and from an intrinsic property of electrons known as spin.

30.1 The Biot–Savart Law

Shortly after Oersted's discovery in 1819 that a compass needle is deflected by a current-carrying conductor, Jean-Baptiste Biot (1774–1862) and Félix Savart

(1791–1841) performed quantitative experiments on the force exerted by an electric current on a nearby magnet. From their experimental results, Biot and Savart arrived at a mathematical expression that gives the magnetic field at some point in space in terms of the current that produces the field. That expression is based on the following experimental observations for the magnetic field $d\vec{\mathbf{B}}$ at a point P associated with a length element $d\vec{\mathbf{s}}$ of a wire carrying a steady current I (Fig. 30.1):

- The vector $d\vec{\mathbf{B}}$ is perpendicular both to $d\vec{\mathbf{s}}$ (which points in the direction of the current) and to the unit vector $\hat{\mathbf{r}}$ directed from $d\vec{\mathbf{s}}$ toward P.
- The magnitude of $d\vec{\mathbf{B}}$ is inversely proportional to r^2, where r is the distance from $d\vec{\mathbf{s}}$ to P.
- The magnitude of $d\vec{\mathbf{B}}$ is proportional to the current and to the magnitude ds of the length element $d\vec{\mathbf{s}}$.
- The magnitude of $d\vec{\mathbf{B}}$ is proportional to $\sin\theta$, where θ is the angle between the vectors $d\vec{\mathbf{s}}$ and $\hat{\mathbf{r}}$.

These observations are summarized in the mathematical expression known today as the **Biot–Savart law:**

Biot–Savart law ▶

$$d\vec{\mathbf{B}} = \frac{\mu_0}{4\pi}\frac{I\,d\vec{\mathbf{s}} \times \hat{\mathbf{r}}}{r^2} \tag{30.1}$$

where μ_0 is a constant called the **permeability of free space:**

Permeability of free space ▶

$$\mu_0 = 4\pi \times 10^{-7}\ \text{T}\cdot\text{m/A} \tag{30.2}$$

Notice that the field $d\vec{\mathbf{B}}$ in Equation 30.1 is the field created at a point by the current in only a small length element $d\vec{\mathbf{s}}$ of the conductor. To find the *total* magnetic field $\vec{\mathbf{B}}$ created at some point by a current of finite size, we must sum up contributions from all current elements $I\,d\vec{\mathbf{s}}$ that make up the current. That is, we must evaluate $\vec{\mathbf{B}}$ by integrating Equation 30.1:

$$\vec{\mathbf{B}} = \frac{\mu_0 I}{4\pi}\int\frac{d\vec{\mathbf{s}} \times \hat{\mathbf{r}}}{r^2} \tag{30.3}$$

where the integral is taken over the entire current distribution. This expression must be handled with special care because the integrand is a cross product and therefore a vector quantity. We shall see one case of such an integration in Example 30.1.

Although the Biot–Savart law was discussed for a current-carrying wire, it is also valid for a current consisting of charges flowing through space such as the electron beam in a television picture tube. In that case, $d\vec{\mathbf{s}}$ represents the length of a small segment of space in which the charges flow.

Interesting similarities exist between Equation 30.1 for the magnetic field due to a current element and Equation 23.9 for the electric field due to a point charge. The magnitude of the magnetic field varies as the inverse square of the distance from the source, as does the electric field due to a point charge. The directions of the two fields are quite different, however. The electric field created by a point charge is radial, but the magnetic field created by a current element is perpendicular to both the length element $d\vec{\mathbf{s}}$ and the unit vector $\hat{\mathbf{r}}$ as described by the cross product in Equation 30.1. Hence, if the conductor lies in the plane of the page as shown in Figure 30.1, $d\vec{\mathbf{B}}$ points out of the page at P and into the page at P'.

Another difference between electric and magnetic fields is related to the source of the field. An electric field is established by an isolated electric charge. The Biot–Savart law gives the magnetic field of an isolated current element at some point, but such an isolated current element cannot exist the way an isolated electric charge can. A current element *must* be part of an extended current distribution because a complete circuit is needed for charges to flow. Therefore, the Biot–Savart law (Eq. 30.1) is only the first step in a calculation of a magnetic field; it must be followed by an integration over the current distribution as in Equation 30.3.

Figure 30.1 The magnetic field $d\vec{\mathbf{B}}$ at a point due to the current I through a length element $d\vec{\mathbf{s}}$ is given by the Biot–Savart law. The direction of the field is out of the page at P and into the page at P'.

Quick Quiz 30.1   Consider the magnetic field due to the current in the length of wire shown in Figure 30.2. Rank the points A, B, and C in terms of magnitude of the magnetic field that is due to the current in just the length element $d\vec{\mathbf{s}}$ shown from greatest to least.

Figure 30.2   (Quick Quiz 30.1) Where is the magnetic field the greatest?

EXAMPLE 30.1   **Magnetic Field Surrounding a Thin, Straight Conductor**

Consider a thin, straight wire carrying a constant current I and placed along the x axis as shown in Figure 30.3. Determine the magnitude and direction of the magnetic field at point P due to this current.

SOLUTION

Conceptualize   From the Biot–Savart law, we expect that the magnitude of the field is proportional to the current in the wire and decreases as the distance a from the wire to point P increases.

Categorize   We are asked to find the magnetic field due to a simple current distribution, so this example is a typical problem for which the Biot–Savart law is appropriate.

Analyze   Let's start by considering a length element $d\vec{\mathbf{s}}$ located a distance r from P. The direction of the magnetic field at point P due to the current in this element is out of the page because $d\vec{\mathbf{s}} \times \hat{\mathbf{r}}$ is out of the page. In fact, because *all* the current elements $I\,d\vec{\mathbf{s}}$ lie in the plane of the page, they all produce a magnetic field directed out of the page at point P. Therefore, the direction of the magnetic field at point P is out of the page and we need only find the magnitude of the field. We place the origin at O and let point P be along the positive y axis, with $\hat{\mathbf{k}}$ being a unit vector pointing out of the page.

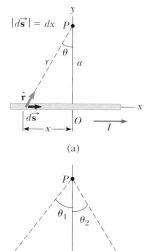

(a)

(b)

Figure 30.3   (Example 30.1) (a) A thin, straight wire carrying a current I. The magnetic field at point P due to the current in each element $d\vec{\mathbf{s}}$ of the wire is out of the page, so the net field at point P is also out of the page. (b) The angles θ_1 and θ_2 used for determining the net field.

Evaluate the cross product in the Biot–Savart law:

$$d\vec{\mathbf{s}} \times \hat{\mathbf{r}} = |d\vec{\mathbf{s}} \times \hat{\mathbf{r}}|\hat{\mathbf{k}} = \left[dx \sin\left(\frac{\pi}{2} - \theta\right) \right]\hat{\mathbf{k}} = (dx \cos\theta)\hat{\mathbf{k}}$$

Substitute into Equation 30.1:

$$(1) \quad d\vec{\mathbf{B}} = (dB)\hat{\mathbf{k}} = \frac{\mu_0 I}{4\pi} \frac{dx \cos\theta}{r^2}\,\hat{\mathbf{k}}$$

From the geometry in Figure 30.3a, express r in terms of θ:

$$(2) \quad r = \frac{a}{\cos\theta}$$

Notice that $\tan\theta = -x/a$ from the right triangle in Figure 30.3a (the negative sign is necessary because $d\vec{\mathbf{s}}$ is located at a negative value of x) and solve for x:

$$x = -a \tan\theta$$

Find the differential dx:

$$(3) \quad dx = -a \sec^2\theta\,d\theta = -\frac{a\,d\theta}{\cos^2\theta}$$

Substitute Equations (2) and (3) into the magnitude of the field from Equation (1):

$$(4) \quad dB = -\frac{\mu_0 I}{4\pi} \frac{(a\,d\theta)\cos\theta\cos^2\theta}{a^2\cos^2\theta} = -\frac{\mu_0 I}{4\pi a}\cos\theta\,d\theta$$

Integrate Equation (4) over all length elements on the wire, where the subtending angles range from θ_1 to θ_2 as defined in Figure 30.3b:

$$B = -\frac{\mu_0 I}{4\pi a} \int_{\theta_1}^{\theta_2} \cos\theta \, d\theta = \frac{\mu_0 I}{4\pi a}(\sin\theta_1 - \sin\theta_2) \quad \text{(30.4)}$$

Finalize We can use this result to find the magnetic field of *any* straight current-carrying wire if we know the geometry and hence the angles θ_1 and θ_2. Consider the special case of an infinitely long, straight wire. If the wire in Figure 30.3b becomes infinitely long, we see that $\theta_1 = \pi/2$ and $\theta_2 = -\pi/2$ for length elements ranging between positions $x = -\infty$ and $x = +\infty$. Because $(\sin\theta_1 - \sin\theta_2) = (\sin\pi/2 - \sin(-\pi/2)) = 2$, Equation 30.4 becomes

$$B = \frac{\mu_0 I}{2\pi a} \quad \text{(30.5)}$$

Equations 30.4 and 30.5 both show that the magnitude of the magnetic field is proportional to the current and decreases with increasing distance from the wire, as expected. Equation 30.5 has the same mathematical form as the expression for the magnitude of the electric field due to a long charged wire (see Eq. 24.7).

Figure 30.4 The right-hand rule for determining the direction of the magnetic field surrounding a long, straight wire carrying a current. Notice that the magnetic field lines form circles around the wire.

The result of Example 30.1 is important because a current in the form of a long, straight wire occurs often. Figure 30.4 is a perspective view of the magnetic field surrounding a long, straight, current-carrying wire. Because of the wire's symmetry, the magnetic field lines are circles concentric with the wire and lie in planes perpendicular to the wire. The magnitude of $\vec{B}$ is constant on any circle of radius a and is given by Equation 30.5. A convenient rule for determining the direction of $\vec{B}$ is to grasp the wire with the right hand, positioning the thumb along the direction of the current. The four fingers wrap in the direction of the magnetic field.

Figure 30.4 also shows that the magnetic field line has no beginning and no end. Rather, it forms a closed loop. That is a major difference between magnetic field lines and electric field lines, which begin on positive charges and end on negative charges. We will explore this feature of magnetic field lines further in Section 30.5.

EXAMPLE 30.2 Magnetic Field Due to a Curved Wire Segment

Calculate the magnetic field at point O for the current-carrying wire segment shown in Figure 30.5. The wire consists of two straight portions and a circular arc of radius a, which subtends an angle θ.

SOLUTION

Conceptualize The magnetic field at O due to the current in the straight segments AA' and CC' is zero because $d\vec{s}$ is parallel to $\hat{r}$ along these paths, which means that $d\vec{s} \times \hat{r} = 0$ for these paths.

Categorize Because we can ignore segments AA' and CC', this example is categorized as an application of the Biot–Savart law to the curved wire segment AC.

Analyze Each length element $d\vec{s}$ along path AC is at the same distance a from O, and the current in each contributes a field element $d\vec{B}$ directed into the page at O. Furthermore, at every point on AC, $d\vec{s}$ is perpendicular to $\hat{r}$; hence, $|d\vec{s} \times \hat{r}| = ds$.

From Equation 30.1, find the magnitude of the field at O due to the current in an element of length ds:

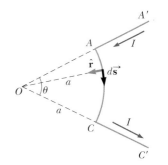

Figure 30.5 (Example 30.2) The magnetic field at O due to the current in the curved segment AC is into the page. The contribution to the field at O due to the current in the two straight segments is zero. The length of the curved segment AC is s.

$$dB = \frac{\mu_0}{4\pi}\frac{I\,ds}{a^2}$$

Integrate this expression over the curved path AC, noting that I and a are constants:

$$B = \frac{\mu_0 I}{4\pi a^2} \int ds = \frac{\mu_0 I}{4\pi a^2} s$$

From the geometry, note that $s = a\theta$ and substitute:

$$B = \frac{\mu_0 I}{4\pi a^2}(a\theta) = \frac{\mu_0 I}{4\pi a}\theta \qquad (30.6)$$

Finalize Equation 30.6 gives the magnitude of the magnetic field at O. The direction of $\vec{B}$ is into the page at O because $d\vec{s} \times \hat{r}$ is into the page for every length element.

What If? What if you were asked to find the magnetic field at the center of a circular wire loop of radius R that carries a current I? Can this question be answered at this point in our understanding of the source of magnetic fields?

Answer Yes, it can. The straight wires in Figure 30.5 do not contribute to the magnetic field. The only contribution is from the curved segment. As the angle θ increases, the curved segment becomes a full circle when $\theta = 2\pi$. Therefore, you can find the magnetic field at the center of a wire loop by letting $\theta = 2\pi$ in Equation 30.6:

$$B = \frac{\mu_0 I}{4\pi a} 2\pi = \frac{\mu_0 I}{2a}$$

This result is a limiting case of a more general result discussed in Example 30.3.

EXAMPLE 30.3   **Magnetic Field on the Axis of a Circular Current Loop**

Consider a circular wire loop of radius a located in the yz plane and carrying a steady current I as in Figure 30.6. Calculate the magnetic field at an axial point P a distance x from the center of the loop.

SOLUTION

Conceptualize Figure 30.6 shows the magnetic field contribution $d\vec{B}$ at P due to a single current element at the top of the ring. This field vector can be resolved into components dB_x parallel to the axis of the ring and $dB_\perp$ perpendicular to the axis. Think about the magnetic field contributions from a current element at the bottom of the loop. Because of the symmetry of the situation, the perpendicular components of the field due to elements at the top and bottom of the ring cancel. This cancellation occurs for all pairs of segments around the ring, so we can ignore the perpendicular component of the field and focus solely on the parallel components, which simply add.

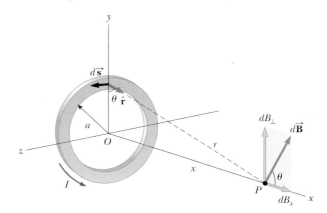

Figure 30.6 (Example 30.3) Geometry for calculating the magnetic field at a point P lying on the axis of a current loop. By symmetry, the total field $\vec{B}$ is along this axis.

Categorize We are asked to find the magnetic field due to a simple current distribution, so this example is a typical problem for which the Biot–Savart law is appropriate.

Analyze In this situation, every length element $d\vec{s}$ is perpendicular to the vector $\hat{r}$ at the location of the element. Therefore, for any element, $|d\vec{s} \times \hat{r}| = (ds)(1)\sin 90° = ds$. Furthermore, all length elements around the loop are at the same distance r from P, where $r^2 = a^2 + x^2$.

Use Equation 30.1 to find the magnitude of $d\vec{B}$ due to the current in any length element $d\vec{s}$:

$$dB = \frac{\mu_0 I}{4\pi} \frac{|d\vec{s} \times \hat{r}|}{r^2} = \frac{\mu_0 I}{4\pi} \frac{ds}{(a^2 + x^2)}$$

Find the x component of the field element:

$$dB_x = \frac{\mu_0 I}{4\pi} \frac{ds}{(a^2 + x^2)} \cos\theta$$

Integrate over the entire loop:

$$B_x = \oint dB_x = \frac{\mu_0 I}{4\pi} \oint \frac{ds \cos \theta}{a^2 + x^2}$$

From the geometry, evaluate $\cos \theta$:

$$\cos \theta = \frac{a}{(a^2 + x^2)^{1/2}}$$

Substitute this expression for $\cos \theta$ into the integral and note that x, a, and θ are all constant:

$$B_x = \frac{\mu_0 I}{4\pi} \oint \frac{ds}{a^2 + x^2} \frac{a}{(a^2 + x^2)^{1/2}} = \frac{\mu_0 I}{4\pi} \frac{a}{(a^2 + x^2)^{3/2}} \oint ds$$

Integrate around the loop:

$$B_x = \frac{\mu_0 I}{4\pi} \frac{a}{(a^2 + x^2)^{3/2}} (2\pi a) = \frac{\mu_0 I a^2}{2(a^2 + x^2)^{3/2}} \quad \text{(30.7)}$$

Finalize To find the magnetic field at the center of the loop, set $x = 0$ in Equation 30.7. At this special point,

$$B = \frac{\mu_0 I}{2a} \quad (\text{at } x = 0) \quad \text{(30.8)}$$

which is consistent with the result of the **What If?** feature of Example 30.2.

The pattern of magnetic field lines for a circular current loop is shown in Figure 30.7a. For clarity, the lines are drawn for only the plane that contains the axis of the loop. The field-line pattern is axially symmetric and looks like the pattern around a bar magnet, which is shown in Figure 30.7c.

What If? What if we consider points on the x axis very far from the loop? How does the magnetic field behave at these distant points?

Answer In this case, in which $x \gg a$, we can neglect the term a^2 in the denominator of Equation 30.7 and obtain

(a)   (b)   (c)

Figure 30.7 (Example 30.3) (a) Magnetic field lines surrounding a current loop. (b) Magnetic field lines surrounding a current loop, displayed with iron filings. (c) Magnetic field lines surrounding a bar magnet. Notice the similarity between this line pattern and that of a current loop.

$$B \approx \frac{\mu_0 I a^2}{2x^3} \quad (\text{for } x \gg a) \quad \text{(30.9)}$$

The magnitude of the magnetic moment μ of the loop is defined as the product of current and loop area (see Eq. 29.15): $\mu = I(\pi a^2)$ for our circular loop. We can express Equation 30.9 as

$$B \approx \frac{\mu_0}{2\pi} \frac{\mu}{x^3} \quad \text{(30.10)}$$

This result is similar in form to the expression for the electric field due to an electric dipole, $E = k_e(p/y^3)$ (see Example 23.5), where $p = 2qa$ is the electric dipole moment as defined in Equation 26.16.

30.2 The Magnetic Force Between Two Parallel Conductors

In Chapter 29, we described the magnetic force that acts on a current-carrying conductor placed in an external magnetic field. Because a current in a conductor sets up its own magnetic field, it is easy to understand that two current-carrying conductors exert magnetic forces on each other. Such forces can be used as the basis for defining the ampere and the coulomb.

Consider two long, straight, parallel wires separated by a distance a and carrying currents I_1 and I_2 in the same direction as in Active Figure 30.8. Let's determine the force exerted on one wire due to the magnetic field set up by the other

ACTIVE FIGURE 30.8

Two parallel wires that each carry a steady current exert a magnetic force on each other. The field $\vec{B}_2$ due to the current in wire 2 exerts a magnetic force of magnitude $F_1 = I_1 \ell B_2$ on wire 1. The force is attractive if the currents are parallel (as shown) and repulsive if the currents are antiparallel.

Sign in at www.thomsonedu.com and go to ThomsonNOW to adjust the currents in the wires and the distance between them to see the effect on the force.

wire. Wire 2, which carries a current I_2 and is identified arbitrarily as the source wire, creates a magnetic field $\vec{B}_2$ at the location of wire 1, the test wire. The direction of $\vec{B}_2$ is perpendicular to wire 1 as shown in Active Figure 30.8. According to Equation 29.10, the magnetic force on a length ℓ of wire 1 is $\vec{F}_1 = I_1 \vec{\ell} \times \vec{B}_2$. Because $\vec{\ell}$ is perpendicular to $\vec{B}_2$ in this situation, the magnitude of $\vec{F}_1$ is $F_1 = I_1 \ell B_2$. Because the magnitude of $\vec{B}_2$ is given by Equation 30.5,

$$F_1 = I_1 \ell B_2 = I_1 \ell \left(\frac{\mu_0 I_2}{2\pi a} \right) = \frac{\mu_0 I_1 I_2}{2\pi a} \ell \qquad (30.11)$$

The direction of $\vec{F}_1$ is toward wire 2 because $\vec{\ell} \times \vec{B}_2$ is in that direction. When the field set up at wire 2 by wire 1 is calculated, the force $\vec{F}_2$ acting on wire 2 is found to be equal in magnitude and opposite in direction to $\vec{F}_1$, which is what we expect because Newton's third law must be obeyed. When the currents are in opposite directions (that is, when one of the currents is reversed in Active Fig. 30.8), the forces are reversed and the wires repel each other. Hence, **parallel conductors carrying currents in the same direction attract each other, and parallel conductors carrying currents in opposite directions repel each other**.

Because the magnitudes of the forces are the same on both wires, we denote the magnitude of the magnetic force between the wires as simply F_B. We can rewrite this magnitude in terms of the force per unit length:

$$\frac{F_B}{\ell} = \frac{\mu_0 I_1 I_2}{2\pi a} \qquad (30.12)$$

The force between two parallel wires is used to define the **ampere** as follows:

When the magnitude of the force per unit length between two long, parallel wires that carry identical currents and are separated by 1 m is 2×10^{-7} N/m, the current in each wire is defined to be 1 A.

◄ Definition of the ampere

The value 2×10^{-7} N/m is obtained from Equation 30.12 with $I_1 = I_2 = 1$ A and $a = 1$ m. Because this definition is based on a force, a mechanical measurement can be used to standardize the ampere. For instance, the National Institute of Standards and Technology uses an instrument called a *current balance* for primary current measurements. The results are then used to standardize other, more conventional instruments such as ammeters.

The SI unit of charge, the **coulomb**, is defined in terms of the ampere: When a conductor carries a steady current of 1 A, the quantity of charge that flows through a cross section of the conductor in 1 s is 1 C.

In deriving Equations 30.11 and 30.12, we assumed that both wires are long compared with their separation distance. In fact, only one wire needs to be long. The equations accurately describe the forces exerted on each other by a long wire and a straight, parallel wire of limited length ℓ.

Quick Quiz 30.2   A loose spiral spring carrying no current is hung from a ceiling. When a switch is thrown so that a current exists in the spring, do the coils (a) move closer together, (b) move farther apart, or (c) not move at all?

EXAMPLE 30.4   **Suspending a Wire**

Two infinitely long, parallel wires are lying on the ground 1.00 cm apart as shown in Figure 30.9a. A third wire, of length 10.0 m and mass 400 g, carries a current of $I_1 = 100$ A and is levitated above the first two wires, at a horizontal position midway between them. The infinitely long wires carry equal currents I_2 in the same direction, but in the direction opposite to that in the levitated wire. What current must the infinitely long wires carry so that the three wires form an equilateral triangle?

(a)   (b)

Figure 30.9 (Example 30.4) (a) Two current-carrying wires lie on the ground and suspend a third wire in the air by magnetic forces. (b) End view. In the situation described in the example, the three wires form an equilateral triangle. The two magnetic forces on the levitated wire are $\vec{\mathbf{F}}_{B,L}$, the force due to the left-hand wire on the ground, and $\vec{\mathbf{F}}_{B,R}$, the force due to the right-hand wire. The gravitational force $\vec{\mathbf{F}}_g$ on the levitated wire is also shown.

SOLUTION

Conceptualize Because the current in the short wire is opposite those in the long wires, the short wire is repelled from both of the others. Imagine the currents in the long wires are increased. The repulsive force becomes stronger, and the levitated wire rises to the point at which the weight of the wire is once again levitated in equilibrium. Figure 30.9b shows the desired situation with the three wires forming an equilateral triangle.

Categorize We model the levitated wire as a particle in equilibrium.

Analyze The horizontal components of the magnetic forces on the levitated wire cancel. The vertical components are both positive and add together.

Find the total magnetic force in the upward direction on the levitated wire:

$$\vec{\mathbf{F}}_B = 2\left(\frac{\mu_0 I_1 I_2}{2\pi a}\,\ell\right)\cos 30.0^\circ \hat{\mathbf{k}} = 0.866\,\frac{\mu_0 I_1 I_2}{\pi a}\,\ell\hat{\mathbf{k}}$$

Find the gravitational force on the levitated wire:

$$\vec{\mathbf{F}}_g = -mg\hat{\mathbf{k}}$$

Apply the particle in equilibrium model by adding the forces and setting the net force equal to zero:

$$\sum \vec{\mathbf{F}} = \vec{\mathbf{F}}_B + \vec{\mathbf{F}}_g = 0.866\,\frac{\mu_0 I_1 I_2}{\pi a}\,\ell\hat{\mathbf{k}} - mg\hat{\mathbf{k}} = 0$$

Solve for the current in the wires on the ground:

$$I_2 = \frac{mg\pi a}{0.866\mu_0 I_1 \ell}$$

Substitute numerical values:

$$I_2 = \frac{(0.400\text{ kg})(9.80\text{ m/s}^2)\pi(0.010\ 0\text{ m})}{0.866(4\pi \times 10^{-7}\text{ T}\cdot\text{m/A})(100\text{ A})(10.0\text{ m})}$$

$$= \boxed{113\text{ A}}$$

Finalize The currents in all wires are on the order of 10^2 A. Such large currents would require specialized equipment. Therefore, this situation would be difficult to establish in practice.

30.3 Ampère's Law

Oersted's 1819 discovery about deflected compass needles demonstrates that a current-carrying conductor produces a magnetic field. Active Figure 30.10a shows how this effect can be demonstrated in the classroom. Several compass needles are placed in a horizontal plane near a long, vertical wire. When no current is present in the wire, all the needles point in the same direction (that of the Earth's magnetic field) as expected. When the wire carries a strong, steady current, the needles all deflect in a direction tangent to the circle as in Active Figure 30.10b. These observations demonstrate that the direction of the magnetic field produced by the current in the wire is consistent with the right-hand rule described in Figure 30.4. When the current is reversed, the needles in Active Figure 30.10b also reverse.

Because the compass needles point in the direction of $\vec{\mathbf{B}}$, we conclude that the lines of $\vec{\mathbf{B}}$ form circles around the wire as discussed in Section 30.1. By symmetry, the magnitude of $\vec{\mathbf{B}}$ is the same everywhere on a circular path centered on the wire and lying in a plane perpendicular to the wire. By varying the current and distance from the wire, we find that B is proportional to the current and inversely proportional to the distance from the wire as described by Equation 30.5.

Now let's evaluate the product $\vec{\mathbf{B}} \cdot d\vec{\mathbf{s}}$ for a small length element $d\vec{\mathbf{s}}$ on the circular path defined by the compass needles and sum the products for all elements over the closed circular path.[1] Along this path, the vectors $d\vec{\mathbf{s}}$ and $\vec{\mathbf{B}}$ are parallel at each point (see Active Fig. 30.10b), so $\vec{\mathbf{B}} \cdot d\vec{\mathbf{s}} = B\,ds$. Furthermore, the magnitude of $\vec{\mathbf{B}}$ is constant on this circle and is given by Equation 30.5. Therefore, the sum of the products $B\,ds$ over the closed path, which is equivalent to the line integral of $\vec{\mathbf{B}} \cdot d\vec{\mathbf{s}}$, is

$$\oint \vec{\mathbf{B}} \cdot d\vec{\mathbf{s}} = B \oint ds = \frac{\mu_0 I}{2\pi r}(2\pi r) = \mu_0 I$$

where $\oint ds = 2\pi r$ is the circumference of the circular path. Although this result was calculated for the special case of a circular path surrounding a wire, it holds for a closed path of *any* shape (an *amperian loop*) surrounding a current that exists in an unbroken circuit. The general case, known as **Ampère's law**, can be stated as follows:

> The line integral of $\vec{\mathbf{B}} \cdot d\vec{\mathbf{s}}$ around any closed path equals $\mu_0 I$, where I is the total steady current passing through any surface bounded by the closed path:
>
> $$\oint \vec{\mathbf{B}} \cdot d\vec{\mathbf{s}} = \mu_0 I \qquad \text{(30.13)}$$

ACTIVE FIGURE 30.10

(a) When no current is present in the wire, all compass needles point in the same direction (toward the Earth's north pole). (b) When the wire carries a strong current, the compass needles deflect in a direction tangent to the circle, which is the direction of the magnetic field created by the current. (c) Circular magnetic field lines surrounding a current-carrying conductor, displayed with iron filings.

Sign in at www.thomsonedu.com and go to ThomsonNOW to change the value of the current and see the effect on the compasses.

---

[1] You may wonder why we would choose to evaluate this scalar product. The origin of Ampère's law is in 19th-century science, in which a "magnetic charge" (the supposed analog to an isolated electric charge) was imagined to be moved around a circular field line. The work done on the charge was related to $\vec{\mathbf{B}} \cdot d\vec{\mathbf{s}}$, just as the work done moving an electric charge in an electric field is related to $\vec{\mathbf{E}} \cdot d\vec{\mathbf{s}}$. Therefore, Ampère's law, a valid and useful principle, arose from an erroneous and abandoned work calculation!

ANDRE-MARIE AMPÈRE
French Physicist (1775–1836)
Ampère is credited with the discovery of electromagnetism, which is the relationship between electric currents and magnetic fields. Ampère's genius, particularly in mathematics, became evident by the time he was 12 years old; his personal life, however, was filled with tragedy. His father, a wealthy city official, was guillotined during the French Revolution, and his wife died young, in 1803. Ampère died at the age of 61 of pneumonia. His judgment of his life is clear from the epitaph he chose for his gravestone: *Tandem Felix* (Happy at Last).

◄   Ampère's law

PITFALL PREVENTION 30.2
Avoiding Problems with Signs

When using Ampère's law, apply the following right-hand rule. Point your thumb in the direction of the current through the amperian loop. Your curled fingers then point in the direction that you should integrate when traversing the loop to avoid having to define the current as negative.

Ampère's law describes the creation of magnetic fields by all continuous current configurations, but at our mathematical level it is useful only for calculating the magnetic field of current configurations having a high degree of symmetry. Its use is similar to that of Gauss's law in calculating electric fields for highly symmetric charge distributions.

Quick Quiz 30.3 Rank the magnitudes of $\oint \vec{\mathbf{B}} \cdot d\vec{\mathbf{s}}$ for the closed paths a through d in Figure 30.11 from least to greatest.

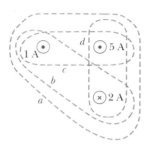

Figure 30.11 (Quick Quiz 30.3) Four closed paths around three current-carrying wires.

Quick Quiz 30.4 Rank the magnitudes of $\oint \vec{\mathbf{B}} \cdot d\vec{\mathbf{s}}$ for the closed paths a through d in Figure 30.12 from least to greatest.

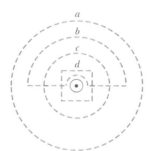

Figure 30.12 (Quick Quiz 30.4) Several closed paths near a single current-carrying wire.

EXAMPLE 30.5 The Magnetic Field Created by a Long Current-Carrying Wire

A long, straight wire of radius R carries a steady current I that is uniformly distributed through the cross section of the wire (Fig. 30.13). Calculate the magnetic field a distance r from the center of the wire in the regions $r \geq R$ and $r < R$.

SOLUTION

Conceptualize Study Figure 30.13 to understand the structure of the wire and the current in the wire. The current creates magnetic fields everywhere, both inside and outside the wire.

Categorize Because the wire has a high degree of symmetry, we categorize this example as an Ampère's law problem. For the $r \geq R$ case, we should arrive at the same result as was obtained in Example 30.1, where we applied the Biot–Savart law to the same situation.

Analyze For the magnetic field exterior to the wire, let us choose for our path of integration circle 1 in Figure 30.13. From symmetry, $\vec{\mathbf{B}}$ must be constant in magnitude and parallel to $d\vec{\mathbf{s}}$ at every point on this circle.

Figure 30.13 (Example 30.5) A long, straight wire of radius R carrying a steady current I uniformly distributed across the cross section of the wire. The magnetic field at any point can be calculated from Ampère's law using a circular path of radius r, concentric with the wire.

Note that the total current passing through the plane of the circle is I and apply Ampère's law:

$$\oint \vec{\mathbf{B}} \cdot d\vec{\mathbf{s}} = B \oint ds = B(2\pi r) = \mu_0 I$$

Solve for B:

$$B = \frac{\mu_0 I}{2\pi r} \quad \text{(for } r \geq R\text{)} \tag{30.14}$$

Now consider the interior of the wire, where $r < R$. Here the current I' passing through the plane of circle 2 is less than the total current I.

Set the ratio of the current I' enclosed by circle 2 to the entire current I equal to the ratio of the area πr^2 enclosed by circle 2 to the cross-sectional area πR^2 of the wire:

$$\frac{I'}{I} = \frac{\pi r^2}{\pi R^2}$$

Solve for I':

$$I' = \frac{r^2}{R^2} I$$

Apply Ampère's law to circle 2:

$$\oint \vec{B} \cdot d\vec{s} = B(2\pi r) = \mu_0 I' = \mu_0 \left(\frac{r^2}{R^2} I\right)$$

Solve for B:

$$B = \left(\frac{\mu_0 I}{2\pi R^2}\right) r \quad \text{(for } r < R\text{)} \tag{30.15}$$

Finalize The magnetic field exterior to the wire is identical in form to Equation 30.5. As is often the case in highly symmetric situations, it is much easier to use Ampère's law than the Biot–Savart law (Example 30.1). The magnetic field interior to the wire is similar in form to the expression for the electric field inside a uniformly charged sphere (see Example 24.3). The magnitude of the magnetic field versus r for this configuration is plotted in Figure 30.14. Inside the wire, $B \to 0$ as $r \to 0$. Furthermore, Equations 30.14 and 30.15 give the same value of the magnetic field at $r = R$, demonstrating that the magnetic field is continuous at the surface of the wire.

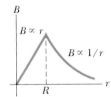

Figure 30.14 (Example 30.5) Magnitude of the magnetic field versus r for the wire shown in Figure 30.13. The field is proportional to r inside the wire and varies as $1/r$ outside the wire.

EXAMPLE 30.6   **The Magnetic Field Created by a Toroid**

A device called a *toroid* (Fig. 30.15) is often used to create an almost uniform magnetic field in some enclosed area. The device consists of a conducting wire wrapped around a ring (a *torus*) made of a nonconducting material. For a toroid having N closely spaced turns of wire, calculate the magnetic field in the region occupied by the torus, a distance r from the center.

SOLUTION

Conceptualize Study Figure 30.15 carefully to understand how the wire is wrapped around the torus. The torus could be a solid material or it could be air, with a stiff wire wrapped into the shape shown in Figure 30.15 to form an empty toroid.

Categorize Because the toroid has a high degree of symmetry, we categorize this example as an Ampère's law problem.

Analyze Consider the circular amperian loop (loop 1) of radius r in the plane of Figure 30.15. By symmetry, the magnitude of the field is constant on this circle and tangent to it, so $\vec{B} \cdot d\vec{s} = B \, ds$. Furthermore, the wire passes through the loop N times, so the total current through the loop is NI.

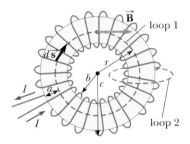

Figure 30.15 (Example 30.6) A toroid consisting of many turns of wire. If the turns are closely spaced, the magnetic field in the interior of the torus (the gold-shaded region) is tangent to the dashed circle (loop 1) and varies as $1/r$. The dimension a is the cross-sectional radius of the torus. The field outside the toroid is very small and can be described by using the amperian loop (loop 2) at the right side, perpendicular to the page.

Apply Ampère's law to loop 1:

$$\oint \vec{\mathbf{B}} \cdot d\vec{\mathbf{s}} = B \oint ds = B(2\pi r) = \mu_0 NI$$

Solve for B:

$$B = \frac{\mu_0 NI}{2\pi r} \tag{30.16}$$

Finalize This result shows that B varies as $1/r$ and hence is *nonuniform* in the region occupied by the torus. If, however, r is very large compared with the cross-sectional radius a of the torus, the field is approximately uniform inside the torus.

For an ideal toroid, in which the turns are closely spaced, the external magnetic field is close to zero, but it is not exactly zero. In Figure 30.15, imagine the radius r of the amperian loop to be either smaller than b or larger than c. In either case, the loop encloses zero net current, so $\oint \vec{\mathbf{B}} \cdot d\vec{\mathbf{s}} = 0$. You might think that this result proves that $\vec{\mathbf{B}} = 0$, but it does not. Consider the amperian loop (loop 2) on the right side of the toroid in Figure 30.15. The plane of this loop is perpendicular to the page, and the toroid passes through the loop. As charges enter the toroid as indicated by the current directions in Figure 30.15, they work their way counterclockwise around the toroid. Therefore, a current passes through the perpendicular amperian loop! This current is small, but not zero. As a result, the toroid acts as a current loop and produces a weak external field of the form shown in Figure 30.7. The reason $\oint \vec{\mathbf{B}} \cdot d\vec{\mathbf{s}} = 0$ for the amperian loops of radius $r < b$ and $r > c$ in the plane of the page is that the field lines are perpendicular to $d\vec{\mathbf{s}}$, *not* because $\vec{\mathbf{B}} = 0$.

30.4 The Magnetic Field of a Solenoid

A **solenoid** is a long wire wound in the form of a helix. With this configuration, a reasonably uniform magnetic field can be produced in the space surrounded by the turns of wire—which we shall call the *interior* of the solenoid—when the solenoid carries a current. When the turns are closely spaced, each can be approximated as a circular loop; the net magnetic field is the vector sum of the fields resulting from all the turns.

Figure 30.16 shows the magnetic field lines surrounding a loosely wound solenoid. The field lines in the interior are nearly parallel to one another, are uniformly distributed, and are close together, indicating that the field in this space is strong and almost uniform.

If the turns are closely spaced and the solenoid is of finite length, the magnetic field lines are as shown in Figure 30.17a. This field line distribution is similar to that surrounding a bar magnet (Fig. 30.17b). Hence, one end of the solenoid behaves like the north pole of a magnet and the opposite end behaves like the south pole. As the length of the solenoid increases, the interior field becomes more uniform and the exterior field becomes weaker. An *ideal solenoid* is

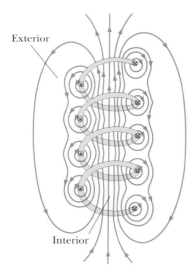

Figure 30.16 The magnetic field lines for a loosely wound solenoid.

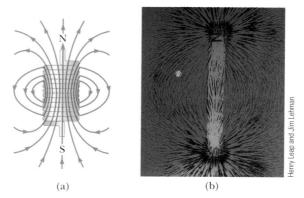

(a) (b)

Figure 30.17 (a) Magnetic field lines for a tightly wound solenoid of finite length, carrying a steady current. The field in the interior space is strong and nearly uniform. Notice that the field lines resemble those of a bar magnet, meaning that the solenoid effectively has north and south poles. (b) The magnetic field pattern of a bar magnet, displayed with small iron filings on a sheet of paper.

approached when the turns are closely spaced and the length is much greater than the radius of the turns. Figure 30.18 shows a longitudinal cross section of part of such a solenoid carrying a current *I*. In this case, the external field is close to zero and the interior field is uniform over a great volume.

Consider the amperian loop (loop 1) perpendicular to the page in Figure 30.18, surrounding the ideal solenoid. This loop encloses a small current as the charges in the wire move coil by coil along the length of the solenoid. Therefore, there is a nonzero magnetic field outside the solenoid. It is a weak field, with circular field lines, like those due to a line of current as in Figure 30.4. For an ideal solenoid, this weak field is the only field external to the solenoid. We could eliminate this field in Figure 30.18 by adding a second layer of turns of wire outside the first layer, with the current carried along the axis of the solenoid in the opposite direction compared with the first layer. Then the net current along the axis is zero.

We can use Ampère's law to obtain a quantitative expression for the interior magnetic field in an ideal solenoid. Because the solenoid is ideal, $\vec{B}$ in the interior space is uniform and parallel to the axis and the magnetic field lines in the exterior space form circles around the solenoid. The planes of these circles are perpendicular to the page. Consider the rectangular path (loop 2) of length ℓ and width w shown in Figure 30.18. Let's apply Ampère's law to this path by evaluating the integral of $\vec{B} \cdot d\vec{s}$ over each side of the rectangle. The contribution along side 3 is zero because the magnetic field lines are perpendicular to the path in this region. The contributions from sides 2 and 4 are both zero, again because $\vec{B}$ is perpendicular to $d\vec{s}$ along these paths, both inside and outside the solenoid. Side 1 gives a contribution to the integral because along this path $\vec{B}$ is uniform and parallel to $d\vec{s}$. The integral over the closed rectangular path is therefore

$$\oint \vec{B} \cdot d\vec{s} = \int_{\text{path 1}} \vec{B} \cdot d\vec{s} = B \int_{\text{path 1}} ds = B\ell$$

The right side of Ampère's law involves the total current *I* through the area bounded by the path of integration. In this case, the total current through the rectangular path equals the current through each turn multiplied by the number of turns. If *N* is the number of turns in the length ℓ, the total current through the rectangle is *NI*. Therefore, Ampère's law applied to this path gives

$$\oint \vec{B} \cdot d\vec{s} = B\ell = \mu_0 NI$$

$$B = \mu_0 \frac{N}{\ell} I = \mu_0 nI \qquad (30.17)$$

where $n = N/\ell$ is the number of turns per unit length.

We also could obtain this result by reconsidering the magnetic field of a toroid (see Example 30.6). If the radius *r* of the torus in Figure 30.15 containing *N* turns is much greater than the toroid's cross-sectional radius *a*, a short section of the toroid approximates a solenoid for which $n = N/2\pi r$. In this limit, Equation 30.16 agrees with Equation 30.17.

Equation 30.17 is valid only for points near the center (that is, far from the ends) of a very long solenoid. As you might expect, the field near each end is smaller than the value given by Equation 30.17. At the very end of a long solenoid, the magnitude of the field is half the magnitude at the center (see Problem 36).

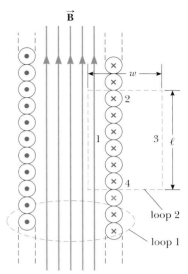

Figure 30.18 Cross-sectional view of an ideal solenoid, where the interior magnetic field is uniform and the exterior field is close to zero. Ampère's law applied to the circular path near the bottom whose plane is perpendicular to the page can be used to show that there is a weak field outside the solenoid. Ampère's law applied to the rectangular dashed path in the plane of the page can be used to calculate the magnitude of the interior field.

◀ Magnetic field inside a solenoid

Quick Quiz 30.5 Consider a solenoid that is very long compared with its radius. Of the following choices, what is the most effective way to increase the magnetic field in the interior of the solenoid? (a) double its length, keeping the number of turns per unit length constant   (b) reduce its radius by half, keeping the number of turns per unit length constant   (c) overwrap the entire solenoid with an additional layer of current-carrying wire

30.5   Gauss's Law in Magnetism

The flux associated with a magnetic field is defined in a manner similar to that used to define electric flux (see Eq. 24.3). Consider an element of area dA on an arbitrarily shaped surface as shown in Figure 30.19. If the magnetic field at this element is $\vec{B}$, the magnetic flux through the element is $\vec{B} \cdot d\vec{A}$, where $d\vec{A}$ is a vector that is perpendicular to the surface and has a magnitude equal to the area dA. Therefore, the total magnetic flux Φ_B through the surface is

Definition of magnetic flux ▶

$$\Phi_B \equiv \int \vec{B} \cdot d\vec{A} \tag{30.18}$$

Consider the special case of a plane of area A in a uniform field $\vec{B}$ that makes an angle θ with $d\vec{A}$. The magnetic flux through the plane in this case is

$$\Phi_B = BA \cos \theta \tag{30.19}$$

If the magnetic field is parallel to the plane as in Active Figure 30.20a, then $\theta = 90°$ and the flux through the plane is zero. If the field is perpendicular to the plane as in Active Figure 30.20b, then $\theta = 0$ and the flux through the plane is BA (the maximum value).

The unit of magnetic flux is $T \cdot m^2$, which is defined as a *weber* (Wb); 1 Wb = $1 \, T \cdot m^2$.

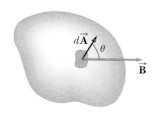

Figure 30.19   The magnetic flux through an area element dA is $\vec{B} \cdot d\vec{A} = B \, dA \cos \theta$, where $d\vec{A}$ is a vector perpendicular to the surface.

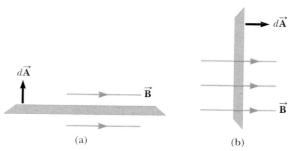

(a)   (b)

ACTIVE FIGURE 30.20

Magnetic flux through a plane lying in a magnetic field. (a) The flux through the plane is zero when the magnetic field is parallel to the plane surface. (b) The flux through the plane is a maximum when the magnetic field is perpendicular to the plane.

Sign in at www.thomsonedu.com and go to ThomsonNOW to rotate the plane and change the value of the field to see the effect on the flux.

EXAMPLE 30.7   **Magnetic Flux Through a Rectangular Loop**

A rectangular loop of width a and length b is located near a long wire carrying a current I (Fig. 30.21). The distance between the wire and the closest side of the loop is c. The wire is parallel to the long side of the loop. Find the total magnetic flux through the loop due to the current in the wire.

SOLUTION

Conceptualize   We know that the magnetic field is a function of distance r from a long wire. Therefore, the magnetic field varies over the area of the rectangular loop.

Figure 30.21   (Example 30.7) The magnetic field due to the wire carrying a current I is not uniform over the rectangular loop.

Categorize   Because the magnetic field varies over the area of the loop, we must integrate over this area to find the total flux.

Analyze Noting that $\vec{\mathbf{B}}$ is parallel to $d\vec{\mathbf{A}}$ at any point within the loop, find the magnetic flux through the rectangular area using Equation 30.18 and incorporate Equation 30.14 for the magnetic field:

$$\Phi_B = \int \vec{\mathbf{B}} \cdot d\vec{\mathbf{A}} = \int B \, dA = \int \frac{\mu_0 I}{2\pi r} \, dA$$

Express the area element (the tan strip in Fig. 30.21) as $dA = b \, dr$ and substitute:

$$\Phi_B = \int \frac{\mu_0 I}{2\pi r} \, b \, dr = \frac{\mu_0 I b}{2\pi} \int \frac{dr}{r}$$

Integrate from $r = c$ to $r = a + c$:

$$\Phi_B = \frac{\mu_0 I b}{2\pi} \int_c^{a+c} \frac{dr}{r} = \frac{\mu_0 I b}{2\pi} \ln r \Big|_c^{a+c}$$

$$= \frac{\mu_0 I b}{2\pi} \ln \left(\frac{a+c}{c} \right) = \boxed{\frac{\mu_0 I b}{2\pi} \ln \left(1 + \frac{a}{c} \right)}$$

Finalize Notice how the flux depends on the size of the loop. Increasing either a or b increases the flux as expected. If c becomes large such that the loop is very far from the wire, the flux approaches zero, also as expected. If c goes to zero, the flux becomes infinite. In principle, this infinite value occurs because the field becomes infinite at $r = 0$ (assuming an infinitesimally thin wire). That will not happen in reality because the thickness of the wire prevents the left edge of the loop from reaching $r = 0$.

In Chapter 24, we found that the electric flux through a closed surface surrounding a net charge is proportional to that charge (Gauss's law). In other words, the number of electric field lines leaving the surface depends only on the net charge within it. This behavior exists because electric field lines originate and terminate on electric charges.

The situation is quite different for magnetic fields, which are continuous and form closed loops. In other words, as illustrated by the magnetic field lines of a current in Figure 30.4 and of a bar magnet in Figure 30.22, magnetic field lines do not begin or end at any point. For any closed surface such as the one outlined by the dashed line in Figure 30.22, the number of lines entering the surface equals the number leaving the surface; therefore, the net magnetic flux is zero. In contrast, for a closed surface surrounding one charge of an electric dipole (Fig. 30.23), the net electric flux is not zero.

Figure 30.22 The magnetic field lines of a bar magnet form closed loops. Notice that the net magnetic flux through a closed surface surrounding one of the poles (or any other closed surface) is zero. (The dashed line represents the intersection of the surface with the page.)

Figure 30.23 The electric field lines surrounding an electric dipole begin on the positive charge and terminate on the negative charge. The electric flux through a closed surface surrounding one of the charges is not zero.

Gauss's law in magnetism states that

Gauss's law in magnetism ▶

the net magnetic flux through any closed surface is always zero:

$$\oint \vec{B} \cdot d\vec{A} = 0 \tag{30.20}$$

This statement represents that isolated magnetic poles (monopoles) have never been detected and perhaps do not exist. Nonetheless, scientists continue the search because certain theories that are otherwise successful in explaining fundamental physical behavior suggest the possible existence of magnetic monopoles.

30.6 Magnetism in Matter

The magnetic field produced by a current in a coil of wire gives us a hint as to what causes certain materials to exhibit strong magnetic properties. Earlier we found that a coil like the one shown in Figure 30.17a has a north pole and a south pole. In general, *any* current loop has a magnetic field and therefore has a magnetic dipole moment, including the atomic-level current loops described in some models of the atom.

The Magnetic Moments of Atoms

Let's begin our discussion with a classical model of the atom in which electrons move in circular orbits around the much more massive nucleus. In this model, an orbiting electron constitutes a tiny current loop (because it is a moving charge) and the magnetic moment of the electron is associated with this orbital motion. Although this model has many deficiencies, some of its predictions are in good agreement with the correct theory, which is expressed in terms of quantum physics.

In our classical model, we assume an electron moves with constant speed v in a circular orbit of radius r about the nucleus as in Figure 30.24. The current I associated with this orbiting electron is its charge e divided by its period T. Using $T = 2\pi/\omega$ and $\omega = v/r$ gives

$$I = \frac{e}{T} = \frac{e\omega}{2\pi} = \frac{ev}{2\pi r}$$

Figure 30.24 An electron moving in the direction of the gray arrow in a circular orbit of radius r has an angular momentum $\vec{L}$ in one direction and a magnetic moment $\vec{\mu}$ in the opposite direction. Because the electron carries a negative charge, the direction of the current due to its motion about the nucleus is opposite the direction of that motion.

The magnitude of the magnetic moment associated with this current loop is given by $\mu = IA$, where $A = \pi r^2$ is the area enclosed by the orbit. Therefore,

$$\mu = IA = \left(\frac{ev}{2\pi r}\right)\pi r^2 = \tfrac{1}{2}evr \tag{30.21}$$

Because the magnitude of the orbital angular momentum of the electron is given by $L = m_e vr$ (Eq. 11.12 with $\phi = 90°$), the magnetic moment can be written as

Orbital magnetic moment ▶

$$\mu = \left(\frac{e}{2m_e}\right)L \tag{30.22}$$

This result demonstrates that **the magnetic moment of the electron is proportional to its orbital angular momentum**. Because the electron is negatively charged, the vectors $\vec{\mu}$ and $\vec{L}$ point in *opposite* directions. Both vectors are perpendicular to the plane of the orbit as indicated in Figure 30.24.

A fundamental outcome of quantum physics is that orbital angular momentum is quantized and is equal to multiples of $\hbar = h/2\pi = 1.05 \times 10^{-34}$ J·s, where h is

Planck's constant (see Chapter 40). The smallest nonzero value of the electron's magnetic moment resulting from its orbital motion is

$$\mu = \sqrt{2}\, \frac{e}{2m_e}\, \hbar \tag{30.23}$$

We shall see in Chapter 42 how expressions such as Equation 30.23 arise.

Because all substances contain electrons, you may wonder why most substances are not magnetic. The main reason is that, in most substances, the magnetic moment of one electron in an atom is canceled by that of another electron orbiting in the opposite direction. The net result is that, for most materials, **the magnetic effect produced by the orbital motion of the electrons is either zero or very small**.

In addition to its orbital magnetic moment, an electron (as well as protons, neutrons, and other particles) has an intrinsic property called **spin** that also contributes to its magnetic moment. Classically, the electron might be viewed as spinning about its axis as shown in Figure 30.25, but you should be very careful with the classical interpretation. The magnitude of the angular momentum $\vec{\mathbf{S}}$ associated with spin is on the same order of magnitude as the magnitude of the angular momentum $\vec{\mathbf{L}}$ due to the orbital motion. The magnitude of the spin angular momentum of an electron predicted by quantum theory is

$$S = \frac{\sqrt{3}}{2}\, \hbar$$

The magnetic moment characteristically associated with the spin of an electron has the value

$$\mu_{\text{spin}} = \frac{e\hbar}{2m_e} \tag{30.24}$$

This combination of constants is called the **Bohr magneton μ_{B}**:

$$\mu_{\text{B}} = \frac{e\hbar}{2m_e} = 9.27 \times 10^{-24}\,\text{J/T} \tag{30.25}$$

Therefore, atomic magnetic moments can be expressed as multiples of the Bohr magneton. (Note that $1\,\text{J/T} = 1\,\text{A} \cdot \text{m}^2$.)

In atoms containing many electrons, the electrons usually pair up with their spins opposite each other; therefore, the spin magnetic moments cancel. Atoms containing an odd number of electrons, however, must have at least one unpaired electron and therefore some spin magnetic moment. The total magnetic moment of an atom is the vector sum of the orbital and spin magnetic moments, and a few examples are given in Table 30.1. Notice that helium and neon have zero moments because their individual spin and orbital moments cancel.

The nucleus of an atom also has a magnetic moment associated with its constituent protons and neutrons. The magnetic moment of a proton or neutron, however, is much smaller than that of an electron and can usually be neglected. We can understand this smaller value by inspecting Equation 30.25 and replacing the mass of the electron with the mass of a proton or a neutron. Because the masses of the proton and neutron are much greater than that of the electron, their magnetic moments are on the order of 10^3 times smaller than that of the electron.

Ferromagnetism

A small number of crystalline substances exhibit strong magnetic effects called **ferromagnetism**. Some examples of ferromagnetic substances are iron, cobalt, nickel, gadolinium, and dysprosium. These substances contain permanent atomic magnetic moments that tend to align parallel to each other even in a weak external magnetic field. Once the moments are aligned, the substance remains magnetized

PITFALL PREVENTION 30.3
The Electron Does Not Spin

The electron is *not* physically spinning. It has an intrinsic angular momentum *as if it were spinning*, but the notion of rotation for a point particle is meaningless. Rotation applies only to a *rigid object*, with an extent in space, as in Chapter 10. Spin angular momentum is actually a relativistic effect.

Figure 30.25 Classical model of a spinning electron. We can adopt this model to remind ourselves that electrons have an intrinsic angular momentum. The model should not be pushed too far, however; it gives an incorrect magnitude for the magnetic moment, incorrect quantum numbers, and too many degrees of freedom.

TABLE 30.1

Magnetic Moments of Some Atoms and Ions

Atom or Ion	Magnetic Moment (10^{-24} J/T)
H	9.27
He	0
Ne	0
Ce^{3+}	19.8
Yb^{3+}	37.1

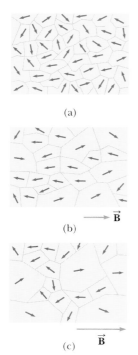

(a)

(b) $\overrightarrow{\mathbf{B}}$

(c) $\overrightarrow{\mathbf{B}}$

Figure 30.26 (a) Random orientation of atomic magnetic dipoles in the domains of an unmagnetized substance. (b) When an external field $\overrightarrow{\mathbf{B}}$ is applied, the domains with components of magnetic moment in the same direction as $\overrightarrow{\mathbf{B}}$ grow larger, giving the sample a net magnetization. (c) As the field is made even stronger, the domains with magnetic moment vectors not aligned with the external field become very small.

after the external field is removed. This permanent alignment is due to a strong coupling between neighboring moments, a coupling that can be understood only in quantum-mechanical terms.

All ferromagnetic materials are made up of microscopic regions called **domains**, regions within which all magnetic moments are aligned. These domains have volumes of about 10^{-12} to 10^{-8} m^3 and contain 10^{17} to 10^{21} atoms. The boundaries between the various domains having different orientations are called **domain walls**. In an unmagnetized sample, the magnetic moments in the domains are randomly oriented so that the net magnetic moment is zero as in Figure 30.26a. When the sample is placed in an external magnetic field $\overrightarrow{\mathbf{B}}$, the size of those domains with magnetic moments aligned with the field grows, which results in a magnetized sample as in Figure 30.26b. As the external field becomes very strong as in Figure 30.26c, the domains in which the magnetic moments are not aligned with the field become very small. When the external field is removed, the sample may retain a net magnetization in the direction of the original field. At ordinary temperatures, thermal agitation is not sufficient to disrupt this preferred orientation of magnetic moments.

Magnetic computer disks store information by alternating the direction of $\overrightarrow{\mathbf{B}}$ for portions of a thin layer of ferromagnetic material. Floppy disks have the layer on a circular sheet of plastic. Hard disks have several rigid platters with magnetic coatings on each side. Audio tapes and videotapes work the same way as floppy disks except that the ferromagnetic material is on a very long strip of plastic. Tiny coils of wire in a recording head are placed close to the magnetic material (which is moving rapidly past the head). Varying the current in the coils creates a magnetic field that magnetizes the recording material. To retrieve the information, the magnetized material is moved past a playback coil. The changing magnetism of the material induces a current in the coil as discussed in Chapter 31. This current is then amplified by audio or video equipment, or it is processed by computer circuitry.

When the temperature of a ferromagnetic substance reaches or exceeds a critical temperature called the **Curie temperature**, the substance loses its residual magnetization. Below the Curie temperature, the magnetic moments are aligned and the substance is ferromagnetic. Above the Curie temperature, the thermal agitation is great enough to cause a random orientation of the moments and the substance becomes paramagnetic. Curie temperatures for several ferromagnetic substances are given in Table 30.2.

Paramagnetism

Paramagnetic substances have a small but positive magnetism resulting from the presence of atoms (or ions) that have permanent magnetic moments. These moments interact only weakly with one another and are randomly oriented in the absence of an external magnetic field. When a paramagnetic substance is placed in an external magnetic field, its atomic moments tend to line up with the field. This alignment process, however, must compete with thermal motion, which tends to randomize the magnetic moment orientations.

Diamagnetism

When an external magnetic field is applied to a diamagnetic substance, a weak magnetic moment is induced in the direction opposite the applied field, causing diamagnetic substances to be weakly repelled by a magnet. Although diamagnetism is present in all matter, its effects are much smaller than those of paramagnetism or ferromagnetism and are evident only when those other effects do not exist.

We can attain some understanding of diamagnetism by considering a classical model of two atomic electrons orbiting the nucleus in opposite directions but with the same speed. The electrons remain in their circular orbits because of the attractive electrostatic force exerted by the positively charged nucleus. Because the magnetic moments of the two electrons are equal in magnitude and opposite in direc-

TABLE 30.2

Curie Temperatures for Several Ferromagnetic Substances

Substance	T_{Curie} (K)
Iron	1 043
Cobalt	1 394
Nickel	631
Gadolinium	317
Fe$_2$O$_3$	893

Leon Lewandowski

High Field Magnet Laboratory, University of Nijmegen, The Netherlands

Photo courtesy of Argonne National Laboratory

(*Left*) Paramagnetism: liquid oxygen, a paramagnetic material, is attracted to the poles of a magnet. (*Right*) Diamagnetism: a frog is levitated in a 16-T magnetic field at the Nijmegen High Field Magnet Laboratory in the Netherlands. The levitation force is exerted on the diamagnetic water molecules in the frog's body. The frog suffered no ill effects from the levitation experience.

Figure 30.27   An illustration of the Meissner effect, shown by this magnet suspended above a cooled ceramic superconductor disk, has become our most visual image of high-temperature superconductivity. Superconductivity is the loss of all resistance to electrical current and is a key to more-efficient energy use. In the Meissner effect, the magnet induces superconducting currents in the disk, which is cooled to $-321°F$ (77 K). The currents create a magnetic force that repels and levitates the disk.

tion, they cancel each other and the magnetic moment of the atom is zero. When an external magnetic field is applied, the electrons experience an additional magnetic force $q\vec{\mathbf{v}} \times \vec{\mathbf{B}}$. This added magnetic force combines with the electrostatic force to increase the orbital speed of the electron whose magnetic moment is antiparallel to the field and to decrease the speed of the electron whose magnetic moment is parallel to the field. As a result, the two magnetic moments of the electrons no longer cancel and the substance acquires a net magnetic moment that is opposite the applied field.

As you recall from Chapter 27, a superconductor is a substance in which the electrical resistance is zero below some critical temperature. Certain types of superconductors also exhibit perfect diamagnetism in the superconducting state. As a result, an applied magnetic field is expelled by the superconductor so that the field is zero in its interior. This phenomenon is known as the **Meissner effect**. If a permanent magnet is brought near a superconductor, the two objects repel each other. This repulsion is illustrated in Figure 30.27, which shows a small permanent magnet levitated above a superconductor maintained at 77 K.

30.7   The Magnetic Field of the Earth

When we speak of a compass magnet having a north pole and a south pole, it is more proper to say that it has a "north-seeking" pole and a "south-seeking" pole. This wording means that one pole of the magnet seeks, or points to, the north geographic pole of the Earth. Because the north pole of a magnet is attracted toward the north geographic pole of the Earth, **the Earth's south magnetic pole is located near the north geographic pole and the Earth's north magnetic pole is located near the south geographic pole**. In fact, the configuration of the Earth's magnetic field, pictured in Figure 30.28 (page 856), is very much like the one that would be achieved by burying a gigantic bar magnet deep in the interior of the Earth.

If a compass needle is suspended in bearings that allow it to rotate in the vertical plane as well as in the horizontal plane, the needle is horizontal with respect to the Earth's surface only near the equator. As the compass is moved northward, the needle rotates so that it points more and more toward the surface of the Earth. Finally, at a point near Hudson Bay in Canada, the north pole of the needle points directly downward. This site, first found in 1832, is considered to be the location of the south magnetic pole of the Earth. It is approximately 1 300 mi from the Earth's geographic North Pole, and its exact position varies slowly with time. Similarly, the north magnetic pole of the Earth is about 1 200 mi away from the Earth's geographic South Pole.

Because of this distance between the north geographic and south magnetic poles, it is only approximately correct to say that a compass needle points north.

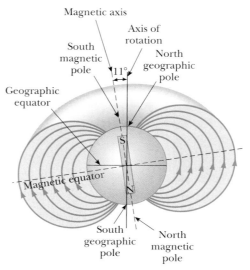

Figure 30.28   The Earth's magnetic field lines. Notice that a south magnetic pole is near the north geographic pole and a north magnetic pole is near the south geographic pole.

The difference between true north, defined as the geographic North Pole, and north indicated by a compass varies from point to point on the Earth. This difference is referred to as *magnetic declination.* For example, along a line through Florida and the Great Lakes, a compass indicates true north, whereas in the state of Washington, it aligns 25° east of true north. Figure 30.29 shows some representative values of the magnetic declination for the contiguous United States.

Although the Earth's magnetic field pattern is similar to the one that would be set up by a bar magnet deep within the Earth, it is easy to understand why the source of this magnetic field cannot be large masses of permanently magnetized material. The Earth does have large deposits of iron ore deep beneath its surface, but the high temperatures in the Earth's core prevent the iron from retaining any permanent magnetization. Scientists consider it more likely that the source of the Earth's magnetic field is convection currents in the Earth's core. Charged ions or electrons circulating in the liquid interior could produce a magnetic field just as a current loop does. There is also strong evidence that the magnitude of a planet's magnetic field is related to the planet's rate of rotation. For example, Jupiter rotates faster than the Earth, and space probes indicate that Jupiter's magnetic field is stronger than the Earth's. Venus, on the other hand, rotates more slowly than the Earth, and its magnetic field is found to be weaker. Investigation into the cause of the Earth's magnetism is ongoing.

It is interesting to point out that that the direction of the Earth's magnetic field has reversed several times during the last million years. Evidence for this reversal is provided by basalt, a type of rock that contains iron and that forms from material spewed forth by volcanic activity on the ocean floor. As the lava cools, it solidifies and retains a picture of the Earth's magnetic field direction. The rocks are dated by other means to provide a timeline for these periodic reversals of the magnetic field.

Figure 30.29   A map of the contiguous United States showing several lines of constant magnetic declination.

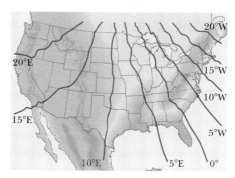

Summary

DEFINITION

The **magnetic flux** Φ_B through a surface is defined by the surface integral

$$\Phi_B \equiv \int \vec{B} \cdot d\vec{A} \qquad (30.18)$$

CONCEPTS AND PRINCIPLES

The **Biot–Savart law** says that the magnetic field $d\vec{B}$ at a point P due to a length element $d\vec{s}$ that carries a steady current I is

$$d\vec{B} = \frac{\mu_0}{4\pi} \frac{I\, d\vec{s} \times \hat{\mathbf{r}}}{r^2} \qquad (30.1)$$

where μ_0 is the **permeability of free space**, r is the distance from the element to the point P, and $\hat{\mathbf{r}}$ is a unit vector pointing from $d\vec{s}$ toward point P. We find the total field at P by integrating this expression over the entire current distribution.

The magnetic force per unit length between two parallel wires separated by a distance a and carrying currents I_1 and I_2 has a magnitude

$$\frac{F_B}{\ell} = \frac{\mu_0 I_1 I_2}{2\pi a} \qquad (30.12)$$

The force is attractive if the currents are in the same direction and repulsive if they are in opposite directions.

Ampère's law says that the line integral of $\vec{B} \cdot d\vec{s}$ around any closed path equals $\mu_0 I$, where I is the total steady current through any surface bounded by the closed path:

$$\oint \vec{B} \cdot d\vec{s} = \mu_0 I \qquad (30.13)$$

The magnitude of the magnetic field at a distance r from a long, straight wire carrying an electric current I is

$$B = \frac{\mu_0 I}{2\pi r} \qquad (30.14)$$

The field lines are circles concentric with the wire.

The magnitudes of the fields inside a toroid and solenoid are

$$B = \frac{\mu_0 NI}{2\pi r} \quad \text{(toroid)} \qquad (30.16)$$

$$B = \mu_0 \frac{N}{\ell} I = \mu_0 n I \quad \text{(solenoid)} \qquad (30.17)$$

where N is the total number of turns.

Gauss's law of magnetism states that the net magnetic flux through any closed surface is zero.

Substances can be classified into one of three categories that describe their magnetic behavior. **Diamagnetic** substances are those in which the magnetic moment is weak and opposite the applied magnetic field. **Paramagnetic** substances are those in which the magnetic moment is weak and in the same direction as the applied magnetic field. In **ferromagnetic** substances, interactions between atoms cause magnetic moments to align and create a strong magnetization that remains after the external field is removed.

Questions

☐ denotes answer available in *Student Solutions Manual/Study Guide;* **O** denotes objective question

1. **O** What creates a magnetic field? Choose every correct answer. (a) a stationary object with electric charge (b) a moving object with electric charge (c) a stationary conductor carrying electric current (d) a difference in electric potential (e) an electric resistor. *Note:* In Chapter 34, we will see that a changing electric field also creates a magnetic field.

2. **O** A long, vertical, metallic wire carries downward electric current. **(i)** What is the direction of the magnetic field it creates at a point 2 cm horizontally east of the center of the wire? (a) north (b) south (c) east (d) west (e) up (f) down **(ii)** What would be the direction of the field if the current consisted of positive charges moving downward, instead of electrons moving upward? Choose from the same possibilities.

3. **O** Suppose you are facing a tall makeup mirror on a vertical wall. Fluorescent tubes framing the mirror carry a clockwise electric current. **(i)** What is the direction of the magnetic field created by that current at a point slightly to the right of the center of the mirror? (a) up (b) down (c) left (d) right (e) horizontally toward you (f) away from you **(ii)** What is the direction of the field the current creates at a point on the wall outside the frame to the right? Choose from the same possibilities.

4. Explain why two parallel wires carrying currents in opposite directions repel each other.

5. **O** In Active Figure 30.8, assume $I_1 = 2$ A and $I_2 = 6$ A. What is the relationship between the magnitude F_1 of the force exerted on wire 1 and the magnitude F_2 of the force exerted on wire 2? (a) $F_1 = 6F_2$ (b) $F_1 = 3F_2$ (c) $F_1 = F_2$ (d) $F_1 = F_2/3$ (e) $F_1 = F_2/6$

6. **O** Answer each question yes or no. (a) Is it possible for each of three stationary charged particles to exert a force of attraction on the other two? (b) Is it possible for each of three stationary charged particles to repel both of the other particles? (c) Is it possible for each of three current-carrying metal wires to attract the other two? (d) Is it possible for each of three current-carrying metal wires to repel both of the other wires? André-Marie Ampère's experiments on electromagnetism are models of logical precision and included observation of the phenomena referred to in this question.

7. Is Ampère's law valid for all closed paths surrounding a conductor? Why is it not useful for calculating $\vec{B}$ for all such paths?

8. Compare Ampère's law with the Biot–Savart law. Which is more generally useful for calculating $\vec{B}$ for a current-carrying conductor?

9. A hollow copper tube carries a current along its length. Why is $\vec{B} = 0$ inside the tube? Is $\vec{B}$ nonzero outside the tube?

10. **O** **(i)** What happens to the magnitude of the magnetic field inside a long solenoid if the current is doubled? (a) It becomes 4 times larger. (b) It becomes twice as large. (c) It is unchanged. (d) It becomes one-half as large. (e) It becomes one-fourth as large. **(ii)** What happens to the field if instead the length of the solenoid is doubled, with the number of turns remaining the same? Choose from the same possibilities. **(iii)** What happens to the field if the number of turns is doubled, with the length remaining the same? Choose from the same possibilities. **(iv)** What happens to the field if the radius is doubled? Choose from the same possibilities.

11. **O** A long solenoid with closely spaced turns carries electric current. Does each turn of wire exert (a) an attractive force on the next adjacent turn, (b) a repulsive force on the next adjacent turn, (c) zero force on the next adjacent turn, or (d) either an attractive or a repulsive force on the next turn, depending on the direction of current in the solenoid?

12. **O** A uniform magnetic field is directed along the *x* axis. For what orientation of a flat, rectangular coil is the flux through the rectangle a maximum? (a) It is a maximum in the *xy* plane. (b) It is a maximum in the *xz* plane. (c) It is a maximum in the *yz* plane. (d) The flux has the same nonzero value for all these orientations. (e) The flux is zero in all cases.

13. The quantity $\oint \vec{B} \cdot d\vec{s}$ in Ampère's law is called *magnetic circulation*. Active Figure 30.10 and Figure 30.13 show paths around which the magnetic circulation was evaluated. Each of these paths encloses an area. What is the magnetic flux through each area? Explain your answer.

14. **O** (a) Two stationary charged particles exert forces of attraction on each other. One of the particles has negative charge. Is the other positive or negative? (b) Is the net electric field at a point halfway between the particles larger, smaller, or the same in magnitude as the field due to one charge by itself? (c) Two straight, vertical, current-carrying wires exert forces of attraction on each other. One of them carries downward current. Does the other wire carry upward or downward current? (d) Is the net magnetic field at a point halfway between the wires larger, smaller, or the same in magnitude as the field due to one wire by itself?

15. **O** Rank the magnitudes of the following magnetic fields from the largest to the smallest, noting any cases of equality. (a) the field 2 cm away from a long, straight wire carrying a current of 3 A (b) the field at the center of a flat, compact, circular coil, 2 cm in radius, with 10 turns, carrying a current of 0.3 A (c) the field at the center of a solenoid 2 cm in radius and 200 cm long, with 1 000 turns, carrying a current of 0.3 A (d) the field at the center of a long, straight metal bar, 2 cm in radius, carrying a current of 300 A (e) a field of 1 mT

16. One pole of a magnet attracts a nail. Will the other pole of the magnet attract the nail? Explain. Explain how a magnet sticks to a refrigerator door.

17. A magnet attracts a piece of iron. The iron can then attract another piece of iron. On the basis of domain alignment, explain what happens in each piece of iron.

18. Why does hitting a magnet with a hammer cause the magnetism to be reduced?

19. Which way would a compass point if you were at the north magnetic pole of the Earth?

20. Figure Q30.20 shows four permanent magnets, each having a hole through its center. Notice that the blue and yellow magnets are levitated above the red ones. (a) How does this levitation occur? (b) What purpose do the rods serve? (c) What can you say about the poles of the magnets from this observation? (d) If the upper magnet were inverted, what do you suppose would happen?

Figure Q30.20

Problems

WebAssign The Problems from this chapter may be assigned online in WebAssign.

ThomsonNOW™ Sign in at **www.thomsonedu.com** and go to ThomsonNOW to assess your understanding of this chapter's topics with additional quizzing and conceptual questions.

1, 2, 3 denotes straightforward, intermediate, challenging; □ denotes full solution available in *Student Solutions Manual/Study Guide;* ▲ denotes coached solution with hints available at **www.thomsonedu.com;** denotes developing symbolic reasoning; ● denotes asking for qualitative reasoning; ▪ denotes computer useful in solving problem

Section 30.1 The Biot–Savart Law

1. In Niels Bohr's 1913 model of the hydrogen atom, an electron circles the proton at a distance of 5.29×10^{-11} m with a speed of 2.19×10^6 m/s. Compute the magnitude of the magnetic field this motion produces at the location of the proton.

2. Calculate the magnitude of the magnetic field at a point 100 cm from a long, thin conductor carrying a current of 1.00 A.

3. (a) A conductor in the shape of a square loop of edge length $\ell = 0.400$ m carries a current $I = 10.0$ A as shown in Figure P30.3. Calculate the magnitude and direction of the magnetic field at the center of the square. (b) **What If?** If this conductor is formed into a single circular turn and carries the same current, what is the value of the magnetic field at the center?

Figure P30.3

4. A conductor consists of a circular loop of radius R and two straight, long sections as shown in Figure P30.4. The wire lies in the plane of the paper and carries a current I. Find an expression for the vector magnetic field at the center of the loop.

Figure P30.4

5. ▲ Determine the magnetic field at a point P located a distance x from the corner of an infinitely long wire bent at a right angle as shown in Figure P30.5. The wire carries a steady current I.

Figure P30.5

6. ▪ Consider a flat, circular current loop of radius R carrying current I. Choose the x axis to be along the axis of the loop, with the origin at the center of the loop. Plot a graph of the ratio of the magnitude of the magnetic field at coordinate x to that at the origin, for $x = 0$ to $x = 5R$. It may be useful to use a programmable calculator or a computer to solve this problem.

7. Two long, straight, parallel wires carry currents that are directed perpendicular to the page as shown in Figure P30.7. Wire 1 carries a current I_1 into the page (in the $-z$ direction) and passes through the x axis at $x = +a$. Wire 2 passes through the x axis at $x = -2a$ and carries an unknown current I_2. The total magnetic field at the origin due to the current-carrying wires has the magnitude $2\mu_0 I_1/(2\pi a)$. The current I_2 can have either of two possible values. (a) Find the value of I_2 with the smaller magnitude, stating it in terms of I_1 and giving its direction. (b) Find the other possible value of I_2.

Figure P30.7

8. A long, straight wire carries current I. A right-angle bend is made in the middle of the wire. The bend forms an arc

of a circle of radius r as shown in Figure P30.8. Determine the magnetic field at the center of the arc.

Figure P30.8

9. One long wire carries current 30.0 A to the left along the x axis. A second long wire carries current 50.0 A to the right along the line ($y = 0.280$ m, $z = 0$). (a) Where in the plane of the two wires is the total magnetic field equal to zero? (b) A particle with a charge of -2.00 μC is moving with a velocity of $150\hat{\mathbf{i}}$ Mm/s along the line ($y = 0.100$ m, $z = 0$). Calculate the vector magnetic force acting on the particle. (c) **What If?** A uniform electric field is applied to allow this particle to pass through this region undeflected. Calculate the required vector electric field.

10. A current path shaped as shown in Figure P30.10 produces a magnetic field at P, the center of the arc. If the arc subtends an angle of 30.0° and the radius of the arc is 0.600 m, what are the magnitude and direction of the field produced at P if the current is 3.00 A?

Figure P30.10

11. Three long, parallel conductors carry currents of $I = 2.00$ A. Figure P30.11 is an end view of the conductors, with each current coming out of the page. Taking $a = 1.00$ cm, determine the magnitude and direction of the magnetic field at points A, B, and C.

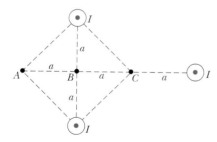

Figure P30.11

12. ● In a long, straight, vertical lightning stroke, electrons move downward and positive ions move upward, to constitute a current of magnitude 20.0 kA. At a location 50.0 m east of the middle of the stroke, a free electron drifts through the air toward the west with a speed of 300 m/s. (a) Find the vector force the lightning stroke exerts on the electron. Make a sketch showing the various vectors involved. Ignore the effect of the Earth's magnetic field. (b) Find the radius of the electron's path. Is it a good

approximation to model the electron as moving in a uniform field? Explain your answer. (c) If it does not collide with any obstacles, how many revolutions will the electron complete during the 60.0-μs duration of the lightning stroke?

13. ● A wire carrying a current I is bent into the shape of an equilateral triangle of side L. (a) Find the magnitude of the magnetic field at the center of the triangle. (b) At a point halfway between the center and any vertex, is the field stronger or weaker than at the center? Give a qualitative argument for your answer.

14. Determine the magnetic field (in terms of I, a, and d) at the origin due to the current loop in Figure P30.14.

Figure P30.14

15. Two long, parallel conductors carry currents $I_1 = 3.00$ A and $I_2 = 3.00$ A, both directed into the page in Figure P30.15. Determine the magnitude and direction of the resultant magnetic field at P.

Figure P30.15

16. The idea that a magnetic field can have therapeutic value has been around for centuries. A rare-earth magnet sold to relieve joint pain is a disk 1.20 mm thick and 3.50 mm in diameter. Its circular flat faces are its north and south poles. Assume it is accurately modeled as a magnetic dipole. Also assume Equation 30.10 describes the magnetic field it produces at all points along its axis. The field is strongest, with the value 40.0 mT, at the center of each flat face. At what distance from the surface is the magnitude of the magnetic field like that of the Earth, with a value of 50.0 μT?

Section 30.2 The Magnetic Force Between Two Parallel Conductors

17. In Figure P30.17, the current in the long, straight wire is $I_1 = 5.00$ A and the wire lies in the plane of the rectangular loop, which carries the current $I_2 = 10.0$ A. The dimensions are $c = 0.100$ m, $a = 0.150$ m, and $\ell = 0.450$ m. Find the magnitude and direction of the net force exerted on the loop by the magnetic field created by the wire.

2 = intermediate;   3 = challenging;   □ = SSM/SG;   ▲ = ThomsonNOW;   ▨ = symbolic reasoning;   ● = qualitative reasoning

Figure P30.17

18. Two long, parallel conductors, separated by 10.0 cm, carry currents in the same direction. The first wire carries current $I_1 = 5.00$ A, and the second carries $I_2 = 8.00$ A. (a) What is the magnitude of the magnetic field created by I_1 at the location of I_2? (b) What is the force per unit length exerted by I_1 on I_2? (c) What is the magnitude of the magnetic field created by I_2 at the location of I_1? (d) What is the force per length exerted by I_2 on I_1?

19. Two long, parallel wires are attracted to each other by a force per unit length of 320 μN/m when they are separated by a vertical distance of 0.500 m. The current in the upper wire is 20.0 A to the right. Determine the location of the line in the plane of the two wires along which the total magnetic field is zero.

20. ● Three long wires (wire 1, wire 2, and wire 3) hang vertically. The distance between wire 1 and wire 2 is 20.0 cm. On the left, wire 1 carries an upward current of 1.50 A. To the right, wire 2 carries a downward current of 4.00 A. Wire 3 is to be located such that when it carries a certain current, each wire experiences no net force. (a) Is this situation possible? Is it possible in more than one way? Describe (b) the position of wire 3 and (c) the magnitude and direction of the current in wire 3.

21. ● The unit of magnetic flux is named for Wilhelm Weber. A practical-size unit of magnetic field is named for Johann Karl Friedrich Gauss. Both were scientists at Göttingen, Germany. Along with their individual accomplishments, together they built a telegraph in 1833. It consisted of a battery and switch, at one end of a transmission line 3 km long, operating an electromagnet at the other end. (André Ampère suggested electrical signaling in 1821; Samuel Morse built a telegraph line between Baltimore and Washington, D.C., in 1844.) Suppose Weber and Gauss's transmission line was as diagrammed in Figure P30.21. Two long, parallel wires, each having a mass per unit length of 40.0 g/m, are supported in a horizontal plane by strings 6.00 cm long. When both wires carry the same current I, the wires repel each other so that the angle θ between the supporting strings is 16.0°. (a) Are the currents in the same direction or in opposite direc-

tions? (b) Find the magnitude of the current. (c) If this apparatus were taken to Mars, would the current required to separate the wires by 16° be larger or smaller than on Earth? Why?

22. ● Two parallel copper conductors are each 0.500 m long. They carry 10.0-A currents in opposite directions. (a) What center-to-center separation must the conductors have if they are to repel each other with a force of 1.00 N? (b) Is this situation physically possible? Explain.

Section 30.3 Ampère's Law

23. ▲ Four long, parallel conductors carry equal currents of $I = 5.00$ A. Figure P30.23 is an end view of the conductors. The current direction is into the page at points A and B (indicated by the crosses) and out of the page at C and D (indicated by the dots). Calculate the magnitude and direction of the magnetic field at point P, located at the center of the square of edge length 0.200 m.

Figure P30.23

24. A long, straight wire lies on a horizontal table and carries a current of 1.20 μA. In a vacuum, a proton moves parallel to the wire (opposite the current) with a constant speed of 2.30×10^4 m/s at a distance d above the wire. Determine the value of d. You may ignore the magnetic field due to the Earth.

25. Figure P30.25 is a cross-sectional view of a coaxial cable. The center conductor is surrounded by a rubber layer, which is surrounded by an outer conductor, which is surrounded by another rubber layer. In a particular application, the current in the inner conductor is 1.00 A out of the page and the current in the outer conductor is 3.00 A into the page. Determine the magnitude and direction of the magnetic field at points a and b.

Figure P30.25

26. The magnetic field 40.0 cm away from a long, straight wire carrying current 2.00 A is 1.00 μT. (a) At what distance is it 0.100 μT? (b) **What If?** At one instant, the two conductors in a long household extension cord carry equal 2.00-A currents in opposite directions. The two wires are 3.00 mm apart. Find the magnetic field 40.0 cm away

Figure P30.21

2 = intermediate;   3 = challenging;   ☐ = SSM/SG;   ▲ = ThomsonNOW;   ▨ = symbolic reasoning;   ● = qualitative reasoning

from the middle of the straight cord, in the plane of the two wires. (c) At what distance is it one-tenth as large? (d) The center wire in a coaxial cable carries current 2.00 A in one direction, and the sheath around it carries current 2.00 A in the opposite direction. What magnetic field does the cable create at points outside?

27. ● ▲ A packed bundle of 100 long, straight, insulated wires forms a cylinder of radius $R = 0.500$ cm. (a) If each wire carries 2.00 A, what are the magnitude and direction of the magnetic force per unit length acting on a wire located 0.200 cm from the center of the bundle? (b) **What If?** Would a wire on the outer edge of the bundle experience a force greater or smaller than the value calculated in part (a)? Give a qualitative argument for your answer.

28. The magnetic coils of a tokamak fusion reactor are in the shape of a toroid having an inner radius of 0.700 m and an outer radius of 1.30 m. The toroid has 900 turns of large-diameter wire, each of which carries a current of 14.0 kA. Find the magnitude of the magnetic field inside the toroid along (a) the inner radius and (b) the outer radius.

29. Consider a column of electric current passing through plasma (ionized gas). Filaments of current within the column are magnetically attracted to one another. They can crowd together to yield a very great current density and a very strong magnetic field in a small region. Sometimes the current can be cut off momentarily by this *pinch effect.* (In a metallic wire, a pinch effect is not important because the current-carrying electrons repel one another with electric forces.) The pinch effect can be demonstrated by making an empty aluminum can carry a large current parallel to its axis. Let R represent the radius of the can and I the upward current, uniformly distributed over its curved wall. Determine the magnetic field (a) just inside the wall and (b) just outside. (c) Determine the pressure on the wall.

30. Niobium metal becomes a superconductor when cooled below 9 K. Its superconductivity is destroyed when the surface magnetic field exceeds 0.100 T. Determine the maximum current a 2.00-mm-diameter niobium wire can carry and remain superconducting, in the absence of any external magnetic field.

31. A long, cylindrical conductor of radius R carries a current I as shown in Figure P30.31. The current density J, however, is not uniform over the cross section of the conductor but is a function of the radius according to $J = br$, where b is a constant. Find an expression for the magnetic field magnitude B (a) at a distance $r_1 < R$ and (b) at a distance $r_2 > R$, measured from the axis.

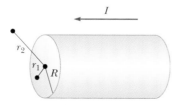

Figure P30.31

32. In Figure P30.32, both currents in the infinitely long wires are 8.00 A in the negative x direction. The wires are separated by the distance $2a = 6.00$ cm. (a) Sketch the

magnetic field pattern in the yz plane. (b) What is the value of the magnetic field at the origin? At ($y = 0$, $z \rightarrow \infty$)? (c) Find the magnetic field at points along the z axis as a function of z. (d) At what distance d along the positive z axis is the magnetic field a maximum? (e) What is this maximum value?

Figure P30.32

33. An infinite sheet of current lying in the yz plane carries a surface current of linear density J_s. The current is in the y direction, and J_s represents the current per unit length measured along the z axis. Figure P30.33 is an edge view of the sheet. Prove that the magnetic field near the sheet is parallel to the sheet and perpendicular to the current direction, with magnitude $\mu_0 J_s/2$. *Suggestion:* Use Ampère's law and evaluate the line integral for a rectangular path around the sheet, represented by the dashed line in Figure P30.33.

Figure P30.33

Section 30.4 The Magnetic Field of a Solenoid

34. ● You are given a certain volume of copper from which you can make copper wire. To insulate the wire, you can have as much enamel as you like. You will use the wire to make a tightly wound solenoid 20 cm long having the greatest possible magnetic field at the center and using a power supply that can deliver a current of 5 A. The solenoid can be wrapped with wire in one or more layers. (a) Should you make the wire long and thin or shorter and thick? Explain. (b) Should you make the solenoid radius small or large? Explain.

35. ▲ What current is required in the windings of a long solenoid that has 1 000 turns uniformly distributed over a length of 0.400 m to produce at the center of the solenoid a magnetic field of magnitude 1.00×10^{-4} T?

36. Consider a solenoid of length ℓ and radius R, containing N closely spaced turns and carrying a steady current I. (a) In terms of these parameters, find the magnetic field at a point along the axis as a function of distance a from the end of the solenoid. (b) Show that as ℓ becomes very long, B approaches $\mu_0 NI/2\ell$ at each end of the solenoid.

2 = intermediate;   3 = challenging;   ☐ = SSM/SG;   ▲ = ThomsonNOW;   = symbolic reasoning;   ● = qualitative reasoning

37. A single-turn square loop of wire, 2.00 cm on each edge, carries a clockwise current of 0.200 A. The loop is inside a solenoid, with the plane of the loop perpendicular to the magnetic field of the solenoid. The solenoid has 30 turns/cm and carries a clockwise current of 15.0 A. Find the force on each side of the loop and the torque acting on the loop.

38. A solenoid 10.0 cm in diameter and 75.0 cm long is made from copper wire of diameter 0.100 cm, with very thin insulation. The wire is wound onto a cardboard tube in a single layer, with adjacent turns touching each other. What power must be delivered to the solenoid if it is to produce a field of 8.00 mT at its center?

Section 30.5 Gauss's Law in Magnetism

39. A cube of edge length $\ell = 2.50$ cm is positioned as shown in Figure P30.39. A uniform magnetic field given by $\vec{B} = (5\hat{i} + 4\hat{j} + 3\hat{k})$ T exists throughout the region. (a) Calculate the magnetic flux through the shaded face. (b) What is the total flux through the six faces?

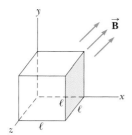

Figure P30.39

40. Consider the hemispherical closed surface in Figure P30.40. The hemisphere is in a uniform magnetic field that makes an angle θ with the vertical. Calculate the magnetic flux through (a) the flat surface S_1 and (b) the hemispherical surface S_2.

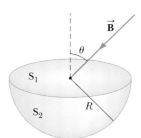

Figure P30.40

41. A solenoid 2.50 cm in diameter and 30.0 cm long has 300 turns and carries 12.0 A. (a) Calculate the flux through the surface of a disk of radius 5.00 cm that is positioned perpendicular to and centered on the axis of the solenoid, as shown in Figure P30.41a. (b) Figure P30.41b shows an enlarged end view of the same solenoid. Calculate the flux through the blue area, which is an annulus with an inner radius of 0.400 cm and an outer radius of 0.800 cm.

(a)   (b)

Figure P30.41

42. *Compare this problem with Problem 65 in Chapter 24.* Consider a magnetic field that is uniform in direction throughout a certain volume. Can it be uniform in magnitude? Must it be uniform in magnitude? Give evidence for your answers.

Section 30.6 Magnetism in Matter

43. At *saturation*, when nearly all the atoms have their magnetic moments aligned, the magnetic field in a sample of iron can be 2.00 T. If each electron contributes a magnetic moment of 9.27×10^{-24} A·m^2 (one Bohr magneton), how many electrons per atom contribute to the saturated field of iron? The number density of atoms in iron is approximately 8.50×10^{28} atoms/m^3.

Section 30.7 The Magnetic Field of the Earth

44. A circular coil of 5 turns and a diameter of 30.0 cm is oriented in a vertical plane with its axis perpendicular to the horizontal component of the Earth's magnetic field. A horizontal compass placed at the center of the coil is made to deflect 45.0° from magnetic north by a current of 0.600 A in the coil. (a) What is the horizontal component of the Earth's magnetic field? (b) The current in the coil is switched off. A "dip needle" is a magnetic compass mounted so that it can rotate in a vertical north–south plane. At this location, a dip needle makes an angle of 13.0° from the vertical. What is the total magnitude of the Earth's magnetic field at this location?

45. The magnetic moment of the Earth is approximately 8.00×10^{22} A·m^2. (a) Imagine that the planetary magnetic field were caused by the complete magnetization of a huge iron deposit. How many unpaired electrons would participate? (b) At two unpaired electrons per iron atom, how many kilograms of iron would compose the deposit? Iron has a density of 7 900 kg/m^3 and approximately 8.50×10^{28} iron atoms/m^3.

46. ● A particular location on the Earth's surface is characterized by a value of gravitational field, a value of magnetic field, and a value of atmospheric pressure. (a) Which of these quantities are vectors and which are scalars? (b) Determine a value for each quantity at your current location. Include the direction of each vector quantity. State your sources. (c) Which of the quantities have separate causes from which of the others?

Additional Problems

47. A very long, thin strip of metal of width w carries a current I along its length as shown in Figure P30.47. Find the magnetic field at the point P in the diagram. The point P is in the plane of the strip at distance b away from it.

2 = intermediate;   3 = challenging;   □ = SSM/SG;   ▲ = ThomsonNOW;   ▨ = symbolic reasoning;   ● = qualitative reasoning

Figure P30.47

48. The magnitude of the Earth's magnetic field at either pole is approximately 7.00×10^{-5} T. Suppose the field fades away, before its next reversal. Scouts, sailors, and conservative politicians around the world join together in a program to replace the field. One plan is to use a current loop around the equator, without relying on magnetization of any materials inside the Earth. Determine the current that would generate such a field if this plan were carried out. Take the radius of the Earth as $R_E = 6.37 \times 10^6$ m.

49. A thin copper bar of length $\ell = 10.0$ cm is supported horizontally by two (nonmagnetic) contacts. The bar carries current $I_1 = 100$ A in the $-x$ direction as shown in Figure P30.49. At a distance $h = 0.500$ cm below one end of the bar, a long, straight wire carries a current $I_2 = 200$ A in the z direction. Determine the magnetic force exerted on the bar.

Figure P30.49

50. Suppose you install a compass on the center of the dashboard of a car. Compute an order-of-magnitude estimate for the magnetic field at this location produced by the current when you switch on the headlights. How does this estimate compare with the Earth's magnetic field? You may suppose the dashboard is made mostly of plastic.

51. ▲ A nonconducting ring of radius 10.0 cm is uniformly charged with a total positive charge 10.0 μC. The ring rotates at a constant angular speed 20.0 rad/s about an axis through its center, perpendicular to the plane of the ring. What is the magnitude of the magnetic field on the axis of the ring 5.00 cm from its center?

52. A nonconducting ring of radius R is uniformly charged with a total positive charge q. The ring rotates at a constant angular speed ω about an axis through its center, perpendicular to the plane of the ring. What is the magnitude of the magnetic field on the axis of the ring a distance $R/2$ from its center?

53. Two circular coils of radius R, each with N turns, are perpendicular to a common axis. The coil centers are a dis-

tance R apart. Each coil carries a steady current I in the same direction as shown in Figure P30.53. (a) Show that the magnetic field on the axis at a distance x from the center of one coil is

$$B = \frac{N\mu_0 IR^2}{2}\left[\frac{1}{(R^2 + x^2)^{3/2}} + \frac{1}{(2R^2 + x^2 - 2Rx)^{3/2}}\right]$$

(b) Show that dB/dx and d^2B/dx^2 are both zero at the point midway between the coils. We may then conclude that the magnetic field in the region midway between the coils is uniform. Coils in this configuration are called *Helmholtz coils*.

Figure P30.53 Problems 53 and 54.

54. Two identical, flat, circular coils of wire each have 100 turns and a radius of 0.500 m. The coils are arranged as a set of Helmholtz coils (see Fig. P30.53), parallel and with separation 0.500 m. Each coil carries a current of 10.0 A. Determine the magnitude of the magnetic field at a point on the common axis of the coils and halfway between them.

55. We have seen that a long solenoid produces a uniform magnetic field directed along the axis of a cylindrical region. To produce a uniform magnetic field directed parallel to a *diameter* of a cylindrical region, however, one can use the *saddle coils* illustrated in Figure P30.55. The loops are wrapped over a somewhat flattened tube. Assume the straight sections of wire are very long. The end view of the tube shows how the windings are applied. The overall current distribution is the superposition of two overlapping, circular cylinders of uniformly distributed current, one toward you and one away from you. The current density J is the same for each cylinder. The position of the axis of one cylinder is described by a position vector $\vec{a}$ relative to the other cylinder. Prove that the magnetic field inside the hollow tube is $\mu_0 Ja/2$ downward. *Suggestion:* The use of vector methods simplifies the calculation.

(a)   (b)

Figure P30.55

56. *You may use the result of Problem 33 in solving this problem.* A very large parallel-plate capacitor carries charge with uniform charge per unit area $+\sigma$ on the upper plate and $-\sigma$ on the lower plate. The plates are horizontal and both move horizontally with speed v to the right. (a) What is the magnetic field between the plates? (b) What is the magnetic field close to the plates but outside of the capacitor? (c) What is the magnitude and direction of the magnetic force per unit area on the upper plate? (d) At what extrapolated speed v will the magnetic force on a plate balance the electric force on the plate? Calculate this speed numerically.

57. ● Two circular loops are parallel, coaxial, and almost in contact, 1.00 mm apart (Fig. P30.57). Each loop is 10.0 cm in radius. The top loop carries a clockwise current of 140 A. The bottom loop carries a counterclockwise current of 140 A. (a) Calculate the magnetic force exerted by the bottom loop on the top loop. (b) Suppose a student thinks the first step in solving part (a) is to use Equation 30.7 to find the magnetic field created by one of the loops. How would you argue for or against this idea? *Suggestion:* Think about how one loop looks to a bug perched on the other loop. (c) The upper loop has a mass of 0.021 0 kg. Calculate its acceleration, assuming the only forces acting on it are the force in part (a) and the gravitational force.

140 A

140 A

Figure P30.57

58. What objects experience a force in an electric field? Chapter 23 gives the answer: any object with electric charge, stationary or moving, other than the charged object that created the field. What creates an electric field? Any object with electric charge, stationary or moving, as you studied in Chapter 23. What objects experience a force in a magnetic field? An electric current or a moving electric charge, other than the current or charge that created the field, as discussed in Chapter 29. What creates a magnetic field? An electric current, as you studied in Section 30.1, or a moving electric charge, as shown in this problem. (a) To understand how a moving charge creates a magnetic field, consider a particle with charge q moving with velocity $\vec{\mathbf{v}}$. Define the position vector $\vec{\mathbf{r}} = r\hat{\mathbf{r}}$ leading from the particle to some location. Show that the magnetic field at that location is

$$\vec{\mathbf{B}} = \frac{\mu_0}{4\pi} \frac{q\vec{\mathbf{v}} \times \hat{\mathbf{r}}}{r^2}$$

(b) Find the magnitude of the magnetic field 1.00 mm to the side of a proton moving at 2.00×10^7 m/s. (c) Find the magnetic force on a second proton at this point, moving with the same speed in the opposite direction. (d) Find the electric force on the second proton.

59. The chapter-opening photograph shows a rail gun. Rail guns have been suggested for launching projectiles into space without chemical rockets and for ground-to-air antimissile weapons of war. A tabletop model rail gun

(Fig. P30.59) consists of two long, parallel, horizontal rails 3.50 cm apart, bridged by a bar BD of mass 3.00 g. The bar is originally at rest at the midpoint of the rails and is free to slide without friction. When the switch is closed, electric current is quickly established in the circuit $ABCDEA$. The rails and bar have low electric resistance, and the current is limited to a constant 24.0 A by the power supply. (a) Find the magnitude of the magnetic field 1.75 cm from a single very long, straight wire carrying current 24.0 A. (b) Find the magnitude and direction of the magnetic field at point C in the diagram, the midpoint of the bar, immediately after the switch is closed. *Suggestion:* Consider what conclusions you can draw from the Biot–Savart law. (c) At other points along the bar BD, the field is in the same direction as at point C, but is larger in magnitude. Assume the average effective magnetic field along BD is five times larger than the field at C. With this assumption, find the magnitude and direction of the force on the bar. (d) Find the acceleration of the bar when it is in motion. (e) Does the bar move with constant acceleration? (f) Find the velocity of the bar after it has traveled 130 cm to the end of the rails.

Figure P30.59

60. ▪ Fifty turns of insulated wire 0.100 cm in diameter are tightly wound to form a flat spiral. The spiral fills a disk surrounding a circle of radius 5.00 cm and extending to a radius 10.00 cm at the outer edge. Assume the wire carries current I at the center of its cross section. Approximate each turn of wire as a circle. Then a loop of current exists at radius 5.05 cm, another at 5.15 cm, and so on. Numerically calculate the magnetic field at the center of the coil.

61. An infinitely long, straight wire carrying a current I_1 is partially surrounded by a loop as shown in Figure P30.61. The loop has a length L and radius R, and it carries a current I_2. The axis of the loop coincides with the wire. Calculate the force exerted on the loop.

R

L

I_2

I_1

Figure P30.61

62. The magnitude of the force on a magnetic dipole $\vec{\boldsymbol{\mu}}$ aligned with a nonuniform magnetic field in the x direction is $F_x = |\vec{\boldsymbol{\mu}}| \, dB/dx$. Suppose two flat loops of wire each have radius R and carry current I. (a) The loops are arranged coaxially and separated by a variable distance x, large compared with R. Show that the magnetic force

between them varies as $1/x^4$. (b) Evaluate the magnitude of this force, taking $I = 10.0$ A, $R = 0.500$ cm, and $x = 5.00$ cm.

63. A wire is formed into the shape of a square of edge length L (Fig. P30.63). Show that when the current in the loop is I, the magnetic field at point P, a distance x from the center of the square along its axis is

$$B = \frac{\mu_0 I L^2}{2\pi(x^2 + L^2/4)\sqrt{x^2 + L^2/2}}$$

Figure P30.63

64. A wire carrying a current I is bent into the shape of an exponential spiral, $r = e^\theta$, from $\theta = 0$ to $\theta = 2\pi$ as suggested in Figure P30.64. To complete a loop, the ends of the spiral are connected by a straight wire along the x axis. Find the magnitude and direction of $\vec{\mathbf{B}}$ at the origin. *Suggestions:* Use the Biot–Savart law. The angle β between

a radial line and its tangent line at any point on the curve $r = f(\theta)$ is related to the function as follows:

$$\tan \beta = \frac{r}{dr/d\theta}$$

Therefore, in this case $r = e^\theta$, $\tan \beta = 1$, and $\beta = \pi/4$, and the angle between $d\vec{\mathbf{s}}$ and $\hat{\mathbf{r}}$ is $\pi - \beta = 3\pi/4$. Also,

$$ds = \frac{dr}{\sin(\pi/4)} = \sqrt{2}\, dr$$

65. A sphere of radius R has a uniform volume charge density ρ. Determine the magnetic field at the center of the sphere when it rotates as a rigid object with angular speed ω about an axis through its center (Fig. P30.65).

Figure P30.65 Problems 65 and 66.

66. A sphere of radius R has a uniform volume charge density ρ. Determine the magnetic dipole moment of the sphere when it rotates as a rigid body with angular speed ω about an axis through its center (Fig. P30.65).

67. A long, cylindrical conductor of radius a has two cylindrical cavities of diameter a through its entire length as shown in Figure P30.67. A current I is directed out of the page and is uniform through a cross section of the conductor. Find the magnitude and direction of the magnetic field in terms of μ_0, I, r, and a at (a) point P_1 and (b) point P_2.

Figure P30.67

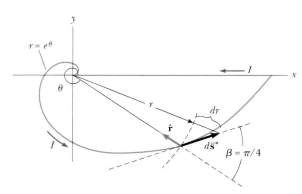

Figure P30.64

Answers to Quick Quizzes

30.1 B, C, A. Point B is closest to the current element. Point C is farther away, and the field is further reduced by the $\sin \theta$ factor in the cross product $d\vec{\mathbf{s}} \times \hat{\mathbf{r}}$. The field at A is zero because $\theta = 0$.

30.2 (a). The coils act like wires carrying parallel currents in the same direction and hence attract one another.

30.3 b, d, a, c. Equation 30.13 indicates that the value of the line integral depends only on the net current through each closed path. Path b encloses 1 A, path d encloses 3 A, path a encloses 4 A, and path c encloses 6 A.

30.4 b, then $a = c = d$. Paths a, c, and d all give the same nonzero value $\mu_0 I$ because the size and shape of the paths do not matter. Path b does not enclose the current; hence, its line integral is zero.

30.5 (c). The magnetic field in a very long solenoid is independent of its length or radius. Overwrapping with an additional layer of wire increases the number of turns per unit length.

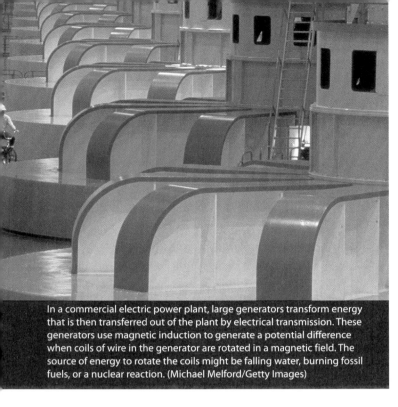

In a commercial electric power plant, large generators transform energy that is then transferred out of the plant by electrical transmission. These generators use magnetic induction to generate a potential difference when coils of wire in the generator are rotated in a magnetic field. The source of energy to rotate the coils might be falling water, burning fossil fuels, or a nuclear reaction. (Michael Melford/Getty Images)

31.1 Faraday's Law of Induction

31.2 Motional emf

31.3 Lenz's Law

31.4 Induced emf and Electric Fields

31.5 Generators and Motors

31.6 Eddy Currents

31 Faraday's Law

So far, our studies in electricity and magnetism have focused on the electric fields produced by stationary charges and the magnetic fields produced by moving charges. This chapter explores the effects produced by magnetic fields that vary in time.

Experiments conducted by Michael Faraday in England in 1831 and independently by Joseph Henry in the United States that same year showed that an emf can be induced in a circuit by a changing magnetic field. The results of these experiments led to a very basic and important law of electromagnetism known as *Faraday's law of induction*. An emf (and therefore a current as well) can be induced in various processes that involve a change in a magnetic flux.

MICHAEL FARADAY
British Physicist and Chemist (1791–1867)
Faraday is often regarded as the greatest experimental scientist of the 1800s. His many contributions to the study of electricity include the invention of the electric motor, electric generator, and transformer as well as the discovery of electromagnetic induction and the laws of electrolysis. Greatly influenced by religion, he refused to work on the development of poison gas for the British military.

31.1 Faraday's Law of Induction

To see how an emf can be induced by a changing magnetic field, consider the experimental results obtained when a loop of wire is connected to a sensitive ammeter as illustrated in Active Figure 31.1 (page 868). When a magnet is moved toward the loop, the reading on the ammeter changes from zero in one direction, arbitrarily shown as negative in Active Figure 31.1a. When the magnet is brought to rest and held stationary relative to the loop (Active Fig. 31.1b), a reading of

ACTIVE FIGURE 31.1

(a) When a magnet is moved toward a loop of wire connected to a sensitive ammeter, the ammeter reading changes from zero, indicating that a current is induced in the loop. (b) When the magnet is held stationary, there is no induced current in the loop, even when the magnet is inside the loop. (c) When the magnet is moved away from the loop, the ammeter reading changes in the opposite direction, indicating that the induced current is opposite that shown in (a). Changing the direction of the magnet's motion changes the direction of the current induced by that motion.

Sign in at www.thomsonedu.com and go to ThomsonNOW to move the magnet and observe the current in the ammeter.

PITFALL PREVENTION 31.1
Induced emf Requires a Change

The *existence* of a magnetic flux through an area is not sufficient to create an induced emf. The magnetic flux must *change* to induce an emf.

zero is observed. When the magnet is moved away from the loop, the reading on the ammeter changes in the opposite direction as shown in Active Figure 31.1c. Finally, when the magnet is held stationary and the loop is moved either toward or away from it, the reading changes from zero. From these observations, we conclude that the loop detects that the magnet is moving relative to it and we relate this detection to a change in magnetic field. Therefore, it seems that a relationship exists between current and changing magnetic field.

These results are quite remarkable because **a current is set up even though no batteries are present in the circuit!** We call such a current an *induced current* and say that it is produced by an *induced emf.*

Now let's describe an experiment conducted by Faraday and illustrated in Active Figure 31.2. A primary coil is wrapped around an iron ring and connected to a switch and a battery. A current in the coil produces a magnetic field when the switch is closed. A secondary coil also is wrapped around the ring and is connected to a sensitive ammeter. No battery is present in the secondary circuit, and the secondary coil is not electrically connected to the primary coil. Any current detected in the secondary circuit must be induced by some external agent.

Initially, you might guess that no current is ever detected in the secondary circuit. Something quite amazing happens when the switch in the primary circuit is either opened or thrown closed, however. At the instant the switch is closed, the ammeter reading changes from zero in one direction and then returns to zero. At the instant the switch is opened, the ammeter changes in the opposite direction and again returns to zero. Finally, the ammeter reads zero when there is either a steady current or no current in the primary circuit. To understand what happens in this experiment, note that when the switch is closed, the current in the primary circuit produces a magnetic field that penetrates the secondary circuit. Furthermore, when the switch is closed, the magnetic field produced by the current in the primary circuit changes from zero to some value over some finite time, and this changing field induces a current in the secondary circuit.

As a result of these observations, Faraday concluded that **an electric current can be induced in a loop by a changing magnetic field.** The induced current exists only while the magnetic field through the loop is changing. Once the magnetic field reaches a steady value, the current in the loop disappears. In effect, the loop behaves as though a source of emf were connected to it for a short time. It is customary to say that **an induced emf is produced in the loop by the changing magnetic field.**

The experiments shown in Active Figures 31.1 and 31.2 have one thing in common: in each case, an emf is induced in a loop when the magnetic flux through the loop changes with time. In general, this emf is directly proportional to the

ACTIVE FIGURE 31.2

Faraday's experiment. When the switch in the primary circuit is closed, the ammeter reading in the secondary circuit changes momentarily. The emf induced in the secondary circuit is caused by the changing magnetic field through the secondary coil.

Sign in at www.thomsonedu.com and go to ThomsonNOW to open and close the switch and observe the current in the ammeter.

time rate of change of the magnetic flux through the loop. This statement can be written mathematically as **Faraday's law of induction:**

$$\mathcal{E} = -\frac{d\Phi_B}{dt} \tag{31.1}$$

 ◄ Faraday's law

where $\Phi_B = \oint \vec{\mathbf{B}} \cdot d\vec{\mathbf{A}}$ is the magnetic flux through the loop. (See Section 30.5.)

If a coil consists of N loops with the same area and Φ_B is the magnetic flux through one loop, an emf is induced in every loop. The loops are in series, so their emfs add; therefore, the total induced emf in the coil is given by

$$\mathcal{E} = -N\frac{d\Phi_B}{dt} \tag{31.2}$$

The negative sign in Equations 31.1 and 31.2 is of important physical significance as discussed in Section 31.3.

Suppose a loop enclosing an area A lies in a uniform magnetic field $\vec{\mathbf{B}}$ as in Figure 31.3. The magnetic flux through the loop is equal to $BA\cos\theta$; hence, the induced emf can be expressed as

$$\mathcal{E} = -\frac{d}{dt}(BA\cos\theta) \tag{31.3}$$

From this expression, we see that an emf can be induced in the circuit in several ways:

- The magnitude of $\vec{\mathbf{B}}$ can change with time.
- The area enclosed by the loop can change with time.
- The angle θ between $\vec{\mathbf{B}}$ and the normal to the loop can change with time.
- Any combination of the above can occur.

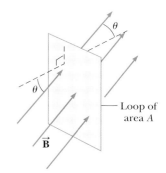

Figure 31.3 A conducting loop that encloses an area A in the presence of a uniform magnetic field $\vec{\mathbf{B}}$. The angle between $\vec{\mathbf{B}}$ and the normal to the loop is θ.

Quick Quiz 31.1 A circular loop of wire is held in a uniform magnetic field, with the plane of the loop perpendicular to the field lines. Which of the following will *not* cause a current to be induced in the loop? (a) crushing the loop (b) rotating the loop about an axis perpendicular to the field lines (c) keeping the orientation of the loop fixed and moving it along the field lines (d) pulling the loop out of the field

---

Some Applications of Faraday's Law

The ground fault interrupter (GFI) is an interesting safety device that protects users of electrical appliances against electric shock. Its operation makes use of Faraday's law. In the GFI shown in Figure 31.4, wire 1 leads from the wall outlet to the appliance to be protected and wire 2 leads from the appliance back to the wall outlet. An iron ring surrounds the two wires, and a sensing coil is wrapped around part of the ring. Because the currents in the wires are in opposite directions and of equal magnitude, there is no magnetic field surrounding the wires and the net magnetic flux through the sensing coil is zero. If the return current in wire 2 changes so that the two currents are not equal, however, circular magnetic field lines exist around the pair of wires. (This can happen if, for example, the appliance becomes wet, enabling current to leak to ground.) Therefore, the net magnetic flux through the sensing coil is no longer zero. Because household current is alternating (meaning that its direction keeps reversing), the magnetic flux through the sensing coil changes with time, inducing an emf in the coil. This induced emf is used to trigger a circuit breaker, which stops the current before it is able to reach a harmful level.

Another interesting application of Faraday's law is the production of sound in an electric guitar. The coil in this case, called the *pickup coil*, is placed near the

Figure 31.4 Essential components of a ground fault interrupter.

Figure 31.5 (a) In an electric guitar, a vibrating magnetized string induces an emf in a pickup coil. (b) The pickups (the circles beneath the metallic strings) of this electric guitar detect the vibrations of the strings and send this information through an amplifier and into speakers. (A switch on the guitar allows the musician to select which set of six pickups is used.)

vibrating guitar string, which is made of a metal that can be magnetized. A permanent magnet inside the coil magnetizes the portion of the string nearest the coil (Fig. 31.5a). When the string vibrates at some frequency, its magnetized segment produces a changing magnetic flux through the coil. The changing flux induces an emf in the coil that is fed to an amplifier. The output of the amplifier is sent to the loudspeakers, which produce the sound waves we hear.

EXAMPLE 31.1 | **Inducing an emf in a Coil**

A coil consists of 200 turns of wire. Each turn is a square of side $d = 18$ cm, and a uniform magnetic field directed perpendicular to the plane of the coil is turned on. If the field changes linearly from 0 to 0.50 T in 0.80 s, what is the magnitude of the induced emf in the coil while the field is changing?

SOLUTION

Conceptualize From the description in the problem, imagine magnetic field lines passing through the coil. Because the magnetic field is changing in magnitude, an emf is induced in the coil.

Categorize We will evaluate the emf using Faraday's law from this section, so we categorize this example as a substitution problem.

Evaluate Equation 31.2 for the situation described here, noting that the magnetic field changes linearly with time:

$$|\mathcal{E}| = N \frac{\Delta \Phi_B}{\Delta t} = N \frac{\Delta(BA)}{\Delta t} = NA \frac{\Delta B}{\Delta t} = Nd^2 \frac{B_f - B_i}{\Delta t}$$

Substitute numerical values:

$$|\mathcal{E}| = (200)(0.18 \text{ m})^2 \frac{(0.50 \text{ T} - 0)}{0.80 \text{ s}} = \boxed{4.0 \text{ V}}$$

What If? What if you were asked to find the magnitude of the induced current in the coil while the field is changing? Can you answer that question?

Answer If the ends of the coil are not connected to a circuit, the answer to this question is easy: the current is zero! (Charges move within the wire of the coil, but they cannot move into or out of the ends of the coil.) For a steady current to exist, the ends of the coil must be connected to an external circuit. Let's assume the coil is connected to a circuit and the total resistance of the coil and the circuit is 2.0 Ω. Then, the current in the coil is

$$I = \frac{\mathcal{E}}{R} = \frac{4.0 \text{ V}}{2.0 \text{ }\Omega} = 2.0 \text{ A}$$

| EXAMPLE 31.2 | **An Exponentially Decaying B Field** |

A loop of wire enclosing an area A is placed in a region where the magnetic field is perpendicular to the plane of the loop. The magnitude of $\vec{B}$ varies in time according to the expression $B = B_{max}e^{-at}$, where a is some constant. That is, at $t = 0$, the field is B_{max}, and for $t > 0$, the field decreases exponentially (Fig. 31.6). Find the induced emf in the loop as a function of time.

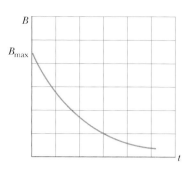

SOLUTION

Conceptualize The physical situation is similar to that in Example 31.1 except for two things: there is only one loop, and the field varies exponentially with time rather than linearly.

Figure 31.6 (Example 31.2) Exponential decrease in the magnitude of the magnetic field with time. The induced emf and induced current vary with time in the same way.

Categorize We will evaluate the emf using Faraday's law from this section, so we categorize this example as a substitution problem.

Evaluate Equation 31.1 for the situation described here:

$$\mathcal{E} = -\frac{d\Phi_B}{dt} = -\frac{d}{dt}(AB_{max}e^{-at}) = -AB_{max}\frac{d}{dt}e^{-at} = \boxed{aAB_{max}e^{-at}}$$

This expression indicates that the induced emf decays exponentially in time. The maximum emf occurs at $t = 0$, where $\mathcal{E}_{max} = aAB_{max}$. The plot of $\mathcal{E}$ versus t is similar to the B-versus-t curve shown in Figure 31.6.

31.2 Motional emf

In Examples 31.1 and 31.2, we considered cases in which an emf is induced in a stationary circuit placed in a magnetic field when the field changes with time. In this section, we describe **motional emf,** the emf induced in a conductor moving through a constant magnetic field.

The straight conductor of length ℓ shown in Figure 31.7 is moving through a uniform magnetic field directed into the page. For simplicity, let's assume the conductor is moving in a direction perpendicular to the field with constant velocity under the influence of some external agent. The electrons in the conductor experience a force $\vec{F}_B = q\vec{v} \times \vec{B}$ that is directed along the length ℓ, perpendicular to both $\vec{v}$ and $\vec{B}$ (Eq. 29.1). Under the influence of this force, the electrons move to the lower end of the conductor and accumulate there, leaving a net positive charge at the upper end. As a result of this charge separation, an electric field $\vec{E}$ is produced inside the conductor. The charges accumulate at both ends until the downward magnetic force qvB on charges remaining in the conductor is balanced by the upward electric force qE. The condition for equilibrium requires that the forces on the electrons balance:

$$qE = qvB \quad \text{or} \quad E = vB$$

The electric field produced in the conductor is related to the potential difference across the ends of the conductor according to the relationship $\Delta V = E\ell$ (Eq. 25.6). Therefore, for the equilibrium condition,

$$\Delta V = E\ell = B\ell v \qquad (31.4)$$

where the upper end of the conductor in Figure 31.7 is at a higher electric potential than the lower end. Therefore, **a potential difference is maintained between the ends of the conductor as long as the conductor continues to move through the uniform magnetic field.** If the direction of the motion is reversed, the polarity of the potential difference is also reversed.

A more interesting situation occurs when the moving conductor is part of a closed conducting path. This situation is particularly useful for illustrating how a

Figure 31.7 A straight electrical conductor of length ℓ moving with a velocity $\vec{v}$ through a uniform magnetic field $\vec{B}$ directed perpendicular to $\vec{v}$. Due to the magnetic force on electrons, the ends of the conductor become oppositely charged, which establishes an electric field in the conductor. In steady state, the electric and magnetic forces on an electron in the wire are balanced.

ACTIVE FIGURE 31.8

(a) A conducting bar sliding with a velocity $\vec{v}$ along two conducting rails under the action of an applied force $\vec{F}_{app}$. The magnetic force $\vec{F}_B$ opposes the motion, and a counterclockwise current I is induced in the loop. (b) The equivalent circuit diagram for the setup shown in (a).

Sign in at www.thomsonedu.com and go to ThomsonNOW to adjust the applied force, the magnetic field, and the resistance to see the effects on the motion of the bar.

changing magnetic flux causes an induced current in a closed circuit. Consider a circuit consisting of a conducting bar of length ℓ sliding along two fixed parallel conducting rails as shown in Active Figure 31.8a. For simplicity, let's assume the bar has zero resistance and the stationary part of the circuit has a resistance R. A uniform and constant magnetic field $\vec{B}$ is applied perpendicular to the plane of the circuit. As the bar is pulled to the right with a velocity $\vec{v}$ under the influence of an applied force $\vec{F}_{app}$, free charges in the bar experience a magnetic force directed along the length of the bar. This force sets up an induced current because the charges are free to move in the closed conducting path. In this case, the rate of change of magnetic flux through the circuit and the corresponding induced motional emf across the moving bar are proportional to the change in area of the circuit.

Because the area enclosed by the circuit at any instant is ℓx, where x is the position of the bar, the magnetic flux through that area is

$$\Phi_B = B\ell x$$

Using Faraday's law and noting that x changes with time at a rate $dx/dt = v$, we find that the induced motional emf is

$$\varepsilon = -\frac{d\Phi_B}{dt} = -\frac{d}{dt}(B\ell x) = -B\ell\frac{dx}{dt}$$

Motional emf ▶
$$\varepsilon = -B\ell v \qquad \textbf{(31.5)}$$

Because the resistance of the circuit is R, the magnitude of the induced current is

$$I = \frac{|\varepsilon|}{R} = \frac{B\ell v}{R} \qquad \textbf{(31.6)}$$

The equivalent circuit diagram for this example is shown in Active Figure 31.8b.

Let's examine the system using energy considerations. Because no battery is in the circuit, you might wonder about the origin of the induced current and the energy delivered to the resistor. We can understand the source of this current and energy by noting that the applied force does work on the conducting bar. Therefore, we model the circuit as a nonisolated system. The movement of the bar through the field causes charges to move along the bar with some average drift velocity; hence, a current is established. The change in energy in the system during some time interval must be equal to the transfer of energy into the system by work, consistent with the general principle of conservation of energy described by Equation 8.2.

Let's verify this mathematically. As the bar moves through the uniform magnetic field $\vec{\mathbf{B}}$, it experiences a magnetic force $\vec{\mathbf{F}}_B$ of magnitude $I\ell B$ (see Section 29.4). Because the bar moves with constant velocity, it is modeled as a particle in equilibrium and the magnetic force must be equal in magnitude and opposite in direction to the applied force, or to the left in Active Figure 31.8a. (If $\vec{\mathbf{F}}_B$ acted in the direction of motion, it would cause the bar to accelerate, violating the principle of conservation of energy.) Using Equation 31.6 and $F_{app} = F_B = I\ell B$, the power delivered by the applied force is

$$\mathcal{P} = F_{app}v = (I\ell B)v = \frac{B^2\ell^2 v^2}{R} = \frac{\mathcal{E}^2}{R} \qquad (31.7)$$

From Equation 27.21, we see that this power input is equal to the rate at which energy is delivered to the resistor.

Quick Quiz 31.2   In Active Figure 31.8a, a given applied force of magnitude F_{app} results in a constant speed v and a power input $\mathcal{P}$. Imagine that the force is increased so that the constant speed of the bar is doubled to $2v$. Under these conditions, what are the new force and the new power input? (a) $2F$ and $2\mathcal{P}$ (b) $4F$ and $2\mathcal{P}$ (c) $2F$ and $4\mathcal{P}$ (d) $4F$ and $4\mathcal{P}$

EXAMPLE 31.3   **Magnetic Force Acting on a Sliding Bar**

The conducting bar illustrated in Figure 31.9 moves on two frictionless, parallel rails in the presence of a uniform magnetic field directed into the page. The bar has mass m, and its length is ℓ. The bar is given an initial velocity $\vec{\mathbf{v}}_i$ to the right and is released at $t = 0$.

(A) Using Newton's laws, find the velocity of the bar as a function of time.

SOLUTION

Conceptualize   As the bar slides to the right in Figure 31.9, a counterclockwise current is established in the circuit consisting of the bar, the rails, and the resistor. The upward current in the bar results in a magnetic force to the left on the bar as shown in the figure. Therefore, the bar must slow down, so our mathematical solution should demonstrate that.

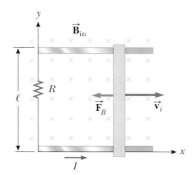

Figure 31.9   (Example 31.3) A conducting bar of length ℓ on two fixed conducting rails is given an initial velocity $\vec{\mathbf{v}}_i$ to the right.

Categorize   The text already categorizes this problem as one that uses Newton's laws. We model the bar as a particle under a net force.

Analyze   From Equation 29.10, the magnetic force is $F_B = -I\ell B$, where the negative sign indicates that the force is to the left. The magnetic force is the *only* horizontal force acting on the bar.

Apply Newton's second law to the bar in the horizontal direction:
$$F_x = ma = m\frac{dv}{dt} = -I\ell B$$

Substitute $I = B\ell v/R$ from Equation 31.6:
$$m\frac{dv}{dt} = -\frac{B^2\ell^2}{R}v$$

Rearrange the equation so that all occurrences of the variable v are on the left and those of t are on the right:
$$\frac{dv}{v} = -\left(\frac{B^2\ell^2}{mR}\right)dt$$

Integrate this equation using the initial condition that $v = v_i$ at $t = 0$ and noting that $(B^2\ell^2/mR)$ is a constant:
$$\int_{v_i}^{v}\frac{dv}{v} = -\frac{B^2\ell^2}{mR}\int_0^t dt$$

$$\ln\left(\frac{v}{v_i}\right) = -\left(\frac{B^2\ell^2}{mR}\right)t$$

Define the constant $\tau = mR/B^2\ell^2$ and solve for the velocity:

$$(1) \quad v = v_i e^{-t/\tau}$$

Finalize This expression for v indicates that the velocity of the bar decreases with time under the action of the magnetic force as expected from our conceptualization of the problem.

(B) Show that the same result is found by using an energy approach.

SOLUTION

Categorize The text of this part of the problem tells us to use an energy approach for the same situation. We model the entire circuit in Figure 31.9 as an isolated system.

Finalize Consider the sliding bar as one system component possessing kinetic energy, which decreases because energy is transferring *out* of the bar by electrical transmission through the rails. The resistor is another system component possessing internal energy, which rises because energy is transferring *into* the resistor. Because energy is not leaving the system, the rate of energy transfer out of the bar equals the rate of energy transfer into the resistor.

Equate the power entering the resistor to that leaving the bar:

$$\mathcal{P}_{resistor} = -\mathcal{P}_{bar}$$

Substitute for the electrical power delivered to the resistor and the time rate of change of kinetic energy for the bar:

$$I^2R = -\frac{d}{dt}\left(\tfrac{1}{2}mv^2\right)$$

Use Equation 31.6 for the current and carry out the derivative:

$$\frac{B^2\ell^2v^2}{R} = -mv\frac{dv}{dt}$$

Rearrange terms:

$$\frac{dv}{v} = -\left(\frac{B^2\ell^2}{mR}\right)dt$$

Finalize This result is the same expression found in part (A).

What If? Suppose you wished to increase the distance through which the bar moves between the time it is initially projected and the time it essentially comes to rest. You can do so by changing one of three variables: v_i, R, or B by a factor of 2 or $\tfrac{1}{2}$. Which variable should you change to maximize the distance, and would you double it or halve it?

Answer Increasing v_i would make the bar move farther. Increasing R would decrease the current and therefore the magnetic force, making the bar move farther. Decreasing B would decrease the magnetic force and make the bar move farther. Which method is most effective, though?

Use Equation (1) to find the distance the bar moves by integration:

$$v = \frac{dx}{dt} = v_i e^{-t/\tau}$$

$$x = \int_0^\infty v_i e^{-t/\tau}\, dt = -v_i\tau e^{-t/\tau}\Big|_0^\infty$$

$$= -v_i\tau(0-1) = v_i\tau = v_i\left(\frac{mR}{B^2\ell^2}\right)$$

This expression shows that doubling v_i or R will double the distance. Changing B by a factor of $\tfrac{1}{2}$, however, causes the distance to be four times as great!

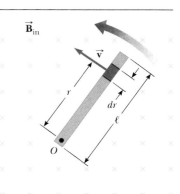

EXAMPLE 31.4   **Motional emf Induced in a Rotating Bar**

A conducting bar of length ℓ rotates with a constant angular speed ω about a pivot at one end. A uniform magnetic field $\vec{B}$ is directed perpendicular to the plane of rotation as shown in Figure 31.10. Find the motional emf induced between the ends of the bar.

SOLUTION

Conceptualize   The rotating bar is different in nature than the sliding bar in Active Figure 31.8. Consider a small segment of the bar, however. It is a short length of conductor moving in a magnetic field and has an emf generated in it. By thinking of each small segment as a source of emf, we see that all segments are in series and the emfs add.

Figure 31.10   (Example 31.4) A conducting bar rotating around a pivot at one end in a uniform magnetic field that is perpendicular to the plane of rotation. A motional emf is induced across the ends of the bar.

Categorize   Based on the conceptualization of the problem, we approach this example as we did Example 31.3, with the added feature that the short segments of the bar are traveling in circular paths.

Analyze   Evaluate the magnitude of the emf induced in a segment of the bar of length dr having a velocity $\vec{v}$ from Equation 31.5:

$$d\mathcal{E} = Bv\,dr$$

Find the total emf between the ends of the bar by adding the emfs induced across all segments:

$$\mathcal{E} = \int Bv\,dr$$

The tangential speed v of an element is related to the angular speed ω through the relationship $v = r\omega$ (Eq. 10.10); use that fact and integrate:

$$\mathcal{E} = B\int v\,dr = B\omega\int_0^\ell r\,dr = \tfrac{1}{2}B\omega\ell^2$$

Finalize   In Equation 31.5 for a sliding bar, we can increase $\mathcal{E}$ by increasing B, ℓ, or v. Increasing any one of these variables by a given factor increases $\mathcal{E}$ by the same factor. Therefore, you would choose whichever of these three variables is most convenient to increase. For the rotating rod, however, there is an advantage to increasing the length of the rod to raise the emf because ℓ is squared. Doubling the length gives four times the emf, whereas doubling the angular speed only doubles the emf.

What If?   Suppose, after reading through this example, you come up with a brilliant idea. A Ferris wheel has radial metallic spokes between the hub and the circular rim. These spokes move in the magnetic field of the Earth, so each spoke acts like the bar in Figure 31.10. You plan to use the emf generated by the rotation of the Ferris wheel to power the lightbulbs on the wheel. Will this idea work?

Answer   Let's estimate the emf that is generated in this situation. We know the magnitude of the magnetic field of the Earth from Table 29.1: $B = 0.5 \times 10^{-4}$ T. A typical spoke on a Ferris wheel might have a length on the order of 10 m. Suppose the period of rotation is on the order of 10 s.

Determine the angular speed of the spoke:

$$\omega = \frac{2\pi}{T} = \frac{2\pi}{10 \text{ s}} = 0.63 \text{ s}^{-1} \sim 1 \text{ s}^{-1}$$

Assume the magnetic field lines of the Earth are horizontal at the location of the Ferris wheel and perpendicular to the spokes. Find the emf generated:

$$\mathcal{E} = \tfrac{1}{2}B\omega\ell^2 = \tfrac{1}{2}(0.5 \times 10^{-4} \text{ T})(1 \text{ s}^{-1})(10 \text{ m})^2$$
$$= 2.5 \times 10^{-3} \text{ V} \sim 1 \text{ mV}$$

This value is a tiny emf, far smaller than that required to operate lightbulbs.

An additional difficulty is related to energy. Even assuming you could find lightbulbs that operate using a potential difference on the order of millivolts, a spoke must be part of a circuit to provide a voltage to the lightbulbs. Consequently, the spoke must carry a current. Because this current-carrying spoke is in a magnetic field, a magnetic force is exerted on the spoke in the direction opposite its direction of motion. As a result, the motor of the Ferris wheel must supply more energy to perform work against this magnetic drag force. The motor must ultimately provide the energy that is operating the lightbulbs, and you have not gained anything for free!

31.3 Lenz's Law

Faraday's law (Eq. 31.1) indicates that the induced emf and the change in flux have opposite algebraic signs. This feature has a very real physical interpretation that has come to be known as **Lenz's law:**[1]

Lenz's law ▶

> The induced current in a loop is in the direction that creates a magnetic field that opposes the change in magnetic flux through the area enclosed by the loop.

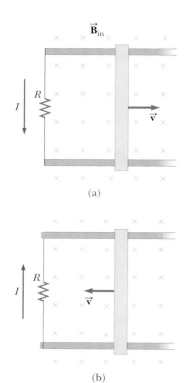

(a)

(b)

Figure 31.11 (a) As the conducting bar slides on the two fixed conducting rails, the magnetic flux due to the external magnetic field into the page through the area enclosed by the loop increases in time. By Lenz's law, the induced current must be counterclockwise to produce a counteracting magnetic field directed out of the page. (b) When the bar moves to the left, the induced current must be clockwise. Why?

That is, the induced current tends to keep the original magnetic flux through the loop from changing. We shall show that this law is a consequence of the law of conservation of energy.

To understand Lenz's law, let's return to the example of a bar moving to the right on two parallel rails in the presence of a uniform magnetic field (the *external* magnetic field; Fig. 31.11a.) As the bar moves to the right, the magnetic flux through the area enclosed by the circuit increases with time because the area increases. Lenz's law states that the induced current must be directed so that the magnetic field it produces opposes the change in the external magnetic flux. Because the magnetic flux due to an external field directed into the page is increasing, the induced current—if it is to oppose this change—must produce a field directed out of the page. Hence, the induced current must be directed counterclockwise when the bar moves to the right. (Use the right-hand rule to verify this direction.) If the bar is moving to the left as in Figure 31.11b, the external magnetic flux through the area enclosed by the loop decreases with time. Because the field is directed into the page, the direction of the induced current must be clockwise if it is to produce a field that also is directed into the page. In either case, the induced current attempts to maintain the original flux through the area enclosed by the current loop.

Let's examine this situation using energy considerations. Suppose the bar is given a slight push to the right. In the preceding analysis, we found that this motion sets up a counterclockwise current in the loop. What happens if we assume the current is clockwise such that the direction of the magnetic force exerted on the bar is to the right? This force would accelerate the rod and increase its velocity, which in turn would cause the area enclosed by the loop to increase more rapidly. The result would be an increase in the induced current, which would cause an increase in the force, which would produce an increase in the current, and so on. In effect, the system would acquire energy with no input of energy. This behavior is clearly inconsistent with all experience and violates the law of conservation of energy. Therefore, the current must be counterclockwise.

Quick Quiz 31.3 Figure 31.12 shows a circular loop of wire falling toward a wire carrying a current to the left. What is the direction of the induced current in the loop of wire? (a) clockwise (b) counterclockwise (c) zero (d) impossible to determine

Figure 31.12 (Quick Quiz 31.3)

[1] Developed by German physicist Heinrich Lenz (1804–1865).

CONCEPTUAL EXAMPLE 31.5   **Application of Lenz's Law**

A magnet is placed near a metal loop as shown in Figure 31.13a.

(A) Find the direction of the induced current in the loop when the magnet is pushed toward the loop.

SOLUTION

As the magnet moves to the right toward the loop, the external magnetic flux through the loop increases with time. To counteract this increase in flux due to a field toward the right, the induced current produces its own magnetic field to the left as illustrated in Figure 31.13b; hence, the induced current is in the direction shown. Knowing that like magnetic poles repel each other, we conclude that the left face of the current loop acts like a north pole and the right face acts like a south pole.

(B) Find the direction of the induced current in the loop when the magnet is pulled away from the loop.

SOLUTION

If the magnet moves to the left as in Figure 31.13c, its flux through the area enclosed by the loop decreases in time. Now the induced current in the loop is in the direction shown in Figure 31.13d because this current direction produces a magnetic field in the same direction as the external field. In this case, the left face of the loop is a south pole and the right face is a north pole.

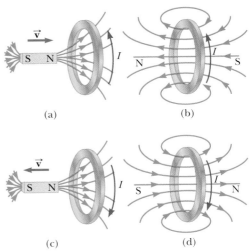

Figure 31.13 (Conceptual Example 31.5) (a) When the magnet is moved toward the stationary conducting loop, a current is induced in the direction shown. The magnetic field lines shown are those due to the bar magnet. (b) This induced current produces its own magnetic field directed to the left that counteracts the increasing external flux. The magnetic field lines shown are those due to the induced current in the ring. (c) When the magnet is moved away from the stationary conducting loop, a current is induced in the direction shown. The magnetic field lines shown are those due to the bar magnet. (d) This induced current produces a magnetic field directed to the right and so counteracts the decreasing external flux. The magnetic field lines shown are those due to the induced current in the ring.

CONCEPTUAL EXAMPLE 31.6   **A Loop Moving Through a Magnetic Field**

A rectangular metallic loop of dimensions ℓ and w and resistance R moves with constant speed v to the right, as in Figure 31.14a. The loop passes through a uniform magnetic field $\vec{B}$ directed into the page and extending a distance $3w$ along the x axis. Define x as the position of the right side of the loop along the x axis.

(A) Plot as a function of x the magnetic flux through the area enclosed by the loop.

SOLUTION

Figure 31.14b shows the flux through the area enclosed by the loop as a function of x. Before the loop enters the field, the flux through the loop is zero. As the loop enters the field, the flux increases linearly with position until the left edge of the loop is just inside the field. Finally, the flux through the loop decreases linearly to zero as the loop leaves the field.

(B) Plot as a function of x the induced motional emf in the loop.

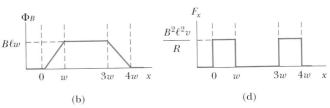

Figure 31.14 (Conceptual Example 31.6) (a) A conducting rectangular loop of width w and length ℓ moving with a velocity $\vec{v}$ through a uniform magnetic field extending a distance $3w$. (b) Magnetic flux through the area enclosed by the loop as a function of loop position. (c) Induced emf as a function of loop position. (d) Applied force required for constant velocity as a function of loop position.

SOLUTION

Before the loop enters the field, no motional emf is induced in it because no field is present (Fig. 31.14c). As the right side of the loop enters the field, the magnetic flux directed into the page increases. Hence, according to Lenz's law, the induced current is counterclockwise because it must produce its own magnetic field directed out of the page. The motional emf $-B\ell v$ (from Eq. 31.5) arises from the magnetic force experienced by charges in the right side of the loop. When the loop is entirely in the field, the change in magnetic flux through the loop is zero; hence, the motional emf vanishes. That happens because once the left side of the loop enters the field, the motional emf induced in it cancels the motional emf present in the right side of the loop. As the right side of the loop leaves the field, the flux through the loop begins to decrease, a clockwise current is induced, and the induced emf is $B\ell v$. As soon as the left side leaves the field, the emf decreases to zero.

(C) Plot as a function of x the external applied force necessary to counter the magnetic force and keep v constant.

SOLUTION

The external force that must be applied to the loop to maintain this motion is plotted in Figure 31.14d. Before the loop enters the field, no magnetic force acts on it; hence, the applied force must be zero if v is constant. When the right side of the loop enters the field, the applied force necessary to maintain constant speed must be equal in magnitude and opposite in direction to the magnetic force exerted on that side. When the loop is entirely in the field, the flux through the loop is not changing with time. Hence, the net emf induced in the loop is zero and the current also is zero. Therefore, no external force is needed to maintain the motion. Finally, as the right side leaves the field, the applied force must be equal in magnitude and opposite in direction to the magnetic force acting on the left side of the loop.

From this analysis, we conclude that power is supplied only when the loop is either entering or leaving the field. Furthermore, this example shows that the motional emf induced in the loop can be zero even when there is motion through the field! A motional emf is induced *only* when the magnetic flux through the loop *changes in time*.

31.4 Induced emf and Electric Fields

We have seen that a changing magnetic flux induces an emf and a current in a conducting loop. In our study of electricity, we related a current to an electric field that applies electric forces on charged particles. In the same way, we can relate an induced current in a conducting loop to an electric field by claiming that **an electric field is created in the conductor as a result of the changing magnetic flux.**

We also noted in our study of electricity that the existence of an electric field is independent of the presence of any test charges. This independence suggests that even in the absence of a conducting loop, a changing magnetic field generates an electric field in empty space.

This induced electric field is *nonconservative,* unlike the electrostatic field produced by stationary charges. To illustrate this point, consider a conducting loop of radius r situated in a uniform magnetic field that is perpendicular to the plane of the loop as in Figure 31.15. If the magnetic field changes with time, an emf $\mathcal{E} = -d\Phi_B/dt$ is, according to Faraday's law (Eq. 31.1), induced in the loop. The induction of a current in the loop implies the presence of an induced electric field $\vec{E}$, which must be tangent to the loop because that is the direction in which the charges in the wire move in response to the electric force. The work done by the electric field in moving a test charge q once around the loop is equal to $q\mathcal{E}$. Because the electric force acting on the charge is $q\vec{E}$, the work done by the electric field in moving the charge once around the loop is $qE(2\pi r)$, where $2\pi r$ is the circumference of the loop. These two expressions for the work done must be equal; therefore,

$$q\mathcal{E} = qE(2\pi r)$$

$$E = \frac{\mathcal{E}}{2\pi r}$$

Using this result along with Equation 31.1 and that $\Phi_B = BA = B\pi r^2$ for a circular loop, the induced electric field can be expressed as

$$E = -\frac{1}{2\pi r}\frac{d\Phi_B}{dt} = -\frac{r}{2}\frac{dB}{dt} \tag{31.8}$$

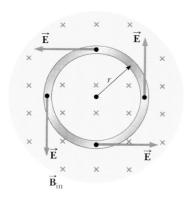

Figure 31.15 A conducting loop of radius r in a uniform magnetic field perpendicular to the plane of the loop. If **B** changes in time, an electric field is induced in a direction tangent to the circumference of the loop.

PITFALL PREVENTION 31.2

Induced Electric Fields

The changing magnetic field does *not* need to exist at the location of the induced electric field. In Figure 31.15, even a loop outside the region of magnetic field experiences an induced electric field.

If the time variation of the magnetic field is specified, the induced electric field can be calculated from Equation 31.8.

The emf for any closed path can be expressed as the line integral of $\vec{E} \cdot d\vec{s}$ over that path: $\mathcal{E} = \oint \vec{E} \cdot d\vec{s}$. In more general cases, E may not be constant and the path may not be a circle. Hence, Faraday's law of induction, $\mathcal{E} = -d\Phi_B/dt$, can be written in the general form

$$\oint \vec{E} \cdot d\vec{s} = -\frac{d\Phi_B}{dt} \qquad (31.9)$$

◀ Faraday's law in general form

The induced electric field $\vec{E}$ in Equation 31.9 is a nonconservative field that is generated by a changing magnetic field. The field $\vec{E}$ that satisfies Equation 31.9 cannot possibly be an electrostatic field because were the field electrostatic and hence conservative, the line integral of $\vec{E} \cdot d\vec{s}$ over a closed loop would be zero (Section 25.1), which would be in contradiction to Equation 31.9.

EXAMPLE 31.7	**Electric Field Induced by a Changing Magnetic Field in a Solenoid**

A long solenoid of radius R has n turns of wire per unit length and carries a time-varying current that varies sinusoidally as $I = I_{max} \cos \omega t$, where I_{max} is the maximum current and ω is the angular frequency of the alternating current source (Fig. 31.16).

(A) Determine the magnitude of the induced electric field outside the solenoid at a distance $r > R$ from its long central axis.

SOLUTION

Conceptualize Figure 31.16 shows the physical situation. As the current in the coil changes, imagine a changing magnetic field at all points in space as well as an induced electric field.

Figure 31.16 (Example 31.7) A long solenoid carrying a time-varying current given by $I = I_{max} \cos \omega t$. An electric field is induced both inside and outside the solenoid.

Categorize Because the current varies in time, the magnetic field is changing, leading to an induced electric field as opposed to the electrostatic electric fields due to stationary electric charges.

Analyze First consider an external point and take the path for the line integral to be a circle of radius r centered on the solenoid as illustrated in Figure 31.16.

Evaluate the right side of Equation 31.9, noting that $\vec{B}$ is perpendicular to the circle bounded by the path of integration and that this magnetic field exists only inside the solenoid:

$$(1) \quad -\frac{d\Phi_B}{dt} = -\frac{d}{dt}(B\pi R^2) = -\pi R^2 \frac{dB}{dt}$$

Evaluate the magnetic field in the solenoid from Equation 30.17:

$$(2) \quad B = \mu_0 nI = \mu_0 nI_{max} \cos \omega t$$

Substitute Equation (2) into Equation (1):

$$(3) \quad -\frac{d\Phi_B}{dt} = -\pi R^2 \mu_0 nI_{max} \frac{d}{dt}(\cos \omega t) = \pi R^2 \mu_0 nI_{max} \omega \sin \omega t$$

Evaluate the left side of Equation 31.9, noting that the magnitude of $\vec{E}$ is constant on the path of integration and $\vec{E}$ is tangent to it:

$$(4) \quad \oint \vec{E} \cdot d\vec{s} = E(2\pi r)$$

Substitute Equations (3) and (4) into Equation 31.9:

$$E(2\pi r) = \pi R^2 \mu_0 nI_{max} \omega \sin \omega t$$

Solve for the magnitude of the electric field:
$$E = \frac{\mu_0 n I_{max} \omega R^2}{2r} \sin \omega t \quad \text{(for } r > R)$$

Finalize   This result shows that the amplitude of the electric field outside the solenoid falls off as $1/r$ and varies sinusoidally with time. As we will learn in Chapter 34, the time-varying electric field creates an additional contribution to the magnetic field. The magnetic field can be somewhat stronger than we first stated, both inside and outside the solenoid. The correction to the magnetic field is small if the angular frequency ω is small. At high frequencies, however, a new phenomenon can dominate: The electric and magnetic fields, each re-creating the other, constitute an electromagnetic wave radiated by the solenoid as we will study in Chapter 34.

(B) What is the magnitude of the induced electric field inside the solenoid, a distance r from its axis?

SOLUTION

Analyze   For an interior point $(r < R)$, the magnetic flux through an integration loop is given by $\Phi_B = B\pi r^2$.

Evaluate the right side of Equation 31.9:
$$(5) \quad -\frac{d\Phi_B}{dt} = -\frac{d}{dt}\left(B\pi r^2\right) = -\pi r^2 \frac{dB}{dt}$$

Substitute Equation (2) into Equation (5):
$$(6) \quad -\frac{d\Phi_B}{dt} = -\pi r^2 \mu_0 n I_{max} \frac{d}{dt}(\cos \omega t) = \pi r^2 \mu_0 n I_{max} \omega \sin \omega t$$

Substitute Equations (4) and (6) into Equation 31.9:
$$E(2\pi r) = \pi r^2 \mu_0 n I_{max} \omega \sin \omega t$$

Solve for the magnitude of the electric field:
$$E = \frac{\mu_0 n I_{max} \omega}{2} r \sin \omega t \quad \text{(for } r < R)$$

Finalize   This result shows that the amplitude of the electric field induced inside the solenoid by the changing magnetic flux through the solenoid increases linearly with r and varies sinusoidally with time.

31.5   Generators and Motors

Electric generators take in energy by work and transfer it out by electrical transmission. To understand how they operate, let us consider the **alternating-current (AC) generator.** In its simplest form, it consists of a loop of wire rotated by some external means in a magnetic field (Active Fig. 31.17a).

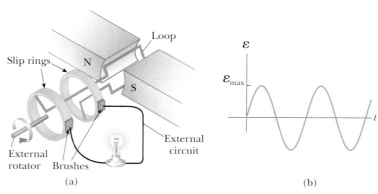

(a)   (b)

ACTIVE FIGURE 31.17

(a) Schematic diagram of an AC generator. An emf is induced in a loop that rotates in a magnetic field.
(b) The alternating emf induced in the loop plotted as a function of time.

Sign in at www.thomsonedu.com and go to ThomsonNOW to adjust the speed of rotation and the strength of the field to see the effects on the emf generated.

In commercial power plants, the energy required to rotate the loop can be derived from a variety of sources. For example, in a hydroelectric plant, falling water directed against the blades of a turbine produces the rotary motion; in a coal-fired plant, the energy released by burning coal is used to convert water to steam, and this steam is directed against the turbine blades.

As a loop rotates in a magnetic field, the magnetic flux through the area enclosed by the loop changes with time, and this change induces an emf and a current in the loop according to Faraday's law. The ends of the loop are connected to slip rings that rotate with the loop. Connections from these slip rings, which act as output terminals of the generator, to the external circuit are made by stationary metallic brushes in contact with the slip rings.

Instead of a single turn, suppose a coil with N turns (a more practical situation), with the same area A, rotates in a magnetic field with a constant angular speed ω. If θ is the angle between the magnetic field and the normal to the plane of the coil as in Figure 31.18, the magnetic flux through the coil at any time t is

$$\Phi_B = BA \cos \theta = BA \cos \omega t$$

where we have used the relationship $\theta = \omega t$ between angular position and angular speed (see Eq. 10.3). (We have set the clock so that $t = 0$ when $\theta = 0$.) Hence, the induced emf in the coil is

$$\mathcal{E} = -N\frac{d\Phi_B}{dt} = -NAB\frac{d}{dt}(\cos \omega t) = NAB\omega \sin \omega t \qquad \textbf{(31.10)}$$

This result shows that the emf varies sinusoidally with time as plotted in Active Figure 31.17b. Equation 31.10 shows that the maximum emf has the value

$$\mathcal{E}_{\text{max}} = NAB\omega \qquad \textbf{(31.11)}$$

which occurs when $\omega t = 90°$ or $270°$. In other words, $\mathcal{E} = \mathcal{E}_{\text{max}}$ when the magnetic field is in the plane of the coil and the time rate of change of flux is a maximum. Furthermore, the emf is zero when $\omega t = 0$ or $180°$, that is, when $\vec{B}$ is perpendicular to the plane of the coil and the time rate of change of flux is zero.

The frequency for commercial generators in the United States and Canada is 60 Hz, whereas in some European countries it is 50 Hz. (Recall that $\omega = 2\pi f$, where f is the frequency in hertz.)

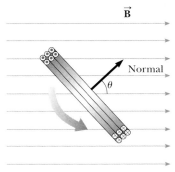

Figure 31.18   A loop enclosing an area A and containing N turns, rotating with constant angular speed ω in a magnetic field. The emf induced in the loop varies sinusoidally in time.

Quick Quiz 31.4   In an AC generator, a coil with N turns of wire spins in a magnetic field. Of the following choices, which does *not* cause an increase in the emf generated in the coil? (a) replacing the coil wire with one of lower resistance (b) spinning the coil faster   (c) increasing the magnetic field   (d) increasing the number of turns of wire on the coil

---

EXAMPLE 31.8	**emf Induced in a Generator**

The coil in an AC generator consists of 8 turns of wire, each of area $A = 0.090\,0$ m², and the total resistance of the wire is 12.0 Ω. The coil rotates in a 0.500-T magnetic field at a constant frequency of 60.0 Hz.

(A) Find the maximum induced emf in the coil.

SOLUTION

Conceptualize   Study Active Figure 31.17 to make sure you understand the operation of an AC generator.

Categorize   We evaluate parameters using equations developed in this section, so we categorize this example as a substitution problem.

Use Equation 31.11 to find the maximum induced emf: $\mathcal{E}_{\text{max}} = NAB\omega = NAB(2\pi f)$

Substitute numerical values:

$$\mathcal{E}_{max} = 8(0.090\ 0\ m^2)(0.500\ T)(2\pi)(60.0\ Hz) = \boxed{136\ V}$$

(B) What is the maximum induced current in the coil when the output terminals are connected to a low-resistance conductor?

SOLUTION

Use Equation 27.7 and the result to part (A):

$$I_{max} = \frac{\mathcal{E}_{max}}{R} = \frac{136\ V}{12.0\ \Omega} = \boxed{11.3\ A}$$

The **direct-current (DC) generator** is illustrated in Active Figure 31.19a. Such generators are used, for instance, in older cars to charge the storage batteries. The components are essentially the same as those of the AC generator except that the contacts to the rotating coil are made using a split ring called a *commutator*.

In this configuration, the output voltage always has the same polarity and pulsates with time as shown in Active Figure 31.19b. We can understand why by noting that the contacts to the split ring reverse their roles every half cycle. At the same time, the polarity of the induced emf reverses; hence, the polarity of the split ring (which is the same as the polarity of the output voltage) remains the same.

A pulsating DC current is not suitable for most applications. To obtain a steadier DC current, commercial DC generators use many coils and commutators distributed so that the sinusoidal pulses from the various coils are out of phase. When these pulses are superimposed, the DC output is almost free of fluctuations.

A **motor** is a device into which energy is transferred by electrical transmission while energy is transferred out by work. A motor is essentially a generator operating in reverse. Instead of generating a current by rotating a coil, a current is supplied to the coil by a battery and the torque acting on the current-carrying coil (Section 29.5) causes it to rotate.

Useful mechanical work can be done by attaching the rotating coil to some external device. As the coil rotates in a magnetic field, however, the changing magnetic flux induces an emf in the coil; this induced emf always acts to reduce the current in the coil. If that were not the case, Lenz's law would be violated. The back emf increases in magnitude as the rotational speed of the coil increases. (The phrase *back emf* is used to indicate an emf that tends to reduce the supplied current.) Because the voltage available to supply current equals the difference between the supply voltage and the back emf, the current in the rotating coil is limited by the back emf.

When a motor is turned on, there is initially no back emf, and the current is very large because it is limited only by the resistance of the coil. As the coil begins

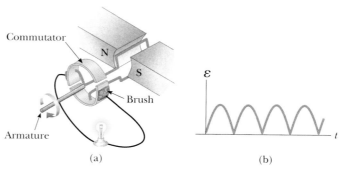

(a)   (b)

ACTIVE FIGURE 31.19

(a) Schematic diagram of a DC generator. (b) The magnitude of the emf varies in time, but the polarity never changes.

Sign in at www.thomsonedu.com and go to ThomsonNOW to adjust the speed of rotation and the strength of the field to see the effects on the emf generated.

to rotate, the induced back emf opposes the applied voltage and the current in the coil decreases. If the mechanical load increases, the motor slows down, which causes the back emf to decrease. This reduction in the back emf increases the current in the coil and therefore also increases the power needed from the external voltage source. For this reason, the power requirements for running a motor are greater for heavy loads than for light ones. If the motor is allowed to run under no mechanical load, the back emf reduces the current to a value just large enough to overcome energy losses due to internal energy and friction. If a very heavy load jams the motor so that it cannot rotate, the lack of a back emf can lead to dangerously high current in the motor's wire. This dangerous situation is explored in the **What If?** section of Example 31.9.

A modern application of motors in automobiles is seen in the development of *hybrid drive systems*. In these automobiles, a gasoline engine and an electric motor are combined to increase the fuel economy of the vehicle and reduce its emissions. Figure 31.20 shows the engine compartment of a Toyota Prius, one of a small number of hybrids available in the United States. In this automobile, power to the wheels can come from either the gasoline engine or the electric motor. In normal driving, the electric motor accelerates the vehicle from rest until it is moving at a speed of about 15 mi/h (24 km/h). During this acceleration period, the engine is not running, so gasoline is not used and there is no emission. At higher speeds, the motor and engine work together so that the engine always operates at or near its most efficient speed. The result is a significantly higher gasoline mileage than that obtained by a traditional gasoline-powered automobile. When a hybrid vehicle brakes, the motor acts as a generator and returns some of the vehicle's kinetic energy back to the battery as stored energy. In a normal vehicle, this kinetic energy is simply lost as it is transformed to internal energy in the brakes and roadway.

Figure 31.20 The engine compartment of the Toyota Prius, a hybrid vehicle.

| EXAMPLE 31.9 | **The Induced Current in a Motor** |

A motor contains a coil with a total resistance of 10 Ω and is supplied by a voltage of 120 V. When the motor is running at its maximum speed, the back emf is 70 V.

(A) Find the current in the coil at the instant the motor is turned on.

SOLUTION

Conceptualize Think about the motor just after it is turned on. It has not yet moved, so there is no back emf generated. As a result, the current in the motor is high. After the motor begins to turn, a back emf is generated and the current decreases.

Categorize We need to combine our new understanding about motors with the relationship between current, voltage, and resistance.

Analyze Evaluate the current in the coil from Equation 27.7 with no back emf generated:

$$I = \frac{\mathcal{E}}{R} = \frac{120\ \text{V}}{10\ \Omega} = \boxed{12\ \text{A}}$$

(B) Find the current in the coil when the motor has reached maximum speed.

SOLUTION

Evaluate the current in the coil with the maximum back emf generated:

$$I = \frac{\mathcal{E} - \mathcal{E}_{\text{back}}}{R} = \frac{120\ \text{V} - 70\ \text{V}}{10\ \Omega} = \frac{50\ \text{V}}{10\ \Omega} = \boxed{5.0\ \text{A}}$$

Finalize The current drawn by the motor when operating at its maximum speed is significantly less than that drawn before it begins to turn.

What If?   Suppose this motor is in a circular saw. When you are operating the saw, the blade becomes jammed in a piece of wood and the motor cannot turn. By what percentage does the power input to the motor increase when it is jammed?

Answer   You may have everyday experiences with motors becoming warm when they are prevented from turning. That is due to the increased power input to the motor. The higher rate of energy transfer results in an increase in the internal energy of the coil, an undesirable effect.

Set up the ratio of power input to the motor when jammed, which is that calculated in part (A), to that when it is not jammed, part (B):

$$\frac{\mathscr{P}_{\text{jammed}}}{\mathscr{P}_{\text{not jammed}}} = \frac{I_A^2 R}{I_B^2 R} = \frac{I_A^2}{I_B^2}$$

Substituting numerical values gives

$$\frac{\mathscr{P}_{\text{jammed}}}{\mathscr{P}_{\text{not jammed}}} = \frac{(12 \text{ A})^2}{(5.0 \text{ A})^2} = 5.76$$

which represents a 476% increase in the input power! Such a high power input can cause the coil to become so hot that it is damaged.

31.6   Eddy Currents

As we have seen, an emf and a current are induced in a circuit by a changing magnetic flux. In the same manner, circulating currents called **eddy currents** are induced in bulk pieces of metal moving through a magnetic field. This phenomenon can be demonstrated by allowing a flat copper or aluminum plate attached at the end of a rigid bar to swing back and forth through a magnetic field (Fig. 31.21). As the plate enters the field, the changing magnetic flux induces an emf in the plate, which in turn causes the free electrons in the plate to move, producing the swirling eddy currents. According to Lenz's law, the direction of the eddy currents is such that they create magnetic fields that oppose the change that causes the currents. For this reason, the eddy currents must produce effective magnetic poles on the plate, which are repelled by the poles of the magnet; this situation gives rise to a repulsive force that opposes the motion of the plate. (If the opposite were true, the plate would accelerate and its energy would increase after each swing, in violation of the law of conservation of energy.)

As indicated in Active Figure 31.22a, with $\vec{\mathbf{B}}$ directed into the page, the induced eddy current is counterclockwise as the swinging plate enters the field at position 1 because the flux due to the external magnetic field into the page through the plate is increasing. Hence, by Lenz's law, the induced current must provide its own magnetic field out of the page. The opposite is true as the plate leaves the field at position 2, where the current is clockwise. Because the induced eddy current always produces a magnetic retarding force $\vec{\mathbf{F}}_B$ when the plate enters or leaves the field, the swinging plate eventually comes to rest.

If slots are cut in the plate as shown in Active Figure 31.22b, the eddy currents and the corresponding retarding force are greatly reduced. We can understand this reduction in force by realizing that the cuts in the plate prevent the formation of any large current loops.

The braking systems on many subway and rapid-transit cars make use of electromagnetic induction and eddy currents. An electromagnet attached to the train is positioned near the steel rails. (An electromagnet is essentially a solenoid with an iron core.) The braking action occurs when a large current is passed through the electromagnet. The relative motion of the magnet and rails induces eddy currents in the rails, and the direction of these currents produces a drag force on the moving train. Because the eddy currents decrease steadily in magnitude as the train slows

Figure 31.21   Formation of eddy currents in a conducting plate moving through a magnetic field. As the plate enters or leaves the field, the changing magnetic flux induces an emf, which causes eddy currents in the plate.

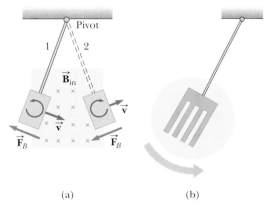

(a)   (b)

ACTIVE FIGURE 31.22

(a) As the conducting plate enters the field (position 1), the eddy currents are counterclockwise. As the plate leaves the field (position 2), the currents are clockwise. In either case, the force on the plate is opposite the velocity and eventually the plate comes to rest. (b) When slots are cut in the conducting plate, the eddy currents are reduced and the plate swings more freely through the magnetic field.

Sign in at www.thomsonedu.com and go to ThomsonNOW to choose to let a solid or a slotted plate swing through the magnetic field and observe the effect.

down, the braking effect is quite smooth. As a safety measure, some power tools use eddy currents to stop rapidly spinning blades once the device is turned off.

Eddy currents are often undesirable because they represent a transformation of mechanical energy to internal energy. To reduce this energy loss, conducting parts are often laminated; that is, they are built up in thin layers separated by a nonconducting material such as lacquer or a metal oxide. This layered structure prevents large current loops and effectively confines the currents to small loops in individual layers. Such a laminated structure is used in transformer cores (see Section 33.8) and motors to minimize eddy currents and thereby increase the efficiency of these devices.

Quick Quiz 31.5   In an equal-arm balance from the early 20th century (Fig. 31.23), an aluminum sheet hangs from one of the arms and passes between the poles of a magnet, causing the oscillations of the balance to decay rapidly. In the absence of such magnetic braking, the oscillation might continue for a long time, and the experimenter would have to wait to take a reading. Why do the oscillations decay? (a) because the aluminum sheet is attracted to the magnet   (b) because currents in the aluminum sheet set up a magnetic field that opposes the oscillations   (c) because aluminum is paramagnetic.

Figure 31.23   (Quick Quiz 31.5) In an old-fashioned equal-arm balance, an aluminum sheet hangs between the poles of a magnet.

Summary

CONCEPTS AND PRINCIPLES

Faraday's law of induction states that the emf induced in a loop is directly proportional to the time rate of change of magnetic flux through the loop, or

$$\mathcal{E} = -\frac{d\Phi_B}{dt} \qquad (31.1)$$

where $\Phi_B = \oint \vec{B} \cdot d\vec{A}$ is the magnetic flux through the loop.

When a conducting bar of length ℓ moves at a velocity $\vec{v}$ through a magnetic field $\vec{B}$, where $\vec{B}$ is perpendicular to the bar and to $\vec{v}$, the **motional emf** induced in the bar is

$$\mathcal{E} = -B\ell v \qquad (31.5)$$

Lenz's law states that the induced current and induced emf in a conductor are in such a direction as to set up a magnetic field that opposes the change that produced them.

A general form of **Faraday's law of induction** is

$$\oint \vec{E} \cdot d\vec{s} = -\frac{d\Phi_B}{dt} \qquad (31.9)$$

where $\vec{E}$ is the nonconservative electric field that is produced by the changing magnetic flux.

Questions

□ denotes answer available in *Student Solutions Manual/Study Guide;* **O** denotes objective question

1. What is the difference between magnetic flux and magnetic field?

2. O Figure Q31.2 is a graph of the magnetic flux through a certain coil of wire as a function of time, during an interval while the radius of the coil is increased, the coil is rotated through 1.5 revolutions, and the external source of the magnetic field is turned off, in that order. Rank the electromotive force induced in the coil at the instants marked A through F from the largest positive value to the largest-magnitude negative value. In your ranking, note any cases of equality and also any instants when the emf is zero.

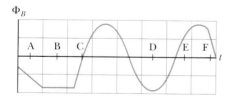

Figure Q31.2

3. O A flat coil of wire is placed in a uniform magnetic field that is in the *y* direction. **(i)** The magnetic flux through the coil is a maximum if the coil is (a) in the *xy* plane (b) in either the *xy* or the *yz* plane (c) in the *xz* plane (d) in any orientation, because it is a constant **(ii)** For what orientation is the flux zero? Choose the best answer from the same possibilities.

4. O A square, flat coil of wire is pulled at constant velocity through a region of uniform magnetic field directed perpendicular to the plane of the coil as shown in Figure

Q31.4. (i) Is current induced in the coil? (a) yes, clockwise (b) yes, counterclockwise (c) no **(ii)** Does charge separation occur in the coil? (a) yes, with the top positive (b) yes, with the top negative (c) no

Figure Q31.4 Questions 4 and 6.

5. The bar in Figure Q31.5 moves on rails to the right with a velocity $\vec{v}$, and the uniform, constant magnetic field is directed out of the page. Why is the induced current clockwise? If the bar were moving to the left, what would be the direction of the induced current?

Figure Q31.5 Questions 5 and 6.

6. O (i) As the square coil of wire in Figure Q31.4 moves perpendicular to the field, is an external force required to keep it moving with constant speed? **(ii)** Answer the same question for the bar in Figure Q31.5. **(iii)** Answer the same question for the bar in Figure Q31.6.

Figure Q31.6

7. In a hydroelectric dam, how is energy produced that is then transferred out by electrical transmission? That is, how is the energy of motion of the water converted to energy that is transmitted by AC electricity?

8. A piece of aluminum is dropped vertically downward between the poles of an electromagnet. Does the magnetic field affect the velocity of the aluminum?

9. O What happens to the amplitude of the induced emf when the rate of rotation of a generator coil is doubled? (a) It becomes 4 times larger. (b) It becomes 2 times larger. (c) It is unchanged. (d) It becomes $\frac{1}{2}$ as large. (e) It becomes $\frac{1}{4}$ as large.

10. When the switch in Figure Q31.10a is closed, a current is set up in the coil and the metal ring springs upward (Fig. Q31.10b). Explain this behavior.

Figure Q31.10   Questions 10 and 11.

11. Assume the battery in Figure Q31.10a is replaced by an AC source and the switch is held closed. If held down, the metal ring on top of the solenoid becomes hot. Why?

12. O A bar magnet is held in a vertical orientation above a loop of wire that lies in a horizontal plane as shown in Figure Q31.12. The south pole of the magnet is on the bottom end, closest to the loop of wire. The magnet is dropped toward the loop. **(i)** While the magnet is falling toward the loop, what is the direction of current in the resistor? (a) to the left   (b) to the right   (c) there is no current   (d) both to the left and to the right   (e) downward   **(ii)** After the magnet has passed through the loop and moves away from it, what is the direction of current in the resistor? Choose from the same possibilities. **(iii)** Now assume the magnet, producing a symmetrical field, is held in a horizontal orientation and then dropped. While it is approaching the loop, what is the direction of current in the resistor? Choose from the same possibilities.

Figure Q31.12

13. O What is the direction of the current in the resistor in Figure Q31.13 **(i)** at an instant immediately after the switch is thrown closed, **(ii)** after the switch has been closed for several seconds, and **(iii)** at an instant after the switch has then been thrown open? Choose each answer from these possibilities: (a) left (b) right   (c) both left and right   (d) The current is zero.

Figure Q31.13

14. In Section 7.7, we defined conservative and nonconservative forces. In Chapter 23, we stated that an electric charge creates an electric field that produces a conservative force. Argue now that induction creates an electric field that produces a nonconservative force.

Problems

WebAssign The Problems from this chapter may be assigned online in WebAssign.

ThomsonNOW   Sign in at **www.thomsonedu.com** and go to ThomsonNOW to assess your understanding of this chapter's topics with additional quizzing and conceptual questions.

1, 2, 3 denotes straightforward, intermediate, challenging;   □ denotes full solution available in *Student Solutions Manual/Study Guide;*   ▲ denotes coached solution with hints available at **www.thomsonedu.com;**   denotes developing symbolic reasoning;   ● denotes asking for qualitative reasoning;   ▪ denotes computer useful in solving problem

Section 31.1 Faraday's Law of Induction

Section 31.3 Lenz's Law

1. *Transcranial magnetic stimulation* is a noninvasive technique used to stimulate regions of the human brain. A small coil

is placed on the scalp, and a brief burst of current in the coil produces a rapidly changing magnetic field inside the brain. The induced emf can stimulate neuronal activity. (a) One such device generates an upward magnetic field within the brain that rises from zero to 1.50 T in 120 ms. Determine the induced emf around a horizontal circle of tissue of radius 1.60 mm. (b) **What If?** The field next changes to 0.500 T downward in 80.0 ms. How does the emf induced in this process compare with that in part (a)?

2. A flat loop of wire consisting of a single turn of cross-sectional area 8.00 cm² is perpendicular to a magnetic field that increases uniformly in magnitude from 0.500 T to 2.50 T in 1.00 s. What is the resulting induced current if the loop has a resistance of 2.00 Ω?

3. A 25-turn circular coil of wire has diameter 1.00 m. It is placed with its axis along the direction of the Earth's magnetic field of 50.0 μT and then in 0.200 s is flipped 180°. An average emf of what magnitude is generated in the coil?

4. ● Your physics teacher asks you to help her set up a demonstration of Faraday's law for the class. As shown in Figure P31.4, the apparatus consists of a strong, permanent magnet producing a field of 110 mT between its poles, a 12-turn coil of radius 2.10 cm cemented onto a wood frame with a handle, some flexible connecting wires, and an ammeter. The idea is to pull the coil out of the center of the magnetic field as quickly as you can and read the average current registered on the meter. The equivalent resistance of the coil, leads, and meter is 2.30 Ω. You can flip the coil out of the field in about 180 ms. The ammeter has a full-scale sensitivity of 1 000 μA. (a) Is this meter sensitive enough to show the induced current clearly? Explain your reasoning. (b) Does the meter in the diagram register a positive or a negative current? Explain your reasoning.

Figure P31.4

$\Delta t = 0.18$ s

5. A rectangular loop of area A is placed in a region where the magnetic field is perpendicular to the plane of the loop. The magnitude of the field is allowed to vary in time according to $B = B_{max}e^{-t/\tau}$, where B_{max} and τ are constants. The field has the constant value B_{max} for $t < 0$. (a) Use Faraday's law to show that the emf induced in the loop is given by

$$\mathcal{E} = \frac{AB_{max}}{\tau}e^{-t/\tau}$$

(b) Obtain a numerical value for $\mathcal{E}$ at $t = 4.00$ s when $A = 0.160$ m², $B_{max} = 0.350$ T, and $\tau = 2.00$ s. (c) For the values of A, B_{max}, and τ given in part (b), what is the maximum value of $\mathcal{E}$?

6. To monitor the breathing of a hospital patient, a thin belt is girded around the patient's chest. The belt is a 200-turn coil. When the patient inhales, the area encircled by the coil increases by 39.0 cm². The magnitude of the Earth's magnetic field is 50.0 μT and makes an angle of 28.0° with the plane of the coil. Assuming a patient takes 1.80 s to inhale, find the average induced emf in the coil during this time interval.

7. ▲ A strong electromagnet produces a uniform magnetic field of 1.60 T over a cross-sectional area of 0.200 m². A coil having 200 turns and a total resistance of 20.0 Ω is placed around the electromagnet. The current in the electromagnet is then smoothly reduced until it reaches zero in 20.0 ms. What is the current induced in the coil?

8. A loop of wire in the shape of a rectangle of width w and length L and a long, straight wire carrying a current I lie on a tabletop as shown in Figure P31.8. (a) Determine the magnetic flux through the loop due to the current I. (b) Suppose the current is changing with time according to $I = a + bt$, where a and b are constants. Determine the emf that is induced in the loop if $b = 10.0$ A/s, $h = 1.00$ cm, $w = 10.0$ cm, and $L = 100$ cm. What is the direction of the induced current in the rectangle?

Figure P31.8 Problems 8 and 67.

9. ▲ An aluminum ring of radius 5.00 cm and resistance 3.00×10^{-4} Ω is placed around one end of a long air-core solenoid with 1 000 turns per meter and radius 3.00 cm as shown in Figure P31.9. Assume the axial component of the field produced by the solenoid is one-half as strong over the area of the end of the solenoid as at the center of the solenoid. Also assume the solenoid produces negligible field outside its cross-sectional area. The current in the solenoid is increasing at a rate of 270 A/s. (a) What is the induced current in the ring? At the center of the ring, what are (b) the magnitude and (c) the direction of the magnetic field produced by the induced current in the ring?

5.00 cm   3.00 cm

Figure P31.9 Problems 9 and 10.

10. An aluminum ring of radius r_1 and resistance R is placed around one end of a long air-core solenoid with n turns per meter and smaller radius r_2 as shown in Figure P31.9. Assume the axial component of the field produced by the solenoid over the area of the end of the solenoid is one-

2 = intermediate;   3 = challenging;   ☐ = SSM/SG;   ▲ = ThomsonNOW;   = symbolic reasoning;   ● = qualitative reasoning

half as strong as at the center of the solenoid. Also assume the solenoid produces negligible field outside its cross-sectional area. The current in the solenoid is increasing at a rate of $\Delta I / \Delta t$. (a) What is the induced current in the ring? (b) At the center of the ring, what is the magnetic field produced by the induced current in the ring? (c) What is the direction of this field?

11. A coil of 15 turns and radius 10.0 cm surrounds a long solenoid of radius 2.00 cm and 1.00×10^3 turns/meter (Fig. P31.11). The current in the solenoid changes as $I = (5.00 \text{ A}) \sin(120t)$. Find the induced emf in the 15-turn coil as a function of time.

Figure P31.11

12. Two circular coils lie in the same plane. The following equation describes the emf induced in the smaller coil by a changing current in the larger coil. (a) Calculate this emf. (b) Write the statement of a problem, including data, for which the equation gives the solution.

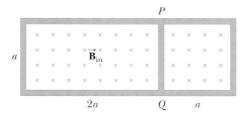

$$\mathcal{E} = -20 \frac{d}{dt} \left[\frac{130 \left(4\pi \times 10^{-7} \frac{\text{T} \cdot \text{m}}{\text{A}} \right) \left(3 \text{ A} - \frac{(6 \text{ A})t}{13 \times 10^{-6} \text{ s}} \right)}{2(0.40 \text{ m})} \pi (0.03 \text{ m})^2 \cos 0° \right]$$

13. Find the current through section PQ of length $a = 65.0$ cm in Figure P31.13. The circuit is located in a magnetic field whose magnitude varies with time according to the expression $B = (1.00 \times 10^{-3} \text{ T/s})t$. Assume the resistance per length of the wire is 0.100 Ω/m.

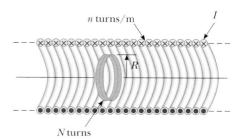

Figure P31.13

14. A 30-turn circular coil of radius 4.00 cm and resistance 1.00 Ω is placed in a magnetic field directed perpendicular to the plane of the coil. The magnitude of the magnetic field varies in time according to the expression $B = 0.010\ 0t + 0.040\ 0t^2$, where t is in seconds and B is in teslas. Calculate the induced emf in the coil at $t = 5.00$ s.

15. A long solenoid has $n = 400$ turns per meter and carries a current given by $I = (30.0 \text{ A})(1 - e^{-1.60t})$. Inside the solenoid and coaxial with it is a coil that has a radius of 6.00 cm and consists of a total of $N = 250$ turns of fine wire (Fig. P31.15). What emf is induced in the coil by the changing current?

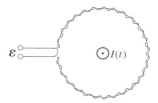

Figure P31.15

16. ● When a wire carries an AC current with a known frequency, you can use a *Rogowski coil* to determine the amplitude I_{max} of the current without disconnecting the wire to shunt the current though a meter. The Rogowski coil, shown in Figure P31.16, simply clips around the wire. It consists of a toroidal conductor wrapped around a circular return cord. Let n represent the number of turns in the toroid per unit distance along it. Let A represent the cross-sectional area of the toroid. Let $I(t) = I_{\text{max}} \sin \omega t$ represent the current to be measured. (a) Show that the amplitude of the emf induced in the Rogowski coil is $\mathcal{E}_{\text{max}} = \mu_0 nA\omega\, I_{\text{max}}$. (b) Explain why the wire carrying the unknown current need not be at the center of the Rogowski coil and why the coil will not respond to nearby currents that it does not enclose.

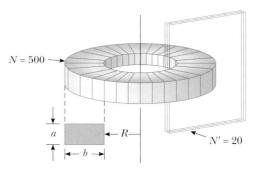

Figure P31.16

17. A coil formed by wrapping 50 turns of wire in the shape of a square is positioned in a magnetic field so that the normal to the plane of the coil makes an angle of 30.0° with the direction of the field. When the magnetic field is increased uniformly from 200 μT to 600 μT in 0.400 s, an emf of magnitude 80.0 mV is induced in the coil. What is the total length of the wire?

18. A toroid having a rectangular cross section ($a = 2.00$ cm by $b = 3.00$ cm) and inner radius $R = 4.00$ cm consists of 500 turns of wire that carries a sinusoidal current $I = I_{\text{max}} \sin \omega t$, with $I_{\text{max}} = 50.0$ A and a frequency $f = \omega/2\pi = 60.0$ Hz. A coil that consists of 20 turns of wire links with the toroid as shown in Figure P31.18. Determine the emf induced in the coil as a function of time.

Figure P31.18

2 = intermediate;   3 = challenging;   ☐ = SSM/SG;   ▲ = ThomsonNOW;   = symbolic reasoning;   ● = qualitative reasoning

19. A piece of insulated wire is shaped into a figure 8 as shown in Figure P31.19. The radius of the upper circle is 5.00 cm and that of the lower circle is 9.00 cm. The wire has a uniform resistance per unit length of 3.00 Ω/m. A uniform magnetic field is applied perpendicular to the plane of the two circles, in the direction shown. The magnetic field is increasing at a constant rate of 2.00 T/s. Find the magnitude and direction of the induced current in the wire.

Figure P31.19

Section 31.2 Motional emf

Section 31.3 Lenz's Law

Problem 61 in Chapter 29 can be assigned with this section.

20. An automobile has a vertical radio antenna 1.20 m long. The automobile travels at 65.0 km/h on a horizontal road where the Earth's magnetic field is 50.0 μT directed toward the north and downward at an angle of 65.0° below the horizontal. (a) Specify the direction the automobile should move so as to generate the maximum motional emf in the antenna, with the top of the antenna positive relative to the bottom. (b) Calculate the magnitude of this induced emf.

21. ● A small airplane with a wingspan of 14.0 m is flying due north at a speed of 70.0 m/s over a region where the vertical component of the Earth's magnetic field is 1.20 μT downward. (a) What potential difference is developed between the wingtips? Which wingtip is at higher potential? (b) **What If?** How would the answer change if the plane turned to fly due east? (c) Can this emf be used to power a light in the passenger compartment? Explain your answer.

22. Consider the arrangement shown in Figure P31.22. Assume $R = 6.00$ Ω, $\ell = 1.20$ m, and a uniform 2.50-T magnetic field is directed into the page. At what speed should the bar be moved to produce a current of 0.500 A in the resistor?

Figure P31.22 Problems 22, 23, and 24.

23. Figure P31.22 shows a top view of a bar that can slide without friction. The resistor is 6.00 Ω, and a 2.50-T magnetic field is directed perpendicularly downward, into the paper. Let $\ell = 1.20$ m. (a) Calculate the applied force

required to move the bar to the right at a constant speed of 2.00 m/s. (b) At what rate is energy delivered to the resistor?

24. ● A conducting rod of length ℓ moves on two horizontal, frictionless rails as shown in Figure P31.22. If a constant force of 1.00 N moves the bar at 2.00 m/s through a magnetic field $\vec{B}$ that is directed into the page, (a) what is the current in the 8.00-Ω resistor R? (b) What is the rate at which energy is delivered to the resistor? (c) What is the mechanical power delivered by the force $\vec{F}_{app}$? (d) Explain the relationship between the quantities computed in parts (b) and (c).

25. The *homopolar generator*, also called the *Faraday disk*, is a low-voltage, high-current electric generator. It consists of a rotating conducting disk with one stationary brush (a sliding electrical contact) at its axle and another at a point on its circumference as shown in Figure P31.25. A magnetic field is applied perpendicular to the plane of the disk. Assume the field is 0.900 T, the angular speed is 3 200 rev/min, and the radius of the disk is 0.400 m. Find the emf generated between the brushes. When superconducting coils are used to produce a large magnetic field, a homopolar generator can have a power output of several megawatts. Such a generator is useful, for example, in purifying metals by electrolysis. If a voltage is applied to the output terminals of the generator, it runs in reverse as a *homopolar motor* capable of providing great torque, useful in ship propulsion.

Figure P31.25

26. **Review problem.** As he starts to restring his acoustic guitar, a student attaches a single string, with linear density 3.00×10^{-3} kg/m, between two fixed points 64.0 cm apart, applies tension 267 N, and is distracted by a video game. His roommate attaches voltmeter leads to the ends of the metallic string and places a magnet across the string as shown in Figure P31.26. The magnet does not touch the string, but produces a uniform field of 4.50 mT

Figure P31.26

2 = intermediate;   3 = challenging;   ☐ = SSM/SG;   ▲ = ThomsonNOW;   ▒ = symbolic reasoning;   ● = qualitative reasoning

over a 2.00-cm length at the center of the string and neg-ligible field elsewhere. Strumming the string sets it vibrat-ing at its fundamental (lowest) frequency. The section of the string in the magnetic field moves perpendicular to the field with a uniform amplitude of 1.50 cm. Find (a) the frequency and (b) the amplitude of the electromotive force induced between the ends of the string.

27. A helicopter has blades of length 3.00 m, extending out from a central hub and rotating at 2.00 rev/s. If the verti-cal component of the Earth's magnetic field is 50.0 μT, what is the emf induced between the blade tip and the center hub?

28. Use Lenz's law to answer the following questions concern-ing the direction of induced currents. (a) What is the direction of the induced current in resistor R in Figure P31.28a when the bar magnet is moved to the left? (b) What is the direction of the current induced in the resistor R immediately after the switch S in Figure P31.28b is closed? (c) What is the direction of the induced current in R when the current I in Figure P31.28c decreases rapidly to zero? (d) A copper bar is moved to the right while its axis is maintained in a direc-tion perpendicular to a magnetic field as shown in Figure P31.28d. If the top of the bar becomes positive relative to the bottom, what is the direction of the magnetic field?

Figure P31.28

29. A rectangular coil with resistance R has N turns, each of length ℓ and width w as shown in Figure P31.29. The coil moves into a uniform magnetic field $\vec{B}$ with constant velocity $\vec{v}$. What are the magnitude and direction of the total magnetic force on the coil (a) as it enters the mag-netic field, (b) as it moves within the field, and (c) as it leaves the field?

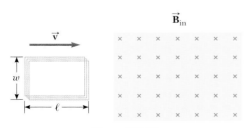

Figure P31.29

30. ● In Figure P31.30, the bar magnet is moved toward the loop. Is $V_a - V_b$ positive, negative, or zero? Explain your reasoning.

Figure P31.30

31. Two parallel rails with negligible resistance are 10.0 cm apart and are connected by a 5.00-Ω resistor. The circuit also contains two metal rods having resistances of 10.0 Ω and 15.0 Ω sliding along the rails (Fig. P31.31). The rods are pulled away from the resistor at constant speeds of 4.00 m/s and 2.00 m/s, respectively. A uniform magnetic field of magnitude 0.010 0 T is applied perpendicular to the plane of the rails. Determine the current in the 5.00-Ω resistor.

Figure P31.31

Section 31.4 Induced emf and Electric Fields

32. For the situation shown in Figure P31.32, the magnetic field changes with time according to the expression $B = (2.00t^3 - 4.00t^2 + 0.800)$ T, and $r_2 = 2R = 5.00$ cm. (a) Calculate the magnitude and direction of the force exerted on an electron located at point P_2 when $t = 2.00$ s. (b) At what instant is this force equal to zero?

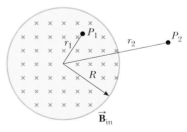

Figure P31.32   Problems 32 and 33.

33. A magnetic field directed into the page changes with time according to $B = (0.030\ 0t^2 + 1.40)$ T, where t is in sec-onds. The field has a circular cross section of radius $R = 2.50$ cm (Fig. P31.32). What are the magnitude and direc-tion of the electric field at point P_1 when $t = 3.00$ s and $r_1 = 0.020\ 0$ m?

2 = intermediate;   3 = challenging;   ☐ = SSM/SG;   ▲ = ThomsonNOW;   = symbolic reasoning;   ● = qualitative reasoning

34. A long solenoid with 1 000 turns per meter and radius 2.00 cm carries an oscillating current given by $I = (5.00 \text{ A}) \sin(100\pi t)$. What is the electric field induced at a radius $r = 1.00$ cm from the axis of the solenoid? What is the direction of this electric field when the current is increasing counterclockwise in the coil?

Section 31.5 Generators and Motors

> Problems 40 and 54 in Chapter 29 can be assigned with this section.

35. ▲ A coil of area 0.100 m² is rotating at 60.0 rev/s with the axis of rotation perpendicular to a 0.200-T magnetic field. (a) If the coil has 1 000 turns, what is the maximum emf generated in it? (b) What is the orientation of the coil with respect to the magnetic field when the maximum induced emf occurs?

36. In a 250-turn automobile alternator, the magnetic flux in each turn is $\Phi_B = (2.50 \times 10^{-4} \text{ Wb}) \cos \omega t$, where ω is the angular speed of the alternator. The alternator is geared to rotate three times for each engine revolution. When the engine is running at an angular speed of 1 000 rev/min, determine (a) the induced emf in the alternator as a function of time and (b) the maximum emf in the alternator.

37. A long solenoid, with its axis along the x axis, consists of 200 turns per meter of wire that carries a steady current of 15.0 A. A coil is formed by wrapping 30 turns of thin wire around a circular frame that has a radius of 8.00 cm. The coil is placed inside the solenoid and mounted on an axis that is a diameter of the coil and coincides with the y axis. The coil is then rotated with an angular speed of 4.00π rad/s. The plane of the coil is in the yz plane at $t = 0$. Determine the emf generated in the coil as a function of time.

38. A bar magnet is spun at constant angular speed ω around an axis as shown in Figure P31.38. A stationary, flat, rectangular conducting loop surrounds the magnet, and at $t = 0$, the magnet is oriented as shown. Make a qualitative graph of the induced current in the loop as a function of time, plotting counterclockwise currents as positive and clockwise currents as negative.

Figure P31.38

39. A motor in normal operation carries a direct current of 0.850 A when connected to a 120-V power supply. The resistance of the motor windings is 11.8 Ω. While in normal operation, (a) what is the back emf generated by the motor? (b) At what rate is internal energy produced in the windings? (c) **What If?** Suppose a malfunction stops the motor shaft from rotating. At what rate will internal energy be produced in the windings in this case? (Most motors have a thermal switch that will turn off the motor to prevent overheating when this stalling occurs.)

40. The rotating loop in an AC generator is a square 10.0 cm on each side. It is rotated at 60.0 Hz in a uniform field of 0.800 T. Calculate (a) the flux through the loop as a function of time, (b) the emf induced in the loop, (c) the current induced in the loop for a loop resistance of 1.00 Ω, (d) the power delivered to the loop, and (e) the torque that must be exerted to rotate the loop.

Section 31.6 Eddy Currents

41. ● Figure P31.41 represents an electromagnetic brake that uses eddy currents. An electromagnet hangs from a railroad car near one rail. To stop the car, a large current is sent through the coils of the electromagnet. The moving electromagnet induces eddy currents in the rails, whose fields oppose the change in the electromagnet's field. The magnetic fields of the eddy currents exert force on the current in the electromagnet, thereby slowing the car. The direction of the car's motion and the direction of the current in the electromagnet are shown correctly in the picture. Determine which of the eddy currents shown on the rails is correct. Explain your answer.

Figure P31.41

42. An *induction furnace* uses electromagnetic induction to produce eddy currents in a conductor, thereby raising the conductor's temperature. Commercial units operate at frequencies ranging from 60 Hz to about 1 MHz and deliver powers from a few watts to several megawatts. Induction heating can be used for warming a metal pan on a kitchen stove. It can be used to avoid oxidation and contamination of the metal when welding in a vacuum enclosure. At high frequencies, induced currents occur only near the surface of the conductor, in a phenomenon called the "skin effect." By creating an induced current for a short time interval at an appropriately high frequency, one can heat a sample down to a controlled depth. For example, the surface of a farm tiller can be tempered to make it hard and brittle for effective cutting while keeping the interior metal soft and ductile to resist breakage.

To explore induction heating, consider a flat conducting disk of radius R, thickness b, and resistivity ρ. A sinusoidal magnetic field $B_{max} \cos \omega t$ is applied perpendicular to the disk. Assume the field is uniform in space and the frequency is so low that the skin effect is not important. Also assume the eddy currents occur in circles concentric with the disk. (a) Calculate the average power delivered

2 = intermediate; 3 = challenging; □ = SSM/SG; ▲ = ThomsonNOW; = symbolic reasoning; ● = qualitative reasoning

to the disk. (b) **What If?** By what factor does the power change when the amplitude of the field doubles? (c) When the frequency doubles? (d) When the radius of the disk doubles?

43. ▲ ● A conducting rectangular loop of mass M, resistance R, and dimensions w by ℓ falls from rest into a magnetic field $\vec{B}$ as shown in Figure P31.43. During the time interval before the top edge of the loop reaches the field, the loop approaches a terminal speed v_T. (a) Show that

$$v_T = \frac{MgR}{B^2w^2}$$

(b) Why is v_T proportional to R? (c) Why is it inversely proportional to B^2?

Figure P31.43

Additional Problems

44. ● Consider the apparatus shown in Figure P31.44 in which a conducting bar can be moved along two rails connected to a lightbulb. The whole system is immersed in a magnetic field of 0.400 T perpendicularly into the page. The vertical distance between the horizontal rails is 0.800 m. The resistance of the lightbulb is 48.0 Ω, assumed to be constant. The bar and rails have negligible resistance. The bar is moved toward the right by a constant force of magnitude 0.600 N. (a) What is the direction of the induced current in the circuit? (b) If the speed of the bar is 15.0 m/s at a particular instant, what is the value of the induced current? (c) Argue that the constant force causes the speed of the bar to increase and approach a certain terminal speed. Find the value of this maximum speed. (d) What power is delivered to the lightbulb when the bar is moving at its terminal speed? (e) We have assumed the resistance of the lightbulb is constant. In reality, as the power delivered to the lightbulb increases, the filament temperature increases and the resistance increases. Explain conceptually (not algebraically) whether the terminal speed found in part (c) changes if the resistance increases. If the terminal speed changes, does it increase or decrease? (f) With the assumption that the resistance of the lightbulb increases as the current increases, explain mathematically whether the power found in part (d)

changes because of the increasing resistance. If it changes, is the actual power larger or smaller than the value previously found?

45. A guitar's steel string vibrates (Fig. 31.5a). The component of magnetic field perpendicular to the area of a pickup coil nearby is given by

$$B = 50.0 \text{ mT} + (3.20 \text{ mT}) \sin (1\,046\pi t)$$

The circular pickup coil has 30 turns and radius 2.70 mm. Find the emf induced in the coil as a function of time.

46. Strong magnetic fields are used in such medical procedures as magnetic resonance imaging, or MRI. A technician wearing a brass bracelet enclosing area 0.005 00 m² places her hand in a solenoid whose magnetic field is 5.00 T directed perpendicular to the plane of the bracelet. The electrical resistance around the bracelet's circumference is 0.020 0 Ω. An unexpected power failure causes the field to drop to 1.50 T in a time interval of 20.0 ms. Find (a) the current induced in the bracelet and (b) the power delivered to the bracelet. *Note:* As this problem implies, you should not wear any metal objects when working in regions of strong magnetic fields.

47. Figure P31.47 is a graph of the induced emf versus time for a coil of N turns rotating with angular speed ω in a uniform magnetic field directed perpendicular to the coil's axis of rotation. **What If?** Copy this sketch (on a larger scale) and on the same set of axes show the graph of emf versus t (a) if the number of turns in the coil is doubled, (b) if instead the angular speed is doubled, and (c) if the angular speed is doubled while the number of turns in the coil is halved.

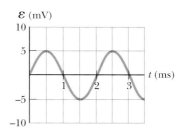

Figure P31.47

48. Two infinitely long solenoids (seen in cross section) pass through a circuit as shown in Figure P31.48. The magnitude of $\vec{B}$ inside each is the same and is increasing at the rate of 100 T/s. What is the current in each resistor?

Figure P31.44

Figure P31.48

2 = intermediate;   3 = challenging;   □ = SSM/SG;   ▲ = ThomsonNOW;   = symbolic reasoning;   ● = qualitative reasoning

49. A conducting rod of length $\ell = 35.0$ cm is free to slide on two parallel conducting bars as shown in Figure P31.49. Two resistors $R_1 = 2.00\ \Omega$ and $R_2 = 5.00\ \Omega$ are connected across the ends of the bars to form a loop. A constant magnetic field $B = 2.50$ T is directed perpendicularly into the page. An external agent pulls the rod to the left with a constant speed of $v = 8.00$ m/s. Find (a) the currents in both resistors, (b) the total power delivered to the resistance of the circuit, and (c) the magnitude of the applied force that is needed to move the rod with this constant velocity.

Figure P31.49

50. A bar of mass m, length d, and resistance R slides without friction in a horizontal plane, moving on parallel rails as shown in Figure P31.50. A battery that maintains a constant emf $\mathcal{E}$ is connected between the rails, and a constant magnetic field $\vec{B}$ is directed perpendicularly to the plane of the page. Assuming the bar starts from rest, show that at time t it moves with a speed

$$v = \frac{\mathcal{E}}{Bd}\left(1 - e^{-B^2 d^2 t / mR}\right)$$

Figure P31.50

51. Suppose you wrap wire onto the core from a roll of cellophane tape to make a coil. How can you use a bar magnet to produce an induced voltage in the coil? What is the order of magnitude of the emf you generate? State the quantities you take as data and their values.

52. Magnetic field values are often determined by using a device known as a *search coil*. This technique depends on the measurement of the total charge passing through a coil in a time interval during which the magnetic flux linking the windings changes either because of the coil's motion or because of a change in the value of B. (a) Show that as the flux through the coil changes from Φ_1 to Φ_2, the charge transferred through the coil is given by $Q = N(\Phi_2 - \Phi_1)/R$, where R is the resistance of the coil and a sensitive ammeter connected across it and N is the number of turns. (b) As a specific example, calculate B when a 100-turn coil of resistance 200 Ω and cross-sectional area 40.0 cm^2 produces the following results. A total charge of 5.00×10^{-4} C passes through the coil when it is rotated in a uniform field from a position where the plane of the coil is perpendicular to the field to a position where the coil's plane is parallel to the field.

53. The plane of a square loop of wire with edge length $a = 0.200$ m is perpendicular to the Earth's magnetic field at a point where $B = 15.0\ \mu$T as shown in Figure P31.53. The total resistance of the loop and the wires connecting it to a sensitive ammeter is 0.500 Ω. If the loop is suddenly collapsed by horizontal forces as shown, what total charge passes through the ammeter?

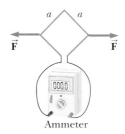

Figure P31.53

54. **Review problem**. A particle with a mass of 2.00×10^{-16} kg and a charge of 30.0 nC starts from rest, is accelerated by a strong electric field, and is fired from a small source inside a region of uniform constant magnetic field 0.600 T. The velocity of the particle is perpendicular to the field. The circular orbit of the particle encloses a magnetic flux of 15.0 μWb. (a) Calculate the speed of the particle. (b) Calculate the potential difference through which the particle accelerated inside the source.

55. ● In Figure P31.55, the rolling axle, 1.50 m long, is pushed along horizontal rails at a constant speed $v = 3.00$ m/s. A resistor $R = 0.400\ \Omega$ is connected to the rails at points a and b, directly opposite each other. The wheels make good electrical contact with the rails, so the axle, rails, and R form a closed-loop circuit. The only significant resistance in the circuit is R. A uniform magnetic field $B = 0.080\ 0$ T is vertically downward. (a) Find the induced current I in the resistor. (b) What horizontal force F is required to keep the axle rolling at constant speed? (c) Which end of the resistor, a or b, is at the higher electric potential? (d) **What If?** After the axle rolls past the resistor, does the current in R reverse direction? Explain your answer.

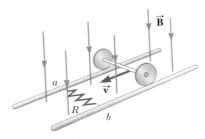

Figure P31.55

56. A conducting rod moves with a constant velocity $\vec{v}$ in a direction perpendicular to a long, straight wire carrying a current I as shown in Figure P31.56. Show that the magnitude of the emf generated between the ends of the rod is

$$|\mathcal{E}| = \frac{\mu_0 v I \ell}{2 \pi r}$$

In this case, note that the emf decreases with increasing r as you might expect.

---

2 = intermediate; 3 = challenging; ☐ = SSM/SG; ▲ = ThomsonNOW; = symbolic reasoning; ● = qualitative reasoning

Figure P31.56

57. ● In Figure P31.57, a uniform magnetic field decreases at a constant rate $dB/dt = -K$, where K is a positive constant. A circular loop of wire of radius a containing a resistance R and a capacitance C is placed with its plane normal to the field. (a) Find the charge Q on the capacitor when it is fully charged. (b) Which plate is at the higher potential? (c) Discuss the force that causes the separation of charges.

Figure P31.57

58. ● Figure P31.58 shows a compact, circular coil with 220 turns and radius 12.0 cm immersed in a uniform magnetic field parallel to the axis of the coil. The rate of change of the field has the constant magnitude 20.0 mT/s. (a) The following question cannot be answered with the information given. *Is the coil carrying clockwise or counterclockwise current?* What additional information is necessary to answer that question? (b) The coil overheats if more than 160 W of power is delivered to it. What resistance would the coil have at this critical point? To run cooler, should it have lower or higher resistance?

Figure P31.58

59. A rectangular coil of 60 turns, dimensions 0.100 m by 0.200 m and total resistance 10.0 Ω, rotates with angular speed 30.0 rad/s about the y axis in a region where a 1.00-T magnetic field is directed along the x axis. The rotation is initiated so that the plane of the coil is perpendicular to the direction of $\vec{\mathbf{B}}$ at $t = 0$. Calculate (a) the maximum induced emf in the coil, (b) the maximum rate of change of magnetic flux through the coil, (c) the induced emf at $t = 0.050\ 0$ s, and (d) the torque exerted by the magnetic field on the coil at the instant when the emf is a maximum.

60. A small, circular washer of radius 0.500 cm is held directly below a long, straight wire carrying a current of 10.0 A. The washer is located 0.500 m above the top of a table (Fig. P31.60). (a) If the washer is dropped from rest, what is the magnitude of the average induced emf in the washer over the time interval between its release and the moment it hits the tabletop? Assume the magnetic field is nearly constant over the area of the washer and equal to the magnetic field at the center of the washer. (b) What is the direction of the induced current in the washer?

Figure P31.60

61. A conducting rod of length ℓ moves with velocity $\vec{\mathbf{v}}$ parallel to a long wire carrying a steady current I. The axis of the rod is maintained perpendicular to the wire with the near end a distance r away as shown in Figure P31.61. Show that the magnitude of the emf induced in the rod is

$$|\mathcal{E}| = \frac{\mu_0 I v}{2\pi} \ln\left(1 + \frac{\ell}{r}\right)$$

Figure P31.61

62. A rectangular loop of dimensions ℓ and w moves with a constant velocity $\vec{\mathbf{v}}$ away from a long wire that carries a current I in the plane of the loop (Fig. P31.62). The total resistance of the loop is R. Derive an expression that gives the current in the loop at the instant the near side is a distance r from the wire.

Figure P31.62

63. The magnetic flux through a metal ring varies with time t according to $\Phi_B = 3(at^3 - bt^2)$ T·m², with $a = 2.00$ s^{-3} and $b = 6.00$ s^{-2}. The resistance of the ring is 3.00 Ω. Determine the maximum current induced in the ring during the interval from $t = 0$ to $t = 2.00$ s.

64. **Review problem.** The bar of mass m in Figure P31.64 is pulled horizontally across parallel, frictionless rails by a massless string that passes over a light, frictionless pulley and is attached to a suspended object of mass M. The uniform magnetic field has a magnitude B, and the distance between the rails is ℓ. The only significant electrical resistance is the load resistor R shown connecting the rails at one end. Derive an expression that gives the horizontal speed of the bar as a function of time, assuming the suspended object is released with the bar at rest at $t = 0$.

Figure P31.64

65. A *betatron* is a device that accelerates electrons to energies in the MeV range by means of electromagnetic induction. Electrons in a vacuum chamber are held in a circular orbit by a magnetic field perpendicular to the orbital plane. The magnetic field is gradually increased to induce an electric field around the orbit. (a) Show that the electric field is in the correct direction to make the electrons speed up. (b) Assume the radius of the orbit remains con-

stant. Show that the average magnetic field over the area enclosed by the orbit must be twice as large as the magnetic field at the circle's circumference.

66. A thin wire 30.0 cm long is held parallel to and 80.0 cm above a long, thin wire carrying 200 A and resting on the horizontal floor (Fig. P31.66). The 30.0-cm wire is released at the instant $t = 0$ and falls, remaining parallel to the current-carrying wire as it falls. Assume the falling wire accelerates at 9.80 m/s². (a) Derive an equation for the emf induced in it as a function of time. (b) What is the minimum value of the emf? (c) What is the maximum value? (d) What is the induced emf 0.300 s after the wire is released?

Figure P31.66

67. ▲ A long, straight wire carries a current that is given by $I = I_{max} \sin(\omega t + \phi)$. The wire lies in the plane of a rectangular coil of N turns of wire, as shown in Figure P31.8. The quantities I_{max}, ω, and ϕ are all constants. Determine the emf induced in the coil by the magnetic field created by the current in the straight wire. Assume $I_{max} = 50.0$ A, $\omega = 200\pi$ s⁻¹, $N = 100$, $h = w = 5.00$ cm, and $L = 20.0$ cm.

Answers to Quick Quizzes

31.1 (c). In all cases except this one, there is a change in the magnetic flux through the loop.

31.2 (c). The force on the wire is of magnitude $F_{app} = F_B = I\ell B$, with I given by Equation 31.6. Therefore, the force is proportional to the speed and the force doubles. Because $\mathcal{P} = F_{app}v$, the doubling of the force *and* the speed results in the power being four times as large.

31.3 (b). At the position of the loop, the magnetic field lines due to the wire point into the page. The loop is entering a region of stronger magnetic field as it drops toward the wire, so the flux is increasing. The induced current must set up a magnetic field that opposes this increase. To do so, it creates a magnetic field directed out of the page. By the right-hand rule for current loops, a counterclockwise current in the loop is required.

31.4 (a). Although reducing the resistance may increase the current the generator provides to a load, it does not alter the emf. Equation 31.11 shows that the emf depends on ω, B, and N, so all other choices increase the emf.

31.5 (b). When the aluminum sheet moves between the poles of the magnet, eddy currents are established in the aluminum. According to Lenz's law, these currents are in a direction so as to oppose the original change, which is the movement of the aluminum sheet in the magnetic field. The same principle is used in common laboratory triple-beam balances. See if you can find the magnet and the aluminum sheet the next time you use a triple-beam balance.

32.1 Self-Induction and Inductance

32.2 *RL* Circuits

32.3 Energy in a Magnetic Field

32.4 Mutual Inductance

32.5 Oscillations in an *LC* Circuit

32.6 The *RLC* Circuit

A treasure hunter uses a metal detector to search for buried objects at a beach. At the end of the metal detector is a coil of wire that is part of a circuit. When the coil comes near a metal object, the inductance of the coil is affected and the current in the circuit changes. This change triggers a signal in the earphones worn by the treasure hunter. We investigate inductance in this chapter. (Stone/Getty Images)

32 Inductance

In Chapter 31, we saw that an emf and a current are induced in a loop of wire when the magnetic flux through the area enclosed by the loop changes with time. This phenomenon of electromagnetic induction has some practical consequences. In this chapter, we first describe an effect known as *self-induction,* in which a time-varying current in a circuit produces an induced emf opposing the emf that initially set up the time-varying current. Self-induction is the basis of the *inductor,* an electrical circuit element. We discuss the energy stored in the magnetic field of an inductor and the energy density associated with the magnetic field.

Next, we study how an emf is induced in a coil as a result of a changing magnetic flux produced by a second coil, which is the basic principle of *mutual induction.* Finally, we examine the characteristics of circuits that contain inductors, resistors, and capacitors in various combinations.

32.1 Self-Induction and Inductance

In this chapter, we need to distinguish carefully between emfs and currents that are caused by physical sources such as batteries and those that are induced by changing magnetic fields. When we use a term (such as *emf* or *current*) without an adjective, we are describing the parameters associated with a physical source. We

JOSEPH HENRY
American Physicist (1797–1878)
Henry became the first director of the Smithsonian Institution and first president of the Academy of Natural Science. He improved the design of the electromagnet and constructed one of the first motors. He also discovered the phenomenon of self-induction, but he failed to publish his findings. The unit of inductance, the henry, is named in his honor.

Figure 32.1 After the switch is closed, the current produces a magnetic flux through the area enclosed by the loop. As the current increases toward its equilibrium value, this magnetic flux changes in time and induces an emf in the loop.

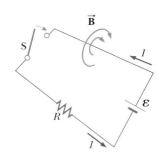

use the adjective *induced* to describe those emfs and currents caused by a changing magnetic field.

Consider a circuit consisting of a switch, a resistor, and a source of emf as shown in Figure 32.1. The circuit diagram is represented in perspective to show the orientations of some of the magnetic field lines due to the current in the circuit. When the switch is thrown to its closed position, the current does not immediately jump from zero to its maximum value $\mathcal{E}/R$. Faraday's law of electromagnetic induction (Eq. 31.1) can be used to describe this effect as follows. As the current increases with time, the magnetic flux through the circuit loop due to this current also increases with time. This increasing flux creates an induced emf in the circuit. The direction of the induced emf is such that it would cause an induced current in the loop (if the loop did not already carry a current), which would establish a magnetic field opposing the change in the original magnetic field. Therefore, the direction of the induced emf is opposite the direction of the emf of the battery, which results in a gradual rather than instantaneous increase in the current to its final equilibrium value. Because of the direction of the induced emf, it is also called a *back emf*, similar to that in a motor as discussed in Chapter 31. This effect is called **self-induction** because the changing flux through the circuit and the resultant induced emf arise from the circuit itself. The emf $\mathcal{E}_L$ set up in this case is called a **self-induced emf.**

To obtain a quantitative description of self-induction, recall from Faraday's law that the induced emf is equal to the negative of the time rate of change of the magnetic flux. The magnetic flux is proportional to the magnetic field, which in turn is proportional to the current in the circuit. Therefore, **a self-induced emf is always proportional to the time rate of change of the current.** For any loop of wire, we can write this proportionality as

$$\mathcal{E}_L = -L\frac{dI}{dt} \tag{32.1}$$

where L is a proportionality constant—called the **inductance** of the loop—that depends on the geometry of the loop and other physical characteristics. If we consider a closely spaced coil of N turns (a toroid or an ideal solenoid) carrying a current I and containing N turns, Faraday's law tells us that $\mathcal{E}_L = -N\, d\Phi_B/dt$. Combining this expression with Equation 32.1 gives

Inductance of an *N*-turn coil ▶

$$L = \frac{N\Phi_B}{I} \tag{32.2}$$

where it is assumed the same magnetic flux passes through each turn and L is the inductance of the entire coil.

From Equation 32.1, we can also write the inductance as the ratio

Inductance ▶

$$L = -\frac{\mathcal{E}_L}{dI/dt} \tag{32.3}$$

Recall that resistance is a measure of the opposition to current ($R = \Delta V/I$); in comparison, Equation 32.3 shows us that inductance is a measure of the opposition to a *change* in current.

The SI unit of inductance is the **henry** (H), which as we can see from Equation 32.3 is 1 volt-second per ampere: $1\,\text{H} = 1\,\text{V}\cdot\text{s/A}$.

As shown in Example 32.1, the inductance of a coil depends on its geometry. This dependence is analogous to the capacitance of a capacitor depending on the geometry of its plates as we found in Chapter 26. Inductance calculations can be quite difficult to perform for complicated geometries, but the examples below involve simple situations for which inductances are easily evaluated.

Quick Quiz 32.1   A coil with zero resistance has its ends labeled a and b. The potential at a is higher than at b. Which of the following could be consistent with this situation? (a) The current is constant and is directed from a to b. (b) The current is constant and is directed from b to a. (c) The current is increasing and is directed from a to b. (d) The current is decreasing and is directed from a to b. (e) The current is increasing and is directed from b to a. (f) The current is decreasing and is directed from b to a.

EXAMPLE 32.1 | **Inductance of a Solenoid**

Consider a uniformly wound solenoid having N turns and length ℓ. Assume ℓ is much longer than the radius of the windings and the core of the solenoid is air.

(A) Find the inductance of the solenoid.

SOLUTION

Conceptualize   The magnetic field lines from each turn of the solenoid pass through all the turns, so an induced emf in each coil opposes changes in the current.

Categorize   Because the solenoid is long, we can use the results for an ideal solenoid obtained in Chapter 30.

Analyze   Find the magnetic flux through each turn of area A in the solenoid, using the expression for the magnetic field from Equation 30.17:

$$\Phi_B = BA = \mu_0 n I A = \mu_0 \frac{N}{\ell} I A$$

Substitute this expression into Equation 32.2:

$$L = \frac{N\Phi_B}{I} = \mu_0 \frac{N^2}{\ell} A \qquad (32.4)$$

(B) Calculate the inductance of the solenoid if it contains 300 turns, its length is 25.0 cm, and its cross-sectional area is 4.00 cm².

SOLUTION

Substitute numerical values into Equation 32.4:

$$L = (4\pi \times 10^{-7}\,\text{T}\cdot\text{m/A})\,\frac{(300)^2}{25.0 \times 10^{-2}\,\text{m}}\,(4.00 \times 10^{-4}\,\text{m}^2)$$

$$= 1.81 \times 10^{-4}\,\text{T}\cdot\text{m}^2/\text{A} = \boxed{0.181\,\text{mH}}$$

(C) Calculate the self-induced emf in the solenoid if the current it carries decreases at the rate of 50.0 A/s.

SOLUTION

Substitute $dI/dt = -50.0$ A/s into Equation 32.1:

$$\varepsilon_L = -L\frac{dI}{dt} = -(1.81 \times 10^{-4}\,\text{H})(-50.0\,\text{A/s})$$

$$= \boxed{9.05\,\text{mV}}$$

Finalize The result for part (A) shows that L depends on geometry and is proportional to the square of the number of turns. Because $N = n\ell$, we can also express the result in the form

$$L = \mu_0 \frac{(n\ell)^2}{\ell} A = \mu_0 n^2 A\ell = \mu_0 n^2 V \tag{32.5}$$

where $V = A\ell$ is the interior volume of the solenoid.

32.2 *RL* Circuits

If a circuit contains a coil such as a solenoid, the inductance of the coil prevents the current in the circuit from increasing or decreasing instantaneously. A circuit element that has a large inductance is called an **inductor** and has the circuit symbol ——. We always assume the inductance of the remainder of a circuit is negligible compared with that of the inductor. Keep in mind, however, that even a circuit without a coil has some inductance that can affect the circuit's behavior.

Because the inductance of an inductor results in a back emf, **an inductor in a circuit opposes changes in the current in that circuit.** The inductor attempts to keep the current the same as it was before the change occurred. If the battery voltage in the circuit is increased so that the current rises, the inductor opposes this change and the rise is not instantaneous. If the battery voltage is decreased, the inductor causes a slow drop in the current rather than an immediate drop. Therefore, the inductor causes the circuit to be "sluggish" as it reacts to changes in the voltage.

Consider the circuit shown in Active Figure 32.2, which contains a battery of negligible internal resistance. This circuit is an **RL circuit** because the elements connected to the battery are a resistor and an inductor. The curved lines on switch S_2 suggest this switch can never be open; it is always set to either a or b. (If the switch is connected to neither a nor b, any current in the circuit suddenly stops.) Suppose S_2 is set to a and switch S_1 is open for $t < 0$ and then thrown closed at $t = 0$. The current in the circuit begins to increase, and a back emf (Eq. 32.1) that opposes the increasing current is induced in the inductor.

With this point in mind, let's apply Kirchhoff's loop rule to this circuit, traversing the circuit in the clockwise direction:

$$\mathcal{E} - IR - L\frac{dI}{dt} = 0 \tag{32.6}$$

where IR is the voltage drop across the resistor. (Kirchhoff's rules were developed for circuits with steady currents, but they can also be applied to a circuit in which the current is changing if we imagine them to represent the circuit at one *instant* of time.) Now let's find a solution to this differential equation, which is similar to that for the *RC* circuit (see Section 28.4).

A mathematical solution of Equation 32.6 represents the current in the circuit as a function of time. To find this solution, we change variables for convenience, letting $x = (\mathcal{E}/R) - I$, so $dx = -dI$. With these substitutions, Equation 32.6 becomes

$$x + \frac{L}{R}\frac{dx}{dt} = 0$$

Rearranging and integrating this last expression gives

$$\int_{x_0}^{x} \frac{dx}{x} = -\frac{R}{L}\int_0^t dt$$

$$\ln\frac{x}{x_0} = -\frac{R}{L}t$$

where x_0 is the value of x at time $t = 0$. Taking the antilogarithm of this result gives

$$x = x_0 e^{-Rt/L}$$

ACTIVE FIGURE 32.2

An *RL* circuit. When switch S_2 is in position a, the battery is in the circuit. When switch S_1 is thrown closed, the current increases and an emf that opposes the increasing current is induced in the inductor. When the switch is thrown to position b, the battery is no longer part of the circuit and the current decreases. The switch is designed so that it is never open, which would cause the current to stop.

Sign in at www.thomsonedu.com and go to ThomsonNOW to adjust the values of R and L and see the effect on the current. A graphical display as in Active Figure 32.3 is available.

Because $I = 0$ at $t = 0$, note from the definition of x that $x_0 = \mathcal{E}/R$. Hence, this last expression is equivalent to

$$\frac{\mathcal{E}}{R} - I = \frac{\mathcal{E}}{R}\,e^{-Rt/L}$$

$$I = \frac{\mathcal{E}}{R}\left(1 - e^{-Rt/L}\right)$$

This expression shows how the inductor affects the current. The current does not increase instantly to its final equilibrium value when the switch is closed, but instead increases according to an exponential function. If the inductance is removed from the circuit, which corresponds to letting L approach zero, the exponential term becomes zero and there is no time dependence of the current in this case; the current increases instantaneously to its final equilibrium value in the absence of the inductance.

We can also write this expression as

$$I = \frac{\mathcal{E}}{R}\left(1 - e^{-t/\tau}\right) \tag{32.7}$$

where the constant τ is the **time constant** of the *RL* circuit:

$$\tau = \frac{L}{R} \tag{32.8}$$

Physically, τ is the time interval required for the current in the circuit to reach $(1 - e^{-1}) = 0.632 = 63.2\%$ of its final value $\mathcal{E}/R$. The time constant is a useful parameter for comparing the time responses of various circuits.

Active Figure 32.3 shows a graph of the current versus time in the *RL* circuit. Notice that the equilibrium value of the current, which occurs as t approaches infinity, is $\mathcal{E}/R$. That can be seen by setting dI/dt equal to zero in Equation 32.6 and solving for the current I. (At equilibrium, the change in the current is zero.) Therefore, the current initially increases very rapidly and then gradually approaches the equilibrium value $\mathcal{E}/R$ as t approaches infinity.

Let's also investigate the time rate of change of the current. Taking the first time derivative of Equation 32.7 gives

$$\frac{dI}{dt} = \frac{\mathcal{E}}{L}\,e^{-t/\tau} \tag{32.9}$$

This result shows that the time rate of change of the current is a maximum (equal to $\mathcal{E}/L$) at $t = 0$ and falls off exponentially to zero as t approaches infinity (Fig. 32.4).

Now consider the *RL* circuit in Active Figure 32.2 again. Suppose switch S_2 has been set at position a long enough (and switch S_1 remains closed) to allow the current to reach its equilibrium value $\mathcal{E}/R$. In this situation, the circuit is described by the outer loop in Active Figure 32.2. If S_2 is thrown from a to b, the circuit is now described by only the right-hand loop in Active Figure 32.2. Therefore, the battery has been eliminated from the circuit. Setting $\mathcal{E} = 0$ in Equation 32.6 gives

$$IR + L\frac{dI}{dt} = 0$$

It is left as a problem (Problem 10) to show that the solution of this differential equation is

$$I = \frac{\mathcal{E}}{R}\,e^{-t/\tau} = I_i e^{-t/\tau} \tag{32.10}$$

where $\mathcal{E}$ is the emf of the battery and $I_i = \mathcal{E}/R$ is the initial current at the instant the switch is thrown to b.

If the circuit did not contain an inductor, the current would immediately decrease to zero when the battery is removed. When the inductor is present, it

ACTIVE FIGURE 32.3

Plot of the current versus time for the *RL* circuit shown in Active Figure 32.2. When switch S_1 is thrown closed at $t = 0$, the current increases toward its maximum value $\mathcal{E}/R$. The time constant τ is the time interval required for I to reach 63.2% of its maximum value.

Sign in at www.thomsonedu.com and go to ThomsonNOW to observe this graph develop after switch S_1 in Active Figure 32.2 is thrown closed.

◄   Time constant of an *RL* circuit

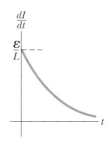

Figure 32.4 Plot of dI/dt versus time for the *RL* circuit shown in Active Figure 32.2. The time rate of change of current is a maximum at $t = 0$, which is the instant at which switch S_1 is thrown closed. The rate decreases exponentially with time as I increases toward its maximum value.

ACTIVE FIGURE 32.5

Current versus time for the right-hand loop of the circuit shown in Active Figure 32.2. For $t < 0$, switch S_2 is at position a. At $t = 0$, the switch is thrown to position b and the current has its maximum value $\mathcal{E}/R$.

Sign in at www.thomsonedu.com and go to Thomson-NOW to observe this graph develop after the switch in Active Figure 32.2 is thrown to position b.

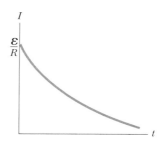

opposes the decrease in the current and causes the current to decrease exponentially. A graph of the current in the circuit versus time (Active Fig. 32.5) shows that the current is continuously decreasing with time.

Quick Quiz 32.2 Consider the circuit in Active Figure 32.2 with S_1 open and S_2 at position a. Switch S_1 is now thrown closed. **(i)** At the instant it is closed, across which circuit element is the voltage equal to the emf of the battery? (a) the resistor (b) the inductor (c) both the inductor and resistor **(ii)** After a very long time, across which circuit element is the voltage equal to the emf of the battery? Choose from among the same answers.

EXAMPLE 32.2 | **Time Constant of an _RL_ Circuit**

Consider the circuit in Active Figure 32.2 again. Suppose the circuit elements have the following values: $\mathcal{E} = 12.0$ V, $R = 6.00$ Ω, and $L = 30.0$ mH.

(A) Find the time constant of the circuit.

SOLUTION

Conceptualize You should understand the behavior of this circuit from the discussion in this section.

Categorize We evaluate the results using equations developed in this section, so this example is a substitution problem.

Evaluate the time constant from Equation 32.8:

$$\tau = \frac{L}{R} = \frac{30.0 \times 10^{-3} \text{ H}}{6.00 \text{ Ω}} = 5.00 \text{ ms}$$

(B) Switch S_2 is at position a, and switch S_1 is thrown closed at $t = 0$. Calculate the current in the circuit at $t = 2.00$ ms.

SOLUTION

Evaluate the current at $t = 2.00$ ms from Equation 32.7:

$$I = \frac{\mathcal{E}}{R}(1 - e^{-t/\tau}) = \frac{12.0 \text{ V}}{6.00 \text{ Ω}}(1 - e^{-2.00 \text{ ms}/5.00 \text{ ms}}) = 2.00 \text{ A } (1 - e^{-0.400})$$

$$= 0.659 \text{ A}$$

(C) Compare the potential difference across the resistor with that across the inductor.

SOLUTION

At the instant the switch is closed, there is no current and therefore no potential difference across the resistor. At this instant, the battery voltage appears entirely across the inductor in the form of a back emf of 12.0 V as the inductor tries to maintain the zero-current condition. (The top end of the inductor in Active Fig. 32.2 is at a higher electric potential than the bottom end.) As time passes, the emf across the inductor decreases and the current in the resistor (and hence the voltage across it) increases as shown in Figure 32.6. The sum of the two voltages at all times is 12.0 V.

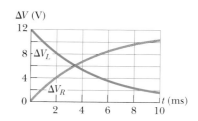

Figure 32.6 (Example 32.2) The time behavior of the voltages across the resistor and inductor in Active Figure 32.2 given the values provided in this example.

What If? In Figure 32.6, the voltages across the resistor and inductor are equal at 3.4 ms. What if you wanted to delay the condition in which the voltages are equal to some later instant, such as $t = 10.0$ ms? Which parameter, L or R, would require the least adjustment, in terms of a percentage change, to achieve that?

Answer Figure 32.6 shows that the voltages are equal when the voltage across the inductor has fallen to half its original value. Therefore, the time interval required for the voltages to become equal is the *half-life* $t_{1/2}$ of the decay. We introduced the half-life in the **What If?** section of Example 28.10 to describe the exponential decay in RC circuits, where $t_{1/2} = 0.693\tau$.

From the desired half-life of 10.0 ms, use the result from Example 28.10 to find the time constant of the circuit:

$$\tau = \frac{t_{1/2}}{0.693} = \frac{10.0 \text{ ms}}{0.693} = 14.4 \text{ ms}$$

Hold L fixed and find the value of R that gives this time constant:

$$\tau = \frac{L}{R} \quad \rightarrow \quad R = \frac{L}{\tau} = \frac{30.0 \times 10^{-3} \text{ H}}{14.4 \text{ ms}} = 2.08 \ \Omega$$

Now hold R fixed and find the appropriate value of L:

$$\tau = \frac{L}{R} \quad \rightarrow \quad L = \tau R = (14.4 \text{ ms})(6.00 \ \Omega) = 86.4 \times 10^{-3} \text{ H}$$

The change in R corresponds to a 65% decrease compared with the initial resistance. The change in L represents a 188% increase in inductance! Therefore, a much smaller percentage adjustment in R can achieve the desired effect than would an adjustment in L.

32.3 Energy in a Magnetic Field

A battery in a circuit containing an inductor must provide more energy than in a circuit without the inductor. Part of the energy supplied by the battery appears as internal energy in the resistance in the circuit, and the remaining energy is stored in the magnetic field of the inductor. Multiplying each term in Equation 32.6 by I and rearranging the expression gives

$$I\mathcal{E} = I^2 R + LI \frac{dI}{dt} \tag{32.11}$$

Recognizing $I\mathcal{E}$ as the rate at which energy is supplied by the battery and $I^2 R$ as the rate at which energy is delivered to the resistor, we see that $LI(dI/dt)$ must represent the rate at which energy is being stored in the inductor. If U is the energy stored in the inductor at any time, we can write the rate dU/dt at which energy is stored as

$$\frac{dU}{dt} = LI \frac{dI}{dt}$$

To find the total energy stored in the inductor at any instant, let's rewrite this expression as $dU = LI \, dI$ and integrate:

$$U = \int dU = \int_0^I LI \, dI = L \int_0^I I \, dI$$

$$U = \tfrac{1}{2} LI^2 \tag{32.12}$$

where L is constant and has been removed from the integral. Equation 32.12 represents the energy stored in the magnetic field of the inductor when the current is I. It is similar in form to Equation 26.11 for the energy stored in the electric field of a capacitor, $U = \tfrac{1}{2} C (\Delta V)^2$. In either case, energy is required to establish a field.

PITFALL PREVENTION 32.1
Capacitors, Resistors, and Inductors Store Energy Differently

Different energy-storage mechanisms are at work in capacitors, inductors, and resistors. A charged capacitor stores energy as electrical potential energy. An inductor stores energy as what we could call magnetic potential energy when it carries current. Energy delivered to a resistor is transformed to internal energy.

◀ Energy stored in an inductor

We can also determine the energy density of a magnetic field. For simplicity, consider a solenoid whose inductance is given by Equation 32.5:

$$L = \mu_0 n^2 V$$

The magnetic field of a solenoid is given by Equation 30.17:

$$B = \mu_0 n I$$

Substituting the expression for L and $I = B/\mu_0 n$ into Equation 32.12 gives

$$U = \tfrac{1}{2} L I^2 = \tfrac{1}{2} \mu_0 n^2 V \left(\frac{B}{\mu_0 n} \right)^2 = \frac{B^2}{2\mu_0} V \tag{32.13}$$

The magnetic energy density, or the energy stored per unit volume in the magnetic field of the inductor, is

Magnetic energy density ▶

$$u_B = \frac{U}{V} = \frac{B^2}{2\mu_0} \tag{32.14}$$

Although this expression was derived for the special case of a solenoid, it is valid for any region of space in which a magnetic field exists. Equation 32.14 is similar in form to Equation 26.13 for the energy per unit volume stored in an electric field, $u_E = \tfrac{1}{2}\epsilon_0 E^2$. In both cases, the energy density is proportional to the square of the field magnitude.

Quick Quiz 32.3 You are performing an experiment that requires the highest-possible magnetic energy density in the interior of a very long current-carrying solenoid. Which of the following adjustments increases the energy density? (More than one choice may be correct.) (a) increasing the number of turns per unit length on the solenoid   (b) increasing the cross-sectional area of the solenoid   (c) increasing only the length of the solenoid while keeping the number of turns per unit length fixed   (d) increasing the current in the solenoid

---

EXAMPLE 32.3   **What Happens to the Energy in the Inductor?**

Consider once again the *RL* circuit shown in Active Figure 32.2, with switch S$_2$ at position *a* and the current having reached its steady-state value. When S$_2$ is thrown to position *b*, the current in the right-hand loop decays exponentially with time according to the expression $I = I_i e^{-t/\tau}$, where $I_i = \mathcal{E}/R$ is the initial current in the circuit and $\tau = L/R$ is the time constant. Show that all the energy initially stored in the magnetic field of the inductor appears as internal energy in the resistor as the current decays to zero.

SOLUTION

Conceptualize Before S$_2$ is thrown to *b*, energy is being delivered at a constant rate to the resistor from the battery and energy is stored in the magnetic field of the inductor. After $t = 0$, when S$_2$ is thrown to *b*, the battery can no longer provide energy and energy is delivered to the resistor only from the inductor.

Categorize We model the right-hand loop of the circuit as an isolated system so that energy is transferred between components of the system but does not leave the system.

Analyze The energy in the magnetic field of the inductor at any time is U. The rate dU/dt at which energy leaves the inductor and is delivered to the resistor is equal to $I^2 R$, where I is the instantaneous current.

Substitute the current given by Equation 32.10 into $dU/dt = I^2 R$:

$$\frac{dU}{dt} = I^2 R = (I_i e^{-Rt/L})^2 R = I_i^2 R e^{-2Rt/L}$$

Solve for dU and integrate this expression over the limits $t = 0$ to $t \to \infty$:

$$U = \int_0^\infty I_i^2 R e^{-2Rt/L} \, dt = I_i^2 R \int_0^\infty e^{-2Rt/L} \, dt$$

The value of the definite integral can be shown to be $L/2R$ (see Problem 26). Use this result to evaluate U:

$$U = I_i^2 R \left(\frac{L}{2R} \right) = \tfrac{1}{2} L I_i^2$$

Finalize This result is equal to the initial energy stored in the magnetic field of the inductor, given by Equation 32.12, as we set out to prove.

EXAMPLE 32.4	**The Coaxial Cable**

Coaxial cables are often used to connect electrical devices, such as your stereo system, and in receiving signals in television cable systems. Model a long coaxial cable as two thin, concentric, cylindrical conducting shells of radii a and b and length ℓ as in Figure 32.7. The conducting shells carry the same current I in opposite directions. Calculate the inductance L of this cable.

SOLUTION

Conceptualize Consider Figure 32.7. Although we do not have a visible coil in this geometry, imagine a thin, radial slice of the coaxial cable such as the light gold rectangle in Figure 32.7. If the inner and outer conductors are connected at the ends of the cable (above and below the figure), this slice represents one large conducting loop. The current in the loop sets up a magnetic field between the inner and outer conductors that passes through this loop. If the current changes, the magnetic field changes and the induced emf opposes the original change in the current in the conductors.

Figure 32.7 (Example 32.4) Section of a long coaxial cable. The inner and outer conductors carry equal currents in opposite directions.

Categorize We categorize this situation as one in which we must return to the fundamental definition of inductance, Equation 32.2.

Analyze We must find the magnetic flux through the light gold rectangle in Figure 32.7. Ampère's law (see Section 30.3) tells us that the magnetic field in the region between the shells is due to the inner conductor and that its magnitude is $B = \mu_0 I / 2\pi r$, where r is measured from the common center of the shells. The magnetic field is zero outside the outer shell ($r > b$) because the net current passing through the area enclosed by a circular path surrounding the cable is zero; hence, from Ampère's law, $\oint \vec{B} \cdot d\vec{s} = 0$. The magnetic field is zero inside the inner shell because the shell is hollow and no current is present within a radius $r < a$.

The magnetic field is perpendicular to the light gold rectangle of length ℓ and width $b - a$, the cross section of interest. Because the magnetic field varies with radial position across this rectangle, we must use calculus to find the total magnetic flux.

Divide the light gold rectangle into strips of width dr such as the darker strip in Figure 32.7. Evaluate the magnetic flux through such a strip:

$$\Phi_B = \int B \, dA = \int B\ell \, dr$$

Substitute for the magnetic field and integrate over the entire light gold rectangle:

$$\Phi_B = \int_a^b \frac{\mu_0 I}{2\pi r} \ell \, dr = \frac{\mu_0 I \ell}{2\pi} \int_a^b \frac{dr}{r} = \frac{\mu_0 I \ell}{2\pi} \ln \left(\frac{b}{a} \right)$$

Use Equation 32.2 to find the inductance of the cable:

$$L = \frac{\Phi_B}{I} = \frac{\mu_0 \ell}{2\pi} \ln \left(\frac{b}{a} \right)$$

Finalize The inductance increases if ℓ increases, if b increases, or if a decreases. This result is consistent with our conceptualization: any of these changes increases the size of the loop represented by our radial slice and through which the magnetic field passes, increasing the inductance.

32.4   Mutual Inductance

Very often, the magnetic flux through the area enclosed by a circuit varies with time because of time-varying currents in nearby circuits. This condition induces an emf through a process known as *mutual induction,* so named because it depends on the interaction of two circuits.

Consider the two closely wound coils of wire shown in cross-sectional view in Figure 32.8. The current I_1 in coil 1, which has N_1 turns, creates a magnetic field. Some of the magnetic field lines pass through coil 2, which has N_2 turns. The magnetic flux caused by the current in coil 1 and passing through coil 2 is represented by Φ_{12}. In analogy to Equation 32.2, we can identify the **mutual inductance** M_{12} of coil 2 with respect to coil 1:

▶ Definition of mutual inductance

$$M_{12} = \frac{N_2 \Phi_{12}}{I_1} \qquad (32.15)$$

Mutual inductance depends on the geometry of both circuits and on their orientation with respect to each other. As the circuit separation distance increases, the mutual inductance decreases because the flux linking the circuits decreases.

If the current I_1 varies with time, we see from Faraday's law and Equation 32.15 that the emf induced by coil 1 in coil 2 is

$$\mathcal{E}_2 = -N_2 \frac{d\Phi_{12}}{dt} = -N_2 \frac{d}{dt}\left(\frac{M_{12}I_1}{N_2}\right) = -M_{12}\frac{dI_1}{dt} \qquad (32.16)$$

In the preceding discussion, it was assumed the current is in coil 1. Let's also imagine a current I_2 in coil 2. The preceding discussion can be repeated to show that there is a mutual inductance M_{21}. If the current I_2 varies with time, the emf induced by coil 2 in coil 1 is

$$\mathcal{E}_1 = -M_{21}\frac{dI_2}{dt} \qquad (32.17)$$

In mutual induction, the emf induced in one coil is always proportional to the rate at which the current in the other coil is changing. Although the proportionality constants M_{12} and M_{21} have been treated separately, it can be shown that they are equal. Therefore, with $M_{12} = M_{21} = M$, Equations 32.16 and 32.17 become

$$\mathcal{E}_2 = -M\frac{dI_1}{dt} \quad \text{and} \quad \mathcal{E}_1 = -M\frac{dI_2}{dt}$$

These two equations are similar in form to Equation 32.1 for the self-induced emf $\mathcal{E} = -L\,(dI/dt)$. The unit of mutual inductance is the henry.

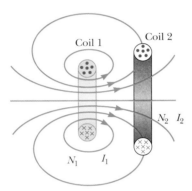

Figure 32.8 A cross-sectional view of two adjacent coils. A current in coil 1 sets up a magnetic field, and some of the magnetic field lines pass through coil 2.

Quick Quiz 32.4   In Figure 32.8, coil 1 is moved closer to coil 2, with the orientation of both coils remaining fixed. Because of this movement, the mutual induction of the two coils (a) increases, (b) decreases, or (c) is unaffected.

EXAMPLE 32.5   **"Wireless" Battery Charger**

An electric toothbrush has a base designed to hold the toothbrush handle when not in use. As shown in Figure 32.9a, the handle has a cylindrical hole that fits loosely over a matching cylinder on the base. When the handle is placed on the base, a changing current in a solenoid inside the base cylinder induces a current in a coil inside the handle. This induced current charges the battery in the handle.

Figure 32.9 (Example 32.5) (a) This electric toothbrush uses the mutual induction of solenoids as part of its battery-charging system. (b) A coil of N_H turns wrapped around the center of a solenoid of N_B turns.

(a)

(b)

We can model the base as a solenoid of length ℓ with N_B turns (Fig. 32.9b), carrying a current I, and having a cross-sectional area A. The handle coil contains N_H turns and completely surrounds the base coil. Find the mutual inductance of the system.

SOLUTION

Conceptualize   Be sure you can identify the two coils in the situation and understand that a changing current in one coil induces a current in the second coil.

Categorize   We will evaluate the result using concepts discussed in this section, so we categorize this example as a substitution problem.

Use Equation 30.17 to express the magnetic field in the interior of the base solenoid:

$$B = \mu_0 \frac{N_B}{\ell} I$$

Find the mutual inductance, noting that the magnetic flux Φ_{BH} through the handle's coil caused by the magnetic field of the base coil is BA:

$$M = \frac{N_H \Phi_{BH}}{I} = \frac{N_H BA}{I} = \mu_0 \frac{N_B N_H}{\ell} A$$

Wireless charging is used in a number of other "cordless" devices. One significant example is the inductive charging used by some manufacturers of electric cars that avoids direct metal-to-metal contact between the car and the charging apparatus.

32.5   Oscillations in an *LC* Circuit

When a capacitor is connected to an inductor as illustrated in Figure 32.10, the combination is an ***LC* circuit.** If the capacitor is initially charged and the switch is then closed, both the current in the circuit and the charge on the capacitor oscillate between maximum positive and negative values. If the resistance of the circuit is zero, no energy is transformed to internal energy. In the following analysis, the resistance in the circuit is neglected. We also assume an idealized situation in which energy is not radiated away from the circuit. This radiation mechanism is discussed in Chapter 34.

Assume the capacitor has an initial charge Q_{max} (the maximum charge) and the switch is open for $t < 0$ and then closed at $t = 0$. Let's investigate what happens from an energy viewpoint.

When the capacitor is fully charged, the energy U in the circuit is stored in the capacitor's electric field and is equal to $Q_{max}^2/2C$ (Eq. 26.11). At this time, the current in the circuit is zero; therefore, no energy is stored in the inductor. After the switch is closed, the rate at which charges leave or enter the capacitor plates (which is also the rate at which the charge on the capacitor changes) is equal to the current in the circuit. After the switch is closed and the capacitor begins to discharge, the energy stored in its electric field decreases. The capacitor's discharge represents a current in the circuit, and some energy is now stored in the magnetic field of the inductor. Therefore, energy is transferred from the electric field of the capacitor to the magnetic field of the inductor. When the capacitor is fully discharged, it stores no energy. At this time, the current reaches its maximum value and all the energy in the circuit is stored in the inductor. The current continues in the same direction, decreasing in magnitude, with the capacitor eventually becoming fully charged again but with the polarity of its plates now opposite the initial polarity. This process is followed by another discharge until the circuit returns to its original state of maximum charge Q_{max} and the plate polarity shown in Figure 32.10. The energy continues to oscillate between inductor and capacitor.

The oscillations of the *LC* circuit are an electromagnetic analog to the mechanical oscillations of the block–spring system studied in Chapter 15. Much of what

Figure 32.10   A simple *LC* circuit. The capacitor has an initial charge Q_{max}, and the switch is open for $t < 0$ and then closed at $t = 0$.

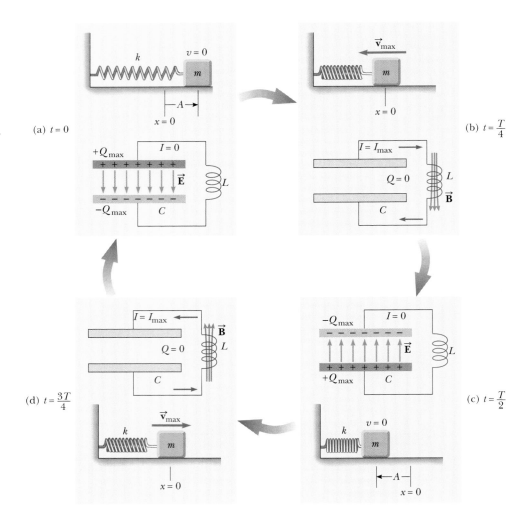

was discussed there is applicable to LC oscillations. For example, we investigated the effect of driving a mechanical oscillator with an external force, which leads to the phenomenon of *resonance*. The same phenomenon is observed in the LC circuit. (See Section 33.7.)

A representation of the energy transfer in an LC circuit is shown in Active Figure 32.11. As mentioned, the behavior of the circuit is analogous to that of the oscillating block–spring system studied in Chapter 15. The potential energy $\frac{1}{2}kx^2$ stored in a stretched spring is analogous to the potential energy $Q_{max}^2/2C$ stored in the capacitor. The kinetic energy $\frac{1}{2}mv^2$ of the moving block is analogous to the magnetic energy $\frac{1}{2}LI^2$ stored in the inductor, which requires the presence of moving charges. In Active Figure 32.11a, all the energy is stored as electric potential energy in the capacitor at $t = 0$ (because $I = 0$), just as all the energy in a block–spring system is initially stored as potential energy in the spring if it is stretched and released at $t = 0$. In Active Figure 32.11b, all the energy is stored as magnetic energy $\frac{1}{2}LI_{max}^2$ in the inductor, where I_{max} is the maximum current. Active Figures 32.11c and 32.11d show subsequent quarter-cycle situations in which the energy is all electric or all magnetic. At intermediate points, part of the energy is electric and part is magnetic.

Consider some arbitrary time t after the switch is closed so that the capacitor has a charge $Q < Q_{max}$ and the current is $I < I_{max}$. At this time, both circuit elements store energy, but the sum of the two energies must equal the total initial energy U stored in the fully charged capacitor at $t = 0$:

Total energy stored in ▶
an LC circuit

$$U = U_C + U_L = \frac{Q^2}{2C} + \frac{1}{2}LI^2 \qquad (32.18)$$

Because we have assumed the circuit resistance to be zero and we ignore electromagnetic radiation, no energy is transformed to internal energy and none is transferred out of the system of the circuit. Therefore, *the total energy of the system must remain constant in time.* We describe the constant energy of the system mathematically by setting $dU/dt = 0$. Therefore, by differentiating Equation 32.18 with respect to time while noting that Q and I vary with time gives

$$\frac{dU}{dt} = \frac{d}{dt}\left(\frac{Q^2}{2C} + \tfrac{1}{2}LI^2\right) = \frac{Q}{C}\frac{dQ}{dt} + LI\frac{dI}{dt} = 0 \tag{32.19}$$

We can reduce this result to a differential equation in one variable by remembering that the current in the circuit is equal to the rate at which the charge on the capacitor changes: $I = dQ/dt$. It then follows that $dI/dt = d^2Q/dt^2$. Substitution of these relationships into Equation 32.19 gives

$$\frac{Q}{C} + L\frac{d^2Q}{dt^2} = 0$$

$$\frac{d^2Q}{dt^2} = -\frac{1}{LC}Q \tag{32.20}$$

Let's solve for Q by noting that this expression is of the same form as the analogous Equations 15.3 and 15.5 for a block–spring system:

$$\frac{d^2x}{dt^2} = -\frac{k}{m}x = -\omega^2 x$$

where k is the spring constant, m is the mass of the block, and $\omega = \sqrt{k/m}$. The solution of this mechanical equation has the general form (Eq. 15.6):

$$x = A\cos(\omega t + \phi)$$

where A is the amplitude of the simple harmonic motion (the maximum value of x), ω is the angular frequency of this motion, and ϕ is the phase constant; the values of A and ϕ depend on the initial conditions. Because Equation 32.20 is of the same mathematical form as the differential equation of the simple harmonic oscillator, it has the solution

$$Q = Q_{max}\cos(\omega t + \phi) \tag{32.21}$$

◀ Charge as a function of time for an ideal *LC* circuit

where Q_{max} is the maximum charge of the capacitor and the angular frequency ω is

$$\omega = \frac{1}{\sqrt{LC}} \tag{32.22}$$

◀ Angular frequency of oscillation in an *LC* circuit

Note that the angular frequency of the oscillations depends solely on the inductance and capacitance of the circuit. Equation 32.22 gives the *natural frequency* of oscillation of the *LC* circuit.

Because Q varies sinusoidally with time, the current in the circuit also varies sinusoidally. We can show that by differentiating Equation 32.21 with respect to time:

$$I = \frac{dQ}{dt} = -\omega Q_{max}\sin(\omega t + \phi) \tag{32.23}$$

◀ Current as a function of time for an ideal *LC* current

To determine the value of the phase angle ϕ, let's examine the initial conditions, which in our situation require that at $t = 0$, $I = 0$, and $Q = Q_{max}$. Setting $I = 0$ at $t = 0$ in Equation 32.23 gives

$$0 = -\omega Q_{max}\sin\phi$$

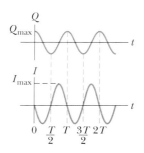

ACTIVE FIGURE 32.12

Graphs of charge versus time and current versus time for a resistanceless, nonradiating *LC* circuit. Notice that *Q* and *I* are 90° out of phase with each other.

Sign in at www.thomsonedu.com and go to ThomsonNOW to observe this graph develop for the *LC* circuit in Active Figure 32.11.

which shows that $\phi = 0$. This value for ϕ also is consistent with Equation 32.21 and the condition that $Q = Q_{max}$ at $t = 0$. Therefore, in our case, the expressions for Q and I are

$$Q = Q_{max} \cos \omega t \qquad (32.24)$$

$$I = -\omega Q_{max} \sin \omega t = -I_{max} \sin \omega t \qquad (32.25)$$

Graphs of Q versus t and I versus t are shown in Active Figure 32.12. The charge on the capacitor oscillates between the extreme values Q_{max} and $-Q_{max}$, and the current oscillates between I_{max} and $-I_{max}$. Furthermore, the current is 90° out of phase with the charge. That is, when the charge is a maximum, the current is zero, and when the charge is zero, the current has its maximum value.

Let's return to the energy discussion of the *LC* circuit. Substituting Equations 32.24 and 32.25 in Equation 32.18, we find that the total energy is

$$U = U_C + U_L = \frac{Q_{max}^2}{2C} \cos^2 \omega t + \tfrac{1}{2} L I_{max}^2 \sin^2 \omega t \qquad (32.26)$$

This expression contains all the features described qualitatively at the beginning of this section. It shows that the energy of the *LC* circuit continuously oscillates between energy stored in the capacitor's electric field and energy stored in the inductor's magnetic field. When the energy stored in the capacitor has its maximum value $Q_{max}^2/2C$, the energy stored in the inductor is zero. When the energy stored in the inductor has its maximum value $\tfrac{1}{2}L I_{max}^2$, the energy stored in the capacitor is zero.

Plots of the time variations of U_C and U_L are shown in Figure 32.13. The sum $U_C + U_L$ is a constant and is equal to the total energy $Q_{max}^2/2C$, or $\tfrac{1}{2}L I_{max}^2$. Analytical verification is straightforward. The amplitudes of the two graphs in Figure 32.13 must be equal because the maximum energy stored in the capacitor (when $I = 0$) must equal the maximum energy stored in the inductor (when $Q = 0$). This equality is expressed mathematically as

$$\frac{Q_{max}^2}{2C} = \frac{L I_{max}^2}{2}$$

Using this expression in Equation 32.26 for the total energy gives

$$U = \frac{Q_{max}^2}{2C} (\cos^2 \omega t + \sin^2 \omega t) = \frac{Q_{max}^2}{2C} \qquad (32.27)$$

because $\cos^2 \omega t + \sin^2 \omega t = 1$.

In our idealized situation, the oscillations in the circuit persist indefinitely; the total energy U of the circuit, however, remains constant only if energy transfers and transformations are neglected. In actual circuits, there is always some resistance and some energy is therefore transformed to internal energy. We mentioned at the beginning of this section that we are also ignoring radiation from the circuit. In reality, radiation is inevitable in this type of circuit, and the total energy in the circuit continuously decreases as a result of this process.

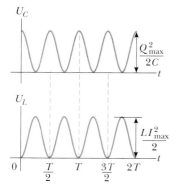

Figure 32.13 Plots of U_C versus t and U_L versus t for a resistanceless, nonradiating *LC* circuit. The sum of the two curves is a constant and is equal to the total energy stored in the circuit.

Quick Quiz 32.5 (i) At an instant of time during the oscillations of an *LC* circuit, the current is at its maximum value. At this instant, what happens to the voltage across the capacitor? (a) It is different from that across the inductor. (b) It is zero. (c) It has its maximum value. (d) It is impossible to determine. (ii) At an instant of time during the oscillations of an *LC* circuit, the current is momentarily zero. From the same choices, describe the voltage across the capacitor at this instant.

EXAMPLE 32.6   **Oscillations in an *LC* Circuit**

In Figure 32.14, the battery has an emf of 12.0 V, the inductance is 2.81 mH, and the capacitance is 9.00 pF. The switch has been set to position *a* for a long time so that the capacitor is charged. The switch is then thrown to position *b*, removing the battery from the circuit and connecting the capacitor directly across the inductor.

(A) Find the frequency of oscillation of the circuit.

SOLUTION

Conceptualize   When the switch is thrown to position *b*, the active part of the circuit is the right-hand loop, which is an *LC* circuit.

Categorize   We use equations developed in this section, so we categorize this example as a substitution problem.

Figure 32.14   (Example 32.6) First the capacitor is fully charged with the switch set to position *a*. Then, the switch is thrown to position *b* and the battery is no longer in the circuit.

Use Equation 32.22 to find the frequency:

$$f = \frac{\omega}{2\pi} = \frac{1}{2\pi\sqrt{LC}}$$

Substitute numerical values:

$$f = \frac{1}{2\pi[(2.81 \times 10^{-3}\ \text{H})(9.00 \times 10^{-12}\ \text{F})]^{1/2}} = \boxed{1.00 \times 10^{6}\ \text{Hz}}$$

(B) What are the maximum values of charge on the capacitor and current in the circuit?

SOLUTION

Find the initial charge on the capacitor, which equals the maximum charge:

$$Q_{\text{max}} = C\,\Delta V = (9.00 \times 10^{-12}\ \text{F})(12.0\ \text{V}) = \boxed{1.08 \times 10^{-10}\ \text{C}}$$

Use Equation 32.25 to find the maximum current from the maximum charge:

$$I_{\text{max}} = \omega Q_{\text{max}} = 2\pi f Q_{\text{max}} = (2\pi \times 10^{6}\ \text{s}^{-1})(1.08 \times 10^{-10}\ \text{C})$$
$$= \boxed{6.79 \times 10^{-4}\ \text{A}}$$

32.6   The *RLC* Circuit

Let's now turn our attention to a more realistic circuit consisting of a resistor, an inductor, and a capacitor connected in series as shown in Active Figure 32.15. We assume the resistance of the resistor represents all the resistance in the circuit. Suppose the switch is at position *a* so that the capacitor has an initial charge Q_{max}. The switch is now thrown to position *b*. After this instant, the total energy stored in the capacitor and inductor at any time is given by Equation 32.18. This total energy, however, is no longer constant as it was in the *LC* circuit because the resistor causes transformation to internal energy. (We continue to ignore electromagnetic radiation from the circuit in this discussion.) Because the rate of energy transformation to internal energy within a resistor is $I^2 R$,

$$\frac{dU}{dt} = -I^2 R$$

where the negative sign signifies that the energy *U* of the circuit is decreasing in time. Substituting this result into Equation 32.19 gives

$$LI\frac{dI}{dt} + \frac{Q}{C}\frac{dQ}{dt} = -I^2 R \qquad \textbf{(32.28)}$$

ACTIVE FIGURE 32.15

A series *RLC* circuit. The switch is set to position *a*, and the capacitor is charged. The switch is then thrown to position *b*.

Sign in at www.thomsonedu.com and go to ThomsonNOW to adjust the values of *R*, *L*, and *C* and see the effect on the decaying charge on the capacitor. A graphical display as in Active Figure 32.16a is available, as is an energy bar graph.

To convert this equation into a form that allows us to compare the electrical oscillations with their mechanical analog, we first use $I = dQ/dt$ and move all terms to the left-hand side to obtain

$$LI\frac{d^2Q}{dt^2} + I^2R + \frac{Q}{C}I = 0$$

Now divide through by I:

$$L\frac{d^2Q}{dt^2} + IR + \frac{Q}{C} = 0$$

$$L\frac{d^2Q}{dt^2} + R\frac{dQ}{dt} + \frac{Q}{C} = 0 \tag{32.29}$$

The *RLC* circuit is analogous to the damped harmonic oscillator discussed in Section 15.6 and illustrated in Figure 15.20. The equation of motion for a damped block–spring system is, from Equation 15.31,

$$m\frac{d^2x}{dt^2} + b\frac{dx}{dt} + kx = 0 \tag{32.30}$$

Comparing Equations 32.29 and 32.30, we see that Q corresponds to the position x of the block at any instant, L to the mass m of the block, R to the damping coefficient b, and C to $1/k$, where k is the force constant of the spring. These and other relationships are listed in Table 32.1.

TABLE 32.1

Analogies Between Electrical and Mechanical Systems

Electric Circuit		One-Dimensional Mechanical System
Charge	$Q \leftrightarrow x$	Position
Current	$I \leftrightarrow v_x$	Velocity
Potential difference	$\Delta V \leftrightarrow F_x$	Force
Resistance	$R \leftrightarrow b$	Viscous damping coefficient
Capacitance	$C \leftrightarrow 1/k$	(k = spring constant)
Inductance	$L \leftrightarrow m$	Mass
Current = time derivative of charge	$I = \dfrac{dQ}{dt} \leftrightarrow v_x = \dfrac{dx}{dt}$	Velocity = time derivative of position
Rate of change of current = second time derivative of charge	$\dfrac{dI}{dt} = \dfrac{d^2Q}{dt^2} \leftrightarrow a_x = \dfrac{dv_x}{dt} = \dfrac{d^2x}{dt^2}$	Acceleration = second time derivative of position
Energy in inductor	$U_L = \frac{1}{2}LI^2 \leftrightarrow K = \frac{1}{2}mv^2$	Kinetic energy of moving object
Energy in capacitor	$U_C = \frac{1}{2}\dfrac{Q^2}{C} \leftrightarrow U = \frac{1}{2}kx^2$	Potential energy stored in a spring
Rate of energy loss due to resistance	$I^2R \leftrightarrow bv^2$	Rate of energy loss due to friction
RLC circuit	$L\dfrac{d^2Q}{dt^2} + R\dfrac{dQ}{dt} + \dfrac{Q}{C} = 0 \leftrightarrow m\dfrac{d^2x}{dt^2} + b\dfrac{dx}{dt} + kx = 0$	Damped object on a spring

Because the analytical solution of Equation 32.29 is cumbersome, we give only a qualitative description of the circuit behavior. In the simplest case, when $R = 0$, Equation 32.29 reduces to that of a simple LC circuit as expected, and the charge and the current oscillate sinusoidally in time. This situation is equivalent to removing all damping in the mechanical oscillator.

When R is small, a situation that is analogous to light damping in the mechanical oscillator, the solution of Equation 32.29 is

$$Q = Q_{max}e^{-Rt/2L}\cos\omega_d t \qquad (32.31)$$

where ω_d, the angular frequency at which the circuit oscillates, is given by

$$\omega_d = \left[\frac{1}{LC} - \left(\frac{R}{2L}\right)^2\right]^{1/2} \qquad (32.32)$$

That is, the value of the charge on the capacitor undergoes a damped harmonic oscillation in analogy with a block–spring system moving in a viscous medium. Equation 32.32 shows that when $R \ll \sqrt{4L/C}$ (so that the second term in the brackets is much smaller than the first), the frequency ω_d of the damped oscillator is close to that of the undamped oscillator, $1/\sqrt{LC}$. Because $I = dQ/dt$, it follows that the current also undergoes damped harmonic oscillation. A plot of the charge versus time for the damped oscillator is shown in Active Figure 32.16a and an oscilloscope trace for a real RLC circuit is shown in Active Figure 32.16b. The maximum value of Q decreases after each oscillation, just as the amplitude of a damped block–spring system decreases in time.

For larger values of R, the oscillations damp out more rapidly; in fact, there exists a critical resistance value $R_c = \sqrt{4L/C}$ above which no oscillations occur. A system with $R = R_c$ is said to be *critically damped*. When R exceeds R_c, the system is said to be *overdamped*.

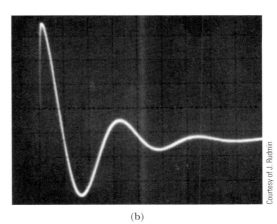

(a)   (b)

ACTIVE FIGURE 32.16

(a) Charge versus time for a damped *RLC* circuit. The charge decays in this way when $R < \sqrt{4L/C}$. The Q-versus-t curve represents a plot of Equation 32.31. (b) Oscilloscope pattern showing the decay in the oscillations of an *RLC* circuit.

Sign in at www.thomsonedu.com and go to ThomsonNOW to observe this graph develop for the damped *RLC* circuit in Active Figure 32.15.

Summary

CONCEPTS AND PRINCIPLES

When the current in a loop of wire changes with time, an emf is induced in the loop according to Faraday's law. The **self-induced emf** is

$$\mathcal{E}_L = -L\frac{dI}{dt} \qquad (32.1)$$

where L is the **inductance** of the loop. Inductance is a measure of how much opposition a loop offers to a change in the current in the loop. Inductance has the SI unit of **henry** (H), where $1\ \text{H} = 1\ \text{V}\cdot\text{s/A}$.

The inductance of any coil is

$$L = \frac{N\Phi_B}{I} \qquad (32.2)$$

where N is the total number of turns and Φ_B is the magnetic flux through the coil. The inductance of a device depends on its geometry. For example, the inductance of an air-core solenoid is

$$L = \mu_0 \frac{N^2}{\ell} A \qquad (32.4)$$

where ℓ is the length of the solenoid and A is the cross-sectional area.

If a resistor and inductor are connected in series to a battery of emf $\mathcal{E}$ at time $t = 0$, the current in the circuit varies in time according to the expression

$$I = \frac{\mathcal{E}}{R}\left(1 - e^{-t/\tau}\right) \qquad (32.7)$$

where $\tau = L/R$ is the **time constant** of the RL circuit. If we replace the battery in the circuit by a resistanceless wire, the current decays exponentially with time according to the expression

$$I = \frac{\mathcal{E}}{R} e^{-t/\tau} \qquad (32.10)$$

where $\mathcal{E}/R$ is the initial current in the circuit.

The energy stored in the magnetic field of an inductor carrying a current I is

$$U = \tfrac{1}{2}LI^2 \qquad (32.12)$$

This energy is the magnetic counterpart to the energy stored in the electric field of a charged capacitor.

The energy density at a point where the magnetic field is B is

$$u_B = \frac{B^2}{2\mu_0} \qquad (32.14)$$

The **mutual inductance** of a system of two coils is

$$M_{12} = \frac{N_2\Phi_{12}}{I_1} = M_{21} = \frac{N_1\Phi_{21}}{I_2} = M \qquad (32.15)$$

This mutual inductance allows us to relate the induced emf in a coil to the changing source current in a nearby coil using the relationships

$$\mathcal{E}_2 = -M_{12}\frac{dI_1}{dt} \quad \text{and} \quad \mathcal{E}_1 = -M_{21}\frac{dI_2}{dt} \qquad (32.16, 32.17)$$

In an LC circuit that has zero resistance and does not radiate electromagnetically (an idealization), the values of the charge on the capacitor and the current in the circuit vary sinusoidally in time at an angular frequency given by

$$\omega = \frac{1}{\sqrt{LC}} \qquad (32.22)$$

The energy in an LC circuit continuously transfers between energy stored in the capacitor and energy stored in the inductor.

In an RLC circuit with small resistance, the charge on the capacitor varies with time according to

$$Q = Q_{\max} e^{-Rt/2L} \cos \omega_d t \qquad (32.31)$$

where

$$\omega_d = \left[\frac{1}{LC} - \left(\frac{R}{2L}\right)^2\right]^{1/2} \qquad (32.32)$$

Questions

□ denotes answer available in *Student Solutions Manual/Study Guide;* **O** denotes objective question

1. The current in a circuit containing a coil, a resistor, and a battery has reached a constant value. Does the coil have an inductance? Does the coil affect the value of the current?

2. What parameters affect the inductance of a coil? Does the inductance of a coil depend on the current in the coil?

3. O Initially, an inductor with no resistance carries a constant current. Then the current is brought to a new constant value twice as large. *After* this change, what has happened to the emf in the inductor? (a) It is larger than before the change by a factor of 4. (b) It is larger by a factor of 2. (c) It has the same nonzero value. (d) It continues to be zero. (e) It has decreased.

4. O A long, fine wire is wound into a coil with inductance 5 mH. The coil is connected across the terminals of a battery, and the current is measured a few seconds after the connection is made. The wire is unwound and wound again into a different coil with $L = 10$ mH. This second coil is connected across the same battery, and the current is measured in the same way. Compared with the current in the first coil, is the current in the second coil (a) four times as large, (b) twice as large, (c) unchanged, (d) half as large, or (e) one-fourth as large?

5. O Two solenoidal coils, A and B, are wound using equal lengths of the same kind of wire. The length of the axis of each coil is large compared with its diameter. The axial length of coil A is twice as large as that of coil B, and coil A has twice as many turns as coil B. What is the ratio of the inductance of coil A to that of coil B? (a) 8 (b) 4 (c) 2 (d) 1 (e) $\frac{1}{2}$ (f) $\frac{1}{4}$ (g) $\frac{1}{8}$

6. A switch controls the current in a circuit that has a large inductance. Is a spark (Fig. Q32.6) more likely to be produced at the switch when the switch is being closed, when it is being opened, or doesn't it matter? The electric arc can melt and oxidize the contact surfaces, resulting in high resistivity of the contacts and eventual destruction of the switch. Before electronic ignitions were invented, distributor contact points in automobiles had to be replaced regularly. Switches in power distribution networks and switches controlling large motors, generators, and electromagnets can suffer from arcing and can be very dangerous to operate.

Figure Q32.6

7. O In Figure Q32.7, the switch is left in position *a* for a long time interval and is then quickly thrown to position *b*. Rank the magnitudes of the voltages across the four cir-

cuit elements a short time thereafter from the largest to the smallest.

Figure Q32.7

8. Consider the four circuits shown in Figure Q32.8, each consisting of a battery, a switch, a lightbulb, a resistor, and either a capacitor or an inductor. Assume the capacitor has a large capacitance and the inductor has a large inductance but no resistance. The lightbulb has high efficiency, glowing whenever it carries electric current. **(i)** Describe what the lightbulb does in each of circuits (a), (b), (c), and (d) after the switch is thrown closed. **(ii)** Describe what the lightbulb does in each circuit after, having been closed for a long time interval, the switch is thrown open.

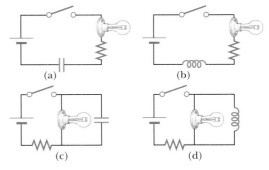

Figure Q32.8

9. O *Don't do this; it's dangerous and illegal.* Suppose a criminal wants to steal energy from the electric company by placing a flat, rectangular coil of wire close to, but not touching, one long, straight, horizontal wire in a transmission line. The long, straight wire carries a sinusoidally varying current. Which of the following statements is true? (a) The method works best if the coil is in a vertical plane surrounding the straight wire. (b) The method works best if the coil is in a vertical plane with the two long sides of the rectangle parallel to the long wire and equally far from it. (c) The method works best if the coil and the long wire are in the same horizontal plane with one long side of the rectangle close to the wire. (d) The method works for any orientation of the coil. (e) The method cannot work without contact between the coil and the long wire.

10. Consider this thesis: "Joseph Henry, America's first professional physicist, caused the most recent basic change in the human view of the Universe when he discovered self-induction during a school vacation at the Albany Academy about 1830. Before that time, one could think of the Universe as composed of only one thing: matter. The energy that temporarily maintains the current after a battery is removed from a coil, on the other hand, is not energy that belongs to any chunk of matter. It is energy in the massless magnetic field surrounding the coil. With Henry's discovery, Nature forced us to admit that the Universe consists of fields as well as matter." Argue for or against the statement. In your view, what makes up the Universe?

11. **O** If the current in an inductor is doubled, by what factor is the stored energy multiplied? (a) 4   (b) 2   (c) 1   (d) $\frac{1}{2}$ (e) $\frac{1}{4}$

12. **O** A solenoidal inductor for a printed circuit board is being redesigned. To save weight, the number of turns is reduced by one-half with the geometric dimensions kept the same. By how much must the current change if the energy stored in the inductor is to remain the same? (a) It must be four times larger (b) It must be two times larger (c) It must be larger by a factor of $\sqrt{2}$. (d) It should be left the same. (e) It should be one-half as large. (f) No change in the current can compensate for the reduction in the number of turns.

13. Discuss the similarities between the energy stored in the electric field of a charged capacitor and the energy stored in the magnetic field of a current-carrying coil.

14. The open switch in Figure Q32.14 is thrown closed at $t =$ 0. Before the switch is closed, the capacitor is uncharged and all currents are zero. Determine the currents in L, C, and R and the potential differences across L, C, and R (a) at the instant after the switch is closed and (b) long after it is closed.

Figure Q32.14

15. **O** The centers of two circular loops are separated by a fixed distance. **(i)** For what relative orientation of the loops is their mutual inductance a maximum? (a) coaxial and lying in parallel planes   (b) lying in the same plane (c) lying in perpendicular planes, with the center of one on the axis of the other   (d) The orientation makes no difference.   **(ii)** For what relative orientation is their mutual inductance a minimum? Choose from the same possibilities.

16. In the LC circuit shown in Figure 32.10, the charge on the capacitor is sometimes zero, but at such instants the current in the circuit is not zero. How is this behavior possible?

17. How can you tell whether an RLC circuit is overdamped or underdamped?

18. Can an object exert a force on itself? When a coil induces an emf in itself, does it exert a force on itself?

Problems

WebAssign The Problems from this chapter may be assigned online in WebAssign.

ThomsonNOW Sign in at **www.thomsonedu.com** and go to ThomsonNOW to assess your understanding of this chapter's topics with additional quizzing and conceptual questions.

1, 2, 3 denotes straightforward, intermediate, challenging;   ☐ denotes full solution available in *Student Solutions Manual/Study Guide;*   ▲ denotes coached solution with hints available at **www.thomsonedu.com;**   denotes developing symbolic reasoning;   ● denotes asking for qualitative reasoning;   ▪ denotes computer useful in solving problem

Section 32.1 Self-Induction and Inductance

1. A 2.00-H inductor carries a steady current of 0.500 A. When the switch in the circuit is opened, the current is effectively zero after 10.0 ms. What is the average induced emf in the inductor during this time interval?

2. A coiled telephone cord forms a spiral having 70 turns, a diameter of 1.30 cm and an unstretched length of 60.0 cm. Determine the inductance of one conductor in the unstretched cord.

3. ▲ A 10.0-mH inductor carries a current $I = I_{max} \sin \omega t$, with $I_{max} = 5.00$ A and $\omega/2\pi = 60.0$ Hz. What is the self-induced emf as a function of time?

4. An emf of 24.0 mV is induced in a 500-turn coil at an instant when the current is 4.00 A and is changing at the rate of 10.0 A/s. What is the magnetic flux through each turn of the coil?

5. An inductor in the form of a solenoid contains 420 turns, is 16.0 cm in length, and has a cross-sectional area of 3.00 cm². What uniform rate of decrease of current through the inductor induces an emf of 175 μV?

6. The current in a 90.0-mH inductor changes with time as $I = 1.00t^2 - 6.00t$ (in SI units). Find the magnitude of the induced emf at (a) $t = 1.00$ s and (b) $t = 4.00$ s. (c) At what time is the emf zero?

7. ● A 40.0-mA current is carried by a uniformly wound air-core solenoid with 450 turns, a 15.0-mm diameter, and 12.0-cm length. Compute (a) the magnetic field inside the solenoid, (b) the magnetic flux through each turn, and (c) the inductance of the solenoid. (d) **What If?** If the current were different, which of these quantities would change?

8. A toroid has a major radius R and a minor radius r and is tightly wound with N turns of wire as shown in Figure P32.8. If $R \gg r$, the magnetic field in the region enclosed by the wire of the torus, of cross-sectional area $A = \pi r^2$, is essentially the same as the magnetic field of a solenoid that has been bent into a large circle of radius R. Modeling the field as the uniform field of a long solenoid, show that the inductance of such a toroid is approximately

$$L \approx \frac{\mu_0 N^2 A}{2\pi R}$$

(An exact expression of the inductance of a toroid with a rectangular cross section is derived in Problem 57.)

Figure P32.8

9. A self-induced emf in a solenoid of inductance L changes in time as $\mathcal{E} = \mathcal{E}_0 e^{-kt}$. Find the total charge that passes through the solenoid, assuming the charge is finite.

Section 32.2 *RL* Circuits

10. Show that $I = I_i e^{-t/\tau}$ is a solution of the differential equation

$$IR + L\frac{dI}{dt} = 0$$

where I_i is the current at $t = 0$ and $\tau = L/R$.

11. A 12.0-V battery is connected into a series circuit containing a 10.0-Ω resistor and a 2.00-H inductor. In what time interval will the current reach (a) 50.0% and (b) 90.0% of its final value?

12. ● In the circuit diagrammed in Figure P32.12, take $\mathcal{E} = 12.0$ V and $R = 24.0~\Omega$. Assume the switch is open for $t < 0$ and is closed at $t = 0$. On a single set of axes, sketch graphs of the current in the circuit as a function of time for $t \geq 0$, assuming (a) the inductance in the circuit is essentially zero, (b) the inductance has an intermediate value, and (c) the inductance has a very large value. Label the initial and final values of the current.

Figure P32.12   Problems 12, 13, 14, and 15.

13. Consider the circuit in Figure P32.12, taking $\mathcal{E} = 6.00$ V, $L = 8.00$ mH, and $R = 4.00~\Omega$. (a) What is the inductive time constant of the circuit? (b) Calculate the current in the circuit 250 μs after the switch is closed. (c) What is the value of the final steady-state current? (d) After what time interval does the current reach 80.0% of its maximum value?

14. In the circuit shown in Figure P32.12, let $L = 7.00$ H, $R = 9.00~\Omega$, and $\mathcal{E} = 120$ V. What is the self-induced emf 0.200 s after the switch is closed?

15. ▲ For the *RL* circuit shown in Figure P32.12, let the inductance be 3.00 H, the resistance 8.00 Ω, and the battery emf 36.0 V. (a) Calculate the ratio of the potential difference across the resistor to the emf across the inductor when the current is 2.00 A. (b) Calculate the emf across the inductor when the current is 4.50 A.

16. A 12.0-V battery is connected in series with a resistor and an inductor. The circuit has a time constant of 500 μs, and the maximum current is 200 mA. What is the value of the inductance of the inductor?

17. An inductor that has an inductance of 15.0 H and a resistance of 30.0 Ω is connected across a 100-V battery. What is the rate of increase of the current (a) at $t = 0$ and (b) at $t = 1.50$ s?

18. The switch in Figure P32.18 is open for $t < 0$ and is then thrown closed at time $t = 0$. Find the current in the inductor and the current in the switch as functions of time thereafter.

Figure P32.18   Problems 18 and 52.

19. A series *RL* circuit with $L = 3.00$ H and a series *RC* circuit with $C = 3.00~\mu$F have equal time constants. If the two circuits contain the same resistance R, (a) what is the value of R and (b) what is the time constant?

20. A current pulse is fed to the partial circuit shown in Figure P32.20. The current begins at zero, becomes 10.0 A between $t = 0$ and $t = 200~\mu$s, and then is zero once again. Determine the current in the inductor as a function of time.

Figure P32.20

21. ▲ A 140-mH inductor and a 4.90-Ω resistor are connected with a switch to a 6.00-V battery as shown in Figure P32.21. (a) After the switch is thrown to a (connecting the battery), what time interval elapses before the current reaches 220 mA? (b) What is the current in the inductor 10.0 s after the switch is closed? (c) Now the switch is

2 = intermediate;   3 = challenging;   ☐ = SSM/SG;   ▲ = ThomsonNOW;   = symbolic reasoning;   ● = qualitative reasoning

quickly thrown from *a* to *b*. What time interval elapses before the current falls to 160 mA?

Figure P32.21

22. ● Two ideal inductors, L_1 and L_2, have *zero* internal resistance and are far apart, so their magnetic fields do not influence each other. (a) Assuming these inductors are connected in series, show that they are equivalent to a single ideal inductor having $L_{eq} = L_1 + L_2$. (b) Assuming these same two inductors are connected in parallel, show that they are equivalent to a single ideal inductor having $1/L_{eq} = 1/L_1 + 1/L_2$. (c) **What If?** Now consider two inductors L_1 and L_2 that have *nonzero* internal resistances R_1 and R_2, respectively. Assume they are still far apart so that their mutual inductance is zero. Assuming these inductors are connected in series, show that they are equivalent to a single inductor having $L_{eq} = L_1 + L_2$ and $R_{eq} = R_1 + R_2$. (d) If these same inductors are now connected in parallel, is it necessarily true that they are equivalent to a single ideal inductor having $1/L_{eq} = 1/L_1 + 1/L_2$ and $1/R_{eq} = 1/R_1 + 1/R_2$? Explain your answer.

Section 32.3 Energy in a Magnetic Field

23. An air-core solenoid with 68 turns is 8.00 cm long and has a diameter of 1.20 cm. How much energy is stored in its magnetic field when it carries a current of 0.770 A?

24. The magnetic field inside a superconducting solenoid is 4.50 T. The solenoid has an inner diameter of 6.20 cm and a length of 26.0 cm. Determine (a) the magnetic energy density in the field and (b) the energy stored in the magnetic field within the solenoid.

25. ▲ On a clear day at a certain location, a 100-V/m vertical electric field exists near the Earth's surface. At the same place, the Earth's magnetic field has a magnitude of 0.500×10^{-4} T. Compute the energy densities of the two fields.

26. Complete the calculation in Example 32.3 by proving that

$$\int_0^\infty e^{-2Rt/L}\, dt = \frac{L}{2R}$$

27. ● A flat coil of wire has an inductance of 40.0 mH and a resistance of 5.00 Ω. It is connected to a 22.0-V battery at the instant $t = 0$. Consider the moment when the current is 3.00 A. (a) At what rate is energy being delivered by the battery? (b) What is the power being delivered to the resistor? (c) At what rate is energy being stored in the magnetic field of the coil? (d) What is the relationship among these three power values? Is this relationship true at other instants as well? Explain the relationship at the moment immediately after $t = 0$ and at a moment several seconds later.

28. A 10.0-V battery, a 5.00-Ω resistor, and a 10.0-H inductor are connected in series. After the current in the circuit has reached its maximum value, calculate (a) the power being supplied by the battery, (b) the power being delivered to the resistor, (c) the power being delivered to the inductor, and (d) the energy stored in the magnetic field of the inductor.

29. Assume the magnitude of the magnetic field outside a sphere of radius R is $B = B_0(R/r)^2$, where B_0 is a constant. Determine the total energy stored in the magnetic field outside the sphere and evaluate your result for $B_0 = 5.00 \times 10^{-5}$ T and $R = 6.00 \times 10^6$ m, values appropriate for the Earth's magnetic field.

Section 32.4 Mutual Inductance

30. Two coils are close to each other. The first coil carries a current given by $I(t) = (5.00\ \text{A})e^{-0.025\,0t}\sin(377t)$. At $t = 0.800$ s, the emf measured across the second coil is -3.20 V. What is the mutual inductance of the coils?

31. Two coils, held in fixed positions, have a mutual inductance of 100 μH. What is the peak emf in one coil when a sinusoidal current given by $I(t) = (10.0\ \text{A})\sin(1\,000t)$ is in the other coil?

32. On a printed circuit board, a relatively long, straight conductor and a conducting rectangular loop lie in the same plane as shown in Figure P31.8 in Chapter 31. Taking $h = 0.400$ mm, $w = 1.30$ mm, and $L = 2.70$ mm, find their mutual inductance.

33. Two solenoids A and B, spaced close to each other and sharing the same cylindrical axis, have 400 and 700 turns, respectively. A current of 3.50 A in coil A produces an average flux of 300 μWb through each turn of A and a flux of 90.0 μWb through each turn of B. (a) Calculate the mutual inductance of the two solenoids. (b) What is the inductance of A? (c) What emf is induced in B when the current in A increases at the rate of 0.500 A/s?

34. ● A solenoid has N_1 turns, radius R_1, and length ℓ. It is so long that its magnetic field is uniform nearly everywhere inside it and is nearly zero outside. A second solenoid has N_2 turns, radius $R_2 < R_1$, and the same length. It lies inside the first solenoid, with their axes parallel. (a) Assume solenoid 1 carries variable current I. Compute the mutual inductance characterizing the emf induced in solenoid 2. (b) Now assume solenoid 2 carries current I. Compute the mutual inductance to which the emf in solenoid 1 is proportional. (c) State how the results of parts (a) and (b) compare with each other.

35. A large coil of radius R_1 and having N_1 turns is coaxial with a small coil of radius R_2 and having N_2 turns. The centers of the coils are separated by a distance x that is much larger than R_2. What is the mutual inductance of the coils? *Suggestion:* John von Neumann proved that the same answer must result from considering the flux through the first coil of the magnetic field produced by the second coil or from considering the flux through the second coil of the magnetic field produced by the first coil. In this problem, it is easy to calculate the flux through the small coil, but it is difficult to calculate the flux through the large coil because to do so, you would have to know the magnetic field away from the axis.

36. Two inductors having inductances L_1 and L_2 are connected in parallel as shown in Figure P32.36a. The mutual inductance between the two inductors is M. Deter-

mine the equivalent inductance L_{eq} for the system (Fig. P32.36b).

(a)   (b)

Figure P32.36

Section 32.5 Oscillations in an *LC* Circuit

37. A 1.00-μF capacitor is charged by a 40.0-V power supply. The fully charged capacitor is then discharged through a 10.0-mH inductor. Find the maximum current in the resulting oscillations.

38. An *LC* circuit consists of a 20.0-mH inductor and a 0.500-μF capacitor. If the maximum instantaneous current is 0.100 A, what is the greatest potential difference across the capacitor?

39. In the circuit of Figure P32.39, the battery emf is 50.0 V, the resistance is 250 Ω, and the capacitance is 0.500 μF. The switch S is closed for a long time interval, and zero potential difference is measured across the capacitor. After the switch is opened, the potential difference across the capacitor reaches a maximum value of 150 V. What is the value of the inductance?

Figure P32.39

40. An *LC* circuit like the one in Figure 32.10 contains an 82.0-mH inductor and a 17.0-μF capacitor that initially carries a 180-μC charge. The switch is open for $t < 0$ and then thrown closed at $t = 0$. (a) Find the frequency (in hertz) of the resulting oscillations. At $t = 1.00$ ms, find (b) the charge on the capacitor and (c) the current in the circuit.

41. A fixed inductance $L = 1.05 \mu$H is used in series with a variable capacitor in the tuning section of a radiotelephone on a ship. What capacitance tunes the circuit to the signal from a transmitter broadcasting at 6.30 MHz?

42. The switch in Figure P32.42 is connected to point a for a long time interval. After the switch is thrown to point b, what are (a) the frequency of oscillation of the *LC* circuit, (b) the maximum charge that appears on the capacitor,

Figure P32.42

(c) the maximum current in the inductor, and (d) the total energy the circuit possesses at $t = 3.00$ s?

43. ▲ An *LC* circuit like that in Figure 32.10 consists of a 3.30-H inductor and an 840-pF capacitor that initially carries a 105-μC charge. The switch is open for $t < 0$ and then thrown closed at $t = 0$. Compute the following quantities at $t = 2.00$ ms: (a) the energy stored in the capacitor, (b) the energy stored in the inductor, and (c) the total energy in the circuit.

Section 32.6 The *RLC* Circuit

44. In Active Figure 32.15, let $R = 7.60$ Ω, $L = 2.20$ mH, and $C = 1.80$ μF. (a) Calculate the frequency of the damped oscillation of the circuit. (b) What is the critical resistance?

45. Consider an *LC* circuit in which $L = 500$ mH and $C = 0.100 \mu$F. (a) What is the resonance frequency ω_0? (b) If a resistance of 1.00 kΩ is introduced into this circuit, what is the frequency of the (damped) oscillations? (c) What is the percent difference between the two frequencies?

46. Show that Equation 32.28 in the text is Kirchhoff's loop rule as applied to the circuit in Active Figure 32.15.

47. Electrical oscillations are initiated in a series circuit containing a capacitance C, inductance L, and resistance R. (a) If $R \ll \sqrt{4L/C}$ (weak damping), what time interval elapses before the amplitude of the current oscillation falls to 50.0% of its initial value? (b) Over what time interval does the energy decrease to 50.0% of its initial value?

Additional Problems

48. **Review problem.** This problem extends the reasoning of Section 26.4, Problem 29 in Chapter 26, Problem 33 in Chapter 30, and Section 32.3. (a) Consider a capacitor with vacuum between its large, closely spaced, oppositely charged parallel plates. Show that the force on one plate can be accounted for by thinking of the electric field between the plates as exerting a "negative pressure" equal to the energy density of the electric field. (b) Consider two infinite plane sheets carrying electric currents in opposite directions with equal linear current densities J_s. Calculate the force per area acting on one sheet due to the magnetic field, of magnitude $\mu_0 J_s/2$, created by the other sheet. (c) Calculate the net magnetic field between the sheets and the field outside of the volume between them. (d) Calculate the energy density in the magnetic field between the sheets. (e) Show that the force on one sheet can be accounted for by thinking of the magnetic field between the sheets as exerting a positive pressure equal to its energy density. This result for magnetic pressure applies to all current configurations, not only to sheets of current.

49. A 1.00-mH inductor and a 1.00-μF capacitor are connected in series. The current in the circuit is described by $I = 20.0t$, where t is in seconds and I is in amperes. The capacitor initially has no charge. Determine (a) the voltage across the inductor as a function of time, (b) the voltage across the capacitor as a function of time, and (c) the time when the energy stored in the capacitor first exceeds that in the inductor.

50. An inductor having inductance L and a capacitor having capacitance C are connected in series. The current in the circuit increases linearly in time as described by $I = Kt$,

2 = intermediate;   3 = challenging;   ☐ = SSM/SG;   ▲ = ThomsonNOW;   = symbolic reasoning;   ● = qualitative reasoning

where K is a constant. The capacitor is initially uncharged. Determine (a) the voltage across the inductor as a function of time, (b) the voltage across the capacitor as a function of time, and (c) the time when the energy stored in the capacitor first exceeds that in the inductor.

51. A capacitor in a series LC circuit has an initial charge Q and is being discharged. Find, in terms of L and C, the flux through each of the N turns in the coil when the charge on the capacitor is $Q/2$.

52. ● In the circuit diagrammed in Figure P32.18, assume that the switch has been closed for a long time interval and is opened at $t = 0$. (a) Before the switch is opened, does the inductor behave as an open circuit, a short circuit, a resistor of some particular resistance, or none of these choices? What current does the inductor carry? (b) How much energy is stored in the inductor for $t < 0$? (c) After the switch is opened, what happens to the energy previously stored in the inductor? (d) Sketch a graph of the current in the inductor for $t \geq 0$. Label the initial and final values and the time constant.

53. ● At the moment $t = 0$, a 24.0-V battery is connected to a 5.00-mH coil and a 6.00-Ω resistor. (a) Immediately thereafter, how does the potential difference across the resistor compare to the emf across the coil? (b) Answer the same question about the circuit several seconds later. (c) Is there an instant at which these two voltages are equal in magnitude? If so, when? Is there more than one such instant? (d) After a 4.00-A current is established in the resistor and coil, the battery is suddenly replaced by a short circuit. Answer questions (a), (b), and (c) again with reference to this new circuit.

54. When the current in the portion of the circuit shown in Figure P32.54 is 2.00 A and increases at a rate of 0.500 A/s, the measured potential difference is $\Delta V_{ab} = 9.00$ V. When the current is 2.00 A and decreases at the rate of 0.500 A/s, the measured potential difference is $\Delta V_{ab} = 5.00$ V. Calculate the values of L and R.

Figure P32.54

55. A time-varying current I is sent through a 50.0-mH inductor as shown in Figure P32.55. Make a graph of the potential at point b relative to the potential at point a.

Figure P32.55

56. ● Consider a series circuit consisting of a 500-μF capacitor, a 32.0-mH inductor, and a resistor R. Explain what you can say about the angular frequency of oscillations for (a) $R = 0$, (b) $R = 4.00$ Ω, (c) $R = 15.0$ Ω, and (d) $R = 17.0$ Ω. Relate the mathematical description of the angu-

lar frequency to the experimentally measurable angular frequency.

57. ● The toroid in Figure P32.57 consists of N turns and has a rectangular cross section. Its inner and outer radii are a and b, respectively. (a) Show that the inductance of the toroid is

$$L = \frac{\mu_0 N^2 h}{2\pi} \ln \frac{b}{a}$$

(b) Using this result, compute the inductance of a 500-turn toroid for which $a = 10.0$ cm, $b = 12.0$ cm, and $h = 1.00$ cm. (c) **What If?** In Problem 8, an approximate equation for the inductance of a toroid with $R \gg r$ was derived. To get a feel for the accuracy of that result, use the expression in Problem 8 to compute the approximate inductance of the toroid described in part (b). How does that result compare with the answer to part (b)?

Figure P32.57

58. (a) A flat, circular coil does not actually produce a uniform magnetic field in the area it encloses. Nevertheless, estimate the inductance of a flat, compact, circular coil, with radius R and N turns, by assuming the field at its center is uniform over its area. (b) A circuit on a laboratory table consists of a 1.5-volt battery, a 270-Ω resistor, a switch, and three 30-cm-long patch cords connecting them. Suppose the circuit is arranged to be circular. Think of it as a flat coil with one turn. Compute the order of magnitude of its inductance and (c) of the time constant describing how fast the current increases when you close the switch.

59. At $t = 0$, the open switch in Figure P32.59 is thrown closed. Using Kirchhoff's rules for the instantaneous currents and voltages in this two-loop circuit, show that the current in the inductor at time $t > 0$ is

$$I(t) = \frac{\mathcal{E}}{R_1}\left[1 - e^{-(R'/L)t}\right]$$

where $R' = R_1 R_2/(R_1 + R_2)$.

Figure P32.59

60. A wire of nonmagnetic material, with radius R, carries current uniformly distributed over its cross section. The total current carried by the wire is I. Show that the magnetic energy per unit length inside the wire is $\mu_0 I^2/16\pi$.

61. In Figure P32.61, the switch is closed for $t < 0$ and steady-state conditions are established. The switch is opened at $t = 0$. (a) Find the initial emf $\mathcal{E}_0$ across L immediately after $t = 0$. Which end of the coil, a or b, is at the higher voltage? (b) Make freehand graphs of the currents in R_1 and in R_2 as a function of time, treating the steady-state directions as positive. Show values before and after $t = 0$. (c) At what moment after $t = 0$ does the current in R_2 have the value 2.00 mA?

Figure P32.61

62. ● The lead-in wires from a television antenna are often constructed in the form of two parallel wires (Fig. P32.62). The two wires carry currents of equal magnitude in opposite directions. Assume the wires carry the current uniformly distributed over their surfaces and no magnetic field exists inside the wires. (a) Why does this configuration of conductors have an inductance? (b) What constitutes the flux loop for this configuration? (c) Show that the inductance of a length x of this type of lead-in is

$$L = \frac{\mu_0 x}{\pi} \ln\left(\frac{w - a}{a}\right)$$

where w is the center-to-center separation of the wires and a is their radius.

Figure P32.62

63. To prevent damage from arcing in an electric motor, a discharge resistor is sometimes placed in parallel with the armature. If the motor is suddenly unplugged while running, this resistor limits the voltage that appears across the armature coils. Consider a 12.0-V DC motor with an armature that has a resistance of 7.50 Ω and an inductance of 450 mH. Assume the magnitude of the self-induced emf in the armature coils is 10.0 V when the motor is running at normal speed. (The equivalent circuit for the armature is shown in Fig. P32.63.) Calculate the maximum resistance R that limits the voltage across the armature to 80.0 V when the motor is unplugged.

Figure P32.63

Review problems. Problems 64 through 67 apply ideas from this and earlier chapters to some properties of superconductors, which were introduced in Section 27.5.

64. *The resistance of a superconductor.* In an experiment carried out by S. C. Collins between 1955 and 1958, a current was maintained in a superconducting lead ring for 2.50 yr with no observed loss. If the inductance of the ring were 3.14×10^{-8} H and the sensitivity of the experiment were 1 part in 10^9, what was the maximum resistance of the ring? *Suggestion:* Treat the ring as an RL circuit carrying decaying current and recall that $e^{-x} \approx 1 - x$ for small x.

65. A novel method of storing energy has been proposed. A huge underground superconducting coil, 1.00 km in diameter, would be fabricated. It would carry a maximum current of 50.0 kA through each winding of a 150-turn Nb_3Sn solenoid. (a) If the inductance of this huge coil were 50.0 H, what would be the total energy stored? (b) What would be the compressive force per meter length acting between two adjacent windings 0.250 m apart?

66. *Superconducting power transmission.* The use of superconductors has been proposed for power transmission lines. A single coaxial cable (Fig. P32.66) could carry 1.00×10^3 MW (the output of a large power plant) at 200 kV, DC, over a distance of 1 000 km without loss. An inner wire of radius 2.00 cm, made from the superconductor Nb_3Sn, carries the current I in one direction. A surrounding superconducting cylinder of radius 5.00 cm would carry the return current I. In such a system, what is the magnetic field (a) at the surface of the inner conductor and (b) at the inner surface of the outer conductor? (c) How much energy would be stored in the space between the conductors in a 1 000-km superconducting line? (d) What is the pressure exerted on the outer conductor?

Figure P32.66

67. ● *The Meissner effect.* Compare this problem with Problem 57 in Chapter 26, pertaining to the force attracting a perfect dielectric into a strong electric field. A fundamental property of a type I superconducting material is *perfect* *diamagnetism*, or demonstration of the *Meissner effect*, illustrated in Figure 30.27 in Section 30.6 and described as follows. The superconducting material has $\vec{\mathbf{B}} = 0$ everywhere inside it. If a sample of the material is placed into an externally produced magnetic field or is cooled to become superconducting while it is in a magnetic field, electric currents appear on the surface of the sample. The currents have precisely the strength and orientation required to make the total magnetic field be zero throughout the interior of the sample. This problem will help you to understand the magnetic force that can then act on the superconducting sample.

A vertical solenoid with a length of 120 cm and a diameter of 2.50 cm consists of 1 400 turns of copper wire carrying a counterclockwise current of 2.00 A as shown in Figure P32.67a. (a) Find the magnetic field in the vacuum inside the solenoid. (b) Find the energy density of the magnetic field, noting that the units J/m^3 of energy density are the same as the units N/m^2 of pressure. (c) Now a superconducting bar 2.20 cm in diameter is inserted partway into the solenoid. Its upper end is far outside the solenoid, where the magnetic field is negligible. The lower end of the bar is deep inside the solenoid. Explain how you identify the direction required for the current on the curved surface of the bar so that the total magnetic field is zero within the bar. The field created by the supercurrents is sketched in Figure P32.67b, and the total field is sketched in Figure P32.67c. (d) The field of the solenoid exerts a force on the current in the superconductor. Explain how you determine the direction of the force on the bar. (e) Calculate the magnitude of the force by multiplying the energy density of the solenoid field times the area of the bottom end of the superconducting bar.

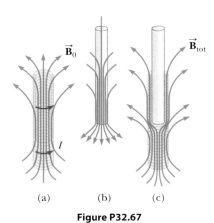

Figure P32.67

Answers to Quick Quizzes

32.1 (c), (f). For the constant current in statements (a) and (b), there is no voltage across the resistanceless inductor. In statement (c), if the current increases, the emf induced in the inductor is in the opposite direction, from b to a, making a higher in potential than b. Similarly, in statement (f), the decreasing current induces an emf in the same direction as the current, from b to a, again making the potential higher at a than at b.

32.2 (i), (b). As the switch is closed, there is no current, so there is no voltage across the resistor. (ii), (a). After a long time, the current has reached its final value and the inductor has no further effect on the circuit.

32.3 (a), (d). Because the energy density depends on the magnitude of the magnetic field, you must increase the magnetic field to increase the energy density. For a solenoid, $B = \mu_0 nI$, where n is the number of turns per unit length. In choice (a), increasing n increases the magnetic field. In choice (b), the change in cross-sectional area has no effect on the magnetic field. In choice (c), increasing the length but keeping n fixed has no effect on the magnetic field. Increasing the current in choice (d) increases the magnetic field in the solenoid.

32.4 (a). M increases because the magnetic flux through coil 2 increases.

32.5 (i), (b). If the current is at its maximum value, the charge on the capacitor is zero. (ii), (c). If the current is zero, this moment is the instant at which the capacitor is fully charged and the current is about to reverse direction.

These large transformers are used to increase the voltage at a power plant for distribution of energy by electrical transmission to the power grid. Voltages can be changed relatively easily because power is distributed by alternating current rather than direct current. (Lester Lefkowitz/Getty Images)

33.1 AC Sources

33.2 Resistors in an AC Circuit

33.3 Inductors in an AC Circuit

33.4 Capacitors in an AC Circuit

33.5 The *RLC* Series Circuit

33.6 Power in an AC Circuit

33.7 Resonance in a Series *RLC* Circuit

33.8 The Transformer and Power Transmission

33.9 Rectifiers and Filters

33 Alternating Current Circuits

In this chapter, we describe alternating-current (AC) circuits. Every time you turn on a television set, a stereo, or any of a multitude of other electrical appliances in a home, you are calling on alternating currents to provide the power to operate them. We begin our study by investigating the characteristics of simple series circuits that contain resistors, inductors, and capacitors and that are driven by a sinusoidal voltage. The primary aim of this chapter can be summarized as follows: if an AC source applies an alternating voltage to a series circuit containing resistors, inductors, and capacitors, we want to know the amplitude and time characteristics of the alternating current. We conclude this chapter with two sections concerning transformers, power transmission, and electrical filters.

33.1 AC Sources

An AC circuit consists of circuit elements and a power source that provides an alternating voltage Δv. This time-varying voltage from the source is described by

$$\Delta v = \Delta V_{\text{max}} \sin \omega t$$

where ΔV_{max} is the maximum output voltage of the source, or the **voltage amplitude.** There are various possibilities for AC sources, including generators as discussed in Section 31.5 and electrical oscillators. In a home, each electrical outlet

PITFALL PREVENTION 33.1
Time-Varying Values

We use lowercase symbols Δv and i to indicate the instantaneous values of time-varying voltages and currents. Capital letters represent fixed values of voltage and current such as ΔV_{max} and I_{max}.

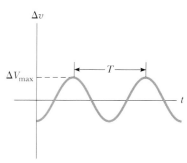

Figure 33.1 The voltage supplied by an AC source is sinusoidal with a period T.

serves as an AC source. Because the output voltage of an AC source varies sinusoidally with time, the voltage is positive during one half of the cycle and negative during the other half as in Figure 33.1. Likewise, the current in any circuit driven by an AC source is an alternating current that also varies sinusoidally with time.

From Equation 15.12, the angular frequency of the AC voltage is

$$\omega = 2\pi f = \frac{2\pi}{T}$$

where f is the frequency of the source and T is the period. The source determines the frequency of the current in any circuit connected to it. Commercial electric-power plants in the United States use a frequency of 60 Hz, which corresponds to an angular frequency of 377 rad/s.

33.2 Resistors in an AC Circuit

Consider a simple AC circuit consisting of a resistor and an AC source 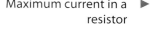 as shown in Active Figure 33.2. At any instant, the algebraic sum of the voltages around a closed loop in a circuit must be zero (Kirchhoff's loop rule). Therefore, $\Delta v + \Delta v_R = 0$ or, using Equation 27.7 for the voltage across the resistor,

$$\Delta v - i_R R = 0$$

If we rearrange this expression and substitute $\Delta V_{max} \sin \omega t$ for Δv, the instantaneous current in the resistor is

$$i_R = \frac{\Delta v}{R} = \frac{\Delta V_{max}}{R} \sin \omega t = I_{max} \sin \omega t \qquad (33.1)$$

where I_{max} is the maximum current:

◄ **Maximum current in a resistor**

$$I_{max} = \frac{\Delta V_{max}}{R} \qquad (33.2)$$

Equation 33.1 shows that the instantaneous voltage across the resistor is

◄ **Voltage across a resistor**

$$\Delta v_R = i_R R = I_{max} R \sin \omega t \qquad (33.3)$$

A plot of voltage and current versus time for this circuit is shown in Active Figure 33.3a. At point a, the current has a maximum value in one direction, arbitrarily

$$\Delta v = \Delta V_{max} \sin \omega t$$

ACTIVE FIGURE 33.2

A circuit consisting of a resistor of resistance R connected to an AC source, designated by the symbol

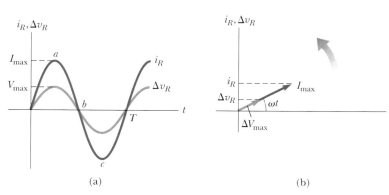

(a)   (b)

ACTIVE FIGURE 33.3

(a) Plots of the instantaneous current i_R and instantaneous voltage Δv_R across a resistor as functions of time. The current is in phase with the voltage, which means that the current is zero when the voltage is zero, maximum when the voltage is maximum, and minimum when the voltage is minimum. At time $t = T$, one cycle of the time-varying voltage and current has been completed. (b) Phasor diagram for the resistive circuit showing that the current is in phase with the voltage.

called the positive direction. Between points a and b, the current is decreasing in magnitude but is still in the positive direction. At point b, the current is momentarily zero; it then begins to increase in the negative direction between points b and c. At point c, the current has reached its maximum value in the negative direction.

The current and voltage are in step with each other because they vary identically with time. Because i_R and Δv_R both vary as $\sin \omega t$ and reach their maximum values at the same time as shown in Active Figure 33.3a, they are said to be **in phase,** similar to the way that two waves can be in phase as discussed in our study of wave motion in Chapter 18. Therefore, **for a sinusoidal applied voltage, the current in a resistor is always in phase with the voltage across the resistor.** For resistors in AC circuits, there are no new concepts to learn. Resistors behave essentially the same way in both DC and AC circuits. That, however, is not the case for capacitors and inductors.

To simplify our analysis of circuits containing two or more elements, we use a graphical representation called a *phasor diagram*. A **phasor** is a vector whose length is proportional to the maximum value of the variable it represents (ΔV_{max} for voltage and I_{max} for current in this discussion). The phasor rotates counterclockwise at an angular speed equal to the angular frequency associated with the variable. The projection of the phasor onto the vertical axis represents the instantaneous value of the quantity it represents.

Active Figure 33.3b shows voltage and current phasors for the circuit of Active Figure 33.2 at some instant of time. The projections of the phasor arrows onto the vertical axis are determined by a sine function of the angle of the phasor with respect to the horizontal axis. For example, the projection of the current phasor in Active Figure 33.3b is $I_{max} \sin \omega t$. Notice that this expression is the same as Equation 33.1. Therefore, the projections of phasors represent current values that vary sinusoidally in time. We can do the same with time-varying voltages. The advantage of this approach is that the phase relationships among currents and voltages can be represented as vector additions of phasors using the vector addition techniques discussed in Chapter 3.

In the case of the single-loop resistive circuit of Active Figure 33.2, the current and voltage phasors lie along the same line in Active Figure 33.3b because i_R and Δv_R are in phase. The current and voltage in circuits containing capacitors and inductors have different phase relationships.

Quick Quiz 33.1 Consider the voltage phasor in Figure 33.4, shown at three instants of time. **(i)** Choose the part of the figure, (a), (b), or (c), that represents the instant of time at which the instantaneous value of the voltage has the largest magnitude. **(ii)** Choose the part of the figure that represents the instant of time at which the instantaneous value of the voltage has the smallest magnitude.

For the simple resistive circuit in Active Figure 33.2, notice that **the average value of the current over one cycle is zero.** That is, the current is maintained in the positive direction for the same amount of time and at the same magnitude as it is maintained in the negative direction. The direction of the current, however, has no effect on the behavior of the resistor. We can understand this concept by realizing that collisions between electrons and the fixed atoms of the resistor result in an increase in the resistor's temperature. Although this temperature increase depends on the magnitude of the current, it is independent of the current's direction.

We can make this discussion quantitative by recalling that the rate at which energy is delivered to a resistor is the power $\mathcal{P} = i^2 R$, where i is the instantaneous current in the resistor. Because this rate is proportional to the square of the current, it makes no difference whether the current is direct or alternating, that is, whether the sign associated with the current is positive or negative. The temperature increase produced by an alternating current having a maximum value I_{max}, however, is not the same as that produced by a direct current equal to I_{max} because the alternating current has this maximum value for only an instant during each cycle (Fig. 33.5a, page 926). What is of importance in an AC circuit is an average

PITFALL PREVENTION 33.2
A Phasor Is Like a Graph

An alternating voltage can be presented in different representations. One graphical representation is shown in Figure 33.1 in which the voltage is drawn in rectangular coordinates, with voltage on the vertical axis and time on the horizontal axis. Active Figure 33.3b shows another graphical representation. The phase space in which the phasor is drawn is similar to polar coordinate graph paper. The radial coordinate represents the amplitude of the voltage. The angular coordinate is the phase angle. The vertical-axis coordinate of the tip of the phasor represents the instantaneous value of the voltage. The horizontal coordinate represents nothing at all. As shown in Active Figure 33.3b, alternating currents can also be represented by phasors.

To help with this discussion of phasors, review Section 15.4, where we represented the simple harmonic motion of a real object by the projection of an imaginary object's uniform circular motion onto a coordinate axis. A phasor is a direct analog to this representation.

(a)

(b)

(c)

Figure 33.4 (Quick Quiz 33.1) A voltage phasor is shown at three instants of time, (a), (b), and (c).

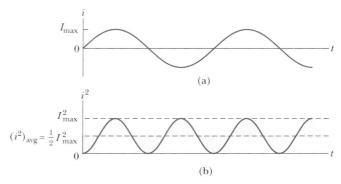

Figure 33.5 (a) Graph of the current in a resistor as a function of time. (b) Graph of the current squared in a resistor as a function of time. Notice that the gray shaded regions *under* the curve and *above* the dashed line for $\frac{1}{2}I_{max}^2$ have the same area as the gray shaded regions *above* the curve and *below* the dashed line for $\frac{1}{2}I_{max}^2$. Therefore, the average value of i^2 is $\frac{1}{2}I_{max}^2$. In general, the average value of $\sin^2 \omega t$ or $\cos^2 \omega t$ over one cycle is $\frac{1}{2}$.

value of current, referred to as the **rms current.** As we learned in Section 21.1, the notation *rms* stands for *root-mean-square*, which in this case means the square root of the mean (average) value of the square of the current: $I_{rms} = \sqrt{(i^2)_{avg}}$. Because i^2 varies as $\sin^2 \omega t$ and because the average value of i^2 is $\frac{1}{2}I_{max}^2$ (see Fig. 33.5b), the rms current is

rms current ▶

$$I_{rms} = \frac{I_{max}}{\sqrt{2}} = 0.707 I_{max} \qquad (33.4)$$

This equation states that an alternating current whose maximum value is 2.00 A delivers to a resistor the same power as a direct current that has a value of $(0.707)(2.00 \text{ A}) = 1.41$ A. The average power delivered to a resistor that carries an alternating current is

Average power delivered ▶
to a resistor

$$\mathcal{P}_{avg} = I_{rms}^2 R$$

Alternating voltage is also best discussed in terms of rms voltage, and the relationship is identical to that for current:

rms voltage ▶

$$\Delta V_{rms} = \frac{\Delta V_{max}}{\sqrt{2}} = 0.707 \, \Delta V_{max} \qquad (33.5)$$

When we speak of measuring a 120-V alternating voltage from an electrical outlet, we are referring to an rms voltage of 120 V. A calculation using Equation 33.5 shows that such an alternating voltage has a maximum value of about 170 V. One reason rms values are often used when discussing alternating currents and voltages is that AC ammeters and voltmeters are designed to read rms values. Furthermore, with rms values, many of the equations we use have the same form as their direct current counterparts.

EXAMPLE 33.1 | **What Is the rms Current?**

The voltage output of an AC source is given by the expression $\Delta v = (200 \text{ V})\sin \omega t$. Find the rms current in the circuit when this source is connected to a 100-Ω resistor.

SOLUTION

Conceptualize Active Figure 33.2 shows the physical situation for this problem.

Categorize We evaluate the current with an equation developed in this section, so we categorize this example as a substitution problem.

Comparing this expression for voltage output with the general form $\Delta v = \Delta V_{max} \sin \omega t$ shows that $\Delta V_{max} = 200$ V. Calculate the rms voltage from Equation 33.5:

$$\Delta V_{rms} = \frac{\Delta V_{max}}{\sqrt{2}} = \frac{200 \text{ V}}{\sqrt{2}} = 141 \text{ V}$$

Find the rms current:

$$I_{rms} = \frac{\Delta V_{rms}}{R} = \frac{141 \text{ V}}{100 \ \Omega} = \boxed{1.41 \text{ A}}$$

33.3   Inductors in an AC Circuit

Now consider an AC circuit consisting only of an inductor connected to the terminals of an AC source as shown in Active Figure 33.6. If $\Delta v_L = \mathcal{E}_L = -L(di_L/dt)$ is the self-induced instantaneous voltage across the inductor (see Eq. 32.1), Kirchhoff's loop rule applied to this circuit gives $\Delta v + \Delta v_L = 0$, or

$$\Delta v - L \frac{di_L}{dt} = 0$$

Substituting $\Delta V_{max} \sin \omega t$ for Δv and rearranging gives

$$\Delta v = L \frac{di_L}{dt} = \Delta V_{max} \sin \omega t \qquad \textbf{(33.6)}$$

Solving this equation for di_L gives

$$di_L = \frac{\Delta V_{max}}{L} \sin \omega t \, dt$$

Integrating this expression[1] gives the instantaneous current i_L in the inductor as a function of time:

$$i_L = \frac{\Delta V_{max}}{L} \int \sin \omega t \, dt = -\frac{\Delta V_{max}}{\omega L} \cos \omega t \qquad \textbf{(33.7)}$$

Using the trigonometric identity $\cos \omega t = -\sin(\omega t - \pi/2)$, we can express Equation 33.7 as

$$i_L = \frac{\Delta V_{max}}{\omega L} \sin\left(\omega t - \frac{\pi}{2}\right) \qquad \textbf{(33.8)} \qquad \blacktriangleleft \quad \text{Current in an inductor}$$

Comparing this result with Equation 33.6 shows that the instantaneous current i_L in the inductor and the instantaneous voltage Δv_L across the inductor are *out* of phase by $\pi/2$ rad = 90°.

A plot of voltage and current versus time is shown in Active Figure 33.7a (page 928). When the current i_L in the inductor is a maximum (point *b* in Active Fig. 33.7a), it is momentarily not changing, so the voltage across the inductor is zero (point *d*). At points such as *a* and *e*, the current is zero and the rate of change of current is at a maximum. Therefore, the voltage across the inductor is also at a maximum (points *c* and *f*). Notice that the voltage reaches its maximum value one quarter of a period before the current reaches its maximum value. Therefore, **for a sinusoidal applied voltage, the current in an inductor always lags behind the voltage across the inductor by 90° (one-quarter cycle in time).**

As with the relationship between current and voltage for a resistor, we can represent this relationship for an inductor with a phasor diagram as in Active Figure 33.7b. The phasors are at 90° to each other, representing the 90° phase difference between current and voltage.

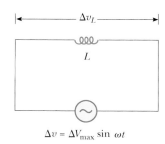

$\Delta v = \Delta V_{max} \sin \omega t$

ACTIVE FIGURE 33.6

A circuit consisting of an inductor of inductance L connected to an AC source.

Sign in at www.thomsonedu.com and go to ThomsonNOW to adjust the inductance, frequency, and maximum voltage. The results can be studied with the graph and the phasor diagram in Active Figure 33.7.

---

[1] We neglect the constant of integration here because it depends on the initial conditions, which are not important for this situation.

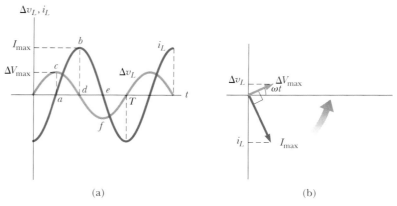

ACTIVE FIGURE 33.7

(a) Plots of the instantaneous current i_L and instantaneous voltage Δv_L across an inductor as functions of time. The current lags behind the voltage by 90°. (b) Phasor diagram for the inductive circuit, showing that the current lags behind the voltage by 90°.

Sign in at www.thomsonedu.com and go to ThomsonNOW to adjust the inductance, frequency, and maximum voltage of the circuit in Active Figure 33.6. The results can be studied with the graph and the phasor diagram in this figure.

Equation 33.7 shows that the current in an inductive circuit reaches its maximum value when $\cos \omega t = \pm 1$:

Maximum current in an inductor ▶

$$I_{max} = \frac{\Delta V_{max}}{\omega L} \qquad (33.9)$$

This expression is similar to the relationship between current, voltage, and resistance in a DC circuit, $I = \Delta V/R$ (Eq. 27.7). Because I_{max} has units of amperes and ΔV_{max} has units of volts, ωL must have units of ohms. Therefore, ωL has the same units as resistance and is related to current and voltage in the same way as resistance. It must behave in a manner similar to resistance in the sense that it represents opposition to the flow of charge. Because ωL depends on the applied frequency ω, the inductor *reacts* differently, in terms of offering opposition to current, for different frequencies. For this reason, we define ωL as the **inductive reactance** X_L:

Inductive reactance ▶

$$X_L \equiv \omega L \qquad (33.10)$$

Therefore, we can write Equation 33.9 as

$$I_{max} = \frac{\Delta V_{max}}{X_L} \qquad (33.11)$$

The expression for the rms current in an inductor is similar to Equation 33.9, with I_{max} replaced by I_{rms} and ΔV_{max} replaced by ΔV_{rms}.

Equation 33.10 indicates that, for a given applied voltage, the inductive reactance increases as the frequency increases. This conclusion is consistent with Faraday's law: the greater the rate of change of current in the inductor, the larger the back emf. The larger back emf translates to an increase in the reactance and a decrease in the current.

Using Equations 33.6 and 33.11, we find that the instantaneous voltage across the inductor is

Voltage across an inductor ▶

$$\Delta v_L = -L \frac{di_L}{dt} = -\Delta V_{max} \sin \omega t = -I_{max} X_L \sin \omega t \qquad (33.12)$$

Quick Quiz 33.2 Consider the AC circuit in Figure 33.8. The frequency of the AC source is adjusted while its voltage amplitude is held constant. When does the lightbulb glow the brightest? (a) It glows brightest at high frequencies. (b) It glows brightest at low frequencies. (c) The brightness is the same at all frequencies.

Figure 33.8 (Quick Quiz 33.2) At what frequencies does the lightbulb glow the brightest?

EXAMPLE 33.2 | **A Purely Inductive AC Circuit**

In a purely inductive AC circuit, $L = 25.0$ mH and the rms voltage is 150 V. Calculate the inductive reactance and rms current in the circuit if the frequency is 60.0 Hz.

SOLUTION

Conceptualize   Active Figure 33.6 shows the physical situation for this problem.

Categorize   We evaluate the reactance and the current from equations developed in this section, so we categorize this example as a substitution problem.

Use Equation 33.10 to find the inductive reactance:

$$X_L = \omega L = 2\pi f L = 2\pi (60.0 \text{ Hz})(25.0 \times 10^{-3} \text{ H})$$
$$= 9.42 \ \Omega$$

Use an rms version of Equation 33.11 to find the rms current:

$$I_{\text{rms}} = \frac{\Delta V_{\text{rms}}}{X_L} = \frac{150 \text{ V}}{9.42 \ \Omega} = 15.9 \text{ A}$$

What If?   If the frequency increases to 6.00 kHz, what happens to the rms current in the circuit?

Answer   If the frequency increases, the inductive reactance also increases because the current is changing at a higher rate. The increase in inductive reactance results in a lower current.
Let's calculate the new inductive reactance and the new rms current:

$$X_L = 2\pi (6.00 \times 10^3 \text{ Hz})(25.0 \times 10^{-3} \text{ H}) = 942 \ \Omega$$

$$I_{\text{rms}} = \frac{150 \text{ V}}{942 \ \Omega} = 0.159 \text{ A}$$

33.4 Capacitors in an AC Circuit

Active Figure 33.9 shows an AC circuit consisting of a capacitor connected across the terminals of an AC source. Kirchhoff's loop rule applied to this circuit gives $\Delta v + \Delta v_C = 0$, or

$$\Delta v - \frac{q}{C} = 0 \tag{33.13}$$

Substituting $\Delta V_{\text{max}} \sin \omega t$ for Δv and rearranging gives

$$q = C \, \Delta V_{\text{max}} \sin \omega t \tag{33.14}$$

where q is the instantaneous charge on the capacitor. Differentiating Equation 33.14 with respect to time gives the instantaneous current in the circuit:

$$i_C = \frac{dq}{dt} = \omega C \, \Delta V_{\text{max}} \cos \omega t \tag{33.15}$$

$\Delta v = \Delta V_{\text{max}} \sin \omega t$

ACTIVE FIGURE 33.9

A circuit consisting of a capacitor of capacitance C connected to an AC source.

Sign in at www.thomsonedu.com and go to ThomsonNOW to adjust the capacitance, frequency, and maximum voltage. The results can be studied with the graph and the phasor diagram in Active Figure 33.10.

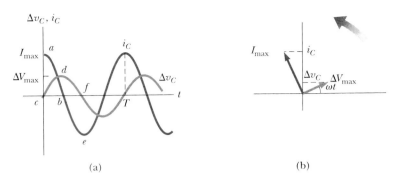

ACTIVE FIGURE 33.10

(a) Plots of the instantaneous current i_C and instantaneous voltage Δv_C across a capacitor as functions of time. The voltage lags behind the current by 90°. (b) Phasor diagram for the capacitive circuit, showing that the current leads the voltage by 90°.

Sign in at www.thomsonedu.com and go to ThomsonNOW to adjust the capacitance, frequency, and maximum voltage of the circuit in Active Figure 33.9. The results can be studied with the graph and the phasor diagram in this figure.

Using the trigonometric identity

$$\cos \omega t = \sin \left(\omega t + \frac{\pi}{2} \right)$$

we can express Equation 33.15 in the alternative form

▶ **Current in a capacitor**

$$i_C = \omega C \, \Delta V_{max} \sin \left(\omega t + \frac{\pi}{2} \right) \tag{33.16}$$

Comparing this expression with $\Delta v = \Delta V_{max} \sin \omega t$ shows that the current is $\pi/2$ rad = 90° out of phase with the voltage across the capacitor. A plot of current and voltage versus time (Active Fig. 33.10a) shows that the current reaches its maximum value one-quarter of a cycle sooner than the voltage reaches its maximum value.

Consider a point such as b where the current is zero at this instant. That occurs when the capacitor reaches its maximum charge so that the voltage across the capacitor is a maximum (point d). At points such as a and e, the current is a maximum, which occurs at those instants when the charge on the capacitor reaches zero and the capacitor begins to recharge with the opposite polarity. When the charge is zero, the voltage across the capacitor is zero (points c and f). Therefore, the current and voltage are out of phase.

As with inductors, we can represent the current and voltage for a capacitor on a phasor diagram. The phasor diagram in Active Figure 33.10b shows that **for a sinusoidally applied voltage, the current always leads the voltage across a capacitor by 90°.**

Equation 33.15 shows that the current in the circuit reaches its maximum value when $\cos \omega t = \pm 1$:

$$I_{max} = \omega C \, \Delta V_{max} = \frac{\Delta V_{max}}{(1/\omega C)} \tag{33.17}$$

As in the case with inductors, this looks like Equation 27.7, so the denominator plays the role of resistance, with units of ohms. We give the combination $1/\omega C$ the symbol X_C, and because this function varies with frequency, we define it as the **capacitive reactance:**

▶ **Capacitive reactance**

$$X_C \equiv \frac{1}{\omega C} \tag{33.18}$$

We can now write Equation 33.17 as

▶ **Maximum current in a capacitor**

$$I_{max} = \frac{\Delta V_{max}}{X_C} \tag{33.19}$$

The rms current is given by an expression similar to Equation 33.19, with I_{max} replaced by I_{rms} and ΔV_{max} replaced by ΔV_{rms}.

Using Equation 33.19, we can express the instantaneous voltage across the capacitor as

$$\Delta v_C = \Delta V_{max} \sin \omega t = I_{max} X_C \sin \omega t \qquad (33.20)$$

◀ Voltage across a capacitor

Equations 33.18 and 33.19 indicate that as the frequency of the voltage source increases, the capacitive reactance decreases and the maximum current therefore increases. The frequency of the current is determined by the frequency of the voltage source driving the circuit. As the frequency approaches zero, the capacitive reactance approaches infinity and the current therefore approaches zero. This conclusion makes sense because the circuit approaches direct current conditions as ω approaches zero and the capacitor represents an open circuit.

Quick Quiz 33.3 Consider the AC circuit in Figure 33.11. The frequency of the AC source is adjusted while its voltage amplitude is held constant. When does the lightbulb glow the brightest? (a) It glows brightest at high frequencies. (b) It glows brightest at low frequencies. (c) The brightness is the same at all frequencies.

Figure 33.11   (Quick Quiz 33.3)

Quick Quiz 33.4 Consider the AC circuit in Figure 33.12. The frequency of the AC source is adjusted while its voltage amplitude is held constant. When does the lightbulb glow the brightest? (a) It glows brightest at high frequencies. (b) It glows brightest at low frequencies. (c) The brightness is the same at all frequencies.

Figure 33.12   (Quick Quiz 33.4)

EXAMPLE 33.3   **A Purely Capacitive AC Circuit**

An 8.00-μF capacitor is connected to the terminals of a 60.0-Hz AC source whose rms voltage is 150 V. Find the capacitive reactance and the rms current in the circuit.

SOLUTION

Conceptualize   Active Figure 33.9 shows the physical situation for this problem.

Categorize   We evaluate the reactance and the current from equations developed in this section, so we categorize this example as a substitution problem.

Use Equation 33.18 to find the capacitive reactance:

$$X_C = \frac{1}{\omega C} = \frac{1}{2\pi f C} = \frac{1}{2\pi (60 \text{ Hz})(8.00 \times 10^{-6}\text{ F})} = 332 \ \Omega$$

Use an rms version of Equation 33.19 to find the rms current:

$$I_{rms} = \frac{\Delta V_{rms}}{X_C} = \frac{150 \text{ V}}{332 \ \Omega} = \boxed{0.452 \text{ A}}$$

What If?   What if the frequency is doubled? What happens to the rms current in the circuit?

Answer   If the frequency increases, the capacitive reactance decreases, which is just the opposite from the case of an inductor. The decrease in capacitive reactance results in an increase in the current.

Let's calculate the new capacitive reactance and the new rms current:

$$X_C = \frac{1}{\omega C} = \frac{1}{2\pi (120 \text{ Hz})(8.00 \times 10^{-6}\text{ F})} = 166 \ \Omega$$

$$I_{rms} = \frac{150 \text{ V}}{166 \ \Omega} = 0.904 \text{ A}$$

(a)

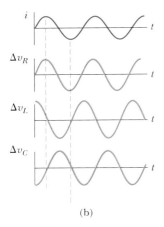

(b)

ACTIVE FIGURE 33.13

(a) A series circuit consisting of a resistor, an inductor, and a capacitor connected to an AC source. (b) Phase relationships for instantaneous voltages in the series *RLC* circuit.

Sign in at www.thomsonedu.com and go to ThomsonNOW to adjust the resistance, inductance, and capacitance. The results can be studied with the graph in this figure and the phasor diagram in Active Figure 33.15.

33.5 The *RLC* Series Circuit

Active Figure 33.13a shows a circuit that contains a resistor, an inductor, and a capacitor connected in series across an alternating voltage source. If the applied voltage varies sinusoidally with time, the instantaneous applied voltage is

$$\Delta v = \Delta V_{max} \sin \omega t$$

while the current varies as

$$i = I_{max} \sin (\omega t - \phi)$$

where ϕ is some **phase angle** between the current and the applied voltage. Based on our discussions of phase in Sections 33.3 and 33.4, we expect that the current will generally not be in phase with the voltage in an *RLC* circuit. Our aim is to determine ϕ and I_{max}. Active Figure 33.13b shows the voltage versus time across each element in the circuit and their phase relationships.

First, because the elements are in series, the current everywhere in the circuit must be the same at any instant. That is, **the current at all points in a series AC circuit has the same amplitude and phase.** Based on the preceding sections, we know that the voltage across each element has a different amplitude and phase. In particular, the voltage across the resistor is in phase with the current, the voltage across the inductor leads the current by 90°, and the voltage across the capacitor lags behind the current by 90°. Using these phase relationships, we can express the instantaneous voltages across the three circuit elements as

$$\Delta v_R = I_{max} R \sin \omega t = \Delta V_R \sin \omega t \tag{33.21}$$

$$\Delta v_L = I_{max} X_L \sin \left(\omega t + \frac{\pi}{2} \right) = \Delta V_L \cos \omega t \tag{33.22}$$

$$\Delta v_C = I_{max} X_C \sin \left(\omega t - \frac{\pi}{2} \right) = -\Delta V_C \cos \omega t \tag{33.23}$$

The sum of these three voltages must equal the voltage from the AC source, but it is important to recognize that because the three voltages have different phase relationships with the current, they cannot be added directly. Figure 33.14 represents the phasors at an instant at which the current in all three elements is momentarily zero. The zero current is represented by the current phasor along the horizontal axis in each part of the figure. Next the voltage phasor is drawn at the appropriate phase angle to the current for each element.

Because phasors are rotating vectors, the voltage phasors in Figure 33.14 can be combined using vector addition as in Active Figure 33.15. In Active Figure 33.15a, the voltage phasors in Figure 33.14 are combined on the same coordinate axes. Active Figure 33.15b shows the vector addition of the voltage phasors. The voltage phasors ΔV_L and ΔV_C are in *opposite* directions along the same line, so we can construct the difference phasor $\Delta V_L - \Delta V_C$, which is perpendicular to the phasor ΔV_R. This diagram shows that the vector sum of the voltage amplitudes ΔV_R, ΔV_L, and ΔV_C equals a phasor whose length is the maximum applied voltage ΔV_{max} and

Figure 33.14 Phase relationships between the voltage and current phasors for (a) a resistor, (b) an inductor, and (c) a capacitor connected in series.

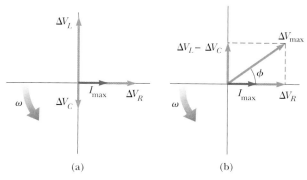

(a)   (b)

ACTIVE FIGURE 33.15

(a) Phasor diagram for the series *RLC* circuit shown in Active Figure 33.13a. The phasor ΔV_R is in phase with the current phasor I_{max}, the phasor ΔV_L leads I_{max} by 90°, and the phasor ΔV_C lags I_{max} by 90°. (b) The inductance and capacitance phasors are added together and then added vectorially to the resistance phasor. The total voltage ΔV_{max} makes an angle ϕ with I_{max}.

Sign in at www.thomsonedu.com and go to ThomsonNOW to adjust the resistance, inductance, and capacitance of the circuit in Active Figure 33.13a. The results can be studied with the graphs in Active Figure 33.13b and the phasor diagram in this figure.

which makes an angle ϕ with the current phasor I_{max}. From the right triangle in Active Figure 33.15b, we see that

$$\Delta V_{max} = \sqrt{\Delta V_R^2 + (\Delta V_L - \Delta V_C)^2} = \sqrt{(I_{max}R)^2 + (I_{max}X_L - I_{max}X_C)^2}$$

$$\Delta V_{max} = I_{max}\sqrt{R^2 + (X_L - X_C)^2}$$

Therefore, we can express the maximum current as

$$I_{max} = \frac{\Delta V_{max}}{\sqrt{R^2 + (X_L - X_C)^2}} \qquad \textbf{(33.24)} \qquad \blacktriangleleft \; \text{Maximum current in an } RLC \text{ circuit}$$

Once again, this expression has the same mathematical form as Equation 27.7. The denominator of the fraction plays the role of resistance and is called the **impedance** *Z* of the circuit:

$$Z \equiv \sqrt{R^2 + (X_L - X_C)^2} \qquad \textbf{(33.25)} \qquad \blacktriangleleft \; \text{Impedance}$$

where impedance also has units of ohms. Therefore, Equation 33.24 can be written in the form

$$I_{max} = \frac{\Delta V_{max}}{Z} \qquad \textbf{(33.26)}$$

Equation 33.26 is the AC equivalent of Equation 27.7. Note that the impedance and therefore the current in an AC circuit depend on the resistance, the inductance, the capacitance, and the frequency (because the reactances are frequency dependent).

From the right triangle in the phasor diagram in Active Figure 33.15b, the phase angle ϕ between the current and the voltage is found as follows:

$$\phi = \tan^{-1}\left(\frac{\Delta V_L - \Delta V_C}{\Delta V_R}\right) = \tan^{-1}\left(\frac{I_{max}X_L - I_{max}X_C}{I_{max}R}\right)$$

$$\phi = \tan^{-1}\left(\frac{X_L - X_C}{R}\right) \qquad \textbf{(33.27)} \qquad \blacktriangleleft \; \text{Phase angle}$$

When $X_L > X_C$ (which occurs at high frequencies), the phase angle is positive, signifying that the current lags the applied voltage as in Active Figure 33.15b. We describe this situation by saying that the circuit is *more inductive than capacitive*. When $X_L < X_C$, the phase angle is negative, signifying that the current leads the applied voltage, and the circuit is *more capacitive than inductive*. When $X_L = X_C$, the phase angle is zero and the circuit is *purely resistive*.

Quick Quiz 33.5 Label each part of Figure 33.16, (a), (b), and (c), as representing $X_L > X_C$, $X_L = X_C$, or $X_L < X_C$.

Figure 33.16 (Quick Quiz 33.5) Match the phasor diagrams to the relationships between the reactances.

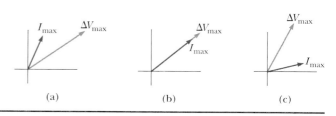

(a)   (b)   (c)

| EXAMPLE 33.4 | **Analyzing a Series *RLC* Circuit** |

A series *RLC* circuit has $R = 425\ \Omega$, $L = 1.25$ H, $C = 3.50\ \mu$F. It is connected to an AC source with $f = 60.0$ Hz and $\Delta V_{max} = 150$ V.

(A) Determine the inductive reactance, the capacitive reactance, and the impedance of the circuit.

SOLUTION

Conceptualize The circuit of interest in this example is shown in Active Figure 33.13a. The current in the combination of the resistor, inductor, and capacitor oscillates at a particular phase angle with respect to the applied voltage.

Categorize The circuit is a simple series *RLC* circuit, so we can use the approach discussed in this section.

Analyze Find the angular frequency:

$$\omega = 2\pi f = 2\pi (60.0\ \text{Hz}) = 377\ \text{s}^{-1}$$

Use Equation 33.10 to find the inductive reactance:

$$X_L = \omega L = (377\ \text{s}^{-1})(1.25\ \text{H}) = \boxed{471\ \Omega}$$

Use Equation 33.18 to find the capacitive reactance:

$$X_C = \frac{1}{\omega C} = \frac{1}{(377\ \text{s}^{-1})(3.50 \times 10^{-6}\ \text{F})} = \boxed{758\ \Omega}$$

Use Equation 33.25 to find the impedance:

$$Z = \sqrt{R^2 + (X_L - X_C)^2}$$
$$= \sqrt{(425\ \Omega)^2 + (471\ \Omega - 758\ \Omega)^2} = \boxed{513\ \Omega}$$

(B) Find the maximum current in the circuit.

SOLUTION

Use Equation 33.26 to find the maximum current:

$$I_{max} = \frac{\Delta V_{max}}{Z} = \frac{150\ \text{V}}{513\ \Omega} = \boxed{0.292\ \text{A}}$$

(C) Find the phase angle between the current and voltage.

SOLUTION

Use Equation 33.27 to calculate the phase angle:

$$\phi = \tan^{-1}\left(\frac{X_L - X_C}{R}\right) = \tan^{-1}\left(\frac{471\ \Omega - 758\ \Omega}{425\ \Omega}\right) = \boxed{-34.0°}$$

(D) Find the maximum voltage across each element.

SOLUTION

Use Equations 33.2, 33.11, and 33.19 to calculate the maximum voltages:

$$\Delta V_R = I_{max} R = (0.292\ \text{A})(425\ \Omega) = \boxed{124\ \text{V}}$$
$$\Delta V_L = I_{max} X_L = (0.292\ \text{A})(471\ \Omega) = \boxed{138\ \text{V}}$$
$$\Delta V_C = I_{max} X_C = (0.292\ \text{A})(758\ \Omega) = \boxed{221\ \text{V}}$$

(E) What replacement value of L should an engineer analyzing the circuit choose such that the current leads the applied voltage by 30.0°? All other values in the circuit stay the same.

SOLUTION

Solve Equation 33.27 for the inductive reactance:

$$X_L = X_C + R \tan \phi$$

Substitute Equations 33.10 and 33.18 into this expression:

$$\omega L = \frac{1}{\omega C} + R \tan \phi$$

Solve for L:

$$L = \frac{1}{\omega}\left(\frac{1}{\omega C} + R \tan \phi\right)$$

Substitute the given values:

$$L = \frac{1}{(377 \text{ s}^{-1})}\left[\frac{1}{(377 \text{ s}^{-1})(3.50 \times 10^{-6} \text{ F})} + (425 \ \Omega) \tan (-30.0°)\right]$$

$$L = \boxed{1.36 \text{ H}}$$

Finalize   Because the capacitive reactance is larger than the inductive reactance, the circuit is more capacitive than inductive. In this case, the phase angle ϕ is negative, so the current leads the applied voltage.

   Using Equations 33.21, 33.22, and 33.23, the instantaneous voltages across the three elements are

$$\Delta v_R = (124 \text{ V}) \sin 377t$$

$$\Delta v_L = (138 \text{ V}) \cos 377t$$

$$\Delta v_C = (-221 \text{ V}) \cos 377t$$

What If?   What if you added up the maximum voltages across the three circuit elements? Is that a physically meaningful quantity?

Answer   The sum of the maximum voltages across the elements is $\Delta V_R + \Delta V_L + \Delta V_C = 483$ V. This sum is much greater than the maximum voltage of the source, 150 V. The sum of the maximum voltages is a meaningless quantity because when sinusoidally varying quantities are added, *both their amplitudes and their phases* must be taken into account. The maximum voltages across the various elements occur at different times. Therefore, the voltages must be added in a way that takes account of the different phases as shown in Active Figure 33.15.

33.6   Power in an AC Circuit

Now let's take an energy approach to analyzing AC circuits and consider the transfer of energy from the AC source to the circuit. The power delivered by a battery to an external DC circuit is equal to the product of the current and the terminal voltage of the battery. Likewise, the instantaneous power delivered by an AC source to a circuit is the product of the current and the applied voltage. For the *RLC* circuit shown in Active Figure 33.13a, we can express the instantaneous power $\mathscr{P}$ as

$$\mathscr{P} = i \Delta v = I_{\max} \sin (\omega t - \phi)\, \Delta V_{\max} \sin \omega t$$

$$\mathscr{P} = I_{\max} \Delta V_{\max} \sin \omega t \sin (\omega t - \phi) \qquad \textbf{(33.28)}$$

This result is a complicated function of time and is therefore not very useful from a practical viewpoint. What is generally of interest is the average power over one or more cycles. Such an average can be computed by first using the trigonometric identity $\sin (\omega t - \phi) = \sin \omega t \cos \phi - \cos \omega t \sin \phi$. Substituting this identity into Equation 33.28 gives

$$\mathscr{P} = I_{\max} \Delta V_{\max} \sin^2 \omega t \cos \phi - I_{\max} \Delta V_{\max} \sin \omega t \cos \omega t \sin \phi \qquad \textbf{(33.29)}$$

Let's now take the time average of $\mathcal{P}$ over one or more cycles, noting that I_{max}, ΔV_{max}, ϕ, and ω are all constants. The time average of the first term on the right of the equal sign in Equation 33.29 involves the average value of $\sin^2 \omega t$, which is $\frac{1}{2}$. The time average of the second term on the right of the equal sign is identically zero because $\sin \omega t \cos \omega t = \frac{1}{2} \sin 2\omega t$, and the average value of $\sin 2\omega t$ is zero. Therefore, we can express the **average power** $\mathcal{P}_{avg}$ as

$$\mathcal{P}_{avg} = \tfrac{1}{2} I_{max} \Delta V_{max} \cos \phi \qquad (33.30)$$

It is convenient to express the average power in terms of the rms current and rms voltage defined by Equations 33.4 and 33.5:

$$\mathcal{P}_{avg} = I_{rms} \Delta V_{rms} \cos \phi \qquad (33.31)$$

Average power delivered ▶
to an *RLC* circuit

where the quantity $\cos \phi$ is called the **power factor.** Active Figure 33.15b shows that the maximum voltage across the resistor is given by $\Delta V_R = \Delta V_{max} \cos \phi = I_{max} R$. Using Equation 33.5 and $\cos \phi = I_{max} R / \Delta V_{max}$, we can express $\mathcal{P}_{avg}$ as

$$\mathcal{P}_{avg} = I_{rms} \Delta V_{rms} \cos \phi = I_{rms} \left(\frac{\Delta V_{max}}{\sqrt{2}} \right) \frac{I_{max} R}{\Delta V_{max}} = I_{rms} \frac{I_{max} R}{\sqrt{2}}$$

Substituting $I_{max} = \sqrt{2} I_{rms}$ from Equation 33.4 gives

$$\mathcal{P}_{avg} = I_{rms}^2 R \qquad (33.32)$$

In words, **the average power delivered by the source is converted to internal energy in the resistor,** just as in the case of a DC circuit. When the load is purely resistive, $\phi = 0$, $\cos \phi = 1$, and, from Equation 33.31, we see that

$$\mathcal{P}_{avg} = I_{rms} \Delta V_{rms}$$

Note that **no power losses are associated with pure capacitors and pure inductors in an AC circuit.** To see why that is true, let's first analyze the power in an AC circuit containing only a source and a capacitor. When the current begins to increase in one direction in an AC circuit, charge begins to accumulate on the capacitor and a voltage appears across it. When this voltage reaches its maximum value, the energy stored in the capacitor as electric potential energy is $\frac{1}{2} C (\Delta V_{max})^2$. This energy storage, however, is only momentary. The capacitor is charged and discharged twice during each cycle: charge is delivered to the capacitor during two quarters of the cycle and is returned to the voltage source during the remaining two quarters. Therefore, **the average power supplied by the source is zero.** In other words, **no power losses occur in a capacitor in an AC circuit.**

Now consider the case of an inductor. When the current in an inductor reaches its maximum value, the energy stored in the inductor is a maximum and is given by $\frac{1}{2} L I_{max}^2$. When the current begins to decrease in the circuit, this stored energy in the inductor returns to the source as the inductor attempts to maintain the current in the circuit.

Equation 33.31 shows that the power delivered by an AC source to any circuit depends on the phase, a result that has many interesting applications. For example, a factory that uses large motors in machines, generators, or transformers has a large inductive load (because of all the windings). To deliver greater power to such devices in the factory without using excessively high voltages, technicians introduce capacitance in the circuits to shift the phase.

Quick Quiz 33.6 An AC source drives an *RLC* circuit with a fixed voltage amplitude. If the driving frequency is ω_1, the circuit is more capacitive than inductive and the phase angle is $-10°$. If the driving frequency is ω_2, the circuit is more inductive than capacitive and the phase angle is $+10°$. At what frequency is the largest amount of power delivered to the circuit? (a) It is largest at ω_1. (b) It is largest at ω_2. (c) The same amount of power is delivered at both frequencies.

EXAMPLE 33.5   **Average Power in an *RLC* Series Circuit**

Calculate the average power delivered to the series *RLC* circuit described in Example 33.4.

SOLUTION

Conceptualize   Consider the circuit in Active Figure 33.13a and imagine energy being delivered to the circuit by the AC source. Review Example 33.4 for other details about this circuit.

Categorize   We find the result by using equations developed in this section, so we categorize this example as a substitution problem.

Use Equation 33.5 and the maximum voltage from Example 33.4 to find the rms voltage from the source:

$$\Delta V_{rms} = \frac{\Delta V_{max}}{\sqrt{2}} = \frac{150 \text{ V}}{\sqrt{2}} = 106 \text{ V}$$

Similarly, find the rms current in the circuit:

$$I_{rms} = \frac{I_{max}}{\sqrt{2}} = \frac{0.292 \text{ A}}{\sqrt{2}} = 0.206 \text{ A}$$

Use Equation 33.31 to find the power delivered by the source:

$$\mathcal{P}_{avg} = I_{rms} V_{rms} \cos \phi = (0.206 \text{ A})(106 \text{ V}) \cos (-34.0°)$$

$$= \boxed{18.1 \text{ W}}$$

33.7   Resonance in a Series *RLC* Circuit

A series *RLC* circuit is said to be **in resonance** when the driving frequency is such that the rms current has its maximum value. In general, the rms current can be written

$$I_{rms} = \frac{\Delta V_{rms}}{Z} \tag{33.33}$$

where *Z* is the impedance. Substituting the expression for *Z* from Equation 33.25 into Equation 33.33 gives

$$I_{rms} = \frac{\Delta V_{rms}}{\sqrt{R^2 + (X_L - X_C)^2}} \tag{33.34}$$

Because the impedance depends on the frequency of the source, the current in the *RLC* circuit also depends on the frequency. The frequency ω_0 at which $X_L - X_C = 0$ is called the **resonance frequency** of the circuit. To find ω_0, we set $X_L = X_C$, which gives $\omega_0 L = 1/\omega_0 C$, or

$$\omega_0 = \frac{1}{\sqrt{LC}} \tag{33.35}$$ ◀ Resonance frequency

This frequency also corresponds to the natural frequency of oscillation of an *LC* circuit (see Section 32.5). Therefore, the rms current in a series *RLC* circuit has its maximum value when the frequency of the applied voltage matches the natural oscillator frequency, which depends only on *L* and *C*. Furthermore, at the resonance frequency, the current is in phase with the applied voltage.

Quick Quiz 33.7   What is the impedance of a series *RLC* circuit at resonance? (a) larger than *R*   (b) less than *R*   (c) equal to *R*   (d) impossible to determine

A plot of rms current versus frequency for a series *RLC* circuit is shown in Active Figure 33.17a. The data assume a constant $\Delta V_{rms} = 5.0$ mV, $L = 5.0$ μH, and $C = 2.0$ nF. The three curves correspond to three values of *R*. In each case,

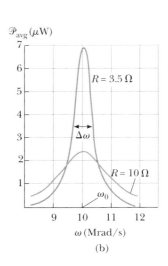

ACTIVE FIGURE 33.17

(a) The rms current versus frequency for a series RLC circuit for three values of R. The current reaches its maximum value at the resonance frequency ω_0. (b) Average power delivered to the circuit versus frequency for the series RLC circuit for two values of R.

Sign in at www.thomsonedu.com and go to ThomsonNOW to adjust the resistance, inductance, and capacitance of the circuit in Active Figure 33.13a. You can then determine the current and power for a given frequency or sweep through the frequencies to generate resonance curves as shown in this figure.

the rms current has its maximum value at the resonance frequency ω_0. Furthermore, the curves become narrower and taller as the resistance decreases.

Equation 33.34 shows that when $R = 0$, the current becomes infinite at resonance. Real circuits, however, always have some resistance, which limits the value of the current to some finite value.

We can also calculate the average power as a function of frequency for a series RLC circuit. Using Equations 33.32, 33.33, and 33.25 gives

$$\mathcal{P}_{\text{avg}} = I_{\text{rms}}^2 R = \frac{(\Delta V_{\text{rms}})^2}{Z^2}\,R = \frac{(\Delta V_{\text{rms}})^2 R}{R^2 + (X_L - X_C)^2} \tag{33.36}$$

Because $X_L = \omega L$, $X_C = 1/\omega C$, and $\omega_0^2 = 1/LC$, the term $(X_L - X_C)^2$ can be expressed as

$$(X_L - X_C)^2 = \left(\omega L - \frac{1}{\omega C}\right)^2 = \frac{L^2}{\omega^2}(\omega^2 - \omega_0^2)^2$$

Using this result in Equation 33.36 gives

◀ Average power as a function of frequency in an *RLC* circuit

$$\mathcal{P}_{\text{avg}} = \frac{(\Delta V_{\text{rms}})^2 R \omega^2}{R^2 \omega^2 + L^2(\omega^2 - \omega_0^2)^2} \tag{33.37}$$

Equation 33.37 shows that **at resonance, when $\omega = \omega_0$, the average power is a maximum** and has the value $(\Delta V_{\text{rms}})^2/R$. Active Figure 33.17b is a plot of average power versus frequency for two values of R in a series RLC circuit. As the resistance is made smaller, the curve becomes sharper in the vicinity of the resonance frequency. This curve sharpness is usually described by a dimensionless parameter known as the **quality factor,**[2] denoted by Q:

◀ Quality factor

$$Q = \frac{\omega_0}{\Delta \omega}$$

---

[2] The quality factor is also defined as the ratio $2\pi E/\Delta E$, where E is the energy stored in the oscillating system and ΔE is the energy decrease per cycle of oscillation due to the resistance.

where $\Delta\omega$ is the width of the curve measured between the two values of ω for which $\mathcal{P}_{avg}$ has one-half its maximum value, called the *half-power points* (see Active Fig. 33.17b.) It is left as a problem (Problem 68) to show that the width at the half-power points has the value $\Delta\omega = R/L$ so that

$$Q = \frac{\omega_0 L}{R} \tag{33.38}$$

A radio's receiving circuit is an important application of a resonant circuit. The radio is tuned to a particular station (which transmits an electromagnetic wave or signal of a specific frequency) by varying a capacitor, which changes the receiving circuit's resonance frequency. When the circuit is driven by the electromagnetic oscillations a radio signal produces in an antenna, the tuner circuit responds with a large amplitude of electrical oscillation only for the station frequency that matches the resonance frequency. Therefore, only the signal from one radio station is passed on to the amplifier and loudspeakers even though signals from all stations are driving the circuit at the same time. Because many signals are often present over a range of frequencies, it is important to design a high-Q circuit to eliminate unwanted signals. In this manner, stations whose frequencies are near but not equal to the resonance frequency give signals at the receiver that are negligibly small relative to the signal that matches the resonance frequency.

EXAMPLE 33.6   **A Resonating Series *RLC* Circuit**

Consider a series *RLC* circuit for which $R = 150\ \Omega$, $L = 20.0$ mH, $\Delta V_{rms} = 20.0$ V, and $\omega = 5\ 000$ s^{-1}. Determine the value of the capacitance for which the current is a maximum.

SOLUTION

Conceptualize   Consider the circuit in Active Figure 33.13a and imagine varying the frequency of the AC source. The current in the circuit has its maximum value at the resonance frequency ω_0.

Categorize   We find the result by using equations developed in this section, so we categorize this example as a substitution problem.

Use Equation 33.35 to solve for the required capacitance in terms of the resonance frequency:

$$\omega_0 = \frac{1}{\sqrt{LC}} \quad \rightarrow \quad C = \frac{1}{\omega_0^2 L}$$

Substitute numerical values:

$$C = \frac{1}{(5.00 \times 10^3\ \text{s}^{-1})^2 (20.0 \times 10^{-3}\ \text{H})} = 2.00\ \mu\text{F}$$

33.8   The Transformer and Power Transmission

As discussed in Section 27.6, it is economical to use a high voltage and a low current to minimize the $I^2 R$ loss in transmission lines when electric power is transmitted over great distances. Consequently, 350-kV lines are common, and in many areas, even higher-voltage (765-kV) lines are used. At the receiving end of such lines, the consumer requires power at a low voltage (for safety and for efficiency in design). In practice, the voltage is decreased to approximately 20 000 V at a distributing station, then to 4 000 V for delivery to residential areas, and finally to 120 V and 240 V at the customer's site. Therefore, a device is needed that can change the alternating voltage and current without causing appreciable changes in the power delivered. The AC transformer is that device.

Figure 33.18 An ideal transformer consists of two coils wound on the same iron core. An alternating voltage Δv_1 is applied to the primary coil, and the output voltage Δv_2 is across the resistor of resistance R_L.

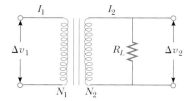

Figure 33.19 Circuit diagram for a transformer.

UPI/CORBIS

NIKOLA TESLA
American Physicist (1856–1943)

Tesla was born in Croatia, but he spent most of his professional life as an inventor in the United States. He was a key figure in the development of alternating-current electricity, high-voltage transformers, and the transport of electrical power using AC transmission lines. Tesla's viewpoint was at odds with the ideas of Thomas Edison, who committed himself to the use of direct current in power transmission. Tesla's AC approach won out.

In its simplest form, the **AC transformer** consists of two coils of wire wound around a core of iron as illustrated in Figure 33.18. (Compare this arrangement to Faraday's experiment in Figure 31.2.) The coil on the left, which is connected to the input alternating voltage source and has N_1 turns, is called the *primary winding* (or the *primary*). The coil on the right, consisting of N_2 turns and connected to a load resistor R_L, is called the *secondary winding* (or the *secondary*). The purposes of the iron core are to increase the magnetic flux through the coil and to provide a medium in which nearly all the magnetic field lines through one coil pass through the other coil. Eddy-current losses are reduced by using a laminated core. Transformation of energy to internal energy in the finite resistance of the coil wires is usually quite small. Typical transformers have power efficiencies from 90% to 99%. In the discussion that follows, let's assume we are working with an *ideal transformer*, one in which the energy losses in the windings and core are zero.

Faraday's law states that the voltage Δv_1 across the primary is

$$\Delta v_1 = -N_1 \frac{d\Phi_B}{dt} \tag{33.39}$$

where Φ_B is the magnetic flux through each turn. If we assume all magnetic field lines remain within the iron core, the flux through each turn of the primary equals the flux through each turn of the secondary. Hence, the voltage across the secondary is

$$\Delta v_2 = -N_2 \frac{d\Phi_B}{dt} \tag{33.40}$$

Solving Equation 33.39 for $d\Phi_B/dt$ and substituting the result into Equation 33.40 gives

$$\Delta v_2 = \frac{N_2}{N_1} \Delta v_1 \tag{33.41}$$

When $N_2 > N_1$, the output voltage Δv_2 exceeds the input voltage Δv_1. This configuration is referred to as a *step-up transformer*. When $N_2 < N_1$, the output voltage is less than the input voltage, and we have a *step-down transformer*.

When the switch in the secondary circuit is closed, a current I_2 is induced in the secondary. (In this discussion, uppercase I and ΔV refer to rms values.) If the load in the secondary circuit is a pure resistance, the induced current is in phase with the induced voltage. The power supplied to the secondary circuit must be provided by the AC source connected to the primary circuit as shown in Figure 33.19. In an ideal transformer where there are no losses, the power $I_1 \Delta V_1$ supplied by the source is equal to the power $I_2 \Delta V_2$ in the secondary circuit. That is,

$$I_1 \Delta V_1 = I_2 \Delta V_2 \tag{33.42}$$

The value of the load resistance R_L determines the value of the secondary current because $I_2 = \Delta V_2/R_L$. Furthermore, the current in the primary is $I_1 = \Delta V_1/R_{eq}$, where

$$R_{eq} = \left(\frac{N_1}{N_2}\right)^2 R_L \tag{33.43}$$

is the equivalent resistance of the load resistance when viewed from the primary side. We see from this analysis that a transformer may be used to match resistances between the primary circuit and the load. In this manner, maximum power transfer can be achieved between a given power source and the load resistance. For example, a transformer connected between the 1-kΩ output of an audio amplifier and an 8-Ω speaker ensures that as much of the audio signal as possible is transferred into the speaker. In stereo terminology, this process is called *impedance matching*.

To operate properly, many common household electronic devices require low voltages. A small transformer that plugs directly into the wall like the one illus-

© Thomson Learning/George Semple

Figure 33.20 The primary winding in this transformer is directly attached to the prongs of the plug. The secondary winding is connected to the power cord on the right, which runs to an electronic device. Many of these power-supply transformers also convert alternating current to direct current.

© Thomson Learning/George Semple

This transformer is smaller than the one in the opening photograph of this chapter. In addition, it is a step-down transformer. It drops the voltage from 4 000 V to 240 V for delivery to a group of residences.

trated in Figure 33.20 can provide the proper voltage. The photograph shows the two windings wrapped around a common iron core that is found inside all these little "black boxes." This particular transformer converts the 120-V AC in the wall socket to 12.5-V AC. (Can you determine the ratio of the numbers of turns in the two coils?) Some black boxes also make use of diodes to convert the alternating current to direct current. (See Section 33.9.)

EXAMPLE 33.7	**The Economics of AC Power**

An electricity-generating station needs to deliver energy at a rate of 20 MW to a city 1.0 km away. A common voltage for commercial power generators is 22 kV, but a step-up transformer is used to boost the voltage to 230 kV before transmission.

(A) If the resistance of the wires is 2.0 Ω and the energy costs are about 10¢/kWh, estimate what it costs the utility company for the energy converted to internal energy in the wires during one day.

SOLUTION

Conceptualize The resistance of the wires is in series with the resistance representing the load (homes and businesses). Therefore, there is a voltage drop in the wires, which means that some of the transmitted energy is converted to internal energy in the wires and never reaches the load.

Categorize This problem involves finding the power delivered to a resistive load in an AC circuit. Let's ignore any capacitive or inductive characteristics of the load and set the power factor equal to 1.

Analyze Calculate I_{rms} in the wires from Equation 33.31:

$$I_{rms} = \frac{\mathcal{P}_{avg}}{\Delta V_{rms}} = \frac{20 \times 10^6 \text{ W}}{230 \times 10^3 \text{ V}} = 87 \text{ A}$$

Determine the rate at which energy is delivered to the resistance in the wires from Equation 33.32:

$$\mathcal{P}_{avg} = I_{rms}^2 R = (87 \text{ A})^2 (2.0 \text{ }\Omega) = 15 \text{ kW}$$

Calculate the energy T_{ET} delivered to the wires over the course of a day:

$$T_{ET} = \mathcal{P}_{avg} \Delta t = (15 \text{ kW})(24 \text{ h}) = 360 \text{ kWh}$$

Find the cost of this energy at a rate of 10¢/kWh:

$$\text{Cost} = (360 \text{ kWh})(\$0.10/\text{kWh}) = \quad \$36$$

(B) Repeat the calculation for the situation in which the power plant delivers the energy at its original voltage of 22 kV.

SOLUTION

Calculate I_{rms} in the wires from Equation 33.31:

$$I_{rms} = \frac{\mathcal{P}_{avg}}{\Delta V_{rms}} = \frac{20 \times 10^6 \text{ W}}{22 \times 10^3 \text{ V}} = 910 \text{ A}$$

From Equation 33.32, determine the rate at which energy is delivered to the resistance in the wires:

$$\mathcal{P}_{avg} = I_{rms}^2 R = (910 \text{ A})^2 (2.0 \text{ }\Omega) = 1.7 \times 10^3 \text{ kW}$$

Calculate the energy delivered to the wires over the course of a day:

$$T_{ET} = \mathcal{P}_{avg} \Delta t = (1.7 \times 10^3 \text{ kW})(24 \text{ h}) = 4.1 \times 10^4 \text{ kWh}$$

Find the cost of this energy at a rate of 10¢/kWh:

$$\text{Cost} = (4.1 \times 10^4 \text{ kWh})(\$0.10/\text{kWh}) = \boxed{\$4.1 \times 10^3}$$

Finalize   Notice the tremendous savings that are possible through the use of transformers and high-voltage transmission lines. Such savings in combination with the efficiency of using alternating current to operate motors led to the universal adoption of alternating current instead of direct current for commercial power grids.

33.9   Rectifiers and Filters

Portable electronic devices such as radios and compact disc players are often powered by direct current supplied by batteries. Many devices come with AC–DC converters such as that shown in Figure 33.20. Such a converter contains a transformer that steps the voltage down from 120 V to, typically, 9 V and a circuit that converts alternating current to direct current. The AC–DC converting process is called **rectification,** and the converting device is called a **rectifier.**

The most important element in a rectifier circuit is a **diode,** a circuit element that conducts current in one direction but not the other. Most diodes used in modern electronics are semiconductor devices. The circuit symbol for a diode is ──▶├──, where the arrow indicates the direction of the current in the diode. A diode has low resistance to current in one direction (the direction of the arrow) and high resistance to current in the opposite direction. To understand how a diode rectifies a current, consider Figure 33.21a, which shows a diode and a resistor connected to the secondary of a transformer. The transformer reduces the voltage from 120-V AC to the lower voltage that is needed for the device having a resistance R (the load resistance). Because the diode conducts current in only one direction, the alternating current in the load resistor is reduced to the form shown by the solid curve in Figure 33.21b. The diode conducts current only when the side of the symbol containing the arrowhead has a positive potential relative to the other side. In this situation, the diode acts as a *half-wave rectifier* because current is present in the circuit only during half of each cycle.

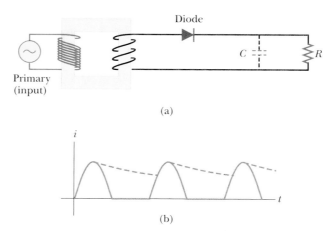

Figure 33.21   (a) A half-wave rectifier with an optional filter capacitor. (b) Current versus time in the resistor. The solid curve represents the current with no filter capacitor, and the dashed curve is the current when the circuit includes the capacitor.

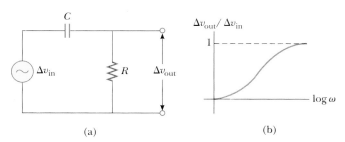

ACTIVE FIGURE 33.22

(a) A simple RC high-pass filter. (b) Ratio of output voltage to input voltage for an RC high-pass filter as a function of the angular frequency of the AC source.

Sign in at www.thomsonedu.com and go to ThomsonNOW to adjust the resistance and capacitance of the circuit in (a). You can then determine the output voltage for a given frequency or sweep through the frequencies to generate a curve like that in (b).

When a capacitor is added to the circuit as shown by the dashed lines and the capacitor symbol in Figure 33.21a, the circuit is a simple DC power supply. The time variation of the current in the load resistor (the dashed curve in Fig. 33.21b) is close to being zero, as determined by the RC time constant of the circuit. As the current in the circuit begins to rise at $t = 0$ in Figure 33.21b, the capacitor charges up. When the current begins to fall, however, the capacitor discharges through the resistor, so the current in the resistor does not fall as quickly as the current from the transformer.

The RC circuit in Figure 33.21a is one example of a **filter circuit**, which is used to smooth out or eliminate a time-varying signal. For example, radios are usually powered by a 60-Hz alternating voltage. After rectification, the voltage still contains a small AC component at 60 Hz (sometimes called *ripple*), which must be filtered. By "filtered," we mean that the 60-Hz ripple must be reduced to a value much less than that of the audio signal to be amplified because without filtering, the resulting audio signal includes an annoying hum at 60 Hz.

We can also design filters that respond differently to different frequencies. Consider the simple series RC circuit shown in Active Figure 33.22a. The input voltage is across the series combination of the two elements. The output is the voltage across the resistor. A plot of the ratio of the output voltage to the input voltage as a function of the logarithm of angular frequency (see Active Fig. 33.22b) shows that at low frequencies, Δv_{out} is much smaller than Δv_{in}, whereas at high frequencies, the two voltages are equal. Because the circuit preferentially passes signals of higher frequency while blocking low-frequency signals, the circuit is called an RC high-pass filter. (See Problem 45 for an analysis of this filter.)

Physically, a high-pass filter works because a capacitor "blocks out" direct current and AC current at low frequencies. At low frequencies, the capacitive reactance is large and much of the applied voltage appears across the capacitor rather than across the output resistor. As the frequency increases, the capacitive reactance drops and more of the applied voltage appears across the resistor.

Now consider the circuit shown in Active Figure 33.23a, where we have interchanged the resistor and capacitor and where the output voltage is taken across the capacitor. At low frequencies, the reactance of the capacitor and the voltage across the capacitor is high. As the frequency increases, the voltage across the capacitor drops. Therefore, this filter is an RC low-pass filter. The ratio of output voltage to input voltage (see Problem 46), plotted as a function of the logarithm of ω in Active Figure 33.23b, shows this behavior.

You may be familiar with crossover networks, which are an important part of the speaker systems for high-fidelity audio systems. These networks use low-pass filters to direct low frequencies to a special type of speaker, the "woofer," which is designed to reproduce the low notes accurately. The high frequencies are sent to the "tweeter" speaker.

(a)

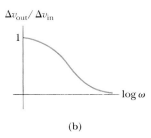

(b)

ACTIVE FIGURE 33.23

(a) A simple RC low-pass filter. (b) Ratio of output voltage to input voltage for an RC low-pass filter as a function of the angular frequency of the AC source.

Sign in at www.thomsonedu.com and go to ThomsonNOW to adjust the resistance and capacitance of the circuit in (a). You can then determine the output voltage for a given frequency or sweep through the frequencies to generate a curve like that in (b).

Summary

DEFINITIONS

In AC circuits that contain inductors and capacitors, it is useful to define the **inductive reactance** X_L and the **capacitive reactance** X_C as

$$X_L \equiv \omega L \qquad (33.10)$$

$$X_C \equiv \frac{1}{\omega C} \qquad (33.18)$$

where ω is the angular frequency of the AC source. The SI unit of reactance is the ohm.

The **impedance** Z of an RLC series AC circuit is

$$Z \equiv \sqrt{R^2 + (X_L - X_C)^2} \qquad (33.25)$$

This expression illustrates that we cannot simply add the resistance and reactances in a circuit. We must account for the applied voltage and current being out of phase, with the **phase angle** ϕ between the current and voltage being

$$\phi = \tan^{-1}\left(\frac{X_L - X_C}{R}\right) \qquad (33.27)$$

The sign of ϕ can be positive or negative, depending on whether X_L is greater or less than X_C. The phase angle is zero when $X_L = X_C$.

CONCEPTS AND PRINCIPLES

The **rms current** and **rms voltage** in an AC circuit in which the voltages and current vary sinusoidally are given by

$$I_{\text{rms}} = \frac{I_{\max}}{\sqrt{2}} = 0.707 I_{\max} \qquad (33.4)$$

$$\Delta V_{\text{rms}} = \frac{\Delta V_{\max}}{\sqrt{2}} = 0.707\,\Delta V_{\max} \qquad (33.5)$$

where $I_{\max}$ and $\Delta V_{\max}$ are the maximum values.

If an AC circuit consists of a source and a resistor, the current is in phase with the voltage. That is, the current and voltage reach their maximum values at the same time.

If an AC circuit consists of a source and an inductor, the current lags the voltage by 90°. That is, the voltage reaches its maximum value one quarter of a period before the current reaches its maximum value.

If an AC circuit consists of a source and a capacitor, the current leads the voltage by 90°. That is, the current reaches its maximum value one quarter of a period before the voltage reaches its maximum value.

The **average power** delivered by the source in an RLC circuit is

$$\mathcal{P}_{\text{avg}} = I_{\text{rms}}\,\Delta V_{\text{rms}} \cos\phi \qquad (33.31)$$

An equivalent expression for the average power is

$$\mathcal{P}_{\text{avg}} = I_{\text{rms}}^2 R \qquad (33.32)$$

The average power delivered by the source results in increasing internal energy in the resistor. No power loss occurs in an ideal inductor or capacitor.

The rms current in a series RLC circuit is

$$I_{\text{rms}} = \frac{\Delta V_{\text{rms}}}{\sqrt{R^2 + (X_L - X_C)^2}} \qquad (33.34)$$

A series RLC circuit is in resonance when the inductive reactance equals the capacitive reactance. When this condition is met, the rms current given by Equation 33.34 has its maximum value. The **resonance frequency** ω_0 of the circuit is

$$\omega_0 = \frac{1}{\sqrt{LC}} \qquad (33.35)$$

The rms current in a series RLC circuit has its maximum value when the frequency of the source equals ω_0, that is, when the "driving" frequency matches the resonance frequency.

AC transformers allow for easy changes in alternating voltage according to

$$\Delta v_2 = \frac{N_2}{N_1}\,\Delta v_1 \qquad (33.41)$$

where N_1 and N_2 are the numbers of windings on the primary and secondary coils, respectively, and Δv_1 and Δv_2 are the voltages on these coils.

Questions

☐ denotes answer available in *Student Solutions Manual/Study Guide;* **O** denotes objective question

1. **O (i)** What is the time average of the "square-wave" potential shown in Figure Q33.1? (a) $\sqrt{2}\,\Delta V_{max}$ (b) ΔV_{max} (c) $\Delta V_{max}/\sqrt{2}$ (d) $\Delta V_{max}/2$ (e) $\Delta V_{max}/4$ **(ii)** What is the rms voltage? Choose from the same possibilities.

Figure Q33.1

2. **O** Do AC ammeters and voltmeters read (a) peak-to-valley, (b) maximum, (c) rms, or (d) average values?

3. **O** A sinusoidally varying potential difference has amplitude 170 V. **(i)** What is its minimum instantaneous value? (a) 240 V (b) 170 V (c) 120 V (d) 0 (e) -120 V (f) -170 V (g) -240 V **(ii)** What is its average value? **(iii)** What is its rms value? Choose from the same possibilities in each case.

4. Why does a capacitor act as a short circuit at high frequencies? Why does it act as an open circuit at low frequencies?

5. Explain how the mnemonic "ELI the ICE man" can be used to recall whether current leads voltage or voltage leads current in *RLC* circuits. Note that E represents emf $\mathcal{E}$.

6. Why is the sum of the maximum voltages across each element in a series *RLC* circuit usually greater than the maximum applied voltage? Doesn't that inequality violate Kirchhoff's loop rule?

7. Does the phase angle depend on frequency? What is the phase angle when the inductive reactance equals the capacitive reactance?

8. **O (i)** When a particular inductor is connected to a source of sinusoidally varying emf with constant amplitude and a frequency of 60 Hz, the rms current is 3 A. What is the rms current if the source frequency is doubled? (a) 12 A (b) 6 A (c) 4.24 A (d) 3 A (e) 2.12 A (f) 1.5 A (g) 0.75 A **(ii)** Repeat part (i) assuming the load is a capacitor instead of an inductor. **(iii)** Repeat part (i) assuming the load is a resistor instead of an inductor.

9. **O** What is the impedance of a series *RLC* circuit at resonance? (a) X_L (b) X_C (c) R (d) $X_L - X_C$ (e) $2X_L$ (f) $\sqrt{2}R$ (g) 0

10. **O** What is the phase angle in a series *RLC* circuit at resonance? (a) $180°$ (b) $90°$ (c) 0 (d) $-90°$ (e) None of these answers is necessarily correct.

11. A certain power supply can be modeled as a source of emf in series with both a resistance of 10 Ω and an inductive reactance of 5 Ω. To obtain maximum power delivered to the load, it is found that the load should have a resistance of $R_L = 10$ Ω, an inductive reactance of zero, and a capacitive reactance of 5 Ω. (a) With this load, is the circuit in resonance? (b) With this load, what fraction of the average power put out by the source of emf is delivered to the load? (c) To increase the fraction of the power delivered to the load, how could the load be changed? You may wish to review Example 28.2 and Problem 4 in Chapter 28 on maximum power transfer in DC circuits.

12. As shown in Figure 7.5, a person pulls a vacuum cleaner at speed v across a horizontal floor, exerting on it a force of magnitude F directed upward at an angle θ with the horizontal. At what rate is the person doing work on the cleaner? State as completely as you can the analogy between power in this situation and in an electric circuit.

13. **O** A circuit containing a generator, a capacitor, an inductor, and a resistor has a high-Q resonance at 1 000 Hz. From greatest to least, rank the following contributions to the impedance of the circuit at that frequency and at lower and higher frequencies, and note any cases of equality in your ranking. (a) X_C at 500 Hz (b) X_C at 1 000 Hz (c) X_C at 1 500 Hz (d) X_L at 500 Hz (e) X_L at 1 000 Hz (f) X_L at 1 500 Hz (g) R at 500 Hz (h) R at 1 000 Hz (i) R at 1 500 Hz

14. Do some research to answer these questions: Who invented the metal detector? Why? Did it work?

15. Will a transformer operate if a battery is used for the input voltage across the primary? Explain.

16. Explain how the quality factor is related to the response characteristics of a radio receiver. Which variable most strongly influences the quality factor?

17. An ice storm breaks a transmission line and interrupts electric power to a town. A homeowner starts a gasoline-powered 120-V generator and clips its output terminals to "hot" and "ground" terminals of the electrical panel for his house. On a power pole down the block is a transformer designed to step down the voltage for household use. It has a ratio of turns N_1/N_2 of 100 to 1. A repairman climbs the pole. What voltage will he encounter on the input side of the transformer? As this question implies, safety precautions must be taken in the use of home generators and during power failures in general.

Problems

WebAssign The Problems from this chapter may be assigned online in WebAssign.

ThomsonNOW   Sign in at **www.thomsonedu.com** and go to ThomsonNOW to assess your understanding of this chapter's topics with additional quizzing and conceptual questions.

1, 2, 3 denotes straightforward, intermediate, challenging;   ☐ denotes full solution available in *Student Solutions Manual/Study Guide;*   ▲ denotes coached solution with hints available at **www.thomsonedu.com;**   denotes developing symbolic reasoning; ● denotes asking for qualitative reasoning;   ▪ denotes computer useful in solving problem

Section 33.1 AC Sources

Section 33.2 Resistors in an AC Circuit

1. The rms output voltage of an AC source is 200 V and the operating frequency is 100 Hz. Write the equation giving the output voltage as a function of time.

2. (a) What is the resistance of a lightbulb that uses an average power of 75.0 W when connected to a 60.0-Hz power source having a maximum voltage of 170 V? (b) **What If?** What is the resistance of a 100-W lightbulb?

3. An AC power supply produces a maximum voltage $\Delta V_{max} = 100$ V. This power supply is connected to a 24.0-Ω resistor, and the current and resistor voltage are measured with an ideal AC ammeter and voltmeter as shown in Figure P33.3. What does each meter read? An ideal ammeter has zero resistance and an ideal voltmeter has infinite resistance.

$\Delta V_{max} = 100$ V

$R = 24.0$ Ω

Figure P33.3

4. In the simple AC circuit shown in Active Figure 33.2, $R = 70.0$ Ω and $\Delta v = \Delta V_{max} \sin \omega t$. (a) If $\Delta v_R = 0.250 \Delta V_{max}$ for the first time at $t = 0.010\ 0$ s, what is the angular frequency of the source? (b) What is the next value of t for which $\Delta v_R = 0.250 \Delta V_{max}$?

5. The current in the circuit shown in Active Figure 33.2 equals 60.0% of the peak current at $t = 7.00$ ms. What is the lowest source frequency that gives this current?

6. An audio amplifier, represented by the AC source and resistor in Figure P33.6, delivers to the speaker alternating voltage at audio frequencies. If the source voltage has an amplitude of 15.0 V, $R = 8.20$ Ω, and the speaker is equivalent to a resistance of 10.4 Ω, what is the time-averaged power transferred to it?

R

Speaker

Figure P33.6

Section 33.3 Inductors in an AC Circuit

7. In a purely inductive AC circuit as shown in Active Figure 33.6, $\Delta V_{max} = 100$ V. (a) The maximum current is 7.50 A at 50.0 Hz. Calculate the inductance L. (b) **What If?** At what angular frequency ω is the maximum current 2.50 A?

8. An inductor has a 54.0-Ω reactance at 60.0 Hz. What is the maximum current if this inductor is connected to a 50.0-Hz source that produces a 100-V rms voltage?

9. ▲ For the circuit shown in Active Figure 33.6, $\Delta V_{max} = 80.0$ V, $\omega = 65.0\pi$ rad/s, and $L = 70.0$ mH. Calculate the current in the inductor at $t = 15.5$ ms.

10. A 20.0-mH inductor is connected to a standard electrical outlet ($\Delta V_{rms} = 120$ V, $f = 60.0$ Hz). Determine the energy stored in the inductor at $t = \frac{1}{180}$ s, assuming this energy is zero at $t = 0$.

11. **Review problem.** Determine the maximum magnetic flux through an inductor connected to a standard electrical outlet ($\Delta V_{rms} = 120$ V, $f = 60.0$ Hz).

Section 33.4 Capacitors in an AC Circuit

12. (a) For what frequencies does a 22.0-μF capacitor have a reactance below 175 Ω? (b) **What If?** What is the reactance of a 44.0-μF capacitor over this same frequency range?

13. What is the maximum current in a 2.20-μF capacitor when it is connected across (a) a North American electrical outlet having $\Delta V_{rms} = 120$ V and $f = 60.0$ Hz, and (b) a European electrical outlet having $\Delta V_{rms} = 240$ V and $f = 50.0$ Hz?

14. A capacitor C is connected to a power supply that operates at a frequency f and produces an rms voltage ΔV. What is the maximum charge that appears on either capacitor plate?

15. What maximum current is delivered by an AC source with $\Delta V_{max} = 48.0$ V and $f = 90.0$ Hz when connected across a 3.70-μF capacitor?

16. A 1.00-mF capacitor is connected to a standard electrical outlet ($\Delta V_{rms} = 120$ V, $f = 60.0$ Hz). Determine the current in the wires at $t = \frac{1}{180}$ s, assuming the energy stored in the capacitor is zero at $t = 0$.

Section 33.5 The *RLC* Series Circuit

17. An inductor ($L = 400$ mH), a capacitor ($C = 4.43$ μF), and a resistor ($R = 500$ Ω) are connected in series. A 50.0-Hz AC source produces a peak current of 250 mA in the circuit. (a) Calculate the required peak voltage ΔV_{max}. (b) Determine the phase angle by which the current leads or lags the applied voltage.

18. At what frequency does the inductive reactance of a 57.0-μH inductor equal the capacitive reactance of a 57.0-μF capacitor?

19. A series AC circuit contains the following components: a 150-Ω resistor, an inductor of 250 mH, a capacitor of 2.00 μF, and a source with $\Delta V_{\text{max}} = 210$ V operating at 50.0 Hz. Calculate the (a) inductive reactance, (b) capacitive reactance, (c) impedance, (d) maximum current, and (e) phase angle between current and source voltage.

20. A sinusoidal voltage $\Delta v(t) = (40.0$ V$) \sin (100t)$ is applied to a series RLC circuit with $L = 160$ mH, $C = 99.0$ μF, and $R = 68.0$ Ω. (a) What is the impedance of the circuit? (b) What is the maximum current? (c) Determine the numerical values for I_{max}, ω, and ϕ in the equation $i(t) = I_{\text{max}} \sin (\omega t - \phi)$.

21. ▲ An RLC circuit consists of a 150-Ω resistor, a 21.0-μF capacitor, and a 460-mH inductor connected in series with a 120-V, 60.0-Hz power supply. (a) What is the phase angle between the current and the applied voltage? (b) Which reaches its maximum earlier, the current or the voltage?

22. Four circuit elements—a capacitor, an inductor, a resistor, and an AC source—are connected together in various ways. First the capacitor is connected to the source, and the rms current is found to be 25.1 mA. The capacitor is disconnected and discharged, and then it is connected in series with the resistor and the source, making the rms current 15.7 mA. The circuit is disconnected and the capacitor discharged. The capacitor is then connected in series with the inductor and the source, making the rms current 68.2 mA. After the circuit is disconnected and the capacitor discharged, all four circuit elements are connected together in a series loop. What is the rms current in the circuit?

23. A person is working near the secondary of a transformer as shown in Figure P33.23. The primary voltage is 120 V at 60.0 Hz. The capacitance C_s, which is the stray capacitance between the hand and the secondary winding, is 20.0 pF. Assuming the person has a body resistance to ground of $R_b = 50.0$ kΩ, determine the rms voltage across the body. *Suggestion:* Model the secondary of the transformer as an AC source.

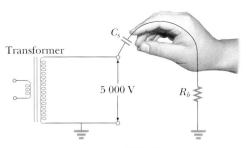

Figure P33.23

24. An AC source with $\Delta V_{\text{max}} = 150$ V and $f = 50.0$ Hz is connected between points a and d in Figure P33.24. Calculate the maximum voltages between (a) points a and b, (b) points b and c, (c) points c and d, and (d) points b and d.

Figure P33.24   Problems 24 and 64.

25. Draw to scale a phasor diagram and determine Z, X_L, X_C, and ϕ for an AC series circuit for which $R = 300$ Ω, $C = 11.0$ μF, $L = 0.200$ H, and $f = (500/\pi)$ Hz.

26. ● In an RLC series circuit that includes a source of alternating current operating at fixed frequency and voltage, the resistance R is equal to the inductive reactance. If the plate separation of the parallel-plate capacitor is reduced to one-half its original value, the current in the circuit doubles. Find the initial capacitive reactance in terms of R. Explain each step in your solution.

Section 33.6 Power in an AC Circuit

27. ▲ An AC voltage of the form $\Delta v = (100$ V$) \sin (1\,000t)$ is applied to a series RLC circuit. Assume the resistance is 400 Ω, the capacitance is 5.00 μF, and the inductance is 0.500 H. Find the average power delivered to the circuit.

28. A series RLC circuit has a resistance of 45.0 Ω and an impedance of 75.0 Ω. What average power is delivered to this circuit when $\Delta V_{\text{rms}} = 210$ V?

29. In a certain series RLC circuit, $I_{\text{rms}} = 9.00$ A, $\Delta V_{\text{rms}} = 180$ V, and the current leads the voltage by 37.0°. (a) What is the total resistance of the circuit? (b) Calculate the reactance of the circuit ($X_L - X_C$).

30. Suppose you manage a factory that uses many electric motors. The motors create a large inductive load to the electric power line as well as a resistive load. The electric company builds an extra-heavy distribution line to supply you with a component of current that is 90° out of phase with the voltage as well as with current in phase with the voltage. The electric company charges you an extra fee for "reactive volt-amps" in addition to the amount you pay for the energy you use. You can avoid the extra fee by installing a capacitor between the power line and your factory. The following problem models this solution.

In an RL circuit, a 120-V (rms), 60.0-Hz source is in series with a 25.0-mH inductor and a 20.0-Ω resistor. What are (a) the rms current and (b) the power factor? (c) What capacitor must be added in series to make the power factor 1? (d) To what value can the supply voltage be reduced if the power supplied is to be the same as before the capacitor was installed?

31. Energy is to be transmitted at the rate of 20.0 kW with only 1.00% loss over a distance of 18.0 km at potential difference ΔV. (a) What is the diameter required for each of the two copper wires in the transmission line? Assume the current density is uniform in the conductors. (b) State how the diameter depends on ΔV. (c) Evaluate the diameter for $\Delta V = 1\,500$ V. (d) If you choose to make the diameter 3.00 mm, what potential difference is required?

32. ● A series circuit consists of an AC generator with an rms voltage of 120 V at a frequency of 60.0 Hz and a magnetic buzzer with a resistance of 100 Ω and an inductance of 100 mH. (a) Find the circuit's power factor. (b) Suppose a higher power factor is desired. Can a power factor of 1.00 be achieved by changing the inductance or any other circuit parameters? (c) Show that a power factor of 1.00 can be attained by inserting a capacitor into the original circuit, and find the value of its capacitance.

33. A diode is a device that allows current to be carried in only one direction (the direction indicated by the arrowhead in its circuit symbol). Find in terms of ΔV and R the

2 = intermediate;   3 = challenging;   □ = SSM/SG;   ▲ = ThomsonNOW;   = symbolic reasoning;   ● = qualitative reasoning

average power delivered to the diode circuit of Figure P33.33.

Figure P33.33

Section 33.7 Resonance in a Series *RLC* Circuit

34. A radar transmitter contains an *LC* circuit oscillating at 1.00×10^{10} Hz. (a) What capacitance resonates with a one-turn loop having an inductance of 400 pH at this frequency? (b) The capacitor has square, parallel plates separated by 1.00 mm of air. What should the edge length of the plates be? (c) What is the common reactance of the loop and capacitor at resonance?

35. An *RLC* circuit is used in a radio to tune into an FM station broadcasting at 99.7 MHz. The resistance in the circuit is 12.0 Ω, and the inductance is 1.40 μH. What capacitance should be used?

36. A series *RLC* circuit has components with the following values: $L = 20.0$ mH, $C = 100$ nF, $R = 20.0$ Ω, and $\Delta V_{max} = 100$ V, with $\Delta v = \Delta V_{max} \sin \omega t$. Find (a) the resonant frequency, (b) the amplitude of the current at the resonant frequency, (c) the *Q* of the circuit, and (d) the amplitude of the voltage across the inductor at resonance.

37. A 10.0-Ω resistor, 10.0-mH inductor, and 100-μF capacitor are connected in series to a 50.0-V (rms) source having variable frequency. Find the energy delivered to the circuit during one period if the operating frequency is twice the resonance frequency.

38. A resistor *R*, inductor *L*, and capacitor *C* are connected in series to an AC source of rms voltage ΔV and variable frequency. Find the energy delivered to the circuit during one period if the operating frequency is twice the resonance frequency.

39. Compute the quality factor for the circuits described in Problems 20 and 21. Which circuit has the sharper resonance?

Section 33.8 The Transformer and Power Transmission

40. A step-down transformer is used for recharging the batteries of portable devices such as tape players. The ratio of turns inside the transformer is 13:1, and the transformer is used with 120-V (rms) household service. If a particular ideal transformer draws 0.350 A from the house outlet, what are (a) the voltage and (b) the current supplied to a tape player from the transformer? (c) How much power is delivered?

41. A transformer has $N_1 = 350$ turns and $N_2 = 2\ 000$ turns. If the input voltage is $\Delta v(t) = (170\ \text{V}) \cos \omega t$, what rms voltage is developed across the secondary coil?

42. A step-up transformer is designed to have an output voltage of $2\ 200$ V (rms) when the primary is connected across a 110-V (rms) source. (a) If the primary winding has 80 turns, how many turns are required on the secondary? (b) If a load resistor across the secondary draws a current of 1.50 A, what is the current in the primary, assuming ideal conditions? (c) **What If?** If the transformer actually has an efficiency of 95.0%, what is the current in the primary when the secondary current is 1.20 A?

43. A transmission line that has a resistance per unit length of 4.50×10^{-4} Ω/m is to be used to transmit 5.00 MW across 400 miles (6.44×10^5 m). The output voltage of the generator is 4.50 kV. (a) What is the line loss if a transformer is used to step up the voltage to 500 kV? (b) What fraction of the input power is lost to the line under these circumstances? (c) **What If?** What difficulties would be encountered in attempting to transmit the 5.00 MW at the generator voltage of 4.50 kV?

Section 33.9 Rectifiers and Filters

44. One particular plug-in power supply for a radio looks similar to the one shown in Figure 33.20 and is marked with the following information: Input 120 V AC 8 W Output 9 V DC 300 mA. Assume these values are accurate to two digits. (a) Find the energy efficiency of the device when the radio is operating. (b) At what rate is energy wasted in the device when the radio is operating? (c) Suppose the input power to the transformer is 8.0 W when the radio is switched off and energy costs \$0.135/kWh from the electric company. Find the cost of having six such transformers around the house, each plugged in for 31 days.

45. Consider the filter circuit shown in Active Figure 33.22a. (a) Show that the ratio of the output voltage to the input voltage is

$$\frac{\Delta v_{out}}{\Delta v_{in}} = \frac{R}{\sqrt{R^2 + \left(\dfrac{1}{\omega C}\right)^2}}$$

(b) What value does this ratio approach as the frequency decreases toward zero? What value does this ratio approach as the frequency increases without limit? (c) At what frequency is the ratio equal to one-half?

46. Consider the filter circuit shown in Active Figure 33.23a. (a) Show that the ratio of the output voltage to the input voltage is

$$\frac{\Delta v_{out}}{\Delta v_{in}} = \frac{1/\omega C}{\sqrt{R^2 + \left(\dfrac{1}{\omega C}\right)^2}}$$

(b) What value does this ratio approach as the frequency decreases toward zero? What value does this ratio approach as the frequency increases without limit? (c) At what frequency is the ratio equal to one-half?

47. ▲ The *RC* high-pass filter shown in Active Figure 33.22a has a resistance $R = 0.500$ Ω. (a) What capacitance gives an output signal that has one-half the amplitude of a 300-Hz input signal? (b) What is the ratio $(\Delta v_{out}/\Delta v_{in})$ for a 600-Hz signal? You may use the result of Problem 45.

48. The *RC* low-pass filter shown in Active Figure 33.23a has a resistance $R = 90.0$ Ω and a capacitance $C = 8.00$ nF. Calculate the ratio $(\Delta v_{out}/\Delta v_{in})$ for an input frequency of (a) 600 Hz and (b) 600 kHz. You may use the result of Problem 46.

49. The resistor in Figure P33.49 represents the midrange speaker in a three-speaker system. Assume its resistance to be constant at 8.00 Ω. The source represents an audio amplifier producing signals of uniform amplitude $\Delta V_{max} = 10.0$ V at all audio frequencies. The inductor and capacitor are to function as a band-pass filter with $\Delta v_{out}/\Delta v_{in} = \frac{1}{2}$ at 200 Hz and at 4 000 Hz. (a) Determine the required values of L and C. (b) Find the maximum value of the ratio $\Delta v_{out}/\Delta v_{in}$. (c) Find the frequency f_0 at which the ratio has its maximum value. (d) Find the phase shift between Δv_{in} and Δv_{out} at 200 Hz, at f_0, and at 4 000 Hz. (e) Find the average power transferred to the speaker at 200 Hz, at f_0, and at 4 000 Hz. (f) Treating the filter as a resonant circuit, find its quality factor.

Figure P33.49

Additional Problems

50. Show that the rms value for the sawtooth voltage shown in Figure P33.50 is $\Delta V_{max}/\sqrt{3}$.

Figure P33.50

51. ● A 400-Ω resistor, an inductor, and a capacitor are in series with a generator. The reactance of the inductor is 700 Ω, and the circuit impedance is 760 Ω. (a) Explain what you can and cannot determine about the reactance of the capacitor. (b) If you find that the source power decreases as you raise the frequency, what do you know about the capacitive reactance in the original circuit? (c) Repeat part (a) assuming the resistance is 200 Ω instead of 400 Ω.

52. ● A capacitor, a coil, and two resistors of equal resistance are arranged in an AC circuit as shown in Figure P33.52. An AC generator provides an emf of 20.0 V (rms) at a frequency of 60.0 Hz. When the double-throw switch S is

Figure P33.52

open as shown in the figure, the rms current is 183 mA. When the switch is closed in position 1, the rms current is 298 mA. When the switch is closed in position 2, the rms current is 137 mA. Determine the values of R, C and L. Is more than one set of values possible? Explain.

53. ▲ A series RLC circuit consists of an 8.00-Ω resistor, a 5.00-μF capacitor, and a 50.0-mH inductor. A variable frequency source applies an emf of 400 V (rms) across the combination. Determine the power delivered to the circuit when the frequency is equal to one-half the resonance frequency.

54. ● A series RLC circuit has resonance angular frequency 2 000 rad/s. When it is operating at some certain frequency, $X_L = 12.0$ Ω and $X_C = 8.00$ Ω. (a) Is this certain frequency higher than, lower than, or the same as the resonance frequency? Explain how you can tell. (b) Explain whether it is possible to determine the values of both L and C. (c) If it is possible, find L and C. If this determination is not possible, give a compact expression for the condition that L and C must satisfy.

55. ● **Review problem.** One insulated conductor from a household extension cord has a mass per length of 19.0 g/m. A section of this conductor is held under tension between two clamps. A subsection is located in a magnetic field of magnitude 15.3 mT directed perpendicular to the length of the cord. When the cord carries an AC current of 9.00 A at a frequency of 60.0 Hz, it vibrates in resonance in its simplest standing-wave vibration state. Determine the relationship that must be satisfied between the separation d of the clamps and the tension T in the cord. Determine one possible combination of values for these variables.

56. Sketch a graph of the phase angle for an RLC series circuit as a function of angular frequency from zero to a frequency much higher than the resonance frequency. Identify the value of ϕ at the resonance angular frequency ω_0. Prove that the slope of the graph of ϕ versus ω at the resonance point is $2Q/\omega_0$.

57. In Figure P33.57, find the rms current delivered by the 45.0-V (rms) power supply when (a) the frequency is very large and (b) the frequency is very small.

Figure P33.57

58. In the circuit shown in Figure P33.58 (page 950), assume all parameters except C are given. (a) Find the current as a function of time. (b) Find the power delivered to the circuit. (c) Find the current as a function of time after *only* switch 1 is opened. (d) After switch 2 is *also* opened, the current and voltage are in phase. Find the capacitance C. (e) Find the impedance of the circuit when both switches are open. (f) Find the maximum energy stored in the capacitor during oscillations. (g) Find the maximum energy stored in the inductor during oscillations.

(h) Now the frequency of the voltage source is doubled. Find the phase difference between the current and the voltage. (i) Find the frequency that makes the inductive reactance one-half the capacitive reactance.

$$\Delta v(t) = \Delta V_{max} \cos \omega t$$

Figure P33.58

59. An 80.0-Ω resistor and a 200-mH inductor are connected in *parallel* across a 100-V (rms), 60.0-Hz source. (a) What is the rms current in the resistor? (b) By what angle does the total current lead or lag behind the voltage?

60. Make an order-of-magnitude estimate of the electric current that the electric company delivers to a town (Fig. P33.60) from a remote generating station. State the data you measure or estimate. If you wish, you may consider a suburban residential community of 20 000 people.

Figure P33.60

61. Consider a series *RLC* circuit having the following circuit parameters: $R = 200 \ \Omega$, $L = 663$ mH, and $C = 26.5 \ \mu$F. The applied voltage has an amplitude of 50.0 V and a frequency of 60.0 Hz. Find the following amplitudes. (a) the current I_{max} and its phase relative to the applied voltage Δv (b) the maximum voltage ΔV_R across the resistor and its phase relative to the current (c) the maximum voltage ΔV_C across the capacitor and its phase relative to the current (d) the maximum voltage ΔV_L across the inductor and its phase relative to the current

62. A voltage $\Delta v = (100$ V$) \sin \omega t$ is applied across a series combination of a 2.00-H inductor, a 10.0-μF capacitor, and a 10.0-Ω resistor. (a) Determine the angular frequency ω_0 at which the power delivered to the resistor is a maximum. (b) Calculate the power delivered at that frequency. (c) Determine the two angular frequencies ω_1 and ω_2 at which the power is one-half the maximum value. *Note:* The Q of the circuit is $\omega_0/(\omega_2 - \omega_1)$.

63. *Impedance matching.* Example 28.2 showed that maximum power is transferred when the internal resistance of a DC source is equal to the resistance of the load. A transformer may be used to provide maximum power transfer between two AC circuits that have different impedances Z_1 and Z_2. (a) Show that the ratio of turns N_1/N_2 needed to meet this condition is

$$\frac{N_1}{N_2} = \sqrt{\frac{Z_1}{Z_2}}$$

(b) Suppose you want to use a transformer as an impedance-matching device between an audio amplifier that has an output impedance of 8.00 kΩ and a speaker that has an input impedance of 8.00 Ω. What should your N_1/N_2 ratio be?

64. ● A power supply with $\Delta V_{rms} = 120$ V is connected between points a and d in Figure P33.24. At what frequency will it deliver a power of 250 W? Explain your answer.

65. Figure P33.65a shows a parallel *RLC* circuit, and the corresponding phasor diagram is given in Figure P33.65b. The instantaneous voltages (and rms voltages) across each of the three circuit elements are the same, and each is in phase with the current in the resistor. The currents in C and L lead or lag the current in the resistor as shown in Figure P33.65b. (a) Show that the rms current delivered by the source is

$$I_{rms} = \Delta V_{rms} \left[\frac{1}{R^2} + \left(\omega C - \frac{1}{\omega L} \right)^2 \right]^{1/2}$$

(b) Show that the phase angle ϕ between ΔV_{rms} and I_{rms} is given by

$$\tan \phi = R \left(\frac{1}{X_C} - \frac{1}{X_L} \right)$$

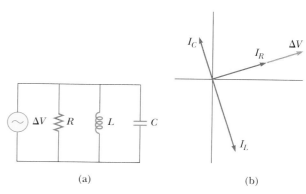

(a)   (b)

Figure P33.65

66. A certain electric circuit is described by the equations

$$\frac{200 \text{ V}}{4.00 \text{ A}} = \sqrt{(35.0 \ \Omega)^2 + \left[\omega(205 \text{ mH}) - \frac{1}{\omega C} \right]^2}$$

$$\omega = 2\pi (100 \text{ Hz})$$

State a problem for which these equations would appear in the solution, giving the data and identifying the unknown. Evaluate the unknown quantity.

67. ■ A series *RLC* circuit is operating at 2 000 Hz. At this frequency, $X_L = X_C = 1$ 884 Ω. The resistance of the circuit is 40.0 Ω. (a) Prepare a table showing the values of X_L, X_C, and Z for $f = 300$, 600, 800, 1 000, 1 500, 2 000, 3 000, 4 000, 6 000, and 10 000 Hz. (b) Plot on the same set of axes X_L, X_C, and Z as a function of ln f.

2 = intermediate;   3 = challenging;   □ = SSM/SG;   ▲ = ThomsonNOW;   = symbolic reasoning;   ● = qualitative reasoning

68. ▪ A series *RLC* circuit with $R = 1.00\ \Omega$, $L = 1.00$ mH, and $C = 1.00$ nF is connected to an AC source delivering 1.00 V (rms). Make a precise graph of the power delivered to the circuit as a function of the frequency and verify that the full width of the resonance peak at half-maximum is $R/2\pi L$.

69. ● Marie Cornu, a physicist at the Polytechnic Institute in Paris, invented phasors in about 1880. This problem helps you see their general utility in representing oscillations. Two mechanical vibrations are represented by the expressions

$$y_1 = (12.0\ \text{cm}) \sin (4.5t)$$

and

$$y_2 = (12.0\ \text{cm}) \sin (4.5t + 70°)$$

Find the amplitude and phase constant of the sum of these functions (a) by using a trigonometric identity (as from Appendix B) and (b) by representing the oscillations as phasors. State the result of comparing the answers to (a) and (b). (c) Phasors make it equally easy to add traveling waves. Find the amplitude and phase constant of the sum of the three waves represented by

$$y_1 = (12.0\ \text{cm}) \sin (15x - 4.5t + 70°)$$

$$y_2 = (15.5\ \text{cm}) \sin (15x - 4.5t - 80°)$$

$$y_3 = (17.0\ \text{cm}) \sin (15x - 4.5t + 160°)$$

Answers to Quick Quizzes

33.1 **(i)**, (c). The phasor in (c) has the largest projection onto the vertical axis. **(ii)**, (b). The phasor in (b) has the smallest-magnitude projection onto the vertical axis.

33.2 (b). For low frequencies, the reactance of the inductor is small, so the current is large. Most of the voltage from the source is across the lightbulb, so the power delivered to it is large.

33.3 (a). For high frequencies, the reactance of the capacitor is small, so the current is large. Most of the voltage from the source is across the lightbulb, so the power delivered to it is large.

33.4 (b). For low frequencies, the reactance of the capacitor is large, so very little current exists in the capacitor branch. The reactance of the inductor is small, so current exists in the inductor branch and the lightbulb glows. As the frequency increases, the inductive reac-tance increases and the capacitive reactance decreases. At high frequencies, more current exists in the capacitor branch than the inductor branch and the lightbulb glows more dimly.

33.5 (a) $X_L < X_C$ (b) $X_L = X_C$ (c) $X_L > X_C$

33.6 (c). The cosine of $-\phi$ is the same as that of $+\phi$, so the $\cos\phi$ factor in Equation 33.31 is the same for both frequencies. The factor ΔV_{rms} is the same because the source voltage is fixed. According to Equation 33.27, changing $+\phi$ to $-\phi$ simply interchanges the values of X_L and X_C. Equation 33.25 tells us that such an interchange does not affect the impedance, so the current I_{rms} in Equation 33.31 is the same for both frequencies.

33.7 (c). At resonance, $X_L = X_C$. According to Equation 33.25, that gives us $Z = R$.

Electromagnetic waves cover a broad spectrum of wavelengths, with waves in various wavelength ranges having distinct properties. These images of the Crab Nebula show different structure for observations made with waves of various wavelengths. The images (clockwise starting from upper left) were taken with x-rays, visible light, radio waves, and infrared waves. (upper left, NASA/CXC/SAO; upper right, Palomar Observatory; lower right, VLA/NRAO; lower left, WM Keck Observatory)

34.1 Displacement Current and the General Form of Ampère's Law

34.2 Maxwell's Equations and Hertz's Discoveries

34.3 Plane Electromagnetic Waves

34.4 Energy Carried by Electromagnetic Waves

34.5 Momentum and Radiation Pressure

34.6 Production of Electromagnetic Waves by an Antenna

34.7 The Spectrum of Electromagnetic Waves

34 Electromagnetic Waves

The waves described in Chapters 16, 17, and 18 are mechanical waves. By definition, the propagation of mechanical disturbances—such as sound waves, water waves, and waves on a string—requires the presence of a medium. This chapter is concerned with the properties of electromagnetic waves, which (unlike mechanical waves) can propagate through empty space.

We begin by considering Maxwell's contributions in modifying Ampère's law, which we studied in Chapter 30. We then discuss Maxwell's equations, which form the theoretical basis of all electromagnetic phenomena. These equations predict the existence of electromagnetic waves that propagate through space at the speed of light c. Heinrich Hertz confirmed Maxwell's prediction when he generated and detected electromagnetic waves in 1887. That discovery has led to many practical communication systems, including radio, television, radar, and optoelectronics.

Next, we learn how electromagnetic waves are generated by oscillating electric charges. The waves radiated from the oscillating charges can be detected at great distances. Furthermore, because electromagnetic waves carry energy and momentum, they can exert pressure on a surface. The chapter concludes with a look at many frequencies covered by electromagnetic waves.

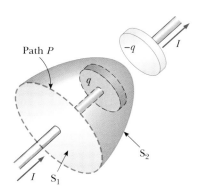

Figure 34.1 Two surfaces S_1 and S_2 near the plate of a capacitor are bounded by the same path P. The conduction current in the wire passes only through S_1, which leads to a contradiction in Ampère's law that is resolved only if one postulates a displacement current through S_2.

JAMES CLERK MAXWELL
Scottish Theoretical Physicist (1831–1879)
Maxwell developed the electromagnetic theory of light and the kinetic theory of gases, and explained the nature of Saturn's rings and color vision. Maxwell's successful interpretation of the electromagnetic field resulted in the field equations that bear his name. Formidable mathematical ability combined with great insight enabled him to lead the way in the study of electromagnetism and kinetic theory. He died of cancer before he was 50.

34.1 Displacement Current and the General Form of Ampère's Law

In Chapter 30, we discussed using Ampère's law (Eq. 30.13) to analyze the magnetic fields created by currents:

$$\oint \vec{\mathbf{B}} \cdot d\vec{\mathbf{s}} = \mu_0 I$$

In this equation, the line integral is over any closed path through which conduction current passes, where conduction current is defined by the expression $I = dq/dt$. (In this section, we use the term *conduction current* to refer to the current carried by charge carriers in the wire to distinguish it from a new type of current we shall introduce shortly.) We now show that **Ampère's law in this form is valid only if any electric fields present are constant in time.** James Clerk Maxwell recognized this limitation and modified Ampère's law to include time-varying electric fields.

Consider a capacitor being charged as illustrated in Figure 34.1. When a conduction current is present, the charge on the positive plate changes but *no conduction current exists in the gap between the plates.* Now consider the two surfaces S_1 and S_2 in Figure 34.1, bounded by the same path P. Ampère's law states that $\oint \vec{\mathbf{B}} \cdot d\vec{\mathbf{s}}$ around this path must equal $\mu_0 I$, where I is the total current through *any* surface bounded by the path P.

When the path P is considered to be the boundary of S_1, $\oint \vec{\mathbf{B}} \cdot d\vec{\mathbf{s}} = \mu_0 I$ because the conduction current I passes through S_1. When the path is considered to be the boundary of S_2, however, $\oint \vec{\mathbf{B}} \cdot d\vec{\mathbf{s}} = 0$ because no conduction current passes through S_2. Therefore, we have a contradictory situation that arises from the discontinuity of the current! Maxwell solved this problem by postulating an additional term on the right side of Ampère's law, which includes a factor called the **displacement current** I_d defined as[1]

$$I_d \equiv \epsilon_0 \frac{d\Phi_E}{dt} \qquad \qquad \textbf{(34.1)}$$

◄ Displacement current

where ϵ_0 is the permittivity of free space (see Section 23.3) and $\Phi_E = \int \vec{\mathbf{E}} \cdot d\vec{\mathbf{A}}$ is the electric flux (see Eq. 24.3) through the surface bounded by the path of integration.

As the capacitor is being charged (or discharged), the changing electric field between the plates may be considered equivalent to a current that acts as a continuation of the conduction current in the wire. When the expression for the displacement current given by Equation 34.1 is added to the conduction current on the right side of Ampère's law, the difficulty represented in Figure 34.1 is resolved. No matter which surface bounded by the path P is chosen, either a conduction

---

[1] *Displacement* in this context does not have the meaning it does in Chapter 2. Despite the inaccurate implications, the word is historically entrenched in the language of physics, so we continue to use it.

current or a displacement current passes through it. With this new term I_d, we can express the general form of Ampère's law (sometimes called the **Ampère–Maxwell law**) as

Ampère–Maxwell law ▶

$$\oint \vec{\mathbf{B}} \cdot d\vec{\mathbf{s}} = \mu_0 (I + I_d) = \mu_0 I + \mu_0 \epsilon_0 \frac{d\Phi_E}{dt} \qquad (34.2)$$

We can understand the meaning of this expression by referring to Figure 34.2. The electric flux through surface S_2 is $\Phi_E = \int \vec{\mathbf{E}} \cdot d\vec{\mathbf{A}} = EA$, where A is the area of the capacitor plates and E is the magnitude of the uniform electric field between the plates. If q is the charge on the plates at any instant, then $E = q/(\epsilon_0 A)$ (see Section 26.2). Therefore, the electric flux through S_2 is

$$\Phi_E = EA = \frac{q}{\epsilon_0}$$

Hence, the displacement current through S_2 is

$$I_d = \epsilon_0 \frac{d\Phi_E}{dt} = \frac{dq}{dt} \qquad (34.3)$$

That is, the displacement current I_d through S_2 is precisely equal to the conduction current I through S_1!

By considering surface S_2, we can identify the displacement current as the source of the magnetic field on the surface boundary. The displacement current has its physical origin in the time-varying electric field. The central point of this formalism is that **magnetic fields are produced both by conduction currents and by time-varying electric fields.** This result was a remarkable example of theoretical work by Maxwell, and it contributed to major advances in the understanding of electromagnetism.

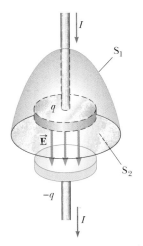

Figure 34.2 Because it exists only in the wires attached to the capacitor plates, the conduction current $I = dq/dt$ passes through S_1 but not through S_2. Only the displacement current $I_d = \epsilon_0 d\Phi_E/dt$ passes through S_2. The two currents must be equal for continuity.

Quick Quiz 34.1 In an RC circuit, the capacitor begins to discharge. **(i)** During the discharge in the region of space between the plates of the capacitor, is there (a) conduction current but no displacement current, (b) displacement current but no conduction current, (c) both conduction and displacement current, or (d) no current of any type? **(ii)** In the same region of space, is there (a) an electric field but no magnetic field, (b) a magnetic field but no electric field, (c) both electric and magnetic fields, or (d) no fields of any type?

EXAMPLE 34.1 | **Displacement Current in a Capacitor**

A sinusoidally varying voltage is applied across an 8.00-μF capacitor. The frequency of the applied voltage is 3.00 kHz, and the voltage amplitude is 30.0 V. Find the displacement current in the capacitor.

SOLUTION

Conceptualize Active Figure 33.9 represents the circuit diagram for this situation. Figure 34.2 shows a close-up of the capacitor and the electric field between the plates.

Categorize We evaluate results using equations discussed in this section, so we categorize this example as a substitution problem.

Evaluate the angular frequency of the source from Equation 15.12:

$$\omega = 2\pi f = 2\pi(3.00 \times 10^3 \text{ Hz}) = 1.88 \times 10^4 \text{ s}^{-1}$$

Use Equation 33.20 to express the voltage across the capacitor as a function of time:

$$\Delta v_C = \Delta V_{\max} \sin \omega t = (30.0 \text{ V}) \sin (1.88 \times 10^4 \, t)$$

Use Equation 34.3 to find the displacement current as a function of time. Note that the charge on the capacitor is $q = C \Delta v_C$:

$$I_d = \frac{dq}{dt} = \frac{d}{dt}(C \Delta v_C) = C\frac{d}{dt}(\Delta v_C)$$

$$= (8.00 \times 10^{-6}\text{ F})\frac{d}{dt}\big[(30.0\text{ V})\sin(1.88 \times 10^4\,t)\big]$$

$$= \boxed{(4.52\text{ A})\cos(1.88 \times 10^4\,t)}$$

34.2 Maxwell's Equations and Hertz's Discoveries

We now present four equations that are regarded as the basis of all electrical and magnetic phenomena. These equations, developed by Maxwell, are as fundamental to electromagnetic phenomena as Newton's laws are to mechanical phenomena. In fact, the theory that Maxwell developed was more far-reaching than even he imagined because it turned out to be in agreement with the special theory of relativity, as Einstein showed in 1905.

Maxwell's equations represent the laws of electricity and magnetism that we have already discussed, but they have additional important consequences. For simplicity, we present **Maxwell's equations** as applied to free space, that is, in the absence of any dielectric or magnetic material. The four equations are

$$\oint \vec{\mathbf{E}} \cdot d\vec{\mathbf{A}} = \frac{q}{\epsilon_0} \qquad (34.4) \qquad \blacktriangleleft \quad \text{Gauss's law}$$

$$\oint \vec{\mathbf{B}} \cdot d\vec{\mathbf{A}} = 0 \qquad (34.5) \qquad \blacktriangleleft \quad \text{Gauss's law in magnetism}$$

$$\oint \vec{\mathbf{E}} \cdot d\vec{\mathbf{s}} = -\frac{d\Phi_B}{dt} \qquad (34.6) \qquad \blacktriangleleft \quad \text{Faraday's law}$$

$$\oint \vec{\mathbf{B}} \cdot d\vec{\mathbf{s}} = \mu_0 I + \epsilon_0\mu_0\frac{d\Phi_E}{dt} \qquad (34.7) \qquad \blacktriangleleft \quad \text{Ampère–Maxwell law}$$

Equation 34.4 is Gauss's law: **the total electric flux through any closed surface equals the net charge inside that surface divided by ϵ_0.** This law relates an electric field to the charge distribution that creates it.

Equation 34.5 is Gauss's law in magnetism, and it states that **the net magnetic flux through a closed surface is zero.** That is, the number of magnetic field lines that enter a closed volume must equal the number that leave that volume, which implies that magnetic field lines cannot begin or end at any point. If they did, it would mean that isolated magnetic monopoles existed at those points. That isolated magnetic monopoles have not been observed in nature can be taken as a confirmation of Equation 34.5.

Equation 34.6 is Faraday's law of induction, which describes the creation of an electric field by a changing magnetic flux. This law states that **the emf, which is the line integral of the electric field around any closed path, equals the rate of change of magnetic flux through any surface bounded by that path.** One consequence of Faraday's law is the current induced in a conducting loop placed in a time-varying magnetic field.

Equation 34.7 is the Ampère–Maxwell law, and it describes the creation of a magnetic field by a changing electric field and by electric current: **the line integral of the magnetic field around any closed path is the sum of μ_0 times the net current through that path and $\epsilon_0\mu_0$ times the rate of change of electric flux through any surface bounded by that path.**

Once the electric and magnetic fields are known at some point in space, the force acting on a particle of charge q can be calculated from the expression

$$\vec{\mathbf{F}} = q\vec{\mathbf{E}} + q\vec{\mathbf{v}} \times \vec{\mathbf{B}} \qquad (34.8)$$

This relationship is called the **Lorentz force law.** (We saw this relationship earlier as Eq. 29.6.) Maxwell's equations, together with this force law, completely describe all classical electromagnetic interactions in a vacuum.

Notice the symmetry of Maxwell's equations. Equations 34.4 and 34.5 are symmetric, apart from the absence of the term for magnetic monopoles in Equation 34.5. Furthermore, Equations 34.6 and 34.7 are symmetric in that the line integrals of $\vec{\mathbf{E}}$ and $\vec{\mathbf{B}}$ around a closed path are related to the rate of change of magnetic flux and electric flux, respectively. Maxwell's equations are of fundamental importance not only to electromagnetism, but to all science. Hertz once wrote, "One cannot escape the feeling that these mathematical formulas have an independent existence and an intelligence of their own, that they are wiser than we are, wiser even than their discoverers, that we get more out of them than we put into them."

In the next section, we show that Equations 34.6 and 34.7 can be combined to obtain a wave equation for both the electric field and the magnetic field. In empty space, where $q = 0$ and $I = 0$, the solution to these two equations shows that the speed at which electromagnetic waves travel equals the measured speed of light. This result led Maxwell to predict that light waves are a form of electromagnetic radiation.

Hertz performed experiments that verified Maxwell's prediction. The experimental apparatus Hertz used to generate and detect electromagnetic waves is shown schematically in Figure 34.3. An induction coil is connected to a transmitter made up of two spherical electrodes separated by a narrow gap. The coil provides short voltage surges to the electrodes, making one positive and the other negative. A spark is generated between the spheres when the electric field near either electrode surpasses the dielectric strength for air (3×10^6 V/m; see Table 26.1). Free electrons in a strong electric field are accelerated and gain enough energy to ionize any molecules they strike. This ionization provides more electrons, which can accelerate and cause further ionizations. As the air in the gap is ionized, it becomes a much better conductor and the discharge between the electrodes exhibits an oscillatory behavior at a very high frequency. From an electric-circuit viewpoint, this experimental apparatus is equivalent to an LC circuit in which the inductance is that of the coil and the capacitance is due to the spherical electrodes.

Because L and C are small in Hertz's apparatus, the frequency of oscillation is high, on the order of 100 MHz. (Recall from Eq. 32.22 that $\omega = 1/\sqrt{LC}$ for an LC circuit.) Electromagnetic waves are radiated at this frequency as a result of the oscillation (and hence acceleration) of free charges in the transmitter circuit. Hertz was able to detect these waves using a single loop of wire with its own spark gap (the receiver). Such a receiver loop, placed several meters from the transmitter, has its own effective inductance, capacitance, and natural frequency of oscillation. In Hertz's experiment, sparks were induced across the gap of the receiving electrodes when the receiver's frequency was adjusted to match that of the transmitter. In this way, Hertz demonstrated that the oscillating current induced in the receiver was produced by electromagnetic waves radiated by the transmitter. His experiment is analogous to the mechanical phenomenon in which a tuning fork responds to acoustic vibrations from an identical tuning fork that is oscillating.

In addition, Hertz showed in a series of experiments that the radiation generated by his spark-gap device exhibited the wave properties of interference, diffraction, reflection, refraction, and polarization, which are all properties exhibited by light as we shall see in Part 5. Therefore, it became evident that the radio-frequency waves Hertz was generating had properties similar to those of light waves and that they differed only in frequency and wavelength. Perhaps his most convincing experiment was the measurement of the speed of this radiation. Waves of known frequency were reflected from a metal sheet and created a standing-wave

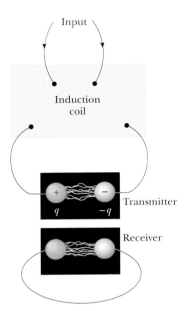

Figure 34.3 Schematic diagram of Hertz's apparatus for generating and detecting electromagnetic waves. The transmitter consists of two spherical electrodes connected to an induction coil, which provides short voltage surges to the spheres, setting up oscillations in the discharge between the electrodes. The receiver is a nearby loop of wire containing a second spark gap.

Hulton-Deutsch Collection/CORBIS

HEINRICH RUDOLF HERTZ
German Physicist (1857–1894)

Hertz made his most important discovery of electromagnetic waves in 1887. After finding that the speed of an electromagnetic wave was the same as that of light, Hertz showed that electromagnetic waves, like light waves, could be reflected, refracted, and diffracted. Hertz died of blood poisoning at the age of 36. During his short life, he made many contributions to science. The hertz, equal to one complete vibration or cycle per second, is named after him.

interference pattern whose nodal points could be detected. The measured distance between the nodal points enabled determination of the wavelength λ. Using the relationship $v = \lambda f$ (Eq. 16.12), Hertz found that v was close to 3×10^8 m/s, the known speed c of visible light.

34.3   Plane Electromagnetic Waves

The properties of electromagnetic waves can be deduced from Maxwell's equations. One approach to deriving these properties is to solve the second-order differential equation obtained from Maxwell's third and fourth equations. A rigorous mathematical treatment of that sort is beyond the scope of this text. To circumvent this problem, let's assume the vectors for the electric field and magnetic field in an electromagnetic wave have a specific space–time behavior that is simple but consistent with Maxwell's equations.

To understand the prediction of electromagnetic waves more fully, let's focus our attention on an electromagnetic wave that travels in the x direction (the *direction of propagation*). For this wave, the electric field $\vec{\mathbf{E}}$ is in the y direction and the magnetic field $\vec{\mathbf{B}}$ is in the z direction as shown in Active Figure 34.4a. Such waves, in which the electric and magnetic fields are restricted to being parallel to a pair of perpendicular axes, are said to be **linearly polarized waves.** Furthermore, let's assume the field magnitudes E and B depend on x and t only, not on the y or z coordinate. Active Figure 34.4b shows a sinusoidal electromagnetic wave, which we discuss below.

Let's also imagine that the source of the electromagnetic waves is such that a wave radiated from *any* position in the yz plane (not only from the origin as might be suggested by Active Fig. 34.4a) propagates in the x direction and all such waves are emitted in phase. If we define a **ray** as the line along which the wave travels, all rays for these waves are parallel. This entire collection of waves is often called a **plane wave.** A surface connecting points of equal phase on all waves is a geometric plane called a **wave front,** as introduced in Chapter 17. In comparison, a point source of radiation sends waves out radially in all directions. A surface connecting points of equal phase for this situation is a sphere, so this wave is called a **spherical wave.**

To generate the prediction of electromagnetic waves, we start with Faraday's law, Equation 34.6:

$$\oint \vec{\mathbf{E}} \cdot d\vec{\mathbf{s}} = -\frac{d\Phi_B}{dt}$$

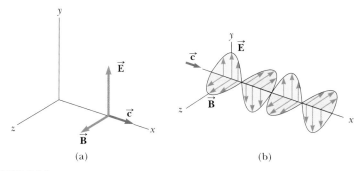

(a)   (b)

ACTIVE FIGURE 34.4

(a) An electromagnetic wave traveling at velocity $\vec{\mathbf{c}}$ in the positive x direction. The wave is shown at an instant of time at which the electric field is along the y direction and has its maximum magnitude and the magnetic field is along the z direction, also with its maximum magnitude. These fields depend only on x and t. (b) Representation of a sinusoidal electromagnetic wave moving in the positive x direction with a speed c.

Sign in at www.thomsonedu.com and go to ThomsonNOW to observe the wave in (b) and the variation of the fields in time. In addition, you can take a "snapshot" of the wave at an instant of time and investigate the electric and magnetic field vectors at that instant.

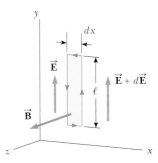

Figure 34.5 At an instant when a plane wave moving in the positive x direction passes through a rectangular path of width dx lying in the xy plane, the electric field in the y direction varies from $\vec{\mathbf{E}}$ to $\vec{\mathbf{E}} + d\vec{\mathbf{E}}$. This spatial variation in $\vec{\mathbf{E}}$ gives rise to a time-varying magnetic field along the z direction, according to Equation 34.11.

Let's again assume the electromagnetic wave is traveling in the x direction, with the electric field $\vec{\mathbf{E}}$ in the positive y direction and the magnetic field $\vec{\mathbf{B}}$ in the positive z direction.

Consider a rectangle of width dx and height ℓ lying in the xy plane as shown in Figure 34.5. To apply Equation 34.6, let's first evaluate the line integral of $\vec{\mathbf{E}} \cdot d\vec{\mathbf{s}}$ around this rectangle. The contributions from the top and bottom of the rectangle are zero because $\vec{\mathbf{E}}$ is perpendicular to $d\vec{\mathbf{s}}$ for these paths. We can express the electric field on the right side of the rectangle as

$$E(x + dx, t) \approx E(x, t) + \frac{dE}{dx}\bigg|_{t \text{ constant}} dx = E(x, t) + \frac{\partial E}{\partial x} dx$$

where $E(x, t)$ is the field on the left side.[2] Therefore, the line integral over this rectangle is approximately

$$\oint \vec{\mathbf{E}} \cdot d\vec{\mathbf{s}} = [E(x + dx, t)]\ell - [E(x, t)]\ell \approx \ell \left(\frac{\partial E}{\partial x} \right) dx \qquad (34.9)$$

Because the magnetic field is in the z direction, the magnetic flux through the rectangle of area $\ell\,dx$ is approximately $\Phi_B = B\ell\,dx$ (assuming dx is very small compared with the wavelength of the wave). Taking the time derivative of the magnetic flux gives

$$\frac{d\Phi_B}{dt} = \ell\,dx \frac{dB}{dt}\bigg|_{x \text{ constant}} = \ell\,dx \frac{\partial B}{\partial t} \qquad (34.10)$$

Substituting Equations 34.9 and 34.10 into Equation 34.6 gives

$$\ell \left(\frac{\partial E}{\partial x} \right) dx = -\ell\,dx \frac{\partial B}{\partial t}$$

$$\frac{\partial E}{\partial x} = -\frac{\partial B}{\partial t} \qquad (34.11)$$

In a similar manner, we can derive a second equation by starting with Maxwell's fourth equation in empty space (Eq. 34.7). In this case, the line integral of $\vec{\mathbf{B}} \cdot d\vec{\mathbf{s}}$ is evaluated around a rectangle lying in the xz plane and having width dx and length ℓ as in Figure 34.6. Noting that the magnitude of the magnetic field changes from $B(x, t)$ to $B(x + dx, t)$ over the width dx and that the direction for taking the line integral is as shown in Figure 34.6, the line integral over this rectangle is found to be approximately

$$\oint \vec{\mathbf{B}} \cdot d\vec{\mathbf{s}} = [B(x, t)]\ell - [B(x + dx, t)]\ell \approx -\ell \left(\frac{\partial B}{\partial x} \right) dx \qquad (34.12)$$

The electric flux through the rectangle is $\Phi_E = E\ell\,dx$, which, when differentiated with respect to time, gives

$$\frac{\partial \Phi_E}{\partial t} = \ell\,dx \frac{\partial E}{\partial t} \qquad (34.13)$$

Substituting Equations 34.12 and 34.13 into Equation 34.7 gives

$$-\ell \left(\frac{\partial B}{\partial x} \right) dx = \mu_0 \epsilon_0 \ell\,dx \left(\frac{\partial E}{\partial t} \right)$$

$$\frac{\partial B}{\partial x} = -\mu_0 \epsilon_0 \frac{\partial E}{\partial t} \qquad (34.14)$$

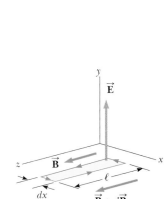

Figure 34.6 At an instant when a plane wave passes through a rectangular path of width dx lying in the xz plane, the magnetic field in the z direction varies from $\vec{\mathbf{B}}$ to $\vec{\mathbf{B}} + d\vec{\mathbf{B}}$. This spatial variation in $\vec{\mathbf{B}}$ gives rise to a time-varying electric field along the y direction, according to Equation 34.14.

[2] Because dE/dx in this equation is expressed as the change in E with x at a given instant t, dE/dx is equivalent to the partial derivative $\partial E/\partial x$. Likewise, dB/dt means the change in B with time at a particular position x; therefore, in Equation 34.10, we can replace dB/dt with $\partial B/\partial t$.

Taking the derivative of Equation 34.11 with respect to x and combining the result with Equation 34.14 gives

$$\frac{\partial^2 E}{\partial x^2} = -\frac{\partial}{\partial x}\left(\frac{\partial B}{\partial t}\right) = -\frac{\partial}{\partial t}\left(\frac{\partial B}{\partial x}\right) = -\frac{\partial}{\partial t}\left(-\mu_0\epsilon_0\frac{\partial E}{\partial t}\right)$$

$$\frac{\partial^2 E}{\partial x^2} = \mu_0\epsilon_0\frac{\partial^2 E}{\partial t^2} \tag{34.15}$$

In the same manner, taking the derivative of Equation 34.14 with respect to x and combining it with Equation 34.11 gives

$$\frac{\partial^2 B}{\partial x^2} = \mu_0\epsilon_0\frac{\partial^2 B}{\partial t^2} \tag{34.16}$$

Equations 34.15 and 34.16 both have the form of the general wave equation[3] with the wave speed v replaced by c, where

$$c = \frac{1}{\sqrt{\mu_0\epsilon_0}} \tag{34.17}$$ ◀ Speed of electromagnetic waves

Let's evaluate this speed numerically:

$$c = \frac{1}{\sqrt{(4\pi \times 10^{-7}\,\text{T}\cdot\text{m/A})(8.854\,19 \times 10^{-12}\,\text{C}^2/\text{N}\cdot\text{m}^2)}}$$

$$= 2.997\,92 \times 10^8\,\text{m/s}$$

Because this speed is precisely the same as the speed of light in empty space, we are led to believe (correctly) that light is an electromagnetic wave.

The simplest solution to Equations 34.15 and 34.16 is a sinusoidal wave for which the field magnitudes E and B vary with x and t according to the expressions

$$E = E_{\text{max}} \cos{(kx - \omega t)} \tag{34.18}$$ ◀ Sinusoidal electric and magnetic fields

$$B = B_{\text{max}} \cos{(kx - \omega t)} \tag{34.19}$$

where E_{max} and B_{max} are the maximum values of the fields. The angular wave number is $k = 2\pi/\lambda$, where λ is the wavelength. The angular frequency is $\omega = 2\pi f$, where f is the wave frequency. The ratio ω/k equals the speed of an electromagnetic wave, c:

$$\frac{\omega}{k} = \frac{2\pi f}{2\pi/\lambda} = \lambda f = c$$

where we have used Equation 16.12, $v = c = \lambda f$, which relates the speed, frequency, and wavelength of any continuous wave. Therefore, for electromagnetic waves, the wavelength and frequency of these waves are related by

$$\lambda = \frac{c}{f} = \frac{3.00 \times 10^8\,\text{m/s}}{f} \tag{34.20}$$

Active Figure 34.4b is a pictorial representation, at one instant, of a sinusoidal, linearly polarized plane wave moving in the positive x direction.

Taking partial derivatives of Equations 34.18 (with respect to x) and 34.19 (with respect to t) gives

$$\frac{\partial E}{\partial x} = -kE_{\text{max}} \sin{(kx - \omega t)}$$

$$\frac{\partial B}{\partial t} = \omega B_{\text{max}} \sin{(kx - \omega t)}$$

---

[3] The general wave equation is of the form $(\partial^2 y/\partial x^2) = (1/v^2)(\partial^2 y/\partial t^2)$, where v is the speed of the wave and y is the wave function. The general wave equation was introduced as Equation 16.27, and we suggest you review Section 16.6.

PITFALL PREVENTION 34.2
$\vec{E}$ Stronger Than $\vec{B}$?

Because the value of c is so large, some students incorrectly interpret Equation 34.21 as meaning that the electric field is much stronger than the magnetic field. Electric and magnetic fields are measured in different units, however, so they cannot be directly compared. In Section 34.4, we find that the electric and magnetic fields contribute equally to the wave's energy.

Substituting these results into Equation 34.11 shows that at any instant,

$$kE_{max} = \omega B_{max}$$

$$\frac{E_{max}}{B_{max}} = \frac{\omega}{k} = c$$

Using these results together with Equations 34.18 and 34.19 gives

$$\frac{E_{max}}{B_{max}} = \frac{E}{B} = c \qquad (34.21)$$

That is, **at every instant, the ratio of the magnitude of the electric field to the magnitude of the magnetic field in an electromagnetic wave equals the speed of light.**

Finally, note that electromagnetic waves obey the superposition principle (which we discussed in Section 18.1 with respect to mechanical waves) because the differential equations involving E and B are linear equations. For example, we can add two waves with the same frequency and polarization simply by adding the magnitudes of the two electric fields algebraically.

Quick Quiz 34.2 What is the phase difference between the sinusoidal oscillations of the electric and magnetic fields in Active Figure 34.4b? (a) 180° (b) 90° (c) 0 (d) impossible to determine

---

EXAMPLE 34.2 **An Electromagnetic Wave**

A sinusoidal electromagnetic wave of frequency 40.0 MHz travels in free space in the x direction as in Figure 34.7.

(A) Determine the wavelength and period of the wave.

Figure 34.7 (Example 34.2) At some instant, a plane electromagnetic wave moving in the x direction has a maximum electric field of 750 N/C in the positive y direction. The corresponding magnetic field at that point has a magnitude E/c and is in the z direction.

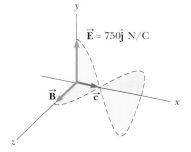

SOLUTION

Conceptualize Imagine the wave in Figure 34.7 moving to the right along the x axis, with the electric and magnetic fields oscillating in phase.

Categorize We evaluate the results using equations developed in this section, so we categorize this example as a substitution problem.

Use Equation 34.20 to find the wavelength of the wave:

$$\lambda = \frac{c}{f} = \frac{3.00 \times 10^8 \text{ m/s}}{40.0 \times 10^6 \text{ Hz}} = 7.50 \text{ m}$$

Find the period T of the wave as the inverse of the frequency:

$$T = \frac{1}{f} = \frac{1}{40.0 \times 10^6 \text{ Hz}} = 2.50 \times 10^{-8} \text{ s}$$

(B) At some point and at some instant, the electric field has its maximum value of 750 N/C and is directed along the y axis. Calculate the magnitude and direction of the magnetic field at this position and time.

SOLUTION

Use Equation 34.21 to find the magnitude of the magnetic field:

$$B_{max} = \frac{E_{max}}{c} = \frac{750 \text{ N/C}}{3.00 \times 10^8 \text{ m/s}} = 2.50 \times 10^{-6} \text{ T}$$

Because $\vec{E}$ and $\vec{B}$ must be perpendicular to each other and perpendicular to the direction of wave propagation (x in this case), we conclude that $\vec{B}$ is in the z direction.

34.4 Energy Carried by Electromagnetic Waves

In our discussion of the nonisolated system model in Section 8.1, we identified electromagnetic radiation as one method of energy transfer across the boundary of a system. The amount of energy transferred by electromagnetic waves is symbolized as T_{ER} in Equation 8.2. The rate of flow of energy in an electromagnetic wave is described by a vector $\vec{S}$, called the **Poynting vector,** which is defined by the expression

$$\vec{S} \equiv \frac{1}{\mu_0}\vec{E} \times \vec{B} \qquad (34.22)$$

◄ Poynting vector

The magnitude of the Poynting vector represents the rate at which energy flows through a unit surface area perpendicular to the direction of wave propagation. Therefore, the magnitude of $\vec{S}$ represents *power per unit area*. The direction of the vector is along the direction of wave propagation (Fig. 34.8). The SI units of $\vec{S}$ are $J/s \cdot m^2 = W/m^2$.

As an example, let's evaluate the magnitude of $\vec{S}$ for a plane electromagnetic wave where $|\vec{E} \times \vec{B}| = EB$. In this case,

$$S = \frac{EB}{\mu_0} \qquad (34.23)$$

Because $B = E/c$, we can also express this result as

$$S = \frac{E^2}{\mu_0 c} = \frac{cB^2}{\mu_0}$$

These equations for S apply at any instant of time and represent the *instantaneous* rate at which energy is passing through a unit area.

What is of greater interest for a sinusoidal plane electromagnetic wave is the time average of S over one or more cycles, which is called the *wave intensity I*. (We discussed the intensity of sound waves in Chapter 17.) When this average is taken, we obtain an expression involving the time average of $\cos^2(kx - \omega t)$, which equals $\frac{1}{2}$. Hence, the average value of S (in other words, the intensity of the wave) is

$$I = S_{avg} = \frac{E_{max}B_{max}}{2\mu_0} = \frac{E_{max}^2}{2\mu_0 c} = \frac{cB_{max}^2}{2\mu_0} \qquad (34.24)$$

◄ Wave intensity

Recall that the energy per unit volume, which is the instantaneous energy density u_E associated with an electric field, is given by Equation 26.13:

$$u_E = \tfrac{1}{2}\epsilon_0 E^2$$

Also recall that the instantaneous energy density u_B associated with a magnetic field is given by Equation 32.14:

$$u_B = \frac{B^2}{2\mu_0}$$

Because E and B vary with time for an electromagnetic wave, the energy densities also vary with time. Using the relationships $B = E/c$ and $c = 1/\sqrt{\epsilon_0\mu_0}$, the expression for u_B becomes

$$u_B = \frac{(E/c)^2}{2\mu_0} = \frac{\epsilon_0\mu_0}{2\mu_0}E^2 = \tfrac{1}{2}\epsilon_0 E^2$$

Comparing this result with the expression for u_E, we see that

$$u_B = u_E = \tfrac{1}{2}\epsilon_0 E^2 = \frac{B^2}{2\mu_0}$$

PITFALL PREVENTION 34.3
An Instantaneous Value

The Poynting vector given by Equation 34.22 is time dependent. Its magnitude varies in time, reaching a maximum value at the same instant as the magnitudes of $\vec{E}$ and $\vec{B}$ do. The *average* rate of energy transfer is given by Equation 34.24.

PITFALL PREVENTION 34.4
Irradiance

In this discussion, intensity is defined in the same way as in Chapter 17 (as power per unit area). In the optics industry, however, power per unit area is called the *irradiance*. Radiant intensity is defined as the power in watts per solid angle (measured in steradians).

Figure 34.8 The Poynting vector $\vec{S}$ for a plane electromagnetic wave is along the direction of wave propagation.

That is, **the instantaneous energy density associated with the magnetic field of an electromagnetic wave equals the instantaneous energy density associated with the electric field.** Hence, in a given volume, the energy is equally shared by the two fields.

The **total instantaneous energy density** u is equal to the sum of the energy densities associated with the electric and magnetic fields:

◀ Total instantaneous energy density of an electromagnetic wave

$$u = u_E + u_B = \epsilon_0 E^2 = \frac{B^2}{\mu_0}$$

When this total instantaneous energy density is averaged over one or more cycles of an electromagnetic wave, we again obtain a factor of $\frac{1}{2}$. Hence, for any electromagnetic wave, the total average energy per unit volume is

◀ Average energy density of an electromagnetic wave

$$u_{avg} = \epsilon_0 (E^2)_{avg} = \tfrac{1}{2}\epsilon_0 E_{max}^2 = \frac{B_{max}^2}{2\mu_0} \tag{34.25}$$

Comparing this result with Equation 34.24 for the average value of S, we see that

$$I = S_{avg} = cu_{avg} \tag{34.26}$$

In other words, **the intensity of an electromagnetic wave equals the average energy density multiplied by the speed of light.**

The Sun delivers about 10^3 W/m^2 of energy to the Earth's surface via electromagnetic radiation. Let's calculate the total power that is incident on the roof of a home. The roof's dimensions are 8.00 m × 20.0 m. We assume the average magnitude of the Poynting vector for solar radiation at the surface of the Earth is $S_{avg} = 1\,000$ W/m^2. This average value represents the power per unit area, or the light intensity. Assuming the radiation is incident normal to the roof, we obtain

$$\mathcal{P}_{avg} = S_{avg}A = (1\,000\text{ W/m}^2)(8.00\text{ m} \times 20.0\text{ m}) = 1.60 \times 10^5\text{ W}$$

This power is large compared to the power requirements of a typical home. If this power were maintained for 24 hours per day and the energy could be absorbed and made available to electrical devices, it would provide more than enough energy for the average home. Solar energy is not easily harnessed, however, and the prospects for large-scale conversion are not as bright as may appear from this calculation. For example, the efficiency of conversion from solar energy is typically 10% for photovoltaic cells, reducing the available power by an order of magnitude. Other considerations reduce the power even further. Depending on location, the radiation is most likely not incident normal to the roof and, even if it is (in locations near the equator), this situation exists for only a short time near the middle of the day. No energy is available for about half of each day during the nighttime hours, and cloudy days further reduce the available energy. Finally, while energy is arriving at a large rate during the middle of the day, some of it must be stored for later use, requiring batteries or other storage devices. All in all, complete solar operation of homes is not currently cost effective for most homes.

Quick Quiz 34.3 An electromagnetic wave propagates in the $-y$ direction. The electric field at a point in space is momentarily oriented in the $+x$ direction. In which direction is the magnetic field at that point momentarily oriented? (a) the $-x$ direction   (b) the $+y$ direction   (c) the $+z$ direction   (d) the $-z$ direction

---

EXAMPLE 34.3   **Fields on the Page**

Estimate the maximum magnitudes of the electric and magnetic fields of the light that is incident on this page because of the visible light coming from your desk lamp. Treat the lightbulb as a point source of electromagnetic radiation that is 5% efficient at transforming energy coming in by electrical transmission to energy leaving by visible light.

SOLUTION

Conceptualize   The filament in your lightbulb emits electromagnetic radiation. The brighter the light, the larger the magnitudes of the electric and magnetic fields.

Categorize   Because the lightbulb is to be treated as a point source, it emits equally in all directions, so the outgoing electromagnetic radiation can be modeled as a spherical wave.

Analyze   Recall from Equation 17.7 that the wave intensity I a distance r from a point source is $I = \mathcal{P}_{avg}/4\pi r^2$, where $\mathcal{P}_{avg}$ is the average power output of the source and $4\pi r^2$ is the area of a sphere of radius r centered on the source.

Set this expression for I equal to the intensity of an electromagnetic wave given by Equation 34.24:

$$I = \frac{\mathcal{P}_{avg}}{4\pi r^2} = \frac{E_{max}^2}{2\mu_0 c}$$

Solve for the electric field magnitude:

$$E_{max} = \sqrt{\frac{\mu_0 c \mathcal{P}_{avg}}{2\pi r^2}}$$

Let's make some assumptions about numbers to enter in this equation. The visible light output of a 60-W lightbulb operating at 5% efficiency is approximately 3.0 W by visible light. (The remaining energy transfers out of the lightbulb by conduction and invisible radiation.) A reasonable distance from the lightbulb to the page might be 0.30 m.

Substitute these values:

$$E_{max} = \sqrt{\frac{(4\pi \times 10^{-7}\ \text{T}\cdot\text{m/A})(3.00 \times 10^8\ \text{m/s})(3.0\ \text{W})}{2\pi (0.30\ \text{m})^2}}$$

$$= \ 45\ \text{V/m}$$

Use Equation 34.21 to find the magnetic field magnitude:

$$B_{max} = \frac{E_{max}}{c} = \frac{45\ \text{V/m}}{3.00 \times 10^8\ \text{m/s}} = \ 1.5 \times 10^{-7}\ \text{T}$$

Finalize   This value of the magnetic field magnitude is two orders of magnitude smaller than the Earth's magnetic field.

34.5   Momentum and Radiation Pressure

Electromagnetic waves transport linear momentum as well as energy. As this momentum is absorbed by some surface, pressure is exerted on the surface. In this discussion, let's assume the electromagnetic wave strikes the surface at normal incidence and transports a total energy T_{ER} to the surface in a time interval Δt. Maxwell showed that if the surface absorbs all the incident energy T_{ER} in this time interval (as does a black body, introduced in Section 20.7), the total momentum $\vec{p}$ transported to the surface has a magnitude

$$p = \frac{T_{ER}}{c} \qquad \text{(complete absorption)} \qquad (34.27)$$

◄ Momentum transported to a perfectly absorbing surface

The pressure exerted on the surface is defined as force per unit area F/A, which when combined with Newton's second law gives

$$P = \frac{F}{A} = \frac{1}{A}\frac{dp}{dt}$$

Substituting Equation 34.27 into this expression for P gives

$$P = \frac{1}{A}\frac{dp}{dt} = \frac{1}{A}\frac{d}{dt}\left(\frac{T_{ER}}{c}\right) = \frac{1}{c}\frac{(dT_{ER}/dt)}{A}$$

PITFALL PREVENTION 34.5
So Many p's

We have p for momentum and P for pressure, and they are both related to $\mathcal{P}$ for power! Be sure to keep all these symbols straight.

We recognize $(dT_{ER}/dt)/A$ as the rate at which energy is arriving at the surface per unit area, which is the magnitude of the Poynting vector. Therefore, the radiation pressure P exerted on the perfectly absorbing surface is

Radiation pressure exerted ▶
on a perfectly absorbing
surface

$$P = \frac{S}{c} \qquad (34.28)$$

If the surface is a perfect reflector (such as a mirror) and incidence is normal, the momentum transported to the surface in a time interval Δt is twice that given by Equation 34.27. That is, the momentum transferred to the surface by the incoming light is $p = T_{ER}/c$ and that transferred by the reflected light also is $p = T_{ER}/c$. Therefore,

$$p = \frac{2T_{ER}}{c} \qquad \text{(complete reflection)} \qquad (34.29)$$

The radiation pressure exerted on a perfectly reflecting surface for normal incidence of the wave is

Radiation pressure exerted ▶
on a perfectly reflecting
surface

$$P = \frac{2S}{c} \qquad (34.30)$$

The pressure on a surface having a reflectivity somewhere between these two extremes has a value between S/c and $2S/c$, depending on the properties of the surface.

Although radiation pressures are very small (about 5×10^{-6} N/m^2 for direct sunlight), NASA is exploring the possibility of *solar sailing* as a low-cost means of sending spacecraft to the planets. Large sheets would experience radiation pressure from sunlight and would be used in much the way canvas sheets are used on earthbound sailboats. In 1973, NASA engineers took advantage of the momentum of the sunlight striking the solar panels of *Mariner 10* to make small course corrections when the spacecraft's fuel supply was running low. (This procedure was carried out when the spacecraft was in the vicinity of Mercury. Would it have worked as well near Neptune?)

Quick Quiz 34.4 To maximize the radiation pressure on the sails of a spacecraft using solar sailing, should the sheets be (a) very black to absorb as much sunlight as possible or (b) very shiny to reflect as much sunlight as possible?

CONCEPTUAL EXAMPLE 34.4 | **Sweeping the Solar System**

A great amount of dust exists in interplanetary space. Although in theory these dust particles can vary in size from molecular size to a much larger size, very little of the dust in our solar system is smaller than about 0.2 μm. Why?

SOLUTION

The dust particles are subject to two significant forces: the gravitational force that draws them toward the Sun and the radiation-pressure force that pushes them away from the Sun. The gravitational force is proportional to the cube of the radius of a spherical dust particle because it is proportional to the mass and therefore to the volume $4\pi r^3/3$ of the particle. The radiation pressure is proportional to the square of the radius because it depends on the planar cross section of the particle. For large particles, the gravitational force is greater than the force from radiation pressure. For particles having radii less than about 0.2 μm, the radiation-pressure force is greater than the gravitational force. As a result, these particles are swept out of our solar system by sunlight.

EXAMPLE 34.5 | **Pressure from a Laser Pointer**

When giving presentations, many people use a laser pointer to direct the attention of the audience to information on a screen. If a 3.0-mW pointer creates a spot on a screen that is 2.0 mm in diameter, determine the radiation pressure on a screen that reflects 70% of the light that strikes it. The power 3.0 mW is a time-averaged value.

SOLUTION

Conceptualize The pressure should not be very large.

Categorize This problem involves a calculation of radiation pressure using an approach like that leading to Equation 34.28 or Equation 34.30, but it is complicated by the 70% reflection.

Analyze We begin by determining the magnitude of the beam's Poynting vector.

Divide the time-averaged power delivered via the electromagnetic wave by the cross-sectional area of the beam:

$$S_{avg} = \frac{\mathcal{P}_{avg}}{A} = \frac{\mathcal{P}_{avg}}{\pi r^2} = \frac{3.0 \times 10^{-3}\ W}{\pi \left(\dfrac{2.0 \times 10^{-3}\ m}{2} \right)^2} = 955\ W/m^2$$

Now let's determine the radiation pressure from the laser beam. Equation 34.30 indicates that a completely reflected beam would apply an average pressure of $P_{avg} = 2S_{avg}/c$. We can model the actual reflection as follows. Imagine that the surface absorbs the beam, resulting in pressure $P_{avg} = S_{avg}/c$. Then the surface emits the beam, resulting in additional pressure $P_{avg} = S_{avg}/c$. If the surface emits only a fraction f of the beam (so that f is the amount of the incident beam reflected), the pressure due to the emitted beam is $P_{avg} = fS_{avg}/c$.

Use this model to find the total pressure on the surface due to absorption and re-emission (reflection):

$$P_{avg} = \frac{S_{avg}}{c} + f\frac{S_{avg}}{c} = (1 + f)\frac{S_{avg}}{c}$$

Evaluate this pressure for a beam that is 70% reflected:

$$P_{avg} = (1 + 0.70)\frac{955\ W/m^2}{3.0 \times 10^8\ m/s} = \boxed{5.4 \times 10^{-6}\ N/m^2}$$

Finalize Consider the magnitude of the Poynting vector, $S_{avg} = 955\ W/m^2$. It is about the same as the intensity of sunlight at the Earth's surface. For this reason, it is not safe to shine the beam of a laser pointer into a person's eyes, which may be more dangerous than looking directly at the Sun. The pressure has an extremely small value, as expected. (Recall from Section 14.2 that atmospheric pressure is approximately $10^5\ N/m^2$.)

What If? What if the laser pointer is moved twice as far away from the screen? Does that affect the radiation pressure on the screen?

Answer Because a laser beam is popularly represented as a beam of light with constant cross section, you might think that the intensity of radiation, and therefore the radiation pressure, is independent of distance from the screen. A laser beam, however, does not have a constant cross section at all distances from the source; rather, there is a small but measurable divergence of the beam. If the laser is moved farther away from the screen, the area of illumination on the screen increases, decreasing the intensity. In turn, the radiation pressure is reduced.

In addition, the doubled distance from the screen results in more loss of energy from the beam due to scattering from air molecules and dust particles as the light travels from the laser to the screen. This energy loss further reduces the radiation pressure on the screen.

34.6   Production of Electromagnetic Waves by an Antenna

Stationary charges and steady currents cannot produce electromagnetic waves. Whenever the current in a wire changes with time, however, the wire emits electromagnetic radiation. **The fundamental mechanism responsible for this radiation is the acceleration of a charged particle. Whenever a charged particle accelerates, it radiates energy.**

Let's consider the production of electromagnetic waves by a *half-wave antenna*. In this arrangement, two conducting rods are connected to a source of alternating voltage (such as an *LC* oscillator) as shown in Figure 34.9 (page 966). The length

Figure 34.9 A half-wave antenna consists of two metal rods connected to an alternating voltage source. This diagram shows $\vec{\mathbf{E}}$ and $\vec{\mathbf{B}}$ at an arbitrary instant when the current is upward. Notice that the electric field lines resemble those of a dipole (shown in Fig. 23.20).

of each rod is equal to one quarter of the wavelength of the radiation emitted when the oscillator operates at frequency f. The oscillator forces charges to accelerate back and forth between the two rods. Figure 34.9 shows the configuration of the electric and magnetic fields at some instant when the current is upward. The separation of charges in the upper and lower portions of the antenna make the electric field lines resemble those of an electric dipole. (As a result, this type of antenna is sometimes called a *dipole antenna*.) Because these charges are continuously oscillating between the two rods, the antenna can be approximated by an oscillating electric dipole. The current representing the movement of charges between the ends of the antenna produces magnetic field lines forming concentric circles around the antenna that are perpendicular to the electric field lines at all points. The magnetic field is zero at all points along the axis of the antenna. Furthermore, $\vec{\mathbf{E}}$ and $\vec{\mathbf{B}}$ are 90° out of phase in time; for example, the current is zero when the charges at the outer ends of the rods are at a maximum.

At the two points where the magnetic field is shown in Figure 34.9, the Poynting vector $\vec{\mathbf{S}}$ is directed radially outward, indicating that energy is flowing away from the antenna at this instant. At later times, the fields and the Poynting vector reverse direction as the current alternates. Because $\vec{\mathbf{E}}$ and $\vec{\mathbf{B}}$ are 90° out of phase at points near the dipole, the net energy flow is zero. From this fact, you might conclude (incorrectly) that no energy is radiated by the dipole.

Energy is indeed radiated, however. Because the dipole fields fall off as $1/r^3$ (as shown in Example 23.5 for the electric field of a static dipole), they are negligible at great distances from the antenna. At these great distances, something else causes a type of radiation different from that close to the antenna. The source of this radiation is the continuous induction of an electric field by the time-varying magnetic field and the induction of a magnetic field by the time-varying electric field, predicted by Equations 34.6 and 34.7. The electric and magnetic fields produced in this manner are in phase with each other and vary as $1/r$. The result is an outward flow of energy at all times.

The angular dependence of the radiation intensity produced by a dipole antenna is shown in Figure 34.10. Notice that the intensity and the power radiated are a maximum in a plane that is perpendicular to the antenna and passing through its midpoint. Furthermore, the power radiated is zero along the antenna's axis. A mathematical solution to Maxwell's equations for the dipole antenna shows that the intensity of the radiation varies as $(\sin^2\theta)/r^2$, where θ is measured from the axis of the antenna.

Electromagnetic waves can also induce currents in a receiving antenna. The response of a dipole receiving antenna at a given position is a maximum when the antenna axis is parallel to the electric field at that point and zero when the axis is perpendicular to the electric field.

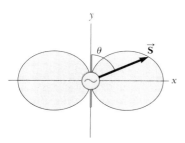

Figure 34.10 Angular dependence of the intensity of radiation produced by an oscillating electric dipole. The distance from the origin to a point on the edge of the gray shape is proportional to the intensity of radiation.

Quick Quiz 34.5 If the antenna in Figure 34.9 represents the source of a distant radio station, what would be the best orientation for your portable radio antenna located to the right of the figure? (a) up–down along the page   (b) left–right along the page   (c) perpendicular to the page

34.7 The Spectrum of Electromagnetic Waves

The various types of electromagnetic waves are listed in Figure 34.11, which shows the **electromagnetic spectrum.** Notice the wide ranges of frequencies and wavelengths. No sharp dividing point exists between one type of wave and the next. Remember that **all forms of the various types of radiation are produced by the same phenomenon, accelerating charges.** The names given to the types of waves are simply a convenient way to describe the region of the spectrum in which they lie.

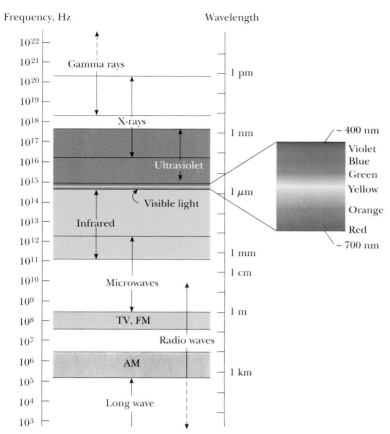

Figure 34.11   The electromagnetic spectrum. Notice the overlap between adjacent wave types. The expanded view to the right shows details of the visible spectrum.

Radio waves, whose wavelengths range from more than 10^4 m to about 0.1 m, are the result of charges accelerating through conducting wires. They are generated by such electronic devices as *LC* oscillators and are used in radio and television communication systems.

Microwaves have wavelengths ranging from approximately 0.3 m to 10^{-4} m and are also generated by electronic devices. Because of their short wavelengths, they are well suited for radar systems and for studying the atomic and molecular properties of matter. Microwave ovens are an interesting domestic application of these waves. It has been suggested that solar energy could be harnessed by beaming microwaves to the Earth from a solar collector in space.

Infrared waves have wavelengths ranging from approximately 10^{-3} m to the longest wavelength of visible light, 7×10^{-7} m. These waves, produced by molecules and room-temperature objects, are readily absorbed by most materials. The infrared (IR) energy absorbed by a substance appears as internal energy because the energy agitates the object's atoms, increasing their vibrational or translational motion, which results in a temperature increase. Infrared radiation has practical and scientific applications in many areas, including physical therapy, IR photography, and vibrational spectroscopy.

Visible light, the most familiar form of electromagnetic waves, is the part of the electromagnetic spectrum the human eye can detect. Light is produced by the rearrangement of electrons in atoms and molecules. The various wavelengths of visible light, which correspond to different colors, range from red ($\lambda \approx 7 \times 10^{-7}$ m) to violet ($\lambda \approx 4 \times 10^{-7}$ m). The sensitivity of the human eye is a function of wavelength, being a maximum at a wavelength of about 5.5×10^{-7} m. With that in mind, why do you suppose tennis balls often have a yellow-green color? Table 34.1 (page 968) provides approximate correspondences between the wavelength of

PITFALL PREVENTION 34.6
"Heat Rays"

Infrared rays are often called "heat rays," but this terminology is a misnomer. Although infrared radiation is used to raise or maintain temperature as in the case of keeping food warm with "heat lamps" at a fast-food restaurant, all wavelengths of electromagnetic radiation carry energy that can cause the temperature of a system to increase. As an example, consider a potato baking in your microwave oven.

TABLE 34.1

Approximate Correspondence Between Wavelengths of Visible Light and Color

Wavelength Range (nm)	Color Description
400–430	Violet
430–485	Blue
485–560	Green
560–590	Yellow
590–625	Orange
625–700	Red

Note: The wavelength ranges here are approximate. Different people will describe colors differently.

Wearing sunglasses that do not block ultraviolet (UV) light is worse for your eyes than wearing no sunglasses at all. The lenses of any sunglasses absorb some visible light, thereby causing the wearer's pupils to dilate. If the glasses do not also block UV light, more damage may be done to the lens of the eye because of the dilated pupils. If you wear no sunglasses at all, your pupils are contracted, you squint, and much less UV light enters your eyes. High-quality sunglasses block nearly all the eye-damaging UV light.

visible light and the color assigned to it by humans. Light is the basis of the science of optics and optical instruments, to be discussed in Chapters 35 through 38.

Ultraviolet waves cover wavelengths ranging from approximately 4×10^{-7} m to 6×10^{-10} m. The Sun is an important source of ultraviolet (UV) light, which is the main cause of sunburn. Sunscreen lotions are transparent to visible light but absorb most UV light. The higher a sunscreen's solar protection factor, or SPF, the greater the percentage of UV light absorbed. Ultraviolet rays have also been implicated in the formation of cataracts, a clouding of the lens inside the eye.

Most of the UV light from the Sun is absorbed by ozone (O_3) molecules in the Earth's upper atmosphere, in a layer called the stratosphere. This ozone shield converts lethal high-energy UV radiation to IR radiation, which in turn warms the stratosphere.

X-rays have wavelengths in the range from approximately 10^{-8} m to 10^{-12} m. The most common source of x-rays is the stopping of high-energy electrons upon bombarding a metal target. X-rays are used as a diagnostic tool in medicine and as a treatment for certain forms of cancer. Because x-rays can damage or destroy living tissues and organisms, care must be taken to avoid unnecessary exposure or overexposure. X-rays are also used in the study of crystal structure because x-ray wavelengths are comparable to the atomic separation distances in solids (about 0.1 nm).

Gamma rays are electromagnetic waves emitted by radioactive nuclei (such as ^{60}Co and ^{137}Cs) and during certain nuclear reactions. High-energy gamma rays are a component of cosmic rays that enter the Earth's atmosphere from space. They have wavelengths ranging from approximately 10^{-10} m to less than 10^{-14} m. Gamma rays are highly penetrating and produce serious damage when absorbed by living tissues. Consequently, those working near such dangerous radiation must be protected with heavily absorbing materials such as thick layers of lead.

Quick Quiz 34.6 In many kitchens, a microwave oven is used to cook food. The frequency of the microwaves is on the order of 10^{10} Hz. Are the wavelengths of these microwaves on the order of (a) kilometers, (b) meters, (c) centimeters, or (d) micrometers?

Quick Quiz 34.7 A radio wave of frequency on the order of 10^5 Hz is used to carry a sound wave with a frequency on the order of 10^3 Hz. Is the wavelength of this radio wave on the order of (a) kilometers, (b) meters, (c) centimeters, or (d) micrometers?

Summary

ThomsonNOW™ Sign in at **www.thomsonedu.com** and go to ThomsonNOW to take a practice test for this chapter.

DEFINITIONS

In a region of space in which there is a changing electric field, there is a **displacement current** defined as

$$I_d \equiv \epsilon_0 \frac{d\Phi_E}{dt} \qquad (34.1)$$

where ϵ_0 is the permittivity of free space (see Section 23.3) and $\Phi_E = \int \vec{E} \cdot d\vec{A}$ is the electric flux.

The rate of flow of energy crossing a unit area by electromagnetic radiation is described by the **Poynting vector** $\vec{S}$, where

$$\vec{S} \equiv \frac{1}{\mu_0} \vec{E} \times \vec{B} \qquad (34.22)$$

(continued)

CONCEPTS AND PRINCIPLES

When used with the **Lorentz force law,** $\vec{F} = q\vec{E} + q\vec{v} \times \vec{B}$, **Maxwell's equations** describe all electromagnetic phenomena:

$$\oint \vec{E} \cdot d\vec{A} = \frac{q}{\epsilon_0} \qquad (34.4)$$

$$\oint \vec{B} \cdot d\vec{A} = 0 \qquad (34.5)$$

$$\oint \vec{E} \cdot d\vec{s} = -\frac{d\Phi_B}{dt} \qquad (34.6)$$

$$\oint \vec{B} \cdot d\vec{s} = \mu_0 I + \epsilon_0 \mu_0 \frac{d\Phi_E}{dt} \qquad (34.7)$$

Electromagnetic waves, which are predicted by Maxwell's equations, have the following properties:

■ The electric field and the magnetic field each satisfy a wave equation. These two wave equations, which can be obtained from Maxwell's third and fourth equations, are

$$\frac{\partial^2 E}{\partial x^2} = \mu_0 \epsilon_0 \frac{\partial^2 E}{\partial t^2} \qquad (34.15)$$

$$\frac{\partial^2 B}{\partial x^2} = \mu_0 \epsilon_0 \frac{\partial^2 B}{\partial t^2} \qquad (34.16)$$

■ The waves travel through a vacuum with the speed of light c, where

$$c = \frac{1}{\sqrt{\mu_0 \epsilon_0}} \qquad (34.17)$$

■ Numerically, the speed of electromagnetic waves in a vacuum is 3.00×10^8 m/s.
■ The electric and magnetic fields are perpendicular to each other and perpendicular to the direction of wave propagation.
■ The instantaneous magnitudes of $\vec{E}$ and $\vec{B}$ in an electromagnetic wave are related by the expression

$$\frac{E}{B} = c \qquad (34.21)$$

■ Electromagnetic waves carry energy.
■ Electromagnetic waves carry momentum.

Because electromagnetic waves carry momentum, they exert pressure on surfaces. If an electromagnetic wave whose Poynting vector is $\vec{S}$ is completely absorbed by a surface upon which it is normally incident, the radiation pressure on that surface is

$$P = \frac{S}{c} \quad \text{(complete absorption)} \qquad (34.28)$$

If the surface totally reflects a normally incident wave, the pressure is doubled.

The electric and magnetic fields of a sinusoidal plane electromagnetic wave propagating in the positive x direction can be written as

$$E = E_{max} \cos(kx - \omega t) \qquad (34.18)$$

$$B = B_{max} \cos(kx - \omega t) \qquad (34.19)$$

where ω is the angular frequency of the wave and k is the angular wave number. These equations represent special solutions to the wave equations for E and B. The wavelength and frequency of electromagnetic waves are related by

$$\lambda = \frac{c}{f} = \frac{3.00 \times 10^8 \text{ m/s}}{f} \qquad (34.20)$$

The average value of the Poynting vector for a plane electromagnetic wave has a magnitude

$$S_{avg} = \frac{E_{max}B_{max}}{2\mu_0} = \frac{E_{max}^2}{2\mu_0 c} = \frac{cB_{max}^2}{2\mu_0} \qquad (34.24)$$

The intensity of a sinusoidal plane electromagnetic wave equals the average value of the Poynting vector taken over one or more cycles.

The electromagnetic spectrum includes waves covering a broad range of wavelengths, from long radio waves at more than 10^4 m to gamma rays at less than 10^{-14} m.

Questions

□ denotes answer available in *Student Solutions Manual/Study Guide;* **O** denotes objective question

1. What new concept did Maxwell's generalized form of Ampère's law include?

2. Do Maxwell's equations allow for the existence of magnetic monopoles? Explain.

3. Radio stations often advertise "instant news." If that means you can hear the news the instant the radio announcer speaks it, is the claim true? What approximate time interval is required for a message to travel from Maine to California by radio waves? Assume the waves can be detected at this range.

4. When light (or other electromagnetic radiation) travels across a given region, what is it that oscillates? What is it that is transported?

5. If a high-frequency current exists in a solenoid containing a metallic core, the core becomes warm due to induction. Explain why the material rises in temperature in this situation.

6. **O** A student working with transmitting apparatus like Heinrich Hertz's wishes to adjust the electrodes to generate electromagnetic waves with a frequency half as large as before. **(i)** How large should she make the effective capacitance of the pair of electrodes? (a) 4 times larger than before   (b) 2 times larger than before   (c) $\frac{1}{2}$ as large as before   (d) $\frac{1}{4}$ as large as before   (e) None of these answers would have the desired effect. **(ii)** After she makes the required adjustment, what will the wavelength of the transmitted wave be? (a) 4 times larger than before   (b) 2 times larger than before   (c) the same as before   (d) $\frac{1}{2}$ as large as before   (e) $\frac{1}{4}$ as large as before   (f) None of these answers is necessarily true.

7. **O** Assume you charge a comb by running it through your hair and then hold the comb next to a bar magnet. Do the electric and magnetic fields produced constitute an electromagnetic wave? (a) Yes they do, necessarily. (b) Yes they do because charged particles are moving inside the bar magnet. (c) They can, but only if the electric field of the comb and the magnetic field of the magnet are perpendicular. (d) They can, but only if both the comb and the magnet are moving. (e) They can, if either the comb or the magnet or both are accelerating.

8. **O** A small source radiates an electromagnetic wave with a single frequency into vacuum, equally in all directions. **(i)** As the wave moves, does its frequency (a) increase, (b) decrease, or (c) stay constant? Answer the same question about **(ii)** its wavelength, **(iii)** its speed, **(iv)** its intensity, and **(v)** the amplitude of its electric field.

9. **O** A plane electromagnetic wave with a single frequency moves in vacuum in the $+x$ direction. Its amplitude is uniform over the yz plane. **(i)** As the wave moves, does its frequency (a) increase, (b) decrease, or (c) stay constant? Answer the same question about **(ii)** its wavelength, **(iii)** its speed, **(iv)** its intensity, and **(v)** the amplitude of its magnetic field.

10. List as many similarities and differences between sound waves and light waves as you can.

11. Describe the physical significance of the Poynting vector.

12. **O** Assume the amplitude of the electric field in a plane electromagnetic wave is E_1 and the amplitude of the magnetic field is B_1. The source of the wave is then adjusted so that the amplitude of the electric field doubles to become $2E_1$. **(i)** What happens to the amplitude of the magnetic field in this process? (a) It becomes 4 times larger. (b) It becomes 2 times larger. (c) It can stay constant. (d) It becomes $\frac{1}{2}$ as large. (e) It becomes $\frac{1}{4}$ as large. (f) None of these answers is necessarily true. **(ii)** What happens to the intensity of the wave? Choose from the same possibilities.

13. **O** A spherical interplanetary dust grain of radius 0.2 μm is at distance r_1 from the Sun. The gravitational force exerted by the Sun on the grain just balances the force due to radiation pressure from the Sun's light. **(i)** Assume the grain is moved to a distance $2r_1$ from the Sun and released. At this location, what is the net force exerted on the grain? (a) toward the Sun (b) away from the Sun (c) zero (d) impossible to determine without knowing the mass of the grain **(ii)** Now assume the grain is moved back to its original location at r_1, compressed so that it crystallizes into a sphere with significantly higher density, and released. In this situation, what is the net force exerted on the grain? (a) toward the Sun (b) away from the Sun (c) zero (d) impossible to determine without knowing the mass of the grain

14. For a given incident energy of an electromagnetic wave, why is the radiation pressure on a perfectly reflecting surface twice as great as that on a perfect absorbing surface?

15. Some television viewers use "rabbit ears" atop their sets (Fig. Q34.15) instead of purchasing cable television service or satellite dishes. Certain orientations of the receiving antenna on a television set give better reception than others. Furthermore, the best orientation varies from station to station. Explain.

Figure Q34.15   Question 15 and Problem 63.

16. What does a radio wave do to the charges in the receiving antenna to provide a signal for your car radio?

17. **O** **(i)** Rank the following kinds of waves according to their wavelength ranges from those with the smallest typical or average wavelength to the largest, noting any cases of equality: (a) gamma rays   (b) infrared light   (c) microwaves   (d) radio waves   (e) ultraviolet light   (f) visible light   (g) x-rays   **(ii)** Rank the kinds of waves according to their frequencies from lowest to highest. **(iii)** Rank the kinds of waves according to their speeds in a vacuum from slowest to fastest.

18. An empty plastic or glass dish being removed from a microwave oven can be cool to the touch, even when food on an adjoining dish is hot. How is this phenomenon possible?

19. Why should an infrared photograph of a person look different from a photograph taken with visible light?

20. Suppose a creature from another planet has eyes that are sensitive to infrared radiation. Describe what the alien would see if it looked around the room you are now in. In particular, what would be bright and what would be dim?

21. A home microwave oven uses electromagnetic waves with a wavelength of about 12.2 cm. Some 2.4-GHz cordless telephones suffer noisy interference when a microwave oven is used nearby. Locate the waves used by both devices on the electromagnetic spectrum. Do you expect them to interfere with each other?

Problems

WebAssign   The Problems from this chapter may be assigned online in WebAssign.

ThomsonNOW   Sign in at **www.thomsonedu.com** and go to ThomsonNOW to assess your understanding of this chapter's topics with additional quizzing and conceptual questions.

1, 2, 3 denotes straightforward, intermediate, challenging;   ☐ denotes full solution available in *Student Solutions Manual/Study Guide;*   ▲ denotes coached solution with hints available at **www.thomsonedu.com;**   denotes developing symbolic reasoning; ● denotes asking for qualitative reasoning;   ▪ denotes computer useful in solving problem

Section 34.1 Displacement Current and the General Form of Ampère's Law

1. A 0.100-A current is charging a capacitor that has square plates 5.00 cm on each side. The plate separation is 4.00 mm. Find (a) the time rate of change of electric flux between the plates and (b) the displacement current between the plates.

2. A 0.200-A current is charging a capacitor that has circular plates 10.0 cm in radius. If the plate separation is 4.00 mm, (a) what is the time rate of increase of electric field between the plates? (b) What is the magnetic field between the plates 5.00 cm from the center?

3. Consider the situation shown in Figure P34.3. An electric field of 300 V/m is confined to a circular area 10.0 cm in diameter and directed outward perpendicular to the plane of the figure. If the field is increasing at a rate of 20.0 V/m · s, what are the direction and magnitude of the magnetic field at the point P, 15.0 cm from the center of the circle?

Figure P34.3

Section 34.2 Maxwell's Equations and Hertz's Discoveries

4. A very long, thin rod carries electric charge with the linear density 35.0 nC/m. It lies along the x axis and moves in the x direction at a speed of 15.0 Mm/s. (a) Find the electric field the rod creates at the point ($x = 0$, $y = 20.0$ cm, $z = 0$). (b) Find the magnetic field it creates at the same point. (c) Find the force exerted on an electron at this point, moving with a velocity of $(240\hat{\mathbf{i}})$ Mm/s.

5. A proton moves through a uniform electric field given by $\vec{\mathbf{E}} = 50.0\hat{\mathbf{j}}$ V/m and a uniform magnetic field $\vec{\mathbf{B}} = (0.200\hat{\mathbf{i}} + 0.300\hat{\mathbf{j}} + 0.400\hat{\mathbf{k}})$ T. Determine the acceleration of the proton when it has a velocity $\vec{\mathbf{v}} = 200\hat{\mathbf{i}}$ m/s.

6. An electron moves through a uniform electric field $\vec{\mathbf{E}} = (2.50\hat{\mathbf{i}} + 5.00\hat{\mathbf{j}})$ V/m and a uniform magnetic field $\vec{\mathbf{B}} = (0.400\hat{\mathbf{k}})$ T. Determine the acceleration of the electron when it has a velocity $\vec{\mathbf{v}} = 10.0\hat{\mathbf{i}}$ m/s.

Section 34.3 Plane Electromagnetic Waves

Note: Assume the medium is vacuum unless specified otherwise.

7. (a) The distance to the North Star, Polaris, is approximately 6.44×10^{18} m. If Polaris were to burn out today, in what year would we see it disappear? (b) What time interval is required for sunlight to reach the Earth? (c) What is the transit time for a microwave radar signal traveling from the Earth to the Moon and back? (d) In what time interval does a radio wave travel once around the Earth in a great circle, close to the planet's surface? (e) How long does it take for light to reach you from a lightning stroke 10.0 km away?

8. The speed of an electromagnetic wave traveling in a transparent nonmagnetic substance is $v = 1/\sqrt{\kappa\mu_0\epsilon_0}$, where κ is the dielectric constant of the substance. Determine the speed of light in water, which has a dielectric constant at optical frequencies of 1.78.

9. ▲ Active Figure 34.4b shows a plane electromagnetic sinusoidal wave propagating in the x direction. Suppose the wavelength is 50.0 m and the electric field vibrates in the xy plane with an amplitude of 22.0 V/m. Calculate (a) the frequency of the wave and (b) the magnitude and direction of $\vec{\mathbf{B}}$ when the electric field has its maximum value in the negative y direction. (c) Write an expression for $\vec{\mathbf{B}}$ with the correct unit vector, with numerical values for B_{max}, k, and ω, and with its magnitude in the form

$$B = B_{\text{max}} \cos(kx - \omega t)$$

10. An electromagnetic wave in vacuum has an electric field amplitude of 220 V/m. Calculate the amplitude of the corresponding magnetic field.

11. In SI units, the electric field in an electromagnetic wave is described by

$$E_y = 100 \sin(1.00 \times 10^7 x - \omega t)$$

2 = intermediate;   3 = challenging;   ☐ = SSM/SG;   ▲ = ThomsonNOW;   = symbolic reasoning;   ● = qualitative reasoning

Find (a) the amplitude of the corresponding magnetic field oscillations, (b) the wavelength λ, and (c) the frequency f.

12. Verify by substitution that the following equations are solutions to Equations 34.15 and 34.16, respectively:

$$E = E_{max} \cos (kx - \omega t)$$

$$B = B_{max} \cos (kx - \omega t)$$

13. Review problem. A standing-wave pattern is set up by radio waves between two metal sheets 2.00 m apart, which is the shortest distance between the plates that produces a standing-wave pattern. What is the frequency of the radio waves?

14. A microwave oven is powered by an electron tube, called a magnetron, that generates electromagnetic waves of frequency 2.45 GHz. The microwaves enter the oven and are reflected by the walls. The standing-wave pattern produced in the oven can cook food unevenly, with hot spots in the food at antinodes and cool spots at nodes, so a turntable is often used to rotate the food and distribute the energy. If a microwave oven intended for use with a turntable is instead used with a cooking dish in a fixed position, the antinodes can appear as burn marks on foods such as carrot strips or cheese. The separation distance between the burns is measured to be 6 cm ± 5%. From these data, calculate the speed of the microwaves.

Section 34.4 Energy Carried by Electromagnetic Waves

15. How much electromagnetic energy per cubic meter is contained in sunlight if the intensity of sunlight at the Earth's surface under a fairly clear sky is 1 000 W/m²?

16. An AM radio station broadcasts isotropically (equally in all directions) with an average power of 4.00 kW. A dipole receiving antenna 65.0 cm long is at a location 4.00 mi from the transmitter. Compute the amplitude of the emf that is induced by this signal between the ends of the receiving antenna.

17. What is the average magnitude of the Poynting vector 5.00 mi from a radio transmitter broadcasting isotropically (equally in all directions) with an average power of 250 kW?

18. ● The power of sunlight reaching each square meter of the Earth's surface on a clear day in the tropics is close to 1 000 W. On a winter day in Manitoba, the power concentration of sunlight can be 100 W/m². Many human activities are described by a power-per-footprint-area on the order of 10^2 W/m² or less. (a) Consider, for example, a family of four paying $80 to the electric company every 30 days for 600 kWh of energy carried by electrical transmission to their house, which has floor dimensions of 13.0 m by 9.50 m. Compute the power-per-area measure of this energy use. (b) Consider a car 2.10 m wide and 4.90 m long traveling at 55.0 mi/h using gasoline having "heat of combustion" 44.0 MJ/kg with fuel economy 25.0 mi/gal. One gallon of gasoline has a mass of 2.54 kg. Find the power-per-area measure of the car's energy use. It can be similar to that of a steel mill in which rocks are melted in blast furnaces. (c) Explain why direct use of solar energy is not practical for a conventional automobile. What are some uses of solar energy that are on their face more practical?

19. ▲ A community plans to build a facility to convert solar radiation to electrical power. The community requires 1.00 MW of power, and the system to be installed has an efficiency of 30.0% (that is, 30.0% of the solar energy incident on the surface is converted to useful energy that can power the community). What must be the effective area of a perfectly absorbing surface used in such an installation, assuming sunlight has a constant intensity of 1 000 W/m²?

20. ● Assuming the antenna of a 10.0-kW radio station radiates spherical electromagnetic waves, compute the maximum value of the magnetic field 5.00 km from the antenna and state how this value compares with the surface magnetic field of the Earth.

21. ▲ The filament of an incandescent lamp has a 150-Ω resistance and carries a direct current of 1.00 A. The filament is 8.00 cm long and 0.900 mm in radius. (a) Calculate the Poynting vector at the surface of the filament, associated with the static electric field producing the current and the current's static magnetic field. (b) Find the magnitudes of the static electric and magnetic fields at the surface of the filament.

22. One of the weapons being considered for the "Star Wars" antimissile system is a laser that could destroy ballistic missiles. When a high-power laser is used in the Earth's atmosphere, the electric field can ionize the air, turning it into a conducting plasma that reflects the laser light. In dry air at 0°C and 1 atm, electric breakdown occurs for fields with amplitudes above about 3.00 MV/m. (a) What laser beam intensity will produce such a field? (b) At this maximum intensity, what power can be delivered in a cylindrical beam of diameter 5.00 mm?

23. In a region of free space, the electric field at an instant of time is $\vec{E} = (80.0\hat{i} + 32.0\hat{j} - 64.0\hat{k})$ N/C and the magnetic field is $\vec{B} = (0.200\hat{i} + 0.080\,0\hat{j} + 0.290\hat{k})\,\mu$T. (a) Show that the two fields are perpendicular to each other. (b) Determine the Poynting vector for these fields.

24. Model the electromagnetic wave in a microwave oven as a plane traveling wave moving to the left, with an intensity of 25.0 kW/m². An oven contains two cubical containers of small mass, each full of water. One has an edge length of 6.00 cm, and the other, 12.0 cm. Energy falls perpendicularly on one face of each container. The water in the smaller container absorbs 70.0% of the energy that falls on it. The water in the larger container absorbs 91.0%. That is, the fraction 0.3 of the incoming microwave energy passes through a 6-cm thickness of water, and the fraction $(0.3)(0.3) = 0.09$ passes through a 12-cm thickness. Find the temperature change of the water in each container over a time interval of 480 s. Assume a negligible amount of energy leaves either container by heat.

25. High-power lasers in factories are used to cut through cloth and metal (Fig. P34.25). One such laser has a beam diameter of 1.00 mm and generates an electric field having an amplitude of 0.700 MV/m at the target. Find (a) the amplitude of the magnetic field produced, (b) the intensity of the laser, and (c) the power delivered by the laser.

26. At one location on the Earth, the rms value of the magnetic field caused by solar radiation is 1.80 μT. From this value, calculate (a) the rms electric field due to solar radiation, (b) the average energy density of the solar component of electromagnetic radiation at this location, and (c) the average magnitude of the Poynting vector for the Sun's radiation.

Figure P34.25

27. Consider a bright star in our night sky. Assume its distance from Earth is 20.0 ly and its power output is 4.00×10^{28} W, about 100 times that of the Sun. (a) Find the intensity of the starlight at the Earth. (b) Find the power of the starlight the Earth intercepts.

Section 34.5 Momentum and Radiation Pressure

28. A possible means of space flight is to place a perfectly reflecting aluminized sheet into orbit around the Earth and then use the light from the Sun to push this "solar sail." Suppose a sail of area 6.00×10^{5} m^2 and mass 6 000 kg is placed in orbit facing the Sun. (a) What force is exerted on the sail? (b) What is the sail's acceleration? (c) What time interval is required for the sail to reach the Moon, 3.84×10^{8} m away? Ignore all gravitational effects, assume the acceleration calculated in part (b) remains constant, and assume a solar intensity of 1 370 W/m^2.

29. A radio wave transmits 25.0 W/m^2 of power per unit area. A flat surface of area A is perpendicular to the direction of propagation of the wave. Calculate the radiation pressure on it, assuming the surface is a perfect absorber.

30. ● A plane electromagnetic wave of intensity 6.00 W/m^2, moving in the x direction, strikes a small pocket mirror, of area 40.0 cm^2, held in the yz plane. (a) What momentum does the wave transfer to the mirror each second? (b) Find the force that the wave exerts on the mirror. (c) Explain the relationship between the answers to parts (a) and (b).

31. ▲ A 15.0-mW helium–neon laser ($\lambda = 632.8$ nm) emits a beam of circular cross section with a diameter of 2.00 mm. (a) Find the maximum electric field in the beam. (b) What total energy is contained in a 1.00-m length of the beam? (c) Find the momentum carried by a 1.00-m length of the beam.

32. ● The intensity of sunlight at the Earth's distance from the Sun is 1 370 W/m^2. (a) Assume the Earth absorbs all the sunlight incident upon it. Find the total force the Sun exerts on the Earth due to radiation pressure. (b) Explain how this force compares with the Sun's gravitational attraction.

33. ● Assume the intensity of solar radiation incident on the upper atmosphere of the Earth is 1 370 W/m^2 and use data from Table 13.2 as necessary. Determine (a) the intensity of solar radiation incident on Mars, (b) the total power incident on Mars, and (c) the radiation force that acts on that planet if it absorbs nearly all the light. (d) State how this force compares with the gravitational attraction exerted by the Sun on Mars. (e) Compare the ratio of the gravitational force to the light-pressure force exerted on

Earth, found in Problem 32, and the ratio of these forces exerted on Mars, found in part (d).

34. A uniform circular disk of mass 24.0 g and radius 40.0 cm hangs vertically from a fixed, frictionless, horizontal hinge at a point on its circumference. A beam of electromagnetic radiation with intensity 10.0 MW/m^2 is incident on the disk in a direction perpendicular to its surface. The disk is perfectly absorbing, and the resulting radiation pressure makes the disk rotate. Find the angle through which the disk rotates as it reaches its new equilibrium position. Assume the radiation is *always* perpendicular to the surface of the disk.

Section 34.6 Production of Electromagnetic Waves by an Antenna

35. Figure 34.9 shows a Hertz antenna (also known as a half-wave antenna because its length is $\lambda/2$). The antenna is located far enough from the ground that reflections do not significantly affect its radiation pattern. Most AM radio stations, however, use a Marconi antenna, which consists of the top half of a Hertz antenna. The lower end of this (quarter-wave) antenna is connected to Earth ground, and the ground itself serves as the missing lower half. What are the heights of the Marconi antennas for radio stations broadcasting at (a) 560 kHz and (b) 1 600 kHz? In the United States, these stations are at 560 and 1 600 on the AM dial.

36. Two handheld radio transceivers with dipole antennas are separated by a large, fixed distance. If the transmitting antenna is vertical, what fraction of the maximum received power will appear in the receiving antenna when it is inclined from the vertical by (a) 15.0°? (b) 45.0°? (c) 90.0°?

37. Two vertical radio-transmitting antennas are separated by half the broadcast wavelength and are driven in phase with each other. In what horizontal directions are (a) the strongest and (b) the weakest signals radiated?

38. **Review problem.** Accelerating charges radiate electromagnetic waves. Calculate the wavelength of radiation produced by a proton moving in a circle of radius R perpendicular to a magnetic field of magnitude B.

39. A large, flat sheet carries a uniformly distributed electric current with current per unit width J_s. Problem 33 in Chapter 30 demonstrated that the current creates a magnetic field on both sides of the sheet, parallel to the sheet and perpendicular to the current, with magnitude $B = \frac{1}{2}\mu_0 J_s$. If the current is in the y direction and oscillates in time according to

$$J_{\max}(\cos \omega t)\hat{\mathbf{j}} = J_{\max}[\cos (-\omega t)]\hat{\mathbf{j}}$$

the sheet radiates an electromagnetic wave as shown in Figure P34.39. The magnetic field of the wave is described

Figure P34.39

---

2 = intermediate;   3 = challenging;   □ = SSM/SG;   ▲ = ThomsonNOW;   ▨ = symbolic reasoning;   ● = qualitative reasoning

by the wave function $\vec{\mathbf{B}} = \frac{1}{2}\mu_0 J_{\max}[\cos(kx - \omega t)]\hat{\mathbf{k}}$. (a) Find the wave function for the electric field in the wave. (b) Find the Poynting vector as a function of x and t. (c) Find the intensity of the wave. (d) **What If?** If the sheet is to emit radiation in each direction (normal to the plane of the sheet) with intensity 570 W/m^2, what maximum value of sinusoidal current density is required?

Section 34.7 The Spectrum of Electromagnetic Waves

40. Classify waves with frequencies of 2 Hz, 2 kHz, 2 MHz, 2 GHz, 2 THz, 2 PHz, 2 EHz, 2 ZHz, and 2 YHz on the electromagnetic spectrum. Classify waves with wavelengths of 2 km, 2 m, 2 mm, 2 μm, 2 nm, 2 pm, 2 fm, and 2 am.

41. What are the wavelengths of electromagnetic waves in free space that have frequencies of (a) 5.00×10^{19} Hz and (b) 4.00×10^9 Hz?

42. Compute an order-of-magnitude estimate for the frequency of an electromagnetic wave with wavelength equal to (a) your height and (b) the thickness of a sheet of paper. How is each wave classified on the electromagnetic spectrum?

43. Twelve VHF television channels (channels 2 through 13) lie in the range of frequencies between 54.0 MHz and 216 MHz. Each channel is assigned a width of 6.0 MHz, with the two ranges 72.0–76.0 MHz and 88.0–174 MHz reserved for non-TV purposes. (Channel 2, for example, lies between 54.0 and 60.0 MHz.) Calculate the broadcast wavelength range for (a) channel 4, (b) channel 6, and (c) channel 8.

44. ● *This just in!* An important news announcement is transmitted by radio waves to people sitting next to their radios 100 km from the station and by sound waves to people sitting across the newsroom 3.00 m from the newscaster. Who receives the news first? Explain. Take the speed of sound in air to be 343 m/s.

45. ● The United States Navy has long proposed the construction of extremely low-frequency (ELF) communication systems. Such waves could penetrate the oceans to reach distant submarines. Calculate the length of a quarter-wavelength antenna for a transmitter generating ELF waves of frequency 75.0 Hz. How practical is this plan?

Additional Problems

46. Write expressions for the electric and magnetic fields of a sinusoidal plane electromagnetic wave having a frequency of 3.00 GHz and traveling in the positive x direction. The amplitude of the electric field is 300 V/m.

47. Assume the intensity of solar radiation incident on the cloud tops of the Earth is 1 370 W/m^2. (a) Calculate the total power radiated by the Sun, taking the average Earth–Sun separation to be 1.496×10^{11} m. (b) Determine the maximum values of the electric and magnetic fields in the sunlight at the Earth's location.

48. The intensity of solar radiation at the top of the Earth's atmosphere is 1 370 W/m^2. Assuming that 60% of the incoming solar energy reaches the Earth's surface and you absorb 50% of the incident energy, make an order-of-magnitude estimate of the amount of solar energy you absorb in a 60-min sunbath.

49. ● You may wish to review Sections 16.5 and 17.3 on the transport of energy by string waves and sound. Active Figure

34.4b is a graphical representation of an electromagnetic wave moving in the x direction. A mathematical representation is given by the two equations $E = E_{\max} \cos(kx - \omega t)$ and $B = B_{\max} \cos(kx - \omega t)$, where $E_{\max}$ is the amplitude of the electric field and $B_{\max}$ is the amplitude of the magnetic field. (a) Sketch a graph of the electric field in this wave at the instant $t = 0$, letting your flat paper represent the xy plane. (b) Compute the energy density u_E in the electric field as a function of x at the instant $t = 0$. Compute (c) the energy density in the magnetic field u_B and (d) the total energy density u as functions of x. Add a curve to your graph representing $u = u_E + u_B$. The energy in a rectangular box of length dx along the x direction and area A parallel to the yz plane is $uA\,dx$. The energy in a "shoebox" of length λ and frontal area A is $E_\lambda = \int_0^\lambda uA\,dx$. (The symbol E_λ for energy in a wavelength imitates the notation of Sections 16.5 and 17.3.) (e) Perform the integration to compute the amount of this energy in terms of A, λ, $E_{\max}$, and universal constants. We may think of the energy transport by the whole wave as a series of these shoeboxes going past as if carried on a conveyor belt. Each cycle passes in a time interval defined as the period $T = 1/f$ of the wave. The power the wave carries through area A is then $\mathcal{P} = T_{\mathrm{ER}}/\Delta t = E_\lambda/T$. The intensity of the wave is $I = \mathcal{P}/A = E_\lambda/AT$. (f) Compute this intensity in terms of $E_{\max}$ and universal constants. Explain how your result compares with that given in Equation 34.24.

50. Consider a small, spherical particle of radius r located in space a distance R from the Sun. (a) Show that the ratio $F_{\mathrm{rad}}/F_{\mathrm{grav}}$ is proportional to $1/r$, where F_{rad} is the force exerted by solar radiation and F_{grav} is the force of gravitational attraction. (b) The result of part (a) means that, for a sufficiently small value of r, the force exerted on the particle by solar radiation exceeds the force of gravitational attraction. Calculate the value of r for which the particle is in equilibrium under the two forces. Assume the particle has a perfectly absorbing surface and a mass density of 1.50 g/cm^3. Let the particle be located 3.75×10^{11} m from the Sun and use 214 W/m^2 as the value of the solar intensity at that point.

51. A dish antenna having a diameter of 20.0 m receives (at normal incidence) a radio signal from a distant source as shown in Figure P34.51. The radio signal is a continuous sinusoidal wave with amplitude $E_{\max} = 0.200\ \mu$V/m. Assume the antenna absorbs all the radiation that falls on the dish. (a) What is the amplitude of the magnetic field in this wave? (b) What is the intensity of the radiation received by this antenna? (c) What is the power received by the antenna? (d) What force is exerted by the radio waves on the antenna?

Figure P34.51

52. ● (a) A stationary charged particle at the origin creates an electric flux of 487 N · m²/C. Find the electric field it creates in the empty space around it as a function of radial distance r away from the particle. (b) A small source at the origin emits an electromagnetic wave with a single frequency into vacuum, equally in all directions, with power 25.0 W. Find the electric field amplitude as a function of radial distance away from the source. Show explicitly how the units in your expression work out. (c) At what distance is the amplitude of the electric field in the wave equal to 3.00 MV/m, representing the dielectric strength of air? (d) As the distance doubles, what happens to the field amplitude? State how this behavior compares with the behavior of the field in part (a).

53. In 1965, Arno Penzias and Robert Wilson discovered the cosmic microwave radiation left over from the Big Bang expansion of the Universe. Suppose the energy density of this background radiation is 4.00×10^{-14} J/m³. Determine the corresponding electric field amplitude.

54. ● A handheld cellular telephone operates in the 860- to 900-MHz band and has a power output of 0.600 W from an antenna 10.0 cm long (Fig. P34.54). (a) Find the average magnitude of the Poynting vector 4.00 cm from the antenna at the location of a typical person's head. Assume the antenna emits energy with cylindrical wave fronts. (The actual radiation from antennas follows a more complicated pattern.) (b) The ANSI/IEEE C95.1-1991 maximum exposure standard is 0.57 mW/cm² for persons living near cellular telephone base stations, who would be continuously exposed to the radiation. State how the answer to part (a) compares with this standard.

Figure P34.54

55. A linearly polarized microwave of wavelength 1.50 cm is directed along the positive x axis. The electric field vector has a maximum value of 175 V/m and vibrates in the xy plane. (a) Assuming the magnetic field component of the wave can be written in the form $B = B_{max} \sin(kx - \omega t)$, give values for B_{max}, k, and ω. Also, determine in which plane the magnetic field vector vibrates. (b) Calculate the average value of the Poynting vector for this wave. (c) What radiation pressure would this wave exert if it were directed at normal incidence onto a perfectly reflecting sheet? (d) What acceleration would be imparted to a 500-g sheet (perfectly reflecting and at normal incidence) with dimensions of 1.00 m × 0.750 m?

56. ● The Earth reflects approximately 38.0% of the incident sunlight from its clouds and surface. (a) Given that the intensity of solar radiation is 1 370 W/m², find the radiation pressure on the Earth, in pascals, at the location where the Sun is straight overhead. (b) State how this quantity compares with normal atmospheric pressure at the Earth's surface, which is 101 kPa.

57. An astronaut, stranded in space 10.0 m from her spacecraft and at rest relative to it, has a mass (including equipment) of 110 kg. Because she has a 100-W light source that forms a directed beam, she considers using the beam as a photon rocket to propel herself continuously toward the spacecraft. (a) Calculate the time interval required for her to reach the spacecraft by this method. (b) **What If?** Suppose she throws the light source away in the direction away from the spacecraft instead. The mass of the light source is 3.00 kg, and, after being thrown, it moves at 12.0 m/s relative to the recoiling astronaut. After what time interval will the astronaut reach the spacecraft?

58. **Review problem.** A 1.00-m-diameter mirror focuses the Sun's rays onto an absorbing plate 2.00 cm in radius, which holds a can containing 1.00 L of water at 20.0°C. (a) If the solar intensity is 1.00 kW/m², what is the intensity on the absorbing plate? (b) What are the maximum magnitudes of the fields $\vec{E}$ and $\vec{B}$? (c) If 40.0% of the energy is absorbed, what time interval is required to bring the water to its boiling point?

59. Lasers have been used to suspend spherical glass beads in the Earth's gravitational field. (a) A black bead has a mass of 1.00 μg and a density of 0.200 g/cm³. Determine the radiation intensity needed to support the bead. (b) If the bead has a radius of 0.200 cm, what power is required for this laser?

60. Lasers have been used to suspend spherical glass beads in the Earth's gravitational field. (a) A black bead has mass m and density ρ. Determine the radiation intensity needed to support the bead. (b) If the bead has radius r, what power is required for this laser?

61. The electromagnetic power radiated by a nonrelativistic particle with charge q moving with acceleration a is

$$\mathcal{P} = \frac{q^2 a^2}{6\pi\epsilon_0 c^3}$$

where ϵ_0 is the permittivity of free space (also called the permittivity of vacuum) and c is the speed of light in vacuum. (a) Show that the right side of this equation has units of watts. (b) An electron is placed in a constant electric field of magnitude 100 N/C. Determine the acceleration of the electron and the electromagnetic power radiated by this electron. (c) **What If?** If a proton is placed in a cyclotron with a radius of 0.500 m and a magnetic field of magnitude 0.350 T, what electromagnetic power does this proton radiate?

62. A plane electromagnetic wave varies sinusoidally at 90.0 MHz as it travels along the $+x$ direction. The peak value of the electric field is 2.00 mV/m, and it is directed along the $\pm y$ direction. (a) Find the wavelength, the period, and the maximum value of the magnetic field. (b) Write expressions in SI units for the space and time variations of the electric field and of the magnetic field. Include both numerical values and subscripts to indicate coordinate directions. (c) Find the average power per unit area that this wave carries through space. (d) Find the average energy density in the radiation (in joules per cubic meter). (e) What radiation pressure would this wave exert upon a perfectly reflecting surface at normal incidence?

63. ▲ **Review problem.** In the absence of cable input or a satellite dish, a television set can use a dipole-receiving antenna for VHF channels and a loop antenna for UHF channels. In Figure Q34.15, the "rabbit ears" form the VHF antenna and the smaller loop of wire is the UHF antenna. The UHF antenna produces an emf from the changing magnetic flux through the loop. The television station broadcasts a signal with a frequency f, and the signal has an electric-field amplitude E_{max} and a magnetic-field amplitude B_{max} at the location of the receiving antenna. (a) Using Faraday's law, derive an expression for the amplitude of the emf that appears in a single-turn, circular loop antenna with a radius r that is small compared with the wavelength of the wave. (b) If the electric field in the signal points vertically, what orientation of the loop gives the best reception?

> *Note:* Section 8.1 introduced electromagnetic radiation as a mode of energy transfer, which was discussed in more detail in Section 20.7. The next three problems use ideas introduced there and in this chapter.

64. ● **Review problem.** Eliza is a black cat with four black kittens. Eliza's mass is 5.50 kg, and each kitten has mass 0.800 kg. One cool night, all five sleep snuggled together on a mat, with their bodies forming one hemisphere. (a) Assuming the purring heap has a uniform density of 990 kg/m³, find the radius of the hemisphere. (b) Find the area of its curved surface. (c) Assume the surface temperature is 31.0°C and the emissivity is 0.970. Find the intensity of radiation emitted by the cats at their curved surface and (d) the radiated power from this surface. (e) You may think of the emitted electromagnetic wave as having a single predominant frequency (of 31.2 THz). Find the amplitude of the electric field just outside the surface of the cozy pile and (f) the amplitude of the magnetic field. (g) Are the sleeping cats charged? Are they current carrying? Are they magnetic? Are they a radiation source? Do they glow in the dark? Give an explanation for your answers so that they do not seem contradictory. (h) **What If?** The next night, the kittens all sleep alone, curling up into separate hemispheres like their mother. Find the total radiated power of the family. (For simplicity, ignore the cats' absorption of radiation from the environment.)

65. **Review problem.** (a) An elderly couple has a solar water heater installed on the roof of their house (Fig. P34.65).

The heater is a flat, closed box with excellent thermal insulation. Its interior is painted black, and its front face is made of insulating glass. Its emissivity for visible light is 0.900, and its emissivity for infrared light is 0.700. Light from the noon Sun is incident perpendicular to the glass with an intensity of 1 000 W/m², and no water enters or leaves the box. Find the steady-state temperature of the box's interior. (b) **What If?** The couple builds an identical box with no water tubes. It lies flat on the ground in front of the house. They use it as a cold frame, where they plant seeds in early spring. Assuming the same noon Sun is at an elevation angle of 50.0°, find the steady-state temperature of the interior of the box when its ventilation slots are tightly closed.

Figure P34.65

66. ● **Review problem.** The study of Creation suggests a Creator with an inordinate fondness for beetles and for small, red stars. A small, red star radiates electromagnetic waves with power 6.00×10^{23} W, which is only 0.159% of the luminosity of the Sun. Consider a spherical planet in a circular orbit around this star. Assume the emissivity of the planet is equal for infrared and for visible light and the planet has a uniform surface temperature. Identify the projected area over which the planet absorbs starlight and the radiating area of the planet. If beetles thrive at a temperature of 310 K, what should the radius of the planet's orbit be?

Answers to Quick Quizzes

34.1 **(i)**, **(b)**. There can be no conduction current because there is no conductor between the plates. There is a time-varying electric field because of the decreasing charge on the plates, and the time-varying electric flux represents a displacement current. **(ii)**, **(c)**. There is a time-varying electric field because of the decreasing charge on the plates. This time-varying electric field produces a magnetic field.

34.2 **(c)**. Active Figure 34.4b shows that the $\vec{\mathbf{B}}$ and $\vec{\mathbf{E}}$ vectors reach their maximum positive values at the same time.

34.3 **(c)**. The $\vec{\mathbf{B}}$ field must be in the $+z$ direction in order that the Poynting vector be directed along the $-y$ direction.

34.4 **(b)**. To maximize the pressure on the sails, they should be perfectly reflective, so the pressure is given by Equation 34.30.

34.5 **(a)**. The best orientation is parallel to the transmitting antenna because that is the orientation of the electric field. The electric field moves electrons in the receiving antenna, thereby inducing a current that is detected and amplified.

34.6 **(c)**. Either Equation 34.20 or Figure 34.11 can be used to find the order of magnitude of the wavelengths.

34.7 **(a)**. Either Equation 34.20 or Figure 34.11 can be used to find the order of magnitude of the wavelength.

PART 5

Light and Optics

Light is basic to almost all life on the Earth. For example, plants convert the energy transferred by sunlight to chemical energy through photosynthesis. In addition, light is the principal means by which we are able to transmit and receive information to and from objects around us and throughout the Universe. Light is a form of electromagnetic radiation and represents energy transfer from the source to the observer.

Many phenomena in our everyday life depend on the properties of light. When you watch a color television or view photos on a computer monitor, you are seeing millions of colors formed from combinations of only three colors that are physically on the screen: red, blue, and green. The blue color of the daytime sky is a result of the optical phenomenon of *scattering* of light by air molecules, as are the red and orange colors of sunrises and sunsets. You see your image in your bathroom mirror in the morning or the images of other cars in your car's rearview mirror when you are driving. These images result from *reflection* of light. If you wear glasses or contact lenses, you are depending on *refraction* of light for clear vision. The colors of a rainbow result from *dispersion* of light as it passes through raindrops hovering in the sky after a rainstorm. If you have ever seen the colored circles of the glory surrounding the shadow of your airplane on clouds as you fly above them, you are seeing an effect that results from *interference* of light. The phenomena mentioned here have been studied by scientists and are well understood.

In the introduction to Chapter 35, we discuss the dual nature of light. In some cases, it is best to model light as a stream of particles; in others, a wave model works better. Chapters 35 through 38 concentrate on those aspects of light that are best understood through the wave model of light. In Part 6, we will investigate the particle nature of light.

The Grand Tetons in western Wyoming are reflected in a smooth lake at sunset. The optical principles we study in this part of the book will explain the nature of the reflected image of the mountains and why the sky appears red. (David Muench/CORBIS)

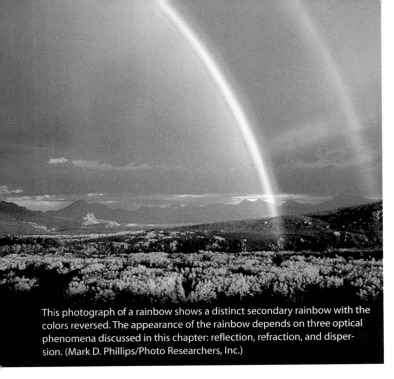

This photograph of a rainbow shows a distinct secondary rainbow with the colors reversed. The appearance of the rainbow depends on three optical phenomena discussed in this chapter: reflection, refraction, and dispersion. (Mark D. Phillips/Photo Researchers, Inc.)

35.1 The Nature of Light

35.2 Measurements of the Speed of Light

35.3 The Ray Approximation in Geometric Optics

35.4 The Wave Under Reflection

35.5 The Wave Under Refraction

35.6 Huygens's Principle

35.7 Dispersion

35.8 Total Internal Reflection

35 The Nature of Light and the Laws of Geometric Optics

This first chapter on optics begins by introducing two historical models for light and discussing early methods for measuring the speed of light. Next we study the fundamental phenomena of geometric optics: reflection of light from a surface and refraction as the light crosses the boundary between two media. We will also study the dispersion of light as it refracts into materials, resulting in visual displays such as the rainbow. Finally, we investigate the phenomenon of total internal reflection, which is the basis for the operation of optical fibers and the burgeoning technology of fiber optics.

Courtesy of Rijksmuseum voor de Geschiedenis der Natuurwetenschappen and Niels Bohr Library

CHRISTIAN HUYGENS
Dutch Physicist and Astronomer
(1629–1695)
Huygens is best known for his contributions to the fields of optics and dynamics. To Huygens, light was a type of vibratory motion, spreading out and producing the sensation of light when impinging on the eye. On the basis of this theory, he deduced the laws of reflection and refraction and explained the phenomenon of double refraction.

35.1 The Nature of Light

Before the beginning of the nineteenth century, light was considered to be a stream of particles that either was emitted by the object being viewed or emanated from the eyes of the viewer. Newton, the chief architect of the particle model of light, held that particles were emitted from a light source and that these particles stimulated the sense of sight upon entering the eye. Using this idea, he was able to explain reflection and refraction.

Most scientists accepted Newton's particle model. During Newton's lifetime, however, another model was proposed, one that argued that light might be some sort of wave motion. In 1678, Dutch physicist and astronomer Christian Huygens showed that a wave model of light could also explain reflection and refraction.

In 1801, Thomas Young (1773–1829) provided the first clear experimental demonstration of the wave nature of light. Young showed that under appropriate conditions light rays interfere with one another. Such behavior could not be explained at that time by a particle model because there was no conceivable way in which two or more particles could come together and cancel one another. Additional developments during the nineteenth century led to the general acceptance of the wave model of light, the most important resulting from the work of Maxwell, who in 1873 asserted that light was a form of high-frequency electromagnetic wave. As discussed in Chapter 34, Hertz provided experimental confirmation of Maxwell's theory in 1887 by producing and detecting electromagnetic waves.

Although the wave model and the classical theory of electricity and magnetism were able to explain most known properties of light, they could not explain some subsequent experiments. The most striking phenomenon is the photoelectric effect, also discovered by Hertz: when light strikes a metal surface, electrons are sometimes ejected from the surface. As one example of the difficulties that arose, experiments showed that the kinetic energy of an ejected electron is independent of the light intensity. This finding contradicted the wave model, which held that a more intense beam of light should add more energy to the electron. Einstein proposed an explanation of the photoelectric effect in 1905 using a model based on the concept of quantization developed by Max Planck (1858–1947) in 1900. The quantization model assumes the energy of a light wave is present in particles called *photons;* hence, the energy is said to be quantized. According to Einstein's theory, the energy of a photon is proportional to the frequency of the electromagnetic wave:

$$E = hf \qquad (35.1)$$

◀ Energy of a photon

where the constant of proportionality $h = 6.63 \times 10^{-34}$ J·s is called *Planck's constant*. We study this theory in Chapter 40.

In view of these developments, light must be regarded as having a dual nature. **Light exhibits the characteristics of a wave in some situations and the characteristics of a particle in other situations.** Light is light, to be sure. The question "Is light a wave or a particle?" is inappropriate, however. Sometimes light acts like a wave, and other times it acts like a particle. In the next few chapters, we investigate the wave nature of light.

35.2   Measurements of the Speed of Light

Light travels at such a high speed (to three digits, $c = 3.00 \times 10^8$ m/s) that early attempts to measure its speed were unsuccessful. Galileo attempted to measure the speed of light by positioning two observers in towers separated by approximately 10 km. Each observer carried a shuttered lantern. One observer would open his lantern first, and then the other would open his lantern at the moment he saw the light from the first lantern. Galileo reasoned that by knowing the transit time of the light beams from one lantern to the other and the distance between the two lanterns, he could obtain the speed. His results were inconclusive. Today, we realize (as Galileo concluded) that it is impossible to measure the speed of light in this manner because the transit time for the light is so much less than the reaction time of the observers.

Roemer's Method

In 1675, Danish astronomer Ole Roemer (1644–1710) made the first successful estimate of the speed of light. Roemer's technique involved astronomical observations of Io, one of the moons of Jupiter. Io has a period of revolution around Jupiter of approximately 42.5 h. The period of revolution of Jupiter around the Sun is about 12 yr; therefore, as the Earth moves through 90° around the Sun, Jupiter revolves through only $(\frac{1}{12})90° = 7.5°$ (Fig. 35.1).

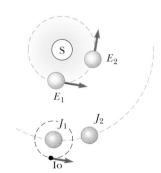

Figure 35.1 Roemer's method for measuring the speed of light. In the time interval during which the Earth travels 90° around the Sun (three months), Jupiter travels only about 7.5° (drawing not to scale).

An observer using the orbital motion of Io as a clock would expect the orbit to have a constant period. After collecting data for more than a year, however, Roemer observed a systematic variation in Io's period. He found that the periods were longer than average when the Earth was receding from Jupiter and shorter than average when the Earth was approaching Jupiter. Roemer attributed this variation in period to the distance between the Earth and Jupiter changing from one observation to the next.

Using Roemer's data, Huygens estimated the lower limit for the speed of light to be approximately 2.3×10^8 m/s. This experiment is important historically because it demonstrated that light does have a finite speed and gave an estimate of this speed.

Fizeau's Method

The first successful method for measuring the speed of light by means of purely terrestrial techniques was developed in 1849 by French physicist Armand H. L. Fizeau (1819–1896). Figure 35.2 represents a simplified diagram of Fizeau's apparatus. The basic procedure is to measure the total time interval during which light travels from some point to a distant mirror and back. If d is the distance between the light source (considered to be at the location of the wheel) and the mirror and if the time interval for one round trip is Δt, the speed of light is $c = 2d/\Delta t$.

To measure the transit time, Fizeau used a rotating toothed wheel, which converts a continuous beam of light into a series of light pulses. The rotation of such a wheel controls what an observer at the light source sees. For example, if the pulse traveling toward the mirror and passing the opening at point A in Figure 35.2 should return to the wheel at the instant tooth B had rotated into position to cover the return path, the pulse would not reach the observer. At a greater rate of rotation, the opening at point C could move into position to allow the reflected pulse to reach the observer. Knowing the distance d, the number of teeth in the wheel, and the angular speed of the wheel, Fizeau arrived at a value of 3.1×10^8 m/s. Similar measurements made by subsequent investigators yielded more precise values for c, which led to the currently accepted value of 2.9979×10^8 m/s.

Toothed wheel   Mirror

Figure 35.2 Fizeau's method for measuring the speed of light using a rotating toothed wheel. The light source is considered to be at the location of the wheel; therefore, the distance d is known.

EXAMPLE 35.1 **Measuring the Speed of Light with Fizeau's Wheel**

Assume Fizeau's wheel has 360 teeth and rotates at 27.5 rev/s when a pulse of light passing through opening A in Figure 35.2 is blocked by tooth B on its return. If the distance to the mirror is 7 500 m, what is the speed of light?

SOLUTION

Conceptualize Imagine a pulse of light passing through opening A in Figure 35.2 and reflecting from the mirror. By the time the pulse arrives back at the wheel, tooth B has rotated into the position previously occupied by opening A.

Categorize We model the wheel as a rigid object under constant angular speed and the pulse of light as a particle under constant speed.

Analyze The wheel has 360 teeth, so it must have 360 openings. Therefore, because the light passes through opening A but is blocked by the tooth immediately adjacent to A, the wheel must rotate through an angular displacement of $\frac{1}{720}$ rev in the time interval during which the light pulse makes its round trip.

Use the rigid object under constant angular speed model to find the time interval for the pulse's round trip:

$$\Delta t = \frac{\Delta \theta}{\omega} = \frac{\frac{1}{720} \text{ rev}}{27.5 \text{ rev/s}} = 5.05 \times 10^{-5} \text{ s}$$

From the particle under constant speed model, find the speed of the pulse of light:

$$c = \frac{2d}{\Delta t} = \frac{2(7\,500 \text{ m})}{5.05 \times 10^{-5} \text{ s}} = 2.97 \times 10^8 \text{ m/s}$$

Finalize This result is very close to the actual value of the speed of light.

35.3 The Ray Approximation in Geometric Optics

The field of **geometric optics** involves the study of the propagation of light. Geometric optics assumes light travels in a fixed direction in a straight line as it passes through a uniform medium and changes its direction when it meets the surface of a different medium or if the optical properties of the medium are nonuniform in either space or time. In our study of geometric optics here and in Chapter 36, we use what is called the **ray approximation.** To understand this approximation, first notice that the rays of a given wave are straight lines perpendicular to the wave fronts as illustrated in Figure 35.3 for a plane wave. In the ray approximation, a wave moving through a medium travels in a straight line in the direction of its rays.

If the wave meets a barrier in which there is a circular opening whose diameter is much larger than the wavelength as in Active Figure 35.4a, the wave emerging from the opening continues to move in a straight line (apart from some small edge effects); hence, the ray approximation is valid. If the diameter of the opening is on the order of the wavelength as in Active Figure 35.4b, the waves spread out from the opening in all directions. This effect, called *diffraction,* will be studied in Chapter 37. Finally, if the opening is much smaller than the wavelength, the opening can be approximated as a point source of waves as shown in Active Fig. 35.4c.

Similar effects are seen when waves encounter an opaque object of dimension d. In that case, when $\lambda \ll d$, the object casts a sharp shadow.

The ray approximation and the assumption that $\lambda \ll d$ are used in this chapter and in Chapter 36, both of which deal with geometric optics. This approximation is very good for the study of mirrors, lenses, prisms, and associated optical instruments such as telescopes, cameras, and eyeglasses.

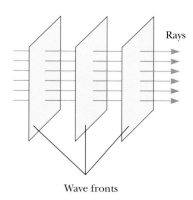

Figure 35.3 A plane wave propagating to the right. Notice that the rays, which always point in the direction of the wave propagation, are straight lines perpendicular to the wave fronts.

35.4 The Wave Under Reflection

We introduced the concept of reflection of waves in a discussion of waves on strings in Section 16.4. As with waves on strings, when a light ray traveling in one medium encounters a boundary with another medium, part of the incident light is reflected. For waves on a one-dimensional string, the reflected wave must necessarily be restricted to a direction along the string. For light waves traveling in three-dimensional space, no such restriction applies and the reflected light waves can be

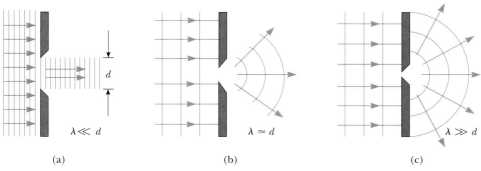

(a)   (b)   (c)

ACTIVE FIGURE 35.4

A plane wave of wavelength λ is incident on a barrier in which there is an opening of diameter d.
(a) When $\lambda \ll d$, the rays continue in a straight-line path and the ray approximation remains valid.
(b) When $\lambda \approx d$, the rays spread out after passing through the opening. (c) When $\lambda \gg d$, the opening behaves as a point source emitting spherical waves.

Sign in at www.thomsonedu.com and go to ThomsonNOW to adjust the size of the opening and observe the effect on the waves passing through.

Courtesy of Henry Leap and Jim Lehman

(a)

(b)

Courtesy of Henry Leap and Jim Lehman

(c)

(d)

Courtesy of Henry Leap and Jim Lehman

Figure 35.5 Schematic representation of (a) specular reflection, where the reflected rays are all parallel to each other, and (b) diffuse reflection, where the reflected rays travel in random directions. (c) and (d) Photographs of specular and diffuse reflection using laser light.

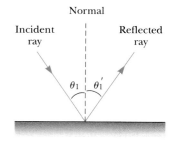

ACTIVE FIGURE 35.6

According to the wave under reflection model, $\theta_1' = \theta_1$. The incident ray, the reflected ray, and the normal all lie in the same plane.

Sign in at www.thomsonedu.com and go to ThomsonNOW to vary the incident angle and see the effect on the reflected ray.

Law of reflection ▶

PITFALL PREVENTION 35.1
Subscript Notation

The subscript 1 refers to parameters for the light in the initial medium. When light travels from one medium to another, we use the subscript 2 for the parameters associated with the light in the new medium. In this discussion, the light stays in the same medium, so we only have to use the subscript 1.

in directions different from the direction of the incident waves. Figure 35.5a shows several rays of a beam of light incident on a smooth, mirror-like, reflecting surface. The reflected rays are parallel to one another as indicated in the figure. The direction of a reflected ray is in the plane perpendicular to the reflecting surface that contains the incident ray. Reflection of light from such a smooth surface is called **specular reflection.** If the reflecting surface is rough as in Figure 35.5b, the surface reflects the rays not as a parallel set but in various directions. Reflection from any rough surface is known as **diffuse reflection.** A surface behaves as a smooth surface as long as the surface variations are much smaller than the wavelength of the incident light.

The difference between these two kinds of reflection explains why it is more difficult to see while driving on a rainy night than on a dry, sunny day. If the road is wet, the smooth surface of the water specularly reflects most of your headlight beams away from your car (and perhaps into the eyes of oncoming drivers). When the road is dry, its rough surface diffusely reflects part of your headlight beam back toward you, allowing you to see the road more clearly. In this book, we restrict our study to specular reflection and use the term *reflection* to mean specular reflection.

Consider a light ray traveling in air and incident at an angle on a flat, smooth surface as shown in Active Figure 35.6. The incident and reflected rays make angles θ_1 and θ_1', respectively, where the angles are measured between the normal and the rays. (The normal is a line drawn perpendicular to the surface at the point where the incident ray strikes the surface.) Experiments and theory show that **the angle of reflection equals the angle of incidence:**

$$\theta_1' = \theta_1 \tag{35.2}$$

This relationship is called the **law of reflection.** Because reflection of waves from an interface between two media is a common phenomenon, we identify an analysis model for this situation: the **wave under reflection.** Equation 35.2 is the mathematical representation of this model.

Quick Quiz 35.1 In the movies, you sometimes see an actor looking in a mirror and you can see his face in the mirror. During the filming of such a scene, what does the actor see in the mirror? (a) his face (b) your face (c) the director's face (d) the movie camera (e) impossible to determine

EXAMPLE 35.2 **The Double-Reflected Light Ray**

Two mirrors make an angle of 120° with each other as illustrated in Figure 35.7a. A ray is incident on mirror M_1 at an angle of 65° to the normal. Find the direction of the ray after it is reflected from mirror M_2.

SOLUTION

Conceptualize Figure 35.7a helps conceptualize this situation. The incoming ray reflects from the first mirror, and the reflected ray is directed toward the second mirror. Therefore, there is a second reflection from the second mirror.

Categorize Because the interactions with both mirrors are simple reflections, we apply the wave under reflection model and some geometry.

Figure 35.7 (Example 35.2) (a) Mirrors M_1 and M_2 make an angle of 120° with each other. (b) The geometry for an arbitrary mirror angle.

Analyze From the law of reflection, the first reflected ray makes an angle of 65° with the normal.

Find the angle the first reflected ray makes with the horizontal:

$$\delta = 90° - 65° = 25°$$

From the triangle made by the first reflected ray and the two mirrors, find the angle the reflected ray makes with M_2:

$$\gamma = 180° - 25° - 120° = 35°$$

Find the angle the first reflected ray makes with the normal to M_2:

$$\theta_{M_2} = 90° - 35° = 55°$$

From the law of reflection, find the angle the second reflected ray makes with the normal to M_2:

$$\theta'_{M_2} = \theta_{M_2} = \boxed{55°}$$

Finalize Let's explore variations in the angle between the mirrors as follows.

What If? If the incoming and outgoing rays in Figure 35.7a are extended behind the mirror, they cross at an angle of 60° and the overall change in direction of the light ray is 120°. This angle is the same as that between the mirrors. What if the angle between the mirrors is changed? Is the overall change in the direction of the light ray always equal to the angle between the mirrors?

Answer Making a general statement based on one data point or one observation is always a dangerous practice! Let's investigate the change in direction for a general situation. Figure 35.7b shows the mirrors at an arbitrary angle ϕ and the incoming light ray striking the mirror at an arbitrary angle θ with respect to the normal to the mirror surface. In accordance with the law of reflection and the sum of the interior angles of a triangle, the angle γ is given by $180° - (90° - \theta) - \phi = 90° + \theta - \phi$.

Consider the triangle highlighted in blue in Figure 35.7b and determine α:

$$\alpha + 2\gamma + 2(90° - \theta) = 180° \quad \rightarrow \quad \alpha = 2(\theta - \gamma)$$

Notice from Figure 35.7b that the change in direction of the light ray is angle β. Use the geometry in the figure to solve for β:

$$\beta = 180° - \alpha = 180° - 2(\theta - \gamma)$$
$$= 180° - 2[\theta - (90° + \theta - \phi)] = 360° - 2\phi$$

Notice that β is not equal to ϕ. For $\phi = 120°$, we obtain $\beta = 120°$, which happens to be the same as the mirror angle; that is true only for this special angle between the mirrors, however. For example, if $\phi = 90°$, we obtain $\beta = 180°$. In that case, the light is reflected straight back to its origin.

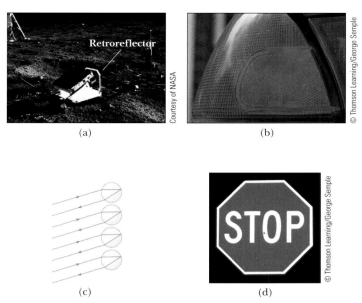

Figure 35.8 Applications of retroreflection. (a) This panel on the Moon reflects a laser beam directly back to its source on the Earth. (b) An automobile taillight has small retroreflectors to ensure that head-light beams are reflected back toward the car that sent them. (c) A light ray hitting a transparent sphere at the proper position is retroreflected. (d) This stop sign appears to glow in headlight beams because its surface is covered with a layer of many tiny retroreflecting spheres. What would you see if the sign had a mirror-like surface?

(a)

(b)

Figure 35.9 (a) An array of mirrors on the surface of a digital micromirror device. Each mirror has an area of approximately 16 μm^2. To provide a sense of scale, the leg of an ant appears in the photograph. (b) A close-up view of two single micromirrors. The mirror on the left is "on," and the one on the right is "off."

If the angle between two mirrors is 90°, the reflected beam returns to the source parallel to its original path as discussed in the **What If?** section of the preceding example. This phenomenon, called *retroreflection,* has many practical applications. If a third mirror is placed perpendicular to the first two so that the three form the corner of a cube, retroreflection works in three dimensions. In 1969, a panel of many small reflectors was placed on the Moon by the *Apollo 11* astronauts (Fig. 35.8a). A laser beam from the Earth is reflected directly back on itself, and its transit time is measured. This information is used to determine the distance to the Moon with an uncertainty of 15 cm. (Imagine how difficult it would be to align a regular flat mirror so that the reflected laser beam would hit a particular location on the Earth!) A more everyday application is found in automobile taillights. Part of the plastic making up the taillight is formed into many tiny cube corners (Fig. 35.8b) so that headlight beams from cars approaching from the rear are reflected back to the drivers. Instead of cube corners, small spherical bumps are sometimes used (Fig. 35.8c). Tiny clear spheres are used in a coating material found on many road signs. Due to retroreflection from these spheres, the stop sign in Figure 35.8d appears much brighter than it would if it were simply a flat, shiny surface. Retrore-flectors are also used for reflective panels on running shoes and running clothing to allow joggers to be seen at night.

Another practical application of the law of reflection is the digital projection of movies, television shows, and computer presentations. A digital projector uses an optical semiconductor chip called a *digital micromirror device.* This device contains an array of tiny mirrors (Fig. 35.9a) that can be individually tilted by means of sig-nals to an address electrode underneath the edge of the mirror. Each mirror cor-responds to a pixel in the projected image. When the pixel corresponding to a given mirror is to be bright, the mirror is in the "on" position and is oriented so as to reflect light from a source illuminating the array to the screen (Fig. 35.9b). When the pixel for this mirror is to be dark, the mirror is "off" and is tilted so that the light is reflected away from the screen. The brightness of the pixel is deter-mined by the total time interval during which the mirror is in the "on" position during the display of one image.

Digital movie projectors use three micromirror devices, one for each of the pri-mary colors red, blue, and green, so that movies can be displayed with up to 35 trillion colors. Because information is stored as binary data, a digital movie does not degrade with time as does film. Furthermore, because the movie is entirely in the form of computer software, it can be delivered to theaters by means of satellites, optical discs, or optical fiber networks.

Several movies have been projected digitally to audiences, and polls show that 85% of viewers describe the image quality as "excellent." The first all-digital movie, from cinematography to postproduction to projection, was *Star Wars Episode II: Attack of the Clones* in 2002.

35.5 The Wave Under Refraction

In addition to the phenomenon of reflection discussed for waves on strings in Section 16.4, we also found that some of the energy of the incident wave transmits into the new medium. Similarly, when a ray of light traveling through a transparent medium encounters a boundary leading into another transparent medium as shown in Active Figure 35.10, part of the energy is reflected and part enters the second medium. As with reflection, the direction of the transmitted wave exhibits an interesting behavior because of the three-dimensional nature of the light waves. The ray that enters the second medium is bent at the boundary and is said to be **refracted.** The incident ray, the reflected ray, and the refracted ray all lie in the same plane. The **angle of refraction,** θ_2 in Active Figure 35.10a, depends on the properties of the two media and on the angle of incidence θ_1 through the relationship

$$\frac{\sin \theta_2}{\sin \theta_1} = \frac{v_2}{v_1} \tag{35.3}$$

where v_1 is the speed of light in the first medium and v_2 is the speed of light in the second medium.

The path of a light ray through a refracting surface is reversible. For example, the ray shown in Active Figure 35.10a travels from point A to point B. If the ray originated at B, it would travel along line BA to reach point A and the reflected ray would point downward and to the left in the glass.

Quick Quiz 35.2   If beam ① is the incoming beam in Active Figure 35.10b, which of the other four red lines are reflected beams and which are refracted beams?

From Equation 35.3, we can infer that when light moves from a material in which its speed is high to a material in which its speed is lower as shown in Active

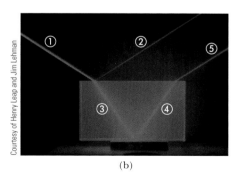

ACTIVE FIGURE 35.10

(a) A ray obliquely incident on an air–glass interface behaves according to the wave under refraction model. The refracted ray is bent toward the normal because $v_2 < v_1$. All rays and the normal lie in the same plane. (b) Light incident on the Lucite block refracts both when it enters the block and when it leaves the block.

Sign in at www.thomsonedu.com and go to ThomsonNOW to vary the incident angle and see the effect on the reflected and refracted rays.

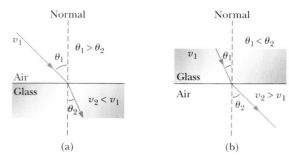

ACTIVE FIGURE 35.11

(a) When the light beam moves from air into glass, the light slows down upon entering the glass and its path is bent toward the normal. (b) When the beam moves from glass into air, the light speeds up upon entering the air and its path is bent away from the normal.

Sign in at www.thomsonedu.com and go to ThomsonNOW to observe light passing through three layers of material. You can vary the incident angle and see the effect on the refracted rays for a variety of values of the index of refraction (defined in Equation 35.4) of the three materials.

Figure 35.12 Light passing from one atom to another in a medium. The dots are electrons, and the vertical arrows represent their oscillations.

Figure 35.13 Overhead view of a barrel rolling from concrete onto grass.

Figure 35.11a, the angle of refraction θ_2 is less than the angle of incidence θ_1 and the ray is bent *toward* the normal. If the ray moves from a material in which light moves slowly to a material in which it moves more rapidly as illustrated in Active Figure 35.11b, θ_2 is greater than θ_1 and the ray is bent *away* from the normal.

The behavior of light as it passes from air into another substance and then re-emerges into air is often a source of confusion to students. When light travels in air, its speed is 3.00×10^8 m/s, but this speed is reduced to approximately 2×10^8 m/s when the light enters a block of glass. When the light re-emerges into air, its speed instantaneously increases to its original value of 3.00×10^8 m/s. This effect is far different from what happens, for example, when a bullet is fired through a block of wood. In that case, the speed of the bullet decreases as it moves through the wood because some of its original energy is used to tear apart the wood fibers. When the bullet enters the air once again, it emerges at a speed lower than it had when it entered the wood.

To see why light behaves as it does, consider Figure 35.12, which represents a beam of light entering a piece of glass from the left. Once inside the glass, the light may encounter an electron bound to an atom, indicated as point A. Let's assume light is absorbed by the atom, which causes the electron to oscillate (a detail represented by the double-headed vertical arrows). The oscillating electron then acts as an antenna and radiates the beam of light toward an atom at B, where the light is again absorbed. The details of these absorptions and radiations are best explained in terms of quantum mechanics (Chapter 42). For now, it is sufficient to think of light passing from one atom to another through the glass. Although light travels from one atom to another at 3.00×10^8 m/s, the absorption and radiation that take place cause the *average* light speed through the material to fall to about 2×10^8 m/s. Once the light emerges into the air, absorption and radiation cease and the light travels at a constant speed of 3.00×10^8 m/s.

A mechanical analog of refraction is shown in Figure 35.13. When the left end of the rolling barrel reaches the grass, it slows down, whereas the right end remains on the concrete and moves at its original speed. This difference in speeds causes the barrel to pivot, which changes the direction of travel.

Index of Refraction

In general, the speed of light in any material is *less* than its speed in vacuum. In fact, *light travels at its maximum speed c in vacuum.* It is convenient to define the **index of refraction** n of a medium to be the ratio

Index of refraction ▶

$$n \equiv \frac{\text{speed of light in vacuum}}{\text{speed of light in a medium}} \equiv \frac{c}{v} \qquad (35.4)$$

TABLE 35.1

Indices of Refraction

Substance	Index of Refraction	Substance	Index of Refraction
Solids at 20°C		*Liquids at 20°C*	
Cubic zirconia	2.20	Benzene	1.501
Diamond (C)	2.419	Carbon disulfide	1.628
Fluorite (CaF_2)	1.434	Carbon tetrachloride	1.461
Fused quartz (SiO_2)	1.458	Ethyl alcohol	1.361
Gallium phosphide	3.50	Glycerin	1.473
Glass, crown	1.52	Water	1.333
Glass, flint	1.66		
Ice (H_2O)	1.309	*Gases at 0°C, 1 atm*	
Polystyrene	1.49	Air	1.000 293
Sodium chloride (NaCl)	1.544	Carbon dioxide	1.000 45

Note: All values are for light having a wavelength of 589 nm in vacuum.

PITFALL PREVENTION 35.2

***n* Is Not an Integer Here**

The symbol n has been used several times as an integer, such as in Chapter 18 to indicate the standing wave mode on a string or in an air column. The index of refraction n is *not* an integer.

This definition shows that the index of refraction is a dimensionless number greater than unity because v is always less than c. Furthermore, n is equal to unity for vacuum. The indices of refraction for various substances are listed in Table 35.1.

As light travels from one medium to another, its frequency does not change but its wavelength does. To see why that is true, consider Figure 35.14. Waves pass an observer at point A in medium 1 with a certain frequency and are incident on the boundary between medium 1 and medium 2. The frequency with which the waves pass an observer at point B in medium 2 must equal the frequency at which they pass point A. If that were not the case, energy would be piling up or disappearing at the boundary. Because there is no mechanism for that to occur, the frequency must be a constant as a light ray passes from one medium into another. Therefore, because the relationship $v = \lambda f$ (Eq. 16.12) must be valid in both media and because $f_1 = f_2 = f$, we see that

$$v_1 = \lambda_1 f \quad \text{and} \quad v_2 = \lambda_2 f \tag{35.5}$$

Because $v_1 \neq v_2$, it follows that $\lambda_1 \neq \lambda_2$ as shown in Figure 35.14.

We can obtain a relationship between index of refraction and wavelength by dividing the first Equation 35.5 by the second and then using Equation 35.4:

$$\frac{\lambda_1}{\lambda_2} = \frac{v_1}{v_2} = \frac{c/n_1}{c/n_2} = \frac{n_2}{n_1} \tag{35.6}$$

This expression gives

$$\lambda_1 n_1 = \lambda_2 n_2$$

If medium 1 is vacuum or, for all practical purposes, air, then $n_1 = 1$. Hence, it follows from Equation 35.6 that the index of refraction of any medium can be expressed as the ratio

$$n = \frac{\lambda}{\lambda_n} \tag{35.7}$$

where λ is the wavelength of light in vacuum and λ_n is the wavelength of light in the medium whose index of refraction is n. From Equation 35.7, we see that because $n > 1$, $\lambda_n < \lambda$.

We are now in a position to express Equation 35.3 in an alternative form. Replacing the v_2/v_1 term in Equation 35.3 with n_1/n_2 from Equation 35.6 gives

$$n_1 \sin \theta_1 = n_2 \sin \theta_2 \tag{35.8}$$

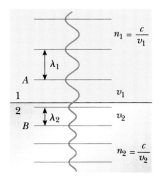

Figure 35.14 As a wave moves from medium 1 to medium 2, its wavelength changes but its frequency remains constant.

PITFALL PREVENTION 35.3

An Inverse Relationship

The index of refraction is *inversely* proportional to the wave speed. As the wave speed v decreases, the index of refraction n increases. Therefore, the higher the index of refraction of a material, the more it *slows down* light from its speed in vacuum. The more the light slows down, the more θ_2 differs from θ_1 in Equation 35.8.

◀   Snell's law of refraction

The experimental discovery of this relationship is usually credited to Willebrord Snell (1591–1626) and it is therefore known as **Snell's law of refraction.** We shall examine this equation further in Section 35.6. Refraction of waves at an interface between two media is a common phenomenon, so we identify an analysis model for this situation: the **wave under refraction.** Equation 35.8 is the mathematical representation of this model for electromagnetic radiation. Other waves, such as seismic waves and sound waves, also exhibit refraction according to this model, and the mathematical representation for these waves is Equation 35.3.

Quick Quiz 35.3   Light passes from a material with index of refraction 1.3 into one with index of refraction 1.2. Compared to the incident ray, what happens to the refracted ray? (a) It bends toward the normal. (b) It is undeflected. (c) It bends away from the normal.

EXAMPLE 35.3   **Angle of Refraction for Glass**

A light ray of wavelength 589 nm traveling through air is incident on a smooth, flat slab of crown glass at an angle of 30.0° to the normal.

(A) Find the angle of refraction.

SOLUTION

Conceptualize   Study Active Figure 35.11a, which illustrates the refraction process occurring in this problem.

Categorize   We evaluate results by using equations developed in this section, so we categorize this example as a substitution problem.

Rearrange Snell's law of refraction to find $\sin \theta_2$:

$$\sin \theta_2 = \frac{n_1}{n_2} \sin \theta_1$$

Substitute the incident angle and, from Table 35.1, $n_1 = 1.00$ for air and $n_2 = 1.52$ for crown glass:

$$\sin \theta_2 = \left(\frac{1.00}{1.52} \right) \sin 30.0° = 0.329$$

Solve for θ_2:

$$\theta_2 = \sin^{-1} (0.329)$$
$$= \boxed{19.2°}$$

(B) Find the speed of this light once it enters the glass.

SOLUTION

Solve Equation 35.4 for the speed of light in the glass:

$$v = \frac{c}{n}$$

Substitute numerical values:

$$v = \frac{3.00 \times 10^8 \text{ m/s}}{1.52} = \boxed{1.97 \times 10^8 \text{ m/s}}$$

(C) What is the wavelength of this light in the glass?

SOLUTION

Use Equation 35.7 to find the wavelength in the glass:

$$\lambda_n = \frac{\lambda}{n} = \frac{589 \text{ nm}}{1.52} = \boxed{388 \text{ nm}}$$

EXAMPLE 35.4   Light Passing Through a Slab

A light beam passes from medium 1 to medium 2, with the latter medium being a thick slab of material whose index of refraction is n_2 (Fig. 35.15). Show that the beam emerging into medium 1 from the other side is parallel to the incident beam.

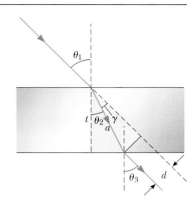

Figure 35.15   (Example 35.4) When light passes through a flat slab of material, the emerging beam is parallel to the incident beam; therefore, $\theta_1 = \theta_3$. The dashed line drawn parallel to the ray coming out the bottom of the slab represents the path the light would take were the slab not there.

SOLUTION

Conceptualize   Follow the path of the light beam as it enters and exits the slab of material in Figure 35.15. The ray bends toward the normal upon entering and away from the normal upon leaving.

Categorize   We evaluate results by using equations developed in this section, so we categorize this example as a substitution problem.

Apply Snell's law of refraction to the upper surface:

$$(1) \quad \sin \theta_2 = \frac{n_1}{n_2} \sin \theta_1$$

Apply Snell's law to the lower surface:

$$(2) \quad \sin \theta_3 = \frac{n_2}{n_1} \sin \theta_2$$

Substitute Equation (1) into Equation (2):

$$\sin \theta_3 = \frac{n_2}{n_1} \left(\frac{n_1}{n_2} \sin \theta_1 \right) = \sin \theta_1$$

Therefore, $\theta_3 = \theta_1$ and the slab does not alter the direction of the beam. It does, however, offset the beam parallel to itself by the distance d shown in Figure 35.15.

What If?   What if the thickness t of the slab is doubled? Does the offset distance d also double?

Answer   Consider the region of the light path within the slab in Figure 35.15. The distance a is the hypotenuse of two right triangles.

Find an expression for a from the gold triangle:

$$a = \frac{t}{\cos \theta_2}$$

Find an expression for d from the blue triangle:

$$d = a \sin \gamma = a \sin (\theta_1 - \theta_2)$$

Combine these equations:

$$d = \frac{t}{\cos \theta_2} \sin (\theta_1 - \theta_2)$$

For a given incident angle θ_1, the refracted angle θ_2 is determined solely by the index of refraction, so the offset distance d is proportional to t. If the thickness doubles, so does the offset distance.

In Example 35.4, the light passes through a slab of material with parallel sides. What happens when light strikes a prism with nonparallel sides as shown in Figure 35.16? In this case, the outgoing ray does not propagate in the same direction as the incoming ray. A ray of single-wavelength light incident on the prism from the left emerges at angle δ from its original direction of travel. This angle δ is called the **angle of deviation**. The **apex angle** Φ of the prism, shown in the figure, is defined as the angle between the surface at which the light enters the prism and the second surface that the light encounters.

Figure 35.16   A prism refracts a single-wavelength light ray through an angle of deviation δ.

EXAMPLE 35.5 **Measuring *n* Using a Prism**

Although we do not prove it here, the minimum angle of deviation δ_{min} for a prism occurs when the angle of incidence θ_1 is such that the refracted ray inside the prism makes the same angle with the normal to the two prism faces[1] as shown in Figure 35.17. Obtain an expression for the index of refraction of the prism material in terms of the minimum angle of deviation and the apex angle Φ.

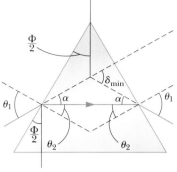

Figure 35.17 (Example 35.5) A light ray passing through a prism at the minimum angle of deviation δ_{min}.

SOLUTION

Conceptualize Study Figure 35.17 carefully and be sure you understand why the light ray comes out of the prism traveling in a different direction.

Categorize In this example, light enters a material through one surface and leaves the material at another surface. Let's apply the wave under refraction model at each surface.

Analyze Consider the geometry in Figure 35.17. The reproduction of the angle $\Phi/2$ at the location of the incoming light ray shows that $\theta_2 = \Phi/2$. The theorem that an exterior angle of any triangle equals the sum of the two opposite interior angles shows that $\delta_{min} = 2\alpha$. The geometry also shows that $\theta_1 = \theta_2 + \alpha$.

Combine these three geometric results:

$$\theta_1 = \theta_2 + \alpha = \frac{\Phi}{2} + \frac{\delta_{min}}{2} = \frac{\Phi + \delta_{min}}{2}$$

Apply the wave under refraction model at the left surface and solve for *n*:

$$(1.00) \sin \theta_1 = n \sin \theta_2 \quad \rightarrow \quad n = \frac{\sin \theta_1}{\sin \theta_2}$$

Substitute for the incident and refracted angles:

$$n = \frac{\sin \left(\dfrac{\Phi + \delta_{min}}{2} \right)}{\sin (\Phi/2)} \tag{35.9}$$

Finalize Knowing the apex angle Φ of the prism and measuring δ_{min}, you can calculate the index of refraction of the prism material. Furthermore, a hollow prism can be used to determine the values of *n* for various liquids filling the prism.

PITFALL PREVENTION 35.4
Of What Use Is Huygens's Principle?

At this point, the importance of Huygens's principle may not be evident. Predicting the position of a future wave front may not seem to be very critical. We will use Huygens's principle in later chapters to explain additional wave phenomena for light, however.

35.6 Huygens's Principle

In this section, we develop the laws of reflection and refraction by using a geometric method proposed by Huygens in 1678. **Huygens's principle** is a geometric construction for using knowledge of an earlier wave front to determine the position of a new wave front at some instant. In Huygens's construction, **all points on a given wave front are taken as point sources for the production of spherical secondary waves, called wavelets, that propagate outward through a medium with speeds characteristic of waves in that medium. After some time interval has passed, the new position of the wave front is the surface tangent to the wavelets.**

First, consider a plane wave moving through free space as shown in Figure 35.18a. At $t = 0$, the wave front is indicated by the plane labeled AA'. In Huygens's construction, each point on this wave front is considered a point source. For clarity, only three points on AA' are shown. With these points as sources for the wavelets, we draw circles, each of radius $c \Delta t$, where c is the speed of light in vacuum and Δt is some time interval during which the wave propagates. The surface drawn tangent to these wavelets is the plane BB', which is the wave front at a later

[1] The details of this proof are available in texts on optics.

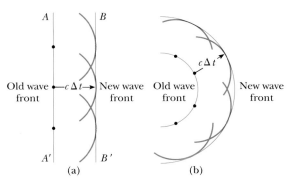

Figure 35.18 Huygens's construction for (a) a plane wave propagating to the right and (b) a spherical wave propagating to the right.

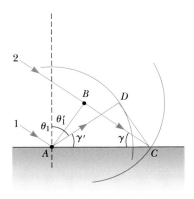

Figure 35.19 Huygens's construction for proving the law of reflection. The instant ray 1 strikes the surface, it sends out a Huygens wavelet from A and ray 2 sends out a Huygens wavelet from B. We choose a radius of the wavelet to be $c\,\Delta t$, where Δt is the time interval for ray 2 to travel from B to C. Triangle ADC is congruent to triangle ABC.

time, and is parallel to AA'. In a similar manner, Figure 35.18b shows Huygens's construction for a spherical wave.

Huygens's Principle Applied to Reflection and Refraction

The laws of reflection and refraction were stated earlier in this chapter without proof. We now derive these laws, using Huygens's principle.

For the law of reflection, refer to Figure 35.19. The line AB represents a plane wave front of the incident light just as ray 1 strikes the surface. At this instant, the wave at A sends out a Huygens wavelet (the red circular arc centered on A). The reflected light propagates toward D. At the same time, the wave at B emits a Huygens wavelet (the red circular arc centered on B) with the light propagating toward C. Figure 35.19 shows these wavelets after a time interval Δt, after which ray 2 strikes the surface. Because both rays 1 and 2 move with the same speed, we must have $AD = BC = c\,\Delta t$.

The remainder of our analysis depends on geometry. Notice that the two triangles ABC and ADC are congruent because they have the same hypotenuse AC and because $AD = BC$. Figure 35.19 shows that

$$\cos \gamma = \frac{BC}{AC} \quad \text{and} \quad \cos \gamma' = \frac{AD}{AC}$$

where $\gamma = 90° - \theta_1$ and $\gamma' = 90° - \theta_1'$. Because $AD = BC$,

$$\cos \gamma = \cos \gamma'$$

Therefore,

$$\gamma = \gamma'$$
$$90° - \theta_1 = 90° - \theta_1'$$

and

$$\theta_1 = \theta_1'$$

which is the law of reflection.

Now let's use Huygens's principle and Figure 35.20 to derive Snell's law of refraction. We focus our attention on the instant ray 1 strikes the surface and the subsequent time interval until ray 2 strikes the surface. During this time interval, the wave at A sends out a Huygens wavelet (the red arc centered on A) and the light refracts toward D. In the same time interval, the wave at B sends out a Huygens wavelet (the red arc centered on B) and the light continues to propagate toward C. Because these two wavelets travel through different media, the radii of the wavelets are different. The radius of the wavelet from A is $AD = v_2\,\Delta t$, where v_2 is the wave speed in the second medium. The radius of the wavelet from B is $BC = v_1\,\Delta t$, where v_1 is the wave speed in the original medium.

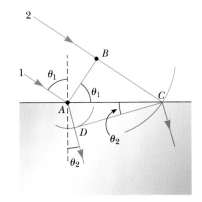

Figure 35.20 Huygens's construction for proving Snell's law of refraction. The instant ray 1 strikes the surface, it sends out a Huygens wavelet from A and ray 2 sends out a Huygens wavelet from B. The two wavelets have different radii because they travel in different media.

From triangles *ABC* and *ADC,* we find that

$$\sin \theta_1 = \frac{BC}{AC} = \frac{v_1 \, \Delta t}{AC} \quad \text{and} \quad \sin \theta_2 = \frac{AD}{AC} = \frac{v_2 \, \Delta t}{AC}$$

Dividing the first equation by the second gives

$$\frac{\sin \theta_1}{\sin \theta_2} = \frac{v_1}{v_2}$$

From Equation 35.4, however, we know that $v_1 = c/n_1$ and $v_2 = c/n_2$. Therefore,

$$\frac{\sin \theta_1}{\sin \theta_2} = \frac{c/n_1}{c/n_2} = \frac{n_2}{n_1}$$

and

$$n_1 \sin \theta_1 = n_2 \sin \theta_2$$

which is Snell's law of refraction.

35.7 Dispersion

An important property of the index of refraction n is that, for a given material, the index varies with the wavelength of the light passing through the material as Figure 35.21 shows. This behavior is called **dispersion.** Because n is a function of wavelength, Snell's law of refraction indicates that light of different wavelengths is refracted at different angles when incident on a material.

Figure 35.21 shows that the index of refraction generally decreases with increasing wavelength. For example, violet light refracts more than red light does when passing into a material.

Now suppose a beam of *white light* (a combination of all visible wavelengths) is incident on a prism as illustrated in Figure 35.22. Clearly, the angle of deviation δ depends on wavelength. The rays that emerge spread out in a series of colors known as the **visible spectrum.** These colors, in order of decreasing wavelength, are red, orange, yellow, green, blue, and violet. Newton showed that each color has a particular angle of deviation and that the colors can be recombined to form the original white light.

Figure 35.21 Variation of index of refraction with vacuum wavelength for three materials.

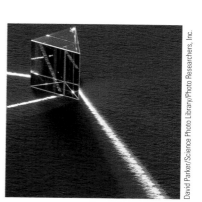

Figure 35.22 White light enters a glass prism at the upper left. A reflected beam of light comes out of the prism below the incoming beam. The beam moving toward the lower right shows distinct colors. Different colors are refracted at different angles because the index of refraction of the glass depends on wavelength. Violet light deviates the most; red light deviates the least.

The dispersion of light into a spectrum is demonstrated most vividly in nature by the formation of a rainbow, which is often seen by an observer positioned between the Sun and a rain shower. To understand how a rainbow is formed, consider Active Figure 35.23. A ray of sunlight (which is white light) passing overhead strikes a drop of water in the atmosphere and is refracted and reflected as follows. It is first refracted at the front surface of the drop, with the violet light deviating the most and the red light the least. At the back surface of the drop, the light is reflected and returns to the front surface, where it again undergoes refraction as it moves from water into air. The rays leave the drop such that the angle between the incident white light and the most intense returning violet ray is 40° and the angle between the incident white light and the most intense returning red ray is 42°. This small angular difference between the returning rays causes us to see a colored bow.

Now suppose an observer is viewing a rainbow as shown in Figure 35.24. If a raindrop high in the sky is being observed, the most intense red light returning from the drop reaches the observer because it is deviated the most; the most intense violet light, however, passes over the observer because it is deviated the least. Hence, the observer sees red light coming from this drop. Similarly, a drop lower in the sky directs the most intense violet light toward the observer and appears violet to the observer. (The most intense red light from this drop passes below the observer's eye and is not seen.) The most intense light from other colors of the spectrum reaches the observer from raindrops lying between these two extreme positions.

The opening photograph for this chapter shows a *double rainbow*. The secondary rainbow is fainter than the primary rainbow, and the colors are reversed. The secondary rainbow arises from light that makes two reflections from the interior surface before exiting the raindrop. In the laboratory, rainbows have been observed in which the light makes more than 30 reflections before exiting the water drop. Because each reflection involves some loss of light due to refraction out of the water drop, the intensity of these higher-order rainbows is small compared with that of the primary rainbow.

Quick Quiz 35.4   In film photography, lenses in a camera use refraction to form an image on a film. Ideally, you want all the colors in the light from the object being photographed to be refracted by the same amount. Of the materials shown in Figure 35.21, which would you choose for a single-element camera lens? (a) crown glass (b) acrylic   (c) fused quartz   (d) impossible to determine

35.8   Total Internal Reflection

An interesting effect called **total internal reflection** can occur when light is directed from a medium having a given index of refraction toward one having a lower index of refraction. Consider Active Figure 35.25a (page 994), in which a light ray travels in medium 1 and meets the boundary between medium 1 and medium 2, where n_1 is greater than n_2. In the figure, labels 1 through 5 indicate various possible directions of the ray consistent with the wave under refraction model. The refracted rays are bent away from the normal because n_1 is greater than n_2. At some particular angle of incidence θ_c, called the **critical angle,** the refracted light ray moves parallel to the boundary so that $\theta_2 = 90°$ (Active Fig. 35.25b). For angles of incidence greater than θ_c, the ray is entirely reflected at the boundary as shown by ray 5 in Active Figure 35.25a.

We can use Snell's law of refraction to find the critical angle. When $\theta_1 = \theta_c$, $\theta_2 = 90°$ and Equation 35.8 gives

$$n_1 \sin \theta_c = n_2 \sin 90° = n_2$$

$$\sin \theta_c = \frac{n_2}{n_1} \quad (\text{for } n_1 > n_2) \qquad (35.10)$$

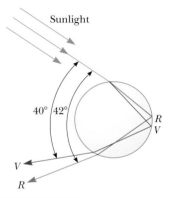

ACTIVE FIGURE 35.23

Path of sunlight through a spherical raindrop. Light following this path contributes to the visible rainbow.

Sign in at www.thomsonedu.com and go to ThomsonNOW to vary the point at which the sunlight enters the raindrop and verify that the angles shown are the maximum angles.

PITFALL PREVENTION 35.5
A Rainbow of Many Light Rays

Pictorial representations such as Active Figure 35.23 are subject to misinterpretation. The figure shows one ray of light entering the raindrop and undergoing reflection and refraction, exiting the raindrop in a range of 40° to 42° from the entering ray. This illustration might be interpreted incorrectly as meaning that *all* light entering the raindrop exits in this small range of angles. In reality, light exits the raindrop over a much larger range of angles, from 0° to 42°. A careful analysis of the reflection and refraction from the spherical raindrop shows that the range of 40° to 42° is where the *highest-intensity light* exits the raindrop.

Figure 35.24   The formation of a rainbow seen by an observer standing with the Sun behind his back.

◀   Critical angle for total internal reflection

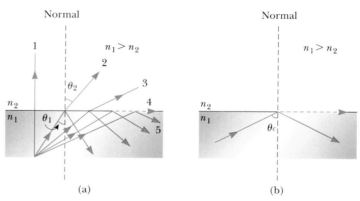

ACTIVE FIGURE 35.25

(a) Rays travel from a medium of index of refraction n_1 into a medium of index of refraction n_2, where $n_2 < n_1$. As the angle of incidence θ_1 increases, the angle of refraction θ_2 increases until θ_2 is 90° (ray 4). The dashed line indicates that no energy actually propagates in this direction. For even larger angles of incidence, total internal reflection occurs (ray 5). (b) The angle of incidence producing an angle of refraction equal to 90° is the critical angle θ_c. At this angle of incidence, all the energy of the incident light is reflected.

Sign in at www.thomsonedu.com and go to ThomsonNOW to vary the incident angle and see the effect on the refracted ray and the distribution of incident energy between the reflected and refracted rays.

This equation can be used only when n_1 is greater than n_2. That is, **total internal reflection occurs only when light is directed from a medium of a given index of refraction toward a medium of lower index of refraction.** If n_1 were less than n_2, Equation 35.10 would give sin $\theta_c > 1$, which is a meaningless result because the sine of an angle can never be greater than unity.

The critical angle for total internal reflection is small when n_1 is considerably greater than n_2. For example, the critical angle for a diamond in air is 24°. Any ray inside the diamond that approaches the surface at an angle greater than 24° is completely reflected back into the crystal. This property, combined with proper faceting, causes diamonds to sparkle. The angles of the facets are cut so that light is "caught" inside the crystal through multiple internal reflections. These multiple reflections give the light a long path through the medium, and substantial dispersion of colors occurs. By the time the light exits through the top surface of the crystal, the rays associated with different colors have been fairly widely separated from one another.

Cubic zirconia also has a high index of refraction and can be made to sparkle very much like a diamond. If a suspect jewel is immersed in corn syrup, the difference in n for the cubic zirconia and that for the corn syrup is small and the critical angle is therefore great. Hence, more rays escape sooner; as a result, the sparkle completely disappears. A real diamond does not lose all its sparkle when placed in corn syrup.

Figure 35.26 (Quick Quiz 35.5) Five nonparallel light rays enter a glass prism from the left.

Quick Quiz 35.5 In Figure 35.26, five light rays enter a glass prism from the left. **(i)** How many of these rays undergo total internal reflection at the slanted surface of the prism? (a) 1 (b) 2 (c) 3 (d) 4 (e) 5 **(ii)** Suppose the prism in Figure 35.26 can be rotated in the plane of the paper. For *all five* rays to experience total internal reflection from the slanted surface, should the prism be rotated (a) clockwise or (b) counterclockwise?

EXAMPLE 35.6   **A View from the Fish's Eye**

Find the critical angle for an air–water boundary. (The index of refraction of water is 1.33.)

SOLUTION

Conceptualize Study Active Figure 35.25 to understand the concept of total internal reflection and the significance of the critical angle.

Categorize   We use concepts developed in this section, so we categorize this example as a substitution problem.

Apply Equation 35.10 to the air–water interface:

$$\sin \theta_c = \frac{n_2}{n_1} = \frac{1.00}{1.33} = 0.752$$

$$\theta_c = \boxed{48.8°}$$

What If?   What if a fish in a still pond looks upward toward the water's surface at different angles relative to the surface as in Figure 35.27? What does it see?

Answer   Because the path of a light ray is reversible, light traveling from medium 2 into medium 1 in Active Figure 35.25a follows the paths shown, but in the *opposite* direction. A fish looking upward toward the water surface as in Figure 35.27 can see out of the water if it looks toward the surface at an angle less than the critical angle. Therefore, when the fish's line of vision makes an angle of $\theta = 40°$ with the normal to the surface, for example, light from above the water reaches the fish's eye. At $\theta = 48.8°$, the critical angle for water, the light has to skim along the water's surface before being refracted to the fish's eye; at this angle, the fish can, in principle, see the entire shore of the pond. At angles greater than the critical angle, the light reaching the fish comes by means of total internal reflection at the surface. Therefore, at $\theta = 60°$, the fish sees a reflection of the bottom of the pond.

Figure 35.27   (Example 35.6) **What If?** A fish looks upward toward the water surface.

Optical Fibers

Another interesting application of total internal reflection is the use of glass or transparent plastic rods to "pipe" light from one place to another. As indicated in Figure 35.28, light is confined to traveling within a rod, even around curves, as the result of successive total internal reflections. Such a light pipe is flexible if thin fibers are used rather than thick rods. A flexible light pipe is called an **optical fiber.** If a bundle of parallel fibers is used to construct an optical transmission line, images can be transferred from one point to another. This technique is used in a sizable industry known as *fiber optics.*

A practical optical fiber consists of a transparent core surrounded by a *cladding,* a material that has a lower index of refraction than the core. The combination may be surrounded by a plastic *jacket* to prevent mechanical damage. Figure 35.29 shows a cutaway view of this construction. Because the index of refraction of the cladding is less than that of the core, light traveling in the core experiences total internal reflection if it arrives at the interface between the core and the cladding at an angle

Figure 35.28   Light travels in a curved transparent rod by multiple internal reflections.

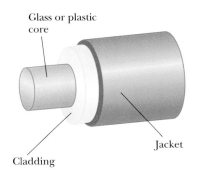

Glass or plastic core

Jacket

Cladding

Figure 35.29   The construction of an optical fiber. Light travels in the core, which is surrounded by a cladding and a protective jacket.

(Left) Strands of glass optical fibers are used to carry voice, video, and data signals in telecommunication networks. *(Right)* A bundle of optical fibers is illuminated by a laser.

of incidence that exceeds the critical angle. In this case, light "bounces" along the core of the optical fiber, losing very little of its intensity as it travels.

Any loss in intensity in an optical fiber is essentially due to reflections from the two ends and absorption by the fiber material. Optical fiber devices are particularly useful for viewing an object at an inaccessible location. For example, physicians often use such devices to examine internal organs of the body or to perform surgery without making large incisions. Optical fiber cables are replacing copper wiring and coaxial cables for telecommunications because the fibers can carry a much greater volume of telephone calls or other forms of communication than electrical wires can.

Summary

DEFINITION

The **index of refraction** n of a medium is defined by the ratio

$$n \equiv \frac{c}{v} \qquad (35.4)$$

where c is the speed of light in a vacuum and v is the speed of light in the medium.

CONCEPTS AND PRINCIPLES

In geometric optics, we use the **ray approximation,** in which a wave travels through a uniform medium in straight lines in the direction of the rays.

Total internal reflection occurs when light travels from a medium of high index of refraction to one of lower index of refraction. The **critical angle** θ_c for which total internal reflection occurs at an interface is given by

$$\sin \theta_c = \frac{n_2}{n_1} \quad (\text{for } n_1 > n_2) \qquad (35.10)$$

ANALYSIS MODELS FOR PROBLEM SOLVING

Wave Under Reflection. The **law of reflection** states that for a light ray (or other type of wave) incident on a smooth surface, the angle of reflection θ_1' equals the angle of incidence θ_1:

$$\theta_1' = \theta_1 \qquad (35.2)$$

Wave Under Refraction. A wave crossing a boundary as it travels from medium 1 to medium 2 is **refracted,** or bent. The angle of refraction θ_2 is related to the incident angle θ_1 by the relationship

$$\frac{\sin \theta_2}{\sin \theta_1} = \frac{v_2}{v_1} \qquad (35.3)$$

where v_1 and v_2 are the speeds of the wave in medium 1 and medium 2, respectively. The incident ray, the reflected ray, the refracted ray, and the normal to the surface all lie in the same plane.

For light waves, **Snell's law of refraction** states that

$$n_1 \sin \theta_1 = n_2 \sin \theta_2 \qquad (35.8)$$

where n_1 and n_2 are the indices of refraction in the two media.

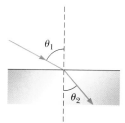

Questions

☐ denotes answer available in *Student Solutions Manual/Study Guide;* **O** denotes objective question

1. Why do astronomers looking at distant galaxies talk about looking backward in time?

2. O What is the order of magnitude of the time interval required for light to travel 10 km as in Galileo's attempt to measure the speed of light? (a) several seconds (b) several milliseconds (c) several microseconds (d) several nanoseconds

3. O In each of the following situations, a wave passes through an opening in an absorbing wall. Rank the situations in order from the one in which the wave is best described by the ray approximation to the one in which the wave coming through the opening spreads out most nearly equally in all directions in the hemisphere beyond the wall. (a) The sound of a low whistle at 1 kHz passes through a doorway 1 m wide. (b) Red light passes through the pupil of your eye. (c) Blue light passes through the pupil of your eye. (d) The wave broadcast by an AM radio station passes through a doorway 1 m wide. (e) An x-ray passes through the space between bones in your elbow joint.

4. The display windows of some department stores are slanted slightly inward at the bottom. This tilt is to decrease the glare from streetlights and the Sun, which would make it difficult for shoppers to see the display inside. Sketch a light ray reflecting from such a window to show how this design works.

5. You take a child for walks around the neighborhood. She loves to listen to echoes from houses when she shouts or when you clap loudly. A house with a large, flat front wall can produce an echo if you stand straight in front of it and reasonably far away. Draw a bird's-eye view of the situation to explain the production of the echo. Shade the area where you can stand to hear the echo. **What If?** The child helps you discover that a house with an L-shaped floor plan can produce echoes if you are standing in a wider range of locations. You can be standing at any reasonably distant location from which you can see the inside corner. Explain the echo in this case and draw another diagram for comparison. **What If?** What if the two wings of the house are not perpendicular? Will you and the child, standing close together, hear echoes? **What If?** What if a rectangular house and its garage have perpendicular walls that would form an inside corner but have a breezeway between them so that the walls do not meet? Will this structure produce strong echoes for peo-

ple in a wide range of locations? Explain your answers with diagrams.

6. The F-117A stealth fighter (Figure Q35.6) is specifically designed to be a *non*retroreflector of radar. What aspects of its design help accomplish this purpose? *Suggestion:* Answer the previous question as preparation for this one. Notice that the bottom of the plane is flat and that all the flat exterior panels meet at odd angles.

Figure Q35.6

7. O A light wave moves between medium 1 and medium 2. Which of the following are correct statements relating its speed, frequency, and wavelength in the two media, the indices of refraction of the media, and the angles of incidence and refraction? Choose all correct statements. (a) $v_1/\sin\theta_1 = v_2/\sin\theta_2$ (b) $\csc\theta_1/n_1 = \csc\theta_2/n_2$ (c) $\lambda_1/\sin\theta_1 = \lambda_2/\sin\theta_2$ (d) $f_1/\sin\theta_1 = f_2/\sin\theta_2$ (e) $n_1/\cos\theta_1 = n_2/\cos\theta_2$

8. Sound waves have much in common with light waves, including the properties of reflection and refraction. Give examples of these phenomena for sound waves.

9. O Consider light traveling from one medium into another with a different index of refraction. (a) Does its wavelength change? (b) Does its frequency change? (c) Does its speed change? (d) Does its direction always change?

10. A laser beam passing through a nonhomogeneous sugar solution follows a curved path. Explain.

11. **O** (a) Can light undergo total internal reflection at a smooth interface between air and water? If so, in which medium must it be traveling originally? (b) Can sound undergo total internal reflection at a smooth interface between air and water? If so, in which medium must it be traveling originally?

12. Explain why a diamond sparkles more than a glass crystal of the same shape and size.

13. Total internal reflection is applied in the periscope of a submarine to let the user "see around corners." In this device, two prisms are arranged as shown in Figure Q35.13 so that an incident beam of light follows the path shown. Parallel tilted, silvered mirrors could be used, but glass prisms with no silvered surfaces give higher light throughput. Propose a reason for the higher efficiency.

Figure Q35.13

14. **O** Suppose you find experimentally that two colors of light, A and B, originally traveling in the same direction in air, are sent through a glass prism, and A changes direction more than B. Which travels more slowly in the prism, A or B? Alternatively, is there insufficient information to determine which moves more slowly?

15. Retroreflection by transparent spheres, mentioned in Section 35.4, can be observed with dewdrops. To do so, look at the shadow of your head where it falls on dewy grass. Compare your observations to the reactions of two other people: Renaissance artist Benvenuto Cellini described the phenomenon and his reaction in his *Autobiography*, at the end of Part One, and American philosopher Henry David Thoreau did the same in *Walden*, "Baker Farm," second paragraph. The optical display around the shadow of your head is called *heiligenschein*, which is German for *holy light*. Try to find a person you know who has seen the heiligenschein. What did that person think about it?

16. How is it possible that a complete circle of a rainbow can sometimes be seen from an airplane? With a stepladder, a lawn sprinkler, and a sunny day, how can you show the complete circle to children?

17. At one restaurant, a worker uses colored chalk to write the daily specials on a blackboard illuminated with a spotlight. At another restaurant, a worker writes with colored grease pencils onto a flat, smooth sheet of transparent acrylic plastic with index of refraction 1.55. The panel hangs in front of a piece of black felt. Small, bright electric lights are installed all along the edges of the sheet, inside an opaque channel. Figure Q35.17 shows a cutaway view of the sign. Explain why viewers at both restaurants see the letters shining against a black background. Explain why the sign at the second restaurant may use less energy from the electric company than the illuminated blackboard at the first restaurant. What would be a good choice for the index of refraction of the material in the grease pencils?

Figure Q35.17

18. **O** The core of an optical fiber transmits light with minimal loss if it is surrounded by what? (a) water (b) diamond (c) air (d) glass (e) fused quartz.

19. Under what conditions is a mirage formed? On a hot day, what are we seeing when we observe "water on the road"?

Problems

WebAssign The Problems from this chapter may be assigned online in WebAssign.

ThomsonNOW™   Sign in at **www.thomsonedu.com** and go to ThomsonNOW to assess your understanding of this chapter's topics with additional quizzing and conceptual questions.

1, 2, 3 denotes straightforward, intermediate, challenging;   ☐ denotes full solution available in *Student Solutions Manual/Study Guide;*   ▲ denotes coached solution with hints available at **www.thomsonedu.com;**   denotes developing symbolic reasoning;   ● denotes asking for qualitative reasoning;   ▀ denotes computer useful in solving problem

Section 35.1 The Nature of Light

Section 35.2 Measurements of the Speed of Light

1. ● The *Apollo 11* astronauts set up a panel of efficient corner-cube retroreflectors on the Moon's surface (Fig. 35.8a). The speed of light can be found by measuring the time interval required for a laser beam to travel from the Earth, reflect from the panel, and return to the Earth. Assume this interval is measured to be 2.51 s at a station where the Moon is at the zenith. What is the measured speed of light? Take the center-to-center distance from the Earth to the Moon to be 3.84×10^8 m. Explain whether it is necessary to consider the sizes of the Earth and the Moon in your calculation.

2. As a result of his observations, Roemer concluded that eclipses of Io by Jupiter were delayed by 22 min during a six-month period as the Earth moved from the point in its orbit where it is closest to Jupiter to the diametrically opposite point where it is farthest from Jupiter. Using 1.50×10^8 km as the average radius of the Earth's orbit around the Sun, calculate the speed of light from these data.

3. In an experiment to measure the speed of light using the apparatus of Fizeau (see Fig. 35.2), the distance between light source and mirror was 11.45 km and the wheel had 720 notches. The experimentally determined value of c was 2.998×10^8 m/s. Calculate the minimum angular speed of the wheel for this experiment.

Section 35.3 The Ray Approximation in Geometric Optics

Section 35.4 The Wave Under Reflection

Section 35.5 The Wave Under Refraction

Note: You may look up indices of refraction in Table 35.1.

4. A dance hall is built without pillars and with a horizontal ceiling 7.20 m above the floor. A mirror is fastened flat against one section of the ceiling. Following an earthquake, the mirror is in place and unbroken. An engineer makes a quick check of whether the ceiling is sagging by directing a vertical beam of laser light up at the mirror and observing its reflection on the floor. (a) Show that if the mirror has rotated to make an angle ϕ with the hori-

zontal, the normal to the mirror makes an angle ϕ with the vertical. (b) Show that the reflected laser light makes an angle 2ϕ with the vertical. (c) Assume the reflected laser light makes a spot on the floor 1.40 cm away from the point vertically below the laser. Find the angle ϕ.

5. The two mirrors illustrated in Figure P35.5 meet at a right angle. The beam of light in the vertical plane P strikes mirror 1 as shown. (a) Determine the distance the reflected light beam travels before striking mirror 2. (b) In what direction does the light beam travel after being reflected from mirror 2?

Figure P35.5

6. Two flat, rectangular mirrors, both perpendicular to a horizontal sheet of paper, are set edge to edge with their reflecting surfaces perpendicular to each other. (a) A light ray in the plane of the paper strikes one of the mirrors at an arbitrary angle of incidence θ_1. Prove that the final direction of the ray, after reflection from both mirrors, is opposite its initial direction. In a clothing store, such a pair of mirrors shows you an image of yourself as others see you, with no apparent right–left reversal. (b) **What If?** Now assume the paper is replaced with a third flat mirror, touching edges with the other two and perpendicular to both. The set of three mirrors is called a *corner-cube reflector.* A ray of light is incident from any direction within the octant of space bounded by the reflecting surfaces. Argue that the ray will reflect once from each mirror and that its final direction will be opposite to its original direction. The *Apollo 11* astronauts placed a panel of corner-cube retroreflectors on the Moon. Analysis of timing data taken with it reveals that

the radius of the Moon's orbit is increasing at the rate of 3.8 cm/yr as it loses kinetic energy because of tidal friction.

7. The distance of a lightbulb from a large plane mirror is twice the distance of a person from the plane mirror. Light from the lightbulb reaches the person by two paths. It travels to the mirror at an angle of incidence θ and reflects from the mirror to the person. It also travels directly to the person without reflecting off the mirror. The total distance traveled by the light in the first case is twice the distance traveled by the light in the second case. Find the value of the angle θ.

8. Two light pulses are emitted simultaneously from a source. Both pulses travel to a detector, but mirrors shunt one pulse along a path that carries it through 6.20 m of ice along the way. Determine the difference in the pulses' times of arrival at the detector.

9. A narrow beam of sodium yellow light, with wavelength 589 nm in vacuum, is incident from air onto a smooth water surface at an angle of incidence of 35.0°. Determine the angle of refraction and the wavelength of the light in water.

10. ● A plane sound wave in air at 20°C, with wavelength 589 mm, is incident on a smooth surface of water at 25°C at an angle of incidence of 3.50°. Determine the angle of refraction for the sound wave and the wavelength of the sound in water. Compare and contrast the behavior of the sound in this problem with the behavior of the light in Problem 9.

11. An underwater scuba diver sees the Sun at an apparent angle of 45.0° above the horizontal. What is the actual elevation angle of the Sun above the horizontal?

12. The wavelength of red helium–neon laser light in air is 632.8 nm. (a) What is its frequency? (b) What is its wavelength in glass that has an index of refraction of 1.50? (c) What is its speed in the glass?

13. A ray of light is incident on a flat surface of a block of crown glass that is surrounded by water. The angle of refraction is 19.6°. Find the angle of reflection.

14. A laser beam with vacuum wavelength 632.8 nm is incident from air onto a block of Lucite as shown in Active Figure 35.10b. The line of sight of the photograph is perpendicular to the plane in which the light moves. Find (a) the speed, (b) the frequency, and (c) the wavelength of the light in the Lucite. *Suggestion:* Use a protractor.

15. Find the speed of light in (a) flint glass, (b) water, and (c) cubic zirconia.

16. A narrow beam of ultrasonic waves reflects off the liver tumor illustrated in Figure P35.16. The speed of the wave is 10.0% less in the liver than in the surrounding medium. Determine the depth of the tumor.

Figure P35.16

17. ▲ A ray of light strikes a flat block of glass ($n = 1.50$) of thickness 2.00 cm at an angle of 30.0° with the normal. Trace the light beam through the glass and find the angles of incidence and refraction at each surface.

18. An opaque cylindrical tank with an open top has a diameter of 3.00 m and is completely filled with water. When the afternoon Sun reaches an angle of 28.0° above the horizon, sunlight ceases to illuminate any part of the bottom of the tank. How deep is the tank?

19. When the light illustrated in Figure P35.19 passes through the glass block, it is shifted laterally by the distance d. Taking $n = 1.50$, find the value of d.

Figure P35.19

20. Find the time interval required for the light to pass through the glass block described in Problem 19.

21. The light beam shown in Figure P35.21 makes an angle of 20.0° with the normal line NN' in the linseed oil. Determine the angles θ and θ'. (The index of refraction of linseed oil is 1.48.)

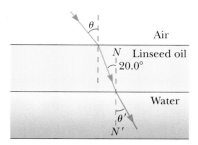

Figure P35.21

22. Three sheets of plastic have unknown indices of refraction. Sheet 1 is placed on top of sheet 2, and a laser beam is directed onto the sheets from above so that it strikes the interface at an angle of 26.5° with the normal. The refracted beam in sheet 2 makes an angle of 31.7° with the normal. The experiment is repeated with sheet 3 on top of sheet 2, and, with the same angle of incidence, the refracted beam makes an angle of 36.7° with the normal. If the experiment is repeated again with sheet 1 on top of sheet 3, what is the expected angle of refraction in sheet 3? Assume the same angle of incidence.

23. ● Light passes from air into flint glass. (a) Is it possible for the component of its velocity perpendicular to the interface to remain constant? Explain your answer. (b) **What If?** Can the component of velocity parallel to the interface remain constant during refraction? Explain your answer.

24. When you look through a window, by what time interval is the light you see delayed by having to go through glass instead of air? Make an order-of-magnitude estimate on the basis of data you specify. By how many wavelengths is it delayed?

25. A prism that has an apex angle of 50.0° is made of cubic zirconia, with $n = 2.20$. What is its angle of minimum deviation?

26. Light of wavelength 700 nm is incident on the face of a fused quartz prism at an angle of 75.0° (with respect to the normal to the surface). The apex angle of the prism is 60.0°. Use the value of n from Figure 35.21 and calculate the angle (a) of refraction at the first surface, (b) of incidence at the second surface, (c) of refraction at the second surface, and (d) between the incident and emerging rays.

27. A triangular glass prism with apex angle $\Phi = 60.0°$ has an index of refraction $n = 1.50$ (Fig. P35.27). What is the smallest angle of incidence θ_1 for which a light ray can emerge from the other side?

Figure P35.27   Problems 27 and 28.

28. A triangular glass prism with apex angle Φ has index of refraction n. (See Fig. P35.27.) What is the smallest angle of incidence θ_1 for which a light ray can emerge from the other side?

29. A triangular glass prism with apex angle 60.0° has an index of refraction of 1.50. (a) Show that if its angle of incidence on the first surface is $\theta_1 = 48.6°$, light will pass symmetrically through the prism as shown in Figure 35.17. (b) Find the angle of deviation $\delta_{\min}$ for $\theta_1 = 48.6°$. (c) **What If?** Find the angle of deviation if the angle of incidence on the first surface is 45.6°. (d) Find the angle of deviation if $\theta_1 = 51.6°$.

Section 35.6 Huygens's Principle

30. The speed of a water wave is described by $v = \sqrt{gd}$, where d is the water depth, assumed to be small compared to the wavelength. Because their speed changes, water waves refract when moving into a region of different depth. Sketch a map of an ocean beach on the eastern side of a landmass. Show contour lines of constant depth under water, assuming reasonably uniform slope. (a) Suppose waves approach the coast from a storm far away to the north–northeast. Demonstrate that the waves move nearly perpendicular to the shoreline when they reach the beach. (b) Sketch a map of a coastline with alternating bays and headlands as suggested in Figure P35.30. Again

Ray Atkeson/Image Archive

Figure P35.30

2 = intermediate;   3 = challenging;   ☐ = SSM/SG;   ▲ = ThomsonNOW;   ▨ = symbolic reasoning;   ● = qualitative reasoning

make a reasonable guess about the shape of contour lines of constant depth. Suppose waves approach the coast, carrying energy with uniform density along originally straight wave fronts. Show that the energy reaching the coast is concentrated at the headlands and has lower intensity in the bays.

Section 35.7 Dispersion

31. ▲ The index of refraction for violet light in silica flint glass is 1.66 and that for red light is 1.62. What is the angular spread of visible light passing through a prism of apex angle 60.0° if the angle of incidence is 50.0°? See Figure P35.31.

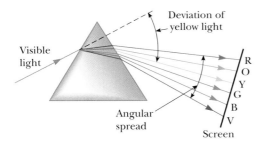

Figure P35.31

32. A narrow, white light beam is incident on a block of fused quartz at an angle of 30.0°. Find the angular spread of the light beam inside the quartz due to dispersion.

Section 35.8 Total Internal Reflection

33. For 589-nm light, calculate the critical angle for the following materials surrounded by air. (a) diamond (b) flint glass (c) ice

34. A glass fiber ($n = 1.50$) is submerged in water ($n = 1.33$). What is the critical angle for light to stay inside the optical fiber?

35. Consider a common mirage formed by superheated air immediately above a roadway. A truck driver whose eyes are 2.00 m above the road, where $n = 1.000\ 3$, looks forward. She perceives the illusion of a patch of water ahead on the road, where her line of sight makes an angle of 1.20° below the horizontal. Find the index of refraction of the air immediately above the road surface. *Suggestion:* Treat this problem as one about total internal reflection.

36. Determine the maximum angle θ for which the light rays incident on the end of the pipe in Figure P35.36 are sub-

ject to total internal reflection along the walls of the pipe. Assume the pipe has an index of refraction of 1.36 and the outside medium is air. Your answer defines the size of the *cone of acceptance* for the light pipe.

Figure P35.36

37. ● An optical fiber has index of refraction n and diameter d. It is surrounded by air. Light is sent into the fiber along its axis as shown in Figure P35.37. (a) Find the smallest outside radius R permitted for a bend in the fiber if no light is to escape. (b) **What If?** Does the result for part (a) predict reasonable behavior as d approaches zero? As n increases? As n approaches 1? (c) Evaluate R assuming the fiber diameter is 100 μm and its index of refraction is 1.40.

Figure P35.37

38. ● A room contains air in which the speed of sound is 343 m/s. The walls of the room are made of concrete in which the speed of sound is 1 850 m/s. (a) Find the critical angle for total internal reflection of sound at the concrete–air boundary. (b) In which medium must the sound be traveling if it is undergo total internal reflection? (c) "A bare concrete wall is a highly efficient mirror for sound." Give evidence for or against this statement.

39. ● Around 1965, engineers at the Toro Company invented a gasoline gauge for small engines diagrammed in Figure P35.39. The gauge has no moving parts. It consists of a flat slab of transparent plastic fitting vertically into a slot in the cap on the gas tank. None of the plastic has a reflective coating. The plastic projects from the horizontal top down nearly to the bottom of the opaque tank. Its lower edge is cut with facets making angles of 45° with the horizontal. A lawn mower operator looks down from above and sees a boundary between bright and dark on the gauge. The location of the boundary, across the width of the plastic, indicates the quantity of gasoline in the tank. Explain how the gauge works. Explain the design requirements, if any, for the index of refraction of the plastic.

Figure P35.39

Additional Problems

40. A digital videodisc records information in a spiral track approximately 1 μm wide. The track consists of a series of pits in the information layer (Fig. P35.40a) that scatter light from a laser beam sharply focused on them. The laser shines in through transparent plastic of thickness $t = 1.20$ mm and index of refraction 1.55 (Fig. P35.40b). Assume the width of the laser beam at the information layer must be $a = 1.00$ μm to read from only one track and not from its neighbors. Assume the width of the beam as it enters the transparent plastic from below is $w = 0.700$ mm. A lens makes the beam converge into a cone with an apex angle $2\theta_1$ before it enters the videodisc. Find the incidence angle θ_1 of the light at the edge of the conical beam. This design is relatively immune to small dust particles degrading the video quality. Particles on the plastic surface would have to be as large as 0.7 mm to obscure the beam.

(a)

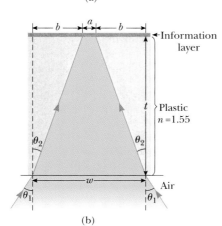

(b)

Figure P35.40

41. ● Figure P35.41a shows a desk ornament globe containing a photograph. The flat photograph is in air, inside a vertical slot located behind a water-filled compartment having the shape of one half of a cylinder. Suppose you are looking at the center of the photograph and then rotate the globe about a vertical axis. You find that the center of the photograph disappears when you rotate the globe beyond a certain maximum angle (Fig. P35.41b). Account for this phenomenon and calculate the maximum angle. Describe what you see when you turn the globe beyond this angle.

(a)   (b)

Figure P35.41

42. ● A light ray enters the atmosphere of a planet and descends vertically to the surface a distance h below. The index of refraction where the light enters the atmosphere is 1.000, and it increases linearly with distance to have the value n at the planet surface. (a) Over what time interval does the light traverse this path? (b) State how this travel time compares with the time interval required in the absence of an atmosphere.

43. A narrow beam of light is incident from air onto the surface of glass with index of refraction 1.56. Find the angle of incidence for which the corresponding angle of refraction is half the angle of incidence. *Suggestion:* You might want to use the trigonometric identity $\sin 2\theta = 2 \sin \theta \cos \theta$.

44. ▼ (a) Consider a horizontal interface between air above and glass of index 1.55 below. Draw a light ray incident from the air at angle of incidence 30.0°. Determine the angles of the reflected and refracted rays and show them on the diagram. (b) **What If?** Now suppose the light ray is incident from the glass at angle of incidence 30.0°. Determine the angles of the reflected and refracted rays and show all three rays on a new diagram. (c) For rays incident from the air onto the air–glass surface, determine and tabulate the angles of reflection and refraction for all the angles of incidence at 10.0° intervals from 0° to 90.0°.

(d) Do the same for light rays coming up to the interface through the glass.

45. ▲ A small light fixture on the bottom of a swimming pool is 1.00 m below the surface. The light emerging from the still water forms a circle on the water surface. What is the diameter of this circle?

46. The walls of a prison cell are perpendicular to the four cardinal compass directions. On the first day of spring, light from the rising Sun enters a rectangular window in the eastern wall. The light traverses 2.37 m horizontally to shine perpendicularly on the wall opposite the window. A young prisoner observes the patch of light moving across this western wall and for the first time forms his own understanding of the rotation of the Earth. (a) With what speed does the illuminated rectangle move? (b) The prisoner holds a small, square mirror flat against the wall at one corner of the rectangle of light. The mirror reflects light back to a spot on the eastern wall close beside the window. With what speed does the smaller square of light move across that wall? (c) Seen from a latitude of 40.0° north, the rising Sun moves through the sky along a line making a 50.0° angle with the southeastern horizon. In what direction does the rectangular patch of light on the western wall of the prisoner's cell move? (d) In what direction does the smaller square of light on the eastern wall move?

47. A hiker stands on an isolated mountain peak near sunset and observes a rainbow caused by water droplets in the air at a distance of 8.00 km along her line of sight. The valley is 2.00 km below the mountain peak and entirely flat. What fraction of the complete circular arc of the rainbow is visible to the hiker? (See Fig. 35.24.)

48. Figure P35.48 shows a top view of a square enclosure. The inner surfaces are plane mirrors. A ray of light enters a small hole in the center of one mirror. (a) At what angle θ must the ray enter if it exits through the hole after being reflected once by each of the other three mirrors? (b) **What If?** Are there other values of θ for which the ray can exit after multiple reflections? If so, sketch one of the ray's paths.

Figure P35.48

49. ▲ A laser beam strikes one end of a slab of material as shown in Figure P35.49. The index of refraction of the slab is 1.48. Determine the number of internal reflections of the beam before it emerges from the opposite end of the slab.

Figure P35.49

50. A 4.00-m-long pole stands vertically in a lake having a depth of 2.00 m. The Sun is 40.0° above the horizontal. Determine the length of the pole's shadow on the bottom of the lake. Take the index of refraction for water to be 1.33.

51. The light beam in Figure P35.51 strikes surface 2 at the critical angle. Determine the angle of incidence θ_1.

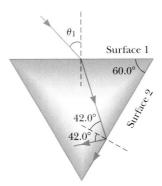

Figure P35.51

52. Builders use a leveling instrument in which the beam from a fixed helium–neon laser reflects in a horizontal plane from a small, flat mirror mounted on a vertical rotating shaft. The light is sufficiently bright and the rotation rate is sufficiently high that the reflected light appears as a horizontal line, wherever it falls on a wall. (a) Assume the mirror is at the center of a circular grain elevator of radius 3.00 m. The mirror spins with constant angular velocity 35.0 rad/s. Find the speed of the spot of laser light on the curved wall. (b) Now assume the spinning mirror is at a perpendicular distance of 3.00 m from point O on a long, flat, vertical wall. When the spot of laser light on the wall is at distance x from point O, what is its speed? (c) What is the minimum value for the speed? What value of x corresponds to it? How does the minimum speed compare with the speed you found in part (a)? (d) What is the maximum speed of the spot on the flat wall? (e) In what time interval does the spot change from its minimum to its maximum speed?

53. ▲ ● A light ray of wavelength 589 nm is incident at an angle θ on the top surface of a block of polystyrene as shown in Figure P35.53. (a) Find the maximum value of θ for which the refracted ray undergoes total internal reflection at the left vertical face of the block. **What If?** Repeat the calculation for the case in which the polystyrene block is immersed in (b) water and (c) carbon disulfide. You will need to explain your answers.

Figure P35.53

54. ● As sunlight enters the Earth's atmosphere, it changes direction due to the small difference between the speeds of light in vacuum and in air. The duration of an *optical* day is defined as the time interval between the instant when the top of the rising Sun is just visible above the horizon and the instant when the top of the Sun just disappears below the horizontal plane. The duration of the *geometric* day is defined as the time interval between the instant when a mathematically straight line between an observer and the top of the Sun just clears the horizon and the instant at which this line just dips below the horizon. (a) Explain which is longer, an optical day or a geometric day. (b) Find the difference between these two time intervals. Model the Earth's atmosphere as uniform, with index of refraction 1.000 293, a sharply defined upper surface, and depth 8 614 m. Assume the observer is at the Earth's equator so that the apparent path of the rising and setting Sun is perpendicular to the horizon.

55. A shallow glass dish is 4.00 cm wide at the bottom as shown in Figure P35.55. When an observer's eye is located as shown, the observer sees the edge of the bottom of the empty dish. When this dish is filled with water, the observer sees the center of the bottom of the dish. Find the height of the dish.

Figure P35.55

56. A ray of light passes from air into water. For its deviation angle $\delta = |\theta_1 - \theta_2|$ to be 10.0°, what must its angle of incidence be?

57. A material having an index of refraction n is surrounded by a vacuum and is in the shape of a quarter circle of radius R (Fig. P35.57). A light ray parallel to the base of the material is incident from the left at a distance L above the base and emerges from the material at the angle θ. Determine an expression for θ.

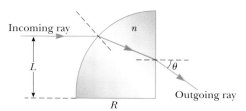

Figure P35.57

58. *Fermat's principle.* Pierre de Fermat (1601–1665) showed that whenever light travels from one point to another, its actual path is the path that requires the smallest time interval. The simplest example is for light propagating in a homogeneous medium. It moves in a straight line because a straight line is the shortest distance between two points. Derive Snell's law of refraction from Fermat's principle. Proceed as follows. In Figure P35.58, a light ray travels from point P in medium 1 to point Q in medium 2. The two points are respectively at perpendicular distances a and b from the interface. The displacement from P to Q has the component d parallel to the interface, and we let x represent the coordinate of the point where the ray enters the second medium. Let $t = 0$ be the instant at which the light starts from P. (a) Show that the time at which the light arrives at Q is

$$t = \frac{r_1}{v_1} + \frac{r_2}{v_2} = \frac{n_1 \sqrt{a^2 + x^2}}{c} + \frac{n_2 \sqrt{b^2 + (d-x)^2}}{c}$$

(b) To obtain the value of x for which t has its minimum value, differentiate t with respect to x and set the derivative equal to zero. Show that the result implies

$$\frac{n_1 x}{\sqrt{a^2 + x^2}} = \frac{n_2 (d-x)}{\sqrt{b^2 + (d-x)^2}}$$

(c) Show that this expression in turn gives Snell's law

$$n_1 \sin \theta_1 = n_2 \sin \theta_2$$

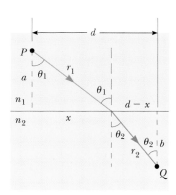

Figure P35.58

2 = intermediate;   3 = challenging;   ☐ = SSM/SG;   ▲ = ThomsonNOW;   ▨ = symbolic reasoning;   ● = qualitative reasoning

59. Refer to Problem 58 for the statement of Fermat's principle of least time. Derive the law of reflection (Eq. 35.2) from Fermat's principle.

60. A transparent cylinder of radius $R = 2.00$ m has a mirrored surface on its right half as shown in Figure P35.60. A light ray traveling in air is incident on the left side of the cylinder. The incident light ray and exiting light ray are parallel, and $d = 2.00$ m. Determine the index of refraction of the material.

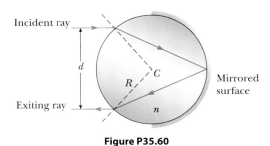

Incident ray

d

R

C

Mirrored surface

Exiting ray

n

Figure P35.60

61. Suppose a luminous sphere of radius R_1 (such as the Sun) is surrounded by a uniform atmosphere of radius R_2 and index of refraction n. When the sphere is viewed from a location far away in vacuum, what is its apparent radius? You will need to distinguish between the two cases (a) $R_2 > nR_1$ and (b) $R_2 < nR_1$.

62. ● A. H. Pfund's method for measuring the index of refraction of glass is illustrated in Figure P35.62. One face of a slab of thickness t is painted white, and a small hole scraped clear at point P serves as a source of diverging rays when the slab is illuminated from below. Ray PBB' strikes the clear surface at the critical angle and is totally reflected as are rays such as PCC'. Rays such as PAA' emerge from the clear surface. On the painted surface, there appears a dark circle of diameter d surrounded by an illuminated region, or halo. (a) Derive an equation for n in terms of the measured quantities d and t. (b) What is

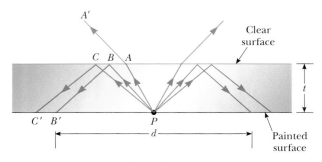

A'

Clear surface

C B A

C' B'

P

t

d

Painted surface

Figure P35.62

the diameter of the dark circle if $n = 1.52$ for a slab 0.600 cm thick? (c) If white light is used, dispersion causes the critical angle to depend on color. Is the inner edge of the white halo tinged with red light or with violet light? Explain.

63. A light ray enters a rectangular block of plastic at an angle $\theta_1 = 45.0°$ and emerges at an angle $\theta_2 = 76.0°$ as shown in Figure P35.63. (a) Determine the index of refraction of the plastic. (b) If the light ray enters the plastic at a point $L = 50.0$ cm from the bottom edge, what time interval is required for the light ray to travel through the plastic?

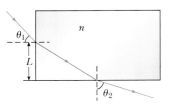

θ_1

n

L

θ_2

Figure P35.63

64. ● ⬛ Students allow a narrow beam of laser light to strike a water surface. They measure the angle of refraction for selected angles of incidence and record the data shown in the accompanying table. Use the data to verify Snell's law of refraction by plotting the sine of the angle of incidence versus the sine of the angle of refraction. Explain what the shape of the graph demonstrates. Use the resulting plot to deduce the index of refraction of water, explaining how you do so.

Angle of Incidence (degrees)	Angle of Refraction (degrees)
10.0	7.5
20.0	15.1
30.0	22.3
40.0	28.7
50.0	35.2
60.0	40.3
70.0	45.3
80.0	47.7

65. **Review problem.** A mirror is often "silvered" with aluminum. By adjusting the thickness of the metallic film, one can make a sheet of glass into a mirror that reflects anything between, say, 3% and 98% of the incident light, transmitting the rest. Prove that it is impossible to construct a "one-way mirror" that would reflect 90% of the electromagnetic waves incident from one side and reflect 10% of those incident from the other side. *Suggestion:* Use Clausius's statement of the second law of thermodynamics.

Answers to Quick Quizzes

35.1 (d). The light rays from the actor's face must reflect from the mirror and into the camera. If these light rays are reversed, light from the camera reflects from the mirror into the eyes of the actor.

35.2 Beams ② and ④ are reflected; beams ③ and ⑤ are refracted.

35.3 (c). Because the light is entering a material in which the index of refraction is lower, the speed of light is higher and the light bends away from the normal.

35.4 (c). An ideal camera lens would have an index of refraction that does not vary with wavelength so that all colors would be bent through the same angle by the lens. Of the three choices, fused quartz has the least variation in n across the visible spectrum. A lens designer can do even better by stacking two lenses of different materials together to make an *achromatic doublet*.

35.5 **(i),** (b). The two bright rays exiting the bottom of the prism on the right in Figure 35.26 result from total internal reflection at the right face of the prism. Notice that there is no refracted light exiting the slanted side for these rays. The light from the other three rays is divided into reflected and refracted parts. **(ii),** (b). Counterclockwise rotation of the prism will cause the rays to strike the slanted side of the prism at a larger angle. When the five rays strike at an angle larger than the critical angle, they all undergo total internal reflection.

The light rays coming from the leaves in the background of this scene did not form a focused image on the film of the camera that took this photograph. Consequently, the background appears very blurry. Light rays passing though the raindrop, however, have been altered so as to form a focused image of the background leaves on the film. In this chapter, we investigate the formation of images as light rays reflect from mirrors and refract through lenses. (Don Hammond/CORBIS)

36.1 Images Formed by Flat Mirrors

36.2 Images Formed by Spherical Mirrors

36.3 Images Formed by Refraction

36.4 Thin Lenses

36.5 Lens Aberrations

36.6 The Camera

36.7 The Eye

36.8 The Simple Magnifier

36.9 The Compound Microscope

36.10 The Telescope

36 Image Formation

This chapter is concerned with the images that result when light rays encounter flat and curved surfaces. Images can be formed by either reflection or refraction, and we can design mirrors and lenses to form images with desired characteristics. We continue to use the ray approximation and assume light travels in straight lines. These two steps lead to valid predictions in the field called *geometric optics*. Subsequent chapters cover interference and diffraction effects, which are the objects of study in the field of *wave optics*.

36.1 Images Formed by Flat Mirrors

Image formation by mirrors can be understood through the analysis of light rays following the wave under reflection model. We begin by considering the simplest possible mirror, the flat mirror. Consider a point source of light placed at O in Figure 36.1, a distance p in front of a flat mirror. The distance p is called the **object distance.** Diverging light rays leave the source and are reflected from the mirror. Upon reflection, the rays continue to diverge. The dashed lines in Figure 36.1 are extensions of the diverging rays back to a point of intersection at I. The diverging rays appear to the viewer to originate at the point I behind the mirror. Point I, which is a distance q behind the mirror, is called the **image** of the object at O. The distance q is called the **image distance.** Regardless of the system under study, images can always be located by extending diverging rays back to a point at which

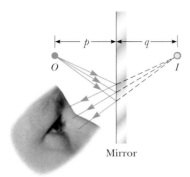

Figure 36.1 An image formed by reflection from a flat mirror. The image point I is located behind the mirror a perpendicular distance q from the mirror (the image distance). The image distance has the same magnitude as the object distance p.

they intersect. **Images are located either at a point from which rays of light** *actually* **diverge or at a point from which they** *appear* **to diverge.**

Images are classified as **real** or **virtual. A real image is formed when light rays pass through and diverge from the image point; a virtual image is formed when the light rays do not pass through the image point but only appear to diverge from that point.** The image formed by the mirror in Figure 36.1 is virtual. The image of an object seen in a flat mirror is *always* virtual. Real images can be displayed on a screen (as at a movie theater), but virtual images cannot be displayed on a screen. We shall see an example of a real image in Section 36.2.

We can use the simple geometry in Active Figure 36.2 to examine the properties of the images of extended objects formed by flat mirrors. Even though there are an infinite number of choices of direction in which light rays could leave each point on the object (represented by a blue arrow), we need to choose only two rays to determine where an image is formed. One of those rays starts at P, follows a path perpendicular to the mirror, and reflects back on itself. The second ray follows the oblique path PR and reflects as shown in Active Figure 36.2 according to the law of reflection. An observer in front of the mirror would extend the two reflected rays back to the point at which they appear to have originated, which is point P' behind the mirror. A continuation of this process for points other than P on the object would result in a virtual image (represented by a yellow arrow) of the entire object behind the mirror. Because triangles PQR and $P'QR$ are congruent, $PQ = P'Q$, so that $|p| = |q|$. Therefore, **the image formed of an object placed in front of a flat mirror is as far behind the mirror as the object is in front of the mirror.**

The geometry in Active Figure 36.2 also reveals that the object height h equals the image height h'. Let us define **lateral magnification** M of an image as follows:

$$M \equiv \frac{\text{image height}}{\text{object height}} = \frac{h'}{h} \qquad (36.1)$$

This general definition of the lateral magnification for an image from any type of mirror is also valid for images formed by lenses, which we study in Section 36.4. For a flat mirror, $M = +1$ for any image because $h' = h$. The positive value of the magnification signifies that the image is upright. (By upright we mean that if the object arrow points upward as in Active Figure 36.2, so does the image arrow.)

A flat mirror produces an image that has an *apparent* left–right reversal. You can see this reversal by standing in front of a mirror and raising your right hand as shown in Figure 36.3. The image you see raises its left hand. Likewise, your hair appears to be parted on the side opposite your real part, and a mole on your right cheek appears to be on your left cheek.

This reversal is not *actually* a left–right reversal. Imagine, for example, lying on your left side on the floor with your body parallel to the mirror surface. Now your head is on the left and your feet are on the right. If you shake your feet, the image does not shake its head! If you raise your right hand, however, the image again raises its left hand. Therefore, the mirror again appears to produce a left–right reversal but in the up–down direction!

The reversal is actually a *front–back reversal,* caused by the light rays going forward toward the mirror and then reflecting back from it. An interesting exercise is to stand in front of a mirror while holding an overhead transparency in front of you so that you can read the writing on the transparency. You will also be able to read the writing on the image of the transparency. You may have had a similar experience if you have attached a transparent decal with words on it to the rear window of your car. If the decal can be read from outside the car, you can also read it when looking into your rearview mirror from inside the car.

Quick Quiz 36.1   You are standing approximately 2 m away from a mirror. The mirror has water spots on its surface. True or False: It is possible for you to see the water spots and your image both in focus at the same time.

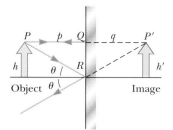

ACTIVE FIGURE 36.2

A geometric construction that is used to locate the image of an object placed in front of a flat mirror. Because the triangles PQR and $P'QR$ are congruent, $|p| = |q|$ and $h = h'$.

Sign in at www.thomsonedu.com and go to ThomsonNOW to move the object and see the effect on the image.

◀ Lateral magnification

PITFALL PREVENTION 36.1
Magnification Does Not Necessarily Imply Enlargement

For optical elements other than flat mirrors, the magnification defined in Equation 36.1 can result in a number with a magnitude larger *or* smaller than 1. Therefore, despite the cultural usage of the word *magnification* to mean *enlargement*, the image could be smaller than the object.

Figure 36.3 The image in the mirror of a person's right hand is reversed front to back, which makes the right hand appear to be a left hand. Notice that the thumb is on the left side of both real hands and on the left side of the image. That the thumb is not on the right side of the image indicates that there is no left-to-right reversal.

CONCEPTUAL EXAMPLE 36.1 **Multiple Images Formed by Two Mirrors**

Two flat mirrors are perpendicular to each other as in Figure 36.4, and an object is placed at point O. In this situation, multiple images are formed. Locate the positions of these images.

SOLUTION

The image of the object is at I_1 in mirror 1 (violet rays) and at I_2 in mirror 2 (blue rays). In addition, a third image is formed at I_3 (brown rays). This third image is the image of I_1 in mirror 2 or, equivalently, the image of I_2 in mirror 1. That is, the image at I_1 (or I_2) serves as the object for I_3. To form this image at I_3, the rays reflect twice after leaving the object at O.

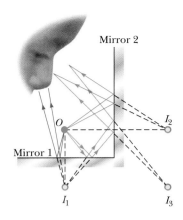

Figure 36.4 (Conceptual Example 36.1) When an object is placed in front of two mutually perpendicular mirrors as shown, three images are formed. Follow the different-colored light rays to understand the formation of each image.

CONCEPTUAL EXAMPLE 36.2 **The Tilting Rearview Mirror**

Most rearview mirrors in cars have a day setting and a night setting. The night setting greatly diminishes the intensity of the image so that lights from trailing vehicles do not temporarily blind the driver. How does such a mirror work?

SOLUTION

Figure 36.5 shows a cross-sectional view of a rearview mirror for each setting. The unit consists of a reflective coating on the back of a wedge of glass. In the day setting (Fig. 36.5a), the light from an object behind the car strikes the glass wedge at point 1. Most of the light enters the wedge, refracting as it crosses the front surface, and reflects from the back surface to return to the front surface, where it is refracted again as it re-enters the air as ray B (for *bright*). In addition, a small portion of the light is reflected at the front surface of the glass as indicated by ray D (for *dim*).

This dim reflected light is responsible for the image observed when the mirror is in the night setting (Fig. 36.5b).

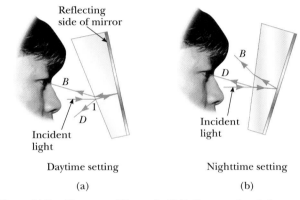

Figure 36.5 (Conceptual Example 36.2) Cross-sectional views of a rearview mirror. (a) With the day setting, the silvered back surface of the mirror reflects a bright ray B into the driver's eyes. (b) With the night setting, the glass of the unsilvered front surface of the mirror reflects a dim ray D into the driver's eyes.

In that case, the wedge is rotated so that the path followed by the bright light (ray B) does not lead to the eye. Instead, the dim light reflected from the front surface of the wedge travels to the eye, and the brightness of trailing headlights does not become a hazard.

36.2 Images Formed by Spherical Mirrors

In the preceding section, we considered images formed from flat mirrors. Now we study images formed by curved mirrors. Although a variety of curvatures are possible, we will restrict our investigation to spherical mirrors. As its name implies, a **spherical mirror** has the shape of a section of a sphere.

Concave Mirrors

We first consider reflection of light from the inner, concave surface of a spherical mirror as shown in Figure 36.6. This type of reflecting surface is called a **concave mirror.** Figure 36.6a shows that the mirror has a radius of curvature R, and its center of curvature is point C. Point V is the center of the spherical section, and a line through C and V is called the **principal axis** of the mirror. Figure 36.6a shows a

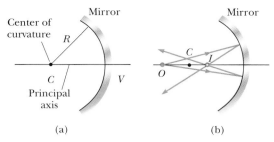

(a)   (b)

Figure 36.6   (a) A concave mirror of radius R. The center of curvature C is located on the principal axis. (b) A point object placed at O in front of a concave spherical mirror of radius R, where O is any point on the principal axis farther than R from the mirror surface, forms a real image at I. If the rays diverge from O at small angles, they all reflect through the same image point.

Figure 36.7   Red, blue, and green light rays are reflected by a curved mirror. Notice that the three colored beams meet at a point.

cross section of a spherical mirror, with its surface represented by the solid, curved black line. (The blue band represents the structural support for the mirrored surface, such as a curved piece of glass on which the silvered surface is deposited.) This type of mirror focuses incoming parallel rays to a point as demonstrated by the colored light rays in Figure 36.7.

Now consider a point source of light placed at point O in Figure 36.6b, where O is any point on the principal axis to the left of C. Two diverging light rays that originate at O are shown. After reflecting from the mirror, these rays converge and cross at the image point I. They then continue to diverge from I as if an object were there. As a result, the image at point I is real.

In this section, we shall consider only rays that diverge from the object and make a small angle with the principal axis. Such rays are called **paraxial rays.** All paraxial rays reflect through the image point as shown in Figure 36.6b. Rays that are far from the principal axis such as those shown in Figure 36.8 converge to other points on the principal axis, producing a blurred image. This effect, called *spherical aberration,* is present to some extent for any spherical mirror and is discussed in Section 36.5.

Figure 36.8   Rays diverging from the object at large angles from the principal axis reflect from a spherical concave mirror to intersect the principal axis at different points, resulting in a blurred image. This condition is called *spherical aberration.*

If the object distance p and radius of curvature R are known, we can use Figure 36.9 to calculate the image distance q. By convention, these distances are measured from point V. Figure 36.9 shows two rays leaving the tip of the object. One of these rays passes through the center of curvature C of the mirror, hitting the mirror perpendicular to the mirror surface and reflecting back on itself. The second ray strikes the mirror at its center (point V) and reflects as shown, obeying the law of reflection. The image of the tip of the arrow is located at the point where these two rays intersect. From the large, gold right triangle in Figure 36.9, we see that $\tan\theta = h/p$, and from the blue right triangle, we see that $\tan\theta = -h'/q$. The negative sign is introduced because the image is inverted, so h' is taken to be negative. Therefore, from Equation 36.1 and these results, we find that the magnification of the image is

$$M = \frac{h'}{h} = -\frac{q}{p} \qquad (36.2)$$

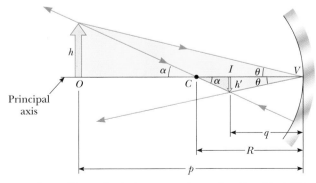

Figure 36.9   The image formed by a spherical concave mirror when the object O lies outside the center of curvature C. This geometric construction is used to derive Equation 36.4.

A satellite-dish antenna is a concave reflector for television signals from a satellite in orbit around the Earth. Because the satellite is so far away, the signals are carried by microwaves that are parallel when they arrive at the dish. These waves reflect from the dish and are focused on the receiver.

Also notice from the green right triangle in Figure 36.9 and the smaller gold right triangle that

$$\tan \alpha = \frac{-h'}{R - q} \quad \text{and} \quad \tan \alpha = \frac{h}{p - R}$$

from which it follows that

$$\frac{h'}{h} = -\frac{R - q}{p - R} \tag{36.3}$$

Comparing Equations 36.2 and 36.3 gives

$$\frac{R - q}{p - R} = \frac{q}{p}$$

Simple algebra reduces this expression to

◀ Mirror equation in terms
of radius of curvature

$$\frac{1}{p} + \frac{1}{q} = \frac{2}{R} \tag{36.4}$$

which is called the *mirror equation*. We present a modified version of this equation shortly.

If the object is very far from the mirror—that is, if p is so much greater than R that p can be said to approach infinity—then $1/p \approx 0$, and Equation 36.4 shows that $q \approx R/2$. That is, when the object is very far from the mirror, the image point is halfway between the center of curvature and the center point on the mirror as shown in Figure 36.10a. The incoming rays from the object are essentially parallel in this figure because the source is assumed to be very far from the mirror. The image point in this special case is called the **focal point** F, and the image distance the **focal length** f, where

◀ Focal length

$$f = \frac{R}{2} \tag{36.5}$$

In Figure 36.7, the colored beams are traveling parallel to the principal axis and the mirror reflects all three beams to the focal point. Notice that the point at which the three beams intersect and the colors add is white.

Because the focal length is a parameter particular to a given mirror, it can be used to compare one mirror with another. Combining Equations 36.4 and 36.5, the **mirror equation** can be expressed in terms of the focal length:

◀ Mirror equation in terms
of focal length

$$\frac{1}{p} + \frac{1}{q} = \frac{1}{f} \tag{36.6}$$

Notice that the focal length of a mirror depends only on the curvature of the mirror and not on the material from which the mirror is made because the formation of the image results from rays reflected from the surface of the material. The situ-

PITFALL PREVENTION 36.2
The *Focal* Point Is Not the *Focus* Point

The focal point *is usually not* the point at which the light rays focus to form an image. The focal point is determined solely by the curvature of the mirror; it does not depend on the location of the object. In general, an image forms at a point different from the focal point of a mirror (or a lens). The *only* exception is when the object is located infinitely far away from the mirror.

(a)

(b)

Henry Leap and Jim Lehman

Figure 36.10   (a) Light rays from a distant object ($p \to \infty$) reflect from a concave mirror through the focal point F. In this case, the image distance $q \approx R/2 = f$, where f is the focal length of the mirror. (b) Reflection of parallel rays from a concave mirror.

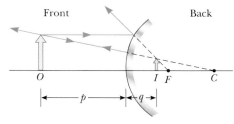

Figure 36.11 Formation of an image by a spherical convex mirror. The image formed by the object is virtual and upright.

ation is different for lenses; in that case, the light actually passes through the material and the focal length depends on the type of material from which the lens is made. (See Section 36.4.)

Convex Mirrors

Figure 36.11 shows the formation of an image by a **convex mirror**, that is, one silvered so that light is reflected from the outer, convex surface. It is sometimes called a **diverging mirror** because the rays from any point on an object diverge after reflection as though they were coming from some point behind the mirror. The image in Figure 36.11 is virtual because the reflected rays only appear to originate at the image point as indicated by the dashed lines. Furthermore, the image is always upright and smaller than the object. This type of mirror is often used in stores to foil shoplifters. A single mirror can be used to survey a large field of view because it forms a smaller image of the interior of the store.

We do not derive any equations for convex spherical mirrors because Equations 36.2, 36.4, and 36.6 can be used for either concave or convex mirrors if we adhere to the following procedure. We will refer to the region in which light rays originate and move toward the mirror as the *front side* of the mirror and the other side as the *back side*. For example, in Figures 36.9 and 36.11, the side to the left of the mirrors is the front side and the side to the right of the mirrors is the back side. Figure 36.12 states the sign conventions for object and image distances, and Table 36.1 summarizes the sign conventions for all quantities. One entry in the table, a *virtual object,* is formally introduced in Section 36.4.

Front, or real side	Back, or virtual, side
p and q positive Incident light →	p and q negative
← Reflected light	No light

Convex or
concave mirror

Figure 36.12 Signs of p and q for convex and concave mirrors.

Ray Diagrams for Mirrors

The positions and sizes of images formed by mirrors can be conveniently determined with *ray diagrams.* These pictorial representations reveal the nature of the image and can be used to check results calculated from the mathematical representation using the mirror and magnification equations. To draw a ray diagram, you must know the position of the object and the locations of the mirror's focal point and center of curvature. You then draw three rays to locate the image as

TABLE 36.1

Sign Conventions for Mirrors

Quantity	Positive When . . .	Negative When . . .
Object location (p)	object is in front of mirror (real object).	object is in back of mirror (virtual object).
Image location (q)	image is in front of mirror (real image).	image is in back of mirror (virtual image).
Image height (h')	image is upright.	image is inverted.
Focal length (f) and radius (R)	mirror is concave.	mirror is convex.
Magnification (M)	image is upright.	image is inverted.

PITFALL PREVENTION 36.3
Watch Your Signs

Success in working mirror problems (as well as problems involving refracting surfaces and thin lenses) is largely determined by proper sign choices when substituting into the equations. The best way to success is to work a multitude of problems on your own.

PITFALL PREVENTION 36.4
Choose a Small Number of Rays

A *huge* number of light rays leave each point on an object (and pass through each point on an image). In a ray diagram, which displays the characteristics of the image, we choose only a few rays that follow simply stated rules. Locating the image by calculation complements the diagram.

shown by the examples in Active Figure 36.13. These rays all start from the same object point and are drawn as follows. You may choose any point on the object; here, let's choose the top of the object for simplicity. For concave mirrors (see Active Figs. 36.13a and 36.13b), draw the following three rays:

- Ray 1 is drawn from the top of the object parallel to the principal axis and is reflected through the focal point F.
- Ray 2 is drawn from the top of the object through the focal point (or as if coming from the focal point if $p < f$) and is reflected parallel to the principal axis.
- Ray 3 is drawn from the top of the object through the center of curvature C and is reflected back on itself.

The intersection of any two of these rays locates the image. The third ray serves as a check of the construction. The image point obtained in this fashion must always agree with the value of q calculated from the mirror equation. With concave mirrors, notice what happens as the object is moved closer to the mirror. The real, inverted image in Active Figure 36.13a moves to the left and becomes larger as the object approaches the focal point. When the object is at the focal point, the image is infinitely far to the left. When the object lies between the focal point and the mirror surface as shown in Active Figure 36.13b, however, the image is to the right, behind the object, and virtual, upright, and enlarged. This latter situation applies when you use a shaving mirror or a makeup mirror, both of which are concave. Your face is closer to the mirror than the focal point, and you see an upright, enlarged image of your face.

ACTIVE FIGURE 36.13

Ray diagrams for spherical mirrors along with corresponding photographs of the images of candles. (a) When the object is located so that the center of curvature lies between the object and a concave mirror surface, the image is real, inverted, and reduced in size. (b) When the object is located between the focal point and a concave mirror surface, the image is virtual, upright, and enlarged. (c) When the object is in front of a convex mirror, the image is virtual, upright, and reduced in size.

Sign in at www.thomsonedu.com and go to ThomsonNOW to move the objects and change the focal length of the mirrors to see the effect on the images.

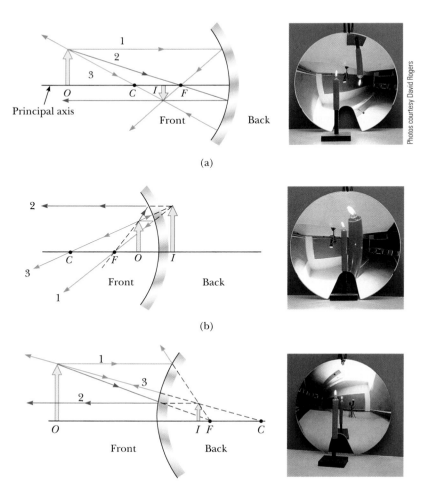

Photos courtesy David Rogers

(a)

(b)

(c)

For convex mirrors (see Active Fig. 36.13c), draw the following three rays:

- Ray 1 is drawn from the top of the object parallel to the principal axis and is reflected *away from* the focal point F.
- Ray 2 is drawn from the top of the object toward the focal point on the back side of the mirror and is reflected parallel to the principal axis.
- Ray 3 is drawn from the top of the object toward the center of curvature C on the back side of the mirror and is reflected back on itself.

In a convex mirror, the image of an object is always virtual, upright, and reduced in size as shown in Active Figure 36.13c. In this case, as the object distance decreases, the virtual image increases in size and moves away from the focal point toward the mirror as the object approaches the mirror. You should construct other diagrams to verify how image position varies with object position.

Quick Quiz 36.2   You wish to start a fire by reflecting sunlight from a mirror onto some paper under a pile of wood. Which would be the best choice for the type of mirror? (a) flat (b) concave (c) convex

Quick Quiz 36.3   Consider the image in the mirror in Figure 36.14. Based on the appearance of this image, would you conclude that (a) the mirror is concave and the image is real, (b) the mirror is concave and the image is virtual, (c) the mirror is convex and the image is real, or (d) the mirror is convex and the image is virtual?

Figure 36.14   (Quick Quiz 36.3) What type of mirror is shown here?

EXAMPLE 36.3   **The Image Formed by a Concave Mirror**

A spherical mirror has a focal length of $+10.0$ cm.

(A) Locate and describe the image for an object distance of 25.0 cm.

SOLUTION

Conceptualize   Because the focal length of the mirror is positive, it is a concave mirror (see Table 36.1). We expect the possibilities of both real and virtual images.

Categorize   Because the object distance in this part of the problem is larger than the focal length, we expect the image to be real. This situation is analogous to that in Active Figure 36.13a.

Analyze   Find the image distance by using Equation 36.6:

$$\frac{1}{q} = \frac{1}{f} - \frac{1}{p}$$

$$\frac{1}{q} = \frac{1}{10.0 \text{ cm}} - \frac{1}{25.0 \text{ cm}}$$

$$q = 16.7 \text{ cm}$$

Find the magnification of the image from Equation 36.2:

$$M = -\frac{q}{p} = -\frac{16.7 \text{ cm}}{25.0 \text{ cm}} = -0.668$$

Finalize   The absolute value of M is less than unity, so the image is smaller than the object, and the negative sign for M tells us that the image is inverted. Because q is positive, the image is located on the front side of the mirror and is real. Look into the bowl of a shiny spoon or stand far away from a shaving mirror to see this image.

(B) Locate and describe the image for an object distance of 10.0 cm.

SOLUTION

Categorize Because the object is at the focal point, we expect the image to be infinitely far away.

Analyze Find the image distance by using Equation 36.6:

$$\frac{1}{q} = \frac{1}{f} - \frac{1}{p}$$

$$\frac{1}{q} = \frac{1}{10.0 \text{ cm}} - \frac{1}{10.0 \text{ cm}}$$

$$q = \boxed{\infty}$$

Finalize This result means that rays originating from an object positioned at the focal point of a mirror are reflected so that the image is formed at an infinite distance from the mirror; that is, the rays travel parallel to one another after reflection. Such is the situation in a flashlight or an automobile headlight, where the bulb filament is placed at the focal point of a reflector, producing a parallel beam of light.

(C) Locate and describe the image for an object distance of 5.00 cm.

SOLUTION

Categorize Because the object distance is smaller than the focal length, we expect the image to be virtual. This situation is analogous to that in Active Figure 36.13b.

Analyze Find the image distance by using Equation 36.6:

$$\frac{1}{q} = \frac{1}{f} - \frac{1}{p}$$

$$\frac{1}{q} = \frac{1}{10.0 \text{ cm}} - \frac{1}{5.00 \text{ cm}}$$

$$q = \boxed{-10.0 \text{ cm}}$$

Find the magnification of the image from Equation 36.2:

$$M = -\frac{q}{p} = -\left(\frac{-10.0 \text{ cm}}{5.00 \text{ cm}}\right) = \boxed{+2.00}$$

Finalize The image is twice as large as the object, and the positive sign for M indicates that the image is upright (see Active Fig. 36.13b). The negative value of the image distance tells us that the image is virtual, as expected. Put your face close to a shaving mirror to see this type of image.

What If? Suppose you set up the candle and mirror apparatus illustrated in Active Figure 36.13a and described here in part (A). While adjusting the apparatus, you accidentally bump the candle and it begins to slide toward the mirror at velocity v_p. How fast does the image of the candle move?

Answer Solve the mirror equation, Equation 36.6, for q:

$$q = \frac{fp}{p - f}$$

Differentiate this equation with respect to time to find the velocity of the image:

$$(1) \quad v_q = \frac{dq}{dt} = \frac{d}{dt}\left(\frac{fp}{p - f}\right) = -\frac{f^2}{(p - f)^2}\frac{dp}{dt} = -\frac{f^2 v_p}{(p - f)^2}$$

Substitute numerical values from part (A):

$$v_q = -\frac{(10.0 \text{ cm})^2 \, v_p}{(25.0 \text{ cm} - 10.0 \text{ cm})^2} = -0.444 v_p$$

Therefore, the speed of the image is less than that of the object in this case.

We can see two interesting behaviors of the function for v_q in Equation (1). First, the velocity is negative regardless of the value of p or f. Therefore, if the object moves toward the mirror, the image moves toward the left in Active Figure 36.13 without regard for the side of the focal point at which the object is located or whether the mirror is con-cave or convex. Second, in the limit of $p \rightarrow 0$, the velocity v_q approaches $-v_p$. As the object moves very close to the mirror, the mirror looks like a plane mirror, the image is as far behind the mirror as the object is in front, and both the object and the image move with the same speed.

EXAMPLE 36.4   **The Image Formed by a Convex Mirror**

An automobile rearview mirror as shown in Figure 36.15 shows an image of a truck located 10.0 m from the mirror. The focal length of the mirror is -0.60 m.

(A) Find the position of the image of the truck.

SOLUTION

Conceptualize   This situation is depicted in Active Figure 36.13c.

Categorize   Because the mirror is convex, we expect it to form an upright, reduced, virtual image for any object position.

Analyze   Find the image distance by using Equation 36.6:

Figure 36.15   (Example 36.4) An approaching truck is seen in a convex mirror on the right side of an automobile. Because the image is reduced in size, the truck appears to be farther away than it actually is. Notice also that the image of the truck is in focus, but the frame of the mirror is not, which demonstrates that the image is not at the same location as the mirror surface.

$$\frac{1}{q} = \frac{1}{f} - \frac{1}{p}$$

$$\frac{1}{q} = \frac{1}{-0.60 \text{ m}} - \frac{1}{10.0 \text{ m}}$$

$$q = \boxed{-0.57 \text{ m}}$$

(B) Find the magnification of the image.

SOLUTION

Analyze   Use Equation 36.2:

$$M = -\frac{q}{p} = -\left(\frac{-0.57 \text{ m}}{10.0 \text{ m}} \right) = \boxed{+0.057}$$

Finalize   The negative value of q in part (A) indicates that the image is virtual, or behind the mirror, as shown in Active Figure 36.13c. The magnification in part (B) indicates that the image is much smaller than the truck and is upright because M is positive. Because of the image's small size, these mirrors carry the inscription, "Objects in this mirror are closer than they appear." Look into your rearview mirror or the back side of a shiny spoon to see an image of this type.

36.3   Images Formed by Refraction

In this section, we describe how images are formed when light rays follow the wave under refraction model at the boundary between two transparent materials. Consider two transparent media having indices of refraction n_1 and n_2, where the boundary between the two media is a spherical surface of radius R (Fig. 36.16). We assume the object at O is in the medium for which the index of refraction is n_1. Let's consider the paraxial rays leaving O. As we shall see, all such rays are refracted at the spherical surface and focus at a single point I, the image point.

Figure 36.17 (page 1018) shows a single ray leaving point O and refracting to point I. Snell's law of refraction applied to this ray gives

$$n_1 \sin \theta_1 = n_2 \sin \theta_2$$

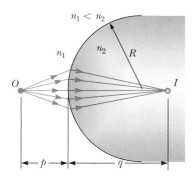

Figure 36.16   An image formed by refraction at a spherical surface. Rays making small angles with the principal axis diverge from a point object at O and are refracted through the image point I.

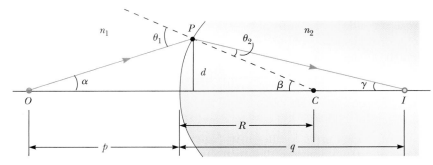

Figure 36.17 Geometry used to derive Equation 36.8, assuming that $n_1 < n_2$.

Because θ_1 and θ_2 are assumed to be small, we can use the small-angle approximation $\sin\theta \approx \theta$ (with angles in radians) and write Snell's law as

$$n_1\theta_1 = n_2\theta_2$$

We know that an exterior angle of any triangle equals the sum of the two opposite interior angles, so applying this rule to triangles OPC and PIC in Figure 36.17 gives

$$\theta_1 = \alpha + \beta$$

$$\beta = \theta_2 + \gamma$$

Combining all three expressions and eliminating θ_1 and θ_2 gives

$$n_1\alpha + n_2\gamma = (n_2 - n_1)\beta \tag{36.7}$$

Figure 36.17 shows three right triangles that have a common vertical leg of length d. For paraxial rays (unlike the relatively large-angle ray shown in Fig. 36.17), the horizontal legs of these triangles are approximately p for the triangle containing angle α, R for the triangle containing angle β, and q for the triangle containing angle γ. In the small-angle approximation, $\tan\theta \approx \theta$, so we can write the approximate relationships from these triangles as follows:

$$\tan\alpha \approx \alpha \approx \frac{d}{p} \qquad \tan\beta \approx \beta \approx \frac{d}{R} \qquad \tan\gamma \approx \gamma \approx \frac{d}{q}$$

Substituting these expressions into Equation 36.7 and dividing through by d gives

◄ Relation between object and image distance for a refracting surface

$$\frac{n_1}{p} + \frac{n_2}{q} = \frac{n_2 - n_1}{R} \tag{36.8}$$

For a fixed object distance p, the image distance q is independent of the angle the ray makes with the axis. This result tells us that all paraxial rays focus at the same point I.

As with mirrors, we must use a sign convention to apply Equation 36.8 to a variety of cases. We define the side of the surface in which light rays originate as the front side. The other side is called the back side. In contrast with mirrors, where real images are formed in front of the reflecting surface, real images are formed by refraction of light rays to the back of the surface. Because of the difference in location of real images, the refraction sign conventions for q and R are opposite the reflection sign conventions. For example, q and R are both positive in Figure 36.17. The sign conventions for spherical refracting surfaces are summarized in Table 36.2.

We derived Equation 36.8 from an assumption that $n_1 < n_2$ in Figure 36.17. This assumption is not necessary, however. Equation 36.8 is valid regardless of which index of refraction is greater.

TABLE 36.2

Sign Conventions for Refracting Surfaces

Quantity	Positive When . . .	Negative When . . .
Object location (p)	object is in front of surface (real object).	object is in back of surface (virtual object).
Image location (q)	image is in back of surface (real image).	image is in front of surface (virtual image).
Image height (h')	image is upright.	image is inverted.
Radius (R)	center of curvature is in back of surface.	center of curvature is in front of surface.

Flat Refracting Surfaces

If a refracting surface is flat, then R is infinite and Equation 36.8 reduces to

$$\frac{n_1}{p} = -\frac{n_2}{q}$$

$$q = -\frac{n_2}{n_1}p \qquad (36.9)$$

From this expression, we see that the sign of q is opposite that of p. Therefore, according to Table 36.2, **the image formed by a flat refracting surface is on the same side of the surface as the object** as illustrated in Active Figure 36.18 for the situation in which the object is in the medium of index n_1 and n_1 is greater than n_2. In this case, a virtual image is formed between the object and the surface. If n_1 is less than n_2, the rays on the back side diverge from one another at lesser angles than those in Active Figure 36.18. As a result, the virtual image is formed to the left of the object.

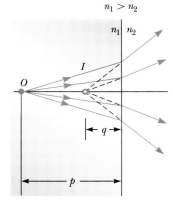

$n_1 > n_2$

ACTIVE FIGURE 36.18

The image formed by a flat refracting surface is virtual and on the same side of the surface as the object. All rays are assumed to be paraxial.

Sign in at www.thomsonedu.com and go to ThomsonNOW to move the object and see the effect on the location of the image.

Quick Quiz 36.4 In Figure 36.16, what happens to the image point I as the object point O is moved to the right from very far away to very close to the refracting surface? (a) It is always to the right of the surface. (b) It is always to the left of the surface. (c) It starts off to the left, and at some position of O, I moves to the right of the surface. (d) It starts off to the right, and at some position of O, I moves to the left of the surface.

Quick Quiz 36.5 In Active Figure 36.18, what happens to the image point I as the object point O moves toward the right-hand surface of the material of index of refraction n_1? (a) It always remains between O and the surface, arriving at the surface just as O does. (b) It moves toward the surface more slowly than O so that eventually O passes I. (c) It approaches the surface and then moves to the right of the surface.

CONCEPTUAL EXAMPLE 36.5   **Let's Go Scuba Diving!**

Objects viewed under water with the naked eye appear blurred and out of focus. A scuba diver using a mask, however, has a clear view of underwater objects. Explain how that works, using the information that the indices of refraction of the cornea, water, and air are 1.376, 1.333, and 1.000 29, respectively.

SOLUTION

Because the cornea and water have almost identical indices of refraction, very little refraction occurs when a person under water views objects with the naked eye. In this case, light rays from an object focus behind the retina, resulting in a blurred image. When a mask is used, however, the air space between the eye and the mask surface provides the normal amount of refraction at the eye–air interface; consequently, the light from the object focuses on the retina.

EXAMPLE 36.6 | **Gaze into the Crystal Ball**

A set of coins is embedded in a spherical plastic paperweight having a radius of 3.0 cm. The index of refraction of the plastic is $n_1 = 1.50$. One coin is located 2.0 cm from the edge of the sphere (Fig. 36.19). Find the position of the image of the coin.

SOLUTION

Conceptualize Because $n_1 > n_2$, where $n_2 = 1.00$ is the index of refraction for air, the rays originating from the coin in Figure 36.19 are refracted away from the normal at the surface and diverge outward.

Categorize Because the light rays originate in one material and then pass through a curved surface into another material, this example involves an image formed by refraction.

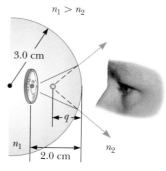

Figure 36.19 (Example 36.6) Light rays from a coin embedded in a plastic sphere form a virtual image between the surface of the object and the sphere surface. Because the object is inside the sphere, the front of the refracting surface is the *interior* of the sphere.

Analyze Apply Equation 36.8, noting from Table 36.2 that R is negative:

$$\frac{n_2}{q} = \frac{n_2 - n_1}{R} - \frac{n_1}{p}$$

$$\frac{1}{q} = \frac{1.00 - 1.50}{-3.0 \text{ cm}} - \frac{1.50}{2.0 \text{ cm}}$$

$$q = -1.7 \text{ cm}$$

Finalize The negative sign for q indicates that the image is in front of the surface; in other words, it is in the same medium as the object as shown in Figure 36.19. Therefore, the image must be virtual. (See Table 36.2.) The coin appears to be closer to the paperweight surface than it actually is.

EXAMPLE 36.7 | **The One That Got Away**

A small fish is swimming at a depth d below the surface of a pond (Fig. 36.20).

(A) What is the apparent depth of the fish as viewed from directly overhead?

SOLUTION

Conceptualize Because $n_1 > n_2$, where $n_2 = 1.00$ is the index of refraction for air, the rays originating from the fish in Figure 36.20a are refracted away from the normal at the surface and diverge outward.

Categorize Because the refracting surface is flat, R is infinite. Hence, we can use Equation 36.9 to determine the location of the image with $p = d$.

(a) (b)

Figure 36.20 (Example 36.7) (a) The apparent depth q of the fish is less than the true depth d. All rays are assumed to be paraxial. (b) Your face appears to the fish to be higher above the surface than it is.

Analyze Use the indices of refraction given in Figure 36.20a in Equation 36.9:

$$q = -\frac{n_2}{n_1} p = -\frac{1.00}{1.33} d = -0.752d$$

Finalize Because q is negative, the image is virtual as indicated by the dashed lines in Figure 36.20a. The apparent depth is approximately three-fourths the actual depth.

(B) If your face is a distance d above the water surface, at what apparent distance above the surface does the fish see your face?

SOLUTION

The light rays from your face are shown in Figure 36.20b.

Conceptualize   Because the rays refract toward the normal, your face appears higher above the surface than it actually is.

Categorize   Because the refracting surface is flat, R is infinite. Hence, we can use Equation 36.9 to determine the location of the image with $p = d$.

Analyze   Use Equation 36.9 to find the image distance:

$$q = -\frac{n_2}{n_1} p = -\frac{1.33}{1.00} d = \boxed{-1.33d}$$

Finalize   The negative sign for q indicates that the image is in the medium from which the light originated, which is the air above the water.

What If?   What if you look more carefully at the fish and measure its apparent *height* from its upper fin to its lower fin? Is the apparent height h' of the fish different from the actual height h?

Answer   Because all points on the fish appear to be fractionally closer to the observer, we expect the height to be smaller. Let the distance d in Figure 36.20a be measured to the top fin, and let the distance to the bottom fin be $d + h$. Then the images of the top and bottom of the fish are located at

$$q_{\text{top}} = -0.752d$$

$$q_{\text{bottom}} = -0.752(d + h)$$

The apparent height h' of the fish is

$$h' = q_{\text{top}} - q_{\text{bottom}} = -0.752d - [-0.752(d + h)] = 0.752h$$

Hence, the fish appears to be approximately three-fourths its actual height.

36.4   Thin Lenses

Lenses are commonly used to form images by refraction in optical instruments such as cameras, telescopes, and microscopes. Let's use what we just learned about images formed by refracting surfaces to help locate the image formed by a lens. Light passing through a lens experiences refraction at two surfaces. The development we shall follow is based on the notion that **the image formed by one refracting surface serves as the object for the second surface.** We shall analyze a thick lens first and then let the thickness of the lens be approximately zero.

Consider a lens having an index of refraction n and two spherical surfaces with radii of curvature R_1 and R_2 as in Figure 36.21 (page 1022). (Notice that R_1 is the radius of curvature of the lens surface the light from the object reaches first and R_2 is the radius of curvature of the other surface of the lens.) An object is placed at point O at a distance p_1 in front of surface 1.

Let's begin with the image formed by surface 1. Using Equation 36.8 and assuming $n_1 = 1$ because the lens is surrounded by air, we find that the image I_1 formed by surface 1 satisfies the equation

$$\frac{1}{p_1} + \frac{n}{q_1} = \frac{n - 1}{R_1} \qquad \textbf{(36.10)}$$

where q_1 is the position of the image formed by surface 1. If the image formed by surface 1 is virtual (Fig. 36.21a), q_1 is negative; it is positive if the image is real (Fig. 36.21b).

Now let's apply Equation 36.8 to surface 2, taking $n_1 = n$ and $n_2 = 1$. (We make this switch in index because the light rays approaching surface 2 are *in the material*

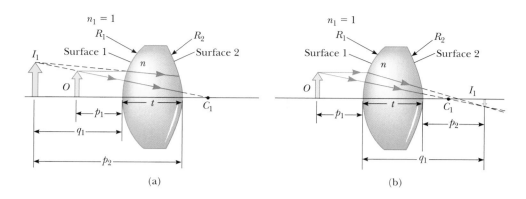

Figure 36.21 To locate the image formed by a lens, we use the virtual image at I_1 formed by surface 1 as the object for the image formed by surface 2. The point C_1 is the center of curvature of surface 1. (a) The image due to surface 1 is virtual, so I_1 is to the left of the surface. (b) The image due to surface 1 is real, so I_1 is to the right of the surface.

of the lens, and this material has index n.) Taking p_2 as the object distance for surface 2 and q_2 as the image distance gives

$$\frac{n}{p_2} + \frac{1}{q_2} = \frac{1-n}{R_2} \tag{36.11}$$

We now introduce mathematically that the image formed by the first surface acts as the object for the second surface. If the image from surface 1 is virtual as in Figure 36.21a, we see that p_2, measured from surface 2, is related to q_1 as $p_2 = -q_1 + t$, where t is the thickness of the lens. Because q_1 is negative, p_2 is a positive number. Figure 36.21b shows the case of the image from surface 1 being real. In this situation, q_1 is positive and $p_2 = -q_1 + t$, where the image from surface 1 acts as a virtual object, so p_2 is negative. Regardless of the type of image from surface 1, the same equation describes the location of the object for surface 2 based on our sign convention. For a *thin* lens (one whose thickness is small compared with the radii of curvature), we can neglect t. In this approximation, $p_2 = -q_1$ for either type of image from surface 1. Hence, Equation 36.11 becomes

$$-\frac{n}{q_1} + \frac{1}{q_2} = \frac{1-n}{R_2} \tag{36.12}$$

Adding Equations 36.10 and 36.12 gives

$$\frac{1}{p_1} + \frac{1}{q_2} = (n-1)\left(\frac{1}{R_1} - \frac{1}{R_2}\right) \tag{36.13}$$

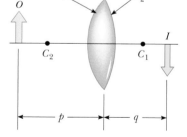

Figure 36.22 Simplified geometry for a thin lens.

For a thin lens, we can omit the subscripts on p_1 and q_2 in Equation 36.13 and call the object distance p and the image distance q, as in Figure 36.22. Hence, we can write Equation 36.13 as

$$\frac{1}{p} + \frac{1}{q} = (n-1)\left(\frac{1}{R_1} - \frac{1}{R_2}\right) \tag{36.14}$$

This expression relates the image distance q of the image formed by a thin lens to the object distance p and to the lens properties (index of refraction and radii of curvature). It is valid only for paraxial rays and only when the lens thickness is much less than R_1 and R_2.

The **focal length** f of a thin lens is the image distance that corresponds to an infinite object distance, just as with mirrors. Letting p approach ∞ and q approach f in Equation 36.14, we see that the inverse of the focal length for a thin lens is

Lens-makers' equation ▶

$$\frac{1}{f} = (n-1)\left(\frac{1}{R_1} - \frac{1}{R_2}\right) \tag{36.15}$$

This relationship is called the **lens-makers' equation** because it can be used to determine the values of R_1 and R_2 needed for a given index of refraction and a desired focal length f. Conversely, if the index of refraction and the radii of curvature of a lens are given, this equation can be used to find the focal length. If the

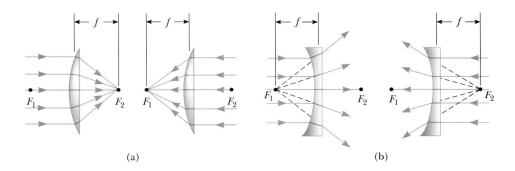

Figure 36.23 Parallel light rays pass through (a) a converging lens and (b) a diverging lens. The focal length is the same for light rays passing through a given lens in either direction. Both focal points F_1 and F_2 are the same distance from the lens.

(a)   (b)

lens is immersed in something other than air, this same equation can be used, with *n* interpreted as the *ratio* of the index of refraction of the lens material to that of the surrounding fluid.

Using Equation 36.15, we can write Equation 36.14 in a form identical to Equation 36.6 for mirrors:

$$\frac{1}{p} + \frac{1}{q} = \frac{1}{f} \qquad (36.16)$$

This equation, called the **thin lens equation,** can be used to relate the image distance and object distance for a thin lens.

Because light can travel in either direction through a lens, each lens has two focal points, one for light rays passing through in one direction and one for rays passing through in the other direction. These two focal points are illustrated in Figure 36.23 for a plano-convex lens (a converging lens) and a plano-concave lens (a diverging lens).

Figure 36.24 is useful for obtaining the signs of *p* and *q*, and Table 36.3 gives the sign conventions for thin lenses. These sign conventions are the *same* as those for refracting surfaces (see Table 36.2).

Various lens shapes are shown in Figure 36.25. Notice that a converging lens is thicker at the center than at the edge, whereas a diverging lens is thinner at the center than at the edge.

Magnification of Images

Consider a thin lens through which light rays from an object pass. As with mirrors (Eq. 36.2), a geometric construction shows that the lateral magnification of the image is

$$M = \frac{h'}{h} = -\frac{q}{p} \qquad (36.17)$$

From this expression, it follows that when *M* is positive, the image is upright and on the same side of the lens as the object. When *M* is negative, the image is inverted and on the side of the lens opposite the object.

> **PITFALL PREVENTION 36.5**
> ## A Lens Has Two Focal Points but Only One Focal Length
>
> A lens has a focal point on each side, front and back. There is only one focal length, however; each of the two focal points is located the same distance from the lens (Fig. 36.23). As a result, the lens forms an image of an object at the same point if it is turned around. In practice, that might not happen because real lenses are not infinitesimally thin.

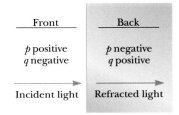

Figure 36.24 A diagram for obtaining the signs of *p* and *q* for a thin lens. (This diagram also applies to a refracting surface.)

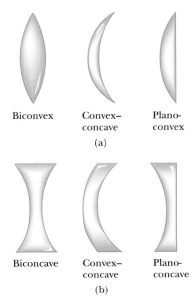

Figure 36.25 Various lens shapes. (a) Converging lenses have a positive focal length and are thickest at the middle. (b) Diverging lenses have a negative focal length and are thickest at the edges.

TABLE 36.3

Sign Conventions for Thin Lenses

Quantity	Positive When . . .	Negative When . . .
Object location (*p*)	object is in front of lens (real object).	object is in back of lens (virtual object).
Image location (*q*)	image is in back of lens (real image).	image is in front of lens (virtual image).
Image height (*h'*)	image is upright.	image is inverted.
R_1 and R_2	center of curvature is in back of lens.	center of curvature is in front of lens.
Focal length (*f*)	a converging lens.	a diverging lens.

Ray Diagrams for Thin Lenses

Ray diagrams are convenient for locating the images formed by thin lenses or systems of lenses. They also help clarify our sign conventions. Active Figure 36.26 shows such diagrams for three single-lens situations.

To locate the image of a *converging* lens (Active Fig. 36.26a and b), the following three rays are drawn from the top of the object:

- Ray 1 is drawn parallel to the principal axis. After being refracted by the lens, this ray passes through the focal point on the back side of the lens.
- Ray 2 is drawn through the center of the lens and continues in a straight line.
- Ray 3 is drawn through the focal point on the front side of the lens (or as if coming from the focal point if $p < f$) and emerges from the lens parallel to the principal axis.

To locate the image of a *diverging* lens (Active Fig. 36.26c), the following three rays are drawn from the top of the object:

- Ray 1 is drawn parallel to the principal axis. After being refracted by the lens, this ray emerges directed away from the focal point on the front side of the lens.
- Ray 2 is drawn through the center of the lens and continues in a straight line.
- Ray 3 is drawn in the direction toward the focal point on the back side of the lens and emerges from the lens parallel to the principal axis.

For the converging lens in Active Figure 36.26a, where the object is to the left of the focal point ($p > f$), the image is real and inverted. When the object is between the focal point and the lens ($p < f$) as in Active Figure 36.26b, the image is virtual and upright. In that case, the lens acts as a magnifying glass, which we study in more detail in Section 36.8. For a diverging lens (Active Fig. 36.26c), the image is always virtual and upright, regardless of where the object is placed. These geometric constructions are reasonably accurate only if the distance between the rays and the principal axis is much less than the radii of the lens surfaces.

Notice that refraction occurs only at the surfaces of the lens. A certain lens design takes advantage of this behavior to produce the *Fresnel lens,* a powerful lens

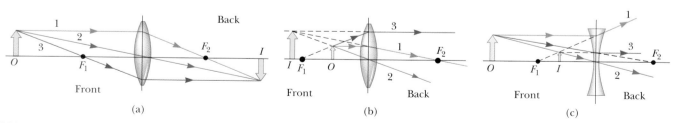

ACTIVE FIGURE 36.26

Ray diagrams for locating the image formed by a thin lens. (a) When the object is in front of and outside the focal point of a converging lens, the image is real, inverted, and on the back side of the lens. (b) When the object is between the focal point and a converging lens, the image is virtual, upright, larger than the object, and on the front side of the lens. (c) When an object is anywhere in front of a diverging lens, the image is virtual, upright, smaller than the object, and on the front side of the lens.

Sign in at www.thomsonedu.com and go to ThomsonNOW to move the objects and change the focal length of the lenses to see the effect on the images.

without great thickness. Because only the surface curvature is important in the refracting qualities of the lens, material in the middle of a Fresnel lens is removed as shown in the cross sections of lenses in Figure 36.27. Because the edges of the curved segments cause some distortion, Fresnel lenses are generally used only in situations in which image quality is less important than reduction of weight. A classroom overhead projector often uses a Fresnel lens; the circular edges between segments of the lens can be seen by looking closely at the light projected onto a screen.

Quick Quiz 36.6 What is the focal length of a pane of window glass? (a) zero (b) infinity (c) the thickness of the glass (d) impossible to determine

Figure 36.27 The Fresnel lens on the left has the same focal length as the thick lens on the right but is made of much less glass.

EXAMPLE 36.8   **Images Formed by a Converging Lens**

A converging lens has a focal length of 10.0 cm.

(A) An object is placed 30.0 cm from the lens. Construct a ray diagram, find the image distance, and describe the image.

SOLUTION

Conceptualize Because the lens is converging, the focal length is positive (see Table 36.3). We expect the possibilities of both real and virtual images.

Categorize Because the object distance is larger than the focal length, we expect the image to be real. The ray diagram for this situation is shown in Figure 36.28a.

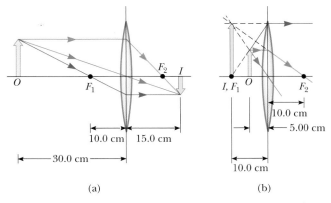

Figure 36.28 (Example 36.8) An image is formed by a converging lens. (a) The object is farther from the lens than the focal point. (b) The object is closer to the lens than the focal point.

Analyze Find the image distance by using Equation 36.16:

$$\frac{1}{q} = \frac{1}{f} - \frac{1}{p}$$

$$\frac{1}{q} = \frac{1}{10.0 \text{ cm}} - \frac{1}{30.0 \text{ cm}}$$

$$q = \boxed{+15.0 \text{ cm}}$$

Find the magnification of the image from Equation 36.17:

$$M = -\frac{q}{p} = -\frac{15.0 \text{ cm}}{30.0 \text{ cm}} = \boxed{-0.500}$$

Finalize The positive sign for the image distance tells us that the image is indeed real and on the back side of the lens. The magnification of the image tells us that the image is reduced in height by one half, and the negative sign for *M* tells us that the image is inverted.

(B) An object is placed 10.0 cm from the lens. Find the image distance and describe the image.

SOLUTION

Categorize Because the object is at the focal point, we expect the image to be infinitely far away.

Analyze Find the image distance by using Equation 36.16:

$$\frac{1}{q} = \frac{1}{f} - \frac{1}{p}$$

$$\frac{1}{q} = \frac{1}{10.0 \text{ cm}} - \frac{1}{10.0 \text{ cm}}$$

$$q = \boxed{\infty}$$

Finalize This result means that rays originating from an object positioned at the focal point of a lens are refracted so that the image is formed at an infinite distance from the lens; that is, the rays travel parallel to one another after refraction.

(C) An object is placed 5.00 cm from the lens. Construct a ray diagram, find the image distance, and describe the image.

SOLUTION

Categorize Because the object distance is smaller than the focal length, we expect the image to be virtual. The ray diagram for this situation is shown in Figure 36.28b.

Analyze Find the image distance by using Equation 36.16:

$$\frac{1}{q} = \frac{1}{f} - \frac{1}{p}$$

$$\frac{1}{q} = \frac{1}{10.0 \text{ cm}} - \frac{1}{5.00 \text{ cm}}$$

$$q = \boxed{-10.0 \text{ cm}}$$

Find the magnification of the image from Equation 36.17:

$$M = -\frac{q}{p} = -\left(\frac{-10.0 \text{ cm}}{5.00 \text{ cm}}\right) = \boxed{+2.00}$$

Finalize The negative image distance tells us that the image is virtual and formed on the side of the lens from which the light is incident, the front side. The image is enlarged, and the positive sign for M tells us that the image is upright.

What If? What if the object moves right up to the lens surface, so that $p \to 0$? Where is the image?

Answer In this case, because $p \ll R$, where R is either of the radii of the surfaces of the lens, the curvature of the lens can be ignored. The lens should appear to have the same effect as a flat piece of material, which suggests that the image is just on the front side of the lens, at $q = 0$. This conclusion can be verified mathematically by rearranging the thin lens equation:

$$\frac{1}{q} = \frac{1}{f} - \frac{1}{p}$$

If we let $p \to 0$, the second term on the right becomes very large compared with the first and we can neglect $1/f$. The equation becomes

$$\frac{1}{q} = -\frac{1}{p} \quad \to \quad q = -p = 0$$

Therefore, q is on the front side of the lens (because it has the opposite sign as p) and right at the lens surface.

EXAMPLE 36.9 **Images Formed by a Diverging Lens**

A diverging lens has a focal length of 10.0 cm.

(A) An object is placed 30.0 cm from the lens. Construct a ray diagram, find the image distance, and describe the image.

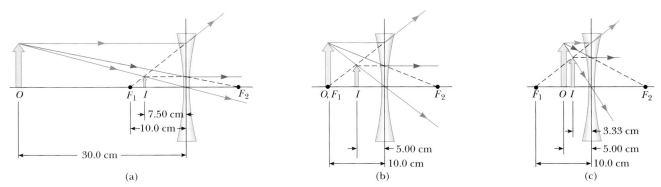

Figure 36.29 (Example 36.9) An image is formed by a diverging lens. (a) The object is farther from the lens than the focal point. (b) The object is at the focal point. (c) The object is closer to the lens than the focal point.

SOLUTION

Conceptualize Because the lens is diverging, the focal length is negative (see Table 36.3). The ray diagram for this situation is shown in Figure 36.29a.

Categorize Because the lens is diverging, we expect it to form an upright, reduced, virtual image for any object position.

Analyze Find the image distance by using Equation 36.16:

$$\frac{1}{q} = \frac{1}{f} - \frac{1}{p}$$

$$\frac{1}{q} = \frac{1}{-10.0 \text{ cm}} - \frac{1}{30.0 \text{ cm}}$$

$$q = \boxed{-7.50 \text{ cm}}$$

Find the magnification of the image from Equation 36.17:

$$M = -\frac{q}{p} = -\left(\frac{-7.50 \text{ cm}}{30.0 \text{ cm}}\right) = \boxed{+0.250}$$

Finalize This result confirms that the image is virtual, smaller than the object, and upright. Look through the diverging lens in a door peephole to see this type of image.

(B) An object is placed 10.0 cm from the lens. Construct a ray diagram, find the image distance, and describe the image.

SOLUTION

The ray diagram for this situation is shown in Figure 36.29b.

Analyze Find the image distance by using Equation 36.16:

$$\frac{1}{q} = \frac{1}{f} - \frac{1}{p}$$

$$\frac{1}{q} = \frac{1}{-10.0 \text{ cm}} - \frac{1}{10.0 \text{ cm}}$$

$$q = \boxed{-5.00 \text{ cm}}$$

Find the magnification of the image from Equation 36.17:

$$M = -\frac{q}{p} = -\left(\frac{-5.00 \text{ cm}}{10.0 \text{ cm}}\right) = \boxed{+0.500}$$

Finalize Notice the difference between this situation and that for a converging lens. For a diverging lens, an object at the focal point does not produce an image infinitely far away.

(C) An object is placed 5.00 cm from the lens. Construct a ray diagram, find the image distance, and describe the image.

SOLUTION

The ray diagram for this situation is shown in Figure 36.29c.

Analyze Find the image distance by using Equation 36.16:

$$\frac{1}{q} = \frac{1}{f} - \frac{1}{p}$$

$$\frac{1}{q} = \frac{1}{-10.0 \text{ cm}} - \frac{1}{5.00 \text{ cm}}$$

$$q = -3.33 \text{ cm}$$

Find the magnification of the image from Equation 36.17:

$$M = -\left(\frac{-3.33 \text{ cm}}{5.00 \text{ cm}}\right) = +0.667$$

Finalize For all three object positions, the image position is negative and the magnification is a positive number smaller than 1, which confirms that the image is virtual, smaller than the object, and upright.

Combination of Thin Lenses

If two thin lenses are used to form an image, the system can be treated in the following manner. First, the image formed by the first lens is located as if the second lens were not present. Then a ray diagram is drawn for the second lens, with the image formed by the first lens now serving as the object for the second lens. The second image formed is the final image of the system. If the image formed by the first lens lies on the back side of the second lens, that image is treated as a **virtual object** for the second lens (that is, in the thin lens equation, p is negative). The same procedure can be extended to a system of three or more lenses. Because the magnification due to the second lens is performed on the magnified image due to the first lens, the **overall magnification of the image due to the combination of lenses is the product of the individual magnifications:**

$$M = M_1 M_2 \tag{36.18}$$

This equation can be used for combinations of any optical elements such as a lens and a mirror. For more than two optical elements, the magnifications due to all elements are multiplied together.

Let's consider the special case of a system of two lenses of focal lengths f_1 and f_2 in contact with each other. If $p_1 = p$ is the object distance for the combination, application of the thin lens equation (Eq. 36.16) to the first lens gives

$$\frac{1}{p} + \frac{1}{q_1} = \frac{1}{f_1}$$

where q_1 is the image distance for the first lens. Treating this image as the object for the second lens, we see that the object distance for the second lens must be $p_2 = -q_1$. (The distances are the same because the lenses are in contact and assumed to be infinitesimally thin. The object distance is negative because the object is virtual.) Therefore, for the second lens,

$$\frac{1}{p_2} + \frac{1}{q_2} = \frac{1}{f_2} \quad \rightarrow \quad -\frac{1}{q_1} + \frac{1}{q} = \frac{1}{f_2}$$

where $q = q_2$ is the final image distance from the second lens, which is the image distance for the combination. Adding the equations for the two lenses eliminates q_1 and gives

$$\frac{1}{p} + \frac{1}{q} = \frac{1}{f_1} + \frac{1}{f_2}$$

If the combination is replaced with a single lens that forms an image at the same location, its focal length must be related to the individual focal lengths by the expression

$$\frac{1}{f} = \frac{1}{f_1} + \frac{1}{f_2}$$

(36.19)   ◀ Focal length for a combination of two thin lenses in contact

Therefore, **two thin lenses in contact with each other are equivalent to a single thin lens having a focal length given by Equation 36.19.**

EXAMPLE 36.10   Where Is the Final Image?

Two thin converging lenses of focal lengths $f_1 = 10.0$ cm and $f_2 = 20.0$ cm are separated by 20.0 cm as illustrated in Figure 36.30. An object is placed 30.0 cm to the left of lens 1. Find the position and the magnification of the final image.

Figure 36.30   (Example 36.10) A combination of two converging lenses. The ray diagram shows the location of the final image due to the combination of lenses. The black dots are the focal points of lens 1 and the red dots are the focal points of lens 2.

SOLUTION

Conceptualize   Imagine light rays passing through the first lens and forming a real image (because $p > f$) in the absence of a second lens. Figure 36.30 shows these light rays forming the inverted image I_1. Once the light rays converge to the image point, they do not stop. They continue through the image point and interact with the second lens. The rays leaving the image point behave in the same way as the rays leaving an object. Therefore, the image of the first lens serves as the object of the second lens.

Categorize   We categorize this problem as one in which the thin lens equation is applied in a stepwise fashion to the two lenses.

Analyze   Find the location of the image formed by lens 1 from the thin lens equation:

$$\frac{1}{q_1} = \frac{1}{f} - \frac{1}{p_1}$$

$$\frac{1}{q_1} = \frac{1}{10.0\text{ cm}} - \frac{1}{30.0\text{ cm}}$$

$$q_1 = +15.0\text{ cm}$$

Find the magnification of the image from Equation 36.17:

$$M_1 = -\frac{q_1}{p_1} = -\frac{15.0\text{ cm}}{30.0\text{ cm}} = -0.500$$

The image formed by this lens acts as the object for the second lens. Therefore, the object distance for the second lens is 20.0 cm − 15.0 cm = 5.00 cm.

Find the location of the image formed by lens 2 from the thin lens equation:

$$\frac{1}{q_2} = \frac{1}{20.0\text{ cm}} - \frac{1}{5.00\text{ cm}}$$

$$q_2 = \boxed{-6.67\text{ cm}}$$

Find the magnification of the image from Equation 36.17:

$$M_2 = -\frac{q_2}{p_2} = -\frac{(-6.67\text{ cm})}{5.00\text{ cm}} = +1.33$$

Find the overall magnification of the system from Equation 36.18:

$$M = M_1 M_2 = (-0.500)(1.33) = \boxed{-0.667}$$

Finalize   The negative sign on the overall magnification indicates that the final image is inverted with respect to the initial object. Because the absolute value of the magnification is less than 1, the final image is smaller than the object. Because q_2 is negative, the final image is on the front, or left, side of lens 2. These conclusions are consistent with the ray diagram in Figure 36.30.

What If?   Suppose you want to create an upright image with this system of two lenses. How must the second lens be moved?

Answer   Because the object is farther from the first lens than the focal length of that lens, the first image is inverted. Consequently, the second lens must invert the image once again so that the final image is upright. An inverted image is only formed by a converging lens if the object is outside the focal point. Therefore, the image formed by the first lens must be to the left of the focal point of the second lens in Figure 36.30. To make that happen, you must move the second lens at least as far away from the first lens as the sum $q_1 + f_2 = 15.0 \text{ cm} + 20.0 \text{ cm} = 35.0$ cm.

36.5   Lens Aberrations

Our analysis of mirrors and lenses assumes rays make small angles with the principal axis and the lenses are thin. In this simple model, all rays leaving a point source focus at a single point, producing a sharp image. Clearly, that is not always true. When the approximations used in this analysis do not hold, imperfect images are formed.

A precise analysis of image formation requires tracing each ray, using Snell's law at each refracting surface and the law of reflection at each reflecting surface. This procedure shows that the rays from a point object do not focus at a single point, with the result that the image is blurred. The departures of actual images from the ideal predicted by our simplified model are called **aberrations.**

Spherical Aberration

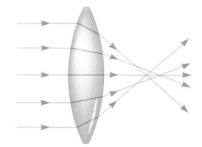

Figure 36.31   Spherical aberration caused by a converging lens. Does a diverging lens cause spherical aberration?

Spherical aberration occurs because the focal points of rays far from the principal axis of a spherical lens (or mirror) are different from the focal points of rays of the same wavelength passing near the axis. Figure 36.31 illustrates spherical aberration for parallel rays passing through a converging lens. Rays passing through points near the center of the lens are imaged farther from the lens than rays passing through points near the edges. Figure 36.8 earlier in the chapter showed a similar situation for a spherical mirror.

Many cameras have an adjustable aperture to control light intensity and reduce spherical aberration. (An aperture is an opening that controls the amount of light passing through the lens.) Sharper images are produced as the aperture size is reduced; with a small aperture, only the central portion of the lens is exposed to the light and therefore a greater percentage of the rays are paraxial. At the same time, however, less light passes through the lens. To compensate for this lower light intensity, a longer exposure time is used.

In the case of mirrors, spherical aberration can be minimized through the use of a parabolic reflecting surface rather than a spherical surface. Parabolic surfaces are not used often, however, because those with high-quality optics are very expensive to make. Parallel light rays incident on a parabolic surface focus at a common point, regardless of their distance from the principal axis. Parabolic reflecting surfaces are used in many astronomical telescopes to enhance image quality.

Chromatic Aberration

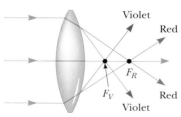

Figure 36.32   Chromatic aberration caused by a converging lens. Rays of different wavelengths focus at different points.

In Chapter 35, we described dispersion, whereby a material's index of refraction varies with wavelength. Because of this phenomenon, violet rays are refracted more than red rays when white light passes through a lens (Fig. 36.32). The figure shows that the focal length of a lens is greater for red light than for violet light. Other wavelengths (not shown in Fig. 36.32) have focal points intermediate between those of red and violet, which causes a blurred image and is called **chromatic aberration.**

Chromatic aberration for a diverging lens also results in a shorter focal length for violet light than for red light, but on the front side of the lens. Chromatic aberration can be greatly reduced by combining a converging lens made of one type of glass and a diverging lens made of another type of glass.

36.6   The Camera

The photographic **camera** is a simple optical instrument whose essential features are shown in Figure 36.33. It consists of a lighttight chamber, a converging lens that produces a real image, and a film behind the lens to receive the image.

Digital cameras are similar to film cameras except that the light does not form an image on photographic film. The image in a digital camera is formed on a *charge-coupled device* (CCD), which digitizes the image, turning it into binary code as we discussed for sound in Section 17.5. (A CCD is described in Section 40.2.) The digital information is then stored on a memory chip for playback on the camera's display screen, or it can be downloaded to a computer. In the discussion that follows, we assume the camera is digital.

A camera is focused by varying the distance between the lens and the CCD. For proper focusing—which is necessary for the formation of sharp images—the lens-to-CCD distance depends on the object distance as well as the focal length of the lens.

The shutter, positioned behind the lens, is a mechanical device that is opened for selected time intervals, called *exposure times.* You can photograph moving objects by using short exposure times or photograph dark scenes (with low light levels) by using long exposure times. If this adjustment were not available, it would be impossible to take stop-action photographs. For example, a rapidly moving vehicle could move enough in the time interval during which the shutter is open to produce a blurred image. Another major cause of blurred images is the movement of the camera while the shutter is open. To prevent such movement, either short exposure times or a tripod should be used, even for stationary objects. Typical shutter speeds (that is, exposure times) are $\frac{1}{30}$ s, $\frac{1}{60}$ s, $\frac{1}{125}$ s, and $\frac{1}{250}$ s. In practice, stationary objects are normally shot with an intermediate shutter speed of $\frac{1}{60}$ s.

The intensity I of the light reaching the CCD is proportional to the area of the lens. Because this area is proportional to the square of the diameter D, it follows that I is also proportional to D^2. Light intensity is a measure of the rate at which energy is received by the CCD per unit area of the image. Because the area of the image is proportional to q^2 and $q \approx f$ (when $p \gg f$, so p can be approximated as infinite), we conclude that the intensity is also proportional to $1/f^2$ and therefore that $I \propto D^2/f^2$.

The ratio f/D is called the ***f*-number** of a lens:

$$f\text{-number} \equiv \frac{f}{D} \tag{36.20}$$

Hence, the intensity of light incident on the CCD varies according to the following proportionality:

$$I \propto \frac{1}{(f/D)^2} \propto \frac{1}{(f\text{-number})^2} \tag{36.21}$$

The *f*-number is often given as a description of the lens's "speed." The lower the *f*-number, the wider the aperture and the higher the rate at which energy from the light exposes the CCD; therefore, a lens with a low *f*-number is a "fast" lens. The conventional notation for an *f*-number is "*f*/" followed by the actual number. For

Figure 36.33   Cross-sectional view of a simple digital camera. The CCD is the light-sensitive component of the camera. In a nondigital camera, the light from the lens falls onto photographic film. In reality, $p \gg q$.

example, "$f/4$" means an f-number of 4; it *does not* mean to divide f by 4! Extremely fast lenses, which have f-numbers as low as approximately $f/1.2$, are expensive because it is very difficult to keep aberrations acceptably small with light rays passing through a large area of the lens. Camera lens systems (that is, combinations of lenses with adjustable apertures) are often marked with multiple f-numbers, usually $f/2.8$, $f/4$, $f/5.6$, $f/8$, $f/11$, and $f/16$. Any one of these settings can be selected by adjusting the aperture, which changes the value of D. Increasing the setting from one f-number to the next higher value (for example, from $f/2.8$ to $f/4$) decreases the area of the aperture by a factor of 2. The lowest f-number setting on a camera lens corresponds to a wide-open aperture and the use of the maximum possible lens area.

Simple cameras usually have a fixed focal length and a fixed aperture size, with an f-number of about $f/11$. This high value for the f-number allows for a large **depth of field,** meaning that objects at a wide range of distances from the lens form reasonably sharp images on the CCD. In other words, the camera does not have to be focused.

Quick Quiz 36.7 A camera can be modeled as a simple converging lens that focuses an image on the CCD, acting as the screen. A camera is initially focused on a distant object. To focus the image of an object close to the camera, must the lens be (a) moved away from the CCD, (b) left where it is, or (c) moved toward the CCD?

EXAMPLE 36.11 **Finding the Correct Exposure Time**

The lens of a digital camera has a focal length of 55 mm and a speed (an f-number) of $f/1.8$. The correct exposure time for this speed under certain conditions is known to be $\frac{1}{500}$ s.

(A) Determine the diameter of the lens.

SOLUTION

Conceptualize Remember that the f-number for a lens relates its focal length to its diameter.

Categorize We evaluate results using equations developed in this section, so we categorize this example as a substitution problem.

Solve Equation 36.20 for D and substitute numerical values:

$$D = \frac{f}{f\text{-number}} = \frac{55 \text{ mm}}{1.8} = 31 \text{ mm}$$

(B) Calculate the correct exposure time if the f-number is changed to $f/4$ under the same lighting conditions.

SOLUTION

The total light energy hitting the CCD is proportional to the product of the intensity and the exposure time. If I is the light intensity reaching the CCD, the energy per unit area received by the CCD in a time interval Δt is proportional to $I \Delta t$. Comparing the two situations, we require that $I_1 \Delta t_1 = I_2 \Delta t_2$, where Δt_1 is the correct exposure time for $f/1.8$ and Δt_2 is the correct exposure time for $f/4$.

Use this result and substitute for I from Equation 36.21:

$$I_1 \Delta t_1 = I_2 \Delta t_2 \quad \rightarrow \quad \frac{\Delta t_1}{(f_1\text{-number})^2} = \frac{\Delta t_2}{(f_2\text{-number})^2}$$

Solve for Δt_2 and substitute numerical values:

$$\Delta t_2 = \left(\frac{f_2\text{-number}}{f_1\text{-number}}\right)^2 \Delta t_1 = \left(\frac{4}{1.8}\right)^2 \left(\tfrac{1}{500} \text{ s}\right) \approx \tfrac{1}{100} \text{ s}$$

As the aperture size is reduced, the exposure time must increase.

36.7   The Eye

Like a camera, a normal eye focuses light and produces a sharp image. The mechanisms by which the eye controls the amount of light admitted and adjusts to produce correctly focused images, however, are far more complex, intricate, and effective than those in even the most sophisticated camera. In all respects, the eye is a physiological wonder.

Figure 36.34 shows the basic parts of the human eye. Light entering the eye passes through a transparent structure called the *cornea* (Fig. 36.35), behind which are a clear liquid (the *aqueous humor*), a variable aperture (the *pupil*, which is an opening in the *iris*), and the *crystalline lens*. Most of the refraction occurs at the outer surface of the eye, where the cornea is covered with a film of tears. Relatively little refraction occurs in the crystalline lens because the aqueous humor in contact with the lens has an average index of refraction close to that of the lens. The iris, which is the colored portion of the eye, is a muscular diaphragm that controls pupil size. The iris regulates the amount of light entering the eye by dilating, or opening, the pupil in low-light conditions and contracting, or closing, the pupil in high-light conditions. The *f*-number range of the human eye is approximately $f/2.8$ to $f/16$.

The cornea–lens system focuses light onto the back surface of the eye, the *retina,* which consists of millions of sensitive receptors called *rods* and *cones.* When stimulated by light, these receptors send impulses via the optic nerve to the brain, where an image is perceived. By this process, a distinct image of an object is observed when the image falls on the retina.

The eye focuses on an object by varying the shape of the pliable crystalline lens through a process called **accommodation.** The lens adjustments take place so swiftly that we are not even aware of the change. Accommodation is limited in that objects very close to the eye produce blurred images. The **near point** is the closest distance for which the lens can accommodate to focus light on the retina. This distance usually increases with age and has an average value of 25 cm. At age 10, the near point of the eye is typically approximately 18 cm. It increases to approximately 25 cm at age 20, to 50 cm at age 40, and to 500 cm or greater at age 60. The **far point** of the eye represents the greatest distance for which the lens of the relaxed eye can focus light on the retina. A person with normal vision can see very distant objects and therefore has a far point that can be approximated as infinity.

Recall that the light leaving the mirror in Figure 36.7 becomes white where it comes together but then diverges into separate colors again. Because nothing but air exists at the point where the rays cross (and hence nothing exists to cause the colors to separate again), seeing white light as a result of a combination of colors must be a visual illusion. In fact, that is the case. Only three types of color-sensitive

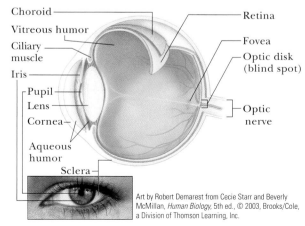

Art by Robert Demarest from Cecie Starr and Beverly McMillan, *Human Biology,* 5th ed., © 2003, Brooks/Cole, a Division of Thomson Learning, Inc.

Figure 36.34   Important parts of the eye.

From Lennart Nilsson, in collaboration with Jan Lindberg. *Behold Man: A Photographic Journey of Discovery Inside the Body,* Boston, Little, Brown & Co., 1974.

Figure 36.35   Close-up photograph of the cornea of the human eye.

Figure 36.36 Approximate color sensitivity of the three types of cones in the retina.

cells are present in the retina. They are called red, green, and blue cones because of the peaks of the color ranges to which they respond (Fig. 36.36). If the red and green cones are stimulated simultaneously (as would be the case if yellow light were shining on them), the brain interprets what is seen as yellow. If all three types of cones are stimulated by the separate colors red, blue, and green as in Figure 36.7, white light is seen. If all three types of cones are stimulated by light that contains *all* colors, such as sunlight, again white light is seen.

Color televisions take advantage of this visual illusion by having only red, green, and blue dots on the screen. With specific combinations of brightness in these three primary colors, our eyes can be made to see any color in the rainbow. Therefore, the yellow lemon you see in a television commercial is not actually yellow, it is red and green! The paper on which this page is printed is made of tiny, matted, translucent fibers that scatter light in all directions, and the resultant mixture of colors appears white to the eye. Snow, clouds, and white hair are not actually white. In fact, there is no such thing as a white pigment. The appearance of these things is a consequence of the scattering of light containing all colors, which we interpret as white.

Conditions of the Eye

When the eye suffers a mismatch between the focusing range of the lens–cornea system and the length of the eye, with the result that light rays from a near object reach the retina before they converge to form an image as shown in Figure 36.37a, the condition is known as **farsightedness** (or *hyperopia*). A farsighted person can usually see faraway objects clearly but not nearby objects. Although the near point of a normal eye is approximately 25 cm, the near point of a farsighted person is much farther away. The refracting power in the cornea and lens is insufficient to focus the light from all but distant objects satisfactorily. The condition can be corrected by placing a converging lens in front of the eye as shown in Figure 36.37b. The lens refracts the incoming rays more toward the principal axis before entering the eye, allowing them to converge and focus on the retina.

A person with **nearsightedness** (or *myopia*), another mismatch condition, can focus on nearby objects but not on faraway objects. The far point of the nearsighted eye is not infinity and may be less than 1 m. The maximum focal length of the nearsighted eye is insufficient to produce a sharp image on the retina, and rays from a distant object converge to a focus in front of the retina. They then continue past that point, diverging before they finally reach the retina and causing blurred vision (Fig. 36.38a). Nearsightedness can be corrected with a diverging lens as shown in Figure 36.38b. The lens refracts the rays away from the principal axis before they enter the eye, allowing them to focus on the retina.

Beginning in middle age, most people lose some of their accommodation ability as their visual muscles weaken and the lens hardens. Unlike farsightedness, which is a mismatch between focusing power and eye length, **presbyopia** (literally, "old-age vision") is due to a reduction in accommodation ability. The cornea and

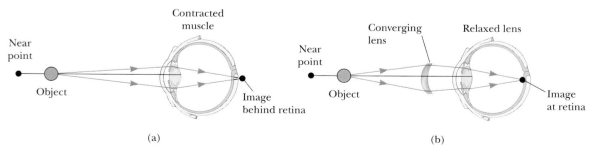

(a)   (b)

Figure 36.37 (a) When a farsighted eye looks at an object located between the near point and the eye, the image point is behind the retina, resulting in blurred vision. The eye muscle contracts to try to bring the object into focus. (b) Farsightedness is corrected with a converging lens.

Figure 36.38 (a) When a nearsighted eye looks at an object that lies beyond the eye's far point, the image is formed in front of the retina, resulting in blurred vision. (b) Nearsightedness is corrected with a diverging lens.

lens do not have sufficient focusing power to bring nearby objects into focus on the retina. The symptoms are the same as those of farsightedness, and the condition can be corrected with converging lenses.

In eyes having a defect known as **astigmatism,** light from a point source produces a line image on the retina. This condition arises when either the cornea, the lens, or both are not perfectly symmetric. Astigmatism can be corrected with lenses that have different curvatures in two mutually perpendicular directions.

Optometrists and ophthalmologists usually prescribe lenses[1] measured in **diopters:** the **power** P of a lens in diopters equals the inverse of the focal length in meters: $P = 1/f$. For example, a converging lens of focal length $+20$ cm has a power of $+5.0$ diopters, and a diverging lens of focal length -40 cm has a power of -2.5 diopters.

Quick Quiz 36.8 Two campers wish to start a fire during the day. One camper is nearsighted, and one is farsighted. Whose glasses should be used to focus the Sun's rays onto some paper to start the fire? (a) either camper (b) the nearsighted camper (c) the farsighted camper

---

36.8 The Simple Magnifier

The simple magnifier, or magnifying glass, consists of a single converging lens. This device increases the apparent size of an object.

Suppose an object is viewed at some distance p from the eye as illustrated in Figure 36.39. The size of the image formed at the retina depends on the angle θ subtended by the object at the eye. As the object moves closer to the eye, θ increases and a larger image is observed. An average normal human eye, however, cannot focus on an object closer than about 25 cm, the near point (Fig. 36.40a, page 1036). Therefore, θ is maximum at the near point.

To further increase the apparent angular size of an object, a converging lens can be placed in front of the eye as in Figure 36.40b, with the object located at point O, immediately inside the focal point of the lens. At this location, the lens forms a virtual, upright, enlarged image. We define **angular magnification** m as the ratio of the angle subtended by an object with a lens in use (angle θ in Fig. 36.40b) to the angle subtended by the object placed at the near point with no lens in use (angle θ_0 in Fig. 36.40a):

$$m \equiv \frac{\theta}{\theta_0} \qquad (36.22)$$

Figure 36.39 The size of the image formed on the retina depends on the angle θ subtended at the eye.

---

[1] The word *lens* comes from *lentil,* the name of an Italian legume. (You may have eaten lentil soup.) Early eyeglasses were called "glass lentils" because the biconvex shape of their lenses resembled the shape of a lentil. The first lenses for farsightedness and presbyopia appeared around 1280; concave eyeglasses for correcting nearsightedness did not appear until more than 100 years later.

Figure 36.40 (a) An object placed at the near point of the eye ($p = 25$ cm) subtends an angle $\theta_0 \approx h/25$ at the eye. (b) An object placed near the focal point of a converging lens produces a magnified image that subtends an angle $\theta \approx h'/25$ at the eye.

(a)   (b)

The angular magnification is a maximum when the image is at the near point of the eye, that is, when $q = -25$ cm. The object distance corresponding to this image distance can be calculated from the thin lens equation:

$$\frac{1}{p} + \frac{1}{-25 \text{ cm}} = \frac{1}{f} \quad \rightarrow \quad p = \frac{25f}{25 + f}$$

where f is the focal length of the magnifier in centimeters. If we make the small-angle approximations

$$\tan \theta_0 \approx \theta_0 \approx \frac{h}{25} \quad \text{and} \quad \tan \theta \approx \theta \approx \frac{h}{p} \tag{36.23}$$

Equation 36.22 becomes

$$m_{\text{max}} = \frac{\theta}{\theta_0} = \frac{h/p}{h/25} = \frac{25}{p} = \frac{25}{25f/(25 + f)}$$

$$m_{\text{max}} = 1 + \frac{25 \text{ cm}}{f} \tag{36.24}$$

Although the eye can focus on an image formed anywhere between the near point and infinity, it is most relaxed when the image is at infinity. For the image formed by the magnifying lens to appear at infinity, the object has to be at the focal point of the lens. In this case, Equations 36.23 become

$$\theta_0 \approx \frac{h}{25} \quad \text{and} \quad \theta \approx \frac{h}{f}$$

and the magnification is

$$m_{\text{min}} = \frac{\theta}{\theta_0} = \frac{25 \text{ cm}}{f} \tag{36.25}$$

George Semple

A simple magnifier, also called a magnifying glass, is used to view an enlarged image of a portion of a map.

With a single lens, it is possible to obtain angular magnifications up to about 4 without serious aberrations. Magnifications up to about 20 can be achieved by using one or two additional lenses to correct for aberrations.

EXAMPLE 36.12   **Magnification of a Lens**

What is the maximum magnification that is possible with a lens having a focal length of 10 cm, and what is the magnification of this lens when the eye is relaxed?

SOLUTION

Conceptualize   Study Figure 36.40b for the situation in which a magnifying glass forms an enlarged image of an object placed inside the focal point. The maximum magnification occurs when the image is located at the near point of the eye. When the eye is relaxed, the image is at infinity.

Categorize   We evaluate results using equations developed in this section, so we categorize this example as a substitution problem.

Evaluate the maximum magnification from Equation 36.24:

$$m_{max} = 1 + \frac{25 \text{ cm}}{f} = 1 + \frac{25 \text{ cm}}{10 \text{ cm}} = \boxed{3.5}$$

Evaluate the minimum magnification, when the eye is relaxed, from Equation 36.25:

$$m_{min} = \frac{25 \text{ cm}}{f} = \frac{25 \text{ cm}}{10 \text{ cm}} = \boxed{2.5}$$

36.9   The Compound Microscope

A simple magnifier provides only limited assistance in inspecting minute details of an object. Greater magnification can be achieved by combining two lenses in a device called a **compound microscope** shown in Active Figure 36.41a. It consists of one lens, the *objective*, that has a very short focal length $f_o < 1$ cm and a second lens, the *eyepiece*, that has a focal length f_e of a few centimeters. The two lenses are separated by a distance L that is much greater than either f_o or f_e. The object, which is placed just outside the focal point of the objective, forms a real, inverted image at I_1, and this image is located at or close to the focal point of the eyepiece. The eyepiece, which serves as a simple magnifier, produces at I_2 a virtual, enlarged image of I_1. The lateral magnification M_1 of the first image is $-q_1/p_1$. Notice from Active Figure 36.41a that q_1 is approximately equal to L and that the object is very close to the focal point of the objective: $p_1 \approx f_o$. Therefore, the lateral magnification by the objective is

$$M_o \approx -\frac{L}{f_o}$$

The angular magnification by the eyepiece for an object (corresponding to the image at I_1) placed at the focal point of the eyepiece is, from Equation 36.25,

$$m_e = \frac{25 \text{ cm}}{f_e}$$

The overall magnification of the image formed by a compound microscope is defined as the product of the lateral and angular magnifications:

$$M = M_o m_e = -\frac{L}{f_o}\left(\frac{25 \text{ cm}}{f_e}\right) \qquad \textbf{(36.26)}$$

The negative sign indicates that the image is inverted.

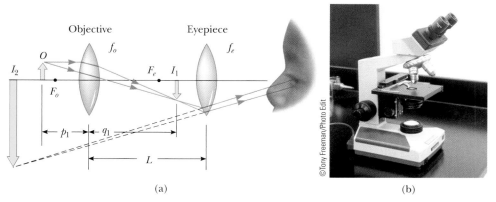

ACTIVE FIGURE 36.41

(a) Diagram of a compound microscope, which consists of an objective lens and an eyepiece lens. (b) A compound microscope. The three-objective turret allows the user to choose from several powers of magnification. Combinations of eyepieces with different focal lengths and different objectives can produce a wide range of magnifications.

Sign in at www.thomsonedu.com and go to ThomsonNOW to adjust the focal lengths of the objective and eyepiece lenses and see the effect on the final image.

The microscope has extended human vision to the point where we can view previously unknown details of incredibly small objects. The capabilities of this instrument have steadily increased with improved techniques for precision grinding of lenses. A question often asked about microscopes is, "If one were extremely patient and careful, would it be possible to construct a microscope that would enable the human eye to see an atom?" The answer is no, as long as light is used to illuminate the object. For an object under an optical microscope (one that uses visible light) to be seen, the object must be at least as large as a wavelength of light. Because the diameter of any atom is many times smaller than the wavelengths of visible light, the mysteries of the atom must be probed using other types of "microscopes."

36.10   The Telescope

Two fundamentally different types of **telescopes** exist; both are designed to aid in viewing distant objects, such as the planets in our solar system. The **refracting telescope** uses a combination of lenses to form an image, and the **reflecting telescope** uses a curved mirror and a lens.

Like the compound microscope, the refracting telescope shown in Active Figure 36.42a has an objective and an eyepiece. The two lenses are arranged so that the objective forms a real, inverted image of a distant object very near the focal point of the eyepiece. Because the object is essentially at infinity, this point at which I_1 forms is the focal point of the objective. The eyepiece then forms, at I_2, an enlarged, inverted image of the image at I_1. To provide the largest possible magnification, the image distance for the eyepiece is infinite. The light rays exit the eyepiece lens parallel to the principal axis, and the image due to the objective lens must form at the focal point of the eyepiece. Hence, the two lenses are separated by a distance $f_o + f_e$, which corresponds to the length of the telescope tube.

The angular magnification of the telescope is given by θ/θ_o, where θ_o is the angle subtended by the object at the objective and θ is the angle subtended by the final image at the viewer's eye. Consider Active Figure 36.42a, in which the object is a very great distance to the left of the figure. The angle θ_o (to the *left* of the objective) subtended by the object at the objective is the same as the angle (to the *right* of the objective) subtended by the first image at the objective. Therefore,

ACTIVE FIGURE 36.42

(a) Lens arrangement in a refracting telescope, with the object at infinity. (b) A refracting telescope.

Sign in at www.thomsonedu.com and go to ThomsonNOW to adjust the focal lengths of the objective and eyepiece lenses and see the effect on the final image.

$$\tan \theta_o \approx \theta_o \approx -\frac{h'}{f_o}$$

where the negative sign indicates that the image is inverted.

The angle θ subtended by the final image at the eye is the same as the angle that a ray coming from the tip of I_1 and traveling parallel to the principal axis makes with the principal axis after it passes through the lens. Therefore,

$$\tan \theta \approx \theta \approx \frac{h'}{f_e}$$

We have not used a negative sign in this equation because the final image is not inverted; the object creating this final image I_2 is I_1, and both it and I_2 point in the same direction. Therefore, the angular magnification of the telescope can be expressed as

$$m = \frac{\theta}{\theta_o} = \frac{h'/f_e}{-h'/f_o} = -\frac{f_o}{f_e} \tag{36.27}$$

This result shows that the angular magnification of a telescope equals the ratio of the objective focal length to the eyepiece focal length. The negative sign indicates that the image is inverted.

When you look through a telescope at such relatively nearby objects as the Moon and the planets, magnification is important. Individual stars in our galaxy, however, are so far away that they always appear as small points of light no matter how great the magnification. To gather as much light as possible, large research telescopes used to study very distant objects must have a large diameter. It is difficult and expensive to manufacture large lenses for refracting telescopes. Another difficulty with large lenses is that their weight leads to sagging, which is an additional source of aberration.

These problems associated with large lenses can be partially overcome by replacing the objective with a concave mirror, which results in a reflecting telescope. Because light is reflected from the mirror and does not pass through a lens, the mirror can have rigid supports on the back side. Such supports eliminate the problem of sagging.

Figure 36.43a shows the design for a typical reflecting telescope. The incoming light rays are reflected by a parabolic mirror at the base. These reflected rays converge toward point A in the figure, where an image would be formed. Before this image is formed, however, a small, flat mirror M reflects the light toward an opening in the tube's side and it passes into an eyepiece. This particular design is said to have a Newtonian focus because Newton developed it. Figure 36.43b shows such a telescope. Notice that the light never passes through glass (except through the

(a)   (b)

Figure 36.43   (a) A Newtonian-focus reflecting telescope. (b) A reflecting telescope. This type of telescope is shorter than that in Figure 36.42b.

small eyepiece) in the reflecting telescope. As a result, problems associated with chromatic aberration are virtually eliminated. The reflecting telescope can be made even shorter by orienting the flat mirror so that it reflects the light back toward the objective mirror and the light enters an eyepiece in a hole in the middle of the mirror.

The largest reflecting telescopes in the world are at the Keck Observatory on Mauna Kea, Hawaii. The site includes two telescopes with diameters of 10 m, each containing 36 hexagonally shaped, computer-controlled mirrors that work together to form a large reflecting surface. Discussions and plans have been initiated for telescopes with different mirrors working together, as at the Keck Observatory, resulting in an effective diameter up to 21 m. In contrast, the largest refracting telescope in the world, at the Yerkes Observatory in Williams Bay, Wisconsin, has a diameter of only 1 m.

Summary

ThomsonNOW™ Sign in at **www.thomsonedu.com** and go to ThomsonNOW to take a practice test for this chapter.

DEFINITIONS

The **lateral magnification** M of the image due to a mirror or lens is defined as the ratio of the image height h' to the object height h. It is equal to the negative of the ratio of the image distance q to the object distance p:

$$M \equiv \frac{\text{image height}}{\text{object height}} = \frac{h'}{h} = -\frac{q}{p} \qquad (36.1, 36.2, 36.17)$$

The **angular magnification** m is the ratio of the angle subtended by an object with a lens in use (angle θ in Fig. 36.40b) to the angle subtended by the object placed at the near point with no lens in use (angle θ_0 in Fig. 36.40a):

$$m \equiv \frac{\theta}{\theta_0} \qquad (36.22)$$

The ratio of the focal length of a camera lens to the diameter of the lens is called the **f-number** of the lens:

$$f\text{-number} \equiv \frac{f}{D} \qquad (36.20)$$

CONCEPTS AND PRINCIPLES

In the paraxial ray approximation, the object distance p and image distance q for a spherical mirror of radius R are related by the **mirror equation:**

$$\frac{1}{p} + \frac{1}{q} = \frac{2}{R} = \frac{1}{f} \qquad (36.4, 36.6)$$

where $f = R/2$ is the **focal length** of the mirror.

An image can be formed by refraction from a spherical surface of radius R. The object and image distances for refraction from such a surface are related by

$$\frac{n_1}{p} + \frac{n_2}{q} = \frac{n_2 - n_1}{R} \qquad (36.8)$$

where the light is incident in the medium for which the index of refraction is n_1 and is refracted in the medium for which the index of refraction is n_2.

(continued)

The inverse of the **focal length** f of a thin lens surrounded by air is given by the **lens-makers' equation:**

$$\frac{1}{f} = (n - 1)\left(\frac{1}{R_1} - \frac{1}{R_2}\right) \qquad (36.15)$$

Converging lenses have positive focal lengths, and **diverging lenses** have negative focal lengths.

For a thin lens, and in the paraxial ray approximation, the object and image distances are related by the **thin lens equation:**

$$\frac{1}{p} + \frac{1}{q} = \frac{1}{f} \qquad (36.16)$$

The maximum magnification of a single lens of focal length f used as a simple magnifier is

$$m_{\text{max}} = 1 + \frac{25 \text{ cm}}{f} \qquad (36.24)$$

The overall magnification of the image formed by a compound microscope is:

$$M = -\frac{L}{f_o}\left(\frac{25 \text{ cm}}{f_e}\right) \qquad (36.26)$$

where f_o and f_e are the focal lengths of the objective and eyepiece lenses, respectively, and L is the distance between the lenses.

The angular magnification of a refracting telescope can be expressed as

$$m = -\frac{f_o}{f_e} \qquad (36.27)$$

where f_o and f_e are the focal lengths of the objective and eyepiece lenses, respectively. The angular magnification of a reflecting telescope is given by the same expression where f_o is the focal length of the objective mirror.

Questions

☐ denotes answer available in *Student Solutions Manual/Study Guide;* **O** denotes objective question

1. Consider a concave spherical mirror with a real object. Is the image always inverted? Is the image always real? Give conditions for your answers.

2. Repeat Question 1 for a convex spherical mirror.

3. **O** **(i)** What is the focal length of a plane mirror? (a) 0 (b) 1 (c) −1 (d) ∞ (e) equal to the mirror height (f) Neither the focal length nor its reciprocal can be defined. **(ii)** What magnification does a plane mirror produce? (a) 0 (b) 1 (c) −1 (d) ∞ (e) Neither the magnification nor its reciprocal can be defined.

4. Do the equations $1/p + 1/q = 1/f$ and $M = -q/p$ apply to the image formed by a flat mirror? Explain your answer.

5. **O** Lulu looks at her image in a makeup mirror. It is enlarged when she is close to the mirror. As she backs away, the image becomes larger, then impossible to iden-

tify when she is 30 cm from the mirror, then upside down when she is beyond 30 cm, and finally small, clear, and upside down when she is much farther from the mirror. **(i)** Is the mirror (a) convex, (b) plane, or (c) concave? **(ii)** What is the magnitude of its focal length? (a) 0 (b) 15 cm (c) 30 cm (d) 60 cm (e) ∞

6. Consider a spherical concave mirror with the object located to the left of the mirror beyond the focal point. Using ray diagrams, show that the image moves to the left as the object approaches the focal point.

7. **O** **(i)** Consider the mirror in Figure 36.11. What are the signs of the following? (a) the object distance (b) the image distance (c) the mirror radius (d) the focal length (e) the object height (f) the image height (g) the magnification **(ii)** Consider the objective lens in Active Figure 36.41a. What are the signs of the following? (a) the object distance (b) the image distance (c) the focal length (d) the object height (e) the image height (f) the magnification **(iii)** Answer the same questions (a) through (f) as in part (ii) for the eyepiece in Active Figure 36.41a.

8. **O** A person spearfishing from a boat sees a stationary fish a few meters away in a direction about 30° below the horizontal. To spear the fish, should the person (a) aim above where he sees the fish, (b) aim precisely at the fish, or (c) aim below the fish? Assume the dense spear does not change direction when it enters the water.

9. **O** A single converging lens can be used to constitute a scale model of each of the following devices in use simply by changing the distance from the lens to a candle representing the object. Rank the cases according to the distance from the object to the lens from the largest to the smallest. (a) a movie projector (b) Batman's signal, used to project an image on clouds high above Gotham City (c) a magnifying glass (d) a burning glass, used to make a sharp image of the Sun on tinder (e) an astronomical refracting telescope, used to make a sharp image of stars on an electronic detector (f) a searchlight, used to produce a beam of parallel rays from a point source. (g) a camera lens, used to photograph a soccer game.

10. In Active Figure 36.26a, assume the blue object arrow is replaced by one that is much taller than the lens. How many rays from the top of the object will strike the lens? How many principal rays can be drawn in a ray diagram?

11. **O** A converging lens in a vertical plane receives light from an object and forms an inverted image on a screen. An opaque card is then placed next to the lens, covering only the upper half of the lens. What happens to the image on the screen? (a) The upper half of the image disappears. (b) The lower half of the image disappears. (c) The entire image disappears. (d) The entire image is still visible, but is dimmer. (e) Half of the image disappears and the rest is dimmer. (f) No change in the image occurs.

12. **O** A converging lens of focal length 8 cm forms a sharp image of an object on a screen. What is the smallest possible distance between the object and the screen? (a) 0 (b) 4 cm (c) 8 cm (d) 16 cm (e) 32 cm (f) ∞

13. Explain this statement: "The focal point of a lens is the location of the image of a point object at infinity." Discuss the notion of infinity in real terms as it applies to object distances. Based on this statement, can you think of a simple method for determining the focal length of a converging lens?

14. Discuss the proper position of a photographic slide relative to the lens in a slide projector. What type of lens must the slide projector have?

15. **O** In this chapter's opening photograph, a water drop functions as a biconvex lens with radii of curvature of small magnitude. What is the location of the image photographed? (a) inside the water drop (b) on the back sur-

face of the drop, farthest from the camera (c) somewhat beyond the back surface of the drop (d) on the front surface of the drop, closest to the camera (e) somewhat closer to the camera than the front surface of the drop

16. Explain why a mirror cannot give rise to chromatic aberration.

17. Can a converging lens be made to diverge light if it is placed into a liquid? **What If?** What about a converging mirror?

18. Explain why a fish in a spherical goldfish bowl appears larger than it really is.

19. Why do some emergency vehicles have the symbol ƎƆИA⅃UBMA written on the front?

20. Lenses used in eyeglasses, whether converging or diverging, are always designed so that the middle of the lens curves away from the eye like the center lenses of Figures 36.25a and 36.25b. Why?

21. **O** The faceplate of a diving mask can be a corrective lens for a diver who does not have perfect vision and who needs essentially the same prescription for both eyes. Then the diver does not have to wear glasses or contact lenses. The proper design allows the person to see clearly both under water and in the air. Normal eyeglasses have lenses with both the front and back surfaces curved. Should the lens of a diving mask be curved (a) on the outer surface only, (b) on the inner surface only, or (c) on both surfaces?

22. In Figures Q36.22a and Q36.22b, which glasses correct nearsightedness and which correct farsightedness?

(a)   (b)

Figure Q36.22 Questions 22 and 23.

23. A child tries on either his hyperopic grandfather's or his myopic brother's glasses and complains, "Everything looks blurry." Why do the eyes of a person wearing glasses not look blurry? (See Figure Q36.22.)

24. In a Jules Verne novel, a piece of ice is shaped to form a magnifying lens to focus sunlight to start a fire. Is that possible?

25. A solar furnace can be constructed by using a concave mirror to reflect and focus sunlight into a furnace enclosure. What factors in the design of the reflecting mirror would guarantee very high temperatures?

26. Figure Q36.26 shows a lithograph by M. C. Escher titled *Hand with Reflection Sphere (Self-Portrait in Spherical Mirror).* Escher said about the work:

> The picture shows a spherical mirror, resting on a left hand. But as a print is the reverse of the original drawing on stone, it was my right hand that you see depicted. (Being left-handed, I needed my left hand to make the drawing.) Such a globe reflection collects almost one's whole surroundings in one disk-shaped image. The whole room, four walls, the floor, and the ceiling, everything, albeit distorted, is compressed into that one small circle. Your own head, or more exactly the point between your eyes, is the absolute center. No matter how you turn or

twist yourself, you can't get out of that central point. You are immovably the focus, the unshakable core, of your world.

Comment on the accuracy of Escher's description.

Figure Q36.26

27. A converging lens of short focal length can take light diverging from a small source and refract it into a beam of parallel rays. A Fresnel lens as shown in Figure 36.27 is used in a lighthouse for this purpose. A concave mirror can take light diverging from a small source and reflect it into a beam of parallel rays. Is it possible to make a Fresnel mirror? Is this idea original, or has it already been done? *Suggestion:* Look at the walls and ceiling of an auditorium.

Problems

WebAssign The Problems from this chapter may be assigned online in WebAssign.

ThomsonNOW™   Sign in at **www.thomsonedu.com** and go to ThomsonNOW to assess your understanding of this chapter's topics with additional quizzing and conceptual questions.

1, 2, 3 denotes straightforward, intermediate, challenging;   ☐ denotes full solution available in *Student Solutions Manual/Study Guide;*   ▲ denotes coached solution with hints available at **www.thomsonedu.com;**   denotes developing symbolic reasoning;   ● denotes asking for qualitative reasoning;   ▪ denotes computer useful in solving problem

Section 36.1 Images Formed by Flat Mirrors

1. Does your bathroom mirror show you older or younger than you actually are? Compute an order-of-magnitude estimate for the age difference based on data you specify.

2. In a church choir loft, two parallel walls are 5.30 m apart. The singers stand against the north wall. The organist faces the south wall, sitting 0.800 m away from it. To enable her to see the choir, a flat mirror 0.600 m wide is mounted on the south wall, straight in front of her. What width of the north wall can the organist see? *Suggestion:* Draw a top-view diagram to justify your answer.

3. Determine the minimum height of a vertical flat mirror in which a person 5 ft 10 in. in height can see his or her full image. (A ray diagram would be helpful.)

4. A person walks into a room that has two flat mirrors on opposite walls. The mirrors produce multiple images of the person. When the person is 5.00 ft from the mirror on the left wall and 10.0 ft from the mirror on the right wall, find the distance from the person to the first three images seen in the mirror on the left.

2 = intermediate;   3 = challenging;   ☐ = SSM/SG;   ▲ = ThomsonNOW;   = symbolic reasoning;   ● = qualitative reasoning

5. A periscope (Fig. P36.5) is useful for viewing objects that cannot be seen directly. It can be used in submarines and when watching golf matches or parades from behind a crowd of people. Suppose the object is a distance p_1 from the upper mirror and the two flat mirrors are separated by a distance h. (a) What is the distance of the final image from the lower mirror? (b) Is the final image real or virtual? (c) Is it upright or inverted? (d) What is its magnification? (e) Does it appear to be left–right reversed?

Figure P36.5

Section 36.2 Images Formed by Spherical Mirrors

6. A concave spherical mirror has a radius of curvature of 20.0 cm. Find the location of the image for object distances of (a) 40.0 cm, (b) 20.0 cm, and (c) 10.0 cm. For each case, state whether the image is real or virtual and upright or inverted. Find the magnification in each case.

7. ▲ A spherical convex mirror has a radius of curvature with a magnitude of 40.0 cm. Determine the position of the virtual image and the magnification for object distances of (a) 30.0 cm and (b) 60.0 cm. (c) Are the images upright or inverted?

8. At an intersection of hospital hallways, a convex mirror is mounted high on a wall to help people avoid collisions. The magnitude of the mirror's radius of curvature is 0.550 m. Locate and describe the image of a patient 10.0 m from the mirror. Determine the magnification of the image.

9. A concave mirror has a radius of curvature of 60.0 cm. Calculate the image position and magnification of an object placed in front of the mirror at distances of (a) 90.0 cm and (b) 20.0 cm. (c) Draw ray diagrams to obtain the image characteristics in each case.

10. A large church has a niche in one wall. On the floor plan, the niche appears as a semicircular indentation of radius 2.50 m. A worshiper stands on the centerline of the niche, 2.00 m out from its deepest point, and whispers a prayer. Where is the sound concentrated after reflection from the back wall of the niche?

11. A dentist uses a mirror to examine a tooth. The tooth is 1.00 cm in front of the mirror, and the image is formed 10.0 cm behind the mirror. Determine (a) the mirror's radius of curvature and (b) the magnification of the image.

12. A certain Christmas tree ornament is a silver sphere having a diameter of 8.50 cm. Determine an object location for which the size of the reflected image is three-fourths the object's size. Use a principal-ray diagram to arrive at a description of the image.

13. (a) A concave mirror forms an inverted image four times larger than the object. Find the focal length of the mirror, assuming the distance between object and image is 0.600 m. (b) A convex mirror forms a virtual image half the size of the object. Assuming the distance between image and object is 20.0 cm, determine the radius of curvature of the mirror.

14. To fit a contact lens to a patient's eye, a *keratometer* can be used to measure the curvature of the eye's front surface, the cornea. This instrument places an illuminated object of known size at a known distance p from the cornea. The cornea reflects some light from the object, forming an image of the object. The magnification M of the image is measured by using a small viewing telescope that allows comparison of the image formed by the cornea with a second calibrated image projected into the field of view by a prism arrangement. Determine the radius of curvature of the cornea for the case $p = 30.0$ cm and $M = 0.013\ 0$.

15. An object 10.0 cm tall is placed at the zero mark of a meterstick. A spherical mirror located at some point on the meterstick creates an image of the object that is upright, 4.00 cm tall, and located at the 42.0-cm mark of the meterstick. (a) Is the mirror convex or concave? (b) Where is the mirror? (c) What is the mirror's focal length?

16. A dedicated sports car enthusiast polishes the inside and outside surfaces of a hubcap that is a section of a sphere. When she looks into one side of the hubcap, she sees an image of her face 30.0 cm in back of the hubcap. She then flips the hubcap over and sees another image of her face 10.0 cm in back of the hubcap. (a) How far is her face from the hubcap? (b) What is the radius of curvature of the hubcap?

17. A spherical mirror is to be used to form, on a screen located 5.00 m from the object, an image five times the size of the object. (a) Describe the type of mirror

required. (b) Where should the mirror be positioned relative to the object?

18. You unconsciously estimate the distance to an object from the angle it subtends in your field of view. This angle θ in radians is related to the linear height of the object h and to the distance d by $\theta = h/d$. Assume you are driving a car and another car, 1.50 m high, is 24.0 m behind you. (a) Suppose your car has a flat passenger-side rearview mirror, 1.55 m from your eyes. How far from your eyes is the image of the car following you? (b) What angle does the image subtend in your field of view? (c) **What If?** Now suppose your car has a convex rearview mirror with a radius of curvature of magnitude 2.00 m (as suggested in Fig. 36.15). How far from your eyes is the image of the car behind you? (d) What angle does the image subtend at your eyes? (e) Based on its angular size, how far away does the following car appear to be?

19. ● **Review problem.** A ball is dropped at $t = 0$ from rest 3.00 m directly above the vertex of a concave mirror that has a radius of curvature of 1.00 m and lies in a horizontal plane. (a) Describe the motion of the ball's image in the mirror. (b) At what instant or instants do the ball and its image coincide?

Section 36.3 Images Formed by Refraction

20. A flint glass plate ($n = 1.66$) rests on the bottom of an aquarium tank. The plate is 8.00 cm thick (vertical dimension) and is covered with a layer of water ($n = 1.33$) 12.0 cm deep. Calculate the apparent thickness of the plate as viewed from straight above the water.

21. A cubical block of ice 50.0 cm on a side is placed over a speck of dust on a level floor. Find the location of the image of the speck as viewed from above. The index of refraction of ice is 1.309.

22. One end of a long glass rod ($n = 1.50$) is formed into a convex surface with a radius of curvature of 6.00 cm. An object is located in air along the axis of the rod. Find the image positions corresponding to object distances of (a) 20.0 cm, (b) 10.0 cm, and (c) 3.00 cm from the end of the rod.

23. A glass sphere ($n = 1.50$) with a radius of 15.0 cm has a tiny air bubble 5.00 cm above its center. The sphere is viewed looking down along the extended radius containing the bubble. What is the apparent depth of the bubble below the surface of the sphere?

24. Figure P36.24 shows a curved surface separating a material with index of refraction n_1 from a material with index n_2. The surface forms an image I of object O. The ray shown in

blue passes through the surface along a radial line. Its angles of incidence and refraction are both zero, so its direction does not change at the surface. For the ray shown in brown, the direction changes according to $n_1 \sin \theta_1 = n_2 \sin \theta_2$. For paraxial rays, we assume θ_1 and θ_2 are small, so we may write $n_1 \tan \theta_1 = n_2 \tan \theta_2$. The magnification is defined as $M = h'/h$. Prove that the magnification is given by $M = -n_1 q/n_2 p$.

Figure P36.24

25. ● As shown in Figure P36.25, a water tank containing lobsters has a curved front made of plastic with uniform thickness and a radius of curvature of magnitude 80.0 cm. Locate and describe the images of lobsters (a) 30.0 cm and (b) 90.0 cm from the base of the front wall. (c) Find the magnification of each image. You may use the result of Problem 24. (d) The lobsters are both 9.00 cm in height. Find the height of each image. (e) Explain why you do not need to know the index of refraction of the plastic to solve this problem.

Figure P36.25

26. A goldfish is swimming at 2.00 cm/s toward the front wall of a rectangular aquarium. What is the apparent speed of the fish measured by an observer looking in from outside the front wall of the tank? The index of refraction of water is 1.33.

Section 36.4 Thin Lenses

27. ▲ The left face of a biconvex lens has a radius of curvature of magnitude 12.0 cm, and the right face has a radius

2 = intermediate;   3 = challenging;   □ = SSM/SG;   ▲ = ThomsonNOW;   = symbolic reasoning;   ● = qualitative reasoning

of curvature of magnitude 18.0 cm. The index of refraction of the glass is 1.44. (a) Calculate the focal length of the lens. (b) **What If?** Calculate the focal length the lens has after it is turned around to interchange the radii of curvature of the two faces.

28. A contact lens is made of plastic with an index of refraction of 1.50. The lens has an outer radius of curvature of +2.00 cm and an inner radius of curvature of +2.50 cm. What is the focal length of the lens?

29. A converging lens has a focal length of 20.0 cm. Locate the image for object distances of (a) 40.0 cm, (b) 20.0 cm, and (c) 10.0 cm. For each case, state whether the image is real or virtual and upright or inverted. Find the magnification in each case.

30. An object located 32.0 cm in front of a lens forms an image on a screen 8.00 cm behind the lens. (a) Find the focal length of the lens. (b) Determine the magnification. (c) Is the lens converging or diverging?

31. ▲ The nickel's image in Figure P36.31 has twice the diameter of the nickel and is 2.84 cm from the lens. Determine the focal length of the lens.

Figure P36.31

32. Suppose an object has thickness dp so that it extends from object distance p to $p + dp$. Prove that the thickness dq of its image is given by $(-q^2/p^2)\,dp$. Then the longitudinal magnification is $dq/dp = -M^2$, where M is the lateral magnification.

33. An object is located 20.0 cm to the left of a diverging lens having a focal length $f = -32.0$ cm. Determine (a) the location and (b) the magnification of the image. (c) Construct a ray diagram for this arrangement.

34. The projection lens in a certain slide projector is a single thin lens. A slide 24.0 mm high is to be projected so that its image fills a screen 1.80 m high. The slide-to-screen

distance is 3.00 m. (a) Determine the focal length of the projection lens. (b) How far from the slide should the lens of the projector be placed so as to form the image on the screen?

35. The use of a lens in a certain situation is described by the equation

$$\frac{1}{p} + \frac{1}{-3.50p} = \frac{1}{7.50 \text{ cm}}$$

Determine (a) the object distance and (b) the image distance. (c) Use a ray diagram to obtain a description of the image. (d) Identify a practical device described by the given equation and write the statement of a problem for which the equation appears in the solution.

36. An antelope is at a distance of 20.0 m from a converging lens of focal length 30.0 cm. The lens forms an image of the animal. If the antelope runs away from the lens at a speed of 5.00 m/s, how fast does the image move? Does the image move toward or away from the lens?

37. ● An object is at a distance d to the left of a flat screen. A converging lens with focal length $f < d/4$ is placed between object and screen. (a) Show that two lens positions exist that form an image on the screen and determine how far these positions are from the object. (b) How do the two images differ from each other?

38. ● In Figure P36.38, a thin converging lens of focal length 14.0 cm forms an image of the square $abcd$, which is $h_c = h_b = 10.0$ cm high and lies between distances of $p_d = 20.0$ cm and $p_a = 30.0$ cm from the lens. (a) Let a', b', c', and d' represent the respective corners of the image. Let q_a represent the image distance for points a' and b', q_d represent the image distance for points c' and d', h'_b represent the distance from point b' to the axis, and h'_c represent the height of c'. Evaluate each of these quantities. Make a sketch of the image. (b) The area of the object is 100 cm². By carrying out the following steps, you will evaluate the area of the image. Let q represent the image distance of any point between a' and d', for which the object distance is p. Let h' represent the distance from the axis to the point at the edge of the image between b' and c' at image distance q. Demonstrate that

$$|h'| = (10 \text{ cm})q\left(\frac{1}{14 \text{ cm}} - \frac{1}{q}\right)$$

(c) Explain why the geometric area of the image is given by

$$\int_{q_a}^{q_d} |h'|\,dq$$

Carry out the integration to find the area of the image.

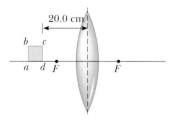

Figure P36.38

39. Figure 36.33 diagrams a cross section of a camera. It has a single lens of focal length 65.0 mm that is to form an image on the CCD at the back of the camera. Suppose the position of the lens has been adjusted to focus the image of a distant object. How far and in what direction must the lens be moved to form a sharp image of an object that is 2.00 m away?

Section 36.5 Lens Aberrations

40. The magnitudes of the radii of curvature are 32.5 cm and 42.5 cm for the two faces of a biconcave lens. The glass has index of refraction 1.53 for violet light and 1.51 for red light. For a very distant object, locate and describe (a) the image formed by violet light and (b) the image formed by red light.

41. Two rays traveling parallel to the principal axis strike a large plano-convex lens having an index of refraction of 1.60 (Fig. P36.41). If the convex face is spherical, a ray near the edge does not pass through the focal point (spherical aberration occurs). Assume this face has a radius of curvature of 20.0 cm and the two rays are at distances $h_1 = 0.500$ cm and $h_2 = 12.0$ cm from the principal axis. Find the difference Δx in the positions where each crosses the principal axis.

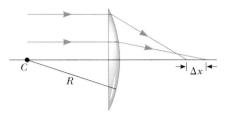

Figure P36.41

Section 36.6 The Camera

42. A camera is being used with a correct exposure at $f/4$ and a shutter speed of $\frac{1}{16}$ s. To photograph a rapidly moving subject, the shutter speed is changed to $\frac{1}{128}$ s. Find the new f-number setting needed to maintain satisfactory exposure.

Section 36.7 The Eye

43. A nearsighted person cannot see objects clearly beyond 25.0 cm (her far point). If she has no astigmatism and contact lenses are prescribed for her, what power and type of lens are required to correct her vision?

44. The accommodation limits for nearsighted Nick's eyes are 18.0 cm and 80.0 cm. When he wears his glasses, he can see faraway objects clearly. At what minimum distance is he able to see objects clearly?

Section 36.8 The Simple Magnifier

Section 36.9 The Compound Microscope

Section 36.10 The Telescope

45. A lens that has a focal length of 5.00 cm is used as a magnifying glass. (a) To obtain maximum magnification, where should the object be placed? (b) What is the magnification?

46. The distance between eyepiece and objective lens in a certain compound microscope is 23.0 cm. The focal length of the eyepiece is 2.50 cm and that of the objective is 0.400 cm. What is the overall magnification of the microscope?

47. The refracting telescope at the Yerkes Observatory has a 1.00-m diameter objective lens of focal length 20.0 m. Assume it is used with an eyepiece of focal length 2.50 cm. (a) Determine the magnification of Mars as seen through this telescope. (b) Are the Martian polar caps right side up or upside down?

48. ● Astronomers often take photographs with the objective lens or mirror of a telescope alone, without an eyepiece. (a) Show that the image size h' for such a telescope is given by $h' = fh/(f - p)$, where h is the object size, f is the objective focal length, and p is the object distance. (b) **What If?** Simplify the expression in part (a) for the case in which the object distance is much greater than objective focal length. (c) The "wingspan" of the International Space Station is 108.6 m, the overall width of its solar panel configuration. Find the width of the image formed by a telescope objective of focal length 4.00 m when the station is orbiting at an altitude of 407 km.

49. A certain telescope has an objective mirror with an aperture diameter of 200 mm and a focal length of 2 000 mm. It captures the image of a nebula on photographic film at its prime focus with an exposure time of 1.50 min. To produce the same light energy per unit area on the film, what is the required exposure time to photograph the same nebula with a smaller telescope that has an objective with a diameter of 60.0 mm and a focal length of 900 mm?

Additional Problems

50. A *zoom lens* system is a combination of lenses that produces a variable magnification of a fixed object as it maintains a fixed image position. The magnification is varied by moving one or more lenses along the axis. Multiple lenses are used in practice to obtain high-quality images, but the effect of zooming in on an object can be demonstrated with a simple two-lens system. An object, two converging lenses, and a screen are mounted on an optical bench. The first lens, which is to the right of the object, has a focal length of 5.00 cm, and the second lens, which is to the right of the first lens, has a focal length of 10.0 cm. The screen is to the right of the second lens. Initially, an object is situated at a distance of 7.50 cm to the left of the first lens, and the image formed on the screen has a magnification of $+1.00$. (a) Find the distance between the object and the screen. (b) Both lenses are now moved along their common axis, while the object and the screen maintain fixed positions, until the image formed on the screen has a magnification of $+3.00$. Find the displacement of each lens from its initial position in part (a). Can the lenses be displaced in more than one way?

51. The distance between an object and its upright image is 20.0 cm. If the magnification is 0.500, what is the focal length of the lens being used to form the image?

52. The distance between an object and its upright image is d. If the magnification is M, what is the focal length of the lens being used to form the image?

53. A real object is located at the zero end of a meterstick. A large concave mirror at the 100-cm end of the meterstick forms an image of the object at the 70.0-cm position. A small convex mirror placed at the 20.0-cm position forms a final image at the 10.0-cm point. What is the radius of curvature of the convex mirror?

54. The lens and mirror in Figure P36.54 have focal lengths of $+80.0$ cm and -50.0 cm, respectively. An object is placed 1.00 m to the left of the lens as shown. Locate the final image, formed by light that has gone through the lens twice. State whether the image is upright or inverted and determine the overall magnification.

Figure P36.54

55. ● An object is originally at the $x_i = 0$ cm position of a meterstick located on the x axis. A converging lens of focal length 26.0 cm is fixed at the position 32.0 cm. Then we gradually slide the object to the position $x_f = 12.0$ cm. Find the location x' of the object's image as a function of the object position x. Describe the pattern of the motion of the image with reference to a graph or a table of values. As the object moves 12 cm to the right, how far does the image move? In what direction or directions?

56. The object in Figure P36.56 is midway between the lens and the mirror. The mirror's radius of curvature is 20.0 cm, and the lens has a focal length of -16.7 cm. Considering only the light that leaves the object and travels first toward the mirror, locate the final image formed by this system. Is this image real or virtual? Is it upright or inverted? What is the overall magnification?

Figure P36.56

57. ● In many applications, it is necessary to expand or decrease the diameter of a beam of parallel rays of light. This change can be made by using a converging lens and a diverging lens in combination. Suppose you have a converging lens of focal length 21.0 cm and a diverging lens of focal length -12.0 cm. How can you arrange these lenses to increase the diameter of a beam of parallel rays? By what factor will the diameter increase?

58. The lens-makers' equation applies to a lens immersed in a liquid if n in the equation is replaced by n_2/n_1. Here n_2 refers to the index of refraction of the lens material and n_1 is that of the medium surrounding the lens. (a) A certain lens has focal length 79.0 cm in air and index of refraction 1.55. Find its focal length in water. (b) A certain mirror has focal length 79.0 cm in air. Find its focal length in water.

59. ▲ A parallel beam of light enters a glass hemisphere perpendicular to the flat face as shown in Figure P36.59. The

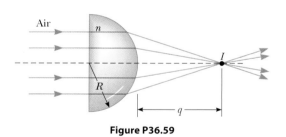

Figure P36.59

2 = intermediate; 3 = challenging; □ = SSM/SG; ▲ = ThomsonNOW; = symbolic reasoning; ● = qualitative reasoning

magnitude of the radius is 6.00 cm, and the index of refraction is 1.560. Determine the point at which the beam is focused. (Assume paraxial rays.)

60. **Review problem.** A spherical lightbulb of diameter 3.20 cm radiates light equally in all directions, with power 4.50 W. (a) Find the light intensity at the surface of the lightbulb. (b) Find the light intensity 7.20 m away from the center of the lightbulb. (c) At this 7.20-m distance, a lens is set up with its axis pointing toward the lightbulb. The lens has a circular face with a diameter 15.0 cm and has a focal length of 35.0 cm. Find the diameter of the image of the lightbulb. (d) Find the light intensity at the image.

61. An object is placed 12.0 cm to the left of a diverging lens of focal length −6.00 cm. A converging lens of focal length 12.0 cm is placed a distance d to the right of the diverging lens. Find the distance d so that the final image is at infinity. Draw a ray diagram for this case.

62. Assume the intensity of sunlight is 1.00 kW/m² at a particular location. A highly reflecting concave mirror is to be pointed toward the Sun to produce a power of at least 350 W at the image. (a) Find the required radius R_a of the circular face area of the mirror. (b) Now suppose the light intensity is to be at least 120 kW/m² at the image. Find the required relationship between R_a and the radius of curvature R of the mirror. The disk of the Sun subtends an angle of 0.533° at the Earth.

63. ▲ The disk of the Sun subtends an angle of 0.533° at the Earth. What are the position and diameter of the solar image formed by a concave spherical mirror with a radius of curvature of 3.00 m?

64. ● Figure P36.64 shows a thin converging lens for which the radii of curvature are $R_1 = 9.00$ cm and $R_2 = -11.0$ cm. The lens is in front of a concave spherical mirror with the radius of curvature $R = 8.00$ cm. (a) Assume its focal points F_1 and F_2 are 5.00 cm from the center of the lens. Determine its index of refraction. (b) The lens and mirror are 20.0 cm apart, and an object is placed 8.00 cm to the left of the lens. Determine the position of the final image and its magnification as seen by the eye in the figure. (c) Is the final image inverted or upright? Explain.

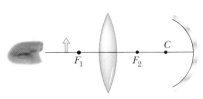

Figure P36.64

65. In a darkened room, a burning candle is placed 1.50 m from a white wall. A lens is placed between candle and wall at a location that causes a larger, inverted image to form on the wall. When the lens is moved 90.0 cm toward the wall, another image of the candle is formed. Find (a) the two object distances that produce the specified images and (b) the focal length of the lens. (c) Characterize the second image.

66. ● A floating strawberry illusion is achieved with two parabolic mirrors, each having a focal length 7.50 cm, facing each other so that their centers are 7.50 cm apart (Fig. P36.66). If a strawberry is placed on the lower mirror, an image of the strawberry is formed at the small opening at the center of the top mirror. Show that the final image is formed at that location and describe its characteristics. *Note:* A very startling effect is to shine a flashlight beam on this image. Even at a glancing angle, the incoming light beam is seemingly reflected from the image! Do you understand why?

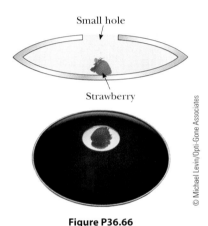

Figure P36.66

67. An object 2.00 cm high is placed 40.0 cm to the left of a converging lens having a focal length of 30.0 cm. A diverging lens with a focal length of −20.0 cm is placed 110 cm to the right of the converging lens. (a) Determine the position and magnification of the final image. (b) Is the image upright or inverted? (c) **What If?** Repeat parts (a) and (b) for the case in which the second lens is a converging lens having a focal length of +20.0 cm.

68. Two lenses made of kinds of glass having different indices of refraction n_1 and n_2 are cemented together to form an *optical doublet*. Optical doublets are often used to correct chromatic aberrations in optical devices. The first lens of a certain doublet has one flat side and one concave side with a radius of curvature of magnitude R. The second lens has two convex sides with radii of curvature also of magnitude R. Show that the doublet can be modeled as a single thin lens with a focal length described by

$$\frac{1}{f} = \frac{2n_2 - n_1 - 1}{R}$$

2 = intermediate;   3 = challenging;   □ = SSM/SG;   ▲ = ThomsonNOW;   ▮ = symbolic reasoning;   ● = qualitative reasoning

Answers to Quick Quizzes

36.1 False. The water spots are 2 m away from you, and your image is 4 m away. You cannot focus your eyes on both at the same time.

36.2 (b). A concave mirror focuses the light from a large area of the mirror onto a small area of the paper, resulting in a very high power input to the paper.

36.3 (b). A convex mirror always forms an image with a magnification less than 1, so the mirror must be concave. In a concave mirror, only virtual images are upright. This particular photograph is of the Hubble Space Telescope primary mirror. The scientists acting as the object for the image are to the left of the photograph and not visible to us.

36.4 (d). When O is far away, the rays refract into the material of index n_2 and converge to form a real image as in Figure 36.16. For certain combinations of R and n_2 as O moves very close to the refracting surface, the incident angle of the rays increases so much that rays are no longer refracted back toward the principal axis. The result is a virtual image as shown in the next column.

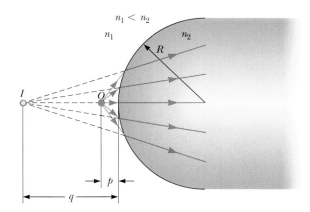

36.5 (a). No matter where O is, the rays refract into the air away from the normal and form a virtual image between O and the surface.

36.6 (b). Because the flat surfaces of the plane have infinite radii of curvature, Equation 36.15 indicates that the focal length is also infinite. Parallel rays striking the plane focus at infinity, which means that they remain parallel after passing through the glass.

36.7 (a). If the object is brought closer to the lens, the image moves farther away from the lens, behind the plane of the CCD. To bring the image back up to the CCD, the lens is moved toward the object and away from the CCD.

36.8 (c). The Sun's rays must converge onto the paper. A far-sighted person wears converging lenses.

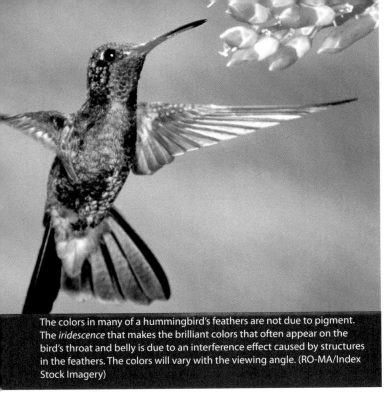

The colors in many of a hummingbird's feathers are not due to pigment. The *iridescence* that makes the brilliant colors that often appear on the bird's throat and belly is due to an interference effect caused by structures in the feathers. The colors will vary with the viewing angle. (RO-MA/Index Stock Imagery)

37.1 Conditions for Interference

37.2 Young's Double-Slit Experiment

37.3 Light Waves in Interference

37.4 Intensity Distribution of the Double-Slit Interference Pattern

37.5 Change of Phase Due to Reflection

37.6 Interference in Thin Films

37.7 The Michelson Interferometer

37 Interference of Light Waves

In Chapter 36, we studied light rays passing through a lens or reflecting from a mirror to describe the formation of images. This discussion completed our study of *geometric optics.* In this chapter and in Chapter 38, we are concerned with *wave optics* or *physical optics,* the study of interference, diffraction, and polarization of light. These phenomena cannot be adequately explained with the ray optics used in Chapters 35 and 36. We now learn how treating light as waves rather than as rays leads to a satisfying description of such phenomena.

37.1 Conditions for Interference

In Chapter 18, we studied the waves in interference model and found that the superposition of two mechanical waves can be constructive or destructive. In constructive interference, the amplitude of the resultant wave is greater than that of either individual wave, whereas in destructive interference, the resultant amplitude is less than that of the larger wave. Light waves also interfere with one another. Fundamentally, all interference associated with light waves arises when the electromagnetic fields that constitute the individual waves combine.

If two lightbulbs are placed side by side so that light from both bulbs combines, no interference effects are observed because the light waves from one bulb are emitted independently of those from the other bulb. The emissions from the two lightbulbs do not maintain a constant phase relationship with each other over time. Light waves from an ordinary source such as a lightbulb undergo random phase changes in time intervals of less than a nanosecond. Therefore, the conditions for constructive interference, destructive interference, or some intermediate state are maintained only for such short time intervals. Because the eye cannot follow such rapid changes, no interference effects are observed. Such light sources are said to be **incoherent.**

To observe interference of waves from two sources, the following conditions must be met:

Conditions for interference ▶

- The sources must be **coherent**; that is, **they must maintain a constant phase with respect to each other.**
- The sources should be **monochromatic**; that is, they should be of a single wavelength.

As an example, single-frequency sound waves emitted by two side-by-side loudspeakers driven by a single amplifier can interfere with each other because the two speakers are coherent. In other words, they respond to the amplifier in the same way at the same time.

(a)

(b)

Figure 37.1 (a) If light waves did not spread out after passing through the slits, no interference would occur. (b) The light waves from the two slits overlap as they spread out, filling what we expect to be shadowed regions with light and producing interference fringes on a screen placed to the right of the slits.

37.2 Young's Double-Slit Experiment

A common method for producing two coherent light sources is to use a monochromatic source to illuminate a barrier containing two small openings, usually in the shape of slits. The light emerging from the two slits is coherent because a single source produces the original light beam and the two slits serve only to separate the original beam into two parts (which, after all, is what is done to the sound signal from two side-by-side loudspeakers). Any random change in the light emitted by the source occurs in both beams at the same time. As a result, interference effects can be observed when the light from the two slits arrives at a viewing screen.

If the light traveled only in its original direction after passing through the slits as shown in Figure 37.1a, the waves would not overlap and no interference pattern would be seen. Instead, as we have discussed in our treatment of Huygens's principle (Section 35.6), the waves spread out from the slits as shown in Figure 37.1b. In other words, the light deviates from a straight-line path and enters the region that

ACTIVE FIGURE 37.2

(a) Schematic diagram of Young's double-slit experiment. Slits S_1 and S_2 behave as coherent sources of light waves that produce an interference pattern on the viewing screen (drawing not to scale). (b) An enlargement of the center of a fringe pattern formed on the viewing screen.

Sign in at www.thomsonedu.com and go to ThomsonNOW to adjust the slit separation and the wavelength of the light and see the effect on the interference pattern.

(a)

(b)

Figure 37.3   An interference pattern involving water waves is produced by two vibrating sources at the water's surface. The pattern is analogous to that observed in Young's double-slit experiment. Notice the regions of constructive (*A*) and destructive (*B*) interference.

would otherwise be shadowed. As noted in Section 35.3, this divergence of light from its initial line of travel is called **diffraction.**

Interference in light waves from two sources was first demonstrated by Thomas Young in 1801. A schematic diagram of the apparatus Young used is shown in Active Figure 37.2a. Plane light waves arrive at a barrier that contains two parallel slits S_1 and S_2. The light from S_1 and S_2 produces on a viewing screen a visible pattern of bright and dark parallel bands called **fringes** (Active Fig. 37.2b). When the light from S_1 and that from S_2 both arrive at a point on the screen such that constructive interference occurs at that location, a bright fringe appears. When the light from the two slits combines destructively at any location on the screen, a dark fringe results. Figure 37.3 is a photograph of an interference pattern produced by two coherent vibrating sources in a water tank.

Figure 37.4 shows some of the ways in which two waves can combine at the screen. In Figure 37.4a, the two waves, which leave the two slits in phase, strike the screen at the central point O. Because both waves travel the same distance, they arrive at O in phase. As a result, constructive interference occurs at this location and a bright fringe is observed. In Figure 37.4b, the two waves also start in phase, but here the lower wave has to travel one wavelength farther than the upper wave to reach point P. Because the lower wave falls behind the upper one by exactly one wavelength, they still arrive in phase at P and a second bright fringe appears at this location. At point R in Figure 37.4c, however, between points O and P, the lower wave has fallen half a wavelength behind the upper wave and a trough of the upper wave overlaps a crest of the lower wave, giving rise to destructive interference at point R. A dark fringe is therefore observed at this location.

(a)   (b)   (c)

Figure 37.4   (a) Constructive interference occurs at point O when the waves combine. (b) Constructive interference also occurs at point P. (c) Destructive interference occurs at point R when the two waves combine because the lower wave falls one-half a wavelength behind the upper wave. (All figures not to scale.)

37.3 Light Waves in Interference

We can describe Young's experiment quantitatively with the help of Figure 37.5. The viewing screen is located a perpendicular distance L from the barrier containing two slits, S_1 and S_2 (Fig. 37.5a). These slits are separated by a distance d, and the source is monochromatic. To reach any arbitrary point P in the upper half of the screen, a wave from the lower slit must travel farther than a wave from the upper slit by a distance $d \sin \theta$ (Fig. 37.5b). This distance is called the **path difference** δ (Greek letter delta). If we assume the rays labeled r_1 and r_2 are parallel, which is approximately true if L is much greater than d, then δ is given by

▶ Path difference

$$\delta = r_2 - r_1 = d \sin \theta \tag{37.1}$$

The value of δ determines whether the two waves are in phase when they arrive at point P. If δ is either zero or some integer multiple of the wavelength, the two waves are in phase at point P and constructive interference results. Therefore, the condition for bright fringes, or **constructive interference,** at point P is

▶ Conditions for constructive interference

$$d \sin \theta_{\text{bright}} = m\lambda \quad (m = 0, \pm 1, \pm 2, \cdots) \tag{37.2}$$

The number m is called the **order number.** For constructive interference, the order number is the same as the number of wavelengths that represents the path difference between the waves from the two slits. The central bright fringe at $\theta_{\text{bright}} = 0$ is called the *zeroth-order maximum.* The first maximum on either side, where $m = \pm 1$, is called the *first-order maximum,* and so forth.

When δ is an odd multiple of $\lambda/2$, the two waves arriving at point P are 180° out of phase and give rise to destructive interference. Therefore, the condition for dark fringes, or **destructive interference,** at point P is

▶ Conditions for destructive interference

$$d \sin \theta_{\text{dark}} = (m + \tfrac{1}{2})\lambda \quad (m = 0, \pm 1, \pm 2, \cdots) \tag{37.3}$$

These equations provide the *angular* positions of the fringes. It is also useful to obtain expressions for the *linear* positions measured along the screen from O to P. From the triangle OPQ in Figure 37.5a, we see that

$$\tan \theta = \frac{y}{L} \tag{37.4}$$

Using this result, the linear positions of bright and dark fringes are given by

$$y_{\text{bright}} = L \tan \theta_{\text{bright}} \tag{37.5}$$

$$y_{\text{dark}} = L \tan \theta_{\text{dark}} \tag{37.6}$$

where θ_{bright} and θ_{dark} are given by Equations 37.2 and 37.3.

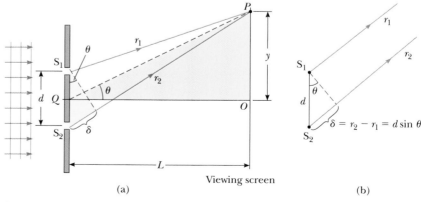

Figure 37.5   (a) Geometric construction for describing Young's double-slit experiment (not to scale). (b) When we assume r_1 is parallel to r_2, the path difference between the two rays is $r_2 - r_1 = d \sin \theta$. For this approximation to be valid, it is essential that $L \gg d$.

When the angles to the fringes are small, the positions of the fringes are linear near the center of the pattern. This can be verified by noting that for small angles, $\tan \theta \approx \sin \theta$, so Equation 37.5 gives the positions of the bright fringes as $y_{bright} = L \sin \theta_{bright}$. Incorporating Equation 37.2 gives

$$y_{bright} = L\left(\frac{m\lambda}{d}\right) \quad \text{(small angles)} \qquad (37.7)$$

This result shows that y_{bright} is linear in the order number m, so the fringes are equally spaced.

As demonstrated in Example 37.1, Young's double-slit experiment provides a method for measuring the wavelength of light. In fact, Young used this technique to do precisely that. In addition, his experiment gave the wave model of light a great deal of credibility. It was inconceivable that particles of light coming through the slits could cancel one another in a way that would explain the dark fringes.

The principles discussed in this section are the basis of the **waves in interference** analysis model. This model was applied to mechanical waves in one dimension in Chapter 18. Here we see the details of applying this model in three dimensions to light.

The faint pastel-colored bows beneath the main rainbow are called *supernumerary bows*. They are formed by interference between rays of light leaving raindrops at angles slightly smaller than the angle of maximum intensity. (See Section 35.7 for a discussion of the rainbow.)

Quick Quiz 37.1   Which of the following causes the fringes in a two-slit interference pattern to move farther apart? (a) decreasing the wavelength of the light (b) decreasing the screen distance L (c) decreasing the slit spacing d (d) immersing the entire apparatus in water

---

EXAMPLE 37.1	**Measuring the Wavelength of a Light Source**

A viewing screen is separated from a double slit by 1.2 m. The distance between the two slits is 0.030 mm. Monochromatic light is directed toward the double slit and forms an interference pattern on the screen. The second-order bright fringe ($m = 2$) is 4.5 cm from the center line on the screen.

(A) Determine the wavelength of the light.

SOLUTION

Conceptualize   Study Figure 37.5 to be sure you understand the phenomenon of interference of light waves.

Categorize   We evaluate results using equations developed in this section, so we categorize this example as a substitution problem.

Solve Equation 37.7 for the wavelength and substitute numerical values:

$$\lambda = \frac{y_{bright}d}{mL} = \frac{(4.5 \times 10^{-2}\ \text{m})(3.0 \times 10^{-5}\ \text{m})}{2(1.2\ \text{m})}$$

$$= 5.6 \times 10^{-7}\ \text{m} = \boxed{560\ \text{nm}}$$

(B) Calculate the distance between adjacent bright fringes.

SOLUTION

Find the distance between adjacent bright fringes from Equation 37.7 and the results of part (A):

$$y_{m+1} - y_m = L\frac{(m+1)\lambda}{d} - L\left(\frac{m\lambda}{d}\right)$$

$$= L\left(\frac{\lambda}{d}\right) = 1.2\ \text{m}\left(\frac{5.6 \times 10^{-7}\ \text{m}}{3.0 \times 10^{-5}\ \text{m}}\right)$$

$$= 2.2 \times 10^{-2}\ \text{m} = \boxed{2.2\ \text{cm}}$$

EXAMPLE 37.2   **Separating Double-Slit Fringes of Two Wavelengths**

A light source emits visible light of two wavelengths: $\lambda = 430$ nm and $\lambda' = 510$ nm. The source is used in a double-slit interference experiment in which $L = 1.50$ m and $d = 0.025\ 0$ mm. Find the separation distance between the third-order bright fringes for the two wavelengths.

SOLUTION

Conceptualize   In Figure 37.5a, imagine light of two wavelengths incident on the slits and forming two interference patterns on the screen.

Categorize   We evaluate results using equations developed in this section, so we categorize this example as a substitution problem.

Use Equation 37.7, with $m = 3$, to find the fringe positions corresponding to these two wavelengths:

$$y_{\text{bright}} = L\left(\frac{m\lambda}{d}\right) = L\left(\frac{3\lambda}{d}\right) = 1.50\ \text{m}\left[\frac{3(430 \times 10^{-9}\ \text{m})}{0.025\ 0 \times 10^{-3}\ \text{m}}\right]$$

$$= 7.74 \times 10^{-2}\ \text{m}$$

$$y'_{\text{bright}} = L\left(\frac{m\lambda'}{d}\right) = L\left(\frac{3\lambda'}{d}\right) = 1.50\ \text{m}\left[\frac{3(510 \times 10^{-9}\ \text{m})}{0.025\ 0 \times 10^{-3}\ \text{m}}\right]$$

$$= 9.18 \times 10^{-2}\ \text{m}$$

Evaluate the separation distance between the two fringes:

$$\Delta y = 9.18 \times 10^{-2}\ \text{m} - 7.74 \times 10^{-2}\ \text{m}$$

$$= 1.44 \times 10^{-2}\ \text{m} = \boxed{1.44\ \text{cm}}$$

What If?   What if we examine the entire interference pattern due to the two wavelengths and look for overlapping fringes? Are there any locations on the screen where the bright fringes from the two wavelengths overlap exactly?

Answer   Find such a location by setting the location of any bright fringe due to λ equal to one due to λ', using Equation 37.7:

$$L\left(\frac{m\lambda}{d}\right) = L\left(\frac{m'\lambda'}{d}\right) \quad \rightarrow \quad \frac{m'}{m} = \frac{\lambda}{\lambda'}$$

Substitute the wavelengths:

$$\frac{m'}{m} = \frac{430\ \text{nm}}{510\ \text{nm}} = \frac{43}{51}$$

Use Equation 37.7 to find the value of y for these fringes:

$$y = 1.50\ \text{m}\left[\frac{51(430 \times 10^{-9}\ \text{m})}{0.025\ 0 \times 10^{-3}\ \text{m}}\right] = 1.32\ \text{m}$$

This value of y is comparable to L, so the small-angle approximation used for Equation 37.7 is *not* valid. This conclusion suggests we should not expect Equation 37.7 to give us the correct result. If you use Equation 37.5, you can show that the bright fringes do indeed overlap when the same condition, $m'/m = \lambda/\lambda'$, is met (see Problem 38). Therefore, the 51st fringe of the 430-nm light does overlap with the 43rd fringe of the 510-nm light, but not at the location of 1.32 m. You are asked to find the correct location as part of Problem 38.

37.4   Intensity Distribution of the Double-Slit Interference Pattern

Notice that the edges of the bright fringes in Active Figure 37.2b are not sharp; rather, there is a gradual change from bright to dark. So far, we have discussed the locations of only the centers of the bright and dark fringes on a distant screen. Let's now direct our attention to the intensity of the light at other points between the positions of maximum constructive and destructive interference. In other words, we now calculate the distribution of light intensity associated with the double-slit interference pattern.

Again, suppose the two slits represent coherent sources of sinusoidal waves such that the two waves from the slits have the same angular frequency ω and are in phase. The total magnitude of the electric field at point P on the screen in Figure 37.6 is the superposition of the two waves. Assuming that the two waves have the same amplitude E_0, we can write the magnitude of the electric field at point P due to each wave separately as

$$E_1 = E_0 \sin \omega t \quad \text{and} \quad E_2 = E_0 \sin (\omega t + \phi) \tag{37.8}$$

Figure 37.6 Construction for analyzing the double-slit interference pattern. A bright fringe, or intensity maximum, is observed at O.

Although the waves are in phase at the slits, *their phase difference ϕ at P depends on the path difference* $\delta = r_2 - r_1 = d \sin \theta$. A path difference of λ (for constructive interference) corresponds to a phase difference of 2π rad. A path difference of δ is the same fraction of λ as the phase difference ϕ is of 2π. We can describe this fraction mathematically with the ratio

$$\frac{\delta}{\lambda} = \frac{\phi}{2\pi}$$

which gives

$$\phi = \frac{2\pi}{\lambda} \delta = \frac{2\pi}{\lambda} d \sin \theta \tag{37.9}$$

◀ Phase difference

This equation shows how the phase difference ϕ depends on the angle θ in Figure 37.5.

Using the superposition principle and Equation 37.8, we obtain the following expression for the magnitude of the resultant electric field at point P:

$$E_P = E_1 + E_2 = E_0[\sin \omega t + \sin (\omega t + \phi)] \tag{37.10}$$

We can simplify this expression by using the trigonometric identity

$$\sin A + \sin B = 2 \sin \left(\frac{A + B}{2}\right) \cos \left(\frac{A - B}{2}\right)$$

Taking $A = \omega t + \phi$ and $B = \omega t$, Equation 37.10 becomes

$$E_P = 2E_0 \cos \left(\frac{\phi}{2}\right) \sin \left(\omega t + \frac{\phi}{2}\right) \tag{37.11}$$

This result indicates that the electric field at point P has the same frequency ω as the light at the slits but that the amplitude of the field is multiplied by the factor $2 \cos (\phi/2)$. To check the consistency of this result, note that if $\phi = 0, 2\pi, 4\pi, \ldots$, the magnitude of the electric field at point P is $2E_0$, corresponding to the condition for maximum constructive interference. These values of ϕ are consistent with Equation 37.2 for constructive interference. Likewise, if $\phi = \pi, 3\pi, 5\pi, \ldots$, the magnitude of the electric field at point P is zero, which is consistent with Equation 37.3 for total destructive interference.

Finally, to obtain an expression for the light intensity at point P, recall from Section 34.4 that *the intensity of a wave is proportional to the square of the resultant electric field magnitude at that point* (Eq. 34.24). Using Equation 37.11, we can therefore express the light intensity at point P as

$$I \propto E_P{}^2 = 4E_0{}^2 \cos^2 \left(\frac{\phi}{2}\right) \sin^2 \left(\omega t + \frac{\phi}{2}\right)$$

Most light-detecting instruments measure time-averaged light intensity, and the time-averaged value of $\sin^2 (\omega t + \phi/2)$ over one cycle is $\frac{1}{2}$. (See Fig. 33.5.) Therefore, we can write the average light intensity at point P as

$$I = I_{\max} \cos^2 \left(\frac{\phi}{2}\right) \tag{37.12}$$

M. Cagnet, M. Françon, J. C. Thrierr, *Atlas of Optical Phenomena*, Berlin, Springer-Verlag, 1962

Figure 37.7 Light intensity versus $d \sin \theta$ for a double-slit interference pattern when the screen is far from the two slits $(L \gg d)$.

where I_{max} is the maximum intensity on the screen and the expression represents the time average. Substituting the value for ϕ given by Equation 37.9 into this expression gives

$$I = I_{max} \cos^2 \left(\frac{\pi d \sin \theta}{\lambda} \right) \tag{37.13}$$

Alternatively, because $\sin \theta \approx y/L$ for small values of θ in Figure 37.5, we can write Equation 37.13 in the form

$$I = I_{max} \cos^2 \left(\frac{\pi d}{\lambda L} y \right) \tag{37.14}$$

Constructive interference, which produces light intensity maxima, occurs when the quantity $\pi \, dy/\lambda L$ is an integral multiple of π, corresponding to $y = (\lambda L/d)\, m$. This result is consistent with Equation 37.7.

A plot of light intensity versus $d \sin \theta$ is given in Figure 37.7. The interference pattern consists of equally spaced fringes of equal intensity. Remember, however, that this result is valid only if the slit-to-screen distance L is much greater than the slit separation and only for small values of θ.

Figure 37.8 shows similar plots of light intensity versus $d \sin \theta$ for light passing through multiple slits. For more than two slits, the pattern contains primary and secondary maxima. For three slits, notice that the primary maxima are nine times more intense than the secondary maxima as measured by the height of the curve because the intensity varies as E^2. For N slits, the intensity of the primary maxima is N^2 times greater than that due to a single slit. As the number of slits increases, the primary maxima increase in intensity and become narrower, while the secondary maxima decrease in intensity relative to the primary maxima. Figure 37.8 also shows that as the number of slits increases, the number of secondary maxima also increases. In fact, the number of secondary maxima is always $N - 2$, where N is the number of slits. In Section 38.4, we shall investigate the pattern for a very large number of slits in a device called a *diffraction grating*.

Quick Quiz 37.2 Using Figure 37.8 as a model, sketch the interference pattern from six slits.

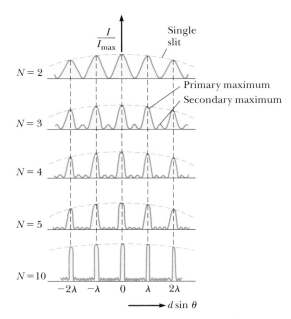

Figure 37.8 Multiple-slit interference patterns. As *N*, the number of slits, is increased, the primary maxima (the tallest peaks in each graph) become narrower but remain fixed in position and the number of secondary maxima increases. For any value of *N*, the decrease in intensity in maxima to the left and right of the central maximum, indicated by the blue dashed arcs, is due to *diffraction patterns* from the individual slits, which are discussed in Chapter 38.

37.5   Change of Phase Due to Reflection

Young's method for producing two coherent light sources involves illuminating a pair of slits with a single source. Another simple, yet ingenious, arrangement for producing an interference pattern with a single light source is known as *Lloyd's mirror*[1] (Fig. 37.9). A point light source S is placed close to a mirror, and a viewing screen is positioned some distance away and perpendicular to the mirror. Light waves can reach point *P* on the screen either directly from S to *P* or by the path involving reflection from the mirror. The reflected ray can be treated as a ray originating from a virtual source S′. As a result, we can think of this arrangement as a double-slit source with the distance between sources S and S′ comparable to length *d* in Figure 37.5. Hence, at observation points far from the source ($L \gg d$), we expect waves from S and S′ to form an interference pattern exactly like the one formed by two real coherent sources. An interference pattern is indeed observed. The positions of the dark and bright fringes, however, are reversed relative to the pattern created by two real coherent sources (Young's experiment). Such a reversal can only occur if the coherent sources S and S′ differ in phase by 180°.

To illustrate further, consider point *P*′, the point where the mirror intersects the screen. This point is equidistant from sources S and S′. If path difference alone were responsible for the phase difference, we would see a bright fringe at *P*′ (because the path difference is zero for this point), corresponding to the central bright fringe of the two-slit interference pattern. Instead, a dark fringe is observed at *P*′. We therefore conclude that a 180° phase change must be produced by reflection from the mirror. In general, **an electromagnetic wave undergoes a phase change of 180° upon reflection from a medium that has a higher index of refraction than the one in which the wave is traveling.**

It is useful to draw an analogy between reflected light waves and the reflections of a transverse wave pulse on a stretched string (Section 16.4). The reflected pulse on a string undergoes a phase change of 180° when reflected from the boundary

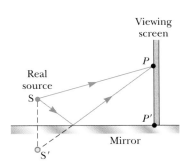

Figure 37.9 Lloyd's mirror. An interference pattern is produced at point *P* on the screen as a result of the combination of the direct ray (blue) and the reflected ray (brown). The reflected ray undergoes a phase change of 180°.

---

[1] Developed in 1834 by Humphrey Lloyd (1800–1881), Professor of Natural and Experimental Philosophy, Trinity College, Dublin.

Figure 37.10 (a) For $n_1 < n_2$, a light ray traveling in medium 1 when reflected from the surface of medium 2 undergoes a 180° phase change. The same thing happens with a reflected pulse traveling along a string fixed at one end. (b) For $n_1 > n_2$, a light ray traveling in medium 1 undergoes no phase change when reflected from the surface of medium 2. The same is true of a reflected wave pulse on a string whose supported end is free to move.

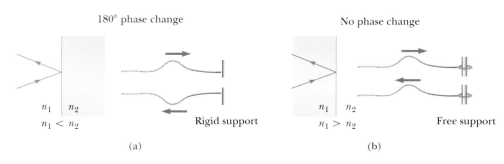

of a denser medium, but no phase change occurs when the pulse is reflected from the boundary of a less dense medium. Similarly, an electromagnetic wave undergoes a 180° phase change when reflected from a boundary leading to an optically denser medium (defined as a medium with a higher index of refraction), but no phase change occurs when the wave is reflected from a boundary leading to a less dense medium. These rules, summarized in Figure 37.10, can be deduced from Maxwell's equations, but the treatment is beyond the scope of this text.

37.6 Interference in Thin Films

Figure 37.11 Interference in light reflected from a thin film is due to a combination of rays 1 and 2 reflected from the upper and lower surfaces of the film. Rays 3 and 4 lead to interference effects for light transmitted through the film.

Interference effects are commonly observed in thin films, such as thin layers of oil on water or the thin surface of a soap bubble. The varied colors observed when white light is incident on such films result from the interference of waves reflected from the two surfaces of the film.

Consider a film of uniform thickness t and index of refraction n as shown in Figure 37.11. Let's assume the light rays traveling in air are nearly normal to the two surfaces of the film. The wavelength of light λ_n in the film (see Section 35.5) is

$$\lambda_n = \frac{\lambda}{n}$$

where λ is the wavelength of the light in free space and n is the index of refraction of the film material.

Reflected ray 1, which is reflected from the upper surface (A) in Figure 37.11, undergoes a phase change of 180° with respect to the incident wave. Reflected ray 2, which is reflected from the lower film surface (B), undergoes no phase change because it is reflected from a medium (air) that has a lower index of refraction. Therefore, ray 1 is 180° out of phase with ray 2, which is equivalent to a path difference of $\lambda_n/2$. We must also consider, however, that ray 2 travels an extra distance $2t$ before the waves recombine in the air above surface A. (Remember that we are considering light rays that are close to normal to the surface. If the rays are not close to normal, the path difference is larger than $2t$.) If $2t = \lambda_n/2$, rays 1 and 2 recombine in phase and the result is constructive interference. In general, the condition for *constructive* interference in thin films is[2]

$$2t = \left(m + \tfrac{1}{2}\right)\lambda_n \quad (m = 0, 1, 2, \dots) \tag{37.15}$$

This condition takes into account two factors: (1) the difference in path length for the two rays (the term $m\lambda_n$) and (2) the 180° phase change upon reflection (the term $\tfrac{1}{2}\lambda_n$). Because $\lambda_n = \lambda/n$, we can write Equation 37.15 as

◀ Conditions for constructive interference in thin films

$$2nt = \left(m + \tfrac{1}{2}\right)\lambda \quad (m = 0, 1, 2, \dots) \tag{37.16}$$

[2] The full interference effect in a thin film requires an analysis of an infinite number of reflections back and forth between the top and bottom surfaces of the film. We focus here only on a single reflection from the bottom of the film, which provides the largest contribution to the interference effect.

If the extra distance $2t$ traveled by ray 2 corresponds to a multiple of λ_n, the two waves combine out of phase and the result is destructive interference. The general equation for *destructive* interference in thin films is

$$2nt = m\lambda \quad (m = 0, 1, 2, \ldots) \tag{37.17}$$

◄ Conditions for destructive interference in thin films

The foregoing conditions for constructive and destructive interference are valid when the medium above the top surface of the film is the same as the medium below the bottom surface or, if there are different media above and below the film, the index of refraction of both is less than n. If the film is placed between two different media, one with $n < n_{\text{film}}$ and the other with $n > n_{\text{film}}$, the conditions for constructive and destructive interference are reversed. In that case, either there is a phase change of 180° for both ray 1 reflecting from surface A and ray 2 reflecting from surface B or there is no phase change for either ray; hence, the net change in relative phase due to the reflections is zero.

Rays 3 and 4 in Figure 37.11 lead to interference effects in the light transmitted through the thin film. The analysis of these effects is similar to that of the reflected light. You are asked to explore the transmitted light in Problems 23, 29, and 30.

PITFALL PREVENTION 37.1
Be Careful with Thin Films

Be sure to include *both* effects—path length and phase change—when analyzing an interference pattern resulting from a thin film. The possible phase change is a new feature we did not need to consider for double-slit interference. Also think carefully about the material on either side of the film. You may have a situation in which there is a 180° phase change at *both* surfaces or at *neither* surface, if there are different materials on either side of the film.

Quick Quiz 37.3   One microscope slide is placed on top of another with their left edges in contact and a human hair under the right edge of the upper slide. As a result, a wedge of air exists between the slides. An interference pattern results when monochromatic light is incident on the wedge. What is at the left edges of the slides? (a) a dark fringe (b) a bright fringe (c) impossible to determine

---

Newton's Rings

Another method for observing interference in light waves is to place a plano-convex lens on top of a flat glass surface as shown in Figure 37.12a. With this arrangement, the air film between the glass surfaces varies in thickness from zero at the point of contact to some value t at point P. If the radius of curvature R of the lens is much greater than the distance r and the system is viewed from above, a pattern of light and dark rings is observed as shown in Figure 37.12b. These circular fringes, discovered by Newton, are called **Newton's rings.**

The interference effect is due to the combination of ray 1, reflected from the flat plate, with ray 2, reflected from the curved surface of the lens. Ray 1 undergoes a phase change of 180° upon reflection (because it is reflected from a medium of higher index of refraction), whereas ray 2 undergoes no phase change (because it is reflected from a medium of lower index of refraction). Hence, the conditions for constructive and destructive interference are given by Equations 37.16 and 37.17, respectively, with $n = 1$ because the film is air. Because there is

(a)

(b)

Courtesy of Bausch and Lomb Optical Company

Figure 37.12   (a) The combination of rays reflected from the flat plate and the curved lens surface gives rise to an interference pattern known as Newton's rings. (b) Photograph of Newton's rings.

(a) A thin film of oil floating on water displays interference, shown by the pattern of colors when white light is incident on the film. Variations in film thickness produce the interesting color pattern. The razor blade gives you an idea of the size of the colored bands. (b) Interference in soap bubbles. The colors are due to interference between light rays reflected from the front and back surfaces of the thin film of soap making up the bubble. The color depends on the thickness of the film, ranging from black, where the film is thinnest, to magenta, where it is thickest.

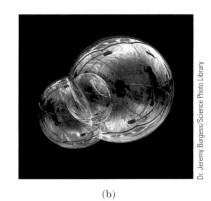

(a) (b)

no path difference and the total phase change is due only to the 180° phase change upon reflection, the contact point at O is dark as seen in Figure 37.12b.

Using the geometry shown in Figure 37.12a, we can obtain expressions for the radii of the bright and dark bands in terms of the radius of curvature R and wavelength λ. For example, the dark rings have radii given by the expression $r \approx \sqrt{m\lambda R/n}$. The details are left as a problem (see Problem 60). We can obtain the wavelength of the light causing the interference pattern by measuring the radii of the rings, provided R is known. Conversely, we can use a known wavelength to obtain R.

One important use of Newton's rings is in the testing of optical lenses. A circular pattern like that pictured in Figure 37.12b is obtained only when the lens is ground to a perfectly symmetric curvature. Variations from such symmetry produce a pattern with fringes that vary from a smooth, circular shape. These variations indicate how the lens must be reground and repolished to remove imperfections.

PROBLEM-SOLVING STRATEGY   **Thin-Film Interference**

The following features should be kept in mind when working thin-film interference problems.

1. *Conceptualize.* Think about what is going on physically in the problem. Identify the light source and the location of the observer.

2. *Categorize.* Confirm that you should use the techniques for thin-film interference by identifying the thin film causing the interference.

3. *Analyze.* The type of interference that occurs is determined by the phase relationship between the portion of the wave reflected at the upper surface of the film and the portion reflected at the lower surface. Phase differences between the two portions of the wave have two causes: differences in the distances traveled by the two portions and phase changes occurring on reflection. *Both* causes must be considered when determining which type of interference occurs. If the media above and below the film both have index of refraction larger than that of the film or if both indices are smaller, use Equation 37.16 for constructive interference and Equation 37.17 for destructive interference. If the film is located between two different media, one with $n < n_{\text{film}}$ and the other with $n > n_{\text{film}}$, reverse these two equations for constructive and destructive interference.

4. *Finalize.* Inspect your final results to see if they make sense physically and are of an appropriate size.

| EXAMPLE 37.3 | **Interference in a Soap Film** |

Calculate the minimum thickness of a soap-bubble film that results in constructive interference in the reflected light if the film is illuminated with light whose wavelength in free space is $\lambda = 600$ nm. The index of refraction of the soap film is 1.33.

SOLUTION

Conceptualize Imagine that the film in Figure 37.11 is soap, with air on both sides.

Categorize We evaluate the result using an equation from this section, so we categorize this example as a substitution problem.

The minimum film thickness for constructive interference in the reflected light corresponds to $m = 0$ in Equation 37.16. Solve this equation for t and substitute numerical values:

$$t = \frac{(0 + \frac{1}{2})\lambda}{2n} = \frac{\lambda}{4n} = \frac{(600 \text{ nm})}{4(1.33)} = \boxed{113 \text{ nm}}$$

What If? What if the film is twice as thick? Does this situation produce constructive interference?

Answer Using Equation 37.16, we can solve for the thicknesses at which constructive interference occurs:

$$t = \left(m + \tfrac{1}{2}\right)\frac{\lambda}{2n} = (2m + 1)\frac{\lambda}{4n} \qquad (m = 0, 1, 2, \dots)$$

The allowed values of m show that constructive interference occurs for *odd* multiples of the thickness corresponding to $m = 0$, $t = 113$ nm. Therefore, constructive interference does *not* occur for a film that is twice as thick.

| EXAMPLE 37.4 | **Nonreflective Coatings for Solar Cells** |

Solar cells—devices that generate electricity when exposed to sunlight—are often coated with a transparent, thin film of silicon monoxide (SiO, $n = 1.45$) to minimize reflective losses from the surface. Suppose a silicon solar cell ($n = 3.5$) is coated with a thin film of silicon monoxide for this purpose (Fig. 37.13a). Determine the minimum film thickness that produces the least reflection at a wavelength of 550 nm, near the center of the visible spectrum.

Figure 37.13 (Example 37.4) (a) Reflective losses from a silicon solar cell are minimized by coating the surface of the cell with a thin film of silicon monoxide. (b) The reflected light from a coated camera lens often has a reddish-violet appearance.

SOLUTION

Conceptualize Figure 37.13a helps us visualize the path of the rays in the SiO film that result in interference in the reflected light.

Categorize Based on the geometry of the SiO layer, we categorize this example as a thin-film interference problem.

Analyze The reflected light is a minimum when rays 1 and 2 in Figure 37.13a meet the condition of destructive interference. In this situation, *both* rays undergo a 180° phase change upon reflection: ray 1 from the upper SiO surface, and ray 2 from the lower SiO surface. The net change in phase due to reflection is therefore zero, and the condition for a reflection minimum requires a path difference of $\lambda_n/2$, where λ_n is the wavelength of the light in SiO. Hence, $2nt = \lambda/2$, where λ is the wavelength in air and n is the index of refraction of SiO.

Solve the equation $2nt = \lambda/2$ for t and substitute numerical values:

$$t = \frac{\lambda}{4n} = \frac{550 \text{ nm}}{4(1.45)} = 94.8 \text{ nm}$$

Finalize   A typical uncoated solar cell has reflective losses as high as 30%, but a coating of SiO can reduce this value to about 10%. This significant decrease in reflective losses increases the cell's efficiency because less reflection means that more sunlight enters the silicon to create charge carriers in the cell. No coating can ever be made perfectly non-reflecting because the required thickness is wavelength-dependent and the incident light covers a wide range of wavelengths.

Glass lenses used in cameras and other optical instruments are usually coated with a transparent thin film to reduce or eliminate unwanted reflection and to enhance the transmission of light through the lenses. The camera lens in Figure 37.13b has several coatings (of different thicknesses) to minimize reflection of light waves having wavelengths near the center of the visible spectrum. As a result, the small amount of light that is reflected by the lens has a greater proportion of the far ends of the spectrum and often appears reddish violet.

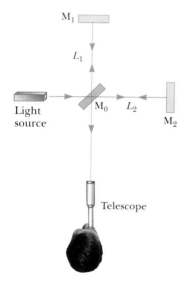

ACTIVE FIGURE 37.14

Diagram of the Michelson interferometer. A single ray of light is split into two rays by mirror M_0, which is called a beam splitter. The path difference between the two rays is varied with the adjustable mirror M_1. As M_1 is moved, an interference pattern changes in the field of view.

Sign in at www.thomsonedu.com and go to ThomsonNOW to move the mirror and see the effect on the interference pattern and use the interferometer to measure the wavelength of light.

37.7   The Michelson Interferometer

The **interferometer,** invented by American physicist A. A. Michelson (1852–1931), splits a light beam into two parts and then recombines the parts to form an interference pattern. The device can be used to measure wavelengths or other lengths with great precision because a large and precisely measurable displacement of one of the mirrors is related to an exactly countable number of wavelengths of light.

A schematic diagram of the interferometer is shown in Active Figure 37.14. A ray of light from a monochromatic source is split into two rays by mirror M_0, which is inclined at 45° to the incident light beam. Mirror M_0, called a *beam splitter,* transmits half the light incident on it and reflects the rest. One ray is reflected from M_0 vertically upward toward mirror M_1, and the second ray is transmitted horizontally through M_0 toward mirror M_2. Hence, the two rays travel separate paths L_1 and L_2. After reflecting from M_1 and M_2, the two rays eventually recombine at M_0 to produce an interference pattern, which can be viewed through a telescope.

The interference condition for the two rays is determined by the difference in their path length. When the two mirrors are exactly perpendicular to each other, the interference pattern is a target pattern of bright and dark circular fringes, similar to Newton's rings. As M_1 is moved, the fringe pattern collapses or expands, depending on the direction in which M_1 is moved. For example, if a dark circle appears at the center of the target pattern (corresponding to destructive interference) and M_1 is then moved a distance $\lambda/4$ toward M_0, the path difference changes by $\lambda/2$. What was a dark circle at the center now becomes a bright circle. As M_1 is moved an additional distance $\lambda/4$ toward M_0, the bright circle becomes a dark circle again. Therefore, the fringe pattern shifts by one-half fringe each time M_1 is moved a distance $\lambda/4$. The wavelength of light is then measured by counting the number of fringe shifts for a given displacement of M_1. If the wavelength is accurately known, mirror displacements can be measured to within a fraction of the wavelength.

We will see an important historical use of the Michelson interferometer in our discussion of relativity in Chapter 39. Modern uses include the following two applications, Fourier transform infrared spectroscopy and the laser interferometer gravitational-wave observatory.

Fourier Transform Infrared Spectroscopy

Spectroscopy is the study of the wavelength distribution of radiation from a sample that can be used to identify the characteristics of atoms or molecules in the sample. Infrared spectroscopy is particularly important to organic chemists when analyzing organic molecules. Traditional spectroscopy involves the use of an optical element, such as a prism (Section 35.5) or a diffraction grating (Section 38.4), which spreads out various wavelengths in a complex optical signal from the sample into different angles. In this way, the various wavelengths of radiation and their intensities in the signal can be determined. These types of devices are limited in their resolution and effectiveness because they must be scanned through the various angular deviations of the radiation.

The technique of *Fourier transform infrared* (FTIR) *spectroscopy* is used to create a higher-resolution spectrum in a time interval of 1 second that may have required 30 minutes with a standard spectrometer. In this technique, the radiation from a sample enters a Michelson interferometer. The movable mirror is swept through the zero-path-difference condition, and the intensity of radiation at the viewing position is recorded. The result is a complex set of data relating light intensity as a function of mirror position, called an *interferogram*. Because there is a relationship between mirror position and light intensity for a given wavelength, the interferogram contains information about all wavelengths in the signal.

In Section 18.8, we discussed Fourier analysis of a waveform. The waveform is a function that contains information about all the individual frequency components that make up the waveform.[3] Equation 18.13 shows how the waveform is generated from the individual frequency components. Similarly, the interferogram can be analyzed by computer, in a process called a *Fourier transform*, to provide all of the wavelength components. This information is the same as that generated by traditional spectroscopy, but the resolution of FTIR spectroscopy is much higher.

Laser Interferometer Gravitational-Wave Observatory

Einstein's general theory of relativity (Section 39.10) predicts the existence of *gravitational waves*. These waves propagate from the site of any gravitational disturbance, which could be periodic and predictable, such as the rotation of a double star around a center of mass, or unpredictable, such as the supernova explosion of a massive star.

In Einstein's theory, gravitation is equivalent to a distortion of space. Therefore, a gravitational disturbance causes an additional distortion that propagates through space in a manner similar to mechanical or electromagnetic waves. When gravitational waves from a disturbance pass by the Earth, they create a distortion of the local space. The laser interferometer gravitational-wave observatory (LIGO) apparatus is designed to detect this distortion. The apparatus employs a Michelson interferometer that uses laser beams with an effective path length of several kilometers. At the end of an arm of the interferometer, a mirror is mounted on a massive pendulum. When a gravitational wave passes by, the pendulum and the attached mirror move and the interference pattern due to the laser beams from the two arms changes.

Two sites for interferometers have been developed in the United States—in Richland, Washington, and in Livingston, Louisiana—to allow coincidence studies of gravitational waves. Figure 37.15 shows the Washington site. The two arms of

LIGO Hanford Observatory

Figure 37.15   The Laser Interferometer Gravitational-Wave Observatory (LIGO) near Richland, Washington. Notice the two perpendicular arms of the Michelson interferometer.

---

[3] In acoustics, it is common to talk about the components of a complex signal in terms of frequency. In optics, it is more common to identify the components by wavelength.

the Michelson interferometer are evident in the photograph. Five data runs have been performed as of 2007. These runs have been coordinated with other gravitational wave detectors, such as GEO in Hannover, Germany, TAMA in Mitaka, Japan, and VIRGO in Cascina, Italy. So far, gravitational waves have not yet been detected, but the data runs have provided critical information for modifications and design features for the next generation of detectors.

Summary

ThomsonNOW™ Sign in at **www.thomsonedu.com** and go to ThomsonNOW to take a practice test for this chapter.

CONCEPTS AND PRINCIPLES

Interference in light waves occurs whenever two or more waves overlap at a given point. An interference pattern is observed if (1) the sources are coherent and (2) the sources have identical wavelengths.

The **intensity** at a point in a double-slit interference pattern is

$$I = I_{max} \cos^2\left(\frac{\pi d \sin\theta}{\lambda}\right) \tag{37.13}$$

where I_{max} is the maximum intensity on the screen and the expression represents the time average.

A wave traveling from a medium of index of refraction n_1 toward a medium of index of refraction n_2 undergoes a 180° phase change upon reflection when $n_2 > n_1$ and undergoes no phase change when $n_2 < n_1$.

The condition for constructive interference in a film of thickness t and index of refraction n surrounded by air is

$$2nt = (m + \tfrac{1}{2})\lambda \quad (m = 0, 1, 2, \dots) \tag{37.16}$$

where λ is the wavelength of the light in free space.

Similarly, the condition for destructive interference in a thin film surrounded by air is

$$2nt = m\lambda \quad (m = 0, 1, 2, \dots) \tag{37.17}$$

ANALYSIS MODEL FOR PROBLEM SOLVING

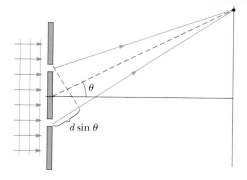

Waves in Interference. Young's double-slit experiment serves as a prototype for interference phenomena involving electromagnetic radiation. In this experiment, two slits separated by a distance d are illuminated by a single-wavelength light source. The condition for bright fringes (**constructive interference**) is

$$d \sin\theta_{bright} = m\lambda \quad (m = 0, \pm1, \pm2, \dots) \tag{37.2}$$

The condition for dark fringes (**destructive interference**) is

$$d \sin\theta_{dark} = (m + \tfrac{1}{2})\lambda \quad (m = 0, \pm1, \pm2, \dots) \tag{37.3}$$

The number m is called the **order number** of the fringe.

Questions

Question 4 in Chapter 18 may be assigned with this chapter.

1. What is the necessary condition on the path length difference between two waves that interfere (a) constructively and (b) destructively?

2. Explain why two flashlights held close together do not produce an interference pattern on a distant screen.

3. O Four trials of Young's double-slit experiment are conducted. (a) In the first trial, blue light passes through two fine slits 400 μm apart and forms an interference pattern on a screen 4 m away. (b) In a second trial, red light passes through the same slits and falls on the same screen. (c) A third trial is performed with red light and the same screen, but with slits 800 μm apart. (d) A final trial is performed with red light, slits 800 μm apart, and a screen 8 m away. **(i)** Rank the trials (a) through (d) from largest to smallest value of the angle between the central maximum and the first-order side maximum. In your ranking, note any cases of equality. **(ii)** Rank the same trials according to the distance between the central maximum and the first-order side maximum on the screen.

4. Suppose you blow smoke into the space between the barrier and the viewing screen in Young's double-slit experiment, shown in Active Figure 37.2a. Would the smoke show evidence of interference within this space? Explain your answer.

5. O Suppose Young's double-slit experiment is performed in air using red light and then the apparatus is immersed in water. What happens to the interference pattern on the screen? (a) It disappears. (b) The bright and dark fringes stay in the same locations, but the contrast is reduced. (c) The bright fringes are closer together. (d) The color shifts toward blue. (e) The bright fringes are farther apart. (f) The bright fringes are in continuous motion. (g) No change happens in the interference pattern.

6. In Young's double-slit experiment, why do we use monochromatic light? If white light is used, how would the pattern change?

7. O Suppose you perform Young's double-slit experiment with the slit separation slightly smaller than the wavelength of the light. As a screen, you use a large half-cylinder with its axis along the midline between the slits. What interference pattern will you see on the interior surface of the cylinder? (a) bright and dark fringes so closely spaced as to be indistinguishable (b) one central bright fringe and two dark fringes only (c) a completely bright screen with no dark fringes (d) one central dark fringe and two bright fringes only (e) a completely dark screen with no bright fringes

8. As a soap bubble evaporates, it appears black immediately before it breaks, as at the top of the circular film shown in Figure Q37.8. Explain this phenomenon in terms of the phase changes that occur on reflection from the two surfaces of the soap film.

Figure Q37.8 Question 8 and Problem 63.

9. O A film of oil on a puddle in a parking lot shows a variety of bright colors in swirled patches. What can you say about the thickness of the oil film? (a) It is much less than the wavelength of visible light. (b) It is of the same order of magnitude as the wavelength of visible light. (c) It is much greater than the wavelength of visible light. (d) It might have any relationship to the wavelength of visible light.

10. O Assume the index of refraction of flint glass is 1.66 and the index of refraction of crown glass is 1.52. **(i)** A film formed by one drop of sassafras oil, on a horizontal surface of a flint glass block, is viewed by reflected light. The film appears brightest at its outer margin, where it is thinnest. A film of the same oil on crown glass appears dark at its outer margin. What can you say about the index of refraction of the oil? (a) It must be less than 1.52. (b) It must be between 1.52 and 1.66. (c) It must be greater than 1.66. (d) None of statements (a) through (c) is necessarily true. **(ii)** Could a very thin film of some other liquid appear bright by reflected light on both of the glass blocks? **(iii)** Could it appear dark on both?

(iv) Could it appear dark on crown glass and bright on flint glass? Experiments described by Thomas Young suggested this question.

11. A lens with outer radius of curvature R and index of refraction n rests on a flat glass plate. The combination is illuminated with white light from above and observed from above. Is there a dark spot or a light spot at the center of the lens? What does it mean if the observed rings are noncircular?

12. Why is the lens on a good-quality camera coated with a thin film?

13. O Green light has a wavelength of 500 nm in air. **(i)** Assume green light is reflected from a mirror with angle of incidence 0°. The incident and reflected waves together constitute a standing wave with what distance from one node to the next antinode? (a) 1 000 nm (b) 500 nm (c) 250 nm (d) 125 nm (e) 62.5 nm **(ii)** The green light is sent into a Michelson interferometer that is adjusted to produce a central bright circle. How far must the interferometer's moving mirror be shifted to change the center of the pattern into a dark circle? Choose from the same possibilities. **(iii)** The light is reflected perpendicularly from a thin film of a plastic with index of refraction 2.00. The film appears bright in the reflected light. How much additional thickness would make the film appear dark?

14. O Using a Michelson interferometer, shown in Active Figure 37.14, you are viewing a dark circle at the center of the interference pattern. As you gradually move the light source toward the central mirror M_0, through a distance $\lambda/2$, what do you see? (a) There is no change in the pattern. (b) The dark circle changes into a bright circle. (c) The dark circle changes into a bright circle and then back into a dark circle. (d) The dark circle changes into a bright circle, then into a dark circle, and then into a bright circle.

Problems

WebAssign The Problems from this chapter may be assigned online in WebAssign.

ThomsonNOW Sign in at **www.thomsonedu.com** and go to ThomsonNOW to assess your understanding of this chapter's topics with additional quizzing and conceptual questions.

1, 2, 3 denotes straightforward, intermediate, challenging; □ denotes full solution available in *Student Solutions Manual/Study Guide;* ▲ denotes coached solution with hints available at **www.thomsonedu.com;** denotes developing symbolic reasoning; ● denotes asking for qualitative reasoning; ☕ denotes computer useful in solving problem

Section 37.1 Conditions for Interference

Section 37.2 Young's Double-Slit Experiment

Section 37.3 Light Waves in Interference

> *Note:* Problems 4, 5, 6, 7, 8, and 10 in Chapter 18 can be assigned with this section.

1. A laser beam ($\lambda = 632.8$ nm) is incident on two slits 0.200 mm apart. How far apart are the bright interference fringes on a screen 5.00 m away from the double slits?

2. A Young's interference experiment is performed with monochromatic light. The separation between the slits is 0.500 mm, and the interference pattern on a screen 3.30 m away shows the first side maximum 3.40 mm from the center of the pattern. What is the wavelength?

3. ▲ Two radio antennas separated by 300 m as shown in Figure P37.3 simultaneously broadcast identical signals at the same wavelength. A radio in a car traveling due north receives the signals. (a) If the car is at the position of the second maximum, what is the wavelength of the signals? (b) How much farther must the car travel to encounter the next minimum in reception? *Note:* Do not use the small-angle approximation in this problem.

Figure P37.3

4. In a location where the speed of sound is 354 m/s, a 2 000-Hz sound wave impinges on two slits 30.0 cm apart. (a) At what angle is the first maximum located? (b) **What If?** If the sound wave is replaced by 3.00-cm microwaves, what slit separation gives the same angle for the first maximum? (c) **What If?** If the slit separation is 1.00 μm, what frequency of light gives the same first maximum angle?

5. ▲ Young's double-slit experiment is performed with 589-nm light and a distance of 2.00 m between the slits and the screen. The tenth interference minimum is observed 7.26 mm from the central maximum. Determine the spacing of the slits.

6. ● Write the statement of a problem, including data, for which the following equations appear in the solution.

$$\lambda = \frac{343 \text{ m/s}}{1\,620/\text{s}}$$

$$(35.0 \text{ cm}) \sin \theta_0 = 0\lambda$$

$$(35.0 \text{ cm}) \sin \theta_{1 \text{ soft}} = 0.5\lambda \qquad (35.0 \text{ cm}) \sin \theta_{1 \text{ loud}} = 1\lambda$$

$$(35.0 \text{ cm}) \sin \theta_{2 \text{ soft}} = 1.5\lambda \qquad (35.0 \text{ cm}) \sin \theta_{2 \text{ loud}} = 2\lambda$$

State the solution to the problem, including values for each quantity that appears as an unknown. State what you can conclude from the last of the set of six equations. Does this equation describe an angle $\theta_{2 \text{ loud}}$ that is larger than 90°?

7. Two narrow, parallel slits separated by 0.250 mm are illuminated by green light ($\lambda = 546.1$ nm). The interference pattern is observed on a screen 1.20 m away from the plane of the slits. Calculate the distance (a) from the central maximum to the first bright region on either side of the central maximum and (b) between the first and second dark bands.

8. A riverside warehouse has two open doors as shown in Figure P37.8. Its walls are lined with sound-absorbing material. A boat on the river sounds its horn. To person A, the sound is loud and clear. To person B, the sound is barely audible. The principal wavelength of the sound

Figure P37.8

waves is 3.00 m. Assuming person B is at the position of the first minimum, determine the distance between the

doors, center to center.

9. Light with wavelength 442 nm passes through a double-slit system that has a slit separation $d = 0.400$ mm. Determine how far away a screen must be placed so that dark fringes appear directly opposite both slits, with only one bright fringe between them.

10. Two slits are separated by 0.320 mm. A beam of 500-nm light strikes the slits, producing an interference pattern. Determine the number of maxima observed in the angular range $-30.0° < \theta < 30.0°$.

11. ● Young's double-slit experiment underlies the *instrument landing system* used to guide aircraft to safe landings when the visibility is poor. Although real systems are more complicated than the example described here, they operate on the same principles. A pilot is trying to align her plane with a runway as suggested in Figure P37.11a. Two radio antennas A_1 and A_2 are positioned adjacent to the runway, separated by 40.0 m. The antennas broadcast unmodulated coherent radio waves at 30.0 MHz. (a) Find the wavelength of the waves. The pilot "locks onto" the strong signal radiated along an interference maximum, and steers the plane to keep the received signal strong. If she has found the central maximum, the plane will have precisely the right heading to land when it reaches the runway. (b) **What If?** Suppose the plane is flying along the first side maximum instead (Fig. P37.11b). How far to the side of the runway centerline will the plane be when it is 2.00 km from the antennas? (c) It is possible to tell the pilot that she is on the wrong maximum by sending out

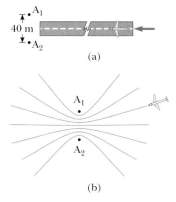

Figure P37.11

two signals from each antenna and equipping the aircraft with a two-channel receiver. The ratio of the two frequencies must not be the ratio of small integers (such as $\frac{3}{4}$). Explain how this two-frequency system would work and why it would not necessarily work if the frequencies were related by an integer ratio.

12. A student holds a laser that emits light of wavelength 633 nm. The beam passes though a pair of slits separated by 0.300 mm, in a glass plate attached to the front of the laser. The beam then falls perpendicularly on a screen, creating an interference pattern on it. The student begins to walk directly toward the screen at 3.00 m/s. The central maximum on the screen is stationary. Find the speed of the first-order maxima on the screen.

13. In Figure 37.5, let L = 1.20 m and d = 0.120 mm and assume the slit system is illuminated with monochromatic 500-nm light. Calculate the phase difference between the two wave fronts arriving at P when (a) θ = 0.500° and (b) y = 5.00 mm. (c) What is the value of θ for which the phase difference is 0.333 rad? (d) What is the value of θ for which the path difference is $\lambda/4$?

14. Coherent light rays of wavelength λ strike a pair of slits separated by distance d at an angle θ_1 as shown in Figure P37.14. Assume an interference maximum is formed at an angle θ_2 a great distance from the slits. Show that $d(\sin \theta_2 - \sin \theta_1) = m\lambda$, where m is an integer.

Figure P37.14

15. ● In a double-slit arrangement of Figure 37.5, d = 0.150 mm, L = 140 cm, λ = 643 nm, and y = 1.80 cm. (a) What is the path difference δ for the rays from the two slits arriving at P? (b) Express this path difference in terms of λ. (c) Does P correspond to a maximum, a minimum, or an intermediate condition? Give evidence for your answer.

Section 37.4 Intensity Distribution of the Double-Slit Interference Pattern

16. The intensity on the screen at a certain point in a double-slit interference pattern is 64.0% of the maximum value. (a) What minimum phase difference (in radians) between

sources produces this result? (b) Express this phase difference as a path difference for 486.1-nm light.

17. ▲ In Figure 37.5, let L = 120 cm and d = 0.250 cm. The slits are illuminated with coherent 600-nm light. Calculate the distance y above the central maximum for which the average intensity on the screen is 75.0% of the maximum.

18. Two slits are separated by 0.180 mm. An interference pattern is formed on a screen 80.0 cm away by 656.3-nm light. Calculate the fraction of the maximum intensity 0.600 cm above the central maximum.

19. ▲ Show that the two waves with wave functions E_1 = 6.00 sin (100πt) and E_2 = 8.00 sin (100πt + $\pi/2$) add to give a wave with the wave function E_R sin (100πt + ϕ). Find the required values for E_R and ϕ.

20. Make a graph of I/I_{max} as a function of θ for the interference pattern produced by the arrangement described in Problem 7. Let θ range over the interval from $-0.3°$ to $+0.3°$.

21. Two narrow, parallel slits separated by 0.850 mm are illuminated by 600-nm light, and the viewing screen is 2.80 m away from the slits. (a) What is the phase difference between the two interfering waves on a screen at a point 2.50 mm from the central bright fringe? (b) What is the ratio of the intensity at this point to the intensity at the center of a bright fringe?

22. ● Monochromatic coherent light of amplitude E_0 and angular frequency ω passes through three parallel slits each separated by a distance d from its neighbor. (a) Show that the time-averaged intensity as a function of the angle θ is

$$I(\theta) = I_{max}\left[1 + 2 \cos\left(\frac{2\pi d \sin \theta}{\lambda}\right)\right]^2$$

(b) Explain how this expression describes both the primary and the secondary maxima. Determine the ratio of the intensities of the primary and secondary maxima.

Section 37.5 Change of Phase Due to Reflection

Section 37.6 Interference in Thin Films

23. ● An oil film (n = 1.45) floating on water is illuminated by white light at normal incidence. The film is 280 nm

thick. Find (a) the color of the light in the visible spectrum most strongly reflected and (b) the color of the light in the spectrum most strongly transmitted. Explain your reasoning.

24. A soap bubble ($n = 1.33$) is floating in air. If the thickness of the bubble wall is 115 nm, what is the wavelength of the light that is most strongly reflected?

25. A thin film of oil ($n = 1.25$) is located on smooth, wet pavement. When viewed perpendicular to the pavement, the film reflects most strongly red light at 640 nm and reflects no green light at 512 nm. How thick is the oil film?

26. A possible means for making an airplane invisible to radar is to coat the plane with an antireflective polymer. If radar waves have a wavelength of 3.00 cm and the index of refraction of the polymer is $n = 1.50$, how thick would you make the coating?

27. A material having an index of refraction of 1.30 is used as an antireflective coating on a piece of glass ($n = 1.50$). What should the minimum thickness of this film be to minimize reflection of 500-nm light?

28. A film of MgF_2 ($n = 1.38$) having thickness 1.00×10^{-5} cm is used to coat a camera lens. Are any wavelengths in the visible spectrum intensified in the reflected light?

29. ● Astronomers observe the chromosphere of the Sun with a filter that passes the red hydrogen spectral line of wavelength 656.3 nm, called the H_α line. The filter consists of a transparent dielectric of thickness d held between two partially aluminized glass plates. The filter is held at a constant temperature. (a) Find the minimum value of d that produces maximum transmission of perpendicular H_α light if the dielectric has an index of refraction of 1.378. (b) **What If?** If the temperature of the filter increases above the normal value, what happens to the transmitted wavelength? (The index of refraction of the filter does not change significantly.) (c) The dielectric will also pass what near-visible wavelength? One of the glass plates is colored red to absorb this light.

30. A beam of 580-nm light passes through two closely spaced glass plates as shown in Figure P37.30. For what minimum nonzero value of the plate separation d is the transmitted light bright?

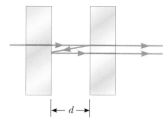

Figure P37.30

31. ▲ An air wedge is formed between two glass plates separated at one edge by a very fine wire as shown in Figure P37.31. When the wedge is illuminated from above by 600-nm light and viewed from above, 30 dark fringes are observed. Calculate the radius of the wire.

Figure P37.31   Problems 31 and 33.

32. When a liquid is introduced into the air space between the lens and the plate in a Newton's-rings apparatus, the diameter of the tenth ring changes from 1.50 to 1.31 cm. Find the index of refraction of the liquid.

33. Two glass plates 10.0 cm long are in contact at one end and separated at the other end by a thread 0.050 0 mm in diameter (Fig. P37.31). Light containing the two wavelengths 400 nm and 600 nm is incident perpendicularly and viewed by reflection. At what distance from the contact point is the next dark fringe?

Section 37.7 The Michelson Interferometer

34. Monochromatic light is beamed into a Michelson interferometer. The movable mirror is displaced 0.382 mm, causing the interferometer pattern to reproduce itself 1 700 times. Determine the wavelength of the light. What color is it?

35. Mirror M_1 in Active Figure 37.14 is moved through a displacement ΔL. During this displacement, 250 fringe reversals (formation of successive dark or bright bands) are counted. The light being used has a wavelength of 632.8 nm. Calculate the displacement ΔL.

36. One leg of a Michelson interferometer contains an evacuated cylinder of length L, having glass plates on each end. A gas is slowly leaked into the cylinder until a pressure of 1 atm is reached. If N bright fringes pass on the screen during this process when light of wavelength λ is used, what is the index of refraction of the gas?

Figure P37.40

Additional Problems

37. In an experiment similar to that of Example 37.1, green light with wavelength 560 nm, sent through a pair of slits 30.0 μm apart, produces bright fringes 2.24 cm apart on a screen 1.20 m away. Calculate the fringe separation for this same arrangement assuming that the apparatus is submerged in a tank containing a sugar solution with index of refraction 1.38.

38. In the **What If?** section of Example 37.2, it was claimed that overlapping fringes in a two-slit interference pattern for two different wavelengths obey the following relationship even for large values of the angle θ:

$$\frac{m'}{m} = \frac{\lambda}{\lambda'}$$

(a) Prove this assertion. (b) Using the data in Example 37.2, find the nonzero value of y on the screen at which the fringes from the two wavelengths first coincide.

39. One radio transmitter A operating at 60.0 MHz is 10.0 m from another similar transmitter B that is 180° out of phase with A. How far must an observer move from A toward B along the line connecting the two transmitters to reach the nearest point where the two beams are in phase?

40. Review problem. This problem extends the result of Problem 10 in Chapter 18. Figure P37.40 shows two adjacent vibrating balls dipping into a pan of water. At distant points, they produce an interference pattern of water waves as shown in Figure 37.3. Let λ represent the wavelength of the ripples. Show that the two sources produce a standing wave along the line segment, of length d, between them. In terms of λ and d, find the number of nodes and the number of antinodes in the standing wave. Find the number of zones of constructive and of destructive interference in the interference pattern far away from the sources. Each line of destructive interference springs from a node in the standing wave, and each line of constructive interference springs from an antinode.

41. Raise your hand and hold it flat. Think of the space between your index finger and your middle finger as one slit and think of the space between middle finger and ring finger as a second slit. (a) Consider the interference resulting from sending coherent visible light perpendicularly through this pair of openings. Compute an order-of-magnitude estimate for the angle between adjacent zones of constructive interference. (b) To make the angles in the interference pattern easy to measure with a plastic protractor, you should use an electromagnetic wave with frequency of what order of magnitude? How is this wave classified on the electromagnetic spectrum?

42. Two coherent waves, coming from sources at different locations, move along the x axis. Their wave functions are

$$E_1 = (860 \text{ V/m}) \sin\left[\frac{2\pi x_1}{650 \text{ nm}} - 2\pi(462 \text{ THz})t + \frac{\pi}{6}\right]$$

and

$$E_2 = (860 \text{ V/m}) \sin\left[\frac{2\pi x_2}{650 \text{ nm}} - 2\pi(462 \text{ THz})t + \frac{\pi}{8}\right]$$

Determine the relationship between x_1 and x_2 that produces constructive interference when the two waves are superposed.

43. In a Young's interference experiment, the two slits are separated by 0.150 mm and the incident light includes two wavelengths: $\lambda_1 = 540$ nm (green) and $\lambda_2 = 450$ nm (blue). The overlapping interference patterns are observed on a screen 1.40 m from the slits. Calculate the minimum distance from the center of the screen to a point where a bright fringe of the green light coincides with a bright fringe of the blue light.

44. In a Young's double-slit experiment using light of wavelength λ, a thin piece of Plexiglas having index of refraction n covers one of the slits. If the center point on the screen is a dark spot instead of a bright spot, what is the minimum thickness of the Plexiglas?

45. **Review problem.** A flat piece of glass is held stationary and horizontal above the flat top end of a 10.0-cm-long vertical metal rod that has its lower end rigidly fixed. The thin film of air between the rod and glass is observed to be bright by reflected light when it is illuminated by light of wavelength 500 nm. As the temperature is slowly increased by 25.0°C, the film changes from bright to dark and back to bright 200 times. What is the coefficient of linear expansion of the metal?

46. A certain crude oil has an index of refraction of 1.25. A ship dumps 1.00 m³ of this oil into the ocean, and the oil spreads into a thin uniform slick. If the film produces a first-order maximum of light of wavelength 500 nm normally incident on it, how much surface area of the ocean does the oil slick cover? Assume the index of refraction of the ocean water is 1.34.

47. Astronomers observe a 60.0-MHz radio source both directly and by reflection from the sea. If the receiving dish is 20.0 m above sea level, what is the angle of the radio source above the horizon at first maximum?

48. Interference effects are produced at point P on a screen as a result of direct rays from a 500-nm source and reflected rays from the mirror as shown in Figure P37.48. Assume the source is 100 m to the left of the screen and 1.00 cm above the mirror. Find the distance y to the first dark band above the mirror.

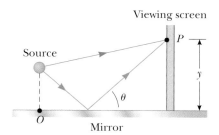

Figure P37.48

49. The waves from a radio station can reach a home receiver by two paths. One is a straight-line path from transmitter to home, a distance of 30.0 km. The second path is by reflection from the ionosphere (a layer of ionized air molecules high in the atmosphere). Assume this reflection takes place at a point midway between receiver and transmitter and the wavelength broadcast by the radio station is 350 m. Find the minimum height of the ionospheric layer that could produce destructive interference between the direct and reflected beams. Assume no phase change occurs on reflection.

50. Many cells are transparent and colorless. Structures of great interest in biology and medicine can be practically invisible to ordinary microscopy. To indicate the size and shape of cell structures, an *interference microscope* reveals a difference in index of refraction as a shift in interference fringes. The idea is exemplified in the following problem. An air wedge is formed between two glass plates in contact along one edge and slightly separated at the opposite edge. When the plates are illuminated with monochromatic light from above, the reflected light has 85 dark fringes. Calculate the number of dark fringes that appear if water ($n = 1.33$) replaces the air between the plates.

51. Measurements are made of the intensity distribution within the central bright fringe in a Young's interference pattern (see Fig. 37.7). At a particular value of y, it is found that $I/I_{max} = 0.810$ when 600-nm light is used. What wavelength of light should be used to reduce the relative intensity at the same location to 64.0% of the maximum intensity?

52. Our discussion of the techniques for determining constructive and destructive interference by reflection from a thin film in air has been confined to rays striking the film at nearly normal incidence. **What If?** Assume a ray is incident at an angle of 30.0° (relative to the normal) on a film with index of refraction 1.38. Calculate the minimum thickness for constructive interference of sodium light with a wavelength of 590 nm.

53. The condition for constructive interference by reflection from a thin film in air as developed in Section 37.6 assumes nearly normal incidence. **What If?** Show that if the light is incident on the film at a nonzero angle ϕ_1 (relative to the normal), the condition for constructive interference is $2nt \cos \theta_2 = (m + \frac{1}{2})\lambda$, where θ_2 is the angle of refraction.

54. ● The quantity δ in Equation 37.1 is called the path difference. Its size in comparison to the wavelength controls the character of the interference between two beams in vacuum by controlling the phase difference between the beams. The analogous quantity nt in Equations 37.16 and 37.17 is called the *optical path length* corresponding to the geometrical distance t. The optical

2 = intermediate; 3 = challenging; ☐ = SSM/SG; ▲ = ThomsonNOW; = symbolic reasoning; ● = qualitative reasoning

path length is proportional to n because a larger index of refraction shortens the wavelength, so more cycles of a wave fit into a particular geometrical distance. (a) Assume a mixture of corn syrup and water is prepared in a tank, with its index of refraction n increasing uniformly from 1.33 at $y = 20.0$ cm at the top to 1.90 at $y = 0$. Write the index of refraction $n(y)$ as a function of y. (b) Compute the optical path length corresponding to the 20-cm height of the tank by calculating

$$\int_0^{20 \text{ cm}} n(y)\,dy$$

(c) Suppose a narrow beam of light is directed into the mixture with its original direction between horizontal and vertically upward. Qualitatively describe its path.

55. (a) Both sides of a uniform film that has index of refraction n and thickness d are in contact with air. For normal incidence of light, an intensity minimum is observed in the reflected light at λ_2 and an intensity maximum is observed at λ_1, where $\lambda_1 > \lambda_2$. Assuming that no intensity minima are observed between λ_1 and λ_2, show that the integer m in Equations 37.16 and 37.17 is given by $m = \lambda_1/2(\lambda_1 - \lambda_2)$. (b) Determine the thickness of the film, assuming $n = 1.40$, $\lambda_1 = 500$ nm, and $\lambda_2 = 370$ nm.

56. Figure P37.56 shows a radio-wave transmitter and a receiver separated by a distance d and both a distance h above the ground. The receiver can receive signals both directly from the transmitter and indirectly from signals that reflect from the ground. Assume the ground is level between the transmitter and receiver and a $180°$ phase shift occurs upon reflection. Determine the longest wavelengths that interfere (a) constructively and (b) destructively.

Figure P37.56

57. Consider the double-slit arrangement shown in Figure P37.57, where the slit separation is d and the distance from the slit to the screen is L. A sheet of transparent plastic having an index of refraction n and thickness t is placed over the upper slit. As a result, the central maximum of the interference pattern moves upward a distance y'. Find y'.

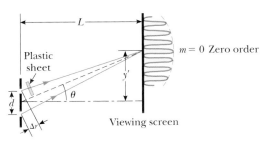

Figure P37.57

58. A piece of transparent material having an index of refraction n is cut into the shape of a wedge as shown in Figure P37.58. The angle of the wedge is small. Monochromatic light of wavelength λ is normally incident from above and is viewed from above. Let h represent the height of the wedge and ℓ its width. Show that bright fringes occur at the positions $x = \lambda\ell(m + \frac{1}{2})/2hn$ and dark fringes occur at the positions $x = \lambda\ell m/2hn$, where $m = 0, 1, 2, \ldots$ and x is measured as shown.

Figure P37.58

59. In a Newton's-rings experiment, a plano-convex glass ($n = 1.52$) lens having diameter 10.0 cm is placed on a flat plate as shown in Figure 37.12a. When 650-nm light is incident normally, 55 bright rings are observed, with the last one precisely on the edge of the lens. (a) What is the radius of curvature of the convex surface of the lens? (b) What is the focal length of the lens?

60. A plano-convex lens has index of refraction n. The curved side of the lens has radius of curvature R and rests on a flat glass surface of the same index of refraction, with a film of index n_{film} between them, as shown in Fig. 37.12a. The lens is illuminated from above by light of wavelength λ. Show that the dark Newton's rings have radii given approximately by

$$r \approx \sqrt{\frac{m\lambda R}{n_{\text{film}}}}$$

where m is an integer and r is much less than R.

61. A plano-concave lens having index of refraction 1.50 is placed on a flat glass plate as shown in Figure P37.61. Its curved surface, with radius of curvature 8.00 m, is on the bottom. The lens is illuminated from above with yellow sodium light of wavelength 589 nm, and a series of concentric bright and dark rings is observed by reflection. The interference pattern has a dark spot at the center, surrounded by 50 dark rings, the largest of which is at the outer edge of the lens. (a) What is the thickness of the air layer at the center of the interference pattern? (b) Calculate the radius of the outermost dark ring. (c) Find the focal length of the lens.

Figure P37.61

62. A plano-convex lens having a radius of curvature of $r = 4.00$ m is placed on a concave glass surface whose radius of curvature is $R = 12.0$ m as shown in Figure P37.62. Determine the radius of the 100th bright ring, assuming 500-nm light is incident normal to the flat surface of the lens.

Figure P37.62

63. Figure Q37.8 shows an unbroken soap film in a circular frame. The film thickness increases from top to bottom, slowly at first and then rapidly. As a simpler model, consider a soap film ($n = 1.33$) contained within a rectangular wire frame. The frame is held vertically so that the film drains downward and forms a wedge with flat faces. The thickness of the film at the top is essentially zero. The film is viewed in reflected white light with near-normal incidence, and the first violet ($\lambda = 420$ nm) interference band is observed 3.00 cm from the top edge of the film. (a) Locate the first red ($\lambda = 680$ nm) interference band. (b) Determine the film thickness at the positions of the violet and red bands. (c) What is the wedge angle of the film?

64. Compact disc (CD) and digital videodisc (DVD) players use interference to generate strong signals from tiny bumps, shown in Figure P35.40. A pit's depth is chosen to be one-quarter of the wavelength of the laser light used to read the disc. Then light reflected from the pit and light reflected from the adjoining flat surface differ in path length traveled by one-half wavelength, interfering destructively at the detector. As the disc rotates, the light intensity drops significantly every time light is reflected from near a pit edge. The space between the leading and trailing edges of a pit determines the time interval between the fluctuations. The series of time intervals is decoded into a series of zeros and ones that carries the stored information. Assume infrared light with a wavelength of 780 nm in vacuum is used in a CD player. The disc is coated with plastic having an index of refraction of 1.50. What should the depth of each pit be? A DVD player uses light of a shorter wavelength, and the pit dimensions are correspondingly smaller. This reduction is one factor resulting in a DVD's greater storage capacity compared with that of a CD.

65. Interference fringes are produced using Lloyd's mirror and a 606-nm source as shown in Figure 37.9. Fringes 1.20 mm apart are formed on a screen 2.00 m from the real source S. Find the vertical distance h of the source above the reflecting surface.

66. Slit 1 of a double slit is wider than slit 2 so that the light from slit 1 has an amplitude 3.00 times that of the light from slit 2. Show that Equation 37.12 is replaced by the equation $I = (4I_{max}/9)(1 + 3\cos^2 \phi/2)$ for this situation.

67. Monochromatic light of wavelength 620 nm passes through a very narrow slit S and then strikes a screen in which are two parallel slits, S_1 and S_2, as shown in Figure P37.67. Slit S_1 is directly in line with S and at a distance of $L = 1.20$ m away from S, whereas S_2 is displaced a distance d to one side. The light is detected at point P on a second screen, equidistant from S_1 and S_2. When either slit S_1 or S_2 is open, equal light intensities are measured at point P. When both slits are open, the intensity is three times larger. Find the minimum possible value for the slit separation d.

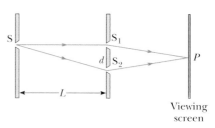

Figure P37.67

Answers to Quick Quizzes

37.1 (c). Equation 37.7 shows that decreasing λ or L will bring the fringes closer together. Immersing the apparatus in water decreases the wavelength so that the fringes move closer together.

37.2 The graph is shown below. The width of the primary maxima is slightly narrower than the $N = 5$ primary width but wider than the $N = 10$ primary width. Because $N = 6$, the secondary maxima are $\frac{1}{36}$ as intense as the primary maxima.

37.3 (a). At the left edge, the air wedge has zero thickness and the only contribution to the interference is the 180° phase shift as the light reflects from the upper surface of the glass slide.

The Hubble Space Telescope does its viewing above the atmosphere and does not suffer from the atmospheric blurring, caused by air turbulence, that plagues ground-based telescopes. Despite this advantage, it does have limitations due to diffraction effects. In this chapter, we show how the wave nature of light limits the ability of any optical system to distinguish between closely spaced objects. (© Denis Scott/CORBIS)

38.1 Introduction to Diffraction Patterns

38.2 Diffraction Patterns from Narrow Slits

38.3 Resolution of Single-Slit and Circular Apertures

38.4 The Diffraction Grating

38.5 Diffraction of X-Rays by Crystals

38.6 Polarization of Light Waves

38 Diffraction Patterns and Polarization

When plane light waves pass through a small aperture in an opaque barrier, the aperture acts as if it were a point source of light, with waves entering the shadow region behind the barrier. This phenomenon, known as diffraction, can be described only with a wave model for light as discussed in Section 35.3. In this chapter, we investigate the features of the *diffraction pattern* that occurs when the light from the aperture is allowed to fall upon a screen.

In Chapter 34, we learned that electromagnetic waves are transverse. That is, the electric and magnetic field vectors associated with electromagnetic waves are perpendicular to the direction of wave propagation. In this chapter, we show that under certain conditions these transverse waves with electric field vectors in all possible transverse directions can be *polarized* in various ways. In other words, only certain directions of the electric field vectors are present in the polarized wave.

38.1 Introduction to Diffraction Patterns

In Section 35.3, we discussed that light of wavelength comparable to or larger than the width of a slit spreads out in all forward directions upon passing through the slit. This phenomenon is called *diffraction*. When light passes through a narrow slit, it spreads beyond the narrow path defined by the slit into regions that would be in shadow if light traveled in straight lines. Other waves, such as sound waves and

Figure 38.1 The diffraction pattern that appears on a screen when light passes through a narrow vertical slit. The pattern consists of a broad central fringe and a series of less intense and narrower side fringes.

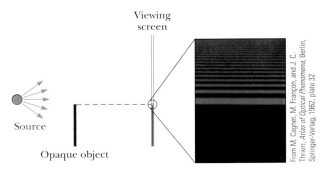

Figure 38.2 Light from a small source passes by the edge of an opaque object and continues on to a screen. A diffraction pattern consisting of bright and dark fringes appears on the screen in the region above the edge of the object.

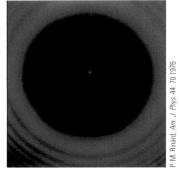

Figure 38.3 Diffraction pattern created by the illumination of a penny, with the penny positioned midway between screen and light source. Note the bright spot at the center.

water waves, also have this property of spreading when passing through apertures or by sharp edges.

You might expect that the light passing through a small opening would simply result in a broad region of light on a screen due to the spreading of the light as it passes through the opening. We find something more interesting, however. A **diffraction pattern** consisting of light and dark areas is observed, somewhat similar to the interference patterns discussed earlier. For example, when a narrow slit is placed between a distant light source (or a laser beam) and a screen, the light produces a diffraction pattern like that shown in Figure 38.1. The pattern consists of a broad, intense central band (called the **central maximum**) flanked by a series of narrower, less intense additional bands (called **side maxima** or **secondary maxima**) and a series of intervening dark bands (or **minima**). Figure 38.2 shows a diffraction pattern associated with light passing by the edge of an object. Again we see bright and dark fringes, which is reminiscent of an interference pattern.

Figure 38.3 shows a diffraction pattern associated with the shadow of a penny. A bright spot occurs at the center, and circular fringes extend outward from the shadow's edge. We can explain the central bright spot by using the wave theory of light, which predicts constructive interference at this point. From the viewpoint of geometric optics (in which light is viewed as rays traveling in straight lines), we expect the center of the shadow to be dark because that part of the viewing screen is completely shielded by the penny.

Shortly before the central bright spot was first observed, one of the supporters of geometric optics, Simeon Poisson, argued that if Augustin Fresnel's wave theory of light were valid, a central bright spot should be observed in the shadow of a circular object illuminated by a point source of light. To Poisson's astonishment, the spot was observed by Dominique Arago shortly thereafter. Therefore, Poisson's prediction reinforced the wave theory rather than disproving it.

PITFALL PREVENTION 38.1
Diffraction Versus Diffraction Pattern

Diffraction refers to the general behavior of waves spreading out as they pass through a slit. We used diffraction in explaining the existence of an interference pattern in Chapter 37. A *diffraction pattern* is actually a misnomer, but is deeply entrenched in the language of physics. The diffraction pattern seen on a screen when a single slit is illuminated is actually another interference pattern. The interference is between parts of the incident light illuminating different regions of the slit.

38.2 Diffraction Patterns from Narrow Slits

Let's consider a common situation, that of light passing through a narrow opening modeled as a slit and projected onto a screen. To simplify our analysis, we assume the observing screen is far from the slit and the rays reaching the screen are approximately parallel. (This situation can also be achieved experimentally by using a converging lens to focus the parallel rays on a nearby screen.) In this model, the pattern on the screen is called a **Fraunhofer diffraction pattern.**[1]

Active Figure 38.4a shows light entering a single slit from the left and diffracting as it propagates toward a screen. Active Figure 38.4b is a photograph of a single-slit Fraunhofer diffraction pattern. A bright fringe is observed along the axis at

[1] If the screen is brought close to the slit (and no lens is used), the pattern is a *Fresnel* diffraction pattern. The Fresnel pattern is more difficult to analyze, so we shall restrict our discussion to Fraunhofer diffraction.

(a)   (b)

ACTIVE FIGURE 38.4

(a) Fraunhofer diffraction pattern of a single slit. The pattern consists of a central bright fringe flanked by much weaker maxima alternating with dark fringes. (Drawing not to scale.) The labels y indicate positions of dark fringes on the screen. (b) Photograph of a single-slit Fraunhofer diffraction pattern.

Sign in at www.thomsonedu.com and go to ThomsonNOW to adjust the slit width and the wavelength of the light to see the effect on the diffraction pattern.

$\theta = 0$, with alternating dark and bright fringes on each side of the central bright fringe.

Until now, we have assumed slits are point sources of light. In this section, we abandon that assumption and see how the finite width of slits is the basis for understanding Fraunhofer diffraction. We can explain some important features of this phenomenon by examining waves coming from various portions of the slit as shown in Figure 38.5. According to Huygens's principle, **each portion of the slit acts as a source of light waves.** Hence, light from one portion of the slit can interfere with light from another portion, and the resultant light intensity on a viewing screen depends on the direction θ. Based on this analysis, we recognize that a diffraction pattern is actually an interference pattern in which the different sources of light are different portions of the single slit!

To analyze the diffraction pattern, let's divide the slit into two halves as shown in Figure 38.5. Keeping in mind that all the waves are in phase as they leave the slit, consider rays 1 and 3. As these two rays travel toward a viewing screen far to the right of the figure, ray 1 travels farther than ray 3 by an amount equal to the path difference $(a/2) \sin \theta$, where a is the width of the slit. Similarly, the path difference between rays 2 and 4 is also $(a/2) \sin \theta$, as is that between rays 3 and 5. If this path difference is exactly half a wavelength (corresponding to a phase difference of $180°$), the pairs of waves cancel each other and destructive interference results. This cancellation occurs for any two rays that originate at points separated by half the slit width because the phase difference between two such points is $180°$. Therefore, waves from the upper half of the slit interfere destructively with waves from the lower half when

$$\frac{a}{2} \sin \theta = \pm \frac{\lambda}{2}$$

or when

$$\sin \theta = \pm \frac{\lambda}{a}$$

Dividing the slit into four equal parts and using similar reasoning, we find that the viewing screen is also dark when

$$\sin \theta = \pm 2 \frac{\lambda}{a}$$

Likewise, dividing the slit into six equal parts shows that darkness occurs on the screen when

$$\sin \theta = \pm 3 \frac{\lambda}{a}$$

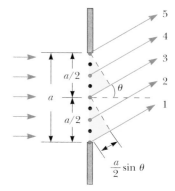

Figure 38.5 Paths of light rays that encounter a narrow slit of width a and diffract toward a screen in the direction described by angle θ. Each portion of the slit acts as a point source of light waves. The path difference between rays 1 and 3, rays 2 and 4, or rays 3 and 5 is $(a/2) \sin \theta$. (Drawing not to scale.)

Therefore, the general condition for destructive interference is

Condition for destructive ▶
interference for a single slit

$$\sin \theta_{\text{dark}} = m\frac{\lambda}{a} \quad m = \pm 1, \pm 2, \pm 3, \ldots \tag{38.1}$$

PITFALL PREVENTION 38.2
Similar Equation Warning!

Equation 38.1 has exactly the same form as Equation 37.2, with d, the slit separation, used in Equation 37.2 and a, the slit width, used in Equation 38.1. Equation 37.2, however, describes the *bright* regions in a two-slit interference pattern, whereas Equation 38.1 describes the *dark* regions in a single-slit diffraction pattern. Furthermore, $m = 0$ does not represent a dark fringe in the diffraction pattern.

This equation gives the values of θ_{dark} for which the diffraction pattern has zero light intensity, that is, when a dark fringe is formed. It tells us nothing, however, about the variation in light intensity along the screen. The general features of the intensity distribution are shown in Active Figure 38.4. A broad, central bright fringe is observed; this fringe is flanked by much weaker bright fringes alternating with dark fringes. The various dark fringes occur at the values of θ_{dark} that satisfy Equation 38.1. Each bright-fringe peak lies approximately halfway between its bordering dark-fringe minima. Notice that the central bright maximum is twice as wide as the secondary maxima.

Quick Quiz 38.1 Suppose the slit width in Active Figure 38.4 is made half as wide. Does the central bright fringe (a) become wider, (b) remain the same, or (c) become narrower?

EXAMPLE 38.1 **Where Are the Dark Fringes?**

Light of wavelength 580 nm is incident on a slit having a width of 0.300 mm. The viewing screen is 2.00 m from the slit. Find the positions of the first dark fringes and the width of the central bright fringe.

SOLUTION

Conceptualize Based on the problem statement, we imagine a single-slit diffraction pattern similar to that in Active Figure 38.4.

Categorize We categorize this example as a straightforward application of our discussion of single-slit diffraction patterns.

Analyze Evaluate Equation 38.1 for the two dark fringes that flank the central bright fringe, which correspond to $m = \pm 1$:

$$\sin \theta_{\text{dark}} = \pm\frac{\lambda}{a} = \pm\frac{5.80 \times 10^{-7}\,\text{m}}{0.300 \times 10^{-3}\,\text{m}} = \pm 1.933 \times 10^{-3}$$

From the triangle in Active Figure 38.4a, notice that $\tan \theta_{\text{dark}} = y_1/L$. Because θ_{dark} is very small, we can use the approximation $\sin \theta_{\text{dark}} \approx \tan \theta_{\text{dark}}$; therefore, $\sin \theta_{\text{dark}} \approx y_1/L$.

Use this result to find the positions of the first minima measured from the central axis:

$$y_1 \approx L \sin \theta_{\text{dark}} = (2.00\,\text{m})(\pm 1.933 \times 10^{-3})$$
$$= \pm 3.866 \times 10^{-3}\,\text{m}$$

Find the width of the central bright fringe:

$$2|y_1| = 7.73 \times 10^{-3}\,\text{m} = \boxed{7.73\,\text{mm}}$$

Finalize Notice that this value is much greater than the width of the slit. Let's explore what happens if we change the slit width.

What If? What if the slit width is increased by an order of magnitude to 3.00 mm? What happens to the diffraction pattern?

Answer Based on Equation 38.1, we expect that the angles at which the dark bands appear will decrease as a increases. Therefore, the diffraction pattern narrows.

From Equation 38.1, find the sines of the angles θ_{dark} for the $m = \pm 1$ dark fringes:

$$\sin \theta_{\text{dark}} = \pm\frac{\lambda}{a} = \pm\frac{5.80 \times 10^{-7}\,\text{m}}{3.00 \times 10^{-3}\,\text{m}} = \pm 1.933 \times 10^{-4}$$

Use this result to find the positions of the first minima measured from the central axis:

$$y_1 \approx L \sin \theta_{\text{dark}} = (2.00 \text{ m})(\pm 1.933 \times 10^{-4})$$
$$= \pm 3.866 \times 10^{-4} \text{ m}$$

Find the width of the central bright fringe:

$$2|y_1| = 7.73 \times 10^{-4} \text{ m} = 0.773 \text{ mm}$$

Notice that this result is *smaller* than the width of the slit. In general, for large values of a, the various maxima and minima are so closely spaced that only a large, central bright area resembling the geometric image of the slit is observed. This concept is very important in the performance of optical instruments such as telescopes.

Intensity of Single-Slit Diffraction Patterns

Analysis of the intensity variation in a diffraction pattern from a single slit of width a shows that the intensity is given by

$$I = I_{\text{max}} \left[\frac{\sin (\pi a \sin \theta / \lambda)}{\pi a \sin \theta / \lambda} \right]^2 \qquad (38.2)$$

◄ Intensity of a single-slit Fraunhofer diffraction pattern

where I_{max} is the intensity at $\theta = 0$ (the central maximum) and λ is the wavelength of light used to illuminate the slit. This expression shows that *minima* occur when

$$\frac{\pi a \sin \theta_{\text{dark}}}{\lambda} = m\pi$$

or

$$\sin \theta_{\text{dark}} = m \frac{\lambda}{a} \quad m = \pm 1, \pm 2, \pm 3, \ldots$$

◄ Condition for intensity minima for a single slit

in agreement with Equation 38.1.

Figure 38.6a represents a plot of Equation 38.2, and Figure 38.6b is a photograph of a single-slit Fraunhofer diffraction pattern. Notice that most of the light intensity is concentrated in the central bright fringe.

Intensity of Two-Slit Diffraction Patterns

When more than one slit is present, we must consider not only diffraction patterns due to the individual slits but also the interference patterns due to the waves coming from different slits. Notice the curved dashed lines in Figure 37.8, which indicate a decrease in intensity of the interference maxima as θ increases. This decrease is due to a diffraction pattern. To determine the effects of both two-slit

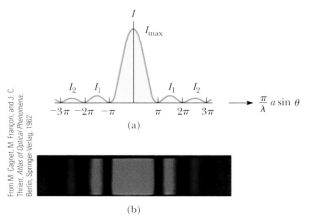

Figure 38.6   (a) A plot of light intensity I versus $(\pi/\lambda) a \sin \theta$ for the single-slit Fraunhofer diffraction pattern. (b) Photograph of a single-slit Fraunhofer diffraction pattern.

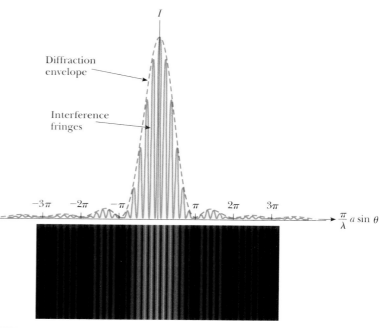

ACTIVE FIGURE 38.7

The combined effects of two-slit and single-slit interference. This pattern is produced when 650-nm light waves pass through two 3.0-μm slits that are 18 μm apart. Notice how the diffraction pattern acts as an "envelope" and controls the intensity of the regularly spaced interference maxima.

Sign in at www.thomsonedu.com and go to ThomsonNOW to adjust the slit width, slit separation, and the wavelength of the light to see the effect on the interference pattern.

interference and a single-slit diffraction pattern from each slit, we combine Equations 37.12 and 38.2:

$$I = I_{max} \cos^2\left(\frac{\pi d \sin\theta}{\lambda}\right)\left[\frac{\sin(\pi a \sin\theta/\lambda)}{\pi a \sin\theta/\lambda}\right]^2 \quad \textbf{(38.3)}$$

Although this expression looks complicated, it merely represents the single-slit diffraction pattern (the factor in square brackets) acting as an "envelope" for a two-slit interference pattern (the cosine-squared factor) as shown in Active Figure 38.7. The broken blue curve in Active Figure 38.7 represents the factor in square brackets in Equation 38.3. The cosine-squared factor by itself would give a series of peaks all with the same height as the highest peak of the brown curve in Active Figure 38.7. Because of the effect of the square-bracket factor, however, these peaks vary in height as shown.

Equation 37.2 indicates the conditions for interference maxima as $d \sin\theta = m\lambda$, where d is the distance between the two slits. Equation 38.1 specifies that the first diffraction minimum occurs when $a \sin\theta = \lambda$, where a is the slit width. Dividing Equation 37.2 by Equation 38.1 (with $m = 1$) allows us to determine which interference maximum coincides with the first diffraction minimum:

$$\frac{d \sin\theta}{a \sin\theta} = \frac{m\lambda}{\lambda}$$

$$\frac{d}{a} = m \quad \textbf{(38.4)}$$

In Active Figure 38.7, $d/a = 18\ \mu\text{m}/3.0\ \mu\text{m} = 6$. Therefore, the sixth interference maximum (if we count the central maximum as $m = 0$) is aligned with the first diffraction minimum and cannot be seen.

Quick Quiz 38.2 Consider the central peak in the diffraction envelope in Active Figure 38.7. Suppose the wavelength of the light is changed to 450 nm. What happens to this central peak? (a) The width of the peak decreases, and the number of interference fringes it encloses decreases. (b) The width of the peak decreases, and the number of interference fringes it encloses increases. (c) The width of the peak decreases, and the number of interference fringes it encloses remains the same. (d) The width of the peak increases, and the number of interference fringes it encloses decreases. (e) The width of the peak increases, and the number of interference fringes it encloses increases. (f) The width of the peak increases, and the number of interference fringes it encloses remains the same. (g) The width of the peak remains the same and the number of interference fringes it encloses decreases. (h) The width of the peak remains the same and the number of interference fringes it encloses increases. (i) The width of the peak remains the same and the number of interference fringes it encloses remains the same.

38.3   Resolution of Single-Slit and Circular Apertures

The ability of optical systems to distinguish between closely spaced objects is limited because of the wave nature of light. To understand this limitation, consider Figure 38.8, which shows two light sources far from a narrow slit of width a. The sources can be two noncoherent point sources S_1 and S_2; for example, they could be two distant stars. If no interference occurred between light passing through different parts of the slit, two distinct bright spots (or images) would be observed on the viewing screen. Because of such interference, however, each source is imaged as a bright central region flanked by weaker bright and dark fringes, a diffraction pattern. What is observed on the screen is the sum of two diffraction patterns: one from S_1 and the other from S_2.

If the two sources are far enough apart to keep their central maxima from overlapping as in Figure 38.8a, their images can be distinguished and are said to be *resolved*. If the sources are close together as in Figure 38.8b, however, the two central maxima overlap and the images are not resolved. To determine whether two images are resolved, the following condition is often used:

> When the central maximum of one image falls on the first minimum of another image, the images are said to be just resolved. This limiting condition of resolution is known as **Rayleigh's criterion.**

From Rayleigh's criterion, we can determine the minimum angular separation $\theta_{\min}$ subtended by the sources at the slit in Figure 38.8 for which the images are just resolved. Equation 38.1 indicates that the first minimum in a single-slit diffraction pattern occurs at the angle for which

$$\sin \theta = \frac{\lambda}{a}$$

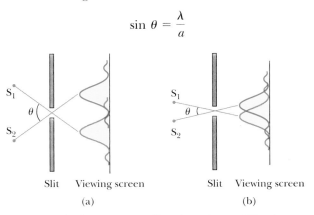

Slit	Viewing screen	Slit	Viewing screen
	(a)		(b)

Figure 38.8   Two point sources far from a narrow slit each produce a diffraction pattern. (a) The angle subtended by the sources at the slit is large enough for the diffraction patterns to be distinguishable. (b) The angle subtended by the sources is so small that their diffraction patterns overlap, and the images are not well resolved. (Notice that the angles are greatly exaggerated. The drawing is not to scale.)

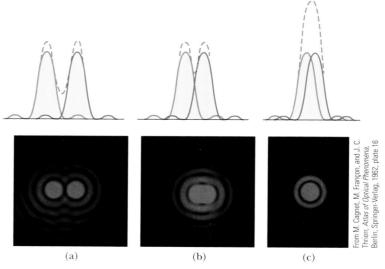

Figure 38.9 Individual diffraction patterns of two point sources (solid curves) and the resultant patterns (dashed curves) for various angular separations of the sources as the light passes through a circular aperture. In each case, the dashed curve is the sum of the two solid curves. (a) The sources are far apart, and the patterns are well resolved. (b) The sources are closer together such that the angular separation just satisfies Rayleigh's criterion, and the patterns are just resolved. (c) The sources are so close together that the patterns are not resolved.

From M. Cagnet, M. Françon, and J. C. Thrier, *Atlas of Optical Phenomena*, Berlin, Springer-Verlag, 1962, plate 16

where a is the width of the slit. According to Rayleigh's criterion, this expression gives the smallest angular separation for which the two images are resolved. Because $\lambda \ll a$ in most situations, $\sin\theta$ is small and we can use the approximation $\sin\theta \approx \theta$. Therefore, the limiting angle of resolution for a slit of width a is

$$\theta_{\min} = \frac{\lambda}{a} \tag{38.5}$$

where $\theta_{\min}$ is expressed in radians. Hence, the angle subtended by the two sources at the slit must be greater than λ/a if the images are to be resolved.

Many optical systems use circular apertures rather than slits. The diffraction pattern of a circular aperture as shown in the photographs of Figure 38.9 consists of a central circular bright disk surrounded by progressively fainter bright and dark rings. Figure 38.9 shows diffraction patterns for three situations in which light from two point sources passes through a circular aperture. When the sources are far apart, their images are well resolved (Fig. 38.9a). When the angular separation of the sources satisfies Rayleigh's criterion, the images are just resolved (Fig. 38.9b). Finally, when the sources are close together, the images are said to be unresolved (Fig. 38.9c).

Analysis shows that the limiting angle of resolution of the circular aperture is

Limiting angle of ▶
resolution for a circular
aperture

$$\theta_{\min} = 1.22 \frac{\lambda}{D} \tag{38.6}$$

where D is the diameter of the aperture. This expression is similar to Equation 38.5 except for the factor 1.22, which arises from a mathematical analysis of diffraction from the circular aperture.

Quick Quiz 38.3 Cat's eyes have pupils that can be modeled as vertical slits. At night, would cats be more successful in resolving (a) headlights on a distant car or (b) vertically separated lights on the mast of a distant boat?

Quick Quiz 38.4 Suppose you are observing a binary star with a telescope and are having difficulty resolving the two stars. You decide to use a colored filter to maximize the resolution. (A filter of a given color transmits only that color of light.) What color filter should you choose? (a) blue (b) green (c) yellow (d) red

---

EXAMPLE 38.2   **Resolution of the Eye**

Light of wavelength 500 nm, near the center of the visible spectrum, enters a human eye. Although pupil diameter varies from person to person, estimate a daytime diameter of 2 mm.

(A) Estimate the limiting angle of resolution for this eye, assuming its resolution is limited only by diffraction.

SOLUTION

Conceptualize   In Figure 38.9, identify the aperture through which the light travels as the pupil of the eye. Light passing through this small aperture causes diffraction patterns to occur on the retina.

Categorize   We evaluate the result using equations developed in this section, so we categorize this example as a substitution problem.

Use Equation 38.6, taking $\lambda = 500$ nm and $D = 2$ mm:

$$\theta_{min} = 1.22 \frac{\lambda}{D} = 1.22 \left(\frac{5.00 \times 10^{-7} \text{ m}}{2 \times 10^{-3} \text{ m}} \right)$$

$$\approx \boxed{3 \times 10^{-4} \text{ rad}} \approx \boxed{1 \text{ min of arc}}$$

(B) Determine the minimum separation distance d between two point sources that the eye can distinguish if the point sources are a distance $L = 25$ cm from the observer (Fig. 38.10).

Figure 38.10   (Example 38.2) Two point sources separated by a distance d as observed by the eye.

SOLUTION

Noting that θ_{min} is small, find d:

$$\sin \theta_{min} \approx \theta_{min} \approx \frac{d}{L} \quad \rightarrow \quad d = L\theta_{min}$$

Substitute numerical values:

$$d = (25 \text{ cm})(3 \times 10^{-4} \text{ rad}) = \boxed{8 \times 10^{-3} \text{ cm}}$$

This result is approximately equal to the thickness of a human hair.

EXAMPLE 38.3   **Resolution of a Telescope**

The Keck telescope at Mauna Kea, Hawaii, has an effective diameter of 10 m. What is its limiting angle of resolution for 600-nm light?

SOLUTION

Conceptualize   In Figure 38.9, identify the aperture through which the light travels as the opening of the telescope. Light passing through this aperture causes diffraction patterns to occur in the final image.

Categorize   We evaluate the result using equations developed in this section, so we categorize this example as a substitution problem.

Use Equation 38.6, taking $\lambda = 6.00 \times 10^{-7}$ m and $D = 10$ m:

$$\theta_{min} = 1.22 \frac{\lambda}{D} = 1.22 \left(\frac{6.00 \times 10^{-7} \text{ m}}{10 \text{ m}} \right)$$

$$= \boxed{7.3 \times 10^{-8} \text{ rad}} \approx \boxed{0.015 \text{ s of arc}}$$

Any two stars that subtend an angle greater than or equal to this value are resolved (if atmospheric conditions are ideal).

What If?   What if we consider radio telescopes? They are much larger in diameter than optical telescopes, but do they have better angular resolutions than optical telescopes? For example, the radio telescope at Arecibo, Puerto Rico, has a diameter of 305 m and is designed to detect radio waves of 0.75-m wavelength. How does its resolution compare with that of the Keck telescope?

Answer   The increase in diameter might suggest that radio telescopes would have better resolution than the Keck telescope, but Equation 38.6 shows that θ_{min} depends on *both* diameter and wavelength. Calculating the minimum angle of resolution for the radio telescope, we find

$$\theta_{min} = 1.22\,\frac{\lambda}{D} = 1.22\left(\frac{0.75 \text{ m}}{305 \text{ m}}\right)$$

$$= 3.0 \times 10^{-3} \text{ rad} \approx 10 \text{ min of arc}$$

This limiting angle of resolution is measured in *minutes* of arc rather than the *seconds* of arc for the optical telescope. Therefore, the change in wavelength more than compensates for the increase in diameter. The limiting angle of resolution for the Arecibo radio telescope is more than 40 000 times larger (that is, *worse*) than the Keck minimum.

The Keck telescope discussed in Example 38.3 can never reach its diffraction limit because the limiting angle of resolution is always set by atmospheric blurring at optical wavelengths. This seeing limit is usually about 1 s of arc and is never smaller than about 0.1 s of arc. The atmospheric blurring is caused by variations in index of refraction with temperature variations in the air. This blurring is one reason for the superiority of photographs from the Hubble Space Telescope, which views celestial objects from an orbital position above the atmosphere.

As an example of the effects of atmospheric blurring, consider telescopic images of Pluto and its moon, Charon. Figure 38.11a, an image taken in 1978, represents the discovery of Charon. In this photograph, taken from an Earth-based telescope, atmospheric turbulence causes the image of Charon to appear only as a bump on the edge of Pluto. In comparison, Figure 38.11b shows a photograph taken with the Hubble Space Telescope. Without the problems of atmospheric turbulence, Pluto and its moon are clearly resolved.

38.4   The Diffraction Grating

The **diffraction grating,** a useful device for analyzing light sources, consists of a large number of equally spaced parallel slits. A *transmission grating* can be made by cutting parallel grooves on a glass plate with a precision ruling machine. The spaces between the grooves are transparent to the light and hence act as separate slits. A *reflection grating* can be made by cutting parallel grooves on the surface of a

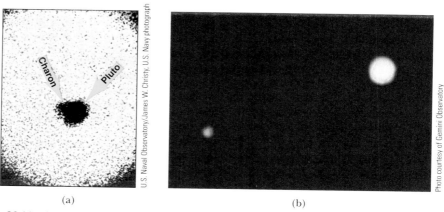

(a)   (b)

Figure 38.11   (a) The photograph on which Charon, the moon of Pluto, was discovered in 1978. From an Earth-based telescope, atmospheric blurring results in Charon appearing only as a subtle bump on the edge of Pluto. (b) A Hubble Space Telescope photo of Pluto and Charon, clearly resolving the two objects.

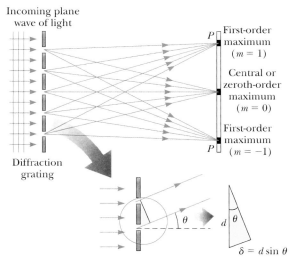

Figure 38.12 Side view of a diffraction grating. The slit separation is d, and the path difference between adjacent slits is $d \sin \theta$.

reflective material. The reflection of light from the spaces between the grooves is specular, and the reflection from the grooves cut into the material is diffuse. Therefore, the spaces between the grooves act as parallel sources of reflected light like the slits in a transmission grating. Current technology can produce gratings that have very small slit spacings. For example, a typical grating ruled with 5 000 grooves/cm has a slit spacing $d = (1/5\ 000)$ cm $= 2.00 \times 10^{-4}$ cm.

A section of a diffraction grating is illustrated in Figure 38.12. A plane wave is incident from the left, normal to the plane of the grating. The pattern observed on the screen far to the right of the grating is the result of the combined effects of interference and diffraction. Each slit produces diffraction, and the diffracted beams interfere with one another to produce the final pattern.

The waves from all slits are in phase as they leave the slits. For an arbitrary direction θ measured from the horizontal, however, the waves must travel different path lengths before reaching the screen. Notice in Figure 38.12 that the path difference δ between rays from any two adjacent slits is equal to $d \sin \theta$. If this path difference equals one wavelength or some integral multiple of a wavelength, waves from all slits are in phase at the screen and a bright fringe is observed. Therefore, the condition for *maxima* in the interference pattern at the angle θ_{bright} is

$$d \sin \theta_{\text{bright}} = m\lambda \quad m = 0, \pm 1, \pm 2, \pm 3, \dots \quad (38.7)$$

We can use this expression to calculate the wavelength if we know the grating spacing d and the angle θ_{bright}. If the incident radiation contains several wavelengths, the mth-order maximum for each wavelength occurs at a specific angle. All wavelengths are seen at $\theta = 0$, corresponding to $m = 0$, the zeroth-order maximum. The first-order maximum ($m = 1$) is observed at an angle that satisfies the relationship $\sin \theta_{\text{bright}} = \lambda/d$, the second-order maximum ($m = 2$) is observed at a larger angle θ_{bright}, and so on. For the small values of d typical in a diffraction grating, the angles θ_{bright} are large, as we see in Example 38.5.

The intensity distribution for a diffraction grating obtained with the use of a monochromatic source is shown in Active Figure 38.13. Notice the sharpness of the principal maxima and the broadness of the dark areas compared with the broad bright fringes characteristic of the two-slit interference pattern (see Fig. 37.7). You should also review Figure 37.8, which shows that the width of the intensity maxima decreases as the number of slits increases. Because the principal maxima are so sharp, they are much brighter than two-slit interference maxima.

Quick Quiz 38.5 Ultraviolet light of wavelength 350 nm is incident on a diffraction grating with slit spacing d and forms an interference pattern on a screen

PITFALL PREVENTION 38.3
A Diffraction Grating Is an Interference Grating

As with *diffraction pattern*, *diffraction grating* is a misnomer, but is deeply entrenched in the language of physics. The diffraction grating depends on diffraction in the same way as the double slit, spreading the light so that light from different slits can interfere. It would be more correct to call it an *interference grating*, but *diffraction grating* is the name in use.

◀ Condition for interference maxima for a grating

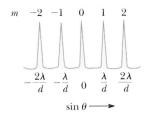

ACTIVE FIGURE 38.13

Intensity versus $\sin \theta$ for a diffraction grating. The zeroth-, first-, and second-order maxima are shown.

Sign in at www.thomsonedu.com and go to ThomsonNOW to choose the number of slits to be illuminated and see the effect on the interference pattern.

a distance L away. The angular positions θ_{bright} of the interference maxima are large. The locations of the bright fringes are marked on the screen. Now red light of wavelength 700 nm is used with a diffraction grating to form another diffraction pattern on the screen. Will the bright fringes of this pattern be located at the marks on the screen if (a) the screen is moved to a distance $2L$ from the grating, (b) the screen is moved to a distance $L/2$ from the grating, (c) the grating is replaced with one of slit spacing $2d$, (d) the grating is replaced with one of slit spacing $d/2$, or (e) nothing is changed?

CONCEPTUAL EXAMPLE 38.4 | **A Compact Disc Is a Diffraction Grating**

Light reflected from the surface of a compact disc is multicolored, as shown in Figure 38.14. The colors and their intensities depend on the orientation of the CD relative to the eye and relative to the light source. Explain how this works.

SOLUTION

The surface of a CD has a spiral grooved track (with adjacent grooves having a separation on the order of 1 μm). Therefore, the surface acts as a reflection grating. The light reflecting from the regions between these closely spaced grooves interferes constructively only in certain directions that depend on the wavelength and the direction of the incident light. Any section of the CD serves as a diffraction grating for white light, sending different colors in different directions. The different colors you see upon viewing one section change when the light source, the CD, or you change position. This change in position causes the angle of incidence or the angle of the diffracted light to be altered.

Figure 38.14 (Conceptual Example 38.4) A compact disc observed under white light. The colors observed in the reflected light and their intensities depend on the orientation of the CD relative to the eye and relative to the light source.

© Kristen Brochmann/Fundamental Photographs

EXAMPLE 38.5 | **The Orders of a Diffraction Grating**

Monochromatic light from a helium–neon laser ($\lambda = 632.8$ nm) is incident normally on a diffraction grating containing 6 000 grooves per centimeter. Find the angles at which the first- and second-order maxima are observed.

SOLUTION

Conceptualize Study Figure 38.12 and imagine that the light coming from the left originates from the helium–neon laser.

Categorize We evaluate results using equations developed in this section, so we categorize this example as a substitution problem.

Calculate the slit separation as the inverse of the number of grooves per centimeter:

$$d = \frac{1}{6\,000} \text{ cm} = 1.667 \times 10^{-4} \text{ cm} = 1\,667 \text{ nm}$$

Solve Equation 38.7 for $\sin \theta$ and substitute numerical values for the first-order maximum ($m = 1$) to find θ_1:

$$\sin \theta_1 = \frac{(1)\lambda}{d} = \frac{632.8 \text{ nm}}{1\,667 \text{ nm}} = 0.379\,7$$

$$\theta_1 = \boxed{22.31°}$$

Repeat for the second-order maximum ($m = 2$):

$$\sin \theta_2 = \frac{(2)\lambda}{d} = \frac{2(632.8 \text{ nm})}{1\,667 \text{ nm}} = 0.759\,4$$

$$\theta_2 = \boxed{49.41°}$$

What If? What if you looked for the third-order maximum? Would you find it?

Answer   For $m = 3$, we find $\sin \theta_3 = 1.139$. Because $\sin \theta$ cannot exceed unity, this result does not represent a realistic solution. Hence, only zeroth-, first-, and second-order maxima can be observed for this situation.

Applications of Diffraction Gratings

A schematic drawing of a simple apparatus used to measure angles in a diffraction pattern is shown in Active Figure 38.15. This apparatus is a *diffraction grating spectrometer*. The light to be analyzed passes through a slit, and a collimated beam of light is incident on the grating. The diffracted light leaves the grating at angles that satisfy Equation 38.7, and a telescope is used to view the image of the slit. The wavelength can be determined by measuring the precise angles at which the images of the slit appear for the various orders.

The spectrometer is a useful tool in *atomic spectroscopy*, in which the light from an atom is analyzed to find the wavelength components. These wavelength components can be used to identify the atom. We shall investigate atomic spectra in Chapter 42 of the extended version of this text.

Another application of diffraction gratings is the *grating light valve* (GLV), which may compete in the near future in video projection with the digital micromirror devices (DMDs) discussed in Section 35.4. A GLV is a silicon microchip fitted with an array of parallel silicon nitride ribbons coated with a thin layer of aluminum (Fig. 38.16). Each ribbon is approximately 20 μm long and 5 μm wide and is separated from the silicon substrate by an air gap on the order of 100 nm. With no voltage applied, all ribbons are at the same level. In this situation, the array of ribbons acts as a flat surface, specularly reflecting incident light.

When a voltage is applied between a ribbon and the electrode on the silicon substrate, an electric force pulls the ribbon downward, closer to the substrate. Alternate ribbons can be pulled down, while those in between remain in an elevated configuration. As a result, the array of ribbons acts as a diffraction grating such that the constructive interference for a particular wavelength of light can be directed toward a screen or other optical display system. By using three such devices—one each for red, blue, and green light—full-color display is possible.

A GLV tends to be simpler to fabricate and higher in resolution than comparable DMDs. On the other hand, DMDs have already made an entry into the market. It will be interesting to watch this technology competition in future years.

Another interesting application of diffraction gratings is **holography,** the production of three-dimensional images of objects. The physics of holography was developed by Dennis Gabor (1900–1979) in 1948 and resulted in the Nobel Prize

ACTIVE FIGURE 38.15

Diagram of a diffraction grating spectrometer. The collimated beam incident on the grating is spread into its various wavelength components with constructive interference for a particular wavelength occurring at the angles θ_{bright} that satisfy the equation $d \sin \theta_{\text{bright}} = m\lambda$, where $m = 0, \pm 1, \pm 2, \dots$.

Sign in at www.thomsonedu.com and go to ThomsonNOW to use the spectrometer and observe constructive interference for various wavelengths.

Silicon Light Machines

Figure 38.16 A small portion of a grating light valve. The alternating reflective ribbons at different levels act as a diffraction grating, offering very high-speed control of the direction of light toward a digital display device.

(a)   (b)

Figure 38.17 In this hologram, a circuit board is shown from two different views. Notice the difference in the appearance of the measuring tape and the view through the magnifying lens in (a) and (b).

in Physics for Gabor in 1971. The requirement of coherent light for holography delayed the realization of holographic images from Gabor's work until the development of lasers in the 1960s. Figure 38.17 shows a hologram and the three-dimensional character of its image. Notice in particular the difference in the view through the magnifying glass in Figures 38.17a and 38.17b.

Figure 38.18 shows how a hologram is made. Light from the laser is split into two parts by a half-silvered mirror at B. One part of the beam reflects off the object to be photographed and strikes an ordinary photographic film. The other half of the beam is diverged by lens L_2, reflects from mirrors M_1 and M_2, and finally strikes the film. The two beams overlap to form an extremely complicated interference pattern on the film. Such an interference pattern can be produced only if the phase relationship of the two waves is constant throughout the exposure of the film. This condition is met by illuminating the scene with light coming through a pinhole or with coherent laser radiation. The hologram records not only the intensity of the light scattered from the object (as in a conventional photograph), but also the phase difference between the reference beam and the beam scattered from the object. Because of this phase difference, an interference pattern is formed that produces an image in which all three-dimensional information available from the perspective of any point on the hologram is preserved.

In a normal photographic image, a lens is used to focus the image so that each point on the object corresponds to a single point on the film. Notice that there is no lens used in Figure 38.18 to focus the light onto the film. Therefore, light from each point on the object reaches *all* points on the film. As a result, each region of the photographic film on which the hologram is recorded contains information about all illuminated points on the object, which leads to a remarkable result: if a small section of the hologram is cut from the film, the complete image can be formed from the small piece! (The quality of the image is reduced, but the entire image is present.)

A hologram is best viewed by allowing coherent light to pass through the developed film as one looks back along the direction from which the beam comes. The interference pattern on the film acts as a diffraction grating. Figure 38.19 shows

Figure 38.18 Experimental arrangement for producing a hologram.

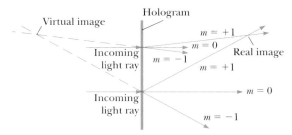

Figure 38.19 Two light rays strike a hologram at normal incidence. For each ray, outgoing rays corresponding to $m = 0$ and $m = \pm 1$ are shown. If the $m = -1$ rays are extended backward, a virtual image of the object photographed in the hologram exists on the front side of the hologram.

two rays of light striking and passing through the film. For each ray, the $m = 0$ and $m = \pm 1$ rays in the diffraction pattern are shown emerging from the right side of the film. The $m = +1$ rays converge to form a real image of the scene, which is not the image that is normally viewed. By extending the light rays corresponding to $m = -1$ behind the film, we see that there is a virtual image located there, with light coming from it in exactly the same way that light came from the actual object when the film was exposed. This image is what one sees when looking through the holographic film.

Holograms are finding a number of applications. You may have a hologram on your credit card. This special type of hologram is called a *rainbow hologram* and is designed to be viewed in reflected white light.

38.5 Diffraction of X-Rays by Crystals

In principle, the wavelength of any electromagnetic wave can be determined if a grating of the proper spacing (on the order of λ) is available. X-rays, discovered by Wilhelm Roentgen (1845–1923) in 1895, are electromagnetic waves of very short wavelength (on the order of 0.1 nm). It would be impossible to construct a grating having such a small spacing by the cutting process described at the beginning of Section 38.4. The atomic spacing in a solid is known to be about 0.1 nm, however. In 1913, Max von Laue (1879–1960) suggested that the regular array of atoms in a crystal could act as a three-dimensional diffraction grating for x-rays. Subsequent experiments confirmed this prediction. The diffraction patterns from crystals are complex because of the three-dimensional nature of the crystal structure. Nevertheless, x-ray diffraction has proved to be an invaluable technique for elucidating these structures and for understanding the structure of matter.

Figure 38.20 shows one experimental arrangement for observing x-ray diffraction from a crystal. A collimated beam of monochromatic x-rays is incident on a crystal. The diffracted beams are very intense in certain directions, corresponding

Figure 38.20 Schematic diagram of the technique used to observe the diffraction of x-rays by a crystal. The array of spots formed on the film is called a Laue pattern.

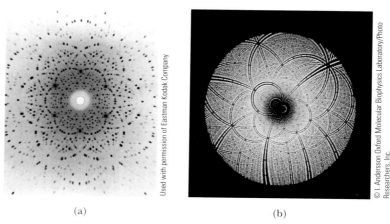

Figure 38.21   (a) A Laue pattern of a single crystal of the mineral beryl (beryllium aluminum silicate). Each dot represents a point of constructive interference. (b) A Laue pattern of the enzyme Rubisco, produced with a wideband x-ray spectrum. This enzyme is present in plants and takes part in the process of photosynthesis. The Laue pattern is used to determine the crystal structure of Rubisco.

to constructive interference from waves reflected from layers of atoms in the crystal. The diffracted beams, which can be detected by a photographic film, form an array of spots known as a *Laue pattern* as in Figure 38.21a. One can deduce the crystalline structure by analyzing the positions and intensities of the various spots in the pattern. Figure 38.21b shows a Laue pattern from a crystalline enzyme, using a wide range of wavelengths so that a swirling pattern results.

The arrangement of atoms in a crystal of sodium chloride (NaCl) is shown in Figure 38.22. Each unit cell (the geometric solid that repeats throughout the crystal) is a cube having an edge length a. A careful examination of the NaCl structure shows that the ions lie in discrete planes (the shaded areas in Fig. 38.22). Now suppose an incident x-ray beam makes an angle θ with one of the planes as in Figure 38.23. The beam can be reflected from both the upper plane and the lower one, but the beam reflected from the lower plane travels farther than the beam reflected from the upper plane. The effective path difference is $2d \sin \theta$. The two beams reinforce each other (constructive interference) when this path difference equals some integer multiple of λ. The same is true for reflection from the entire family of parallel planes. Hence, the condition for *constructive* interference (maxima in the reflected beam) is

> **PITFALL PREVENTION 38.4**
> **Different Angles**
>
> Notice in Figure 38.23 that the angle θ is measured from the reflecting surface rather than from the normal as in the case of the law of reflection in Chapter 35. With slits and diffraction gratings, we also measured the angle θ from the normal to the array of slits. Because of historical tradition, the angle is measured differently in Bragg diffraction, so interpret Equation 38.8 with care.

Bragg's law ▶

$$2d \sin \theta = m\lambda \quad m = 1, 2, 3, \ldots \tag{38.8}$$

This condition is known as **Bragg's law,** after W. L. Bragg (1890–1971), who first derived the relationship. If the wavelength and diffraction angle are measured, Equation 38.8 can be used to calculate the spacing between atomic planes.

Figure 38.22   Crystalline structure of sodium chloride (NaCl). The blue spheres represent Cl^- ions, and the red spheres represent Na^+ ions. The length of the cube edge is $a = 0.562\ 737$ nm.

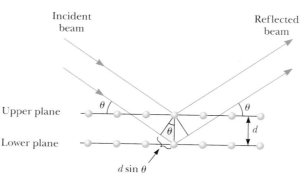

Figure 38.23   A two-dimensional description of the reflection of an x-ray beam from two parallel crystalline planes separated by a distance d. The beam reflected from the lower plane travels farther than the beam reflected from the upper plane by a distance $2d \sin \theta$.

38.6 Polarization of Light Waves

In Chapter 34, we described the transverse nature of light and all other electro-magnetic waves. Polarization, discussed in this section, is firm evidence of this transverse nature.

An ordinary beam of light consists of a large number of waves emitted by the atoms of the light source. Each atom produces a wave having some particular ori-entation of the electric field vector $\vec{E}$, corresponding to the direction of atomic vibration. The *direction of polarization* of each individual wave is defined to be the direction in which the electric field is vibrating. In Figure 38.24, this direction happens to lie along the y axis. An individual electromagnetic wave, however, could have its $\vec{E}$ vector in the yz plane, making any possible angle with the y axis. Because all directions of vibration from a wave source are possible, the resultant electromagnetic wave is a superposition of waves vibrating in many different direc-tions. The result is an **unpolarized** light beam, represented in Figure 38.25a. The direction of wave propagation in this figure is perpendicular to the page. The arrows show a few possible directions of the electric field vectors for the individual waves making up the resultant beam. At any given point and at some instant of time, all these individual electric field vectors add to give one resultant electric field vector.

As noted in Section 34.3, a wave is said to be **linearly polarized** if the resultant electric field $\vec{E}$ vibrates in the same direction *at all times* at a particular point as shown in Figure 38.25b. (Sometimes, such a wave is described as *plane-polarized,* or simply *polarized.*) The plane formed by $\vec{E}$ and the direction of propagation is called the *plane of polarization* of the wave. If the wave in Figure 38.24 represents the resultant of all individual waves, the plane of polarization is the xy plane.

A linearly polarized beam can be obtained from an unpolarized beam by removing all waves from the beam except those whose electric field vectors oscil-late in a single plane. We now discuss four processes for producing polarized light from unpolarized light.

Polarization by Selective Absorption

The most common technique for producing polarized light is to use a material that transmits waves whose electric fields vibrate in a plane parallel to a certain direction and that absorbs waves whose electric fields vibrate in all other direc-tions.

In 1938, E. H. Land (1909–1991) discovered a material, which he called *Polaroid,* that polarizes light through selective absorption. This material is fabri-cated in thin sheets of long-chain hydrocarbons. The sheets are stretched during manufacture so that the long-chain molecules align. After a sheet is dipped into a solution containing iodine, the molecules become good electrical conductors. Conduction takes place primarily along the hydrocarbon chains because electrons can move easily only along the chains. If light whose electric field vector is parallel to the chains is incident on the material, the electric field accelerates electrons along the chains and energy is absorbed from the radiation. Therefore, the light does not pass through the material. Light whose electric field vector is perpendicu-lar to the chains passes through the material because electrons cannot move from one molecule to the next. As a result, when unpolarized light is incident on the material, the exiting light is polarized perpendicular to the molecular chains.

It is common to refer to the direction perpendicular to the molecular chains as the *transmission axis.* In an ideal polarizer, all light with $\vec{E}$ parallel to the transmis-sion axis is transmitted and all light with $\vec{E}$ perpendicular to the transmission axis is absorbed.

Active Figure 38.26 (page 1094) represents an unpolarized light beam incident on a first polarizing sheet, called the *polarizer.* Because the transmission axis is ori-ented vertically in the figure, the light transmitted through this sheet is polarized vertically. A second polarizing sheet, called the *analyzer,* intercepts the beam. In

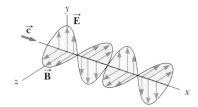

Figure 38.24 Schematic diagram of an electromagnetic wave propagating at velocity $\vec{c}$ in the x direction. The electric field vibrates in the xy plane, and the magnetic field vibrates in the xz plane.

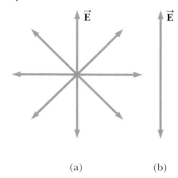

(a)   (b)

Figure 38.25 (a) A representation of an unpolarized light beam viewed along the direction of propagation (perpendicular to the page). The transverse electric field can vibrate in any direction in the plane of the page with equal probability. (b) A linearly polarized light beam with the electric field vibrating in the vertical direction.

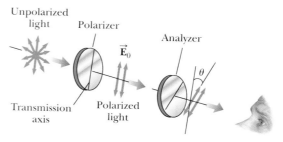

ACTIVE FIGURE 38.26

Two polarizing sheets whose transmission axes make an angle θ with each other. Only a fraction of the polarized light incident on the analyzer is transmitted through it.

Sign in at www.thomsonedu.com and go to ThomsonNOW to rotate the analyzer and see the effect on the transmitted light.

Active Figure 38.26, the analyzer transmission axis is set at an angle θ to the polarizer axis. We call the electric field vector of the first transmitted beam $\vec{E}_0$. The component of $\vec{E}_0$ perpendicular to the analyzer axis is completely absorbed. The component of $\vec{E}_0$ parallel to the analyzer axis, which is transmitted through the analyzer, is $E_0 \cos\theta$. Because the intensity of the transmitted beam varies as the square of its magnitude, we conclude that the intensity I of the (polarized) beam transmitted through the analyzer varies as

Malus's law ▶

$$I = I_{\max}\cos^2\theta \tag{38.9}$$

where $I_{\max}$ is the intensity of the polarized beam incident on the analyzer. This expression, known as **Malus's law**,[2] applies to any two polarizing materials whose transmission axes are at an angle θ to each other. This expression shows that the intensity of the transmitted beam is maximum when the transmission axes are parallel ($\theta = 0$ or $180°$) and is zero (complete absorption by the analyzer) when the transmission axes are perpendicular to each other. This variation in transmitted intensity through a pair of polarizing sheets is illustrated in Figure 38.27.

Polarization by Reflection

When an unpolarized light beam is reflected from a surface, the reflected light may be completely polarized, partially polarized, or unpolarized, depending on the angle of incidence. If the angle of incidence is $0°$, the reflected beam is unpolarized. For other angles of incidence, the reflected light is polarized to some extent, and for one particular angle of incidence, the reflected light is completely polarized. Let's now investigate reflection at that special angle.

Suppose an unpolarized light beam is incident on a surface as in Figure 38.28a. Each individual electric field vector can be resolved into two components: one parallel to the surface (and perpendicular to the page in Fig. 38.28, represented by

Figure 38.27 The intensity of light transmitted through two polarizers depends on the relative orientation of their transmission axes. (a) The transmitted light has maximum intensity when the transmission axes are aligned with each other. (b) The transmitted light has lesser intensity when the transmission axes are at an angle of $45°$ with each other. (c) The transmitted light intensity is a minimum when the transmission axes are perpendicular to each other.

(a)

(b)

(c)

Henry Leap and Jim Lehman

[2] Named after its discoverer, E. L. Malus (1775–1812). Malus discovered that reflected light was polarized by viewing it through a calcite ($CaCO_3$) crystal.

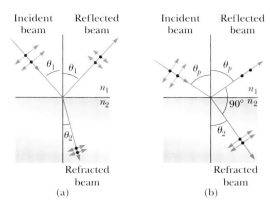

Figure 38.28 (a) When unpolarized light is incident on a reflecting surface, the reflected and refracted beams are partially polarized. (b) The reflected beam is completely polarized when the angle of incidence equals the polarizing angle θ_p, which satisfies the equation $n = \tan \theta_p$. At this incident angle, the reflected and refracted rays are perpendicular to each other.

the dots) and the other (represented by the brown arrows) perpendicular both to the first component and to the direction of propagation. Therefore, the polarization of the entire beam can be described by two electric field components in these directions. It is found that the parallel component reflects more strongly than the perpendicular component, resulting in a partially polarized reflected beam. Furthermore, the refracted beam is also partially polarized.

Now suppose the angle of incidence θ_1 is varied until the angle between the reflected and refracted beams is 90° as in Figure 38.28b. At this particular angle of incidence, the reflected beam is completely polarized (with its electric field vector parallel to the surface) and the refracted beam is still only partially polarized. The angle of incidence at which this polarization occurs is called the **polarizing angle** θ_p.

We can obtain an expression relating the polarizing angle to the index of refraction of the reflecting substance by using Figure 38.28b. From this figure, we see that $\theta_p + 90° + \theta_2 = 180°$; therefore, $\theta_2 = 90° - \theta_p$. Using Snell's law of refraction (Eq. 35.8) gives

$$\frac{n_2}{n_1} = \frac{\sin \theta_1}{\sin \theta_2} = \frac{\sin \theta_p}{\sin \theta_2}$$

Because $\sin \theta_2 = \sin (90° - \theta_p) = \cos \theta_p$, we can write this expression as $n_2/n_1 = \sin \theta_p/\cos \theta_p$, which means that

$$\tan \theta_p = \frac{n_2}{n_1} \tag{38.10}$$

◀ Brewster's law

This expression is called **Brewster's law,** and the polarizing angle θ_p is sometimes called **Brewster's angle,** after its discoverer, David Brewster (1781–1868). Because n varies with wavelength for a given substance, Brewster's angle is also a function of wavelength.

We can understand polarization by reflection by imagining that the electric field in the incident light sets electrons at the surface of the material in Figure 38.28b into oscillation. The component directions of oscillation are (1) parallel to the arrows shown on the refracted beam of light and (2) perpendicular to the page. The oscillating electrons act as antennas radiating light with a polarization parallel to the direction of oscillation. For the oscillations in direction 1, there is no radiation in the perpendicular direction, which is along the reflected ray (see the $\theta = 90°$ direction in Fig. 34.10). For oscillations in direction 2, the electrons radiate light with a polarization perpendicular to the page (the $\theta = 0$ direction in Fig. 34.10). Therefore, the light reflected from the surface at this angle is completely polarized parallel to the surface.

Polarization by reflection is a common phenomenon. Sunlight reflected from water, glass, and snow is partially polarized. If the surface is horizontal, the electric

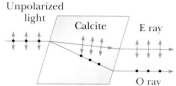

Figure 38.29 Unpolarized light incident at an angle to the optic axis in a calcite crystal splits into an ordinary (O) ray and an extraordinary (E) ray. These two rays are polarized in mutually perpendicular directions. (Drawing not to scale.)

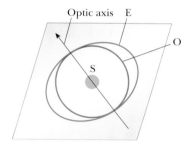

Figure 38.30 A point source S inside a double-refracting crystal produces a spherical wave front corresponding to the ordinary (O) ray and an elliptical wave front corresponding to the extraordinary (E) ray. The two waves propagate with the same velocity along the optic axis.

Figure 38.31 A calcite crystal produces a double image because it is a birefringent (double-refracting) material.

field vector of the reflected light has a strong horizontal component. Sunglasses made of polarizing material reduce the glare of reflected light. The transmission axes of such lenses are oriented vertically so that they absorb the strong horizontal component of the reflected light. If you rotate sunglasses through 90°, they are not as effective at blocking the glare from shiny horizontal surfaces.

Polarization by Double Refraction

Solids can be classified on the basis of internal structure. Those in which the atoms are arranged in a specific order are called *crystalline;* the NaCl structure of Figure 38.22 is one example of a crystalline solid. Those solids in which the atoms are distributed randomly are called *amorphous.* When light travels through an amorphous material such as glass, it travels with a speed that is the same in all directions. That is, glass has a single index of refraction. In certain crystalline materials such as calcite and quartz, however, the speed of light is not the same in all directions. In these materials, the speed of light depends on the direction of propagation *and* on the plane of polarization of the light. Such materials are characterized by two indices of refraction. Hence, they are often referred to as **double-refracting** or **birefringent** materials.

When unpolarized light enters a birefringent material, it may split into an **ordinary (O) ray** and an **extraordinary (E) ray.** These two rays have mutually perpendicular polarizations and travel at different speeds through the material. The two speeds correspond to two indices of refraction, n_O for the ordinary ray and n_E for the extraordinary ray.

There is one direction, called the **optic axis,** along which the ordinary and extraordinary rays have the same speed. If light enters a birefringent material at an angle to the optic axis, however, the different indices of refraction will cause the two polarized rays to split and travel in different directions as shown in Figure 38.29.

The index of refraction n_O for the ordinary ray is the same in all directions. If one could place a point source of light inside the crystal as in Figure 38.30, the ordinary waves would spread out from the source as spheres. The index of refraction n_E varies with the direction of propagation. A point source sends out an extraordinary wave having wave fronts that are elliptical in cross section. The difference in speed for the two rays is a maximum in the direction perpendicular to the optic axis. For example, in calcite, $n_O = 1.658$ at a wavelength of 589.3 nm, and n_E varies from 1.658 along the optic axis to 1.486 perpendicular to the optic axis. Values for n_O and the extreme value of n_E for various double-refracting crystals are given in Table 38.1.

If you place a calcite crystal on a sheet of paper and then look through the crystal at any writing on the paper, you would see two images as shown in Figure 38.31. As can be seen from Figure 38.29, these two images correspond to one formed by the ordinary ray and one formed by the extraordinary ray. If the two images are viewed through a sheet of rotating polarizing glass, they alternately appear and dis-

TABLE 38.1

Indices of Refraction for Some Double-Refracting Crystals at a Wavelength of 589.3 nm

Crystal	n_O	n_E	n_O/n_E
Calcite ($CaCO_3$)	1.658	1.486	1.116
Quartz (SiO_2)	1.544	1.553	0.994
Sodium nitrate ($NaNO_3$)	1.587	1.336	1.188
Sodium sulfite ($NaSO_3$)	1.565	1.515	1.033
Zinc chloride ($ZnCl_2$)	1.687	1.713	0.985
Zinc sulfide (ZnS)	2.356	2.378	0.991

(a)   (b)

Figure 38.32   (a) Strain distribution in a plastic model of a hip replacement used in a medical research laboratory. The pattern is produced when the plastic model is viewed between a polarizer and analyzer oriented perpendicular to each other. (b) A plastic model of an arch structure under load conditions observed between perpendicular polarizers. Such patterns are useful in the optimal design of architectural components.

appear because the ordinary and extraordinary rays are plane-polarized along mutually perpendicular directions.

Some materials such as glass and plastic become birefringent when stressed. Suppose an unstressed piece of plastic is placed between a polarizer and an analyzer so that light passes from polarizer to plastic to analyzer. When the plastic is unstressed and the analyzer axis is perpendicular to the polarizer axis, none of the polarized light passes through the analyzer. In other words, the unstressed plastic has no effect on the light passing through it. If the plastic is stressed, however, regions of greatest stress become birefringent and the polarization of the light passing through the plastic changes. Hence, a series of bright and dark bands is observed in the transmitted light, with the bright bands corresponding to regions of greatest stress.

Engineers often use this technique, called *optical stress analysis,* in designing structures ranging from bridges to small tools. They build a plastic model and analyze it under different load conditions to determine regions of potential weakness and failure under stress. Two examples of plastic models under stress are shown in Figure 38.32.

Polarization by Scattering

When light is incident on any material, the electrons in the material can absorb and reradiate part of the light. Such absorption and reradiation of light by electrons in the gas molecules that make up air is what causes sunlight reaching an observer on the Earth to be partially polarized. You can observe this effect—called **scattering**—by looking directly up at the sky through a pair of sunglasses whose lenses are made of polarizing material. Less light passes through at certain orientations of the lenses than at others.

Figure 38.33 illustrates how sunlight becomes polarized when it is scattered. The phenomenon is similar to that creating completely polarized light upon reflection from a surface at Brewster's angle. An unpolarized beam of sunlight traveling in the horizontal direction (parallel to the ground) strikes a molecule of one of the gases that make up air, setting the electrons of the molecule into vibration. These vibrating charges act like the vibrating charges in an antenna. The horizontal component of the electric field vector in the incident wave results in a horizontal component of the vibration of the charges, and the vertical component of the vector results in a vertical component of vibration. If the observer in Figure 38.33 is looking straight up (perpendicular to the original direction of propagation of the light), the vertical oscillations of the charges send no radiation toward the observer. Therefore, the observer sees light that is completely polarized in the horizontal direction as indicated by the brown arrows. If the observer looks in other directions, the light is partially polarized in the horizontal direction.

Variations in the color of scattered light in the atmosphere can be understood as follows. When light of various wavelengths λ is incident on gas molecules of diameter d, where $d \ll \lambda$, the relative intensity of the scattered light varies as $1/\lambda^4$. The condition $d \ll \lambda$ is satisfied for scattering from oxygen (O_2) and nitrogen (N_2) molecules in the atmosphere, whose diameters are about 0.2 nm. Hence,

Unpolarized
light

Air
molecule

Figure 38.33   The scattering of unpolarized sunlight by air molecules. The scattered light traveling perpendicular to the incident light is plane-polarized because the vertical vibrations of the charges in the air molecule send no light in this direction.

This photograph is a view of a rocket launch from Vandenburg Air Force Base in California. The trail left by the rocket shows the effects of scattering of light by air molecules. The lower portion of the trail appears red due to the scattering of wavelengths at the violet end of the spectrum as the light from the Sun travels through a large portion of the atmosphere to light up the trail. The upper portion of the trail is illuminated by light that has traveled through much less atmosphere and appears white.

short wavelengths (violet light) are scattered more efficiently than long wavelengths (red light). Therefore, when sunlight is scattered by gas molecules in the air, the short-wavelength radiation (violet) is scattered more intensely than the long-wavelength radiation (red).

When you look up into the sky in a direction that is not toward the Sun, you see the scattered light, which is predominantly violet. Your eyes, however, are not very sensitive to violet light. Light of the next color in the spectrum, blue, is scattered with less intensity than violet, but your eyes are far more sensitive to blue light than to violet light. Hence, you see a blue sky. If you look toward the west at sunset (or toward the east at sunrise), you are looking in a direction toward the Sun and are seeing light that has passed through a large distance of air. Most of the blue light has been scattered by the air between you and the Sun. The light that survives this trip through the air to you has had much of its blue component scattered and is therefore heavily weighted toward the red end of the spectrum; as a result, you see the red and orange colors of sunset.

Optical Activity

Many important applications of polarized light involve materials that display **optical activity.** A material is said to be optically active if it rotates the plane of polarization of any light transmitted through the material. The angle through which the light is rotated by a specific material depends on the length of the path through the material and on concentration if the material is in solution. One optically active material is a solution of the common sugar dextrose. A standard method for determining the concentration of sugar solutions is to measure the rotation produced by a fixed length of the solution.

Molecular asymmetry determines whether a material is optically active. For example, some proteins are optically active because of their spiral shape.

The liquid crystal displays found in most calculators have their optical activity changed by the application of electric potential across different parts of the display. Try using a pair of polarizing sunglasses to investigate the polarization used in the display of your calculator.

Quick Quiz 38.6 A polarizer for microwaves can be made as a grid of parallel metal wires approximately 1 cm apart. Is the electric field vector for microwaves transmitted through this polarizer (a) parallel or (b) perpendicular to the metal wires?

Quick Quiz 38.7 You are walking down a long hallway that has many light fixtures in the ceiling and a very shiny, newly waxed floor. When looking at the floor, you see reflections of every light fixture. Now you put on sunglasses that are polarized. Some of the reflections of the light fixtures can no longer be seen. (Try it!) Are the reflections that disappear those (a) nearest to you, (b) farthest from you, or (c) at an intermediate distance from you?

Summary

CONCEPTS AND PRINCIPLES

Diffraction is the deviation of light from a straight-line path when the light passes through an aperture or around an obstacle. Diffraction is due to the wave nature of light.

The **Fraunhofer diffraction pattern** produced by a single slit of width a on a distant screen consists of a central bright fringe and alternating bright and dark fringes of much lower intensities. The angles θ_{dark} at which the diffraction pattern has zero intensity, corresponding to destructive interference, are given by

$$\sin \theta_{dark} = m \frac{\lambda}{a} \quad m = \pm 1, \pm 2, \pm 3, \ldots \quad (38.1)$$

Rayleigh's criterion, which is a limiting condition of resolution, states that two images formed by an aperture are just distinguishable if the central maximum of the diffraction pattern for one image falls on the first minimum of the diffraction pattern for the other image. The limiting angle of resolution for a slit of width a is $\theta_{min} = \lambda/a$, and the limiting angle of resolution for a circular aperture of diameter D is given by $\theta_{min} = 1.22\lambda/D$.

A **diffraction grating** consists of a large number of equally spaced, identical slits. The condition for intensity maxima in the interference pattern of a diffraction grating for normal incidence is

$$d \sin \theta_{bright} = m\lambda \quad m = 0, \pm 1, \pm 2, \pm 3, \ldots \quad (38.7)$$

where d is the spacing between adjacent slits and m is the order number of the intensity maximum.

When polarized light of intensity I_{max} is emitted by a polarizer and then is incident on an analyzer, the light transmitted through the analyzer has an intensity equal to $I_{max} \cos^2 \theta$, where θ is the angle between the polarizer and analyzer transmission axes.

In general, reflected light is partially polarized. Reflected light, however, is completely polarized when the angle of incidence is such that the angle between the reflected and refracted beams is 90°. This angle of incidence, called the **polarizing angle** θ_p, satisfies **Brewster's law:**

$$\tan \theta_p = \frac{n_2}{n_1} \quad (38.10)$$

where n_1 is the index of refraction of the medium in which the light initially travels and n_2 is the index of refraction of the reflecting medium.

Questions

☐ denotes answer available in *Student Solutions Manual/Study Guide;* **O** denotes objective question

1. Why can you hear around corners, but not see around corners?

2. Holding your hand at arm's length, you can readily block sunlight from reaching your eyes. Why can you not block sound from reaching your ears this way?

3. **O** Consider a wave passing through a single slit. What happens to the width of the central maximum of its diffraction pattern as the slit is made half as wide? (a) The central maximum becomes one-fourth as wide. (b) It becomes one-half as wide. (c) Its width does not change. (d) It becomes two times wider. (e) It becomes four times wider.

4. **O** Assume Figure 38.1 was photographed with red light of a single wavelength λ_0. The light passed through a single slit of width a and traveled distance L to the screen where the photograph was made. Consider the width of the central bright fringe, measured between the dark fringes on both sides of it. Rank from largest to smallest the widths of the central fringe in the following situations and note any cases of equality. (a) The experiment is performed as photographed. (b) The experiment is performed with light whose frequency is increased by 50%. (c) The experiment is performed with light whose wavelength is increased by 50%. (Its wavelength is $3\lambda_0/2$.) (d) The experiment is performed with the original light and with a slit of width $2a$. (e) The experiment is performed with the original light and slit, and with distance $2L$ to the screen. (f) The experiment is performed with light of twice the original intensity.

5. **O** In Active Figure 38.4, assume the slit is in a barrier that is opaque to x-rays as well as to visible light. The photograph in Active Figure 38.4b shows the diffraction pattern produced with visible light. What will happen if the experiment is repeated with x-rays as the incoming wave and with no other changes? (a) The diffraction pattern is similar. (b) There is no noticeable diffraction pattern but rather a projected shadow of high intensity on the screen, having the same width as the slit. (c) The central maximum is much wider and the minima occur at larger angles than with visible light. (d) No x-rays reach the screen.

6. **O** Off in the distance, you see the headlights of a car, but they are indistinguishable from the single headlight of a motorcycle. Assume the car's headlights are now switched from low beam to high beam so that the light intensity you receive becomes three times greater. What then happens to your ability to resolve the two light sources? (a) It increases by a factor of 9. (b) It increases by a factor of 3. (c) It remains the same. (d) It becomes one-third as good. (e) It becomes one-ninth as good.

7. A laser beam is incident at a shallow angle on a horizontal machinist's ruler that has a finely calibrated scale. The engraved rulings on the scale give rise to a diffraction pattern on a vertical screen. Discuss how you can use this technique to obtain a measure of the wavelength of the laser light.

8. **O** When you receive a chest x-ray at a hospital, the x-rays pass through a set of parallel ribs in your chest. Do your ribs act as a diffraction grating for x-rays? (a) Yes. They produce diffracted beams that can be observed separately. (b) Not to a measurable extent. The ribs are too far apart. (c) Essentially not. The ribs are too close together. (d) Essentially not. The ribs are too few in number. (e) Absolutely not. X-rays cannot diffract.

9. **O** Certain sunglasses use a polarizing material to reduce the intensity of light reflected as glare from water or automobiles. What orientation should the polarizing filters have to be most effective? (a) The polarizers should absorb light with its electric field horizontal. (b) The polarizers should absorb light with its electric field vertical. (c) The polarizers should absorb both horizontal and vertical electric fields. (d) The polarizers should not absorb either horizontal or vertical electric fields.

10. Is light from the sky polarized? Why is it that clouds seen through Polaroid glasses stand out in bold contrast to the sky?

11. **O** When unpolarized light passes straight through a diffraction grating, does it become polarized? (a) No, it does not. (b) Yes, it does, with the transmission axis parallel to the slits or grooves in the grating. (c) Yes, it does, with the transmission axis perpendicular to the slits or grooves in the grating. (d) It possibly does because an electric field above some threshold is blocked out by the grating if the field is perpendicular to the slits.

12. If a coin is glued to a glass sheet and this arrangement is held in front of a laser beam, the projected shadow has diffraction rings around its edge and a bright spot in the center. How are these effects possible?

13. How could the index of refraction of a flat piece of opaque volcanic glass be determined?

14. A laser produces a beam a few millimeters wide, with uniform intensity across its width. A hair is stretched vertically across the front of the laser to cross the beam. How is the diffraction pattern it produces on a distant screen related to that of a vertical slit equal in width to the hair? How could you determine the width of the hair from measurements of its diffraction pattern?

15. A radio station serves listeners in a city to the northeast of its broadcast site. It broadcasts from three adjacent towers on a mountain ridge, along a line running east and west. Show that by introducing time delays among the signals the individual towers radiate, the station can maximize net intensity in the direction toward the city (and in the opposite direction) and minimize the signal transmitted in other directions. The towers together are said to form a *phased array*.

16. John William Strutt, Lord Rayleigh (1842–1919), is known as the last person to understand all physics and all mathematics. He invented an improved foghorn. To warn ships

of a coastline, a foghorn should radiate sound in a wide horizontal sheet over the ocean's surface. It should not waste energy by broadcasting sound upward. It should not emit sound downward because the water in front of the foghorn would reflect that sound upward. Rayleigh's foghorn trumpet is shown in Figure Q38.16. Is it installed in the correct orientation? Decide whether the long dimension of the rectangular opening should be horizontal or vertical, and argue for your decision.

Figure Q38.16

Problems

WebAssign The Problems from this chapter may be assigned online in WebAssign.

ThomsonNOW™ Sign in at **www.thomsonedu.com** and go to ThomsonNOW to assess your understanding of this chapter's topics with additional quizzing and conceptual questions.

1, 2, 3 denotes straightforward, intermediate, challenging;   □ denotes full solution available in *Student Solutions Manual/Study Guide;*   ▲ denotes coached solution with hints available at **www.thomsonedu.com;**   denotes developing symbolic reasoning;   ● denotes asking for qualitative reasoning;   ▪ denotes computer useful in solving problem

Section 38.2 Diffraction Patterns from Narrow Slits

1. Helium–neon laser light ($\lambda = 632.8$ nm) is sent through a 0.300-mm-wide single slit. What is the width of the central maximum on a screen 1.00 m from the slit?

2. A beam of monochromatic green light is diffracted by a slit of width 0.550 mm. The diffraction pattern forms on a wall 2.06 m beyond the slit. The distance between the positions of zero intensity on both sides of the central bright fringe is 4.10 mm. Calculate the wavelength of the light.

3. ▲ A screen is placed 50.0 cm from a single slit, which is illuminated with 690-nm light. If the distance between the first and third minima in the diffraction pattern is 3.00 mm, what is the width of the slit?

4. Coherent microwaves of wavelength 5.00 cm enter a tall, narrow window in a building otherwise essentially opaque to the microwaves. If the window is 36.0 cm wide, what is the distance from the central maximum to the first-order minimum along a wall 6.50 m from the window?

5. Sound with a frequency 650 Hz from a distant source passes through a doorway 1.10 m wide in a sound-absorbing wall. Find the number and approximate directions of the diffraction-maximum beams radiated into the space beyond.

6. ● A horizontal laser beam of wavelength 632.8 nm has a circular cross section 2.00 mm in diameter. A rectangular aperture is to be placed in the center of the beam so that when the light falls perpendicularly on a wall 4.50 m away, the central maximum fills a rectangle 110 mm wide and 6.00 mm high. The dimensions are measured between the minima bracketing the central maximum. (a) Find the required width and height of the aperture. (b) Is the longer dimension of the central bright patch in the diffraction pattern horizontal or vertical? Is the longer dimension of the aperture horizontal or vertical? Explain the relationship between these two rectangles, using a diagram.

7. A diffraction pattern is formed on a screen 120 cm away from a 0.400-mm-wide slit. Monochromatic 546.1-nm light is used. Calculate the fractional intensity I/I_{max} at a point on the screen 4.10 mm from the center of the principal maximum.

8. **What If?** Assume the light in Figure 38.5 strikes the single slit at an angle β from the perpendicular direction. Show that Equation 38.1, the condition for destructive interference, must be modified to read

$$\sin \theta_{dark} = m\left(\frac{\lambda}{a}\right) - \sin \beta$$

9. Assume light with a wavelength of 650 nm passes through two slits 3.00 μm wide, with their centers 9.00 μm apart. Make a sketch of the combined diffraction and interference pattern in the form of a graph of intensity versus $\phi = (\pi a \sin \theta)/\lambda$. You may use Active Figure 38.7 as a starting point.

10. Coherent light of wavelength 501.5 nm is sent through two parallel slits in a large, flat wall. Each slit is 0.700 μm wide. Their centers are 2.80 μm apart. The light then falls on a semicylindrical screen, with its axis at the midline between the slits. (a) Predict the direction of each interference maximum on the screen as an angle away from the bisector of the line joining the slits. (b) Describe the pattern of light on the screen, specifying the number of bright fringes and the location of each. (c) Find the intensity of light on the screen at the center of each bright fringe, expressed as a fraction of the light intensity I_{max} at the center of the pattern.

Section 38.3 Resolution of Single-Slit and Circular Apertures

11. The pupil of a cat's eye narrows to a vertical slit of width 0.500 mm in daylight. What is the angular resolution for horizontally separated mice? Assume the average wavelength of the light is 500 nm.

12. ● Yellow light of wavelength 589 nm is used to view an object under a microscope. The objective diameter is 9.00 mm. (a) What is the limiting angle of resolution? (b) Suppose it is possible to use visible light of any wavelength. What color should you choose to give the smallest possible angle of resolution, and what is this angle? (c) Suppose water fills the space between the object and the objective. What effect does this change have on the resolving power when 589-nm light is used?

13. ▲ A helium–neon laser emits light that has a wavelength of 632.8 nm. The circular aperture through which the beam emerges has a diameter of 0.500 cm. Estimate the diameter of the beam 10.0 km from the laser.

14. ● Narrow, parallel, glowing gas-filled tubes in a variety of colors form block letters to spell out the name of a night-club. Adjacent tubes are all 2.80 cm apart. The tubes forming one letter are filled with neon and radiate predominantly red light with a wavelength of 640 nm. For

another letter, the tubes emit predominantly blue light at 440 nm. The pupil of a dark-adapted viewer's eye is 5.20 mm in diameter. If she is in a certain range of distances away, the viewer can resolve the separate tubes of one color but not the other. Which color is easier to resolve? State how you decide. The viewer's distance must be in what range for her to resolve the tubes of only one of these two colors?

15. Impressionist painter Georges Seurat created paintings with an enormous number of dots of pure pigment, each of which was approximately 2.00 mm in diameter. The idea was to have colors such as red and green next to each other to form a scintillating canvas (Fig. P38.15). Outside what distance would one be unable to discern individual dots on the canvas? Assume $\lambda = 500$ nm and a pupil diameter of 4.00 mm.

Figure P38.15

16. What are the approximate dimensions of the smallest object on the Earth that astronauts can resolve by eye when they are orbiting 250 km above the Earth? Assume $\lambda = 500$ nm and a pupil diameter of 5.00 mm.

17. A spy satellite can consist of a large-diameter concave mirror forming an image on a digital-camera detector and sending the picture to a ground receiver by radio waves. In effect, it is an astronomical telescope in orbit, looking down instead of up. Can a spy satellite read a license plate? Can it read the date on a dime? Argue for your answers by making an order-of-magnitude calculation, specifying the data you estimate.

2 = intermediate;   3 = challenging;   ☐ = SSM/SG;   ▲ = ThomsonNOW;   = symbolic reasoning;   ● = qualitative reasoning

18. A circular radar antenna on a Coast Guard ship has a diameter of 2.10 m and radiates at a frequency of 15.0 GHz. Two small boats are located 9.00 km away from the ship. How close together could the boats be and still be detected as two objects?

Section 38.4 The Diffraction Grating

Note: In the following problems, assume the light is incident normally on the gratings.

19. White light is spread out into its spectral components by a diffraction grating. If the grating has 2 000 grooves per centimeter, at what angle does red light of wavelength 640 nm appear in first order?

20. Light from an argon laser strikes a diffraction grating that has 5 310 grooves per centimeter. The central and first-order principal maxima are separated by 0.488 m on a wall 1.72 m from the grating. Determine the wavelength of the laser light.

21. ▲ The hydrogen spectrum includes a red line at 656 nm and a blue-violet line at 434 nm. What are the angular separations between these two spectral lines obtained with a diffraction grating that has 4 500 grooves/cm?

22. A helium–neon laser (λ = 632.8 nm) is used to calibrate a diffraction grating. If the first-order maximum occurs at 20.5°, what is the spacing between adjacent grooves in the grating?

23. Three discrete spectral lines occur at angles of 10.09°, 13.71°, and 14.77° in the first-order spectrum of a grating spectrometer. (a) If the grating has 3 660 slits/cm, what are the wavelengths of the light? (b) At what angles are these lines found in the second-order spectrum?

24. Show that whenever white light is passed through a diffraction grating of any spacing size, the violet end of the continuous visible spectrum in third order always overlaps with red light at the other end of the second-order spectrum.

25. A refrigerator shelf is an array of parallel wires with uniform spacing of 1.30 cm between centers. In air at 20°C, ultrasound with a frequency of 37.2 kHz from a distant source falls perpendicularly on the shelf. Find the number of diffracted beams leaving the other side of the shelf. Find the direction of each beam.

26. The laser in a CD player must precisely follow the spiral track, along which the distance between one loop of the spiral and the next is only about 1.25 μm. A feedback mechanism lets the player know if the laser drifts off the track so that the player can steer it back again. Figure P38.26 shows how a diffraction grating is used to provide information to keep the beam on track. The laser light passes through a diffraction grating before it reaches the disk. The strong central maximum of the diffraction pattern is used to read the information in the track of pits. The two first-order side maxima are used for steering. The grating is designed so that the first-order maxima fall on the flat surfaces on both sides of the information track. Both side beams are reflected into their own detectors. As long as both beams are reflecting from smooth nonpitted surfaces, they are detected with constant high intensity. If the main beam wanders off the track, however, one of the side beams begins to strike pits on the information track and the reflected light diminishes. This change is used with an electronic circuit to guide the beam back to the desired location. Assume the laser light has a wavelength of 780 nm and the diffraction grating is positioned 6.90 μm from the disk. Assume the first-order beams are to fall on the disk 0.400 μm on either side of the information track. What should be the number of grooves per millimeter in the grating?

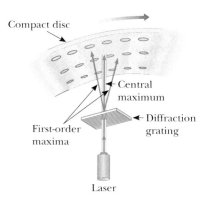

Figure P38.26

27. A grating with 250 grooves/mm is used with an incandescent light source. Assume the visible spectrum to range in

wavelength from 400 to 700 nm. In how many orders can one see (a) the entire visible spectrum and (b) the short-wavelength region?

28. ● A wide beam of laser light with a wavelength of 632.8 nm is directed through several narrow parallel slits, separated by 1.20 mm, and falls on a sheet of photographic film 1.40 m away. The exposure time is chosen so that the film stays unexposed everywhere except at the central region of each bright fringe. (a) Find the distance between these interference maxima. The film is printed as a transparency; it is opaque everywhere except at the exposed lines. Next, the same beam of laser light is directed through the transparency and allowed to fall on a screen 1.40 m beyond. (b) Argue that several narrow parallel bright regions, separated by 1.20 mm, appear on the screen as real images of the original slits. If the screen is removed, light diverges from the images of the original slits with the same reconstructed wave fronts as the original slits produced. *Suggestion:* You may find it useful to draw diagrams similar to Figure 38.12. A similar train of thought, at a soccer game, led Dennis Gabor to invent holography.

29. A diffraction grating has 4 200 rulings/cm. On a screen 2.00 m from the grating, it is found that for a particular order m, the maxima corresponding to two closely spaced wavelengths of sodium (589.0 nm and 589.6 nm) are separated by 1.59 mm. Determine the value of m.

Section 38.5 Diffraction of X-Rays by Crystals

30. Potassium iodide (KI) has the same crystalline structure as NaCl, with atomic planes separated by 0.353 nm. A monochromatic x-ray beam shows a first-order diffraction maximum when the grazing angle is 7.60°. Calculate the x-ray wavelength.

31. ▲ If the interplanar spacing of NaCl is 0.281 nm, what is the predicted angle at which 0.140-nm x-rays are diffracted in a first-order maximum?

32. In water of uniform depth, a wide pier is supported on pilings in several parallel rows 2.80 m apart. Ocean waves of uniform wavelength roll in, moving in a direction that makes an angle of 80.0° with the rows of posts. Find the three longest wavelengths of waves that are strongly reflected by the pilings.

33. ● The atoms in a crystal lie in planes separated by a few tenths of a nanometer. Can they produce a diffraction pattern for visible light as they do for x-rays? Explain your answer with reference to Bragg's law.

Section 38.6 Polarization of Light Waves

Problem 36 in Chapter 34 can be assigned with this section.

34. Unpolarized light passes through two ideal Polaroid sheets. The axis of the first is vertical and that of the second is at 30.0° to the vertical. What fraction of the incident light is transmitted?

35. Plane-polarized light is incident on a single polarizing disk with the direction of $\vec{E}_0$ parallel to the direction of the transmission axis. Through what angle should the disk be rotated so that the intensity in the transmitted beam is reduced by a factor of (a) 3.00, (b) 5.00, and (c) 10.0?

36. The angle of incidence of a light beam onto a reflecting surface is continuously variable. The reflected ray in air is completely polarized when the angle of incidence is 48.0°. What is the index of refraction of the reflecting material?

37. The critical angle for total internal reflection for sapphire surrounded by air is 34.4°. Calculate the polarizing angle for sapphire.

38. For a particular transparent medium surrounded by air, show that the critical angle for total internal reflection and the polarizing angle are related by $\cot \theta_p = \sin \theta_c$.

39. ● **Review problem.** (a) A transparent plate with index of refraction n_2 is immersed in a medium with index n_1. Light traveling in the surrounding medium strikes the top surface of the plate at Brewster's angle. Show that if and only if the surfaces of the plate are parallel, the refracted light strikes the bottom surface of the plate at Brewster's angle for that interface. (b) **What If?** Instead of a plate, consider a prism of index of refraction n_2 separating

media of different refractive indices n_1 and n_3. The light propagates in a plane, containing also the apex angle of the prism. Is there one particular apex angle between the surfaces of the prism for which light can fall on both its surfaces at Brewster's angle as it passes through the prism? If so, determine it.

40. In Figure P38.40, suppose the transmission axes of the left and right polarizing disks are perpendicular to each other. Also, let the center disk be rotated on the common axis with an angular speed ω. Show that if unpolarized light is incident on the left disk with an intensity I_{max}, the intensity of the beam emerging from the right disk is

$$I = \tfrac{1}{16} I_{max}(1 - \cos 4\omega t)$$

This result means that the intensity of the emerging beam is modulated at a rate four times the rate of rotation of the center disk. *Suggestion:* Use the trigonometric identities $\cos^2 \theta = (1 + \cos 2\theta)/2$ and $\sin^2 \theta = (1 - \cos 2\theta)/2$, and recall that $\theta = \omega t$.

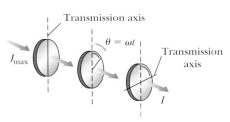

Figure P38.40

41. ● An unpolarized beam of light is incident on a stack of ideal polarizing filters. The axis of the first filter is perpendicular to the axis of the last filter in the stack. Find the fraction by which the transmitted beam's intensity is reduced in the following three cases. (a) Three filters are in the stack, each with its transmission axis at 45.0° relative to the preceding filter. (b) Four filters are in the stack, each with its transmission axis at 30.0° relative to the preceding filter. (c) Seven filters are in the stack, each with its axis at 15.0° relative to the preceding filter. (d) Comment on comparing the answers to parts (a), (b), and (c).

Additional Problems

42. ● Laser light with a wavelength of 632.8 nm is directed through one slit or two slits and allowed to fall on a

screen 2.60 m beyond. Figure P38.42 shows the pattern on the screen, with a centimeter ruler below it. Did the light pass through one slit or two slits? Explain how you can tell. If one slit, find its width. If two slits, find the distance between their centers.

Figure P38.42

43. You use a sequence of ideal polarizing filters, each with its axis making the same angle with the axis of the previous filter, to rotate the plane of polarization of a polarized light beam by a total of 45.0°. You wish to have an intensity reduction no larger than 10.0%. (a) How many polarizers do you need to achieve your goal? (b) What is the angle between adjacent polarizers?

44. ● Figure P38.44 shows a megaphone in use. Construct a theoretical description of how a megaphone works. You may assume the sound of your voice radiates just through the opening of your mouth. Most of the information in speech is carried not in a signal at the fundamental frequency, but in noises and in harmonics, with frequencies of a few thousand hertz. Does your theory allow any prediction that is simple to test?

© Doug Pensinger/Getty Images

Figure P38.44

45. In a single-slit diffraction pattern, (a) find the ratio of the intensity of the first-order side maximum to the intensity of the central maximum. (b) Find the ratio of the intensity of the second-order side maximum to the intensity of the central maximum. You may assume each side maximum is halfway between the adjacent minima.

46. Consider a light wave passing through a slit. Give a mathematical argument that more than 90% of the transmitted energy is in the central maximum of the diffraction pattern. *Suggestions:* Think of the energy as represented by the shaded area in Figure 38.6a. Solve Problem 45 as preparation for this one. You are not expected to calculate the precise percentage, but explain the steps of your reasoning. You may use the identification

$$\frac{1}{1^2} + \frac{1}{3^2} + \frac{1}{5^2} + \cdots = \frac{\pi^2}{8}$$

47. Light from a helium–neon laser ($\lambda = 632.8$ nm) is incident on a single slit. What is the maximum width of the slit for which no diffraction minima are observed?

48. ● Two motorcycles separated laterally by 2.30 m are approaching an observer who is holding a "snooper scope" sensitive to infrared light of wavelength 885 nm. What aperture diameter is required if the motorcycles' headlights are to be resolved at a distance of 12.0 km? Assume the light propagates through perfectly steady and uniform air. Comment on how realistic this assumption is.

49. **Review problem.** A beam of 541-nm light is incident on a diffraction grating that has 400 grooves/mm. (a) Determine the angle of the second-order ray. (b) **What If?** If the entire apparatus is immersed in water, what is the new second-order angle of diffraction? (c) Show that the two diffracted rays of parts (a) and (b) are related through the law of refraction.

50. The *Very Large Array* (VLA) is a set of 27 radio telescope dishes in Catron and Socorro counties, New Mexico (Fig. P38.50). The antennas can be moved apart on railroad tracks, and their combined signals give the resolving power of a synthetic aperture 36.0 km in diameter. (a) If the detectors are tuned to a frequency of 1.40 GHz, what is the angular resolution of the VLA? (b) Clouds of hydrogen radiate at this frequency. What must be the separation distance of two clouds at the center of the galaxy, 26 000 light-years away, if they are to be resolved? (c) **What If?** As the telescope looks up, a circling hawk looks down. Find the angular resolution of the hawk's eye. Assume the hawk is most sensitive to green light having a wavelength of 500 nm and it has a pupil of diameter 12.0 mm. (d) A mouse is on the ground 30.0 m below. By what distance must the mouse's whiskers be separated if the hawk can resolve them?

Figure P38.50

51. A 750-nm light beam hits the flat surface of a certain liquid, and the beam is split into a reflected ray and a refracted ray. If the reflected ray is completely polarized at 36.0°, what is the wavelength of the refracted ray?

52. ● Iridescent peacock feathers are shown in Figure P38.52a. The surface of one microscopic barbule is composed of transparent keratin that supports rods of dark brown melanin in a regular lattice, represented in Figure P38.52b. (Your fingernails are made of keratin, and melanin is the dark pigment giving color to human skin.) In a portion of the feather that can appear turquoise

(a)   (b)

Figure P38.52

(blue-green), assume the melanin rods are uniformly separated by 0.25 μm, with air between them. (a) Explain how this structure can appear blue-green when it contains no blue or green pigment. (b) Explain how it can also appear violet if light falls on it in a different direction. (c) Explain how it can present different colors to your two eyes simultaneously, which is a characteristic of iridescence. (d) A compact disc can appear to be any color of the rainbow. Explain why this portion of the feather cannot appear yellow or red. (e) What could be different about the array of melanin rods in a portion of the feather that does appear to be red?

53. Light of wavelength 500 nm is incident normally on a diffraction grating. If the third-order maximum of the diffraction pattern is observed at 32.0°, (a) what is the number of rulings per centimeter for the grating? (b) Determine the total number of primary maxima that can be observed in this situation.

54. ● Light in air strikes a water surface at the polarizing angle. The part of the beam refracted into the water strikes a submerged slab of material with refractive index n as shown in Figure P38.54. The light reflected from the upper surface of the slab is completely polarized. (a) Find the angle θ between the water surface and the surface of the slab as a function of n. (b) Identify the maximum imaginable value of θ and describe the physical situation to which it corresponds. (c) Identify the minimum imaginable value of θ and describe the physical situation to which it corresponds.

Figure P38.54

55. A beam of bright red light of wavelength 654 nm passes through a diffraction grating. Enclosing the space beyond the grating is a large screen forming one half of a cylinder centered on the grating, with its axis parallel to the slits in the grating. Fifteen bright spots appear on the screen. Find the maximum and minimum possible values for the slit separation in the diffraction grating.

56. A *pinhole camera* has a small circular aperture of diameter D. Light from distant objects passes through the aperture into an otherwise dark box, falling on a screen located a distance L away. If D is too large, the display on the screen will be fuzzy because a bright point in the field of view will send light onto a circle of diameter slightly larger than D. On the other hand, if D is too small, diffraction will blur the display on the screen. The screen shows a reasonably sharp image if the diameter of the central disk of the diffraction pattern, specified by Equation 38.6, is equal to D at the screen. (a) Show that for monochromatic light with plane wave fronts and $L \gg D$, the condition for a sharp view is fulfilled if $D^2 = 2.44\lambda L$. (b) Find the optimum pinhole diameter for 500-nm light projected onto a screen 15.0 cm away.

57. An American standard television picture is composed of approximately 485 horizontal lines of varying light intensity. Assume your ability to resolve the lines is limited only by the Rayleigh criterion and the pupils of your eyes are 5.00 mm in diameter. Calculate the ratio of minimum viewing distance to the vertical dimension of the picture such that you will not be able to resolve the lines. Assume the average wavelength of the light coming from the screen is 550 nm.

58. (a) Light traveling in a medium of index of refraction n_1 is incident at an angle θ on the surface of a medium of index n_2. The angle between reflected and refracted rays is β. Show that

$$\tan \theta = \frac{n_2 \sin \beta}{n_1 - n_2 \cos \beta}$$

Suggestion: Use the identity $\sin (A + B) = \sin A \cos B + \cos A \sin B$. (b) **What If?** Show that this expression for $\tan \theta$ reduces to Brewster's law when $\beta = 90°$, $n_1 = 1$, and $n_2 = n$.

59. Suppose the single slit in Active Figure 38.4 is 6.00 cm wide and in front of a microwave source operating at 7.50 GHz. (a) Calculate the angle subtended by the first minimum in the diffraction pattern. (b) What is the relative intensity I/I_{max} at $\theta = 15.0°$? (c) Assume two such sources, separated laterally by 20.0 cm, are behind the slit. What must the maximum distance between the plane of the sources and the slit be if the diffraction patterns are to be resolved? In this case, the approximation $\sin \theta \approx \tan \theta$ is not valid because of the relatively small value of a/λ.

60. ● (a) Two polarizing sheets are placed together with their transmission axes crossed so that no light is transmitted. A third sheet is inserted between them with its transmission axis at an angle of 45.0° with respect to each of the other axes. Find the fraction of incident unpolarized light intensity transmitted by the three-sheet combination. Assume each polarizing sheet is ideal. (b) A parent is searching an Internet collection of recipes for easy main dishes. She uses a computer to eliminate all the recipes containing meat and then to eliminate all the remaining recipes containing cheese. No recipes from the original database remain after these two sorting processes. With the same database, she next tries a sequence of three selection rounds: eliminating all recipes containing meat, eliminating remaining recipes containing more than a little meat or cheese, and eliminating remaining recipes containing any cheese. Will any recipes remain in the list after these three sorting processes? Compare and contrast the results of the recipe-sorting experiment with the results of the polarization experiment in part (a).

61. The scale of a map is a number of kilometers per centimeter, specifying the distance on the ground that any distance on the map represents. The scale of a spectrum is its *dispersion,* a number of nanometers per centimeter, specifying the change in wavelength that a distance across the spectrum represents. You must know the dispersion if you want to compare one spectrum with another or to make a measurement of, for example, a Doppler shift. Let y represent the position relative to the center of a diffraction pattern projected onto a flat screen at distance L by a diffraction grating with slit spacing d. The dispersion is $d\lambda/dy$. (a) Prove that the dispersion is given by

$$\frac{d\lambda}{dy} = \frac{L^2 d}{m(L^2 + y^2)^{3/2}}$$

(b) Calculate the dispersion in first order for light with a mean wavelength of 550 nm, analyzed with a grating having 8 000 rulings/cm, and projected onto a screen 2.40 m away.

62. Two closely spaced wavelengths of light are incident on a diffraction grating. (a) Starting with Equation 38.7, show that the angular dispersion of the grating is given by

$$\frac{d\theta}{d\lambda} = \frac{m}{d\cos\theta}$$

(b) A square grating 2.00 cm on each side containing 8 000 equally spaced slits is used to analyze the spectrum of mercury. Two closely spaced lines emitted by this element have wavelengths of 579.065 nm and 576.959 nm. What is the angular separation of these two wavelengths in the second-order spectrum?

63. Figure P38.63a is a three-dimensional sketch of a birefringent crystal. The dotted lines illustrate how a thin, parallel-

faced slab of material could be cut from the larger specimen with the crystal's optic axis parallel to the faces of the plate. A section cut from the crystal in this manner is known as a *retardation plate.* When a beam of light is incident on the plate perpendicular to the direction of the optic axis as shown in Figure P38.63b, the O ray and the E ray travel along a single straight line, but with different speeds. (a) Let the thickness of the plate be d. Show that the phase difference between the O ray and the E ray is

$$\theta = \frac{2\pi d}{\lambda}|n_O - n_E|$$

where λ is the wavelength in air. (b) In a particular case, the incident light has a wavelength of 550 nm. Find the minimum value of d for a quartz plate for which $\theta = \pi/2$. Such a plate is called a *quarter-wave plate.* Use values of n_O and n_E from Table 38.1.

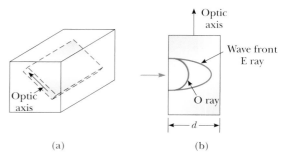

(a)   (b)

Figure P38.63

64. ▀ How much diffraction spreading does a light beam undergo? One quantitative answer is the *full width at half maximum* of the central maximum of the single-slit Fraunhofer diffraction pattern. You can evaluate this angle of spreading in this problem and the next. (a) In Equation 38.2, define $\pi a \sin\theta/\lambda = \phi$ and show that, at the point where $I = 0.5I_{max}$, we must have $\sin\phi = \phi/\sqrt{2}$. (b) Let $y_1 = \sin\phi$ and $y_2 = \phi/\sqrt{2}$. Plot y_1 and y_2 on the same set of axes over a range from $\phi = 1$ rad to $\phi = \pi/2$ rad. Determine ϕ from the point of intersection of the two curves. (c) Then show that if the fraction λ/a is not large, the angular full width at half maximum of the central diffraction maximum is $\Delta\theta = 0.886\lambda/a$.

65. ▀ Another method to solve the transcendental equation $\phi = \sqrt{2}\sin\phi$ in Problem 64 is to guess a first value of ϕ, use a computer or calculator to see how nearly it fits, and continue to update your estimate until the equation balances. How many steps (iterations) does this process take?

horizontal tangent at maxima and also at minima. You will need to solve a transcendental equation.

66. ☕ The diffraction pattern of a single slit is described by the equation

$$I_\theta = I_{max} \frac{\sin^2 \phi}{\phi^2}$$

where $\phi = (\pi a \sin \theta)/\lambda$. The central maximum is at $\phi = 0$, and the side maxima are *approximately* at $\phi = (m + \frac{1}{2})\pi$ for $m = 1, 2, 3, \ldots$. Determine more precisely (a) the location of the first side maximum, where $m = 1$, and (b) the location of the second side maximum. Observe in Figure 38.6a that the graph of intensity versus ϕ has a

67. ☕ Light of wavelength 632.8 nm illuminates a single slit, and a diffraction pattern is formed on a screen 1.00 m from the slit. Using the data in the following table, plot relative intensity versus position. Choose an appropriate value for the slit width a and, on the same graph used for the experimental data, plot the theoretical expression for the relative intensity

$$\frac{I_\theta}{I_{max}} = \frac{\sin^2 \phi}{\phi^2}$$

where $\phi = (\pi a \sin \theta)/\lambda$. What value of a gives the best fit of theory and experiment?

Relative Intensity	Position Relative to Central Maximum (mm)	Relative Intensity	Position Relative to Central Maximum (mm)
1.00	0	0.029	10.5
0.95	0.8	0.013	11.3
0.80	1.6	0.002	12.1
0.60	2.4	0.000 3	12.9
0.39	3.2	0.005	13.7
0.21	4.0	0.012	14.5
0.079	4.8	0.016	15.3
0.014	5.6	0.015	16.1
0.003	6.5	0.010	16.9
0.015	7.3	0.004 4	17.7
0.036	8.1	0.000 6	18.5
0.047	8.9	0.000 3	19.3
0.043	9.7	0.003	20.2

Answers to Quick Quizzes

38.1 (a). Equation 38.1 shows that a decrease in a results in an increase in the angles at which the dark fringes appear.

38.2 (i). In Equation 38.4, the ratio d/a is independent of wavelength, so the number of interference fringes in the central diffraction pattern peak remains the same. Equation 38.1 tells us that a decrease in wavelength causes a decrease in the width of the central peak if I is graphed against y as in Active Figure 38.4. If I is graphed against $(\pi/\lambda) a \sin \theta$ as in Active Figure 38.7, however, the peak width is independent of λ.

38.3 (b). The effective slit width in the vertical direction of the cat's eye is larger than that in the horizontal direction. Therefore, the cat's eye has more resolving power for lights separated in the vertical direction and would be more effective at resolving the mast lights on the boat.

38.4 (a). We would like to reduce the minimum angular separation for two objects below the angle subtended by the two stars in the binary system. That can be done by reducing the wavelength of the light, which in essence makes the aperture larger, relative to the light wavelength, increasing the resolving power. Therefore, we should choose a blue filter.

38.5   (c). Doubling the wavelength makes the pattern wider. Choices (a) and (d) make the pattern even wider. From Equation 38.10, we see that choice (b) causes $\sin \theta$ to be twice as large. Because the small angle approximation cannot be used, however, doubling $\sin \theta$ is not the same as doubling θ, which would translate to a doubling of the position of a maximum along the screen. If we only consider small-angle maxima, choice (b) would work, but it does not work in the large-angle case.

38.6   (b). Electric field vectors parallel to the metal wires cause electrons in the metal to oscillate parallel to the wires. Therefore, the energy from the waves with these electric field vectors is transferred to the metal by accelerating these electrons and is eventually transformed to internal energy through the resistance of the metal. Waves with electric field vectors perpendicular to the metal wires pass through because they are not able to accelerate electrons in the wires.

38.7   (c). At some intermediate distance, the light rays from the fixtures will strike the floor at Brewster's angle and reflect to your eyes. Because this light is polarized horizontally, it will not pass through your polarized sunglasses. Tilting your head to the side will cause the reflections to reappear.

PART 6

Modern Physics

At the end of the 19th century, many scientists believed they had learned most of what there was to know about physics. Newton's laws of motion and theory of universal gravitation, Maxwell's theoretical work in unifying electricity and magnetism, the laws of thermodynamics and kinetic theory, and the principles of optics were highly successful in explaining a variety of phenomena.

At the turn of the 20th century, however, a major revolution shook the world of physics. In 1900, Max Planck provided the basic ideas that led to the formulation of the quantum theory, and in 1905, Albert Einstein formulated his special theory of relativity. The excitement of the times is captured in Einstein's own words: "It was a marvelous time to be alive." Both theories were to have a profound effect on our understanding of nature. Within a few decades, they inspired new developments in the fields of atomic physics, nuclear physics, and condensed-matter physics.

In Chapter 39, we shall introduce the special theory of relativity. The theory provides us with a new and deeper view of physical laws. Although the predictions of this theory often violate our common sense, the theory correctly describes the results of experiments involving speeds near the speed of light. The extended version of this textbook, *Physics for Scientists and Engineers with Modern Physics,* covers the basic concepts of quantum mechanics and their application to atomic and molecular physics, and we introduce solid-state physics, nuclear physics, particle physics, and cosmology.

Even though the physics that was developed during the 20th century has led to a multitude of important technological achievements, the story is still incomplete. Discoveries will continue to evolve during our lifetimes, and many of these discoveries will deepen or refine our understanding of nature and the Universe around us. It is still a "marvelous time to be alive."

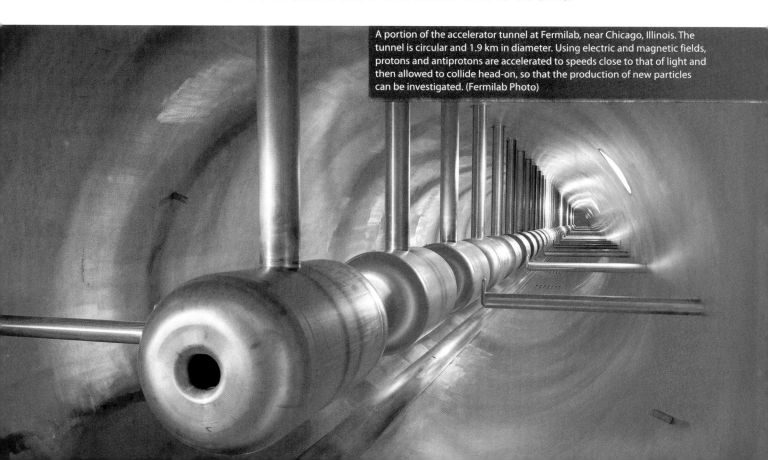

A portion of the accelerator tunnel at Fermilab, near Chicago, Illinois. The tunnel is circular and 1.9 km in diameter. Using electric and magnetic fields, protons and antiprotons are accelerated to speeds close to that of light and then allowed to collide head-on, so that the production of new particles can be investigated. (Fermilab Photo)

39.1 The Principle of Galilean Relativity

39.2 The Michelson–Morley Experiment

39.3 Einstein's Principle of Relativity

39.4 Consequences of the Special Theory of Relativity

39.5 The Lorentz Transformation Equations

39.6 The Lorentz Velocity Transformation Equations

39.7 Relativistic Linear Momentum

39.8 Relativistic Energy

39.9 Mass and Energy

39.10 The General Theory of Relativity

Standing on the shoulders of a giant. David Serway, son of one of the authors, watches over two of his children, Nathan and Kaitlyn, as they frolic in the arms of Albert Einstein's statue at the Einstein memorial in Washington, D.C. It is well known that Einstein, the principal architect of relativity, was very fond of children. (Emily Serway)

39 Relativity

Our everyday experiences and observations involve objects that move at speeds much less than the speed of light. Newtonian mechanics was formulated by observing and describing the motion of such objects, and this formalism is very successful in describing a wide range of phenomena that occur at low speeds. Nonetheless, it fails to describe properly the motion of objects whose speeds approach that of light.

Experimentally, the predictions of Newtonian theory can be tested at high speeds by accelerating electrons or other charged particles through a large electric potential difference. For example, it is possible to accelerate an electron to a speed of $0.99c$ (where c is the speed of light) by using a potential difference of several million volts. According to Newtonian mechanics, if the potential difference is increased by a factor of 4, the electron's kinetic energy is four times greater and its speed should double to $1.98c$. Experiments show, however, that the speed of the electron—as well as the speed of any other object in the Universe—always remains less than the speed of light, regardless of the size of the accelerating voltage. Because it places no upper limit on speed, Newtonian mechanics is contrary to modern experimental results and is clearly a limited theory.

In 1905, at the age of only 26, Einstein published his special theory of relativity. Regarding the theory, Einstein wrote:

The relativity theory arose from necessity, from serious and deep contradictions in the old theory from which there seemed no escape. The strength of the new theory lies in the consistency and simplicity with which it solves all these difficulties.[1]

Although Einstein made many other important contributions to science, the special theory of relativity alone represents one of the greatest intellectual achievements of all time. With this theory, experimental observations can be correctly predicted over the range of speeds from $v = 0$ to speeds approaching the speed of light. At low speeds, Einstein's theory reduces to Newtonian mechanics as a limiting situation. It is important to recognize that Einstein was working on electromagnetism when he developed the special theory of relativity. He was convinced that Maxwell's equations were correct, and to reconcile them with one of his postulates, he was forced into the revolutionary notion of assuming that space and time are not absolute.

This chapter gives an introduction to the special theory of relativity, with emphasis on some of its predictions. In addition to its well-known and essential role in theoretical physics, the special theory of relativity has practical applications, including the design of nuclear power plants and modern global positioning system (GPS) units. These devices do not work if designed in accordance with nonrelativistic principles.

39.1   The Principle of Galilean Relativity

To describe a physical event, we must establish a frame of reference. You should recall from Chapter 5 that an inertial frame of reference is one in which an object is observed to have no acceleration when no forces act on it. Furthermore, any frame moving with constant velocity with respect to an inertial frame must also be an inertial frame.

There is no absolute inertial reference frame. Therefore, the results of an experiment performed in a vehicle moving with uniform velocity must be identical to the results of the same experiment performed in a stationary vehicle. The formal statement of this result is called the **principle of Galilean relativity:**

> The laws of mechanics must be the same in all inertial frames of reference.

◀ Principle of Galilean relativity

Let's consider an observation that illustrates the equivalence of the laws of mechanics in different inertial frames. A pickup truck moves with a constant velocity as shown in Figure 39.1a. If a passenger in the truck throws a ball straight up

(a)   (b)

Figure 39.1   (a) The observer in the truck sees the ball move in a vertical path when thrown upward. (b) The Earth-based observer sees the ball's path as a parabola.

[1] A. Einstein and L. Infeld, *The Evolution of Physics* (New York: Simon and Schuster, 1961).

and if air effects are neglected, the passenger observes that the ball moves in a vertical path. The motion of the ball appears to be precisely the same as if the ball were thrown by a person at rest on the Earth. The law of universal gravitation and the equations of motion under constant acceleration are obeyed whether the truck is at rest or in uniform motion.

Both observers agree on the laws of physics: they each throw a ball straight up, and it rises and falls back into their own hand. Do the observers agree on the path of the ball thrown by the observer in the truck? The observer on the ground sees the path of the ball as a parabola as illustrated in Figure 39.1b, while, as mentioned earlier, the observer in the truck sees the ball move in a vertical path. Furthermore, according to the observer on the ground, the ball has a horizontal component of velocity equal to the velocity of the truck. Although the two observers disagree on certain aspects of the situation, they agree on the validity of Newton's laws and on such classical principles as conservation of energy and conservation of linear momentum. This agreement implies that no mechanical experiment can detect any difference between the two inertial frames. The only thing that can be detected is the relative motion of one frame with respect to the other.

Quick Quiz 39.1 Which observer in Figure 39.1 sees the ball's *correct* path? (a) the observer in the truck   (b) the observer on the ground   (c) both observers

Suppose some physical phenomenon, which we call an *event*, occurs and is observed by an observer at rest in an inertial reference frame. The wording "in a frame" means that the observer is at rest with respect to the origin of that frame. The event's location and time of occurrence can be specified by the four coordinates (x, y, z, t). We would like to be able to transform these coordinates from those of an observer in one inertial frame to those of another observer in a frame moving with uniform relative velocity compared with the first frame.

Consider two inertial frames S and S′ (Fig. 39.2). The S′ frame moves with a constant velocity $\vec{v}$ along the common x and x' axes, where $\vec{v}$ is measured relative to S. We assume the origins of S and S′ coincide at $t = 0$ and an event occurs at point P in space at some instant of time. An observer in S describes the event with space–time coordinates (x, y, z, t), whereas an observer in S′ uses the coordinates (x', y', z', t') to describe the same event. As we see from the geometry in Figure 39.2, the relationships among these various coordinates can be written

$$x' = x - vt \qquad y' = y \qquad z' = z \qquad t' = t \tag{39.1}$$

Figure 39.2 An event occurs at a point P. The event is seen by two observers in inertial frames S and S′, where S′ moves with a velocity $\vec{v}$ relative to S.

◄ Galilean transformation equations

These equations are the **Galilean space–time transformation equations.** Note that time is assumed to be the same in both inertial frames. That is, within the framework of classical mechanics, all clocks run at the same rate, regardless of their velocity, so the time at which an event occurs for an observer in S is the same as the time for the same event in S′. Consequently, the time interval between two successive events should be the same for both observers. Although this assumption may seem obvious, it turns out to be incorrect in situations where v is comparable to the speed of light.

Now suppose a particle moves through a displacement of magnitude dx along the x axis in a time interval dt as measured by an observer in S. It follows from Equations 39.1 that the corresponding displacement dx' measured by an observer in S′ is $dx' = dx - v\,dt$, where frame S′ is moving with speed v in the x direction relative to frame S. Because $dt = dt'$, we find that

$$\frac{dx'}{dt'} = \frac{dx}{dt} - v$$

or

$$u'_x = u_x - v \tag{39.2}$$

where u_x and u'_x are the x components of the velocity of the particle measured by observers in S and S′, respectively. (We use the symbol $\vec{u}$ rather than $\vec{v}$ for particle

PITFALL PREVENTION 39.1

The Relationship Between the S and S′ Frames

Many of the mathematical representations in this chapter are true *only* for the specified relationship between the S and S′ frames. The x and x' axes coincide, except their origins are different. The y and y' axes (and the z and z' axes) are parallel, but they do not coincide due to the displacement of the origin of S′ with respect to that of S. We choose the time $t = 0$ to be the instant at which the origins of the two coordinate systems coincide. If the S′ frame is moving in the positive x direction relative to S, then v is positive; otherwise, it is negative.

velocity because $\vec{\mathbf{v}}$ is already used for the relative velocity of two reference frames.) Equation 39.2 is the **Galilean velocity transformation equation.** It is consistent with our intuitive notion of time and space as well as with our discussions in Section 4.6. As we shall soon see, however, it leads to serious contradictions when applied to electromagnetic waves.

Quick Quiz 39.2   A baseball pitcher with a 90-mi/h fastball throws a ball while standing on a railroad flatcar moving at 110 mi/h. The ball is thrown in the same direction as that of the velocity of the train. If you apply the Galilean velocity transformation equation to this situation, is the speed of the ball relative to the Earth (a) 90 mi/h, (b) 110 mi/h, (c) 20 mi/h, (d) 200 mi/h, or (e) impossible to determine?

The Speed of Light

It is quite natural to ask whether the principle of Galilean relativity also applies to electricity, magnetism, and optics. Experiments indicate that the answer is no. Recall from Chapter 34 that Maxwell showed that the speed of light in free space is $c = 3.00 \times 10^8$ m/s. Physicists of the late 1800s thought light waves moved through a medium called the *ether* and the speed of light was c only in a special, absolute frame at rest with respect to the ether. The Galilean velocity transformation equation was expected to hold for observations of light made by an observer in any frame moving at speed v relative to the absolute ether frame. That is, if light travels along the x axis and an observer moves with velocity $\vec{\mathbf{v}}$ along the x axis, the observer measures the light to have speed $c \pm v$, depending on the directions of travel of the observer and the light.

Because the existence of a preferred, absolute ether frame would show that light is similar to other classical waves and that Newtonian ideas of an absolute frame are true, considerable importance was attached to establishing the existence of the ether frame. Prior to the late 1800s, experiments involving light traveling in media moving at the highest laboratory speeds attainable at that time were not capable of detecting differences as small as that between c and $c \pm v$. Starting in about 1880, scientists decided to use the Earth as the moving frame in an attempt to improve their chances of detecting these small changes in the speed of light.

Observers fixed on the Earth can take the view that they are stationary and that the absolute ether frame containing the medium for light propagation moves past them with speed v. Determining the speed of light under these circumstances is similar to determining the speed of an aircraft traveling in a moving air current, or wind; consequently, we speak of an "ether wind" blowing through our apparatus fixed to the Earth.

A direct method for detecting an ether wind would use an apparatus fixed to the Earth to measure the ether wind's influence on the speed of light. If v is the speed of the ether relative to the Earth, light should have its maximum speed $c + v$ when propagating downwind as in Figure 39.3a. Likewise, the speed of light should have its minimum value $c - v$ when the light is propagating upwind as in Figure 39.3b and an intermediate value $(c^2 - v^2)^{1/2}$ when the light is directed such that it travels perpendicular to the ether wind as in Figure 39.3c. If the Sun is assumed to be at rest in the ether, the velocity of the ether wind would be equal to the orbital velocity of the Earth around the Sun, which has a magnitude of approximately 30 km/s or 3×10^4 m/s. Because $c = 3 \times 10^8$ m/s, it is necessary to detect a change in speed of approximately 1 part in 10^4 for measurements in the upwind or downwind directions. Although such a change is experimentally measurable, all attempts to detect such changes and establish the existence of the ether wind (and hence the absolute frame) proved futile! We shall discuss the classic experimental search for the ether in Section 39.2.

The principle of Galilean relativity refers only to the laws of mechanics. If it is assumed the laws of electricity and magnetism are the same in all inertial frames, a

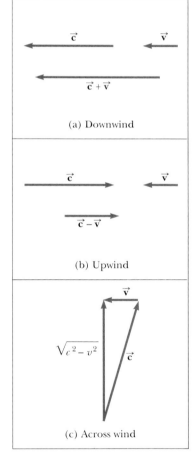

Figure 39.3   If the velocity of the ether wind relative to the Earth is $\vec{\mathbf{v}}$ and the velocity of light relative to the ether is $\vec{\mathbf{c}}$, the speed of light relative to the Earth is (a) $c + v$ in the downwind direction, (b) $c - v$ in the upwind direction, and (c) $(c^2 - v^2)^{1/2}$ in the direction perpendicular to the wind.

paradox concerning the speed of light immediately arises. That can be understood by recognizing that Maxwell's equations imply that the speed of light always has the fixed value 3.00×10^8 m/s in all inertial frames, a result in direct contradiction to what is expected based on the Galilean velocity transformation equation. According to Galilean relativity, the speed of light should *not* be the same in all inertial frames.

To resolve this contradiction in theories, we must conclude that either (1) the laws of electricity and magnetism are not the same in all inertial frames or (2) the Galilean velocity transformation equation is incorrect. If we assume the first alternative, a preferred reference frame in which the speed of light has the value c must exist and the measured speed must be greater or less than this value in any other reference frame, in accordance with the Galilean velocity transformation equation. If we assume the second alternative, we must abandon the notions of absolute time and absolute length that form the basis of the Galilean space–time transformation equations.

39.2 The Michelson–Morley Experiment

The most famous experiment designed to detect small changes in the speed of light was first performed in 1881 by Albert A. Michelson (see Section 37.7) and later repeated under various conditions by Michelson and Edward W. Morley (1838–1923). As we shall see, the outcome of the experiment contradicted the ether hypothesis.

The experiment was designed to determine the velocity of the Earth relative to that of the hypothetical ether. The experimental tool used was the Michelson interferometer, which was discussed in Section 37.7 and is shown again in Active Figure 39.4. Arm 2 is aligned along the direction of the Earth's motion through space. The Earth moving through the ether at speed v is equivalent to the ether flowing past the Earth in the opposite direction with speed v. This ether wind blowing in the direction opposite the direction of the Earth's motion should cause the speed of light measured in the Earth frame to be $c - v$ as the light approaches mirror M_2 and $c + v$ after reflection, where c is the speed of light in the ether frame.

The two light beams reflect from M_1 and M_2 and recombine, and an interference pattern is formed as discussed in Section 37.7. The interference pattern is observed while the interferometer is rotated through an angle of 90°. This rotation interchanges the speed of the ether wind between the arms of the interferometer. The rotation should cause the fringe pattern to shift slightly but measurably. Measurements failed, however, to show any change in the interference pattern! The Michelson–Morley experiment was repeated at different times of the year when the ether wind was expected to change direction and magnitude, but the results were always the same: **no fringe shift of the magnitude required was ever observed.**[2]

The negative results of the Michelson–Morley experiment not only contradicted the ether hypothesis, but also showed that it is impossible to measure the absolute velocity of the Earth with respect to the ether frame. Einstein, however, offered a postulate for his special theory of relativity that places quite a different interpretation on these null results. In later years, when more was known about the nature of light, the idea of an ether that permeates all of space was abandoned. **Light is now understood to be an electromagnetic wave, which requires no medium for its propagation.** As a result, the idea of an ether in which these waves travel became unnecessary.

ACTIVE FIGURE 39.4

According to the ether wind theory, the speed of light should be $c - v$ as the beam approaches mirror M_2 and $c + v$ after reflection.

---

[2] From an Earth-based observer's point of view, changes in the Earth's speed and direction of motion in the course of a year are viewed as ether wind shifts. Even if the speed of the Earth with respect to the ether were zero at some time, six months later the speed of the Earth would be 60 km/s with respect to the ether and as a result a fringe shift should be noticed. No shift has ever been observed, however.

Details of the Michelson–Morley Experiment

To understand the outcome of the Michelson–Morley experiment, let's assume the two arms of the interferometer in Active Figure 39.4 are of equal length L. We shall analyze the situation as if there were an ether wind because that is what Michelson and Morley expected to find. As noted above, the speed of the light beam along arm 2 should be $c - v$ as the beam approaches M_2 and $c + v$ after the beam is reflected. We model a pulse of light as a particle under constant speed. Therefore, the time interval for travel to the right for the pulse is $\Delta t = L/(c - v)$, and the time interval for travel to the left is $\Delta t = L/(c + v)$. The total time interval for the round trip along arm 2 is

$$\Delta t_{\text{arm 2}} = \frac{L}{c + v} + \frac{L}{c - v} = \frac{2Lc}{c^2 - v^2} = \frac{2L}{c}\left(1 - \frac{v^2}{c^2}\right)^{-1}$$

Now consider the light beam traveling along arm 1, perpendicular to the ether wind. Because the speed of the beam relative to the Earth is $(c^2 - v^2)^{1/2}$ in this case (see Fig. 39.3c), the time interval for travel for each half of the trip is $\Delta t = L/(c^2 - v^2)^{1/2}$ and the total time interval for the round trip is

$$\Delta t_{\text{arm 1}} = \frac{2L}{(c^2 - v^2)^{1/2}} = \frac{2L}{c}\left(1 - \frac{v^2}{c^2}\right)^{-1/2}$$

The time difference Δt between the horizontal round trip (arm 2) and the vertical round trip (arm 1) is

$$\Delta t = \Delta t_{\text{arm 2}} - \Delta t_{\text{arm 1}} = \frac{2L}{c}\left[\left(1 - \frac{v^2}{c^2}\right)^{-1} - \left(1 - \frac{v^2}{c^2}\right)^{-1/2}\right]$$

Because $v^2/c^2 \ll 1$, we can simplify this expression by using the following binomial expansion after dropping all terms higher than second order:

$$(1 - x)^n \approx 1 - nx \quad (\text{for } x \ll 1)$$

In our case, $x = v^2/c^2$, and we find that

$$\Delta t = \Delta t_{\text{arm 2}} - \Delta t_{\text{arm 1}} \approx \frac{Lv^2}{c^3} \tag{39.3}$$

This time difference between the two instants at which the reflected beams arrive at the viewing telescope gives rise to a phase difference between the beams, producing an interference pattern when they combine at the position of the telescope. A shift in the interference pattern should be detected when the interferometer is rotated through 90° in a horizontal plane so that the two beams exchange roles. This rotation results in a time difference twice that given by Equation 39.3. Therefore, the path difference that corresponds to this time difference is

$$\Delta d = c(2\,\Delta t) = \frac{2Lv^2}{c^2}$$

Because a change in path length of one wavelength corresponds to a shift of one fringe, the corresponding fringe shift is equal to this path difference divided by the wavelength of the light:

$$\text{Shift} = \frac{2Lv^2}{\lambda c^2} \tag{39.4}$$

In the experiments by Michelson and Morley, each light beam was reflected by mirrors many times to give an effective path length L of approximately 11 m. Using this value, taking v to be equal to 3.0×10^4 m/s (the speed of the Earth around the Sun), and using 500 nm for the wavelength of the light, we expect a fringe shift of

$$\text{Shift} = \frac{2(11 \text{ m})(3.0 \times 10^4 \text{ m/s})^2}{(5.0 \times 10^{-7} \text{ m})(3.0 \times 10^8 \text{ m/s})^2} = 0.44$$

The instrument used by Michelson and Morley could detect shifts as small as 0.01 fringe, but **it detected no shift whatsoever in the fringe pattern!** The experiment has been repeated many times since by different scientists under a wide variety of conditions, and no fringe shift has ever been detected. Therefore, it was concluded that the motion of the Earth with respect to the postulated ether cannot be detected.

Many efforts were made to explain the null results of the Michelson–Morley experiment and to save the ether frame concept and the Galilean velocity transformation equation for light. All proposals resulting from these efforts have been shown to be wrong. No experiment in the history of physics received such valiant efforts to explain the absence of an expected result as did the Michelson–Morley experiment. The stage was set for Einstein, who solved the problem in 1905 with his special theory of relativity.

39.3   Einstein's Principle of Relativity

In the previous section, we noted the impossibility of measuring the speed of the ether with respect to the Earth and the failure of the Galilean velocity transformation equation in the case of light. Einstein proposed a theory that boldly removed these difficulties and at the same time completely altered our notion of space and time.[3] He based his special theory of relativity on two postulates:

1. **The principle of relativity:** The laws of physics must be the same in all inertial reference frames.
2. **The constancy of the speed of light:** The speed of light in vacuum has the same value, $c = 3.00 \times 10^8$ m/s, in all inertial frames, regardless of the velocity of the observer or the velocity of the source emitting the light.

The first postulate asserts that *all* the laws of physics—those dealing with mechanics, electricity and magnetism, optics, thermodynamics, and so on—are the same in all reference frames moving with constant velocity relative to one another. This postulate is a generalization of the principle of Galilean relativity, which refers only to the laws of mechanics. From an experimental point of view, Einstein's principle of relativity means that any kind of experiment (measuring the speed of light, for example) performed in a laboratory at rest must give the same result when performed in a laboratory moving at a constant velocity with respect to the first one. Hence, no preferred inertial reference frame exists, and it is impossible to detect absolute motion.

Note that postulate 2 is required by postulate 1: if the speed of light were not the same in all inertial frames, measurements of different speeds would make it possible to distinguish between inertial frames. As a result, a preferred, absolute frame could be identified, in contradiction to postulate 1.

Although the Michelson–Morley experiment was performed before Einstein published his work on relativity, it is not clear whether or not Einstein was aware of the details of the experiment. Nonetheless, the null result of the experiment can be readily understood within the framework of Einstein's theory. According to his principle of relativity, the premises of the Michelson–Morley experiment were incorrect. In the process of trying to explain the expected results, we stated that when light traveled against the ether wind, its speed was $c - v$, in accordance with the Galilean velocity transformation equation. If the state of motion of the observer or of the source has no influence on the value found for the speed of

ALBERT EINSTEIN
German-American Physicist (1879–1955)
Einstein, one of the greatest physicists of all times, was born in Ulm, Germany. In 1905, at age 26, he published four scientific papers that revolutionized physics. Two of these papers were concerned with what is now considered his most important contribution: the special theory of relativity.

In 1916, Einstein published his work on the general theory of relativity. The most dramatic prediction of this theory is the degree to which light is deflected by a gravitational field. Measurements made by astronomers on bright stars in the vicinity of the eclipsed Sun in 1919 confirmed Einstein's prediction, and Einstein became a world celebrity as a result. Einstein was deeply disturbed by the development of quantum mechanics in the 1920s despite his own role as a scientific revolutionary. In particular, he could never accept the probabilistic view of events in nature that is a central feature of quantum theory. The last few decades of his life were devoted to an unsuccessful search for a unified theory that would combine gravitation and electromagnetism.

AIP Niels Bohr Library

---

[3] A. Einstein, "On the Electrodynamics of Moving Bodies," *Ann. Physik* **17**:891, 1905. For an English translation of this article and other publications by Einstein, see the book by H. Lorentz, A. Einstein, H. Minkowski, and H. Weyl, *The Principle of Relativity* (New York: Dover, 1958).

light, however, one always measures the value to be c. Likewise, the light makes the return trip after reflection from the mirror at speed c, not at speed $c + v$. Therefore, the motion of the Earth does not influence the fringe pattern observed in the Michelson–Morley experiment, and a null result should be expected.

If we accept Einstein's theory of relativity, we must conclude that relative motion is unimportant when measuring the speed of light. At the same time, we must alter our commonsense notion of space and time and be prepared for some surprising consequences. As you read the pages ahead, keep in mind that our commonsense ideas are based on a lifetime of everyday experiences and not on observations of objects moving at hundreds of thousands of kilometers per second. Therefore, these results may seem strange, but that is only because we have no experience with them.

39.4   Consequences of the Special Theory of Relativity

As we examine some of the consequences of relativity in this section, we restrict our discussion to the concepts of simultaneity, time intervals, and lengths, all three of which are quite different in relativistic mechanics from what they are in Newtonian mechanics. In relativistic mechanics, for example, the distance between two points and the time interval between two events depend on the frame of reference in which they are measured.

Simultaneity and the Relativity of Time

A basic premise of Newtonian mechanics is that a universal time scale exists that is the same for all observers. Newton and his followers took simultaneity for granted. In his special theory of relativity, Einstein abandoned this assumption.

Einstein devised the following thought experiment to illustrate this point. A boxcar moves with uniform velocity, and two bolts of lightning strike its ends as illustrated in Figure 39.5a, leaving marks on the boxcar and on the ground. The marks on the boxcar are labeled A' and B', and those on the ground are labeled A and B. An observer O' moving with the boxcar is midway between A' and B', and a ground observer O is midway between A and B. The events recorded by the observers are the striking of the boxcar by the two lightning bolts.

The light signals emitted from A and B at the instant at which the two bolts strike later reach observer O at the same time as indicated in Figure 39.5b. This observer realizes that the signals traveled at the same speed over equal distances and so concludes that the events at A and B occurred simultaneously. Now consider the same events as viewed by observer O'. By the time the signals have reached observer O, observer O' has moved as indicated in Figure 39.5b. Therefore, the signal from B' has already swept past O', but the signal from A' has not yet reached O'. In other words, O' sees the signal from B' before seeing the signal from A'. According to Einstein, *the two observers must find that light travels at the same*

Figure 39.5   (a) Two lightning bolts strike the ends of a moving boxcar. (b) The events appear to be simultaneous to the stationary observer O who is standing midway between A and B. The events do not appear to be simultaneous to observer O', who claims that the front of the car is struck before the rear. Notice in (b) that the leftward-traveling light signal has already passed O', but the rightward-traveling signal has not yet reached O'.

(a)   (b)

speed. Therefore, observer O' concludes that one lightning bolt strikes the front of the boxcar *before* the other one strikes the back.

This thought experiment clearly demonstrates that the two events that appear to be simultaneous to observer O do *not* appear to be simultaneous to observer O'. In other words,

> two events that are simultaneous in one reference frame are in general not simultaneous in a second frame moving relative to the first.

Simultaneity is not an absolute concept but rather one that depends on the state of motion of the observer. Einstein's thought experiment demonstrates that two observers can disagree on the simultaneity of two events. **This disagreement, however, depends on the transit time of light to the observers and therefore does *not* demonstrate the deeper meaning of relativity.** In relativistic analyses of high-speed situations, simultaneity is relative even when the transit time is subtracted out. In fact, in all the relativistic effects that we discuss, we ignore differences caused by the transit time of light to the observers.

Time Dilation

To illustrate that observers in different inertial frames can measure different time intervals between a pair of events, consider a vehicle moving to the right with a speed v such as the boxcar shown in Active Figure 39.6a. A mirror is fixed to the ceiling of the vehicle, and observer O' at rest in the frame attached to the vehicle holds a flashlight a distance d below the mirror. At some instant, the flashlight emits a pulse of light directed toward the mirror (event 1), and at some later time after reflecting from the mirror, the pulse arrives back at the flashlight (event 2). Observer O' carries a clock and uses it to measure the time interval Δt_p between these two events. (The subscript p stands for *proper*, as we shall see in a moment.) We model the pulse of light as a particle under constant speed. Because the light pulse has a speed c, the time interval required for the pulse to travel from O' to the mirror and back is

$$\Delta t_p = \frac{\text{distance traveled}}{\text{speed}} = \frac{2d}{c} \tag{39.5}$$

Now consider the same pair of events as viewed by observer O in a second frame as shown in Active Figure 39.6b. According to this observer, the mirror and the flashlight are moving to the right with a speed v, and as a result, the sequence of

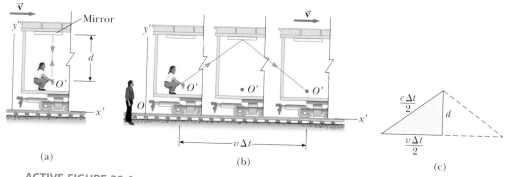

(a)   (b)   (c)

ACTIVE FIGURE 39.6

(a) A mirror is fixed to a moving vehicle, and a light pulse is sent out by observer O' at rest in the vehicle. (b) Relative to a stationary observer O standing alongside the vehicle, the mirror and O' move with a speed v. Notice that what observer O measures for the distance the pulse travels is greater than $2d$. (c) The right triangle for calculating the relationship between Δt and Δt_p.

Sign in at www.thomsonedu.com and go to ThomsonNOW to observe the bouncing of the light pulse for various speeds of the train.

events appears entirely different. By the time the light from the flashlight reaches the mirror, the mirror has moved to the right a distance $v\,\Delta t/2$, where Δt is the time interval required for the light to travel from O' to the mirror and back to O' as measured by O. Observer O concludes that because of the motion of the vehicle, if the light is to hit the mirror, it must leave the flashlight at an angle with respect to the vertical direction. Comparing Active Figure 39.6a with Active Figure 39.6b, we see that the light must travel farther in part (b) than in part (a). (Notice that neither observer "knows" that he or she is moving. Each is at rest in his or her own inertial frame.)

According to the second postulate of the special theory of relativity, both observers must measure c for the speed of light. Because the light travels farther according to O, the time interval Δt measured by O is longer than the time interval Δt_p measured by O'. To obtain a relationship between these two time intervals, let's use the right triangle shown in Active Figure 39.6c. The Pythagorean theorem gives

$$\left(\frac{c\,\Delta t}{2}\right)^2 = \left(\frac{v\,\Delta t}{2}\right)^2 + d^2$$

Solving for Δt gives

$$\Delta t = \frac{2d}{\sqrt{c^2 - v^2}} = \frac{2d}{c\sqrt{1 - \dfrac{v^2}{c^2}}} \tag{39.6}$$

Because $\Delta t_p = 2d/c$, we can express this result as

$$\Delta t = \frac{\Delta t_p}{\sqrt{1 - \dfrac{v^2}{c^2}}} = \gamma \Delta t_p \tag{39.7}$$ ◄ Time dilation

where

$$\gamma = \frac{1}{\sqrt{1 - \dfrac{v^2}{c^2}}} \tag{39.8}$$

Because γ is always greater than unity, Equation 39.7 shows that **the time interval Δt measured by an observer moving with respect to a clock is longer than the time interval Δt_p measured by an observer at rest with respect to the clock. This effect is known as time dilation.**

Time dilation is not observed in our everyday lives, which can be understood by considering the factor γ. This factor deviates significantly from a value of 1 only for very high speeds as shown in Figure 39.7 and Table 39.1. For example, for a speed of $0.1c$, the value of γ is 1.005. Therefore, there is a time dilation of only 0.5% at one-tenth the speed of light. Speeds encountered on an everyday basis are far slower than $0.1c$, so we do not experience time dilation in normal situations.

TABLE 39.1

Approximate Values for γ at Various Speeds

v/c	γ
0.001 0	1.000 000 5
0.010	1.000 05
0.10	1.005
0.20	1.021
0.30	1.048
0.40	1.091
0.50	1.155
0.60	1.250
0.70	1.400
0.80	1.667
0.90	2.294
0.92	2.552
0.94	2.931
0.96	3.571
0.98	5.025
0.99	7.089
0.995	10.01
0.999	22.37

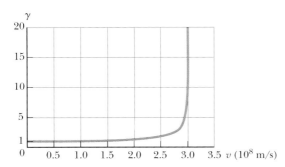

Figure 39.7   Graph of γ versus v. As the speed approaches that of light, γ increases rapidly.

The time interval Δt_p in Equations 39.5 and 39.7 is called the **proper time interval.** (Einstein used the German term *Eigenzeit*, which means "own-time.") In general, **the proper time interval is the time interval between two events measured by an observer who sees the events occur at the same point in space.**

If a clock is moving with respect to you, the time interval between ticks of the moving clock is observed to be longer than the time interval between ticks of an identical clock in your reference frame. Therefore, it is often said that a moving clock is measured to run more slowly than a clock in your reference frame by a factor γ. We can generalize this result by stating that all physical processes, including mechanical, chemical, and biological ones, are measured to slow down when those processes occur in a frame moving with respect to the observer. For example, the heartbeat of an astronaut moving through space keeps time with a clock inside the spacecraft. Both the astronaut's clock and heartbeat are measured to slow down relative to a clock back on the Earth (although the astronaut would have no sensation of life slowing down in the spacecraft).

Quick Quiz 39.3 Suppose the observer O' on the train in Active Figure 39.6 aims her flashlight at the far wall of the boxcar and turns it on and off, sending a pulse of light toward the far wall. Both O' and O measure the time interval between when the pulse leaves the flashlight and when it hits the far wall. Which observer measures the proper time interval between these two events? (a) O' (b) O (c) both observers   (d) neither observer

Quick Quiz 39.4   A crew on a spacecraft watches a movie that is two hours long. The spacecraft is moving at high speed through space. Does an Earth-based observer watching the movie screen on the spacecraft through a powerful telescope measure the duration of the movie to be (a) longer than, (b) shorter than, or (c) equal to two hours?

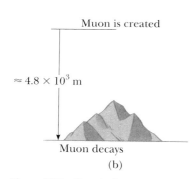

Figure 39.8 Travel of muons according to an Earth-based observer. (a) Without relativistic considerations, muons created in the atmosphere and traveling downward with a speed of 0.99c travel only about 6.6 $\times$ 10^2 m before decaying with an average lifetime of 2.2 μs. Therefore, very few muons reach the surface of the Earth. (b) With relativistic considerations, the muon's lifetime is dilated according to an observer on the Earth. Hence, according to this observer, the muon can travel about 4.8 $\times$ 10^3 m before decaying. The result is many of them arriving at the surface.

Time dilation is a very real phenomenon that has been verified by various experiments involving natural clocks. One experiment reported by J. C. Hafele and R. E. Keating provided direct evidence of time dilation.[4] Time intervals measured with four cesium atomic clocks in jet flight were compared with time intervals measured by Earth-based reference atomic clocks. To compare these results with theory, many factors had to be considered, including periods of speeding up and slowing down relative to the Earth, variations in direction of travel, and the weaker gravitational field experienced by the flying clocks than that experienced by the Earth-based clock. The results were in good agreement with the predictions of the special theory of relativity and were explained in terms of the relative motion between the Earth and the jet aircraft. In their paper, Hafele and Keating stated that "relative to the atomic time scale of the U.S. Naval Observatory, the flying clocks lost 59 $\pm$ 10 ns during the eastward trip and gained 273 $\pm$ 7 ns during the westward trip."

Another interesting example of time dilation involves the observation of *muons,* unstable elementary particles that have a charge equal to that of the electron and a mass 207 times that of the electron. Muons can be produced by the collision of cosmic radiation with atoms high in the atmosphere. Slow-moving muons in the laboratory have a lifetime that is measured to be the proper time interval $\Delta t_p = 2.2 \mu$s. If we take 2.2 μs as the average lifetime of a muon and assume their speed is close to the speed of light, we find that these particles can travel a distance of approximately $(3.0 \times 10^8 \text{ m/s})(2.2 \times 10^{-6} \text{ s}) \approx 6.6 \times 10^2$ m before they decay (Fig. 39.8a). Hence, they are unlikely to reach the surface of the Earth from high in the atmosphere where they are produced. Experiments show, however, that a large number of muons *do* reach the surface. The phenomenon of time dilation explains this effect. As measured by an observer on the Earth, the muons have a dilated lifetime

[4] J. C. Hafele and R. E. Keating, "Around the World Atomic Clocks: Relativistic Time Gains Observed," *Science* **177**:168, 1972.

equal to $\gamma \Delta t_p$. For example, for $v = 0.99c$, $\gamma \approx 7.1$, and $\gamma \Delta t_p \approx 16 \mu s$. Hence, the average distance traveled by the muons in this time interval as measured by an observer on the Earth is approximately $(0.99)(3.0 \times 10^8 \text{ m/s})(16 \times 10^{-6} \text{ s}) \approx 4.8 \times 10^3$ m as indicated in Figure 39.8b.

In 1976, at the laboratory of the European Council for Nuclear Research in Geneva, muons injected into a large storage ring reached speeds of approximately $0.999\,4c$. Electrons produced by the decaying muons were detected by counters around the ring, enabling scientists to measure the decay rate and hence the muon lifetime. The lifetime of the moving muons was measured to be approximately 30 times as long as that of the stationary muon, in agreement with the prediction of relativity to within two parts in a thousand.

| EXAMPLE 39.1 | **What Is the Period of the Pendulum?** |

The period of a pendulum is measured to be 3.00 s in the reference frame of the pendulum. What is the period when measured by an observer moving at a speed of $0.960c$ relative to the pendulum?

SOLUTION

Conceptualize   Let's change frames of reference. Instead of the observer moving at $0.960c$, we can take the equivalent point of view that the observer is at rest and the pendulum is moving at $0.960c$ past the stationary observer. Hence, the pendulum is an example of a clock moving at high speed with respect to an observer.

Categorize   Based on the Conceptualize step, we can categorize this problem as one involving time dilation.

Analyze   The proper time interval, measured in the rest frame of the pendulum, is $\Delta t_p = 3.00$ s.

Use Equation 39.7 to find the dilated time interval:

$$\Delta t = \gamma \Delta t_p = \frac{1}{\sqrt{1 - \frac{(0.960c)^2}{c^2}}} \Delta t_p = \frac{1}{\sqrt{1 - 0.921\,6}} \Delta t_p$$

$$= 3.57(3.00 \text{ s}) = \boxed{10.7 \text{ s}}$$

Finalize   This result shows that a moving pendulum is indeed measured to take longer to complete a period than a pendulum at rest does. The period increases by a factor of $\gamma = 3.57$.

What If?   What if the speed of the observer increases by 4.00%? Does the dilated time interval increase by 4.00%?

Answer   Based on the highly nonlinear behavior of γ as a function of v in Figure 39.7, we would guess that the increase in Δt would be different from 4.00%.

Find the new speed if it increases by 4.00%:

$$v_{\text{new}} = (1.040\,0)(0.960c) = 0.998\,4c$$

Perform the time dilation calculation again:

$$\Delta t = \gamma \Delta t_p = \frac{1}{\sqrt{1 - \frac{(0.998\,4c)^2}{c^2}}} \Delta t_p = \frac{1}{\sqrt{1 - 0.996\,8}} \Delta t_p$$

$$= 17.68(3.00 \text{ s}) = 53.1 \text{ s}$$

Therefore, the 4.00% increase in speed results in almost a 400% increase in the dilated time!

EXAMPLE 39.2 | **How Long Was Your Trip?**

Suppose you are driving your car on a business trip and are traveling at 30 m/s. Your boss, who is waiting at your destination, expects the trip to take 5.0 h. When you arrive late, your excuse is that clock in your car registered the passage of 5.0 h but that you were driving fast and so your clock ran more slowly than the clock in your boss's office. If your car clock actually did indicate a 5.0-h trip, how much time passed on your boss's clock, which was at rest on the Earth?

SOLUTION

Conceptualize The observer is your boss standing stationary on the Earth. The clock is in your car, moving at 30 m/s with respect to your boss.

Categorize The speed of 30 m/s suggests we might categorize this problem as one in which we use classical concepts and equations. Based on the problem statement that the moving clock runs more slowly than a stationary clock however, we categorize this problem as one involving time dilation.

Analyze The proper time interval, measured in the rest frame of the car, is $\Delta t_p = 5.0$ h.

Use Equation 39.8 to evaluate γ:

$$\gamma = \frac{1}{\sqrt{1 - \dfrac{v^2}{c^2}}} = \frac{1}{\sqrt{1 - \dfrac{(3.0 \times 10^1 \text{ m/s})^2}{(3.0 \times 10^8 \text{ m/s})^2}}} = \frac{1}{\sqrt{1 - 10^{-14}}}$$

If you try to determine this value on your calculator, you will probably obtain $\gamma = 1$. Instead, perform a binomial expansion:

$$\gamma = (1 - 10^{-14})^{-1/2} \approx 1 + \tfrac{1}{2}(10^{-14}) = 1 + 5.0 \times 10^{-15}$$

Use Equation 39.7 to find the dilated time interval measured by your boss:

$$\Delta t = \gamma \, \Delta t_p = (1 + 5.0 \times 10^{-15})(5.0 \text{ h})$$
$$= 5.0 \text{ h} + 2.5 \times 10^{-14} \text{ h} = \boxed{5.0 \text{ h} + 0.090 \text{ ns}}$$

Finalize Your boss's clock would be only 0.090 ns ahead of your car clock. You might want to think of another excuse!

The Twin Paradox

An intriguing consequence of time dilation is the *twin paradox* (Fig. 39.9). Consider an experiment involving a set of twins named Speedo and Goslo. When they are 20 years old, Speedo, the more adventuresome of the two, sets out on an epic journey from the Earth to Planet X, located 20 lightyears away. One lightyear (ly) is the distance light travels through free space in 1 year. Furthermore, Speedo's spacecraft is capable of reaching a speed of $0.95c$ relative to the inertial frame of his twin brother back home on the Earth. After reaching Planet X, Speedo becomes homesick and immediately returns to the Earth at the same speed $0.95c$.

Figure 39.9 (a) As one twin leaves his brother on the Earth, both are the same age. (b) When Speedo returns from his journey to Planet X, he is younger than his twin Goslo.

(a)   (b)

Upon his return, Speedo is shocked to discover that Goslo has aged 42 years and is now 62 years old. Speedo, on the other hand, has aged only 13 years.

The paradox is *not* that the twins have aged at different rates. Here is the paradox. From Goslo's frame of reference, he was at rest while his brother traveled at a high speed away from him and then came back. According to Speedo, however, he himself remained stationary while Goslo and the Earth raced away from him and then headed back. Therefore, we might expect Speedo to claim that Goslo ages more slowly than himself. The situation appears to be symmetrical from either twin's point of view. Which twin *actually* ages more slowly?

The situation is actually not symmetrical. Consider a third observer moving at a constant speed relative to Goslo. According to the third observer, Goslo never changes inertial frames. Goslo's speed relative to the third observer is always the same. The third observer notes, however, that Speedo accelerates during his journey when he slows down and starts moving back toward the Earth, *changing reference frames in the process.* From the third observer's perspective, there is something very different about the motion of Goslo when compared to Speedo. Therefore, there is no paradox: only Goslo, who is always in a single inertial frame, can make correct predictions based on special relativity. Goslo finds that instead of aging 42 years, Speedo ages only $(1 - v^2/c^2)^{1/2}(42 \text{ years}) = 13 \text{ years}$. Of these 13 years, Speedo spends 6.5 years traveling to Planet X and 6.5 years returning.

Quick Quiz 39.5 Suppose astronauts are paid according to the amount of time they spend traveling in space. After a long voyage traveling at a speed approaching c, would a crew rather be paid according to (a) an Earth-based clock, (b) their spacecraft's clock, or (c) either clock?

Length Contraction

The measured distance between two points in space also depends on the frame of reference of the observer. **The proper length L_p of an object is the length measured by someone at rest relative to the object.** The length of an object measured by someone in a reference frame that is moving with respect to the object is always less than the proper length. This effect is known as **length contraction.**

To understand length contraction, consider a spacecraft traveling with a speed v from one star to another. There are two observers: one on the Earth and the other in the spacecraft. The observer at rest on the Earth (and also assumed to be at rest with respect to the two stars) measures the distance between the stars to be the proper length L_p. According to this observer, the time interval required for the spacecraft to complete the voyage is $\Delta t = L_p/v$. The passages of the two stars by the spacecraft occur at the same position for the space traveler. Therefore, the space traveler measures the proper time interval Δt_p. Because of time dilation, the proper time interval is related to the Earth-measured time interval by $\Delta t_p = \Delta t/\gamma$. Because the space traveler reaches the second star in the time Δt_p, he or she concludes that the distance L between the stars is

$$L = v \, \Delta t_p = v \, \frac{\Delta t}{\gamma}$$

Because the proper length is $L_p = v \, \Delta t$, we see that

$$L = \frac{L_p}{\gamma} = L_p \sqrt{1 - \frac{v^2}{c^2}} \qquad (39.9) \qquad \blacktriangleleft \text{ Length contraction}$$

where $\sqrt{1 - v^2/c^2}$ is a factor less than unity. **If an object has a proper length L_p when it is measured by an observer at rest with respect to the object, its length L when it moves with speed v in a direction parallel to its length is measured to be shorter according to** $L = L_p \sqrt{1 - v^2/c^2} = L_p/\gamma$.

PITFALL PREVENTION 39.4
The Proper Length

As with the proper time interval, it is *very* important in relativistic calculations to correctly identify the observer who measures the proper length. The proper length between two points in space is always the length measured by an observer at rest with respect to the points. Often, the proper time interval and the proper length are *not* measured by the same observer.

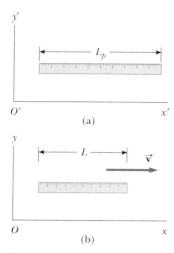

ACTIVE FIGURE 39.10

(a) A meterstick measured by an observer in a frame attached to the stick (that is, both have the same velocity) has its proper length L_p. (b) The meterstick measured by an observer in a frame in which the stick has a velocity $\vec{v}$ relative to the frame is measured to be shorter than its proper length L_p by a factor $(1 - v^2/c^2)^{1/2}$.

Sign in at www.thomsonedu.com and go to ThomsonNOW to view the meterstick from the points of view of two observers and compare the measured length of the stick.

For example, suppose a meterstick moves past a stationary Earth-based observer with speed v as in Active Figure 39.10. The length of the meterstick as measured by an observer in a frame attached to the stick is the proper length L_p shown in Active Figure 39.10a. The length of the stick L measured by the Earth observer is shorter than L_p by the factor $(1 - v^2/c^2)^{1/2}$ as suggested in Active Figure 39.10b. Notice that **length contraction takes place only along the direction of motion.**

The proper length and the proper time interval are defined differently. The proper length is measured by an observer for whom the end points of the length remain fixed in space. The proper time interval is measured by someone for whom the two events take place at the same position in space. As an example of this point, let's return to the decaying muons moving at speeds close to the speed of light. An observer in the muon's reference frame measures the proper lifetime, whereas an Earth-based observer measures the proper length (the distance between the creation point and the decay point in Fig. 39.8b). In the muon's reference frame, there is no time dilation, but the distance of travel to the surface is shorter when measured in this frame. Likewise, in the Earth observer's reference frame, there is time dilation, but the distance of travel is measured to be the proper length. Therefore, when calculations on the muon are performed in both frames, the outcome of the experiment in one frame is the same as the outcome in the other frame: more muons reach the surface than would be predicted without relativistic effects.

Quick Quiz 39.6 You are packing for a trip to another star. During the journey, you will be traveling at 0.99c. You are trying to decide whether you should buy smaller sizes of your clothing, because you will be thinner on your trip due to length contraction. You also plan to save money by reserving a smaller cabin to sleep in because you will be shorter when you lie down. Should you (a) buy smaller sizes of clothing, (b) reserve a smaller cabin, (c) do neither of these things, or (d) do both of these things?

Quick Quiz 39.7 You are observing a spacecraft moving away from you. You measure it to be shorter than when it was at rest on the ground next to you. You also see a clock through the spacecraft window, and you observe that the passage of time on the clock is measured to be slower than that of the watch on your wrist. Compared with when the spacecraft was on the ground, what do you measure if the spacecraft turns around and comes *toward* you at the same speed? (a) The spacecraft is measured to be longer, and the clock runs faster. (b) The spacecraft is measured to be longer, and the clock runs slower. (c) The spacecraft is measured to be shorter, and the clock runs faster. (d) The spacecraft is measured to be shorter, and the clock runs slower.

Space–Time Graphs

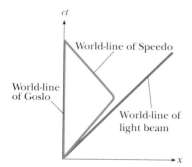

Figure 39.11 The twin paradox on a space–time graph. The twin who stays on the Earth has a world-line along the ct axis (green). The path of the traveling twin through space–time is represented by a world-line that changes direction (blue).

It is sometimes helpful to represent a physical situation with a *space–time graph,* in which ct is the ordinate and position x is the abscissa. The twin paradox is displayed in such a graph in Figure 39.11 from Goslo's point of view. A path through space–time is called a **world-line.** At the origin, the world-lines of Speedo (blue) and Goslo (green) coincide because the twins are in the same location at the same time. After Speedo leaves on his trip, his world-line diverges from that of his brother. Goslo's world-line is vertical because he remains fixed in location. At Goslo and Speedo's reunion, the two world-lines again come together. It would be impossible for Speedo to have a world-line that crossed the path of a light beam that left the Earth when he did. To do so would require him to have a speed greater than c (which, as shown in Sections 39.6 and 39.7, is not possible).

World-lines for light beams are diagonal lines on space–time graphs, typically drawn at 45° to the right or left of vertical (assuming that the x and ct axes have the same scales), depending on whether the light beam is traveling in the direction of increasing or decreasing x. These two world-lines mean that all possible future events for Goslo and Speedo lie within two 45° lines extending from the origin. Either twin's presence at an event outside this "light cone" would require that twin to move at a speed greater than c, which we have said is not possible. Also, the only past events that Goslo and Speedo could have experienced occur between two similar 45° world-lines that approach the origin from below the x axis.

EXAMPLE 39.3   **A Voyage to Sirius**

An astronaut takes a trip to Sirius, which is located a distance of 8 lightyears from the Earth. The astronaut measures the time of the one-way journey to be 6 years. If the spaceship moves at a constant speed of $0.8c$, how can the 8-ly distance be reconciled with the 6-year trip time measured by the astronaut?

SOLUTION

Conceptualize   An observer on the Earth measures light to require 8 years to travel from Earth to Sirius. The astronaut measures a time interval of only 6 years. Is the astronaut traveling faster than light?

Categorize   Because the astronaut is measuring a length of space between Earth and Sirius that is in motion with respect to her, we categorize this example as a length contraction problem.

Analyze   The distance of 8 ly represents the proper length from the Earth to Sirius measured by an observer on the Earth seeing both objects nearly at rest.

Calculate the contracted length measured by the astronaut using Equation 39.9:

$$L = \frac{8 \text{ ly}}{\gamma} = (8 \text{ ly})\sqrt{1 - \frac{v^2}{c^2}} = (8 \text{ ly})\sqrt{1 - \frac{(0.8c)^2}{c^2}} = 5 \text{ ly}$$

Use the particle under constant speed model to find the travel time measured on the astronaut's clock:

$$\Delta t = \frac{L}{v} = \frac{5 \text{ ly}}{0.8c} = \frac{5 \text{ ly}}{0.8(1 \text{ ly/yr})} = 6 \text{ yr}$$

Finalize   Note that we have used the value for the speed of light as $c = 1$ ly/yr. The trip takes a time interval shorter than 8 years for the astronaut because, to her, the distance between the Earth and Sirius is measured to be shorter.

What If?   What if this trip is observed with a very powerful telescope by a technician in Mission Control on the Earth? At what time will this technician *see* that the astronaut has arrived at Sirius?

Answer   The time interval the technician measures for the astronaut to arrive is

$$\Delta t = \frac{L_p}{v} = \frac{8 \text{ ly}}{0.8c} = 10 \text{ yr}$$

For the technician to *see* the arrival, the light from the scene of the arrival must travel back to the Earth and enter the telescope. This travel requires a time interval of

$$\Delta t = \frac{L_p}{v} = \frac{8 \text{ ly}}{c} = 8 \text{ yr}$$

Therefore, the technician sees the arrival after 10 yr + 8 yr = 18 yr. If the astronaut immediately turns around and comes back home, she arrives, according to the technician, 20 years after leaving, only 2 years *after the technician saw her arrive*! In addition, the astronaut would have aged by only 12 years.

EXAMPLE 39.4   **The Pole-in-the-Barn Paradox**

The twin paradox, discussed earlier, is a classic "paradox" in relativity. Another classic "paradox" is as follows. Suppose a runner moving at $0.75c$ carries a horizontal pole 15 m long toward a barn that is 10 m long. The barn has front and rear doors that are initially open. An observer on the ground can instantly and simultaneously close and open the two doors by remote control. When the runner and the pole are inside the barn, the ground observer closes and then opens both doors so that the runner and pole are momentarily captured inside the barn and then proceed to exit the barn from the back door. Do both the runner and the ground observer agree that the runner makes it safely through the barn?

SOLUTION

Conceptualize   From your everyday experience, you would be surprised to see a 15-m pole fit inside a 10-m barn.

Categorize   The pole is in motion with respect to the ground observer so that the observer measures its length to be contracted, whereas the stationary barn has a proper length of 10 m. We categorize this example as a length contraction problem.

Analyze   Use Equation 39.9 to find the contracted length of the pole according to the ground observer:

$$L_{\text{pole}} = L_p\sqrt{1 - \frac{v^2}{c^2}} = (15\ \text{m})\sqrt{1 - (0.75)^2} = 9.9\ \text{m}$$

Therefore, the ground observer measures the pole to be slightly shorter than the barn and there is no problem with momentarily capturing the pole inside it. The "paradox" arises when we consider the runner's point of view.

Use Equation 39.9 to find the contracted length of the barn according to the running observer:

$$L_{\text{barn}} = L_p\sqrt{1 - \frac{v^2}{c^2}} = (10\ \text{m})\sqrt{1 - (0.75)^2} = 6.6\ \text{m}$$

Because the pole is in the rest frame of the runner, the runner measures it to have its proper length of 15 m. How can a 15-m pole fit inside a 6.6-m barn? Although this question is the classic one that is often asked, it is not the question we have asked because it is not the important one. We asked, *"Does the runner make it safely through the barn?"*

The resolution of the "paradox" lies in the relativity of simultaneity. The closing of the two doors is measured to be simultaneous by the ground observer. Because the doors are at different positions, however, they do not close simultaneously as measured by the runner. The rear door closes and then opens first, allowing the leading end of the pole to exit. The front door of the barn does not close until the trailing end of the pole passes by.

We can analyze this "paradox" using a space–time graph. Figure 39.12a is a space–time graph from the ground observer's point of view. We choose $x = 0$ as the position of the front door of the barn and $t = 0$ as the instant at which the leading end of the pole is located at the front door of the barn. The world-lines for the two doors of the barn are separated by 10 m and are vertical because the barn is not moving relative to this observer. For the pole, we follow two tilted world-lines, one for each end of the moving pole. These world-lines are 9.9 m apart horizontally, which is the con-

Figure 39.12   (Example 39.4) Space–time graphs for the pole-in-the-barn paradox. (a) From the ground observer's point of view, the world-lines for the front and back doors of the barn are vertical lines. The world-lines for the ends of the pole are tilted and are 9.9 m apart horizontally. The front door of the barn is at $x = 0$, and the leading end of the pole enters the front door at $t = 0$. The entire pole is inside the barn at the time indicated by the dashed line. (b) From the runner's point of view, the world-lines for the ends of the pole are vertical. The barn is moving in the negative direction, so the world-lines for the front and back doors are tilted to the left. The leading end of the pole exits the back door before the trailing end arrives at the front door.

(a)

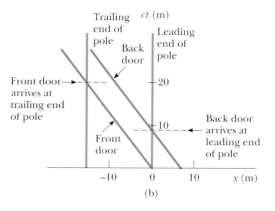
(b)

tracted length seen by the ground observer. As seen in Figure 39.12a, the pole is entirely within the barn at one instant.

Figure 39.12b shows the space–time graph according to the runner. Here, the world-lines for the pole are separated by 15 m and are vertical because the pole is at rest in the runner's frame of reference. The barn is hurtling *toward* the runner, so the world-lines for the front and rear doors of the barn are tilted to the left. The world-lines for the barn are separated by 6.6 m, the contracted length as seen by the runner. The leading end of the pole leaves the rear door of the barn long before the trailing end of the pole enters the barn. Therefore, the opening of the rear door occurs before the closing of the front door.

From the ground observer's point of view, use the particle under constant velocity model to find the time after $t = 0$ at which the trailing end of the pole enters the barn:	(1) $t = \dfrac{\Delta x}{v} = \dfrac{9.9 \text{ m}}{0.75c} = \dfrac{13.2 \text{ m}}{c}$
From the runner's point of view, use the particle under constant velocity model to find the time at which the leading end of the pole leaves the barn:	(2) $t = \dfrac{\Delta x}{v} = \dfrac{6.6 \text{ m}}{0.75c} = \dfrac{8.8 \text{ m}}{c}$
Find the time at which the trailing end of the pole enters the front door of the barn:	(3) $t = \dfrac{\Delta x}{v} = \dfrac{15 \text{ m}}{0.75c} = \dfrac{20 \text{ m}}{c}$

Finalize   From Equation (1), the pole should be completely inside the barn at a time corresponding to $ct = 13.2$ m. This situation is consistent with the point on the ct axis in Figure 39.12a where the pole is inside the barn. From Equation (2), the leading end of the pole leaves the barn at $ct = 8.8$ m. This situation is consistent with the point on the ct axis in Figure 39.12b where the back door of the barn arrives at the leading end of the pole. Equation (3) gives $ct = 20$ m, which agrees with the instant shown in Figure 39.12b at which the front door of the barn arrives at the trailing end of the pole.

The Relativistic Doppler Effect

Another important consequence of time dilation is the shift in frequency observed for light emitted by atoms in motion as opposed to light emitted by atoms at rest. This phenomenon, known as the Doppler effect, was introduced in Chapter 17 as it pertains to sound waves. In the case of sound, the motion of the source with respect to the medium of propagation can be distinguished from the motion of the observer with respect to the medium. Light waves must be analyzed differently, however, because they require no medium of propagation and no method exists for distinguishing the motion of a light source from the motion of the observer.

If a light source and an observer approach each other with a relative speed v, the frequency f_{obs} measured by the observer is

$$f_{obs} = \frac{\sqrt{1 + v/c}}{\sqrt{1 - v/c}} f_{source} \qquad (39.10)$$

where f_{source} is the frequency of the source measured in its rest frame. This relativistic Doppler shift equation, unlike the Doppler shift equation for sound, depends only on the relative speed v of the source and observer and holds for relative speeds as great as c. As you might expect, the equation predicts that $f_{obs} > f_{source}$ when the source and observer approach each other. We obtain the expression for the case in which the source and observer recede from each other by substituting negative values for v in Equation 39.10.

The most spectacular and dramatic use of the relativistic Doppler effect is the measurement of shifts in the frequency of light emitted by a moving astronomical object such as a galaxy. Light emitted by atoms and normally found in the extreme violet region of the spectrum is shifted toward the red end of the spectrum for atoms in other galaxies, indicating that these galaxies are *receding* from us. American astronomer Edwin Hubble (1889–1953) performed extensive measurements of this *red shift* to confirm that most galaxies are moving away from us, indicating that the Universe is expanding.

39.5 The Lorentz Transformation Equations

Figure 39.13 Events occur at points *P* and *Q* and are observed by an observer at rest in the S frame and another in the S' frame, which is moving to the right with a speed *v*.

Suppose two events occur at points *P* and *Q* and are reported by two observers, one at rest in a frame S and another in a frame S' that is moving to the right with speed *v* as in Figure 39.13. The observer in S reports the events with space–time coordinates (x, y, z, t) and the observer in S' reports the same events using the coordinates (x', y', z', t'). Equation 39.1 predicts that the distance between the two points in space at which the events occur does not depend on motion of the observer: $\Delta x = \Delta x'$. Because this prediction is contradictory to the notion of length contraction, the Galilean transformation is not valid when *v* approaches the speed of light. In this section, we present the correct transformation equations that apply for all speeds in the range $0 < v < c$.

The equations that are valid for all speeds and that enable us to transform coordinates from S to S' are the **Lorentz transformation equations:**

◀ Lorentz transformation for S → S'

$$x' = \gamma(x - vt) \qquad y' = y \qquad z' = z \qquad t' = \gamma\left(t - \frac{v}{c^2}x\right) \tag{39.11}$$

These transformation equations were developed by Hendrik A. Lorentz (1853–1928) in 1890 in connection with electromagnetism. It was Einstein, however, who recognized their physical significance and took the bold step of interpreting them within the framework of the special theory of relativity.

Notice the difference between the Galilean and Lorentz time equations. In the Galilean case, $t = t'$. In the Lorentz case, however, the value for t' assigned to an event by an observer O' in the S' frame in Figure 39.13 depends both on the time t and on the coordinate x as measured by an observer O in the S frame, which is consistent with the notion that an event is characterized by four space–time coordinates (x, y, z, t). In other words, in relativity, space and time are *not* separate concepts but rather are closely interwoven with each other.

If you wish to transform coordinates in the S' frame to coordinates in the S frame, simply replace *v* by $-v$ and interchange the primed and unprimed coordinates in Equations 39.11:

◀ Inverse Lorentz transformation for S' → S

$$x = \gamma(x' + vt') \qquad y = y' \qquad z = z' \qquad t = \gamma\left(t' + \frac{v}{c^2}x'\right) \tag{39.12}$$

When $v \ll c$, the Lorentz transformation equations should reduce to the Galilean equations. As *v* approaches zero, $v/c \ll 1$; therefore, $\gamma \to 1$ and Equations 39.11 indeed reduce to the Galilean space–time transformation equations in Equation 39.1.

In many situations, we would like to know the difference in coordinates between two events or the time interval between two events as seen by observers O and O'. From Equations 39.11 and 39.12, we can express the differences between the four variables x, x', t, and t' in the form

$$\left.\begin{array}{l} \Delta x' = \gamma(\Delta x - v\,\Delta t) \\ \Delta t' = \gamma\left(\Delta t - \dfrac{v}{c^2}\,\Delta x\right) \end{array}\right\} \text{S} \to \text{S}' \tag{39.13}$$

$$\left.\begin{array}{l} \Delta x = \gamma(\Delta x' + v\,\Delta t') \\ \Delta t = \gamma\left(\Delta t' + \dfrac{v}{c^2}\,\Delta x'\right) \end{array}\right\} \text{S}' \to \text{S} \tag{39.14}$$

where $\Delta x' = x_2' - x_1'$ and $\Delta t' = t_2' - t_1'$ are the differences measured by observer O' and $\Delta x = x_2 - x_1$ and $\Delta t = t_2 - t_1$ are the differences measured by observer O. (We have not included the expressions for relating the y and z coordinates because they are unaffected by motion along the x direction.[5])

EXAMPLE 39.5   **Simultaneity and Time Dilation Revisited**

(A) Use the Lorentz transformation equations in difference form to show that simultaneity is not an absolute concept.

SOLUTION

Conceptualize   Imagine two events that are simultaneous and separated in space such that $\Delta t' = 0$ and $\Delta x' \neq 0$ according to an observer O' who is moving with speed v relative to O.

Categorize   The statement of the problem tells us to categorize this example as one involving the Lorentz transformation.

Analyze   From the expression for Δt given in Equation 39.14, find the time interval Δt measured by observer O:

$$\Delta t = \gamma\left(\Delta t' + \frac{v}{c^2}\Delta x'\right) = \gamma\left(0 + \frac{v}{c^2}\Delta x'\right) = \gamma\frac{v}{c^2}\Delta x'$$

Finalize   The time interval for the same two events as measured by O is nonzero, so the events do not appear to be simultaneous to O.

(B) Use the Lorentz transformation equations in difference form to show that a moving clock is measured to run more slowly than a clock that is at rest with respect to an observer.

SOLUTION

Conceptualize   Imagine that observer O' carries a clock that he uses to measure a time interval $\Delta t'$. He finds that two events occur at the same place in his reference frame ($\Delta x' = 0$) but at different times ($\Delta t' \neq 0$). Observer O' is moving with speed v relative to O.

Categorize   The statement of the problem tells us to categorize this example as one involving the Lorentz transformation.

Analyze   From the expression for Δt given in Equation 39.14, find the time interval Δt measured by observer O:

$$\Delta t = \gamma\left(\Delta t' + \frac{v}{c^2}\Delta x'\right) = \gamma\left(\Delta t' + \frac{v}{c^2}(0)\right) = \gamma\Delta t'$$

Finalize   This result is the equation for time dilation found earlier (Eq. 39.7), where $\Delta t' = \Delta t_p$ is the proper time interval measured by the clock carried by observer O'. Therefore, O measures the moving clock to run slow.

39.6   The Lorentz Velocity Transformation Equations

Suppose two observers in relative motion with respect to each other are both observing an object's motion. Previously, we defined an event as occurring at an instant of time. Now let's interpret the "event" as the object's motion. We know that the Galilean velocity transformation (Eq. 39.2) is valid for low speeds. How do the observers' measurements of the velocity of the object relate to each other if the speed of the object is close to that of light? Once again, S' is our frame moving

---

[5] Although relative motion of the two frames along the x axis does not change the y and z coordinates of an object, it does change the y and z velocity components of an object moving in either frame as noted in Section 39.6.

at a speed v relative to S. Suppose an object has a velocity component u'_x measured in the S′ frame, where

$$u'_x = \frac{dx'}{dt'} \qquad (39.15)$$

Using Equation 39.11, we have

$$dx' = \gamma(dx - v\, dt)$$

$$dt' = \gamma\left(dt - \frac{v}{c^2}\, dx\right)$$

Substituting these values into Equation 39.15 gives

$$u'_x = \frac{dx - v\, dt}{dt - \frac{v}{c^2}\, dx} = \frac{\frac{dx}{dt} - v}{1 - \frac{v}{c^2}\frac{dx}{dt}}$$

The term dx/dt, however, is simply the velocity component u_x of the object measured by an observer in S, so this expression becomes

Lorentz velocity transformation for S → S′ ▶

$$u'_x = \frac{u_x - v}{1 - \frac{u_x v}{c^2}} \qquad (39.16)$$

If the object has velocity components along the y and z axes, the components as measured by an observer in S′ are

$$u'_y = \frac{u_y}{\gamma\left(1 - \frac{u_x v}{c^2}\right)} \quad \text{and} \quad u'_z = \frac{u_z}{\gamma\left(1 - \frac{u_x v}{c^2}\right)} \qquad (39.17)$$

Notice that u'_y and u'_z do not contain the parameter v in the numerator because the relative velocity is along the x axis.

When v is much smaller than c (the nonrelativistic case), the denominator of Equation 39.16 approaches unity and so $u'_x \approx u_x - v$, which is the Galilean velocity transformation equation. In another extreme, when $u_x = c$, Equation 39.16 becomes

$$u'_x = \frac{c - v}{1 - \frac{cv}{c^2}} = \frac{c\left(1 - \frac{v}{c}\right)}{1 - \frac{v}{c}} = c$$

This result shows that a speed measured as c by an observer in S is also measured as c by an observer in S′, independent of the relative motion of S and S′. This conclusion is consistent with Einstein's second postulate: the speed of light must be c relative to all inertial reference frames. Furthermore, we find that the speed of an object can never be measured as larger than c. That is, the speed of light is the ultimate speed. We shall return to this point later.

To obtain u_x in terms of u'_x, we replace v by $-v$ in Equation 39.16 and interchange the roles of u_x and u'_x:

$$u_x = \frac{u'_x + v}{1 + \frac{u'_x v}{c^2}} \qquad (39.18)$$

PITFALL PREVENTION 39.5

What Can the Observers Agree On?

We have seen several measurements that the two observers O and O' do *not* agree on: (1) the time interval between events that take place in the same position in one of the frames, (2) the distance between two points that remain fixed in one of their frames, (3) the velocity components of a moving particle, and (4) whether two events occurring at different locations in both frames are simultaneous or not. The two observers *can* agree on (1) their relative speed of motion v with respect to each other, (2) the speed c of any ray of light, and (3) the simultaneity of two events which take place at the same position *and* time in some frame.

Quick Quiz 39.8 You are driving on a freeway at a relativistic speed. **(i)** Straight ahead of you, a technician standing on the ground turns on a searchlight and a beam of light moves exactly vertically upward as seen by the technician. As you observe the beam of light, do you measure the magnitude of the vertical component of its velocity as (a) equal to c, (b) greater than c, or (c) less than c? **(ii)** If the technician aims the searchlight directly at you instead of upward, do you measure the magnitude of the horizontal component of its velocity as (a) equal to c, (b) greater than c, or (c) less than c?

EXAMPLE 39.6   Relative Velocity of Two Spacecraft

Two spacecraft A and B are moving in opposite directions as shown in Figure 39.14. An observer on the Earth measures the speed of spacecraft A to be $0.750c$ and the speed of spacecraft B to be $0.850c$. Find the velocity of spacecraft B as observed by the crew on spacecraft A.

SOLUTION

Conceptualize   There are two observers, one on the Earth and one on spacecraft A. The event is the motion of spacecraft B.

Figure 39.14   (Example 39.6) Two spacecraft A and B move in opposite directions. The speed of spacecraft B relative to spacecraft A is *less* than c and is obtained from the relativistic velocity transformation equation.

Categorize   Because the problem asks to find an observed velocity, we categorize this example as one requiring the Lorentz velocity transformation.

Analyze   The Earth-based observer at rest in the S frame makes two measurements, one of each spacecraft. We want to find the velocity of spacecraft B as measured by the crew on spacecraft A. Therefore, $u_x = -0.850c$. The velocity of spacecraft A is also the velocity of the observer at rest in spacecraft A (the S′ frame) relative to the observer at rest on the Earth. Therefore, $v = 0.750c$.

Obtain the velocity u'_x of spacecraft B relative to spacecraft A using Equation 39.16:

$$u'_x = \frac{u_x - v}{1 - \dfrac{u_x v}{c^2}} = \frac{-0.850c - 0.750c}{1 - \dfrac{(-0.850c)(0.750c)}{c^2}} = \boxed{-0.977c}$$

Finalize   The negative sign indicates that spacecraft B is moving in the negative x direction as observed by the crew on spacecraft A. Is that consistent with your expectation from Figure 39.14? Notice that the speed is less than c. That is, an object whose speed is less than c in one frame of reference must have a speed less than c in any other frame. (Had you used the Galilean velocity transformation equation in this example, you would have found that $u'_x = u_x - v = -0.850c - 0.750c = -1.60c$, which is impossible. The Galilean transformation equation does not work in relativistic situations.)

What If?   What if the two spacecraft pass each other? What is their relative speed now?

Answer   The calculation using Equation 39.16 involves only the velocities of the two spacecraft and does not depend on their locations. After they pass each other, they have the same velocities, so the velocity of spacecraft B as observed by the crew on spacecraft A is the same, $-0.977c$. The only difference after they pass is that spacecraft B is receding from spacecraft A, whereas it was approaching spacecraft A before it passed.

EXAMPLE 39.7   Relativistic Leaders of the Pack

Two motorcycle pack leaders named David and Emily are racing at relativistic speeds along perpendicular paths as shown in Figure 39.15. How fast does Emily recede as seen by David over his right shoulder?

Figure 39.15   (Example 39.7) David moves east with a speed $0.75c$ relative to the police officer, and Emily travels south at a speed $0.90c$ relative to the officer.

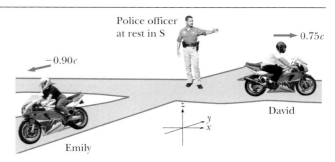

SOLUTION

Conceptualize The two observers are David and the police officer in Figure 39.15. The event is the motion of Emily. Figure 39.15 represents the situation as seen by the police officer at rest in frame S. Frame S′ moves along with David.

Categorize Because the problem asks to find an observed velocity, we categorize this problem as one requiring the Lorentz velocity transformation. The motion takes place in two dimensions.

Analyze Identify the velocity components for David and Emily according to the police officer:

David: $v_x = v = 0.75c$ $v_y = 0$

Emily: $u_x = 0$ $u_y = -0.90c$

Using Equations 39.16 and 39.17, calculate u_x' and u_y' for Emily as measured by David:

$$u_x' = \frac{u_x - v}{1 - \frac{u_x v}{c^2}} = \frac{0 - 0.75c}{1 - \frac{(0)(0.75c)}{c^2}} = -0.75c$$

$$u_y' = \frac{u_y}{\gamma\left(1 - \frac{u_x v}{c^2}\right)} = \frac{\sqrt{1 - \frac{(0.75c)^2}{c^2}}\,(-0.90c)}{\left(1 - \frac{(0)(0.75c)}{c^2}\right)} = -0.60c$$

Using the Pythagorean theorem, find the speed of Emily as measured by David:

$$u' = \sqrt{(u_x')^2 + (u_y')^2} = \sqrt{(-0.75c)^2 + (-0.60c)^2} = \boxed{0.96c}$$

Finalize This speed is less than c, as required by the special theory of relativity.

PITFALL PREVENTION 39.6
Watch Out for "Relativistic Mass"

Some older treatments of relativity maintained the conservation of momentum principle at high speeds by using a model in which a particle's mass increases with speed. You might still encounter this notion of "relativistic mass" in your outside reading, especially in older books. Be aware that this notion is no longer widely accepted; today, mass is considered as *invariant*, independent of speed. The mass of an object in all frames is considered to be the mass as measured by an observer at rest with respect to the object.

39.7 Relativistic Linear Momentum

To describe the motion of particles within the framework of the special theory of relativity properly, you must replace the Galilean transformation equations by the Lorentz transformation equations. Because the laws of physics must remain unchanged under the Lorentz transformation, we must generalize Newton's laws and the definitions of linear momentum and energy to conform to the Lorentz transformation equations and the principle of relativity. These generalized definitions should reduce to the classical (nonrelativistic) definitions for $v \ll c$.

First, recall from the isolated system model that when two particles (or objects that can be modeled as particles) collide, the total momentum of the isolated system of the two particles remains constant. Suppose we observe this collision in a reference frame S and confirm that the momentum of the system is conserved. Now imagine that the momenta of the particles are measured by an observer in a second reference frame S′ moving with velocity $\vec{v}$ relative to the first frame. Using the Lorentz velocity transformation equation and the classical definition of linear momentum, $\vec{p} = m\vec{u}$ (where $\vec{u}$ is the velocity of a particle), we find that linear momentum is *not* measured to be conserved by the observer in S′. Because the laws of physics are the same in all inertial frames, however, linear momentum of the system must be conserved in all frames. We have a contradiction. In view of this contradiction and assuming the Lorentz velocity transformation equation is correct, we must modify the definition of linear momentum so that the momentum of an isolated system is conserved for all observers. For any particle, the correct relativistic equation for linear momentum that satisfies this condition is

◀ Definition of relativistic linear momentum

$$\vec{p} \equiv \frac{m\vec{u}}{\sqrt{1 - \frac{u^2}{c^2}}} = \gamma m\vec{u} \qquad (39.19)$$

where m is the mass of the particle and $\vec{u}$ is the velocity of the particle. When u is much less than c, $\gamma = (1 - u^2/c^2)^{-1/2}$ approaches unity and $\vec{p}$ approaches $m\vec{u}$. Therefore, the relativistic equation for $\vec{p}$ reduces to the classical expression when u is much smaller than c, as it should.

The relativistic force $\vec{F}$ acting on a particle whose linear momentum is $\vec{p}$ is defined as

$$\vec{F} \equiv \frac{d\vec{p}}{dt} \qquad (39.20)$$

where $\vec{p}$ is given by Equation 39.19. This expression, which is the relativistic form of Newton's second law, is reasonable because it preserves classical mechanics in the limit of low velocities and is consistent with conservation of linear momentum for an isolated system ($\vec{F}_{ext} = 0$) both relativistically and classically.

It is left as an end-of-chapter problem (Problem 61) to show that under relativistic conditions, the acceleration $\vec{a}$ of a particle decreases under the action of a constant force, in which case $a \propto (1 - u^2/c^2)^{3/2}$. This proportionality shows that as the particle's speed approaches c, the acceleration caused by any finite force approaches zero. Hence, it is impossible to accelerate a particle from rest to a speed $u \geq c$. This argument reinforces that the speed of light is the ultimate speed, the speed limit of the Universe. It is the maximum possible speed for energy transfer and for information transfer. Any object with mass must move at a lower speed.

EXAMPLE 39.8   **Linear Momentum of an Electron**

An electron, which has a mass of 9.11×10^{-31} kg, moves with a speed of $0.750c$. Find the magnitude of its relativistic momentum and compare this value with the momentum calculated from the classical expression.

SOLUTION

Conceptualize   Imagine an electron moving with high speed. The electron carries momentum, but the magnitude of its momentum is not given by $p = mu$ because the speed is relativistic.

Categorize   We categorize this example as a substitution problem involving a relativistic equation.

Use Equation 39.19 with $u = 0.750c$ to find the momentum:

$$p = \frac{m_e u}{\sqrt{1 - \dfrac{u^2}{c^2}}}$$

$$p = \frac{(9.11 \times 10^{-31} \text{ kg})(0.750)(3.00 \times 10^8 \text{ m/s})}{\sqrt{1 - \dfrac{(0.750c)^2}{c^2}}}$$

$$= 3.10 \times 10^{-22} \text{ kg} \cdot \text{m/s}$$

The classical expression (used incorrectly here) gives $p_{classical} = m_e u = 2.05 \times 10^{-22}$ kg·m/s. Hence, the correct relativistic result is 50% greater than the classical result!

39.8   Relativistic Energy

We have seen that the definition of linear momentum requires generalization to make it compatible with Einstein's postulates. This conclusion implies that the definition of kinetic energy must most likely be modified also.

To derive the relativistic form of the work–kinetic energy theorem, imagine a particle moving in one dimension along the x axis. A force in the x direction causes the momentum of the particle to change according to Equation 39.20. In what follows, we assume the particle is accelerated from rest to some final speed u. The work done by the force F on the particle is

$$W = \int_{x_1}^{x_2} F \, dx = \int_{x_1}^{x_2} \frac{dp}{dt} \, dx \qquad (39.21)$$

To perform this integration and find the work done on the particle and the relativistic kinetic energy as a function of u, we first evaluate dp/dt:

$$\frac{dp}{dt} = \frac{d}{dt} \frac{mu}{\sqrt{1 - \dfrac{u^2}{c^2}}} = \frac{m}{\left(1 - \dfrac{u^2}{c^2}\right)^{3/2}} \frac{du}{dt}$$

Substituting this expression for dp/dt and $dx = u \, dt$ into Equation 39.21 gives

$$W = \int_0^t \frac{m}{\left(1 - \dfrac{u^2}{c^2}\right)^{3/2}} \frac{du}{dt}(u \, dt) = m \int_0^u \frac{u}{\left(1 - \dfrac{u^2}{c^2}\right)^{3/2}} \, du$$

where we use the limits 0 and u in the integral because the integration variable has been changed from t to u. Evaluating the integral gives

$$W = \frac{mc^2}{\sqrt{1 - \dfrac{u^2}{c^2}}} - mc^2 \qquad (39.22)$$

Recall from Chapter 7 that the work done by a force acting on a system consisting of a single particle equals the change in kinetic energy of the particle. Because we assumed the initial speed of the particle is zero, its initial kinetic energy is zero. Therefore, the work W in Equation 39.22 is equivalent to the relativistic kinetic energy K:

◀ Relativistic kinetic energy

$$K = \frac{mc^2}{\sqrt{1 - \dfrac{u^2}{c^2}}} - mc^2 = \gamma mc^2 - mc^2 = (\gamma - 1)mc^2 \qquad (39.23)$$

This equation is routinely confirmed by experiments using high-energy particle accelerators.

At low speeds, where $u/c \ll 1$, Equation 39.23 should reduce to the classical expression $K = \frac{1}{2}mu^2$. We can check that by using the binomial expansion $(1 - \beta^2)^{-1/2} \approx 1 + \frac{1}{2}\beta^2 + \cdots$ for $\beta \ll 1$, where the higher-order powers of β are neglected in the expansion. (In treatments of relativity, β is a common symbol used to represent u/c or v/c.) In our case, $\beta = u/c$, so

$$\gamma = \frac{1}{\sqrt{1 - \dfrac{u^2}{c^2}}} = \left(1 - \frac{u^2}{c^2}\right)^{-1/2} \approx 1 + \frac{1}{2}\frac{u^2}{c^2}$$

Substituting this result into Equation 39.23 gives

$$K \approx \left[\left(1 + \frac{1}{2}\frac{u^2}{c^2}\right) - 1\right]mc^2 = \frac{1}{2}mu^2 \quad \text{(for } u/c \ll 1)$$

which is the classical expression for kinetic energy. A graph comparing the relativistic and nonrelativistic expressions is given in Figure 39.16. In the relativistic case, the particle speed never exceeds c, regardless of the kinetic energy. The two curves are in good agreement when $u \ll c$.

Figure 39.16 A graph comparing relativistic and nonrelativistic kinetic energy of a moving particle. The energies are plotted as a function of particle speed u. In the relativistic case, u is always less than c.

The constant term mc^2 in Equation 39.23, which is independent of the speed of the particle, is called the **rest energy** E_R of the particle:

$$E_R = mc^2 \qquad (39.24)$$

◀ Rest energy

Equation 39.24 shows that **mass is a form of energy,** where c^2 is simply a constant conversion factor. This expression also shows that a small mass corresponds to an enormous amount of energy, a concept fundamental to nuclear and elementary-particle physics.

The term γmc^2 in Equation 39.23, which depends on the particle speed, is the sum of the kinetic and rest energies. It is called the **total energy** E:

$$\text{Total energy} = \text{kinetic energy} + \text{rest energy}$$

$$E = K + mc^2 \qquad (39.25)$$

or

$$E = \frac{mc^2}{\sqrt{1 - \dfrac{u^2}{c^2}}} = \gamma mc^2 \qquad (39.26)$$

◀ Total energy of a relativistic particle

In many situations, the linear momentum or energy of a particle rather than its speed is measured. It is therefore useful to have an expression relating the total energy E to the relativistic linear momentum p, which is accomplished by using the expressions $E = \gamma mc^2$ and $p = \gamma mu$. By squaring these equations and subtracting, we can eliminate u (Problem 37). The result, after some algebra, is[6]

$$E^2 = p^2c^2 + (mc^2)^2 \qquad (39.27)$$

◀ Energy–momentum relationship for a relativistic particle

When the particle is at rest, $p = 0$, so $E = E_R = mc^2$.

In Section 35.1, we introduced the concept of a particle of light, called a **photon.** For particles that have zero mass, such as photons, we set $m = 0$ in Equation 39.27 and find that

$$E = pc \qquad (39.28)$$

This equation is an exact expression relating total energy and linear momentum for photons, which always travel at the speed of light (in vacuum).

Finally, because the mass m of a particle is independent of its motion, m must have the same value in all reference frames. For this reason, m is often called the **invariant mass.** On the other hand, because the total energy and linear momentum of a particle both depend on velocity, these quantities depend on the reference frame in which they are measured.

When dealing with subatomic particles, it is convenient to express their energy in electron volts (Section 25.1) because the particles are usually given this energy by acceleration through a potential difference. The conversion factor, as you recall from Equation 25.5, is

$$1 \text{ eV} = 1.60 \times 10^{-19} \text{ J}$$

For example, the mass of an electron is 9.11×10^{-31} kg. Hence, the rest energy of the electron is

$$m_e c^2 = (9.109 \times 10^{-31} \text{ kg})(2.998 \times 10^8 \text{ m/s})^2 = 8.187 \times 10^{-14} \text{ J}$$

$$= (8.187 \times 10^{-14} \text{ J})(1 \text{ eV}/1.602 \times 10^{-19} \text{ J}) = 0.511 \text{ MeV}$$

Quick Quiz 39.9   The following *pairs* of energies—particle 1: E, $2E$; particle 2: E, $3E$; particle 3: $2E$, $4E$—represent the rest energy and total energy of three different particles. Rank the particles from greatest to least according to their (a) mass, (b) kinetic energy, and (c) speed.

---

[6] One way to remember this relationship is to draw a right triangle having a hypotenuse of length E and legs of lengths pc and mc^2.

EXAMPLE 39.9 **The Energy of a Speedy Proton**

(A) Find the rest energy of a proton in units of electron volts.

SOLUTION

Conceptualize Even if the proton is not moving, it has energy associated with its mass. If it moves, the proton possesses more energy, with the total energy being the sum of its rest energy and its kinetic energy.

Categorize The phrase "rest energy" suggests we must take a relativistic rather than a classical approach to this problem.

Analyze Use Equation 39.24 to find the rest energy:

$$E_R = m_p c^2 = (1.673 \times 10^{-27}\ \text{kg})(2.998 \times 10^8\ \text{m/s})^2$$

$$= (1.504 \times 10^{-10}\ \text{J})\left(\frac{1.00\ \text{eV}}{1.602 \times 10^{-19}\ \text{J}}\right) = \boxed{938\ \text{MeV}}$$

(B) If the total energy of a proton is three times its rest energy, what is the speed of the proton?

SOLUTION

Use Equation 39.26 to relate the total energy of the proton to the rest energy:

$$E = 3m_p c^2 = \frac{m_p c^2}{\sqrt{1 - \dfrac{u^2}{c^2}}} \rightarrow 3 = \frac{1}{\sqrt{1 - \dfrac{u^2}{c^2}}}$$

Solve for u:

$$\left(1 - \frac{u^2}{c^2}\right) = \tfrac{1}{9} \rightarrow \frac{u^2}{c^2} = \tfrac{8}{9}$$

$$u = \frac{\sqrt{8}}{3}\, c = 0.943c = \boxed{2.83 \times 10^8\ \text{m/s}}$$

(C) Determine the kinetic energy of the proton in units of electron volts.

SOLUTION

Use Equation 39.25 to find the kinetic energy of the proton:

$$K = E - m_p c^2 = 3m_p c^2 - m_p c^2 = 2m_p c^2$$

$$= 2(938\ \text{MeV}) = \boxed{1.88 \times 10^3\ \text{MeV}}$$

(D) What is the proton's momentum?

SOLUTION

Use Equation 39.27 to calculate the momentum:

$$E^2 = p^2 c^2 + (m_p c^2)^2 = (3m_p c^2)^2$$

$$p^2 c^2 = 9(m_p c^2)^2 - (m_p c^2)^2 = 8(m_p c^2)^2$$

$$p = \sqrt{8}\,\frac{m_p c^2}{c} = \sqrt{8}\,\frac{(938\ \text{MeV})}{c} = \boxed{2.65 \times 10^3\ \text{MeV}/c}$$

Finalize The unit of momentum in part (D) is written MeV/c, which is a common unit in particle physics. For comparison, you might want to solve this example using classical equations.

What If? In classical physics, if the momentum of a particle doubles, the kinetic energy increases by a factor of 4. What happens to the kinetic energy of the proton in this example if its momentum doubles?

Answer Based on what we have seen so far in relativity, it is likely you would predict that its kinetic energy does not increase by a factor of 4.

Find the new doubled momentum:

$$p_{\text{new}} = 2\left(\sqrt{8}\,\frac{m_p c^2}{c}\right) = 4\sqrt{2}\,\frac{m_p c^2}{c}$$

Use this result in Equation 39.27 to find the new total energy:

$$E_{\text{new}}^2 = p_{\text{new}}^2 c^2 + (m_p c^2)^2$$

$$E_{\text{new}}^2 = \left(4\sqrt{2}\,\frac{m_p c^2}{c}\right)^2 c^2 + (m_p c^2)^2 = 33(m_p c^2)^2$$

$$E_{\text{new}} = \sqrt{33}(m_p c^2) = 5.7\,m_p c^2$$

Use Equation 39.25 to find the new kinetic energy:

$$K_{\text{new}} = E_{\text{new}} - m_p c^2 = 5.7\,m_p c^2 - m_p c^2 = 4.7\,m_p c^2$$

This value is a little more than twice the kinetic energy found in part (C), not four times. In general, the factor by which the kinetic energy increases if the momentum doubles depends on the initial momentum, but it approaches 4 as the momentum approaches zero. In this latter situation, classical physics correctly describes the situation.

39.9   Mass and Energy

Equation 39.26, $E = \gamma mc^2$, represents the total energy of a particle. This important equation suggests that even when a particle is at rest ($\gamma = 1$), it still possesses enormous energy through its mass. The clearest experimental proof of the equivalence of mass and energy occurs in nuclear and elementary-particle interactions in which the conversion of mass into kinetic energy takes place. Consequently, we cannot use the principle of conservation of energy in relativistic situations as it was outlined in Chapter 8. We must modify the principle by including rest energy as another form of energy storage.

This concept is important in atomic and nuclear processes, in which the change in mass is a relatively large fraction of the initial mass. In a conventional nuclear reactor, for example, the uranium nucleus undergoes *fission*, a reaction that results in several lighter fragments having considerable kinetic energy. In the case of ^{235}U, which is used as fuel in nuclear power plants, the fragments are two lighter nuclei and a few neutrons. The total mass of the fragments is less than that of the ^{235}U by an amount Δm. The corresponding energy Δmc^2 associated with this mass difference is exactly equal to the total kinetic energy of the fragments. The kinetic energy is absorbed as the fragments move through water, raising the internal energy of the water. This internal energy is used to produce steam for the generation of electricity.

Next, consider a basic *fusion* reaction in which two deuterium atoms combine to form one helium atom. The decrease in mass that results from the creation of one helium atom from two deuterium atoms is $\Delta m = 4.25 \times 10^{-29}$ kg. Hence, the corresponding energy that results from one fusion reaction is $\Delta mc^2 = 3.83 \times 10^{-12}$ J $= 23.9$ MeV. To appreciate the magnitude of this result, consider that if only 1 g of deuterium were converted to helium, the energy released would be on the order of 10^{12} J! In 2007's cost of electrical energy, this energy would be worth approximately \$30 000. We shall present more details of these nuclear processes in Chapter 45 of the extended version of this textbook.

EXAMPLE 39.10   **Mass Change in a Radioactive Decay**

The ^{216}Po nucleus is unstable and exhibits radioactivity (Chapter 44). It decays to ^{212}Pb by emitting an alpha particle, which is a helium nucleus, ^{4}He. The relevant masses are $m_i = m(^{216}\text{Po}) = 216.001\ 905$ u, and $m_f = m(^{212}\text{Pb}) + m(^4\text{He}) = 211.991\ 888$ u $+ 4.002\ 603$ u.

(A) Find the mass change of the system in this decay.

SOLUTION

Conceptualize The initial system is the ^{216}Po nucleus. Imagine the mass of the system decreasing during the decay and transforming to kinetic energy of the alpha particle and the ^{212}Pb nucleus after the decay.

Categorize We use concepts discussed in this section, so we categorize this example as a substitution problem.

Calculate the mass change:

$$\Delta m = 216.001\ 905\ u - (211.991\ 888\ u + 4.002\ 603\ u)$$

$$= 0.007\ 414\ u = \boxed{1.23 \times 10^{-29}\ kg}$$

(B) Find the energy this mass change represents.

SOLUTION

Use Equation 39.24 to find the energy associated with this mass change:

$$E = \Delta mc^2 = (1.23 \times 10^{-29}\ kg)(3.00 \times 10^8\ m/s)^2$$

$$= 1.11 \times 10^{-12}\ J = \boxed{6.92\ MeV}$$

39.10 The General Theory of Relativity

Up to this point, we have sidestepped a curious puzzle. Mass has two seemingly different properties: a *gravitational attraction* for other masses and an *inertial* property that represents a resistance to acceleration. To designate these two attributes, we use the subscripts g and i and write

Gravitational property: $F_g = m_g g$

Inertial property: $\sum F = m_i a$

The value for the gravitational constant G was chosen to make the magnitudes of m_g and m_i numerically equal. Regardless of how G is chosen, however, the strict proportionality of m_g and m_i has been established experimentally to an extremely high degree: a few parts in 10^{12}. Therefore, it appears that gravitational mass and inertial mass may indeed be exactly proportional.

Why, though? They seem to involve two entirely different concepts: a force of mutual gravitational attraction between two masses and the resistance of a single mass to being accelerated. This question, which puzzled Newton and many other physicists over the years, was answered by Einstein in 1916 when he published his theory of gravitation, known as the *general theory of relativity*. Because it is a mathematically complex theory, we offer merely a hint of its elegance and insight.

In Einstein's view, the dual behavior of mass was evidence for a very intimate and basic connection between the two behaviors. He pointed out that no mechanical experiment (such as dropping an object) could distinguish between the two situations illustrated in Figures 39.17a and 39.17b. In Figure 39.17a, a person standing in an elevator on the surface of a planet feels pressed into the floor due to the gravitational force. If he releases his briefcase, he observes it moving toward the floor with acceleration $\vec{g} = -g\hat{j}$. In Figure 39.17b, the person is in an elevator in empty space accelerating upward with $\vec{a}_{el} = +g\hat{j}$. The person feels pressed into the floor with the same force as in Figure 39.17a. If he releases his briefcase, he observes it moving toward the floor with acceleration g, exactly as in the previous situation. In each situation, an object released by the observer undergoes a downward acceleration of magnitude g relative to the floor. In Figure 39.17a, the person is at rest in an inertial frame in a gravitational field due to the planet. In Figure 39.17b, the person is in a noninertial frame accelerating in gravity-free space. Einstein's claim is that these two situations are completely equivalent.

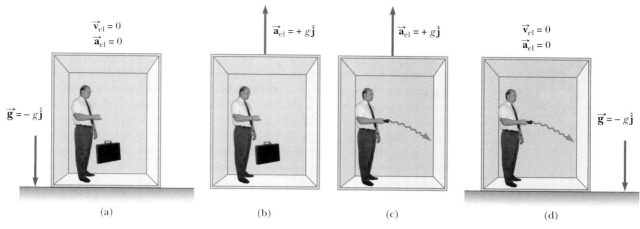

Figure 39.17   (a) The observer is at rest in an elevator in a uniform gravitational field $\vec{\mathbf{g}} = -g\hat{\mathbf{j}}$, directed downward. The observer drops his brief-case, which moves downward with acceleration g. (b) The observer is in a region where gravity is negligible, but the elevator moves upward with an acceleration $\vec{\mathbf{a}}_{el} = +g\hat{\mathbf{j}}$. The observer releases his briefcase, which moves downward (according to the observer) with acceleration g relative to the floor of the elevator. According to Einstein, the frames of reference in (a) and (b) are equivalent in every way. No local experiment can distinguish any difference between the two frames. (c) In the accelerating frame, a ray of light would appear to bend downward due to the acceleration. (d) If (a) and (b) are truly equivalent, as Einstein proposed, (c) suggests that a ray of light would bend downward in a gravitational field.

Einstein carried this idea further and proposed that *no* experiment, mechanical or otherwise, could distinguish between the two situations. This extension to include all phenomena (not just mechanical ones) has interesting consequences. For example, suppose a light pulse is sent horizontally across the elevator as in Figure 39.17c, in which the elevator is accelerating upward in empty space. From the point of view of an observer in an inertial frame outside the elevator, the light travels in a straight line while the floor of the elevator accelerates upward. According to the observer on the elevator, however, the trajectory of the light pulse bends downward as the floor of the elevator (and the observer) accelerates upward. Therefore, based on the equality of parts (a) and (b) of the figure, Einstein proposed that **a beam of light should also be bent downward by a gravitational field** as in Figure 39.17d. Experiments have verified the effect, although the bending is small. A laser aimed at the horizon falls less than 1 cm after traveling 6 000 km. (No such bending is predicted in Newton's theory of gravitation.)

Einstein's **general theory of relativity** has two postulates:

■ All the laws of nature have the same form for observers in any frame of reference, whether accelerated or not.
■ In the vicinity of any point, a gravitational field is equivalent to an accelerated frame of reference in gravity-free space (the **principle of equivalence**).

◀ Postulates of the general theory of relativity

One interesting effect predicted by the general theory is that time is altered by gravity. A clock in the presence of gravity runs slower than one located where gravity is negligible. Consequently, the frequencies of radiation emitted by atoms in the presence of a strong gravitational field are *redshifted* to lower frequencies when compared with the same emissions in the presence of a weak field. This gravitational redshift has been detected in spectral lines emitted by atoms in massive stars. It has also been verified on the Earth by comparing the frequencies of gamma rays emitted from nuclei separated vertically by about 20 m.

The second postulate suggests a gravitational field may be "transformed away" at any point if we choose an appropriate accelerated frame of reference, a freely falling one. Einstein developed an ingenious method of describing the acceleration necessary to make the gravitational field "disappear." He specified a concept, the *curvature of space–time*, that describes the gravitational effect at every point. In fact, the curvature of space–time completely replaces Newton's gravitational theory. According to Einstein, there is no such thing as a gravitational force. Rather, the presence of a mass causes a curvature of space–time in the vicinity of the mass,

Einstein's cross. The four bright spots are images of the same galaxy that have been bent around a massive object located between the galaxy and the Earth. The massive object acts like a lens, causing the rays of light that were diverging from the distant galaxy to converge on the Earth. (If the intervening massive object had a uniform mass distribution, we would see a bright ring instead of four spots.)

Courtesy of NASA

and this curvature dictates the space–time path that all freely moving objects must follow.

As an example of the effects of curved space–time, imagine two travelers moving on parallel paths a few meters apart on the surface of the Earth and maintaining an exact northward heading along two longitude lines. As they observe each other near the equator, they will claim that their paths are exactly parallel. As they approach the North Pole, however, they notice that they are moving closer together and will meet at the North Pole. Therefore, they claim that they moved along parallel paths, but moved toward each other, *as if there were an attractive force between them.* The travelers make this conclusion based on their everyday experience of moving on flat surfaces. From our mental representation, however, we realize they are walking on a curved surface, and it is the geometry of the curved surface, rather than an attractive force, that causes them to converge. In a similar way, general relativity replaces the notion of forces with the movement of objects through curved space–time.

One prediction of the general theory of relativity is that a light ray passing near the Sun should be deflected in the curved space–time created by the Sun's mass. This prediction was confirmed when astronomers detected the bending of starlight near the Sun during a total solar eclipse that occurred shortly after World War I (Fig. 39.18). When this discovery was announced, Einstein became an international celebrity.

If the concentration of mass becomes very great as is believed to occur when a large star exhausts its nuclear fuel and collapses to a very small volume, a **black hole** may form. Here, the curvature of space–time is so extreme that within a certain distance from the center of the black hole all matter and light become trapped as discussed in Section 13.6.

Figure 39.18 Deflection of starlight passing near the Sun. Because of this effect, the Sun or some other remote object can act as a *gravitational lens*. In his general theory of relativity, Einstein calculated that starlight just grazing the Sun's surface should be deflected by an angle of 1.75 s of arc.

Summary

DEFINITIONS

The relativistic expression for the **linear momentum** of a particle moving with a velocity $\vec{u}$ is

$$\vec{p} \equiv \frac{m\vec{u}}{\sqrt{1 - \dfrac{u^2}{c^2}}} = \gamma m \vec{u} \qquad (39.19)$$

The relativistic force $\vec{F}$ acting on a particle whose linear momentum is $\vec{p}$ is defined as

$$\vec{F} \equiv \frac{d\vec{p}}{dt} \qquad (39.20)$$

(continued)

CONCEPTS AND PRINCIPLES

The two basic postulates of the special theory of relativity are as follows:

- The laws of physics must be the same in all inertial reference frames.
- The speed of light in vacuum has the same value, $c = 3.00 \times 10^8$ m/s, in all inertial frames, regardless of the velocity of the observer or the velocity of the source emitting the light.

Three consequences of the special theory of relativity are as follows:

- Events that are measured to be simultaneous for one observer are not necessarily measured to be simultaneous for another observer who is in motion relative to the first.
- Clocks in motion relative to an observer are measured to run slower by a factor $\gamma = (1 - v^2/c^2)^{-1/2}$. This phenomenon is known as **time dilation.**
- The length of objects in motion are measured to be contracted in the direction of motion by a factor $1/\gamma = (1 - v^2/c^2)^{1/2}$. This phenomenon is known as **length contraction.**

To satisfy the postulates of special relativity, the Galilean transformation equations must be replaced by the **Lorentz transformation equations:**

$$x' = \gamma(x - vt) \quad y' = y \quad z' = z \quad t' = \gamma\left(t - \frac{v}{c^2}x\right) \quad \textbf{(39.11)}$$

where $\gamma = (1 - v^2/c^2)^{-1/2}$ and the S' frame moves in the x direction relative to the S frame.

The relativistic form of the **Lorentz velocity transformation equation** is

$$u'_x = \frac{u_x - v}{1 - \frac{u_x v}{c^2}} \quad \textbf{(39.16)}$$

where u'_x is the x component of the velocity of an object as measured in the S' frame and u_x is its component as measured in the S frame.

The relativistic expression for the **kinetic energy** of a particle is

$$K = \frac{mc^2}{\sqrt{1 - \frac{u^2}{c^2}}} - mc^2 = (\gamma - 1)mc^2 \quad \textbf{(39.23)}$$

The constant term mc^2 in Equation 39.23 is called the **rest energy** E_R of the particle:

$$E_R = mc^2 \quad \textbf{(39.24)}$$

The total energy E of a particle is given by

$$E = \frac{mc^2}{\sqrt{1 - \frac{u^2}{c^2}}} = \gamma mc^2 \quad \textbf{(39.26)}$$

The relativistic linear momentum of a particle is related to its total energy through the equation

$$E^2 = p^2c^2 + (mc^2)^2 \quad \textbf{(39.27)}$$

Questions

☐ denotes answer available in *Student Solutions Manual/Study Guide;* **O** denotes objective question

1. The speed of light in water is 230 Mm/s. Suppose an electron is moving through water at 250 Mm/s. Does that motion violate the principle of relativity?

2. **O** You measure the volume of a cube at rest to be V_0. You then measure the volume of the same cube as it passes you in a direction parallel to one side of the cube. The speed of the cube is $0.98c$, so $\gamma \approx 5$. Is the volume you measure close to (a) $V_0/125$, (b) $V_0/25$, (c) $V_0/5$, (d) V_0, (e) $5V_0$, (f) $25V_0$, or (g) $125V_0$?

3. **O** A spacecraft built in the shape of a sphere moves past an observer on the Earth with a speed of $0.5c$. What shape does the observer measure for the spacecraft as it goes by? (a) a sphere (b) a cigar shape, elongated along the direction of motion (c) a round pillow shape, flattened along the direction of motion (d) a conical shape, pointing in the direction of motion

4. **O** A spacecraft zooms past the Earth with a constant velocity. An observer on the Earth measures that an

undamaged clock on the spacecraft is ticking at one-third the rate of an identical clock on the Earth. What does an observer on the spacecraft measure about the Earth clock's ticking rate? (a) It runs more than three times faster than his own clock. (b) It runs three times faster than his own. (c) It runs at the same rate as his own. (d) It runs at approximately half the rate of his own. (e) It runs at one-third the rate of his own. (f) It runs at less than one-third the rate of his own.

5. Explain why, when defining the length of a rod, it is necessary to specify that the positions of the ends of the rod are to be measured simultaneously.

6. O Two identical clocks are set side by side and synchronized. One remains on the Earth. The other is put into orbit around the Earth moving toward the east. (i) As measured by an observer on the Earth, while in rapid motion does the orbiting clock (a) run faster than the Earth-based clock, (b) run at the same rate, or (c) run slower? (ii) The orbiting clock is returned to its original location and brought to rest relative to the Earth. Thereafter, (a) its reading lags farther and farther behind the Earth-based clock, (b) it lags behind the Earth-based clock by a constant amount, (c) it is synchronized with the Earth-based clock, (d) it is ahead of the Earth-based clock by a constant amount, or (e) it gets farther and farther ahead of the Earth-based clock.

7. A train is approaching you at very high speed as you stand next to the tracks. Just as an observer on the train passes you, you both begin to play the same Beethoven symphony on portable CD players. (a) According to you, whose CD player finishes the symphony first? (b) **What If?** According to the observer on the train, whose CD player finishes the symphony first? (c) Whose CD player actually finishes the symphony first?

8. List some ways our day-to-day lives would change if the speed of light were only 50 m/s.

9. How is acceleration indicated on a space–time graph?

10. Explain how the Doppler effect with microwaves is used to determine the speed of an automobile.

11. In several cases, a nearby star has been found to have a large planet orbiting about it, although light from the planet could not be seen separately from the starlight. Using the ideas of a system rotating about its center of mass and of the Doppler shift for light, explain how an astronomer could determine the presence of the invisible planet.

12. A particle is moving at a speed less than $c/2$. If the speed of the particle is doubled, what happens to its momentum?

13. O Rank the following particles according to the magnitudes of their momentum from the largest to the smallest. If any have equal amounts of momentum, or zero momentum, display that fact in your ranking. (a) a 1-MeV photon (b) a proton with kinetic energy $K = 1$ MeV (c) an electron with $K = 1$ MeV (d) a grain of dust with $K = 1$ MeV = 160 fJ

14. Give a physical argument that shows it is impossible to accelerate an object of mass m to the speed of light, even with a continuous force acting on it.

15. O (i) Does the speed of an electron have an upper limit? (a) yes, the speed of light c (b) yes, with another value (c) no (ii) Does the magnitude of an electron's momentum have an upper limit? (a) yes, $m_e c$ (b) yes, with another value (c) no (iii) Does the electron's kinetic energy have an upper limit? (a) yes, $m_e c^2$ (b) yes, $\frac{1}{2}m_e c^2$ (c) yes, with another value (d) no

16. O A distant astronomical object (a quasar) is moving away from us at half the speed of light. What is the speed of the light we receive from this quasar? (a) greater than c (b) c (c) between $c/2$ and c (d) $c/2$ (e) between 0 and $c/2$ (f) 0

17. "Newtonian mechanics correctly describes objects moving at ordinary speeds and relativistic mechanics correctly describes objects moving very fast." "Relativistic mechanics must make a smooth transition as it reduces to Newtonian mechanics in a case in which the speed of an object becomes small compared with the speed of light." Argue for or against each of these two statements.

18. Two cards have straight edges. Suppose the top edge of one card crosses the bottom edge of another card at a small angle as shown in Figure Q39.18a. A person slides the cards together at a moderately high speed. In what direction does the intersection point of the edges move? Show that the intersection point can move at a speed greater than the speed of light.
A small flashlight is suspended in a horizontal plane and set into rapid rotation. Show that the spot of light it produces on a distant screen can move across the screen at a speed greater than the speed of light. (If you use a laser pointer as shown in Figure Q39.18b, make sure the direct laser light cannot enter a person's eyes.) Argue that these experiments do not invalidate the principle that no material, no energy, and no information can move faster than light moves in a vacuum.

(a)   (b)

Figure Q39.18

19. With regard to reference frames, how does general relativity differ from special relativity?

20. Two identical clocks are in the same house, one upstairs in a bedroom and the other downstairs in the kitchen. Which clock runs slower? Explain.

Problems

WebAssign The Problems from this chapter may be assigned online in WebAssign.

ThomsonNOW Sign in at **www.thomsonedu.com** and go to ThomsonNOW to assess your understanding of this chapter's topics with additional quizzing and conceptual questions.

1, 2, 3 denotes straightforward, intermediate, challenging;   ☐ denotes full solution available in *Student Solutions Manual/Study Guide*;   ▲ denotes coached solution with hints available at **www.thomsonedu.com;**   denotes developing symbolic reasoning;   ● denotes asking for qualitative reasoning;   ▪ denotes computer useful in solving problem

Section 39.1 The Principle of Galilean Relativity

1. In a laboratory frame of reference, an observer notes that Newton's second law is valid. Show that it is also valid for an observer moving at a constant speed, small compared with the speed of light, relative to the laboratory frame.

2. Show that Newton's second law is *not* valid in a reference frame moving past the laboratory frame of Problem 1 with a constant acceleration.

3. A 2 000-kg car moving at 20.0 m/s collides and locks together with a 1 500-kg car at rest at a stop sign. Show that momentum is conserved in a reference frame moving at 10.0 m/s in the direction of the moving car.

Section 39.2 The Michelson–Morley Experiment

Section 39.3 Einstein's Principle of Relativity

Section 39.4 Consequences of the Special Theory of Relativity

Problem 37 in Chapter 4 can be assigned with this section.

4. How fast must a meterstick be moving if its length is measured to shrink to 0.500 m?

5. At what speed does a clock move if it is measured to run at a rate one-half the rate of a clock at rest with respect to an observer?

6. An astronaut is traveling in a space vehicle moving at $0.500c$ relative to the Earth. The astronaut measures her pulse rate at 75.0 beats per minute. Signals generated by the astronaut's pulse are radioed to the Earth when the vehicle is moving in a direction perpendicular to the line that connects the vehicle with an observer on the Earth. (a) What pulse rate does the Earth-based observer measure? (b) **What If?** What would be the pulse rate if the speed of the space vehicle were increased to $0.990c$?

7. An atomic clock moves at 1 000 km/h for 1.00 h as measured by an identical clock on the Earth. At the end of the 1.00-h interval, how many nanoseconds slow will the moving clock be compared with the Earth clock?

8. A muon formed high in the Earth's atmosphere travels at speed $v = 0.990c$ for a distance of 4.60 km before it decays into an electron, a neutrino, and an antineutrino ($\mu^- \rightarrow e^- + \nu + \bar{\nu}$). (a) For what time interval does the muon live as measured in its reference frame? (b) How far does the Earth travel as measured in the frame of the muon?

2 = intermediate;   3 = challenging;   ☐ = SSM/SG;   ▲ = ThomsonNOW;   ▪ = symbolic reasoning;   ● = qualitative reasoning

9. ▲ A spacecraft with a proper length of 300 m takes 0.750 μs to pass an Earth-based observer. Determine the speed of the spacecraft as measured by the Earth observer.

10. (a) An object of proper length L_p takes a time interval Δt to pass an Earth-based observer. Determine the speed of the object as measured by the Earth observer. (b) A column of tanks, 300 m long, takes 75.0 s to pass a child waiting at a street corner on her way to school. Determine the speed of the armored vehicles. (c) Show that the answer to part (a) includes the answer to Problem 9 as a special case and includes the answer to part (b) as another special case.

11. ● **Review problem.** In 1963, astronaut Gordon Cooper orbited the Earth 22 times. The press stated that for each orbit, he aged 2 millionths of a second less than he would have had he remained on the Earth. (a) Assuming that he was 160 km above the Earth in a circular orbit, determine the difference in elapsed time between someone on the Earth and the orbiting astronaut for the 22 orbits. You may use the approximation

$$\frac{1}{\sqrt{1-x}} \approx 1 + \frac{x}{2}$$

for small x. (b) Did the press report accurate information? Explain.

12. For what value of v does $\gamma = 1.010\ 0$? Observe that for speeds lower than this value, time dilation and length contraction are effects amounting to less than 1%.

13. A friend passes by you in a spacecraft traveling at a high speed. He tells you that his spacecraft is 20.0 m long and that the identically constructed spacecraft you are sitting in is 19.0 m long. According to your observations, (a) how long is your spacecraft, (b) how long is your friend's spacecraft, and (c) what is the speed of your friend's spacecraft?

14. The identical twins Speedo and Goslo join a migration from the Earth to Planet X. It is 20.0 ly away in a reference frame in which both planets are at rest. The twins, of the same age, depart at the same moment on different spacecraft. Speedo's spacecraft travels steadily at $0.950c$ and Goslo's at $0.750c$. Calculate the age difference between the twins after Goslo's spacecraft lands on Planet X. Which twin is older?

15. **Review problem.** An alien civilization occupies a brown dwarf, nearly stationary relative to the Sun, several lightyears away. The extraterrestrials have come to love

original broadcasts of *I Love Lucy*, on television channel 2, at carrier frequency 57.0 MHz. Their line of sight to us is in the plane of the Earth's orbit. Find the difference between the highest and lowest frequencies they receive due to the Earth's orbital motion around the Sun.

16. Police radar detects the speed of a car (Fig. P39.16) as follows. Microwaves of a precisely known frequency are broadcast toward the car. The moving car reflects the microwaves with a Doppler shift. The reflected waves are received and combined with an attenuated version of the transmitted wave. Beats occur between the two microwave signals. The beat frequency is measured. (a) For an electromagnetic wave reflected back to its source from a mirror approaching at speed v, show that the reflected wave has frequency

$$f = f_{source}\frac{c+v}{c-v}$$

where f_{source} is the source frequency. (b) When v is much less than c, the beat frequency is much smaller than the transmitted frequency. In this case, use the approximation $f + f_{source} \approx 2f_{source}$ and show that the beat frequency can be written as $f_{beat} = 2v/\lambda$. (c) What beat frequency is measured for a car speed of 30.0 m/s if the microwaves have frequency 10.0 GHz? (d) If the beat frequency measurement is accurate to ±5 Hz, how accurate is the speed measurement?

Trent Steffler/David R. Frazier Photolibrary

Figure P39.16

17. *The redshift.* A light source recedes from an observer with a speed v_{source} that is small compared with c. (a) Show that the fractional shift in the measured wavelength is given by the approximate expression

$$\frac{\Delta\lambda}{\lambda} \approx \frac{v_{source}}{c}$$

This phenomenon is known as the redshift because the visible light is shifted toward the red. (b) Spectroscopic measurements of light at $\lambda = 397$ nm coming from a galaxy in Ursa Major reveal a redshift of 20.0 nm. What is the recessional speed of the galaxy?

2 = intermediate;   3 = challenging;   □ = SSM/SG;   ▲ = ThomsonNOW;   = symbolic reasoning;   ● = qualitative reasoning

18. A physicist drives through a stop light. When he is pulled over, he tells the police officer that the Doppler shift made the red light of wavelength 650 nm appear green to him, with a wavelength of 520 nm. The police officer writes out a traffic citation for speeding. How fast was the physicist traveling, according to his own testimony?

Section 39.5 The Lorentz Transformation Equations

19. Suzanne observes two light pulses to be emitted from the same location, but separated in time by 3.00 μs. Mark observes the emission of the same two pulses to be separated in time by 9.00 μs. (a) How fast is Mark moving relative to Suzanne? (b) According to Mark, what is the separation in space of the two pulses?

20. A moving rod is observed to have a length of 2.00 m and to be oriented at an angle of 30.0° with respect to the direction of motion as shown in Figure P39.20. The rod has a speed of 0.995c. (a) What is the proper length of the rod? (b) What is the orientation angle in the proper frame?

Figure P39.20

21. An observer in reference frame S measures two events to be simultaneous. Event A occurs at the point (50.0 m, 0, 0) at the instant 9:00:00 Universal time on January 15, 2008. Event B occurs at the point (150 m, 0, 0) at the same moment. A second observer, moving past with a velocity of $0.800c\hat{\mathbf{i}}$, also observes the two events. In her reference frame S′, which event occurred first and what time interval elapsed between the events?

22. A red light flashes at position $x_R = 3.00$ m and time $t_R = 1.00 \times 10^{-9}$ s, and a blue light flashes at $x_B = 5.00$ m and $t_B = 9.00 \times 10^{-9}$ s, all measured in the S reference frame. Reference frame S′ moves uniformly to the right and has its origin at the same point as S at $t = t' = 0$. Both flashes are observed to occur at the same place in S′. (a) Find the relative speed between S and S′. (b) Find the location of the two flashes in frame S′. (c) At what time does the red flash occur in the S′ frame?

Section 39.6 The Lorentz Velocity Transformation Equations

23. ▲ Two jets of material from the center of a radio galaxy are ejected in opposite directions. Both jets move at 0.750c relative to the galaxy. Determine the speed of one jet relative to the other.

24. A Klingon spacecraft moves away from the Earth at a speed of 0.800c (Fig. P39.24). The starship *Enterprise* pursues at a speed of 0.900c relative to the Earth. Observers on the Earth measure the *Enterprise* to be overtaking the Klingon craft at a relative speed of 0.100c. With what speed is the *Enterprise* overtaking the Klingon craft as measured by the crew of the *Enterprise*?

Figure P39.24

Section 39.7 Relativistic Linear Momentum

25. Calculate the momentum of an electron moving with a speed of (a) 0.010 0c, (b) 0.500c, and (c) 0.900c.

26. The nonrelativistic expression for the momentum of a particle, $p = mu$, agrees with experiment if $u \ll c$. For what speed does the use of this equation give an error in the momentum of (a) 1.00% and (b) 10.0%?

27. A golf ball travels with a speed of 90.0 m/s. By what fraction does its relativistic momentum magnitude p differ from its classical value mu? That is, calculate the ratio $(p - mu)/mu$.

28. The speed limit on a certain roadway is 90.0 km/h. Suppose speeding fines are made proportional to the amount by which a vehicle's momentum exceeds the momentum it would have when traveling at the speed limit. The fine for driving at 190 km/h (that is, 100 km/h over the speed limit) is \$80.0. What then will be the fine for traveling (a) at 1 090 km/h? (b) At 1 000 000 090 km/h?

29. ▲ An unstable particle at rest spontaneously breaks into two fragments of unequal mass. The mass of the first fragment is 2.50×10^{-28} kg, and that of the other is 1.67×10^{-27} kg. If the lighter fragment has a speed of $0.893c$ after the breakup, what is the speed of the heavier fragment?

Section 39.8 Relativistic Energy

30. An electron has a kinetic energy five times greater than its rest energy. Find its (a) total energy and (b) speed.

31. A proton in a high-energy accelerator moves with a speed of $c/2$. Use the work–kinetic energy theorem to find the work required to increase its speed to (a) $0.750c$ and (b) $0.995c$.

32. Show that for any object moving at less than one-tenth the speed of light, the relativistic kinetic energy agrees with the result of the classical equation $K = \frac{1}{2}mu^2$ to within less than 1%. Therefore, for most purposes, the classical equation is good enough to describe these objects.

33. Find the momentum of a proton in MeV/c units assuming its total energy is twice its rest energy.

34. ● (a) Find the kinetic energy of a 78.0-kg spacecraft launched out of the solar system with speed 106 km/s by using the classical equation $K = \frac{1}{2}mu^2$. (b) **What If?** Calculate its kinetic energy using the relativistic equation. (c) Explain the result of comparing the results of parts (a) and (b).

35. ▲ A proton moves at $0.950c$. Calculate its (a) rest energy, (b) total energy, and (c) kinetic energy.

36. An unstable particle with a mass of 3.34×10^{-27} kg is initially at rest. The particle decays into two fragments that fly off along the x axis with velocity components $0.987c$ and $-0.868c$. Find the masses of the fragments. *Suggestion:* Use conservation of both energy and momentum.

37. Show that the energy–momentum relationship $E^2 = p^2c^2 + (mc^2)^2$ follows from the expressions $E = \gamma mc^2$ and $p = \gamma mu$.

38. In a typical color television picture tube, the electrons are accelerated from rest through a potential difference of 25 000 V. (a) What speed do the electrons have when they strike the screen? (b) What is their kinetic energy in joules?

39. The rest energy of an electron is 0.511 MeV. The rest energy of a proton is 938 MeV. Assume both particles have kinetic energies of 2.00 MeV. Find the speed of (a) the electron and (b) the proton. (c) By how much does the speed of the electron exceed that of the proton? (d) Repeat the calculations assuming both particles have kinetic energies of 2 000 MeV.

40. Consider electrons accelerated to an energy of 20.0 GeV in the 3.00-km-long Stanford Linear Accelerator. (a) What is the γ factor for the electrons? (b) What is the electrons' speed? (c) How long does the accelerator appear to the electrons?

41. A pion at rest ($m_\pi = 273m_e$) decays to a muon ($m_\mu = 207m_e$) and an antineutrino ($m_{\bar{\nu}} \approx 0$). The reaction is written $\pi^- \rightarrow \mu^- + \bar{\nu}$. Find the kinetic energy of the muon and the energy of the antineutrino in electron volts. *Suggestion:* Use conservation of both energy and momentum for the decay process.

42. Consider a car moving at highway speed u. Is its actual kinetic energy larger or smaller than $\frac{1}{2}mu^2$? Make an order-of-magnitude estimate of the amount by which its actual kinetic energy differs from $\frac{1}{2}mu^2$. In your solution, state the quantities you take as data and the values you measure or estimate for them. You may find Appendix B.5 useful.

Section 39.9 Mass and Energy

43. ● When 1.00 g of hydrogen combines with 8.00 g of oxygen, 9.00 g of water is formed. During this chemical reaction, 2.86×10^5 J of energy is released. Is the mass of the water larger or smaller than the mass of the reactants? What is the difference in mass? Explain whether the change in mass is likely to be detectable.

44. In a nuclear power plant, the fuel rods last 3 yr before they are replaced. If a plant with rated thermal power 1.00 GW operates at 80.0% capacity for 3.00 yr, what is the loss of mass of the fuel?

45. The power output of the Sun is 3.85×10^{26} W. How much mass is converted to energy in the Sun each second?

46. A gamma ray (a high-energy photon) can produce an electron (e⁻) and a positron (e⁺) when it enters the electric field of a heavy nucleus: $\gamma \rightarrow e^+ + e^-$. What minimum gamma-ray energy is required to accomplish this task? *Note:* The masses of the electron and the positron are equal.

Section 39.10 The General Theory of Relativity

47. An Earth satellite used in the global positioning system (GPS) moves in a circular orbit with period 11 h 58 min. (a) Determine the radius of its orbit. (b) Determine its speed. (c) The satellite contains an oscillator producing the principal nonmilitary GPS signal. Its frequency is 1 575.42 MHz in the reference frame of the satellite. When it is received on the Earth's surface, what is the fractional change in this frequency due to time dilation as described by special relativity? (d) The gravitational "blueshift" of the frequency according to general relativity is a separate effect. It is called a blueshift to indicate a change to a higher frequency. The magnitude of that fractional change is given by

$$\frac{\Delta f}{f} = \frac{\Delta U_g}{mc^2}$$

where ΔU_g is the change in gravitational potential energy of an object–Earth system when the object of mass m is moved between the two points where the signal is observed. Calculate this fractional change in frequency. (e) What is the overall fractional change in frequency? Superposed on both of these relativistic effects is a Doppler shift that is generally much larger. It can be a redshift or a blueshift, depending on the motion of a particular satellite relative to a GPS receiver (Fig. P39.47).

Figure P39.47

Photo courtesy of Gamin Ltd.

Additional Problems

48. *Houston, we've got a problem.* An astronaut wishes to visit the Andromeda galaxy, making a one-way trip that will take 30.0 yr in the spacecraft's frame of reference. Assume the galaxy is 2.00×10^6 ly away and the astronaut's speed is constant. (a) How fast must he travel relative to the Earth? (b) What will be the kinetic energy of his 1 000-metric-ton spacecraft? (c) What is the cost of this energy if it is purchased at a typical consumer price for electric energy of $0.130/kWh?

49. ▲ The cosmic rays of highest energy are protons that have kinetic energy on the order of 10^{13} MeV. (a) How long would it take a proton of this energy to travel across the Milky Way galaxy, having a diameter $\sim 10^5$ ly, as measured in the proton's frame? (b) From the point of view of the proton, how many kilometers across is the galaxy?

50. An electron has a speed of $0.750c$. (a) Find the speed of a proton that has the same kinetic energy as the electron. (b) **What If?** Find the speed of a proton that has the same momentum as the electron.

51. ● The equation

$$K = \left(\frac{1}{\sqrt{1 - u^2/c^2}} - 1 \right) mc^2$$

gives the kinetic energy of a particle moving at speed u. (a) Solve the equation for u. (b) From the equation for u, identify the minimum possible value of speed and the corresponding kinetic energy. (c) Identify the maximum possible speed and the corresponding kinetic energy. (d) Differentiate the equation for u with respect to time to obtain an equation describing the acceleration of a particle as a function of its kinetic energy and the power input to the particle. (e) Observe that for a nonrelativistic particle we have $u = (2K/m)^{1/2}$ and that differentiating this equation with respect to time gives $a = \mathcal{P}/(2mK)^{1/2}$. State the limiting form of the expression in part (d) at low energy. State how it compares with the nonrelativistic expression. (f) State the limiting form of the expression in part (d) at high energy. (g) Consider a particle with constant input power. Explain how the answer to part (f) helps account for the answer to part (c).

52. Ted and Mary are playing a game of catch in frame S′, which is moving at $0.600c$ with respect to frame S, while Jim, at rest in frame S, watches the action (Fig. P39.52). Ted throws the ball to Mary at $0.800c$ (according to Ted), and their separation (measured in S′) is 1.80×10^{12} m. (a) According to Mary, how fast is the ball moving? (b) According to Mary, what time interval is required for the ball to reach her? (c) According to Jim, how far apart are Ted and Mary and how fast is the ball moving? (d) According to Jim, what time interval is required for the ball to reach Mary?

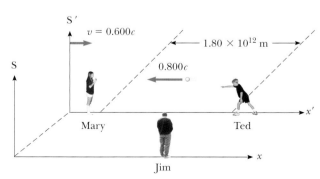

Figure P39.52

2 = intermediate;   3 = challenging;   □ = SSM/SG;   ▲ = ThomsonNOW;   = symbolic reasoning;   ● = qualitative reasoning

53. The net nuclear fusion reaction inside the Sun can be written as $4^1H \rightarrow {}^4He + E$. The rest energy of each hydrogen atom is 938.78 MeV, and the rest energy of the helium-4 atom is 3 728.4 MeV. Calculate the percentage of the starting mass that is transformed to other forms of energy.

54. An object disintegrates into two fragments. One fragment has mass 1.00 MeV/c^2 and momentum 1.75 MeV/c in the positive x direction. The other fragment has mass 1.50 MeV/c^2 and momentum 2.00 MeV/c in the positive y direction. Find the (a) mass and (b) speed of the original object.

55. Spacecraft I, containing students taking a physics exam, approaches the Earth with a speed of 0.600c (relative to the Earth), while spacecraft II, containing professors proctoring the exam, moves at 0.280c (relative to the Earth) directly toward the students. If the professors stop the exam after 50.0 min have passed on their clock, for what time interval does the exam last as measured by (a) the students and (b) an observer on the Earth?

56. Review problem. An electron is traveling through water at a speed 10.0% faster than the speed of light in water. Determine the electron's (a) total energy, (b) kinetic energy, and (c) momentum. The electron gives off Cerenkov radiation, the electromagnetic equivalent of a bow wave or a sonic boom. (d) Find the angle between the shock wave and the electron's direction of motion. Around the core of a nuclear reactor shielded by a large pool of water, Cerenkov radiation appears as a blue glow.

57. An alien spaceship traveling at 0.600c toward the Earth launches a landing craft with an advance guard of purchasing agents and environmental educators. The landing craft travels in the same direction with a speed of 0.800c relative to the mother ship. As observed on the Earth, the spaceship is 0.200 ly from the Earth when the landing craft is launched. (a) What speed do the Earth-based observers measure for the approaching landing craft? (b) What is the distance to the Earth at the moment of the landing craft's launch as observed by the aliens? (c) What travel time is required for the landing craft to reach the Earth as observed by the aliens on the mother ship? (d) If the landing craft has a mass of 4.00×10^5 kg, what is its kinetic energy as observed in the Earth reference frame?

58. ● *Speed of light in a moving medium.* The motion of a transparent medium influences the speed of light. This effect was first observed by Fizeau in 1851. Consider a light beam in water that moves with speed v in a horizontal pipe. Assume the light travels in the same direction as the water. The speed of light with respect to the water is c/n, where $n = 1.33$ is the index of refraction of water. (a) Use the velocity transformation equation to show that the speed of the light measured in the laboratory frame is

$$u = \frac{c}{n}\left(\frac{1 + nv/c}{1 + v/nc}\right)$$

(b) Show that for $v \ll c$, the expression from part (a) becomes, to a good approximation,

$$u \approx \frac{c}{n} + v - \frac{v}{n^2}$$

Argue for or against the view that we should expect the result to be $u = (c/n) + v$ according to the Galilean transformation and that the presence of the term $-v/n^2$ represents a relativistic effect appearing even at "nonrelativistic" speeds. (c) Evaluate u in the limit as the speed of the water approaches c.

59. A supertrain (proper length 100 m) travels at a speed of 0.950c as it passes through a tunnel (proper length 50.0 m). As seen by a trackside observer, is the train ever completely within the tunnel? If so, how much space is there to spare?

60. Imagine that the entire Sun collapses to a sphere of radius R_g such that the work required to remove a small mass m from the surface would be equal to its rest energy mc^2. This radius is called the *gravitational radius* for the Sun. Find R_g. The ultimate fate of very massive stars is thought to be collapsing beyond their gravitational radii into black holes.

61. ● A particle with electric charge q moves along a straight line in a uniform electric field $\vec{E}$ with a speed of u. The electric force exerted on the charge is $q\vec{E}$. The motion and the electric field are both in the x direction. (a) Show that the acceleration of the particle in the x direction is given by

$$a = \frac{du}{dt} = \frac{qE}{m}\left(1 - \frac{u^2}{c^2}\right)^{3/2}$$

(b) Discuss the significance of the dependence of the acceleration on the speed. (c) **What If?** If the particle starts from rest at $x = 0$ at $t = 0$, how would you proceed to find the speed of the particle and its position at time t?

62. An observer in a coasting spacecraft moves toward a mirror at speed v relative to the reference frame labeled by S in Figure P39.62. The mirror is stationary with respect to S. A light pulse emitted by the spacecraft travels toward the mirror and is reflected back to the spacecraft. The front of the spacecraft is a distance d from the mirror (as measured by observers in S) at the moment the light pulse leaves the spacecraft. What is the total travel time of the pulse as measured by observers in (a) the S frame and (b) the front of the spacecraft?

2 = intermediate;   3 = challenging;   □ = SSM/SG;   ▲ = ThomsonNOW;   = symbolic reasoning;   ● = qualitative reasoning

Figure P39.62

63. ● Massive stars ending their lives in supernova explosions produce the nuclei of all the atoms in the bottom half of the periodic table, by fusion of smaller nuclei. This problem roughly models that process. A particle of mass m moving along the x axis with a velocity component $+u$ collides head-on and sticks to a particle of mass $m/3$ moving along the x axis with the velocity component $-u$. (a) What is the mass M of the resulting particle? (b) Evaluate the expression from part (a) in the limit $u \to 0$. Explain whether the result agrees with what you should expect from nonrelativistic physics.

64. The creation and study of new elementary particles is an important part of contemporary physics. Especially interesting is the discovery of a very massive particle. To create a particle of mass M requires an energy Mc^2. With enough energy, an exotic particle can be created by allowing a fast-moving particle of ordinary matter, such as a proton, to collide with a similar target particle. Consider a perfectly inelastic collision between two protons: an incident proton with mass m_p, kinetic energy K, and momentum magnitude p joins with an originally stationary target proton to form a single product particle of mass M. You might think that the creation of a new product particle, nine times more massive than in a previous experiment, would require only nine times more energy for the incident proton. Unfortunately, not all the kinetic energy of the incoming proton is available to create the product particle because conservation of momentum requires that the system as a whole still must have some kinetic energy after the collision. Only a fraction of the energy of the incident particle is therefore available to create a new particle. In this problem, you must determine how the energy available for particle creation depends on the energy of the moving proton. Show that the energy available to create a product particle is given by

$$Mc^2 = 2m_p c^2 \sqrt{1 + \frac{K}{2m_p c^2}}$$

This result shows that when the kinetic energy K of the incident proton is large compared with its rest energy $m_p c^2$, then M approaches $(2m_p K)^{1/2}/c$. Therefore, if the energy of the incoming proton is increased by a factor of 9, the mass you can create increases only by a factor of 3. This disappointing result is the main reason that most modern accelerators such as those at CERN (in Europe), at Fermilab (near Chicago), at SLAC (at Stanford), and at DESY (in Germany) use *colliding beams*. Here the total momentum of a pair of interacting particles can be zero. The center of mass can be at rest after the collision, so, in principle, all the initial kinetic energy can be used for particle creation, according to

$$Mc^2 = 2mc^2 + K = 2mc^2 \left(1 + \frac{K}{2mc^2} \right)$$

where K is the total kinetic energy of two identical colliding particles. Here if $K \gg mc^2$, we have M directly proportional to K as we would desire. These machines are difficult to build and to operate, but they open new vistas in physics.

65. ● Suppose our Sun is about to explode. In an effort to escape, we depart in a spacecraft at $v = 0.800c$ and head toward the star Tau Ceti, 12.0 ly away. When we reach the midpoint of our journey from the Earth, we see our Sun explode, and, unfortunately, at the same instant we see Tau Ceti explode as well. (a) In the spacecraft's frame of reference, should we conclude that the two explosions occurred simultaneously? If not, which occurred first? (b) **What If?** In a frame of reference in which the Sun and Tau Ceti are at rest, did they explode simultaneously? If not, which exploded first?

66. ☕ Prepare a graph of the relativistic kinetic energy and the classical kinetic energy, both as a function of speed, for an object with a mass of your choice. At what speed does the classical kinetic energy underestimate the experimental value by 1%? By 5%? By 50%?

67. A ^{57}Fe nucleus at rest emits a 14.0-keV photon. Use conservation of energy and momentum to deduce the kinetic energy of the recoiling nucleus in electron volts. Use $Mc^2 = 8.60 \times 10^{-9}$ J for the final state of the ^{57}Fe nucleus.

Answers to Quick Quizzes

39.1 (c). Although the observers' measurements differ, both are correct.

39.2 (d). The Galilean velocity transformation gives us $u_x = u'_x + v = 90$ mi/h $+ 110$ mi/h $= 200$ mi/h.

39.3 (d). The two events (the pulse leaving the flashlight and the pulse hitting the far wall) take place at different locations for both observers, so neither measures the proper time interval.

2 = intermediate;   3 = challenging;   ☐ = SSM/SG;   ▲ = ThomsonNOW;   = symbolic reasoning;   ● = qualitative reasoning

39.4 (a). The two events are the beginning and the end of the movie, both of which take place at rest with respect to the spacecraft crew. Therefore, the crew measures the proper time interval of 2 h. Any observer in motion with respect to the spacecraft, which includes the observer on Earth, will measure a longer time interval due to time dilation.

39.5 (a). If their on-duty time is based on clocks that remain on the Earth, the astronauts will have larger paychecks. A shorter time interval will have passed for the astronauts in their frame of reference than for their employer back on the Earth.

39.6 (c). Both your body and your sleeping cabin are at rest in your reference frame; therefore, they will have their proper length according to you. There will be no change in measured lengths of objects, including yourself, within your spacecraft.

39.7 (d). Time dilation and length contraction depend only on the relative speed of one observer relative to another, not on whether the observers are receding or approaching each other.

39.8 **(i),** (c). Because of your motion toward the source of the light, the light beam has a horizontal component of velocity as measured by you. The magnitude of the vector sum of the horizontal and vertical component vectors must be equal to c, so the magnitude of the vertical component must be smaller than c. **(ii),** (a). In this case, there is only a horizontal component of the velocity of the light and you must measure a speed of c.

39.9 (a) $m_3 > m_2 = m_1$; the rest energy of particle 3 is $2E$, whereas it is E for particles 1 and 2. (b) $K_3 = K_2 > K_1$; the kinetic energy is the difference between the total energy and the rest energy. The kinetic energy is $4E - 2E = 2E$ for particle 3, $3E - E = 2E$ for particle 2, and $2E - E = E$ for particle 1. (c) $u_2 > u_3 = u_1$; from Equation 39.26, $E = \gamma E_R$. Solving for the square of the particle speed u, we find that $u^2 = c^2(1 - (E_R/E)^2)$. Therefore, the particle with the smallest ratio of rest energy to total energy will have the largest speed. Particles 1 and 3 have the same ratio as each other, and the ratio of particle 2 is smaller.

This lightbulb filament glows with an orange color. Why? Classical physics is unable to explain the experimentally observed wavelength distribution of electromagnetic radiation from a hot object. A theory proposed in 1900 and describing the radiation from such objects represents the dawn of quantum physics. (Steve Cole/Getty Images)

40.1 Blackbody Radiation and Planck's Hypothesis

40.2 The Photoelectric Effect

40.3 The Compton Effect

40.4 Photons and Electromagnetic Waves

40.5 The Wave Properties of Particles

40.6 The Quantum Particle

40.7 The Double-Slit Experiment Revisited

40.8 The Uncertainty Principle

40 Introduction to Quantum Physics

In Chapter 39, we discussed that Newtonian mechanics must be replaced by Einstein's special theory of relativity when dealing with particle speeds comparable to the speed of light. As the 20th century progressed, many experimental and theoretical problems were resolved by the special theory of relativity. For many other problems, however, neither relativity nor classical physics could provide a theoretical answer. Attempts to apply the laws of classical physics to explain the behavior of matter on the atomic scale were consistently unsuccessful. For example, the emission of discrete wavelengths of light from atoms in a high-temperature gas could not be explained within the framework of classical physics.

As physicists sought new ways to solve these puzzles, another revolution took place in physics between 1900 and 1930. A new theory called *quantum mechanics* was highly successful in explaining the behavior of particles of microscopic size. Like the special theory of relativity, the quantum theory requires a modification of our ideas concerning the physical world.

The first explanation of a phenomenon using quantum theory was introduced by Max Planck. Many subsequent mathematical developments and interpretations were made by a number of distinguished physicists, including Einstein, Bohr, de Broglie, Schrödinger, and Heisenberg. Despite the great success of the quantum theory, Einstein frequently played the role of its critic, especially with regard to the manner in which the theory was interpreted.

PITFALL PREVENTION 40.1
Expect to Be Challenged

If the discussions of quantum physics in this and subsequent chapters seem strange and confusing to you, it's because your whole life experience has taken place in the macroscopic world, where quantum effects are not evident.

Because an extensive study of quantum theory is beyond the scope of this book, this chapter is simply an introduction to its underlying principles.

40.1 Blackbody Radiation and Planck's Hypothesis

An object at any temperature emits electromagnetic waves in the form of **thermal radiation** from its surface as discussed in Section 20.7. The characteristics of this radiation depend on the temperature and properties of the object's surface. Careful study shows that the radiation consists of a continuous distribution of wavelengths from all portions of the electromagnetic spectrum. If the object is at room temperature, the wavelengths of thermal radiation are mainly in the infrared region and hence the radiation is not detected by the human eye. As the surface temperature of the object increases, the object eventually begins to glow visibly red. At sufficiently high temperatures, the glowing object appears white, as in the hot tungsten filament of a lightbulb.

From a classical viewpoint, thermal radiation originates from accelerated charged particles in the atoms near the surface of the object; those charged particles emit radiation much as small antennas do. The thermally agitated particles can have a distribution of energies, which accounts for the continuous spectrum of radiation emitted by the object. By the end of the 19th century, however, it became apparent that the classical theory of thermal radiation was inadequate. The basic problem was in understanding the observed distribution of wavelengths in the radiation emitted by a black body. As defined in Section 20.7, a **black body** is an ideal system that absorbs all radiation incident on it. The electromagnetic radiation emitted by the black body is called **blackbody radiation.**

A good approximation of a black body is a small hole leading to the inside of a hollow object as shown in Figure 40.1. Any radiation incident on the hole from outside the cavity enters the hole and is reflected a number of times on the interior walls of the cavity; hence, the hole acts as a perfect absorber. The nature of the radiation leaving the cavity through the hole depends only on the temperature of the cavity walls and not on the material of which the walls are made. The spaces between lumps of hot charcoal (Fig. 40.2) emit light that is very much like blackbody radiation.

The radiation emitted by oscillators in the cavity walls experiences boundary conditions. As the radiation reflects from the cavity's walls, standing electromagnetic waves are established within the three-dimensional interior of the cavity. Many standing-wave modes are possible, and the distribution of the energy in the cavity among these modes determines the wavelength distribution of the radiation leaving the cavity through the hole.

The wavelength distribution of radiation from cavities was studied experimentally in the late 19th century. Active Figure 40.3 shows how the intensity of blackbody radiation varies with temperature and wavelength. The following two consistent experimental findings were seen as especially significant:

1. **The total power of the emitted radiation increases with temperature.** We discussed this behavior briefly in Chapter 20, where we introduced **Stefan's law:**

Stefan's law   ▶

$$\mathcal{P} = \sigma A e T^4 \qquad (40.1)$$

where $\mathcal{P}$ is the power in watts radiated at all wavelengths from the surface of an object, σ is the Stefan–Boltzmann constant, equal to 5.670×10^{-8} W/m² · K⁴, A is the surface area of the object in square meters, e is the emissivity of the surface, and T is the surface temperature in kelvins. For a black body, the emissivity is $e = 1$ exactly.

2. **The peak of the wavelength distribution shifts to shorter wavelengths as the temperature increases.** This behavior is described by the following relationship, called **Wien's displacement law:**

Figure 40.1 The opening to the cavity inside a hollow object is a good approximation of a black body. Light entering the small opening strikes the interior walls, where some is absorbed and some is reflected at a random angle. The cavity walls reradiate at wavelengths corresponding to their temperature, producing standing waves in the cavity. Some of the energy from these standing waves can leave through the opening.

Figure 40.2 The glow emanating from the spaces between these hot charcoal briquettes is, to a close approximation, blackbody radiation. The color of the light depends only on the temperature of the briquettes.

$$\lambda_{\max} T = 2.898 \times 10^{-3} \text{ m} \cdot \text{K}$$ **(40.2)**   ◄ Wien's displacement law

where $\lambda_{\max}$ is the wavelength at which the curve peaks and T is the absolute temperature of the surface of the object emitting the radiation. The wavelength at the curve's peak is inversely proportional to the absolute temperature; that is, as the temperature increases, the peak is "displaced" to shorter wavelengths (Active Fig. 40.3).

Wien's displacement law is consistent with the behavior of the object mentioned at the beginning of this section. At room temperature, the object does not appear to glow because the peak is in the infrared region of the electromagnetic spectrum. At higher temperatures, it glows red because the peak is in the near infrared with some radiation at the red end of the visible spectrum, and at still higher temperatures, it glows white because the peak is in the visible so that all colors are emitted.

ACTIVE FIGURE 40.3

Intensity of blackbody radiation versus wavelength at three temperatures. The amount of radiation emitted (the area under a curve) increases with increasing temperature. The visible range of wavelengths is between 0.4 μm and 0.7 μm. Therefore, the 4 000-K curve has a peak that is near the visible range and that represents an object that would glow with a yellowish-white appearance. At approximately 6 000 K, the peak is in the center of the visible wavelengths and the object appears white.

Sign in at www.thomsonedu.com and go to ThomsonNOW to adjust the temperature of the black body and study the radiation emitted from it.

Quick Quiz 40.1   Figure 40.4 shows two stars in the constellation Orion. Betelgeuse appears to glow red, whereas Rigel looks blue in color. Which star has a higher surface temperature? (a) Betelgeuse   (b) Rigel   (c) both the same   (d) impossible to determine

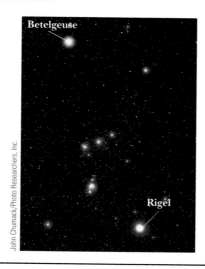

Figure 40.4   (Quick Quiz 40.1) Which star is hotter, Betelgeuse or Rigel?

A successful theory for blackbody radiation must predict the shape of the curves in Active Figure 40.3, the temperature dependence expressed in Stefan's law, and the shift of the peak with temperature described by Wien's displacement law. Early attempts to use classical ideas to explain the shapes of the curves in Active Figure 40.3 failed.

Let's consider one of these early attempts. To describe the distribution of energy from a black body, we define $I(\lambda, T) \, d\lambda$ to be the intensity, or power per unit area, emitted in the wavelength interval $d\lambda$. The result of a calculation based on a classical theory of blackbody radiation known as the **Rayleigh–Jeans law** is

$$I(\lambda, T) = \frac{2\pi c k_{\text{B}} T}{\lambda^4}$$ **(40.3)**   ◄ Rayleigh–Jeans law

where k_{B} is Boltzmann's constant. The black body is modeled as the hole leading into a cavity supporting many modes of oscillation of the electromagnetic field caused by accelerated charges in the cavity walls, resulting in the emission of electromagnetic waves at all wavelengths. In the classical theory used to derive Equation 40.3, the average energy for each wavelength of the standing-wave modes is assumed to be proportional to $k_{\text{B}} T$, based on the theorem of equipartition of energy discussed in Section 21.1.

Figure 40.5 Comparison of experimental results and the curve predicted by the Rayleigh–Jeans law for the distribution of blackbody radiation.

© Bettmann/CORBIS

MAX PLANCK
German Physicist (1858–1947)

Planck introduced the concept of "quantum of action" (Planck's constant, h) in an attempt to explain the spectral distribution of blackbody radiation, which laid the foundations for quantum theory. In 1918, he was awarded the Nobel Prize in Physics for this discovery of the quantized nature of energy.

PITFALL PREVENTION 40.2
n Is Again an Integer

In the preceding chapters on optics, we used the symbol n for the index of refraction, which was not an integer. Here we are again using n as we did in Chapter 18 to indicate the standing-wave mode on a string or in an air column. In quantum physics, n is often used as an integer quantum number to identify a particular quantum state of a system.

An experimental plot of the blackbody radiation spectrum, together with the theoretical prediction of the Rayleigh–Jeans law, is shown in Figure 40.5. At long wavelengths, the Rayleigh–Jeans law is in reasonable agreement with experimental data, but at short wavelengths, major disagreement is apparent.

As λ approaches zero, the function $I(\lambda, T)$ given by Equation 40.3 approaches infinity. Hence, according to classical theory, not only should short wavelengths predominate in a blackbody spectrum, but also the energy emitted by any black body should become infinite in the limit of zero wavelength. In contrast to this prediction, the experimental data plotted in Figure 40.5 show that as λ approaches zero, $I(\lambda, T)$ also approaches zero. This mismatch of theory and experiment was so disconcerting that scientists called it the *ultraviolet catastrophe*. (This "catastrophe"—infinite energy—occurs as the wavelength approaches zero; the word *ultraviolet* was applied because ultraviolet wavelengths are short.)

In 1900, Max Planck developed a theory of blackbody radiation that leads to an equation for $I(\lambda, T)$ that is in complete agreement with experimental results at all wavelengths. Planck assumed the cavity radiation came from atomic oscillators in the cavity walls in Figure 40.1. Planck made two bold and controversial assumptions concerning the nature of the oscillators in the cavity walls:

- The energy of an oscillator can have only certain *discrete* values E_n:

$$E_n = nhf \qquad \textbf{(40.4)}$$

where n is a positive integer called a **quantum number**,[1] f is the oscillator's frequency, and h is a parameter Planck introduced that is now called **Planck's constant**. Because the energy of each oscillator can have only discrete values given by Equation 40.4, we say the energy is **quantized.** Each discrete energy value corresponds to a different **quantum state,** represented by the quantum number n. When the oscillator is in the $n = 1$ quantum state, its energy is hf; when it is in the $n = 2$ quantum state, its energy is $2hf$; and so on.

- The oscillators emit or absorb energy when making a transition from one quantum state to another. The entire energy difference between the initial and final states in the transition is emitted or absorbed as a single quantum of radiation. If the transition is from one state to a lower adjacent state—say, from the $n = 3$ state to the $n = 2$ state—Equation 40.4 shows that the amount of energy emitted by the oscillator and carried by the quantum of radiation is

$$E = hf \qquad \textbf{(40.5)}$$

An oscillator emits or absorbs energy only when it changes quantum states. If it remains in one quantum state, no energy is absorbed or emitted. Figure 40.6 is an **energy-level diagram** showing the quantized energy levels and allowed transitions proposed by Planck. This important semigraphical representation is used often in quantum physics.[2] The vertical axis is linear in energy, and the allowed energy levels are represented as horizontal lines. The quantized system can have only the energies represented by the horizontal lines.

The key point in Planck's theory is the radical assumption of quantized energy states. This development—a clear deviation from classical physics—marked the birth of the quantum theory.

In the Rayleigh–Jeans model, the average energy associated with a particular wavelength of standing waves in the cavity is the same for all wavelengths and is equal to $k_B T$. Planck used the same classical ideas as in the Rayleigh–Jeans model to arrive at the energy density as a product of constants and the average energy for a given wavelength, but the average energy is not given by the equipartition theo-

---

[1] A quantum number is generally an integer (although half-integer quantum numbers can occur) that describes an allowed state of a system, such as the values of n describing the normal modes of oscillation of a string fixed at both ends, as discussed in Section 18.3.

[2] We first saw an energy-level diagram in Section 21.4.

rem. A wave's average energy is the average energy difference between levels of the oscillator, *weighted according to the probability of the wave being emitted*. This weighting is based on the occupation of higher-energy states as described by the Boltzmann distribution law, which was discussed in Section 21.5. According to this law, the probability of a state being occupied is proportional to the factor $e^{-E/k_B T}$, where E is the energy of the state.

At low frequencies, the energy levels are close together as on the right in Active Figure 40.7, and many of the energy states are excited because the Boltzmann factor $e^{-E/k_B T}$ is relatively large for these states. Therefore, there are many contributions to the outgoing radiation, although each contribution has very low energy. Now, consider high-frequency radiation, that is, radiation with short wavelength. To obtain this radiation, the allowed energies are very far apart as on the left in Active Figure 40.7. The probability of thermal agitation exciting these high energy levels is small because of the small value of the Boltzmann factor for large values of E. At high frequencies, the low probability of excitation results in very little contribution to the total energy, even though each quantum is of large energy. This low probability "turns the curve over" and brings it down to zero again at short wavelengths.

Using this approach, Planck generated a theoretical expression for the wavelength distribution that agreed remarkably well with the experimental curves in Active Figure 40.3:

$$I(\lambda, T) = \frac{2\pi h c^2}{\lambda^5 \left(e^{hc/\lambda k_B T} - 1\right)}$$ **(40.6)**   ◀ Planck's wavelength distribution function

This function includes the parameter h, which Planck adjusted so that his curve matched the experimental data at all wavelengths. The value of this parameter is found to be independent of the material of which the black body is made and independent of the temperature; it is a fundamental constant of nature. The value of h, Planck's constant, which was first introduced in Chapter 35, is

$$h = 6.626 \times 10^{-34}\,\text{J}\cdot\text{s}$$ **(40.7)**   ◀ Planck's constant

At long wavelengths, Equation 40.6 reduces to the Rayleigh–Jeans expression, Equation 40.3 (see Problem 13), and at short wavelengths, it predicts an exponential decrease in $I(\lambda, T)$ with decreasing wavelength, in agreement with experimental results.

When Planck presented his theory, most scientists (including Planck!) did not consider the quantum concept to be realistic. They believed it was a mathematical

Figure 40.6 Allowed energy levels for an oscillator with frequency f. Allowed transitions are indicated by the double-headed arrows.

ACTIVE FIGURE 40.7

In Planck's model, the average energy associated with a given wavelength is the product of the energy of a transition and a factor related to the probability of the transition occurring. As the energy levels move farther apart at shorter wavelengths (higher energy), the probability of excitation decreases, as does the probability of a transition from the excited state.

Sign in at www.thomsonedu.com and go to ThomsonNOW to investigate the energy levels and observe the emission of radiation of different wavelengths.

Figure 40.8 An ear thermometer measures a patient's temperature by detecting the intensity of infrared radiation leaving the eardrum.

trick that happened to predict the correct results. Hence, Planck and others continued to search for a more "rational" explanation of blackbody radiation. Subsequent developments, however, showed that a theory based on the quantum concept (rather than on classical concepts) had to be used to explain not only blackbody radiation but also a number of other phenomena at the atomic level.

In 1905, Einstein rederived Planck's results by assuming the cavity oscillations of the electromagnetic field were themselves quantized. In other words, he proposed that quantization is a fundamental property of light and other electromagnetic radiation, which led to the concept of photons as shall be discussed in Section 40.2. Critical to the success of the quantum or photon theory was the relation between energy and frequency, which classical theory completely failed to predict.

You may have had your body temperature measured at the doctor's office by an *ear thermometer*, which can read your temperature in a matter of seconds (Fig. 40.8). In a fraction of a second, this type of thermometer measures the amount of infrared radiation emitted by the eardrum. It then converts the amount of radiation into a temperature reading. This thermometer is very sensitive because temperature is raised to the fourth power in Stefan's law. Suppose you have a fever 1°C above normal. Because absolute temperatures are found by adding 273 to Celsius temperatures, the ratio of your fever temperature to normal body temperature of 37°C is

$$\frac{T_{\text{fever}}}{T_{\text{normal}}} = \frac{38°\text{C} + 273°\text{C}}{37°\text{C} + 273°\text{C}} = 1.003\ 2$$

which is only a 0.32% increase in temperature. The increase in radiated power, however, is proportional to the fourth power of temperature, so

$$\frac{\mathscr{P}_{\text{fever}}}{\mathscr{P}_{\text{normal}}} = \left(\frac{38°\text{C} + 273°\text{C}}{37°\text{C} + 273°\text{C}}\right)^4 = 1.013$$

The result is a 1.3% increase in radiated power, which is easily measured by modern infrared radiation sensors.

EXAMPLE 40.1 | **Thermal Radiation from Different Objects**

(A) Find the peak wavelength of the blackbody radiation emitted by the human body when the skin temperature is 35°C.

SOLUTION

Conceptualize Thermal radiation is emitted from the surface of any object. The peak wavelength is related to the surface temperature through Wien's displacement law (Eq. 40.2).

Categorize We evaluate results using an equation developed in this section, so we categorize this example as a substitution problem.

Solve Equation 40.2 for λ_{max}:

$$(1) \quad \lambda_{\text{max}} = \frac{2.898 \times 10^{-3}\ \text{m} \cdot \text{K}}{T}$$

Substitute the surface temperature:

$$\lambda_{\text{max}} = \frac{2.898 \times 10^{-3}\ \text{m} \cdot \text{K}}{308\ \text{K}} = 9.4\ \mu\text{m}$$

This radiation is in the infrared region of the spectrum and is invisible to the human eye. Some animals (pit vipers, for instance) are able to detect radiation of this wavelength and therefore can locate warm-blooded prey even in the dark.

(B) Find the peak wavelength of the blackbody radiation emitted by the tungsten filament of a lightbulb, which operates at 2 000 K.

SOLUTION

Substitute the filament temperature into Equation (1):

$$\lambda_{max} = \frac{2.898 \times 10^{-3} \text{ m} \cdot \text{K}}{2\,000 \text{ K}} = \boxed{1.4 \ \mu\text{m}}$$

This radiation is also in the infrared, meaning that most of the energy emitted by a lightbulb is not visible to us.

(C) Find the peak wavelength of the blackbody radiation emitted by the Sun, which has a surface temperature of approximately 5 800 K.

SOLUTION

Substitute the surface temperature into Equation (1):

$$\lambda_{max} = \frac{2.898 \times 10^{-3} \text{ m} \cdot \text{K}}{5\,800 \text{ K}} = \boxed{0.50 \ \mu\text{m}}$$

This radiation is near the center of the visible spectrum, near the color of a yellow-green tennis ball. Because it is the most prevalent color in sunlight, our eyes have evolved to be most sensitive to light of approximately this wavelength.

EXAMPLE 40.2   **The Quantized Oscillator**

A 2.0-kg block is attached to a massless spring that has a force constant of $k = 25$ N/m. The spring is stretched 0.40 m from its equilibrium position and released from rest.

(A) Find the total energy of the system and the frequency of oscillation according to classical calculations.

SOLUTION

Conceptualize We understand the details of the block's motion from our study of simple harmonic motion in Chapter 15.

Categorize The phrase "according to classical calculations" tells us to categorize this part of the problem as a classical analysis of the oscillator. We model the block as a particle in simple harmonic motion.

Analyze Based on the way that the block is set into motion, its amplitude is 0.40 m.

Evaluate the total energy of the block–spring system using Equation 15.21:

$$E = \tfrac{1}{2}kA^2 = \tfrac{1}{2}(25 \text{ N/m})(0.40 \text{ m})^2 = \boxed{2.0 \text{ J}}$$

Evaluate the frequency of oscillation from Equation 15.14:

$$f = \frac{1}{2\pi}\sqrt{\frac{k}{m}} = \frac{1}{2\pi}\sqrt{\frac{25 \text{ N/m}}{2.0 \text{ kg}}} = \boxed{0.56 \text{ Hz}}$$

(B) Assuming the energy of the oscillator is quantized, find the quantum number n for the system oscillating with this amplitude.

SOLUTION

Categorize This part of the problem is categorized as a quantum analysis of the oscillator. We model the block–spring system as a Planck oscillator.

Analyze Solve Equation 40.4 for the quantum number n:

$$n = \frac{E_n}{hf}$$

Substitute numerical values:

$$n = \frac{2.0 \text{ J}}{(6.626 \times 10^{-34} \text{ J} \cdot \text{s})(0.56 \text{ Hz})} = \boxed{5.4 \times 10^{33}}$$

Finalize Notice that 5.4×10^{33} is a very large quantum number, which is typical for macroscopic systems. Changes between quantum states for the oscillator are explored next.

What If?   Suppose the oscillator makes a transition from the $n = 5.4 \times 10^{33}$ state to the state corresponding to $n = 5.4 \times 10^{33} - 1$. By how much does the energy of the oscillator change in this one-quantum change?

Answer   From Equation 40.5, the energy carried away due to the transition between states differing in n by 1 is

$$E = hf = (6.626 \times 10^{-34}\,\text{J} \cdot \text{s})(0.56\,\text{Hz}) = 3.7 \times 10^{-34}\,\text{J}$$

This energy change due to a one-quantum change is fractionally equal to $3.7 \times 10^{-34}\,\text{J}/2.0\,\text{J}$, or on the order of one part in 10^{34}! It is such a small fraction of the total energy of the oscillator that it cannot be detected. Therefore, even though the energy of a macroscopic block–spring system is quantized and does indeed decrease by small quantum jumps, our senses perceive the decrease as continuous. Quantum effects become important and detectable only on the submicroscopic level of atoms and molecules.

ACTIVE FIGURE 40.9

A circuit diagram for studying the photoelectric effect. When light strikes plate E (the emitter), photoelectrons are ejected from the plate. Electrons moving from plate E to plate C (the collector) constitute a current in the circuit.

Sign in at www.thomsonedu.com and go to ThomsonNOW to observe the motion of electrons for various frequencies of light and plate voltages.

ACTIVE FIGURE 40.10

Photoelectric current versus applied potential difference for two light intensities. The current increases with intensity but reaches a saturation level for large values of ΔV. At voltages equal to or more negative than $-\Delta V_s$, where ΔV_s is the stopping potential, the current is zero.

Sign in at www.thomsonedu.com and go to ThomsonNOW to sweep through the voltage range and observe the current curve for different intensities of radiation.

40.2   The Photoelectric Effect

Blackbody radiation was the first phenomenon to be explained with a quantum model. In the latter part of the 19th century, at the same time that data were taken on thermal radiation, experiments showed that light incident on certain metallic surfaces causes electrons to be emitted from those surfaces. This phenomenon, which was first discussed in Section 35.1, is known as the **photoelectric effect,** and the emitted electrons are called **photoelectrons.**[3]

Active Figure 40.9 is a diagram of an apparatus for studying the photoelectric effect. An evacuated glass or quartz tube contains a metallic plate E (the emitter) connected to the negative terminal of a battery and another metallic plate C (the collector) that is connected to the positive terminal of the battery. When the tube is kept in the dark, the ammeter reads zero, indicating no current in the circuit. However, when plate E is illuminated by light having an appropriate wavelength, a current is detected by the ammeter, indicating a flow of charges across the gap between plates E and C. This current arises from photoelectrons emitted from plate E and collected at plate C.

Active Figure 40.10 is a plot of photoelectric current versus potential difference ΔV applied between plates E and C for two light intensities. At large values of ΔV, the current reaches a maximum value; all the electrons emitted from E are collected at C, and the current cannot increase further. In addition, the maximum current increases as the intensity of the incident light increases, as you might expect, and more electrons are ejected by the higher-intensity light. Finally, when ΔV is negative—that is, when the battery in the circuit is reversed to make plate E positive and plate C negative—the current drops because many of the photoelectrons emitted from E are repelled by the now negative plate C. In this situation, only those photoelectrons having a kinetic energy greater than $e|\Delta V|$ reach plate C, where e is the magnitude of the charge on the electron. When ΔV is equal to or more negative than $-\Delta V_s$, where ΔV_s is the **stopping potential,** no photoelectrons reach C and the current is zero.

Let's model the combination of the electric field between the plates and an electron ejected from plate E as an isolated system. Suppose this electron stops just as it reaches plate C. Because the system is isolated, the total mechanical energy of the system must be conserved, so we have

$$K_1 + U_1 = K_2 + U_2$$

where configuration 1 refers to the instant the electron leaves the metal with kinetic energy K_1 and configuration 2 is when the electron stops just before touching plate C. If we define the electric potential energy of the system in configuration 1 to be zero, we have

$$K_1 + 0 = 0 + (-e)(-\Delta V)$$

---

[3] Photoelectrons are not different from other electrons. They are given this name solely because of their ejection from a metal by light in the photoelectric effect.

Now suppose the potential difference ΔV is increased in the negative direction just until the current is zero. In this case, the electron that stops immediately before reaching plate C has the maximum possible kinetic energy upon leaving the metal surface and ΔV equals the stopping potential ΔV_s. The previous equation can then be written as

$$K_{max} = e\Delta V_s \qquad (40.8)$$

This equation allows us to measure K_{max} experimentally by determining the voltage ΔV_s at which the current drops to zero.

Several features of the photoelectric effect are listed below. For each feature, we compare the predictions made by a classical approach, using the wave model for light, with the experimental results.

1. Dependence of photoelectron kinetic energy on light intensity

   Classical prediction: Electrons should absorb energy continuously from the electromagnetic waves. As the light intensity incident on a metal is increased, energy should be transferred into the metal at a higher rate and the electrons should be ejected with more kinetic energy.

   Experimental result: The maximum kinetic energy of photoelectrons is *independent* of light intensity as shown in Active Figure 40.10 with both curves falling to zero at the *same* negative voltage. (According to Equation 40.8, the maximum kinetic energy is proportional to the stopping potential.)

2. Time interval between incidence of light and ejection of photoelectrons

   Classical prediction: At low light intensities, a measurable time interval should pass between the instant the light is turned on and the time an electron is ejected from the metal. This time interval is required for the electron to absorb the incident radiation before it acquires enough energy to escape from the metal.

   Experimental result: Electrons are emitted from the surface of the metal almost *instantaneously* (less than 10^{-9} s after the surface is illuminated), even at very low light intensities.

3. Dependence of ejection of electrons on light frequency

   Classical prediction: Electrons should be ejected from the metal at any incident light frequency, as long as the light intensity is high enough, because energy is transferred to the metal regardless of the incident light frequency.

   Experimental result: No electrons are emitted if the incident light frequency falls below some **cutoff frequency** f_c, whose value is characteristic of the material being illuminated. No electrons are ejected below this cutoff frequency *regardless* of the light intensity.

4. Dependence of photoelectron kinetic energy on light frequency

   Classical prediction: There should be *no* relationship between the frequency of the light and the electron kinetic energy. The kinetic energy should be related to the intensity of the light.

   Experimental result: The maximum kinetic energy of the photoelectrons increases with increasing light frequency.

For these features, experimental results contradict *all four* classical predictions. A successful explanation of the photoelectric effect was given by Einstein in 1905, the same year he published his special theory of relativity. As part of a general paper on electromagnetic radiation, for which he received a Nobel Prize in Physics in 1921, Einstein extended Planck's concept of quantization to electromagnetic waves as mentioned in Section 40.1. Einstein assumed light (or any other electromagnetic wave) of frequency f can be considered a stream of quanta, regardless of the source of the radiation. Today we call these quanta **photons.** Each photon has an energy E given by Equation 40.5, $E = hf$, and each moves in a vacuum at the speed of light c, where $c = 3.00 \times 10^8$ m/s.

Quick Quiz 40.2 While standing outdoors one evening, you are exposed to the following four types of electromagnetic radiation: yellow light from a sodium street lamp, radio waves from an AM radio station, radio waves from an FM radio station, and microwaves from an antenna of a communications system. Rank these types of waves in terms of increasing photon energy, lowest first.

In Einstein's model of the photoelectric effect, a photon of the incident light gives *all* its energy hf to a *single* electron in the metal. Therefore, the absorption of energy by the electrons is not a continuous absorption process as envisioned in the wave model; rather, it is a discontinuous process in which energy is delivered to the electrons in discrete bundles. The energy transfer is accomplished via a one photon–one electron event.[4]

Electrons ejected from the surface of the metal and not making collisions with other metal atoms before escaping possess the maximum kinetic energy K_{max}. According to Einstein, the maximum kinetic energy for these liberated electrons is

◀ Photoelectric effect equation

$$K_{max} = hf - \phi \tag{40.9}$$

where ϕ is called the **work function** of the metal. **The work function represents the minimum energy with which an electron is bound in the metal** and is on the order of a few electron volts. Table 40.1 lists work functions for various metals.

We can understand Equation 40.9 by rearranging it as follows:

$$K_{max} + \phi = hf$$

In this form, Einstein's equation is equivalent to Equation 8.2 applied to the non-isolated system of the electron and the metal. Here, K_{max} is the change ΔK in kinetic energy of the electron, assuming it begins at rest; ϕ is the change ΔU in potential energy of the system, assuming the potential energy is defined to be zero when the electron is within the metal; and hf is the transfer of energy into the system by electromagnetic radiation (T_{ER}).

Using the photon model of light, one can explain the observed features of the photoelectric effect that could not be understood using classical concepts:

1. Dependence of photoelectron kinetic energy on light intensity

    Equation 40.9 shows that K_{max} is independent of the light intensity. The maximum kinetic energy of any one electron, which equals $hf - \phi$, depends only on the light frequency and the work function. If the light intensity is doubled, the number of photons arriving per unit time is doubled, which doubles the rate at which photoelectrons are emitted. The maximum kinetic energy of any one photoelectron, however, is unchanged.

2. Time interval between incidence of light and ejection of photoelectrons

    Near-instantaneous emission of electrons is consistent with the photon model of light. The incident energy appears in small packets, and there is a one-to-one interaction between photons and electrons. If the incident light has very low intensity, there are very few photons arriving per unit time interval; each photon, however, can have sufficient energy to eject an electron immediately.

3. Dependence of ejection of electrons on light frequency

    Because the photon must have energy greater than the work function ϕ to eject an electron, the photoelectric effect cannot be observed below a certain cutoff frequency. If the energy of an incoming photon does not satisfy this requirement, an electron cannot be ejected from the surface, regardless of light intensity.

TABLE 40.1

Work Functions of Selected Metals

Metal	ϕ (eV)
Na	2.46
Al	4.08
Cu	4.70
Zn	4.31
Ag	4.73
Pt	6.35
Pb	4.14
Fe	4.50

Note: Values are typical for metals listed. Actual values may vary depending on whether the metal is a single crystal or polycrystalline. Values may also depend on the face from which electrons are ejected from crystalline metals. Furthermore, different experimental procedures may produce differing values.

---

[4] In principle, two photons could combine to provide an electron with their combined energy. That is highly improbable, however, without the high intensity of radiation available from very strong lasers.

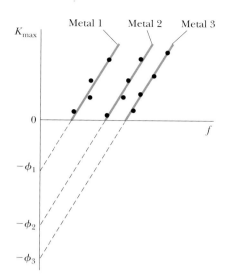

ACTIVE FIGURE 40.11

A plot of K_{max} for photoelectrons versus frequency of incident light in a typical photoelectric effect experiment. Photons with frequency less than the cutoff frequency for a given metal do not have sufficient energy to eject an electron from the metal.

Sign in at www.thomsonedu.com and go to ThomsonNOW to sweep through the frequency range and observe the curve for different target metals.

4. Dependence of photoelectron kinetic energy on light frequency

A photon of higher frequency carries more energy and therefore ejects a higher-energy photoelectron than does a photon of lower frequency.

Einstein's model predicts a linear relationship (Eq. 40.9) between the maximum electron kinetic energy K_{max} and the light frequency f. Experimental observation of a linear relationship between K_{max} and f would be a final confirmation of Einstein's theory. Indeed, such a linear relationship is observed as sketched in Active Figure 40.11, and the slope of the lines in such a plot is Planck's constant h. The intercept on the horizontal axis gives the cutoff frequency below which no photoelectrons are emitted. The cutoff frequency is related to the work function through the relationship $f_c = \phi/h$. The cutoff frequency corresponds to a **cutoff wavelength** λ_c, where

$$\lambda_c = \frac{c}{f_c} = \frac{c}{\phi/h} = \frac{hc}{\phi} \qquad (40.10)$$

◀ Cutoff wavelength

and c is the speed of light. Wavelengths greater than λ_c incident on a material having a work function ϕ do not result in the emission of photoelectrons.

The combination hc in Equation 40.10 often occurs when relating a photon's energy to its wavelength. A common shortcut when solving problems is to express this combination in useful units according to the following approximation:

$$hc = 1\,240 \text{ eV} \cdot \text{nm}$$

One of the first practical uses of the photoelectric effect was as the detector in a camera's light meter. Light reflected from the object to be photographed strikes a photoelectric surface in the meter, causing it to emit photoelectrons that then pass through a sensitive ammeter. The magnitude of the current in the ammeter depends on the light intensity.

The phototube, another early application of the photoelectric effect, acts much like a switch in an electric circuit. It produces a current in the circuit when light of sufficiently high frequency falls on a metal plate in the phototube, but produces no current in the dark. Phototubes were used in burglar alarms and in the detection of the soundtrack on motion picture film. Modern semiconductor devices have now replaced older devices based on the photoelectric effect.

Today, the photoelectric effect is used in the operation of photomultiplier tubes. Figure 40.12 shows the structure of such a device. A photon striking the photocathode ejects an electron by means of the photoelectric effect. This electron accelerates across the potential difference between the photocathode and the first *dynode*, shown as being at +200 V relative to the photocathode in Figure

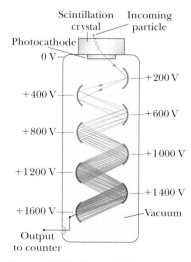

Figure 40.12 The multiplication of electrons in a photomultiplier tube.

40.12. This high-energy electron strikes the dynode and ejects several more electrons. The same process is repeated through a series of dynodes at ever higher potentials until an electrical pulse is produced as millions of electrons strike the last dynode. The tube is therefore called a *multiplier*: one photon at the input has resulted in millions of electrons at the output.

The photomultiplier tube is used in nuclear detectors to detect photons produced by the interaction of energetic charged particles or gamma rays with certain materials. It is also used in astronomy in a technique called *photoelectric photometry*. In that technique, the light collected by a telescope from a single star is allowed to fall on a photomultiplier tube for a time interval. The tube measures the total energy transferred by light during the time interval, which can then be converted to a luminosity of the star.

The photomultiplier tube is being replaced in many astronomical observations with a *charge-coupled device* (CCD), which is the same device used in a digital camera (Section 36.6). In a CCD, an array of pixels is formed on the silicon surface of an integrated circuit (Section 43.7). When the surface is exposed to light from an astronomical scene through a telescope or a terrestrial scene through a digital camera, electrons generated by the photoelectric effect are caught in "traps" beneath the surface. The number of electrons is related to the intensity of the light striking the surface. A signal processor measures the number of electrons associated with each pixel and converts this information into a digital code that a computer can use to reconstruct and display the scene.

The *electron bombardment CCD camera* allows higher sensitivity than a conventional CCD. In this device, electrons ejected from a photocathode by the photoelectric effect are accelerated through a high voltage before striking a CCD array. The higher energy of the electrons results in a very sensitive detector of low-intensity radiation.

Quick Quiz 40.3 Consider one of the curves in Active Figure 40.10. Suppose the intensity of the incident light is held fixed but its frequency is increased. Does the stopping potential in Active Figure 40.10 (a) remain fixed, (b) move to the right, or (c) move to the left?

Quick Quiz 40.4 Suppose classical physicists had the idea of plotting K_{max} versus f as in Active Figure 40.11. Draw a graph of what the expected plot would look like, based on the wave model for light.

EXAMPLE 40.3	The Photoelectric Effect for Sodium

A sodium surface is illuminated with light having a wavelength of 300 nm. The work function for sodium metal is 2.46 eV.

(A) Find the maximum kinetic energy of the ejected photoelectrons.

SOLUTION

Conceptualize Imagine a photon striking the metal surface and ejecting an electron. The electron with the maximum energy is one near the surface that experiences no interactions with other particles in the metal that would reduce its energy on its way out of the metal.

Categorize We evaluate the results using equations developed in this section, so we categorize this example as a substitution problem.

Find the energy of each photon in the illuminating light beam from Equation 40.5:

$$E = hf = \frac{hc}{\lambda} = \frac{1\,240 \text{ eV} \cdot \text{nm}}{300 \text{ nm}} = 4.13 \text{ eV}$$

From Equation 40.9, find the maximum kinetic energy of an electron:

$$K_{max} = hf - \phi = 4.13 \text{ eV} - 2.46 \text{ V} = 1.67 \text{ eV}$$

(B) Find the cutoff wavelength λ_c for sodium.

SOLUTION

Calculate λ_c using Equation 40.10:

$$\lambda_c = \frac{hc}{\phi} = \frac{1\,240 \text{ eV} \cdot \text{nm}}{2.46 \text{ eV}} = \boxed{504 \text{ nm}}$$

40.3   The Compton Effect

In 1919, Einstein concluded that a photon of energy E travels in a single direction and carries a momentum equal to $E/c = hf/c$. In 1923, Arthur Holly Compton (1892–1962) and Peter Debye (1884–1966) independently carried Einstein's idea of photon momentum further.

Prior to 1922, Compton and his coworkers had accumulated evidence showing that the classical wave theory of light failed to explain the scattering of x-rays from electrons. According to classical theory, electromagnetic waves of frequency f_0 incident on electrons should have two effects: (1) radiation pressure (see Section 34.5) should cause the electrons to accelerate in the direction of propagation of the waves, and (2) the oscillating electric field of the incident radiation should set the electrons into oscillation at the apparent frequency f', where f' is the frequency in the frame of the moving electrons. This apparent frequency is different from the frequency f_0 of the incident radiation because of the Doppler effect (see Section 17.4). Each electron first absorbs radiation as a moving particle and then reradiates as a moving particle, thereby exhibiting two Doppler shifts in the frequency of radiation.

Because different electrons move at different speeds after the interaction, depending on the amount of energy absorbed from the electromagnetic waves, the scattered wave frequency at a given angle to the incoming radiation should show a distribution of Doppler-shifted values. Contrary to this prediction, Compton's experiments showed that at a given angle only *one* frequency of radiation is observed. Compton and his coworkers explained these experiments by treating photons not as waves but rather as point-like particles having energy hf and momentum hf/c and by assuming the energy and momentum of the isolated system of the colliding photon–electron pair are conserved. Compton adopted a particle model for something that was well known as a wave, and today this scattering phenomenon is known as the **Compton effect.** Figure 40.13 shows the quantum picture of the collision between an individual x-ray photon and an electron. In the quantum model, the electron is scattered through an angle ϕ with respect to this direction as in a billiard-ball type of collision. (The symbol ϕ used here is an angle and is not to be confused with the work function, which was discussed in the preceding section.)

Figure 40.14 is a schematic diagram of the apparatus used by Compton. The x-rays, scattered from a graphite target, were analyzed with a rotating crystal spectrometer, and the intensity was measured with an ionization chamber that generated

ARTHUR HOLLY COMPTON
American Physicist (1892–1962)
Compton was born in Wooster, Ohio, and attended Wooster College and Princeton University. He became the director of the laboratory at the University of Chicago, where experimental work concerned with sustained nuclear chain reactions was conducted. This work was of central importance to the construction of the first nuclear weapon. His discovery of the Compton effect led to his sharing of the 1927 Nobel Prize in Physics with Charles Wilson.

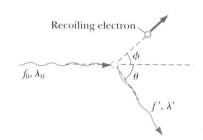

Figure 40.13 The quantum model for x-ray scattering from an electron. The collision of the photon with the electron displays the particle-like nature of the photon.

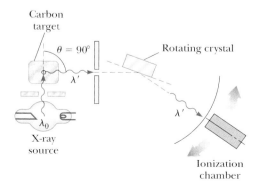

Figure 40.14 Schematic diagram of Compton's apparatus. The wavelength was measured with a rotating crystal spectrometer using graphite (carbon) as the target.

a current proportional to the intensity. The incident beam consisted of monochromatic x-rays of wavelength $\lambda_0 = 0.071$ nm. The experimental intensity-versus-wavelength plots observed by Compton for four scattering angles (corresponding to θ in Fig. 40.13) are shown in Figure 40.15. The graphs for the three nonzero angles show two peaks, one at λ_0 and one at $\lambda' > \lambda_0$. The shifted peak at λ' is caused by the scattering of x-rays from free electrons, which was predicted by Compton to depend on scattering angle as

Compton shift equation ▶

$$\lambda' - \lambda_0 = \frac{h}{m_e c} (1 - \cos \theta) \qquad (40.11)$$

where m_e is the mass of the electron. This expression is known as the **Compton shift equation.** The factor $h/m_e c$, called the **Compton wavelength** of the electron, has a currently accepted value of

Compton wavelength ▶

$$\lambda_C = \frac{h}{m_e c} = 0.002\ 43 \text{ nm}$$

The unshifted peak at λ_0 in Figure 40.15 is caused by x-rays scattered from electrons tightly bound to the target atoms. This unshifted peak also is predicted by Equation 40.11 if the electron mass is replaced with the mass of a carbon atom, which is approximately 23 000 times the mass of the electron. Therefore, there is a wavelength shift for scattering from an electron bound to an atom, but it is so small that it was undetectable in Compton's experiment.

Compton's measurements were in excellent agreement with the predictions of Equation 40.11. These results were the first to convince many physicists of the fundamental validity of quantum theory.

Quick Quiz 40.5 For any given scattering angle θ, Equation 40.11 gives the same value for the Compton shift for any wavelength. Keeping that in mind, for which of the following types of radiation is the fractional shift in wavelength at a given scattering angle the largest? (a) radio waves (b) microwaves (c) visible light (d) x-rays

Derivation of the Compton Shift Equation

We can derive the Compton shift equation by assuming the photon behaves like a particle and collides elastically with a free electron initially at rest as shown in Figure 40.13. The photon is treated as a particle having energy $E = hf = hc/\lambda$ and zero rest energy. We apply the isolated system model to the photon and the electron. In the scattering process, the total energy and total linear momentum of the system are conserved. Applying the principle of conservation of energy to this process gives

$$\frac{hc}{\lambda_0} = \frac{hc}{\lambda'} + K_e$$

where hc/λ_0 is the energy of the incident photon, hc/λ' is the energy of the scattered photon, and K_e is the kinetic energy of the recoiling electron. Because the electron may recoil at a speed comparable to that of light, we must use the relativistic expression $K_e = (\gamma - 1) m_e c^2$ (Eq. 39.23). Therefore,

$$\frac{hc}{\lambda_0} = \frac{hc}{\lambda'} + (\gamma - 1) m_e c^2 \qquad (40.12)$$

where $\gamma = 1/\sqrt{1 - (u^2/c^2)}$ and u is the speed of the electron.

Next, let's apply the law of conservation of momentum to this collision, noting that the x and y components of momentum are each conserved independently. Equation 39.28 shows that the momentum of a photon has a magnitude $p = E/c$, and we know from Equation 40.5 that $E = hf$. Therefore, $p = hf/c$. Substituting λf for c (Eq. 16.12) in this expression gives $p = h/\lambda$. Because the relativistic expres-

$\theta = 0°$
Primary beam

$\theta = 45°$

$\theta = 90°$

$\theta = 135°$

Figure 40.15 Scattered x-ray intensity versus wavelength for Compton scattering at $\theta = 0°$, $45°$, $90°$, and $135°$.

sion for the momentum of the recoiling electron is $p_e = \gamma m_e u$ (Eq. 39.19), we obtain the following expressions for the x and y components of linear momentum, where the angles are as described in Figure 40.13:

$$x \text{ component: } \quad \frac{h}{\lambda_0} = \frac{h}{\lambda'} \cos\theta + \gamma m_e u \cos\phi \qquad \textbf{(40.13)}$$

$$y \text{ component: } \quad 0 = \frac{h}{\lambda'} \sin\theta - \gamma m_e u \sin\phi \qquad \textbf{(40.14)}$$

Eliminating u and ϕ from Equations 40.12 through 40.14 gives a single expression that relates the remaining three variables (λ', λ_0, and θ). After some algebra (see Problem 59), we obtain Equation 40.11.

EXAMPLE 40.4   **Compton Scattering at 45°**

X-rays of wavelength $\lambda_0 = 0.200\,000$ nm are scattered from a block of material. The scattered x-rays are observed at an angle of 45.0° to the incident beam. Calculate their wavelength.

SOLUTION

Conceptualize   Imagine the process in Figure 40.13, with the photon scattered at 45° to its original direction.

Categorize   We evaluate the result using an equation developed in this section, so we categorize this example as a substitution problem.

Solve Equation 40.11 for the wavelength of the scattered x-ray:

$$(1) \quad \lambda' = \lambda_0 + \frac{h(1 - \cos\theta)}{m_e c}$$

Substitute numerical values:

$$\lambda' = 0.200\,000 \times 10^{-9} \text{ m} + \frac{(6.626 \times 10^{-34}\,\text{J}\cdot\text{s})(1 - \cos 45.0°)}{(9.11 \times 10^{-31}\,\text{kg})(3.00 \times 10^8\,\text{m/s})}$$

$$= 0.200\,000 \times 10^{-9} \text{ m} + 7.10 \times 10^{-13} \text{ m} = \boxed{0.200\,710 \text{ nm}}$$

What If?   What if the detector is moved so that scattered x-rays are detected at an angle larger than 45°? Does the wavelength of the scattered x-rays increase or decrease as the angle θ increases?

Answer   In Equation (1), if the angle θ increases, $\cos\theta$ decreases. Consequently, the factor $(1 - \cos\theta)$ increases. Therefore, the scattered wavelength increases.

We could also apply an energy argument to achieve this same result. As the scattering angle increases, more energy is transferred from the incident photon to the electron. As a result, the energy of the scattered photon decreases with increasing scattering angle. Because $E = hf$, the frequency of the scattered photon decreases, and because $\lambda = c/f$, the wavelength increases.

40.4   Photons and Electromagnetic Waves

Phenomena such as the photoelectric effect and the Compton effect offer iron-clad evidence that when light (or other forms of electromagnetic radiation) and matter interact, the light behaves as if it were composed of particles having energy hf and momentum h/λ. How can light be considered a photon (in other words, a particle) when we know it is a wave? On the one hand, we describe light in terms of photons having energy and momentum. On the other hand, light and other electromagnetic waves exhibit interference and diffraction effects, which are consistent only with a wave interpretation.

Which model is correct? Is light a wave or a particle? The answer depends on the phenomenon being observed. Some experiments can be explained either better or solely with the photon model, whereas others are explained either better or solely with the wave model. The end result is that **we must accept both models and admit that the true nature of light is not describable in terms of any single classical picture.** The same light beam that can eject photoelectrons from a metal (meaning that the beam consists of photons) can also be diffracted by a grating (meaning that the beam is a wave). In other words, **the particle model and the wave model of light complement each other.**

The success of the particle model of light in explaining the photoelectric effect and the Compton effect raises many other questions. If light is a particle, what is the meaning of the "frequency" and "wavelength" of the particle, and which of these two properties determines its energy and momentum? Is light *simultaneously* a wave and a particle? Although photons have no rest energy (a nonobservable quantity because a photon cannot be at rest), is there a simple expression for the *effective mass* of a moving photon? If photons have effective mass, do they experience gravitational attraction? What is the spatial extent of a photon, and how does an electron absorb or scatter one photon? Some of these questions can be answered, but others demand a view of atomic processes that is too pictorial and literal. Many of them stem from classical analogies such as colliding billiard balls and ocean waves breaking on a seashore. Quantum mechanics gives light a more flexible nature by treating the particle model and the wave model of light as both necessary and complementary. Neither model can be used exclusively to describe all properties of light. A complete understanding of the observed behavior of light can be attained only if the two models are combined in a complementary manner.

40.5 The Wave Properties of Particles

LOUIS DE BROGLIE
French Physicist (1892–1987)
De Broglie was born in Dieppe, France. At the Sorbonne in Paris, he studied history in preparation for what he hoped would be a career in the diplomatic service. The world of science is lucky he changed his career path to become a theoretical physicist. De Broglie was awarded the Nobel Prize in Physics in 1929 for his prediction of the wave nature of electrons.

Students introduced to the dual nature of light often find the concept difficult to accept. In the world around us, we are accustomed to regarding such things as baseballs solely as particles and other things such as sound waves solely as forms of wave motion. Every large-scale observation can be interpreted by considering either a wave explanation or a particle explanation, but in the world of photons and electrons, such distinctions are not as sharply drawn. Even more disconcerting is that, under certain conditions, the things we unambiguously call "particles" exhibit wave characteristics.

In his 1923 doctoral dissertation, Louis de Broglie postulated that **because photons have both wave and particle characteristics, perhaps all forms of matter have both properties.** This highly revolutionary idea had no experimental confirmation at the time. According to de Broglie, electrons, just like light, have a dual particle–wave nature.

In Section 40.3, we found that the momentum of a photon can be expressed as

$$p = \frac{h}{\lambda}$$

This equation shows that the photon wavelength can be specified by its momentum: $\lambda = h/p$. De Broglie suggested that material particles of momentum p have a characteristic wavelength that is given by the same expression, $\lambda = h/p$. Because the magnitude of the momentum of a particle of mass m and speed u is $p = mu$, the **de Broglie wavelength** of that particle is[5]

---

[5] The de Broglie wavelength for a particle moving at *any* speed u is $\lambda = h/\gamma mu$, where $\gamma = [1 - (u^2/c^2)]^{-1/2}$.

$$\lambda = \frac{h}{p} = \frac{h}{mu} \qquad\qquad (40.15) \qquad\qquad \blacktriangleleft \quad \text{de Broglie wavelength}$$

Furthermore, in analogy with photons, de Broglie postulated that particles obey the Einstein relation $E = hf$, where E is the total energy of the particle. The frequency of a particle is then

$$f = \frac{E}{h} \qquad\qquad (40.16)$$

The dual nature of matter is apparent in Equations 40.15 and 40.16 because each contains both particle quantities (p and E) and wave quantities (λ and f).

The Davisson–Germer Experiment

De Broglie's 1923 proposal that matter exhibits both wave and particle properties was regarded as pure speculation. If particles such as electrons had wave properties, under the correct conditions they should exhibit diffraction effects. Only three years later, C. J. Davisson (1881–1958) and L. H. Germer (1896–1971) succeeded in measuring the wavelength of electrons. Their important discovery provided the first experimental confirmation of the matter waves proposed by de Broglie.

Interestingly, the intent of the initial Davisson–Germer experiment was not to confirm the de Broglie hypothesis. In fact, their discovery was made by accident (as is often the case). The experiment involved the scattering of low-energy electrons (approximately 54 eV) from a nickel target in a vacuum. During one experiment, the nickel surface was badly oxidized because of an accidental break in the vacuum system. After the target was heated in a flowing stream of hydrogen to remove the oxide coating, electrons scattered by it exhibited intensity maxima and minima at specific angles. The experimenters finally realized that the nickel had formed large crystalline regions upon heating and that the regularly spaced planes of atoms in these regions served as a diffraction grating for electrons. (See the discussion of diffraction of x-rays by crystals in Section 38.5.)

Shortly thereafter, Davisson and Germer performed more extensive diffraction measurements on electrons scattered from single-crystal targets. Their results showed conclusively the wave nature of electrons and confirmed the de Broglie relationship $p = h/\lambda$. In the same year, G. P. Thomson (1892–1975) of Scotland also observed electron diffraction patterns by passing electrons through very thin gold foils. Diffraction patterns have since been observed in the scattering of helium atoms, hydrogen atoms, and neutrons. Hence, the wave nature of particles has been established in various ways.

The problem of understanding the dual nature of matter and radiation is conceptually difficult because the two models seem to contradict each other. This problem as it applies to light was discussed earlier. The **principle of complementarity** states that the **wave and particle models of either matter or radiation complement each other.** Neither model can be used exclusively to describe matter or radiation adequately. Because humans tend to generate mental images based on their experiences from the everyday world (baseballs, water waves, and so forth), we use both descriptions in a complementary manner to explain any given set of data from the quantum world.

PITFALL PREVENTION 40.3
What's Waving?

If particles have wave properties, what's waving? You are familiar with waves on strings, which are very concrete. Sound waves are more abstract, but you are likely comfortable with them. Electromagnetic waves are even more abstract, but at least they can be described in terms of physical variables and electric and magnetic fields. In contrast, waves associated with particles are completely abstract and cannot be associated with a physical variable. In Chapter 41, we describe the wave associated with a particle in terms of probability.

Quick Quiz 40.6 An electron and a proton both moving at nonrelativistic speeds have the same de Broglie wavelength. Which of the following quantities are also the same for the two particles? (a) speed   (b) kinetic energy   (c) momentum   (d) frequency

EXAMPLE 40.5 | **Wavelengths for Microscopic and Macroscopic Objects**

(A) Calculate the de Broglie wavelength for an electron ($m_e = 9.11 \times 10^{-31}$ kg) moving at 1.00×10^7 m/s.

SOLUTION

Conceptualize Imagine the electron moving through space. From a classical viewpoint, it is a particle under constant velocity. From the quantum viewpoint, the electron has a wavelength associated with it.

Categorize We evaluate the result using an equation developed in this section, so we categorize this example as a substitution problem.

Evaluate the wavelength using Equation 40.15:

$$\lambda = \frac{h}{m_e u} = \frac{6.63 \times 10^{-34} \, \text{J} \cdot \text{s}}{(9.11 \times 10^{-31} \, \text{kg})(1.00 \times 10^7 \, \text{m/s})} = \boxed{7.28 \times 10^{-11} \, \text{m}}$$

(B) A rock of mass 50 g is thrown with a speed of 40 m/s. What is its de Broglie wavelength?

SOLUTION

Evaluate the de Broglie wavelength using Equation 40.15:

$$\lambda = \frac{h}{mu} = \frac{6.63 \times 10^{-34} \, \text{J} \cdot \text{s}}{(50 \times 10^{-3} \, \text{kg})(40 \, \text{m/s})} = \boxed{3.32 \times 10^{-34} \, \text{m}}$$

This wavelength is much smaller than any aperture through which the rock could possibly pass. Hence, we could not observe diffraction effects, and as a result, the wave properties of large-scale objects cannot be observed.

The Electron Microscope

A practical device that relies on the wave characteristics of electrons is the **electron microscope.** A *transmission* electron microscope, used for viewing flat, thin samples, is shown in Figure 40.16. In many respects, it is similar to an optical micro-

Electron gun
Cathode
Anode
Electromagnetic lens
Electromagnetic condenser lens
Screen
Visual transmission

Vacuum
Core
Coil
Electron beam
Specimen goes here
Specimen chamber door
Projector lens
Photo chamber

(a)

© David Parker/Photo Researchers, Inc.

(b)

Figure 40.16 (a) Diagram of a transmission electron microscope for viewing a thinly sectioned sample. The "lenses" that control the electron beam are magnetic deflection coils. (b) An electron microscope.

scope; the electron microscope, however, has a much greater resolving power because it can accelerate electrons to very high kinetic energies, giving them very short wavelengths. No microscope can resolve details that are significantly smaller than the wavelength of the waves used to illuminate the object. Typically, the wavelengths of electrons are approximately 100 times shorter than those of the visible light used in optical microscopes. As a result, an electron microscope with ideal lenses would be able to distinguish details approximately 100 times smaller than those distinguished by an optical microscope. (Electromagnetic radiation of the same wavelength as the electrons in an electron microscope is in the x-ray region of the spectrum.)

The electron beam in an electron microscope is controlled by electrostatic or magnetic deflection, which acts on the electrons to focus the beam and form an image. Rather than examining the image through an eyepiece as in an optical microscope, the viewer looks at an image formed on a monitor or other type of display screen. Figure 40.17 shows the amazing detail available with an electron microscope.

Figure 40.17 A color-enhanced electron microscope photograph shows significant detail of a storage mite, *Lepidoglyphus destructor*. The mite is so small, with a maximum length of 0.75 mm, that ordinary microscopes do not reveal minute anatomical details.

40.6 The Quantum Particle

Because in the past we considered the particle and wave models to be distinct, the discussions presented in previous sections may be quite disturbing. The notion that both light and material particles have both particle and wave properties does not fit with this distinction. Experimental evidence shows, however, that this conclusion is exactly what we must accept. The recognition of this dual nature leads to a new model, the **quantum particle,** which is a combination of the particle model introduced in Chapter 2 and the wave model discussed in Chapter 16. In this new model, entities have both particle and wave characteristics, and we must choose one appropriate behavior—particle or wave—to understand a particular phenomenon.

In this section, we shall explore this model in a way that might make you more comfortable with this idea. We shall do so by demonstrating that an entity that exhibits properties of a particle can be constructed from waves.

Let's first recall some characteristics of ideal particles and ideal waves. An ideal particle has zero size. Therefore, an essential feature of a particle is that it is *localized* in space. An ideal wave has a single frequency and is infinitely long as suggested by Figure 40.18a. Therefore, an ideal wave is *unlocalized* in space. A localized entity can be built from infinitely long waves as follows. Imagine drawing one wave along the x axis, with one of its crests located at x = 0, as at the top of Figure 40.18b. Now draw a second wave, of the same amplitude but a different frequency, with one of its crests also at x = 0. As a result of the superposition of these two waves, *beats* exist as the waves are alternately in phase and out of phase. (Beats were discussed in Section 18.7.) The bottom curve in Figure 40.18b shows the results of superposing these two waves.

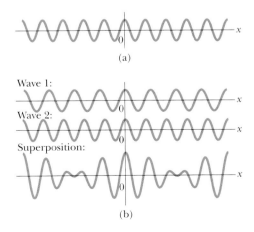

(a)

Wave 1:

Wave 2:

Superposition:

(b)

Figure 40.18 (a) An idealized wave of an exact single frequency is the same throughout space and time. (b) If two ideal waves with slightly different frequencies are combined, beats result (Section 18.7). The regions of space at which there is constructive interference are different from those at which there is destructive interference.

ACTIVE FIGURE 40.19

If a large number of waves are combined, the result is a wave packet, which represents a particle.

Sign in at www.thomsonedu.com and go to ThomsonNOW to choose the number of waves to add together and observe the resulting wave packet.

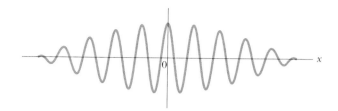

Notice that we have already introduced some localization by superposing the two waves. A single wave has the same amplitude everywhere in space; no point in space is any different from any other point. By adding a second wave, however, there is something different about the in-phase points compared with the out-of-phase points.

Now imagine that more and more waves are added to our original two, each new wave having a new frequency. Each new wave is added so that one of its crests is at $x = 0$ with the result that all the waves add constructively at $x = 0$. When we add a large number of waves, the probability of a positive value of a wave function at any point $x \neq 0$ is equal to the probability of a negative value, and there is destructive interference *everywhere* except near $x = 0$, where all of the crests are superposed. The result is shown in Active Figure 40.19. The small region of constructive interference is called a **wave packet.** This localized region of space is different from all other regions. We can identify the wave packet as a particle because it has the localized nature of a particle! The location of the wave packet corresponds to the particle's position.

The localized nature of this entity is the *only* characteristic of a particle that was generated with this process. We have not addressed how the wave packet might achieve such particle characteristics as mass, electric charge, and spin. Therefore, you may not be completely convinced that we have built a particle. As further evidence that the wave packet can represent the particle, let's show that the wave packet has another characteristic of a particle.

To simplify the mathematical representation, we return to our combination of two waves. Consider two waves with equal amplitudes but different frequencies f_1 and f_2. We can represent the waves mathematically as

$$y_1 = A \cos (k_1 x - \omega_1 t) \quad \text{and} \quad y_2 = A \cos (k_2 x - \omega_2 t)$$

where, as in Chapter 16, $k = 2\pi/\lambda$ and $\omega = 2\pi f$. Using the superposition principle, let's add the waves:

$$y = y_1 + y_2 = A \cos (k_1 x - \omega_1 t) + A \cos (k_2 x - \omega_2 t)$$

It is convenient to write this expression in a form that uses the trigonometric identity

$$\cos a + \cos b = 2 \cos \left(\frac{a - b}{2} \right) \cos \left(\frac{a + b}{2} \right)$$

Letting $a = k_1 x - \omega_1 t$ and $b = k_2 x - \omega_2 t$ gives

$$y = 2A \cos \left[\frac{(k_1 x - \omega_1 t) - (k_2 x - \omega_2 t)}{2} \right] \cos \left[\frac{(k_1 x - \omega_1 t) + (k_2 x - \omega_2 t)}{2} \right]$$

$$y = \left[2A \cos \left(\frac{\Delta k}{2} x - \frac{\Delta \omega}{2} t \right) \right] \cos \left(\frac{k_1 + k_2}{2} x - \frac{\omega_1 + \omega_2}{2} t \right) \quad \textbf{(40.17)}$$

where $\Delta k = k_1 - k_2$ and $\Delta \omega = \omega_1 - \omega_2$. The second cosine factor represents a wave with a wave number and frequency that are equal to the averages of the values for the individual waves.

In Equation 40.17, the factor in square brackets represents the envelope of the wave as shown by the blue curve in Active Figure 40.20. This factor also has the mathematical form of a wave. **This envelope of the combination can travel through space with a different speed than the individual waves.** As an extreme example of

$$2A \cos\left(\frac{\Delta k}{2}x - \frac{\Delta\omega}{2}t\right)$$

ACTIVE FIGURE 40.20

The beat pattern of Figure 40.18b, with an envelope function (blue curve) superimposed.

Sign in at www.thomsonedu.com and go to ThomsonNOW to observe the movement of the waves and the envelope.

this possibility, imagine combining two identical waves moving in opposite directions. The two waves move with the same speed, but the envelope has a speed of *zero* because we have built a standing wave, which we studied in Section 18.2.

For an individual wave, the speed is given by Equation 16.11,

$$v_{\text{phase}} = \frac{\omega}{k} \tag{40.18}$$

◄ Phase speed of a wave in a wave packet

This speed is called the **phase speed** because it is the rate of advance of a crest on a single wave, which is a point of fixed phase. Equation 40.18 can be interpreted as follows: the phase speed of a wave is the ratio of the coefficient of the time variable t to the coefficient of the space variable x in the equation for the wave, $y = A \cos (kx - \omega t)$.

The factor in brackets in Equation 40.17 is of the form of a wave, so it moves with a speed given by this same ratio:

$$v_g = \frac{\text{coefficient of time variable } t}{\text{coefficient of space variable } x} = \frac{(\Delta\omega/2)}{(\Delta k/2)} = \frac{\Delta\omega}{\Delta k}$$

The subscript g on the speed indicates that it is commonly called the **group speed,** or the speed of the wave packet (the *group* of waves) we have built. We have generated this expression for a simple addition of two waves. When a large number of waves are superposed to form a wave packet, this ratio becomes a derivative:

$$v_g = \frac{d\omega}{dk} \tag{40.19}$$

◄ Group speed of a wave packet

Multiplying the numerator and the denominator by $\hbar$, where $\hbar = h/2\pi$, gives

$$v_g = \frac{\hbar \, d\omega}{\hbar \, dk} = \frac{d(\hbar\omega)}{d(\hbar k)} \tag{40.20}$$

Let's look at the terms in the parentheses of Equation 40.20 separately. For the numerator,

$$\hbar\omega = \frac{h}{2\pi}(2\pi f) = hf = E$$

For the denominator,

$$\hbar k = \frac{h}{2\pi}\left(\frac{2\pi}{\lambda}\right) = \frac{h}{\lambda} = p$$

Therefore, Equation 40.20 can be written as

$$v_g = \frac{d(\hbar\omega)}{d(\hbar k)} = \frac{dE}{dp} \tag{40.21}$$

Because we are exploring the possibility that the envelope of the combined waves represents the particle, consider a free particle moving with a speed u that is small compared with the speed of light. The energy of the particle is its kinetic energy:

$$E = \tfrac{1}{2}mu^2 = \frac{p^2}{2m}$$

Differentiating this equation with respect to p gives

$$v_g = \frac{dE}{dp} = \frac{d}{dp}\left(\frac{p^2}{2m}\right) = \frac{1}{2m}(2p) = u \qquad (40.22)$$

Therefore, the group speed of the wave packet is identical to the speed of the particle that it is modeled to represent, giving us further confidence that the wave packet is a reasonable way to build a particle.

Quick Quiz 40.7 As an analogy to wave packets, consider an "automobile packet" that occurs near the scene of an accident on a freeway. The phase speed is analogous to the speed of individual automobiles as they move through the backup caused by the accident. The group speed can be identified as the speed of the leading edge of the packet of cars. For the automobile packet, is the group speed (a) the same as the phase speed, (b) less than the phase speed, or (c) greater than the phase speed?

40.7 The Double-Slit Experiment Revisited

Wave–particle duality is now a firmly accepted concept reinforced by experimental results, including those of the Davisson–Germer experiment. As with the postulates of special relativity, however, this concept often leads to clashes with familiar thought patterns we hold from everyday experience.

One way to crystallize our ideas about the electron's wave–particle duality is through an experiment in which electrons are fired at a double slit. Consider a parallel beam of mono-energetic electrons incident on a double slit as in Figure 40.21. Let's assume the slit widths are small compared with the electron wavelength so that we need not worry about diffraction maxima and minima as discussed for light in Section 38.2. An electron detector screen is positioned far from the slits at a distance much greater than d, the separation distance of the slits. If the detector screen collects electrons for a long enough time, we find a typical wave interference pattern for the counts per minute, or probability of arrival of electrons. Such an interference pattern would not be expected if the electrons behaved as classical particles, giving clear evidence that electrons are interfering, a distinct wave-like behavior.

If we measure the angles θ at which the maximum intensity of electrons arrives at the detector screen in Figure 40.21, we find they are described by exactly the same equation (Eq. 37.2) as that for light, $d \sin \theta = m\lambda$, where m is the order number and λ is the electron wavelength. Therefore, the dual nature of the electron is

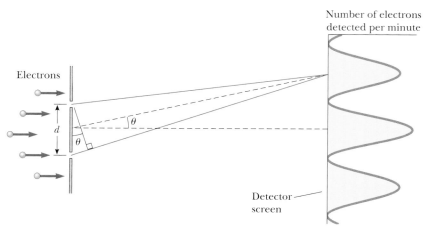

Figure 40.21 Electron interference. The slit separation d is much greater than the individual slit widths and much less than the distance between the slit and the detector screen.

clearly shown in this experiment: **the electrons are detected as particles at a localized spot on the detector screen at some instant of time, but the probability of arrival at that spot is determined by finding the intensity of two interfering waves.**

At extremely low electron beam intensities, one electron at a time arrives at the double slit. It is tempting to assume the electron goes through either slit 1 or slit 2. You might argue that there are no interference effects because there is not a second electron going through the other slit to interfere with the first. This assumption places too much emphasis on the particle model of the electron, however. The interference pattern is still observed if the time interval for the measurement is sufficiently long for many electrons to arrive at the detector screen! This situation is illustrated by the computer-simulated patterns in Active Figure 40.22 where the interference pattern becomes clearer as the number of electrons reaching the detector screen increases. Hence, our assumption that the electron is localized and goes through only one slit when both slits are open must be wrong (a painful conclusion!).

To interpret these results, we are forced to conclude that **an electron interacts with both slits simultaneously.** If you try to determine experimentally which slit the electron goes through, the act of measuring destroys the interference pattern. It is impossible to determine which slit the electron goes through. In effect, we can say only that **the electron passes through both slits!** The same arguments apply to photons.

If we restrict ourselves to a pure particle model, it is an uncomfortable notion that the electron can be present at both slits at once. From the quantum particle model, however, the particle can be considered to be built from waves that exist throughout space. Therefore, the wave components of the electron are present at both slits at the same time, and this model leads to a more comfortable interpretation of this experiment.

40.8 The Uncertainty Principle

Whenever one measures the position or velocity of a particle at any instant, experimental uncertainties are built into the measurements. According to classical mechanics, there is no fundamental barrier to an ultimate refinement of the apparatus or experimental procedures. In other words, it is possible, in principle, to make such measurements with arbitrarily small uncertainty. Quantum theory predicts, however, that **it is fundamentally impossible to make simultaneous measurements of a particle's position and momentum with infinite accuracy.**

In 1927, Werner Heisenberg (1901–1976) introduced this notion, which is now known as the **Heisenberg uncertainty principle:**

If a measurement of the position of a particle is made with uncertainty Δx and a simultaneous measurement of its x component of momentum is made with uncertainty Δp_x, the product of the two uncertainties can never be smaller than $\hbar/2$:

$$\Delta x \, \Delta p_x \geq \frac{\hbar}{2} \tag{40.23}$$

That is, **it is physically impossible to measure simultaneously the exact position and exact momentum of a particle.** Heisenberg was careful to point out that the inescapable uncertainties Δx and Δp_x do not arise from imperfections in practical measuring instruments. Rather, **the uncertainties arise from the quantum structure of matter.**

To understand the uncertainty principle, imagine that a particle has a single wavelength that is known *exactly*. According to the de Broglie relation, $\lambda = h/p$, we would therefore know the momentum to be precisely $p = h/\lambda$. In reality, a

(a) After 28 electrons

(b) After 1 000 electrons

(c) After 10 000 electrons

(d) Two-slit electron pattern

From C. Jönsson, *Zeitschrift für Physik* **161**:454, 1961; used with permission

ACTIVE FIGURE 40.22

(a)–(c) Computer-simulated interference patterns for a beam of electrons incident on a double slit. (d) Photograph of a double-slit interference pattern produced by electrons.

Sign in at www.thomsonedu.com and go to ThomsonNOW to watch the interference pattern develop over time and see how it is destroyed by the action of keeping track of which slit an electron goes through.

◄ Heisenberg uncertainty principle

WERNER HEISENBERG
German Theoretical Physicist (1901–1976)
Heisenberg obtained his Ph.D. in 1923 at the University of Munich. While other physicists tried to develop physical models of quantum phenomena, Heisenberg developed an abstract mathematical model called *matrix mechanics*. The more widely accepted physical models were shown to be equivalent to matrix mechanics. Heisenberg made many other significant contributions to physics, including his famous uncertainty principle for which he received a Nobel Prize in Physics in 1932, the prediction of two forms of molecular hydrogen, and theoretical models of the nucleus.

PITFALL PREVENTION 40.4
The Uncertainty Principle

Some students incorrectly interpret the uncertainty principle as meaning that a measurement interferes with the system. For example, if an electron is observed in a hypothetical experiment using an optical microscope, the photon used to see the electron collides with it and makes it move, giving it an uncertainty in momentum. This scenario does *not* represent the basis of the uncertainty principle. The uncertainty principle is independent of the measurement process and is based on the wave nature of matter.

single-wavelength wave would exist throughout space. Any region along this wave is the same as any other region (Fig. 40.18a). If we were to ask, Where is the particle this wave represents?, no special location in space along the wave could be identified with the particle; all points along the wave are the same. Therefore, we have *infinite* uncertainty in the position of the particle, and we know nothing about its location. Perfect knowledge of the particle's momentum has cost us all information about its location.

In comparison, now consider a particle whose momentum is uncertain so that it has a range of possible values of momentum. According to the de Broglie relation, the result is a range of wavelengths. Therefore, the particle is not represented by a single wavelength, but rather by a combination of wavelengths within this range. This combination forms a wave packet as we discussed in Section 40.6 and illustrated in Active Figure 40.19. If you were asked to determine the location of the particle, you could only say that it is somewhere in the region defined by the wave packet because there is a distinct difference between this region and the rest of space. Therefore, by losing some information about the momentum of the particle, we have gained information about its position.

If you were to lose *all* information about the momentum, you would be adding together waves of all possible wavelengths, resulting in a wave packet of zero length. Therefore, if you know nothing about the momentum, you know exactly where the particle is.

The mathematical form of the uncertainty principle states that the product of the uncertainties in position and momentum is always larger than some minimum value. This value can be calculated from the types of arguments discussed above, which result in the value of $\hbar/2$ in Equation 40.23.

Another form of the uncertainty principle can be generated by reconsidering Active Figure 40.19. Imagine that the horizontal axis is time rather than spatial position x. We can then make the same arguments that were made about knowledge of wavelength and position in the time domain. The corresponding variables would be frequency and time. Because frequency is related to the energy of the particle by $E = hf$, the uncertainty principle in this form is

$$\Delta E \, \Delta t \geq \frac{\hbar}{2} \tag{40.24}$$

The form of the uncertainty principle given in Equation 40.24 suggests that energy conservation can appear to be violated by an amount ΔE as long as it is only for a short time interval Δt consistent with that equation. We shall use this notion to estimate the rest energies of particles in Chapter 46.

Quick Quiz 40.8 A particle's location is measured and specified as being exactly at $x = 0$, with *zero* uncertainty in the x direction. How does that location affect the uncertainty of its velocity component in the y direction? (a) It does not affect it. (b) It makes it infinite. (c) It makes it zero.

EXAMPLE 40.6 **Locating an Electron**

The speed of an electron is measured to be 5.00×10^3 m/s to an accuracy of 0.003 00%. Find the minimum uncertainty in determining the position of this electron.

SOLUTION

Conceptualize The fractional value given for the accuracy of the electron's speed can be interpreted as the fractional uncertainty in its momentum. This uncertainty corresponds to a minimum uncertainty in the electron's position through the uncertainty principle.

Categorize We evaluate the result using concepts developed in this section, so we categorize this example as a substitution problem.

Assume the electron is moving along the x axis and find the x component of its momentum:

$$p_x = mu_x = (9.11 \times 10^{-31} \text{ kg})(5.00 \times 10^3 \text{ m/s}) = 4.56 \times 10^{-27} \text{ kg} \cdot \text{m/s}$$

Find the uncertainty in p_x as 0.003 00% of this value:

$$\Delta p_x = (0.000\ 030\ 0)(4.56 \times 10^{-27} \text{ kg} \cdot \text{m/s}) = 1.37 \times 10^{-31} \text{ kg} \cdot \text{m/s}$$

Solve Equation 40.23 for the uncertainty in the electron's position:

$$\Delta x \geq \frac{\hbar}{2\Delta p_x} = \frac{1.055 \times 10^{-34} \text{ J} \cdot \text{s}}{2(1.37 \times 10^{-31} \text{ kg} \cdot \text{m/s})} = 0.386 \text{ mm}$$

EXAMPLE 40.7   **The Line Width of Atomic Emissions**

Atoms have quantized energy levels similar to those of Planck's oscillators, although the energy levels of an atom are usually not evenly spaced. When an atom makes a transition between states, energy is emitted in the form of a photon. Although an excited atom can radiate at any time from $t = 0$ to $t = \infty$, the average time interval after excitation during which an atom radiates is called the **lifetime** τ. If $\tau = 1.0 \times 10^{-8}$ s, use the uncertainty principle to compute the line width Δf produced by this finite lifetime.

SOLUTION

Conceptualize   The lifetime τ given for the excited state can be interpreted as the uncertainty Δt in the time at which the transition occurs. This uncertainty corresponds to a minimum uncertainty in the frequency of the radiated photon through the uncertainty principle.

Categorize   We evaluate the result using concepts developed in this section, so we categorize this example as a substitution problem.

Use Equation 40.5 to relate the uncertainty in the photon's frequency to the uncertainty in its energy:

$$E = hf \quad \rightarrow \quad \Delta E = h\Delta f \quad \rightarrow \quad \Delta f = \frac{\Delta E}{h}$$

Use Equation 40.24 to substitute for the uncertainty in the photon's energy, giving the minimum value of Δf:

$$\Delta f \geq \frac{1}{h}\frac{\hbar}{2\Delta t} = \frac{1}{h}\frac{(h/2\pi)}{2\Delta t} = \frac{1}{4\pi\Delta t} = \frac{1}{4\pi\tau}$$

Substitute for the lifetime of the excited state:

$$\Delta f \geq \frac{1}{4\pi(1.0 \times 10^{-8} \text{ s})} = 8.0 \times 10^6 \text{ Hz}$$

What If?   What if this same lifetime were associated with a transition that emits a radio wave rather than a visible light wave from an atom? Is the fractional line width $\Delta f/f$ larger or smaller than for the visible light?

Answer   Because we are assuming the same lifetime for both transitions, Δf is independent of the frequency of radiation. Radio waves have lower frequencies than light waves, so the ratio $\Delta f/f$ will be larger for the radio waves. Assuming a light-wave frequency f of 6.00×10^{14} Hz, the fractional line width is

$$\frac{\Delta f}{f} = \frac{8.0 \times 10^6 \text{ Hz}}{6.00 \times 10^{14} \text{ Hz}} = 1.3 \times 10^{-8}$$

This narrow fractional line width can be measured with a sensitive interferometer. Usually, however, temperature and pressure effects overshadow the natural line width and broaden the line through mechanisms associated with the Doppler effect and collisions.

Assuming a radio-wave frequency f of 94.7×10^6 Hz, the fractional line width is

$$\frac{\Delta f}{f} = \frac{8.0 \times 10^6 \text{ Hz}}{94.7 \times 10^6 \text{ Hz}} = 8.4 \times 10^{-2}$$

Therefore, for the radio wave, this same absolute line width corresponds to a fractional line width of more than 8%.

Summary

CONCEPTS AND PRINCIPLES

The characteristics of **blackbody radiation** cannot be explained using classical concepts. Planck introduced the quantum concept and Planck's constant h when he assumed atomic oscillators existing only in discrete energy states were responsible for this radiation. In Planck's model, radiation is emitted in single quantized packets whenever an oscillator makes a transition between discrete energy states. The energy of a packet is

$$E = hf \qquad (40.5)$$

where f is the frequency of the oscillator. Einstein successfully extended Planck's quantum hypothesis to the standing waves of electromagnetic radiation in a cavity used in the blackbody radiation model.

The **photoelectric effect** is a process whereby electrons are ejected from a metal surface when light is incident on that surface. In Einstein's model, light is viewed as a stream of particles, or **photons,** each having energy $E = hf$, where h is Planck's constant and f is the frequency. The maximum kinetic energy of the ejected photoelectron is

$$K_{max} = hf - \phi \qquad (40.9)$$

where ϕ is the **work function** of the metal.

X-rays are scattered at various angles by electrons in a target. In such a scattering event, a shift in wavelength is observed for the scattered x-rays, a phenomenon known as the **Compton effect.** Classical physics does not predict the correct behavior in this effect. If the x-ray is treated as a photon, conservation of energy and linear momentum applied to the photon–electron collisions yields, for the Compton shift,

$$\lambda' - \lambda_0 = \frac{h}{m_e c}(1 - \cos\theta) \qquad (40.11)$$

where m_e is the mass of the electron, c is the speed of light, and θ is the scattering angle.

Light has a dual nature in that it has both wave and particle characteristics. Some experiments can be explained either better or solely by the particle model, whereas others can be explained either better or solely by the wave model.

Every object of mass m and momentum $p = mu$ has wave properties, with a **de Broglie wavelength** given by

$$\lambda = \frac{h}{p} = \frac{h}{mu} \qquad (40.15)$$

By combining a large number of waves, a single region of constructive interference, called a **wave packet,** can be created. The wave packet carries the characteristic of localization like a particle does, but it has wave properties because it is built from waves. For an individual wave in the wave packet, the **phase speed** is

$$v_{phase} = \frac{\omega}{k} \qquad (40.18)$$

For the wave packet as a whole, the **group speed** is

$$v_g = \frac{d\omega}{dk} \qquad (40.19)$$

For a wave packet representing a particle, the group speed can be shown to be the same as the speed of the particle.

The **Heisenberg uncertainty principle** states that if a measurement of the position of a particle is made with uncertainty Δx and a simultaneous measurement of its linear momentum is made with uncertainty Δp_x, the product of the two uncertainties is restricted to

$$\Delta x \, \Delta p_x \geq \frac{\hbar}{2} \qquad (40.23)$$

Another form of the uncertainty principle relates measurements of energy and time:

$$\Delta E \, \Delta t \geq \frac{\hbar}{2} \qquad (40.24)$$

Questions

☐ denotes answer available in *Student Solutions Manual/Study Guide;* **O** denotes objective question

1. The classical model of blackbody radiation given by the Rayleigh–Jeans law has two major flaws. Identify them and explain how Planck's law deals with them.

2. All objects radiate energy. Why, then, are we not able to see all objects in a dark room?

3. **O** In a certain experiment, a filament in an evacuated light-bulb carries a current I_1 and you measure the spectrum of light emitted by the filament, which behaves as a black body at temperature T_1. The wavelength emitted with highest intensity (symbolized by λ_{max}) has the value λ_1. You then increase the potential difference across the filament by a factor of 8, and the current increases by a factor of 2. **(i)** After this change, what is the new value of the temperature of the filament? (a) $16T_1$ (b) $8T_1$ (c) $4T_1$ (d) $2T_1$ (e) still T_1 **(ii)** What is the new value of the wavelength emitted with highest intensity? (a) $4\lambda_1$ (b) $2\lambda_1$ (c) $\sqrt{2}\lambda_1$ (d) λ_1 (e) $\lambda_1/\sqrt{2}$ (f) $\lambda_1/2$ (g) $\lambda_1/4$

4. If the photoelectric effect is observed for one metal, can you conclude that the effect will also be observed for another metal under the same conditions? Explain.

5. What does the slope of the lines in Active Figure 40.11 represent? What does the y intercept represent? How would such graphs for different metals compare with one another?

6. Why does the existence of a cutoff frequency in the photoelectric effect favor a particle theory for light over a wave theory?

7. In the photoelectric effect, explain why the stopping potential depends on the frequency of light but not on the intensity.

8. Which has more energy, a photon of ultraviolet radiation or a photon of yellow light?

9. **O** Which of the following is most likely to cause sunburn by delivering more energy to individual molecules in skin cells? (a) infrared light   (b) visible light   (c) ultraviolet light   (d) microwaves   (e) Choices (a) through (d) are equally likely.

10. How does the Compton effect differ from the photoelectric effect?

11. **O** An x-ray photon is scattered by an originally stationary electron. What happens to the frequency of the scattered photon relative to the frequency of the incident photon? Is the frequency of the scattered photon (a) lower, (b) higher, or (c) unchanged?

12. Suppose a photograph were made of a person's face using only a few photons. Would the result be simply a very faint image of the face? Explain your answer.

13. **O** Consider (a) an electron, (b) a photon, and (c) a proton, all moving in vacuum. Choose all correct answers for each question. **(i)** Which of the three possess rest energy? **(ii)** Which have charge? **(iii)** Which carry energy? **(iv)** Which carry momentum? **(v)** Which move at the speed of light? **(vi)** Which have a wavelength characterizing their motion?

14. Is light a wave or a particle? Support your answer by citing specific experimental evidence.

15. Is an electron a wave or a particle? Support your answer by citing some experimental results.

16. Why was the demonstration of electron diffraction by Davisson and Germer an important experiment?

17. **O** An electron and a proton, moving in opposite directions, are accelerated from rest through the same potential difference. Which particle has the longer wavelength? (a) The electron does. (b) The proton does. (c) Both are the same. (d) Neither has a wavelength

18. If matter has a wave nature, why is this wave-like characteristic not observable in our daily experiences?

19. **O** Rank the wavelengths of the following quantum particles from the largest to the smallest. If any have equal wavelengths, display the equality in your ranking. (a) a photon with energy 3 eV   (b) an electron with kinetic energy 3 eV   (c) a proton with kinetic energy 3 eV   (d) a photon with energy 0.3 eV   (e) a photon with momentum 3 eV/c = 1.6 × 10^{-27} kg · m/s   (f) an electron with momentum 3 eV/c (g) a proton with momentum 3 eV/c

20. In describing the passage of electrons through a slit and arriving at a screen, physicist Richard Feynman said that "electrons arrive in lumps, like particles, but the probability of arrival of these lumps is determined as the intensity of the waves would be. It is in this sense that the electron behaves sometimes like a particle and sometimes like a wave." Elaborate on this point in your own words. For further discussion, see R. Feynman, *The Character of Physical Law* (Cambridge, MA: MIT Press, 1980), chap. 6.

21. Why is an electron microscope more suitable than an optical microscope for "seeing" objects less than 1 μm in size?

22. **O** Both an electron and a proton are accelerated to the same speed, and the experimental uncertainty in the speed is the same for the two particles. The positions of the two particles are also measured. Is the minimum possible uncertainty in the electron's position (a) less than the minimum possible uncertainty in the proton's position, (b) the same as that for the proton, (c) more than that for the proton, or (d) impossible to tell from the given information?

23. Shown in the photograph opening Chapter 37 and in Figure P38.52a in Chapter 38, *iridescence* is the phenomenon that gives shining colors to the feathers of peacocks, hummingbirds, resplendent quetzals, and even ducks and grackles. Without pigments, it colors Morpho butterflies, Urania moths, some beetles and flies, rainbow trout, and mother-of-pearl in abalone shells. Iridescent colors change as you turn an object. They can look different to your two eyes so that the object appears to have metallic luster. Iridescent colors were first described in print not by an artist or biologist, but by a physicist, Isaac Newton. They are produced by a wide variety of intricate structures in different species; Problem 52 in Chapter 38 describes those in a peacock feather. These structures were all unknown until the invention of the electron microscope. Explain why light microscopes cannot reveal them.

24. *Blacker than black, brighter than white.* (a) Take a large, closed, empty cardboard box. Cut a slot a few millimeters wide in one side. Use black pens, markers, and black material to make some stripes next to the slot as shown in Figure Q40.24a. Inspect the slot and the stripes with care and choose which is blackest; the figure may not show enough contrast to reveal which it is. Explain why it is blackest. (b) Locate an intricately shaped compact fluorescent light fixture. Look at it through dark glasses and describe where it appears brightest. Explain why it is brightest there. Figure Q40.24b shows two such light fixtures held near each other. *Suggestion:* Gustav Kirchhoff, professor at Heidelberg and master of the obvious, gave the same answer to part (a) as you likely will. His answer to part (b) would begin as follows. When electromagnetic radiation falls on its surface, an object reflects some fraction r of the energy and absorbs the rest. Whether the fraction reflected is 0.8 or 0.001, the fraction absorbed is $a = 1 - r$. Suppose the object and its surroundings are at the same temperature. The energy the object absorbs joins its fund of internal energy, but the second law of thermodynamics implies that the absorbed energy cannot raise the object's temperature. It does not produce a tem-

perature increase because the object's energy budget has one more term: energy radiated. . . .

You still have to make the observations and answer questions (a) and (b), but you can incorporate some of Kirchhoff's ideas into your answer if you wish.

(a)

(b)

Figure Q40.24

Problems

WebAssign The Problems from this chapter may be assigned online in WebAssign.

ThomsonNOW™ Sign in at **www.thomsonedu.com** and go to ThomsonNOW to assess your understanding of this chapter's topics with additional quizzing and conceptual questions.

1, 2, 3 denotes straightforward, intermediate, challenging; ☐ denotes full solution available in *Student Solutions Manual/Study Guide;* ▲ denotes coached solution with hints available at **www.thomsonedu.com;** denotes developing symbolic reasoning; ● denotes asking for qualitative reasoning; ▀ denotes computer useful in solving problem

Section 40.1 Blackbody Radiation and Planck's Hypothesis

1. The human eye is most sensitive to 560-nm light. What is the temperature of a black body that would radiate most intensely at this wavelength?

2. ● (a) Model the tungsten filament of a lightbulb as a blackbody at temperature 2 900 K. Determine the wavelength of light it emits most strongly. (b) Explain why the answer to part (a) suggests that more energy from the lightbulb goes into infrared radiation than into visible light.

3. ● Figure P40.3 shows the spectrum of light emitted by a firefly. Determine the temperature of a black body that

Figure P40.3

would emit radiation peaked at the same wavelength. Based on your result, explain whether firefly radiation is blackbody radiation.

4. (a) Lightning produces a maximum air temperature on the order of 10^4 K, whereas (b) a nuclear explosion produces a temperature on the order of 10^7 K. Use Wien's displacement law to find the order of magnitude of the wavelength of the thermally produced photons radiated with greatest intensity by each of these sources. Name the part of the electromagnetic spectrum where you would expect each to radiate most strongly.

5. A black body at 7 500 K consists of an opening of diameter 0.050 0 mm, looking into an oven. Find the number of photons per second escaping the hole and having wavelengths between 500 nm and 501 nm.

6. Consider a black body of surface area 20.0 cm^2 and temperature 5 000 K. (a) How much power does it radiate? (b) At what wavelength does it radiate most intensely? Find the spectral power per wavelength at (c) this wavelength and at wavelengths of (d) 1.00 nm (an x- or gamma ray), (e) 5.00 nm (ultraviolet light or an x-ray), (f) 400 nm (at the boundary between UV and visible light), (g) 700 nm (at the boundary between visible and infrared light), (h) 1.00 mm (infrared light or a microwave), and (i) 10.0 cm (a microwave or radio wave).

(j) Approximately how much power does the object radiate as visible light?

7. The radius of our Sun is 6.96×10^8 m, and its total power output is 3.85×10^{26} W. (a) Assuming the Sun's surface emits as a black body, calculate its surface temperature. (b) Using the result of part (a), find λ_{max} for the Sun.

8. The average threshold of dark-adapted (scotopic) vision is 4.00×10^{-11} W/m^2 at a central wavelength of 500 nm. If light having this intensity and wavelength enters the eye and the pupil is open to its maximum diameter of 8.50 mm, how many photons per second enter the eye?

9. ▲ Calculate the energy, in electron volts, of a photon whose frequency is (a) 620 THz, (b) 3.10 GHz, and (c) 46.0 MHz. (d) Determine the corresponding wavelengths for these photons and state the classification of each on the electromagnetic spectrum.

10. A simple pendulum has a length of 1.00 m and a mass of 1.00 kg. The amplitude of oscillations of the pendulum is 3.00 cm. Calculate the quantum number for the pendulum.

11. An FM radio transmitter has a power output of 150 kW and operates at a frequency of 99.7 MHz. How many photons per second does the transmitter emit?

12. **Review problem.** This problem is about how strongly matter is coupled to radiation, the subject with which quantum mechanics began. For a simple model, consider a solid iron sphere 2.00 cm in radius. Assume its temperature is always uniform throughout its volume. (a) Find the mass of the sphere. (b) Assume the sphere is at 20°C and has emissivity 0.860. Find the power with which it radiates electromagnetic waves. (c) If it were alone in the Universe, at what rate would the sphere's temperature be changing? (d) Assume Wien's law describes the sphere. Find the wavelength λ_{max} of electromagnetic radiation it emits most strongly. Although it emits a spectrum of waves having all different wavelengths, assume its power output is carried by photons of wavelength λ_{max}. Find (e) the energy of one photon and (f) the number of photons it emits each second. *Note:* The answer to part (f) gives an indication of how fast the object is emitting and also absorbing photons when it is in thermal equilibrium with its surroundings at 20°C.

13. Show that at long wavelengths, Planck's radiation law (Eq. 40.6) reduces to the Rayleigh–Jeans law (Eq. 40.3).

Section 40.2 The Photoelectric Effect

14. Molybdenum has a work function of 4.20 eV. (a) Find the cutoff wavelength and cutoff frequency for the photoelectric effect. (b) What is the stopping potential if the incident light has a wavelength of 180 nm?

15. Two light sources are used in a photoelectric experiment to determine the work function for a particular metal surface. When green light from a mercury lamp ($\lambda = 546.1$ nm) is used, a stopping potential of 0.376 V reduces the photocurrent to zero. (a) Based on this measurement, what is the work function for this metal? (b) What stopping potential would be observed when using the yellow light from a helium discharge tube ($\lambda = 587.5$ nm)?

16. Electrons are ejected from a metallic surface with speeds ranging up to 4.60×10^5 m/s when light with a wavelength of 625 nm is used. (a) What is the work function

of the surface? (b) What is the cutoff frequency for this surface?

17. ● Lithium, beryllium, and mercury have work functions of 2.30 eV, 3.90 eV, and 4.50 eV, respectively. Light with a wavelength of 400 nm is incident on each of these metals. (a) Determine which of these metals exhibit the photoelectric effect. Explain your reasoning. (b) Find the maximum kinetic energy for the photoelectrons in each case.

18. ● From the scattering of sunlight, J. J. Thomson calculated the classical radius of the electron as having the value 2.82×10^{-15} m. Sunlight with an intensity of 500 W/m^2 falls on a disk with this radius. Calculate the time interval required to accumulate 1.00 eV of energy. Assume light is a classical wave and the light striking the disk is completely absorbed. Explain how your result compares with the observation that photoelectrons are emitted promptly (within 10^{-9} s).

19. **Review problem.** An isolated copper sphere of radius 5.00 cm, initially uncharged, is illuminated by ultraviolet light of wavelength 200 nm. What charge does the photoelectric effect induce on the sphere? The work function for copper is 4.70 eV.

20. ● **Review problem.** A light source emitting radiation at 7.00×10^{14} Hz is incapable of ejecting photoelectrons from a certain metal. In an attempt to use this source to eject photoelectrons from the metal, the source is given a velocity toward the metal. (a) Explain how this procedure can produce photoelectrons. (b) When the speed of the light source is equal to $0.280c$, photoelectrons just begin to be ejected from the metal. What is the work function of the metal? (c) When the speed of the light source is increased to $0.900c$, determine the maximum kinetic energy of the photoelectrons.

Section 40.3 The Compton Effect

21. Calculate the energy and momentum of a photon of wavelength 700 nm.

22. ● X-rays with a wavelength of 120.0 pm undergo Compton scattering. (a) Find the wavelengths of the photons scattered at angles of 30.0°, 60.0°, 90.0°, 120°, 150°, and 180°. (b) Find the energy of the scattered electron in each case. (c) Which of the scattering angles provides the electron with the greatest energy? Explain whether you could answer this question without doing any calculations.

23. ▲ A 0.001 60-nm photon scatters from a free electron. For what (photon) scattering angle does the recoiling electron have kinetic energy equal to the energy of the scattered photon?

24. X-rays having an energy of 300 keV undergo Compton scattering from a target. The scattered rays are detected at 37.0° relative to the incident rays. Find (a) the Compton shift at this angle, (b) the energy of the scattered x-ray, and (c) the energy of the recoiling electron.

25. A 0.880-MeV photon is scattered by a free electron initially at rest such that the scattering angle of the scattered electron is equal to that of the scattered photon ($\theta = \phi$ in Fig. 40.13). (a) Determine the angles θ and ϕ. (b) Determine the energy and momentum of the scattered photon. (c) Determine the kinetic energy and momentum of the scattered electron.

2 = intermediate;   3 = challenging;   ☐ = SSM/SG;   ▲ = ThomsonNOW;   = symbolic reasoning;   ● = qualitative reasoning

26. A photon having energy E_0 is scattered by a free electron initially at rest such that the scattering angle of the scattered electron is equal to that of the scattered photon ($\theta = \phi$ in Fig. 40.13). (a) Determine the angles θ and ϕ. (b) Determine the energy and momentum of the scattered photon. (c) Determine the kinetic energy and momentum of the scattered electron.

27. After a 0.800-nm x-ray photon scatters from a free electron, the electron recoils at 1.40×10^6 m/s. (a) What is the Compton shift in the photon's wavelength? (b) Through what angle is the photon scattered?

28. In a Compton scattering experiment, an x-ray photon scatters through an angle of 17.4° from a free electron that is initially at rest. The electron recoils with a speed of 2 180 km/s. Calculate (a) the wavelength of the incident photon and (b) the angle through which the electron scatters.

29. ● In a Compton scattering experiment, a photon is scattered through an angle of 90.0° and the electron is set into motion in a direction at an angle of 20.0° to the original direction of the photon. Explain whether this information is sufficient to determine uniquely the wavelength of the scattered photon. If it is, find this wavelength.

30. A photon having wavelength λ scatters off a free electron at A (Fig. P40.30), producing a second photon having wavelength λ'. This photon then scatters off another free electron at B, producing a third photon having wavelength λ'' and moving in a direction directly opposite the original photon as shown in the figure. Determine the numerical value of $\Delta\lambda = \lambda'' - \lambda$.

Figure P40.30

31. Find the maximum fractional energy loss for a 0.511-MeV gamma ray that is Compton scattered from (a) a free electron and (b) a free proton.

Section 40.4 Photons and Electromagnetic Waves

32. ● An electromagnetic wave is called *ionizing radiation* if its photon energy is larger than, say, 10.0 eV so that a single photon has enough energy to break apart an atom. With reference to Figure 34.11, explain what region or regions of the electromagnetic spectrum fit this definition of ionizing radiation and what do not.

33. **Review problem.** A helium–neon laser produces a beam of diameter 1.75 mm, delivering 2.00×10^{18} photons/s. Each photon has a wavelength of 633 nm. (a) Calculate the amplitudes of the electric and magnetic fields inside the beam. (b) If the beam shines perpendicularly onto a perfectly reflecting surface, what force does it exert on the surface? (c) If the beam is absorbed by a block of ice at 0°C for 1.50 h, what mass of ice is melted?

Section 40.5 The Wave Properties of Particles

34. Calculate the de Broglie wavelength for a proton moving with a speed of 1.00×10^6 m/s.

35. Calculate the de Broglie wavelength for an electron that has kinetic energy (a) 50.0 eV and (b) 50.0 keV.

36. (a) An electron has a kinetic energy of 3.00 eV. Find its wavelength. (b) **What If?** A photon has energy 3.00 eV. Find its wavelength.

37. ● ▲ The nucleus of an atom is on the order of 10^{-14} m in diameter. For an electron to be confined to a nucleus, its de Broglie wavelength would have to be on this order of magnitude or smaller. (a) What would be the kinetic energy of an electron confined to this region? (b) Make an order-of-magnitude estimate of the electric potential energy of a system of an electron inside an atomic nucleus. Would you expect to find an electron in a nucleus? Explain.

38. In the Davisson–Germer experiment, 54.0-eV electrons were diffracted from a nickel lattice. If the first maximum in the diffraction pattern was observed at $\phi = 50.0°$ (Fig. P40.38), what was the lattice spacing a between the vertical rows of atoms in the figure? (It is not the same as the spacing between the horizontal rows of atoms.)

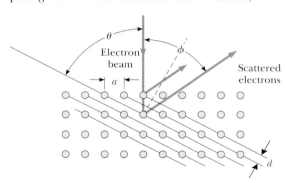

Figure P40.38

39. ● (a) Show that the frequency f and wavelength λ of a freely moving quantum particle with mass are related by the expression

$$\left(\frac{f}{c}\right)^2 = \frac{1}{\lambda^2} + \frac{1}{\lambda_C{}^2}$$

where $\lambda_C = h/mc$ is the Compton wavelength of the particle. (b) Is it ever possible for a particle having nonzero mass to have the same wavelength *and* frequency as a photon? Explain.

40. A photon has an energy equal to the kinetic energy of an electron with speed u, which may be close to the speed of light. (a) Calculate the ratio of the wavelength of the photon to the wavelength of the electron. (b) Evaluate the ratio for the particle speed $u = 0.900c$. (c) **What If?** What would happen to the answer if the material particle were a proton instead of an electron? (d) Evaluate the ratio for the particle speed $u = 0.001\,00c$. (e) What value does the ratio of the wavelengths approach at high particle speeds? (f) At low particle speeds?

41. The resolving power of a microscope depends on the wavelength used. If you wanted to "see" an atom, a resolution of approximately 1.00×10^{-11} m would be required. (a) If electrons are used (in an electron microscope), what minimum kinetic energy is required for the electrons? (b) **What If?** If photons are used, what minimum photon energy is needed to obtain the required resolution?

42. ● After learning about de Broglie's hypothesis that material particles of momentum p move as waves with wavelength $\lambda = h/p$, an 80.0-kg student has grown concerned about being diffracted when passing through a 75.0-cm-wide doorway. Assume significant diffraction occurs when the width of the diffraction aperture is less that 10.0 times the wavelength of the wave being diffracted. (a) Determine the maximum speed at which the student can pass through the doorway if he is to be significantly diffracted. (b) With that speed, over what time interval does the student pass through the doorway if it is in a wall 15.0 cm thick? State how your answer compares with the age of the Universe, which is about 4×10^{17} s. (c) Explain whether this student should worry about being diffracted.

43. ● Robert Hofstadter won the 1961 Nobel Prize in Physics for his pioneering work in studying the scattering of 20-GeV electrons from nuclei. (a) What is the γ factor for an electron with total energy 20.0 GeV, defined by $\gamma = 1/\sqrt{1 - u^2/c^2}$? (b) Find the momentum of the electron. (c) Find the wavelength of the electron. State how it compares with the diameter of an atomic nucleus, typically on the order of 10^{-14} m.

Section 40.6 The Quantum Particle

44. Consider a freely moving quantum particle with mass m and speed u. Its energy is $E = K = \frac{1}{2}mu^2$. Determine the phase speed of the quantum wave representing the particle and show that it is different from the speed at which the particle transports mass and energy.

45. For a free relativistic quantum particle moving with speed u, the total energy is $E = hf = \hbar\omega = \sqrt{p^2c^2 + m^2c^4}$ and the momentum is $p = h/\lambda = \hbar k = \gamma mu$. For the quantum wave representing the particle, the group speed is $v_g = d\omega/dk$. Prove that the group speed of the wave is the same as the speed of the particle.

Section 40.7 The Double-Slit Experiment Revisited

46. A modified oscilloscope is used to perform an electron interference experiment. Electrons are incident on a pair of narrow slits 0.060 0 μm apart. The bright bands in the interference pattern are separated by 0.400 mm on a screen 20.0 cm from the slits. Determine the potential difference through which the electrons were accelerated to give this pattern.

47. ▲ Neutrons traveling at 0.400 m/s are directed through a pair of slits having a 1.00-mm separation. An array of detectors is placed 10.0 m from the slits. (a) What is the de Broglie wavelength of the neutrons? (b) How far off axis is the first zero-intensity point on the detector array? (c) When a neutron reaches a detector, can we say which slit the neutron passed through? Explain.

48. In a certain vacuum tube, electrons evaporate from a hot cathode at a slow, steady rate and accelerate from rest through a potential difference of 45.0 V. Then they travel 28.0 cm as they pass through an array of slits and fall on a screen to produce an interference pattern. If the beam current is below a certain value, only one electron at a time will be in flight in the tube. What is this value? In this situation, the interference pattern still appears, showing that each individual electron can interfere with itself.

Section 40.8 The Uncertainty Principle

49. ▲ An electron ($m_e = 9.11 \times 10^{-31}$ kg) and a bullet ($m = 0.020\ 0$ kg) each have a velocity with a magnitude of 500 m/s, accurate to within 0.010 0%. Within what limits could we determine the position of the objects along the direction of the velocity?

50. Suppose Fuzzy, a quantum–mechanical duck, lives in a world in which $h = 2\pi$ J · s. Fuzzy has a mass of 2.00 kg and is initially known to be within a pond 1.00 m wide. (a) What is the minimum uncertainty in the component of the duck's velocity parallel to the width of the pond? (b) Assuming that this uncertainty in speed prevails for 5.00 s, determine the uncertainty in Fuzzy's position after this time interval.

51. ● An air rifle is used to shoot 1.00-g particles at 100 m/s through a hole of diameter 2.00 mm. How far from the rifle must an observer be to see the beam spread by 1.00 cm because of the uncertainty principle? State how this answer compares with the diameter of the visible part of the Universe, 2×10^{26} m.

52. Use the uncertainty principle to show that if an electron were confined inside an atomic nucleus of diameter 2×10^{-15} m, it would have to be moving relativistically, whereas a proton confined to the same nucleus can be moving nonrelativistically.

53. A woman on a ladder drops small pellets toward a point target on the floor. (a) Show that according to the uncertainty principle, the average miss distance must be at least

$$\Delta x_f = \left(\frac{2\hbar}{m}\right)^{1/2}\left(\frac{2H}{g}\right)^{1/4}$$

where m is the mass of each pellet and H is the initial height of each pellet above the floor. Assume the spread in impact points is given by $\Delta x_f = \Delta x_i + (\Delta v_x)t$. (b) If $H = 2.00$ m and $m = 0.500$ g, what is Δx_f?

Additional Problems

54. **Review problem.** Design an incandescent lamp filament. Specify the length and radius a tungsten wire can have to radiate electromagnetic waves with power 75.0 W when its ends are connected across a 120-V power supply. Assume its constant operating temperature is 2 900 K and its emissivity is 0.450. Also assume it takes in energy only by electric transmission and loses energy only by electromagnetic radiation. From Table 27.2, you may take the resistivity of tungsten at 2 900 K as

$$5.6 \times 10^{-8}\ \Omega \cdot \text{m}[1 + (4.5 \times 10^{-3}/°\text{C})(2\ 607°\text{C})]$$

$$= 7.13 \times 10^{-7}\ \Omega \cdot \text{m}$$

55. ▲ The accompanying table shows data obtained in a photoelectric experiment. (a) Using these data, make a graph similar to Active Figure 40.11 that plots as a straight line. From the graph, determine (b) an experimental value for

Planck's constant (in joule-seconds) and (c) the work function (in electron volts) for the surface. Two significant figures for each answer are sufficient.

Wavelength (nm)	Maximum Kinetic Energy of Photoelectrons (eV)
588	0.67
505	0.98
445	1.35
399	1.63

56. Figure P40.56 shows the stopping potential versus the incident photon frequency for the photoelectric effect for sodium. Use the graph to find (a) the work function of sodium, (b) the ratio h/e, and (c) the cutoff wavelength. The data are taken from R. A. Millikan, *Physical Review* **7**:362 (1916).

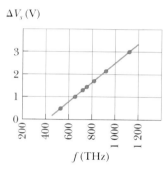

Figure P40.56

57. **Review problem.** Photons of wavelength λ are incident on a metal. The most energetic electrons ejected from the metal are bent into a circular arc of radius R by a magnetic field having a magnitude B. What is the work function of the metal?

58. ● Ultraviolet light with a single wavelength and with intensity 550 W/m² is incident normally on the surface of a metal that has a work function of 3.44 eV. Photoelectrons are emitted with a maximum speed of 420 km/s. (a) Find the maximum possible rate of photoelectron emission from 1 cm² of the surface by imagining that every photon produces one photoelectron. (b) Find the electric current these electrons constitute. (c) How do you suppose the actual current compares with this maximum possible current?

59. Derive the equation for the Compton shift (Eq. 40.11) from Equations 40.12 through 40.14.

60. Show that a photon cannot transfer all of its energy to a free electron. *Suggestion:* Note that system energy and momentum must be conserved.

61. The total power per unit area radiated by a black body at a temperature T is the area under the $I(\lambda, T)$-versus-λ curve as shown in Active Figure 40.3. (a) Show that this power per unit area is

$$\int_0^\infty I(\lambda, T)\,d\lambda = \sigma T^4$$

where $I(\lambda, T)$ is given by Planck's radiation law and σ is a constant independent of T. This result is known as Stefan's law. (See Section 20.7.) To carry out the integration,

you should make the change of variable $x = hc/\lambda k_B T$ and use

$$\int_0^\infty \frac{x^3\,dx}{e^x - 1} = \frac{\pi^4}{15}$$

(b) Show that the Stefan–Boltzmann constant σ has the value

$$\sigma = \frac{2\pi^5 k_B^{\,4}}{15 c^2 h^3} = 5.67 \times 10^{-8}\ \text{W/m}^2 \cdot \text{K}^4$$

62. ■ Derive Wien's displacement law from Planck's law. Proceed as follows. In Active Figure 40.3, notice that the wavelength at which a black body radiates with greatest intensity is the wavelength for which the graph of $I(\lambda, T)$ versus λ has a horizontal tangent. From Equation 40.6, evaluate the derivative $dI/d\lambda$. Set it equal to zero. Solve the resulting transcendental equation numerically to prove $hc/\lambda_{\max} k_B T = 4.965\ldots$ or $\lambda_{\max} T = hc/4.965 k_B$. Evaluate the constant as precisely as possible and compare it with Wien's experimental value.

63. ● The neutron has a mass of 1.67×10^{-27} kg. Neutrons emitted in nuclear reactions can be slowed down by collisions with matter. They are referred to as thermal neutrons after they come into thermal equilibrium with the environment. The average kinetic energy $(3k_B T/2)$ of a thermal neutron is approximately 0.04 eV. Calculate the de Broglie wavelength of a neutron with a kinetic energy of 0.040 0 eV. How does your answer compare with the characteristic atomic spacing in a crystal? Explain whether you expect thermal neutrons to exhibit diffraction effects when scattered by a crystal.

64. Johnny Jumper's favorite trick is to step out of his 16th-story window and fall 50.0 m into a pool. A news reporter takes a picture of 75.0-kg Johnny just before he makes a splash, using an exposure time of 5.00 ms. Find (a) Johnny's de Broglie wavelength at this moment, (b) the uncertainty of his kinetic energy measurement during such an interval of time, and (c) the percent error caused by such an uncertainty.

65. Show that the ratio of the Compton wavelength λ_C to the de Broglie wavelength $\lambda = h/p$ for a relativistic electron is

$$\frac{\lambda_C}{\lambda} = \left[\left(\frac{E}{m_e c^2}\right)^2 - 1\right]^{1/2}$$

where E is the total energy of the electron and m_e is its mass.

66. A photon of initial energy E_0 undergoes Compton scattering at an angle θ from a free electron (mass m_e) initially at rest. Using relativistic equations for energy and momentum conservation, derive the following relationship for the final energy E' of the scattered photon:

$$E' = \frac{E_0}{1 + \left(\dfrac{E_0}{m_e c^2}\right)(1 - \cos\theta)}$$

67. A π^0 meson is an unstable particle produced in high-energy particle collisions. Its rest energy is approximately 135 MeV, and it exists for a lifetime of only 8.70×10^{-17} s before decaying into two gamma rays. Using the uncertainty principle, estimate the fractional uncertainty $\Delta m/m$ in its mass determination.

68. A photon with wavelength λ_0 moves toward a free electron that is moving with speed u in the same direction as the photon (Fig. P40.68a). The photon scatters at an angle θ (Fig. P40.68b). Show that the wavelength of the scattered photon is

$$\lambda' = \lambda_0\left(\frac{1 - (u/c)\cos\theta}{1 - (u/c)}\right) + \frac{h}{m_e c}\sqrt{\frac{1 + (u/c)}{1 - (u/c)}}(1 - \cos\theta)$$

(a)   (b)

Figure P40.68

Answers to Quick Quizzes

40.1 (b). A very hot star has its peak in the blackbody intensity distribution curve at wavelengths shorter than the visible. As a result, more blue light is emitted than red light.

40.2 AM radio, FM radio, microwaves, sodium light. The order of photon energy will be the same as the order of frequency. See Figure 34.11 for a pictorial representation of electromagnetic radiation in order of frequency.

40.3 (c). When the frequency is increased, the photons each carry more energy, so a stopping potential larger in magnitude is required for the current to fall to zero.

40.4 Classical physics predicts that light of sufficient intensity causes emission of photoelectrons, independent of frequency and without a cutoff frequency. Also, the greater the intensity, the larger the maximum kinetic energy of the electrons, with some time delay in emission at low intensities. Therefore, the classical expectation (which did not match experiment) yields a graph like the following drawing:

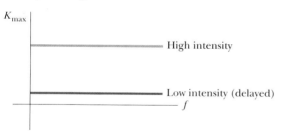

40.5 (d). The shift $\Delta\lambda$ is independent of λ. Therefore, the largest fractional shift will correspond to the smallest wavelength.

40.6 (c). According to Equation 40.15, two particles with the same de Broglie wavelength have the same momentum $p = mu$. If the electron and proton have the same momentum, they cannot have the same speed because of the difference in their masses. For the same reason, because $K = p^2/2m$, they cannot have the same kinetic energy. Because the particles have different kinetic energies, Equation 40.16 tells us that the particles do not have the same frequency.

40.7 (b). The group speed is zero because the leading edge of the packet remains fixed at the location of the accident.

40.8 (a). The uncertainty principle relates uncertainty in position and velocity along the same axis. The zero uncertainty in position along the x axis results in infinite uncertainty in its velocity component in the x direction, but it is unrelated to the y direction.

A *quantum corral* shows two aspects of current technological advances in physics. The first aspect involves control over individual atoms. This corral is formed by positioning iron atoms in a stadium-shaped ring on a copper surface. The second aspect is the ability to image the individual atoms with a scanning tunneling microscope. The corral can be used to study the quantized states of electrons trapped in a small region. (Courtesy of IBM Research, Almaden Research Center. Unauthorized use prohibited.)

41.1 An Interpretation of Quantum Mechanics

41.2 The Quantum Particle Under Boundary Conditions

41.3 The Schrödinger Equation

41.4 A Particle in a Well of Finite Height

41.5 Tunneling Through a Potential Energy Barrier

41.6 Applications of Tunneling

41.7 The Simple Harmonic Oscillator

41 Quantum Mechanics

In this chapter, we introduce *quantum mechanics*, **an extremely successful theory** for explaining the behavior of microscopic particles. This theory, developed in the 1920s by Erwin Schrödinger, Werner Heisenberg, and others, enables us to understand a host of phenomena involving atoms, molecules, nuclei, and solids. The discussion in this chapter follows from the quantum particle model that was developed in Chapter 40 and incorporates some of the features of the waves under boundary conditions model that was explored in Chapter 18. We also discuss practical applications of quantum mechanics, including the scanning tunneling microscope and nanoscale devices that may be used in future quantum computers. Finally, we shall return to the simple harmonic oscillator that was introduced in Chapter 15 and examine it from a quantum mechanical point of view.

41.1 An Interpretation of Quantum Mechanics

In Chapter 40, we introduced some new and strange ideas. In particular, we concluded on the basis of experimental evidence that both matter and electromagnetic radiation are sometimes best modeled as particles and sometimes as waves, depending on the phenomenon being observed. We can improve our understanding of quantum physics by making another connection between particles and waves using the notion of probability, a concept that was introduced in Chapter 40.

We begin by discussing electromagnetic radiation using the particle model. The probability per unit volume of finding a photon in a given region of space at an instant of time is proportional to the number N of photons per unit volume at that time:

$$\frac{\text{Probability}}{V} \propto \frac{N}{V}$$

The number of photons per unit volume is proportional to the intensity of the radiation:

$$\frac{N}{V} \propto I$$

Now, let's form a connection between the particle model and the wave model by recalling that the intensity of electromagnetic radiation is proportional to the square of the electric field amplitude E for the electromagnetic wave (Eq. 34.24):

$$I \propto E^2$$

Equating the beginning and the end of this series of proportionalities gives

$$\frac{\text{Probability}}{V} \propto E^2 \qquad \textbf{(41.1)}$$

Therefore, for electromagnetic radiation, the probability per unit volume of finding a particle associated with this radiation (the photon) is proportional to the square of the amplitude of the associated electromagnetic wave.

Recognizing the wave–particle duality of both electromagnetic radiation and matter, we should suspect a parallel proportionality for a material particle: the probability per unit volume of finding the particle is proportional to the square of the amplitude of a wave representing the particle. In Chapter 40, we learned that there is a de Broglie wave associated with every particle. The amplitude of the de Broglie wave associated with a particle is not a measurable quantity because the wave function representing a particle is generally a complex function as we discuss below. In contrast, the electric field for an electromagnetic wave is a real function. The matter analog to Equation 41.1 relates the square of the amplitude of the wave to the probability per unit volume of finding the particle. Hence, the amplitude of the wave associated with the particle is called the **probability amplitude,** or the **wave function,** and it has the symbol Ψ.

In general, the complete wave function Ψ for a system depends on the positions of all the particles in the system and on time; therefore, it can be written $\Psi(\vec{\mathbf{r}}_1, \vec{\mathbf{r}}_2, \vec{\mathbf{r}}_3, \ldots, \vec{\mathbf{r}}_j, \ldots, t)$, where $\vec{\mathbf{r}}_j$ is the position vector of the jth particle in the system. For many systems of interest, including all those we study in this text, the wave function Ψ is mathematically separable in space and time and can be written as a product of a space function ψ for one particle of the system and a complex time function:[1]

$$\Psi(\vec{\mathbf{r}}_1, \vec{\mathbf{r}}_2, \vec{\mathbf{r}}_3, \ldots, \vec{\mathbf{r}}_j, \ldots, t) = \psi(\vec{\mathbf{r}}_j)e^{-i\omega t} \qquad \textbf{(41.2)}$$

◀ Space- and time-dependent wave function Ψ

where $\omega \ (= 2\pi f)$ is the angular frequency of the wave function and $i = \sqrt{-1}$.

For any system in which the potential energy is time-independent and depends only on the positions of particles within the system, the important information about the system is contained within the space part of the wave function. The time

---

[1] The standard form of a complex number is $a + ib$. The notation $e^{i\theta}$ is equivalent to the standard form as follows:

$$e^{i\theta} = \cos\theta + i\sin\theta$$

Therefore, the notation $e^{-i\omega t}$ in Equation 41.2 is equivalent to $\cos(-\omega t) + i\sin(-\omega t) = \cos\omega t - i\sin\omega t$.

part is simply the factor $e^{-i\omega t}$. Therefore, an understanding of ψ is the critical aspect of a given problem.

The wave function ψ is often complex-valued. The absolute square $|\psi|^2 = \psi^*\psi$, where ψ^* is the complex conjugate[2] of ψ, is always real and positive and is proportional to the probability per unit volume of finding a particle at a given point at some instant. The wave function contains within it all the information that can be known about the particle.

Although ψ cannot be measured, we can measure the real quantity $|\psi|^2$, which can be interpreted as follows. If ψ represents a single particle, then $|\psi|^2$—called the **probability density**—is the relative probability per unit volume that the particle will be found at any given point in the volume. This interpretation can also be stated in the following manner. If dV is a small volume element surrounding some point, the probability of finding the particle in that volume element is

◀ Probability density $|\psi|^2$

$$P(x, y, z)\,dV = |\psi|^2 dV \tag{41.3}$$

This probabilistic interpretation of the wave function was first suggested by Max Born (1882–1970) in 1928. In 1926, Erwin Schrödinger proposed a wave equation that describes the manner in which the wave function changes in space and time. The *Schrödinger wave equation*, which we shall examine in Section 41.3, represents a key element in the theory of quantum mechanics.

The concepts of quantum mechanics, strange as they sometimes may seem, developed from classical ideas. In fact, when the techniques of quantum mechanics are applied to macroscopic systems, the results are essentially identical to those of classical physics. This blending of the two approaches occurs when the de Broglie wavelength is small compared with the dimensions of the system. The situation is similar to the agreement between relativistic mechanics and classical mechanics when $v \ll c$.

In Section 40.5, we found that the de Broglie equation relates the momentum of a particle to its wavelength through the relation $p = h/\lambda$. If an ideal free particle has a precisely known momentum p_x, its wave function is an infinitely long sinusoidal wave of wavelength $\lambda = h/p_x$ and the particle has equal probability of being at any point along the x axis (Fig. 40.18a). The wave function ψ for such a free particle moving along the x axis can be written as

◀ Wave function for
a free particle

$$\psi(x) = Ae^{ikx} \tag{41.4}$$

where A is a constant amplitude and $k = 2\pi/\lambda$ is the angular wave number (Eq. 16.8) of the wave representing the particle.[3]

One-Dimensional Wave Functions and Expectation Values

This section discusses only one-dimensional systems, where the particle must be located along the x axis, so the probability $|\psi|^2\,dV$ in Equation 41.3 is modified to become $|\psi|^2\,dx$. The probability that the particle will be found in the infinitesimal interval dx around the point x is

$$P(x)\,dx = |\psi|^2\,dx \tag{41.5}$$

Although it is not possible to specify the position of a particle with complete certainty, it is possible through $|\psi|^2$ to specify the probability of observing it in a

---

[2] For a complex number $z = a + ib$, the complex conjugate is found by changing i to $-i$: $z^* = a - ib$. The product of a complex number and its complex conjugate is always real and positive. That is, $z^*z = (a - ib)(a + ib) = a^2 - (ib)^2 = a^2 - (i)^2 b^2 = a^2 + b^2$.

[3] For the free particle, the full wave function, based on Equation 41.2, is

$$\Psi(x, t) = Ae^{ikx}e^{-i\omega t} = Ae^{i(kx - \omega t)} = A[\cos(kx - \omega t) + i\sin(kx - \omega t)]$$

The real part of this wave function has the same form as the waves we added together to form wave packets in Section 40.6.

region surrounding a given point x. **The probability of finding the particle in the arbitrary interval** $a \leq x \leq b$ **is**

$$P_{ab} = \int_a^b |\psi|^2 \, dx \qquad (41.6)$$

The probability P_{ab} is the area under the curve of $|\psi|^2$ versus x between the points $x = a$ and $x = b$ as in Figure 41.1.

Experimentally, there is a finite probability of finding a particle in an interval near some point at some instant. The value of that probability must lie between the limits 0 and 1. For example, if the probability is 0.30, there is a 30% chance of finding the particle in the interval.

Because the particle must be somewhere along the x axis, the sum of the probabilities over all values of x must be 1:

$$\int_{-\infty}^{\infty} |\psi|^2 \, dx = 1 \qquad (41.7)$$

Any wave function satisfying Equation 41.7 is said to be **normalized.** Normalization is simply a statement that the particle exists at some point in space.

Once the wave function for a particle is known, it is possible to calculate the average position at which you would expect to find the particle after many measurements. This average position is called the **expectation value** of x and is defined by the equation

$$\langle x \rangle \equiv \int_{-\infty}^{\infty} \psi^* x \psi \, dx \qquad (41.8)$$

(Brackets, $\langle \ldots \rangle$, are used to denote expectation values.) Furthermore, one can find the expectation value of any function $f(x)$ associated with the particle by using the following equation:[4]

$$\langle f(x) \rangle \equiv \int_{-\infty}^{\infty} \psi^* f(x) \psi \, dx \qquad (41.9)$$

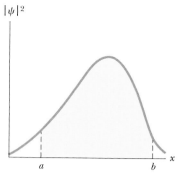

Figure 41.1   The probability of a particle being in the interval $a \leq x \leq b$ is the area under the probability density curve from a to b.

◀ Normalization condition on ψ

◀ Expectation value for position x

◀ Expectation value for a function $f(x)$

Quick Quiz 41.1   Consider the wave function for the free particle, Equation 41.4. At what value of x is the particle most likely to be found at a given time? (a) at $x = 0$ (b) at small nonzero values of x (c) at large values of x (d) anywhere along the x axis

---

EXAMPLE 41.1   **A Wave Function for a Particle**

Consider a particle whose wave function is graphed in Figure 41.2 and is given by

$$\psi(x) = Ae^{-ax^2}$$

(A) What is the value of A if this wave function is normalized?

SOLUTION

Conceptualize   The particle is not a free particle because the wave function is not a sinusoidal function. Figure 41.2 indicates that the particle is constrained to remain close to $x = 0$ at all times. Think of a physical system in which the particle

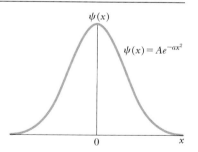

Figure 41.2   (Example 41.1) A symmetric wave function for a particle, given by $\psi(x) = Ae^{-ax^2}$.

---

[4] Expectation values are analogous to "weighted averages," in which each possible value of a function is multiplied by the probability of the occurrence of that value before summing over all possible values. We write the expectation value as $\int_{-\infty}^{\infty} \psi^* f(x) \psi \, dx$ rather than $\int_{-\infty}^{\infty} f(x) |\psi|^2 \, dx$ because $f(x)$ may be represented by an operator (such as a derivative) rather than a simple multiplicative function in more advanced treatments of quantum mechanics. In these situations, the operator is applied only to ψ and not to ψ^*.

always stays close to a given point. Examples of such systems are a block on a spring, a marble at the bottom of a bowl, and the bob of a simple pendulum.

Categorize   Because the statement of the problem describes the wave nature of a particle, this example requires a quantum approach rather than a classical approach.

Analyze   Apply the normalization condition, Equation 41.7, to the wave function:

$$\int_{-\infty}^{\infty} |\psi|^2 \, dx = \int_{-\infty}^{\infty} (Ae^{-ax^2})^2 \, dx = A^2 \int_{-\infty}^{\infty} e^{-2ax^2} \, dx = 1$$

Express the integral as the sum of two integrals:

$$(1) \quad A^2 \int_{-\infty}^{\infty} e^{-2ax^2} \, dx = A^2 \left(\int_{0}^{\infty} e^{-2ax^2} \, dx + \int_{-\infty}^{0} e^{-2ax^2} \, dx \right) = 1$$

Change the integration variable from x to $-x$ in the second integral:

$$\int_{-\infty}^{0} e^{-2ax^2} \, dx = \int_{\infty}^{0} e^{-2a(-x)^2} (-dx) = -\int_{\infty}^{0} e^{-2ax^2} \, dx$$

Reverse the order of the limits, which introduces a negative sign:

$$-\int_{\infty}^{0} e^{-2ax^2} \, dx = \int_{0}^{\infty} e^{-2ax^2} \, dx$$

Substitute this expression for the second integral in Equation (1):

$$A^2 \left(\int_{0}^{\infty} e^{-2ax^2} \, dx + \int_{0}^{\infty} e^{-2ax^2} \, dx \right) = 1$$

$$(2) \qquad 2A^2 \int_{0}^{\infty} e^{-2ax^2} \, dx = 1$$

Evaluate the integral with the help of Table B.6 in Appendix B:

$$\int_{0}^{\infty} e^{-2ax^2} \, dx = \tfrac{1}{2} \sqrt{\frac{\pi}{2a}}$$

Substitute this result into Equation (2) and solve for A:

$$2A^2 \left(\tfrac{1}{2} \sqrt{\frac{\pi}{2a}} \right) = 1 \quad \rightarrow \quad A = \boxed{\left(\frac{2a}{\pi} \right)^{1/4}}$$

(B) What is the expectation value of x for this particle?

SOLUTION

Evaluate the expectation value using Equation 41.8:

$$\langle x \rangle \equiv \int_{-\infty}^{\infty} \psi^* x \psi \, dx = \int_{-\infty}^{\infty} (Ae^{-ax^2}) x (Ae^{-ax^2}) \, dx$$

$$= A^2 \int_{-\infty}^{\infty} x e^{-2ax^2} \, dx$$

As in part (A), express the integral as a sum of two integrals:

$$(3) \qquad \langle x \rangle = A^2 \left(\int_{0}^{\infty} x e^{-2ax^2} \, dx + \int_{-\infty}^{0} x e^{-2ax^2} \, dx \right)$$

Change the integration variable from x to $-x$ in the second integral:

$$\int_{-\infty}^{0} x e^{-2ax^2} \, dx = \int_{\infty}^{0} -x e^{-2a(-x)^2} (-dx) = \int_{\infty}^{0} x e^{-2ax^2} \, dx$$

Reverse the order of the limits, which introduces a negative sign:

$$\int_{\infty}^{0} x e^{-2ax^2} \, dx = -\int_{0}^{\infty} x e^{-2ax^2} \, dx$$

Substitute this expression for the second integral in Equation (3):

$$\langle x \rangle = A^2 \left(\int_0^\infty x e^{-2ax^2}\, dx - \int_0^\infty x e^{-2ax^2}\, dx \right) = 0$$

Finalize Given the symmetry of the wave function around $x = 0$ in Figure 41.2, it is not surprising that the average position of the particle is at $x = 0$. In Section 41.7, we show that the wave function studied in this example represents the lowest-energy state of the quantum harmonic oscillator.

41.2   The Quantum Particle Under Boundary Conditions

The free particle discussed in Section 41.1 has no boundary conditions; it can be anywhere in space. The particle in Example 41.1 is not a free particle. Figure 41.2 shows that the particle is always restricted to positions near $x = 0$. In this section, we shall investigate the effects of restrictions on the motion of a quantum particle.

A Particle in a Box

We begin by applying some of the ideas we have developed to a simple physical problem, a particle confined to a one-dimensional region of space, called the *particle-in-a-box* problem (even though the "box" is one-dimensional!). From a classical viewpoint, if a particle is bouncing elastically back and forth along the x axis between two impenetrable walls separated by a distance L as in Figure 41.3a, it can be modeled as a particle under constant speed. If the speed of the particle is u, the magnitude of its momentum mu remains constant as does its kinetic energy. (Recall that in Chapter 39 we used u for particle speed to distinguish it from v, the speed of a reference frame.) Classical physics places no restrictions on the values of a particle's momentum and energy. The quantum-mechanical approach to this problem is quite different and requires that we find the appropriate wave function consistent with the conditions of the situation.

Because the walls are impenetrable, there is zero probability of finding the particle outside the box, so the wave function $\psi(x)$ must be zero for $x < 0$ and $x > L$. To be a mathematically well-behaved function, $\psi(x)$ must be continuous in space. There must be no discontinuous jumps in the value of the wave function at any point.[5] Therefore, if ψ is zero outside the walls, it must also be zero *at* the walls; that is, $\psi(0) = 0$ and $\psi(L) = 0$. Only those wave functions that satisfy these boundary conditions are allowed.

Figure 41.3b, a graphical representation of the particle-in-a-box problem, shows the potential energy of the particle–environment system as a function of the position of the particle. As long as the particle is inside the box, the potential energy of the system does not depend on the location of the particle and we can choose its constant value to be zero. Outside the box, we must ensure that the wave function is zero. We can do so by defining the system's potential energy as infinitely large if the particle were outside the box. Therefore, the only way a particle could be outside the box is if the system has an infinite amount of energy, which is impossible.

The wave function for a particle in the box can be expressed as a real sinusoidal function:[6]

$$\psi(x) = A \sin\left(\frac{2\pi x}{\lambda}\right) \tag{41.10}$$

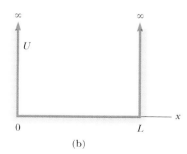

(a)

(b)

Figure 41.3 (a) A particle of mass m and speed u, confined to bouncing between two impenetrable walls separated by a distance L. (b) The potential energy function for the system.

---

[5] If the wave function were not continuous at a point, the derivative of the wave function at that point would be infinite. This result leads to difficulties in the Schrödinger equation, for which the wave function is a solution as discussed in Section 41.3.

[6] We shall show this result explicitly in Section 41.3.

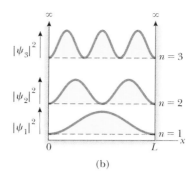

(a)   (b)

ACTIVE FIGURE 41.4

The first three allowed states for a particle confined to a one-dimensional box. The states are shown superimposed on the potential energy function of Figure 41.3b. (a) The wave functions ψ for $n = 1, 2$, and 3. (b) The probability densities $|\psi|^2$ for $n = 1, 2$, and 3. The wave functions and probability densities are plotted vertically from separate axes that are offset vertically for clarity. The positions of these axes on the potential energy function suggest the relative energies of the states.

Sign in at www.thomsonedu.com and go to ThomsonNOW to measure the probability of a particle being between two points for the three quantum states in the figure.

where λ is the de Broglie wavelength associated with the particle. This wave function must satisfy the boundary conditions at the walls. The boundary condition $\psi(0) = 0$ is satisfied already because the sine function is zero when $x = 0$. The boundary condition $\psi(L) = 0$ gives

$$\psi(L) = 0 = A \sin\left(\frac{2\pi L}{\lambda}\right)$$

which can only be true if

$$\frac{2\pi L}{\lambda} = n\pi \quad \rightarrow \quad \lambda = \frac{2L}{n} \tag{41.11}$$

where $n = 1, 2, 3, \ldots$. Therefore, only certain wavelengths for the particle are allowed! Each of the allowed wavelengths corresponds to a quantum state for the system, and n is the quantum number. Incorporating Equation 41.11 in Equation 41.10 gives

◀ **Wave functions for a particle in a box**

$$\psi(x) = A \sin\left(\frac{2\pi x}{2L/n}\right) = A \sin\left(\frac{n\pi x}{L}\right) \tag{41.12}$$

Normalizing this wave function shows that $A = \sqrt{2/L}$. (See Problem 15.) Therefore, the normalized wave function for the particle in a box is

◀ **Normalized wave function for a particle in a box**

$$\psi_n(x) = \sqrt{\frac{2}{L}} \sin\left(\frac{n\pi x}{L}\right) \tag{41.13}$$

Active Figures 41.4a and b are graphical representations of ψ versus x and $|\psi|^2$ versus x for $n = 1, 2$, and 3 for the particle in a box.[7] Although ψ can be positive or negative, $|\psi|^2$ is always positive. Because $|\psi|^2$ represents a probability density, a negative value for $|\psi|^2$ would be meaningless.

Further inspection of Active Figure 41.4b shows that $|\psi|^2$ is zero at the boundaries, satisfying our boundary conditions. In addition, $|\psi|^2$ is zero at other points, depending on the value of n. For $n = 2$, $|\psi|^2 = 0$ at $x = L/2$; for $n = 3$, $|\psi|^2 = 0$ at $x = L/3$ and at $x = 2L/3$. The number of zero points increases by one each time the quantum number increases by one.

---

[7] Note that $n = 0$ is not allowed because, according to Equation 41.12, the wave function would be $\psi = 0$, which is not a physically reasonable wave function. For example, it cannot be normalized because $\int_{-\infty}^{\infty} |\psi|^2 dx = \int_{-\infty}^{\infty} (0) dx = 0$, but Equation 41.7 tells us that this integral must equal 1.

Because the wavelengths of the particle are restricted by the condition $\lambda = 2L/n$, the magnitude of the momentum of the particle is also restricted to specific values, which can be found from the expression for the de Broglie wavelength, Equation 40.15:

$$p = \frac{h}{\lambda} = \frac{h}{2L/n} = \frac{nh}{2L}$$

We have chosen the potential energy of the system to be zero when the particle is inside the box. Therefore, the energy of the system is simply the kinetic energy of the particle and the allowed values are given by

$$E_n = \tfrac{1}{2}mu^2 = \frac{p^2}{2m} = \frac{(nh/2L)^2}{2m}$$

$$E_n = \left(\frac{h^2}{8mL^2}\right)n^2 \quad n = 1, 2, 3, \ldots \qquad (41.14)$$

This expression shows that **the energy of the particle is quantized.** The lowest allowed energy corresponds to the **ground state,** which is the lowest energy state for any system. For the particle in a box, the ground state corresponds to $n = 1$, for which $E_1 = h^2/8mL^2$. Because $E_n = n^2E_1$, the **excited states** corresponding to $n = 2, 3, 4, \ldots$ have energies given by $4E_1, 9E_1, 16E_1, \ldots$.

Active Figure 41.5 is an energy-level diagram describing the energy values of the allowed states. Because the lowest energy of the particle in a box is not zero, then, **according to quantum mechanics, the particle can never be at rest.** The smallest energy it can have, corresponding to $n = 1$, is called the **ground-state energy.** This result contradicts the classical viewpoint, in which $E = 0$ is an acceptable state, as are *all* positive values of E.

Quick Quiz 41.2   Consider an electron, a proton, and an alpha particle (a helium nucleus), each trapped separately in identical boxes. **(i)** Which particle corresponds to the highest ground-state energy? (a) the electron   (b) the proton   (c) the alpha particle   (d) The ground-state energy is the same in all three cases. **(ii)** Which particle has the longest wavelength when the system is in the ground state? (a) the electron   (b) the proton   (c) the alpha particle   (d) All three particles have the same wavelength.

Quick Quiz 41.3   A particle is in a box of length L. Suddenly, the length of the box is increased to $2L$. What happens to the energy levels shown in Active Figure 41.5? (a) Nothing; they are unaffected. (b) They move farther apart. (c) They move closer together.

PITFALL PREVENTION 41.2
Reminder: Energy Belongs to a System

We often refer to the energy of a particle in commonly used language. As in Pitfall Prevention 41.1, we are actually describing the energy of the *system* of the particle and whatever environment is establishing the impenetrable walls. For the particle in a box, the only type of energy is kinetic energy belonging to the particle, which is the origin of the common description.

◄   Quantized energies for a particle in a box

ACTIVE FIGURE 41.5

Energy-level diagram for a particle confined to a one-dimensional box of length L. The lowest allowed energy is $E_1 = h^2/8mL^2$.

Sign in at www.thomsonedu.com and go to ThomsonNOW to adjust the length of the box and the mass of the particle and see the effect on the energy levels.

EXAMPLE 41.2   **Microscopic and Macroscopic Particles in Boxes**

(A) An electron is confined between two impenetrable walls 0.200 nm apart. Determine the energy levels for the states $n = 1, 2,$ and 3.

SOLUTION

Conceptualize   In Figure 41.3a, imagine that the particle is an electron and the walls are very close together.

Categorize   We evaluate the energy levels using an equation developed in this section, so we categorize this example as a substitution problem.

Use Equation 41.14 for the $n = 1$ state:

$$E_1 = \frac{h^2}{8m_eL^2}(1)^2 = \frac{(6.63 \times 10^{-34}\,\text{J}\cdot\text{s})^2}{8(9.11 \times 10^{-31}\,\text{kg})(2.00 \times 10^{-10}\,\text{m})^2}$$

$$= 1.51 \times 10^{-18}\,\text{J} = \boxed{9.42\ \text{eV}}$$

Using $E_n = n^2 E_1$, find the energies of the $n = 2$ and $n = 3$ states:

$$E_2 = (2)^2 E_1 = 4(9.42 \text{ eV}) = \boxed{37.7 \text{ eV}}$$

$$E_3 = (3)^2 E_1 = 9(9.42 \text{ eV}) = \boxed{84.8 \text{ eV}}$$

(B) Find the speed of the electron in the $n = 1$ state.

SOLUTION

Solve the classical expression for kinetic energy for the particle speed:

$$K = \tfrac{1}{2} m_e u^2 \quad \rightarrow \quad u = \sqrt{\frac{2K}{m_e}}$$

Recognize that the kinetic energy of the particle is equal to the system energy and substitute E_n for K:

$$(1) \quad u = \sqrt{\frac{2E_n}{m_e}}$$

Substitute numerical values from part (A):

$$u = \sqrt{\frac{2(1.51 \times 10^{-18} \text{ J})}{9.11 \times 10^{-31} \text{ kg}}} = \boxed{1.82 \times 10^6 \text{ m/s}}$$

Simply placing the electron in the box results in a *minimum* speed of the electron equal to 0.6% of the speed of light!

(C) A 0.500-kg baseball is confined between two rigid walls of a stadium that can be modeled as a box of length 100 m. Calculate the minimum speed of the baseball.

SOLUTION

Conceptualize   In Figure 41.3a, imagine that the particle is a baseball and the walls are those of the stadium.

Categorize   This part of the example is a substitution problem in which we apply a quantum approach to a macroscopic object.

Use Equation 41.14 for the $n = 1$ state:

$$E_1 = \frac{h^2}{8mL^2} (1)^2 = \frac{(6.63 \times 10^{-34} \text{ J} \cdot \text{s})^2}{8(0.500 \text{ kg})(100 \text{ m})^2} = 1.10 \times 10^{-71} \text{ J}$$

Use Equation (1) to find the speed:

$$u = \sqrt{\frac{2(1.10 \times 10^{-71} \text{ J})}{0.500 \text{ kg}}} = 6.63 \times 10^{-36} \text{ m/s}$$

This speed is so small that the object can be considered to be at rest, which is what one would expect for the minimum speed of a macroscopic object.

What If?   What if a sharp line drive is hit so that the baseball is moving with a speed of 150 m/s? What is the quantum number of the state in which the baseball now resides?

Answer   We expect the quantum number to be very large because the baseball is a macroscopic object.

Evaluate the kinetic energy of the baseball:

$$\tfrac{1}{2} mu^2 = \tfrac{1}{2}(0.500 \text{ kg})(150 \text{ m/s})^2 = 5.62 \times 10^3 \text{ J}$$

From Equation 41.14, calculate the quantum number n:

$$n = \sqrt{\frac{8mL^2 E_n}{h^2}} = \sqrt{\frac{8(0.500 \text{ kg})(100 \text{ m})^2(5.62 \times 10^3 \text{ J})}{(6.63 \times 10^{-34} \text{ J} \cdot \text{s})^2}} = 2.26 \times 10^{37}$$

This result is a tremendously large quantum number. As the baseball pushes air out of the way, hits the ground, and rolls to a stop, it moves through more than 10^{37} quantum states. These states are so close together in energy that we cannot observe the transitions from one state to the next. Rather, we see what appears to be a smooth variation in the speed of the ball. The quantum nature of the universe is simply not evident in the motion of macroscopic objects.

EXAMPLE 41.3   **The Expectation Values for the Particle in a Box**

A particle of mass m is confined to a one-dimensional box between $x = 0$ and $x = L$. Find the expectation value of the position x of the particle in the state characterized by quantum number n.

SOLUTION

Conceptualize   Active Figure 41.4b shows that the probability for the particle to be at a given location varies with position within the box. Can you predict what the expectation value of x will be from the symmetry of the wave functions?

Categorize   The statement of the example categorizes the problem for us: we focus on a quantum particle in a box and on the calculation of its expectation value of x.

Analyze   In Equation 41.8, the integration from $-\infty$ to ∞ reduces to the limits 0 to L because $\psi = 0$ everywhere except in the box.

Substitute Equation 41.13 into Equation 41.8 to find the expectation value for x:

$$\langle x \rangle = \int_{-\infty}^{\infty} \psi_n^* x \psi_n \, dx = \int_0^L x \left[\sqrt{\frac{2}{L}} \sin \left(\frac{n\pi x}{L} \right) \right]^2 dx$$

$$= \frac{2}{L} \int_0^L x \sin^2 \left(\frac{n\pi x}{L} \right) dx$$

Evaluate the integral by consulting an integral table or by mathematical integration:[8]

$$\langle x \rangle = \frac{2}{L} \left[\frac{x^2}{4} - \frac{x \sin \left(2 \frac{n\pi x}{L} \right)}{4 \frac{n\pi}{L}} - \frac{\cos \left(2 \frac{n\pi x}{L} \right)}{8 \left(\frac{n\pi}{L} \right)^2} \right]_0^L$$

$$= \frac{2}{L} \left[\frac{L^2}{4} \right] = \boxed{\frac{L}{2}}$$

Finalize   This result shows that the expectation value of x is at the center of the box for all values of n, which you would expect from the symmetry of the square of the wave functions (the probability density) about the center (Active Fig. 41.4b).

The $n = 2$ wave function in Active Figure 41.4b has a value of zero at the midpoint of the box. Can the expectation value of the particle be at a position at which the particle has zero probability of existing? Remember that the expectation value is the *average* position. Therefore, the particle is as likely to be found to the right of the midpoint as to the left, so its average position is at the midpoint even though its probability of being there is zero. As an analogy, consider a group of students for whom the average final examination score is 50%. There is no requirement that some student achieve a score of exactly 50% for the average of all students to be 50%.

Boundary Conditions on Particles in General

The discussion of the particle in a box is very similar to the discussion in Chapter 18 of standing waves on strings:

- Because the ends of the string must be nodes, the wave functions for allowed waves must be zero at the boundaries of the string. Because the particle in a box cannot exist outside the box, the allowed wave functions for the particle must be zero at the boundaries.
- The boundary conditions on the string waves lead to quantized wavelengths and frequencies of the waves. The boundary conditions on the wave function for the particle in a box lead to quantized wavelengths and frequencies of the particle.

[8] To integrate this function, first replace $\sin^2 (n\pi x/L)$ with $\frac{1}{2}(1 - \cos 2n\pi x/L)$ (refer to Table B.3 in Appendix B), which allows $\langle x \rangle$ to be expressed as two integrals. The second integral can then be evaluated by partial integration (Section B.7 in Appendix B).

In quantum mechanics, it is very common for particles to be subject to boundary conditions. We therefore introduce a new analysis model, the **quantum particle under boundary conditions.** In many ways, this model is similar to the waves under boundary conditions model studied in Section 18.3. In fact, the allowed wavelengths for the wave function of a particle in a box (Eq. 41.11) are identical in form to the allowed wavelengths for mechanical waves on a string fixed at both ends (Eq. 18.4).

The quantum particle under boundary conditions model *differs* in some ways from the waves under boundary conditions model:

- In most cases of quantum particles, the wave function is *not* a simple sinusoidal function like the wave function for waves on strings. Furthermore, the wave function for a quantum particle may be a complex function.
- For a quantum particle, frequency is related to energy through $E = hf$, so the quantized frequencies lead to quantized energies.
- There may be no stationary "nodes" associated with the wave function of a quantum particle under boundary conditions. Systems more complicated than the particle in a box have more complicated wave functions, and some boundary conditions may not lead to zeroes of the wave function at fixed points.

In general, **an interaction of a quantum particle with its environment represents one or more boundary conditions, and, if the interaction restricts the particle to a finite region of space, results in quantization of the energy of the system.** Boundary conditions on quantum wave functions are related to the coordinates describing the problem. For the particle in a box, the wave function must be zero at two values of *x*. In the case of a three-dimensional system such as the hydrogen atom we shall discuss in Chapter 42, the problem is best presented in *spherical coordinates*. These coordinates, an extension of the plane polar coordinates introduced in Section 3.1, consist of a radial coordinate *r* and two angular coordinates. The generation of the wave function and application of the boundary conditions for the hydrogen atom are beyond the scope of this book. We shall, however, examine the behavior of some of the hydrogen-atom wave functions in Chapter 42.

Boundary conditions on wave functions that exist for all values of *x* require that the wave function approach zero as $x \rightarrow \infty$ (so that the wave function can be normalized) and remain finite as $x \rightarrow 0$. One boundary condition on any angular parts of wave functions is that adding 2π radians to the angle must return the wave function to the same value because an addition of 2π results in the same angular position.

Quick Quiz 41.4   Which of the following exhibit quantized energy levels? (a) an atom in a crystal   (b) an electron and a proton in a hydrogen atom   (c) a proton in the nucleus of a heavy atom   (d) all of the above   (e) none of the above

---

41.3   The Schrödinger Equation

In Section 34.3, we discussed a wave equation for electromagnetic radiation that follows from Maxwell's equations. The waves associated with particles also satisfy a wave equation. The wave equation for material particles is different from that associated with photons because material particles have a nonzero rest energy. The appropriate wave equation was developed by Schrödinger in 1926. In analyzing the behavior of a quantum system, the approach is to determine a solution to this equation and then apply the appropriate boundary conditions to the solution. This process yields the allowed wave functions and energy levels of the system under consideration. Proper manipulation of the wave function then enables one to calculate all measurable features of the system.

The Schrödinger equation as it applies to a particle of mass m confined to moving along the x axis and interacting with its environment through a potential energy function $U(x)$ is

$$-\frac{\hbar^2}{2m}\frac{d^2\psi}{dx^2} + U\psi = E\psi \qquad (41.15)$$

◀ Time-independent Schrödinger equation

where E is a constant equal to the total energy of the system (the particle and its environment). Because this equation is independent of time, it is commonly referred to as the **time-independent Schrödinger equation.** (We shall not discuss the time-dependent Schrödinger equation in this book.)

The Schrödinger equation is consistent with the principle of conservation of mechanical energy of a system. Problem 25 shows, both for a free particle and a particle in a box, that the first term in the Schrödinger equation reduces to the kinetic energy of the particle multiplied by the wave function. Therefore, Equation 41.15 indicates that the total energy of the system is the sum of the kinetic energy and the potential energy and that the total energy is a constant: $K + U = E =$ constant.

In principle, if the potential energy function U for a system is known, one can solve Equation 41.15 and obtain the wave functions and energies for the allowed states of the system. Because U may be discontinuous in position, it may be necessary to obtain solutions to the equation for different regions of the x axis. Solutions to the Schrödinger equation in different regions must join smoothly at the boundaries, so $\psi(x)$ must be *continuous*. In addition, $d\psi/dx$ must also be continuous for finite values of the potential energy.[9]

The task of solving the Schrödinger equation may be very difficult, depending on the form of the potential energy function. As it turns out, the Schrödinger equation is extremely successful in explaining the behavior of atomic and nuclear systems, whereas classical physics fails to explain this behavior. Furthermore, when quantum mechanics is applied to macroscopic objects, the results agree with classical physics.

AIP Emilio Segrè Visual Archives

ERWIN SCHRÖDINGER
Austrian Theoretical Physicist (1887–1961)
Schrödinger is best known as one of the creators of quantum mechanics. His approach to quantum mechanics was demonstrated to be mathematically equivalent to the more abstract matrix mechanics developed by Heisenberg. Schrödinger also produced important papers in the fields of statistical mechanics, color vision, and general relativity.

The Particle in a Box Revisited

To see how the quantum particle under boundary conditions model is applied to a problem, let's return to our particle in a one-dimensional box of length L (see Fig. 41.3) and analyze it with the Schrödinger equation. Figure 41.3b is the potential-energy diagram that describes this problem. Potential-energy diagrams are a useful representation for understanding and solving problems with the Schrödinger equation.

Because of the shape of the curve in Figure 41.3b, the particle in a box is sometimes said to be in a **square well**,[10] where a **well** is an upward-facing region of the curve in a potential-energy diagram. (A downward-facing region is called a *barrier*, which we investigate in Section 41.5.) Figure 41.3b shows an infinite square well.

In the region $0 < x < L$, where $U = 0$, we can express the Schrödinger equation in the form

$$\frac{d^2\psi}{dx^2} = -\frac{2mE}{\hbar^2}\psi = -k^2\psi \qquad (41.16)$$

where

$$k = \frac{\sqrt{2mE}}{\hbar}$$

PITFALL PREVENTION 41.3
Potential Wells

A potential well such as that in Figure 41.3b is a graphical representation of energy, not a pictorial representation. so you would not see this shape if you were able to observe the situation. A particle moves *only horizontally* at a fixed vertical position in a potential-energy diagram, representing the conserved energy of the system of the particle and its environment.

---

[9] If $d\psi/dx$ were not continuous, we would not be able to evaluate $d^2\psi/dx^2$ in Equation 41.15 at the point of discontinuity.

[10] It is called a square well even if it has a rectangular shape in a potential energy diagram.

The solution to Equation 41.16 is a function ψ whose second derivative is the negative of the same function multiplied by a constant k^2. Both the sine and cosine functions satisfy this requirement. Therefore, the most general solution to the equation is a linear combination of both solutions:

$$\psi(x) = A \sin kx + B \cos kx$$

where A and B are constants that are determined by the boundary and normalization conditions.

The first boundary condition on the wave function is that $\psi(0) = 0$:

$$\psi(0) = A \sin 0 + B \cos 0 = 0 + B = 0$$

which means that $B = 0$. Therefore, our solution reduces to

$$\psi(x) = A \sin kx$$

The second boundary condition, $\psi(L) = 0$, when applied to the reduced solution gives

$$\psi(L) = A \sin kL = 0$$

This equation could be satisfied by setting $A = 0$, but that would mean that $\psi = 0$ everywhere, which is not a valid wave function. The boundary condition is satisfied if kL is an integral multiple of π, that is, if $kL = n\pi$, where n is an integer. Substituting $k = \sqrt{2mE}/\hbar$ into this expression gives

$$kL = \frac{\sqrt{2mE}}{\hbar} L = n\pi$$

Each value of the integer n corresponds to a quantized energy that we call E_n. Solving for the allowed energies E_n gives

$$E_n = \left(\frac{h^2}{8mL^2} \right) n^2 \tag{41.17}$$

which are identical to the allowed energies in Equation 41.14.

Substituting the values of k in the wave function, the allowed wave functions $\psi_n(x)$ are given by

$$\psi_n(x) = A \sin \left(\frac{n\pi x}{L} \right) \tag{41.18}$$

which is the wave function (Eq. 41.12) used in our initial discussion of the particle in a box.

41.4 A Particle in a Well of Finite Height

Now consider a particle in a *finite* potential well, that is, a system having a potential energy that is zero when the particle is in the region $0 < x < L$ and a finite value U when the particle is outside this region as in Figure 41.6. Classically, if the total energy E of the system is less than U, the particle is permanently bound in the potential well. If the particle were outside the well, its kinetic energy would have to be negative, which is an impossibility. According to quantum mechanics, however, **a finite probability exists that the particle can be found outside the well even if $E < U$.** That is, the wave function ψ is generally nonzero outside the well—regions I and III in Figure 41.6—so the probability density $|\psi|^2$ is also nonzero in these regions. Although this notion may be uncomfortable to accept, the uncertainty principle indicates that the energy of the system is uncertain. This uncertainty allows the particle to be outside the well as long as the apparent violation of conservation of energy does not exist in any measurable way.

In region II, where $U = 0$, the allowed wave functions are again sinusoidal because they represent solutions of Equation 41.16. The boundary conditions,

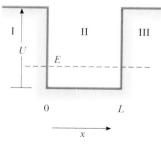

Figure 41.6 Potential energy diagram of a well of finite height U and length L. A particle is trapped in the well. The total energy E of the particle–well system is less than U.

however, no longer require that ψ be zero at the ends of the well, as was the case with the infinite square well.

The Schrödinger equation for regions I and III may be written

$$\frac{d^2\psi}{dx^2} = \frac{2m(U - E)}{\hbar^2}\psi \qquad (41.19)$$

Because $U > E$, the coefficient of ψ on the right-hand side is necessarily positive. Therefore, we can express Equation 41.19 as

$$\frac{d^2\psi}{dx^2} = C^2\psi \qquad (41.20)$$

where $C^2 = 2m(U - E)/\hbar^2$ is a positive constant in regions I and III. As you can verify by substitution, the general solution of Equation 41.20 is

$$\psi = Ae^{Cx} + Be^{-Cx} \qquad (41.21)$$

where A and B are constants.

We can use this general solution as a starting point for determining the appropriate solution for regions I and III. The solution must remain finite as $x \to \pm\infty$. Therefore, in region I, where $x < 0$, the function ψ cannot contain the term Be^{-Cx}. This requirement is handled by taking $B = 0$ in this region to avoid an infinite value for ψ for large negative values of x. Likewise, in region III, where $x > L$, the function ψ cannot contain the term Ae^{Cx}. This requirement is handled by taking $A = 0$ in this region to avoid an infinite value for ψ for large positive x values. Hence, the solutions in regions I and III are

$$\psi_{\mathrm{I}} = Ae^{Cx} \qquad \text{for } x < 0$$

$$\psi_{\mathrm{III}} = Be^{-Cx} \qquad \text{for } x > L$$

In region II, the wave function is sinusoidal and has the general form

$$\psi_{\mathrm{II}}(x) = F \sin kx + G \cos kx$$

where F and G are constants.

These results show that the wave functions outside the potential well (where classical physics forbids the presence of the particle) decay exponentially with distance. At large negative x values, ψ_{I} approaches zero; at large positive x values, ψ_{III} approaches zero. These functions, together with the sinusoidal solution in region II, are shown in Active Figure 41.7a for the first three energy states. In evaluating the complete wave function, we impose the following boundary conditions:

$$\psi_{\mathrm{I}} = \psi_{\mathrm{II}} \quad \text{and} \quad \frac{d\psi_{\mathrm{I}}}{dx} = \frac{d\psi_{\mathrm{II}}}{dx} \quad \text{at } x = 0$$

$$\psi_{\mathrm{II}} = \psi_{\mathrm{III}} \quad \text{and} \quad \frac{d\psi_{\mathrm{II}}}{dx} = \frac{d\psi_{\mathrm{III}}}{dx} \quad \text{at } x = L$$

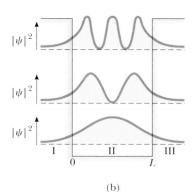

(a)   (b)

ACTIVE FIGURE 41.7

(a) Wave functions ψ and (b) probability densities $|\psi|^2$ for the lowest three energy states for a particle in a potential well of finite height. The states are shown superimposed on the potential energy function of Figure 41.6. The wave functions and probability densities are plotted vertically from separate axes that are offset vertically for clarity. The positions of these axes on the potential energy function suggest the relative energies of the states.

Sign in at www.thomsonedu.com and go to ThomsonNOW to adjust the length of the box and see the effect on the quantized states.

These four boundary conditions and the normalization condition (Eq. 41.7) are sufficient to determine the four constants A, B, F, and G and the allowed values of the energy E. Active Figure 41.7b plots the probability densities for these states. In each case, the wave functions inside and outside the potential well join smoothly at the boundaries.

The notion of trapping particles in potential wells is used in the burgeoning field of **nanotechnology,** which refers to the design and application of devices having dimensions ranging from 1 to 100 nm. The fabrication of these devices often involves manipulating single atoms or small groups of atoms to form structures such as the quantum corral shown in the opening photograph of this chapter.

One area of nanotechnology of interest to researchers is the **quantum dot,** a small region that is grown in a silicon crystal and acts as a potential well. This region can trap electrons into states with quantized energies. The wave functions for a particle in a quantum dot look similar to those in Active Figure 41.7a if L is on the order of nanometers. The storage of binary information using quantum dots is an active field of research. A simple binary scheme would involve associating a one with a quantum dot containing an electron and a zero with an empty dot. Other schemes involve cells of multiple dots such that arrangements of electrons among the dots correspond to ones and zeroes. Several research laboratories are studying the properties and potential applications of quantum dots. Information should be forthcoming from these laboratories at a steady rate in the next few years.

41.5 Tunneling Through a Potential Energy Barrier

Figure 41.8 Wave function ψ for a particle incident from the left on a barrier of height U and width L. The wave function is sinusoidal in regions I and III, but is exponentially decaying in region II. The wave function is plotted vertically from an axis positioned at the energy of the particle.

Consider the potential energy function shown in Figure 41.8. In this situation, the potential energy has a constant value of U in the region of width L and is zero in all other regions.[11] A potential energy function of this shape is called a **square barrier,** and U is called the **barrier height.** A very interesting and peculiar phenomenon occurs when a moving particle encounters such a barrier of finite height and width. Suppose a particle of energy $E < U$ is incident on the barrier from the left (Fig. 41.8). Classically, the particle is reflected by the barrier. If the particle were located in region II, its kinetic energy would be negative, which is not classically allowed. Consequently, region II and therefore region III are both classically *forbidden* to the particle incident from the left. According to quantum mechanics, however, **all regions are accessible to the particle, regardless of its energy.** (Although all regions are accessible, the probability of the particle being in a classically forbidden region is very low.) According to the uncertainty principle, the particle could be within the barrier as long as the time interval during which it is in the barrier is short and consistent with Equation 40.24. If the barrier is relatively narrow, this short time interval can allow the particle to move through the barrier.

Let's approach this situation using a mathematical representation. The Schrödinger equation has valid solutions in all three regions. The solutions in regions I and III are sinusoidal like Equation 41.12. In region II, the solution is exponential like Equation 41.21. Applying the boundary conditions that the wave functions in the three regions and their derivatives must join smoothly at the boundaries, a full solution, such as the one represented by the curve in Figure 41.8, can be found. Because the probability of locating the particle is proportional to $|\psi|^2$, the probability of finding the particle beyond the barrier in region III is nonzero. This result is in complete disagreement with classical physics. The movement of the particle to the far side of the barrier is called **tunneling** or **barrier penetration.**

The probability of tunneling can be described with a **transmission coefficient** T and a **reflection coefficient** R. The transmission coefficient represents the probability that the particle penetrates to the other side of the barrier, and the reflection

PITFALL PREVENTION 41.4
"Height" on an Energy Diagram

The word *height* (as in *barrier height*) refers to an energy in discussions of barriers in potential energy diagrams. For example, we might say the height of the barrier is 10 eV. On the other hand, the barrier *width* refers to the traditional usage of such a word and is an actual physical length measurement between the two locations of the vertical sides of the barrier.

[11] It is common in physics to refer to L as the *length* of a well but the *width* of a barrier.

coefficient is the probability that the particle is reflected by the barrier. Because the incident particle is either reflected or transmitted, we require that $T + R = 1$. An approximate expression for the transmission coefficient that is obtained in the case of $T \ll 1$ (a very wide barrier or a very high barrier, that is, $U \gg E$) is

$$T \approx e^{-2CL} \tag{41.22}$$

where

$$C = \frac{\sqrt{2m(U - E)}}{\hbar} \tag{41.23}$$

This quantum model of barrier penetration and specifically Equation 41.22 show that T can be nonzero. That the phenomenon of tunneling is observed experimentally provides further confidence in the principles of quantum physics.

Quick Quiz 41.5   Which of the following changes would increase the probability of transmission of a particle through a potential barrier? (You may choose more than one answer.) (a) decreasing the width of the barrier   (b) increasing the width of the barrier   (c) decreasing the height of the barrier   (d) increasing the height of the barrier   (e) decreasing the kinetic energy of the incident particle (f) increasing the kinetic energy of the incident particle

---

EXAMPLE 41.4   **Transmission Coefficient for an Electron**

A 30-eV electron is incident on a square barrier of height 40 eV.

(A) What is the probability that the electron tunnels through the barrier if its width is 1.0 nm?

SOLUTION

Conceptualize   Because the particle energy is smaller than the height of the potential barrier, we expect the electron to reflect from the barrier with a probability of 100% according to classical physics. Because of the tunneling phenomenon, however, there is a finite probability that the particle can appear on the other side of the barrier.

Categorize   We evaluate the probability using an equation developed in this section, so we categorize this example as a substitution problem.

Evaluate the quantity $U - E$ that appears in Equation 41.23:

$$U - E = 40 \text{ eV} - 30 \text{ eV} = 10 \text{ eV} \left(\frac{1.6 \times 10^{-19} \text{ J}}{1 \text{ eV}}\right) = 1.6 \times 10^{-18} \text{ J}$$

Evaluate the quantity $2CL$ using Equation 41.23:

$$(1) \quad 2CL = 2 \frac{\sqrt{2(9.11 \times 10^{-31} \text{ kg})(1.6 \times 10^{-18} \text{ J})}}{1.055 \times 10^{-34} \text{ J} \cdot \text{s}} (1.0 \times 10^{-9} \text{ m}) = 32.4$$

From Equation 41.22, find the probability of tunneling through the barrier:

$$T \approx e^{-2CL} = e^{-32.4} = 8.5 \times 10^{-15}$$

(B) What is the probability that the electron tunnels through the barrier if its width is 0.10 nm?

SOLUTION

In this case, the width L in Equation (1) is one-tenth as large, so evaluate the new value of $2CL$:

$$2CL = (0.1)(32.4) = 3.24$$

From Equation 41.22, find the new probability of tunneling through the barrier:

$$T \approx e^{-2CL} = e^{-3.24} = 0.039$$

In part (A), the electron has approximately 1 chance in 10^{14} of tunneling through the barrier. In part (B), however, the electron has a much higher probability (3.9%) of penetrating the barrier. Therefore, reducing the width of the barrier by only one order of magnitude increases the probability of tunneling by about 12 orders of magnitude!

41.6 Applications of Tunneling

As we have seen, tunneling is a quantum phenomenon, a manifestation of the wave nature of matter. Many examples exist (on the atomic and nuclear scales) for which tunneling is very important.

Alpha Decay

One form of radioactive decay is the emission of alpha particles (the nuclei of helium atoms) by unstable, heavy nuclei (Chapter 44). To escape from the nucleus, an alpha particle must penetrate a barrier whose height is several times larger than the energy of the nucleus–alpha particle system. The barrier results from a combination of the attractive nuclear force (discussed in Chapter 44) and the Coulomb repulsion (discussed in Chapter 23) between the alpha particle and the rest of the nucleus. Occasionally, an alpha particle tunnels through the barrier, which explains the basic mechanism for this type of decay and the large variations in the mean lifetimes of various radioactive nuclei.

Figure 41.8 shows the wave function of a particle tunneling through a barrier in one dimension. A similar wave function having spherical symmetry describes the barrier penetration of an alpha particle leaving a radioactive nucleus. The wave function exists both inside and outside the nucleus, and its amplitude is constant in time. In this way, the wave function correctly describes the small but constant probability that the nucleus will decay. The moment of decay cannot be predicted. In general, quantum mechanics implies that the future is indeterminate. This feature is in contrast to classical mechanics, from which the trajectory of an object can be calculated to arbitrarily high precision from precise knowledge of its initial position and velocity and of the forces exerted on it. Do not think that the future is undetermined simply because we have incomplete information about the present. The wave function contains all the information about the state of a system. Sometimes precise predictions can be made, such as the energy of a bound system, but sometimes only probabilities can be calculated about the future. The fundamental laws of nature are probabilistic. Therefore, it appears that Einstein's famous statement about quantum mechanics, "God does not roll dice," was wrong.

A radiation detector can be used to show that a nucleus decays by emitting a particle at a particular moment and in a particular direction. To point out the contrast between this experimental result and the wave function describing it, Schrödinger imagined a box containing a cat, a radioactive sample, a radiation counter, and a vial of poison. When a nucleus in the sample decays, the counter triggers the administration of lethal poison to the cat. Quantum mechanics correctly predicts the probability of finding the cat dead when the box is opened. Before the box is opened, does the cat have a wave function describing it as fractionally dead, with some chance of being alive?

This question is under continuing investigation, never with actual cats but sometimes with interference experiments building upon the experiment described in Section 40.7. Does the act of measurement change the system from a probabilistic to a definite state? When a particle emitted by a radioactive nucleus is detected at one particular location, does the wave function describing the particle drop instantaneously to zero everywhere else in the Universe? (Einstein called such a state change a "spooky action at a distance.") Is there a fundamental difference between a quantum system and a macroscopic system? The answers to these questions are unknown.

Nuclear Fusion

The basic reaction that powers the Sun and, indirectly, almost everything else in the solar system is fusion, which we shall study in Chapter 45. In one step of the process that occurs at the core of the Sun, protons must approach one another to within such a small distance that they fuse and form a deuterium nucleus. (See Section 45.4.) According to classical physics, these protons cannot overcome and penetrate the barrier caused by their mutual electrical repulsion. Quantum mechanically, however, the protons are able to tunnel through the barrier and fuse together.

Scanning Tunneling Microscopes

The scanning tunneling microscope (STM) enables scientists to obtain highly detailed images of surfaces at resolutions comparable to the size of a *single atom*. Figure 41.9, showing the surface of a piece of graphite, demonstrates what STMs can do. What makes this image so remarkable is that its resolution is approximately 0.2 nm. For an optical microscope, the resolution is limited by the wavelength of the light used to make the image. Therefore, an optical microscope has a resolution no better than 200 nm, about half the wavelength of visible light, and so could never show the detail displayed in Figure 41.9.

Scanning tunneling microscopes achieve such high resolution by using the basic idea shown in Figure 41.10. An electrically conducting probe with a very sharp tip is brought near the surface to be studied. The empty space between tip and surface represents the "barrier" we have been discussing, and the tip and surface are the two walls of the "potential well." Because electrons obey quantum rules rather than Newtonian rules, they can "tunnel" across the barrier of empty space. If a voltage is applied between surface and tip, electrons in the atoms of the surface material can tunnel preferentially from surface to tip to produce a tunneling current. In this way, the tip samples the distribution of electrons immediately above the surface.

In the empty space between tip and surface, the electron wave function falls off exponentially (see region II in Fig. 41.8 and Example 41.4). For tip-to-surface distances $z > 1$ nm (that is, beyond a few atomic diameters), essentially no tunneling takes place. This exponential behavior causes the current of electrons tunneling from surface to tip to depend very strongly on z. By monitoring the tunneling current as the tip is scanned over the surface, scientists obtain a sensitive measure of the topography of the electron distribution on the surface. The result of this scan is used to make images like that in Figure 41.9. In this way, the STM can measure the height of surface features to within 0.001 nm, approximately 1/100 of an atomic diameter!

You can appreciate the sensitivity of STMs by examining Figure 41.9. Of the six carbon atoms in each ring, three appear lower than the other three. In fact, all six atoms are at the same height, but all have slightly different electron distributions. The three atoms that appear lower are bonded to other carbon atoms directly beneath them in the underlying atomic layer; as a result, their electron distributions, which are responsible for the bonding, extend downward beneath the surface. The atoms in the surface layer that appear higher do not lie directly over subsurface atoms and hence are not bonded to any underlying atoms. For these higher-appearing atoms, the electron distribution extends upward into the space above the surface. Because STMs map the topography of the electron distribution, this extra electron density makes these atoms appear higher in Figure 41.9.

The STM has one serious limitation: Its operation depends on the electrical conductivity of the sample and the tip. Unfortunately, most materials are not electrically conductive at their surfaces. Even metals, which are usually excellent electrical conductors, are covered with nonconductive oxides. A newer microscope, the atomic force microscope, or AFM, overcomes this limitation.

Resonant Tunneling Devices

Let's expand on the quantum-dot discussion in Section 41.4 by exploring the **resonant tunneling device.** Active Figure 41.11a (page 1204) shows the physical

Figure 41.9   The surface of graphite as "viewed" with a scanning tunneling microscope. This type of microscope enables scientists to see details with a lateral resolution of about 0.2 nm and a vertical resolution of 0.001 nm. The contours seen here represent the ring-like arrangement of individual carbon atoms on the crystal surface.

Figure 41.10   Schematic view of a scanning tunneling microscope. A scan of the tip over the sample can reveal surface contours down to the atomic level. An STM image is composed of a series of scans displaced laterally from one another. (Based on a drawing from P. K. Hansma, V. B. Elings, O. Marti, and C. Bracker, *Science* **242**: 209, 1988. © 1988 by the AAAS.)

ACTIVE FIGURE 41.11

(a) The physical structure of a resonant tunneling device. (b) A potential energy diagram showing the double barrier representing the walls of the quantum dot. (c) A voltage is applied across the device. The distortion of the potential energy curve causes one of the states in the quantum dot to resonate with the incident electron energy.

Sign in at www.thomsonedu.com and go to ThomsonNOW to vary the voltage across the resonant tunneling device.

construction of such a device. The island of gallium arsenide in the center is a quantum dot located between two barriers formed from the thin extensions of aluminum arsenide. Active Figure 41.11b shows both the potential barriers encountered by electrons incident from the left and the quantized energy levels in the quantum dot. This situation differs from the one shown in Figure 41.8 in that there are quantized energy levels on the right of the first barrier. In Figure 41.8, an electron that tunnels through the barrier is considered a free particle and can have any energy. In contrast, in Active Figure 41.11b, as the electron with the energy shown encounters the first barrier it has no energy levels available on the right side of the barrier, which greatly reduces the probability of tunneling.

Active Figure 41.11c shows the effect of applying a voltage: the potential decreases with position as we move to the right across the device. The deformation of the potential barrier results in an energy level in the quantum dot coinciding with the energy of the incident electrons. This "resonance" of energies gives the device its name. When the voltage is applied, the probability of tunneling increases tremendously and the device carries current. In this manner, the device can be used as a very fast switch on a nanotechnological scale.

Resonant Tunneling Transistors

Figure 41.12a shows the addition of a gate electrode at the top of the resonant tunneling device over the quantum dot. This electrode turns the device into a **resonant tunneling transistor.** The basic function of a transistor is amplification, converting a small varying voltage into a large varying voltage. Figure 41.12b, representing the potential-energy diagram for the tunneling transistor, has a slope at the bottom of the quantum dot due to the differing voltages at the source and

Figure 41.12 (a) The addition of a gate electrode to the structure in Active Figure 41.11 converts it to a resonant tunneling transistor. (b) A potential energy diagram showing the double barrier representing the walls of the quantum dot. (c) A voltage is applied to the gate electrode. The potential in the region of the quantum dot drops, along with the quantized energy levels.

drain electrodes. In this configuration, there is no resonance between the electron energies outside the quantum dot and the quantized energies within the dot. By applying a small voltage to the gate electrode as in Figure 41.12c, the quantized energies can be brought into resonance with the electron energy outside the well and resonant tunneling occurs. The resulting current causes a voltage across an external resistor that is much larger than that of the gate voltage; hence, the device amplifies the input signal to the gate electrode.

41.7   The Simple Harmonic Oscillator

Consider a particle that is subject to a linear restoring force $F = -kx$, where k is a constant and x is the position of the particle relative to equilibrium ($x = 0$). The classical motion of a particle subject to such a force is simple harmonic motion, which was discussed in Chapter 15. The potential energy of the system is, from Equation 15.20,

$$U = \tfrac{1}{2}kx^2 = \tfrac{1}{2}m\omega^2 x^2$$

where the angular frequency of vibration is $\omega = \sqrt{k/m}$. Classically, if the particle is displaced from its equilibrium position and released, it oscillates between the points $x = -A$ and $x = A$, where A is the amplitude of the motion. Furthermore, its total energy E is, from Equation 15.21,

$$E = K + U = \tfrac{1}{2}kA^2 = \tfrac{1}{2}m\omega^2 A^2$$

In the classical model, any value of E is allowed, including $E = 0$, which is the total energy when the particle is at rest at $x = 0$.

Let's investigate how the simple harmonic oscillator is treated from a quantum point of view. The Schrödinger equation for this problem is obtained by substituting $U = \tfrac{1}{2}m\omega^2 x^2$ into Equation 41.15:

$$-\frac{\hbar^2}{2m}\frac{d^2\psi}{dx^2} + \tfrac{1}{2}m\omega^2 x^2\psi = E\psi \qquad (41.24)$$

The mathematical technique for solving this equation is beyond the level of this book; nonetheless, it is instructive to guess at a solution. We take as our guess the following wave function:

$$\psi = Be^{-Cx^2} \qquad (41.25)$$

Substituting this function into Equation 41.24 shows that it is a satisfactory solution to the Schrödinger equation, provided that

$$C = \frac{m\omega}{2\hbar} \quad \text{and} \quad E = \tfrac{1}{2}\hbar\omega$$

It turns out that the solution we have guessed corresponds to the ground state of the system, which has an energy $\tfrac{1}{2}\hbar\omega$. Because $C = m\omega/2\hbar$, it follows from Equation 41.25 that the wave function for this state is

$$\psi = Be^{-(m\omega/2\hbar)x^2} \qquad (41.26)$$

◀ Wave function for the ground state of a simple harmonic oscillator

where B is a constant to be determined from the normalization condition. This result is but one solution to Equation 41.24. The remaining solutions that describe the excited states are more complicated, but all solutions include the exponential factor e^{-Cx^2}.

The energy levels of a harmonic oscillator are quantized as we would expect because the oscillating particle is bound to stay near $x = 0$. The energy of a state having an arbitrary quantum number n is

$$E_n = (n + \tfrac{1}{2})\hbar\omega \quad n = 0, 1, 2, \ldots \qquad (41.27)$$

◀ Quantized energies for a simple harmonic oscillator

The state $n = 0$ corresponds to the ground state, whose energy is $E_0 = \tfrac{1}{2}\hbar\omega$; the state $n = 1$ corresponds to the first excited state, whose energy is $E_1 = \tfrac{3}{2}\hbar\omega$; and so

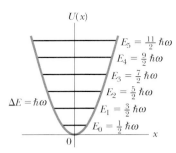

Figure 41.13 Energy-level diagram for a simple harmonic oscillator, superimposed on the potential energy function. The levels are equally spaced, with separation $\hbar\omega$. The ground-state energy is $E_0 = \frac{1}{2}\hbar\omega$.

on. The energy-level diagram for this system is shown in Figure 41.13. The separations between adjacent levels are equal and given by

$$\Delta E = \hbar\omega \tag{41.28}$$

Notice that the energy levels for the harmonic oscillator in Figure 41.13 are equally spaced, just as Planck proposed for the oscillators in the walls of the cavity that was used in the model for blackbody radiation in Section 40.1. Planck's Equation 40.4 for the energy levels of the oscillators differs from Equation 41.27 only in the term $\frac{1}{2}$ added to n. This additional term does not affect the energy emitted in a transition, given by Equation 40.5, which is equivalent to Equation 41.28. That Planck generated these concepts without the benefit of the Schrödinger equation is testimony to his genius.

EXAMPLE 41.5 | **Molar Specific Heat of Hydrogen Gas**

In Figure 21.7 (Section 21.4), which shows the molar specific heat of hydrogen as a function of temperature, vibration does not contribute to the molar specific heat at room temperature. Explain why, modeling the hydrogen molecule as a simple harmonic oscillator. The effective spring constant for the bond in the hydrogen molecule is 573 N/m.

SOLUTION

Conceptualize Imagine the only mode of vibration available to a diatomic molecule. This mode (shown in Fig. 21.6c) consists of the two atoms always moving in opposite directions with equal speeds.

Categorize We categorize this example as a quantum harmonic oscillator problem, with the molecule modeled as a two-particle system.

Analyze The motion of the particles relative to the center of mass can be analyzed by considering the oscillation of a single particle with reduced mass μ. (See Problem 39.)

Use the result of Problem 39 to evaluate the reduced mass of the hydrogen molecule, in which the masses of the two particles are the same:

$$\mu = \frac{m_1 m_2}{m_1 + m_2} = \frac{m^2}{2m} = \frac{1}{2}m$$

Using Equation 41.28, calculate the energy necessary to excite the molecule from its ground vibrational state to its first excited vibrational state:

$$\Delta E = \hbar\omega = \hbar\sqrt{\frac{k}{\mu}} = \hbar\sqrt{\frac{k}{\frac{1}{2}m}} = \hbar\sqrt{\frac{2k}{m}}$$

Substitute numerical values, noting that m is the mass of a hydrogen atom:

$$\Delta E = (1.055 \times 10^{-34}\,\text{J}\cdot\text{s})\sqrt{\frac{2(573\,\text{N/m})}{1.67 \times 10^{-27}\,\text{kg}}} = 8.74 \times 10^{-20}\,\text{J}$$

Set this energy equal to $\frac{3}{2}k_\text{B}T$ from Equation 21.4 and find the temperature at which the average molecular translational kinetic energy is equal to that required to excite the first vibrational state of the molecule:

$$\frac{3}{2}k_\text{B}T = 8.74 \times 10^{-20}\,\text{J}$$

$$T = \frac{2}{3}\left(\frac{8.74 \times 10^{-20}\,\text{J}}{k_\text{B}}\right) = \frac{2}{3}\left(\frac{8.74 \times 10^{-20}\,\text{J}}{1.38 \times 10^{-23}\,\text{J/K}}\right) = 4.22 \times 10^3\,\text{K}$$

Finalize   The temperature of the gas must be more than 4 000 K for the translational kinetic energy to be comparable to the energy required to excite the first vibrational state. This excitation energy must come from collisions between molecules, so if the molecules do not have sufficient translational kinetic energy, they cannot be excited to the first vibrational state and vibration does not contribute to the molar specific heat. Hence, the curve in Figure 21.7 does not rise to a value corresponding to the contribution of vibration until the hydrogen gas has been raised to thousands of kelvins.

Figure 21.7 shows that rotational energy levels must be more closely spaced in energy than vibrational levels because they are excited at a lower temperature than the vibrational levels. The translational energy levels are those of a particle in a three-dimensional box, where the box is the container holding the gas. These levels are given by an expression similar to Equation 41.14. Because the box is macroscopic in size, L is very large and the energy levels are very close together. In fact, they are so close together that translational energy levels are excited at a fraction of a kelvin.

Summary

ThomsonNOW™   Sign in at **www.thomsonedu.com** and go to ThomsonNOW to take a practice test for this chapter.

DEFINITIONS

The **wave function** Ψ for a system is a mathematical function that can be written as a product of a space function ψ for one particle of the system and a complex time function:

$$\Psi(\vec{r}_1, \vec{r}_2, \vec{r}_3, \ldots, \vec{r}_j, \ldots, t) = \psi(\vec{r}_j)e^{-i\omega t} \qquad (41.2)$$

where $\omega \ (= 2\pi f)$ is the angular frequency of the wave function and $i = \sqrt{-1}$. The wave function contains within it all the information that can be known about the particle.

The measured position x of a particle, averaged over many trials, is called the **expectation value** of x and is defined by

$$\langle x \rangle \equiv \int_{-\infty}^{\infty} \psi^* x \psi \, dx \qquad (41.8)$$

CONCEPTS AND PRINCIPLES

In quantum mechanics, a particle in a system can be represented by a wave function $\psi(x, y, z)$. The probability per unit volume (or probability density) that a particle will be found at a point is $|\psi|^2 = \psi^* \psi$, where ψ^* is the complex conjugate of ψ. If the particle is confined to moving along the x axis, the probability that it is located in an interval dx is $|\psi|^2 \, dx$. Furthermore, the sum of all these probabilities over all values of x must be 1:

$$\int_{-\infty}^{\infty} |\psi|^2 dx = 1 \qquad (41.7)$$

This expression is called the **normalization condition.**

If a particle of mass m is confined to moving in a one-dimensional box of length L whose walls are impenetrable, then ψ must be zero at the walls and outside the box. The wave functions for this system are given by

$$\psi(x) = A \sin\left(\frac{n\pi x}{L}\right) \quad n = 1, 2, 3, \ldots \quad (41.12)$$

where A is the maximum value of ψ. The allowed states of a particle in a box have quantized energies given by

$$E_n = \left(\frac{h^2}{8mL^2}\right)n^2 \quad n = 1, 2, 3, \ldots \quad (41.14)$$

The wave function for a system must satisfy the **Schrödinger equation.** The time-independent Schrödinger equation for a particle confined to moving along the x axis is

$$-\frac{\hbar^2}{2m}\frac{d^2\psi}{dx^2} + U\psi = E\psi \qquad (41.15)$$

where U is the potential energy of the system and E is the total energy.

(*continued*)

ANALYSIS MODEL FOR PROBLEM SOLVING

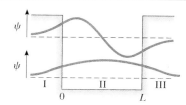

Quantum Particle Under Boundary Conditions. An interaction of a quantum particle with its environment represents one or more boundary conditions. If the interaction restricts the particle to a finite region of space, the energy of the system is quantized. All wave functions must satisfy the following four boundary conditions: (1) $\psi(x)$ must remain finite as x approaches 0, (2) $\psi(x)$ must approach zero as x approaches $\pm\infty$, (3) $\psi(x)$ must be continuous for all values of x, and (4) $d\psi/dx$ must be continuous for all finite values of $U(x)$. If the solution to Equation 41.15 is piecewise, conditions (3) and (4) must be applied at the boundaries between regions of x in which Equation 41.15 has been solved.

Questions

☐ denotes answer available in *Student Solutions Manual/Study Guide;* **O** denotes objective question

1. What is the significance of the wave function ψ?

2. **O** The probability of finding a certain quantum particle in the section of the x axis between $x = 4$ nm and $x = 7$ nm is 48%. The particle's wave function $\psi(x)$ is constant over this range. (i) What numerical value can be attributed to $\psi(x)$? (a) 0.48 (b) 0.16 (c) 0.12 (d) 0.69 (e) 0.40 (f) 0.35 (ii) What is its unit? (a) nm (b) $(\text{nm})^{1/2}$ (c) $(\text{nm})^{1/3}$ (d) $(\text{nm})^{-1}$ (e) $(\text{nm})^{-1/2}$ (f) $(\text{nm})^{-1/3}$

3. **O** (i) Is each one of the following statements (a) through (j) true or false for a photon? (ii) Is each one of the statements true or false for an electron? Make a list of your answers.
   (a) It is a quantum particle, behaving in some experiments like a classical particle and in some experiments like a classical wave.
   (b) Its rest energy is zero.
   (c) Its rest energy is nonzero.
   (d) It carries energy in its motion.
   (e) It carries momentum in its motion.
   (f) Its motion is described by a wave function that has a wavelength and satisfies a wave equation.
   (g) For one-dimensional motion, the wave equation is

$$\frac{\partial^2 E}{\partial x^2} = \mu_0 \epsilon_0 \frac{\partial^2 E}{\partial t^2}$$

   where the wave function E is the electric field magnitude, with a similar equation for $B = E/c$.
   (h) For one-dimensional motion, the wave equation is

$$-\frac{\hbar^2}{2m}\frac{d^2\psi}{dx^2} + U\psi = E\psi$$

   where E is the constant energy.

   (i) The intensity of the wave is proportional to the square of its wave function.
   (j) The intensity of the wave is measured from the rate at which the quantum particles bombard a detector.

4. **O** A quantum particle of mass m_1 is in a square well with infinitely high walls and length 3 nm. Rank the situations (a) through (e) according to the particle's energy from highest to lowest, noting any cases of equality. (a) The particle of mass m_1 is in the ground state of the well. (b) The same particle is in the $n = 2$ excited state of the same well. (c) A particle with mass $2m_1$ is in the ground state of the same well. (d) A particle of mass m_1 is in the ground state of a well of length 6 nm. (e) A particle of mass m_1 is in the ground state of a well of length 3 nm, and the uncertainty principle has become inoperative; that is, Planck's constant has been reduced to zero.

5. For a quantum particle in a box, the probability density at certain points is zero as seen in Active Figure 41.4b. Does this value imply that the particle cannot move across these points? Explain.

6. Discuss the relationship between ground-state energy and the uncertainty principle.

7. **O** Two square wells have the same length. Well 1 has walls of finite height, and well 2 has walls of infinite height. Both wells contain identical quantum particles, one in each well. (i) Is the wavelength of the ground-state wave function (a) greater for well 1, (b) greater for well 2, or (c) equal for both wells? (ii) Is the magnitude of the ground-state momentum (a) greater for well 1, (b) greater for well 2, or (c) equal for both wells? (iii) Is the ground-state energy of the particle (a) greater for well 1, (b) greater for well 2, or (c) equal for both wells?

8. What is the Schrödinger equation? How is it useful in describing quantum phenomena?

9. **O** A beam of quantum particles with kinetic energy 2 eV is reflected from a potential barrier of small width and original height 3 eV. How does the fraction of the particles that are reflected change as the barrier height is reduced to 2.01 eV? (a) It increases. (b) It decreases. (c) It stays constant at zero. (d) It stays constant at 1. (e) It stays constant with some other value.

10. **O** Suppose a tunneling current in an electronic device goes through a potential-energy barrier. The tunneling current is small because the width of the barrier is large and the barrier is high. To increase the current most effectively, what should you do? (a) Reduce the width of the barrier. (b) Reduce the height of the barrier. (c) Either choice (a) or choice (b) is equally effective. (d) Neither choice (a) nor choice (b) increases the current.

11. A philosopher once said that "it is necessary for the very existence of science that the same conditions always produce the same results." In view of what has been discussed in this chapter, present an argument showing that this statement is false. How might the statement be reworded to make it true?

12. In quantum mechanics, it is possible for the energy E of a particle to be less than the potential energy, but classically this condition is not possible. Explain.

13. **O** Unlike the idealized diagram of Figure 41.10, a typical tip used for a scanning tunneling microscope is rather jagged on the atomic scale, with several irregularly spaced points. For such a tip, does most of the tunneling current occur between the sample and (a) all the points of the tip equally, (b) the most centrally located point, (c) the point closest to the sample, or (d) the point farthest from the sample?

Problems

WebAssign The Problems from this chapter may be assigned online in WebAssign.

ThomsonNOW™ Sign in at **www.thomsonedu.com** and go to ThomsonNOW to assess your understanding of this chapter's topics with additional quizzing and conceptual questions.

1, 2, 3 denotes straightforward, intermediate, challenging; ☐ denotes full solution available in *Student Solutions Manual/Study Guide;* ▲ denotes coached solution with hints available at **www.thomsonedu.com;** denotes developing symbolic reasoning; ● denotes asking for qualitative reasoning; ▀ denotes computer useful in solving problem

Section 41.1 An Interpretation of Quantum Mechanics

1. ▲ A free electron has a wave function

$$\psi(x) = Ae^{i(5.00 \times 10^{10}x)}$$

where x is in meters. Find (a) its de Broglie wavelength, (b) its momentum, and (c) its kinetic energy in electron volts.

2. The wave function for a quantum particle is

$$\psi(x) = \sqrt{\frac{a}{\pi(x^2 + a^2)}}$$

for $a > 0$ and $-\infty < x < +\infty$. Determine the probability that the particle is located somewhere between $x = -a$ and $x = +a$.

Section 41.2 The Quantum Particle Under Boundary Conditions

3. An electron is confined to a one-dimensional region in which its ground-state ($n = 1$) energy is 2.00 eV. (a) What is the length L of the region? (b) What energy input is required to promote the electron to its first excited state?

4. An electron that has an energy of approximately 6 eV moves between infinitely high walls 1.00 nm apart. Find (a) the quantum number n for the energy state that the electron occupies and (b) the precise energy of the electron.

5. ▲ An electron is contained in a one-dimensional box of length 0.100 nm. (a) Draw an energy-level diagram for the electron for levels up to $n = 4$. (b) Find the wavelengths of all photons that can be emitted by the electron in making downward transitions that could eventually carry it from the $n = 4$ state to the $n = 1$ state.

6. A bead of mass 5.00 g slides on a horizontal wire 20.0 cm long. Its speed is 0.100 nm/yr, so it is apparently at rest. Treat this system as a quantum particle in a one-dimensional well with infinitely high walls. Calculate the quantum number of the state described.

7. A ruby laser emits 694.3-nm light. Assume light of this wavelength is due to a transition of an electron in a box from its $n = 2$ state to its $n = 1$ state. Find the length of the box.

8. A laser emits light of wavelength λ. Assume this light is due to a transition of an electron in a box from its $n = 2$ state to its $n = 1$ state. Find the length of the box.

9. ● The nuclear potential energy that binds protons and neutrons in a nucleus is often approximated by a square well. Imagine a proton confined in an infinitely high square well of length 10.0 fm, a typical nuclear diameter. Calculate the wavelength and energy associated with the photon emitted when the proton moves from the $n = 2$ state to the ground state. Identify the region of the electromagnetic spectrum to which this wavelength belongs.

10. ● A proton is confined to move in a one-dimensional box of length 0.200 nm. (a) Find the lowest possible energy of the proton. (b) **What If?** What is the lowest possible energy of an electron confined to the same box? (c) How do you account for the great difference in your results for parts (a) and (b)?

11. ● Use the quantum-particle-in-a-box model to calculate the first three energy levels of a neutron trapped in an

atomic nucleus of diameter 20.0 fm. Explain whether the energy-level differences have a realistic order of magnitude.

12. A photon with wavelength λ is absorbed by an electron confined to a box. As a result, the electron moves from state $n = 1$ to $n = 4$. (a) Find the length of the box. (b) What is the wavelength of the photon emitted in the transition of that electron from the state $n = 4$ to the state $n = 2$?

13. ● For a quantum particle of mass m in the ground state of a square well with length L and infinitely high walls, the uncertainty in position is $\Delta x \approx L$. (a) Use the uncertainty principle to estimate the uncertainty in its momentum. (b) Because the particle stays inside the box, its average momentum must be zero. Its average squared momentum is then $\langle p^2 \rangle \approx (\Delta p)^2$. Estimate the energy of the particle. State how the result compares with the actual ground-state energy.

14. ● A quantum particle in an infinitely deep square well has a wave function given by

$$\psi_2(x) = \sqrt{\frac{2}{L}} \sin\left(\frac{2\pi x}{L}\right)$$

for $0 \leq x \leq L$ and zero otherwise. (a) Determine the expectation value of x. (b) Determine the probability of finding the particle near $L/2$ by calculating the probability that the particle lies in the range $0.490L \leq x \leq 0.510L$. (c) **What If?** Determine the probability of finding the particle near $L/4$ by calculating the probability that the particle lies in the range $0.240L \leq x \leq 0.260L$. (d) Argue that the result of part (a) does not contradict the results of parts (b) and (c).

15. The wave function for a quantum particle confined to moving in a one-dimensional box is

$$\psi(x) = A \sin\left(\frac{n\pi x}{L}\right)$$

Use the normalization condition on ψ to show that

$$A = \sqrt{\frac{2}{L}}$$

Suggestion: Because the length of the box is L, the wave function is zero for $x < 0$ and for $x > L$, so the normalization condition (Eq. 41.7) reduces to

$$\int_0^L |\psi|^2 \, dx = 1$$

16. An electron is trapped in an infinitely deep potential well 0.300 nm in length. (a) If the electron is in its ground state, what is the probability of finding it within 0.100 nm of the left-hand wall? (b) Identify the classical probability of finding the electron in this interval and state how it compares with the answer to part (a). (c) Repeat parts (a) and (b) assuming the particle is in the 99th energy state.

17. ● An electron in an infinitely deep square well has a wave function that is given by

$$\psi_2(x) = \sqrt{\frac{2}{L}} \sin\left(\frac{2\pi x}{L}\right)$$

for $0 \leq x \leq L$ and is zero otherwise. What are the most probable positions of the electron? Explain how you identify them.

18. ⬛ A quantum particle is in the $n = 1$ state of an infinitely deep square well with walls at $x = 0$ and $x = L$. Let ℓ be an arbitrary value of x between $x = 0$ and $x = L$. (a) Find an expression for the probability, as a function of ℓ, that the particle will be found between $x = 0$ and $x = \ell$. (b) Sketch the probability as a function of the variable ℓ/L. Choose values of ℓ/L ranging from 0 to 1.00 in steps of 0.100. (c) Explain why the probability function must have particular values at $\ell/L = 0$ and at $\ell/L = 1$. (d) Find the value of ℓ for which the probability of finding the particle between $x = 0$ and $x = \ell$ is twice the probability of finding the particle between $x = \ell$ and $x = L$. You can solve the transcendental equation for ℓ/L numerically.

19. ● A quantum particle in an infinitely deep square well has a wave function

$$\psi_1(x) = \sqrt{\frac{2}{L}} \sin\left(\frac{\pi x}{L}\right)$$

for $0 \leq x \leq L$ and is zero otherwise. (a) Determine the probability of finding the particle between $x = 0$ and $x = L/3$. (b) Use the result of this calculation and a symmetry argument to find the probability of finding the particle between $x = L/3$ and $x = 2L/3$. Do not re-evaluate the integral. (c) **What If?** State how the result of part (a) compares with the classical probability.

Section 41.3 The Schrödinger Equation

20. The wave function of a quantum particle is

$$\psi(x) = A \cos(kx) + B \sin(kx)$$

where A, B, and k are constants. Show that ψ is a solution of the Schrödinger equation (Eq. 41.15), assuming the particle is free ($U = 0$), and find the corresponding energy E of the particle.

21. Show that the wave function $\psi = Ae^{i(kx-\omega t)}$ is a solution to the Schrödinger equation (Eq. 41.15), where $k = 2\pi/\lambda$ and $U = 0$.

22. In a region of space, a quantum particle with zero total energy has a wave function

$$\psi(x) = Axe^{-x^2/L^2}$$

(a) Find the potential energy U as a function of x. (b) Make a sketch of $U(x)$ versus x.

23. A quantum particle of mass m moves in a potential well of length $2L$. Its potential energy is infinite for $x < -L$ and for $x > +L$. Inside the region $-L < x < L$, its potential energy is given by

$$U(x) = -\frac{\hbar^2 x^2}{mL^2(L^2 - x^2)}$$

In addition, the particle is in a stationary state that is described by the wave function $\psi(x) = A(1 - x^2/L^2)$ for $-L < x < +L$ and by $\psi(x) = 0$ elsewhere. (a) Determine the energy of the particle in terms of $\hbar$, m, and L. *Suggestion:* Use the Schrödinger equation, Eq. 41.15. (b) Show that

$$A = \left(\frac{15}{16L}\right)^{1/2}$$

(c) Determine the probability that the particle is located between $x = -L/3$ and $x = +L/3$.

24. Consider a quantum particle moving in a one-dimensional box for which the walls are at $x = -L/2$ and $x = L/2$. (a) Write the wave functions and probability densities for $n = 1$, $n = 2$, and $n = 3$. (b) Sketch the wave functions and probability densities. *Suggestion:* Make an analogy to the case of a particle in a box for which the walls are at $x = 0$ and $x = L$.

25. Prove that the first term in the Schrödinger equation, $-(\hbar^2/2m)(d^2\psi/dx^2)$, reduces to the kinetic energy of the quantum particle multiplied by the wave function (a) for a freely moving particle, with the wave function given by Equation 41.4, and (b) for a particle in a box, with the wave function given by Equation 41.13.

Section 41.4 A Particle in a Well of Finite Height

26. Sketch the wave function $\psi(x)$ and the probability density $|\psi(x)|^2$ for the $n = 4$ state of a quantum particle in a finite potential well. (See Active Fig. 41.7.)

27. Suppose a quantum particle is trapped in its ground state in a box that has infinitely high walls (see Active Fig. 41.4a). Now suppose the left-hand wall is suddenly lowered to a finite height and width. (a) Qualitatively sketch the wave function for the particle a short time later. (b) If the box has a length L, what is the wavelength of the wave that penetrates the left-hand wall?

Section 41.5 Tunneling Through a Potential Energy Barrier

28. An electron having total energy $E = 4.50$ eV approaches a rectangular energy barrier with $U = 5.00$ eV and $L = 950$ pm as shown in Figure P41.28. Classically, the electron cannot pass through the barrier because $E < U$. Quantum-mechanically, however, the probability of tunneling is not zero. Calculate this probability, which is the transmission coefficient.

Figure P41.28 Problems 28, 29, and 31.

29. **What If?** In Problem 28, by how much would the width L of the potential barrier have to be increased for the chance of an incident 4.50-eV electron tunneling through the barrier to be one in a million?

30. Calculate the transmission probability for quantum-mechanical tunneling in each of the following cases. (a) An electron with an energy deficit of $U - E = 0.010\ 0$ eV is incident on a barrier of width $L = 0.100$ nm. (b) An electron with an energy deficit of 1.00 eV is incident on the same barrier. (c) An alpha particle (mass 6.65×10^{-27} kg) with an energy deficit of 1.00 MeV is incident on a barrier of width 1.00 fm. (d) An 8.00-kg bowling ball with an energy deficit of 1.00 J is incident on a barrier of width 2.00 cm.

31. An electron with kinetic energy $E = 5.00$ eV is incident on a barrier with thickness $L = 0.200$ nm and height $U = 10.0$ eV (Fig. P41.28). What is the probability that the electron (a) tunnels through the barrier? (b) Is reflected?

32. An electron has a kinetic energy of 12.0 eV. The electron is incident upon a rectangular barrier of height 20.0 eV and thickness 1.00 nm. By what factor would the electron's probability of tunneling through the barrier increase if the electron absorbed all the energy of a photon of green light (with wavelength 546 nm) at the instant it reached the barrier?

Section 41.6 Applications of Tunneling

33. A scanning tunneling microscope (STM) can precisely determine the depths of surface features because the current through its tip is very sensitive to differences in the width of the gap between the tip and the sample surface. Assume the electron wave function falls off exponentially in this direction with a decay length of 0.100 nm, that is, with $C = 10.0/\text{nm}$. Determine the ratio of the current when the STM tip is 0.500 nm above a surface feature to the current when the tip is 0.515 nm above the surface.

34. The design criterion for a typical scanning tunneling microscope specifies that it must be able to detect, on the sample below its tip, surface features that differ in height by only 0.002 00 nm. To achieve this resolution, what percentage change in electron transmission must the electronics of the STM be able to detect? Assume the electron transmission coefficient is e^{-2CL} with $C = 10.0/\text{nm}$.

Section 41.7 The Simple Harmonic Oscillator

Note: Problem 41 in Chapter 16 can be assigned with this section.

35. Show that Equation 41.26 is a solution of Equation 41.24 with energy $E = \frac{1}{2}\hbar\omega$.

36. A one-dimensional harmonic oscillator wave function is

$$\psi = Axe^{-bx^2}$$

(a) Show that ψ satisfies Equation 41.24. (b) Find b and the total energy E. (c) Is this wave function for the ground state or for the first excited state?

37. A quantum simple harmonic oscillator consists of an electron bound by a restoring force proportional to its position relative to a certain equilibrium point. The proportionality constant is 8.99 N/m. What is the longest wavelength of light that can excite the oscillator?

38. (a) Normalize the wave function for the ground state of a simple harmonic oscillator. That is, apply Equation 41.7 to Equation 41.26 and find the required value for the constant B in terms of m, ω, and fundamental constants. (b) Determine the probability of finding the oscillator in a narrow interval $-\delta/2 < x < \delta/2$ around its equilibrium position.

39. Two particles with masses m_1 and m_2 are joined by a light spring of force constant k. They vibrate along a straight line with their center of mass fixed. (a) Show that the total energy

$$\tfrac{1}{2}m_1u_1^2 + \tfrac{1}{2}m_2u_2^2 + \tfrac{1}{2}kx^2$$

can be written as $\frac{1}{2}\mu u^2 + \frac{1}{2}kx^2$, where $u = |u_1| + |u_2|$ is the *relative* speed of the particles and $\mu = m_1 m_2/(m_1 + m_2)$ is the reduced mass of the system. This result demonstrates

2 = intermediate;   3 = challenging;   □ = SSM/SG;   ▲ = ThomsonNOW;   = symbolic reasoning;   ● = qualitative reasoning

that the pair of freely vibrating particles can be precisely modeled as a single particle vibrating on one end of a spring that has its other end fixed. (b) Differentiate the equation

$$\tfrac{1}{2}\mu u^2 + \tfrac{1}{2}kx^2 = \text{constant}$$

with respect to x. Proceed to show that the system executes simple harmonic motion. Find its frequency.

40. The total energy of a particle–spring system in which the particle moves with simple harmonic motion along the x axis is

$$E = \frac{p_x{}^2}{2m} + \frac{kx^2}{2}$$

where p_x is the momentum of the quantum particle and k is the spring constant. (a) Using the uncertainty principle, show that this expression can also be written as

$$E \geq \frac{p_x{}^2}{2m} + \frac{k\hbar^2}{8p_x{}^2}$$

(b) Show that the minimum energy of the harmonic oscillator is

$$E_{\min} = K + U = \tfrac{1}{4}\hbar\sqrt{\frac{k}{m}} + \frac{\hbar\omega}{4} = \frac{\hbar\omega}{2}$$

Additional Problems

41. A marble rolls back and forth across a shoebox at a constant speed of 0.8 m/s. Make an order-of-magnitude estimate of the probability of its escaping through the wall of the box by quantum tunneling. State the quantities you take as data and the values you measure or estimate for them.

42. A particle of mass 2.00×10^{-28} kg is confined to a one-dimensional box of length 1.00×10^{-10} m. For $n = 1$, what are (a) the particle's wavelength, (b) its momentum, and (c) its ground-state energy?

43. ▲ ● An electron is represented by the time-independent wave function

$$\psi(x) = \begin{cases} Ae^{-\alpha x} & \text{for } x > 0 \\ Ae^{+\alpha x} & \text{for } x < 0 \end{cases}$$

(a) Sketch the wave function as a function of x. (b) Sketch the probability density representing the likelihood that the electron is found between x and $x + dx$. (c) Only an infinite value of potential energy could produce the discontinuity in the derivative of the wave function at $x = 0$. Aside from this feature, argue that $\psi(x)$ can be a physically reasonable wave function. (d) Normalize the wave function. (e) Determine the probability of finding the electron somewhere in the range of

$$x_1 = -\frac{1}{2\alpha} \quad \text{to} \quad x_2 = \frac{1}{2\alpha}$$

44. Prove that assuming $n = 0$ for a quantum particle in an infinitely deep potential well leads to a violation of the uncertainty principle $\Delta p_x \, \Delta x \geq \hbar/2$.

45. ▲ An electron in an infinitely deep potential well has a ground-state energy of 0.300 eV. (a) Show that the photon emitted in a transition from the $n = 3$ state to the

$n = 1$ state has a wavelength of 517 nm, which makes it green visible light. (b) Find the wavelength and the spectral region for each of the other five transitions that take place among the four lowest energy levels.

46. Particles incident from the left in Figure P41.46 are confronted with a step in potential energy. The step has a height U at $x = 0$. The particles have energy $E > U$. Classically, all the particles would continue moving forward with reduced speed. According to quantum mechanics, however, a fraction of the particles are reflected at the step. (a) Prove that the reflection coefficient R for this case is

$$R = \frac{(k_1 - k_2)^2}{(k_1 + k_2)^2}$$

where $k_1 = 2\pi/\lambda_1$ and $k_2 = 2\pi/\lambda_2$ are the wave numbers for the incident and transmitted particles. Proceed as follows. Show that the wave function $\psi_1 = Ae^{ik_1x} + Be^{-ik_1x}$ satisfies the Schrödinger equation in region 1, for $x < 0$. Here Ae^{ik_1x} represents the incident beam and Be^{-ik_1x} represents the reflected particles. Show that $\psi_2 = Ce^{ik_2x}$ satisfies the Schrödinger equation in region 2, for $x > 0$. Impose the boundary conditions $\psi_1 = \psi_2$ and $d\psi_1/dx = d\psi_2/dx$, at $x = 0$, to find the relationship between B and A. Then evaluate $R = B^2/A^2$. (b) A particle that has kinetic energy $E = 7.00$ eV is incident from a region where the potential energy is zero onto one in which $U = 5.00$ eV. Find its probability of being reflected and its probability of being transmitted.

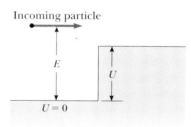

Figure P41.46   Problems 46 and 47.

47. Particles incident from the left in Figure P41.46 are confronted with a step in potential energy. The step has a height U at $x = 0$, and the particles have energy $E = 2U$. Classically, all the particles would pass into the region of higher potential energy at the right. According to quantum mechanics, however, a fraction of the particles are reflected at the barrier. Use the result of Problem 46 to determine the fraction of the incident particles that are reflected. This situation is analogous to the partial reflection and transmission of light striking an interface between two different media.

48. An electron is trapped in a quantum dot. The quantum dot may be modeled as a one-dimensional, rigid-walled box of length 1.00 nm. (a) Sketch the wave functions and probability densities for the $n = 1$ and $n = 2$ states. (b) For the $n = 1$ state, calculate the probability of finding the electron between $x_1 = 0.150$ nm and $x_2 = 0.350$ nm, where $x = 0$ is the left side of the box. (c) Repeat part (b) for the $n = 2$ state. (d) Calculate the energies in electron

volts of the $n = 1$ and $n = 2$ states. *Suggestion:* For parts (b) and (c), use Equation 41.6 and note that

$$\int \sin^2 ax \, dx = \tfrac{1}{2}x - \frac{1}{4a} \sin 2ax$$

49. An atom in an excited state 1.80 eV above the ground state remains in that excited state 2.00 μs before moving to the ground state. Find (a) the frequency and (b) the wavelength of the emitted photon. (c) Find the approximate uncertainty in energy of the photon.

50. An electron is confined to move in the xy plane in a rectangle whose dimensions are L_x and L_y. That is, the electron is trapped in a two-dimensional potential well having lengths of L_x and L_y. In this situation, the allowed energies of the electron depend on two quantum numbers n_x and n_y and are given by

$$E = \frac{h^2}{8m_e}\left(\frac{n_x^{\,2}}{L_x^{\,2}} + \frac{n_y^{\,2}}{L_y^{\,2}}\right)$$

(a) Assuming $L_x = L_y = L$, find the energies of the lowest four energy levels for the electron. (b) Construct an energy-level diagram for the electron and determine the energy difference between the second excited state and the ground state.

51. For a quantum particle described by a wave function $\psi(x)$, the expectation value of a physical quantity $f(x)$ associated with the particle is defined by

$$\langle f(x)\rangle \equiv \int_{-\infty}^{\infty} \psi^* f(x)\psi \, dx$$

For a particle in an infinitely deep one-dimensional box extending from $x = 0$ to $x = L$, show that

$$\langle x^2\rangle = \frac{L^2}{3} - \frac{L^2}{2n^2\pi^2}$$

52. A quantum particle is described by the wave function

$$\psi(x) = \begin{cases} A\cos\left(\dfrac{2\pi x}{L}\right) & \text{for } -\dfrac{L}{4} \le x \le \dfrac{L}{4} \\ 0 & \text{for other values of } x \end{cases}$$

(a) Determine the normalization constant A. (b) What is the probability that the particle will be found between $x = 0$ and $x = L/8$ if its position is measured? *Suggestion:* Use Equation 41.6.

53. A quantum particle has a wave function

$$\psi(x) = \begin{cases} \sqrt{\dfrac{2}{a}}\, e^{-x/a} & \text{for } x > 0 \\ 0 & \text{for } x < 0 \end{cases}$$

(a) Find and sketch the probability density. (b) Find the probability that the particle will be at any point where $x < 0$. (c) Show that ψ is normalized and then find the probability of finding the particle between $x = 0$ and $x = a$.

54. A quantum particle of mass m is placed in a one-dimensional box of length L. **What If?** Assume the box is so small that the particle's motion is *relativistic* and $K = $

$p^2/2m$ is not valid. (a) Derive an expression for the kinetic energy levels of the particle. (b) Assume the particle is an electron in a box of length $L = 1.00 \times 10^{-12}$ m. Find its lowest possible kinetic energy. By what percent is the nonrelativistic equation in error? *Suggestion:* See Equation 39.23.

55. ● Consider a "crystal" consisting of two nuclei and two electrons as shown in Figure P41.55. (a) Taking into account all the pairs of interactions, find the potential energy of the system as a function of d. (b) Assuming the electrons to be restricted to a one-dimensional box of length $3d$, find the minimum kinetic energy of the two electrons. (c) Find the value of d for which the total energy is a minimum. (d) State how this value of d compares with the spacing of atoms in lithium, which has a density of 0.530 g/cm³ and an atomic mass of 7 u. This type of calculation can be used to estimate the density of crystals and certain stars.

Figure P41.55

56. *The simple harmonic oscillator excited.* The wave function

$$\psi(x) = Bxe^{-(m\omega/2\hbar)x^2}$$

is a solution to the simple harmonic oscillator problem. (a) Find the energy of this state. (b) At what position are you least likely to find the particle? (c) At what positions are you most likely to find the particle? (d) Determine the value of B required to normalize the wave function. (e) **What If?** Determine the classical probability of finding the particle in an interval of small length δ centered at the position $x = 2(\hbar/m\omega)^{1/2}$. (f) What is the actual probability of finding the particle in this interval?

57. *Normalization of wave functions.* (a) Find the normalization constant A for a wave function made up of the two lowest states of a quantum particle in a box:

$$\psi(x) = A\left[\sin\left(\frac{\pi x}{L}\right) + 4\sin\left(\frac{2\pi x}{L}\right)\right]$$

(b) A particle is described in the space $-a \le x \le a$ by the wave function

$$\psi(x) = A\cos\left(\frac{\pi x}{2a}\right) + B\sin\left(\frac{\pi x}{a}\right)$$

Determine the relationship between the values of A and B required for normalization. *Suggestion:* Use the identity $\sin 2\theta = 2\sin\theta\cos\theta$.

58. The normalized wave functions for the ground state, $\psi_0(x)$, and the first excited state, $\psi_1(x)$, of a quantum harmonic oscillator are

$$\psi_0(x) = \left(\frac{a}{\pi}\right)^{1/4} e^{-ax^2/2} \qquad \psi_1(x) = \left(\frac{4a^3}{\pi}\right)^{1/4} x e^{-ax^2/2}$$

where $a = m\omega/\hbar$. A mixed state, $\psi_{01}(x)$, is constructed from these states:

$$\psi_{01}(x) = \frac{1}{\sqrt{2}} \left[\psi_0(x) + \psi_1(x) \right]$$

The symbol $\langle q \rangle_s$ denotes the expectation value of the quantity q for the state $\psi_s(x)$. Calculate the following expectation values: (a) $\langle x \rangle_0$ (b) $\langle x \rangle_1$ (c) $\langle x \rangle_{01}$

59. A two-slit electron diffraction experiment is done with slits of *unequal* widths. When only slit 1 is open, the num-

ber of electrons reaching the screen per second is 25.0 times the number of electrons reaching the screen per second when only slit 2 is open. When both slits are open, an interference pattern results in which the destructive interference is not complete. Find the ratio of the probability of an electron arriving at an interference maximum to the probability of an electron arriving at an adjacent interference minimum. *Suggestion:* Use the superposition principle.

Answers to Quick Quizzes

41.1 (d). The probability density for this wave function is $|\psi|^2 = \psi^*\psi = (Ae^{-ikx})(Ae^{ikx}) = A^2$, which is independent of x. Consequently, the particle is equally likely to be found at any value of x, which is consistent with the uncertainty principle. If the wavelength is known precisely (based on a specific value of k in Eq. 41.4), we have no knowledge of the position of the particle.

41.2 (i), (a). In Equation 41.14, setting $n = 1$ for the ground-state energy shows that the energy is inversely proportional to the particle mass. (ii), (d). The wavelength is determined by the length L of the well.

41.3 (c). According to Equation 41.14, if L is increased, all quantized energies become smaller. Therefore, the

energy levels move closer together. As L becomes macroscopic, the energy levels are so close together that the quantized behavior cannot be observed.

41.4 (d). The particles in all three parts (a), (b), and (c) are part of a bound system.

41.5 (a), (c), (f). Decreasing the barrier height and increasing the particle energy both reduce the value of C in Equation 41.23, increasing the transmission coefficient in Equation 41.22. Decreasing the width L of the barrier increases the transmission coefficient in Equation 41.22.

42.1 Atomic Spectra of Gases

42.2 Early Models of the Atom

42.3 Bohr's Model of the Hydrogen Atom

42.4 The Quantum Model of the Hydrogen Atom

42.5 The Wave Functions for Hydrogen

42.6 Physical Interpretation of the Quantum Numbers

42.7 The Exclusion Principle and the Periodic Table

42.8 More on Atomic Spectra: Visible and X-Ray

42.9 Spontaneous and Stimulated Transitions

42.10 Lasers

This street in the Ginza district in Tokyo displays many signs formed from neon lamps of varying bright colors. The light from these lamps has its origin in transitions between quantized energy states in the atoms contained in the lamps. In this chapter we investigate those transitions. (© Ken Straiton/Corbis)

42 Atomic Physics

In Chapter 41, we introduced some basic concepts and techniques used in quantum mechanics along with their applications to various one-dimensional systems. In this chapter, we apply quantum mechanics to atomic systems. A large portion of the chapter is focused on the application of quantum mechanics to the study of the hydrogen atom. Understanding the hydrogen atom, the simplest atomic system, is important for several reasons:

- The hydrogen atom is the only atomic system that can be solved exactly.

- Much of what was learned in the 20th century about the hydrogen atom, with its single electron, can be extended to such single-electron ions as He^+ and Li^{2+}.

- The hydrogen atom is an ideal system for performing precise tests of theory against experiment and for improving our overall understanding of atomic structure.

- The quantum numbers that are used to characterize the allowed states of hydrogen can also be used to investigate more complex atoms, and such a description enables us to understand the periodic table of the elements. This understanding is one of the greatest triumphs of quantum mechanics.

- The basic ideas about atomic structure must be well understood before we attempt to deal with the complexities of molecular structures and the electronic structure of solids.

The full mathematical solution of the Schrödinger equation applied to the hydrogen atom gives a complete and beautiful description of the atom's properties. Because the mathematical procedures involved are beyond the scope of this text, however, many details are omitted. The solutions for some states of hydrogen are discussed, together with the quantum numbers used to characterize various allowed states. We also discuss the physical significance of the quantum numbers and the effect of a magnetic field on certain quantum states.

A new physical idea, the *exclusion principle,* is presented in this chapter. This principle is extremely important for understanding the properties of multielectron atoms and the arrangement of elements in the periodic table.

Finally, we apply our knowledge of atomic structure to describe the mechanisms involved in the production of x-rays and in the operation of a laser.

42.1 Atomic Spectra of Gases

As pointed out in Section 40.1, all objects emit thermal radiation characterized by a continuous distribution of wavelengths. In sharp contrast to this continuous-distribution spectrum is the discrete **line spectrum** observed when a low-pressure gas is subject to an electric discharge. (Electric discharge occurs when the gas is subjected to a potential difference that creates an electric field greater than the dielectric strength of the gas.) Observation and analysis of these spectral lines is called **emission spectroscopy.**

When the light from a gas discharge is examined using a spectrometer (see Active Fig. 38.15), it is found to consist of a few bright lines of color on a generally dark background. This discrete line spectrum contrasts sharply with the continuous rainbow of colors seen when a glowing solid is viewed through the same instrument. Figure 42.1a shows that the wavelengths contained in a given line spectrum are characteristic of the element emitting the light. The simplest line spectrum is that for atomic hydrogen, and we describe this spectrum in detail. Because no two elements have the same line spectrum, this phenomenon represents a practical and sensitive technique for identifying the elements present in unknown samples.

Another form of spectroscopy very useful in analyzing substances is **absorption spectroscopy.** An absorption spectrum is obtained by passing white light from a continuous source through a gas or a dilute solution of the element being analyzed. The absorption spectrum consists of a series of dark lines superimposed on

Figure 42.1 (a) Emission line spectra for hydrogen, mercury, and neon. (b) The absorption spectrum for hydrogen. Notice that the dark absorption lines occur at the same wavelengths as the hydrogen emission lines in (a). (K. W. Whitten, R. E. Davis, M. L. Peck, and G. G. Stanley, *General Chemistry*, 7th ed., Belmont, CA, Brooks/Cole, 2004.)

the continuous spectrum of the light source as shown in Figure 42.1b for atomic hydrogen.

The absorption spectrum of an element has many practical applications. For example, the continuous spectrum of radiation emitted by the Sun must pass through the cooler gases of the solar atmosphere. The various absorption lines observed in the solar spectrum have been used to identify elements in the solar atmosphere. In early studies of the solar spectrum, experimenters found some lines that did not correspond to any known element. A new element had been discovered! The new element was named helium, after the Greek word for Sun, *helios*. Helium was subsequently isolated from subterranean gas on the Earth.

Using this technique, scientists have examined the light from stars other than our Sun and have never detected elements other than those present on the Earth. Absorption spectroscopy has also been useful in analyzing heavy-metal contamination of the food chain. For example, the first determination of high levels of mercury in tuna was made with the use of atomic absorption spectroscopy.

The discrete emissions of light from gas discharges are used in "neon" signs such as those in the opening photograph of this chapter. Neon, the first gas used in these types of signs and the gas after which these signs are named, emits strongly in the red region. As a result, a glass tube filled with neon gas emits bright red light when an applied voltage causes a continuous discharge. Early signs used different gases to provide different colors, although the brightness of these signs was generally very low. Many present-day "neon" signs contain mercury vapor, which emits strongly in the ultraviolet range of the electromagnetic spectrum. The inside of a present-day sign's glass tube is coated with a material that emits a particular color when it absorbs ultraviolet radiation from the mercury. The color of the light from the tube results from the particular material chosen. A household fluorescent light operates in the same manner, with a white-emitting material coating the inside of the glass tube.

From 1860 to 1885, scientists accumulated a great deal of data on atomic emissions using spectroscopic measurements. In 1885, a Swiss schoolteacher, Johann Jacob Balmer (1825–1898), found an empirical equation that correctly predicted the wavelengths of four visible emission lines of hydrogen: H_α(red), H_β (blue-green), H_γ (blue-violet), and H_δ (violet). Figure 42.2 shows these and other lines (in the ultraviolet) in the emission spectrum of hydrogen. The four visible lines occur at the wavelengths 656.3 nm, 486.1 nm, 434.1 nm, and 410.2 nm. The complete set of lines is called the **Balmer series.** The wavelengths of these lines can be described by the following equation, which is a modification made by Johannes Rydberg (1854–1919) of Balmer's original equation:

$$\frac{1}{\lambda} = R_H\left(\frac{1}{2^2} - \frac{1}{n^2}\right) \quad n = 3, 4, 5, \ldots \tag{42.1}$$

◀ Balmer series

where R_H is a constant now called the **Rydberg constant** with a value of $1.097\,373\,2 \times 10^7$ m^{-1}. The integer values of n from 3 to 6 give the four visible lines from 656.3 nm (red) down to 410.2 nm (violet). Equation 42.1 also describes the ultraviolet spectral lines in the Balmer series if n is carried out beyond $n = 6$. The **series limit** is the shortest wavelength in the series and corresponds to $n \to \infty$, with a wavelength of 364.6 nm as in Figure 42.2. The measured spectral lines agree with the empirical equation, Equation 42.1, to within 0.1%.

Other lines in the spectrum of hydrogen were found following Balmer's discovery. These spectra are called the Lyman, Paschen, and Brackett series after their discoverers. The wavelengths of the lines in these series can be calculated through the use of the following empirical equations:

$$\frac{1}{\lambda} = R_H\left(1 - \frac{1}{n^2}\right) \quad n = 2, 3, 4, \ldots \tag{42.2}$$

◀ Lyman series

$$\frac{1}{\lambda} = R_H\left(\frac{1}{3^2} - \frac{1}{n^2}\right) \quad n = 4, 5, 6, \ldots \tag{42.3}$$

◀ Paschen series

λ (nm)

364.6   410.2   434.1   486.1   656.3

Figure 42.2 The Balmer series of spectral lines for atomic hydrogen, with several lines marked with the wavelength in nanometers. (The horizontal wavelength axis is not to scale.) The line labeled 364.6 is the shortest-wavelength line and is in the ultraviolet region of the electromagnetic spectrum. The other labeled lines are in the visible region.

Brackett series ▶

$$\frac{1}{\lambda} = R_{\mathrm{H}}\left(\frac{1}{4^2} - \frac{1}{n^2}\right) \quad n = 5, 6, 7, \ldots \qquad (42.4)$$

No theoretical basis existed for these equations; they simply worked. The same constant R_{H} appears in each equation, and all equations involve small integers. In Section 42.3, we shall discuss the remarkable achievement of a theory for the hydrogen atom that provided an explanation for these equations.

42.2   Early Models of the Atom

The model of the atom in the days of Newton was a tiny, hard, indestructible sphere. Although this model provided a good basis for the kinetic theory of gases, new models had to be devised when experiments revealed the electrical nature of atoms. In 1897, J. J. Thomson established the charge-to-mass ratio for electrons. (See Fig. 29.14 in Section 29.3.) The following year, he suggested a model that describes the atom as a region in which positive charge is spread out in space with electrons embedded throughout the region, much like the seeds in a watermelon or raisins in thick pudding (Fig. 42.3). The atom as a whole would then be electrically neutral.

In 1911, Ernest Rutherford (1871–1937) and his students Hans Geiger and Ernest Marsden performed a critical experiment that showed that Thomson's model could not be correct. In this experiment, a beam of positively charged alpha particles (helium nuclei) was projected into a thin metallic foil such as the target in Figure 42.4a. Most of the particles passed through the foil as if it were empty space, but some of the results of the experiment were astounding. Many of the particles deflected from their original direction of travel were scattered through *large* angles. Some particles were even deflected backward, completely reversing their direction of travel! When Geiger informed Rutherford that some alpha particles were scattered backward, Rutherford wrote, "It was quite the most incredible event that has ever happened to me in my life. It was almost as incredible as if you fired a 15-inch [artillery] shell at a piece of tissue paper and it came back and hit you."

Such large deflections were not expected on the basis of Thomson's model. According to that model, the positive charge of an atom in the foil is spread out over such a great volume (the entire atom) that there is no concentration of positive charge strong enough to cause any large-angle deflections of the positively charged alpha particles. Furthermore, the electrons are so much less massive than the alpha particles that they would not cause large-angle scattering either. Rutherford explained his astonishing results by developing a new atomic model, one that

JOSEPH JOHN THOMSON
English physicist (1856–1940)
The recipient of a Nobel Prize in Physics in 1906, Thomson is usually considered the discoverer of the electron. He opened up the field of subatomic particle physics with his extensive work on the deflection of cathode rays (electrons) in an electric field.

Stock Montage, Inc.

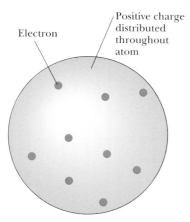

Figure 42.3   Thomson's model of the atom: negatively charged electrons in a volume of continuous positive charge.

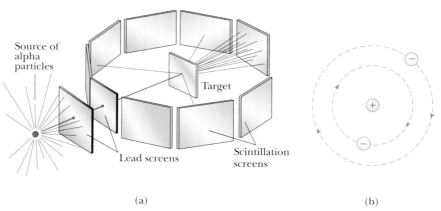

(a)   (b)

Figure 42.4   (a) Rutherford's technique for observing the scattering of alpha particles from a thin foil target. The source is a naturally occurring radioactive substance, such as radium. (b) Rutherford's planetary model of the atom.

assumed the positive charge in the atom was concentrated in a region that was small relative to the size of the atom. He called this concentration of positive charge the **nucleus** of the atom. Any electrons belonging to the atom were assumed to be in the relatively large volume outside the nucleus. To explain why these electrons were not pulled into the nucleus by the attractive electric force, Rutherford modeled them as moving in orbits around the nucleus in the same manner as the planets orbit the Sun (Fig. 42.4b). For this reason, this model is often referred to as the planetary model of the atom.

Two basic difficulties exist with Rutherford's planetary model. As we saw in Section 42.1, an atom emits (and absorbs) certain characteristic frequencies of electromagnetic radiation and no others, but the Rutherford model cannot explain this phenomenon. A second difficulty is that Rutherford's electrons are undergoing a centripetal acceleration. According to Maxwell's theory of electromagnetism, centripetally accelerated charges revolving with frequency f should radiate electromagnetic waves of frequency f. Unfortunately, this classical model leads to a prediction of self-destruction when applied to the atom. As the electron radiates, energy is carried away from the atom, the radius of the electron's orbit steadily decreases, and its frequency of revolution increases. This process would lead to an ever-increasing frequency of emitted radiation and an ultimate collapse of the atom as the electron plunges into the nucleus (Fig. 42.5).

Figure 42.5 The classical model of the nuclear atom. Because the accelerating electron radiates energy, the orbit decays until the electron falls into the nucleus.

42.3 Bohr's Model of the Hydrogen Atom

Given the situation described at the end of Section 42.2, the stage was set for Niels Bohr in 1913 when he presented a new model of the hydrogen atom that circumvented the difficulties of Rutherford's planetary model. Bohr applied Planck's ideas of quantized energy levels (Section 40.1) to orbiting atomic electrons. Bohr's theory was historically important to the development of quantum physics, and it appeared to explain the spectral line series described by Equations 42.1 through 42.4. Although Bohr's model is now considered obsolete and has been completely replaced by a probabilistic quantum-mechanical theory, we can use this model to develop the notions of energy quantization and angular momentum quantization as applied to atomic-sized systems.

Bohr combined ideas from Planck's original quantum theory, Einstein's concept of the photon, Rutherford's planetary model of the atom, and Newtonian mechanics to arrive at a semiclassical model based on some revolutionary postulates. The basic ideas of the Bohr theory as it applies to the hydrogen atom are as follows:

1. The electron moves in circular orbits around the proton under the influence of the electric force of attraction as shown in Figure 42.6.
2. Only certain electron orbits are stable. When in one of these **stationary states,** as Bohr called them, the electron does not emit energy in the form of radiation. Hence, the total energy of the atom remains constant and classical mechanics can be used to describe the electron's motion. Bohr's model claims that the centripetally accelerated electron does not continuously emit radiation, losing energy and eventually spiraling into the nucleus, as predicted by classical physics in the form of Rutherford's planetary model.
3. The atom emits radiation when the electron makes a transition from a more energetic initial orbit to a lower-energy orbit. This transition cannot be visualized or treated classically. In particular, the frequency f of the photon emitted in the transition is related to the change in the atom's energy and is not equal to the frequency of the electron's orbital motion. The frequency of the emitted radiation is found from the energy-conservation expression

$$E_i - E_f = hf \tag{42.5}$$

where E_i is the energy of the initial state, E_f is the energy of the final state, and $E_i > E_f$. In addition, energy of an incident photon can be absorbed by

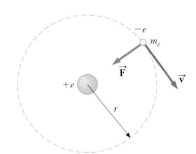

Figure 42.6 Diagram representing Bohr's model of the hydrogen atom. The orbiting electron is allowed to be only in specific orbits of discrete radii.

NIELS BOHR
Danish Physicist (1885–1962)
Bohr was an active participant in the early development of quantum mechanics and provided much of its philosophical framework. During the 1920s and 1930s, he headed the Institute for Advanced Studies in Copenhagen. The institute was a magnet for many of the world's best physicists and provided a forum for the exchange of ideas. When Bohr visited the United States in 1939 to attend a scientific conference, he brought news that the fission of uranium had been observed by Hahn and Strassman in Berlin. The results were the foundations of the nuclear weapon developed in the United States during World War II. Bohr was awarded the 1922 Nobel Prize in Physics for his investigation of the structure of atoms and the radiation emanating from them.

the atom, but only if the photon has an energy that exactly matches the difference in energy between an allowed state of the atom and a higher-energy state. Upon absorption, the photon disappears and the atom makes a transition to the higher-energy state.

4. The size of an allowed electron orbit is determined by a condition imposed on the electron's orbital angular momentum: the allowed orbits are those for which the electron's orbital angular momentum about the nucleus is quantized and equal to an integral multiple of $\hbar = h/2\pi$,

$$m_e v r = n\hbar \qquad n = 1, 2, 3, \ldots \tag{42.6}$$

where m_e is the electron mass, v is the electron's speed in its orbit, and r is the orbital radius.

Assumption 3 implies qualitatively the existence of a characteristic discrete emission line spectrum *and also* a corresponding absorption line spectrum of the kind shown in Figure 42.1 for hydrogen. Using these four assumptions, we can calculate the allowed energy levels and find quantitative values of the emission wavelengths of the hydrogen atom.

The electric potential energy of the system shown in Figure 42.6 is given by Equation 25.13, $U = k_e q_1 q_2 / r = -k_e e^2 / r$, where k_e is the Coulomb constant and the negative sign arises from the charge $-e$ on the electron. Therefore, the *total* energy of the atom, which consists of the electron's kinetic energy and the system's potential energy, is

$$E = K + U = \tfrac{1}{2} m_e v^2 - k_e \frac{e^2}{r} \tag{42.7}$$

The electron is a particle in uniform circular motion, so the electric force $k_e e^2 / r^2$ exerted on the electron must equal the product of its mass and its centripetal acceleration ($a_c = v^2/r$):

$$\frac{k_e e^2}{r^2} = \frac{m_e v^2}{r}$$

$$v^2 = \frac{k_e e^2}{m_e r} \tag{42.8}$$

From Equation 42.8, we find that the kinetic energy of the electron is

$$K = \tfrac{1}{2} m_e v^2 = \frac{k_e e^2}{2r}$$

Substituting this value of K into Equation 42.7 gives the following expression for the total energy of the atom:[1]

$$E = -\frac{k_e e^2}{2r} \tag{42.9}$$

Because the total energy is *negative,* which indicates a bound electron–proton system, energy in the amount of $k_e e^2 / 2r$ must be added to the atom to remove the electron and make the total energy of the system zero.

We can obtain an expression for r, the radius of the allowed orbits, by solving Equation 42.6 for v^2 and equating it to Equation 42.8:

$$v^2 = \frac{n^2 \hbar^2}{m_e^2 r^2} = \frac{k_e e^2}{m_e r}$$

$$r_n = \frac{n^2 \hbar^2}{m_e k_e e^2} \qquad n = 1, 2, 3, \ldots \tag{42.10}$$

[1] Compare Equation 42.9 with its gravitational counterpart, Equation 13.18.

Equation 42.10 shows that the radii of the allowed orbits have discrete values: they are quantized. The result is based on the *assumption* that the electron can exist only in certain allowed orbits determined by the integer n.

The orbit with the smallest radius, called the **Bohr radius** a_0, corresponds to $n = 1$ and has the value

$$a_0 = \frac{\hbar^2}{m_e k_e e^2} = 0.052\ 9\ \text{nm} \qquad \text{(42.11)}$$

◄ Bohr radius

Substituting Equation 42.11 into Equation 42.10 gives a general expression for the radius of any orbit in the hydrogen atom:

$$r_n = n^2 a_0 = n^2 (0.052\ 9\ \text{nm}) \qquad \text{(42.12)}$$

◄ Radii of Bohr orbits in hydrogen

Bohr's theory predicts a value for the radius of a hydrogen atom on the right order of magnitude, based on experimental measurements. This result was a striking triumph for Bohr's theory. The first three Bohr orbits are shown to scale in Active Figure 42.7.

The quantization of orbit radii leads to energy quantization. Substituting $r_n = n^2 a_0$ into Equation 42.9 gives

$$E_n = -\frac{k_e e^2}{2a_0}\left(\frac{1}{n^2}\right) \qquad n = 1, 2, 3, \dots \qquad \text{(42.13)}$$

◄ Allowed energies of the Bohr hydrogen atom

Inserting numerical values into this expression, we find that

$$E_n = -\frac{13.606\ \text{eV}}{n^2} \qquad n = 1, 2, 3, \dots \qquad \text{(42.14)}$$

Only energies satisfying this equation are permitted. The lowest allowed energy level, the ground state, has $n = 1$ and energy $E_1 = -13.606$ eV. The next energy level, the first excited state, has $n = 2$ and energy $E_2 = E_1/2^2 = -3.401$ eV. Active Figure 42.8 is an energy-level diagram showing the energies of these discrete energy states and the corresponding quantum numbers n. The uppermost level corresponds to $n = \infty$ (or $r = \infty$) and $E = 0$.

Notice how the allowed energies of the hydrogen atom differ from those of the particle in a box. The particle-in-a-box energies (Eq. 41.14) increase as n^2, so they become farther apart in energy as n increases. On the other hand, the energies of the hydrogen atom (Eq. 42.14) vary inversely with n^2, so their separation in energy becomes smaller as n increases. The separation between energy levels approaches zero as n approaches infinity and the energy approaches zero.

Zero energy represents the boundary between a bound system of an electron and a proton and an unbound system. If the energy of the atom is raised from that of the ground state to any energy larger than zero, the atom is **ionized.** The minimum energy required to ionize the atom in its ground state is called the **ionization energy.** As can be seen from Active Figure 42.8, the ionization energy for hydrogen

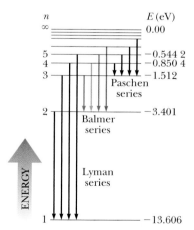

ACTIVE FIGURE 42.8

An energy-level diagram for the hydrogen atom. Quantum numbers are given on the left, and energies (in electron volts) are given on the right. Vertical arrows represent the four lowest-energy transitions for each of the spectral series shown. The colored arrows for the Balmer series indicate that this series results in visible light.

Sign in at **www.thomsonedu.com** and go to ThomsonNOW to choose the initial and final states of the hydrogen atom and observe the transitions in this figure and in Active Figure 42.7.

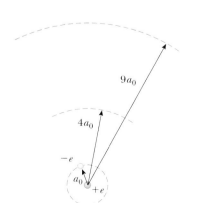

ACTIVE FIGURE 42.7

The first three circular orbits predicted by the Bohr model of the hydrogen atom.

Sign in at **www.thomsonedu.com** and go to ThomsonNOW to choose the initial and final states of the hydrogen atom and observe the transitions in this figure and in Active Figure 42.8.

in the ground state, based on Bohr's calculation, is 13.6 eV. This finding constituted another major achievement for the Bohr theory because the ionization energy for hydrogen had already been measured to be 13.6 eV.

Equations 42.5 and 42.13 can be used to calculate the frequency of the photon emitted when the electron makes a transition from an outer orbit to an inner orbit:

Frequency of a photon ▶
emitted from hydrogen

$$f = \frac{E_i - E_f}{h} = \frac{k_e e^2}{2a_0 h}\left(\frac{1}{n_f^2} - \frac{1}{n_i^2}\right) \tag{42.15}$$

Because the quantity measured experimentally is wavelength, it is convenient to use $c = f\lambda$ to express Equation 42.15 in terms of wavelength:

$$\frac{1}{\lambda} = \frac{f}{c} = \frac{k_e e^2}{2a_0 h c}\left(\frac{1}{n_f^2} - \frac{1}{n_i^2}\right) \tag{42.16}$$

Remarkably, this expression, which is purely theoretical, is *identical* to the general form of the empirical relationships discovered by Balmer and Rydberg and given by Equations 42.1 to 42.4:

$$\frac{1}{\lambda} = R_H\left(\frac{1}{n_f^2} - \frac{1}{n_i^2}\right) \tag{42.17}$$

provided the constant $k_e e^2/2a_0 h c$ is equal to the experimentally determined Rydberg constant. Soon after Bohr demonstrated that these two quantities agree to within approximately 1%, this work was recognized as the crowning achievement of his new quantum theory of the hydrogen atom. Furthermore, Bohr showed that all the spectral series for hydrogen have a natural interpretation in his theory. The different series correspond to transitions to different final states characterized by the quantum number n_f. Active Figure 42.8 shows the origin of these spectral series as transitions between energy levels.

Bohr extended his model for hydrogen to other elements in which all but one electron had been removed. These systems have the same structure as the hydrogen atom except that the nuclear charge is larger. Ionized elements such as He^+, Li^{2+}, and Be^{3+} were suspected to exist in hot stellar atmospheres, where atomic collisions frequently have enough energy to completely remove one or more atomic electrons. Bohr showed that many mysterious lines observed in the spectra of the Sun and several other stars could not be due to hydrogen but were correctly predicted by his theory if attributed to singly ionized helium. In general, the number of protons in the nucleus of an atom is called the **atomic number** of the element and is given the symbol Z. To describe a single electron orbiting a fixed nucleus of charge $+Ze$, Bohr's theory gives

$$r_n = (n^2)\frac{a_0}{Z} \tag{42.18}$$

PITFALL PREVENTION 42.2
The Bohr Model Is Great, But . . .

The Bohr model correctly predicts the ionization energy and general features of the spectrum for hydrogen, but it cannot account for the spectra of more complex atoms and is unable to predict many subtle spectral details of hydrogen and other simple atoms. Scattering experiments show that the electron in a hydrogen atom does not move in a flat circle around the nucleus. Instead, the atom is spherical. The ground-state angular momentum of the atom is zero and not $\hbar$.

$$E_n = -\frac{k_e e^2}{2a_0}\left(\frac{Z^2}{n^2}\right) \quad n = 1, 2, 3, \ldots \tag{42.19}$$

Although the Bohr theory was triumphant in its agreement with some experimental results on the hydrogen atom, it suffered from some difficulties. One of the first indications that the Bohr theory needed to be modified arose when improved spectroscopic techniques were used to examine the spectral lines of hydrogen. It was found that many of the lines in the Balmer and other series were not single lines at all. Instead, each was a group of lines spaced very close together. An additional difficulty arose when it was observed that in some situations certain single spectral lines were split into three closely spaced lines when the atoms were placed in a strong magnetic field. Efforts to explain these and other deviations from the Bohr model led to modifications in the theory and ultimately to a replacement theory that will be discussed in Section 42.4.

Bohr's Correspondence Principle

In our study of relativity, we found that Newtonian mechanics is a special case of relativistic mechanics and is usable only for speeds much less than c. Similarly, **quantum physics agrees with classical physics when the difference between quantized levels becomes vanishingly small**. This principle, first set forth by Bohr, is called the **correspondence principle**.[2]

For example, consider an electron orbiting the hydrogen atom with $n > 10\ 000$. For such large values of n, the energy differences between adjacent levels approach zero; therefore, the levels are nearly continuous. Consequently, the classical model is reasonably accurate in describing the system for large values of n. According to the classical picture, the frequency of the light emitted by the atom is equal to the frequency of revolution of the electron in its orbit about the nucleus. Calculations show that for $n > 10\ 000$, this frequency is different from that predicted by quantum mechanics by less than 0.015%.

Quick Quiz 42.1   A hydrogen atom is in its ground state. Incident on the atom are many photons each having an energy of 10.5 eV. What is the result? (a) The atom is excited to a higher allowed state. (b) The atom is ionized. (c) The photons pass by the atom without interaction.

Quick Quiz 42.2   A hydrogen atom makes a transition from the $n = 3$ level to the $n = 2$ level. It then makes a transition from the $n = 2$ level to the $n = 1$ level. Which transition results in emission of the longest-wavelength photon? (a) the first transition   (b) the second transition   (c) neither transition because the wavelengths are the same for both

---

EXAMPLE 42.1	**Electronic Transitions in Hydrogen**

(A) The electron in a hydrogen atom makes a transition from the $n = 2$ energy level to the ground level ($n = 1$). Find the wavelength and frequency of the emitted photon.

SOLUTION

Conceptualize   Imagine the electron in a circular orbit about the nucleus as in the Bohr model in Figure 42.6. When the electron makes a transition to a lower stationary state, it emits a photon with a given frequency.

Categorize   We evaluate the results using equations developed in this section, so we categorize this example as a substitution problem.

Use Equation 42.17 to obtain λ, with $n_i = 2$ and $n_f = 1$:

$$\frac{1}{\lambda} = R_{\text{H}}\left(\frac{1}{1^2} - \frac{1}{2^2}\right) = \frac{3R_{\text{H}}}{4}$$

$$\lambda = \frac{4}{3R_{\text{H}}} = \frac{4}{3(1.097 \times 10^7\ \text{m}^{-1})} = 1.22 \times 10^{-7}\ \text{m} = \boxed{122\ \text{nm}}$$

Use Equation 34.20 to find the frequency of the photon:

$$f = \frac{c}{\lambda} = \frac{3.00 \times 10^8\ \text{m/s}}{1.22 \times 10^{-7}\ \text{m}} = \boxed{2.47 \times 10^{15}\ \text{Hz}}$$

(B) In interstellar space, highly excited hydrogen atoms called Rydberg atoms have been observed. Find the wavelength to which radio astronomers must tune to detect signals from electrons dropping from the $n = 273$ level to the $n = 272$ level.

---

[2] In reality, the correspondence principle is the starting point for Bohr's postulate 4 on angular momentum quantization. To see how postulate 4 arises from the correspondence principle, see J. W. Jewett, *Physics Begins with Another M . . . Mysteries, Magic, Myth, and Modern Physics* (Boston: Allyn & Bacon, 1996), pp. 353–356.

SOLUTION

Use Equation 42.17, this time with $n_i = 273$ and $n_f = 272$:

$$\frac{1}{\lambda} = R_H\left(\frac{1}{n_f^2} - \frac{1}{n_i^2}\right) = R_H\left(\frac{1}{(272)^2} - \frac{1}{(273)^2}\right) = 9.88 \times 10^{-8} R_H$$

Solve for λ:

$$\lambda = \frac{1}{9.88 \times 10^{-8} R_H} = \frac{1}{(9.88 \times 10^{-8})(1.097 \times 10^7 \text{ m}^{-1})} = 0.922 \text{ m}$$

(C) What is the radius of the electron orbit for a Rydberg atom for which $n = 273$?

SOLUTION

Use Equation 42.12 to find the radius of the orbit:

$$r_{273} = (273)^2(0.052\ 9 \text{ nm}) = 3.94\ \mu\text{m}$$

This radius is large enough that the atom is on the verge of becoming macroscopic!

(D) How fast is the electron moving in a Rydberg atom for which $n = 273$?

SOLUTION

Solve Equation 42.8 for the electron's speed:

$$v = \sqrt{\frac{k_e e^2}{m_e r}} = \sqrt{\frac{(8.99 \times 10^9 \text{ N} \cdot \text{m}^2/\text{C}^2)(1.60 \times 10^{-19} \text{ C})^2}{(9.11 \times 10^{-31} \text{ kg})(3.94 \times 10^{-6} \text{ m})}}$$

$$= 8.02 \times 10^3 \text{ m/s}$$

What If? What if radiation from the Rydberg atom in part (B) is treated classically? What is the wavelength of radiation emitted by the atom in the $n = 273$ level?

Answer Classically, the frequency of the emitted radiation is that of the rotation of the electron around the nucleus.

Calculate this frequency using the period defined in Equation 4.15:

$$f = \frac{1}{T} = \frac{v}{2\pi r}$$

Substitute the radius and speed from parts (C) and (D):

$$f = \frac{v}{2\pi r} = \frac{8.02 \times 10^3 \text{ m/s}}{2\pi(3.94 \times 10^{-6} \text{ m})} = 3.24 \times 10^8 \text{ Hz}$$

Find the wavelength of the radiation from Equation 34.20:

$$\lambda = \frac{c}{f} = \frac{3.00 \times 10^8 \text{ m/s}}{3.24 \times 10^8 \text{ Hz}} = 0.926 \text{ m}$$

This value is less than 0.5% different from the wavelength calculated in part (B). As indicated in the discussion of Bohr's correspondence principle, this difference becomes even smaller for higher values of n.

42.4 The Quantum Model of the Hydrogen Atom

In the preceding section, we described how the Bohr model views the electron as a particle orbiting the nucleus in nonradiating, quantized energy levels. This approach leads to an analysis that combines both classical and quantum concepts. Although the model demonstrates excellent agreement with some experimental results, it cannot explain others. These difficulties are removed when a full quantum model involving the Schrödinger equation is used to describe the hydrogen atom.

The formal procedure for solving the problem of the hydrogen atom is to substitute the appropriate potential energy function into the Schrödinger equation, find solutions to the equation, and apply boundary conditions as we did for the particle in a box in Chapter 41. The potential energy function for the hydrogen atom is that due to the electrical interaction between the electron and the proton:

$$U(r) = -k_e \frac{e^2}{r} \qquad \textbf{(42.20)}$$

where r is the radial distance from the proton (situated at $r = 0$) to the electron and $k_e = 8.99 \times 10^9 \ \text{N} \cdot \text{m}^2/\text{C}^2$ is the Coulomb constant.

The mathematics for the hydrogen atom problem is more complicated than that for the particle in a box because the atom is three-dimensional and U depends on the radial coordinate r. If the time-independent Schrödinger equation (Eq. 41.15) is extended to three-dimensional rectangular coordinates, the result is

$$-\frac{\hbar^2}{2m}\left(\frac{\partial^2 \psi}{\partial x^2} + \frac{\partial^2 \psi}{\partial y^2} + \frac{\partial^2 \psi}{\partial z^2}\right) + U\psi = E\psi$$

It is easier to solve this equation for the hydrogen atom if rectangular coordinates are converted to spherical polar coordinates, an extension of the plane polar coordinates introduced in Section 3.1. In spherical polar coordinates, a point in space is represented by the three variables r, θ, and ϕ, where r is the radial distance from the origin, $r = \sqrt{x^2 + y^2 + z^2}$. With the point represented at the end of a position vector $\vec{r}$ as shown in Figure 42.9, the angular coordinate θ specifies its angular position relative to the z axis. Once that position vector is projected onto the xy plane, the angular coordinate ϕ specifies the projection's (and therefore the point's) angular position relative to the x axis.

The conversion of the three-dimensional time-independent Schrödinger equation for $\psi(x, y, z)$ to the equivalent form for $\psi(r, \theta, \phi)$ is straightforward but very tedious, so we omit the details.[3] In Chapter 41, we separated the time dependence from the space dependence in the general wave function Ψ. In this case of the hydrogen atom, the three space variables in $\psi(x, y, z)$ can be similarly separated by writing the wave function as a product of functions of each single variable:

$$\psi(r, \theta, \phi) = R(r)f(\theta)g(\phi)$$

In this way, Schrödinger's equation, which is a three-dimensional partial differential equation, can be transformed into three separate ordinary differential equations: one for $R(r)$, one for $f(\theta)$, and one for $g(\phi)$. Each of these functions is subject to boundary conditions. For example, $R(r)$ must remain finite as $r \to 0$ and $r \to \infty$; furthermore, $g(\phi)$ must have the same value as $g(\phi + 2\pi)$.

The potential energy function given in Equation 42.20 depends *only* on the radial coordinate r and not on either of the angular coordinates; therefore, it appears only in the equation for $R(r)$. As a result, the equations for θ and ϕ are independent of the particular system and their solutions are valid for *any* system exhibiting rotation.

When the full set of boundary conditions is applied to all three functions, three different quantum numbers are found for each allowed state of the hydrogen atom, one for each of the separate differential equations. These quantum numbers are restricted to integer values and correspond to the three independent degrees of freedom (three space dimensions).

The first quantum number, associated with the radial function $R(r)$ of the full wave function, is called the **principal quantum number** and is assigned the symbol n. The differential equation for $R(r)$ leads to functions giving the probability of finding the electron at a certain radial distance from the nucleus. In Section 42.5,

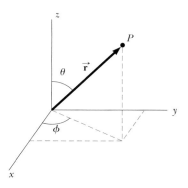

Figure 42.9 A point P in space is located by means of a position vector $\vec{r}$. In Cartesian coordinates, the components of this vector are x, y, and z. In spherical polar coordinates, the point is described by r, the distance from the origin; θ, the angle between $\vec{r}$ and the z axis; and ϕ, the angle between the x axis and a projection of $\vec{r}$ onto the xy plane.

---

[3] Descriptions of the solutions to the Schrödinger equation for the hydrogen atom are available in modern physics textbooks such as R. A. Serway, C. Moses, and C. A. Moyer, *Modern Physics*, 3rd ed. (Belmont, CA: Brooks/Cole, 2005).

we will describe two of these radial wave functions. The energies of the allowed states for the hydrogen atom are found to be related to n as follows:

◄ Allowed energies of the
quantum hydrogen atom

$$E_n = -\left(\frac{k_e e^2}{2a_0}\right)\frac{1}{n^2} = -\frac{13.606 \text{ eV}}{n^2} \quad n = 1, 2, 3, \ldots \quad (42.21)$$

PITFALL PREVENTION 42.3
Energy Depends on n Only for Hydrogen

The implication in Equation 42.21 that the energy depends only on the quantum number n is true only for the hydrogen atom. For more complicated atoms, we will use the same quantum numbers developed here for hydrogen. The energy levels for these atoms depend primarily on n, but they also depend to a lesser degree on other quantum numbers.

This result is in exact agreement with that obtained in the Bohr theory (Eqs. 42.13 and 42.14)! This agreement is remarkable because the Bohr theory and the full quantum theory arrive at the result from completely different starting points.

The **orbital quantum number,** symbolized ℓ, comes from the differential equation for $f(\theta)$ and is associated with the orbital angular momentum of the electron. The **orbital magnetic quantum number** m_ℓ arises from the differential equation for $g(\phi)$. Both ℓ and m_ℓ are integers. We will expand our discussion of these two quantum numbers in Section 42.6, where we also introduce a fourth (nonintegral) quantum number, resulting from a relativistic treatment of the hydrogen atom.

The application of boundary conditions on the three parts of the full wave function leads to important relationships among the three quantum numbers as well as certain restrictions on their values:

◄ Restrictions on the values
of hydrogen-atom
quantum numbers

The values of n are integers that can range from 1 to ∞.

The values of ℓ are integers that can range from 0 to $n - 1$.

The values of m_ℓ are integers that can range from $-\ell$ to ℓ.

PITFALL PREVENTION 42.4
Quantum Numbers Describe a System

It is common to assign the quantum numbers to an electron. Remember, however, that these quantum numbers arise from the Schrödinger equation, which involves a potential energy function for the *system* of the electron and the nucleus. Therefore, it is more *proper* to assign the quantum numbers to the atom, but it is more *popular* to assign them to an electron. We follow this latter usage because it is so common.

For example, if $n = 1$, only $\ell = 0$ and $m_\ell = 0$ are permitted. If $n = 2$, then ℓ may be 0 or 1; if $\ell = 0$, then $m_\ell = 0$; but if $\ell = 1$, then m_ℓ may be 1, 0, or -1. Table 42.1 summarizes the rules for determining the allowed values of ℓ and m_ℓ for a given n.

For historical reasons, **all states having the same principal quantum number are said to form a shell.** Shells are identified by the letters K, L, M, . . ., which designate the states for which $n = 1, 2, 3, \ldots$. Likewise, **all states having the same values of n and ℓ are said to form a subshell.** The letters[4] s, p, d, f, g, h, . . . are used to designate the subshells for which $\ell = 0, 1, 2, 3, \ldots$. The state designated by $3p$, for example, has the quantum numbers $n = 3$ and $\ell = 1$; the $2s$ state has the quantum numbers $n = 2$ and $\ell = 0$. These notations are summarized in Table 42.2.

States that violate the rules given in Table 42.1 do not exist. (They do not satisfy the boundary conditions on the wave function.) For instance, the $2d$ state, which would have $n = 2$ and $\ell = 2$, cannot exist because the highest allowed value of ℓ is $n - 1$, which in this case is 1. Therefore, for $n = 2$, the $2s$ and $2p$ states are allowed but $2d$, $2f$, . . . are not. For $n = 3$, the allowed subshells are $3s$, $3p$, and $3d$.

TABLE 42.1

Three Quantum Numbers for the Hydrogen Atom

Quantum Number	Name	Allowed Values	Number of Allowed States
n	Principal quantum number	$1, 2, 3, \ldots$	Any number
ℓ	Orbital quantum number	$0, 1, 2, \ldots, n - 1$	n
m_ℓ	Orbital magnetic quantum number	$-\ell, -\ell + 1, \ldots, 0, \ldots, \ell - 1, \ell$	$2\ell + 1$

[4] The first four of these letters come from early classifications of spectral lines: sharp, principal, diffuse, and fundamental. The remaining letters are in alphabetical order.

TABLE 42.2

Atomic Shell and Subshell Notations

n	Shell Symbol	ℓ	Subshell Symbol
1	K	0	s
2	L	1	p
3	M	2	d
4	N	3	f
5	O	4	g
6	P	5	h

Quick Quiz 42.3   How many possible subshells are there for the $n = 4$ level of hydrogen? (a) 5   (b) 4   (c) 3   (d) 2   (e) 1

Quick Quiz 42.4   When the principal quantum number is $n = 5$, how many different values of (a) ℓ and (b) m_ℓ are possible?

EXAMPLE 42.2   **The $n = 2$ Level of Hydrogen**

For a hydrogen atom, determine the allowed states corresponding to the principal quantum number $n = 2$ and calculate the energies of these states.

SOLUTION

Conceptualize   Think about the atom in the $n = 2$ quantum state. There is only one such state in the Bohr theory, but our discussion of the quantum theory allows for more states because of the possible values of ℓ and m_ℓ.

Categorize   We evaluate the results using rules discussed in this section, so we categorize this example as a substitution problem.

From Table 42.1, we find that when $n = 2$, ℓ can be 0 or 1. Find the possible values of m_ℓ from Table 42.1:

$$\ell = 0 \quad \rightarrow \quad m_\ell = 0$$

$$\ell = 1 \quad \rightarrow \quad m_\ell = -1, 0, \text{ or } 1$$

Hence, we have one state, designated as the $2s$ state, that is associated with the quantum numbers $n = 2$, $\ell = 0$, and $m_\ell = 0$, and we have three states, designated as $2p$ states, for which the quantum numbers are $n = 2$, $\ell = 1$, $m_\ell = -1$; $n = 2$, $\ell = 1$, $m_\ell = 0$; and $n = 2$, $\ell = 1$, $m_\ell = 1$.

Find the energy for all four of these states with $n = 2$ from Equation 42.21:

$$E_2 = -\frac{13.606 \text{ eV}}{2^2} = \boxed{-3.401 \text{ eV}}$$

42.5   The Wave Functions for Hydrogen

Because the potential energy of the hydrogen atom depends only on the radial distance r between nucleus and electron, some of the allowed states for this atom can be represented by wave functions that depend only on r. For these states, $f(\theta)$ and $g(\phi)$ are constants. The simplest wave function for hydrogen is the one that describes the $1s$ state and is designated $\psi_{1s}(r)$:

$$\psi_{1s}(r) = \frac{1}{\sqrt{\pi a_0^3}}\, e^{-r/a_0}$$

(42.22)   ◀ Wave function for hydrogen in its ground state

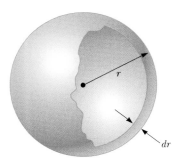

Figure 42.10 A spherical shell of radius r and thickness dr has a volume equal to $4\pi r^2\,dr$.

where a_0 is the Bohr radius. (In Problem 19, you can verify that this function satisfies the Schrödinger equation.) Note that ψ_{1s} approaches zero as r approaches ∞ and is normalized as presented (see Eq. 41.7). Furthermore, because ψ_{1s} depends only on r, it is *spherically symmetric*. This symmetry exists for all s states.

Recall that the probability of finding a particle in any region is equal to an integral of the probability density $|\psi|^2$ for the particle over the region. The probability density for the $1s$ state is

$$|\psi_{1s}|^2 = \left(\frac{1}{\pi a_0{}^3}\right)e^{-2r/a_0} \tag{42.23}$$

Because we imagine the nucleus to be fixed in space at $r = 0$, we can assign this probability density to the question of locating the electron. According to Equation 41.3, the probability of finding the electron in a volume element dV is $|\psi|^2\,dV$. It is convenient to define the *radial probability density function* $P(r)$ as the probability per unit radial length of finding the electron in a spherical shell of radius r and thickness dr. Therefore, $P(r)\,dr$ is the probability of finding the electron in this shell. The volume dV of such an infinitesimally thin shell equals its surface area $4\pi r^2$ multiplied by the shell thickness dr (Fig. 42.10), so we can write this probability as

$$P(r)\,dr = |\psi|^2\,dV = |\psi|^2 4\pi r^2\,dr$$

Therefore, the radial probability density function is

$$P(r) = 4\pi r^2 |\psi|^2 \tag{42.24}$$

Substituting Equation 42.23 into Equation 42.24 gives the radial probability density function for the hydrogen atom in its ground state:

► **Radial probability density for the 1s state of hydrogen**

$$P_{1s}(r) = \left(\frac{4r^2}{a_0{}^3}\right)e^{-2r/a_0} \tag{42.25}$$

A plot of the function $P_{1s}(r)$ versus r is presented in Figure 42.11a. The peak of the curve corresponds to the most probable value of r for this particular state. We show in Example 42.3 that this peak occurs at the Bohr radius, the radial position of the electron when the hydrogen atom is in its ground state in the Bohr theory, another remarkable agreement between the Bohr theory and the quantum theory.

According to quantum mechanics, the atom has no sharply defined boundary as suggested by the Bohr theory. The probability distribution in Figure 42.11a suggests that the charge of the electron can be modeled as being extended throughout a region of space, commonly referred to as an *electron cloud*. Figure 42.11b shows the probability density of the electron in a hydrogen atom in the $1s$ state as a function of position in the xy plane. The darkness of the blue color corresponds to the value of the probability density. The darkest portion of the distribution appears at $r = a_0$, corresponding to the most probable value of r for the electron.

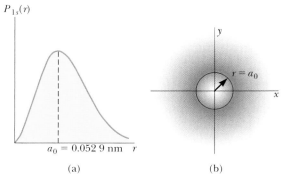

(a) (b)

Figure 42.11 (a) The probability of finding the electron as a function of distance from the nucleus for the hydrogen atom in the $1s$ (ground) state. The probability has its maximum value when r equals the Bohr radius a_0. (b) The cross section in the xy plane of the spherical electronic charge distribution for the hydrogen atom in its $1s$ state.

EXAMPLE 42.3	The Ground State of Hydrogen

(A) Calculate the most probable value of r for an electron in the ground state of the hydrogen atom.

SOLUTION

Conceptualize Do not imagine the electron in orbit around the proton as in the Bohr theory of the hydrogen atom. Instead, imagine the charge of the electron spread out in space around the proton.

Categorize Because the statement of the problem asks for the "most probable value of r," we categorize this example as a problem in which the quantum approach is used. (In the Bohr atom, the electron moves in an orbit with an *exact* value of r.)

Analyze The most probable value of r corresponds to the maximum in the plot of $P_{1s}(r)$ versus r. We can evaluate the most probable value of r by setting $dP_{1s}/dr = 0$ and solving for r.

Differentiate Equation 42.25 and set the result equal to zero:

$$\frac{dP_{1s}}{dr} = \frac{d}{dr}\left[\left(\frac{4r^2}{a_0{}^3}\right)e^{-2r/a_0}\right] = 0$$

$$e^{-2r/a_0}\frac{d}{dr}(r^2) + r^2\frac{d}{dr}(e^{-2r/a_0}) = 0$$

$$2re^{-2r/a_0} + r^2(-2/a_0)e^{-2r/a_0} = 0$$

$$(1)\quad 2r[1-(r/a_0)]e^{-2r/a_0} = 0$$

Set the bracketed expression equal to zero and solve for r:

$$1 - \frac{r}{a_0} = 0 \quad\rightarrow\quad r = \boxed{a_0}$$

Finalize The most probable value of r is the Bohr radius! Equation (1) is also satisfied at $r = 0$ and as $r \to \infty$. These points are locations of the *minimum* probability, which is equal to zero as seen in Figure 42.11a.

(B) Calculate the probability that the electron in the ground state of hydrogen will be found outside the first Bohr radius.

SOLUTION

Analyze The probability is found by integrating the radial probability density function $P_{1s}(r)$ for this state from the Bohr radius a_0 to ∞.

Set up this integral using Equation 42.25:

$$P = \int_{a_0}^{\infty} P_{1s}(r)\,dr = \frac{4}{a_0{}^3}\int_{a_0}^{\infty} r^2 e^{-2r/a_0}\,dr$$

Put the integral in dimensionless form by changing variables from r to $z = 2r/a_0$, noting that $z = 2$ when $r = a_0$ and that $dr = (a_0/2)\,dz$:

$$P = \frac{4}{a_0{}^3}\int_{a_0}^{\infty}\left(\frac{za_0}{2}\right)^2 e^{-z}\left(\frac{a_0}{2}\right)dz = \tfrac{1}{2}\int_{2}^{\infty} z^2 e^{-z}\,dz$$

Evaluate the integral using partial integration (see Appendix B.7):

$$P = -\tfrac{1}{2}(z^2 + 2z + 2)e^{-z}\Big|_{2}^{\infty}$$

Evaluate between the limits:

$$P = -\tfrac{1}{2}(0) - [-\tfrac{1}{2}(4+4+2)e^{-2}] = 5e^{-2} = \boxed{0.677 \text{ or } 67.7\%}$$

Finalize This probability is larger than 50%. The reason for this value is the asymmetry in the radial probability density function (Fig. 42.11a), which has more area to the right of the peak than to the left.

What If? What if you were asked for the *average* value of r for the electron in the ground state rather than the most probable value?

Answer   The average value of r is the same as the expectation value for r.

Use Equation 42.25 to evaluate the average value of r:

$$r_{\text{avg}} = \langle r \rangle = \int_0^\infty rP(r)\,dr = \int_0^\infty r\left(\frac{4r^2}{a_0{}^3}\right)e^{-2r/a_0}\,dr$$

$$= \left(\frac{4}{a_0{}^3}\right)\int_0^\infty r^3 e^{-2r/a_0}\,dr$$

Evaluate the integral with the help of the first integral listed in Table B.6 in Appendix B:

$$r_{\text{avg}} = \left(\frac{4}{a_0{}^3}\right)\left(\frac{3!}{(2/a_0)^4}\right) = \tfrac{3}{2}a_0$$

Again, the average value is larger than the most probable value because of the asymmetry in the wave function as seen in Figure 42.11a.

Wave function for ▶
hydrogen in the 2s state

The next-simplest wave function for the hydrogen atom is the one corresponding to the 2s state ($n = 2$, $\ell = 0$). The normalized wave function for this state is

$$\psi_{2s}(r) = \frac{1}{4\sqrt{2\pi}}\left(\frac{1}{a_0}\right)^{3/2}\left(2 - \frac{r}{a_0}\right)e^{-r/2a_0} \qquad (42.26)$$

Again notice that ψ_{2s} depends only on r and is spherically symmetric. The energy corresponding to this state is $E_2 = -(13.606/4)$ eV $= -3.401$ eV. This energy level represents the first excited state of hydrogen. A plot of the radial probability density function for this state in comparison to the 1s state is shown in Active Figure 42.12. The plot for the 2s state has two peaks. In this case, the most probable value corresponds to that value of r that has the highest value of P ($\approx 5a_0$). An electron in the 2s state would be much farther from the nucleus (on the average) than an electron in the 1s state.

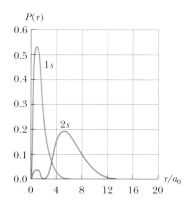

ACTIVE FIGURE 42.12

The radial probability density function versus r/a_0 for the 1s and 2s states of the hydrogen atom.

Sign in at www.thomsonedu.com and go to ThomsonNOW to choose values of r/a_0 and find the probability that the electron is located between two values.

42.6   Physical Interpretation of the Quantum Numbers

The principal quantum number n of a particular state in the hydrogen atom determines the energy of the atom according to Equation 42.21. Now let's see what the other quantum numbers in our atomic model correspond to physically.

The Orbital Quantum Number ℓ

We begin this discussion by returning briefly to the Bohr model of the atom. If the electron moves in a circle of radius r, the magnitude of its angular momentum relative to the center of the circle is $L = m_e vr$. The direction of $\vec{\mathbf{L}}$ is perpendicular to the plane of the circle and is given by a right-hand rule. According to classical physics, the magnitude L of the orbital angular momentum can have any value. The Bohr model of hydrogen, however, postulates that the magnitude of the angular momentum of the electron is restricted to multiples of $\hbar$; that is, $L = n\hbar$. This model must be modified because it predicts (incorrectly) that the ground state of hydrogen has one unit of angular momentum. Furthermore, if L is taken to be zero in the Bohr model, the electron must be pictured as a particle oscillating along a straight line through the nucleus, which is a physically unacceptable situation.

These difficulties are resolved with the quantum-mechanical model of the atom, although we must give up the convenient mental representation of an electron orbiting in a well-defined circular path. Despite the absence of this representation, the atom does indeed possess an angular momentum and it is still called orbital angular momentum. According to quantum mechanics, an atom in a state whose principal quantum number is n can take on the following *discrete* values of the magnitude of the orbital angular momentum:[5]

$$L = \sqrt{\ell(\ell + 1)}\,\hbar \qquad \ell = 0, 1, 2, \ldots, n - 1 \qquad \text{(42.27)}$$

◀ Allowed values of L

Given these allowed values of ℓ, we see that $L = 0$ (corresponding to $\ell = 0$) is an acceptable value of the magnitude of the angular momentum. That L can be zero in this model serves to point out the inherent difficulties in any attempt to describe results based on quantum mechanics in terms of a purely particle-like (classical) model. In the quantum-mechanical interpretation, the electron cloud for the $L = 0$ state is spherically symmetric and has no fundamental rotation axis.

The Orbital Magnetic Quantum Number m_ℓ

Because angular momentum is a vector, its direction must be specified. Recall from Chapter 29 that a current loop has a corresponding magnetic moment $\vec{\mu} = I\vec{A}$ (Eq. 29.15), where I is the current in the loop and $\vec{A}$ is a vector perpendicular to the loop whose magnitude is the area of the loop. Such a moment placed in a magnetic field $\vec{B}$ interacts with the field. Suppose a weak magnetic field applied along the z axis defines a direction in space. According to classical physics, the energy of the loop–field system depends on the direction of the magnetic moment of the loop with respect to the magnetic field as described by Equation 29.18, $U = -\vec{\mu} \cdot \vec{B}$. Any energy between $-\mu B$ and $+\mu B$ is allowed by classical physics.

In the Bohr theory, the circulating electron represents a current loop. In the quantum-mechanical approach to the hydrogen atom, we abandon the circular orbit viewpoint of the Bohr theory, but the atom still possesses an orbital angular momentum. Therefore, there is some sense of rotation of the electron around the nucleus and a magnetic moment is present due to this angular momentum.

As mentioned in Section 42.3, spectral lines from some atoms are observed to split into groups of three closely spaced lines when the atoms are placed in a magnetic field. Suppose the hydrogen atom is located in a magnetic field. According to quantum mechanics, **there are discrete directions allowed for the magnetic moment vector $\vec{\mu}$ with respect to the magnetic field vector $\vec{B}$.** This situation is very different from that in classical physics, in which all directions are allowed.

Because the magnetic moment $\vec{\mu}$ of the atom can be related[6] to the angular momentum vector $\vec{L}$, the discrete directions of $\vec{\mu}$ translate to the direction of $\vec{L}$ being quantized. This quantization means that L_z (the projection of $\vec{L}$ along the z axis) can have only discrete values. The orbital magnetic quantum number m_ℓ specifies the allowed values of the z component of the orbital angular momentum according to the expression[7]

$$L_z = m_\ell \hbar \qquad \text{(42.28)}$$

◀ Allowed values of L_z

The quantization of the possible orientations of $\vec{L}$ with respect to an external magnetic field is often referred to as **space quantization.**

Let's look at the possible orientations of $\vec{L}$ for a given value of ℓ. Recall that m_ℓ can have values ranging from $-\ell$ to ℓ. If $\ell = 0$, then $m_\ell = 0$ and $L_z = 0$. If $\ell = 1$, the possible values of m_ℓ are -1, 0, and 1, so L_z may be $-\hbar$, 0, or $\hbar$. If $\ell = 2$, then m_ℓ can be -2, -1, 0, 1, or 2, corresponding to L_z values of $-2\hbar$, $-\hbar$, 0, $\hbar$, or $2\hbar$, and so on.

---

[5] Equation 42.27 is a direct result of the mathematical solution of the Schrödinger equation and the application of angular boundary conditions. This development, however, is beyond the scope of this book.

[6] See Equation 30.22 for this relationship as derived from a classical viewpoint. Quantum mechanics arrives at the same result.

[7] As with Equation 42.27, the relationship expressed in Equation 42.28 arises from the solution to the Schrödinger equation and application of boundary conditions.

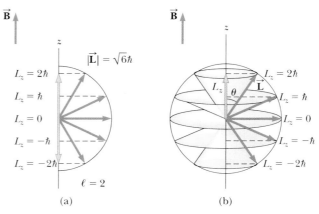

Figure 42.13   A vector model for $\ell = 2$. (a) The allowed projections on the z axis of the orbital angular momentum $\vec{L}$. (b) The orbital angular momentum vector $\vec{L}$ lies on the surface of a cone.

Figure 42.13a shows a **vector model** that describes space quantization for the case $\ell = 2$. Notice that $\vec{L}$ can never be aligned parallel or antiparallel to $\vec{B}$ because the maximum value of L_z is $\ell\hbar$, which is less than the magnitude of the angular momentum $L = \sqrt{\ell(\ell + 1)}\hbar$. The angular momentum vector $\vec{L}$ is allowed to be perpendicular to $\vec{B}$, which corresponds to the case of $L_z = 0$ and $\ell = 0$.

The vector $\vec{L}$ does not point in one specific direction even though its z component is fixed. If $\vec{L}$ were known exactly, all three components L_x, L_y, and L_z would be specified, which is inconsistent with the uncertainty principle. How can the magnitude and z component of a vector be specified, but the vector not be completely specified? To answer, imagine that L_x and L_y are completely unspecified so that $\vec{L}$ lies anywhere on the surface of a cone that makes an angle θ with the z axis as shown in Figure 42.13b. From the figure, we see that θ is also quantized and that its values are specified through the relationship

$$\cos\theta = \frac{L_z}{L} = \frac{m_\ell}{\sqrt{\ell(\ell + 1)}} \tag{42.29}$$

If the atom is placed in a magnetic field, the energy $U = -\vec{\mu} \cdot \vec{B}$ is additional energy for the atom–field system beyond that described in Equation 42.21. Because the directions of $\vec{\mu}$ are quantized, there are discrete total energies for the atom corresponding to different values of m_ℓ. Figure 42.14a shows a transition between two atomic levels in the absence of a magnetic field. In Figure 42.14b, a magnetic field is applied and the upper level, with $\ell = 1$, splits into three levels

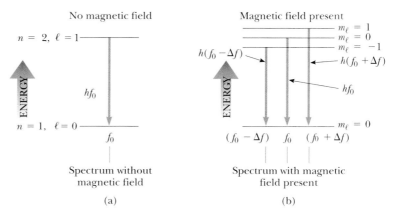

Figure 42.14   The Zeeman effect. (a) Energy levels for the ground and first excited states of a hydrogen atom. When $\vec{B} = 0$, only a single spectral line at f_0 is observed. (b) When the atom is immersed in a magnetic field $\vec{B}$, the state with $\ell = 1$ splits into three states. Atoms in the excited states decay to the ground state with the emission of photons with three different energies, giving rise to emission lines at f_0, $f_0 + \Delta f$, and $f_0 - \Delta f$, where Δf is the frequency shift of the emission caused by the magnetic field.

corresponding to the different directions of $\vec{\mu}$. There are now three possible transitions from the $\ell = 1$ subshell to the $\ell = 0$ subshell. Therefore, in a collection of atoms, there are atoms in all three states and the single spectral line in Figure 42.14a splits into three spectral lines. This phenomenon is called the *Zeeman effect*.

The Zeeman effect can be used to measure extraterrestrial magnetic fields. For example, the splitting of spectral lines in light from hydrogen atoms in the surface of the Sun can be used to calculate the magnitude of the magnetic field at that location. The Zeeman effect is one of many phenomena that cannot be explained with the Bohr model but are successfully explained by the quantum model of the atom.

EXAMPLE 42.4   **Space Quantization for Hydrogen**

Consider the hydrogen atom in the $\ell = 3$ state. Calculate the magnitude of $\vec{L}$, the allowed values of L_z, and the corresponding angles θ that $\vec{L}$ makes with the z axis.

SOLUTION

Conceptualize   Consider Figure 42.13, which is a vector model for $\ell = 2$. Draw such a vector model for $\ell = 3$ to help with this problem.

Categorize   We evaluate results using equations developed in this section, so we categorize this example as a substitution problem.

Calculate the magnitude of the orbital angular momentum using Equation 42.27:
$$L = \sqrt{\ell(\ell + 1)}\hbar = \sqrt{3(3 + 1)}\hbar = 2\sqrt{3}\hbar$$

Calculate the allowed values of L_z using Equation 42.28 with $m_\ell = -3, -2, -1, 0, 1, 2,$ and 3:
$$L_z = -3\hbar, -2\hbar, -\hbar, 0, \hbar, 2\hbar, 3\hbar$$

Calculate the allowed values of $\cos \theta$ using Equation 42.29:
$$\cos \theta = \frac{\pm 3}{2\sqrt{3}} = \pm 0.866 \qquad \cos \theta = \frac{\pm 2}{2\sqrt{3}} = \pm 0.577$$
$$\cos \theta = \frac{\pm 1}{2\sqrt{3}} = \pm 0.289 \qquad \cos \theta = \frac{0}{2\sqrt{3}} = 0$$

Find the angles corresponding to these values of $\cos \theta$:
$$\theta = 30.0°, 54.7°, 73.2°, 90.0°, 107°, 125°, 150°$$

What If?   What if the value of ℓ is an arbitrary integer? For an arbitrary value of ℓ, how many values of m_ℓ are allowed?

Answer   For a given value of ℓ, the values of m_ℓ range from $-\ell$ to $+\ell$ in steps of 1. Therefore, there are 2ℓ nonzero values of m_ℓ (specifically, $\pm 1, \pm 2, \ldots, \pm \ell$). In addition, one more value of $m_\ell = 0$ is possible, for a total of $2\ell + 1$ values of m_ℓ. This result is critical in understanding the results of the Stern–Gerlach experiment described below with regard to spin.

The Spin Magnetic Quantum Number m_s

The three quantum numbers n, ℓ, and m_ℓ discussed so far are generated by applying boundary conditions to solutions of the Schrödinger equation, and we can assign a physical interpretation to each quantum number. Let's now consider **electron spin,** which does *not* come from the Schrödinger equation.

Wolfgang Pauli and Niels Bohr watch a spinning top. The spin of the electron is analogous to the spin of the top but is different in many ways.

Courtesy of AIP Niels Bohr Library, Margarethe Bohr Collection

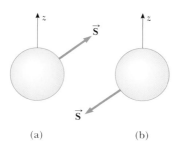

(a)   (b)

Figure 42.15 The spin of an electron can be either (a) up or (b) down relative to a specified z axis. The spin can never be aligned with the axis.

PITFALL PREVENTION 42.5
The Electron Is Not Spinning

Although the concept of a spinning electron is conceptually useful, it should not be taken literally. The spin of the Earth is a mechanical rotation. On the other hand, electron spin is a purely quantum effect that gives the electron an angular momentum as if it were physically spinning.

In Example 42.2, we found four quantum states corresponding to $n = 2$. In reality, however, eight such states occur. The additional four states can be explained by requiring a fourth quantum number for each state, the **spin magnetic quantum number** m_s.

The need for this new quantum number arises because of an unusual feature observed in the spectra of certain gases, such as sodium vapor. Close examination of one prominent line in the emission spectrum of sodium reveals that the line is, in fact, two closely spaced lines called a *doublet*.[8] The wavelengths of these lines occur in the yellow region of the electromagnetic spectrum at 589.0 nm and 589.6 nm. In 1925, when this doublet was first observed, it could not be explained with the existing atomic theory. To resolve this dilemma, Samuel Goudsmit (1902–1978) and George Uhlenbeck (1900–1988), following a suggestion made by Austrian physicist Wolfgang Pauli, proposed the spin quantum number.

To describe this new quantum number, it is convenient (but technically incorrect) to imagine the electron spinning about its axis as it orbits the nucleus as described in Section 30.6. As illustrated in Figure 42.15, only two directions exist for the electron spin. If the direction of spin is as shown in Figure 42.15a, the electron is said to have *spin up*. If the direction of spin is as shown in Figure 42.15b, the electron is said to have *spin down*. In the presence of a magnetic field, the energy of the electron is slightly different for the two spin directions. This energy difference accounts for the sodium doublet.

The classical description of electron spin—as resulting from a spinning electron—is incorrect. More recent theory indicates that the electron is a point particle, without spatial extent. Therefore, the electron cannot be considered to be spinning. Despite this conceptual difficulty, all experimental evidence supports the idea that an electron does have some intrinsic angular momentum that can be described by the spin magnetic quantum number. Paul Dirac (1902–1984) showed that this fourth quantum number originates in the relativistic properties of the electron.

In 1921, Otto Stern (1888–1969) and Walter Gerlach (1889–1979) performed an experiment that demonstrated space quantization. Their results, however, were not in quantitative agreement with the atomic theory that existed at that time. In their experiment, a beam of silver atoms sent through a nonuniform magnetic field was split into two discrete components (Fig. 42.16). Stern and Gerlach repeated the experiment using other atoms, and in each case the beam split into two or more components. The classical argument is as follows. If the z direction is chosen to be the direction of the maximum nonuniformity of $\vec{B}$, the net magnetic force on the atoms is along the z axis and is proportional to the component of the magnetic moment $\vec{\mu}$ of the atom in the z direction. Classically, $\vec{\mu}$ can have any orientation, so the deflected beam should be spread out continuously. According to quantum mechanics, however, the deflected beam has an integral number of discrete components and the number of components determines the number of possible values of μ_z. Therefore, because the Stern–Gerlach experiment showed split beams, space quantization was at least qualitatively verified.

For the moment, let's assume the magnetic moment $\vec{\mu}$ of the atom is due to the orbital angular momentum. Because μ_z is proportional to m_ℓ, the number of possible values of μ_z is $2\ell + 1$ as found in the **What If?** section of Example 42.4. Furthermore, because ℓ is an integer, the number of values of μ_z is always odd. This prediction is not consistent with Stern and Gerlach's observation of two components (an *even* number) in the deflected beam of silver atoms. Hence, either quantum mechanics is incorrect or the model is in need of refinement.

In 1927, T. E. Phipps and J. B. Taylor repeated the Stern–Gerlach experiment using a beam of hydrogen atoms. Their experiment was important because it

---

[8] This phenomenon is a Zeeman effect for spin and is identical in nature to the Zeeman effect for orbital angular momentum discussed before Example 42.4 except that no external magnetic field is required. The magnetic field for this Zeeman effect is internal to the atom and arises from the relative motion of the electron and the nucleus.

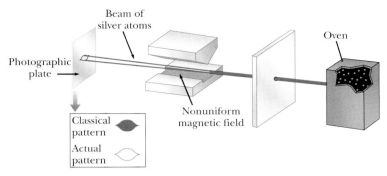

Figure 42.16   The technique used by Stern and Gerlach to verify space quantization. A beam of silver atoms is split in two by a nonuniform magnetic field.

involved an atom containing a single electron in its ground state, for which the quantum theory makes reliable predictions. Recall that $\ell = 0$ for hydrogen in its ground state, so $m_\ell = 0$. Therefore, we would not expect the beam to be deflected by the magnetic field because the magnetic moment $\vec{\boldsymbol{\mu}}$ of the atom is zero. The beam in the Phipps–Taylor experiment, however, was again split into two components. On the basis of that result, we must conclude that something other than the electron's orbital motion is contributing to the atomic magnetic moment.

As we learned earlier, Goudsmit and Uhlenbeck had proposed that the electron has an intrinsic angular momentum, spin, apart from its orbital angular momentum. In other words, the total angular momentum of the electron in a particular electronic state contains both an orbital contribution $\vec{\mathbf{L}}$ and a spin contribution $\vec{\mathbf{S}}$. The Phipps–Taylor result confirmed the hypothesis of Goudsmit and Uhlenbeck.

In 1929, Dirac used the relativistic form of the total energy of a system to solve the relativistic wave equation for the electron in a potential well. His analysis confirmed the fundamental nature of electron spin. (Spin, like mass and charge, is an *intrinsic* property of a particle, independent of its surroundings.) Furthermore, the analysis showed that electron spin[9] can be described by a single quantum number s, whose value can be only $s = \frac{1}{2}$. The spin angular momentum of the electron *never changes*. This notion contradicts classical laws, which dictate that a rotating charge slows down in the presence of an applied magnetic field because of the Faraday emf that accompanies the changing field. Furthermore, if the electron is viewed as a spinning ball of charge subject to classical laws, parts of the electron near its surface would be rotating with speeds exceeding the speed of light. Therefore, the classical picture must not be pressed too far; ultimately, spin of an electron is a quantum entity defying any simple classical description.

Because spin is a form of angular momentum, it must follow the same quantum rules as orbital angular momentum. In accordance with Equation 42.27, the magnitude of the **spin angular momentum $\vec{\mathbf{S}}$** for the electron is

$$S = \sqrt{s(s+1)}\,\hbar = \frac{\sqrt{3}}{2}\,\hbar \qquad (42.30)$$

◄ Magnitude of the spin angular momentum of an electron

Like orbital angular momentum $\vec{\mathbf{L}}$, spin angular momentum $\vec{\mathbf{S}}$ exhibits space quantization as described in Figure 42.17. It can have two orientations relative to a z axis, specified by the **spin magnetic quantum number** $m_s = \pm\frac{1}{2}$. Similar to Equation 42.28 for orbital angular momentum, the z component of spin angular momentum is

$$S_z = m_s\hbar = \pm\tfrac{1}{2}\hbar \qquad (42.31)$$

◄ Allowed values of S_z

[9] Scientists often use the word *spin* when referring to the spin angular momentum quantum number. For example, it is common to say, "The electron has a spin of one half."

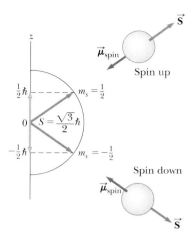

Figure 42.17 Spin angular momentum $\vec{S}$ exhibits space quantization. This figure shows the two allowed orientations of the spin angular momentum vector $\vec{S}$ and the spin magnetic moment $\vec{\mu}_{\text{spin}}$ for a spin-$\frac{1}{2}$ particle, such as the electron.

The two values $\pm\hbar/2$ for S_z correspond to the two possible orientations for $\vec{S}$ shown in Figure 42.17. The value $m_s = +\frac{1}{2}$ refers to the spin-up case, and $m_s = -\frac{1}{2}$ refers to the spin-down case. Notice that Equations 42.30 and 42.31 do not allow the spin vector to lie along the z axis. The actual direction of $\vec{S}$ is at a relatively large angle with respect to the z axis as shown in Figures 42.15 and 42.17.

The spin magnetic moment $\vec{\mu}_{\text{spin}}$ of the electron is related to its spin angular momentum $\vec{S}$ by the expression

$$\vec{\mu}_{\text{spin}} = -\frac{e}{m_e}\vec{S} \tag{42.32}$$

where e is the electronic charge and m_e is the mass of the electron. Because $S_z = \pm\frac{1}{2}\hbar$, the z component of the spin magnetic moment can have the values

$$\mu_{\text{spin},z} = \pm\frac{e\hbar}{2m_e} \tag{42.33}$$

As we learned in Section 30.6, the quantity $e\hbar/2m_e$ is the Bohr magneton $\mu_B = 9.27 \times 10^{-24}$ J/T. The ratio of magnetic moment to angular momentum is twice as great for spin angular momentum (Eq. 42.32) as it is for orbital angular momentum (Eq. 30.22). The factor of 2 is explained in a relativistic treatment first carried out by Dirac.

Today, physicists explain the Stern–Gerlach and Phipps–Taylor experiments as follows. The observed magnetic moments for both silver and hydrogen are due to spin angular momentum only, with no contribution from orbital angular momentum. A single-electron atom such as hydrogen has its electron spin quantized in the magnetic field in such a way that the z component of spin angular momentum is either $\frac{1}{2}\hbar$ or $-\frac{1}{2}\hbar$, corresponding to $m_s = \pm\frac{1}{2}$. Electrons with spin $+\frac{1}{2}$ are deflected downward, and those with spin $-\frac{1}{2}$ are deflected upward.

The Stern–Gerlach experiment provided two important results. First, it verified the concept of space quantization. Second, it showed that spin angular momentum exists, even though this property was not recognized until four years after the experiments were performed.

As mentioned earlier, there are eight quantum states corresponding to $n = 2$ in the hydrogen atom, not four as found in Example 42.2. Each of the four states in Example 42.2 is actually two states because of the two possible values of m_s. Table 42.3 shows the quantum states corresponding to these eight states.

TABLE 42.3

Quantum Numbers for the $n = 2$ State of Hydrogen

n	ℓ	m_ℓ	m_s	Subshell	Shell	Number of States in Subshell
2	0	0	$\frac{1}{2}$	2s	L	2
2	0	0	$-\frac{1}{2}$			
2	1	1	$\frac{1}{2}$			
2	1	1	$-\frac{1}{2}$			
2	1	0	$\frac{1}{2}$	2p	L	6
2	1	0	$-\frac{1}{2}$			
2	1	-1	$\frac{1}{2}$			
2	1	-1	$-\frac{1}{2}$			

42.7 The Exclusion Principle and the Periodic Table

We have found that the state of a hydrogen atom is specified by four quantum numbers: n, ℓ, m_ℓ, and m_s. As it turns out, the number of states available to other atoms may also be predicted by this same set of quantum numbers. In fact, these four quantum numbers can be used to describe all the electronic states of an atom, regardless of the number of electrons in its structure.

For our discussion of atoms with many electrons, it is often easiest to assign the quantum numbers to the electrons in the atom as opposed to the entire atom. An obvious question that arises here is, "How many electrons can be in a particular quantum state?" Pauli answered this important question in 1925, in a statement known as the **exclusion principle**:

> No two electrons can ever be in the same quantum state; therefore, no two electrons in the same atom can have the same set of quantum numbers.

If this principle were not valid, an atom could radiate energy until every electron in the atom is in the lowest possible energy state and therefore the chemical behavior of the elements would be grossly modified. Nature as we know it would not exist.

In reality, we can view the electronic structure of complex atoms as a succession of filled levels increasing in energy. As a general rule, the order of filling of an atom's subshells is as follows. Once a subshell is filled, the next electron goes into the lowest-energy vacant subshell. We can understand this behavior by recognizing that if the atom were not in the lowest energy state available to it, it would radiate energy until it reached this state.

Before we discuss the electronic configuration of various elements, it is convenient to define an *orbital* as the atomic state characterized by the quantum numbers n, ℓ, and m_ℓ. The exclusion principle tells us that **only two electrons can be present in any orbital**. One of these electrons has a spin magnetic quantum number $m_s = +\frac{1}{2}$, and the other has $m_s = -\frac{1}{2}$. Because each orbital is limited to two electrons, the number of electrons that can occupy the various shells is also limited.

Table 42.4 shows the allowed quantum states for an atom up to $n = 3$. The arrows pointing upward indicate an electron described by $m_s = +\frac{1}{2}$, and those pointing downward indicate that $m_s = -\frac{1}{2}$. The $n = 1$ shell can accommodate only two electrons because $m_\ell = 0$ means that only one orbital is allowed. (The three quantum numbers describing this orbital are $n = 1$, $\ell = 0$, and $m_\ell = 0$.) The $n = 2$ shell has two subshells, one for $\ell = 0$ and one for $\ell = 1$. The $\ell = 0$ subshell is limited to two electrons because $m_\ell = 0$. The $\ell = 1$ subshell has three allowed

WOLFGANG PAULI
Austrian Theoretical Physicist (1900–1958)
An extremely talented theoretician who made important contributions in many areas of modern physics, Pauli gained public recognition at the age of 21 with a masterful review article on relativity that is still considered one of the finest and most comprehensive introductions to the subject. His other major contributions were the discovery of the exclusion principle, the explanation of the connection between particle spin and statistics, theories of relativistic quantum electrodynamics, the neutrino hypothesis, and the hypothesis of nuclear spin.

TABLE 42.4

Allowed Quantum States for an Atom Up to $n = 3$

Shell	n	1	2			3								
Subshell	ℓ	0	0	1		0	1			2				
Orbital	m_ℓ	0	0	1	0	−1	0	1	0	−1	2	1	0	−1	−2
	m_s	↑↓	↑↓	↑↓	↑↓	↑↓	↑↓	↑↓	↑↓	↑↓	↑↓	↑↓	↑↓	↑↓	↑↓

orbitals, corresponding to $m_\ell = 1$, 0, and −1. Because each orbital can accommodate two electrons, the $\ell = 1$ subshell can hold six electrons. Therefore, the $n = 2$ shell can contain eight electrons as shown in Table 42.3. The $n = 3$ shell has three subshells ($\ell = 0$, 1, 2) and nine orbitals, accommodating up to 18 electrons. In general, each shell can accommodate up to $2n^2$ electrons.

The exclusion principle can be illustrated by examining the electronic arrangement in a few of the lighter atoms. The atomic number Z of any element is the number of protons in the nucleus of an atom of that element. A neutral atom of that element has Z electrons. Hydrogen ($Z = 1$) has only one electron, which, in the ground state of the atom, can be described by either of two sets of quantum numbers n, ℓ, m_ℓ, m_s: 1, 0, 0, $\frac{1}{2}$ or 1, 0, 0, $-\frac{1}{2}$. This electronic configuration is often written $1s^1$. The notation $1s$ refers to a state for which $n = 1$ and $\ell = 0$, and the superscript indicates that one electron is present in the s subshell.

Helium ($Z = 2$) has two electrons. In the ground state, their quantum numbers are 1, 0, 0, $\frac{1}{2}$ and 1, 0, 0, $-\frac{1}{2}$. No other possible combinations of quantum numbers exist for this level, and we say that the K shell is filled. This electronic configuration is written $1s^2$.

Lithium ($Z = 3$) has three electrons. In the ground state, two of them are in the $1s$ subshell. The third is in the $2s$ subshell because this subshell is slightly lower in energy than the $2p$ subshell.[10] Hence, the electronic configuration for lithium is $1s^2 2s^1$.

The electronic configurations of lithium and the next several elements are provided in Figure 42.18. The electronic configuration of beryllium ($Z = 4$), with its four electrons, is $1s^2 2s^2$, and boron ($Z = 5$) has a configuration of $1s^2 2s^2 2p^1$. The $2p$ electron in boron may be described by any of the six equally probable sets of quantum numbers listed in Table 42.3. In Figure 42.18, we show this electron in the leftmost $2p$ box with spin up, but it is equally likely to be in any $2p$ box with spin either up or down.

Carbon ($Z = 6$) has six electrons, giving rise to a question concerning how to assign the two $2p$ electrons. Do they go into the same orbital with paired spins (↑ ↓), or do they occupy different orbitals with unpaired spins (↑ ↑)? Experimental data show that the most stable configuration (that is, the one with the lowest energy) is the latter, in which the spins are unpaired. Hence, the two $2p$ electrons in carbon and the three $2p$ electrons in nitrogen ($Z = 7$) have unpaired spins as Figure 42.18 shows. The general rule that governs such situations, called **Hund's rule,** states that

> when an atom has orbitals of equal energy, the order in which they are filled by electrons is such that a maximum number of electrons have unpaired spins.

Some exceptions to this rule occur in elements having subshells that are close to being filled or half-filled.

---

[10] To a first approximation, energy depends only on the quantum number n, as we have discussed. Because of the effect of the electronic charge shielding the nuclear charge, however, energy depends on ℓ also in multielectron atoms. We shall discuss these shielding effects in Section 42.8.

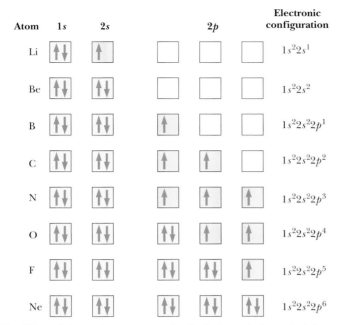

Atom	$1s$	$2s$		$2p$		Electronic configuration
Li	↑↓	↑				$1s^2 2s^1$
Be	↑↓	↑↓				$1s^2 2s^2$
B	↑↓	↑↓	↑			$1s^2 2s^2 2p^1$
C	↑↓	↑↓	↑	↑		$1s^2 2s^2 2p^2$
N	↑↓	↑↓	↑	↑	↑	$1s^2 2s^2 2p^3$
O	↑↓	↑↓	↑↓	↑	↑	$1s^2 2s^2 2p^4$
F	↑↓	↑↓	↑↓	↑↓	↑	$1s^2 2s^2 2p^5$
Ne	↑↓	↑↓	↑↓	↑↓	↑↓	$1s^2 2s^2 2p^6$

Figure 42.18   The filling of electronic states must obey both the exclusion principle and Hund's rule.

In 1871, long before quantum mechanics was developed, the Russian chemist Dmitri Mendeleev (1834–1907) made an early attempt at finding some order among the chemical elements. He was trying to organize the elements for the table of contents of a book he was writing. He arranged the atoms in a table similar to that shown in Figure 42.19 (page 1240), according to their atomic masses and chemical similarities. The first table Mendeleev proposed contained many blank spaces, and he boldly stated that the gaps were there only because the elements had not yet been discovered. By noting the columns in which some missing elements should be located, he was able to make rough predictions about their chemical properties. Within 20 years of this announcement, most of these elements were indeed discovered.

The elements in the **periodic table** (Fig. 42.19) are arranged so that all those in a column have similar chemical properties. For example, consider the elements in the last column, which are all gases at room temperature: He (helium), Ne (neon), Ar (argon), Kr (krypton), Xe (xenon), and Rn (radon). The outstanding characteristic of all these elements is that they do not normally take part in chemical reactions; that is, they do not readily join with other atoms to form molecules. They are therefore called *noble gases*.

We can partially understand this behavior by looking at the electronic configurations in Figure 42.19. The chemical behavior of an element depends on the outermost shell that contains electrons. The electronic configuration for helium is $1s^2$, and the $n = 1$ shell (which is the outermost shell because it is the only shell) is filled. Also, the energy of the atom in this configuration is considerably lower than the energy for the configuration in which an electron is in the next available level, the $2s$ subshell. Next, look at the electronic configuration for neon, $1s^2 2s^2 2p^6$. Again, the outermost shell ($n = 2$ in this case) is filled and a wide gap in energy occurs between the filled $2p$ subshell and the next available one, the $3s$ subshell. Argon has the configuration $1s^2 2s^2 2p^6 3s^2 3p^6$. Here, it is only the $3p$ subshell that is filled, but again a wide gap in energy occurs between the filled $3p$ subshell and the next available one, the $3d$ subshell. This pattern continues through all the noble gases. Krypton has a filled $4p$ subshell, xenon a filled $5p$ subshell, and radon a filled $6p$ subshell.

The column to the left of the noble gases in the periodic table consists of a group of elements called the *halogens*: fluorine, chlorine, bromine, iodine, and

Group I	Group II	Transition elements										Group III	Group VI	Group V	Group VI	Group VII	Group 0
H 1 $1s^1$																H 1 $1s^1$	He 2 $1s^2$
Li 3 $2s^1$	Be 4 $2s^2$											B 5 $2p^1$	C 6 $2p^2$	N 7 $2p^3$	O 8 $2p^4$	F 9 $2p^5$	Ne 10 $2p^6$
Na 11 $3s^1$	Mg 12 $3s^2$											Al 13 $3p^1$	Si 14 $3p^2$	P 15 $3p^3$	S 16 $3p^4$	Cl 17 $3p^5$	Ar 18 $3p^6$
K 19 $4s^1$	Ca 20 $4s^2$	Sc 21 $3d^1 4s^2$	Ti 22 $3d^2 4s^2$	V 23 $3d^3 4s^2$	Cr 24 $3d^5 4s^1$	Mn 25 $3d^5 4s^2$	Fe 26 $3d^6 4s^2$	Co 27 $3d^7 4s^2$	Ni 28 $3d^8 4s^2$	Cu 29 $3d^{10} 4s^1$	Zn 30 $3d^{10} 4s^2$	Ga 31 $4p^1$	Ge 32 $4p^2$	As 33 $4p^3$	Se 34 $4p^4$	Br 35 $4p^5$	Kr 36 $4p^6$
Rb 37 $5s^1$	Sr 38 $5s^2$	Y 39 $4d^1 5s^2$	Zr 40 $4d^2 5s^2$	Nb 41 $4d^4 5s^1$	Mo 42 $4d^5 5s^1$	Tc 43 $4d^5 5s^2$	Ru 44 $4d^7 5s^1$	Rh 45 $4d^8 5s^1$	Pd 46 $4d^{10}$	Ag 47 $4d^{10} 5s^1$	Cd 48 $4d^{10} 5s^2$	In 49 $5p^1$	Sn 50 $5p^2$	Sb 51 $5p^3$	Te 52 $5p^4$	I 53 $5p^5$	Xe 54 $5p^6$
Cs 55 $6s^1$	Ba 56 $6s^2$	57–71*	Hf 72 $5d^2 6s^2$	Ta 73 $5d^3 6s^2$	W 74 $5d^4 6s^2$	Re 75 $5d^5 6s^2$	Os 76 $5d^6 6s^2$	Ir 77 $5d^7 6s^2$	Pt 78 $5d^9 6s^1$	Au 79 $5d^{10} 6s^1$	Hg 80 $5d^{10} 6s^2$	Tl 81 $6p^1$	Pb 82 $6p^2$	Bi 83 $6p^3$	Po 84 $6p^4$	At 85 $6p^5$	Rn 86 $6p^6$
Fr 87 $7s^1$	Ra 88 $7s^2$	89–103**	Rf 104 $6d^2 7s^2$	Db 105 $6d^3 7s^2$	Sg 106 $6d^4 7s^2$	Bh 107 $6d^5 7s^2$	Hs 108 $6d^6 7s^2$	Mt 109 $6d^7 7s^2$	Ds 110 $6d^9 7s^1$	Rg 111	112		114				

*Lanthanide series

La 57 $5d^1 6s^2$	Ce 58 $5d^1 4f^1 6s^2$	Pr 59 $4f^3 6s^2$	Nd 60 $4f^4 6s^2$	Pm 61 $4f^5 6s^2$	Sm 62 $4f^6 6s^2$	Eu 63 $4f^7 6s^2$	Gd 64 $5d^1 4f^7 6s^2$	Tb 65 $5d^1 4f^8 6s^2$	Dy 66 $4f^{10} 6s^2$	Ho 67 $4f^{11} 6s^2$	Er 68 $4f^{12} 6s^2$	Tm 69 $4f^{13} 6s^2$	Yb 70 $4f^{14} 6s^2$	Lu 71 $5d^1 4f^{14} 6s^2$

**Actinide series

Ac 89 $6d^1 7s^2$	Th 90 $6d^2 7s^2$	Pa 91 $5f^2 6d^1 7s^2$	U 92 $5f^3 6d^1 7s^2$	Np 93 $5f^4 6d^1 7s^2$	Pu 94 $5f^6 7s^2$	Am 95 $5f^7 7s^2$	Cm 96 $5f^7 6d^1 7s^2$	Bk 97 $5f^8 6d^1 7s^2$	Cf 98 $5f^{10} 6d^0 7s^2$	Es 99 $5f^{11} 6d^0 7s^2$	Fm 100 $5f^{12} 6d^0 7s^2$	Md 101 $5f^{13} 6d^0 7s^2$	No 102 $5f^{14} 6d^0 7s^2$	Lr 103 $5f^{14} 7s^2 7p^1$

Figure 42.19 The periodic table of the elements is an organized tabular representation of the elements that shows their periodic chemical behavior. Elements in a given column have similar chemical behavior. This table shows the element name, the atomic number, and the electron configuration. A more complete periodic table is available in Appendix C.

astatine. At room temperature, fluorine and chlorine are gases, bromine is a liquid, and iodine and astatine are solids. In each of these atoms, the outer subshell is one electron short of being filled. As a result, the halogens are chemically very active, readily accepting an electron from another atom to form a closed shell. The halogens tend to form strong ionic bonds with atoms at the other side of the periodic table. (We shall discuss ionic bonds in Chapter 43.) In a halogen lightbulb, bromine or iodine atoms combine with tungsten atoms evaporated from the filament and return them to the filament, resulting in a longer-lasting lightbulb. In addition, the filament can be operated at a higher temperature than in ordinary lightbulbs, giving a brighter and whiter light.

At the left side of the periodic table, the Group I elements consist of hydrogen and the *alkali metals*: lithium, sodium, potassium, rubidium, cesium, and francium. Each of these atoms contains one electron in a subshell outside of a closed subshell. Therefore, these elements easily form positive ions because the lone electron is bound with a relatively low energy and is easily removed. Therefore, the alkali metal atoms are chemically active and form very strong bonds with halogen atoms. For example, table salt, NaCl, is a combination of an alkali metal and a halogen. Because the outer electron is weakly bound, pure alkali metals tend to be good electrical conductors. Because of their high chemical activity, however, they are not generally found in nature in pure form.

It is interesting to plot ionization energy versus atomic number Z as in Figure 42.20. Notice the pattern of $\Delta Z = 2, 8, 8, 18, 18, 32$ for the various peaks. This pattern follows from the exclusion principle and helps explain why the elements repeat their chemical properties in groups. For example, the peaks at $Z = 2, 10,$ 18, and 36 correspond to the noble gases helium, neon, argon, and krypton,

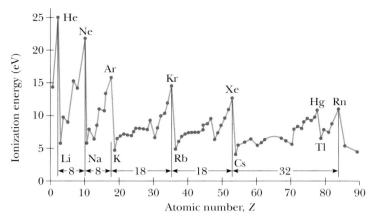

Figure 42.20   Ionization energy of the elements versus atomic number.

respectively, which, as we have mentioned, all have filled outermost shells. These elements have relatively high ionization energies and similar chemical behavior.

42.8   More on Atomic Spectra: Visible and X-Ray

In Section 42.1, we discussed the observation and early interpretation of visible spectral lines from gases. These spectral lines have their origin in transitions between quantized atomic states. We shall investigate these transitions more deeply in these final three sections of this chapter.

A modified energy-level diagram for hydrogen is shown in Figure 42.21. In this diagram, the allowed values of ℓ for each shell are separated horizontally. Figure 42.21 shows only those states up to $\ell = 2$; the shells from $n = 4$ upward would have more sets of states to the right, which are not shown. Transitions for which ℓ does not change are very unlikely to occur and are called *forbidden transitions.* (Such transitions actually can occur, but their probability is very low relative to the probability of "allowed" transitions.) The various diagonal lines represent allowed transitions between stationary states. Whenever an atom makes a transition from a higher energy state to a lower one, a photon of light is emitted. The frequency of this photon is $f = \Delta E / h$, where ΔE is the energy difference between the two states and h is Planck's constant. The **selection rules** for the *allowed transitions* are

$$\Delta \ell = \pm 1 \quad \text{and} \quad \Delta m_\ell = 0, \pm 1 \qquad \textbf{(42.34)}$$

Because the orbital angular momentum of an atom changes when a photon is emitted or absorbed (that is, as a result of a transition between states) and because angular momentum of the isolated atom–photon system must be conserved, we conclude that **the photon involved in the process must carry angular momentum**. In fact, the photon has an angular momentum equivalent to that of a particle having a spin of 1. Therefore, a photon has energy, linear momentum, and angular momentum.

Recall from Equation 42.19 that the allowed energies for one-electron atoms and ions, such as hydrogen and He^+, are

$$E_n = -\frac{k_e e^2}{2a_0} \left(\frac{Z^2}{n^2} \right) = -\frac{(13.6 \text{ eV}) Z^2}{n^2} \qquad \textbf{(42.35)}$$

This equation was developed from the Bohr theory, but it serves as a good first approximation in quantum theory as well. For multielectron atoms, the positive nuclear charge Ze is largely shielded by the negative charge of the inner-shell electrons. Therefore, the outer electrons interact with a net charge that is smaller than

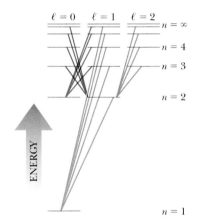

Figure 42.21   Some allowed electronic transitions for hydrogen, represented by the colored lines. These transitions must obey the selection rule $\Delta \ell = \pm 1$.

◀   Selection rules for allowed atomic transitions

Figure 42.22 The x-ray spectrum of a metal target consists of a broad continuous spectrum (*bremsstrahlung*) plus a number of sharp lines, which are due to *characteristic x-rays*. The data shown were obtained when 37-keV electrons bombarded a molybdenum target.

the nuclear charge. The expression for the allowed energies for multielectron atoms has the same form as Equation 42.35 with Z replaced by an effective atomic number Z_{eff}:

$$E_n = -\frac{(13.6 \text{ eV}) Z_{\text{eff}}^2}{n^2} \qquad (42.36)$$

where Z_{eff} depends on n and ℓ.

X-Ray Spectra

X-rays are emitted when high-energy electrons or any other charged particles bombard a metal target. The x-ray spectrum typically consists of a broad continuous band containing a series of sharp lines as shown in Figure 42.22. In Section 34.6, we mentioned that an accelerated electric charge emits electromagnetic radiation. The x-rays in Figure 42.22 are the result of the slowing down of high-energy electrons as they strike the target. It may take several interactions with the atoms of the target before the electron loses all its kinetic energy. The amount of kinetic energy lost in any given interaction can vary from zero up to the entire kinetic energy of the electron. Therefore, the wavelength of radiation from these interactions lies in a continuous range from some minimum value up to infinity. It is this general slowing down of the electrons that provides the continuous curve in Figure 42.22, which shows the cutoff of x-rays below a minimum wavelength value that depends on the kinetic energy of the incoming electrons. X-ray radiation with its origin in the slowing down of electrons is called **bremsstrahlung,** the German word for "braking radiation."

The discrete lines in Figure 42.22, called **characteristic x-rays** and discovered in 1908, have a different origin. Their origin remained unexplained until the details of atomic structure were understood. The first step in the production of characteristic x-rays occurs when a bombarding electron collides with a target atom. The electron must have sufficient energy to remove an inner-shell electron from the atom. The vacancy created in the shell is filled when an electron in a higher level drops down into the level containing the vacancy. The time interval for that to happen is very short, less than 10^{-9} s. This transition is accompanied by the emission of a photon whose energy equals the difference in energy between the two levels. Typically, the energy of such transitions is greater than 1 000 eV and the emitted x-ray photons have wavelengths in the range of 0.01 nm to 1 nm.

Let's assume the incoming electron has dislodged an atomic electron from the innermost shell, the K shell. If the vacancy is filled by an electron dropping from the next higher shell—the L shell—the photon emitted has an energy corresponding to the K_α characteristic x-ray line on the curve of Figure 42.22. In this notation, K refers to the final level of the electron and the subscript α, as the *first* letter of the Greek alphabet, refers to the initial level as the *first* one above the final level. Figure 42.23 shows this transition as well as others discussed below. If the vacancy in the K shell is filled by an electron dropping from the M shell, the K_β line in Figure 42.22 is produced.

Other characteristic x-ray lines are formed when electrons drop from upper levels to vacancies other than those in the K shell. For example, L lines are produced when vacancies in the L shell are filled by electrons dropping from higher shells. An L_α line is produced as an electron drops from the M shell to the L shell, and an L_β line is produced by a transition from the N shell to the L shell.

Although multielectron atoms cannot be analyzed exactly with either the Bohr model or the Schrödinger equation, we can apply Gauss's law from Chapter 24 to make some surprisingly accurate estimates of expected x-ray energies and wavelengths. Consider an atom of atomic number Z in which one of the two electrons in the K shell has been ejected. Imagine drawing a gaussian sphere immediately inside the most probable radius of the L electrons. The electric field at the position of the L electrons is a combination of the fields created by the nucleus, the single K electron, the other L electrons, and the outer electrons. The wave func-

Figure 42.23 Transitions between higher and lower atomic energy levels that give rise to x-ray photons from heavy atoms when they are bombarded with high-energy electrons.

tions of the outer electrons are such that the electrons have a very high probability of being farther from the nucleus than the L electrons are. Therefore, the outer electrons are much more likely to be outside the gaussian surface than inside and, on average, do not contribute significantly to the electric field at the position of the L electrons. The effective charge inside the gaussian surface is the positive nuclear charge and one negative charge due to the single K electron. Ignoring the interactions between L electrons, a single L electron behaves as if it experiences an electric field due to a charge $(Z - 1)e$ enclosed by the gaussian surface. The nuclear charge is shielded by the electron in the K shell such that Z_{eff} in Equation 42.36 is $Z - 1$. For higher-level shells, the nuclear charge is shielded by electrons in all of the inner shells.

We can now use Equation 42.36 to estimate the energy associated with an electron in the L shell:

$$E_L = -(Z - 1)^2 \frac{13.6 \text{ eV}}{2^2}$$

After the atom makes the transition, there are two electrons in the K shell. We can approximate the energy associated with one of these electrons as that of a one-electron atom. (In reality, the nuclear charge is reduced somewhat by the negative charge of the other electron, but let's ignore this effect.) Therefore,

$$E_K \approx -Z^2(13.6 \text{ eV}) \qquad \textbf{(42.37)}$$

As Example 42.5 shows, the energy of the atom with an electron in an M shell can be estimated in a similar fashion. Taking the energy difference between the initial and final levels, we can then calculate the energy and wavelength of the emitted photon.

In 1914, Henry G. J. Moseley (1887–1915) plotted $\sqrt{1/\lambda}$ versus the Z values for a number of elements where λ is the wavelength of the K_α line of each element. He found that the plot is a straight line as in Figure 42.24, which is consistent with rough calculations of the energy levels given by Equation 42.37. From this plot, Moseley determined the Z values of elements that had not yet been discovered and produced a periodic table in excellent agreement with the known chemical properties of the elements. Until that experiment, atomic numbers had been merely placeholders for the elements that appeared in the periodic table, the elements being ordered according to mass.

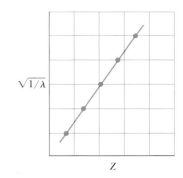

Figure 42.24 A Moseley plot of $\sqrt{1/\lambda}$ versus Z, where λ is the wavelength of the K_α x-ray line of the element of atomic number Z.

Quick Quiz 42.5   In an x-ray tube, as you increase the energy of the electrons striking the metal target, do the wavelengths of the characteristic x-rays (a) increase, (b) decrease, or (c) do not change?

Quick Quiz 42.6   True or False: It is possible for an x-ray spectrum to show the continuous spectrum of x-rays without the presence of the characteristic x-rays.

EXAMPLE 42.5   **Estimating the Energy of an X-Ray**

Estimate the energy of the characteristic x-ray emitted from a tungsten target when an electron drops from an M shell ($n = 3$ state) to a vacancy in the K shell ($n = 1$ state). The atomic number for tungsten is $Z = 74$.

SOLUTION

Conceptualize   Imagine an accelerated electron striking a tungsten atom and ejecting an electron from the K shell. Subsequently, an electron in the M shell drops down to fill the vacancy and the energy difference between the states is emitted as an x-ray photon.

Categorize   We estimate the results using equations developed in this section, so we categorize this example as a substitution problem.

Use Equation 42.37 and $Z = 74$ for tungsten to estimate the energy associated with the electron in the K shell:

$$E_K \approx -(74)^2(13.6 \text{ eV}) = -7.4 \times 10^4 \text{ eV}$$

Use Equation 42.36 and that nine electrons shield the nuclear charge (eight electrons in the $n = 2$ state and one electron in the $n = 1$ state) to estimate the energy of the M shell:

$$E_M \approx -\frac{(13.6 \text{ eV})(74 - 9)^2}{(3)^2} \approx -6.4 \times 10^3 \text{ eV}$$

Find the energy of the emitted x-ray photon:

$$hf = E_M - E_K \approx -6.4 \times 10^3 \text{ eV} - (-7.4 \times 10^4 \text{ eV})$$

$$\approx 6.8 \times 10^4 \text{ eV} = \boxed{68 \text{ keV}}$$

Consultation of x-ray tables shows that the M–K transition energies in tungsten vary from 66.9 keV to 67.7 keV, where the range of energies is due to slightly different energy values for states of different ℓ. Therefore, our estimate differs from the midpoint of this experimentally measured range by approximately 1%.

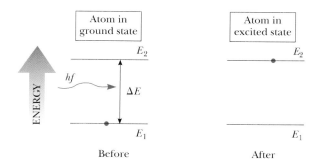

42.9 Spontaneous and Stimulated Transitions

We have seen that an atom absorbs and emits electromagnetic radiation only at frequencies that correspond to the energy differences between allowed states. Let's now examine more details of these processes. Consider an atom having the allowed energy levels labeled E_1, E_2, E_3, ... When radiation is incident on the atom, only those photons whose energy hf matches the energy separation ΔE between two energy levels can be absorbed by the atom as represented in Active Figure 42.25. This process is called **stimulated absorption** because the photon stimulates the atom to make the upward transition. At ordinary temperatures, most of the atoms in a sample are in the ground state. If a vessel containing many atoms of a gaseous element is illuminated with radiation of all possible photon frequencies (that is, a continuous spectrum), only those photons having energy $E_2 - E_1$, $E_3 - E_1$, $E_4 - E_1$, and so on are absorbed by the atoms. As a result of this absorption, some of the atoms are raised to excited states.

Once an atom is in an excited state, the excited atom can make a transition back to a lower energy level, emitting a photon in the process as in Active Figure 42.26. This process is known as **spontaneous emission** because it happens naturally, without requiring an event to trigger the transition. Typically, an atom remains in an excited state for only about 10^{-8} s.

In addition to spontaneous emission, **stimulated emission** occurs. Suppose an atom is in an excited state E_2 as in Active Figure 42.27. If the excited state is a *metastable state*—that is, if its lifetime is much longer than the typical 10^{-8} s lifetime of excited states—the time interval until spontaneous emission occurs is relatively long. Let's imagine that during that interval a photon of energy $hf = E_2 - E_1$ is incident on the atom. One possibility is that the photon energy is sufficient for the

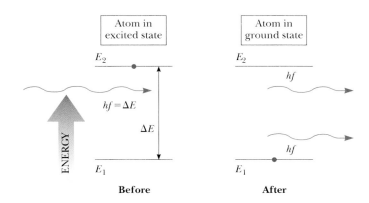

Before   **After**

ACTIVE FIGURE 42.27

Stimulated emission of a photon by an incoming photon of energy $hf = E_2 - E_1$. Initially, the atom is in the excited state. The incoming photon stimulates the atom to emit a second photon of energy given by $hf = E_2 - E_1$.

Sign in at www.thomsonedu.com and go to ThomsonNOW to adjust the energy difference between states and observe stimulated emission.

photon to ionize the atom. Another possibility is that the interaction between the incoming photon and the atom causes the atom to return to the ground state[11] and thereby emit a second photon with energy $hf = E_2 - E_1$. In this process, the incident photon is not absorbed; therefore, after the stimulated emission, two photons with identical energy exist: the incident photon and the emitted photon. The two are in phase and travel in the same direction, which is an important consideration in lasers, discussed next.

42.10   Lasers

In this section, we explore the nature of laser light and a variety of applications of lasers in our technological society. The primary properties of laser light that make it useful in these technological applications are the following:

- Laser light is coherent. The individual rays of light in a laser beam maintain a fixed phase relationship with each other.
- Laser light is monochromatic. Light in a laser beam has a very narrow range of wavelengths.
- Laser light has a small angle of divergence. The beam spreads out very little, even over large distances.

To understand the origin of these properties, let's combine our knowledge of atomic energy levels from this chapter with some special requirements for the atoms that emit laser light.

We have described how an incident photon can cause atomic energy transitions either upward (stimulated absorption) or downward (stimulated emission). The two processes are equally probable. When light is incident on a collection of atoms, a net absorption of energy usually occurs because when the system is in thermal equilibrium, many more atoms are in the ground state than in excited states. If the situation can be inverted so that more atoms are in an excited state than in the ground state, however, a net emission of photons can result. Such a condition is called **population inversion.**

Population inversion is, in fact, the fundamental principle involved in the operation of a **laser** (an acronym for *light amplification by stimulated emission of radiation*). The full name indicates one of the requirements for laser light: to achieve laser action, the process of stimulated emission must occur.

Suppose an atom is in the excited state E_2 as in Active Figure 42.27 and a photon with energy $hf = E_2 - E_1$ is incident on it. As described in Section 42.9, the

---

[11] This phenomenon is fundamentally due to *resonance*. The incoming photon has a frequency and drives the system of the atom at that frequency. Because the driving frequency matches that associated with a transition between states—one of the natural frequencies of the atom—there is a large response: the atom makes the transition.

Figure 42.28 Schematic diagram of a laser design. The tube contains the atoms that are the active medium. An external source of energy (for example, an optical or electrical device) "pumps" the atoms to excited states. The parallel end mirrors confine the photons to the tube, but mirror 2 is only partially reflective.

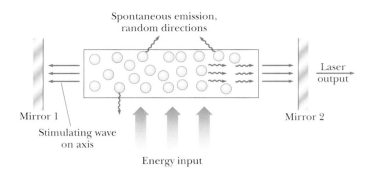

incoming photon can stimulate the excited atom to return to the ground state and thereby emit a second photon having the same energy hf and traveling in the same direction. The incident photon is not absorbed, so after the stimulated emission, there are two identical photons: the incident photon and the emitted photon. The emitted photon is in phase with the incident photon. These photons can stimulate other atoms to emit photons in a chain of similar processes. The many photons produced in this fashion are the source of the intense, coherent light in a laser.

For the stimulated emission to result in laser light, there must be a buildup of photons in the system. The following three conditions must be satisfied to achieve this buildup:

- The system must be in a state of population inversion: there must be more atoms in an excited state than in the ground state. That must be true because the number of photons emitted must be greater than the number absorbed.
- The excited state of the system must be a *metastable state*, meaning that its lifetime must be long compared with the usually short lifetimes of excited states, which are typically 10^{-8} s. In this case, the population inversion can be established and stimulated emission is likely to occur before spontaneous emission.
- The emitted photons must be confined in the system long enough to enable them to stimulate further emission from other excited atoms. That is achieved by using reflecting mirrors at the ends of the system. One end is made totally reflecting, and the other is partially reflecting. A fraction of the light intensity passes through the partially reflecting end, forming the beam of laser light (Fig. 42.28).

One device that exhibits stimulated emission of radiation is the helium–neon gas laser. Figure 42.29 is an energy-level diagram for the neon atom in this system. The mixture of helium and neon is confined to a glass tube that is sealed at the ends by mirrors. A voltage applied across the tube causes electrons to sweep through the tube, colliding with the atoms of the gases and raising them into excited states. Neon atoms are excited to state E_3^* through this process (the asterisk indicates a metastable state) and also as a result of collisions with excited helium atoms. Stimulated emission occurs, causing neon atoms to make transitions to state E_2. Neighboring excited atoms are also stimulated. The result is the production of coherent light at a wavelength of 632.8 nm.

Applications

Since the development of the first laser in 1960, tremendous growth has occurred in laser technology. Lasers that cover wavelengths in the infrared, visible, and ultraviolet regions are now available. Applications include surgical "welding" of detached retinas, precision surveying and length measurement, precision cutting of metals and other materials (such as the fabric in Fig. 42.30), and telephone communication along optical fibers. These and other applications are possible because of the unique characteristics of laser light. In addition to being highly

Figure 42.29 Energy-level diagram for a neon atom in a helium–neon laser. The atom emits 632.8-nm photons through stimulated emission in the transition $E_3^* - E_2$. That is the source of coherent light in the laser.

monochromatic, laser light is also highly directional and can be sharply focused to produce regions of extremely intense light energy (with energy densities 10^{12} times the density in the flame of a typical cutting torch).

Lasers are used in precision long-range distance measurement (range finding). In recent years, it has become important in astronomy and geophysics to measure as precisely as possible the distances from various points on the surface of the Earth to a point on the Moon's surface. To facilitate these measurements, the *Apollo* astronauts set up a 0.5-m square of reflector prisms on the Moon, which enables laser pulses directed from an Earth-based station to be retroreflected to the same station (see Fig. 35.8a). Using the known speed of light and the measured round-trip travel time of a laser pulse, the Earth–Moon distance can be determined to a precision of better than 10 cm.

Because various laser wavelengths can be absorbed in specific biological tissues, lasers have a number of medical applications. For example, certain laser procedures have greatly reduced blindness in patients with glaucoma and diabetes. Glaucoma is a widespread eye condition characterized by a high fluid pressure in the eye, a condition that can lead to destruction of the optic nerve. A simple laser operation (iridectomy) can "burn" open a tiny hole in a clogged membrane, relieving the destructive pressure. A serious side effect of diabetes is neovascularization, the proliferation of weak blood vessels, which often leak blood. When neovascularization occurs in the retina, vision deteriorates (diabetic retinopathy) and finally is destroyed. Today, it is possible to direct the green light from an argon ion laser through the clear eye lens and eye fluid, focus on the retina edges, and photocoagulate the leaky vessels. Even people who have only minor vision defects such as nearsightedness are benefiting from the use of lasers to reshape the cornea, changing its focal length and reducing the need for eyeglasses.

Laser surgery is now an everyday occurrence at hospitals and medical clinics around the world. Infrared light at 10 μm from a carbon dioxide laser can cut through muscle tissue, primarily by vaporizing the water contained in cellular material. Laser power of approximately 100 W is required in this technique. The advantage of the "laser knife" over conventional methods is that laser radiation cuts tissue and coagulates blood at the same time, leading to a substantial reduction in blood loss. In addition, the technique virtually eliminates cell migration, an important consideration when tumors are being removed.

A laser beam can be trapped in fine optical fiber light guides (endoscopes) by means of total internal reflection. An endoscope can be introduced through natural orifices, conducted around internal organs, and directed to specific interior body locations, eliminating the need for invasive surgery. For example, bleeding in the gastrointestinal tract can be optically cauterized by endoscopes inserted through the patient's mouth.

In biological and medical research, it is often important to isolate and collect unusual cells for study and growth. A laser cell separator exploits the tagging of specific cells with fluorescent dyes. All cells are then dropped from a tiny charged nozzle and laser-scanned for the dye tag. If triggered by the correct light-emitting tag, a small voltage applied to parallel plates deflects the falling electrically charged cell into a collection beaker.

An exciting area of research and technological applications arose in the 1990s with the development of *laser trapping* of atoms. One scheme, called *optical molasses* and developed by Steven Chu of Stanford University and his colleagues, involves focusing six laser beams onto a small region in which atoms are to be trapped. Each pair of lasers is along one of the *x*, *y*, and *z* axes and emits light in opposite directions (Fig. 42.31). The frequency of the laser light is tuned to be slightly below the absorption frequency of the subject atom. Imagine that an atom has been placed into the trap region and moves along the positive *x* axis toward the laser that is emitting light toward it (the rightmost laser in Fig. 42.31). Because the atom is moving, the light from the laser appears Doppler-shifted upward in frequency in the reference frame of the atom. Therefore, a match between the Doppler-shifted laser frequency and the absorption frequency of the atom exists

Figure 42.30 This robot carrying laser scissors, which can cut up to 50 layers of fabric at a time, is one of the many applications of laser technology.

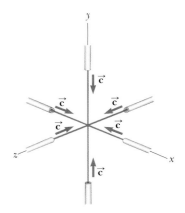

Figure 42.31 An optical trap for atoms is formed at the intersection point of six counterpropagating laser beams along mutually perpendicular axes. The frequency of the laser light is tuned to be immediately below that for absorption by the trapped atoms. If an atom moves away from the trap, it absorbs the Doppler-shifted laser light and the momentum of the light pushes the atom back into the trap.

Figure 42.32 A staff member of the National Institute of Standards and Technology views a sample of trapped sodium atoms (the small yellow dot in the center of the vacuum chamber) cooled to a temperature of less than 1 mK.

and the atom absorbs photons.[12] The momentum carried by these photons results in the atom being pushed back to the center of the trap. By incorporating six lasers, the atoms are pushed back into the trap regardless of which way they move along any axis.

In 1986, Chu developed *optical tweezers*, a device that uses a single tightly focused laser beam to trap and manipulate small particles. In combination with microscopes, optical tweezers have opened up many new possibilities for biologists. Optical tweezers have been used to manipulate live bacteria without damage, move chromosomes within a cell nucleus, and measure the elastic properties of a single DNA molecule. Chu shared the 1997 Nobel Prize in Physics with two of his colleagues for the development of the techniques of optical trapping.

An extension of laser trapping, *laser cooling*, is possible because the normal high speeds of the atoms are reduced when they are restricted to the region of the trap. As a result, the temperature of the collection of atoms can be reduced to a few microkelvins. The technique of laser cooling allows scientists to study the behavior of atoms at extremely low temperatures (Fig. 42.32).

[12] The laser light traveling in the same direction as the atom is Doppler-shifted further downward in frequency, so there is no absorption. Therefore, the atom is not pushed out of the trap by the diametrically opposed laser.

Summary

ThomsonNOW™ Sign in at **www.thomsonedu.com** and go to ThomsonNOW to take a practice test for this chapter.

CONCEPTS AND PRINCIPLES

The wavelengths of spectral lines from hydrogen, called the **Balmer series,** can be described by the equation

$$\frac{1}{\lambda} = R_{\mathrm{H}}\left(\frac{1}{2^2} - \frac{1}{n^2}\right) \quad n = 3, 4, 5, \ldots$$

(42.1)

where R_{H} is the **Rydberg constant.** The spectral lines corresponding to values of n from 3 to 6 are in the visible range of the electromagnetic spectrum. Values of n higher than 6 correspond to spectral lines in the ultraviolet region of the spectrum.

The Bohr model of the atom is successful in describing the spectra of atomic hydrogen and hydrogen-like ions. One basic assumption of the model is that the electron can exist only in discrete orbits such that the angular momentum of the electron is an integral multiple of $h/2\pi = \hbar$. When we assume circular orbits and a simple Coulomb attraction between electron and proton, the energies of the quantum states for hydrogen are calculated to be

$$E_n = -\frac{k_e e^2}{2a_0}\left(\frac{1}{n^2}\right) \quad n = 1, 2, 3, \ldots \quad \textbf{(42.13)}$$

where n is an integer called the **quantum number,** k_e is the Coulomb constant, e is the electronic charge, and $a_0 = 0.052\ 9$ nm is the **Bohr radius.**

If the electron in a hydrogen atom makes a transition from an orbit whose quantum number is n_i to one whose quantum number is n_f, where $n_f < n_i$, a photon is emitted by the atom. The frequency of this photon is

$$f = \frac{k_e e^2}{2a_0 h}\left(\frac{1}{n_f^{\,2}} - \frac{1}{n_i^{\,2}}\right) \quad \textbf{(42.15)}$$

(continued)

Quantum mechanics can be applied to the hydrogen atom by the use of the potential energy function $U(r) = -k_e e^2/r$ in the Schrödinger equation. The solution to this equation yields wave functions for allowed states and allowed energies:

$$E_n = -\left(\frac{k_e e^2}{2a_0}\right)\frac{1}{n^2} = -\frac{13.606 \text{ eV}}{n^2} \quad n = 1, 2, 3, \ldots \quad \textbf{(42.21)}$$

where n is the **principal quantum number.** The allowed wave functions depend on three quantum numbers: n, ℓ, and m_ℓ, where ℓ is the **orbital quantum number** and m_ℓ is the **orbital magnetic quantum number.** The restrictions on the quantum numbers are

$$n = 1, 2, 3, \ldots$$
$$\ell = 0, 1, 2, \ldots, n-1$$
$$m_\ell = -\ell, -\ell+1, \ldots, \ell-1, \ell$$

All states having the same principal quantum number n form a **shell,** identified by the letters K, L, M, . . . (corresponding to $n = 1, 2, 3, \ldots$). All states having the same values of n and ℓ form a **subshell,** designated by the letters s, p, d, f, . . . (corresponding to $\ell = 0, 1, 2, 3, \ldots$).

An atom in a state characterized by a specific value of n can have the following values of L, the magnitude of the atom's orbital angular momentum $\vec{L}$:

$$L = \sqrt{\ell(\ell + 1)}\,\hbar$$
$$\ell = 0, 1, 2, \ldots, n-1 \quad \textbf{(42.27)}$$

The allowed values of the projection of $\vec{L}$ along the z axis are

$$L_z = m_\ell \hbar \quad \textbf{(42.28)}$$

Only discrete values of L_z are allowed as determined by the restrictions on m_ℓ. This quantization of L_z is referred to as **space quantization.**

The electron has an intrinsic angular momentum called the **spin angular momentum.** Electron spin can be described by a single quantum number $s = \frac{1}{2}$. To describe a quantum state completely, it is necessary to include a fourth quantum number m_s, called the **spin magnetic quantum number.** This quantum number can have only two values, $\pm\frac{1}{2}$. The magnitude of the spin angular momentum is

$$S = \frac{\sqrt{3}}{2}\,\hbar \quad \textbf{(42.30)}$$

and the z component of $\vec{S}$ is

$$S_z = m_s\hbar = \pm\tfrac{1}{2}\hbar \quad \textbf{(42.31)}$$

That is, the spin angular momentum is also quantized in space, as specified by the spin magnetic quantum number $m_s = \pm\frac{1}{2}$.

The **exclusion principle** states that **no two electrons in an atom can be in the same quantum state.** In other words, no two electrons can have the same set of quantum numbers n, ℓ, m_ℓ, and m_s. Using this principle, the electronic configurations of the elements can be determined. This principle serves as a basis for understanding atomic structure and the chemical properties of the elements.

The magnetic moment $\vec{\mu}_{\text{spin}}$ associated with the spin angular momentum of an electron is

$$\vec{\mu}_{\text{spin}} = -\frac{e}{m_e}\vec{S} \quad \textbf{(42.32)}$$

The z component of $\vec{\mu}_{\text{spin}}$ can have the values

$$\mu_{\text{spin},z} = \pm\frac{e\hbar}{2m_e} \quad \textbf{(42.33)}$$

The x-ray spectrum of a metal target consists of a set of sharp characteristic lines superimposed on a broad continuous spectrum. **Bremsstrahlung** is x-radiation with its origin in the slowing down of high-energy electrons as they encounter the target. **Characteristic x-rays** are emitted by atoms when an electron undergoes a transition from an outer shell to a vacancy in an inner shell.

Atomic transitions can be described with three processes: **stimulated absorption,** in which an incoming photon raises the atom to a higher energy state; **spontaneous emission,** in which the atom makes a transition to a lower energy state, emitting a photon; and **stimulated emission,** in which an incident photon causes an excited atom to make a downward transition, emitting a photon identical to the incident one.

Questions

denotes answer available in *Student Solutions Manual/Study Guide;* **O** denotes objective question

1. Suppose the electron in the hydrogen atom obeyed classical mechanics rather than quantum mechanics. Why should a gas of such hypothetical atoms emit a continuous spectrum rather than the observed line spectrum?

2. **O** (a) Can a hydrogen atom in the ground state absorb a photon of energy less than 13.6 eV? (b) Can this atom absorb a photon of energy greater than 13.6 eV?

3. **O** Compare this question with Quick Quiz 42.1. When an electron collides with an atom, it can transfer all or some of its energy to the atom. A hydrogen atom is in its ground state. Incident on the atom are several electrons each having a kinetic energy of 10.5 eV. What is the result? (a) The atom is excited to a higher allowed state. (b) The atom is ionized. (c) The electrons pass by the atom without interaction.

4. **O** Let $-E$ represent the energy of a hydrogen atom. **(i)** What is the kinetic energy of the electron? (a) $2E$ (b) E (c) $E/2$ (d) 0 (e) $-E/2$ (f) $-E$ (g) $-2E$ **(ii)** What is the potential energy of the atom? Choose from the same possibilities (a) through (g).

5. **O** **(i)** Rank the following transitions for a hydrogen atom from the transition with the greatest gain in energy to that with the greatest loss, showing any cases of equality: (a) $n_i = 2$; $n_f = 5$ (b) $n_i = 5$; $n_f = 3$ (c) $n_i = 7$; $n_f = 4$ (d) $n_i = 4$; $n_f = 7$ **(ii)** Rank the same transitions according to the wavelength of the photon absorbed or emitted by an otherwise isolated atom from greatest wavelength to smallest.

6. **O** (a) In the hydrogen atom, can the quantum number n increase without limit? (b) Can the frequency of possible lines in the spectrum of hydrogen increase without limit? (c) Can the wavelength of possible lines in the spectrum of hydrogen increase without limit?

7. According to Bohr's model of the hydrogen atom, what is the uncertainty in the radial coordinate of the electron? What is the uncertainty in the radial component of the velocity of the electron? In what way does the model violate the uncertainty principle?

8. **O** Consider the quantum numbers (a) n, (b) ℓ, (c) m_ℓ, and (d) m_s. **(i)** Which of these quantum numbers are fractional as opposed to being integers? **(ii)** Which can be negative as opposed to being always positive? **(iii)** Which can be zero?

9. Why are three quantum numbers needed to describe the state of a one-electron atom (ignoring spin)?

10. Compare the Bohr theory and the Schrödinger treatment of the hydrogen atom. Comment on the total energy and orbital angular momentum.

11. Could the Stern–Gerlach experiment be performed with ions rather than neutral atoms? Explain.

12. Why is a *nonuniform* magnetic field used in the Stern–Gerlach experiment?

13. Discuss some consequences of the exclusion principle.

14. An energy of about 21 eV is required to excite an electron in a helium atom from the $1s$ state to the $2s$ state. The same transition for the He$^+$ ion requires approximately twice as much energy. Explain.

15. Why do lithium, potassium, and sodium exhibit similar chemical properties?

16. It is easy to understand how two electrons (one spin up, one spin down) fill the $n = 1$ or K shell for a helium atom. How is it possible that eight more electrons can fit into the $n = 2$ shell, filling the K and L shells for a neon atom?

17. **O** **(i)** What is the principal quantum number of the initial state of an atom as it emits an M_β line in an x-ray spectrum? (a) 1 (b) 2 (c) 3 (d) 4 (e) 5 (f) none of these answers **(ii)** What is the principal quantum number of the final state for this transition? Choose from the same possibilities.

18. Does the intensity of light from a laser fall off as $1/r^2$? Explain.

19. Why is stimulated emission so important in the operation of a laser?

20. (a) "As soon as I define a particular direction as the z axis, precisely one half of the electrons in this part of the Universe have their magnetic moment vectors oriented at 54.735 61° to that axis, and all the rest have their magnetic moments at 125.264 39°." Argue for or against this statement. (b) "The Universe is not simply stranger than we suppose; it is stranger than we *can* suppose." Argue for or against this statement.

21. A message reads, *"All your base are belong to us!"* Argue for or against the view that a scientific discovery is like a communication from an utterly alien source, in need of interpretation and susceptible to misunderstanding. Argue for or against the view that the human mind is not necessarily well adapted to understand the Universe. Argue for or against the view that education in science is the best preparation for life in a rapidly changing world.

Problems

1, 2, 3 denotes straightforward, intermediate, challenging;   ☐ denotes full solution available in *Student Solutions Manual/Study Guide;* ▲ denotes coached solution with hints available at **www.thomsonedu.com;**   denotes developing symbolic reasoning; ● denotes asking for qualitative reasoning; ▀ denotes computer useful in solving problem

Section 42.1 Atomic Spectra of Gases

1. (a) What value of n_i is associated with the 94.96-nm spectral line in the Lyman series of hydrogen? (b) **What If?** Could this wavelength be associated with the Paschen series? The Balmer series?

2. (a) An isolated atom of a certain element emits light of wavelength 520 nm when the atom falls from its fifth excited state into its second excited state. The atom emits a photon of wavelength 410 nm when it drops from its sixth excited state into its second excited state. Find the wavelength of the light radiated when the atom makes a transition from its sixth to its fifth excited state. (b) Solve the same problem again in symbolic terms. Letting λ_{BA} represent the wavelength emitted in the transition B to A and λ_{CA} represent the shorter wavelength emitted in the transition C to A, find λ_{CB}. This problem exemplifies the *Ritz combination principle,* an empirical rule formulated in 1908.

Section 42.2 Early Models of the Atom

3. ▲ According to classical physics, a charge e moving with an acceleration a radiates at a rate

$$\frac{dE}{dt} = -\frac{1}{6\pi\epsilon_0} \frac{e^2 a^2}{c^3}$$

(a) Show that an electron in a classical hydrogen atom (see Fig. 42.5) spirals into the nucleus at a rate

$$\frac{dr}{dt} = -\frac{e^4}{12\pi^2 \epsilon_0^{\ 2} r^2 m_e^{\ 2} c^3}$$

(b) Find the time interval over which the electron reaches $r = 0$, starting from $r_0 = 2.00 \times 10^{-10}$ m.

4. Review problem. In the Rutherford scattering experiment, 4.00-MeV alpha particles (^{4}He nuclei containing 2 protons and 2 neutrons) scatter off gold nuclei (containing 79 protons and 118 neutrons). Assume a particular alpha particle makes a direct head-on collision with the gold nucleus and scatters backward at 180°. Determine (a) the distance of closest approach of the alpha particle to the gold nucleus and (b) the maximum force exerted on the alpha particle. Assume the gold nucleus remains fixed throughout the entire process.

Section 42.3 Bohr's Model of the Hydrogen Atom

5. The Balmer series for the hydrogen atom corresponds to electronic transitions that terminate in the state with quantum number $n = 2$ as shown in Figure 42.2 and Active Figure 42.8. (a) Consider the photon of longest wavelength. Determine its energy and wavelength. (b) Con-

sider the spectral line of shortest wavelength. Find its photon energy and wavelength.

6. For a hydrogen atom in its ground state, use the Bohr model to compute (a) the orbital speed of the electron, (b) the kinetic energy of the electron, and (c) the electric potential energy of the atom.

7. ▲ A hydrogen atom is in its first excited state ($n = 2$). Using the Bohr theory of the atom, calculate (a) the radius of the orbit, (b) the linear momentum of the electron, (c) the angular momentum of the electron, (d) the kinetic energy of the electron, (e) the potential energy of the system, and (f) the total energy of the system.

8. ● A monochromatic beam of light is absorbed by a collection of ground-state hydrogen atoms in such a way that six different wavelengths are observed when the hydrogen relaxes back to the ground state. (a) What is the wavelength of the incident beam? Explain the steps in your solution. (b) What is the longest wavelength in the emission spectrum of these atoms? To what portion of the electromagnetic spectrum and to what series does it belong? (c) What is the shortest wavelength? To what series does it belong?

9. (a) Construct an energy-level diagram for the He$^+$ ion, for which $Z = 2$. (b) What is the ionization energy for He$^+$?

10. A photon with energy 2.28 eV is barely capable of causing a photoelectric effect when it strikes a sodium plate. Suppose the photon is instead absorbed by hydrogen. Find (a) the minimum n for a hydrogen atom that can be ionized by such a photon and (b) the speed of the released electron far from the nucleus.

11. The positron is the antiparticle to the electron. It has the same mass and a positive electric charge of the same magnitude as that of the electron. Positronium is a hydrogen-like atom consisting of a positron and an electron revolving around each other. Using the Bohr model, find the allowed distances between the two particles and the allowed energies of the system.

12. ● An electron is in the nth Bohr orbit of the hydrogen atom. (a) Show that the period of the electron is $T = t_0 n^3$ and determine the numerical value of t_0. (b) On average, an electron remains in the $n = 2$ orbit for approximately 10 μs before it jumps down to the $n = 1$ (ground-state) orbit. How many revolutions does the electron make in the excited state? (c) Define the period of one revolution as an electron year, analogous to an Earth year being the period of the Earth's motion around the Sun. Explain whether we should think of the electron in the $n = 2$ orbit as "living for a long time."

Section 42.4 The Quantum Model of the Hydrogen Atom

13. A general expression for the energy levels of one-electron atoms and ions is

$$E_n = -\frac{\mu k_e^2 q_1^2 q_2^2}{2\hbar^2 n^2}$$

Here μ is the reduced mass of the atom, given by $\mu = m_1 m_2 / (m_1 + m_2)$, where m_1 is the mass of the electron and m_2 is the mass of the nucleus; k_e is the Coulomb constant; and q_1 and q_2 are the charges of the electron and the nucleus, respectively. The wavelength for the $n = 3$ to $n = 2$ transition of the hydrogen atom is 656.3 nm (visible red light). **What If?** What are the wavelengths for this same transition in (a) positronium, which consists of an electron and a positron, and (b) singly ionized helium? *Note:* A positron is a positively charged electron.

14. Ordinary hydrogen gas is a mixture of two kinds of atoms (isotopes) containing either one- or two-particle nuclei. These isotopes are hydrogen-1 with a proton nucleus, and hydrogen-2, called deuterium, with a deuteron nucleus. A deuteron is one proton and one neutron bound together. Hydrogen-1 and deuterium have identical chemical properties, but they can be separated via an ultracentrifuge or by other methods. Their emission spectra show lines of the same colors at very slightly different wavelengths. (a) Use the equation given in Problem 13 to show that the difference in wavelength between the hydrogen-1 and deuterium spectral lines associated with a particular electron transition is given by

$$\lambda_{\rm H} - \lambda_{\rm D} = \left(1 - \frac{\mu_{\rm H}}{\mu_{\rm D}}\right)\lambda_{\rm H}$$

(b) Evaluate the wavelength difference for the Balmer alpha line of hydrogen, with wavelength 656.3 nm, emitted by an atom making a transition from an $n = 3$ state to an $n = 2$ state. Harold Urey observed this wavelength difference in 1931 and so confirmed his discovery of deuterium.

15. ● An electron of momentum p is at a distance r from a stationary proton. The electron has kinetic energy $K = p^2 / 2m_e$. The atom has potential energy $U = -k_e e^2 / r$ and total energy $E = K + U$. If the electron is bound to the proton to form a hydrogen atom, its average position is at the proton but the uncertainty in its position is approximately equal to the radius r of its orbit. The electron's average vector momentum is zero, but its average squared momentum is approximately equal to the squared uncertainty in its momentum as given by the uncertainty principle. Treating the atom as a one-dimensional system, (a) estimate the uncertainty in the electron's momentum in terms of r. (b) Estimate the electron's kinetic, potential, and total energies in terms of r. (c) The actual value of r is the one that *minimizes the total energy*, resulting in a stable atom. Find that value of r and the resulting total energy. State how your answers compare with the predictions of the Bohr theory.

Section 42.5 The Wave Functions for Hydrogen

16. Plot the wave function $\psi_{1s}(r)$ versus r (see Eq. 42.22) and the radial probability density function $P_{1s}(r)$ versus r (see

Eq. 42.25) for hydrogen. Let r range from 0 to $1.5a_0$, where a_0 is the Bohr radius.

17. The ground-state wave function for the electron in a hydrogen atom is

$$\psi_{1s}(r) = \frac{1}{\sqrt{\pi a_0{}^3}} e^{-r/a_0}$$

where r is the radial coordinate of the electron and a_0 is the Bohr radius. (a) Show that the wave function as given is normalized. (b) Find the probability of locating the electron between $r_1 = a_0/2$ and $r_2 = 3a_0/2$.

18. The wave function for an electron in the $2p$ state of hydrogen is

$$\psi_{2p} = \frac{1}{\sqrt{3}(2a_0)^{3/2}} \frac{r}{a_0} e^{-r/2a_0}$$

What is the most likely distance from the nucleus to find an electron in the $2p$ state?

19. ▲ For a spherically symmetric state of a hydrogen atom, the Schrödinger equation in spherical coordinates is

$$-\frac{\hbar^2}{2m_e}\left(\frac{d^2\psi}{dr^2} + \frac{2}{r}\frac{d\psi}{dr}\right) - \frac{k_e e^2}{r}\psi = E\psi$$

Show that the 1s wave function for an electron in hydrogen,

$$\psi_{1s}(r) = \frac{1}{\sqrt{\pi a_0{}^3}} e^{-r/a_0}$$

satisfies the Schrödinger equation.

20. In an experiment, electrons are fired at a sample of neutral hydrogen atoms and observations are made of how the incident particles scatter. A large set of trials can be thought of as containing 1 000 observations of the electron in the ground state of a hydrogen atom being momentarily at a distance $a_0/2$ from the nucleus. In this set of trials, how many times is the atomic electron observed at a distance $2a_0$ from the nucleus?

Section 42.6 Physical Interpretation of the Quantum Numbers

21. List the possible sets of quantum numbers for the hydrogen atom associated with (a) the $3d$ subshell and (b) the $3p$ subshell.

22. Calculate the orbital angular momentum for a hydrogen atom in (a) the $4d$ state and (b) the $6f$ state.

23. (a) State a problem for which the following equation appears in the solution:

$$4.714 \times 10^{-34}\,{\rm J}\cdot{\rm s} = \frac{\sqrt{\ell(\ell+1)}(6.626 \times 10^{-34}\,{\rm J}\cdot{\rm s})}{2\pi}$$

(b) Solve the equation for the unknown appearing in it and state the name of this quantity.

24. A hydrogen atom is in its fifth excited state, with principal quantum number 6. The atom emits a photon with a wavelength of 1 090 nm. Determine the maximum possible orbital angular momentum of the atom after emission.

25. ▲ How many sets of quantum numbers are possible for a hydrogen atom for which (a) $n = 1$, (b) $n = 2$, (c) $n = 3$, (d) $n = 4$, and (e) $n = 5$? Check your results to show that they agree with the general rule that the number of sets of quantum numbers for a shell is equal to $2n^2$.

26. Find all possible values of L, L_z, and θ for a hydrogen atom in a $3d$ state.

27. ● (a) Find the mass density of a proton, modeling it as a solid sphere of radius 1.00×10^{-15} m. (b) **What If?** Consider a classical model of an electron as a uniform solid sphere with the same density as the proton. Find its radius. (c) Imagine that this electron possesses spin angular momentum $I\omega = \hbar/2$ because of classical rotation about the z axis. Determine the speed of a point on the equator of the electron. (d) State how this speed compares to the speed of light.

28. An electron is in the N shell. Determine the maximum value the z component of its angular momentum could have.

29. The ρ^- meson has a charge of $-e$, a spin quantum number of 1, and a mass 1 507 times that of the electron. The possible values for its spin magnetic quantum number are -1, 0, and 1. **What If?** Imagine that the electrons in atoms were replaced by ρ^- mesons. List the possible sets of quantum numbers for ρ^- mesons in the $3d$ subshell.

Section 42.7 The Exclusion Principle and the Periodic Table

30. (a) What is the electronic configuration for the ground state of oxygen ($Z = 8$)? (b) Write out a set of possible values for the quantum numbers n, ℓ, m_ℓ, and m_s for each electron in oxygen.

31. ● As we go down the periodic table, which subshell is filled first, the $3d$ or the $4s$ subshell? Which electronic configuration has a lower energy, $[\text{Ar}]3d^44s^2$ or $[\text{Ar}]3d^54s^1$? Which has the greater number of unpaired spins? Identify this element and discuss Hund's rule in this case. *Note:* The notation $[\text{Ar}]$ represents the filled configuration for argon.

32. Devise a table similar to that shown in Figure 42.18 for atoms containing 11 through 19 electrons. Use Hund's rule and educated guesswork.

33. A certain element has its outermost electron in a $3p$ subshell. It has valence $+3$ because it has three more electrons than a certain noble gas. What element is it?

34. Two electrons in the same atom both have $n = 3$ and $\ell = 1$. (a) List the quantum numbers for the possible states of the atom. (b) **What If?** How many states would be possible if the exclusion principle were inoperative?

35. ● ▲ (a) Scanning through Figure 42.19 in order of increasing atomic number, notice that the electrons usually fill the subshells in such a way that those subshells with the lowest values of $n + \ell$ are filled first. If two subshells have the same value of $n + \ell$, the one with the lower value of n is generally filled first. Using these two rules, write the order in which the subshells are filled through $n + \ell = 7$. (b) Predict the chemical valence for the elements that have atomic numbers 15, 47, and 86, and compare your predictions with the actual valences (which may be found in a chemistry textbook).

36. For a neutral atom of element 110, what would be the probable ground-state electronic configuration?

37. **Review problem.** For an electron with magnetic moment $\vec{\boldsymbol{\mu}}_s$ in a magnetic field $\vec{\mathbf{B}}$, Section 29.5 showed the follow-

ing. The electron-field system can be in a higher energy state with the z component of the electron's magnetic moment opposite to the field or a lower energy state with the z component of the magnetic moment in the direction of the field. The difference in energy between the two states is $2\mu_B B$.

Under high resolution, many spectral lines are observed to be doublets. The most famous of these are the two yellow lines in the spectrum of sodium (the D lines), with wavelengths of 588.995 nm and 589.592 nm. Their existence was explained in 1925 by Goudsmit and Uhlenbeck, who postulated that an electron has intrinsic spin angular momentum. When the sodium atom is excited with its outermost electron in a $3p$ state, the orbital motion of the outermost electron creates a magnetic field. The atom's energy is somewhat different depending on whether the electron is itself spin-up or spin-down in this field. Then the photon energy the atom radiates as it falls back into its ground state depends on the energy of the excited state. Calculate the magnitude of the internal magnetic field mediating this so-called spin-orbit coupling.

Section 42.8 More on Atomic Spectra: Visible and X-Ray

38. (a) Determine the possible values of the quantum numbers ℓ and m_ℓ for the He$^+$ ion in the state corresponding to $n = 3$. (b) What is the energy of this state?

39. ● The $3p$ level of sodium has an energy of -3.0 eV, and the $3d$ level has an energy of -1.5 eV. Determine Z_{eff} for each of these states. Explain the difference.

40. ● In x-ray production, electrons are accelerated through a high voltage and then decelerated by striking a target. (a) To make possible the production of x-rays of wavelength λ, what is the minimum potential difference ΔV through which the electrons must be accelerated? (b) State how the required potential difference depends on the wavelength. (c) Explain whether your result agrees with the information in Figure 42.22. (d) Does the relationship from part (a) apply to other kinds of electromagnetic radiation besides x-rays? What does the potential difference approach as λ goes to zero? What does the potential difference approach as λ increases without limit?

41. Use the method illustrated in Example 42.5 to calculate the wavelength of the x-ray emitted from a molybdenum target ($Z = 42$) when an electron moves from the L shell ($n = 2$) to the K shell ($n = 1$).

42. The K series of the discrete x-ray spectrum of tungsten contains wavelengths of 0.018 5 nm, 0.020 9 nm, and 0.021 5 nm. The K-shell ionization energy is 69.5 keV. Determine the ionization energies of the L, M, and N shells. Draw a diagram of the transitions.

43. The wavelength of characteristic x-rays in the K$_\beta$ line from a particular source is 0.152 nm. Determine the material in the target.

Section 42.9 Spontaneous and Stimulated Transitions

Section 42.10 Lasers

44. Figure P42.44 shows portions of the energy-level diagrams of the helium and neon atoms. An electrical discharge excites the He atom from its ground state to its excited state of 20.61 eV. The excited He atom collides with a Ne atom in its ground state and excites this atom to the state at 20.66 eV. Lasing action takes place for electron transitions from E_3^* to E_2 in the Ne atoms. From the data in the figure, show that the wavelength of the red He–Ne laser light is approximately 633 nm.

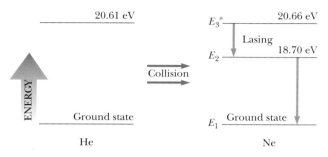

Figure P42.44

45. The carbon dioxide laser is one of the most powerful developed. The energy difference between the two laser levels is 0.117 eV. Determine the frequency and wavelength of the radiation emitted by this laser. In what portion of the electromagnetic spectrum is this radiation?

46. A neodymium-yttrium-aluminum garnet laser used in eye surgery emits a 3.00-mJ pulse in 1.00 ns, focused to a spot 30.0 μm in diameter on the retina. (a) Find (in SI units) the power per unit area at the retina. (In the optics industry, this quantity is called the *irradiance*.) (b) What energy is delivered to an area of molecular size, taken as a circular area 0.600 nm in diameter?

47. ▲ A ruby laser delivers a 10.0-ns pulse of 1.00 MW average power. If the photons have a wavelength of 694.3 nm, how many are contained in the pulse?

48. ● The number N of atoms in a particular state is called the population of that state. This number depends on the energy of that state and the temperature. In thermal equilibrium, the population of atoms in a state of energy E_n is given by a Boltzmann distribution expression

$$N = N_g e^{-(E_n - E_g)/k_B T}$$

where N_g is the population of the ground state of energy E_g and T is the absolute temperature. For simplicity, assume each energy level has only one quantum state associated with it. (a) Before the power is switched on, the neon atoms in a laser are in thermal equilibrium at 27.0°C. Find the equilibrium ratio of the populations of the states E_3^* and E_2 shown in Figure 42.29. Lasers operate by a clever artificial production of a "population inversion" between the upper and lower atomic energy states involved in the lasing transition. This term means that more atoms are in the upper excited state than in the lower one. Consider the helium–neon laser transition at 632.8 nm. Assume 2% more atoms occur in the upper state than in the lower. (b) To demonstrate how unnatural such a situation is, find the temperature for which the Boltzmann distribution describes a 2.00% population inversion. (c) Why does such a situation not occur naturally?

49. **Review problem.** A helium–neon laser can produce a green laser beam instead of a red one. Refer to Figure 42.29, which omits some energy levels between E_2 and E_1.

After a population inversion is established, neon atoms make a variety of downward transitions in falling from the state labeled E_3^* down eventually to level E_1. The atoms emit both red light with a wavelength of 632.8 nm and green light with a wavelength of 543 nm in a competing transition. Assume the atoms are in a cavity between mirrors designed to reflect the green light with high efficiency but to allow the red light to leave the cavity immediately. Then stimulated emission can lead to the buildup of a collimated beam of green light between the mirrors having a greater intensity than that of the red light. To constitute the radiated laser beam, a small fraction of the green light is permitted to escape by transmission through one mirror. The mirrors forming the resonant cavity can be made of layers of silicon dioxide and titanium dioxide. (a) How thick a layer of silicon dioxide, between layers of titanium dioxide, would minimize reflection of the red light? (b) What should be the thickness of a similar but separate layer of silicon dioxide to maximize reflection of the green light?

Additional Problems

50. ● As the Earth moves around the Sun, its orbits are quantized. (a) Follow the steps of Bohr's analysis of the hydrogen atom to show that the allowed radii of the Earth's orbit are given by

$$r = \frac{n^2 \hbar^2}{G M_S M_E^2}$$

where n is an integer quantum number, M_S is the mass of the Sun, and M_E is the mass of the Earth. (b) Calculate the numerical value of n. (c) Find the distance between the orbit for quantum number n and the next orbit out from the Sun corresponding to the quantum number $n + 1$. Discuss the significance of your results.

51. **LENINGRAD, 1930**—Four years after publication of the Schrödinger equation, Lev Davidovich Landau, age 23, solved the equation for a charged particle moving in a uniform magnetic field. A single electron moving perpendicular to a field $\vec{B}$ can be considered as a model atom without a nucleus or as the irreducible quantum limit of the cyclotron. Landau proved its energy is quantized in uniform steps of $e\hbar B/m_e$.

CAMBRIDGE, MA, 1999—Gerald Gabrielse trapped a single electron in an evacuated centimeter-size metal can cooled to a temperature of 80 mK. In a magnetic field of magnitude 5.26 T, the electron circulated for hours in its lowest energy level, generating a measurable signal as it moved. (a) Evaluate the size of a quantum jump in the electron's energy. (b) For comparison, evaluate $k_B T$ as a measure of the energy available to the electron in blackbody radiation from the walls of its container. (c) Microwave radiation was introduced to excite the electron. Calculate the frequency and wavelength of the photon that the electron absorbed as it jumped to its second energy level. Measurement of the resonant absorption frequency verified the theory and permitted precise determination of properties of the electron.

52. ⬛ Example 42.3 calculates the most probable value and the average value for the radial coordinate r of the electron in the ground state of a hydrogen atom. **What If?** For comparison with these modal and mean values, find the

median value of r. Proceed as follows. (a) Derive an expression for the probability, as a function of r, that the electron in the ground state of hydrogen will be found outside a sphere of radius r centered on the nucleus. (b) Make a graph of the probability as a function of r/a_0. Choose values of r/a_0 ranging from 0 to 4.00 in steps of 0.250. (c) Find the value of r for which the probability of finding the electron outside a sphere of radius r is equal to the probability of finding the electron inside this sphere. You must solve a transcendental equation numerically, and your graph is a good starting point.

53. *An example of the correspondence principle.* Use Bohr's model of the hydrogen atom to show that when the electron moves from the state n to the state $n - 1$, the frequency of the emitted light is

$$ f = \left(\frac{2\pi^2 m_e k_e^2 e^4}{h^3 n^2} \right) \frac{2n - 1}{(n - 1)^2} $$

Show that as $n \to \infty$, this expression varies as $1/n^3$ and reduces to the classical frequency one expects the atom to emit. *Suggestion:* To calculate the classical frequency, note that the frequency of revolution is $v/2\pi r$, where v is the speed of the electron and r is given by Eq. 42.10.

54. **Review problem.** (a) How much energy is required to cause an electron in hydrogen to move from the $n = 1$ state to the $n = 2$ state? (b) Suppose the atom gains this energy through collisions among hydrogen atoms at a high temperature. At what temperature would the average atomic kinetic energy $3k_B T/2$ be great enough to excite the electron? Here k_B is Boltzmann's constant.

55. ● Astronomers observe a series of spectral lines in the light from a distant galaxy. On the hypothesis that the lines form the Lyman series for a (new?!) one-electron atom, they start to construct the energy-level diagram shown in Figure P42.55, which gives the wavelengths of the first four lines and the short-wavelength limit of this series. Based on this information, calculate (a) the energies of the ground state and first four excited states for this one-electron atom and (b) the wavelengths of the first three lines and the short-wavelength limit in the Balmer series for this atom. (c) Show that the wavelengths of the first four lines and the short-wavelength limit of the Lyman series for the hydrogen atom are all 60.0% of the wavelengths for the Lyman series in the one-electron

atom described in part (b). (d) Based on this observation, explain why this atom could be hydrogen.

56. (a) For a hydrogen atom making a transition from the $n = 4$ state to the $n = 2$ state, determine the wavelength of the photon created in the process. (b) Assuming that the atom was initially at rest, determine the recoil speed of the hydrogen atom when it emits this photon.

57. Suppose a hydrogen atom is in the 2s state, with its wave function given by Equation 42.26. Taking $r = a_0$, calculate values for (a) $\psi_{2s}(a_0)$, (b) $|\psi_{2s}(a_0)|^2$, and (c) $P_{2s}(a_0)$.

58. An elementary theorem in statistics states that the root-mean-square uncertainty in a quantity r is given by $\Delta r = \sqrt{\langle r^2 \rangle - \langle r \rangle^2}$. Evaluate the uncertainty in the radial position of the electron in the ground state of the hydrogen atom. Use the average value of r found in Example 42.3: $\langle r \rangle = 3a_0/2$. The average value of the squared distance between the electron and the proton is given by

$$ \langle r^2 \rangle = \int_{\text{all space}} |\psi|^2 \, r^2 \, dV = \int_0^\infty P(r) \, r^2 \, dr $$

59. A pulsed ruby laser emits light at 694.3 nm. For a 14.0-ps pulse containing 3.00 J of energy, find (a) the physical length of the pulse as it travels through space and (b) the number of photons in it. (c) The beam has a circular cross section of diameter 0.600 cm. Find the number of photons per cubic millimeter.

60. A pulsed laser emits light of wavelength λ. For a pulse of duration Δt having energy E, find (a) the physical length of the pulse as it travels through space and (b) the number of photons in it. (c) The beam has a circular cross section having diameter d. Find the number of photons per unit volume.

61. Assume three identical uncharged particles of mass m and spin $\frac{1}{2}$ are contained in a one-dimensional box of length L. What is the ground-state energy of this system?

62. The force on a magnetic moment μ_z in a nonuniform magnetic field B_z is given by $F_z = \mu_z(dB_z/dz)$. If a beam of silver atoms travels a horizontal distance of 1.00 m through such a field and each atom has a speed of 100 m/s, how strong must be the field gradient dB_z/dz to deflect the beam 1.00 mm?

63. (a) Show that the most probable radial position for an electron in the 2s state of hydrogen is $r = 5.236a_0$. (b) Show that the wave function given by Equation 42.26 is normalized.

64. **Review problem.** Steven Chu, Claude Cohen-Tannoudji, and William Phillips received the 1997 Nobel Prize in Physics for "the development of methods to cool and trap atoms with laser light." One part of their work was with a beam of atoms (mass $\sim 10^{-25}$ kg) that move at a speed on the order of 1 km/s, similar to the speed of molecules in air at room temperature. An intense laser light beam tuned to a visible atomic transition (assume 500 nm) is directed straight into the atomic beam; that is, the atomic beam and light beam are traveling in opposite directions. An atom in the ground state immediately absorbs a photon. Total system momentum is conserved in the absorption process. After a lifetime on the order of 10^{-8} s, the excited atom radiates by spontaneous emission. It has an equal probability of emitting a photon in any direction. Therefore, the average "recoil" of the atom is zero over many absorption and emission cycles. (a) Estimate the

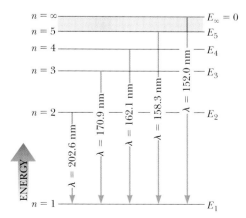

Figure P42.55

average deceleration of the atomic beam. (b) What is the order of magnitude of the distance over which the atoms in the beam are brought to a halt?

65. An electron in chromium moves from the $n = 2$ state to the $n = 1$ state without emitting a photon. Instead, the excess energy is transferred to an outer electron (one in the $n = 4$ state), which is then ejected by the atom. In this Auger (pronounced "ohjay") process, the ejected electron is referred to as an Auger electron. Use the Bohr theory to find the kinetic energy of the Auger electron.

66. Suppose the ionization energy of an atom is 4.10 eV. In the spectrum of this same atom, we observe emission lines with wavelengths 310 nm, 400 nm, and 1 377.8 nm. Use this information to construct the energy-level diagram with the fewest levels. Assume the higher levels are closer together.

67. For hydrogen in the 1s state, what is the probability of finding the electron farther than $2.50a_0$ from the nucleus?

68. All atoms have the same size, to an order of magnitude. (a) To demonstrate this fact, estimate the diameters for aluminum (with molar mass 27.0 g/mol and density 2.70 g/cm³) and uranium (molar mass 238 g/mol and density 18.9 g/cm³). (b) What do the results of part (a) imply about the wave functions for inner-shell electrons as we progress to higher and higher atomic mass atoms? *Suggestion:* The molar volume is approximately D^3N_A, where D is the atomic diameter and N_A is Avogadro's number.

69. In the technique known as electron spin resonance (ESR), a sample containing unpaired electrons is placed in a magnetic field. Consider the simplest situation, in which only one electron is present and therefore only two energy states are possible, corresponding to $m_s = \pm\frac{1}{2}$. In ESR, the absorption of a photon causes the electron's spin magnetic moment to flip from the lower energy state to the higher energy state. According to Section 29.5, the change in energy is $2\mu_B B$. (The lower energy state corresponds to the case in which the z component of the magnetic moment $\vec{\mu}_{spin}$ is aligned with the magnetic field, and the higher energy state corresponds to the case in which the z component of $\vec{\mu}_{spin}$ is aligned opposite to the field.) What is the photon frequency required to excite an ESR transition in a 0.350-T magnetic field?

Answers to Quick Quizzes

42.1 (c). Because the energy of 10.5 eV does not correspond to raising the atom from the ground state to an allowed excited state, there is no interaction between the photon and the atom.

42.2 (a). The longest-wavelength photon is associated with the lowest energy transition, which is $n = 3$ to $n = 2$.

42.3 (b). The number of subshells is the same as the number of allowed values of ℓ. The allowed values of ℓ for $n = 4$ are $\ell = 0, 1, 2$, and 3, so there are four subshells.

42.4 (a). Five values (0, 1, 2, 3, 4) of ℓ and (b) nine different values $(-4, -3, -2, -1, 0, 1, 2, 3, 4)$ of m_ℓ as follows:

ℓ	m_ℓ
0	0
1	$-1, 0, 1$
2	$-2, -1, 0, 1, 2$
3	$-3, -2, -1, 0, 1, 2, 3$
4	$-4, -3, -2, -1, 0, 1, 2, 3, 4$

42.5 (c). The wavelengths of the characteristic x-rays are determined by the separation between energy levels in the atoms of the target, which is unrelated to the energy with which electrons are fired at the target. The only dependence is that the incoming electrons must have enough energy to eject an atomic electron from an inner shell.

42.6 True. If the electrons arrive at the target with very low energy, atomic electrons cannot be ejected and characteristic x-rays do not appear. Because the incoming electrons experience accelerations, the continuous spectrum appears.

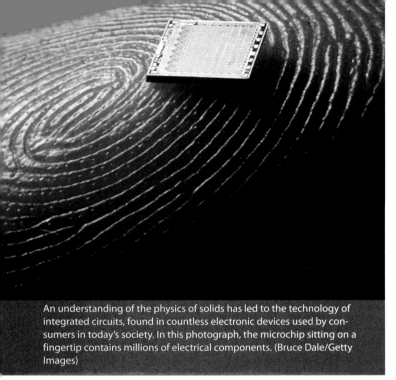

An understanding of the physics of solids has led to the technology of integrated circuits, found in countless electronic devices used by consumers in today's society. In this photograph, the microchip sitting on a fingertip contains millions of electrical components. (Bruce Dale/Getty Images)

43.1 Molecular Bonds

43.2 Energy States and Spectra of Molecules

43.3 Bonding in Solids

43.4 Free-Electron Theory of Metals

43.5 Band Theory of Solids

43.6 Electrical Conduction in Metals, Insulators, and Semiconductors

43.7 Semiconductor Devices

43.8 Superconductivity

43 Molecules and Solids

The most random atomic arrangement, that of a gas, was well understood in the 1800s as discussed in Chapter 21. In a crystalline solid, the atoms are not randomly arranged; rather, they form a regular array. The symmetry of the arrangement of atoms both stimulated and allowed rapid progress in the field of solid-state physics in the 20th century. Recently, our understanding of liquids and amorphous solids has advanced. (In an amorphous solid such as glass or paraffin, the atoms do not form a regular array.) The recent interest in the physics of low-cost amorphous materials has been driven by their use in such devices as solar cells, memory elements, and fiber-optic waveguides.

We begin this chapter by studying the aggregates of atoms known as molecules. We describe the bonding mechanisms in molecules, the various modes of molecular excitation, and the radiation emitted or absorbed by molecules. Next, we show how molecules combine to form solids. Then, by examining their energy-level structure, we explain the differences between insulating, conducting, semiconducting, and superconducting materials. The chapter also includes discussions of semiconducting junctions and several semiconductor devices.

43.1   Molecular Bonds

The bonding mechanisms in a molecule are fundamentally due to electric forces between atoms (or ions). The forces between atoms in the system of a molecule are related to a potential energy function. A stable molecule is expected at a configuration for which the potential energy function has its minimum value. (See Section 7.9.)

A potential energy function that can be used to model a molecule should account for two known features of molecular bonding:

1. The force between atoms is repulsive at very small separation distances. When two atoms are brought close to each other, some of their electron shells overlap, resulting in repulsion between the shells. This repulsion is partly electrostatic in origin and partly the result of the exclusion principle. Because all electrons must obey the exclusion principle, some electrons in the overlapping shells are forced into higher energy states and the system energy increases as if a repulsive force existed between the atoms.
2. At somewhat larger separations, the force between atoms is attractive. If that were not true, the atoms in a molecule would not be bound together.

Taking into account these two features, the potential energy for a system of two atoms can be represented by an expression of the form

$$U(r) = -\frac{A}{r^n} + \frac{B}{r^m} \tag{43.1}$$

where r is the internuclear separation distance between the two atoms and n and m are small integers. The parameter A is associated with the attractive force and B with the repulsive force. Example 7.9 gives one common model for such a potential energy function, the Lennard–Jones potential.

Potential energy versus internuclear separation distance for a two-atom system is graphed in Figure 43.1. At large separation distances between the two atoms, the slope of the curve is positive, corresponding to a net attractive force. At the equilibrium separation distance, the attractive and repulsive forces just balance. At this point, the potential energy has its minimum value and the slope of the curve is zero.

A complete description of the bonding mechanisms in molecules is highly complex because bonding involves the mutual interactions of many particles. In this section, we discuss only some simplified models.

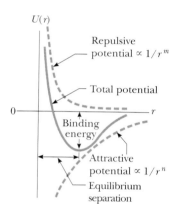

Figure 43.1 Total potential energy as a function of internuclear separation distance for a system of two atoms.

*Labels in figure:* $U(r)$; Repulsive potential $\propto 1/r^m$; Total potential; Binding energy; Attractive potential $\propto 1/r^n$; Equilibrium separation; 0; r

Ionic Bonding

When two atoms combine in such a way that one or more outer electrons are transferred from one atom to the other, the bond formed is called an **ionic bond.** Ionic bonds are fundamentally caused by the Coulomb attraction between oppositely charged ions.

A familiar example of an ionically bonded solid is sodium chloride, NaCl, which is common table salt. Sodium, which has the electronic configuration $1s^2 2s^2 2p^6 3s^1$, is ionized relatively easily, giving up its $3s$ electron to form a Na^+ ion. The energy required to ionize the atom to form Na^+ is 5.1 eV. Chlorine, which has the electronic configuration $1s^2 2s^2 2p^5$, is one electron short of the filled-shell structure of argon. If we compare the energy of a system of a free electron and a Cl atom with one in which the electron joins the atom to make the Cl^- ion, we find that the energy of the ion is lower. When the electron makes a transition from the $E = 0$ state to the negative energy state associated with the available shell in the atom, energy is released. This amount of energy is called the **electron affinity** of the atom. For chlorine, the electron affinity is 3.6 eV. Therefore, the energy required to form Na^+ and Cl^- from isolated atoms is $5.1 - 3.6 = 1.5$ eV. It costs 5.1 eV to remove the electron from the Na atom, but 3.6 eV of it is gained back when that electron is allowed to join with the Cl atom.

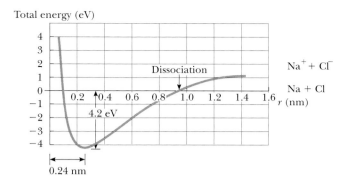

Figure 43.2 Total energy versus internuclear separation distance for Na^+ and Cl^- ions. The horizontal axis is labeled $Na + Cl$ because we define zero energy as that for the system of neutral sodium and chlorine *atoms*. The asymptote of the curve for large values of r is marked $Na^+ + Cl^-$ because that is the energy of the system of sodium and chloride *ions*.

Now imagine that these two charged ions interact with one another to form a NaCl "molecule."[1] The total energy of the NaCl molecule versus internuclear separation distance is graphed in Figure 43.2. At very large separation distances, the energy of the system of ions is 1.5 eV as calculated above. The total energy has a minimum value of -4.2 eV at the equilibrium separation distance, which is approximately 0.24 nm. Hence, the energy required to break the Na^+–Cl^- bond and form neutral sodium and chlorine atoms, called the **dissociation energy,** is 4.2 eV. The energy of the molecule is lower than that of the system of two neutral atoms. Consequently, it is **energetically favorable** for the molecule to form: if a lower energy state of a system exists, the system tends to emit energy to achieve this lower energy state. The system of neutral sodium and chlorine atoms can reduce its total energy by transferring energy out of the system (by electromagnetic radiation, for example) and forming the NaCl molecule.

Covalent Bonding

A **covalent bond** between two atoms is one in which electrons supplied by either one or both atoms are shared by the two atoms. Many diatomic molecules—such as H_2, F_2, and CO—owe their stability to covalent bonds. The bond between two hydrogen atoms can be described by using atomic wave functions. The ground-state wave function for a hydrogen atom (Chapter 42) is

$$\psi_{1s}(r) = \frac{1}{\sqrt{\pi a_0{}^3}}\, e^{-r/a_0}$$

This wave function is graphed in Active Figure 43.3a for two hydrogen atoms that are far apart. There is very little overlap of the wave functions $\psi_1(r)$ for atom 1, located at $r = 0$, and $\psi_2(r)$ for atom 2, located some distance away. Suppose now the two atoms are brought close together. As that happens, their wave functions overlap and form the compound wave function $\psi_1(r) + \psi_2(r)$ shown in Active Figure 43.3b. Notice that the probability amplitude is larger between the atoms than it is on either side of the combination of atoms. As a result, the probability is higher that the electrons associated with the atoms will be located between the atoms than on the outer regions of the system. Consequently, the average position of negative charge in the system is halfway between the atoms. This scenario can be modeled as if there were a fixed negative charge between the atoms, exerting attractive Coulomb forces on both nuclei. Therefore, there is an overall attractive force between the atoms, resulting in a covalent bond.

Because of the exclusion principle, the two electrons in the ground state of H_2 must have antiparallel spins. Also because of the exclusion principle, if a third H atom is brought near the H_2 molecule, the third electron would have to occupy a

PITFALL PREVENTION 43.1
Ionic and Covalent Bonds

In practice, these descriptions of ionic and covalent bonds represent extreme ends of a spectrum of bonds involving electron transfer. In a real bond, the electron may not be *completely* transferred as in an ionic bond or *equally* shared as in a covalent bond. Therefore, real bonds lie somewhere between these extremes.

ACTIVE FIGURE 43.3

Ground-state wave functions $\psi_1(r)$ and $\psi_2(r)$ for two atoms making a covalent bond. (a) The atoms are far apart, and their wave functions overlap minimally. (b) The atoms are close together, forming a composite wave function $\psi_1(r) + \psi_2(r)$ for the system. The probability amplitude for an electron to be between the atoms is high.

Sign in at www.thomsonedu.com and go to ThomsonNOW to move the individual wave functions and observe the composite wave function.

---

[1] NaCl does not exist as an isolated molecule. In the solid state, NaCl forms a crystalline array of ions as described in Section 43.3. In the liquid state or in solution with water, the Na^+ and Cl^- ions dissociate and are free to move relative to each other.

higher energy level, which is not an energetically favorable situation. For this reason, the H_3 molecule is not stable and does not form.

Van der Waals Bonding

Ionic and covalent bonds occur between atoms to form molecules or ionic solids, so they can be described as bonds *within* molecules. Two additional types of bonds, van der Waals bonds and hydrogen bonds, can occur *between* molecules.

You might think that two neutral molecules would not interact by means of the electric force because they each have zero net charge. They are attracted to each other, however, by weak electrostatic forces called **van der Waals forces.** Likewise, atoms that do not form ionic or covalent bonds are attracted to each other by van der Waals forces. Noble gas atoms, for example, because of their filled shell structure, do not generally form molecules or bond to each other to form a liquid. Because of van der Waals forces, however, at sufficiently low temperatures at which thermal excitations are negligible, noble gases first condense to liquids and then solidify. (The exception is helium, which does not solidify at atmospheric pressure.)

The van der Waals force results from the following situation. While being electrically neutral, a molecule has a charge distribution with positive and negative centers at different positions in the molecule. As a result, the molecule may act as an electric dipole. (See Section 23.4.) Because of the dipole electric fields, two molecules can interact such that there is an attractive force between them.

There are three types of van der Waals forces. The first type, called the *dipole–dipole force*, is an interaction between two molecules each having a permanent electric dipole moment. For example, polar molecules such as HCl have permanent electric dipole moments and attract other polar molecules.

The second type, the *dipole–induced dipole force*, results when a polar molecule having a permanent electric dipole moment induces a dipole moment in a nonpolar molecule. In this case, the electric field of the polar molecule creates the dipole moment in the nonpolar molecule, which then results in an attractive force between the molecules.

The third type is called the *dispersion force*, an attractive force that occurs between two nonpolar molecules. In this case, although the average dipole moment of a nonpolar molecule is zero, the average of the square of the dipole moment is nonzero because of charge fluctuations. Two nonpolar molecules near each other tend to have dipole moments that are correlated in time so as to produce an attractive van der Waals force.

Hydrogen Bonding

Because hydrogen has only one electron, it is expected to form a covalent bond with only one other atom within a molecule. A hydrogen atom in a given molecule can also form a second type of bond between molecules called a **hydrogen bond.** Let's use the water molecule H_2O as an example. In the two covalent bonds in this molecule, the electrons from the hydrogen atoms are more likely to be found near the oxygen atom than near the hydrogen atoms, leaving essentially bare protons at the positions of the hydrogen atoms. This unshielded positive charge can be attracted to the negative end of another polar molecule. Because the proton is unshielded by electrons, the negative end of the other molecule can come very close to the proton to form a bond strong enough to form a solid crystalline structure, such as that of ordinary ice. The bonds within a water molecule are covalent, but the bonds between water molecules in ice are hydrogen bonds.

The hydrogen bond is relatively weak compared with other chemical bonds and can be broken with an input energy of approximately 0.1 eV. Because of this weakness, ice melts at the low temperature of 0°C. Even though this bond is weak, however, hydrogen bonding is a critical mechanism responsible for the linking of biological molecules and polymers. For example, in the case of the DNA (deoxyribonucleic acid) molecule, which has a double-helix structure (Fig. 43.4), hydrogen

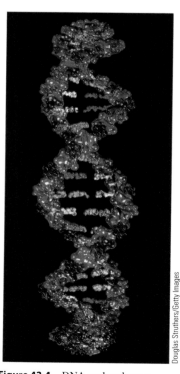

Figure 43.4 DNA molecules are held together by hydrogen bonds.

bonds formed by the sharing of a proton between two atoms create linkages between the turns of the helix.

Quick Quiz 43.1   For each of the following atoms or molecules, identify the most likely type of bonding that occurs between the atoms or between the molecules. Choose from the following list: ionic, covalent, van der Waals, hydrogen. (a) atoms of krypton   (b) potassium and chlorine atoms   (c) hydrogen fluoride (HF) molecules   (d) chlorine and oxygen atoms in a hypochlorite ion (ClO^-)

43.2   Energy States and Spectra of Molecules

Consider an individual molecule in the gaseous phase of a substance. The energy E of the molecule can be divided into four categories: (1) electronic energy, due to the interactions between the molecule's electrons and nuclei; (2) translational energy, due to the motion of the molecule's center of mass through space; (3) rotational energy, due to the rotation of the molecule about its center of mass; and (4) vibrational energy, due to the vibration of the molecule's constituent atoms:

$$E = E_{el} + E_{trans} + E_{rot} + E_{vib}$$

◀ Total energy of a molecule

We explored the roles of translational, rotational, and vibrational energy of molecules in determining the molar specific heats of gases in Sections 21.2 and 21.4. Because the translational energy is unrelated to internal structure, this molecular energy is unimportant in interpreting molecular spectra. The electronic energy of a molecule is very complex because it involves the interaction of many charged particles, but various techniques have been developed to approximate its values. Although the electronic energies can be studied, significant information about a molecule can be determined by analyzing its quantized rotational and vibrational energy states. Transitions between these states give spectral lines in the microwave and infrared regions of the electromagnetic spectrum, respectively.

Rotational Motion of Molecules

Let's consider the rotation of a molecule around its center of mass, confining our discussion to the diatomic molecule (Active Fig. 43.5a, page 1262) but noting that the same ideas can be extended to polyatomic molecules. A diatomic molecule aligned along an x axis has only two rotational degrees of freedom, corresponding to rotations about the y and z axes passing through the molecule's center of mass. If ω is the angular frequency of rotation about one of these axes, the rotational kinetic energy of the molecule about that axis can be expressed as

$$E_{rot} = \tfrac{1}{2}I\omega^2 \qquad (43.2)$$

In this equation, I is the moment of inertia of the molecule about its center of mass, given by

$$I = \left(\frac{m_1 m_2}{m_1 + m_2}\right)r^2 = \mu r^2 \qquad (43.3)$$

◀ Moment of inertia for a diatomic molecule

where m_1 and m_2 are the masses of the atoms that form the molecule, r is the atomic separation, and μ is the **reduced mass** of the molecule (see Example 41.5 and Problem 39 in Chapter 41):

$$\mu = \frac{m_1 m_2}{m_1 + m_2} \qquad (43.4)$$

◀ Reduced mass of a diatomic molecule

The magnitude of the molecule's angular momentum about its center of mass is $L = I\omega$, which classically can have any value. Quantum mechanics, however,

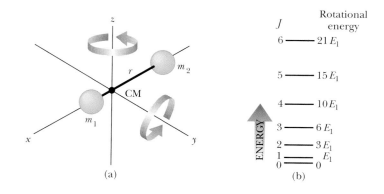

restricts the molecule to certain quantized rotational frequencies such that the angular momentum of the molecule has the values[2]

▶ Allowed values of rotational angular momentum

$$L = \sqrt{J(J+1)}\,\hbar \qquad J = 0, 1, 2, \dots \qquad \textbf{(43.5)}$$

where J is an integer called the **rotational quantum number.** Combining Equations 43.5 and 43.2, we obtain an expression for the allowed values of the rotational kinetic energy of the molecule:

$$E_{\text{rot}} = \tfrac{1}{2}I\omega^2 = \frac{1}{2I}(I\omega)^2 = \frac{L^2}{2I} = \frac{(\sqrt{J(J+1)}\,\hbar)^2}{2I}$$

▶ Allowed values of rotational energy

$$E_{\text{rot}} = E_J = \frac{\hbar^2}{2I}J(J+1) \qquad J = 0, 1, 2, \dots \qquad \textbf{(43.6)}$$

The allowed rotational energies of a diatomic molecule are plotted in Active Figure 43.5b. As the quantum number J goes up, the states become farther apart as displayed earlier for rotational energy levels in Figure 21.8.

For most molecules, transitions between adjacent rotational energy levels result in radiation that lies in the microwave range of frequencies ($f \sim 10^{11}$ Hz). When a molecule absorbs a microwave photon, the molecule jumps from a lower rotational energy level to a higher one. The allowed rotational transitions of linear molecules are regulated by the selection rule $\Delta J = \pm 1$. Given this selection rule, all absorption lines in the spectrum of a linear molecule correspond to energy separations equal to $E_J - E_{J-1}$, where $J = 1, 2, 3, \dots$. From Equation 43.6, we see that the energies of the absorbed photons are given by

$$E_{\text{photon}} = \Delta E_{\text{rot}} = E_J - E_{J-1} = \frac{\hbar^2}{2I}[J(J+1) - (J-1)J]$$

▶ Energy of a photon absorbed in a transition between adjacent rotational levels

$$E_{\text{photon}} = \frac{\hbar^2}{I}J = \frac{h^2}{4\pi^2 I}J \qquad J = 1, 2, 3, \dots \qquad \textbf{(43.7)}$$

where J is the rotational quantum number of the higher energy state. Because $E_{\text{photon}} = hf$, where f is the frequency of the absorbed photon, we see that the allowed frequency for the transition $J = 0$ to $J = 1$ is $f_1 = h/4\pi^2 I$. The frequency corresponding to the $J = 1$ to $J = 2$ transition is $2f_1$, and so on. These predictions are in excellent agreement with the observed frequencies.

Quick Quiz 43.2 A gas of identical diatomic molecules absorbs electromagnetic radiation over a wide range of frequencies. Molecule 1 is in the $J = 0$ rotation state and makes a transition to the $J = 1$ state. Molecule 2 is in the $J = 2$ state and

[2] Equation 43.5 is similar to Equation 42.27 for orbital angular momentum in an atom. The relationship between the magnitude of the angular momentum of a system and the associated quantum number is the same as it is in these equations for *any* system that exhibits rotation as long as the potential energy function for the system is spherically symmetric.

makes a transition to the $J = 3$ state. Is the ratio of the frequency of the photon that excited molecule 2 to that of the photon that excited molecule 1 equal to (a) 1, (b) 2, (c) 3, (d) 4, or (e) impossible to determine?

EXAMPLE 43.1   **Rotation of the CO Molecule**

The $J = 0$ to $J = 1$ rotational transition of the CO molecule occurs at a frequency of 1.15×10^{11} Hz.

(A) Use this information to calculate the moment of inertia of the molecule.

SOLUTION

Conceptualize   Imagine that the two atoms in Active Figure 43.5a are carbon and oxygen. The center of mass of the molecule is not midway between the atoms because of the difference in masses of the C and O atoms.

Categorize   The statement of the problem tells us to categorize this example as one involving a quantum-mechanical treatment and to restrict our investigation to the rotational motion of a diatomic molecule.

Analyze   Use Equation 43.7 to find the energy of a photon that excites the molecule from the $J = 0$ to the $J = 1$ rotational level:

$$E_{\text{photon}} = \frac{h^2}{4\pi^2 I}(1) = \frac{h^2}{4\pi^2 I}$$

Equate this energy to $E = hf$ for the absorbed photon and solve for I:

$$\frac{h^2}{4\pi^2 I} = hf \;\; \rightarrow \;\; I = \frac{h}{4\pi^2 f}$$

Substitute the frequency given in the problem statement:

$$I = \frac{6.626 \times 10^{-34}\,\text{J} \cdot \text{s}}{4\pi^2 (1.15 \times 10^{11}\,\text{s}^{-1})} = 1.46 \times 10^{-46}\,\text{kg} \cdot \text{m}^2$$

(B) Calculate the bond length of the molecule.

SOLUTION

Find the reduced mass μ of the CO molecule:

$$\mu = \frac{m_1 m_2}{m_1 + m_2} = \frac{(12\,\text{u})(16\,\text{u})}{12\,\text{u} + 16\,\text{u}} = 6.86\,\text{u}$$

$$= (6.86\,\text{u})\left(\frac{1.66 \times 10^{-27}\,\text{kg}}{1\,\text{u}}\right) = 1.14 \times 10^{-26}\,\text{kg}$$

Solve Equation 43.3 for r and substitute for the reduced mass and the moment of inertia from part (A):

$$r = \sqrt{\frac{I}{\mu}} = \sqrt{\frac{1.46 \times 10^{-46}\,\text{kg} \cdot \text{m}^2}{1.14 \times 10^{-26}\,\text{kg}}}$$

$$= 1.13 \times 10^{-10}\,\text{m} = 0.113\,\text{nm}$$

Finalize   The moment of inertia of the molecule and the separation distance between the atoms are both very small, as expected for a microscopic system.

What If?   What if another photon of frequency 1.15×10^{11} Hz is incident on the CO molecule while that molecule is in the $J = 1$ state? What happens?

Answer   Because the rotational quantum states are not equally spaced in energy, the $J = 1$ to $J = 2$ transition does not have the same energy as the $J = 0$ to $J = 1$ transition. Therefore, the molecule will *not* be excited to the $J = 2$ state. Two possibilities exist. The photon could pass by the molecule with no interaction, or the photon could induce a stimulated emission, similar to that for atoms and discussed in Section 42.9. In this case, the molecule makes a transition back to the $J = 0$ state and the original photon and a second identical photon leave the scene of the interaction.

Vibrational Motion of Molecules

If we consider a molecule to be a flexible structure in which the atoms are bonded together by "effective springs" as shown in Active Figure 43.6a, we can model the molecule as a simple harmonic oscillator as long as the atoms in the molecule are not too far from their equilibrium positions. Recall from Section 15.3 that the potential energy function for a simple harmonic oscillator is parabolic, varying as the square of the displacement from equilibrium. (See Eq. 15.20.) Active Figure 43.6b shows a plot of potential energy versus atomic separation for a diatomic molecule, where r_0 is the equilibrium atomic separation. For separations close to r_0, the shape of the potential energy curve closely resembles a parabola.

According to classical mechanics, the frequency of vibration for the system shown in Active Figure 43.6a is given by Equation 15.14:

$$f = \frac{1}{2\pi}\sqrt{\frac{k}{\mu}} \tag{43.8}$$

where k is the effective spring constant and μ is the reduced mass given by Equation 43.4.

Quantum mechanics predicts that a molecule vibrates in quantized states as described in Section 41.7. The vibrational motion and quantized vibrational energy can be altered if the molecule acquires energy of the proper value to cause a transition between quantized vibrational states. As discussed in Section 41.7, the allowed vibrational energies are

$$E_{\text{vib}} = (v + \tfrac{1}{2})hf \qquad v = 0, 1, 2, \ldots \tag{43.9}$$

where v is an integer called the **vibrational quantum number**. (We used n in Section 41.7 for a general harmonic oscillator, but v is often used for the quantum number when discussing molecular vibrations.) If the system is in the lowest vibrational state, for which $v = 0$, its ground-state energy is $\tfrac{1}{2}hf$. In the first excited vibrational state, $v = 1$ and the energy is $\tfrac{3}{2}hf$, and so on.

Substituting Equation 43.8 into Equation 43.9 gives the following expression for the allowed vibrational energies:

Allowed values of ▶
vibrational energy

$$E_{\text{vib}} = (v + \tfrac{1}{2})\frac{h}{2\pi}\sqrt{\frac{k}{\mu}} \qquad v = 0, 1, 2, \ldots \tag{43.10}$$

The selection rule for the allowed vibrational transitions is $\Delta v = \pm 1$. Transitions between vibrational levels are caused by absorption of photons in the infrared region of the spectrum. The energy of an absorbed photon is equal to the energy difference between any two successive vibrational levels. Therefore, the photon energy is given by

$$E_{\text{photon}} = \Delta E_{\text{vib}} = \frac{h}{2\pi}\sqrt{\frac{k}{\mu}} \tag{43.11}$$

ACTIVE FIGURE 43.6

(a) Effective-spring model of a diatomic molecule. The vibration is along the molecular axis. (b) Plot of the potential energy of a diatomic molecule versus atomic separation distance, where r_0 is the equilibrium separation distance of the atoms. Compare with Figure 15.11.

Sign in at www.thomsonedu.com and go to ThomsonNOW to adjust the spring constant and choose the initial vibrational energy state of the molecule. Then observe transitions of the molecule to lower energy states in Active Figure 43.7.

(a)

(b)

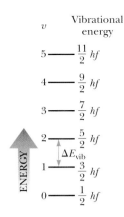

Vibrational
energy

v

$5 \longrightarrow \frac{11}{2} hf$

$4 \longrightarrow \frac{9}{2} hf$

$3 \longrightarrow \frac{7}{2} hf$

$2 \longrightarrow \frac{5}{2} hf$

ΔE_{vib}

$1 \longrightarrow \frac{3}{2} hf$

$0 \longrightarrow \frac{1}{2} hf$

ENERGY

ACTIVE FIGURE 43.7

Allowed vibrational energies of a diatomic molecule, where f is the frequency of vibration of the molecule, given by Equation 43.8. The spacings between adjacent vibrational levels are equal if the molecule behaves as a harmonic oscillator.

Sign in at www.thomsonedu.com and go to ThomsonNOW to adjust the spring constant in Active Figure 43.6 and choose the initial vibrational energy state of the molecule. Then observe transitions of the molecule to other energy states.

The vibrational energies of a diatomic molecule are plotted in Active Figure 43.7. At ordinary temperatures, most molecules have vibrational energies corresponding to the $v = 0$ state because the spacing between vibrational states is much greater than $k_B T$, where k_B is Boltzmann's constant and T is the temperature.

Quick Quiz 43.3   A gas of identical diatomic molecules absorbs electromagnetic radiation over a wide range of frequencies. Molecule 1, initially in the $v = 0$ vibrational state, makes a transition to the $v = 1$ state. Molecule 2, initially in the $v = 2$ state, makes a transition to the $v = 3$ state. What is the ratio of the frequency of the photon that excited molecule 2 to that of the photon that excited molecule 1? (a) 1   (b) 2   (c) 3   (d) 4   (e) impossible to determine

---

EXAMPLE 43.2	**Vibration of the CO Molecule**

The frequency of the photon that causes the $v = 0$ to $v = 1$ transition in the CO molecule is 6.42×10^{13} Hz. We ignore any changes in the rotational energy for this example.

(A) Calculate the force constant k for this molecule.

SOLUTION

Conceptualize   Imagine that the two atoms in Active Figure 43.6a are carbon and oxygen. As the molecule vibrates, a given point on the imaginary spring is at rest. This point is not midway between the atoms because of the difference in masses of the C and O atoms.

Categorize   The statement of the problem tells us to categorize this example as one involving a quantum-mechanical treatment and to restrict our investigation to the vibrational motion of a diatomic molecule.

Analyze   Set Equation 43.11 equal to the photon energy hf and solve for the force constant:

$$\frac{h}{2\pi}\sqrt{\frac{k}{\mu}} = hf \quad \rightarrow \quad k = 4\pi^2 \mu f^2$$

Substitute the frequency given in the problem statement and the reduced mass from Example 43.1:

$$k = 4\pi^2 (1.14 \times 10^{-26} \text{ kg})(6.42 \times 10^{13} \text{ s}^{-1})^2 = \boxed{1.85 \times 10^3 \text{ N/m}}$$

(B) What is the classical amplitude A of vibration for this molecule in the $v = 0$ vibrational state?

SOLUTION

Equate the maximum elastic potential energy $\frac{1}{2}kA^2$ in the molecule (Eq. 15.21) to the vibrational energy given by Equation 43.10 with $v = 0$ and solve for A:

$$\tfrac{1}{2}kA^2 = \frac{h}{4\pi}\sqrt{\frac{k}{\mu}} \quad \rightarrow \quad A = \sqrt{\frac{h}{2\pi}}\left(\frac{1}{\mu k}\right)^{1/4}$$

Substitute the value for k from part (A) and the value for μ:

$$A = \sqrt{\frac{6.626 \times 10^{-34}\,\text{J}\cdot\text{s}}{2\pi}\left[\frac{1}{(1.14 \times 10^{-26}\,\text{kg})(1.85 \times 10^{3}\,\text{N/m})}\right]^{1/4}}$$

$$= 4.79 \times 10^{-12}\,\text{m} = \boxed{0.004\ 79\ \text{nm}}$$

Finalize Comparing this result with the bond length of 0.113 nm we calculated in Example 43.1 shows that the classical amplitude of vibration is approximately 4% of the bond length.

Molecular Spectra

In general, a molecule vibrates and rotates simultaneously. To a first approximation, these motions are independent of each other, so the total energy of the molecule for these motions is the sum of Equations 43.6 and 43.9:

$$E = (v + \tfrac{1}{2})hf + \frac{\hbar^{2}}{2I}\,J(J+1) \tag{43.12}$$

The energy levels of any molecule can be calculated from this expression, and each level is indexed by the two quantum numbers v and J. From these calculations, an energy-level diagram like the one shown in Active Figure 43.8a can be constructed. For each allowed value of the vibrational quantum number v, there is a complete set of rotational levels corresponding to $J = 0, 1, 2, \ldots$. The energy separation between successive rotational levels is much smaller than the separation between successive vibrational levels. As noted earlier, most molecules at ordinary temperatures are in the $v = 0$ vibrational state; these molecules can be in various rotational states as Active Figure 43.8a shows.

When a molecule absorbs a photon with the appropriate energy, the vibrational quantum number v increases by one unit while the rotational quantum number J either increases or decreases by one unit as can be seen in Active Figure 43.8. Therefore, the molecular absorption spectrum consists of two groups of lines: one group to the right of center and satisfying the selection rules $\Delta J = +1$ and $\Delta v = +1$; the other group to the left of center and satisfying the selection rules $\Delta J = -1$ and $\Delta v = +1$.

The energies of the absorbed photons can be calculated from Equation 43.12:

$$E_{\text{photon}} = \Delta E = hf + \frac{\hbar^{2}}{I}\,(J+1) \quad J = 0, 1, 2, \ldots \quad (\Delta J = +1) \tag{43.13}$$

$$E_{\text{photon}} = \Delta E = hf - \frac{\hbar^{2}}{I}\,J \quad\quad J = 1, 2, 3, \ldots \quad (\Delta J = -1) \tag{43.14}$$

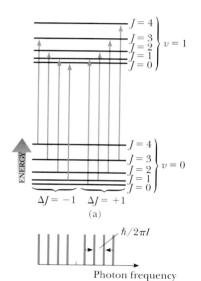

ACTIVE FIGURE 43.8

(a) Absorptive transitions between the $v = 0$ and $v = 1$ vibrational states of a diatomic molecule. The transitions obey the selection rule $\Delta J = \pm 1$ and fall into two sequences, those for $\Delta J = +1$ and those for $\Delta J = -1$. The transition energies are given by Equations 43.13 and 43.14. Compare the energy levels in this figure with those in Figure 21.8. (b) Expected lines in the absorption spectrum of a molecule. The lines to the right of the center mark correspond to transitions in which J changes by $+1$; the lines to the left of the center mark correspond to transitions for which J changes by -1. These same lines appear in the emission spectrum.

Sign in at www.thomsonedu.com and go to ThomsonNOW to adjust the spring constant and the moment of inertia of the molecule and observe the effect on the energy levels and the spectral lines.

where J is the rotational quantum number of the *initial* state. Equation 43.13 generates the series of equally spaced lines *higher* than the frequency f, whereas Equation 43.14 generates the series *lower* than this frequency. Adjacent lines are separated in frequency by the fundamental unit $\hbar/2\pi I$. Active Figure 43.8b shows the expected frequencies in the absorption spectrum of the molecule; these same frequencies appear in the emission spectrum.

The experimental absorption spectrum of the HCl molecule shown in Figure 43.9 follows this pattern very well and reinforces our model. One peculiarity is apparent, however: each line is split into a doublet. This doubling occurs because two chlorine isotopes (see Section 44.1) were present in the sample used to obtain this spectrum. Because the isotopes have different masses, the two HCl molecules have different values of I.

The intensity of the spectral lines in Figure 43.9 follows an interesting pattern, rising first as one moves away from the central gap (at about 8.65×10^{13} Hz, corresponding to the forbidden $J = 0$ to $J = 0$ transition) and then falling. This intensity is determined by a product of two functions of J. The first function corresponds to the number of available states for a given value of J. This function is $2J + 1$, corresponding to the number of values of m_J, the molecular rotation ana-

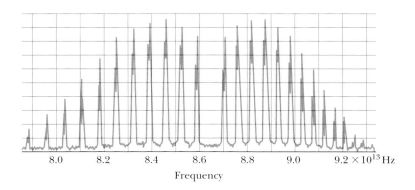

Figure 43.9 Absorption spectrum of the HCl molecule. Each line is split into a doublet because the sample contains two chlorine isotopes that have different masses and therefore different moments of inertia.

log to m_ℓ for atomic states. For example, the $J = 2$ state has five substates with five values of m_J ($m_J = -2, -1, 0, 1, 2$), whereas the $J = 1$ state has only three substates ($m_J = -1, 0, 1$). Therefore, on average and without regard for the second function described below, five-thirds as many molecules make the transition from the $J = 2$ state as from the $J = 1$ state.

The second function determining the envelope of the intensity of the spectral lines is the Boltzmann factor, introduced in Section 21.5. The number of molecules in an excited rotational state is given by

$$n = n_0 e^{-\hbar^2 J(J+1)/(2Ik_B T)}$$

where n_0 is the number of molecules in the $J = 0$ state.

Multiplying these factors together indicates that the intensity of spectral lines should be described by a function of J as follows:

$$I \propto (2J + 1)e^{-\hbar^2 J(J+1)/(2Ik_B T)} \tag{43.15}$$

◀ Intensity variation in the vibration–rotation spectrum of a molecule

The factor $(2J + 1)$ increases with J while the exponential second factor decreases. The product of the two factors gives a behavior that closely describes the envelope of the spectral lines in Figure 43.9.

The excitation of rotational and vibrational energy levels is an important consideration in current models of global warming. Most of the absorption lines for CO_2 are in the infrared portion of the spectrum. Therefore, visible light from the Sun is not absorbed by atmospheric CO_2 but instead strikes the Earth's surface, warming it. In turn, the surface of the Earth, being at a much lower temperature than the Sun, emits thermal radiation that peaks in the infrared portion of the electromagnetic spectrum (Section 40.1). This infrared radiation is absorbed by the CO_2 molecules in the air instead of radiating out into space. Atmospheric CO_2 acts like a one-way valve for energy from the Sun and is responsible, along with some other atmospheric molecules, for raising the temperature of the Earth's surface above its value in the absence of an atmosphere. This phenomenon is commonly called the "greenhouse effect." The burning of fossil fuels in today's industrialized society adds more CO_2 to the atmosphere. This addition of CO_2 increases the absorption of infrared radiation, raising the Earth's temperature further. In turn, this increase in temperature causes substantial climatic changes. The increased temperature also results in melting of ice from Arctic ice sheets, raising sea levels worldwide and endangering coastal population centers. In February, 2007, the Intergovernmental Panel on Climate Change of the United Nations issued "Climate Change 2007: The Physical Science Basis." This survey, based on the work of over 2 500 scientists from more than 130 countries, contains chilling new statements that global warming is clearly linked to human activity and that the global warming issue is no longer a matter of debate.

CONCEPTUAL EXAMPLE 43.3   **Comparing Figures 43.8 and 43.9**

In Active Figure 43.8a, the transitions indicated correspond to spectral lines that are equally spaced as shown in Active Figure 43.8b. The actual spectrum in Figure 43.9, however, shows lines that move closer together as the frequency increases. Why does the spacing of the actual spectral lines differ from the diagram in Active Figure 43.8?

SOLUTION

In Active Figure 43.8, we modeled the rotating diatomic molecule as a rigid object (Chapter 10). In reality, however, as the molecule rotates faster and faster, the effective spring in Active Figure 43.6a stretches and provides the increased force associated with the larger centripetal acceleration of each atom. As the molecule stretches along its length, its moment of inertia I increases. Therefore, the rotational part of the energy expression in Equation 43.12 has an extra dependence on J in the moment of inertia I. Because the increasing moment of inertia is in the denominator, as J increases, the energies do not increase as rapidly with J as indicated in Equation 43.12. With each higher energy level being lower than indicated by Equation 43.12, the energy associated with a transition to that level is smaller, as is the frequency of the absorbed photon, destroying the even spacing of the spectral lines and giving the uneven spacing seen in Figure 43.9.

(a)

(b)

(c)

Figure 43.10   (a) Crystalline structure of NaCl. (b) Each positive sodium ion (orange spheres) is surrounded by six negative chloride ions (blue spheres). (c) Each chloride ion is surrounded by six sodium ions.

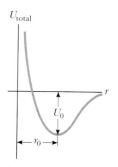

Figure 43.11   Total potential energy versus ion separation distance for an ionic solid, where U_0 is the ionic cohesive energy and r_0 is the equilibrium separation distance between ions.

43.3   Bonding in Solids

A crystalline solid consists of a large number of atoms arranged in a regular array, forming a periodic structure. The ions in the NaCl crystal are ionically bonded, as already noted, and the carbon atoms in diamond form covalent bonds with one another. The metallic bond described at the end of this section is responsible for the cohesion of copper, silver, sodium, and other solid metals.

Ionic Solids

Many crystals are formed by ionic bonding, in which the dominant interaction between ions is the Coulomb force. Consider the NaCl crystal in Figure 43.10. Each Na^+ ion has six nearest-neighbor Cl^- ions, and each Cl^- ion has six nearest-neighbor Na^+ ions. Each Na^+ ion is attracted to its six Cl^- neighbors. The corresponding potential energy is $-6k_e e^2/r$, where k_e is the Coulomb constant and r is the separation distance between each Na^+ and Cl^-. In addition, there are 12 next-nearest-neighbor Na^+ ions at a distance of $\sqrt{2}r$ from the Na^+ ion, and these 12 positive ions exert weaker repulsive forces on the central Na^+. Furthermore, beyond these 12 Na^+ ions are more Cl^- ions that exert an attractive force, and so on. The net effect of all these interactions is a resultant negative electric potential energy

$$U_{\text{attractive}} = -\alpha k_e \frac{e^2}{r} \qquad (43.16)$$

where α is a dimensionless number known as the **Madelung constant.** The value of α depends only on the particular crystalline structure of the solid. For example, $\alpha = 1.747\,6$ for the NaCl structure. When the constituent ions of a crystal are brought close together, a repulsive force exists because of electrostatic forces and the exclusion principle as discussed in Section 43.1. The potential energy term B/r^m in Equation 43.1 accounts for this repulsive force. We do not include neighbors other than nearest neighbors here because the repulsive forces occur only for ions that are very close together. (Electron shells must overlap for exclusion-principle effects to become important.) Therefore, we can express the total potential energy of the crystal as

$$U_{\text{total}} = -\alpha k_e \frac{e^2}{r} + \frac{B}{r^m} \qquad (43.17)$$

where m in this expression is some small integer.

A plot of total potential energy versus ion separation distance is shown in Figure 43.11. The potential energy has its minimum value U_0 at the equilibrium separation, when $r = r_0$. It is left as a problem (Problem 57) to show that

$$U_0 = -\alpha k_e \frac{e^2}{r_0}\left(1 - \frac{1}{m}\right) \qquad (43.18)$$

This minimum energy U_0 is called the **ionic cohesive energy** of the solid, and its absolute value represents the energy required to separate the solid into a collection of isolated positive and negative ions. Its value for NaCl is -7.84 eV per ion pair.

To calculate the **atomic cohesive energy,** which is the binding energy relative to the energy of the neutral atoms, 5.14 eV must be added to the ionic cohesive energy value to account for the transition from Na^+ to Na and 3.62 eV must be subtracted to account for the conversion of Cl^- to Cl. Therefore, the atomic cohesive energy of NaCl is

$$-7.84 \text{ eV} + 5.14 \text{ eV} - 3.62 \text{ eV} = -6.32 \text{ eV}$$

In other words, 6.32 eV of energy per ion pair is needed to separate the solid into isolated neutral atoms of Na and Cl.

Ionic crystals form relatively stable, hard crystals. They are poor electrical conductors because they contain no free electrons; each electron in the solid is bound tightly to one of the ions, so it is not sufficiently mobile to carry current. Ionic crystals have high melting points; for example, the melting point of NaCl is 801°C. Ionic crystals are transparent to visible radiation because the shells formed by the electrons in ionic solids are so tightly bound that visible radiation does not possess sufficient energy to promote electrons to the next allowed shell. Infrared radiation is absorbed strongly because the vibrations of the ions have natural resonant frequencies in the low-energy infrared region.

 ◀ Properties of ionic crystals

Covalent Solids

Solid carbon, in the form of diamond, is a crystal whose atoms are covalently bonded. Because atomic carbon has the electronic configuration $1s^2 2s^2 2p^2$, it is four electrons short of filling its $n = 2$ shell, which can accommodate eight electrons. Hence, two carbon atoms have a strong attraction for each other, with a cohesive energy of 7.37 eV. In the diamond structure, each carbon atom is covalently bonded to four other carbon atoms located at four corners of a cube as shown in Figure 43.12a.

The crystalline structure of diamond is shown in Figure 43.12b. Notice that each carbon atom forms covalent bonds with four nearest-neighbor atoms. The basic structure of diamond is called tetrahedral (each carbon atom is at the center of a regular tetrahedron), and the angle between the bonds is 109.5°. Other crystals such as silicon and germanium have the same structure.

Carbon is interesting in that it can form several different types of structures. In addition to the diamond structure, it forms graphite, with completely different properties. In this form, the carbon atoms form flat layers with hexagonal arrays of atoms. A very weak interaction between the layers allows the layers to be removed easily under friction, as occurs in the graphite used in pencil lead.

Carbon atoms can also form a large hollow structure; in this case, the compound is called **buckminsterfullerene** after the famous architect R. Buckminster Fuller, who invented the geodesic dome. The unique shape of this molecule (Fig.

© Thomson Learning/Charles D. Winters

A cylinder of nearly pure crystalline silicon (Si), approximately 25 cm long. Such crystals are cut into wafers and processed to make various semiconductor devices.

(a)

(b)

Figure 43.12   (a) Each carbon atom in a diamond crystal is covalently bonded to four other carbon atoms so that a tetrahedral structure is formed. (b) The crystal structure of diamond, showing the tetrahedral bond arrangement.

Charles D. Winters

Figure 43.13 Computer rendering of a "buckyball," short for the molecule buckminsterfullerene. These nearly spherical molecular structures that look like soccer balls were named for the inventor of the geodesic dome. This form of carbon, C_{60}, was discovered by astrophysicists investigating the carbon gas that exists between stars. Scientists are actively studying the properties and potential uses of buckminster-fullerene and related molecules.

TABLE 43.1

Atomic Cohesive Energies of Some Covalent Solids

Solid	Cohesive Energy (eV per ion pair)
C (diamond)	7.37
Si	4.63
Ge	3.85
InAs	5.70
SiC	6.15
ZnS	6.32
CuCl	9.24

43.13) provides a "cage" to hold other atoms or molecules. Related structures, called "buckytubes" because of their long, narrow cylindrical arrangements of carbon atoms, may provide the basis for extremely strong, yet lightweight materials.

The atomic cohesive energies of some covalent solids are given in Table 43.1. The large energies account for the hardness of covalent solids. Diamond is particularly hard and has an extremely high melting point (about 4 000 K). Covalently bonded solids are usually very hard, have high bond energies and high melting points, and are good electrical insulators.

Metallic Solids

Metallic bonds are generally weaker than ionic or covalent bonds. The outer electrons in the atoms of a metal are relatively free to move throughout the material, and the number of such mobile electrons in a metal is large. The metallic structure can be viewed as a "sea" or a "gas" of nearly free electrons surrounding a lattice of positive ions (Fig. 43.14). The bonding mechanism in a metal is the attractive force between the entire collection of positive ions and the electron gas. Metals have a cohesive energy in the range of 1 to 3 eV per atom, which is less than the cohesive energies of ionic or covalent solids.

Light interacts strongly with the free electrons in metals. Hence, visible light is absorbed and re-emitted quite close to the surface of a metal, which accounts for the shiny nature of metal surfaces. In addition to the high electrical conductivity of metals produced by the free electrons, the nondirectional nature of the metallic bond allows many different types of metal atoms to be dissolved in a host metal in varying amounts. The resulting *solid solutions*, or *alloys*, may be designed to have particular properties, such as tensile strength, ductility, electrical and thermal conductivity, and resistance to corrosion.

Because the bonding in metals is between all the electrons and all the positive ions, metals tend to bend when stressed. This bending is in contrast to nonmetallic solids, which tend to fracture when stressed. Fracturing results because bonding in nonmetallic solids is primarily with nearest-neighbor ions or atoms. When the distortion causes sufficient stress between some set of nearest neighbors, fracture occurs.

Metal ion

Electron gas

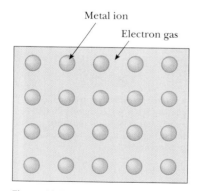

Figure 43.14 Highly schematic diagram of a metal. The blue area represents the electron gas, and the orange circles represent the positive metal ions.

43.4 Free-Electron Theory of Metals

In Section 27.3, we described a classical free-electron theory of electrical conduction in metals that led to Ohm's law. According to this theory, a metal is modeled as a classical gas of conduction electrons moving through a fixed lattice of ions.

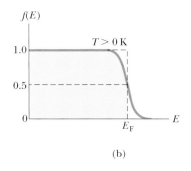

(a)   (b)

ACTIVE FIGURE 43.15

Plot of the Fermi–Dirac distribution function $f(E)$ versus energy at (a) $T = 0$ K and (b) $T > 0$ K. The energy E_F is the Fermi energy.

Sign in at www.thomsonedu.com and go to ThomsonNOW to adjust the temperature and observe the effect on the Fermi–Dirac distribution function.

Although this theory predicts the correct functional form of Ohm's law, it does not predict the correct values of electrical and thermal conductivities.

A quantum-based free-electron theory of metals remedies the shortcomings of the classical model by taking into account the wave nature of the electrons. In this model, the outer-shell electrons are free to move through the metal but are trapped within a three-dimensional box formed by the metal surfaces. Therefore, each electron is represented as a particle in a box. As discussed in Section 41.2, particles in a box are restricted to quantized energy levels.

Statistical physics can be applied to a collection of particles in an effort to relate microscopic properties to macroscopic properties as we saw with kinetic theory of gases in Chapter 21. In the case of electrons, it is necessary to use *quantum statistics*, with the requirement that each state of the system can be occupied by only two electrons (one with spin up and the other with spin down) as a consequence of the exclusion principle. The probability that a particular state having energy E is occupied by one of the electrons in a solid is

$$f(E) = \frac{1}{e^{(E - E_F)/k_B T} + 1} \qquad (43.19)$$

◄ Fermi–Dirac distribution function

where $f(E)$ is called the **Fermi–Dirac distribution function** and E_F is called the **Fermi energy.** A plot of $f(E)$ versus E at $T = 0$ K is shown in Active Figure 43.15a. Notice that $f(E) = 1$ for $E < E_F$ and $f(E) = 0$ for $E > E_F$. That is, at 0 K, all states having energies less than the Fermi energy are occupied and all states having energies greater than the Fermi energy are vacant. A plot of $f(E)$ versus E at some temperature $T > 0$ K is shown in Active Figure 43.15b. This curve shows that as T increases, the distribution rounds off slightly. Because of thermal excitation, states near and below E_F lose population and states near and above E_F gain population. The Fermi energy E_F also depends on temperature, but the dependence is weak in metals.

Let's now follow up on our discussion of the particle in a box in Chapter 41 to generalize the results to a three-dimensional box. Recall that if a particle of mass m is confined to move in a one-dimensional box of length L, the allowed states have quantized energy levels given by Equation 41.14:

$$E_n = \frac{h^2}{8mL^2} n^2 = \frac{\hbar^2 \pi^2}{2mL^2} n^2 \quad n = 1, 2, 3, \ldots$$

Now imagine a piece of metal in the shape of a solid cube of sides L and volume L^3 and focus on one electron that is free to move anywhere in this volume. Therefore, the electron is modeled as a particle in a three-dimensional box. In this model, we require that $\psi(x, y, z) = 0$ at the boundaries of the metal. It can be shown (see Problem 38) that the energy for such an electron is

$$E = \frac{\hbar^2 \pi^2}{2m_e L^2} (n_x^2 + n_y^2 + n_z^2) \qquad (43.20)$$

where m_e is the mass of the electron and n_x, n_y, and n_z are quantum numbers. As we expect, the energies are quantized, and each allowed value of the energy is

Figure 43.16 Plot of the electron distribution function versus energy in a metal at (a) $T = 0$ K and (b) $T = 300$ K. The Fermi energy E_F is 3 eV.

characterized by this set of three quantum numbers (one for each degree of freedom) and the spin quantum number m_s. For example, the ground state, corresponding to $n_x = n_y = n_z = 1$, has an energy equal to $3\hbar^2\pi^2/2m_eL^2$ and can be occupied by two electrons, corresponding to spin-up and spin-down.

Because of the macroscopic size L of the box, the energy levels for the electrons are very close together. As a result, we can treat the quantum numbers as continuous variables. Under this assumption, the number of allowed states per unit volume that have energies between E and $E + dE$ is

$$g(E)\,dE = \frac{8\sqrt{2}\pi m_e^{3/2}}{h^3} E^{1/2}\,dE \qquad (43.21)$$

(See Example 43.5.) The function $g(E)$ is called the **density-of-states function**.

If a metal is in thermal equilibrium, the number of electrons per unit volume $N(E)\,dE$ that have energy between E and $E + dE$ is equal to the product of the number of allowed states and the probability that a state is occupied; that is, $N(E)\,dE = g(E)f(E)\,dE$:

$$N(E)\,dE = \left(\frac{8\sqrt{2}\pi m_e^{3/2}}{h^3} E^{1/2}\right)\left(\frac{1}{e^{(E-E_F)/k_BT} + 1}\right)dE \qquad (43.22)$$

Plots of $N(E)$ versus E for two temperatures are given in Figure 43.16.

If n_e is the total number of electrons per unit volume, we require that

$$n_e = \int_0^\infty N(E)\,dE = \frac{8\sqrt{2}\pi m_e^{3/2}}{h^3}\int_0^\infty \frac{E^{1/2}\,dE}{e^{(E-E_F)/k_BT} + 1} \qquad (43.23)$$

We can use this condition to calculate the Fermi energy. At $T = 0$ K, the Fermi–Dirac distribution function $f(E) = 1$ for $E < E_F$ and $f(E) = 0$ for $E > E_F$. Therefore, at $T = 0$ K, Equation 43.23 becomes

$$n_e = \frac{8\sqrt{2}\pi m_e^{3/2}}{h^3}\int_0^{E_F} E^{1/2}\,dE = \tfrac{2}{3}\frac{8\sqrt{2}\pi m_e^{3/2}}{h^3} E_F^{3/2} \qquad (43.24)$$

Solving for the Fermi energy at 0 K gives

▶ Fermi energy at $T = 0$ K

$$E_F(0) = \frac{h^2}{2m_e}\left(\frac{3n_e}{8\pi}\right)^{2/3} \qquad (43.25)$$

The order of magnitude of the Fermi energy for metals is approximately 5 eV. Representative values for various metals are given in Table 43.2. It is left as a problem (Problem 35) to show that the average energy of a free electron in a metal at 0 K is

$$E_{avg} = \tfrac{3}{5}E_F \qquad (43.26)$$

TABLE 43.2

Calculated Values of the Fermi Energy for Metals at 300 K Based on the Free-Electron Theory

Metal	Electron Concentration (m^{-3})	Fermi Energy (eV)
Li	4.70×10^{28}	4.72
Na	2.65×10^{28}	3.23
K	1.40×10^{28}	2.12
Cu	8.46×10^{28}	7.05
Ag	5.85×10^{28}	5.48
Au	5.90×10^{28}	5.53

In summary, we can consider a metal to be a system comprising a very large number of energy levels available to the free electrons. These electrons fill the levels in accordance with the Pauli exclusion principle, beginning with $E = 0$ and ending with E_F. At $T = 0$ K, all levels below the Fermi energy are filled and all levels above the Fermi energy are empty. At 300 K, a small fraction of the free electrons are excited above the Fermi energy.

EXAMPLE 43.4   **The Fermi Energy of Gold**

Each atom of gold (Au) contributes one free electron to the metal. Compute the Fermi energy for gold.

SOLUTION

Conceptualize   Imagine electrons filling available levels at $T = 0$ K in gold until the solid is neutral. The highest energy filled is the Fermi energy.

Categorize   We evaluate the result using a result from this section, so we categorize this example as a substitution problem.

Substitute the concentration of free electrons in gold from Table 43.2 into Equation 43.25 to calculate the Fermi energy at 0 K:

$$E_F(0) = \frac{(6.626 \times 10^{-34}\,\text{J}\cdot\text{s})^2}{2(9.11 \times 10^{-31}\,\text{kg})}\left[\frac{3(5.90 \times 10^{28}\,\text{m}^{-3})}{8\pi}\right]^{2/3}$$

$$= 8.85 \times 10^{-19}\,\text{J} = 5.53\,\text{eV}$$

EXAMPLE 43.5   **Deriving Equation 43.21**

Based on the allowed states of a particle in a three-dimensional box, derive Equation 43.21.

SOLUTION

Conceptualize   Imagine a particle confined to a three-dimensional box, subject to boundary conditions in three dimensions.

Categorize   We categorize this problem as that of a quantum system in which the energies of the particle are quantized. Further-

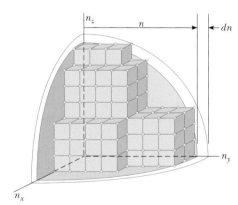

Figure 43.17 (Example 43.5) The allowed states of particles in a three-dimensional box can be represented by dots (blue circles) in a quantum number space. This space is not traditional space in which a location is specified by coordinates x, y, and z; rather, it is a space in which allowed states can be specified by coordinates representing the quantum numbers. The dots representing the allowed states are located at integer values of n_x, n_y, and n_z and are therefore at the corners of cubes with sides of "length" 1. The number of allowed states having energies between E and $E + dE$ corresponds to the number of dots in the spherical shell of radius n and thickness dn.

more, we can base the solution to the problem on our understanding of the particle in a one-dimensional box.

Analyze   As noted previously, the allowed states of the particle in a three-dimensional box are described by three quantum numbers n_x, n_y, and n_z. Imagine a three-dimensional *quantum number space* whose axes represent n_x, n_y, and n_z. The allowed states in this space can be represented as dots located at integral values of the three quantum numbers as in Figure 43.17.

Defining $E_0 = \hbar^2\pi^2/2m_eL^2$ and $n = (E/E_0)^{1/2}$, rewrite Equation 43.20:

$$(1)\quad n_x{}^2 + n_y{}^2 + n_z{}^2 = \frac{2m_eL^2}{\hbar^2\pi^2}E = \frac{E}{E_0} = n^2$$

In the quantum number space, Equation (1) is the equation of a sphere of radius n. Therefore, the number of allowed states having energies between E and $E + dE$ is equal to the number of points in a spherical shell of radius n and thickness dn.

Find the "volume" of this shell, which represents the total number of states $G(E)\,dE$:

$$(2) \quad G(E)\,dE = \tfrac{1}{8}(4\pi n^2\,dn) = \tfrac{1}{2}\pi n^2\,dn$$

We have taken one-eighth of the total volume because we are restricted to the octant of a three-dimensional space in which all three quantum numbers are positive.

Replace n in Equation (2) with its equivalent in terms of E using the relation $n^2 = E/E_0$ from Equation (1):

$$G(E)\,dE = \tfrac{1}{2}\pi \left(\frac{E}{E_0}\right)d\left[\left(\frac{E}{E_0}\right)^{1/2}\right] = \tfrac{1}{2}\pi\,\frac{E}{(E_0)^{3/2}}\,d\left[(E)^{1/2}\right]$$

Evaluate the differential:

$$G(E)\,dE = \tfrac{1}{2}\pi\left[\frac{E}{(E_0)^{3/2}}\right]\left(\tfrac{1}{2}E^{-1/2}\,dE\right) = \tfrac{1}{4}\pi E_0^{-3/2}E^{1/2}\,dE$$

Substitute for E_0 from its definition above:

$$G(E)\,dE = \tfrac{1}{4}\pi\left(\frac{\hbar^2\pi^2}{2m_eL^2}\right)^{-3/2}E^{1/2}\,dE$$

$$= \frac{\sqrt{2}}{2}\,\frac{m_e^{3/2}L^3}{\hbar^3\pi^2}\,E^{1/2}\,dE$$

Letting $g(E)$ represent the number of states per unit volume, where L^3 is the volume V of the cubical box in normal space, find $g(E) = G(E)/V$:

$$g(E)\,dE = \frac{G(E)}{V}\,dE = \frac{\sqrt{2}}{2}\,\frac{m_e^{3/2}}{\hbar^3\pi^2}\,E^{1/2}\,dE$$

Substitute $\hbar = h/2\pi$:

$$g(E)\,dE = \frac{4\sqrt{2}\pi m_e^{3/2}}{h^3}\,E^{1/2}\,dE$$

Multiply by 2 for the two possible spin states in each particle-in-a-box state:

$$g(E)\,dE = \frac{8\sqrt{2}\pi m_e^{3/2}}{h^3}\,E^{1/2}\,dE$$

Finalize This result is Equation 43.21, which is what we set out to derive.

43.5 Band Theory of Solids

In Section 43.4, the electrons in a metal were modeled as particles free to move around inside a three-dimensional box and we ignored the influence of the parent atoms. In this section, we make the model more sophisticated by incorporating the contribution of the parent atoms that form the crystal.

Recall from Section 41.1 that the probability density $|\psi|^2$ for a system is physically significant, but the probability amplitude ψ is not. Let's consider as an example an atom that has a single s electron outside of a closed shell. Both of the following wave functions are valid for such an atom with atomic number Z:

$$\psi_s^+(r) = +Af(r)e^{-Zr/na_0} \qquad \psi_s^-(r) = -Af(r)e^{-Zr/na_0}$$

where A is the normalization constant and $f(r)$ is a function[3] of r that varies with the value of n. Choosing either of these wave functions leads to the same value of $|\psi|^2$, so both choices are equivalent. A difference arises, however, when two atoms are combined.

If two identical atoms are very far apart, they do not interact and their electronic energy levels can be considered to be those of isolated atoms. Suppose the two atoms are sodium, each having a lone $3s$ electron that is in a well-defined

[3] The functions $f(r)$ are called *Laguerre polynomials.* They can be found in the quantum treatment of the hydrogen atom in modern physics textbooks.

quantum state. As the two sodium atoms are brought closer together, their wave functions begin to overlap as we discussed for covalent bonding in Section 43.1. The properties of the combined system differ depending on whether the two atoms are combined with wave functions $\psi_s^+(r)$ as in Figure 43.18a or whether one is combined with wave function $\psi_s^+(r)$ and the other with $\psi_s^-(r)$ as in Figure 43.18b. The choice of two atoms with wave function $\psi_s^-(r)$ is physically equivalent to that with two positive wave functions, so we do not consider it separately. When two wave functions $\psi_s^+(r)$ are combined, the result is a composite wave function in which the probability amplitudes add between the atoms. If $\psi_s^+(r)$ combines with $\psi_s^-(r)$, however, the wave functions between the nuclei subtract. Therefore, the composite probability amplitudes for the two possibilities are different. **These two possible combinations of wave functions represent two possible states of the two-atom system.** We interpret these curves as representing the probability amplitude of finding an electron. The positive–positive curve shows some probability of finding the electron at the midpoint between the atoms. The positive–negative function shows no such probability. A state with a high probability of an electron *between* two positive nuclei must have a different energy than a state with a high probability of the electron being elsewhere! Therefore, the states are *split* into two energy levels due to the two ways of combining the wave functions. The energy difference is relatively small, so the two states are close together on an energy scale.

Figure 43.19a shows this splitting effect as a function of separation distance. For large separations r, the electron clouds do not overlap and there is no splitting. As the atoms are brought closer so that r decreases, the electron clouds overlap and we need to consider the system of two atoms.

When a large number of atoms are brought together to form a solid, a similar phenomenon occurs. The individual wave functions can be brought together in various combinations of $\psi_s^+(r)$ and $\psi_s^-(r)$, each possible combination corresponding to a different energy. As the atoms are brought close together, the various isolated-atom energy levels split into multiple energy levels for the composite system. This splitting in levels for five atoms in close proximity is shown in Figure 43.19b. In this case, there are five energy levels corresponding to five different combinations of isolated-atom wave functions.

If we extend this argument to the large number of atoms found in solids (on the order of 10^{23} atoms per cubic centimeter), we obtain a large number of levels of varying energy so closely spaced that they may be regarded as a continuous **band** of energy levels as shown in Figure 43.19c. In the case of sodium, it is customary to refer to the continuous distributions of allowed energy levels as s bands because the bands originate from the s levels of the individual sodium atoms.

Figure 43.20 (page 1276) shows the allowed energy bands of sodium. Notice that energy gaps, corresponding to *forbidden energies*, occur between the allowed bands. In addition, some bands exhibit sufficient spreading in energy that there is an overlap between bands arising from different quantum states ($3s$ and $3p$).

As indicated by the blue-shaded areas in Figure 43.20, the $1s$, $2s$, and $2p$ bands of sodium are each full of electrons because the $1s$, $2s$, and $2p$ states of each atom

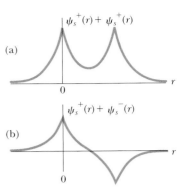

Figure 43.18   The wave functions of two atoms combine to form a composite wave function for the two-atom system when the atoms are close together. (a) Two atoms with wave functions $\psi_s^+(r)$ combine. (b) Two atoms with wave functions $\psi_s^+(r)$ and $\psi_s^-(r)$ combine.

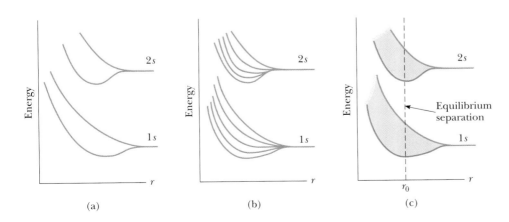

(a)   (b)   (c)

Figure 43.19   (a) Splitting of the $1s$ and $2s$ levels when two sodium atoms are brought together. (b) Splitting of the $1s$ and $2s$ levels when five sodium atoms are brought together. (c) Formation of energy bands when a large number of sodium atoms are assembled to form a solid.

Figure 43.20 Energy bands of a sodium crystal. Notice the energy gaps (white regions) between the allowed bands; electrons cannot occupy states that lie in these gaps. Blue represents energy bands occupied by the sodium electrons when the atom is in its ground state. Gold represents energy bands that are empty.

are full. An energy level in which the orbital angular momentum is ℓ can hold $2(2\ell + 1)$ electrons. The factor 2 arises from the two possible electron spin orientations, and the factor $2\ell + 1$ corresponds to the number of possible orientations of the orbital angular momentum. The capacity of each band for a system of N atoms is $2(2\ell + 1)N$ electrons. Therefore, the $1s$ and $2s$ bands each contain $2N$ electrons ($\ell = 0$), and the $2p$ band contains $6N$ electrons ($\ell = 1$). Because sodium has only one $3s$ electron and there are a total of N atoms in the solid, the $3s$ band contains only N electrons and is partially full as indicated by the blue coloring in Figure 43.20. The $3p$ band, which is the higher region of the overlapping bands, is completely empty (all gold in the figure).

Band theory allows us to build simple models to understand the behavior of conductors, insulators, and semiconductors as well as that of semiconductor devices, as we shall discuss in the following sections.

43.6   Electrical Conduction in Metals, Insulators, and Semiconductors

Good electrical conductors contain a high density of free charge carriers, and the density of free charge carriers in insulators is nearly zero. Semiconductors, first introduced in Section 23.2, are a class of technologically important materials in which charge-carrier densities are intermediate between those of insulators and those of conductors. In this section, we discuss the mechanisms of conduction in these three classes of materials in terms of a model based on energy bands.

Metals

If a material is to be a good electrical conductor, the charge carriers in the material must be free to move in response to an applied electric field. Let's consider the electrons in a metal as the charge carriers. The motion of the electrons in response to an electric field represents an increase in energy of the system (the metal lattice and the free electrons) corresponding to the additional kinetic energy of the moving electrons. Therefore, when an electric field is applied to a conductor, electrons must move upward to an available higher energy state on an energy-level diagram.

Figure 43.21 shows a half-filled band in a metal at $T = 0$ K, where the blue region represents levels filled with electrons. Because electrons obey Fermi–Dirac statistics, all levels below the Fermi energy are filled with electrons and all levels above the Fermi energy are empty. The Fermi energy lies in the band at the highest filled state. At temperatures slightly greater than 0 K, some electrons are thermally excited to levels above E_F, but overall there is little change from the 0 K case. **If a potential difference is applied to the metal, however, electrons having energies near the Fermi energy require only a small amount of additional energy from the applied electric field to reach nearby empty energy states above the Fermi energy.** Therefore, electrons in a metal experiencing only a weak applied electric field are free to move because many empty levels are available close to the occupied energy levels. The model of metals based on band theory demonstrates that metals are excellent electrical conductors.

Figure 43.21 Half-filled band of a metal, an electrical conductor. At $T = 0$ K, the Fermi energy lies in the middle of the band.

Insulators

Now consider the two outermost energy bands of a material in which the lower band is filled with electrons and the higher band is empty at 0 K (Fig. 43.22). The lower, filled band is called the **valence band,** and the upper, empty band is the **conduction band.** (The conduction band is the one that is partially filled in a metal.) It is common to refer to the energy separation between the valence and conduction bands as the **energy gap** E_g of the material. The Fermi energy lies somewhere in the energy gap[4] as shown in Figure 43.22.

Suppose a material has a relatively large energy gap of, for example, approximately 5 eV. At 300 K (room temperature), $k_BT = 0.025$ eV, which is much smaller than the energy gap. At such temperatures, the Fermi–Dirac distribution predicts that very few electrons are thermally excited into the conduction band. There are no available states that lie close in energy and into which electrons can move upward to account for the extra kinetic energy associated with motion through the material in response to an electric field. Consequently, the electrons do not move; the material is an insulator. Although an insulator has many vacant states in its conduction band that can accept electrons, these states are separated from the filled states by a large energy gap. Only a few electrons occupy these states, so the overall electrical conductivity of insulators is very small.

Semiconductors

Semiconductors have the same type of band structure as an insulator, but the energy gap is much smaller, on the order of 1 eV. Table 43.3 shows the energy gaps for some representative materials. The band structure of a semiconductor is shown in Figure 43.23. Because the Fermi level is located near the middle of the gap for a semiconductor and E_g is small, appreciable numbers of electrons are thermally excited from the valence band to the conduction band. Because of the many empty levels above the thermally filled levels in the conduction band, a small applied potential difference can easily raise the energy of the electrons in the conduction band, resulting in a moderate current.

At $T = 0$ K, all electrons in these materials are in the valence band and no energy is available to excite them across the energy gap. Therefore, semiconductors are poor conductors at very low temperatures. Because the thermal excitation of electrons across the narrow gap is more probable at higher temperatures, the conductivity of semiconductors increases rapidly with temperature, contrasting sharply with the conductivity of metals, which decreases slowly with increasing temperature.

Charge carriers in a semiconductor can be negative, positive, or both. When an electron moves from the valence band into the conduction band, it leaves behind

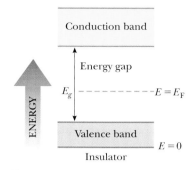

Figure 43.22 An electrical insulator at $T = 0$ K has a filled valence band and an empty conduction band. The Fermi level lies somewhere between these bands in the region known as the energy gap.

TABLE 43.3

Energy-Gap Values for Some Semiconductors

Crystal	E_g (eV)	
	0 K	300 K
Si	1.17	1.14
Ge	0.74	0.67
InP	1.42	1.34
GaP	2.32	2.26
GaAs	1.52	1.42
CdS	2.58	2.42
CdTe	1.61	1.56
ZnO	3.44	3.2
ZnS	3.91	3.6

Figure 43.23 Band structure of a semiconductor at ordinary temperatures ($T \approx 300$ K). The energy gap is much smaller than in an insulator, and some electrons from the valence band occupy states in the conduction band.

[4] We defined the Fermi energy as the energy of the highest filled state at $T = 0$, which might suggest that the Fermi energy should be at the top of the valence band in Figure 43.22. A more sophisticated general treatment of the Fermi energy, however, shows that it is located at that energy at which the probability of occupation is one-half (see Active Fig. 43.15b). According to this definition, the Fermi energy lies in the energy gap between the bands.

Figure 43.24 Movement of charges (holes and electrons) in an intrinsic semiconductor. The electrons move in the direction opposite the direction of the external electric field, and the holes move in the direction of the field.

a vacant site, called a **hole,** in the otherwise filled valence band. This hole (electron-deficient site) acts as a charge carrier in the sense that a free electron from a nearby site can transfer into the hole. Whenever an electron does so, it creates a new hole at the site it abandoned. Therefore, the net effect can be viewed as the hole migrating through the material in the direction opposite the direction of electron movement. The hole behaves as if it were a particle with a positive charge $+e$.

A pure semiconductor crystal containing only one element or one compound is called an **intrinsic semiconductor.** In these semiconductors, there are equal numbers of conduction electrons and holes. Such combinations of charges are called **electron–hole pairs.** In the presence of an external electric field, the holes move in the direction of the field and the conduction electrons move in the direction opposite the field (Fig. 43.24). Because the electrons and holes have opposite signs, both motions correspond to a current in the same direction.

Quick Quiz 43.4   Consider the data on three materials given in the table.

Material	Conduction Band	E_g
A	Empty	1.2 eV
B	Half full	1.2 eV
C	Empty	8.0 eV

Identify these materials as a conductor, an insulator, or a semiconductor.

Doped Semiconductors

When impurities are added to a semiconductor, both the band structure of the semiconductor and its resistivity are modified. The process of adding impurities, called **doping,** is important in controlling the conductivity of semiconductors. For example, when an atom containing five outer-shell electrons, such as arsenic, is added to a Group IV semiconductor, four of the electrons form covalent bonds with atoms of the semiconductor and one is left over (Fig. 43.25a). This extra electron is nearly free of its parent atom and can be modeled as having an energy level that lies in the energy gap, immediately below the conduction band (Fig. 43.25b). Such a pentavalent atom in effect donates an electron to the structure and hence is referred to as a **donor atom.** Because the spacing between the energy level of the electron of the donor atom and the bottom of the conduction band is very small (typically, approximately 0.05 eV), only a small amount of thermal excitation is needed to cause this electron to move into the conduction band. (Recall that the average energy of an electron at room temperature is approximately $k_B T \approx 0.025$ eV.) Semiconductors doped with donor atoms are called **n-type semiconductors** because the majority of charge carriers are electrons, which are *nega*tively charged.

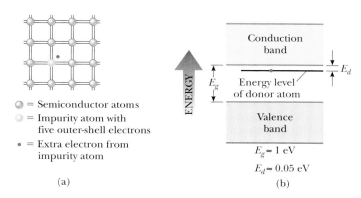

Figure 43.25 (a) Two-dimensional representation of a semiconductor consisting of Group IV atoms (gray) and an impurity atom (yellow) that has five outer-shell electrons. Each double line between atoms represents a covalent bond in which two electrons are shared. (b) Energy-band diagram for a semiconductor in which the nearly free electron of the impurity atom lies in the energy gap, immediately below the bottom of the conduction band. A small amount of energy can excite the electron into the conduction band.

If a Group IV semiconductor is doped with atoms containing three outer-shell electrons, such as indium and aluminum, the three electrons form covalent bonds with neighboring semiconductor atoms, leaving an electron deficiency—a hole—where the fourth bond would be if an impurity-atom electron were available to form it (Fig. 43.26a). This situation can be modeled by placing an energy level in the energy gap, immediately above the valence band, as in Figure 43.26b. An electron from the valence band has enough energy at room temperature to fill this impurity level, leaving behind a hole in the valence band. This hole can carry current in the presence of an electric field. Because a trivalent atom accepts an electron from the valence band, such impurities are referred to as **acceptor atoms.** A semiconductor doped with trivalent (acceptor) impurities is known as a *p*-type semiconductor because the majority of charge carriers are *p*ositively charged holes.

When conduction in a semiconductor is the result of acceptor or donor impurities, the material is called an **extrinsic semiconductor.** The typical range of doping densities for extrinsic semiconductors is 10^{13} to 10^{19} cm^{-3}, whereas the electron density in a typical semiconductor is roughly 10^{21} cm^{-3}.

43.7 Semiconductor Devices

The electronics of the first half of the 20th century was based on vacuum tubes, in which electrons pass through empty space between a cathode and an anode. We have seen vacuum tubes in Figure 29.5 (the television picture tube), Figure 29.9 (circular electron beam), Figure 29.14 (Thomson's apparatus for measuring e/m_e for the electron), and Figure 40.9 (photoelectric effect apparatus).

The transistor was invented in 1948, leading to a shift away from vacuum tubes and toward semiconductors as the basis of electronic devices. This phase of electronics has been under way for several decades. As discussed in Chapter 41, there may be a new phase of electronics in the near future using nanotechnological devices employing quantum dots and other nanoscale structures.

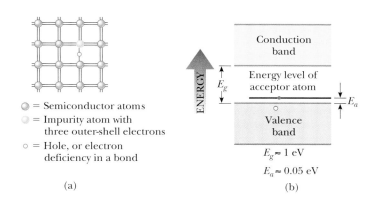

Figure 43.26 (a) Two-dimensional representation of a semiconductor consisting of Group IV atoms (gray) and an impurity atom (yellow) having three outer-shell electrons. The single line between the impurity atom and the semiconductor atom below it represents that there is only one electron shared in this bond. (b) Energy-band diagram for a semiconductor in which the energy level associated with the trivalent impurity atom lies in the energy gap, immediately above the top of the valence band. This diagram shows an electron excited into the energy level of the acceptor atom, leaving a hole in the valence band.

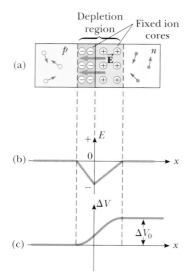

Figure 43.27 (a) Physical arrangement of a p–n junction. (b) Internal electric field magnitude versus x for the p–n junction. (c) Internal electric potential difference ΔV versus x for the p–n junction. The potential difference ΔV_0 represents the potential difference across the junction in the absence of an applied electric field.

In this section, we discuss electronic devices based on semiconductors, which are still in wide use and will be for many years to come.

The Junction Diode

A fundamental unit of a semiconductor device is formed when a p-type semiconductor is joined to an n-type semiconductor to form a **p–n junction**. A **junction diode** is a device that is based on a single p–n junction. The role of a diode of any type is to pass current in one direction but not the other. Therefore, it acts as a one-way valve for current.

The p–n junction shown in Figure 43.27a consists of three distinct regions: a p region, an n region, and a small area that extends several micrometers to either side of the interface, called a *depletion region*.

The depletion region may be visualized as arising when the two halves of the junction are brought together. The mobile n-side donor electrons nearest the junction (deep-blue area in Fig. 43.27a) diffuse to the p side and fill holes located there, leaving behind immobile positive ions. While this process occurs, we can model the holes that are being filled as diffusing to the n side, leaving behind a region (brown area in Fig. 43.27a) of fixed negative ions.

Because the two sides of the depletion region each carry a net charge, an internal electric field on the order of 10^4 to 10^6 V/cm exists in the depletion region (see Fig. 43.27b). This field produces an electric force on any remaining mobile charge carriers that sweeps them out of the depletion region. The depletion region is so named because it is depleted of mobile charge carriers. This internal electric field creates an internal potential difference ΔV_0 that prevents further diffusion of holes and electrons across the junction and thereby ensures zero current in the junction when no potential difference is applied.

The operation of the junction as a diode is easiest to understand in terms of the potential difference graph shown in Figure 43.27c. If a voltage ΔV is applied to the junction such that the p side is connected to the positive terminal of a voltage source as shown in Figure 43.28a, the internal potential difference ΔV_0 across the junction decreases; the decrease results in a current that increases exponentially with increasing forward voltage, or *forward bias*. For *reverse bias* (where the n side of the junction is connected to the positive terminal of a voltage source), the internal potential difference ΔV_0 increases with increasing reverse bias; the increase results in a very small reverse current that quickly reaches a saturation value I_0. The current–voltage relationship for an ideal diode is

$$I = I_0\left(e^{e\Delta V/k_{\mathrm{B}}T} - 1\right) \tag{43.27}$$

where the first e is the base of the natural logarithm, the second e represents the magnitude of the electron charge, k_{B} is Boltzmann's constant, and T is the absolute temperature. Figure 43.28b shows a circuit diagram for a diode under forward bias, and Figure 43.28c shows an I–ΔV plot characteristic of a real p–n junction, demonstrating the diode behavior.

Figure 43.28 (a) A p–n junction under forward bias. (b) The circuit diagram for a diode under forward bias, showing a battery with an adjustable voltage. Both positive and negative voltages can be applied to the diode to study its nonlinear behavior. (c) The characteristic curve for a real p–n junction.

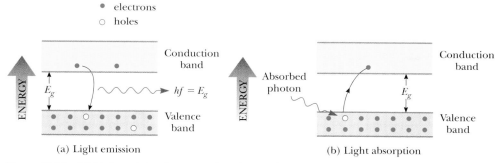

Figure 43.29   (a) Light emission from a semiconductor. (b) Light absorption by a semiconductor.

Light-Emitting and Light-Absorbing Diodes

Light-emitting diodes (LEDs) and semiconductor lasers are common examples of devices that depend on the behavior of semiconductors. LEDs are used in traffic signals, in electronic displays, and as indicator lights for electronic equipment. Semiconductor lasers are often used for pointers in presentations and in compact disc and DVD playback equipment.

Light emission and absorption in semiconductors is similar to light emission and absorption by gaseous atoms except that in the discussion of semiconductors we must incorporate the concept of energy bands rather than the discrete energy levels in single atoms. As shown in Figure 43.29a, an electron excited electrically into the conduction band can easily recombine with a hole (especially if the electron is injected into a p region). As this recombination takes place, a photon of energy E_g is emitted. With proper design of the semiconductor and the associated plastic envelope or mirrors, the light from a large number of these transitions serves as the source of an LED or a semiconductor laser.

Conversely, an electron in the valence band may absorb an incoming photon of light and be promoted to the conduction band, leaving a hole behind (Fig. 43.29b). This absorbed energy can be used to operate an electrical circuit. One device that operates on this principle is the photovoltaic solar cell, which appears in many handheld calculators. Arrays of solar cells are used in generating electric power in space vehicles and in remote areas on the Earth.

EXAMPLE 43.6   **Where's the Remote?**

Estimate the band gap of the semiconductor in the infrared LED of a typical television remote control.

SOLUTION

Conceptualize   Imagine electrons in Figure 43.29a falling from the conduction band to the valence band, emitting infrared photons in the process.

Categorize   We use concepts discussed in this section, so we categorize this example as a substitution problem.

In Chapter 34, we learned that the wavelength of infrared light ranges from 700 nm to 1 mm. Let's pick a number that is easy to work with, such as 1 000 nm (which is not a bad estimate because remote controls typically operate in the range of 880 to 950 nm.)

Estimate the energy hf of the photons from the remote control:

$$E = hf = \frac{hc}{\lambda} = \frac{1\ 240\ \text{eV} \cdot \text{nm}}{1\ 000\ \text{nm}} = 1.2\ \text{eV}$$

This value corresponds to an energy gap E_g of approximately 1.2 eV in the LED's semiconductor.

Figure 43.30 (a) The structure of a metal-oxide-semiconductor field-effect transistor (MOSFET). (b) A source–drain voltage is applied, with the result that current exists in the circuit. (c) A gate voltage is applied. The gate voltage can be used to control the source–drain current so that the MOSFET acts as an amplifier.

The Transistor

The invention of the transistor by John Bardeen (1908–1991), Walter Brattain (1902–1987), and William Shockley (1910–1989) in 1948 totally revolutionized the world of electronics. For this work, these three men shared the Nobel Prize in Physics in 1956. By 1960, the transistor had replaced the vacuum tube in many electronic applications. The advent of the transistor created a multitrillion-dollar industry that produces such popular devices as MP3 players, handheld calculators, computers, television receivers, wireless telephones, and electronic games.

A **junction transistor** consists of a semiconducting material in which a very narrow *n* region is sandwiched between two *p* regions or a *p* region is sandwiched between two *n* regions. In either case, the transistor is formed from two *p–n* junctions. These types of transistors were used widely in the early days of semiconductor electronics.

During the 1960s, the electronics industry converted many electronic applications from the junction transistor to the **field-effect transistor,** which is much easier to manufacture and just as effective. Figure 43.30a shows the structure of a very common device, the **MOSFET,** or **metal-oxide-semiconductor field-effect transistor.** You are likely using millions of MOSFET devices when you are working on your computer.

There are three metal connections (the M in MOSFET) to the transistor: the *source, drain,* and *gate.* The source and drain are connected to *n*-type semiconductor regions (the S in MOSFET) at either end of the structure. These regions are connected by a narrow channel of additional *n*-type material, the *n* channel. The source and drain regions and the *n* channel are embedded in a *p*-type substrate material, which forms a depletion region, as in the junction diode, along the bottom of the *n* channel. (Depletion regions also exist at the junctions underneath the source and drain regions, but we will ignore them because the operation of the device depends primarily on the behavior in the channel.)

The gate is separated from the *n* channel by a layer of insulating silicon dioxide (the O in MOSFET, for oxide). Therefore, it does not make electrical contact with the rest of the semiconducting material.

Imagine that a voltage source ΔV_{SD} is applied across the source and drain as shown in Figure 43.30b. In this situation, electrons flow through the upper region of the *n* channel. Electrons cannot flow through the depletion region in the lower part of the *n* channel because this region is depleted of charge carriers. Now a second voltage ΔV_{SG} is applied across the source and gate as in Figure 43.30c. The positive potential on the gate electrode results in an electric field below the gate that is directed downward in the *n* channel (the field in "field-effect"). This electric field exerts upward forces on electrons in the region below the gate, causing them to move into the *n* channel. Consequently, the depletion region becomes smaller, widening the area through which there is current between the top of the *n* channel and the depletion region. As the area becomes wider, the current increases.

If a varying voltage, such as that generated from music stored on a compact disc, is applied to the gate, the area through which the source–drain current exists varies in size according to the varying gate voltage. A small variation in gate voltage results in a large variation in current and a correspondingly large voltage across the resistor in Figure 43.30c. Therefore, the MOSFET acts as a voltage amplifier. A circuit consisting of a chain of such transistors can result in a very small initial signal from a microphone being amplified enough to drive powerful speakers at an outdoor concert.

The Integrated Circuit

Invented independently by Jack Kilby (b. 1923, Nobel Prize in Physics, 2000) at Texas Instruments in late 1958 and by Robert Noyce (1927–1990) at Fairchild Camera and Instrument in early 1959, the integrated circuit has been justly called "the most remarkable technology ever to hit mankind." Kilby's first device is shown in Figure 43.31. Integrated circuits have indeed started a "second industrial revolution" and are found at the heart of computers, watches, cameras, automobiles, aircraft, robots, space vehicles, and all sorts of communication and switching networks.

In simplest terms, an **integrated circuit** is a collection of interconnected transistors, diodes, resistors, and capacitors fabricated on a single piece of silicon known as a *chip*. Contemporary electronic devices often contain many integrated circuits (Fig. 43.32). State-of-the-art chips easily contain several million components within a 1-cm^2 area, and the number of components per square inch has increased steadily since the integrated circuit was invented. Figure 43.33 illustrates the dramatic advances made in chip technology since Intel introduced the first microprocessor in 1971.

Integrated circuits were invented partly to solve the interconnection problem spawned by the transistor. In the era of vacuum tubes, power and size considerations of individual components set modest limits on the number of components that could be interconnected in a given circuit. With the advent of the tiny, low-power, highly reliable transistor, design limits on the number of components disappeared and were replaced by the problem of wiring together hundreds of thousands of components. The magnitude of this problem can be appreciated when we consider that second-generation computers (consisting of discrete transistors rather than integrated circuits) contained several hundred thousand components requiring more than a million joints that had to be hand-soldered and tested.

In addition to solving the interconnection problem, integrated circuits possess the advantages of miniaturization and fast response, two attributes critical for high-speed computers. Because the response time of a circuit depends on the time interval required for electrical signals traveling at the speed of light to pass from one component to another, miniaturization and close packing of components result in fast response times.

Figure 43.31 Jack Kilby's first integrated circuit, tested on September 12, 1958.

Figure 43.32 Integrated circuits are prevalent in many electronic devices. All the flat circuit elements with black-topped surfaces in this photograph are integrated circuits.

43.8   Superconductivity

We learned in Section 27.5 that there is a class of metals and compounds known as **superconductors** whose electrical resistance decreases to virtually zero below a certain temperature T_c called the *critical temperature* (Table 27.3). Let's now look at these amazing materials in greater detail, using what we know about the properties of solids to help us understand the behavior of superconductors.

Let's start by examining the Meissner effect, introduced in Section 30.6 as the exclusion of magnetic flux from the interior of superconductors. The Meissner

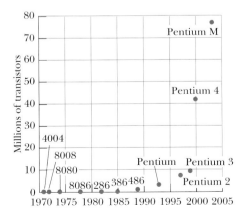

Figure 43.33 Dramatic advances in chip technology related to computer microchips manufactured by Intel, shown by a plot of the number of transistors on a single computer chip versus year of manufacture.

$T > T_c$ $T < T_c$

(a) (b)

Figure 43.34 A superconductor in the form of a long cylinder in the presence of an external magnetic field. (a) At temperatures above T_c, the field lines penetrate the cylinder because it is in its normal state. (b) When the cylinder is cooled to $T < T_c$ and becomes superconducting, magnetic flux is excluded from its interior by the induction of surface currents.

effect is illustrated in Figure 43.34 for a superconducting material in the shape of a long cylinder. Notice that the magnetic field penetrates the cylinder when its temperature is greater than T_c (Fig. 43.34a). As the temperature is lowered to below T_c, however, the field lines are spontaneously expelled from the interior of the superconductor (Fig. 43.34b). Therefore, a superconductor is more than a perfect conductor (resistivity $\rho = 0$); it is also a perfect diamagnet ($\vec{\mathbf{B}} = 0$). The property that $\vec{\mathbf{B}} = 0$ in the interior of a superconductor is as fundamental as the property of zero resistance. If the magnitude of the applied magnetic field exceeds a critical value B_c, defined as the value of B that destroys a material's superconducting properties, the field again penetrates the sample.

Because a superconductor is a perfect diamagnet, it repels a permanent magnet. In fact, one can perform a demonstration of the Meissner effect by floating a small permanent magnet above a superconductor and achieving magnetic levitation as seen in Figure 30.27 in Chapter 30.

Recall from our study of electricity that a good conductor expels static electric fields by moving charges to its surface. In effect, the surface charges produce an electric field that exactly cancels the externally applied field inside the conductor. In a similar manner, a superconductor expels magnetic fields by forming surface currents. To see why that happens, consider again the superconductor shown in Figure 43.34. Let's assume the sample is initially at a temperature $T > T_c$ as illustrated in Figure 43.34a so that the magnetic field penetrates the cylinder. As the cylinder is cooled to a temperature $T < T_c$, the field is expelled as shown in Figure 43.34b. Surface currents induced on the superconductor's surface produce a magnetic field that exactly cancels the externally applied field inside the superconductor. As you would expect, the surface currents disappear when the external magnetic field is removed.

A successful theory for superconductivity in metals was published in 1957 by J. Bardeen, L. N. Cooper, and J. R. Schrieffer; it is generally called BCS theory, based on the first letters of their last names. This theory led to a Nobel Prize in Physics for the three scientists in 1972. In this theory, two electrons can interact via distortions in the array of lattice ions so that there is a net attractive force between the electrons.[5] As a result, the two electrons are bound into an entity called a *Cooper pair*, which behaves like a particle with integral spin. Particles with integral spin are called *bosons*. (As noted in Pitfall Prevention 42.6, *fermions* make up another class of particles, those with half-integral spin.) An important feature of bosons is that they do not obey the Pauli exclusion principle. Consequently, at very low temperatures, it is possible for all bosons in a collection of such particles to be in the lowest quantum state. The entire collection of Cooper pairs in the metal is described by a single wave function. Above the energy level associated with this wave function is an energy gap equal to the binding energy of a Cooper pair. Under the action of an applied electric field, the Cooper pairs experience an electric force and move through the metal. A random scattering event of a Cooper pair from a lattice ion would represent resistance to the electric current. Such a collision would change the energy of the Cooper pair because some energy would be transferred to the lattice ion. There are no available energy levels below that of the Cooper pair (it is already in the lowest state), however, and none available above because of the energy gap. As a result, collisions do not occur and there is no resistance to the movement of Cooper pairs.

An important development in physics that elicited much excitement in the scientific community was the discovery of high-temperature copper oxide–based superconductors. The excitement began with a 1986 publication by J. Georg Bednorz (b. 1950) and K. Alex Müller (b. 1927), scientists at the IBM Zurich Research

---

[5] A highly simplified explanation of this attraction between electrons is as follows. The attractive Coulomb force between one electron and the surrounding positively charged lattice ions causes the ions to move inward slightly toward the electron. As a result, there is a higher concentration of positive charge in this region than elsewhere in the lattice. A second electron is attracted to the higher concentration of positive charge.

Laboratory in Switzerland. In their seminal paper,[6] Bednorz and Müller reported strong evidence for superconductivity at 30 K in an oxide of barium, lanthanum, and copper. They were awarded the Nobel Prize in Physics in 1987 for their remarkable discovery. Shortly thereafter, a new family of compounds was open for investigation and research activity in the field of superconductivity proceeded vigorously. In early 1987, groups at the University of Alabama at Huntsville and the University of Houston announced superconductivity at approximately 92 K in an oxide of yttrium, barium, and copper ($YBa_2Cu_3O_7$). Later that year, teams of scientists from Japan and the United States reported superconductivity at 105 K in an oxide of bismuth, strontium, calcium, and copper. More recently, scientists have reported superconductivity at temperatures as high as 150 K in an oxide containing mercury. Today, one cannot rule out the possibility of room-temperature superconductivity, and the mechanisms responsible for the behavior of high-temperature superconductors are still under investigation. The search for novel superconducting materials continues both for scientific reasons and because practical applications become more probable and widespread as the critical temperature is raised.

Although BCS theory was very successful in explaining superconductivity in metals, there is currently no widely accepted theory for high-temperature superconductivity. It remains an area of active research.

Summary

ThomsonNOW™ Sign in at **www.thomsonedu.com** and go to ThomsonNOW to take a practice test for this chapter.

CONCEPTS AND PRINCIPLES

Two or more atoms combine to form molecules because of a net attractive force between the atoms. The mechanisms responsible for molecular bonding can be classified as follows:

- **Ionic bonds** form primarily because of the Coulomb attraction between oppositely charged ions. Sodium chloride (NaCl) is one example.
- **Covalent bonds** form when the constituent atoms of a molecule share electrons. For example, the two electrons of the H_2 molecule are equally shared between the two nuclei.
- **Van der Waals bonds** are weak electrostatic bonds between molecules or between atoms that do not form ionic or covalent bonds. These bonds are responsible for the condensation of noble gas atoms and nonpolar molecules into the liquid phase.
- **Hydrogen bonds** form between the center of positive charge in a polar molecule that includes one or more hydrogen atoms and the center of negative charge in another polar molecule.

The allowed values of the rotational energy of a diatomic molecule are

$$E_{rot} = E_J = \frac{\hbar^2}{2I} J(J+1) \quad J = 0, 1, 2, \dots \quad \textbf{(43.6)}$$

where I is the moment of inertia of the molecule and J is an integer called the **rotational quantum number.** The selection rule for transitions between rotational states is $\Delta J = \pm 1$.

The allowed values of the vibrational energy of a diatomic molecule are

$$E_{vib} = (v + \tfrac{1}{2}) \frac{h}{2\pi} \sqrt{\frac{k}{\mu}} \quad v = 0, 1, 2, \dots \quad \textbf{(43.10)}$$

where v is the **vibrational quantum number,** k is the force constant of the "effective spring" bonding the molecule, and μ is the **reduced mass** of the molecule. The selection rule for allowed vibrational transitions is $\Delta v = \pm 1$, and the energy difference between any two adjacent levels is the same, regardless of which two levels are involved.

(continued)

[6] J. G. Bednorz and K. A. Müller, *Z. Phys. B* **64:**189, 1986.

Bonding mechanisms in solids can be classified in a manner similar to the schemes for molecules. For example, the Na^+ and Cl^- ions in NaCl form **ionic bonds,** whereas the carbon atoms in diamond form **covalent bonds.** The **metallic bond** is characterized by a net attractive force between positive ion cores and the mobile free electrons of a metal.

In the **free-electron theory of metals,** the free electrons fill the quantized levels in accordance with the Pauli exclusion principle. The number of states per unit volume available to the conduction electrons having energies between E and $E + dE$ is

$$N(E)\,dE = \left(\frac{8\sqrt{2}\pi m_e^{3/2}}{h^3}E^{1/2}\right)\left(\frac{1}{e^{(E-E_F)/k_B T} + 1}\right)dE \qquad \textbf{(43.22)}$$

where E_F is the **Fermi energy.** At $T = 0$ K, all levels below E_F are filled, all levels above E_F are empty and

$$E_F(0) = \frac{h^2}{2m_e}\left(\frac{3n_e}{8\pi}\right)^{2/3} \qquad \textbf{(43.25)}$$

where n_e is the total number of conduction electrons per unit volume. Only those electrons having energies near E_F can contribute to the electrical conductivity of the metal.

In a crystalline solid, the energy levels of the system form a set of **bands.** Electrons occupy the lowest energy states, with no more than one electron per state. Energy gaps are present between the bands of allowed states.

A **semiconductor** is a material having an energy gap of approximately 1 eV and a valence band that is filled at $T = 0$ K. Because of the small energy gap, a significant number of electrons can be thermally excited from the valence band into the conduction band. The band structures and electrical properties of a Group IV semiconductor can be modified by the addition of either donor atoms containing five outer-shell electrons or acceptor atoms containing three outer-shell electrons. A semiconductor **doped** with donor impurity atoms is called an **n-type semiconductor,** and one doped with acceptor impurity atoms is called a **p-type semiconductor.**

Questions

☐ denotes answer available in *Student Solutions Manual/Study Guide;* **O** denotes objective question

Note: Questions 6 and 8 in Chapter 27 can be assigned with this chapter.

1. Discuss models for the different types of bonds that form stable molecules.

2. Discuss the three major forms of excitation of a molecule (other than translational motion) and the relative energies associated with these three forms.

3. **O** Consider a typical material composed of covalently bonded diatomic molecules. Rank the following energies from the largest in magnitude to the smallest in magnitude. (a) the latent heat of fusion per molecule   (b) the molecular binding energy   (c) the energy of the first excited state of molecular rotation   (d) the energy of the first excited state of molecular vibration

4. How can the analysis of the rotational spectrum of a molecule lead to an estimate of the size of that molecule?

5. **O** An infrared absorption spectrum like that in Figure 43.9 is taken when the sample of HCl is at a much higher temperature. Compared with Figure 43.9, in this new spectrum are the highest absorption peaks (a) at the same frequencies, (b) farther from the gap, or (c) closer to the gap?

6. **O** What kind of bonding likely holds the atoms together in the following solids **(i), (ii),** and **(iii)**? Choose your answers from these possibilities: (a) ionic bonding, (b) covalent bonding, and (c) metallic bonding. **(i)** The crystal is transparent, brittle, and soluble in water. It is a poor conductor of electricity. **(ii)** The crystal is opaque, brittle, very hard, and a good electric insulator. **(iii)** The solid is opaque, shiny, flexible, and a good electric conductor.

7. **O** The Fermi energy for silver is 5.48 eV. In a piece of solid silver, free-electron energy levels are measured near 2 eV and near 6 eV. **(i)** Near which of these energies are the energy levels closer together? (a) 2 eV   (b) 6 eV   (c) The spacing is the same. **(ii)** Near which of these energies are more electrons occupying energy levels? (a) 2 eV   (b) 6 eV   (c) The number of electrons is the same.

8. Discuss the differences in the band structures of metals, insulators, and semiconductors. How does the band-structure model enable you to understand the electrical properties of these materials better?

9. **O** As discussed in Chapter 27, the conductivity of metals decreases with increasing temperature due to electron collisions with vibrating atoms. In contrast, the conductivity of semiconductors increases with increasing tempera-

ture. What property of a semiconductor is responsible for this behavior? (a) Atomic vibrations decrease as temperature increases. (b) The number of conduction electrons and the number of holes increase steeply with increasing temperature. (c) The energy gap decreases with increasing temperature. (d) Electrons do not collide with atoms in a semiconductor.

10. When a photon is absorbed by a semiconductor, an electron–hole pair is created. Give a physical explanation of this statement using the energy-band model as the basis for your description.

11. Pentavalent atoms such as arsenic are donor atoms in a semiconductor such as silicon, whereas trivalent atoms such as indium are acceptors. Inspect the periodic table in Appendix C and determine what other elements might make good donors or acceptors.

12. What essential assumptions are made in the free-electron theory of metals? How does the energy-band model differ from the free-electron theory in describing the properties of metals?

13. How do the vibrational and rotational levels of heavy hydrogen (D_2) molecules compare with those of H_2 molecules?

14. The energies of photons of visible light range between 1.8 and 3.1 eV. Does that explain why silicon, with an energy gap of 1.1 eV (see Table 43.3), appears opaque, whereas diamond, with an energy gap of 5.5 eV, appears transparent?

15. **O** (i) Should you expect an *n*-type doped semiconductor to have (a) higher, (b) lower, or (c) the same conductivity as an intrinsic (pure) semiconductor? **(ii)** Should you expect a *p*-type doped semiconductor to have (a) higher, (b) lower, or (c) the same conductivity as an intrinsic (pure) semiconductor?

16. **O** Is each one of the following statements true or false for a superconductor below its critical temperature? (a) It can carry infinite current. (b) It must carry some nonzero current. (c) Its interior electric field must be zero. (d) Its internal magnetic field must be zero. (e) No internal energy appears when it carries electric current.

Problems

WebAssign The Problems from this chapter may be assigned online in WebAssign.

ThomsonNOW Sign in at **www.thomsonedu.com** and go to ThomsonNOW to assess your understanding of this chapter's topics with additional quizzing and conceptual questions.

1, 2, 3 denotes straightforward, intermediate, challenging; ☐ denotes full solution available in *Student Solutions Manual/Study Guide*; ▲ denotes coached solution with hints available at **www.thomsonedu.com**; denotes developing symbolic reasoning; ● denotes asking for qualitative reasoning; ▪ denotes computer useful in solving problem

Section 43.1 Molecular Bonds

1. ▲ **Review problem.** A K^+ ion and a Cl^- ion are separated by a distance of 5.00×10^{-10} m. Assuming the two ions act like charged particles, determine (a) the force each ion exerts on the other and (b) the potential energy of the two-ion system in electron volts.

2. Potassium chloride is an ionically bonded molecule, sold as a salt substitute for use in a low-sodium diet. The electron affinity of chlorine is 3.6 eV. An energy input of 0.7 eV is required to form separate K^+ and Cl^- ions from separate K and Cl atoms. What is the ionization energy of K?

3. One description of the potential energy of a diatomic molecule is given by the Lennard–Jones potential,

$$U = \frac{A}{r^{12}} - \frac{B}{r^6}$$

where A and B are constants. Find, in terms of A and B, (a) the value r_0 at which the energy is a minimum and (b) the energy E required to break up a diatomic molecule. (c) Evaluate r_0 in meters and E in electron volts for the H_2 molecule, taking $A = 0.124 \times 10^{-120}$ eV · m^{12} and $B = 1.488 \times 10^{-60}$ eV · m^6. *Note:* Although this potential is widely used for modeling, it is known to have serious defects. For example, its behavior at both small and large values of r is significantly in error.

4. In the potassium iodide molecule, assume the K and I atoms bond ionically by the transfer of one electron from K to I. (a) The ionization energy of K is 4.34 eV, and the

electron affinity of I is 3.06 eV. What energy is needed to transfer an electron from K to I, to form K^+ and I^- ions from neutral atoms? This quantity is sometimes called the activation energy E_a. (b) A model potential energy function for the KI molecule is the Lennard–Jones potential:

$$U(r) = 4\epsilon \left[\left(\frac{\sigma}{r} \right)^{12} - \left(\frac{\sigma}{r} \right)^6 \right] + E_a$$

where r is the internuclear separation distance and σ and ϵ are adjustable parameters. The E_a term is added to ensure the correct asymptotic behavior at large r. At the equilibrium separation distance, $r = r_0 = 0.305$ nm, $U(r)$ is a minimum, and $dU/dr = 0$. Now $U(r_0)$ is the negative of the dissociation energy: $U(r_0) = -3.37$ eV. Evaluate σ and ϵ. (c) Calculate the force needed to break up a KI molecule. (d) Calculate the force constant for small oscillations about $r = r_0$. *Suggestion:* Set $r = r_0 + s$, where $s/r_0 \ll 1$, and expand $U(r)$ in powers of s/r_0 up to second-order terms.

5. A van der Waals dispersion force between helium atoms produces a very shallow potential well, with a depth on the order of 1 meV. At approximately what temperature would you expect helium to condense?

Section 43.2 Energy States and Spectra of Molecules

6. Assume the distance between the protons in the H_2 molecule is 0.750×10^{-10} m. (a) Find the energy of the first rotational state, with $J = 1$. (b) Find the wavelength of radiation emitted in the transition from $J = 1$ to $J = 0$.

7. The cesium iodide (CsI) molecule has an atomic separation of 0.127 nm. (a) Determine the energy of the lowest excited rotational state and the frequency of the photon absorbed in the $J = 0$ to $J = 1$ transition. (b) **What If?** What would be the fractional change in this frequency if the estimate of the atomic separation is off by 10%?

8. ● The photon frequency that would be absorbed by the NO molecule in a transition from vibration state $v = 0$ to $v = 1$, with no change in rotation state, is 56.3 THz. The bond between the atoms has an effective spring constant of 1 530 N/m. Use this information to calculate the reduced mass of the NO molecule. Then compute a value for μ using Equation 43.4. State how the two results compare.

9. ▲ An HCl molecule is excited to its first rotational energy level, corresponding to $J = 1$. If the distance between its nuclei is 0.127 5 nm, what is the angular speed of the molecule about its center of mass?

10. The CO molecule makes a transition from the $J = 1$ to the $J = 2$ rotational state when it absorbs a photon of frequency 2.30×10^{11} Hz. Find the moment of inertia of this molecule from these data.

11. A diatomic molecule consists of two atoms having masses m_1 and m_2 and separated by a distance r. Show that the moment of inertia about an axis through the center of mass of the molecule is given by Equation 43.3, $I = \mu r^2$.

12. (a) Calculate the moment of inertia of an NaCl molecule about its center of mass. The atoms are separated by a distance $r = 0.28$ nm. (b) Calculate the wavelength of radiation emitted when an NaCl molecule undergoes a transition from the $J = 2$ state to the $J = 1$ state.

13. The rotational spectrum of the HCl molecule contains lines with wavelengths of 0.060 4, 0.069 0, 0.080 4, 0.096 4, and 0.120 4 mm. What is the moment of inertia of the molecule?

14. ● The effective spring constant describing the potential energy of the HI molecule is 320 N/m. That for the HF molecule is 970 N/m. Calculate the minimum amplitude of vibration for (a) the HI molecule and (b) the HF molecule. (c) Which molecule has the weaker bond? Explain how you can tell.

15. ▲ Taking the effective force constant of a vibrating HCl molecule as $k = 480$ N/m, find the energy difference between the ground state and the first excited vibrational level.

16. The nuclei of the O_2 molecule are separated by 1.20×10^{-10} m. The mass of each oxygen atom in the molecule is 2.66×10^{-26} kg. (a) Determine the rotational energies of an oxygen molecule in electron volts for the levels corresponding to $J = 0$, 1, and 2. (b) The effective force constant k between the atoms in the oxygen molecule is 1 177 N/m. Determine the vibrational energies (in electron volts) corresponding to $v = 0$, 1, and 2.

17. Figure P43.17 is a model of a benzene molecule. All atoms lie in a plane, and the carbon atoms form a regular hexagon, as do the hydrogen atoms. The carbon atoms are 0.110 nm apart center to center. Determine the allowed energies of rotation about an axis perpendicular to the plane of the paper through the center point O. Hydrogen and carbon atoms have masses of 1.67×10^{-27} kg and 1.99×10^{-26} kg, respectively.

Figure P43.17

18. (a) Calculate the longest wavelength in the rotational spectrum of HCl. Take the Cl atom to be the isotope ^{35}Cl. The equilibrium separation of the H and Cl atoms is 0.127 46 nm. The atomic mass of the H atom is 1.007 825 u, and that of the ^{35}Cl atom is 34.968 853 u. (b) **What If?** Repeat the calculation in part (a), but take the Cl atom to be the isotope ^{37}Cl, which has atomic mass 36.965 903 u. The equilibrium separation distance is the same as in part (a). (c) Naturally occurring chlorine contains approximately three parts of ^{35}Cl to one part of ^{37}Cl. Because of the two different Cl masses, each line in the microwave rotational spectrum of HCl is split into a doublet. Calculate the doublet separation for the longest wavelength. (Figure 43.9 shows the doublets in the infrared vibrational spectrum).

19. Calculate the moment of inertia of an HCl molecule from its infrared absorption spectrum shown in Figure 43.9.

20. An H_2 molecule is in its vibrational and rotational ground states. It absorbs a photon of wavelength 2.211 2 μm and jumps to the $v = 1$, $J = 1$ energy level. It then drops to the $v = 0$, $J = 2$ energy level while emitting a photon of wavelength 2.405 4 μm. Calculate (a) the moment of inertia of the H_2 molecule about an axis through its center of mass and perpendicular to the H–H bond, (b) the vibrational frequency of the H_2 molecule, and (c) the equilibrium separation distance for this molecule.

21. Photons of what frequencies can be spontaneously emitted by CO molecules in the state with $v = 1$ and $J = 0$?

22. Most of the mass of an atom is in its nucleus. Model the mass distribution in a diatomic molecule as two spheres, each of radius 2.00×10^{-15} m and mass 1.00×10^{-26} kg, located at points along the x axis in Active Figure 43.5a, and separated by 2.00×10^{-10} m. Rotation about the axis joining the nuclei in the diatomic molecule is ordinarily ignored because the first excited state would have an energy that is too high to access. To see why, calculate the ratio of the energy of the first excited state for rotation about the x axis to the energy of the first excited state for rotation about the y axis.

Section 43.3 Bonding in Solids

23. Use Equation 43.18 to calculate the ionic cohesive energy for NaCl. Take $\alpha = 1.747\,6$, $r_0 = 0.281$ nm, and $m = 8$.

24. Use a magnifying glass to look at the table salt that comes out of a salt shaker. Compare what you see with Figure 43.10a. The distance between a sodium ion and a nearest-neighbor chlorine ion is 0.261 nm. (a) Make an order-of-

magnitude estimate of the number N of atoms in a typical grain of salt. (b) **What If?** Suppose you had a number of grains of salt equal to this number N. What would be the volume of this quantity of salt?

25. Consider a one-dimensional chain of alternating positive and negative ions. Show that the potential energy associated with one of the ions and its interactions with the rest of this hypothetical crystal is

$$U(r) = -k_e\alpha\frac{e^2}{r}$$

where the Madelung constant is $\alpha = 2\ln 2$ and r is the interionic spacing. *Suggestion:* Use the series expansion for $\ln(1 + x)$.

26. The distance between the K^+ and Cl^- ions in a KCl crystal is 0.314 nm. Calculate the distances from one K^+ ion to its nearest-neighbor K^+ ions, to its second-nearest-neighbor K^+ ions, and to its third-nearest-neighbor K^+ ions.

Section 43.4 Free-Electron Theory of Metals

Section 43.5 Band Theory of Solids

27. The Fermi energy for silver is 5.48 eV. Silver has a density of 10.6×10^3 kg/m^3 and an atomic mass of 108 u. Use this information to show that silver has one free electron per atom.

28. ● (a) State what the Fermi energy depends on according to the free-electron theory of metals and how the Fermi energy depends on that quantity. (b) Show that Equation 43.25 can be expressed as $E_F = (3.65 \times 10^{-19})n_e^{2/3}$ eV, where E_F is in electron volts when n_e is in electrons per cubic meter. (c) According to Table 43.2, by what factor does the free-electron concentration in copper exceed that in potassium? Which of these metals has the larger Fermi energy? By what factor is the Fermi energy larger? Explain whether this behavior is predicted by Equation 43.25.

29. ● (a) Find the typical speed of a conduction electron in copper, taking its kinetic energy as equal to the Fermi energy, 7.05 eV. (b) State how this speed compares with a drift speed of 0.1 mm/s.

30. For copper at 300 K, calculate the probability that a state with an energy equal to 99.0% of the Fermi energy is occupied.

31. The Fermi energy of copper at 300 K is 7.05 eV. (a) What is the average energy of a conduction electron in copper at 300 K? (b) At what temperature would the average energy of a molecule in an ideal gas be equal to the energy calculated in part (a)?

32. Sodium is a monovalent metal having a density of 0.971 g/cm^3 and a molar mass of 23.0 g/mol. Use this information to calculate (a) the density of charge carriers and (b) the Fermi energy of sodium.

33. ▲ Calculate the energy of a conduction electron in silver at 800 K, assuming the probability of finding an electron in that state is 0.950. The Fermi energy of silver is 5.48 eV at this temperature.

34. When solid silver starts to melt, what is the approximate fraction of the conduction electrons that are thermally excited above the Fermi level?

35. Show that the average kinetic energy of a conduction electron in a metal at 0 K is $E_{avg} = \frac{3}{5}E_F$. *Suggestion:* In general, the average kinetic energy is

$$E_{avg} = \frac{1}{n_e}\int EN(E)\,dE$$

where n_e is the density of particles, $N(E)\,dE$ is given by Equation 43.22, and the integral is over all possible values of the energy.

36. Consider a cube of gold 1.00 mm on an edge. Calculate the approximate number of conduction electrons in this cube whose energies lie in the range 4.000 to 4.025 eV.

37. ● (a) Consider a system of electrons confined to a three-dimensional box. Calculate the ratio of the number of allowed energy levels at 8.50 eV to the number at 7.00 eV. (b) **What If?** Copper has a Fermi energy of 7.0 eV at 300 K. Calculate the ratio of the number of occupied levels in copper at an energy of 8.50 eV to the number at the Fermi energy. State how your answer compares with that obtained in part (a).

38. **Review problem.** An electron moves in a three-dimensional box of edge length L and volume L^3. The wave function of the particle is $\psi = A \sin(k_x x) \sin(k_y y) \sin(k_z z)$. Show that its energy is given by Equation 43.20,

$$E = \frac{\hbar^2\pi^2}{2m_e L^2}(n_x{}^2 + n_y{}^2 + n_z{}^2)$$

where the quantum numbers (n_x, n_y, n_z) are integers ≥ 1. *Suggestions:* The Schrödinger equation in three dimensions may be written

$$\frac{\hbar^2}{2m}\left(\frac{\partial^2\psi}{\partial x^2} + \frac{\partial^2\psi}{\partial y^2} + \frac{\partial^2\psi}{\partial z^2}\right) = (U - E)\psi$$

To confine the electron inside the box, take $U = 0$ inside and $U = \infty$ outside.

Section 43.6 Electrical Conduction in Metals, Insulators, and Semiconductors

39. The energy gap for silicon at 300 K is 1.14 eV. (a) Find the lowest-frequency photon that can promote an electron from the valence band to the conduction band. (b) What is the wavelength of this photon?

40. Light from a hydrogen discharge tube is incident on a CdS crystal. Which spectral lines from the Balmer series are absorbed and which are transmitted?

41. ▲ ● Most solar radiation has a wavelength of 1 μm or less. What energy gap should the material in a solar cell have if it is to absorb this radiation? Is silicon appropriate (see Table 43.3)? Explain your answer.

42. A light-emitting diode (LED) made of the semiconductor GaAsP emits red light ($\lambda = 650$ nm). Determine the energy-band gap E_g in the semiconductor.

43. You are asked to build a scientific instrument that is thermally isolated from its surroundings. You wish to use an external laser to raise the temperature of a target inside the instrument. (It might be a calorimeter, but these design criteria could apply to other devices as well.) Because you know that diamond is transparent and a

good thermal insulator, you decide to use a diamond window in the apparatus. Diamond has an energy gap of 5.5 eV between its valence and conduction bands. What is the shortest laser wavelength you can use to warm the sample inside?

44. **Review problem.** Silicon is a semiconductor widely used in computer chips and other electronic devices. Its most important properties result from doping it with impurities so as to control its electrical conductivity. Phosphorus, which is adjacent to silicon in the periodic table, has five outer valence electrons, compared with four for silicon. When a phosphorus atom is substituted for a silicon atom in a crystal, four of the phosphorus valence electrons form bonds with neighboring atoms and the remaining electron is much more loosely bound. You can model the electron as free to move through the crystal lattice. The phosphorus nucleus has one more positive charge than does the silicon nucleus, however, so the extra electron provided by the phosphorus atom is attracted to this single nuclear charge $+e$. The energy levels of the extra electron are similar to those of the electron in the Bohr hydrogen atom with two important exceptions. First, the Coulomb attraction between the electron and the positive charge on the phosphorus nucleus is reduced by a factor of $1/\kappa$ from what it would be in free space (see Eq. 26.21), so the orbit radii are greatly increased. Here κ is the dielectric constant of the crystal, with a value of 11.7 in silicon. Second, the influence of the periodic electric potential of the lattice causes the electron to move as if it had an effective mass m^*, which is quite different from the mass m_e of a free electron. You can use the Bohr model of hydrogen to obtain fairly accurate values for the allowed energy levels of the extra electron. These energy levels, called donor states, play an important role in semiconductor devices. Assume $m^* = 0.220m_e$. Calculate the energy and the radius for an extra electron in the first Bohr orbit around a donor atom in silicon.

Section 43.7 Semiconductor Devices

45. (a) For what value of the bias voltage ΔV in Equation 43.27 does $I = 9.00I_0$? (b) **What If?** What if $I = -0.900I_0$? Assume $T = 300$ K.

46. The diode shown in Figure 43.28 is connected in series with a battery and a 150-Ω resistor. What battery emf is required for a current of 25.0 mA?

47. You put a diode in a microelectronic circuit to protect the system in case an untrained person installs the battery backward. In the correct forward-bias situation, the current is 200 mA with a potential difference of 100 mV across the diode at room temperature (300 K). If the battery were reversed, what would the magnitude of the current through the diode be?

48. ● A diode, a resistor, and a battery are connected in a series circuit. The diode is at a temperature for which $k_BT = 25.0$ meV, and the saturation value I_0 of the current is 1.00 μA. The resistance of the resistor is 745 Ω, and the battery maintains a constant potential difference between its terminals of 2.42 V. (a) Find graphically the current in the loop. Proceed as follows. On the same axes, draw graphs of the diode current I_D and the current in the wire I_W versus the voltage across the diode ΔV. Choose values of ΔV ranging from 0 to 0.250 V in steps of 0.025 V. Determine the value of ΔV at the intersection of the two graph lines and calculate the corresponding currents I_D and I_W. Explain whether they agree. (b) Find the ohmic resistance of the diode, defined as the ratio $\Delta V/I_D$. (c) Find the dynamic resistance of the diode, which is defined as the derivative $d(\Delta V)/dI_D$.

Section 43.8 Superconductivity

Note: Problem 30 in Chapter 30 and Problems 64 through 67 in Chapter 32 can also be assigned with this section.

49. Determine the current generated in a superconducting ring of niobium metal 2.00 cm in diameter when a 0.020 0-T magnetic field directed perpendicular to the ring is suddenly decreased to zero. The inductance of the ring is 3.10×10^{-8} H.

50. ● *A convincing demonstration of zero resistance.* A direct and relatively simple demonstration of zero DC resistance can be carried out using the four-point probe method. The probe shown in Figure P43.50 consists of a disk of $YBa_2Cu_3O_7$ (a high-T_c superconductor) to which four wires are attached by indium solder or some other suitable contact material. Current is maintained through the sample by applying a DC voltage between points a and b, and it is measured with a DC ammeter. The current can be varied with the variable resistance R. The potential difference ΔV_{cd} between c and d is measured with a digital voltmeter. When the probe is immersed in liquid nitrogen, the sample quickly cools to 77 K, below the critical temperature of the material, 92 K. The current remains approximately constant, but ΔV_{cd} *drops abruptly to zero.* (a) Explain this observation on the basis of what you know about superconductors. (b) The data in the accompanying table represent actual values of ΔV_{cd} for different values of I taken on the sample at room temperature in the senior author's laboratory. A 6-V battery in series with a variable resistor R supplied the current. The values of R ranged from 10 Ω to 100 Ω. Make an I–ΔV plot of the data and determine whether the sample behaves in a linear manner. From the data, obtain a value for the DC resistance of the sample at room temperature. (c) At room temperature, it was found that $\Delta V_{cd} = 2.234$ mV for $I = 100.3$ mA, but after the sample was cooled to 77 K, $\Delta V_{cd} = 0$ and $I = 98.1$ mA. What do you think might have caused the slight decrease in current?

Current Versus Potential Difference ΔV_{cd} Measured in a Bulk Ceramic Sample of $YBa_2Cu_3O_{7-\delta}$ at Room Temperature	
I (mA)	ΔV_{cd} (mV)
57.8	1.356
61.5	1.441
68.3	1.602
76.8	1.802
87.5	2.053
102.2	2.398
123.7	2.904
155	3.61

2 = intermediate;   3 = challenging;   □ = SSM/SG;   ▲ = ThomsonNOW;   = symbolic reasoning;   ● = qualitative reasoning

Figure P43.50

51. A thin rod of superconducting material 2.50 cm long is placed into a 0.540-T magnetic field with its cylindrical axis along the magnetic field lines. (a) Sketch the directions of the applied field and the induced surface current. (b) Find the magnitude of the surface current on the curved surface of the rod.

Additional Problems

52. The effective spring constant associated with bonding in the N_2 molecule is 2 297 N/m. The nitrogen atoms each have a mass of 2.32×10^{-26} kg, and their nuclei are 0.120 nm apart. Assume the molecule is rigid and in the ground vibrational state. Calculate the J value of the rotational state that has the same energy as the first excited vibrational state.

53. As you will learn in Chapter 44, carbon-14 (^{14}C) is an isotope of carbon. It has the same chemical properties and electronic structure as the much more abundant isotope carbon-12 (^{12}C), but it has different nuclear properties. Its mass is 14 u, greater than that of carbon-12 because of the two extra neutrons in the carbon-14 nucleus. Assume the CO molecular potential energy is the same for both isotopes of carbon and the examples in Section 43.2 contain accurate data and results for carbon monoxide with carbon-12 atoms. (a) What is the vibrational frequency of ^{14}CO? (b) What is the moment of inertia of ^{14}CO? (c) What wavelengths of light can be absorbed by ^{14}CO in the ($v = 0$, $J = 10$) state that cause it to end up in the $v = 1$ level?

54. Under pressure, liquid helium can solidify as each atom bonds with four others, and each bond has an average energy of 1.74×10^{-23} J. Find the latent heat of fusion for helium in joules per gram. (The molar mass of He is 4.00 g/mol.)

55. The hydrogen molecule comes apart (dissociates) when it is excited internally by 4.5 eV. Assuming this molecule behaves like a harmonic oscillator having classical angular frequency $\omega = 8.28 \times 10^{14}$ rad/s, find the highest vibrational quantum number for a state below the 4.5-eV dissociation energy.

56. The dissociation energy of ground-state molecular hydrogen is 4.48 eV, but it only takes 3.96 eV to dissociate it when it starts in the first excited vibrational state with $J = 0$. Using this information, determine the depth of the H_2 molecular potential-energy function.

57. Show that the ionic cohesive energy of an ionically bonded solid is given by Equation 43.18. *Suggestion:* Start with Equation 43.17 and note that $dU/dr = 0$ at $r = r_0$.

58. (a) Show that the force exerted on an ion in an ionic solid can by written as

$$F = -k_e\alpha\frac{e^2}{r^2}\left[1 - \left(\frac{r_0}{r}\right)^{m-1}\right]$$

where α is the Madelung constant and r_0 is the equilibrium separation. *Suggestion:* Start with Equation 43.17 and note that $F = -dU/dr = 0$ at $r = r_0$. (b) Imagine that an ion in the solid is displaced a small distance x from r_0. Show that the ion experiences a restoring force $F = -Kx$, where

$$K = \frac{k_e\alpha e^2}{r_0^3}(m - 1)$$

(c) Use the result of part (b) to find the frequency of vibration of a Na^+ ion in NaCl. Take $m = 8$ and use the value $\alpha = 1.747\ 6$.

59. A particle moves in one-dimensional motion through a field for which the potential energy of the particle–field system is

$$U(x) = \frac{A}{x^3} - \frac{B}{x}$$

where $A = 0.150$ eV $\cdot$ nm^3 and $B = 3.68$ eV $\cdot$ nm. The general shape of this function is shown in Figure 43.11, where x replaces r. (a) Find the static equilibrium position x_0 of the particle. (b) Determine the depth U_0 of this potential well. (c) In moving along the x axis, what maximum force toward the negative x direction does the particle experience?

60. A particle of mass m moves in one-dimensional motion through a field for which the potential energy of the particle–field system is

$$U(x) = \frac{A}{x^3} - \frac{B}{x}$$

where A and B are constants with appropriate units. The general shape of this function is shown in Figure 43.11, where x replaces r. (a) Find the static equilibrium position x_0 of the particle in terms of m, A, and B. (b) Determine the depth U_0 of this potential well. (c) In moving along the x axis, what maximum force toward the negative x direction does the particle experience?

61. As an alternative to Equation 43.1, another useful model for the potential energy of a diatomic molecule is the Morse potential

$$U(r) = B[e^{-a(r-r_0)} - 1]^2$$

where B, a, and r_0 are parameters used to adjust the shape of the potential and its depth. (a) What is the equilibrium separation of the nuclei? (b) What is the depth of the potential well, defined as the difference in energy between the potential's minimum value and its asymptote as r approaches infinity? (c) If μ is the reduced mass of the system of two nuclei, what is the vibrational frequency of the diatomic molecule in its ground state? (Assume the potential is nearly parabolic about the well minimum). (d) What amount of energy needs to be supplied to the ground-state molecule to separate the two nuclei to infinity?

62. ■ The Fermi–Dirac distribution function can be written as

$$f(E) = \frac{1}{e^{(E-E_F)/k_B T} + 1} = \frac{1}{e^{(E/E_F - 1)T_F/T} + 1}$$

where T_F is the *Fermi temperature*, defined according to

$$k_B T_F \equiv E_F$$

Write a spreadsheet to calculate and plot $f(E)$ versus E/E_F at a fixed temperature T. Examine the curves obtained for $T = 0.1T_F$, $0.2T_F$, and $0.5T_F$.

Answers to Quick Quizzes

43.1 (a) van der Waals   (b) ionic   (c) hydrogen   (d) covalent

43.2 (c). Equation 43.7 shows that the energy spacing between adjacent states is proportional to J, the quantum number of the higher state in the transition. Because the frequency of the absorbed photon is proportional to the energy separation of the states, the frequencies are in the same ratio as the energy separations.

43.3 (a). This situation is similar to Quick Quiz 43.2 except that the vibrational states are all separated by the same energy difference.

43.4 A: semiconductor; B: conductor; C: insulator

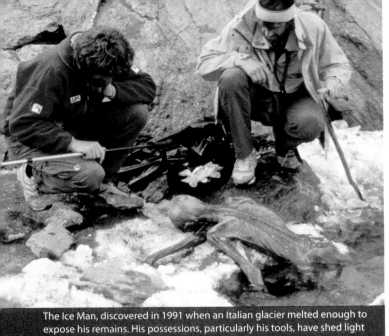

The Ice Man, discovered in 1991 when an Italian glacier melted enough to expose his remains. His possessions, particularly his tools, have shed light on the way people lived in the Bronze Age. Radioactivity was used to determine how long ago this person lived. (Paul Hanny/Gamma Liaison)

44.1 Some Properties of Nuclei

44.2 Nuclear Binding Energy

44.3 Nuclear Models

44.4 Radioactivity

44.5 The Decay Processes

44.6 Natural Radioactivity

44.7 Nuclear Reactions

44.8 Nuclear Magnetic Resonance and Magnetic Resonance Imaging

44 Nuclear Structure

The year 1896 marks the birth of nuclear physics when French physicist Antoine-Henri Becquerel (1852–1908) discovered radioactivity in uranium compounds. This discovery prompted scientists to investigate the details of radioactivity and, ultimately, the structure of the nucleus. Pioneering work by Ernest Rutherford showed that the radiation emitted from radioactive substances is of three types: alpha, beta, and gamma rays, classified according to the nature of their electric charge and their ability to penetrate matter and ionize air. Later experiments showed that alpha rays are helium nuclei, beta rays are electrons, and gamma rays are high-energy photons.

In 1911, Rutherford, Hans Geiger, and Ernest Marsden performed the alpha-particle scattering experiments described in Section 42.2. These experiments established that the nucleus of an atom can be modeled as a point mass and point charge and that most of the atomic mass is contained in the nucleus. Subsequent studies revealed the presence of a new type of force, the short-range nuclear force, which is predominant at particle separation distances less than approximately 10^{-14} m and is zero for large distances.

In this chapter, we discuss the properties and structure of the atomic nucleus. We start by describing the basic properties of nuclei, followed by a discussion of nuclear forces and binding energy, nuclear models, and the phenomenon of radioactivity. Finally, we explore nuclear reactions and the various processes by which nuclei decay.

44.1   Some Properties of Nuclei

All nuclei are composed of two types of particles: protons and neutrons. The only exception is the ordinary hydrogen nucleus, which is a single proton. We describe the atomic nucleus by the number of protons and neutrons it contains, using the following quantities:

- the **atomic number** Z, which equals the number of protons in the nucleus (sometimes called the *charge number*)
- the **neutron number** N, which equals the number of neutrons in the nucleus
- the **mass number** $A = Z + N$, which equals the number of **nucleons** (neutrons plus protons) in the nucleus

PITFALL PREVENTION 44.1
Mass Number Is Not Atomic Mass

The mass number A should not be confused with the atomic mass. Mass number is an integer specific to an isotope and has no units; it is simply a count of the number of nucleons. Atomic mass has units and is generally not an integer because it is an average of the masses of a given element's naturally occurring isotopes.

In representing nuclei, it is convenient to use the symbol $^A_Z X$ to show how many protons and neutrons are present, where X represents the chemical symbol of the element. For example, $^{56}_{26}Fe$ (iron) has mass number 56 and atomic number 26; therefore, it contains 26 protons and 30 neutrons. When no confusion is likely to arise, we omit the subscript Z because the chemical symbol can always be used to determine Z. Therefore, $^{56}_{26}Fe$ is the same as ^{56}Fe and can also be expressed as "iron-56."

The nuclei of all atoms of a particular element contain the same number of protons but often contain different numbers of neutrons. Nuclei related in this way are called **isotopes.** The isotopes of an element have the same Z value but different N and A values.

The natural abundance of isotopes can differ substantially. For example $^{11}_6C$, $^{12}_6C$, $^{13}_6C$, and $^{14}_6C$ are four isotopes of carbon. The natural abundance of the $^{12}_6C$ isotope is approximately 98.9%, whereas that of the $^{13}_6C$ isotope is only about 1.1%. Some isotopes, such as $^{11}_6C$ and $^{14}_6C$, do not occur naturally but can be produced by nuclear reactions in the laboratory or by cosmic rays.

Even the simplest element, hydrogen, has isotopes: ^1_1H, the ordinary hydrogen nucleus; ^2_1H, deuterium; and ^3_1H, tritium.

Quick Quiz 44.1   For each part of this Quick Quiz, choose from the following answers: (a) protons   (b) neutrons   (c) nucleons. **(i)** The three nuclei ^{12}C, ^{13}N, and ^{14}O have the same number of what type of particle? **(ii)** The three nuclei ^{12}N, ^{13}N, and ^{14}N have the same number of what type of particle? **(iii)** The three nuclei ^{14}C, ^{14}N, and ^{14}O have the same number of what type of particle?

Charge and Mass

The proton carries a single positive charge e, equal in magnitude to the charge $-e$ on the electron ($e = 1.6 \times 10^{-19}$ C). The neutron is electrically neutral as its name implies. Because the neutron has no charge, it was difficult to detect with early experimental apparatus and techniques. Today, neutrons are easily detected with devices such as plastic scintillators.

Nuclear masses can be measured with great precision using a mass spectrometer (see Section 29.3) and by the analysis of nuclear reactions. The proton is approximately 1 836 times as massive as the electron, and the masses of the proton and the neutron are almost equal. The **atomic mass unit** u is defined in such a way that

TABLE 44.1

Masses of Selected Particles in Various Units

Particle	kg	Mass u	MeV/c^2
Proton	$1.672\ 62 \times 10^{-27}$	1.007 276	938.28
Neutron	$1.674\ 93 \times 10^{-27}$	1.008 665	939.57
Electron	$9.109\ 39 \times 10^{-31}$	$5.485\ 79 \times 10^{-4}$	0.510 999
$^1_1$H atom	$1.673\ 53 \times 10^{-27}$	1.007 825	938.783
$^4_2$He nucleus	$6.644\ 66 \times 10^{-27}$	4.001 506	3 727.38
$^{12}_6$C atom	$1.992\ 65 \times 10^{-27}$	12.000 000	11 177.9

the mass of one atom of the isotope ^{12}C is exactly 12 u, where 1 u is equal to $1.660\ 539 \times 10^{-27}$ kg. According to this definition, the proton and neutron each have a mass of approximately 1 u and the electron has a mass that is only a small fraction of this value. The masses of these particles and others important to the phenomena discussed in this chapter are given in Table 44.1.

You might wonder how six protons and six neutrons, each having a mass larger than 1 u, can be combined with six electrons to form a carbon-12 atom having a mass of exactly 12 u. The bound system of ^{12}C has a lower rest energy (Section 39.8) than that of six separate protons and six separate neutrons. According to Equation 39.24, $E_R = mc^2$, this lower rest energy corresponds to a smaller mass for the bound system. The difference in mass accounts for the binding energy when the particles are combined to form the nucleus. We shall discuss this point in more detail in Section 44.2.

It is often convenient to express the atomic mass unit in terms of its *rest-energy equivalent*. For one atomic mass unit,

$$E_R = mc^2 = (1.660\ 539 \times 10^{-27}\ \text{kg})(2.997\ 92 \times 10^8\ \text{m/s})^2 = 931.494\ \text{MeV}$$

where we have used the conversion 1 eV = $1.602\ 177 \times 10^{-19}$ J.

Based on the rest-energy expression in Equation 39.24, nuclear physicists often express mass in terms of the unit MeV/c^2.

EXAMPLE 44.1   **The Atomic Mass Unit**

Use Avogadro's number to show that 1 u = 1.66×10^{-27} kg.

SOLUTION

Conceptualize   From the definition of the mole given in Section 19.5, we know that exactly 12 g (= 1 mol) of ^{12}C contains Avogadro's number of atoms.

Categorize   We evaluate the atomic mass unit that was introduced in this section, so we categorize this example as a substitution problem.

Find the mass m of one ^{12}C atom:
$$m = \frac{0.012\ \text{kg}}{6.02 \times 10^{23}\ \text{atoms}} = 1.99 \times 10^{-26}\ \text{kg}$$

Because one atom of ^{12}C is defined to have a mass of 12.0 u, divide by 12.0 to find the mass equivalent to 1 u:
$$1\ \text{u} = \frac{1.99 \times 10^{-26}\ \text{kg}}{12.0} = 1.66 \times 10^{-27}\ \text{kg}$$

ACTIVE FIGURE 44.1

An alpha particle on a head-on collision course with a nucleus of charge Ze. Because of the Coulomb repulsion between the charges of the same sign, the alpha particle approaches to a distance d from the nucleus, called the distance of closest approach.

Sign in at www.thomsonedu.com and go to ThomsonNOW to adjust the atomic number of the target nucleus and the kinetic energy of the alpha particle. Then observe the approach of the alpha particle toward the nucleus.

The Size and Structure of Nuclei

In Rutherford's scattering experiments, positively charged nuclei of helium atoms (alpha particles) were directed at a thin piece of metallic foil. As the alpha particles moved through the foil, they often passed near a metal nucleus. Because of the positive charge on both the incident particles and the nuclei, the particles were deflected from their straight-line paths by the Coulomb repulsive force.

Rutherford used conservation of energy for an isolated system to find an expression for the separation distance d at which an alpha particle approaching a nucleus head-on is turned around by Coulomb repulsion. In such a head-on collision, the kinetic energy of the incoming particle must be converted completely to electric potential energy of the alpha particle–nucleus system when the particle stops momentarily at the point of closest approach (the final configuration of the system) before moving back along the same path (Active Fig. 44.1). Applying the conservation of energy principle to the system gives

$$K_i + U_i = K_f + U_f$$

$$\tfrac{1}{2}mv^2 + 0 = 0 + k_e \frac{q_1 q_2}{d}$$

where m is the mass of the alpha particle and v is its initial speed. Solving for d gives

$$d = 2k_e \frac{q_1 q_2}{mv^2} = 2k_e \frac{(2e)(Ze)}{mv^2} = 4k_e \frac{Ze^2}{mv^2}$$

where Z is the atomic number of the target nucleus. From this expression, Rutherford found that the alpha particles approached nuclei to within 3.2×10^{-14} m when the foil was made of gold. Therefore, the radius of the gold nucleus must be less than this value. From the results of his scattering experiments, Rutherford concluded that the positive charge in an atom is concentrated in a small sphere, which he called the nucleus, whose radius is no greater than approximately 10^{-14} m.

Because such small lengths are common in nuclear physics, an often-used convenient length unit is the femtometer (fm), which is sometimes called the **fermi** and is defined as

$$1 \text{ fm} \equiv 10^{-15} \text{ m}$$

In the early 1920s, it was known that the nucleus of an atom contains Z protons and has a mass nearly equivalent to that of A protons, where on average $A \approx 2Z$ for lighter nuclei ($Z \leq 20$) and $A > 2Z$ for heavier nuclei. To account for the nuclear mass, Rutherford proposed that each nucleus must also contain $A - Z$ neutral particles that he called neutrons. In 1932, British physicist James Chadwick (1891–1974) discovered the neutron, and he was awarded the Nobel Prize in Physics in 1935 for this important work.

Since the time of Rutherford's scattering experiments, a multitude of other experiments have shown that most nuclei are approximately spherical and have an average radius given by

Figure 44.2 A nucleus can be modeled as a cluster of tightly packed spheres, where each sphere is a nucleon.

Nuclear radius ▶

$$r = r_0 A^{1/3} \tag{44.1}$$

where r_0 is a constant equal to 1.2×10^{-15} m and A is the mass number. Because the volume of a sphere is proportional to the cube of its radius, it follows from Equation 44.1 that the volume of a nucleus (assumed to be spherical) is directly proportional to A, the total number of nucleons. This proportionality suggests that *all nuclei have nearly the same density*. When nucleons combine to form a nucleus, they combine as though they were tightly packed spheres (Fig. 44.2). This fact has led to an analogy between the nucleus and a drop of liquid, in which the density of the drop is independent of its size. We shall discuss the liquid-drop model of the nucleus in Section 44.3.

EXAMPLE 44.2   **The Volume and Density of a Nucleus**

Consider a nucleus of mass number A.

(A) Find an approximate expression for the mass of the nucleus.

SOLUTION

Categorize Let's assume A is large enough that we can imagine the nucleus to be spherical.

Analyze The mass of the proton is approximately equal to that of the neutron. Therefore, if the mass of one of these particles is m, the mass of the nucleus is approximately Am.

(B) Find an expression for the volume of this nucleus in terms of A.

SOLUTION

Assume the nucleus is spherical and use Equation 44.1:

$$(1) \quad V_{\text{nucleus}} = \tfrac{4}{3}\pi r^3 = \tfrac{4}{3}\pi r_0^{\,3} A$$

(C) Find a numerical value for the density of this nucleus.

SOLUTION

Use Equation 1.1 and substitute Equation (1):

$$\rho = \frac{m_{\text{nucleus}}}{V_{\text{nucleus}}} = \frac{Am}{\tfrac{4}{3}\pi r_0^{\,3} A} = \frac{3m}{4\pi r_0^{\,3}}$$

Substitute numerical values:

$$\rho = \frac{3(1.67 \times 10^{-27}\ \text{kg})}{4\pi (1.2 \times 10^{-15}\ \text{m})^3} = 2.3 \times 10^{17}\ \text{kg/m}^3$$

Finalize The nuclear density is approximately 2.3×10^{14} times the density of water ($\rho_{\text{water}} = 1.0 \times 10^3\ \text{kg/m}^3$).

What If? What if the Earth could be compressed until it had this density? How large would it be?

Answer Because this density is so large, we predict that an Earth of this density would be very small.

Use Equation 1.1 and the mass of the Earth to find the volume of the compressed Earth:

$$V = \frac{M_E}{\rho} = \frac{5.98 \times 10^{24}\ \text{kg}}{2.3 \times 10^{17}\ \text{kg/m}^3} = 2.6 \times 10^7\ \text{m}^3$$

From this volume, find the radius:

$$V = \tfrac{4}{3}\pi r^3 \quad \rightarrow \quad r = \left(\frac{3V}{4\pi}\right)^{1/3} = \left[\frac{3(2.6 \times 10^7\ \text{m}^3)}{4\pi}\right]^{1/3}$$

$$= 1.8 \times 10^2\ \text{m}$$

An Earth of this radius is indeed a small Earth!

Nuclear Stability

You might expect that the very large repulsive Coulomb forces between the close-packed protons in a nucleus should cause the nucleus to fly apart. Because that does not happen, there must be a counteracting attractive force. The **nuclear force** is a very short range (about 2 fm) attractive force that acts between all nuclear particles. The protons attract each other by means of the nuclear force, and, at the same time, they repel each other through the Coulomb force. The nuclear force also acts between pairs of neutrons and between neutrons and protons. The nuclear force dominates the Coulomb repulsive force within the nucleus (at short ranges), so stable nuclei can exist.

Figure 44.3 (a) Potential energy versus separation distance for a neutron–proton system. (b) Potential energy versus separation distance for a proton–proton system. The difference in the two curves is due to the large Coulomb repulsion in the case of the proton–proton interaction. To display the difference in the curves on this scale, the height of the peak for the proton–proton curve has been exaggerated by a factor of 10.

The nuclear force is independent of charge. In other words, the forces associated with the proton–proton, proton–neutron, and neutron–neutron interactions are the same, apart from the additional repulsive Coulomb force for the proton–proton interaction.

Evidence for the limited range of nuclear forces comes from scattering experiments and from studies of nuclear binding energies. The short range of the nuclear force is shown in the neutron–proton (n–p) potential energy plot of Figure 44.3a obtained by scattering neutrons from a target containing hydrogen. The depth of the n–p potential energy well is 40 to 50 MeV, and there is a strong repulsive component that prevents the nucleons from approaching much closer than 0.4 fm.

The nuclear force does not affect electrons, enabling energetic electrons to serve as point-like probes of nuclei. The charge independence of the nuclear force also means that the main difference between the n–p and p–p interactions is that the p–p potential energy consists of a *superposition* of nuclear and Coulomb interactions as shown in Figure 44.3b. At distances less than 2 fm, both p–p and n–p potential energies are nearly identical, but for distances of 2 fm or greater, the p–p potential has a positive energy barrier with a maximum at 4 fm.

The existence of the nuclear force results in approximately 270 stable nuclei; hundreds of other nuclei have been observed, but they are unstable. A plot of neutron number N versus atomic number Z for a number of stable nuclei is given in Figure 44.4. The stable nuclei are represented by the blue dots, which lie in a narrow range called the *line of stability*. Notice that the light stable nuclei contain an equal number of protons and neutrons; that is, $N = Z$. Also notice that in heavy stable nuclei, the number of neutrons exceeds the number of protons: above $Z = 20$, the line of stability deviates upward from the line representing $N = Z$. This deviation can be understood by recognizing that as the number of protons increases, the strength of the Coulomb force increases, which tends to break the nucleus apart. As a result, more neutrons are needed to keep the nucleus stable because neutrons experience only the attractive nuclear force. Eventually, the repulsive Coulomb forces between protons cannot be compensated by the addi-

Figure 44.4 Neutron number N versus atomic number Z for stable nuclei (blue dots). These nuclei lie in a narrow band called the line of stability. The dashed line corresponds to the condition $N = Z$.

tion of more neutrons. This point occurs at $Z = 83$, meaning that elements that contain more than 83 protons do not have stable nuclei.

44.2   Nuclear Binding Energy

As mentioned in the discussion of ^{12}C in Section 44.1, the total mass of a nucleus is less than the sum of the masses of its individual nucleons. Therefore, **the rest energy of the bound system (the nucleus) is less than the combined rest energy of the separated nucleons.** This difference in energy is called the **binding energy** of the nucleus and can be interpreted as the energy that must be added to a nucleus to break it apart into its components. Therefore, to separate a nucleus into protons and neutrons, energy must be delivered to the system.

Conservation of energy and the Einstein mass–energy equivalence relationship show that the binding energy E_b of any nucleus of mass M_A is

$$E_b \, (\mathrm{MeV}) = \left[ZM(\mathrm{H}) + Nm_n - M(^A_Z\mathrm{X}) \right] \times 931.494 \, \mathrm{MeV/u} \qquad \textbf{(44.2)}$$

◀ Binding energy of a nucleus

where $M(\mathrm{H})$ is the atomic mass of the neutral hydrogen atom, $M(^A_Z\mathrm{X})$ represents the atomic mass of an atom of the isotope $^A_Z\mathrm{X}$, m_n is the mass of the neutron, and the masses are all in atomic mass units. The mass of the Z electrons included in $M(\mathrm{H})$ cancels with the mass of the Z electrons included in the term $M(^A_Z\mathrm{X})$ within a small difference associated with the atomic binding energy of the electrons. Because atomic binding energies are typically several electron volts and nuclear binding energies are several million electron volts, this difference is negligible.

A plot of binding energy per nucleon E_b/A as a function of mass number A for various stable nuclei is shown in Figure 44.5. Notice that the curve in Figure 44.5 peaks in the vicinity of $A = 60$. That is, nuclei having mass numbers either greater or less than 60 are not as strongly bound as those near the middle of the periodic table. The decrease in binding energy per nucleon for $A > 60$ implies that energy is released when a heavy nucleus splits, or *fissions*, into two lighter nuclei. Energy is released in fission because the nucleons in each product nucleus are more tightly

PITFALL PREVENTION 44.2
Binding Energy

When separate nucleons are combined to form a nucleus, the energy of the system is reduced. Therefore, the change in energy is negative. The absolute value of this change is called the binding energy. This difference in sign may be confusing. For example, an *increase* in binding energy corresponds to a *decrease* in the energy of the system.

Figure 44.5   Binding energy per nucleon versus mass number for nuclei that lie along the line of stability in Figure 44.4. Some representative nuclei appear as blue dots with labels. (Nuclei to the right of ^{208}Pb are unstable. The curve represents the binding energy for the most stable isotopes.)

bound to one another than are the nucleons in the original nucleus. The important process of fission and a second important process of *fusion*, in which energy is released as light nuclei combine, shall be considered in detail in Chapter 45.

Another important feature of Figure 44.5 is that the binding energy per nucleon is approximately constant at around 8 MeV per nucleon for all nuclei with $A > 50$. For these nuclei, the nuclear forces are said to be *saturated*, meaning that in the closely packed structure shown in Figure 44.2, a particular nucleon can form attractive bonds with only a limited number of other nucleons.

Figure 44.5 provides insight into fundamental questions about the origin of the chemical elements. In the early life of the Universe, the only elements that existed were hydrogen and helium. Clouds of cosmic gas coalesced under gravitational forces to form stars. As a star ages, it produces heavier elements from the lighter elements contained within it, beginning by fusing hydrogen atoms to form helium. This process continues as the star becomes older, generating atoms having larger and larger atomic numbers, up to the peak of the curve shown in Figure 44.5.

The nuclide $^{62}_{28}\text{Ni}$ has the largest binding energy per nucleon of 8.794 5 MeV. It takes additional energy to create elements with mass numbers larger than 62 because of their lower binding energies per nucleon. This energy comes from the supernova explosion that occurs at the end of some large stars' lives. Therefore, all the heavy atoms in your body were produced from the explosions of ancient stars. You are literally made of stardust!

44.3   Nuclear Models

The details of the nuclear force are still an area of active research. Several nuclear models have been proposed that are useful in understanding general features of nuclear experimental data and the mechanisms responsible for binding energy. Two such models, the liquid-drop model and the shell model, are discussed below.

Liquid-Drop Model

In 1936, Bohr proposed treating nucleons like molecules in a drop of liquid. In this **liquid-drop model,** the nucleons interact strongly with one another and undergo frequent collisions as they jiggle around within the nucleus. This jiggling motion is analogous to the thermally agitated motion of molecules in a drop of liquid.

Four major effects influence the binding energy of the nucleus in the liquid-drop model:

- **The volume effect.** Figure 44.5 shows that for $A > 50$, the binding energy per nucleon is approximately constant, which indicates that the nuclear force on a given nucleon is due only to a few nearest neighbors and not to all the other nucleons in the nucleus. On average, then, the binding energy associated with the nuclear force for each nucleon is the same in all nuclei: that associated with an interaction with a few neighbors. This property indicates that the total binding energy of the nucleus is proportional to A and therefore proportional to the nuclear volume. The contribution to the binding energy of the entire nucleus is $C_1 A$, where C_1 is an adjustable constant that can be determined by fitting the prediction of the model to experimental results.
- **The surface effect.** Because nucleons on the surface of the drop have fewer neighbors than those in the interior, surface nucleons reduce the binding energy by an amount proportional to their number. Because the number of surface nucleons is proportional to the surface area $4\pi r^2$ of the nucleus (modeled as a sphere) and because $r^2 \propto A^{2/3}$ (Eq. 44.1), the surface term can be expressed as $-C_2 A^{2/3}$, where C_2 is a second adjustable constant.

- **The Coulomb repulsion effect.** Each proton repels every other proton in the nucleus. The corresponding potential energy per pair of interacting protons is $k_e e^2/r$, where k_e is the Coulomb constant. The total electric potential energy is equivalent to the work required to assemble Z protons, initially infinitely far apart, into a sphere of volume V. This energy is proportional to the number of proton pairs $Z(Z-1)/2$ and inversely proportional to the nuclear radius. Consequently, the reduction in binding energy that results from the Coulomb effect is $-C_3 Z(Z-1)/A^{1/3}$, where C_3 is yet another adjustable constant.

- **The symmetry effect.** Another effect that lowers the binding energy is related to the symmetry of the nucleus in terms of values of N and Z. For small values of A, stable nuclei tend to have $N \approx Z$. Any large asymmetry between N and Z for light nuclei reduces the binding energy and makes the nucleus less stable. For larger A, the value of N for stable nuclei is naturally larger than Z. This effect can be described by a binding-energy term of the form $-C_4(N-Z)^2/A$, where C_4 is another adjustable constant.[1] For small A, any large asymmetry between values of N and Z makes this term relatively large and reduces the binding energy. For large A, this term is small and has little effect on the overall binding energy.

Adding these contributions gives the following expression for the total binding energy:

$$E_b = C_1 A - C_2 A^{2/3} - C_3 \frac{Z(Z-1)}{A^{1/3}} - C_4 \frac{(N-Z)^2}{A} \qquad (44.3)$$

◀ Semiempirical binding-energy formula

This equation, often referred to as the **semiempirical binding-energy formula,** contains four constants that are adjusted to fit the theoretical expression to experimental data. For nuclei having $A \geq 15$, the constants have the values

$$C_1 = 15.7 \text{ MeV} \qquad C_2 = 17.8 \text{ MeV}$$

$$C_3 = 0.71 \text{ MeV} \qquad C_4 = 23.6 \text{ MeV}$$

Equation 44.3, together with these constants, fits the known nuclear mass values very well. The liquid-drop model does not, however, account for some finer details of nuclear structure, such as stability rules and angular momentum. Equation 44.3 is a *theoretical* equation for the binding energy, based on the liquid-drop model, whereas binding energies calculated from Equation 44.2 are *experimental* values based on mass measurements.

EXAMPLE 44.3   **Applying the Semiempirical Binding-Energy Formula**

The nucleus ^{64}Zn has a tabulated binding energy of 559.09 MeV. Use the semiempirical binding-energy formula to generate a theoretical estimate of the binding energy for this nucleus.

SOLUTION

Conceptualize   Imagine bringing the separate protons and neutrons together to form a ^{64}Zn nucleus. The rest energy of the nucleus is smaller than the rest energy of the individual particles. The difference in rest energy is the binding energy.

Categorize   From the text of the problem, we know to apply the liquid-drop model. This example is a substitution problem.

---

[1] The liquid-drop model *describes* that heavy nuclei have $N > Z$. The shell model, as we shall see shortly, *explains* why that is true with a physical argument.

For the ^{64}Zn nucleus, $Z = 30$, $N = 34$, and $A = 64$. Evaluate the four terms of the semiempirical binding-energy formula:

$$C_1 A = (15.7 \text{ MeV})(64) = 1\,005 \text{ MeV}$$

$$C_2 A^{2/3} = (17.8 \text{ MeV})(64)^{2/3} = 285 \text{ MeV}$$

$$C_3 \frac{Z(Z-1)}{A^{1/3}} = (0.71 \text{ MeV}) \frac{(30)(29)}{(64)^{1/3}} = 154 \text{ MeV}$$

$$C_4 \frac{(N-Z)^2}{A} = (23.6 \text{ MeV}) \frac{(34-30)^2}{64} = 5.90 \text{ MeV}$$

Substitute these values into Equation 44.3: $E_b = 1\,005 \text{ MeV} - 285 \text{ MeV} - 154 \text{ MeV} - 5.90 \text{ MeV} = \boxed{560 \text{ MeV}}$

This value differs from the tabulated value by less than 0.2%. Notice how the sizes of the terms decrease from the first to the fourth term. The fourth term is particularly small for this nucleus, which does not have an excessive number of neutrons.

The Shell Model

The liquid-drop model describes the general behavior of nuclear binding energies relatively well. When the binding energies are studied more closely, however, we find the following features:

- Most stable nuclei have an even value of A. Furthermore, only eight stable nuclei have odd values for both Z and N.
- Figure 44.6 shows a graph of the difference between the binding energy per nucleon calculated by Equation 44.3 and the measured binding energy. There is evidence for regularly spaced peaks in the data that are not described by the semiempirical binding-energy formula. The peaks occur at values of N or Z that have become known as **magic numbers:**

Magic numbers ▶

$$Z \text{ or } N = 2, 8, 20, 28, 50, 82 \tag{44.4}$$

- High-precision studies of nuclear radii show deviations from the simple expression in Equation 44.1. Graphs of experimental data show peaks in the curve of radius versus N at values of N equal to the magic numbers.
- A group of *isotones* is a collection of nuclei having the same value of N and varying values of Z. When the number of stable isotones is graphed as function of N, there are peaks in the graph, again at the magic numbers in Equation 44.4.

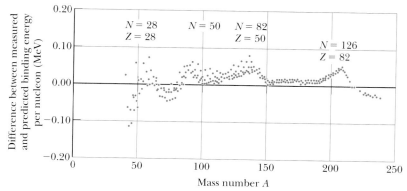

Figure 44.6   The difference between measured binding energies and those calculated from the liquid-drop model as a function of A. The appearance of regular peaks in the experimental data suggests behavior that is not predicted in the liquid-drop model. (Adapted from R. A. Dunlap, *The Physics of Nuclei and Particles*, Brooks/Cole, Belmont CA, 2004.)

- Several other nuclear measurements show anomalous behavior at the magic numbers.[2]

These peaks in graphs of experimental data are reminiscent of the peaks in Figure 42.20 for the ionization energy of atoms, which arose because of the shell structure of the atom. The **shell model** of the nucleus, also called the **independent-particle model,** was developed independently by two German scientists: Maria Goeppert-Mayer in 1949 and Hans Jensen in 1950. Goeppert-Mayer and Jensen shared the 1963 Nobel Prize in Physics for their work. In this model, each nucleon is assumed to exist in a shell, similar to an atomic shell for an electron. The nucleons exist in quantized energy states, and there are few collisions between nucleons. Obviously, the assumptions of this model differ greatly from those made in the liquid-drop model.

The quantized states occupied by the nucleons can be described by a set of quantum numbers. Because both the proton and the neutron have spin $\frac{1}{2}$, the exclusion principle can be applied to describe the allowed states (as it was for electrons in Chapter 42). That is, each state can contain only two protons (or two neutrons) having *opposite* spins (Fig. 44.7). The protons have a set of allowed states, and these differ from those of the neutrons because the two species move in different potential wells. The proton energy levels are farther apart than the neutron levels because the protons experience a superposition of the Coulomb force and the nuclear force, whereas the neutrons experience only the nuclear force.

One factor influencing the observed characteristics of nuclear ground states is *nuclear spin–orbit* effects. The spin–orbit interaction between the spin of an electron and its orbital motion in an atom gives rise to the sodium doublet discussed in Section 42.6 and is magnetic in origin. In contrast, the spin–orbit effect for nucleons in a nucleus is due to the nuclear force. It is much stronger than in the atomic case, and it has opposite sign. When these effects are taken into account, the shell model is able to account for the observed magic numbers.

The shell model helps us understand why nuclei containing an even number of protons and neutrons are more stable than other nuclei. (There are 160 stable even–even isotopes.) Any particular state is filled when it contains two protons (or two neutrons) having opposite spins. An extra proton or neutron can be added to the nucleus only at the expense of increasing the energy of the nucleus. This increase in energy leads to a nucleus that is less stable than the original nucleus. A careful inspection of the stable nuclei shows that the majority have a special stability when their nucleons combine in pairs, which results in a total angular momentum of zero.

The shell model also helps us understand why nuclei tend to have more neutrons than protons. As in Figure 44.7, the proton energy levels are higher than those for neutrons due to the extra energy associated with Coulomb repulsion. This effect becomes more pronounced as Z increases. Consequently, as Z increases and higher states are filled, a proton level for a given quantum number will be much higher in energy than the neutron level for the same quantum number. In fact, it will be even higher in energy than neutron levels for higher quantum numbers. Hence, it is more energetically favorable for the nucleus to form with neutrons in the lower energy levels rather than protons in the higher energy levels, so the number of neutrons is greater than the number of protons.

More sophisticated models of the nucleus have been and continue to be developed. For example, the *collective model* combines features of the liquid-drop and shell models. The development of theoretical models of the nucleus continues to be an active area of research.

MARIA GOEPPERT-MAYER
German Scientist (1906–1972)
Goeppert-Mayer was born and educated in Germany. She is best known for her development of the shell model (independent-particle model) of the nucleus, published in 1950. A similar model was simultaneously developed by Hans Jensen, another German scientist. Goeppert-Mayer and Jensen were awarded the Nobel Prize in Physics in 1963 for their extraordinary work in understanding the structure of the nucleus.

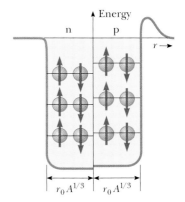

Figure 44.7 A square potential well containing 12 nucleons. The orange circles represent protons, and the green circles represent neutrons. The energy levels for the protons are slightly higher than those for the neutrons because of the electric potential energy associated with the system of protons. The difference in the levels increases as Z increases. Notice that only two nucleons having opposite spins can occupy a given level, as required by the exclusion principle.

[2] For further details, see chapter 5 of R. A. Dunlap, *The Physics of Nuclei and Particles* (Belmont, CA: Brooks/Cole, 2004).

MARIE CURIE
Polish Scientist (1867–1934)

In 1903, Marie Curie shared the Nobel Prize in Physics with her husband, Pierre, and with Becquerel for their studies of radioactive substances. In 1911, she was awarded a Nobel Prize in Chemistry for the discovery of radium and polonium. "I persist in believing that the ideas that then guided us are the only ones which can lead to true social progress. We cannot hope to build a better world without improving the individual. Toward this end, each of us must work toward his own highest development, accepting at the same time his share of responsibility in the general life of humanity."

FPG International

PITFALL PREVENTION 44.3
Rays or Particles?

Early in the history of nuclear physics, the term *radiation* was used to describe the emanations from radioactive nuclei. We now know that alpha radiation and beta radiation involve the emission of particles with nonzero rest energy. Even though they are not examples of electromagnetic radiation, the use of the term *radiation* for all three types of emission is deeply entrenched in our language and in the physics community.

44.4 Radioactivity

In 1896, Becquerel accidentally discovered that uranyl potassium sulfate crystals emit an invisible radiation that can darken a photographic plate when the plate is covered to exclude light. After a series of experiments, he concluded that the radiation emitted by the crystals was of a new type, one that requires no external stimulation and was so penetrating that it could darken protected photographic plates and ionize gases. This process of spontaneous emission of radiation by uranium was soon to be called **radioactivity.**

Subsequent experiments by other scientists showed that other substances were more powerfully radioactive. The most significant early investigations of this type were conducted by Marie and Pierre Curie. After several years of careful and laborious chemical separation processes on tons of pitchblende, a radioactive ore, the Curies reported the discovery of two previously unknown elements, both radioactive, named polonium and radium. Additional experiments, including Rutherford's famous work on alpha-particle scattering, suggested that radioactivity is the result of the *decay,* or disintegration, of unstable nuclei.

Three types of radioactive decay occur in radioactive substances: alpha (α) decay, in which the emitted particles are ^{4}He nuclei; beta (β) decay, in which the emitted particles are either electrons or positrons; and gamma (γ) decay, in which the emitted particles are high-energy photons. A **positron** is a particle like the electron in all respects except that the positron has a charge of $+e$. (The positron is the *antiparticle* of the electron; see Section 46.2.) The symbol e^- is used to designate an electron, and e^+ designates a positron.

We can distinguish among these three forms of radiation by using the scheme described in Figure 44.8. The radiation from radioactive samples that emit all three types of particles is directed into a region in which there is a magnetic field. The radiation beam splits into three components, two bending in opposite directions and the third experiencing no change in direction. This simple observation shows that the radiation of the undeflected beam carries no charge (the gamma ray), the component deflected upward corresponds to positively charged particles (alpha particles), and the component deflected downward corresponds to negatively charged particles (e^-). If the beam includes a positron (e^+), it is deflected upward like the alpha particle, but it follows a different trajectory due to its smaller mass.

The three types of radiation have quite different penetrating powers. Alpha particles barely penetrate a sheet of paper, beta particles can penetrate a few millimeters of aluminum, and gamma rays can penetrate several centimeters of lead.

The decay process is probabilistic in nature and can be described with statistical calculations for a radioactive substance of macroscopic size containing a large number of radioactive nuclei. For such large numbers, the rate at which a particular decay process occurs in a sample is proportional to the number of radioactive nuclei present (that is, the number of nuclei that have not yet decayed). If N is the

Figure 44.8 The radiation from radioactive sources can be separated into three components by using a magnetic field to deflect the charged particles. The detector array at the right records the events. The gamma ray is not deflected by the magnetic field.

ACTIVE FIGURE 44.9

Plot of the exponential decay of radioactive nuclei. The vertical axis represents the number of undecayed radioactive nuclei present at any time t, and the horizontal axis is time. The time $T_{1/2}$ is the half-life of the sample.

Sign in at www.thomsonedu.com and go to ThomsonNOW to observe the decay curves for nuclei with varying half-lives.

number of undecayed radioactive nuclei present at some instant, the rate of change of N with time is

$$\frac{dN}{dt} = -\lambda N \tag{44.5}$$

where λ, called the **decay constant,** is the probability of decay per nucleus per second. The negative sign indicates that dN/dt is negative; that is, N decreases in time.

Equation 44.5 can be written in the form

$$\frac{dN}{N} = -\lambda dt$$

which, upon integration, gives

$$N = N_0 e^{-\lambda t} \tag{44.6}$$

where the constant N_0 represents the number of undecayed radioactive nuclei at $t = 0$. Equation 44.6 shows that the number of undecayed radioactive nuclei in a sample decreases exponentially with time. The plot of N versus t shown in Active Figure 44.9 illustrates the exponential nature of the decay.

The **decay rate** R, which is the number of decays per second, can be obtained by combining Equations 44.5 and 44.6:

$$R = \left| \frac{dN}{dt} \right| = \lambda N = \lambda N_0 e^{-\lambda t} = R_0 e^{-\lambda t} \tag{44.7}$$

where $R_0 = \lambda N_0$ is the decay rate at $t = 0$. The decay rate R of a sample is often referred to as its **activity.** Note that both N and R decrease exponentially with time.

Another parameter useful in characterizing nuclear decay is the **half-life** $T_{1/2}$:

> The **half-life** of a radioactive substance is the time interval during which half of a given number of radioactive nuclei decay.

To find an expression for the half-life, we first set $N = N_0/2$ and $t = T_{1/2}$ in Equation 44.6 to give

$$\frac{N_0}{2} = N_0 e^{-\lambda T_{1/2}}$$

Canceling the N_0 factors and then taking the reciprocal of both sides, we obtain $e^{\lambda T_{1/2}} = 2$. Taking the natural logarithm of both sides gives

$$T_{1/2} = \frac{\ln 2}{\lambda} = \frac{0.693}{\lambda} \tag{44.8}$$

After a time interval equal to one half-life, there are $N_0/2$ radioactive nuclei remaining (by definition); after two half-lives, half of these remaining nuclei have decayed and $N_0/4$ radioactive nuclei are left; after three half-lives, $N_0/8$ are left;

PITFALL PREVENTION 44.4
Notation Warning

In Section 44.1, we introduced the symbol N as an integer representing the number of neutrons in a nucleus. In this discussion, the symbol N represents the number of undecayed nuclei in a radioactive sample remaining after some time interval. As you read further, be sure to consider the context to determine the appropriate meaning for the symbol N.

◄ Exponential behavior of the number of undecayed nuclei

◄ Exponential behavior of the decay rate

PITFALL PREVENTION 44.5
Half-life

It is *not* true that all the original nuclei have decayed after two half-lives! In one half-life, half of the original nuclei will decay. In the second half-life, half of those remaining will decay, leaving $\frac{1}{4}$ of the original number.

◄ Half-life

and so on. In general, after n half-lives, the number of undecayed radioactive nuclei remaining is $N_0(\frac{1}{2})^n$.

A frequently used unit of activity is the **curie** (Ci), defined as

The curie ▶

$$1 \text{ Ci} \equiv 3.7 \times 10^{10} \text{ decays/s}$$

This value was originally selected because it is the approximate activity of 1 g of radium. The SI unit of activity is the **becquerel** (Bq):

The becquerel ▶

$$1 \text{ Bq} \equiv 1 \text{ decay/s}$$

Therefore, $1 \text{ Ci} = 3.7 \times 10^{10}$ Bq. The curie is a rather large unit, and the more frequently used activity units are the millicurie and the microcurie.

Quick Quiz 44.2 On your birthday, you measure the activity of a sample of ^{210}Bi, which has a half-life of 5.01 days. The activity you measure is 1.000 μCi. What is the activity of this sample on your next birthday? (a) 1.000 μCi (b) 0 (c) $\sim$ 0.2 μCi (d) $\sim$ 0.01 μCi (e) $\sim 10^{-22}$ μCi

---

EXAMPLE 44.4   **How Many Nuclei Are Left?**

The isotope carbon-14, $^{14}_{6}$C, is radioactive and has a half-life of 5 730 years. If you start with a sample of 1 000 carbon-14 nuclei, how many nuclei will still be undecayed in 25 000 years?

SOLUTION

Conceptualize   The time interval of 25 000 years is much longer than the half-life, so only a small fraction of the originally undecayed nuclei will remain.

Categorize   The text of the problem allows us to categorize this example as a radioactive decay problem.

Analyze   Divide the time interval by the half-life to determine the number of half-lives:

$$n = \frac{25\ 000 \text{ yr}}{5\ 730 \text{ yr}} = 4.363$$

Determine how many undecayed nuclei are left after 4.363 half-lives:

$$N = N_0(\tfrac{1}{2})^n = 1\ 000(\tfrac{1}{2})^{4.363} = \boxed{49}$$

Finalize   As we have mentioned, radioactive decay is a probabilistic process and accurate statistical predictions are possible only with a very large number of atoms. The original sample in this example contains only 1 000 nuclei, which is certainly not a very large number. Therefore, if you counted the number of undecayed nuclei remaining after 25 000 years, it might not be exactly 49.

EXAMPLE 44.5   **The Activity of Carbon**

At time $t = 0$, a radioactive sample contains 3.50 μg of pure $^{11}_{6}$C, which has a half-life of 20.4 min.

(A) Determine the number N_0 of nuclei in the sample at $t = 0$.

SOLUTION

Conceptualize   The half-life is relatively short, so the number of undecayed nuclei drops rapidly. The molar mass of $^{11}_{6}$C is approximately 11.0 g/mol.

Categorize   We evaluate results using equations developed in this section, so we categorize this example as a substitution problem.

Find the number of moles in 3.50 μg of pure $^{11}_{6}$C:

$$n = \frac{3.50 \times 10^{-6}\,\text{g}}{11.0\,\text{g/mol}} = 3.18 \times 10^{-7}\,\text{mol}$$

Find the number of undecayed nuclei in this amount of pure $^{11}_{6}$C:

$$N_0 = 3.18 \times 10^{-7}\,\text{mol}(6.02 \times 10^{23}\,\text{nuclei/mol}) = \boxed{1.92 \times 10^{17}\,\text{nuclei}}$$

(B) What is the activity of the sample initially and after 8.00 h?

SOLUTION

Find the decay constant:

$$\lambda = \frac{0.693}{T_{1/2}} = \frac{0.693}{20.4\,\text{min}}\left(\frac{1\,\text{min}}{60\,\text{s}}\right) = 5.66 \times 10^{-4}\,\text{s}^{-1}$$

Find the initial activity of the sample:

$$R_0 = \lambda N_0 = (5.66 \times 10^{-4}\,\text{s}^{-1})(1.92 \times 10^{17}) = \boxed{1.08 \times 10^{14}\,\text{Bq}}$$

Use Equation 44.7 to find the activity at $t = 8.00\,\text{h} = 2.88 \times 10^4\,\text{s}$:

$$R = R_0 e^{-\lambda t} = (1.08 \times 10^{14}\,\text{Bq})e^{-(5.66\times10^{-4}\,\text{s}^{-1})(2.88\times10^4\,\text{s})} = \boxed{8.96 \times 10^6\,\text{Bq}}$$

EXAMPLE 44.6 **A Radioactive Isotope of Iodine**

A sample of the isotope ^{131}I, which has a half-life of 8.04 days, has an activity of 5.0 mCi at the time of shipment. Upon receipt of the sample at a medical laboratory, the activity is 2.1 mCi. How much time has elapsed between the two measurements?

SOLUTION

Conceptualize   The sample is continuously decaying as it is in transit. The decrease in the activity is 58% during the time interval between shipment and receipt, so we expect the elapsed time to be greater than the half-life of 8.04 d.

Categorize   The stated activity corresponds to many decays per second, so N is large and we can categorize this problem as one in which we can use our statistical analysis of radioactivity.

Analyze   Solve Equation 44.7 for the ratio of the final activity to the initial activity:

$$\frac{R}{R_0} = e^{-\lambda t}$$

Take the natural logarithm of both sides:

$$\ln\left(\frac{R}{R_0}\right) = -\lambda t$$

Solve for the time t:

$$(1)\quad t = -\frac{1}{\lambda}\ln\left(\frac{R}{R_0}\right)$$

Use Equation 44.8 to find λ:

$$(2)\quad \lambda = \frac{0.693}{T_{1/2}} = \frac{0.693}{8.04\,\text{d}} = 8.62 \times 10^{-2}\,\text{d}^{-1}$$

Substitute Equation (2) into Equation (1):

$$t = -\left(\frac{1}{8.62 \times 10^{-2}\,\text{d}^{-1}}\right)\ln\left(\frac{2.1\,\text{mCi}}{5.0\,\text{mCi}}\right) = \boxed{10\,\text{d}}$$

Finalize   This result is indeed greater than the half-life, as expected. This example demonstrates the difficulty in shipping radioactive samples with short half-lives. If the shipment is delayed by several days, only a small fraction of the sample might remain upon receipt. This difficulty can be addressed by shipping a combination of isotopes in which the desired isotope is the product of a decay occurring within the sample. It is possible for the desired isotope to be in *equilibrium*, in which case it is created at the same rate as it decays. Therefore, the amount of the desired isotope remains constant during the shipping process and subsequent storage. When needed, the desired isotope can be separated from the rest of the sample; its decay from the initial activity begins at this point rather than upon shipment.

- Beta (electron)
- Stable
- Beta (positron) or electron capture
- Alpha

ACTIVE FIGURE 44.10

A close-up view of the line of stability in Figure 44.4 from $Z = 65$ to $Z = 80$. The blue dots represent stable nuclei as in Figure 44.4. The other colored dots represent unstable isotopes above and below the line of stability, with the color of the dot indicating the primary means of decay.

Sign in at www.thomsonedu.com and go to ThomsonNOW to study the decay modes and decay energies by clicking on any of the colored dots.

Before decay

After decay

ACTIVE FIGURE 44.11

The alpha decay of radium-226. The radium nucleus is initially at rest. After the decay, the radon nucleus has kinetic energy K_{Rn} and momentum $\vec{p}_{Rn}$ and the alpha particle has kinetic energy K_α and momentum $\vec{p}_\alpha$.

Sign in at www.thomsonedu.com and go to ThomsonNOW to observe the decay of radium-226. For a large number of decays, observe the development of the graph in Active Figure 44.14b (page 1313).

44.5 The Decay Processes

As we stated in Section 44.4, a radioactive nucleus spontaneously decays by one of three processes: alpha decay, beta decay, or gamma decay. Active Figure 44.10 shows a close-up view of a portion of Figure 44.4 from $Z = 65$ to $Z = 80$. The blue circles are the stable nuclei seen in Figure 44.4. In addition, unstable nuclei above and below the line of stability for each value of Z are shown. Above the line of stability, the red circles show unstable nuclei that are neutron-rich and undergo a beta decay process in which an electron is emitted. Below the blue circles are green circles corresponding to proton-rich unstable nuclei that primarily undergo a beta-decay process in which a positron is emitted or a competing process called electron capture. Beta decay and electron capture are described in more detail below. Further below the line of stability (with a few exceptions) are yellow circles that represent very proton-rich nuclei for which the primary decay mechanism is alpha decay, which we discuss first.

Alpha Decay

A nucleus emitting an alpha particle ($^4_2$He) loses two protons and two neutrons. Therefore, the atomic number Z decreases by 2, the mass number A decreases by 4, and the neutron number decreases by 2. The decay can be written

$$^A_Z X \rightarrow \ ^{A-4}_{Z-2} Y + ^4_2 He \tag{44.9}$$

where X is called the **parent nucleus** and Y the **daughter nucleus.** As a general rule in any decay expression such as this one, (1) the sum of the mass numbers A must be the same on both sides of the decay and (2) the sum of the atomic numbers Z must be the same on both sides of the decay. As examples, ^{238}U and ^{226}Ra are both alpha emitters and decay according to the schemes

$$^{238}_{92} U \rightarrow \ ^{234}_{90} Th + ^4_2 He \tag{44.10}$$

$$^{226}_{88} Ra \rightarrow \ ^{222}_{86} Rn + ^4_2 He \tag{44.11}$$

The decay of ^{226}Ra is shown in Active Figure 44.11.

When the nucleus of one element changes into the nucleus of another as happens in alpha decay, the process is called **spontaneous decay.** In any spontaneous decay, relativistic energy and momentum of the isolated parent nucleus must be conserved. If we call M_X the mass of the parent nucleus, M_Y the mass of the daughter nucleus, and M_α the mass of the alpha particle, we can define the **disintegration energy** Q of the system as

$$Q = (M_X - M_Y - M_\alpha)c^2 \tag{44.12}$$

The energy Q is in joules when the masses are in kilograms and c is the speed of light, 3.00×10^8 m/s. When the masses are expressed in atomic mass units u, however, Q can be calculated in MeV using the expression

$$Q = (M_X - M_Y - M_\alpha) \times 931.494 \text{ MeV/u} \tag{44.13}$$

Table 44.2 (page 1310) contains information on selected isotopes, including masses of neutral atoms that can be used in Equation 44.13 and similar equations.

The disintegration energy Q appears in the form of kinetic energy in the daughter nucleus and the alpha particle and is sometimes referred to as the Q value of the nuclear decay. Consider the case of the ^{226}Ra decay described in Active Figure 44.11. If the parent nucleus is at rest before the decay, the total kinetic energy of the products is 4.87 MeV. (See Example 44.7.) Most of this kinetic energy is associated with the alpha particle because this particle is much less massive than the daughter nucleus ^{222}Rn. That is, because momentum must be conserved, the lighter alpha particle recoils with a much higher speed than does the daughter nucleus. Generally, less massive particles carry off most of the energy in nuclear decays.

Experimental observations of alpha-particle energies show a number of discrete energies rather than a single energy because the daughter nucleus may be left in an excited quantum state after the decay. As a result, not all the disintegration energy is available as kinetic energy of the alpha particle and daughter nucleus. The emission of an alpha particle is followed by one or more gamma-ray photons (see below) as the excited nucleus decays to the ground state. The observed discrete alpha-particle energies represent evidence of the quantized nature of the nucleus and allow a determination of the energies of the quantum states.

If one assumes ^{238}U (or any other alpha emitter) decays by emitting either a proton or a neutron, the mass of the decay products would exceed that of the parent nucleus, corresponding to a negative Q value. A negative Q value indicates that such a proposed decay does not occur spontaneously.

PITFALL PREVENTION 44.6
Another Q

We have seen the symbol Q before, but this use is a brand new meaning for this symbol: the disintegration energy. In this context, it is not heat, charge, or quality factor for a resonance, for which we have used Q before.

Quick Quiz 44.3 Which of the following is the correct daughter nucleus associated with the alpha decay of $^{157}_{72}$Hf? (a) $^{153}_{72}$Hf (b) $^{153}_{70}$Yb (c) $^{157}_{70}$Yb

---

EXAMPLE 44.7   **The Energy Liberated When Radium Decays**

The ^{226}Ra nucleus undergoes alpha decay according to Equation 44.11. Calculate the Q value for this process. From Table 44.2, the masses are 226.025 403 u for ^{226}Ra, 222.017 570 u for ^{222}Rn, and 4.002 603 u for $^4_2$He.

SOLUTION

Conceptualize Study Active Figure 44.11 to understand the process of alpha decay in this nucleus.

Categorize We use an equation developed in this section, so we categorize this example as a substitution problem.

Evaluate Q using Equation 44.13:

$$Q = (M_X - M_Y - M_\alpha) \times 931.494 \text{ MeV/u}$$

$$= (226.025\ 403 \text{ u} - 222.017\ 570 \text{ u} - 4.002\ 603 \text{ u}) \times 931.494 \text{ MeV/u}$$

$$= (0.005\ 230 \text{ u}) \times 931.494 \text{ MeV/u} = \boxed{4.87 \text{ MeV}}$$

What If? Suppose you measured the kinetic energy of the alpha particle from this decay. Would you measure 4.87 MeV?

Answer The value of 4.87 MeV is the disintegration energy for the decay. It includes the kinetic energy of both the alpha particle and the daughter nucleus after the decay. Therefore, the kinetic energy of the alpha particle would be *less* than 4.87 MeV.

Let's determine this kinetic energy mathematically. The parent nucleus is an isolated system that decays into an alpha particle and a daughter nucleus. Therefore, momentum must be conserved for the system.

Set up a conservation of momentum equation, noting that the initial momentum of the system is zero:

$$(1) \quad 0 = M_Y v_Y - M_\alpha v_\alpha$$

Set the disintegration energy equal to the sum of the kinetic energies of the alpha particle and the daughter nucleus (assuming the daughter nucleus is left in the ground state):

$$(2) \quad Q = \tfrac{1}{2} M_\alpha v_\alpha^2 + \tfrac{1}{2} M_Y v_Y^2$$

Solve Equation (1) for v_Y and substitute into Equation (2):

$$Q = \tfrac{1}{2} M_\alpha v_\alpha^2 + \tfrac{1}{2} M_Y \left(\frac{M_\alpha v_\alpha}{M_Y}\right)^2 = \tfrac{1}{2} M_\alpha v_\alpha^2 \left(1 + \frac{M_\alpha}{M_Y}\right)$$

$$= K_\alpha \left(\frac{M_Y + M_\alpha}{M_Y}\right)$$

Solve for the kinetic energy of the alpha particle:

$$K_\alpha = Q\left(\frac{M_Y}{M_Y + M_\alpha}\right)$$

Evaluate this kinetic energy for the specific decay of ^{226}Ra that we are exploring in this example:

$$K_\alpha = (4.87 \text{ MeV})\left(\frac{222}{222 + 4}\right) = 4.79 \text{ MeV}$$

TABLE 44.2

Chemical and Nuclear Information for Selected Isotopes

Atomic Number Z	Element	Chemical Symbol	Mass Number A (* means radioactive)	Mass of Neutral Atom (u)	Percent Abundance	Half-life, if Radioactive $T_{1/2}$
−1	electron	e^-	0	0.000 549		
0	neutron	n	1*	1.008 665		614 s
1	hydrogen	^{1}H = p	1	1.007 825	99.988 5	
	[deuterium	^{2}H = D]	2	2.014 102	0.011 5	
	[tritium	^{3}H = T]	3*	3.016 049		12.33 yr
2	helium	He	3	3.016 029	0.000 137	
	[alpha particle	$\alpha = {}^4$He]	4	4.002 603	99.999 863	
			6*	6.018 888		0.81 s
3	lithium	Li	6	6.015 122	7.5	
			7	7.016 004	92.5	
4	beryllium	Be	7*	7.016 929		53.3 d
			9	9.012 182	100	
5	boron	B	10	10.012 937	19.9	
			11	11.009 306	80.1	
6	carbon	C	11*	11.011 434		20.4 min
			12	12.000 000	98.93	
			13	13.003 355	1.07	
			14*	14.003 242		5 730 yr
7	nitrogen	N	13*	13.005 739		9.96 min
			14	14.003 074	99.632	
			15	15.000 109	0.368	
8	oxygen	O	14*	14.008 595		70.6 s
			15*	15.003 065		122 s
			16	15.994 915	99.757	
			17	16.999 132	0.038	
			18	17.999 160	0.205	
9	fluorine	F	18*	18.000 938		109.8 min
			19	18.998 403	100	
10	neon	Ne	20	19.992 440	90.48	
11	sodium	Na	23	22.989 770	100	
12	magnesium	Mg	23*	22.994 125		11.3 s
			24	23.985 042	78.99	
13	aluminum	Al	27	26.981 539	100	
14	silicon	Si	27*	26.986 705		4.2 s
15	phosphorus	P	30*	29.978 314		2.50 min
			31	30.973 762	100	
			32*	31.973 907		14.26 d
16	sulfur	S	32	31.972 071	94.93	
19	potassium	K	39	38.963 707	93.258 1	
			40*	39.963 999	0.011 7	1.28×10^9 yr
20	calcium	Ca	40	39.962 591	96.941	
			42	41.958 618	0.647	
			43	42.958 767	0.135	
25	manganese	Mn	55	54.938 050	100	
26	iron	Fe	56	55.934 942	91.754	
			57	56.935 399	2.119	

(continued)

TABLE 44.2

Chemical and Nuclear Information for Selected Isotopes (*Continued*)

Atomic Number Z	Element	Chemical Symbol	Mass Number A (* means radioactive)	Mass of Neutral Atom (u)	Percent Abundance	Half-life, if Radioactive $T_{1/2}$
27	cobalt	Co	57*	56.936 296		270 d
			59	58.933 200	100	
			60*	59.933 822		5.27 yr
28	nickel	Ni	58	57.935 348	68.076 9	
			60	59.930 790	26.223 1	
29	copper	Cu	63	62.929 601	69.17	
			64*	63.929 599		12.7 h
			65	64.927 794	30.83	
30	zinc	Zn	64	63.929 147	48.63	
37	rubidium	Rb	87*	86.909 184	27.83	
38	strontium	Sr	87	86.908 880	7.00	
			88	87.905 614	82.58	
			90*	89.907 738		29.1 yr
41	niobium	Nb	93	92.906 378	100	
42	molybdenum	Mo	94	93.905 088	9.25	
44	ruthenium	Ru	98	97.905 287	1.87	
55	cesium	Cs	137*	136.907 074		30 yr
56	barium	Ba	137	136.905 821	11.232	
58	cerium	Ce	140	139.905 434	88.450	
59	praseodymium	Pr	141	140.907 648	100	
60	neodymium	Nd	144*	143.910 083	23.8	2.3×10^{15} yr
61	promethium	Pm	145*	144.912 744		17.7 yr
79	gold	Au	197	196.966 552	100	
80	mercury	Hg	198	197.966 752	9.97	
			202	201.970 626	29.86	
82	lead	Pb	206	205.974 449	24.1	
			207	206.975 881	22.1	
			208	207.976 636	52.4	
			214*	213.999 798		26.8 min
83	bismuth	Bi	209	208.980 383	100	
84	polonium	Po	210*	209.982 857		138.38 d
			216*	216.001 905		0.145 s
			218*	218.008 966		3.10 min
86	radon	Rn	220*	220.011 384		55.6 s
			222*	222.017 570		3.823 d
88	radium	Ra	226*	226.025 403		1 600 yr
90	thorium	Th	232*	232.038 050	100	1.40×10^{10} yr
			234*	234.043 596		24.1 d
92	uranium	U	235*	235.043 923	0.720 0	7.04×10^8 yr
			236*	236.045 562		2.34×10^7 yr
			238*	238.050 783	99.274 5	4.47×10^9 yr
93	neptunium	Np	236*	236.046 560		1.15×10^5 yr
			237*	237.048 167		2.14×10^6 yr
94	plutonium	Pu	239*	239.052 156		24 120 yr

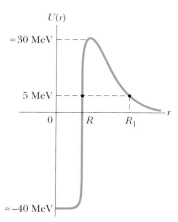

Figure 44.12 Potential energy versus separation distance for a system consisting of an alpha particle and a daughter nucleus. Classically, the energy of the alpha particle is not sufficiently large to overcome the energy barrier, so the particle should not be able to escape from the nucleus. In reality, the alpha particle does escape by tunneling through the barrier.

To understand the mechanism of alpha decay, let's model the parent nucleus as a system consisting of (1) the alpha particle, already formed as an entity within the nucleus, and (2) the daughter nucleus that will result when the alpha particle is emitted. Figure 44.12 shows a plot of potential energy versus separation distance r between the alpha particle and the daughter nucleus, where the distance marked R is the range of the nuclear force. The curve represents the combined effects of (1) the repulsive Coulomb force, which gives the positive part of the curve for $r > R$, and (2) the attractive nuclear force, which causes the curve to be negative for $r < R$. As shown in Example 44.7, a typical disintegration energy Q is approximately 5 MeV, which is the approximate kinetic energy of the alpha particle, represented by the lower dashed line in Figure 44.12.

According to classical physics, the alpha particle is trapped in a potential well. How, then, does it ever escape from the nucleus? The answer to this question was first provided by George Gamow (1904–1968) in 1928 and independently by R. W. Gurney and E. U. Condon in 1929, using quantum mechanics. In the view of quantum mechanics, there is always some probability that a particle can tunnel through a barrier (Section 41.5). That is exactly how we can describe alpha decay: the alpha particle tunnels through the barrier in Figure 44.12, escaping the nucleus. Furthermore, this model agrees with the observation that higher-energy alpha particles come from nuclei with shorter half-lives. For higher-energy alpha particles in Figure 44.12, the barrier is narrower and the probability is higher that tunneling occurs. The higher probability translates to a shorter half-life.

As an example, consider the decays of ^{238}U and ^{226}Ra in Equations 44.10 and 44.11, along with the corresponding half-lives and alpha-particle energies:

$$^{238}\text{U}: \quad T_{1/2} = 4.47 \times 10^9 \text{ yr} \qquad K_\alpha = 4.20 \text{ MeV}$$

$$^{226}\text{Ra}: \quad T_{1/2} = 1.60 \times 10^3 \text{ yr} \qquad K_\alpha = 4.79 \text{ MeV}$$

Notice that a relatively small difference in alpha-particle energy is associated with a tremendous difference of six orders of magnitude in the half-life. The origin of this effect can be understood as follows. Figure 44.12 shows that the curve below an alpha-particle energy of 5 MeV has a slope with a relatively small magnitude. Therefore, a small difference in energy on the vertical axis has a relatively large effect on the width of the potential barrier. Second, recall Equation 41.22, which describes the exponential dependence of the probability of transmission on the barrier width. These two factors combine to give the very sensitive relationship between half-life and alpha-particle energy that the data above suggest.

A life-saving application of alpha decay is the household smoke detector, shown in Figure 44.13. The detector consists of an ionization chamber, a sensitive current detector, and an alarm. A weak radioactive source (usually $^{241}_{95}$Am) ionizes the air in the chamber of the detector, creating charged particles. A voltage is maintained between the plates inside the chamber, setting up a small but detectable current in the external circuit due to the ions acting as charge carriers between the plates. As

(a)

(b)

Figure 44.13 (a) A smoke detector uses alpha decay to determine whether smoke is in the air. The alpha source is in the black cylinder at the right. (b) Smoke entering the chamber reduces the detected current, causing the alarm to sound.

long as the current is maintained, the alarm is deactivated. If smoke drifts into the chamber, however, the ions become attached to the smoke particles. These heavier particles do not drift as readily as do the lighter ions, which causes a decrease in the detector current. The external circuit senses this decrease in current and sets off the alarm.

Beta Decay

When a radioactive nucleus undergoes beta decay, the daughter nucleus contains the same number of nucleons as the parent nucleus but the atomic number is changed by 1, which means that the number of protons changes:

$$^A_Z X \rightarrow \ ^A_{Z+1} Y + e^- \quad \text{(incomplete expression)} \tag{44.14}$$

$$^A_Z X \rightarrow \ ^A_{Z-1} Y + e^+ \quad \text{(incomplete expression)} \tag{44.15}$$

where, as mentioned in Section 44.4, e^- designates an electron and e^+ designates a positron, with *beta particle* being the general term referring to either. *Beta decay is not described completely by these expressions.* We shall give reasons for this statement shortly.

As with alpha decay, the nucleon number and total charge are both conserved in beta decays. Because A does not change but Z does, we conclude that in beta decay, either a neutron changes to a proton (Eq. 44.14) or a proton changes to a neutron (Eq. 44.15). Note that the electron or positron emitted in these decays is not present beforehand in the nucleus; it is created in the process of the decay from the rest energy of the decaying nucleus. Two typical beta-decay processes are

$$^{14}_{6} C \rightarrow \ ^{14}_{7} N + e^- \quad \text{(incomplete expression)} \tag{44.16}$$

$$^{12}_{7} N \rightarrow \ ^{12}_{6} C + e^+ \quad \text{(incomplete expression)} \tag{44.17}$$

Let's consider the energy of the system undergoing beta decay before and after the decay. As with alpha decay, energy of the isolated system must be conserved. Experimentally, it is found that beta particles from a single type of nucleus are emitted over a continuous range of energies (Active Fig. 44.14a), as opposed to alpha decay, in which the alpha particles are emitted with discrete energies (Active Fig. 44.14b). The kinetic energy of the system after the decay is equal to the decrease in rest energy of the system, that is, the Q value. Because all decaying nuclei in the sample have the same initial mass, however, *the Q value must be the same for each decay.* So, why do the emitted particles have the range of kinetic energies shown in Active Figure 44.14a? The law of conservation of energy seems to be violated! It becomes worse: further analysis of the decay processes described by Equations 44.14 and 44.15 shows that the laws of conservation of angular momentum (spin) and linear momentum are also violated!

After a great deal of experimental and theoretical study, Pauli in 1930 proposed that a third particle must be present in the decay products to carry away the "missing" energy and momentum. Fermi later named this particle the **neutrino** (little neutral one) because it had to be electrically neutral and have little or no mass. Although it eluded detection for many years, the neutrino (symbol ν, Greek nu) was finally detected experimentally in 1956 by Frederick Reines, who received the Nobel Prize in Physics for this work in 1995. The neutrino has the following properties:

- It has zero electric charge.
- Its mass is either zero (in which case it travels at the speed of light) or very small; much recent persuasive experimental evidence suggests that the neutrino mass is not zero. Current experiments place the upper bound of the mass of the neutrino at approximately 7 eV/c^2.
- It has a spin of $\frac{1}{2}$, which allows the law of conservation of angular momentum to be satisfied in beta decay.
- It interacts very weakly with matter and is therefore very difficult to detect.

ACTIVE FIGURE 44.14

(a) Distribution of beta particle energies in a typical beta decay. All energies are observed up to a maximum value. (b) In contrast, the energies of alpha particles from an alpha decay are discrete.

Sign in at www.thomsonedu.com and go to ThomsonNOW to observe the development of these graphs for the decays in Active Figures 44.11 and 44.15.

◀ Properties of the neutrino

We can now write the beta-decay processes (Eqs. 44.14 and 44.15) in their correct and complete form:

Beta decay processes ▶

$$^A_Z X \rightarrow \ _{Z+1}^{\ A}Y + e^- + \bar{\nu} \quad \text{(complete expression)} \quad \textbf{(44.18)}$$

$$^A_Z X \rightarrow \ _{Z-1}^{\ A}Y + e^+ + \nu \quad \text{(complete expression)} \quad \textbf{(44.19)}$$

as well as those for carbon-14 and nitrogen-12 (Eqs. 44.16 and 44.17):

$$^{14}_6 C \rightarrow \ ^{14}_7 N + e^- + \bar{\nu} \quad \text{(complete expression)} \quad \textbf{(44.20)}$$

$$^{12}_7 N \rightarrow \ ^{12}_6 C + e^+ + \nu \quad \text{(complete expression)} \quad \textbf{(44.21)}$$

where the symbol $\bar{\nu}$ represents the **antineutrino,** the antiparticle to the neutrino. We shall discuss antiparticles further in Chapter 46. For now, it suffices to say that **a neutrino is emitted in positron decay and an antineutrino is emitted in electron decay.** As with alpha decay, the decays listed above are analyzed by applying conservation laws, but relativistic expressions must be used for beta particles because their kinetic energy is large (typically 1 MeV) compared with their rest energy of 0.511 MeV. Active Figure 44.15 shows a pictorial representation of the decays described by Equations 44.20 and 44.21.

In Equation 44.18, the number of protons has increased by one and the number of neutrons has decreased by one. We can write the fundamental process of e^- decay in terms of a neutron changing into a proton as follows:

$$n \rightarrow p + e^- + \bar{\nu} \quad \textbf{(44.22)}$$

The electron and the antineutrino are ejected from the nucleus, with the net result that there is one more proton and one fewer neutron, consistent with the changes in Z and $A - Z$. A similar process occurs in e^+ decay, with a proton changing into a neutron, a positron, and a neutrino. This latter process can only occur within the nucleus, with the result that the nuclear mass decreases. It cannot occur for an isolated proton because its mass is less than that of the neutron.

A process that competes with e^+ decay is **electron capture,** which occurs when a parent nucleus captures one of its own orbital electrons and emits a neutrino. The final product after decay is a nucleus whose charge is $Z - 1$:

Electron capture ▶

$$^A_Z X + \ ^{\ 0}_{-1} e \rightarrow \ _{Z-1}^{\ A}Y + \nu \quad \textbf{(44.23)}$$

In most cases, it is a K-shell electron that is captured and the process is therefore referred to as **K capture.** One example is the capture of an electron by $^7_4$Be:

$$^7_4 Be + \ ^{\ 0}_{-1} e \rightarrow \ ^7_3 Li + \nu$$

PITFALL PREVENTION 44.7
Mass Number of the Electron

An alternative notation for an electron in Equation 44.23 is the symbol $^{\ 0}_{-1} e$, which does not imply that the electron has zero rest energy. The mass of the electron is so much smaller than that of the lightest nucleon, however, that we approximate it as zero in the context of nuclear decays and reactions.

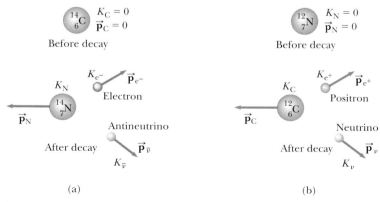

ACTIVE FIGURE 44.15

(a) The beta decay of carbon-14. The final products of the decay are the nitrogen-14 nucleus, an electron, and an antineutrino. (b) The beta decay of nitrogen-12. The final products of the decay are the carbon-12 nucleus, a positron, and a neutrino.

Sign in at www.thomsonedu.com and go to ThomsonNOW to observe the decay of carbon-14. For a large number of decays, observe the development of the graph in Active Figure 44.14a.

Because the neutrino is very difficult to detect, electron capture is usually observed by the x-rays given off as higher-shell electrons cascade downward to fill the vacancy created in the K shell.

Finally, we specify Q values for the beta-decay processes. The Q values for e^- decay and electron capture are given by $Q = (M_X - M_Y) c^2$, where M_X and M_Y are the masses of neutral atoms. In e^- decay, the parent nucleus experiences an increase in atomic number and, for the atom to become neutral, an electron must be absorbed by the atom. If the neutral parent atom and an electron (which will eventually combine with the daughter to form a neutral atom) is the initial system and the final system is the neutral daughter atom and the beta-ejected electron, the system contains a free electron both before and after the decay. Therefore, in subtracting the initial and final masses of the system, this electron mass cancels. The Q values for e^+ decay are given by $Q = (M_X - M_Y - 2m_e) c^2$. The extra term $-2m_ec^2$ in this expression is necessary because the atomic number of the parent decreases by one when the daughter is formed. After it is formed by the decay, the daughter atom sheds one electron to form a neutral atom. Therefore, the final products are the daughter atom, the shed electron, and the ejected positron.

These relationships are useful in determining whether or not a process is energetically possible. For example, the expression for proposed e^+ decay for a particular parent nucleus may turn out to be negative. In that case, this decay does not occur. The expression for electron capture for this parent nucleus, however, may give a positive number, so electron capture can occur even though e^+ decay is not possible. Such is the case for the decay of ^7_4Be shown above.

Quick Quiz 44.4   Which of the following is the correct daughter nucleus associated with the beta decay of $^{184}_{72}\text{Hf}$? (a) $^{183}_{72}\text{Hf}$ (b) $^{183}_{73}\text{Ta}$ (c) $^{184}_{73}\text{Ta}$

Carbon Dating

The beta decay of ^{14}C (Eq. 44.20) is commonly used to date organic samples. Cosmic rays in the upper atmosphere cause nuclear reactions (Section 44.7) that create ^{14}C. The ratio of ^{14}C to ^{12}C in the carbon dioxide molecules of our atmosphere has a constant value of approximately 1.3×10^{-12}. The carbon atoms in all living organisms have this same $^{14}\text{C}/^{12}\text{C}$ ratio because the organisms continuously exchange carbon dioxide with their surroundings. When an organism dies, however, it no longer absorbs ^{14}C from the atmosphere, and so the $^{14}\text{C}/^{12}\text{C}$ ratio decreases as the ^{14}C decays with a half-life of 5 730 yr. It is therefore possible to measure the age of a material by measuring its ^{14}C activity. Using this technique, scientists have been able to identify samples of wood, charcoal, bone, and shell as having lived from 1 000 to 25 000 years ago. This knowledge has helped us reconstruct the history of living organisms—including humans—during this time span.

A particularly interesting example is the dating of the Dead Sea Scrolls. This group of manuscripts was discovered by a shepherd in 1947. Translation showed them to be religious documents, including most of the books of the Old Testament. Because of their historical and religious significance, scholars wanted to know their age. Carbon dating applied to the material in which they were wrapped established their age at approximately 1 950 yr.

CONCEPTUAL EXAMPLE 44.8   **The Age of Ice Man**

In 1991, a German tourist discovered the well-preserved remains of a man, now called the "Ice Man," trapped in a glacier in the Italian Alps. (See the photograph at the opening of this chapter.) Radioactive dating with ^{14}C revealed that this person was alive approximately 5 300 years ago. Why did scientists date a sample of the Ice Man using ^{14}C rather than ^{11}C, which is a beta emitter having a half-life of 20.4 min?

SOLUTION

Because ^{14}C has a half-life of 5 730 yr, the fraction of ^{14}C nuclei remaining after one half-life is high enough to allow accurate measurements of changes in the sample's activity. Because ^{11}C has a very short half-life, it is not useful; its activity decreases to a vanishingly small value over the age of the sample, making it impossible to detect.

An isotope used to date a sample must be present in a known amount in the sample when it is formed. As a general rule, the isotope chosen to date a sample should also have a half-life that is on the same order of magni-

tude as the age of the sample. If the half-life is much less than the age of the sample, there won't be enough activity left to measure because almost all the original radioactive nuclei will have decayed. If the half-life is much greater than the age of the sample, the amount of decay that has taken place since the sample died will be too small to measure. For example, if you have a specimen estimated to have died 50 years ago, neither ^{14}C (5 730 yr) nor ^{11}C (20 min) is suitable. If you know your sample contains hydrogen, however, you can measure the activity of ^{3}H (tritium), a beta emitter that has a half-life of 12.3 yr.

EXAMPLE 44.9 | **Radioactive Dating**

A piece of charcoal containing 25.0 g of carbon is found in some ruins of an ancient city. The sample shows a ^{14}C activity R of 250 decays/min. How long has the tree from which this charcoal came been dead?

SOLUTION

Conceptualize Because the charcoal was found in ancient ruins, we expect the current activity to be smaller than the initial activity. If we can determine the initial activity, we can find out how long the wood has been dead.

Categorize The text of the question helps us categorize this example as a carbon dating problem.

Analyze Calculate the decay constant for ^{14}C:

$$\lambda = \frac{0.693}{T_{1/2}} = \frac{0.693}{(5\ 730\ \text{yr})(3.16 \times 10^{7}\ \text{s/yr})}$$

$$= 3.83 \times 10^{-12}\ \text{s}^{-1}$$

Calculate the number of moles in 25.0 g of carbon:

$$n = \frac{25.0\ \text{g}}{12.0\ \text{g/mol}} = 2.083\ \text{mol}$$

Find the number of ^{12}C nuclei in 25.0 g of carbon:

$$N(^{12}C) = 2.083\ \text{mol}(6.02 \times 10^{23}\ \text{nuclei/mol}) = 1.25 \times 10^{24}\ \text{nuclei}$$

Knowing that the ratio of ^{14}C to ^{12}C in the live sample was 1.3×10^{-12}, find the number of ^{14}C nuclei in 25.0 g *before* decay:

$$N_0(^{14}C) = (1.3 \times 10^{-12})(1.25 \times 10^{24}) = 1.63 \times 10^{12}\ \text{nuclei}$$

Find the initial activity of the sample:

$$R_0 = \lambda N_0 = (3.83 \times 10^{-12}\ \text{s}^{-1})(1.63 \times 10^{12}\ \text{nuclei})$$

$$= 6.24\ \text{decays/s} = 374\ \text{decays/min}$$

Solve Equation 44.7 for t:

$$e^{-\lambda t} = \frac{R}{R_0} \quad \rightarrow \quad -\lambda t = \ln\left(\frac{R}{R_0}\right) \quad \rightarrow \quad t = -\frac{1}{\lambda}\ln\left(\frac{R}{R_0}\right)$$

Evaluate t using $R = 250$ decays/min and $R_0 = 368$ decays/min:

$$t = -\frac{1}{3.83 \times 10^{-12}\ \text{s}^{-1}}\ln\left(\frac{250\ \text{min}^{-1}}{374\ \text{min}^{-1}}\right)$$

$$= 1.06 \times 10^{11}\ \text{s} = \boxed{3.3 \times 10^{3}\ \text{yr}}$$

Finalize Note that the time interval found here is on the same order of magnitude as the half-life, so ^{14}C is a valid isotope to use for this sample, as discussed in Conceptual Example 44.8.

Gamma Decay

Very often, a nucleus that undergoes radioactive decay is left in an excited energy state. The nucleus can then undergo a second decay to a lower energy state, perhaps to the ground state, by emitting a high-energy photon:

$$_{Z}^{A}X^* \rightarrow \; _{Z}^{A}X + \gamma \qquad (44.24)$$

◀ Gamma decay

where X* indicates a nucleus in an excited state. The typical half-life of an excited nuclear state is 10^{-10} s. Photons emitted in such a de-excitation process are called gamma rays. Such photons have very high energy (1 MeV to 1 GeV) relative to the energy of visible light (approximately 1 eV). Recall from Section 42.3 that the energy of a photon emitted or absorbed by an atom equals the difference in energy between the two electronic states involved in the transition. Similarly, a gamma-ray photon has an energy hf that equals the energy difference ΔE between two nuclear energy levels. When a nucleus decays by emitting a gamma ray, the only change in the nucleus is that it ends up in a lower energy state. There are no changes in Z, N, or A.

A nucleus may reach an excited state as the result of a violent collision with another particle. More common, however, is for a nucleus to be in an excited state after it has undergone alpha or beta decay. The following sequence of events represents a typical situation in which gamma decay occurs:

$$_{5}^{12}\text{B} \rightarrow \; _{6}^{12}\text{C}^* + e^- + \overline{\nu} \qquad (44.25)$$

$$_{6}^{12}\text{C}^* \rightarrow \; _{6}^{12}\text{C} + \gamma \qquad (44.26)$$

Figure 44.16 shows the decay scheme for ^{12}B, which undergoes beta decay to either of two levels of ^{12}C. It can either (1) decay directly to the ground state of ^{12}C by emitting a 13.4-MeV electron or (2) undergo beta decay to an excited state of ^{12}C* followed by gamma decay to the ground state. The latter process results in the emission of a 9.0-MeV electron and a 4.4-MeV photon.

The various pathways by which a radioactive nucleus can undergo decay are summarized in Table 44.3.

Figure 44.16 An energy-level diagram showing the initial nuclear state of a ^{12}B nucleus and two possible lower-energy states of the ^{12}C nucleus. The beta decay of the ^{12}B nucleus can result in either of two situations: the ^{12}C nucleus is in the ground state or in the excited state, in which case the nucleus is denoted as ^{12}C*. In the latter case, the beta decay to ^{12}C* is followed by a gamma decay to ^{12}C as the excited nucleus makes a transition to the ground state.

44.6 Natural Radioactivity

Radioactive nuclei are generally classified into two groups: (1) unstable nuclei found in nature, which give rise to **natural radioactivity,** and (2) unstable nuclei produced in the laboratory through nuclear reactions, which exhibit **artificial radioactivity.**

As Table 44.4 shows, there are three series of naturally occurring radioactive nuclei. Each series starts with a specific long-lived radioactive isotope whose half-life exceeds that of any of its unstable descendants. The three natural series begin with the isotopes ^{238}U, ^{235}U, and ^{232}Th, and the corresponding stable end products are three isotopes of lead: ^{206}Pb, ^{207}Pb, and ^{208}Pb. The fourth series in Table 44.4 begins with ^{237}Np and has as its stable end product ^{209}Bi. The element ^{237}Np is a

TABLE 44.3

Various Decay Pathways

Alpha decay	$_{Z}^{A}X \rightarrow \; _{Z-2}^{A-4}Y + _{2}^{4}\text{He}$
Beta decay (e^-)	$_{Z}^{A}X \rightarrow \; _{Z+1}^{A}Y + e^- + \overline{\nu}$
Beta decay (e^+)	$_{Z}^{A}X \rightarrow \; _{Z-1}^{A}Y + e^+ + \nu$
Electron capture	$_{Z}^{A}X + _{-1}^{0}e \rightarrow \; _{Z-1}^{A}Y + \nu$
Gamma decay	$_{Z}^{A}X^* \rightarrow \; _{Z}^{A}X + \gamma$

TABLE 44.4

The Four Radioactive Series

Series		Starting Isotope	Half-life (years)	Stable End Product
Uranium		$_{92}^{238}$U	4.47×10^9	$_{82}^{206}$Pb
Actinium	Natural	$_{92}^{235}$U	7.04×10^8	$_{82}^{207}$Pb
Thorium		$_{90}^{232}$Th	1.41×10^{10}	$_{82}^{208}$Pb
Neptunium		$_{93}^{237}$Np	2.14×10^6	$_{83}^{209}$Bi

Figure 44.17 Successive decays for the ^{232}Th series.

transuranic element (one having an atomic number greater than that of uranium) not found in nature. This element has a half-life of "only" 2.14×10^6 years.

Figure 44.17 shows the successive decays for the ^{232}Th series. First, ^{232}Th undergoes alpha decay to ^{228}Ra. Next, ^{228}Ra undergoes two successive beta decays to ^{228}Th. The series continues and finally branches when it reaches ^{212}Bi. At this point, there are two decay possibilities. The sequence shown in Figure 44.17 is characterized by a mass-number decrease of either 4 (for alpha decays) or 0 (for beta or gamma decays). The two uranium series are more complex than the ^{232}Th series. In addition, several naturally occurring radioactive isotopes, such as ^{14}C and ^{40}K, are not part of any decay series.

Because of these radioactive series, our environment is constantly replenished with radioactive elements that would otherwise have disappeared long ago. For example, because the Solar System is approximately 5×10^9 years old, the supply of ^{226}Ra (whose half-life is only 1 600 years) would have been depleted by radioactive decay long ago if it were not for the radioactive series starting with ^{238}U.

44.7   Nuclear Reactions

We have studied radioactivity, which is a spontaneous process in which the structure of a nucleus changes. It is also possible to stimulate changes in the structure of nuclei by bombarding them with energetic particles. Such collisions, which change the identity of the target nuclei, are called **nuclear reactions.** Rutherford was the first to observe them, in 1919, using naturally occurring radioactive sources for the bombarding particles. Since then, thousands of nuclear reactions have been observed following the development of charged-particle accelerators in the 1930s. With today's advanced technology in particle accelerators and particle detectors, it is possible to achieve particle energies of at least 1 000 GeV = 1 TeV. These high-energy particles are used to create new particles whose properties are helping to solve the mysteries of the nucleus.

Consider a reaction in which a target nucleus X is bombarded by a particle a, resulting in a daughter nucleus Y and an outgoing particle b:

Nuclear reaction ▶
$$a + X \;\rightarrow\; Y + b \tag{44.27}$$

Sometimes this reaction is written in the more compact form

$$X(a, b)Y$$

In Section 44.5, the Q value, or disintegration energy, of a radioactive decay was defined as the rest energy transformed to kinetic energy as a result of the decay process. Likewise, we define the **reaction energy** Q associated with a nuclear reaction as *the total change in rest energy resulting from the reaction:*

Reaction energy Q ▶
$$Q = (M_a + M_X - M_Y - M_b)c^2 \tag{44.28}$$

As an example, consider the reaction ^{7}Li$(p, \alpha)^4$He. The notation p indicates a proton, which is a hydrogen nucleus. Therefore, we can write this reaction in the expanded form

$$^1_1\text{H} + {}^7_3\text{Li} \;\rightarrow\; {}^4_2\text{He} + {}^4_2\text{He}$$

The Q value for this reaction is 17.3 MeV. A reaction such as this one, for which Q is positive, is called **exothermic.** A reaction for which Q is negative is called **endothermic.** To satisfy conservation of momentum, an endothermic reaction does not occur unless the bombarding particle has a kinetic energy greater than Q. (See Problem 49.) The minimum energy necessary for such a reaction to occur is called the **threshold energy.**

If particles a and b in a nuclear reaction are identical so that X and Y are also necessarily identical, the reaction is called a **scattering event.** If the kinetic energy of the system (a and X) before the event is the same as that of the system (b and

Y) after the event, it is classified as *elastic scattering*. If the kinetic energy of the system after the event is less than that before the event, the reaction is described as *inelastic scattering*. In this case, the target nucleus has been raised to an excited state by the event, which accounts for the difference in energy. The final system now consists of b and an excited nucleus Y*, and eventually it will become b, Y, and γ, where γ is the gamma-ray photon that is emitted when the system returns to the ground state. This elastic and inelastic terminology is identical to that used in describing collisions between macroscopic objects as discussed in Section 9.3.

In addition to energy and momentum, the total charge and total number of nucleons must be conserved in any nuclear reaction. For example, consider the reaction $^{19}\text{F}(\text{p}, \alpha)^{16}\text{O}$, which has a Q value of 8.11 MeV. We can show this reaction more completely as

$$^{1}_{1}\text{H} + ^{19}_{9}\text{F} \quad \rightarrow \quad ^{16}_{8}\text{O} + ^{4}_{2}\text{He} \tag{44.29}$$

The total number of nucleons before the reaction $(1 + 19 = 20)$ is equal to the total number after the reaction $(16 + 4 = 20)$. Furthermore, the total charge is the same before $(1 + 9)$ and after $(8 + 2)$ the reaction.

44.8 Nuclear Magnetic Resonance and Magnetic Resonance Imaging

In this section, we describe an important application of nuclear physics in medicine called magnetic resonance imaging. To understand this application, we first discuss the spin angular momentum of the nucleus. This discussion has parallels with the discussion of spin for atomic electrons.

In Chapter 42, we discussed that the electron has an intrinsic angular momentum, called spin. Nuclei also have spin because their component particles—neutrons and protons—each have spin $\frac{1}{2}$ as well as orbital angular momentum within the nucleus. All types of angular momentum obey the quantum rules that were outlined for orbital and spin angular momentum in Chapter 42. In particular, two quantum numbers associated with the angular momentum determine the allowed values of the magnitude of the angular momentum vector and its direction in space. The magnitude of the nuclear angular momentum is $\sqrt{I(I + 1)}\hbar$, where I is called the **nuclear spin quantum number** and may be an integer or a half-integer, depending on how the individual proton and neutron spins combine. The quantum number I is the analog to ℓ for the electron in an atom as discussed in Section 42.6. Furthermore, there is a quantum number m_I that is the analog to m_ℓ, in that the allowed projections of the nuclear spin angular momentum vector on the z axis are $m_I\hbar$. The values of m_I range from $-I$ to $+I$ in steps of 1. (In fact, for *any* type of spin with a quantum number S, there is a quantum number m_S that ranges in value from $-S$ to $+S$ in steps of 1.) Therefore, the maximum value of the z component of the spin angular momentum vector is $I\hbar$. Figure 44.18 is a vector model (see Section 42.6) illustrating the possible orientations of the nuclear spin vector and its projections along the z axis for the case in which $I = \frac{3}{2}$.

Nuclear spin has an associated nuclear magnetic moment, similar to that of the electron. The spin magnetic moment of a nucleus is measured in terms of the **nuclear magneton** μ_n, a unit of moment defined as

$$\mu_n \equiv \frac{e\hbar}{2m_p} = 5.05 \times 10^{-27} \, \text{J/T} \tag{44.30}$$

◄ Nuclear magneton

where m_p is the mass of the proton. This definition is analogous to that of the Bohr magneton μ_B, which corresponds to the spin magnetic moment of a free electron (see Section 42.6). Note that μ_n is smaller than μ_B ($= 9.274 \times 10^{-24} \, \text{J/T}$) by a factor of 1 836 because of the large difference between the proton mass and the electron mass.

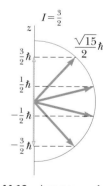

Figure 44.18 A vector model showing possible orientations of the nuclear spin angular momentum vector and its projections along the z axis for the case $I = \frac{3}{2}$.

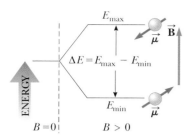

Figure 44.19 A nucleus with spin $\frac{1}{2}$ can occupy one of two energy states when placed in an external magnetic field. The lower-energy state E_{min} corresponds to the case where the spin is aligned with the field as much as possible according to quantum mechanics, and the higher-energy state E_{max} corresponds to the case where the spin is opposite the field as much as possible.

Figure 44.20 Experimental arrangement for nuclear magnetic resonance. The radio-frequency magnetic field created by the coil surrounding the sample and provided by the variable-frequency oscillator is perpendicular to the constant magnetic field created by the electromagnet. When the nuclei in the sample meet the resonance condition, the nuclei absorb energy from the radio-frequency field of the coil; this absorption changes the characteristics of the circuit in which the coil is included. Most modern NMR spectrometers use superconducting magnets at fixed field strengths and operate at frequencies of approximately 200 MHz.

Figure 44.21 A color-enhanced MRI scan of a human brain.

The magnetic moment of a free proton is $2.792\ 8\mu_n$. Unfortunately, there is no general theory of nuclear magnetism that explains this value. The neutron also has a magnetic moment, which has a value of $-1.913\ 5\mu_n$. The negative sign indicates that this moment is opposite the spin angular momentum of the neutron. The existence of a magnetic moment for the neutron is surprising in view of the neutron being uncharged. That suggests that the neutron is not a fundamental particle but rather has an underlying structure consisting of charged constituents. We shall explore this structure in Chapter 46.

The potential energy associated with a magnetic dipole moment $\vec{\mu}$ in an external magnetic field $\vec{B}$ is given by $-\vec{\mu} \cdot \vec{B}$ (Eq. 29.18). When the magnetic moment $\vec{\mu}$ is lined up with the field as closely as quantum physics allows, the potential energy of the dipole–field system has its minimum value E_{min}. When $\vec{\mu}$ is as antiparallel to the field as possible, the potential energy has its maximum value E_{max}. In general, there are other energy states between these values corresponding to the quantized directions of the magnetic moment with respect to the field. For a nucleus with spin $\frac{1}{2}$, there are only two allowed states, with energies E_{min} and E_{max}. These two energy states are shown in Figure 44.19.

It is possible to observe transitions between these two spin states using a technique called **NMR,** for **nuclear magnetic resonance.** A constant magnetic field ($\vec{B}$ in Fig. 44.19) is introduced to define a z axis and split the energies of the spin states. A second, weaker, oscillating magnetic field is then applied perpendicular to $\vec{B}$, creating a cloud of radio-frequency photons around the sample. When the frequency of the oscillating field is adjusted so that the photon energy matches the energy difference between the spin states, there is a net absorption of photons by the nuclei that can be detected electronically.

Figure 44.20 is a simplified diagram of the apparatus used in nuclear magnetic resonance. The energy absorbed by the nuclei is supplied by the generator producing the oscillating magnetic field. Nuclear magnetic resonance and a related technique called *electron spin resonance* are extremely important methods for studying nuclear and atomic systems and the ways in which these systems interact with their surroundings.

A widely used medical diagnostic technique called **MRI,** for **magnetic resonance imaging,** is based on nuclear magnetic resonance. Because nearly two-thirds of the atoms in the human body are hydrogen (which gives a strong NMR signal), MRI works exceptionally well for viewing internal tissues. The patient is placed inside a large solenoid that supplies a magnetic field that is constant in time but whose magnitude varies spatially across the body. Because of the variation in the field, hydrogen atoms in different parts of the body have different energy splittings between spin states, so the resonance signal can be used to provide information about the positions of the protons. A computer is used to analyze the position information to provide data for constructing a final image. Contrast in the final image among different types of tissues is created by computer analysis of the time intervals for the nuclei to return to the lower-energy spin state between pulses of radio-frequency photons. Contrast can be enhanced with the use of contrast agents such as gadolinium compounds or iron oxide nanoparticles taken orally or injected intravenously. An MRI scan showing incredible detail in internal body structure is shown in Figure 44.21.

The main advantage of MRI over other imaging techniques is that it causes minimal cellular damage. The photons associated with the radio-frequency signals used in MRI have energies of only about 10^{-7} eV. Because molecular bond strengths are much larger (approximately 1 eV), the radio-frequency radiation causes little cellular damage. In comparison, x-rays have energies ranging from 10^4 to 10^6 eV and can cause considerable cellular damage. Therefore, despite some individuals' fears of the word *nuclear* associated with MRI, the radio-frequency radiation involved is overwhelmingly safer than the x-rays that these individuals might accept more readily. A disadvantage of MRI is that the equipment required to conduct the procedure is very expensive, so MRI images are costly.

The magnetic field produced by the solenoid is sufficient to lift a car, and the radio signal is about the same magnitude as that from a small commercial broadcasting station. Although MRI is inherently safe in normal use, the strong magnetic field of the solenoid requires diligent care to ensure that no ferromagnetic materials are located in the room near the MRI apparatus. Several accidents have occurred, such as a 2000 incident in which a gun pulled from a police officer's hand discharged upon striking the machine.

Summary

DEFINITIONS

A nucleus is represented by the symbol $^A_Z X$, where A is the **mass number** (the total number of nucleons) and Z is the **atomic number** (the total number of protons). The total number of neutrons in a nucleus is the **neutron number** N, where $A = N + Z$. Nuclei having the same Z value but different A and N values are **isotopes** of each other.

The magnetic moment of a nucleus is measured in terms of the **nuclear magneton** μ_n, where

$$\mu_n \equiv \frac{e\hbar}{2m_p} = 5.05 \times 10^{-27}\,\text{J/T} \quad \textbf{(44.30)}$$

CONCEPTS AND PRINCIPLES

Assuming that nuclei are spherical, their radius is given by

$$r = r_0 A^{1/3} \quad \textbf{(44.1)}$$

where $r_0 = 1.2$ fm.

Nuclei are stable because of the **nuclear force** between nucleons. This short-range force dominates the Coulomb repulsive force at distances of less than about 2 fm and is independent of charge. Light stable nuclei have equal numbers of protons and neutrons. Heavy stable nuclei have more neutrons than protons. The most stable nuclei have Z and N values that are both even.

The difference between the sum of the masses of a group of separate nucleons and the mass of the compound nucleus containing these nucleons, when multiplied by c^2, gives the **binding energy** E_b of the nucleus. The binding energy of the nucleus of mass M_A can be calculated using the expression

$$E_b\,(\text{MeV}) =$$
$$[ZM(\text{H}) + Nm_n - M(^A_Z X)] \times 931.494\,\text{MeV/u}$$
$$\textbf{(44.2)}$$

where $M(\text{H})$ is the atomic mass of the neutral hydrogen atom, $M(^A_Z X)$ represents the atomic mass of an atom of the isotope $^A_Z X$, and m_n is the mass of the neutron.

The **liquid-drop model** of nuclear structure treats the nucleons as molecules in a drop of liquid. The four main contributions influencing binding energy are the volume effect, the surface effect, the Coulomb repulsion effect, and the symmetry effect. Summing such contributions results in the **semiempirical binding-energy formula:**

$$E_b = C_1 A - C_2 A^{2/3} - C_3 \frac{Z(Z-1)}{A^{1/3}} - C_4 \frac{(N-Z)^2}{A}$$
$$\textbf{(44.3)}$$

The **shell model**, or **independent-particle model**, assumes each nucleon exists in a shell and can only have discrete energy values. The stability of certain nuclei can be explained with this model.

(continued)

A radioactive substance decays by **alpha decay, beta decay,** or **gamma decay.** An alpha particle is the ^{4}He nucleus, a beta particle is either an electron (e^-) or a positron (e^+), and a gamma particle is a high-energy photon.

If a radioactive material contains N_0 radioactive nuclei at $t = 0$, the number N of nuclei remaining after a time t has elapsed is

$$N = N_0 e^{-\lambda t} \tag{44.6}$$

where λ is the **decay constant,** a number equal to the probability per second that a nucleus will decay. The **decay rate,** or **activity,** of a radioactive substance is

$$R = \left| \frac{dN}{dt} \right| = R_0 e^{-\lambda t} \tag{44.7}$$

where $R_0 = \lambda N_0$ is the activity at $t = 0$. The **half-life** $T_{1/2}$ is the time interval required for half of a given number of radioactive nuclei to decay, where

$$T_{1/2} = \frac{0.693}{\lambda} \tag{44.8}$$

In alpha decay, a helium nucleus is ejected from the parent nucleus with a discrete set of kinetic energies. A nucleus undergoing beta decay emits either an electron (e^-) and an antineutrino ($\bar{\nu}$) or a positron (e^+) and a neutrino (ν). The electron or positron is ejected with a range of energies. In **electron capture,** the nucleus of an atom absorbs one of its own electrons and emits a neutrino. In gamma decay, a nucleus in an excited state decays to its ground state and emits a gamma ray.

Nuclear reactions can occur when a target nucleus X is bombarded by a particle a, resulting in a daughter nucleus Y and an outgoing particle b:

$$a + X \rightarrow Y + b \tag{44.27}$$

The mass-energy conversion in such a reaction, called the **reaction energy** Q, is

$$Q = (M_a + M_X - M_Y - M_b)c^2 \tag{44.28}$$

Questions

□ denotes answer available in *Student Solutions Manual/Study Guide;* **O** denotes objective question

1. In Rutherford's experiment, assume an alpha particle is headed directly toward the nucleus of an atom. Why doesn't the alpha particle make physical contact with the nucleus?

2. **O** Figure 44.17 shows that the nucleus ^{212}Bi can decay both by emitting an alpha particle and by emitting an electron (with a different probability). As formed naturally, this isotope also emits gamma rays. Let W represent such a nucleus. Let X represent the daughter nucleus produced by its alpha decay, Y the daughter nucleus after beta decay, and Z the nucleus after gamma emission. (a) Rank the nuclei W, X, Y, and Z according to their mass numbers from the largest to the smallest. Note in your ranking any cases of equality. (b) Rank the nuclei according to their atomic numbers. (c) Rank the nuclei according to the numbers of neutrons they contain.

3. Explain why nuclei that are well off the line of stability in Figure 44.4 tend to be unstable.

4. Why are very heavy nuclei unstable?

5. **O** Consider two heavy nuclei X and Y having similar mass numbers. If X has the higher binding energy, which nucleus tends to be more unstable?

6. Why do nearly all the naturally occurring isotopes lie above the $N = Z$ line in Figure 44.4?

7. **O** (i) To predict the behavior of a nucleus in a fission reaction, which model would be more appropriate, (a) the liquid-drop model or (b) the shell model? (ii) Which model would be more successful in predicting the magnetic moment of a given nucleus? (iii) Which could better explain the gamma-ray spectrum of an excited nucleus?

8. "If no more people were to be born, the law of population growth would strongly resemble the radioactive decay law." Discuss this statement.

9. **O** Two samples of the same radioactive nuclide are prepared. Sample G has twice the initial activity of sample H. (i) How does the half-life of G compare with the half-life of H? (a) It is two times larger. (b) It is the same. (c) It is half as large. (ii) After each has passed through five half-lives, how do their activities compare? (a) G has more than twice the activity of H. (b) G has twice the activity of H. (c) G and H have the same activity. (d) G has lower activity than H.

10. *Do two halves make a whole?* What fraction of a radioactive sample has decayed after two half-lives have elapsed?

11. The radioactive nucleus $^{226}_{88}$Ra has a half-life of approximately 1.6×10^3 years. Being that the Solar System is approximately 5 billion years old, why do we still find this nucleus in nature (Fig. Q44.11)?

© 1990 Richard Megna, Fundamental Photographs

Figure Q44.11   Paint on the hands and numbers of this antique watch contains a small amount of natural radium mixed with a phosphorescent material. The decay of the radium causes the phosphor to glow continuously.

12. **O** A free neutron undergoes beta decay by emitting an electron with a half-life of 614 s. Can a free proton undergo a similar decay? (a) yes, the same decay (b) yes, but by emitting a positron (c) yes, but with a very different half-life (d) no

13. If a nucleus such as ^{226}Ra initially at rest undergoes alpha decay, which has more kinetic energy after the decay, the alpha particle or the daughter nucleus? Explain your answer.

14. Can a nucleus emit alpha particles that have different energies? Explain.

15. **O** Which of the following quantities represents the reaction energy of a nuclear reaction?
    (a) (final mass − initial mass)$/c^2$
    (b) (initial mass − final mass)$/c^2$
    (c) (final mass − initial mass)c^2
    (d) (initial mass − final mass)c^2
    (e) none of these choices

16. Suppose it could be shown that the cosmic-ray intensity at the Earth's surface was much greater 10 000 years ago. How would this difference affect what we accept as valid carbon-dated values of the age of ancient samples of once-living matter? Explain your answer.

17. How many values of I_z are possible for $I = \frac{5}{2}$? For $I = 3$?

18. **O** In nuclear magnetic resonance, how does increasing the value of the constant magnetic field change the frequency of the radio-frequency field that excites a particular transition? (a) The frequency is proportional to the square of the constant field. (b) The frequency is directly proportional to the constant field. (c) The frequency is independent of the constant field. (d) The frequency is inversely proportional to the constant field. (e) The frequency is proportional to the reciprocal of the square of the constant field.

19. Do all natural events have causes? Is the Universe intelligible? Give reasons for your answers. *Note:* You may wish to consider again Question 16 in Chapter 6 on whether the future is determinate.

Problems

Web**Assign**   The Problems from this chapter may be assigned online in WebAssign.

Thomson**NOW**   Sign in at **www.thomsonedu.com** and go to ThomsonNOW to assess your understanding of this chapter's topics with additional quizzing and conceptual questions.

1, 2, 3 denotes straightforward, intermediate, challenging;   □ denotes full solution available in *Student Solutions Manual/Study Guide;*   ▲ denotes coached solution with hints available at **www.thomsonedu.com;**   denotes developing symbolic reasoning;   ● denotes asking for qualitative reasoning;   ⚑ denotes computer useful in solving problem

Section 44.1 Some Properties of Nuclei

1. What is the order of magnitude of the number of protons in your body? Of the number of neutrons? Of the number of electrons?

2. **Review problem.** Singly ionized carbon is accelerated through 1 000 V and passed into a mass spectrometer to determine the isotopes present (see Chapter 29). The magnitude of the magnetic field in the spectrometer is 0.200 T. (a) Determine the orbit radii for the ^{12}C and the ^{13}C isotopes as they pass through the field. (b) Show that the ratio of radii may be written in the form

$$\frac{r_1}{r_2} = \sqrt{\frac{m_1}{m_2}}$$

and verify that your radii in part (a) agree with this equation.

3. (a) What fraction of the space in a tank of hydrogen gas at 0°C and 1 atm is occupied by the hydrogen molecules themselves? Assume each hydrogen atom is a sphere with diameter 0.100 nm and a hydrogen molecule consists of two such spheres in contact. (b) What fraction of the space within one hydrogen atom is occupied by its nucleus, of radius 1.20 fm?

4. ● In a Rutherford scattering experiment, alpha particles having kinetic energy of 7.70 MeV are fired toward a gold nucleus. (a) Use energy conservation to determine the distance of closest approach between the alpha particle and gold nucleus. Assume the nucleus remains at rest. (b) **What If?** Calculate the de Broglie wavelength for the 7.70-MeV alpha particle and compare it with the distance obtained in part (a). (c) Based on this comparison, why is it proper to treat the alpha particle as a particle and not as a wave in the Rutherford scattering experiment?

5. (a) Use energy methods to calculate the distance of closest approach for a head-on collision between an alpha particle having an initial energy of 0.500 MeV and a gold nucleus (^{197}Au) at rest. Assume the gold nucleus remains at rest during the collision. (b) What minimum initial speed must the alpha particle have to get as close as 300 fm?

2 = intermediate;   3 = challenging;   □ = SSM/SG;   ▲ = ThomsonNOW;   = symbolic reasoning;   ● = qualitative reasoning

6. Find the radius of (a) a nucleus of $^4_2$He and (b) a nucleus of $^{238}_{92}$U.

7. A star ending its life with a mass of two times the Sun's mass is expected to collapse, combining its protons and electrons to form a neutron star. Such a star could be thought of as a gigantic atomic nucleus. If a star of mass $2 \times 1.99 \times 10^{30}$ kg collapsed into neutrons ($m_n = 1.67 \times 10^{-27}$ kg), what would its radius be? Assume $r = r_0 A^{1/3}$.

8. Review problem. What would be the gravitational force exerted by each of two golf balls on the other if they were made of nuclear matter? Assume each ball has a 4.30-cm diameter and they are 1.00 m apart.

Section 44.2 Nuclear Binding Energy

9. Calculate the binding energy per nucleon for (a) ^{2}H, (b) ^{4}He, (c) ^{56}Fe, and (d) ^{238}U.

10. ● The iron isotope ^{56}Fe is near the peak of the stability curve. That is the fundamental reason that iron is more common in the Universe than heavier elements, as the spectra of the Sun and of many other stars reveal. Show that ^{56}Fe has a higher binding energy per nucleon than its neighbors ^{55}Mn and ^{59}Co. State how your results compare with Figure 44.5.

11. ▲ Nuclei having the same mass numbers are called *isobars*. The isotope $^{139}_{57}$La is stable. A radioactive isobar, $^{139}_{59}$Pr, is located below the line of stable nuclei in Figure 44.4 and decays by e^+ emission. Another radioactive isobar of $^{139}_{57}$La, $^{139}_{55}$Cs, decays by e^- emission and is located above the line of stable nuclei in Figure 44.4. (a) Which of these three isobars has the highest neutron-to-proton ratio? (b) Which has the greatest binding energy per nucleon? (c) Which do you expect to be heavier, $^{139}_{59}$Pr or $^{139}_{55}$Cs?

12. ● Two nuclei having the same mass number are called *isobars*. Calculate the difference in binding energy per nucleon for the isobars $^{23}_{11}$Na and $^{23}_{12}$Mg. How do you account for the difference?

13. ▲ A pair of nuclei for which $Z_1 = N_2$ and $Z_2 = N_1$ are called *mirror isobars* (the atomic and neutron numbers are interchanged). Binding-energy measurements on these nuclei can be used to obtain evidence of the charge independence of nuclear forces (that is, proton–proton, proton–neutron, and neutron–neutron nuclear forces are equal). Calculate the difference in binding energy for the two mirror isobars $^{15}_8$O and $^{15}_7$N. The electric repulsion among eight protons rather than seven accounts for the difference.

14. ● The energy required to construct a uniformly charged sphere of total charge Q and radius R is $U = 3k_eQ^2/5R$, where k_e is the Coulomb constant (see Problem 64). Assume a ^{40}Ca nucleus contains 20 protons uniformly distributed in a spherical volume. (a) How much energy is required to counter their electrical repulsion according to the above equation? *Suggestion:* First calculate the radius of a ^{40}Ca nucleus. (b) Calculate the binding energy of ^{40}Ca. (c) Explain what you can conclude from comparing the result of part (b) with that of part (a).

15. Calculate the minimum energy required to remove a neutron from the $^{43}_{20}$Ca nucleus.

Section 44.3 Nuclear Models

16. ● (a) In the liquid-drop model of nuclear structure, why does the surface-effect term $-C_2 A^{2/3}$ have a negative sign? (b) **What If?** The binding energy of the nucleus increases as the volume-to-surface ratio increases. Calculate this ratio for both spherical and cubical shapes and explain which is more plausible for nuclei.

17. Using the graph in Figure 44.5, estimate how much energy is released when a nucleus of mass number 200 fissions into two nuclei each of mass number 100.

18. (a) Use the semiempirical binding-energy formula to compute the binding energy for $^{56}_{26}$Fe. (b) What percentage is contributed to the binding energy by each of the four terms?

Section 44.4 Radioactivity

19. A sample of radioactive material contains 1.00×10^{15} atoms and has an activity of 6.00×10^{11} Bq. What is its half-life?

20. The half-life of ^{131}I is 8.04 days. On a certain day, the activity of an iodine-131 sample is 6.40 mCi. What is its activity 40.2 days later?

21. A freshly prepared sample of a certain radioactive isotope has an activity of 10.0 mCi. After 4.00 h, its activity is 8.00 mCi. (a) Find the decay constant and half-life. (b) How many atoms of the isotope were contained in the freshly prepared sample? (c) What is the sample's activity 30.0 h after it is prepared?

22. From the equation expressing the law of radioactive decay, derive the following useful formulas for the decay constant and the half-life, in terms of the time interval Δt during which the decay rate decreases from R_0 to R:

$$\lambda = \frac{1}{\Delta t} \ln\left(\frac{R_0}{R}\right) \qquad T_{1/2} = \frac{(\ln 2)\Delta t}{\ln(R_0/R)}$$

23. The radioactive isotope ^{198}Au has a half-life of 64.8 h. A sample containing this isotope has an initial activity ($t = 0$) of 40.0 μCi. Calculate the number of nuclei that decay in the time interval between $t_1 = 10.0$ h and $t_2 = 12.0$ h.

24. A radioactive nucleus has half-life $T_{1/2}$. A sample containing these nuclei has initial activity R_0. Calculate the number of nuclei that decay during the interval between the times t_1 and t_2.

25. Consider a radioactive sample. Determine the ratio of the number of nuclei decaying during the first half of its half-life to the number of nuclei decaying during the second half of its half-life.

26. ● (a) The daughter nucleus formed in radioactive decay is often radioactive. Let N_{10} represent the number of parent nuclei at time $t = 0$, $N_1(t)$ the number of parent nuclei at time t, and λ_1 the decay constant of the parent. Suppose the number of daughter nuclei at time $t = 0$ is zero. Let $N_2(t)$ be the number of daughter nuclei at time t and let λ_2 be the decay constant of the daughter. Show that $N_2(t)$ satisfies the differential equation

$$\frac{dN_2}{dt} = \lambda_1 N_1 - \lambda_2 N_2$$

(b) Verify by substitution that this differential equation has the solution

$$N_2(t) = \frac{N_{10}\lambda_1}{\lambda_1 - \lambda_2}\left(e^{-\lambda_2 t} - e^{-\lambda_1 t}\right)$$

2 = intermediate;   3 = challenging;   ☐ = SSM/SG;   ▲ = ThomsonNOW;   = symbolic reasoning;   ● = qualitative reasoning

This equation is the law of successive radioactive decays. (c) ^{218}Po decays into ^{214}Pb with a half-life of 3.10 min, and ^{214}Pb decays into ^{214}Bi with a half-life of 26.8 min. On the same axes, plot graphs of $N_1(t)$ for ^{218}Po and $N_2(t)$ for ^{214}Pb. Let $N_{10} = 1\,000$ nuclei and choose values of t from 0 to 36 min in 2-min intervals. The curve for ^{214}Pb at first rises to a maximum and then starts to decay. At what instant t_m is the number of ^{214}Pb nuclei a maximum? (d) By applying the condition for a maximum $dN_2/dt = 0$, derive a symbolic equation for t_m in terms of λ_1 and λ_2. Explain whether the value obtained in part (c) agrees with this equation.

27. In an experiment on the transport of nutrients in the root structure of a plant, two radioactive nuclides X and Y are used. Initially, 2.50 times more nuclei of type X are present than of type Y. Precisely three days later, there are 4.20 times more nuclei of type X than of type Y. Isotope Y has a half-life of 1.60 d. What is the half-life of isotope X?

Section 44.5 The Decay Processes

28. Identify the missing nuclide or particle (X):
    (a) $X \rightarrow {}^{65}_{28}Ni + \gamma$
    (b) ${}^{215}_{84}Po \rightarrow X + \alpha$
    (c) $X \rightarrow {}^{55}_{26}Fe + e^+ + \nu$
    (d) ${}^{109}_{48}Cd + X \rightarrow {}^{109}_{47}Ag + \nu$
    (e) ${}^{14}_{7}N + {}^{4}_{2}He \rightarrow X + {}^{17}_{8}O$

29. Find the energy released in the alpha decay
$$^{238}_{92}U \rightarrow {}^{234}_{90}Th + {}^{4}_{2}He$$
    You will find Table 44.2 useful.

30. A living specimen in equilibrium with the atmosphere contains one atom of ^{14}C (half-life = 5 730 yr) for every 7.7×10^{11} stable carbon atoms. An archeological sample of wood (cellulose, $C_{12}H_{22}O_{11}$) contains 21.0 mg of carbon. When the sample is placed inside a shielded beta counter with 88.0% counting efficiency, 837 counts are accumulated in one week. Assuming the cosmic-ray flux and the Earth's atmosphere have not changed appreciably since the sample was formed, find the age of the sample.

31. ● A sample consists of 1.00×10^6 radioactive nuclei with a half-life of 10.0 h. No other nuclei are present at time $t = 0$. The stable daughter nuclei accumulate in the sample as time goes on. (a) Derive an equation giving the number of daughter nuclei N_d as a function of time. (b) Sketch or describe a graph of the number of daughter nuclei as a function of time. What are the maximum and minimum numbers of daughter nuclei, and when do they occur? What are the maximum and minimum rates of change in the number of daughter nuclei, and when do they occur?

32. A ^{3}H nucleus beta decays into ^{3}He by creating an electron and an antineutrino according to the reaction
$$^{3}_{1}H \rightarrow {}^{3}_{2}He + e^- + \overline{\nu}$$
    The symbols in this reaction refer to nuclei. Write the reaction referring to neutral atoms by adding one electron to both sides. Then use Table 44.2 to determine the total energy released in this reaction.

33. The nucleus $^{15}_{8}O$ decays by electron capture. The nuclear reaction is written
$$^{15}_{8}O + e^- \rightarrow {}^{15}_{7}N + \nu$$

(a) Write the process going on for a single particle within the nucleus. (b) Write the decay process referring to neutral atoms. (c) Determine the energy of the neutrino. Disregard the daughter's recoil.

34. Determine which decays can occur spontaneously:
    (a) ${}^{40}_{20}Ca \rightarrow e^+ + {}^{40}_{19}K$
    (b) ${}^{98}_{44}Ru \rightarrow {}^{4}_{2}He + {}^{94}_{42}Mo$
    (c) ${}^{144}_{60}Nd \rightarrow {}^{4}_{2}He + {}^{140}_{58}Ce$

Section 44.6 Natural Radioactivity

35. Enter the correct isotope symbol in each open square in Figure P44.35, which shows the sequences of decays in the natural radioactive series starting with the long-lived isotope uranium-235 and ending with the stable nucleus lead-207.

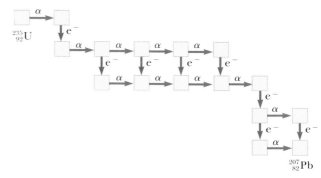

Figure P44.35

36. A rock sample contains traces of ^{238}U, ^{235}U, ^{232}Th, ^{208}Pb, ^{207}Pb, and ^{206}Pb. Analysis shows that the ratio of the amount of ^{238}U to ^{206}Pb is 1.164. (a) Assuming the rock originally contained no lead, determine the age of the rock. (b) What should be the ratios of ^{235}U to ^{207}Pb and of ^{232}Th to ^{208}Pb so that they would yield the same age for the rock? Ignore the minute amounts of the intermediate decay products in the decay chains. *Note:* This form of multiple dating gives reliable geological dates.

37. *Indoor air pollution.* Uranium is naturally present in rock and soil. At one step in its series of radioactive decays, ^{238}U produces the chemically inert gas radon-222, with a half-life of 3.82 days. The radon seeps out of the ground to mix into the atmosphere, typically making open air radioactive with activity 0.3 pCi/L. In homes, ^{222}Rn can be a serious pollutant, accumulating to reach much higher activities in enclosed spaces. If the radon radioactivity exceeds 4 pCi/L, the Environmental Protection Agency suggests taking action to reduce it, such as by reducing infiltration of air from the ground. (a) Convert the activity 4 pCi/L to units of becquerels per cubic meter. (b) How many ^{222}Rn atoms are in one cubic meter of air displaying this activity? (c) What fraction of the mass of the air does the radon constitute?

38. ● The most common isotope of radon is ^{222}Rn, which has half-life 3.82 days. (a) What fraction of the nuclei that were on the Earth one week ago are now undecayed? (b) Of those that existed one year ago? (c) In view of these results, explain why radon remains a problem, contributing significantly to our background radiation exposure.

2 = intermediate;   3 = challenging;   □ = SSM/SG;   ▲ = ThomsonNOW;   = symbolic reasoning;   ● = qualitative reasoning

Section 44.7 Nuclear Reactions

39. Identify the unknown nuclei and particles X and X′ in the following nuclear reactions:
    (a) $X + {}^{4}_{2}He \rightarrow {}^{24}_{12}Mg + {}^{1}_{0}n$
    (b) ${}^{235}_{92}U + {}^{1}_{0}n \rightarrow {}^{90}_{38}Sr + X + 2{}^{1}_{0}n$
    (c) $2{}^{1}_{1}H \rightarrow {}^{2}_{1}H + X + X'$

40. After determining that the Sun has existed for hundreds of millions of years, but before the discovery of nuclear physics, scientists could not explain why the Sun has continued to burn for such a long time interval. For example, if it were a coal fire, it would have burned up in about 3 000 yr. Assume the Sun, whose mass is 1.99×10^{30} kg, originally consisted entirely of hydrogen and its total power output is 3.85×10^{26} W. (a) Assuming the energy-generating mechanism of the Sun is the fusion of hydrogen into helium via the net reaction

    $$4({}^{1}_{1}H) + 2(e^{-}) \rightarrow {}^{4}_{2}He + 2\nu + \gamma$$

    calculate the energy (in joules) given off by this reaction. (b) Determine how many hydrogen atoms constitute the Sun. Take the mass of one hydrogen atom to be 1.67×10^{-27} kg. (c) If the total power output remains constant, after what time interval will all the hydrogen be converted into helium, making the Sun die? The actual projected lifetime of the Sun is about 10 billion years because only the hydrogen in a relatively small core is available as a fuel. Only in the core are temperatures and densities high enough for the fusion reaction to be self-sustaining.

41. ▲ Natural gold has only one isotope, ${}^{197}_{79}Au$. If natural gold is irradiated by a flux of slow neutrons, electrons are emitted. (a) Write the reaction equation. (b) Calculate the maximum energy of the emitted electrons.

42. A beam of 6.61-MeV protons is incident on a target of ${}^{27}_{13}Al$. Those that collide produce the reaction

    $$p + {}^{27}_{13}Al \rightarrow {}^{27}_{14}Si + n$$

    Ignoring any recoil of the product nucleus, determine the kinetic energy of the emerging neutrons. You may use Table 44.2.

43. The following reactions are observed:

    $${}^{9}_{4}Be + n \rightarrow {}^{10}_{4}Be + \gamma \qquad Q = 6.812 \text{ MeV}$$
    $${}^{9}_{4}Be + \gamma \rightarrow {}^{8}_{4}Be + n \qquad Q = -1.665 \text{ MeV}$$

    Using the mass of ${}^{9}Be$ from Table 44.2, calculate the masses of ${}^{8}Be$ and ${}^{10}Be$ in unified mass units to four decimal places.

44. (a) Suppose ${}^{10}_{5}B$ is struck by an alpha particle, releasing a proton and a product nucleus in the reaction. What is the product nucleus? (b) An alpha particle and a product nucleus are produced when ${}^{13}_{6}C$ is struck by a proton. What is the product nucleus?

Section 44.8 Nuclear Magnetic Resonance and Magnetic Resonance Imaging

45. The radio frequency at which a nucleus displays resonance absorption between spin states is called the Larmor frequency and is given by

    $$f = \frac{\Delta E}{h} = \frac{2\mu B}{h}$$

Calculate the Larmor frequency for (a) free neutrons in a magnetic field of 1.00 T, (b) free protons in a magnetic field of 1.00 T, and (c) free protons in the Earth's magnetic field at a location where the magnitude of the field is 50.0 μT.

46. Construct a diagram like that of Figure 44.18 for the cases when I equals (a) $\frac{5}{2}$ and (b) 4.

Additional Problems

47. (a) One method of producing neutrons for experimental use is bombardment of light nuclei with alpha particles. In the method used by James Chadwick in 1932, alpha particles emitted by polonium are incident on beryllium nuclei:

    $${}^{4}_{2}He + {}^{9}_{4}Be \rightarrow {}^{12}_{6}C + {}^{1}_{0}n$$

    What is the Q value? (b) Neutrons are also often produced by small-particle accelerators. In one design, deuterons accelerated in a Van de Graaff generator bombard other deuterium nuclei:

    $${}^{2}_{1}H + {}^{2}_{1}H \rightarrow {}^{3}_{2}He + {}^{1}_{0}n$$

    Is this reaction exothermic or endothermic? Calculate its Q value.

48. As part of his discovery of the neutron in 1932, James Chadwick determined the mass of the newly identified particle by firing a beam of fast neutrons, all having the same speed, at two different targets and measuring the maximum recoil speeds of the target nuclei. The maximum speeds arise when an elastic head-on collision occurs between a neutron and a stationary target nucleus. (a) Represent the masses and final speeds of the two target nuclei as m_1, v_1, m_2, and v_2 and assume Newtonian mechanics applies. Show that the neutron mass can be calculated from the equation

    $$m_n = \frac{m_1 v_1 - m_2 v_2}{v_2 - v_1}$$

    (b) Chadwick directed a beam of neutrons (produced from a nuclear reaction) on paraffin, which contains hydrogen. The maximum speed of the protons ejected was found to be 3.3×10^7 m/s. Because the velocity of the neutrons could not be determined directly, a second experiment was performed using neutrons from the same source and nitrogen nuclei as the target. The maximum recoil speed of the nitrogen nuclei was found to be 4.7×10^6 m/s. The masses of a proton and a nitrogen nucleus were taken as 1 u and 14 u, respectively. What was Chadwick's value for the neutron mass?

49. When the nuclear reaction represented by Equation 44.27 is endothermic, the reaction energy Q is negative. For the reaction to proceed, the incoming particle must have a minimum energy called the threshold energy, E_{th}. Some fraction of the energy of the incident particle is transferred to the compound nucleus to conserve momentum. Therefore, E_{th} must be greater than the magnitude of Q. (a) Show that

    $$E_{th} = -Q\left(1 + \frac{M_a}{M_X}\right)$$

2 = intermediate;   3 = challenging;   ☐ = SSM/SG;   ▲ = ThomsonNOW;   ▨ = symbolic reasoning;   ● = qualitative reasoning

(b) Calculate the threshold energy of the incident alpha particle in the reaction

$$^{4}_{2}\text{He} + ^{14}_{7}\text{N} \rightarrow ^{17}_{8}\text{O} + ^{1}_{1}\text{H}$$

50. **Review problem.** (a) Is the mass of a hydrogen atom in its ground state larger or smaller than the sum of the masses of a proton and an electron? (b) What is the mass difference? (c) How large is the difference as a percentage of the total mass? (d) Is it large enough to affect the value of the atomic mass listed to six decimal places in Table 44.2?

51. Write the statement of a problem for which the following equation appears in the solution. Determine the value of the unknown in the equation and identify its meaning.

10.012 937 u + 4.002 603 u

$$= 13.003\ 355\ \text{u} + 1.007\ 825\ \text{u} + Q/c^2$$

52. A by-product of some fission reactors is the isotope $^{239}_{94}\text{Pu}$, an alpha emitter having a half-life of 24 120 yr:

$$^{239}_{94}\text{Pu} \rightarrow ^{235}_{92}\text{U} + \alpha$$

Consider a sample of 1.00 kg of pure $^{239}_{94}\text{Pu}$ at $t = 0$. Calculate (a) the number of $^{239}_{94}\text{Pu}$ nuclei present at $t = 0$ and (b) the initial activity in the sample. (c) **What If?** For what times interval does the sample have to be stored if a "safe" activity level is 0.100 Bq?

53. (a) Can ^{57}Co decay by e^+ emission? Explain. You may use Table 44.2. (b) **What If?** Can ^{14}C decay by e^- emission? Explain. (c) If either answer is yes, what is the range of kinetic energies available for the beta particle?

54. (a) Find the radius of the $^{12}_{6}\text{C}$ nucleus. (b) Find the force of repulsion between a proton at the surface of a $^{12}_{6}\text{C}$ nucleus and the remaining five protons. (c) How much work (in MeV) has to be done to overcome this electric repulsion in transporting the last proton from a large distance up to the surface of the nucleus? (d) Repeat parts (a), (b), and (c) for $^{238}_{92}\text{U}$.

55. ● (a) Why is the beta decay $p \rightarrow n + e^+ + \nu$ forbidden for a free proton? (b) **What If?** Why is the same reaction possible if the proton is bound in a nucleus? For example, the following reaction occurs:

$$^{13}_{7}\text{N} \rightarrow ^{13}_{6}\text{C} + e^+ + \nu$$

(c) How much energy is released in the reaction given in part (b)? *Suggestion:* Add seven electrons to both sides of the reaction to write it for neutral atoms. You may use Table 44.2.

56. ▼ The activity of a radioactive sample was measured over 12 h, with the net count rates shown in the table.

Time (h)	Counting Rate (counts/min)
1.00	3 100
2.00	2 450
4.00	1 480
6.00	910
8.00	545
10.0	330
12.0	200

(a) Plot the logarithm of counting rate as a function of time. (b) Determine the decay constant and half-life of the radioactive nuclei in the sample. (c) What counting rate would you expect for the sample at $t = 0$? (d) Assuming the efficiency of the counting instrument is 10.0%, calculate the number of radioactive atoms in the sample at $t = 0$.

57. After the sudden release of radioactivity from the Chernobyl nuclear reactor accident in 1986, the radioactivity of milk in Poland rose to 2 000 Bq/L due to iodine-131 present in the grass eaten by dairy cattle. Radioactive iodine, with half-life 8.04 days, is particularly hazardous because the thyroid gland concentrates iodine. The Chernobyl accident caused a measurable increase in thyroid cancers among children in Belarus. (a) For comparison, find the activity of milk due to potassium. Assume 1 liter of milk contains 2.00 g of potassium, of which 0.011 7% is the isotope ^{40}K with a half-life 1.28×10^9 yr. (b) After what elapsed time would the activity due to iodine fall below that due to potassium?

58. When, after a reaction or disturbance of any kind, a nucleus is left in an excited state, it can return to its normal (ground) state by emission of a gamma-ray photon (or several photons). This process is illustrated by Equation 44.24. The emitting nucleus must recoil to conserve both energy and momentum. (a) Show that the recoil energy of the nucleus is

$$E_r = \frac{(\Delta E)^2}{2Mc^2}$$

where ΔE is the difference in energy between the excited and ground states of a nucleus of mass M. (b) Calculate the recoil energy of the ^{57}Fe nucleus when it decays by gamma emission from the 14.4-keV excited state. For this calculation, take the mass to be 57 u. *Suggestions:* When writing the equation for conservation of energy, use $(Mv)^2/2M$ for the kinetic energy of the recoiling nucleus. Also, assume $hf \ll Mc^2$ and use the binomial expansion.

59. A theory of nuclear astrophysics proposes that all the elements heavier than iron are formed in supernova explosions ending the lives of massive stars. Assume the amounts of ^{235}U and ^{238}U were equal at the time of the explosion. How long ago did the star(s) explode that released the elements that formed our Earth? The present $^{235}\text{U}/^{238}\text{U}$ ratio is 0.007 25. The half-lives of ^{235}U and ^{238}U are 0.704×10^9 yr and 4.47×10^9 yr.

60. Europeans named a certain direction in the sky as between the horns of Taurus the Bull. On the day they named as July 4, 1054, a brilliant light appeared there. Europeans left no surviving record of the supernova, which could be seen in daylight for some days. As it faded, it remained visible for years, dimming for a time with the 77.1-day half-life of the radioactive cobalt-56 that had been created in the explosion. (a) The remains of the star now form the Crab nebula (see the photographs opening Chapter 34). In it, the cobalt-56 has now decreased to what fraction of its original activity? (b) Suppose an American, of the people called the Anasazi, made a charcoal drawing of the supernova. The carbon-14 in the charcoal has now decayed to what fraction of its original activity?

2 = intermediate;   3 = challenging;   □ = SSM/SG;   ▲ = ThomsonNOW;   ⬛ = symbolic reasoning;   ● = qualitative reasoning

61. Review problem. Consider the Bohr model of the hydrogen atom, with the electron in the ground state. The magnetic field at the nucleus produced by the orbiting electron has a value of 12.5 T. (See Problem 1 in Chapter 30.) The proton can have its magnetic moment aligned in either of two directions perpendicular to the plane of the electron's orbit. The interaction of the proton's magnetic moment with the electron's magnetic field causes a difference in energy between the states with the two different orientations of the proton's magnetic moment. Find that energy difference in eV.

62. *Student determination of the half-life of* ^{137}*Ba.* The radioactive barium isotope ^{137}Ba has a relatively short half-life and can be easily extracted from a solution containing its parent cesium (^{137}Cs). This barium isotope is commonly used in an undergraduate laboratory exercise for demonstrating the radioactive decay law. Undergraduate students using modest experimental equipment took the data presented in Figure P44.62. Determine the half-life for the decay of ^{137}Ba using their data.

63. Free neutrons have a characteristic half-life of 10.4 min. What fraction of a group of free neutrons with kinetic energy 0.040 0 eV decays before traveling a distance of 10.0 km?

64. Review problem. Consider a model of the nucleus in which the positive charge (Ze) is uniformly distributed throughout a sphere of radius R. By integrating the energy density $\frac{1}{2}\epsilon_0 E^2$ over all space, show that the electric potential energy may be written

$$U = \frac{3Z^2 e^2}{20\pi\epsilon_0 R} = \frac{3k_r Z^2 e^2}{5R}$$

Problem 62 in Chapter 25 derived the same result by a different method.

65. ● In a piece of rock from the Moon, the ^{87}Rb content is assayed to be 1.82×10^{10} atoms per gram of material and the ^{87}Sr content is found to be 1.07×10^9 atoms per gram. (a) Calculate the age of the rock. (b) **What If?** Could the material in the rock actually be much older? What assumption is implicit in using the radioactive dating method? The relevant decay is ^{87}Rb $\rightarrow$ ^{87}Sr + e$^-$ + $\bar{\nu}$. The half-life of the decay is 4.75×10^{10} yr.

66. The ground state of $^{93}_{43}$Tc (molar mass 92.910 2 g/mol) decays by electron capture and e$^+$ emission to energy levels of the daughter (molar mass 92.906 8 g/mol in ground state) at 2.44 MeV, 2.03 MeV, 1.48 MeV, and 1.35 MeV. (a) For which of these levels are electron capture and e$^+$ decay allowed? (b) Identify the daughter and sketch the decay scheme, assuming all excited states de-excite by direct γ decay to the ground state.

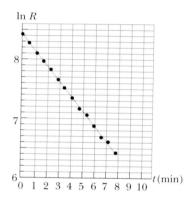

Figure P44.62

Answers to Quick Quizzes

44.1 **(i),** (b). The value of $N = A - Z$ is the same for all three nuclei. **(ii),** (a). The value of Z is the same for all three nuclei because they are all nuclei of nitrogen. **(iii),** (c). The value of A is the same for all three nuclei as seen by the unchanging superscript.

44.2 (e). A year of 365 days is equivalent to 365 d/5.01 d $\approx$ 73 half-lives. Therefore, the activity will be reduced after one year to approximately $(\frac{1}{2})^{73}$ (1.000 μCi) $\sim 10^{-22}$ μCi.

44.3 (b). In alpha decay, the atomic number decreases by two and the atomic mass number decreases by four.

44.4 (c). In e$^-$ decay, the atomic number increases by one and the atomic mass number stays fixed. None of the choices is consistent with e$^+$ decay, so we assume the decay must be by e$^-$.

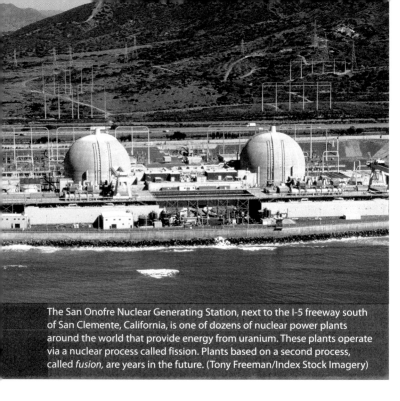

45.1 Interactions Involving Neutrons

45.2 Nuclear Fission

45.3 Nuclear Reactors

45.4 Nuclear Fusion

45.5 Radiation Damage

45.6 Radiation Detectors

45.7 Uses of Radiation

The San Onofre Nuclear Generating Station, next to the I-5 freeway south of San Clemente, California, is one of dozens of nuclear power plants around the world that provide energy from uranium. These plants operate via a nuclear process called fission. Plants based on a second process, called *fusion*, are years in the future. (Tony Freeman/Index Stock Imagery)

45 Applications of Nuclear Physics

In this chapter, we study two means for deriving energy from nuclear reactions: fission, in which a large nucleus splits into two smaller nuclei, and fusion, in which two small nuclei fuse to form a larger one. In both cases, the released energy can be used either constructively (as in electric power plants) or destructively (as in nuclear weapons). We also examine the ways in which radiation interacts with matter and look at several devices used to detect radiation. The chapter concludes with a discussion of some industrial and biological applications of radiation.

45.1 Interactions Involving Neutrons

Nuclear fission is the process that occurs in present-day nuclear reactors and ultimately results in energy supplied to a community by electrical transmission. Nuclear fusion is an area of active research, but it has not yet been commercially developed for the supply of energy. We will discuss fission first and then explore fusion in Section 45.4.

To understand nuclear fission and the physics of nuclear reactors, we must first understand how neutrons interact with nuclei. Because of their charge neutrality, neutrons are not subject to Coulomb forces and as a result do not interact electrically with electrons or the nucleus. Therefore, neutrons can easily penetrate deep into an atom and collide with the nucleus.

A fast neutron (energy greater than approximately 1 MeV) traveling through matter undergoes many collisions with nuclei. In each collision, the neutron gives up some of its kinetic energy to a nucleus. For fast neutrons in some materials, elastic collisions dominate. Materials for which that occurs are called **moderators** because they slow down (or moderate) the originally energetic neutrons very effectively. Moderator nuclei should be of low mass so that a large amount of kinetic energy is transferred to them in elastic collisions. For this reason, materials that are abundant in hydrogen, such as paraffin and water, are good moderators for neutrons.

Eventually, most neutrons bombarding a moderator become **thermal neutrons,** which means they are in thermal equilibrium with the moderator material. Their average kinetic energy at room temperature is, from Equation 21.4,

$$K_{\text{avg}} = \tfrac{3}{2}k_B T \approx \tfrac{3}{2}(1.38 \times 10^{-23}\,\text{J/K})(300\,\text{K}) = 6.21 \times 10^{-21}\,\text{J} \approx 0.04\,\text{eV}$$

which corresponds to a neutron root-mean-square speed of approximately 2 800 m/s. Thermal neutrons have a distribution of speeds, just as the molecules in a container of gas do (see Chapter 21). High-energy neutrons, those with energy of several MeV, *thermalize* (that is, their average energy reaches K_{avg}) in less than 1 ms when they are incident on a moderator.

Once the neutrons have thermalized and the energy of a particular neutron is sufficiently low, there is a high probability the neutron will be captured by a nucleus, an event that is accompanied by the emission of a gamma ray. This **neutron capture** reaction can be written

Neutron capture reaction ▶

$$^{1}_{0}\text{n} + {}^{A}_{Z}\text{X} \;\rightarrow\; {}^{A+1}_{Z}\text{X*} \;\rightarrow\; {}^{A+1}_{Z}\text{X} + \gamma \tag{45.1}$$

Once the neutron is captured, the nucleus ${}^{A+1}_{Z}\text{X*}$ is in an excited state for a very short time before it undergoes gamma decay. The product nucleus ${}^{A+1}_{Z}\text{X}$ is usually radioactive and decays by beta emission.

The neutron-capture rate for neutrons passing through any sample depends on the type of atoms in the sample and on the energy of the incident neutrons. The interaction of neutrons with matter increases with decreasing neutron energy because a slow neutron spends a larger time interval in the vicinity of target nuclei.

PITFALL PREVENTION 45.1
Binding Energy Reminder

Remember from Chapter 44 that binding energy is the absolute value of the system energy and is related to the system mass. Therefore, when considering Figure 44.5, imagine flipping it upside down for a curve representing system mass. In a fission reaction, the system mass decreases. This decrease in mass appears in the system as kinetic energy of the fission products.

45.2 Nuclear Fission

As mentioned in Section 44.2, nuclear **fission** occurs when a heavy nucleus, such as ${}^{235}\text{U}$, splits into two smaller nuclei. Fission is initiated when a heavy nucleus captures a thermal neutron as described by the first step in Equation 45.1. The absorption of the neutron creates a nucleus that is unstable and can change to a lower-energy configuration by splitting into two smaller nuclei. In such a reaction, the combined mass of the daughter nuclei is less than the mass of the parent nucleus, and the difference in mass is called the **mass defect.** Multiplying the mass defect by c^2 gives the numerical value of the released energy. Energy is released because the binding energy per nucleon of the daughter nuclei is approximately 1 MeV greater than that of the parent nucleus (see Fig. 44.5).

Nuclear fission was first observed in 1938 by Otto Hahn (1879–1968) and Fritz Strassman (1902–1980) following some basic studies by Fermi. After bombarding uranium with neutrons, Hahn and Strassman discovered among the reaction products two medium-mass elements, barium and lanthanum. Shortly thereafter, Lise Meitner (1878–1968) and her nephew Otto Frisch (1904–1979) explained what had happened. After absorbing a neutron, the uranium nucleus had split into two nearly equal fragments plus several neutrons. Such an occurrence was of considerable interest to physicists attempting to understand the nucleus, but it was to have even more far-reaching consequences. Measurements showed that approximately 200 MeV of energy was released in each fission event, and this fact was to affect the course of history.

The fission of ^{235}U by thermal neutrons can be represented by the reaction

$$^{1}_{0}n + {}^{235}_{92}U \rightarrow {}^{236}_{92}U^* \rightarrow X + Y + neutrons \qquad (45.2)$$

◀ **General fission reaction for ^{235}U**

where ^{236}U* is an intermediate excited state that lasts for approximately 10^{-12} s before splitting into medium-mass nuclei X and Y, which are called **fission fragments.** In any fission reaction, there are many combinations of X and Y that satisfy the requirements of conservation of energy and charge. In the case of uranium, for example, approximately 90 daughter nuclei can be formed.

Fission also results in the production of several neutrons, typically two or three. On average, approximately 2.5 neutrons are released per event. A typical fission reaction for uranium is

$$^{1}_{0}n + {}^{235}_{92}U \rightarrow {}^{141}_{56}Ba + {}^{92}_{36}Kr + 3({}^{1}_{0}n) \qquad (45.3)$$

Figure 45.1 shows a pictorial representation of the fission event in Equation 45.3.

Figure 45.2 is a graph of the distribution of fission products versus mass number A. The most probable products have mass numbers $A \approx 140$ and $A \approx 95$. Suppose these products are $^{140}_{53}I$ (with 87 neutrons) and $^{95}_{39}Y$ (with 56 neutrons). If these nuclei are located on the graph of Figure 44.4, it is seen that both are well above the line of stability. Because these fragments are very unstable owing to their unusually high number of neutrons, they almost instantaneously release two or three neutrons.

Let's estimate the disintegration energy Q released in a typical fission process. From Figure 44.5, we see that the binding energy per nucleon is approximately 7.2 MeV for heavy nuclei ($A \approx 240$) and approximately 8.2 MeV for nuclei of intermediate mass. The amount of energy released is 8.2 MeV $-$ 7.2 MeV $=$ 1 MeV per nucleon. Because there are a total of 235 nucleons in $^{235}_{92}U$, the energy released per fission event is approximately 235 MeV, a large amount of energy relative to the amount released in chemical processes. For example, the energy released in the combustion of one molecule of octane used in gasoline engines is about one millionth of the energy released in a single fission event!

Figure 45.1 A nuclear fission event. Before the event, a slow neutron approaches a ^{235}U nucleus. After the event, there are two lighter nuclei and three neutrons.

Quick Quiz 45.1 When a nucleus undergoes fission, the two daughter nuclei are generally radioactive. By which process are they most likely to decay? (a) alpha decay   (b) beta decay (e$^-$)   (c) beta decay (e$^+$)

Quick Quiz 45.2 Which of the following are possible fission reactions?

(a) $^{1}_{0}n + {}^{235}_{92}U \rightarrow {}^{140}_{54}Xe + {}^{94}_{38}Sr + 2({}^{1}_{0}n)$

(b) $^{1}_{0}n + {}^{235}_{92}U \rightarrow {}^{132}_{50}Sn + {}^{101}_{42}Mo + 3({}^{1}_{0}n)$

(c) $^{1}_{0}n + {}^{239}_{94}Pu \rightarrow {}^{137}_{53}I + {}^{97}_{41}Nb + 3({}^{1}_{0}n)$

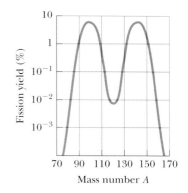

Figure 45.2 Distribution of fission products versus mass number for the fission of ^{235}U bombarded with thermal neutrons. Notice that the vertical axis is logarithmic.

EXAMPLE 45.1   **The Energy Released in the Fission of ^{235}U**

Calculate the energy released when 1.00 kg of ^{235}U fissions, taking the disintegration energy per event to be $Q = $ 208 MeV.

SOLUTION

Conceptualize Imagine a nucleus of ^{235}U absorbing a neutron and then splitting into two smaller nuclei and several neutrons as in Figure 45.1.

Categorize The problem statement tells us to categorize this example as one involving an energy analysis of nuclear fission.

Analyze Because $A = 235$ for uranium, one mole of this isotope has a mass of 235 g.

From the given mass of ^{235}U, find the number of nuclei in our sample:

$$N = \left(\frac{1.00 \times 10^3 \text{ g}}{235 \text{ g}}\right)(6.02 \times 10^{23} \text{ nuclei})$$

$$= 2.56 \times 10^{24} \text{ nuclei}$$

Find the total energy released when all nuclei undergo fission:

$$E = NQ = (2.56 \times 10^{24} \text{ nuclei})(208 \text{ MeV/nucleus})$$

$$= \boxed{5.32 \times 10^{26} \text{ MeV}}$$

Finalize Convert this energy to kWh:

$$E = (5.32 \times 10^{26} \text{ MeV})\left(\frac{1.60 \times 10^{-13} \text{ J}}{\text{MeV}}\right)\left(\frac{1 \text{ kWh}}{3.60 \times 10^6 \text{ J}}\right) = 2.37 \times 10^7 \text{ kWh}$$

which is enough energy to keep a 100-W lightbulb operating for 30 000 years! If the available fission energy in 1 kg of ^{235}U were suddenly released, it would be equivalent to detonating about 20 000 tons of TNT.

45.3 Nuclear Reactors

In Section 45.2, we learned that when ^{235}U fissions, one incoming neutron results in an average of 2.5 neutrons emitted per event. These neutrons can trigger other nuclei to fission. Because more neutrons are produced by the event than are absorbed, there is the possibility of an ever-building chain reaction (Active Fig. 45.3). Calculations show that if the chain reaction is not controlled (that is, if it does not proceed slowly), it can result in a violent explosion, with the sudden release of an enormous amount of energy. When the reaction is controlled, however, the energy released can be put to constructive use. In the United States, for example, nearly 20% of the electricity generated each year comes from nuclear power plants, and nuclear power is used extensively in many other countries, including France, Japan, and Germany.

A nuclear reactor is a system designed to maintain what is called a **self-sustained chain reaction.** This important process was first achieved in 1942 by Enrico Fermi and his team at the University of Chicago, using naturally occurring uranium as

ACTIVE FIGURE 45.3

A nuclear chain reaction initiated by the capture of a neutron. Uranium nuclei are shown in magenta, neutrons in green, and daughter nuclei in orange.

Sign in at www.thomsonedu.com and go to ThomsonNOW to observe the chain reaction.

Courtesy of Chicago Historical Society

Figure 45.4 Artist's rendition of the world's first nuclear reactor. Because of wartime secrecy, there are few photographs of the completed reactor, which was composed of layers of moderating graphite interspersed with uranium. A self-sustained chain reaction was first achieved on December 2, 1942. Word of the success was telephoned immediately to Washington, D.C., with this message: "The Italian navigator has landed in the New World and found the natives very friendly." The historic event took place in an improvised laboratory in the racquet court under the stands of the University of Chicago's Stagg Field, and the Italian navigator was Enrico Fermi.

the fuel.[1] In the first nuclear reactor (Fig. 45.4), Fermi placed bricks of graphite (carbon) between the fuel elements. Carbon nuclei are about 12 times more massive than neutrons, but after several collisions with carbon nuclei, a neutron is slowed sufficiently to increase its likelihood of fission with ^{235}U. In this design, carbon is the moderator; most modern reactors use water as the moderator.

Most reactors in operation today also use uranium as fuel. Naturally occurring uranium contains only 0.7% of the ^{235}U isotope, however, with the remaining 99.3% being ^{238}U. This fact is important to the operation of a reactor because ^{238}U almost never fissions. Instead, it tends to absorb neutrons without a subsequent fission event, producing neptunium and plutonium. For this reason, reactor fuels must be artificially *enriched* to contain at least a few percent ^{235}U.

To achieve a self-sustained chain reaction, an average of one neutron emitted in each ^{235}U fission must be captured by another ^{235}U nucleus and cause that nucleus to undergo fission. A useful parameter for describing the level of reactor operation is the **reproduction constant** K, defined as **the average number of neutrons from each fission event that cause another fission event.** As we have seen, K has an average value of 2.5 in the uncontrolled fission of uranium.

A self-sustained and controlled chain reaction is achieved when $K = 1$. Under this condition, the reactor is said to be **critical.** When $K < 1$, the reactor is subcritical and the reaction dies out. When $K > 1$, the reactor is supercritical and a runaway reaction occurs. In a nuclear reactor used to furnish power to a utility company, it is necessary to maintain a value of K close to 1. If K rises above this value, the internal energy produced in the reaction could melt the reactor.

Several types of reactor systems allow the kinetic energy of fission fragments to be transformed to other types of energy and eventually transferred out of the reactor plant by electrical transmission. The most common reactor in use in the United States is the pressurized-water reactor (Fig. 45.5, page 1334). We shall examine this type because its main parts are common to all reactor designs. Fission events in the uranium **fuel elements** in the reactor core raise the temperature of the water contained in the primary loop, which is maintained at high pressure to keep the water from boiling. (This water also serves as the moderator to slow down the neutrons released in the fission events with energy of approximately 2 MeV.) The hot water is pumped through a heat exchanger, where the internal energy of the water is transferred by conduction to the water contained in the secondary loop. The hot water in the secondary loop is converted to steam, which does work to drive a turbine–generator system to create electric power. The water in the secondary loop is isolated from the water in the primary loop to avoid contamination of the secondary water and the steam by radioactive nuclei from the reactor core.

National Accelerator Laboratory

ENRICO FERMI
Italian Physicist (1901–1954)
Fermi was awarded the Nobel Prize in Physics in 1938 for producing transuranic elements by neutron irradiation and for his discovery of nuclear reactions brought about by thermal neutrons. He made many other outstanding contributions to physics, including his theory of beta decay, the free-electron theory of metals, and the development of the world's first fission reactor in 1942. Fermi was truly a gifted theoretical and experimental physicist. He was also well known for his ability to present physics in a clear and exciting manner. "Whatever Nature has in store for mankind, unpleasant as it may be, men must accept, for ignorance is never better than knowledge."

---

[1] Although Fermi's reactor was the first manufactured nuclear reactor, there is evidence that a natural fission reaction may have sustained itself for perhaps hundreds of thousands of years in a deposit of uranium in Gabon, West Africa. See G. Cowan, "A Natural Fission Reactor," *Scientific American* **235**(5): 36, 1976.

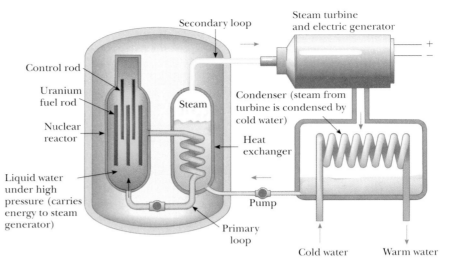

Figure 45.5   Main components of a pressurized-water nuclear reactor.

In any reactor, a fraction of the neutrons produced in fission leak out of the uranium fuel elements before inducing other fission events. If the fraction leaking out is too large, the reactor will not operate. The percentage lost is large if the fuel elements are very small because leakage is a function of the ratio of surface area to volume. Therefore, a critical feature of the reactor design is an optimal surface area–to–volume ratio of the fuel elements.

Control of Power Level

Safety is of critical importance in the operation of a nuclear reactor. The reproduction constant K must not be allowed to rise above 1, lest a runaway reaction occur. Consequently, reactor design must include a means of controlling the value of K.

The basic design of a nuclear reactor core is shown in Figure 45.6. The fuel elements consist of uranium that has been enriched in the ^{235}U isotope. To control the power level, **control rods** are inserted into the reactor core. These rods are made of materials such as cadmium that are very efficient in absorbing neutrons. By adjusting the number and position of the control rods in the reactor core, the K value can be varied and any power level within the design range of the reactor can be achieved.

Quick Quiz 45.3   To reduce the value of the reproduction constant K, do you (a) push the control rods deeper into the core or (b) pull the control rods farther out of the core?

Figure 45.6   Cross section of a reactor core showing the control rods, fuel elements containing enriched fuel, and moderating material, all surrounded by a radiation shield.

Safety and Waste Disposal

The 1979 near-disaster at a nuclear power plant at Three Mile Island in Pennsylvania and the 1986 accident at the Chernobyl reactor in Ukraine rightfully focused attention on reactor safety. The Three Mile Island accident was the result of inadequate control-room instrumentation and poor emergency-response training. There were no injuries or detectable health effects from the event, even though more than one-third of the fuel melted.

Unfortunately, at Chernobyl the activity of the materials released immediately after the accident totaled approximately 1.2×10^{19} Bq and resulted in the evacuation of 135 000 people. Thirty individuals died during the accident or shortly thereafter, and data from the Ukraine Radiological Institute suggest that more than 2 500 deaths could be attributed to the Chernobyl accident. In the period 1986–1997, there was a tenfold increase in the number of children contracting thyroid cancer from the ingestion of radioactive iodine in milk from cows that ate contaminated grass. One conclusion of an international conference studying the Ukraine accident was that the main causes of the Chernobyl accident were the coincidence of severe deficiencies in the reactor physical design and a violation of safety procedures. Most of these deficiencies have since been addressed at plants of similar design in Russia and neighboring countries of the former Soviet Union.

Commercial reactors achieve safety through careful design and rigid operating protocol, and only when these variables are compromised do reactors pose a danger. Radiation exposure and the potential health risks associated with such exposure are controlled by three layers of containment. The fuel and radioactive fission products are contained inside the reactor vessel. Should this vessel rupture, the reactor building acts as a second containment structure to prevent radioactive material from contaminating the environment. Finally, the reactor facilities must be in a remote location to protect the general public from exposure should radiation escape the reactor building.

A continuing concern about nuclear fission reactors is the safe disposal of radioactive material when the reactor core is replaced. This waste material contains long-lived, highly radioactive isotopes and must be stored over long time intervals in such a way that there is no chance of environmental contamination. At present, sealing radioactive wastes in waterproof containers and burying them in deep geologic repositories seems to be the most promising solution.

Transport of reactor fuel and reactor wastes poses additional safety risks. Accidents during transport of nuclear fuel could expose the public to harmful levels of radiation. The U.S. Department of Energy requires stringent crash tests of all containers used to transport nuclear materials. Container manufacturers must demonstrate that their containers will not rupture even in high-speed collisions.

Despite these risks, there are advantages to the use of nuclear power to be weighed against the risks. For example, nuclear power plants do not produce air pollution and greenhouse gases as do fossil fuel plants, and the supply of uranium on the Earth is predicted to last longer than the supply of fossil fuels. For each source of energy—whether nuclear, hydroelectric, fossil fuel, wind, solar or other—the risks must be weighed against the benefits and the availability of the energy source.

45.4   Nuclear Fusion

In Chapter 44, we found that the binding energy for light nuclei ($A < 20$) is much smaller than the binding energy for heavier nuclei, which suggests a process that is the reverse of fission. As mentioned in Section 39.9, when two light nuclei combine to form a heavier nucleus, the process is called nuclear **fusion.** Because the mass of the final nucleus is less than the combined masses of the original nuclei, there is a loss of mass accompanied by a release of energy.

Two examples of such energy-liberating fusion reactions are as follows:

$$^1_1\text{H} + ^1_1\text{H} \;\rightarrow\; ^2_1\text{H} + \text{e}^+ + \nu$$

$$^1_1\text{H} + ^2_1\text{H} \;\rightarrow\; ^3_2\text{He} + \gamma$$

These reactions occur in the core of a star and are responsible for the outpouring of energy from the star. The second reaction is followed by either hydrogen–helium fusion or helium–helium fusion:

$$^1_1\text{H} + ^3_2\text{He} \;\rightarrow\; ^4_2\text{He} + \text{e}^+ + \nu$$

$$^3_2\text{He} + ^3_2\text{He} \;\rightarrow\; ^4_2\text{He} + ^1_1\text{H} + ^1_1\text{H}$$

These fusion reactions are the basic reactions in the **proton–proton cycle,** believed to be one of the basic cycles by which energy is generated in the Sun and other stars that contain an abundance of hydrogen. Most of the energy production takes place in the Sun's interior, where the temperature is approximately 1.5×10^7 K.

PITFALL PREVENTION 45.2
Fission and Fusion

The words *fission* and *fusion* sound similar, but they correspond to different processes. Consider the binding-energy curve in Figure 44.5. There are two directions from which you can approach the peak of the curve so that energy is released: combining two light nuclei, or fusion, and separating a heavy nucleus into two lighter nuclei, or fission.

Because such high temperatures are required to drive these reactions, they are called **thermonuclear fusion reactions.** All the reactions in the proton–proton cycle are exothermic. An overview of the cycle is that four protons combine to generate an alpha particle, positrons, gamma rays, and neutrinos.

Quick Quiz 45.4   In the core of a star, hydrogen nuclei combine in fusion reactions. Once the hydrogen has been exhausted, fusion of helium nuclei can occur. If the star is sufficiently massive, fusion of heavier and heavier nuclei can occur once the helium is used up. Consider a fusion reaction involving two nuclei with the same value of A. For this reaction to be exothermic, which of the following values of A are impossible? (a) 12   (b) 20   (c) 28   (d) 64

EXAMPLE 45.2   **Energy Released in Fusion**

Find the total energy released in the fusion reactions in the proton–proton cycle.

SOLUTION

Conceptualize   The net nuclear result of the proton–proton cycle is to fuse four protons to form an alpha particle. Study the reactions above for the proton–proton cycle to be sure you understand how four protons become an alpha particle.

Categorize   We use concepts discussed in this section, so we categorize this example as a substitution problem.

Find the initial mass of four protons using atomic masses from Table 44.2:

$$4(1.007\ 825\ u) = 4.031\ 300\ u$$

Find the change in mass of the system as this value minus the mass of the resultant alpha particle:

$$4.031\ 300\ u - 4.002\ 603\ u = 0.028\ 697\ u$$

Convert this mass change into energy units:

$$E = 0.028\ 697\ u \times 931.494\ \text{MeV/u} = \boxed{26.7\ \text{MeV}}$$

This energy is shared among the alpha particle and other particles such as positrons, gamma rays, and neutrinos.

Terrestrial Fusion Reactions

The enormous amount of energy released in fusion reactions suggests the possibility of harnessing this energy for useful purposes. A great deal of effort is currently under way to develop a sustained and controllable thermonuclear reactor, a fusion power reactor. Controlled fusion is often called the ultimate energy source because of the availability of its fuel source: water. For example, if deuterium were used as the fuel, 0.12 g of it could be extracted from 1 gal of water at a cost of about four cents. This amount of deuterium would release approximately 10^{10} J if all nuclei underwent fusion. By comparison, 1 gal of gasoline releases approximately 10^8 J upon burning and costs far more than four cents.

An additional advantage of fusion reactors is that comparatively few radioactive by-products are formed. For the proton–proton cycle, for instance, the end product is safe, nonradioactive helium. Unfortunately, a thermonuclear reactor that can deliver a net power output spread over a reasonable time interval is not yet a reality, and many difficulties must be resolved before a successful device is constructed.

The Sun's energy is based in part on a set of reactions in which hydrogen is converted to helium. The proton–proton interaction is not suitable for use in a fusion reactor, however, because the event requires very high temperatures and densities. The process works in the Sun only because of the extremely high density of protons in the Sun's interior.

The reactions that appear most promising for a fusion power reactor involve deuterium ($^2_1$H) and tritium ($^3_1$H):

$$^2_1\text{H} + ^2_1\text{H} \rightarrow ^3_2\text{He} + ^1_0\text{n} \qquad Q = 3.27 \text{ MeV}$$

$$^2_1\text{H} + ^2_1\text{H} \rightarrow ^3_1\text{H} + ^1_1\text{H} \qquad Q = 4.03 \text{ MeV} \qquad \textbf{(45.4)}$$

$$^2_1\text{H} + ^3_1\text{H} \rightarrow ^4_2\text{He} + ^1_0\text{n} \qquad Q = 17.59 \text{ MeV}$$

As noted earlier, deuterium is available in almost unlimited quantities from our lakes and oceans and is very inexpensive to extract. Tritium, however, is radioactive ($T_{1/2} = 12.3$ yr) and undergoes beta decay to ^{3}He. For this reason, tritium does not occur naturally to any great extent and must be artificially produced.

One major problem in obtaining energy from nuclear fusion is that the Coulomb repulsive force between two nuclei, which carry positive charges, must be overcome before they can fuse. Figure 45.7 is a graph of potential energy as a function of the separation distance between two deuterons (deuterium nuclei, each having charge $+e$). The potential energy is positive in the region $r > R$, where the Coulomb repulsive force dominates ($R \approx 1$ fm), and negative in the region $r < R$, where the nuclear force dominates. The fundamental problem then is to give the two nuclei enough kinetic energy to overcome this repulsive force. This requirement can be accomplished by raising the fuel to extremely high temperatures (to approximately 10^8 K, far greater than the interior temperature of the Sun). At these high temperatures, the atoms are ionized and the system consists of a collection of electrons and nuclei, commonly referred to as a *plasma*.

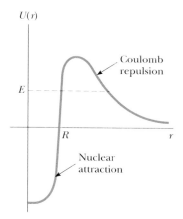

Figure 45.7 Potential energy as a function of separation distance between two deuterons. The Coulomb repulsive force is dominant at long range, and the nuclear force is dominant at short range, where R is on the order of 1 fm. If we neglect tunneling, the two deuterons require an energy E greater than the height of the barrier to undergo fusion.

EXAMPLE 45.3   **The Fusion of Two Deuterons**

For the nuclear force to overcome the repulsive Coulomb force, the separation distance between two deuterons must be approximately 1.0×10^{-14} m.

(A) Calculate the height of the potential barrier due to the repulsive force.

SOLUTION

Conceptualize Imagine moving two deuterons toward each other. As they move closer together, the Coulomb repulsion force becomes stronger. Work must be done on the system to push against this force, and this work appears in the system of two deuterons as electric potential energy.

Categorize We categorize this problem as one involving the electric potential energy of a system of two charged particles.

Analyze Evaluate the potential energy associated with two charges separated by a distance r (Eq. 25.13) for two deuterons:

$$U = k_e \frac{q_1 q_2}{r} = k_e \frac{(+e)^2}{r} = (8.99 \times 10^9 \text{ N} \cdot \text{m}^2/\text{C}^2) \frac{(1.60 \times 10^{-19} \text{ C})^2}{1.0 \times 10^{-14} \text{ m}}$$

$$= 2.3 \times 10^{-14} \text{ J} = \boxed{0.14 \text{ MeV}}$$

(B) Estimate the temperature required for a deuteron to overcome the potential barrier, assuming an energy of $\frac{3}{2} k_\text{B} T$ per deuteron (where k_B is Boltzmann's constant).

SOLUTION

Because the total Coulomb energy of the pair is 0.14 MeV, the Coulomb energy per deuteron is equal to 0.07 MeV $= 1.1 \times 10^{-14}$ J.

Set this energy equal to the average energy per deuteron:

$$\tfrac{3}{2} k_\text{B} T = 1.1 \times 10^{-14} \text{ J}$$

Solve for T:

$$T = \frac{2(1.1 \times 10^{-14} \text{ J})}{3(1.38 \times 10^{-23} \text{ J/K})} = \boxed{5.6 \times 10^8 \text{ K}}$$

(C) Find the energy released in the deuterium–deuterium reaction

$$^2_1H + ^2_1H \rightarrow ^3_1H + ^1_1H$$

SOLUTION

The mass of a single deuterium atom is equal to 2.014 102 u. Therefore, the total mass of the system before the reaction is 4.028 204 u.

Find the sum of the masses after the reaction: $3.016\ 049\ u + 1.007\ 825\ u = 4.023\ 874\ u$

Find the change in mass and convert to energy units: $4.028\ 204\ u - 4.023\ 874\ u$

$$= 0.004\ 33\ u$$

$$= 0.004\ 33\ u \times 931.494\ MeV/u = \boxed{4.03\ MeV}$$

Finalize   The calculated temperature in part (B) is too high because the particles in the plasma have a Maxwellian speed distribution (Section 21.5) and therefore some fusion reactions are caused by particles in the high-energy tail of this distribution. Furthermore, even those particles that do not have enough energy to overcome the barrier have some probability of tunneling through. When these effects are taken into account, a temperature of "only" 4×10^8 K appears adequate to fuse two deuterons in a plasma. In part (C), notice that the energy value is consistent with that already given in Equation 45.4.

What If?   Suppose the tritium resulting from the reaction in part (C) reacts with another deuterium in the reaction

$$^2_1H + ^3_1H \rightarrow ^4_2He + ^1_0n$$

How much energy is released in the sequence of two reactions?

Answer   The overall effect of the sequence of two reactions is that three deuterium nuclei have combined to form a helium nucleus, a hydrogen nucleus, and a neutron. The initial mass is $3(2.014\ 102\ u) = 6.042\ 306\ u$. After the reaction, the sum of the masses is $4.002\ 603\ u + 1.007\ 825\ u + 1.008\ 665 = 6.019\ 093\ u$. The excess mass is equal to 0.023 213 u, equivalent to an energy of 21.6 MeV. Notice that this value is the sum of the Q values for the second and third reactions in Equation 45.4.

The temperature at which the power generation rate in any fusion reaction exceeds the loss rate is called the **critical ignition temperature** T_{ignit}. This temperature for the deuterium–deuterium (D–D) reaction is 4×10^8 K. From the relationship $E \approx \frac{3}{2}k_BT$, the ignition temperature is equivalent to approximately 52 keV. The critical ignition temperature for the deuterium–tritium (D–T) reaction is approximately 4.5×10^7 K, or only 6 keV. A plot of the power $\mathscr{P}_{gen}$ generated by fusion versus temperature for the two reactions is shown in Figure 45.8. The straight green line represents the power $\mathscr{P}_{lost}$ lost via the radiation mechanism known as bremsstrahlung (Section 42.8). In this principal mechanism of energy loss, radiation (primarily x-rays) is emitted as the result of electron–ion collisions within the plasma. The intersections of the $\mathscr{P}_{lost}$ line with the $\mathscr{P}_{gen}$ curves give the critical ignition temperatures.

In addition to the high-temperature requirements, two other critical parameters determine whether or not a thermonuclear reactor is successful: the **ion density** n and **confinement time** τ, which is the time interval during which energy injected into the plasma remains within the plasma. British physicist J. D. Lawson has shown that both the ion density and confinement time must be large enough to ensure that more fusion energy is released than the amount required to raise the temperature of the plasma. For a given value of n, the probability of fusion between two particles increases as τ increases. For a given value of τ, the collision rate between nuclei increases as n increases. The product $n\tau$ is referred to as the **Lawson number** of a reaction. A graph of the value of $n\tau$ necessary to achieve a net energy output for the D–T and D–D reactions at different temperatures is

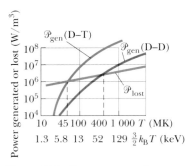

Figure 45.8   Power generated versus temperature for deuterium–deuterium (D–D) and deuterium–tritium (D–T) fusion. The green line represents power lost as a function of temperature. When the generation rate exceeds the loss rate, ignition takes place.

shown in Figure 45.9. In particular, **Lawson's criterion** states that a net energy output is possible for values of $n\tau$ that meet the following conditions:

$$n\tau \geq 10^{14} \text{ s/cm}^3 \quad \text{(D–T)} \qquad \textbf{(45.5)}$$

$$n\tau \geq 10^{16} \text{ s/cm}^3 \quad \text{(D–D)}$$

These values represent the minima of the curves in Figure 45.9.

Lawson's criterion was arrived at by comparing the energy required to raise the temperature of a given plasma with the energy generated by the fusion process.[2] The energy E_{in} required to raise the temperature of the plasma is proportional to the ion density n, which we can express as $E_{in} = C_1 n$, where C_1 is some constant. The energy generated by the fusion process is proportional to $n^2\tau$, or $E_{gen} = C_2 n^2 \tau$. This dependence may be understood by realizing that the fusion energy released is proportional to both the rate at which interacting ions collide ($\propto n^2$) and the confinement time τ. Net energy is produced when $E_{gen} > E_{in}$. When the constants C_1 and C_2 are calculated for different reactions, the condition that $E_{gen} \geq E_{in}$ leads to Lawson's criterion.

Current efforts are aimed at meeting Lawson's criterion at temperatures exceeding T_{ignit}. Although the minimum required plasma densities have been achieved, the problem of confinement time is more difficult. The two basic techniques under investigation for solving this problem are magnetic confinement and inertial confinement.

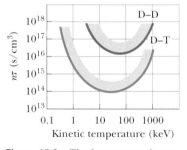

Figure 45.9 The Lawson number $n\tau$ at which net energy output is possible versus temperature for the D–T and D–D fusion reactions. The regions above the colored curves represent favorable conditions for fusion.

Magnetic Confinement

Many fusion-related plasma experiments use **magnetic confinement** to contain the plasma. A toroidal device called a **tokamak,** first developed in Russia, is shown in Figure 45.10a. A combination of two magnetic fields is used to confine and stabilize the plasma: (1) a strong toroidal field produced by the current in the toroidal windings surrounding a doughnut-shaped vacuum chamber and (2) a weaker "poloidal" field produced by the toroidal current. In addition to confining the

Figure 45.10 (a) Diagram of a tokamak used in the magnetic confinement scheme. (b) Interior view of the closed Tokamak Fusion Test Reactor (TFTR) vacuum vessel at the Princeton Plasma Physics Laboratory. (c) The National Spherical Torus Experiment (NSTX) that began operation in March 1999.

[2] Lawson's criterion neglects the energy needed to set up the strong magnetic field used to confine the hot plasma in a magnetic confinement approach. This energy is expected to be about 20 times greater than the energy required to raise the temperature of the plasma. It is therefore necessary either to have a magnetic energy recovery system or to use superconducting magnets.

plasma, the toroidal current is used to raise its temperature. The resultant helical magnetic field lines spiral around the plasma and keep it from touching the walls of the vacuum chamber. (If the plasma touches the walls, its temperature is reduced and heavy impurities sputtered from the walls "poison" it, leading to large power losses.)

One major breakthrough in magnetic confinement in the 1980s was in the area of auxiliary energy input to reach ignition temperatures. Experiments have shown that injecting a beam of energetic neutral particles into the plasma is a very efficient method of raising it to ignition temperatures. Radio-frequency energy input will probably be needed for reactor-size plasmas.

When it was in operation from 1982 to 1997, the Tokamak Fusion Test Reactor (TFTR, Fig. 45.10b) at Princeton University reported central ion temperatures of 510 million degrees Celsius, more than 30 times greater than the temperature at the center of the Sun. The $n\tau$ values in the TFTR for the D–T reaction were well above 10^{13} s/cm^3 and close to the value required by Lawson's criterion. In 1991, reaction rates of 6×10^{17} D–T fusions per second were reached in the JET tokamak at Abington, England.

One of the new generation of fusion experiments is the National Spherical Torus Experiment (NSTX) at the Princeton Plasma Physics Laboratory and shown in Figure 45.10c. This reactor was brought on line in February 1999 and has been running fusion experiments since then. Rather than the doughnut-shaped plasma of a tokamak, the NSTX produces a spherical plasma that has a hole through its center. The major advantage of the spherical configuration is its ability to confine the plasma at a higher pressure in a given magnetic field. This approach could lead to development of smaller, more economical fusion reactors.

An international collaborative effort involving the United States, the European Union, Japan, China, Korea, India, and Russia is currently under way to build a fusion reactor called ITER. This acronym stands for International Thermonuclear Experimental Reactor, although recently the emphasis has shifted to interpreting "iter" in terms of its Latin meaning, "the way." One reason proposed for this change is to avoid public misunderstanding and negative connotations toward the word *thermonuclear*. This facility will address the remaining technological and scientific issues concerning the feasibility of fusion power. The design is completed, and Cadarache, France, was chosen in June 2005 as the reactor site. Construction will require about 10 years, with fusion operation projected to begin in 2016. If the planned device works as expected, the Lawson number for ITER will be about six times greater than the current record holder, the JT-60 U tokamak in Japan. ITER is expected to produce 1.5 GW of power, and the energy content of the alpha particles inside the reactor will be so intense that they will sustain the fusion reaction, allowing the auxiliary energy sources to be turned off once the reaction is initiated.

EXAMPLE 45.4 | **Inside a Fusion Reactor**

In 1998, the JT-60U tokamak in Japan operated with a D–T plasma density of 4.8×10^{13} cm^{-3} at a temperature (in energy units) of 24.1 keV. It confined this plasma inside a magnetic field for 1.1 s.

(A) Do these data meet Lawson's criterion?

SOLUTION

Conceptualize With the help of the third of Equations 45.4, imagine many such reactions occurring in a plasma of high temperature and high density.

Categorize We use the concept of the Lawson number discussed in this section, so we categorize this example as a substitution problem.

Evaluate the Lawson number for the JT-60U: $n\tau = (4.8 \times 10^{13}\text{ cm}^{-3})(1.1\text{ s}) = 5.3 \times 10^{13}\text{ s/cm}^3$

This value is close to meeting Lawson's criterion of 10^{14} s/cm^3 for a D–T plasma given in Equation 45.5. In fact, scientists recorded a power gain of 1.25, indicating that the reactor operated slightly past the break-even point and produced more energy than it required to maintain the plasma.

(B) How does the plasma density compare with the density of atoms in an ideal gas when the gas is under standard conditions ($T = 0°$C and $P = 1$ atm)?

SOLUTION

Find the density of atoms in a sample of ideal gas by evaluating N_A/V_{mol}, where N_A is Avogadro's number and V_{mol} is the molar volume of an ideal gas under standard conditions, 2.24×10^{-2} m^3/mol:

$$\frac{N_A}{V_{mol}} = \frac{6.02 \times 10^{23} \text{ atoms/mol}}{2.24 \times 10^{-2} \text{ m}^3/\text{mol}} = 2.7 \times 10^{25} \text{ atoms/m}^3$$

$$= 2.7 \times 10^{19} \text{ atoms/cm}^3$$

This value is more than 500 000 times greater than the plasma density in the reactor.

Inertial Confinement

The second technique for confining a plasma, called **inertial confinement**, makes use of a D–T target that has a very high particle density. In this scheme, the confinement time is very short (typically 10^{-11} to 10^{-9} s), and, because of their own inertia, the particles do not have a chance to move appreciably from their initial positions. Therefore, Lawson's criterion can be satisfied by combining a high particle density with a short confinement time.

Laser fusion is the most common form of inertial confinement. A small D–T pellet, approximately 1 mm in diameter, is struck simultaneously by several focused, high-intensity laser beams, resulting in a large pulse of input energy that causes the surface of the fuel pellet to evaporate (Fig. 45.11). The escaping particles exert a third-law reaction force on the core of the pellet, resulting in a strong, inwardly moving compressive shock wave. This shock wave increases the pressure and density of the core and produces a corresponding increase in temperature. When the temperature of the core reaches ignition temperature, fusion reactions occur.

One of the leading laser fusion laboratories in the United States is the Omega facility at the University of Rochester in New York. This facility focuses 24 laser beams on the target. Currently under construction at the Lawrence Livermore National Laboratory in Livermore, California, is the National Ignition Facility. The research apparatus there will include 192 laser beams that can be focused on a deuterium–tritium pellet. Construction is expected to be final in 2009, with fusion ignition tests planned in 2010.

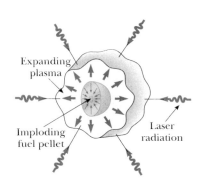

Figure 45.11 In inertial confinement, a D–T fuel pellet fuses when struck by several high-intensity laser beams simultaneously.

Fusion Reactor Design

In the D–T fusion reaction

$$^2_1\text{H} + ^3_1\text{H} \rightarrow ^4_2\text{He} + ^1_0\text{n} \qquad Q = 17.59 \text{ MeV}$$

the alpha particle carries 20% of the energy and the neutron carries 80%, or approximately 14 MeV. A diagram of the deuterium–tritium fusion reaction is shown in Active Figure 45.12. Because the alpha particles are charged, they are primarily absorbed by the plasma, causing the plasma's temperature to increase. In contrast, the 14-MeV neutrons, being electrically neutral, pass through the plasma and are absorbed by a surrounding blanket material, where their large kinetic energy is extracted and used to generate electric power.

One scheme is to use molten lithium metal as the neutron-absorbing material and to circulate the lithium in a closed heat-exchange loop, thereby producing steam and driving turbines as in a conventional power plant. Figure 45.13 (page 1342) shows a diagram of such a reactor. It is estimated that a blanket of lithium

ACTIVE FIGURE 45.12

Deuterium–tritium fusion. Eighty percent of the energy released is in the 14-MeV neutron.

Sign in at www.thomsonedu.com and go to ThomsonNOW to observe several fusion reactions and measure the energy released.

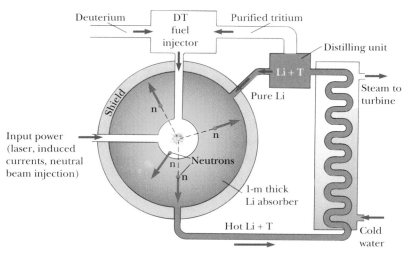

Figure 45.13 Diagram of a fusion reactor.

approximately 1 m thick will capture nearly 100% of the neutrons from the fusion of a small D–T pellet.

The capture of neutrons by lithium is described by the reaction

$$\frac{1}{0}n + \frac{6}{3}Li \rightarrow \frac{3}{1}H + \frac{4}{2}He$$

where the kinetic energies of the charged tritium $\frac{3}{1}H$ and alpha particle are converted to internal energy in the molten lithium. An extra advantage of using lithium as the energy-transfer medium is that the tritium produced can be separated from the lithium and returned as fuel to the reactor.

Advantages and Problems of Fusion

If fusion power can ever be harnessed, it will offer several advantages over fission-generated power: (1) low cost and abundance of fuel (deuterium), (2) impossibility of runaway accidents, and (3) decreased radiation hazard. Some of the anticipated problems and disadvantages include (1) scarcity of lithium, (2) limited supply of helium, which is needed for cooling the superconducting magnets used to produce strong confining fields, and (3) structural damage and induced radioactivity caused by neutron bombardment. If such problems and the engineering design factors can be resolved, nuclear fusion may become a feasible source of energy by the middle of the twenty-first century.

45.5 Radiation Damage

In Chapter 34, we learned that electromagnetic radiation is all around us in the form of radio waves, microwaves, light waves, and so on. In this section, we describe forms of radiation that can cause severe damage as they pass through matter, such as radiation resulting from radioactive processes and radiation in the form of energetic particles such as neutrons and protons.

The degree and type of damage depend on several factors, including the type and energy of the radiation and the properties of the matter. The metals used in nuclear reactor structures can be severely weakened by high fluxes of energetic neutrons because these high fluxes often lead to metal fatigue. The damage in such situations is in the form of atomic displacements, often resulting in major alterations in the properties of the material.

Radiation damage in biological organisms is primarily due to ionization effects in cells. A cell's normal operation may be disrupted when highly reactive ions are formed as the result of ionizing radiation. For example, hydrogen and the hydroxyl radical OH^- produced from water molecules can induce chemical reactions that may break bonds in proteins and other vital molecules. Furthermore, the ionizing radiation may affect vital molecules directly by removing electrons from their structure. Large doses of radiation are especially dangerous because damage to a great number of molecules in a cell may cause the cell to die. Although the death of a single cell is usually not a problem, the death of many cells may result in irreversible damage to the organism. Cells that divide rapidly, such as those of the digestive tract, reproductive organs, and hair follicles, are especially susceptible. In addition, cells that survive the radiation may become defective. These defective cells can produce more defective cells and can lead to cancer.

In biological systems, it is common to separate radiation damage into two categories: somatic damage and genetic damage. *Somatic damage* is that associated with any body cell except the reproductive cells. Somatic damage can lead to cancer or can seriously alter the characteristics of specific organisms. *Genetic damage* affects only reproductive cells. Damage to the genes in reproductive cells can lead to defective offspring. It is important to be aware of the effect of diagnostic treatments, such as x-rays and other forms of radiation exposure, and to balance the significant benefits of treatment with the damaging effects.

Damage caused by radiation also depends on the radiation's penetrating power. Alpha particles cause extensive damage, but penetrate only to a shallow depth in a material due to the strong interaction with other charged particles. Neutrons do not interact via the electric force and hence penetrate deeper, causing significant damage. Gamma rays are high-energy photons that can cause severe damage, but often pass through matter without interaction.

Several units have been used historically to quantify the amount, or dose, of any radiation that interacts with a substance.

> The **roentgen** (R) is that amount of ionizing radiation that produces an electric charge of 3.33×10^{-10} C in 1 cm^3 of air under standard conditions.

Equivalently, the roentgen is that amount of radiation that increases the energy of 1 kg of air by 8.76×10^{-3} J.

For most applications, the roentgen has been replaced by the rad (an acronym for *radiation absorbed dose*):

> One **rad** is that amount of radiation that increases the energy of 1 kg of absorbing material by 1×10^{-2} J.

Although the rad is a perfectly good physical unit, it is not the best unit for measuring the degree of biological damage produced by radiation because damage depends not only on the dose but also on the type of the radiation. For example, a given dose of alpha particles causes about ten times more biological damage than an equal dose of x-rays. The **RBE** (relative biological effectiveness) factor for a given type of radiation is **the number of rads of x-radiation or gamma radiation that produces the same biological damage as 1 rad of the radiation being used.** The RBE factors for different types of radiation are given in Table 45.1 (page 1344). The values are only approximate because they vary with particle energy and with the form of the damage. The RBE factor should be considered only a first-approximation guide to the actual effects of radiation.

Finally, the **rem** (radiation equivalent in man) is the product of the dose in rad and the RBE factor:

$$\text{Dose in rem} \equiv \text{dose in rad} \times \text{RBE} \qquad (45.6)$$ ◄ Radiation dose in rem

TABLE 45.1

RBE Factors for Several Types of Radiation

Radiation	RBE Factor
X-rays and gamma rays	1.0
Beta particles	1.0–1.7
Alpha particles	10–20
Thermal neutrons	4–5
Fast neutrons and protons	10
Heavy ions	20

Note: RBE = relative biological effectiveness.

According to this definition, 1 rem of any two types of radiation produces the same amount of biological damage. Table 45.1 shows that a dose of 1 rad of fast neutrons represents an effective dose of 10 rem, but 1 rad of gamma radiation is equivalent to a dose of only 1 rem.

Low-level radiation from natural sources such as cosmic rays and radioactive rocks and soil delivers to each of us a dose of approximately 0.13 rem/yr. This radiation, called *background radiation,* varies with geography, with the main factors being altitude (exposure to cosmic rays) and geology (radon gas released by some rock formations, deposits of naturally radioactive minerals).

The upper limit of radiation dose rate recommended by the U.S. government (apart from background radiation) is approximately 0.5 rem/yr. Many occupations involve much higher radiation exposures, so an upper limit of 5 rem/yr has been set for combined whole-body exposure. Higher upper limits are permissible for certain parts of the body, such as the hands and the forearms. A dose of 400 to 500 rem results in a mortality rate of approximately 50% (which means that half the people exposed to this radiation level die). The most dangerous form of exposure for most people is either ingestion or inhalation of radioactive isotopes, especially isotopes of those elements the body retains and concentrates, such as ^{90}Sr.

This discussion has focused on measurements of radiation dosage in units such as rads and rems because these units are still widely used. They have, however, been formally replaced with new SI units. The rad has been replaced with the *gray* (Gy), equal to 100 rad, and the rem has been replaced with the *sievert* (Sv), equal to 100 rem. Table 45.2 summarizes the older and the current SI units of radiation dosage.

45.6 Radiation Detectors

Particles passing through matter interact with the matter in several ways. The particle can, for example, ionize atoms, scatter from atoms, or be absorbed by atoms. Radiation detectors exploit these interactions to allow a measurement of the particle's energy, momentum, or charge and sometimes the very existence of the parti-

TABLE 45.2

Units for Radiation Dosage

Quantity	SI Unit	Symbol	Relation to Other SI units	Older Unit	Conversion
Absorbed dose	gray	Gy	= 1 J/kg	rad	1 Gy = 100 rad
Dose equivalent	sievert	Sv	= 1 J/kg	rem	1 Sv = 100 rem

(a)   (b)

Figure 45.14 (a) Artificially colored bubble-chamber photograph showing tracks of particles that have passed through the chamber. (b) This research scientist is studying a photograph of particle tracks made in a bubble chamber at Fermilab. The curved tracks are produced by charged particles moving through the chamber in the presence of an applied magnetic field. Negatively charged particles deflect in one direction, and positively charged particles deflect in the opposite direction.

cle if it is otherwise difficult to detect. Various devices have been developed for detecting radiation. These devices are used for a variety of purposes, including medical diagnoses, radioactive dating measurements, measuring background radiation, and measuring the mass, energy, and momentum of particles created in high-energy nuclear reactions.

In the early part of the 20th century, detectors were much simpler than those used today. We discuss three of these early detectors first. A **photographic emulsion** is the simplest example of a detector. A charged particle ionizes the atoms in an emulsion layer. The particle's path corresponds to a family of points at which chemical changes have occurred in the emulsion. When the emulsion is developed, the particle's track becomes visible. A **cloud chamber** contains a gas that has been supercooled to slightly below its usual condensation point. An energetic particle passing through ionizes the gas along the particle's path. The ions serve as centers for condensation of the supercooled gas. The particle's track can be seen with the naked eye and can be photographed. A magnetic field can be applied to determine the charges of the particles as well as their momentum and energy. A device called a **bubble chamber** uses a liquid (usually liquid hydrogen) maintained near its boiling point. Ions produced by incoming charged particles leave bubble tracks, which can be photographed (Fig. 45.14). Because the density of the detecting medium in a bubble chamber is much higher than the density of the gas in a cloud chamber, the bubble chamber has a much higher sensitivity.

More contemporary detectors involve more sophisticated processes. In an **ion chamber** (Fig. 45.15), electron–ion pairs are generated as radiation passes through a gas and produces an electrical signal. Two plates in the chamber are connected to a voltage supply and thereby maintained at different electric potentials. The positive plate attracts the electrons, and the negative plate attracts positive ions, causing a current pulse that is proportional to the number of electron–ion pairs produced when a particle passes through the chamber. When an ion chamber is

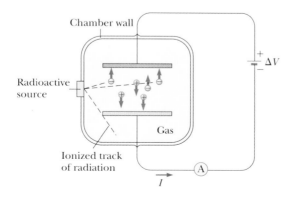

Figure 45.15 Simplified diagram of an ion chamber. The radioactive source creates electrons and positive ions that are collected by the charged plates. The current set up in the external circuit is proportional to a radioactive particle's kinetic energy if the particle stops in the chamber.

Figure 45.16 (a) Diagram of a Geiger counter. The voltage between the wire electrode and the metallic tube is usually approximately 1 000 V. (b) A scientist uses a Geiger counter to make a measurement.

used both to detect the presence of a particle and to measure its energy, it is called a **proportional counter.**

The **Geiger counter** (Fig. 45.16) is perhaps the most common form of ion chamber used to detect radioactivity. It can be considered the prototype of all counters that use the ionization of a medium as the basic detection process. A Geiger counter consists of a thin wire electrode aligned along the central axis of a cylindrical metallic tube filled with a gas at low pressure. The wire is maintained at a high positive electric potential (approximately 10^3 V) relative to the tube. When a high-energy particle resulting, for example, from a radioactive decay enters the tube through a thin window at one end, some of the gas atoms are ionized. The electrons removed from these atoms are attracted toward the wire electrode, and, in the process, they ionize other atoms in their path. This sequential ionization results in an *avalanche* of electrons that produces a current pulse. After the pulse has been amplified, it can either be used to trigger an electronic counter or be delivered to a loudspeaker that clicks each time a particle is detected. Although a Geiger counter easily detects the presence of a particle, the energy lost by the particle in the counter is *not* proportional to the current pulse produced. Therefore, a Geiger counter cannot be used to measure the energy of a particle.

A **semiconductor-diode detector** is essentially a reverse-bias p–n junction. Recall from Section 43.7 that a p–n junction passes current readily when forward-biased and prohibits a current when reverse-biased. As an energetic particle passes through the junction, electrons are excited into the conduction band and holes are formed in the valence band. The internal electric field sweeps the electrons toward the positive (n) side of the junction and the holes toward the negative (p) side. This movement of electrons and holes creates a pulse of current that is measured with an electronic counter. In a typical device, the duration of the pulse is 10^{-8} s.

A **scintillation counter** usually uses a solid or liquid material whose atoms are easily excited by radiation. The excited atoms then emit photons when they return to their ground state. Common materials used as scintillators are transparent crystals of sodium iodide and certain plastics. If the scintillator material is attached to a photomultiplier tube (Section 40.2), the photons emitted by the scintillator can be detected and an electrical signal produced.

Both the scintillator and the semiconductor-diode detector are much more sensitive than a Geiger counter mainly because of the higher density of the detecting medium. Both measure the total energy deposited in the detector, which can be very useful in particle identification. In addition, if the particle stops in the detector, both instruments can be used to measure the total particle energy.

Track detectors are devices used to view the tracks of charged particles directly. High-energy particles produced in particle accelerators may have energies ranging from 10^9 to 10^{12} eV. Therefore, they often cannot be stopped and cannot have their energy measured with the detectors already mentioned. Instead, the energy and momentum of these energetic particles are found from the curvature of their path in a magnetic field of known magnitude and direction.

A **spark chamber** is a counting device that consists of an array of conducting parallel plates and is capable of recording a three-dimensional track record. Even-numbered plates are grounded, and odd-numbered plates are maintained at a high electric potential (approximately 10 kV). The spaces between the plates contain an inert gas at atmospheric pressure. When a charged particle passes through the chamber, gas atoms are ionized, resulting in a current surge and visible sparks along the particle path. These sparks may be photographed or electronically detected and the data sent to a computer for path reconstruction and determination of particle mass, momentum, and energy.

Newer versions of the spark chamber have been developed. A **drift chamber** has thousands of high-voltage wires arrayed through the space of the detector, which is filled with gas. The result is an array of thousands of proportional counters. When a charged particle passes through the detector, it ionizes gas molecules and the ejected electrons drift toward the high-voltage wires, creating an electrical signal upon arrival. A computer detects the signals and reconstructs the path through the detector. A large-volume, sophisticated drift chamber that has provided significant results in studying particles formed in collisions of atoms is the Solenoidal Tracker at RHIC (STAR). (The acronym RHIC stands for Relativistic Heavy Ion Collider, a facility at Brookhaven National Laboratory that began operation in 2000.) This type of drift chamber is called a **time projection chamber.** A photograph of the STAR detector is shown in Figure 45.17.

Figure 45.17 The STAR detector at the Relativistic Heavy Ion Collider at Brookhaven National Laboratory.

45.7 Uses of Radiation

Nuclear physics applications are extremely widespread in manufacturing, medicine, and biology. In this section, we present a few of these applications and the underlying theories supporting them.

Tracing

Radioactive tracers are used to track chemicals participating in various reactions. One of the most valuable uses of radioactive tracers is in medicine. For example, iodine, a nutrient needed by the human body, is obtained largely through the intake of iodized salt and seafood. To evaluate the performance of the thyroid, the patient drinks a very small amount of radioactive sodium iodide containing ^{131}I, an artificially produced isotope of iodine (the natural, nonradioactive isotope is ^{127}I). The amount of iodine in the thyroid gland is determined as a function of time by measuring the radiation intensity at the neck area. How much of the isotope ^{131}I remains in the thyroid is a measure of how well that gland is functioning.

A second medical application is indicated in Figure 45.18. A solution containing radioactive sodium is injected into a vein in the leg, and the time at which the radioisotope arrives at another part of the body is detected with a radiation counter. The elapsed time is a good indication of the presence or absence of constrictions in the circulatory system.

Tracers are also useful in agricultural research. Suppose the best method of fertilizing a plant is to be determined. A certain element in a fertilizer, such as nitrogen, can be *tagged* (identified) with one of its radioactive isotopes. The fertilizer is then sprayed on one group of plants, sprinkled on the ground for a second group, and raked into the soil for a third. A Geiger counter is then used to track the nitrogen through each of the three groups.

Tracing techniques are as wide ranging as human ingenuity can devise. Today, applications range from checking how teeth absorb fluoride to monitoring how cleansers contaminate food-processing equipment to studying deterioration inside an automobile engine. In this last case, a radioactive material is used in the manufacture of the car's piston rings and the oil is checked for radioactivity to determine the amount of wear on the rings.

Figure 45.18   A tracer technique for determining the condition of the human circulatory system.

Materials Analysis

For centuries, a standard method of identifying the elements in a sample of material has been chemical analysis, which involves determining how the material reacts with various chemicals. A second method is spectral analysis, which works because each element, when excited, emits its own characteristic set of electromagnetic wavelengths. These methods are now supplemented by a third technique, **neutron activation analysis.** A disadvantage of both chemical and spectral methods is that a fairly large sample of the material must be destroyed for the analysis. In addition, extremely small quantities of an element may go undetected by either method. Neutron activation analysis has an advantage over chemical analysis and spectral analysis in both respects.

When a material is irradiated with neutrons, nuclei in the material absorb the neutrons and are changed to different isotopes, most of which are radioactive. For example, ^{65}Cu absorbs a neutron to become ^{66}Cu, which undergoes beta decay:

$$^1_0n + {}^{65}_{29}\text{Cu} \rightarrow {}^{66}_{29}\text{Cu} \rightarrow {}^{66}_{30}\text{Zn} + e^- + \bar{\nu}$$

The presence of the copper can be deduced because it is known that ^{66}Cu has a half-life of 5.1 min and decays with the emission of beta particles having maximum energies of 2.63 and 1.59 MeV. Also emitted in the decay of ^{66}Cu is a 1.04-MeV gamma ray. By examining the radiation emitted by a substance after it has been exposed to neutron irradiation, one can detect extremely small amounts of an element in that substance.

Neutron activation analysis is used routinely in a number of industries. In commercial aviation, for example, it is used to check airline luggage for hidden explosives. One nonroutine use is of historical interest. Napoleon died on the island of St. Helena in 1821, supposedly of natural causes. Over the years, suspicion has existed that his death was not all that natural. After his death, his head was shaved and locks of his hair were sold as souvenirs. In 1961, the amount of arsenic in a sample of this hair was measured by neutron activation analysis, and an unusually large quantity of arsenic was found. (Activation analysis is so sensitive that very small pieces of a single hair could be analyzed.) Results showed that the arsenic was fed to him irregularly. In fact, the arsenic concentration pattern corresponded to the fluctuations in the severity of Napoleon's illness as determined from historical records.

Art historians use neutron activation analysis to detect forgeries. The pigments used in paints have changed throughout history, and old and new pigments react differently to neutron activation. The method can even reveal hidden works of art behind existing paintings because an older, hidden layer of paint reacts differently than the surface layer to neutron activation.

Radiation Therapy

Radiation causes much damage to rapidly dividing cells. Therefore, it is useful in cancer treatment because tumor cells divide extremely rapidly. Several mechanisms can be used to deliver radiation to a tumor. In some cases, a narrow beam of x-rays or radiation from a source such as ^{60}Co is used as shown in Figure 45.19. In other situations, thin radioactive needles called *seeds* are implanted in the cancerous tissue. The radioactive isotope ^{131}I is used to treat cancer of the thyroid.

Food Preservation

Radiation is finding increasing use as a means of preserving food because exposure to high levels of radiation can destroy or incapacitate bacteria and mold spores (Fig. 45.20). Techniques include exposing foods to gamma rays, high-energy electron beams, and x-rays. Food preserved by such exposure can be placed in a sealed container (to keep out new spoiling agents) and stored for long periods of time.

Figure 45.19 This large machine is being set to deliver a dose of radiation from ^{60}Co in an effort to destroy a cancerous tumor. Cancer cells are especially susceptible to this type of therapy because they tend to divide more often than cells of healthy tissue nearby.

Figure 45.20 The strawberries on the left are untreated and have become moldy. The unspoiled strawberries on the right have been irradiated. The radiation has killed or incapacitated the mold spores that have spoiled the strawberries on the left.

There is little or no evidence of adverse effect on the taste or nutritional value of food from irradiation. The safety of irradiated foods has been endorsed by the World Health Organization, the Centers for Disease Control and Prevention, the U.S. Department of Agriculture, and the Food and Drug Administration.

Summary

ThomsonNOW Sign in at **www.thomsonedu.com** and go to ThomsonNOW to take a practice test for this chapter.

CONCEPTS AND PRINCIPLES

The probability that neutrons are captured as they move through matter generally increases with decreasing neutron energy. A **thermal neutron** is a slow-moving neutron that has a high probability of being captured by a nucleus in a **neutron capture event:**

$$_{0}^{1}n + {}_{Z}^{A}X \rightarrow {}_{Z}^{A+1}X^* \rightarrow {}_{Z}^{A+1}X + \gamma \qquad (45.1)$$

where ${}_{Z}^{A+1}X^*$ is an excited intermediate nucleus that rapidly emits a photon.

Nuclear fission occurs when a very heavy nucleus, such as ^{235}U, splits into two smaller **fission fragments.** Thermal neutrons can create fission in ^{235}U:

$$_{0}^{1}n + {}_{92}^{235}U \rightarrow {}_{92}^{236}U^* \rightarrow X + Y + \text{neutrons} \qquad (45.2)$$

where ^{236}U* is an intermediate excited state and X and Y are the fission fragments. On average, 2.5 neutrons are released per fission event. The fragments then undergo a series of beta and gamma decays to various stable isotopes. The energy released per fission event is approximately 200 MeV.

The **reproduction constant** K is the average number of neutrons released from each fission event that cause another event. In a fission reactor, it is necessary to maintain $K \approx 1$. The value of K is affected by such factors as reactor geometry, mean neutron energy, and probability of neutron capture.

In **nuclear fusion,** two light nuclei fuse to form a heavier nucleus and release energy. The major obstacle in obtaining useful energy from fusion is the large Coulomb repulsive force between the charged nuclei at small separation distances. The temperature required to produce fusion is on the order of 10^8 K, and at this temperature, all matter occurs as a plasma.

In a fusion reactor, the plasma temperature must reach the **critical ignition temperature,** the temperature at which the power generated by the fusion reactions exceeds the power lost in the system. The most promising fusion reaction is the D–T reaction, which has a critical ignition temperature of approximately 4.5×10^7 K. Two critical parameters in fusion reactor design are **ion density** n and **confinement time** τ, the time interval during which the interacting particles must be maintained at $T > T_{\text{ignit}}$. **Lawson's criterion** states that for the D–T reaction, $n\tau \geq 10^{14}$ s/cm^3.

Questions

☐ denotes answer available in *Student Solutions Manual/Study Guide;* **O** denotes objective question

1. Why is water a better shield against neutrons than lead or steel?

2. If a nucleus captures a slow-moving neutron, the product is left in a highly excited state, with an energy approximately 8 MeV above the ground state. Explain the source of the excitation energy.

3. **O** If the moderator were suddenly removed from a nuclear reactor in an electric generating station, what is the most likely consequence? (a) The reactor would go supercritical, and a runaway reaction would occur. (b) The nuclear reaction would proceed in the same way, but the reactor would overheat. (c) The reactor would become subcritical, and the reaction would die out. (d) No change would occur in the reactor's operation.

4. Discuss the advantages and disadvantages of fission reactors from the point of view of safety, pollution, and resources. Make a comparison with power generated from the burning of fossil fuels.

5. **O** On August 6, 1945, the United States dropped a nuclear bomb on Hiroshima. The Americans called it "little boy," and the Japanese called it the "original child bomb." The energy released was approximately 5×10^{13} J, equivalent to that from 12 000 metric tons of TNT. What is the order of magnitude of the mass converted into energy in this explosion? (a) 1 mg (b) 1 g (c) 1 kg (d) 1 000 kg (e) 10 000 metric tons

6. **O** A certain $^{235}_{92}$U nucleus absorbs a neutron and fissions into the products $^{137}_{53}$I and $^{96}_{39}$Y together with how many neutrons? (a) 0 (b) 1 (c) 2 (d) 3 (e) 4 (f) an indeterminate number

7. Why would a fusion reactor produce less radioactive waste than a fission reactor?

8. Lawson's criterion states that the product of ion density and confinement time must exceed a certain number before a break-even fusion reaction can occur. Why should these two parameters determine the outcome?

9. Discuss the similarities and differences between fusion and fission.

10. Discuss the advantages and disadvantages of fusion power from the viewpoint of safety, pollution, and resources.

11. **O** You may use Figure 44.5, the curve of binding energy, to answer this question. Three nuclear reactions take place, each involving 108 nucleons: (1) eighteen ^{6}Li nuclei fuse in pairs to form nine ^{12}C nuclei, (2) four nuclei each with 27 nucleons fuse in pairs to form two nuclei with 54 nucleons, and (3) one nucleus with 108 nucleons fissions to form two nuclei with 54 nucleons. Rank these three reactions according to the reaction energy from the largest positive Q value (representing energy output) to the largest negative value (representing energy input). Also include $Q = 0$ in your ranking to make clear which of the reactions put out energy and which absorb energy. Note any cases of equality in your ranking.

12. **O** In Figure 45.14, the particle tracks in the bubble chamber are generally spirals rather than sections of circles.

What is the primary reason for this shape? (a) The magnetic field is not perpendicular to the velocity of the particles. (b) The magnetic field is not uniform in space. (c) The forces on the particles increase with time. (d) The speeds of the particles decrease with time.

13. **O** Choose all correct answers. In the operation of a Geiger counter, the amplitude of the current pulse is (a) proportional to the kinetic energy of the particle producing the pulse, (b) proportional to the number of particles entering the tube to produce the pulse, (c) proportional to the RBE factor of the type of particle producing the pulse, or (d) independent of all these factors.

14. The design of a photomultiplier tube (Fig. 40.12) might suggest that any number of dynodes may be used to amplify a weak signal. What factors do you suppose would limit the amplification in this device?

15. **O** If an alpha particle and an electron have the same kinetic energy, which undergoes the greater deflection when passed through a magnetic field? (a) The alpha particle does. (b) The electron does. (c) They undergo the same deflection. (d) Neither is deflected.

16. **O** Working with radioactive materials at a laboratory over one year, (a) Tom received 1 rem of alpha radiation, (b) Karen received 1 rad of fast neutrons, (c) Paul received 1 rad of thermal neutrons as a whole-body dose, and (d) Ingrid received 1 rad of thermal neutrons to her hands only. Rank these four doses according to the likely amount of biological damage from the greatest to the least, noting any cases of equality.

17. *And swift, and swift past comprehension*
Turn round Earth's beauty and her might.
The heavens blaze in alternation
With deep and chill and rainy night.
In mighty currents foams the ocean
Up from the rocks' abyssal base,
With rock and sea torn into motion
In ever-swift celestial race.
Corrosive, choking smoke is spraying.
Above infernos, lava flies.
A perilous bridge, the land is swaying
Between them and the gaping skies.
And tempests bluster in a contest
From sea to land, from land to sea.
In rage they forge a chain around us
Of primal meaning, energy.
There flames a lightning disaster
Before the thunder, in its way.
But all Your servants honor, Master,
The gentle order of Your day.

Johann Wolfgang von Goethe wrote the song of the archangels in *Faust* half a century before the law of conservation of energy was recognized. Students often find it useful to list several "forms of energy," from kinetic to nuclear. Argue for or against the view that these lines of poetry make an obvious or oblique reference to every form of energy and energy transfer.

Problems

1, 2, 3 denotes straightforward, intermediate, challenging;   ☐ denotes full solution available in *Student Solutions Manual/Study Guide;*   ▲ denotes coached solution with hints available at **www.thomsonedu.com;**   denotes developing symbolic reasoning; ● denotes asking for qualitative reasoning;   ▪ denotes computer useful in solving problem

Section 45.2 Nuclear Fission

Note: Problem 47 in Chapter 25 and Problems 17 and 40 in Chapter 44 can be assigned with this chapter.

1. Burning one metric ton (1 000 kg) of coal can yield an energy of 3.30×10^{10} J. Fission of one nucleus of uranium-235 yields an average of approximately 208 MeV. What mass of uranium produces the same energy as one ton of coal?

2. Find the energy released in the fission reaction

$$_{0}^{1}n + {}_{92}^{235}U \; \rightarrow \; {}_{40}^{98}Zr + {}_{52}^{135}Te + 3({}_{0}^{1}n)$$

The atomic masses of the fission products are ${}_{40}^{98}Zr$, 97.912 7 u; and ${}_{52}^{135}Te$, 134.916 5 u.

3. Strontium-90 is a particularly dangerous fission product of ${}^{235}U$ because it is radioactive and it substitutes for calcium in bones. What other direct fission products would accompany it in the neutron-induced fission of ${}^{235}U$? *Note:* This reaction may release two, three, or four free neutrons.

4. (a) The following fission reaction is typical of those occurring in a nuclear electric generating station:

$$_{0}^{1}n + {}_{92}^{235}U \; \rightarrow \; {}_{56}^{141}Ba + {}_{36}^{92}Kr + 3({}_{0}^{1}n)$$

Find the energy released. The masses of the products are

$$M({}_{56}^{141}Ba) = 140.914 \, 4 \text{ u} \quad M({}_{36}^{92}Kr) = 91.926 \, 2 \text{ u}$$

(b) What fraction of the initial mass of the system is transformed?

5. ▲ List the nuclear reactions required to produce ${}^{233}U$ from ${}^{232}Th$ under fast neutron bombardment.

6. ● A reaction that has been considered as a source of energy is the absorption of a proton by a boron-11 nucleus to produce three alpha particles:

$$_{1}^{1}H + {}_{5}^{11}B \; \rightarrow \; 3({}_{2}^{4}He)$$

This reaction is an attractive possibility because boron is easily obtained from the Earth's crust. A disadvantage is that the protons and boron nuclei must have large kinetic energies for the reaction to take place. This requirement contrasts to the initiation of uranium fission by slow neutrons. (a) How much energy is released in each reaction? (b) Why must the reactant particles have high kinetic energies?

7. **Review problem.** Suppose enriched uranium containing 3.40% of the fissionable isotope ${}_{92}^{235}U$ is used as fuel for a ship. The water exerts an average friction force of magnitude 1.00×10^{5} N on the ship. How far can the ship travel per kilogram of fuel? Assume the energy released

per fission event is 208 MeV and the ship's engine has an efficiency of 20.0%.

8. A typical nuclear fission power plant produces approximately 1.00 GW of electrical power. Assume the plant has an overall efficiency of 40.0% and each fission produces 200 MeV of energy. Calculate the mass of ${}^{235}U$ consumed each day.

Section 45.3 Nuclear Reactors

9. ▲ It has been estimated that on the order of 10^{9} tons of natural uranium is available at concentrations exceeding 100 parts per million, of which 0.7% is the fissionable isotope ${}^{235}U$. Assume all the world's energy use (7×10^{12} J/s) were supplied by ${}^{235}U$ fission in conventional nuclear reactors, releasing 208 MeV for each reaction. How long would the supply last? The estimate of uranium supply is taken from K. S. Deffeyes and I. D. MacGregor, "World Uranium Resources," *Scientific American* **242**(1): 66, 1980.

10. ● To minimize neutron leakage from a reactor, the surface area-to-volume ratio should be a minimum. For a given volume V, calculate this ratio for (a) a sphere, (b) a cube, and (c) a parallelepiped of dimensions $a \times a \times 2a$. (d) Which of these shapes would have minimum leakage? Which would have maximum leakage? Explain your answers.

11. If the reproduction constant is 1.000 25 for a chain reaction in a fission reactor and the average time interval between successive fissions is 1.20 ms, by what factor does the reaction rate increase in one minute?

12. A large nuclear power reactor produces approximately 3 000 MW of power in its core. Three months after a reactor is shut down, the core power from radioactive byproducts is 10.0 MW. Assuming each emission delivers 1.00 MeV of energy to the power, find the activity in becquerels three months after the reactor is shut down.

13. ● A particle cannot generally be localized to distances much smaller than its de Broglie wavelength. This fact can be taken to mean that a slow neutron appears to be larger to a target particle than does a fast neutron in the sense that the slow neutron has probabilities of being found over a larger volume of space. For a thermal neutron at room temperature of 300 K, find (a) the linear momentum and (b) the de Broglie wavelength. State how this effective size compares with both nuclear and atomic dimensions.

14. The probability of a nuclear reaction increases dramatically when the incident particle is given energy above the "Coulomb barrier," which is the electric potential energy of the two nuclei when their surfaces barely touch. Compute

the Coulomb barrier for the absorption of an alpha particle by a gold nucleus.

Section 45.4 Nuclear Fusion

15. (a) Consider a fusion generator built to create 3.00 GW of power. Determine the rate of fuel burning in grams per hour if the D–T reaction is used. (b) Do the same for the D–D reaction assuming the reaction products are split evenly between (n, ^{3}He) and (p, ^{3}H).

16. ● Two nuclei having atomic numbers Z_1 and Z_2 approach each other with a total energy E. (a) When they are far apart, they interact only by electric repulsion. If they approach to a distance of 1.00×10^{-14} m, the nuclear force suddenly takes over to make them fuse. Find the minimum value of E, in terms of Z_1 and Z_2, required to produce fusion. (b) State how E depends on the atomic numbers. (c) If $Z_1 + Z_2$ is to have a certain target value such as 60, would it be energetically favorable to take $Z_1 = 1$ and $Z_2 = 59$, or $Z_1 = Z_2 = 30$, or what? Explain your answer. (d) Evaluate from your expression the minimum energy for fusion for the D–D and D–T reactions (the first and third reactions in Eq. 45.4).

17. ● **Review problem.** Consider the deuterium–tritium fusion reaction with the tritium nucleus at rest:

$$^2_1H + ^3_1H \rightarrow ^4_2He + ^1_0n$$

(a) Suppose the reactant nuclei will spontaneously fuse if their surfaces touch. From Equation 44.1, determine the required distance of closest approach between their centers. (b) What is the electric potential energy (in eV) at this distance? (c) Suppose the deuteron is fired straight at an originally stationary tritium nucleus with just enough energy to reach the required distance of closest approach. What is the common speed of the deuterium and tritium nuclei, in terms of the initial deuteron speed v_i, as they touch? *Note:* At this point, the two nuclei have a common velocity equal to the center-of-mass velocity. (d) Use energy methods to find the minimum initial deuteron energy required to achieve fusion. (e) Why does the fusion reaction actually occur at much lower deuteron energies than the energy calculated in part (d)?

18. Of all the hydrogen in the oceans, 0.030 0% of the mass is deuterium. The oceans have a volume of 317 million mi^3. (a) If nuclear fusion were controlled and all the deuterium in the oceans were fused to $^4_2$He, how many joules of energy would be released? (b) **What If?** World power consumption is approximately 7.00×10^{12} W. If consumption were 100 times greater, how many years would the energy calculated in part (a) last?

19. ▲ To understand why plasma containment is necessary, consider the rate at which an unconfined plasma would be lost. (a) Estimate the rms speed of deuterons in a plasma at 4.00×10^8 K. (b) **What If?** Estimate the order of magnitude of the time interval during which such a plasma would remain in a 10-cm cube if no steps were taken to contain it.

20. It has been suggested that fusion reactors are safe from explosion because the plasma never contains enough energy to do much damage. (a) In 1992, the TFTR reactor achieved an ion temperature of 4.0×10^8 K, an ion density of 2.0×10^{13} cm^{-3}, and a confinement time of

1.4 s. Calculate the amount of energy stored in the plasma of the TFTR reactor. (b) How many kilograms of water could be boiled away by this much energy? The plasma volume of the TFTR reactor is approximately 50 m^3.

21. **Review problem.** To confine a stable plasma, the magnetic energy density in the magnetic field (Eq. 32.14) must exceed the pressure $2nk_BT$ of the plasma by a factor of at least 10. In this problem, assume a confinement time $\tau = 1.00$ s. (a) Using Lawson's criterion, determine the ion density required for the D–T reaction. (b) From the ignition-temperature criterion, determine the required plasma pressure. (c) Determine the magnitude of the magnetic field required to contain the plasma.

22. One old prediction for the future was to have a fusion reactor supply energy to dissociate the molecules in garbage into separate atoms and then to ionize the atoms. This material could be put through a giant mass spectrometer so that trash would be a new source of isotopically pure elements, the mine of the future. Assuming an average atomic mass of 56 and an average charge of 26 (a high estimate, considering all the organic materials) at a beam current of 1.00 MA, how long would it take to process 1.00 metric ton of trash?

Section 45.5 Radiation Damage

23. A small building has become accidentally contaminated with radioactivity. The longest-lived material in the building is strontium-90. ($^{90}_{38}$Sr has an atomic mass 89.907 7 u, and its half-life is 29.1 yr. It is particularly dangerous because it substitutes for calcium in bones.) Assume the building initially contained 5.00 kg of this substance uniformly distributed throughout the building and the safe level is defined as less than 10.0 decays/min (which is small compared to background radiation). How long will the building be unsafe?

24. **Review problem.** A particular radioactive source produces 100 mrad of 2-MeV gamma rays per hour at a distance of 1.00 m from the source. (a) How long could a person stand at this distance before accumulating an intolerable dose of 1 rem? (b) **What If?** Assuming the radioactive source is a point source, at what distance would a person receive a dose of 10.0 mrad/h?

25. ● Assume an x-ray technician takes an average of eight x-rays per day and receives a dose of 5 rem/yr as a result. (a) Estimate the dose in rem per x-ray taken. (b) Explain how the technician's exposure compares with low-level background radiation.

26. *Lead shielding.* When gamma rays are incident on matter, the intensity of the gamma rays passing through the material varies with depth x as $I(x) = I_0e^{-\mu x}$, where I_0 is the intensity of the radiation at the surface of the material and μ is the absorption coefficient. For 0.400-MeV gamma rays in lead, the absorption coefficient is 1.59 cm^{-1}. (a) Determine the "half-thickness" for lead, that is, the thickness of lead that would absorb half the incident gamma rays. (b) What thickness reduces the radiation by a factor of 10^4?

27. ▲ A "clever" technician decides to warm some water for his coffee with an x-ray machine. If the machine produces 10.0 rad/s, over what time interval will the temperature of an insulated cup of water rise by 50.0°C?

28. **Review problem.** The danger to the body from a high dose of gamma rays is not due to the amount of energy absorbed; rather, it is due to the ionizing nature of the radiation. As an illustration, calculate the rise in body temperature that results if a "lethal" dose of 1 000 rad is absorbed strictly as internal energy. Take the specific heat of living tissue as 4 186 J/kg · °C.

29. Technetium-99 is used in certain medical diagnostic procedures. Assume 1.00×10^{-8} g of ^{99}Tc is injected into a 60.0-kg patient and half of the 0.140-MeV gamma rays are absorbed in the body. Determine the total radiation dose received by the patient.

30. To destroy a cancerous tumor, a dose of gamma radiation with a total energy of 2.12 J is to be delivered in 30.0 days from implanted sealed capsules containing palladium-103. Assume this isotope has a half-life of 17.0 d and emits gamma rays of energy 21.0 keV, which are entirely absorbed within the tumor. (a) Find the initial activity of the set of capsules. (b) Find the total mass of radioactive palladium that these "seeds" should contain.

31. Strontium-90 from the testing of nuclear bombs can still be found in the atmosphere. Each decay of ^{90}Sr releases 1.1 MeV of energy into the bones of a person who has had strontium replace his or her body's calcium. Assume a 70.0-kg person receives 1.00 μg of ^{90}Sr from contaminated milk. Calculate the absorbed dose rate (in joules per kilogram) in one year. Take the half-life of ^{90}Sr to be 29.1 yr.

Section 45.6 Radiation Detectors

32. Assume a photomultiplier tube (Fig. 40.12) has seven dynodes with potentials of 100, 200, 300, . . . , 700 V. The average energy required to free an electron from the dynode surface is 10.0 eV. Assume only one electron is incident and the tube functions with 100% efficiency. (a) How many electrons are freed at the first dynode? (b) How many electrons are collected at the last dynode? (c) What is the energy available to the counter for each electron?

33. In a Geiger tube, the voltage between the electrodes is typically 1.00 kV and the current pulse discharges a 5.00-pF capacitor. (a) What is the energy amplification of this device for a 0.500-MeV electron? (b) How many electrons participate in the avalanche caused by the single initial electron?

34. ● (a) Your grandmother recounts to you how, as young children, your father, aunts, and uncles made the screen door slam continually as they ran between the house and the back yard. The time interval between one slam and the next varied randomly, but the average slamming rate remained constant at 38.0/h from dawn to dusk every summer day. If the slamming rate suddenly dropped to zero, the children would have found a nest of baby field mice or gotten into some other mischief requiring adult intervention. Approximately how long after the last screen-door slam would a prudent and attentive parent wait before leaving her or his tasks to see about the children? Explain your reasoning. (b) A student wishes to measure the half-life of a radioactive substance using a small sample. Consecutive clicks of her Geiger counter are randomly spaced in time. The counter registers 372 counts during one 5.00-min interval and 337 counts

during the next 5.00 min. The average background rate is 15 counts per minute. Find the most probable value for the half-life. (c) Estimate the uncertainty in the half-life determination. Explain your reasoning.

Section 45.7 Uses of Radiation

35. During the manufacture of a steel engine component, radioactive iron (^{59}Fe) is included in the total mass of 0.200 kg. The component is placed in a test engine when the activity due to this isotope is 20.0 μCi. After a 1 000-h test period, some of the lubricating oil is removed from the engine and found to contain enough ^{59}Fe to produce 800 disintegrations/min/L of oil. The total volume of oil in the engine is 6.50 L. Calculate the total mass worn from the engine component per hour of operation. The half-life of ^{59}Fe is 45.1 d.

36. You want to find out how many atoms of the isotope ^{65}Cu are in a small sample of material. You bombard the sample with neutrons to ensure that on the order of 1% of these copper nuclei absorb a neutron. After activation, you turn off the neutron flux and then use a highly efficient detector to monitor the gamma radiation that comes out of the sample. Assume half of the ^{66}Cu nuclei emit a 1.04-MeV gamma ray in their decay. (The other half of the activated nuclei decay directly to the ground state of ^{66}Ni.) If after 10 min (two half-lives) you have detected 10^4 MeV of photon energy at 1.04 MeV, (a) approximately how many ^{65}Cu atoms are in the sample? (b) Assume the sample contains natural copper. Refer to the isotopic abundances listed in Table 44.2 and estimate the total mass of copper in the sample.

37. *Neutron activation analysis* is a method for chemical analysis at the level of isotopes. When a sample is irradiated by neutrons, radioactive atoms are produced continuously and then decay according to their characteristic half-lives. (a) Assume one species of radioactive nuclei is produced at a constant rate R and its decay is described by the conventional radioactive decay law. Assuming irradiation begins at time $t = 0$, show that the number of radioactive atoms accumulated at time t is

$$N = \frac{R}{\lambda}\left(1 - e^{-\lambda t}\right)$$

(b) What is the maximum number of radioactive atoms that can be produced?

38. *A thickness gauge.* When gamma rays are incident on matter, the intensity of the gamma rays passing through the material varies with depth x as $I(x) = I_0 e^{-\mu x}$, where I_0 is the intensity of the radiation at the surface of the material and μ is the absorption coefficient. For low-energy gamma rays in steel, take the absorption coefficient to be 0.720 mm^{-1}. (a) Determine the "half-thickness" for steel, that is, the thickness of steel that would absorb half the incident gamma rays. (b) In a steel mill, the thickness of sheet steel passing into a roller is measured by monitoring the intensity of gamma radiation reaching a detector below the rapidly moving metal from a small source immediately above the metal. If the thickness of the sheet changes from 0.800 mm to 0.700 mm, by what percentage does the gamma-ray intensity change?

2 = intermediate;   3 = challenging;   □ = SSM/SG;   ▲ = ThomsonNOW;   = symbolic reasoning;   ● = qualitative reasoning

Additional Problems

39. Carbon detonations are powerful nuclear reactions that temporarily tear apart the cores inside massive stars late in their lives. These blasts are produced by carbon fusion, which requires a temperature of approximately 6×10^8 K to overcome the strong Coulomb repulsion between carbon nuclei. (a) Estimate the repulsive energy barrier to fusion, using the temperature required for carbon fusion. (In other words, what is the average kinetic energy of a carbon nucleus at 6×10^8 K?) (b) Calculate the energy (in MeV) released in each of these "carbon-burning" reactions:

$$^{12}\text{C} + {}^{12}\text{C} \rightarrow {}^{20}\text{Ne} + {}^4\text{He}$$

$$^{12}\text{C} + {}^{12}\text{C} \rightarrow {}^{24}\text{Mg} + \gamma$$

(c) Calculate the energy in kilowatt-hours given off when 2.00 kg of carbon completely fuses according to the first reaction.

40. **Review problem.** Consider a nucleus at rest, which then spontaneously splits into two fragments of masses m_1 and m_2. Show that the fraction of the total kinetic energy carried by fragment m_1 is

$$\frac{K_1}{K_{\text{tot}}} = \frac{m_2}{m_1 + m_2}$$

and the fraction carried by m_2 is

$$\frac{K_2}{K_{\text{tot}}} = \frac{m_1}{m_1 + m_2}$$

assuming relativistic corrections can be ignored. *Note:* If the parent nucleus was moving before the decay, the fission products still share the kinetic energy as shown as long as all velocities are measured in the center-of-mass frame of reference, in which the total momentum of the system is zero.

41. A stationary $^{236}_{92}\text{U}$ nucleus fissions spontaneously into two primary fragments, $^{87}_{35}\text{Br}$ and $^{149}_{57}\text{La}$. (a) Calculate the disintegration energy. The required atomic masses are 86.920 711 u for $^{87}_{35}\text{Br}$, 148.934 370 u for $^{149}_{57}\text{La}$, and 236.045 562 u for $^{236}_{92}\text{U}$. (b) How is the disintegration energy split between the two primary fragments? You may use the result of Problem 40. (c) Calculate the speed of each fragment immediately after the fission.

42. ● Explain how the fractional energy loss in a typical ^{235}U fission reaction compares with the fractional energy loss in D–T fusion.

43. ● The half-life of tritium is 12.3 yr. If the TFTR fusion reactor contained 50.0 m^3 of tritium at a density equal to 2.00×10^{14} ions/cm^3, how many curies of tritium were in the plasma? State how this value compares with a fission inventory (the estimated supply of fissionable material) of 4×10^{10} Ci.

44. A fission reactor is hit by a missile, and 5.00×10^6 Ci of ^{90}Sr, with half-life 27.7 yr, evaporates into the air. The strontium falls out over an area of 10^4 km^2. After what time interval will the activity of the ^{90}Sr reach the agriculturally "safe" level of 2.00 μCi/m^2?

45. **Review problem.** A nuclear power plant operates by using the energy released in nuclear fission to convert 20°C water into 400°C steam. How much water could theoretically be converted to steam by the complete fissioning of 1.00 g of ^{235}U at 200 MeV/fission?

46. **Review problem.** A nuclear power plant operates by using the energy released in nuclear fission to convert liquid water at T_c into steam at T_h. How much water could theoretically be converted to steam by the complete fissioning of a mass m of ^{235}U at 200 MeV/fission?

47. ● Consider a 1.00-kg sample of natural uranium composed primarily of ^{238}U, a smaller amount (0.720% by mass) of ^{235}U, and a trace (0.005%) of ^{234}U, which has a half-life of 2.44×10^5 yr. (a) Find the activity in curies due to each of the isotopes. (b) What fraction of the total activity is due to each isotope? (c) Explain whether the activity of this sample is dangerous.

48. **Review problem.** The first nuclear bomb was a fissioning mass of plutonium-239 exploded in the Trinity test before dawn on July 16, 1945, at Alamogordo, New Mexico. Enrico Fermi was 14 km away, lying on the ground facing away from the bomb. After the whole sky had flashed with unbelievable brightness, Fermi stood up and began dropping bits of paper to the ground. They first fell at his feet in the calm and silent air. As the shock wave passed, about 40 s after the explosion, the paper then in flight jumped approximately 5 cm away from ground zero. (a) Assume the shock wave in air propagated equally in all directions without absorption. Find the change in volume of a sphere of radius 14 km as it expands by 5 cm. (b) Find the work $P \, \Delta V$ done by the air in this sphere on the next layer of air farther from the center. (c) Assume the shock wave carried on the order of one-tenth of the energy of the explosion. Make an order-of-magnitude estimate of the bomb yield. (d) One ton of exploding TNT releases 4.2 GJ of energy. What was the order of magnitude of the energy of the Trinity test in equivalent tons of TNT? The dawn revealed the mushroom cloud. Fermi's immediate knowledge of the bomb yield agreed with that determined days later by analysis of elaborate measurements.

49. ● Approximately 1 of every 3 300 water molecules contains one deuterium atom. (a) If all the deuterium nuclei in 1 L of water are fused in pairs according to the D–D fusion reaction $^2\text{H} + {}^2\text{H} \rightarrow {}^3\text{He} + \text{n} + 3.27$ MeV, how much energy in joules is liberated? (b) **What If?** Burning gasoline produces approximately 3.40×10^7 J/L. State how the energy obtainable from the fusion of the deuterium in 1 L of water compares with the energy liberated from the burning of 1 L of gasoline.

50. **Review problem.** A very slow neutron (with speed approximately equal to zero) can initiate the reaction

$$^1_0\text{n} + {}^{10}_5\text{B} \rightarrow {}^7_3\text{Li} + {}^4_2\text{He}$$

The alpha particle moves away with speed 9.25×10^6 m/s. Calculate the kinetic energy of the lithium nucleus. Use nonrelativistic equations.

51. A certain nuclear plant generates internal energy at a rate of 3.065 GW and transfers energy out of the plant by electrical transmission at a rate of 1.000 GW. Of the wasted energy, 3.0% is ejected to the atmosphere and the remainder is passed into a river. A state law requires that the river water be warmed by no more than 3.50°C when it is returned to the river. (a) Determine the amount of cooling water necessary (in kilograms per hour and cubic meters per hour) to cool the plant. (b) Assume fission

generates 7.80×10^{10} J/g of ^{235}U. Determine the rate of fuel burning (in kilograms per hour) of ^{235}U.

52. The alpha-emitter polonium-210 ($^{210}_{84}$Po) is used in a nuclear energy source on a spacecraft (Fig. P45.52). Determine the initial power output of the source. Assume it contains 0.155 kg of ^{210}Po and the efficiency for conversion of radioactive decay energy to energy transferred by electrical transmission is 1.00%.

Courtesy of NASA Ames

Figure P45.52 The Pioneer 10 spacecraft leaves the Solar System. It carries radioactive power supplies at the ends of two booms. Solar panels would not work far from the Sun.

53. ● Natural uranium must be processed to produce uranium enriched in ^{235}U for bombs and power plants. The processing yields a large quantity of nearly pure ^{238}U as a by-product, called "depleted uranium." Because of its high mass density, ^{238}U is used in armor-piercing artillery shells. (a) Find the edge dimension of a 70.0-kg cube of ^{238}U ($\rho = 18.7 \times 10^3$ kg/m^3). (b) The isotope ^{238}U has a long half-life of 4.47×10^9 yr. As soon as one nucleus decays, a relatively rapid series of 14 steps begins that together constitutes the net reaction

$$^{238}_{92}U \rightarrow 8(^4_2He) + 6(^0_{-1}e) + ^{206}_{82}Pb + 6\bar{\nu} + Q_{net}$$

Find the net decay energy. (Refer to Table 44.2.) (c) Argue that a radioactive sample with decay rate R and decay energy Q has power output $\mathcal{P} = QR$. (d) Consider an artillery shell with a jacket of 70.0 kg of ^{238}U. Find its power output due to the radioactivity of the uranium and its daughters. Assume the shell is old enough that the daughters have reached steady-state amounts. Express the power in joules per year. (e) **What If?** A 17-year-old soldier of mass 70.0 kg works in an arsenal where many such artillery shells are stored. Assume his radiation exposure is limited to 5.00 rem per year. Find the rate in joules per year at which he can absorb energy of radiation. Assume an average RBE factor of 1.10.

54. A 2.0-MeV neutron is emitted in a fission reactor. If it loses half its kinetic energy in each collision with a moderator atom, how many collisions does it undergo as it becomes a thermal neutron, with energy 0.039 eV?

55. ▲ Assume a deuteron and a triton are at rest when they fuse according to the reaction

$$^2_1H + ^3_1H \rightarrow ^4_2He + ^1_0n + 17.6 \text{ MeV}$$

Determine the kinetic energy acquired by the neutron.

56. A sealed capsule containing the radiopharmaceutical phosphorus-32 ($^{32}_{15}$P), an e$^-$ (beta) emitter, is implanted into a patient's tumor. The average kinetic energy of the beta particles is 700 keV. The initial activity is 5.22 MBq. Determine the absorbed dose during a 10.0-day period. Assume the beta particles are completely absorbed in 100 g of tissue. *Suggestion:* Find the number of beta particles emitted.

57. ● (a) Calculate the energy (in kilowatt-hours) released if 1.00 kg of ^{239}Pu undergoes complete fission and the energy released per fission event is 200 MeV. (b) Calculate the energy (in electron volts) released in the deuterium–tritium fusion reaction

$$^2_1H + ^3_1H \rightarrow ^4_2He + ^1_0n$$

(c) Calculate the energy (in kilowatt-hours) released if 1.00 kg of deuterium undergoes fusion according to this reaction. (d) **What If?** Calculate the energy (in kilowatt-hours) released by the combustion of 1.00 kg of coal if each C + O$_2$ → CO$_2$ reaction yields 4.20 eV. (e) List advantages and disadvantages of each of these methods of energy generation.

58. The Sun radiates energy at the rate of 3.85×10^{26} W. Suppose the net reaction

$$4(^1_1H) + 2(^0_{-1}e) \rightarrow ^4_2He + 2\nu + \gamma$$

accounts for all the energy released. Calculate the number of protons fused per second.

59. Consider the two nuclear reactions

$$\text{(I)} \quad A + B \rightarrow C + E$$
$$\text{(II)} \quad C + D \rightarrow F + G$$

(a) Show that the net disintegration energy for these two reactions ($Q_{net} = Q_I + Q_{II}$) is identical to the disintegration energy for the net reaction

$$A + B + D \rightarrow E + F + G$$

(b) One chain of reactions in the proton–proton cycle in the Sun's core is

$$^1_1H + ^1_1H \rightarrow ^2_1H + ^0_1e + \nu$$
$$^0_1e + ^0_{-1}e \rightarrow 2\gamma$$
$$^1_1H + ^2_1H \rightarrow ^3_2He + \gamma$$
$$^1_1H + ^3_2He \rightarrow ^4_2He + ^0_1e + \nu$$
$$^0_1e + ^0_{-1}e \rightarrow 2\gamma$$

Based on part (a), what is Q_{net} for this sequence?

60. Suppose the target in a laser fusion reactor is a sphere of solid hydrogen that has a diameter of 1.50×10^{-4} m and a density of 0.200 g/cm^3. Assume half of the nuclei are ^{2}H and half are ^{3}H. (a) If 1.00% of a 200-kJ laser pulse is delivered to this sphere, what temperature does the sphere reach? (b) If all the hydrogen "burns" according to the D–T reaction, how many joules of energy are released?

61. ● In addition to the proton–proton cycle described in the chapter text, the carbon cycle, first proposed by Hans Bethe in 1939, is another cycle by which energy is released in stars as hydrogen is converted to helium. The

2 = intermediate;   3 = challenging;   □ = SSM/SG;   ▲ = ThomsonNOW;   = symbolic reasoning;   ● = qualitative reasoning

carbon cycle requires higher temperatures than the proton–proton cycle. The series of reactions is

$$^{12}\text{C} + {^1}\text{H} \rightarrow {^{13}}\text{N} + \gamma$$

$$^{13}\text{N} \rightarrow {^{13}}\text{C} + e^+ + \nu$$

$$e^+ + e^- \rightarrow 2\gamma$$

$$^{13}\text{C} + {^1}\text{H} \rightarrow {^{14}}\text{N} + \gamma$$

$$^{14}\text{N} + {^1}\text{H} \rightarrow {^{15}}\text{O} + \gamma$$

$$^{15}\text{O} \rightarrow {^{15}}\text{N} + e^+ + \nu$$

$$e^+ + e^- \rightarrow 2\gamma$$

$$^{15}\text{N} + {^1}\text{H} \rightarrow {^{12}}\text{C} + {^4}\text{He}$$

(a) Assuming the proton–proton cycle requires a temperature of 1.5×10^7 K, estimate by proportion the temperature required for the carbon cycle. (b) Calculate the Q value for each step in the carbon cycle and the overall energy released. (c) Do you think the energy carried off by the neutrinos is deposited in the star? Explain.

62. When photons pass through matter, the intensity I of the beam (measured in watts per square meter) decreases exponentially according to

$$I = I_0 e^{-\mu x}$$

where I is the intensity of the beam that just passed through a thickness x of material and I_0 is the intensity of the incident beam. The constant μ is known as the *linear absorption coefficient*, and its value depends on the absorbing material and the wavelength of the photon beam. This wavelength (or energy) dependence allows us to filter out unwanted wavelengths from a broad-spectrum x-ray beam. (a) Two x-ray beams of wavelengths λ_1 and λ_2 and equal incident intensities pass through the same metal plate. Show that the ratio of the emergent beam intensities is

$$\frac{I_2}{I_1} = e^{-(\mu_2 - \mu_1)x}$$

(b) Compute the ratio of intensities emerging from an aluminum plate 1.00 mm thick if the incident beam contains equal intensities of 50 pm and 100 pm x-rays. The values of μ for aluminum at these two wavelengths are

$\mu_1 = 5.4$ cm^{-1} at 50 pm and $\mu_2 = 41.0$ cm^{-1} at 100 pm. (c) Repeat part (b) for an aluminum plate 10.0 mm thick.

63. *To build a bomb.* (a) At time $t = 0$, a sample of uranium is exposed to a neutron source that causes N_0 nuclei to undergo fission. The sample is in a supercritical state, with a reproduction constant $K > 1$. A chain reaction occurs that proliferates fission throughout the mass of uranium. The chain reaction can be thought of as a succession of *generations*. The N_0 fissions produced initially are the zeroth generation of fissions. From this generation, $N_0 K$ neutrons go off to produce fission of new uranium nuclei. The $N_0 K$ fissions that occur subsequently are the first generation of fissions, and from this generation, $N_0 K^2$ neutrons go in search of uranium nuclei in which to cause fission. The subsequent $N_0 K^2$ fissions are the second generation of fissions. This process can continue until all the uranium nuclei have fissioned. Show that the cumulative total of fissions N that have occurred up to and including the nth generation after the zeroth generation is given by

$$N = N_0 \left(\frac{K^{n+1} - 1}{K - 1} \right)$$

(b) Consider a hypothetical uranium bomb made from 5.50 kg of isotopically pure ^{235}U. The chain reaction has a reproduction constant of 1.10 and starts with a zeroth generation of 1.00×10^{20} fissions. The average time interval between one fission generation and the next is 10.0 ns. How long after the zeroth generation does it take the uranium in this bomb to fission completely? (c) Assume the bulk modulus of uranium is 150 GPa. Find the speed of sound in uranium. You may ignore the density difference between ^{235}U and natural uranium. (d) Find the time interval required for a compressional wave to cross the radius of a 5.50-kg sphere of uranium. This time interval indicates how quickly the motion of explosion begins. (e) Fission must occur in a time interval that is short compared with that in part (d); otherwise, most of the uranium will disperse in small chunks without having fissioned. Can the bomb considered in part (b) release the explosive energy of all its uranium? If so, how much energy does it release in equivalent tons of TNT? Assume one ton of TNT releases 4.20 GJ and each uranium fission releases 200 MeV of energy.

Answers to Quick Quizzes

45.1 (b). According to Figure 44.4, the ratio N/Z increases with increasing Z. As a result, when a heavy nucleus fissions to two lighter nuclei, the lighter nuclei tend to have too many neutrons for the nucleus to be stable. Beta decay in which electrons are ejected decreases the number of neutrons and increases the number of protons so as to stabilize the nucleus.

45.2 (a) and (b). In both of these cases, the Z and A values balance on the two sides of the equations. In reaction (c), $Z_{\text{left}} = Z_{\text{right}}$, but $A_{\text{left}} \neq A_{\text{right}}$.

45.3 (a). To reduce the value of K, more neutrons need to be absorbed, so a larger volume of the control rods must be inside the reactor core.

45.4 (d). Figure 44.5 shows that the curve representing the binding energy per nucleon peaks at $A \approx 60$. Consequently, combining two nuclei with equal values of $A > 60$ results in an increase in mass, so a fusion reaction will not occur.

A shower of particle tracks from a head-on collision of gold nuclei, each moving with energy 100 GeV. This collision occurred at the Relativistic Heavy Ion Collider (RHIC) at Brookhaven National Laboratory and was recorded with the STAR (Solenoidal Tracker at RHIC) detector. The tracks represent many fundamental particles arising from the energy of the collision. (Courtesy of Brookhaven National Laboratory/RHIC-STAR)

46.1 The Fundamental Forces in Nature

46.2 Positrons and Other Antiparticles

46.3 Mesons and the Beginning of Particle Physics

46.4 Classification of Particles

46.5 Conservation Laws

46.6 Strange Particles and Strangeness

46.7 Finding Patterns in the Particles

46.8 Quarks

46.9 Multicolored Quarks

46.10 The Standard Model

46.11 The Cosmic Connection

46.12 Problems and Perspectives

46 Particle Physics and Cosmology

The word *atom* comes from the Greek *atomos*, which means "indivisible." The early Greeks believed that atoms were the indivisible constituents of matter; that is, they regarded them as elementary particles. After 1932, physicists viewed all matter as consisting of three constituent particles: electrons, protons, and neutrons. Beginning in the 1940s, many "new" particles were discovered in experiments involving high-energy collisions between known particles. The new particles are characteristically very unstable and have very short half-lives, ranging between 10^{-6} s and 10^{-23} s. So far, more than 300 of these particles have been catalogued.

Until the 1960s, physicists were bewildered by the great number and variety of subatomic particles that were being discovered. They wondered whether the particles had no systematic relationship connecting them or whether a pattern was emerging that would provide a better understanding of the elaborate structure in the subatomic world. That the neutron has a magnetic moment despite having zero electric charge (Section 44.8) suggests an underlying structure to the neutron. The periodic table explains how more than 100 elements can be formed from three types of particles (electrons, protons, and neutrons), which suggests there is, perhaps, a means of forming more than 300 subatomic particles from a small number of basic building blocks.

Recall Figure 1.2, which illustrated the various levels of structure in matter. We studied the atomic structure of matter in Chapter 42. In Chapter 44, we investigated the substructure of the atom by describing the structure of the nucleus. As mentioned in Section 1.2, the protons and neutrons in the nucleus, and a host of other exotic particles, are now known to be composed of six different varieties of particles called *quarks*. In this concluding chapter, we examine the current theory of elementary particles, in which all matter is constructed from only two families of particles, quarks and leptons. We also discuss how clarifications of such models might help scientists understand the birth and evolution of the Universe.

46.1 The Fundamental Forces in Nature

As noted in Section 5.1, all natural phenomena can be described by four fundamental forces acting between particles. In order of decreasing strength, they are the nuclear force, the electromagnetic force, the weak force, and the gravitational force.

The nuclear force discussed in Chapter 44 is an attractive force between nucleons. It has a very short range and is negligible for separation distances between nucleons greater than approximately 10^{-15} m (about the size of the nucleus). The electromagnetic force, which binds atoms and molecules together to form ordinary matter, has a strength of approximately 10^{-2} times that of the nuclear force. This long-range force decreases in magnitude as the inverse square of the separation between interacting particles. The weak force is a short-range force that tends to produce instability in certain nuclei. It is responsible for decay processes, and its strength is only about 10^{-5} times that of the nuclear force. Finally, the gravitational force is a long-range force that has a strength of only about 10^{-39} times that of the nuclear force. Although this familiar interaction is the force that holds the planets, stars, and galaxies together, its effect on elementary particles is negligible.

In Section 13.4, we discussed the difficulty early scientists had with the notion of the gravitational force acting at a distance, with no physical contact between the interacting objects. To resolve this difficulty, the concept of the gravitational field was introduced. Similarly, in Chapter 23, we introduced the electric field to describe the electric force acting between charged objects, and we followed that with a discussion of the magnetic field in Chapter 29. In modern physics, the nature of the interaction between particles is carried a step further. These interactions are described in terms of the exchange of entities called **field particles** or **exchange particles**. Field particles are also called **gauge bosons**.[1] The interacting particles continuously emit and absorb field particles. The emission of a field particle by one particle and its absorption by another manifests as a force between the two interacting particles. In the case of the electromagnetic interaction, for instance, the field particles are photons. In the language of modern physics, the electromagnetic force is said to be *mediated* by photons, and photons are the field particles of the electromagnetic field. Likewise, the nuclear force is mediated by field particles called *gluons*. The weak force is mediated by field particles called *W* and *Z bosons*, and the gravitational force is proposed to be mediated by field particles called *gravitons*. These interactions, their ranges, and their relative strengths are summarized in Table 46.1.

PAUL ADRIEN MAURICE DIRAC
British Physicist (1902–1984)
Dirac was instrumental in the understanding of antimatter and the unification of quantum mechanics and relativity. He made many contributions to the development of quantum physics and cosmology. In 1933, Dirac won a Nobel Prize in Physics.

Courtesy of AIP Emilio Segrè Visual Archives

46.2 Positrons and Other Antiparticles

In the 1920s, Paul Dirac developed a relativistic quantum-mechanical description of the electron that successfully explained the origin of the electron's spin and its magnetic moment. His theory had one major problem, however: its relativistic

---

[1] The word *bosons* suggests that the field particles have integral spin as discussed in Section 43.8. The word *gauge* comes from *gauge theory*, which is a sophisticated mathematical analysis that is beyond the scope of this book.

TABLE 46.1
Particle Interactions

Interaction	Relative Strength	Range of Force	Mediating Field Particle	Mass of Field Particle (GeV/c^2)
Nuclear	1	Short (≈ 1 fm)	Gluon	0
Electromagnetic	10^{-2}	∞	Photon	0
Weak	10^{-5}	Short ($\approx 10^{-3}$ fm)	$W^{\pm}$, Z^0 bosons	80.4, 80.4, 91.2
Gravitational	10^{-39}	∞	Graviton	0

wave equation required solutions corresponding to negative energy states, and if negative energy states existed, an electron in a state of positive energy would be expected to make a rapid transition to one of these states, emitting a photon in the process.

Dirac circumvented this difficulty by postulating that all negative energy states are filled. The electrons occupying these negative energy states are collectively called the *Dirac sea.* Electrons in the Dirac sea are not directly observable because the Pauli exclusion principle does not allow them to react to external forces; there are no available states to which an electron can make a transition in response to an external force. Therefore, an electron in such a state acts as an isolated system unless an interaction with the environment is strong enough to excite the electron to a positive energy state. Such an excitation causes one of the negative energy states to be vacant as in Figure 46.1, leaving a hole in the sea of filled states. *The hole can react to external forces and is observable.* The hole reacts in a way similar to that of the electron except that it has a positive charge: it is the *antiparticle* to the electron.

This theory strongly suggested that *an antiparticle exists for every particle,* not only for fermions such as electrons but also for bosons. That has subsequently been verified for *all* particles known today. The antiparticle for a charged particle has the same mass as the particle but opposite charge. For example, the electron's antiparticle (the *positron* mentioned in Section 44.4) has a rest energy of 0.511 MeV and a positive charge of $+1.60 \times 10^{-19}$ C.

Carl Anderson (1905–1991) observed the positron experimentally in 1932 and was awarded a Nobel Prize in Physics in 1936 for this achievement. Anderson discovered the positron while examining tracks created in a cloud chamber by electron-like particles of positive charge. (These early experiments used cosmic rays—mostly energetic protons passing through interstellar space—to initiate high-energy reactions on the order of several GeV.) To discriminate between positive and negative charges, Anderson placed the cloud chamber in a magnetic field, causing moving charges to follow curved paths. He noted that some of the electron-like tracks deflected in a direction corresponding to a positively charged particle.

Since Anderson's discovery, positrons have been observed in a number of experiments. A common source of positrons is **pair production.** In this process, a gamma-ray photon with sufficiently high energy interacts with a nucleus and an electron–positron pair is created from the photon. (The presence of the nucleus allows the principle of conservation of momentum to be satisfied.) Because the total rest energy of the electron–positron pair is $2m_ec^2 = 1.02$ MeV (where m_e is the mass of the electron), the photon must have at least this much energy to create an electron–positron pair. The energy of a photon is converted to rest energy of the electron and positron in accordance with Einstein's relationship $E_R = mc^2$. If the gamma-ray photon has energy in excess of the rest energy of the electron–positron pair, the excess appears as kinetic energy of the two particles. Figure 46.2 (page 1360) shows early observations of tracks of electron–positron pairs in a bubble chamber created by 300-MeV gamma rays striking a lead sheet.

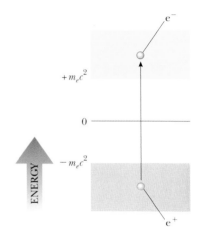

Figure 46.1 Dirac's model for the existence of antielectrons (positrons). The states lower in energy than $-m_ec^2$ are filled with electrons as indicated by the blue coloring. This set of filled states is called the Dirac sea. One of these electrons can make a transition out of its state only if it is provided with energy equal to or larger than $2m_ec^2$. This transition leaves a vacancy in the Dirac sea, which can behave as a particle identical to the electron except for its positive charge.

PITFALL PREVENTION 46.1
Antiparticles

An antiparticle is not identified solely on the basis of opposite charge; even neutral particles have antiparticles, which are defined in terms of other properties, such as spin.

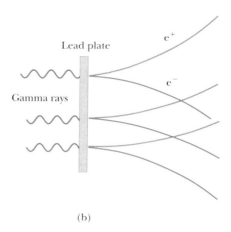

(a)

(b)

Figure 46.2 (a) Bubble-chamber tracks of electron–positron pairs produced by 300-MeV gamma rays striking a lead sheet from the left. (b) The pertinent pair-production events. The positrons deflect upward, and the electrons downward in an applied magnetic field.

Quick Quiz 46.1 Given the identification of the particles in Figure 46.2b, is the direction of the external magnetic field in Figure 46.2a (a) into the page, (b) out of the page, or (c) impossible to determine?

The reverse process can also occur. Under the proper conditions, an electron and a positron can annihilate each other to produce two gamma-ray photons that have a combined energy of at least 1.02 MeV:

$$e^- + e^+ \rightarrow 2\gamma$$

Because the initial momentum of the electron–positron system is approximately zero, the two gamma rays travel in opposite directions after the annihilation, satisfying the principle of conservation of momentum for the system.

Practically every known elementary particle has a distinct antiparticle. Among the exceptions are the photon and the neutral pion (π^0; see Section 46.3). Following the construction of high-energy accelerators in the 1950s, many other antiparticles were revealed. They included the antiproton, discovered by Emilio Segré (1905–1989) and Owen Chamberlain (1920–2006) in 1955, and the antineutron,[2] discovered shortly thereafter.

Electron–positron annihilation is used in the medical diagnostic technique called *positron-emission tomography* (PET). The patient is injected with a glucose solution containing a radioactive substance that decays by positron emission, and the material is carried throughout the body by the blood. A positron emitted during a decay event in one of the radioactive nuclei in the glucose solution annihilates with an electron in the surrounding tissue, resulting in two gamma-ray photons emitted in opposite directions. A gamma detector surrounding the patient pinpoints the source of the photons and, with the assistance of a computer, displays an image of the sites at which the glucose accumulates. (Glucose metabolizes rapidly in cancerous tumors and accumulates at those sites, providing a strong signal for a PET detector system.) The images from a PET scan can indicate a wide variety of disorders in the brain, including Alzheimer's disease (Fig. 46.3). In addition, because glucose metabolizes more rapidly in active areas of the brain, a PET scan can indicate areas of the brain involved in the activities in which the patient is engaging at the time of the scan, such as language use, music, and vision.

---

[2] Antiparticles of charged particles have the opposite charge. Antiparticles for uncharged particles, such as the neutron, are a little more difficult to describe. One basic process that can detect the existence of an antiparticle is pair annihilation. For example, a neutron and an antineutron can annihilate to form two gamma rays. Because the photon and the neutral pion do not have distinct antiparticles, pair annihilation is not observed with either of these particles.

Figure 46.3   PET scans of the brain of a healthy older person *(left)* and that of a patient suffering from Alzheimer's disease *(right)*. Lighter regions contain higher concentrations of radioactive glucose, indicating higher metabolism rates and therefore increased brain activity.

46.3   Mesons and the Beginning of Particle Physics

Physicists in the mid-1930s had a fairly simple view of the structure of matter. The building blocks were the proton, the electron, and the neutron. Three other particles were either known or postulated at the time: the photon, the neutrino, and the positron. Together these six particles were considered the fundamental constituents of matter. With this simple picture, however, no one was able to answer the following important question: the protons in any nucleus should strongly repel one another due to their charges of the same sign, so what is the nature of the force that holds the nucleus together? Scientists recognized that this mysterious force must be much stronger than anything encountered in nature up to that time. This force is the nuclear force discussed in Section 44.1 and examined in historical perspective in the following paragraphs.

The first theory to explain the nature of the nuclear force was proposed in 1935 by Japanese physicist Hideki Yukawa, an effort that earned him a Nobel Prize in Physics in 1949. To understand Yukawa's theory, recall the introduction of field particles in Section 46.1, which stated that each fundamental force is mediated by a field particle exchanged between the interacting particles. Yukawa used this idea to explain the nuclear force, proposing the existence of a new particle whose exchange between nucleons in the nucleus causes the nuclear force. He established that the range of the force is inversely proportional to the mass of this particle and predicted the mass to be approximately 200 times the mass of the electron. (Yukawa's predicted particle is *not* the gluon mentioned in Section 46.1, which is massless and is today considered to be the field particle for the nuclear force.) Because the new particle would have a mass between that of the electron and that of the proton, it was called a **meson** (from the Greek *meso*, "middle").

In efforts to substantiate Yukawa's predictions, physicists began experimental searches for the meson by studying cosmic rays entering the Earth's atmosphere. In 1937, Carl Anderson and his collaborators discovered a particle of mass 106 MeV/c^2, approximately 207 times the mass of the electron. This particle was thought to be Yukawa's meson. Subsequent experiments, however, showed that the particle interacted very weakly with matter and hence could not be the field particle for the nuclear force. That puzzling situation inspired several theoreticians to propose two mesons having slightly different masses equal to approximately 200 times that of the electron, one having been discovered by Anderson and the other, still undiscovered, predicted by Yukawa. This idea was confirmed in 1947 with the discovery of the **pi meson** (π), or simply **pion**. The particle discovered by Anderson in 1937, the one initially thought to be Yukawa's meson, is not really a meson. (We shall discuss the characteristics of mesons in Section 46.4.) Instead, it takes part in the weak and electromagnetic interactions only and is now called the **muon** (μ).

HIDEKI YUKAWA
Japanese Physicist (1907–1981)
Yukawa was awarded the Nobel Prize in Physics in 1949 for predicting the existence of mesons. This photograph of him at work was taken in 1950 in his office at Columbia University. Yukawa came to Columbia in 1949 after spending the early part of his career in Japan.

Figure 46.4 Feynman diagram representing a photon mediating the electromagnetic force between two electrons.

RICHARD FEYNMAN
American Physicist (1918–1988)
Inspired by Dirac, Feynman developed quantum electrodynamics, the theory of the interaction of light and matter on a relativistic and quantum basis. In 1965, Feynman won the Nobel Prize in Physics. The prize was shared by Feynman, Julian Schwinger, and Sin Itiro Tomonaga. Early in Feynman's career, he was a leading member of the team developing the first nuclear weapon in the Manhattan Project. Toward the end of his career, he worked on the commission investigating the 1986 *Challenger* tragedy and demonstrated the effects of cold temperatures on the rubber O-rings used in the space shuttle.

The pion comes in three varieties, corresponding to three charge states: π^+, π^-, and π^0. The π^+ and π^- particles (π^- is the antiparticle of π^+) each have a mass of 139.6 MeV/c^2, and the π^0 mass is 135.0 MeV/c^2. Two muons exist: μ^- and its antiparticle μ^+.

Pions and muons are very unstable particles. For example, the π^-, which has a mean lifetime of 2.6×10^{-8} s, decays to a muon and an antineutrino.[3] The muon, which has a mean lifetime of 2.2 μs, then decays to an electron, a neutrino, and an antineutrino:

$$\pi^- \rightarrow \mu^- + \bar{\nu}$$
$$\mu^- \rightarrow e^- + \nu + \bar{\nu} \qquad \textbf{(46.1)}$$

For chargeless particles (as well as some charged particles, such as the proton), a bar over the symbol indicates an antiparticle, as for the neutrino in beta decay (see Section 44.5). Other antiparticles, such as e^+ and μ^+, use a different notation.

The interaction between two particles can be represented in a simple diagram called a **Feynman diagram,** developed by American physicist Richard P. Feynman. Figure 46.4 is such a diagram for the electromagnetic interaction between two electrons. A Feynman diagram is a qualitative graph of time on the vertical axis versus space on the horizontal axis. It is qualitative in the sense that the actual values of time and space are not important, but the overall appearance of the graph provides a pictorial representation of the process.

In the simple case of the electron–electron interaction in Figure 46.4, a photon (the field particle) mediates the electromagnetic force between the electrons. Notice that the entire interaction is represented in the diagram as occurring at a single point in time. Therefore, the paths of the electrons appear to undergo a discontinuous change in direction at the moment of interaction. The electron paths shown in Figure 46.4 are different from the *actual* paths, which would be curved due to the continuous exchange of large numbers of field particles.

In the electron–electron interaction, the photon, which transfers energy and momentum from one electron to the other, is called a *virtual photon* because it vanishes during the interaction without having been detected. In Chapter 40, we discussed that a photon has energy $E = hf$, where f is its frequency. Consequently, for a system of two electrons initially at rest, the system has energy $2m_ec^2$ before a virtual photon is released and energy $2m_ec^2 + hf$ after the virtual photon is released (plus any kinetic energy of the electron resulting from the emission of the photon). Is that a violation of the law of conservation of energy for an isolated system? No; this process does *not* violate the law of conservation of energy because the virtual photon has a very short lifetime Δt that makes the uncertainty in the energy $\Delta E \approx \hbar/2\,\Delta t$ of the system greater than the photon energy. Therefore, within the constraints of the uncertainty principle, the energy of the system is conserved.

Now consider a pion mediating the nuclear force between a proton and a neutron as in Yukawa's model (Fig. 46.5a). The rest energy E_R of a pion of mass m_π is given by Einstein's equation $E_R = m_\pi c^2$. To conserve energy, as with the photon in Figure 46.4, the uncertainty in the system energy must be greater than the rest energy of the pion: $\Delta E > E_R$. The existence of the pion would violate the law of conservation of energy if the particle existed for a time interval greater than $\Delta t \approx \hbar/2E_R$ (from the uncertainty principle), where E_R is the rest energy of the pion and Δt is the time interval required for the pion to transfer from one nucleon to the other. Therefore,

$$\Delta t \approx \frac{\hbar}{2E_R} = \frac{\hbar}{2m_\pi c^2} \quad \rightarrow \quad m_\pi c^2 = \frac{\hbar}{2\Delta t} \qquad \textbf{(46.2)}$$

---

[3] The antineutrino is another zero-charge particle for which the identification of the antiparticle is more difficult than that for a charged particle. Although the details are beyond the scope of this book, the neutrino and antineutrino can be differentiated by means of the relationship between the linear momentum and the spin angular momentum of the particles.

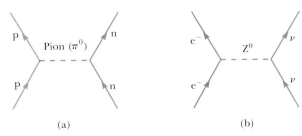

Figure 46.5   (a) Feynman diagram representing a proton and a neutron interacting via the nuclear force with a neutral pion mediating the force. (This model is *not* the current model for nucleon interaction.) (b) Feynman diagram for an electron and a neutrino interacting via the weak force, with a Z^0 boson mediating the force.

Because the pion cannot travel faster than the speed of light, the maximum distance it can travel in a time interval Δt is $d = c \Delta t$. Therefore,

$$m_\pi c^2 \approx \frac{\hbar}{2(d/c)} = \frac{\hbar c}{2d} \tag{46.3}$$

Table 46.1 shows that the range of the nuclear force is approximately 1×10^{-15} m. Using this value for d in Equation 46.3, we estimate the rest energy of the pion to be

$$m_\pi c^2 \approx \frac{(1.055 \times 10^{-34}\,\text{J} \cdot \text{s})(3.00 \times 10^8\,\text{m/s})}{2(1 \times 10^{-15}\,\text{m})}$$

$$= 1.6 \times 10^{-11}\,\text{J} \approx 100\,\text{MeV}$$

Because this result is the same order of magnitude as the observed rest energies of the pions, we have some confidence in the field-particle model.

The concept just described is quite revolutionary. In effect, it says that a system of two nucleons can change into two nucleons plus a pion as long as it returns to its original state in a very short time interval. (Remember that this description is the older historical model, which assumes the pion is the field particle for the nuclear force; the gluon is the actual field particle in current models.) Physicists often say that a nucleon undergoes *fluctuations* as it emits and absorbs field particles. These fluctuations are a consequence of a combination of quantum mechanics (through the uncertainty principle) and special relativity (through Einstein's energy–mass relationship $E_R = mc^2$).

In this section, we discussed the field particles that were originally proposed to mediate the nuclear force (pions) and those that mediate the electromagnetic force (photons). The graviton, the field particle for the gravitational force, has yet to be observed. In 1983, $W^\pm$ and Z^0 particles, which mediate the weak force, were discovered by Italian physicist Carlo Rubbia (b. 1934) and his associates, using a proton–antiproton collider. Rubbia and Simon van der Meer (b. 1925), both at CERN,[4] shared the 1984 Nobel Prize in Physics for the discovery of the $W^\pm$ and Z^0 particles and the development of the proton–antiproton collider. Figure 46.5b shows a Feynman diagram for a weak interaction mediated by a Z^0 boson.

46.4   Classification of Particles

All particles other than field particles can be classified into two broad categories, *hadrons* and *leptons*. The criterion for separating these particles into categories is whether or not they interact via the strong force. The nuclear force between nucleons in a nucleus is a particular manifestation of the strong force, but we will use the term *strong force* to refer to any interaction between particles made up of

---

[4] CERN was originally the Conseil Européen pour la Recherche Nucléaire (European Organization for Nuclear Research); the name has been altered to European Laboratory for Particle Physics, but the CERN acronym has been retained.

PITFALL PREVENTION 46.2
The Nuclear Force and the Strong Force

The nuclear force discussed in Chapter 44 was historically called the strong force. Once the quark theory (Section 46.8) was established, however, the phrase *strong force* was reserved for the force between quarks. We shall follow this convention: the strong force is between quarks or particles built from quarks, and the nuclear force is between nucleons in a nucleus. The nuclear force is a secondary result of the strong force as discussed in Section 46.9. It is sometimes called the *residual strong force*. Because of this historical development of the names for these forces, other books sometimes refer to the nuclear force as the strong force.

TABLE 46.2

Some Particles and Their Properties

Category	Particle Name	Symbol	Anti-particle	Mass (MeV/c^2)	B	L_e	L_μ	L_τ	S	Lifetime(s)	Spin
Leptons	Electron	e^-	e^+	0.511	0	+1	0	0	0	Stable	$\frac{1}{2}$
	Electron–neutrino	ν_e	$\overline{\nu}_e$	$< 7\text{eV}/c^2$	0	+1	0	0	0	Stable	$\frac{1}{2}$
	Muon	μ^-	μ^+	105.7	0	0	+1	0	0	2.20×10^{-6}	$\frac{1}{2}$
	Muon–neutrino	ν_μ	$\overline{\nu}_\mu$	< 0.3	0	0	+1	0	0	Stable	$\frac{1}{2}$
	Tau	τ^-	τ^+	1 784	0	0	0	+1	0	$< 4 \times 10^{-13}$	$\frac{1}{2}$
	Tau–neutrino	ν_τ	$\overline{\nu}_\tau$	< 30	0	0	0	+1	0	Stable	$\frac{1}{2}$
Hadrons											
Mesons	Pion	π^+	π^-	139.6	0	0	0	0	0	2.60×10^{-8}	0
		π^0	Self	135.0	0	0	0	0	0	0.83×10^{-16}	0
	Kaon	K^+	K^-	493.7	0	0	0	0	+1	1.24×10^{-8}.	0
		K_S^0	$\overline{K}_S^0$	497.7	0	0	0	0	+1	0.89×10^{-10}	0
		K_L^0	$\overline{K}_L^0$	497.7	0	0	0	0	+1	5.2×10^{-8}	0
	Eta	η	Self	548.8	0	0	0	0	0	$< 10^{-18}$	0
		η'	Self	958	0	0	0	0	0	2.2×10^{-21}	0
Baryons	Proton	p	$\overline{\text{p}}$	938.3	+1	0	0	0	0	Stable	$\frac{1}{2}$
	Neutron	n	$\overline{\text{n}}$	939.6	+1	0	0	0	0	614	$\frac{1}{2}$
	Lambda	Λ^0	$\overline{\Lambda}^0$	1 115.6	+1	0	0	0	-1	2.6×10^{-10}	$\frac{1}{2}$
	Sigma	Σ^+	$\overline{\Sigma}^-$	1 189.4	+1	0	0	0	-1	0.80×10^{-10}	$\frac{1}{2}$
		Σ^0	$\overline{\Sigma}^0$	1 192.5	+1	0	0	0	-1	6×10^{-20}	$\frac{1}{2}$
		Σ^-	$\overline{\Sigma}^+$	1 197.3	+1	0	0	0	-1	1.5×10^{-10}	$\frac{1}{2}$
	Delta	Δ^{++}	$\overline{\Delta}^{--}$	1 230	+1	0	0	0	0	6×10^{-24}	$\frac{3}{2}$
		Δ^+	$\overline{\Delta}^-$	1 231	+1	0	0	0	0	6×10^{-24}	$\frac{3}{2}$
		Δ^0	$\overline{\Delta}^0$	1 232	+1	0	0	0	0	6×10^{-24}	$\frac{3}{2}$
		Δ^-	$\overline{\Delta}^+$	1 234	+1	0	0	0	0	6×10^{-24}	$\frac{3}{2}$
	Xi	Ξ^0	$\overline{\overline{\Xi}}^0$	1 315	+1	0	0	0	-2	2.9×10^{-10}	$\frac{1}{2}$
		Ξ^-	Ξ^+	1 321	+1	0	0	0	-2	1.64×10^{-10}	$\frac{1}{2}$
	Omega	Ω^-	Ω^+	1 672	+1	0	0	0	-3	0.82×10^{-10}	$\frac{3}{2}$

quarks. (For more detail on quarks and the strong force, see Section 46.8.) Table 46.2 provides a summary of the properties of hadrons and leptons.

Hadrons

Particles that interact through the strong force (as well as through the other fundamental forces) are called **hadrons.** The two classes of hadrons, *mesons* and *baryons,* are distinguished by their masses and spins.

Mesons all have zero or integer spin (0 or 1). As indicated in Section 46.3, the name comes from the expectation that Yukawa's proposed meson mass would lie between the masses of the electron and the proton. Several meson masses do lie in this range, although mesons having masses greater than that of the proton have been found to exist.

All mesons decay finally into electrons, positrons, neutrinos, and photons. The pions are the lightest known mesons and have masses of approximately 1.4×10^2 MeV/c^2, and all three pions—π^+, π^-, and π^0—have a spin of 0. (This spin-0 characteristic indicates that the particle discovered by Anderson in 1937, the muon, is not a meson. The muon has spin $\frac{1}{2}$ and belongs in the *lepton* classification, described below.)

Baryons, the second class of hadrons, have masses equal to or greater than the proton mass (the name *baryon* means "heavy" in Greek), and their spin is always a half-integer value ($\frac{1}{2}$ or $\frac{3}{2}$). Protons and neutrons are baryons, as are many other particles. With the exception of the proton, all baryons decay in such a way that the end products include a proton. For example, the baryon called the Ξ hyperon (Greek letter xi) decays to the Λ^0 baryon (Greek letter lambda) in approximately 10^{-10} s. The Λ^0 then decays to a proton and a π^- in approximately 3×10^{-10} s.

Today it is believed that hadrons are not elementary particles but instead are composed of more elementary units called quarks, per Section 46.8.

Leptons

Leptons (from the Greek *leptos*, meaning "small" or "light") are particles that do not interact by means of the strong force. All leptons have spin $\frac{1}{2}$. Unlike hadrons, which have size and structure, leptons appear to be truly elementary, meaning that they have no structure and are point-like.

Quite unlike the case with hadrons, the number of known leptons is small. Currently, scientists believe that only six leptons exist: the electron, the muon, the tau, and a neutrino associated with each: e^-, μ^-, τ^-, ν_e, ν_μ, and ν_τ. The tau lepton, discovered in 1975, has a mass about twice that of the proton. Direct experimental evidence for the neutrino associated with the tau was announced by the Fermi National Accelerator Laboratory (Fermilab) in July 2000. Each of the six leptons has an antiparticle.

Current studies indicate that neutrinos have a small but nonzero mass. If they do have mass, they cannot travel at the speed of light. In addition, because so many neutrinos exist, their combined mass may be sufficient to cause all the matter in the Universe to eventually collapse into a single point, which might then explode and create a completely new Universe! We shall discuss this possibility in more detail in Section 46.11.

46.5   Conservation Laws

The laws of conservation of energy, linear momentum, angular momentum, and electric charge provide us with a set of rules that all processes must follow. In Chapter 44, we learned that conservation laws are important for understanding why certain radioactive decays and nuclear reactions occur and others do not. In the study of elementary particles, a number of additional conservation laws are important. Although the two described here have no theoretical foundation, they are supported by abundant empirical evidence.

Baryon Number

Experimental results show that whenever a baryon is created in a decay or nuclear reaction, an antibaryon is also created. This scheme can be quantified by assigning every particle a quantum number, the **baryon number,** as follows: $B = +1$ for all baryons, $B = -1$ for all antibaryons, and $B = 0$ for all other particles. (See Table 46.2.) The **law of conservation of baryon number** states that **whenever a nuclear reaction or decay occurs, the sum of the baryon numbers before the process must equal the sum of the baryon numbers after the process.**

◀ Conservation of baryon number

If baryon number is conserved, the proton must be absolutely stable. For example, a decay of the proton to a positron and a neutral pion would satisfy conservation of energy, momentum, and electric charge. Such a decay has never been observed, however. The law of conservation of baryon number would be consistent with the absence of this decay because the proposed decay would involve the loss of a baryon. Based on experimental observations as pointed out in Example 46.2, all we can say at present is that protons have a half-life of at least 10^{33} years (the estimated age of the Universe is only 10^{10} years). Some recent theories, however, predict that the proton is unstable. According to this theory, baryon number is not absolutely conserved.

> **Quick Quiz 46.2**   Consider the decays **(i)** n → $\pi^+ + \pi^- + \mu^+ + \mu^-$ and **(ii)** n → p + π^-. From the following choices, which conservation laws are violated by each decay? (a) energy   (b) electric charge   (c) baryon number   (d) angular momentum   (e) no conservation laws

---

EXAMPLE 46.1   **Checking Baryon Numbers**

Use the law of conservation of baryon number to determine whether each of the following reactions can occur:

(A) p + n → p + p + n + $\bar{p}$

SOLUTION

Conceptualize   Because the mass on the right is larger than the mass on the left, the initial particles must have sufficient kinetic energy that energy conservation is satisfied for this reaction.

Categorize   We use a conservation law developed in this section, so we categorize this example as a substitution problem.

Evaluate the total baryon number for the left side of the reaction:

$$1 + 1 = 2$$

Evaluate the total baryon number for the right side of the reaction:

$$1 + 1 + 1 + (-1) = 2$$

Therefore, baryon number is conserved and the reaction can occur.

(B) p + n → p + p + $\bar{p}$

SOLUTION

Evaluate the total baryon number for the left side of the reaction:

$$1 + 1 = 2$$

Evaluate the total baryon number for the right side of the reaction:

$$1 + 1 + (-1) = 1$$

Because baryon number is not conserved, the reaction cannot occur.

EXAMPLE 46.2   **Detecting Proton Decay**

Measurements taken at the Super Kamiokande neutrino detection facility (Fig. 46.6) indicate that the half-life of protons is at least 10^{33} yr.

(A) Estimate how long we would have to watch, on average, to see a proton in a glass of water decay.

SOLUTION

Conceptualize   Imagine the number of protons in a glass of water. Although this number is huge, the probability of a single proton undergoing decay is small, so we would expect to wait for a long time interval before observing a decay.

Categorize   Because a half-life is provided in the problem, we categorize this problem as one in which we can apply our statistical analysis techniques from Section 44.4.

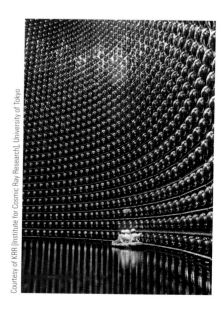

Courtesy of KRR Institute for Cosmic Ray Research, University of Tokyo

Figure 46.6   (Example 46.2) This detector at the Super Kamiokande neutrino facility in Japan is used to study photons and neutrinos. It holds 50 000 metric tons of highly purified water and 13 000 photomultipliers. The photograph was taken while the detector was being filled. Technicians in a raft clean the photodetectors before they are submerged.

Analyze   Let's estimate that a glass contains approximately 250 g of water.

Find the number of molecules of water in the glass:

$$\frac{(250 \text{ g})(6.02 \times 10^{23} \text{ molecules/mol})}{18 \text{ g/mol}} = 8.4 \times 10^{24} \text{ molecules}$$

Each water molecule contains one proton in each of its two hydrogen atoms plus eight protons in its oxygen atom, for a total of ten protons. Therefore, there are 8.4×10^{25} protons in the glass of water.

Find the decay constant (Section 44.4) from Equation 44.8:

$$(1) \quad \lambda = \frac{0.693}{T_{1/2}} = \frac{0.693}{10^{33} \text{ yr}} = 6.9 \times 10^{-34} \text{ yr}^{-1}$$

Find the activity of the protons from Equation 44.7:

$$(2) \quad R = \lambda N = (6.9 \times 10^{-34} \text{ yr}^{-1})(8.4 \times 10^{25})$$

$$= 5.8 \times 10^{-8} \text{ yr}^{-1}$$

Finalize   Equation (1) represents the probability that *one* proton decays in one year. The probability that *any* proton in our glass of water decays in the one-year interval is given by Equation (2). Therefore, we must watch our glass of water for $1/R \approx$ 17 million years! That indeed is a long time interval, as expected.

(B) The Super Kamiokande neutrino facility contains 50 000 metric tons of water. Estimate the average time interval between detected proton decays in this much water if the half-life of a proton is 10^{33} yr.

SOLUTION

Analyze   Find the ratio of the number of molecules in 50 000 metric tons of water to that in the glass of water in part (A), which is the same as the ratio of masses:

$$\frac{N_{\text{Kamiokande}}}{N_{\text{glass}}} = \frac{m_{\text{Kamiokande}}}{m_{\text{glass}}}$$

$$= \frac{50\,000 \text{ metric ton}}{250 \text{ g}}\left(\frac{1\,000 \text{ kg}}{1 \text{ metric ton}}\right)\left(\frac{1\,000 \text{ g}}{1 \text{ kg}}\right)$$

$$= 2.0 \times 10^{8}$$

Find the number of molecules in the Kamiokande facility:

$$N_{\text{Kamiokande}} = (2.0 \times 10^{8})N_{\text{glass}}$$

$$= (2.0 \times 10^{8})(8.4 \times 10^{24} \text{ molecules})$$

$$= 1.7 \times 10^{33} \text{ molecules}$$

Find the decay rate for protons in the facility:

$$R = (6.9 \times 10^{-34} \text{ yr}^{-1})(10 \text{ protons/molecule})(1.7 \times 10^{33} \text{ molecules})$$

$$\approx 12 \text{ yr}^{-1}$$

Finalize   The average time interval between decays is about one-twelfth of a year, or approximately one month. That is much shorter than the time interval in part (A) due to the tremendous amount of water in the detector facility. Despite this rosy prediction of one proton decay per month, a proton decay has never been observed. This suggests that the half-life of the proton may be larger than 10^{33} years or that proton decay simply does not occur.

Lepton Number

There are three conservation laws involving lepton numbers, one for each variety of lepton. The **law of conservation of electron lepton number** states that whenever a nuclear reaction or decay occurs, **the sum of the electron lepton numbers before the process must equal the sum of the electron lepton numbers after the process.**

◀ Conservation of electron lepton number

The electron and the electron neutrino are assigned an electron lepton number $L_e = +1$, and the antileptons e^+ and $\bar{\nu}_e$ are assigned an electron lepton number $L_e = -1$. All other particles have $L_e = 0$. For example, consider the decay of the neutron:

$$n \rightarrow p + e^- + \bar{\nu}_e$$

Before the decay, the electron lepton number is $L_e = 0$; after the decay, it is $0 + 1 + (-1) = 0$. Therefore, electron lepton number is conserved. (Baryon number must also be conserved, of course, and it is: before the decay, $B = +1$, and after the decay, $B = +1 + 0 + 0 = +1$.)

Similarly, when a decay involves muons, the muon lepton number L_μ is conserved. The μ^- and the ν_μ are assigned a muon lepton number $L_\mu = +1$, and the antimuons μ^+ and $\bar{\nu}_\mu$ are assigned a muon lepton number $L_\mu = -1$. All other particles have $L_\mu = 0$.

Finally, tau lepton number L_τ is conserved with similar assignments made for the tau lepton, its neutrino, and their two antiparticles.

Quick Quiz 46.3   Consider the following decay: $\pi^0 \rightarrow \mu^- + e^+ + \nu_\mu$. What conservation laws are violated by this decay? (a) energy   (b) angular momentum   (c) electric charge   (d) baryon number   (e) electron lepton number   (f) muon lepton number   (g) tau lepton number   (h) no conservation laws

Quick Quiz 46.4   Suppose a claim is made that the decay of the neutron is given by $n \rightarrow p + e^-$. What conservation laws are violated by this decay? (a) energy (b) angular momentum   (c) electric charge   (d) baryon number   (e) electron lepton number   (f) muon lepton number   (g) tau lepton number   (h) no conservation laws

---

EXAMPLE 46.3   Checking Lepton Numbers

Use the law of conservation of lepton numbers to determine whether each of the following decay schemes (A) and (B) can occur:

(A) $\mu^- \rightarrow e^- + \bar{\nu}_e + \nu_\mu$

SOLUTION

Conceptualize   Because this decay involves a muon and an electron, L_μ and L_e must each be conserved separately if the decay is to occur.

Categorize   We use a conservation law developed in this section, so we categorize this example as a substitution problem.

Evaluate the lepton numbers before the decay:

$$L_\mu = +1 \qquad L_e = 0$$

Evaluate the total lepton numbers after the decay:

$$L_\mu = 0 + 0 + 1 = +1 \qquad L_e = +1 + (-1) + 0 = 0$$

Therefore, both numbers are conserved and on this basis the decay is possible.

(B) $\pi^+ \rightarrow \mu^+ + \nu_\mu + \nu_e$

SOLUTION

Evaluate the lepton numbers before the decay:

$$L_\mu = 0 \qquad L_e = 0$$

Evaluate the total lepton numbers after the decay:

$$L_\mu = -1 + 1 + 0 = 0 \qquad L_e = 0 + 0 + 1 = 1$$

Therefore, the decay is not possible because electron lepton number is not conserved.

46.6   Strange Particles and Strangeness

Many particles discovered in the 1950s were produced by the interaction of pions with protons and neutrons in the atmosphere. A group of these—the kaon (K), lambda (Λ), and sigma (Σ) particles—exhibited unusual properties both as they were created and as they decayed; hence, they were called *strange particles.*

One unusual property of strange particles is that they are always produced in pairs. For example, when a pion collides with a proton, a highly probable result is the production of two neutral strange particles (Fig. 46.7):

$$\pi^- + p \rightarrow K^0 + \Lambda^0$$

The reaction $\pi^- + p \rightarrow K^0 + n$, where only one of the final particles is strange, never occurs, however, even though no known conservation laws would be violated and even though the energy of the pion is sufficient to initiate the reaction.

The second peculiar feature of strange particles is that although they are produced in reactions involving the strong interaction at a high rate, they do not decay into particles that interact via the strong force at a high rate. Instead, they decay very slowly, which is characteristic of the weak interaction. Their half-lives are in the range 10^{-10} s to 10^{-8} s, whereas most other particles that interact via the strong force have much shorter lifetimes on the order of 10^{-23} s.

To explain these unusual properties of strange particles, a new quantum number S, called **strangeness,** was introduced, together with a conservation law. The strangeness numbers for some particles are given in Table 46.2. The production of strange particles in pairs is explained by assigning $S = +1$ to one of the particles, $S = -1$ to the other, and $S = 0$ to all nonstrange particles. The **law of conservation of strangeness** states that **in a nuclear reaction or decay that occurs via the**

◄ Conservation of strangeness

Figure 46.7   This bubble-chamber photograph shows many events, and the inset is a drawing of identified tracks. The strange particles Λ^0 and K^0 are formed at the bottom as a π^- particle interacts with a proton in the reaction $\pi^- + p \rightarrow \Lambda^0 + K^0$. (Notice that the neutral particles leave no tracks, as indicated by the dashed lines in the inset.) The Λ^0 then decays in the reaction $\Lambda^0 \rightarrow \pi^- + p$ and the K^0 in the reaction $K^0 \rightarrow \pi^+ + \mu^- + \bar{\nu}_\mu$.

strong force, strangeness is conserved; that is, the sum of the strangeness numbers before the process must equal the sum of the strangeness numbers after the process. In processes that occur via the weak interaction, strangeness may not be conserved.

The low decay rate of strange particles can be explained by assuming the strong and electromagnetic interactions obey the law of conservation of strangeness but the weak interaction does not. Because the decay of a strange particle involves the loss of one strange particle, it violates strangeness conservation and hence proceeds slowly via the weak interaction.

EXAMPLE 46.4   **Is Strangeness Conserved?**

(A) Use the law of strangeness conservation to determine whether the reaction $\pi^0 + n \rightarrow K^+ + \Sigma^-$ occurs.

SOLUTION

Conceptualize   Because there are two strange particles on the right side of the reaction and none on the left, we see that we will need to investigate conservation of strangeness.

Categorize   We use a conservation law developed in this section, so we categorize this example as a substitution problem.

Evaluate the strangeness for the left side of the reaction using Table 46.2:

$$S = 0 + 0 = 0$$

Evaluate the strangeness for the right side of the reaction:

$$S = +1 - 1 = 0$$

Therefore, strangeness is conserved and the reaction is allowed.

(B) Show that the reaction $\pi^- + p \rightarrow \pi^- + \Sigma^+$ does not conserve strangeness.

SOLUTION

Evaluate the strangeness for the left side of the reaction:

$$S = 0 + 0 = 0$$

Evaluate the strangeness for the right side of the reaction:

$$S = 0 + (-1) = -1$$

Therefore, strangeness is not conserved.

46.7 Finding Patterns in the Particles

One tool scientists use is the detection of patterns in data, patterns that contribute to our understanding of nature. One of the best examples of this tool's use is the development of the periodic table, which provides a fundamental understanding of the chemical behavior of the elements. As mentioned in the introduction, the periodic table explains how more than 100 elements can be formed from three particles, the electron, the proton, and the neutron. The table of nuclides, part of which is shown in Table 44.2, contains hundreds of nuclides, but all can be built from protons and neutrons.

The number of particles observed by particle physicists is also in the hundreds. Is it possible that a small number of entities exist from which all these can be built? Taking a hint from the success of the periodic table and the table of nuclides, let's explore the historical search for patterns among the particles.

Many classification schemes have been proposed for grouping particles into families. Consider, for instance, the baryons listed in Table 46.2 that have spins of $\frac{1}{2}$: p, n, Λ^0, Σ^+, Σ^0, Σ^-, Ξ^0, and Ξ^-. If we plot strangeness versus charge for these

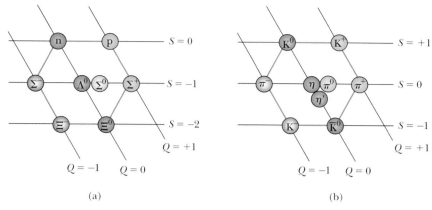

(a)   (b)

Figure 46.8   (a) The hexagonal eightfold-way pattern for the eight spin-$\frac{1}{2}$ baryons. This strangeness-versus-charge plot uses a sloping axis for charge number Q and a horizontal axis for strangeness S. (b) The eightfold-way pattern for the nine spin-zero mesons.

baryons using a sloping coordinate system as in Figure 46.8a, a fascinating pattern is observed: six of the baryons form a hexagon, and the remaining two are at the hexagon's center.

As a second example, consider the following nine spin-zero mesons listed in Table 46.2: π^+, π^0, π^-, K^+, K^0, K^-, η, η', and the antiparticle $\overline{K}^0$. Figure 46.8b is a plot of strangeness versus charge for this family. Again, a hexagonal pattern emerges. In this case, each particle on the perimeter of the hexagon lies opposite its antiparticle and the remaining three (which form their own antiparticles) are at the center of the hexagon. These and related symmetric patterns were developed independently in 1961 by Murray Gell-Mann and Yuval Ne'eman (1925–2006). Gell-Mann called the patterns the **eightfold way,** after the eightfold path to nirvana in Buddhism.

Groups of baryons and mesons can be displayed in many other symmetric patterns within the framework of the eightfold way. For example, the family of spin-$\frac{3}{2}$ baryons known in 1961 contains nine particles arranged in a pattern like that of the pins in a bowling alley as in Figure 46.9. (The particles Σ^{*+}, Σ^{*0}, Σ^{*-}, Ξ^{*0}, and Ξ^{*-} are excited states of the particles Σ^+, Σ^0, Σ^-, Ξ^0, and Ξ^-. In these higher-energy states, the spins of the three quarks (see Section 46.8) making up the particle are aligned so that the total spin of the particle is $\frac{3}{2}$.) When this pattern was proposed, an empty spot occurred in it (at the bottom position), corresponding to a particle that had never been observed. Gell-Mann predicted that the missing particle, which he called the omega minus (Ω^-), should have spin $\frac{3}{2}$, charge -1, strangeness -3, and rest energy of approximately 1 680 MeV. Shortly thereafter,

MURRAY GELL-MANN
American Physicist (b. 1929)
In 1969, Murray Gell-Mann was awarded the Nobel Prize in Physics for his theoretical studies dealing with subatomic particles.

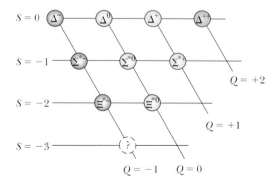

Figure 46.9   The pattern for the higher-mass, spin-$\frac{3}{2}$ baryons known at the time the pattern was proposed. The three Σ^* and two Ξ^* particles are excited states of the corresponding spin-$\frac{1}{2}$ particles in Figure 46.8. These excited states have higher mass and spin $\frac{3}{2}$. The absence of a particle in the bottom position was evidence of a new particle yet to be discovered, the Ω^-.

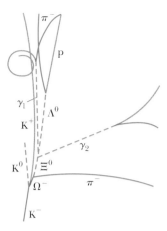

Figure 46.10 Discovery of the Ω^- particle. The photograph on the left shows the original bubble-chamber tracks. The drawing on the right isolates the tracks of the important events. The K^- particle at the bottom collides with a proton to produce the first detected Ω^- particle plus a K^0 and a K^+.

in 1964, scientists at the Brookhaven National Laboratory found the missing particle through careful analyses of bubble-chamber photographs (Fig. 46.10) and confirmed all its predicted properties.

The prediction of the missing particle in the eightfold way has much in common with the prediction of missing elements in the periodic table. Whenever a vacancy occurs in an organized pattern of information, experimentalists have a guide for their investigations.

46.8 Quarks

As mentioned earlier, leptons appear to be truly elementary particles because there are only a few types of them, and experiments indicate that they have no measurable size or internal structure. Hadrons, on the other hand, are complex particles having size and structure. The existence of the strangeness–charge patterns of the eightfold way suggests that hadrons have substructure. Furthermore, hundreds of types of hadrons exist and many decay into other hadrons.

Mesons

π^+

Baryons

p

K$^-$

n

ACTIVE FIGURE 46.11

Quark composition of two mesons and two baryons.

Sign in at www.thomsonedu.com and go to ThomsonNOW to observe the quark composition for the mesons and baryons in Tables 46.4 and 46.5.

The Original Quark Model

In 1963, Gell-Mann and George Zweig (b. 1937) independently proposed a model for the substructure of hadrons. According to their model, all hadrons are composed of two or three elementary constituents called **quarks.** (Gell-Mann borrowed the word *quark* from the passage "Three quarks for Muster Mark" in James Joyce's *Finnegans Wake.* In Zweig's model, he called the constituents "aces.") The model has three types of quarks, designated by the symbols u, d, and s, that are given the arbitrary names **up, down,** and **strange.** The various types of quarks are called **flavors.** Active Figure 46.11 is a pictorial representation of the quark compositions of several hadrons.

An unusual property of quarks is that they carry a fractional electronic charge. The u, d, and s quarks have charges of $+2e/3$, $-e/3$, and $-e/3$, respectively, where e is the elementary charge 1.60×10^{-19} C. These and other properties of quarks and antiquarks are given in Table 46.3. Quarks have spin $\frac{1}{2}$, which means that all quarks are fermions, defined as any particle having half-integral spin, as pointed out in Section 43.8. As Table 46.3 shows, associated with each quark is an antiquark of opposite charge, baryon number, and strangeness.

TABLE 46.3

Properties of Quarks and Antiquarks

Quarks

Name	Symbol	Spin	Charge	Baryon Number	Strangeness	Charm	Bottomness	Topness
Up	u	$\frac{1}{2}$	$+\frac{2}{3}e$	$\frac{1}{3}$	0	0	0	0
Down	d	$\frac{1}{2}$	$-\frac{1}{3}e$	$\frac{1}{3}$	0	0	0	0
Strange	s	$\frac{1}{2}$	$-\frac{1}{3}e$	$\frac{1}{3}$	-1	0	0	0
Charmed	c	$\frac{1}{2}$	$+\frac{2}{3}e$	$\frac{1}{3}$	0	$+1$	0	0
Bottom	b	$\frac{1}{2}$	$-\frac{1}{3}e$	$\frac{1}{3}$	0	0	$+1$	0
Top	t	$\frac{1}{2}$	$+\frac{2}{3}e$	$\frac{1}{3}$	0	0	0	$+1$

Antiquarks

Name	Symbol	Spin	Charge	Baryon Number	Strangeness	Charm	Bottomness	Topness
Anti-up	$\overline{u}$	$\frac{1}{2}$	$-\frac{2}{3}e$	$-\frac{1}{3}$	0	0	0	0
Anti-down	$\overline{d}$	$\frac{1}{2}$	$+\frac{1}{3}e$	$-\frac{1}{3}$	0	0	0	0
Anti-strange	$\overline{s}$	$\frac{1}{2}$	$+\frac{1}{3}e$	$-\frac{1}{3}$	$+1$	0	0	0
Anti-charmed	$\overline{c}$	$\frac{1}{2}$	$-\frac{2}{3}e$	$-\frac{1}{3}$	0	-1	0	0
Anti-bottom	$\overline{b}$	$\frac{1}{2}$	$+\frac{1}{3}e$	$-\frac{1}{3}$	0	0	-1	0
Anti-top	$\overline{t}$	$\frac{1}{2}$	$-\frac{2}{3}e$	$-\frac{1}{3}$	0	0	0	-1

The compositions of all hadrons known when Gell-Mann and Zweig presented their model can be completely specified by three simple rules:

- A meson consists of one quark and one antiquark, giving it a baryon number of 0, as required.
- A baryon consists of three quarks.
- An antibaryon consists of three antiquarks.

The theory put forth by Gell-Mann and Zweig is referred to as the *original quark model*.

Quick Quiz 46.5 Using a coordinate system like that in Figure 46.8, draw an eightfold-way diagram for the three quarks in the original quark model.

Charm and Other Developments

Although the original quark model was highly successful in classifying particles into families, some discrepancies occurred between its predictions and certain experimental decay rates. Consequently, several physicists proposed a fourth quark flavor in 1967. They argued that if four types of leptons exist (as was thought at the time), there should also be four flavors of quarks because of an underlying symmetry in nature. The fourth quark, designated c, was assigned a property called **charm.** A *charmed* quark has charge $+2e/3$, just as the up quark does, but its charm distinguishes it from the other three quarks. This introduces a new quantum number C, representing charm. The new quark has charm $C = +1$, its antiquark has charm of $C = -1$, and all other quarks have $C = 0$. Charm, like strangeness, is conserved in strong and electromagnetic interactions but not in weak interactions.

Evidence that the charmed quark exists began to accumulate in 1974, when a heavy meson called the J/Ψ particle (or simply Ψ, Greek letter psi) was discovered

TABLE 46.4

Quark Composition of Mesons

		Antiquarks								
		$\bar{\mathbf{b}}$		$\bar{\mathbf{c}}$		$\bar{\mathbf{s}}$		$\bar{\mathbf{d}}$		$\bar{\mathbf{u}}$	
Quarks	**b**	Υ ($\bar{b}b$)		B_c^- ($\bar{c}b$)		$\overline{B}_s^0$ ($\bar{s}b$)		$\overline{B}_d^0$ ($\bar{d}b$)		B^- ($\bar{u}b$)	
	c	B_c^+ ($\bar{b}c$)		J/Ψ ($\bar{c}c$)		D_s^+ ($\bar{s}c$)		D^+ ($\bar{d}c$)		D^0 ($\bar{u}c$)	
	s	B_s^0 ($\bar{b}s$)		D_s^- ($\bar{c}s$)		η, η' ($\bar{s}s$)		$\overline{K}^0$ ($\bar{d}s$)		K^- ($\bar{u}s$)	
	d	B_d^0 ($\bar{b}d$)		D^- ($\bar{c}d$)		K^0 ($\bar{s}d$)		π^0, η, η' ($\bar{d}d$)		π^- ($\bar{u}d$)	
	u	B^+ ($\bar{b}u$)		$\overline{D}^0$ ($\bar{c}u$)		K^+ ($\bar{s}u$)		π^+ ($\bar{d}u$)		π^0, η, η' ($\bar{u}u$)	

Note: The top quark does not form mesons because it decays too quickly.

TABLE 46.5

Quark Composition of Several Baryons

Particle	Quark Composition
p	uud
n	udd
Λ^0	uds
Σ^+	uus
Σ^0	uds
Σ^-	dds
Δ^{++}	uuu
Δ^+	uud
Δ^0	udd
Δ^-	ddd
Ξ^0	uss
Ξ^-	dss
Ω^-	sss

Note: Some baryons have the same quark composition, such as the p and the Δ^+ and the n and the Δ^0. In these cases, the Δ particles are considered to be excited states of the proton and neutron.

independently by two groups, one led by Burton Richter (b. 1931) at the Stanford Linear Accelerator (SLAC), and the other led by Samuel Ting (b. 1936) at the Brookhaven National Laboratory. In 1976, Richter and Ting were awarded a Nobel Prize in Physics for this work. The J/Ψ particle does not fit into the three-quark model; instead, it has properties of a combination of the proposed charmed quark and its antiquark ($c\bar{c}$). It is much more massive than the other known mesons ($\sim 3\,100$ MeV/c^2), and its lifetime is much longer than the lifetimes of particles that interact via the strong force. Soon, related mesons were discovered, corresponding to such quark combinations as $\bar{c}d$ and $c\bar{d}$, all of which have great masses and long lifetimes. The existence of these new mesons provided firm evidence for the fourth quark flavor.

In 1975, researchers at Stanford University reported strong evidence for the tau (τ) lepton, mass 1 784 MeV/c^2. This fifth type of lepton led physicists to propose that more flavors of quarks might exist, on the basis of symmetry arguments similar to those leading to the proposal of the charmed quark. These proposals led to more elaborate quark models and the prediction of two new quarks, **top** (t) and **bottom** (b). (Some physicists prefer *truth* and *beauty*.) To distinguish these quarks from the others, quantum numbers called *topness* and *bottomness* (with allowed values +1, 0, −1) were assigned to all quarks and antiquarks (see Table 46.3). In 1977, researchers at the Fermi National Laboratory, under the direction of Leon Lederman (b. 1922), reported the discovery of a very massive new meson Υ (Greek letter upsilon), whose composition is considered to be $b\bar{b}$, providing evidence for the bottom quark. In March 1995, researchers at Fermilab announced the discovery of the top quark (supposedly the last of the quarks to be found), which has a mass of 173 GeV/c^2.

Table 46.4 lists the quark compositions of mesons formed from the up, down, strange, charmed, and bottom quarks. Table 46.5 shows the quark combinations for the baryons listed in Table 46.2. Notice that only two flavors of quarks, u and d, are contained in all hadrons encountered in ordinary matter (protons and neutrons).

Will the discoveries of elementary particles ever end? How many "building blocks" of matter actually exist? At present, physicists believe that the elementary particles in nature are six quarks and six leptons, together with their antiparticles, and the four field particles listed in Table 46.1. Table 46.6 lists the rest energies and charges of the quarks and leptons.

Despite extensive experimental effort, no isolated quark has ever been observed. Physicists now believe that at ordinary temperatures, quarks are permanently confined inside ordinary particles because of an exceptionally strong force that prevents them from escaping, called (appropriately) the **strong force**[5] (which we introduced at the beginning of Section 46.4 and will discuss further in Section 46.10). This force increases with separation distance, similar to the force exerted

---

[5] As a reminder, the original meaning of the term *strong force* was the short-range attractive force between nucleons, which we have called the *nuclear force*. The nuclear force between nucleons is a secondary effect of the strong force between quarks.

TABLE 46.6

The Elementary Particles and Their Rest Energies and Charges

Particle	Rest Energy	Charge
Quarks		
u	360 MeV	$+\frac{2}{3}e$
d	360 MeV	$-\frac{1}{3}e$
s	540 MeV	$-\frac{1}{3}e$
c	1 500 MeV	$+\frac{2}{3}e$
b	5 GeV	$-\frac{1}{3}e$
t	173 GeV	$+\frac{2}{3}e$
Leptons		
e^-	511 keV	$-e$
μ^-	105.7 MeV	$-e$
τ^-	1 784 MeV	$-e$
ν_e	< 7 eV	0
ν_μ	< 0.3 MeV	0
ν_τ	< 30 MeV	0

by a stretched spring. Current efforts are under way to form a **quark–gluon plasma,** a state of matter in which the quarks are freed from neutrons and protons. In 2000, scientists at CERN announced evidence for a quark–gluon plasma formed by colliding lead nuclei. Experiments continue at CERN as well as at the Relativistic Heavy Ion Collider (RHIC) at Brookhaven to verify the production of a quark–gluon plasma.

Quick Quiz 46.6 Doubly charged baryons, such as the Δ^{++}, are known to exist. True or False: Doubly charged mesons also exist.

46.9 Multicolored Quarks

Shortly after the concept of quarks was proposed, scientists recognized that certain particles had quark compositions that violated the exclusion principle. In Section 42.7, we applied the exclusion principle to electrons in atoms. The principle is more general, however, and applies to all particles with half-integral spin ($\frac{1}{2}$, $\frac{3}{2}$, etc.), which are collectively called fermions. Because all quarks are fermions having spin $\frac{1}{2}$, they are expected to follow the exclusion principle. One example of a particle that appears to violate the exclusion principle is the Ω^- (sss) baryon, which contains three strange quarks having parallel spins, giving it a total spin of $\frac{3}{2}$. All three quarks have the same spin quantum number, in violation of the exclusion principle. Other examples of baryons made up of identical quarks having parallel spins are the Δ^{++} (uuu) and the Δ^- (ddd).

To resolve this problem, it was suggested that quarks possess an additional property called **color charge.** This property is similar in many respects to electric charge except that it occurs in six varieties rather than two. The colors assigned to quarks are red, green, and blue, and antiquarks have the colors antired, antigreen, and antiblue. Therefore, the colors red, green, and blue serve as the "quantum numbers" for the color of the quark. To satisfy the exclusion principle, the three quarks in any baryon must all have different colors. Look again at the quarks in the baryons in Active Figure 46.11 and notice the colors. The three colors "neutralize" to white. A quark and an antiquark in a meson must be of a color and the

PITFALL PREVENTION 46.3
Color Charge Is Not Really Color

The description of color for a quark has nothing to do with visual sensation from light. It is simply a convenient name for a property that is analogous to electric charge.

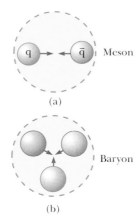

Figure 46.12 (a) A green quark is attracted to an antigreen quark. This forms a meson whose quark structure is (q$\bar{\text{q}}$). (b) Three quarks of different colors attract each other to form a baryon.

(a) Yukawa's pion model

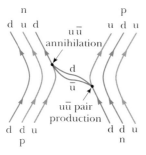

(b) Quark model

Figure 46.13 (a) A nuclear interaction between a proton and a neutron explained in terms of Yukawa's pion-exchange model. Because the pion carries charge, the proton and neutron switch identities. (b) The same interaction, explained in terms of quarks and gluons. The exchanged $\bar{\text{u}}$d quark pair makes up a π^- meson.

corresponding anticolor and will consequently neutralize to white, similar to the way electric charges + and − neutralize to zero net charge. (See the mesons in Active Fig. 46.11.) The apparent violation of the exclusion principle in the Ω^- baryon is removed because the three quarks in the particle have different colors.

The new property of color increases the number of quarks by a factor of 3 because each of the six quarks comes in three colors. Although the concept of color in the quark model was originally conceived to satisfy the exclusion principle, it also provided a better theory for explaining certain experimental results. For example, the modified theory correctly predicts the lifetime of the π^0 meson.

The theory of how quarks interact with each other is called **quantum chromodynamics,** or QCD, to parallel the name *quantum electrodynamics* (the theory of the electrical interaction between light and matter). In QCD, each quark is said to carry a color charge, in analogy to electric charge. The strong force between quarks is often called the **color force.** Therefore, the terms *strong force* and *color force* are used interchangeably.

In Section 46.1, we stated that the nuclear interaction between hadrons is mediated by massless field particles called **gluons.** As mentioned earlier, the nuclear force is actually a secondary effect of the strong force between quarks. The gluons are the mediators of the strong force. When a quark emits or absorbs a gluon, the quark's color may change. For example, a blue quark that emits a gluon may become a red quark and a red quark that absorbs this gluon becomes a blue quark.

The color force between quarks is analogous to the electric force between charges: particles with the same color repel, and those with opposite colors attract. Therefore, two green quarks repel each other, but a green quark is attracted to an antigreen quark. The attraction between quarks of opposite color to form a meson (q$\bar{\text{q}}$) is indicated in Figure 46.12a. Differently colored quarks also attract one another, although with less intensity than the oppositely colored quark and antiquark. For example, a cluster of red, blue, and green quarks all attract one another to form a baryon as in Figure 46.12b. Therefore, every baryon contains three quarks of three different colors.

Although the nuclear force between two colorless hadrons is negligible at large separations, the net strong force between their constituent quarks is not exactly zero at small separations. This residual strong force is the nuclear force that binds protons and neutrons to form nuclei. It is similar to the force between two electric dipoles. Each dipole is electrically neutral. An electric field surrounds the dipoles, however, because of the separation of the positive and negative charges (see Section 23.6). As a result, an electric interaction occurs between the dipoles that is weaker than the force between single charges. In Section 43.1, we explored how this interaction results in the Van der Waals force between neutral molecules.

According to QCD, a more basic explanation of the nuclear force can be given in terms of quarks and gluons. Figure 46.13a shows the nuclear interaction between a neutron and a proton by means of Yukawa's pion, in this case a π^-. This drawing differs from Figure 46.5a, in which the field particle is a π^0; there is no transfer of charge from one nucleon to the other in Figure 46.5a. In Figure 46.13a, the charged pion carries charge from one nucleon to the other, so the nucleons change identities, with the proton becoming a neutron and the neutron becoming a proton.

Let's look at the same interaction from the viewpoint of the quark model, shown in Figure 46.13b. In this Feynman diagram, the proton and neutron are represented by their quark constituents. Each quark in the neutron and proton is continuously emitting and absorbing gluons. The energy of a gluon can result in the creation of quark–antiquark pairs. This process is similar to the creation of electron–positron pairs in pair production, which we investigated in Section 46.2. When the neutron and proton approach to within 1 fm of each other, these gluons and quarks can be exchanged between the two nucleons, and such exchanges produce the nuclear force. Figure 46.13b depicts one possibility for the process shown in Figure 46.13a. A down quark in the neutron on the right emits a gluon.

The energy of the gluon is then transformed to create a $u\bar{u}$ pair. The u quark stays within the nucleon (which has now changed to a proton), and the recoiling d quark and the $\bar{u}$ antiquark are transmitted to the proton on the left side of the diagram. Here the $\bar{u}$ annihilates a u quark within the proton and the d is captured. The net effect is to change a u quark to a d quark, and the proton on the left has changed to a neutron.

As the d quark and $\bar{u}$ antiquark in Figure 46.13b transfer between the nucleons, the d and $\bar{u}$ exchange gluons with each other and can be considered to be bound to each other by means of the strong force. Looking back at Table 46.4, we see that this combination is a π^-, or Yukawa's field particle! Therefore, the quark model of interactions between nucleons is consistent with the pion-exchange model.

46.10   The Standard Model

Scientists now believe there are three classifications of truly elementary particles: leptons, quarks, and field particles. These three types of particles are further classified as either fermions or bosons. Quarks and leptons have spin $\frac{1}{2}$ and hence are fermions, whereas the field particles have integral spin of 1 or higher and are bosons.

Recall from Section 46.1 that the weak force is believed to be mediated by the W^+, W^-, and Z^0 bosons. These particles are said to have *weak charge,* just as quarks have color charge. Therefore, each elementary particle can have mass, electric charge, color charge, and weak charge. Of course, one or more of these could be zero.

In 1979, Sheldon Glashow (b. 1932), Abdus Salam (1926–1996), and Steven Weinberg (b. 1933) won a Nobel Prize in Physics for developing a theory that unifies the electromagnetic and weak interactions. This **electroweak theory** postulates that the weak and electromagnetic interactions have the same strength when the particles involved have very high energies. The two interactions are viewed as different manifestations of a single unifying electroweak interaction. The theory makes many concrete predictions, but perhaps the most spectacular is the prediction of the masses of the W and Z particles at approximately 82 GeV/c^2 and 93 GeV/c^2, respectively. These predictions are close to the masses in Table 46.1 determined by experiment.

The combination of the electroweak theory and QCD for the strong interaction is referred to in high-energy physics as the **Standard Model.** Although the details of the Standard Model are complex, its essential ingredients can be summarized with the help of Figure 46.14. (Although the Standard Model does not include the gravitational force at present, we include gravity in Figure 46.14 because physicists hope to eventually incorporate this force into a unified theory.) This diagram

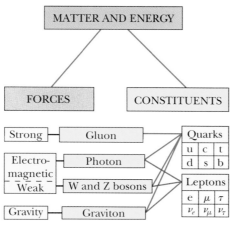

Figure 46.14   The Standard Model of particle physics.

Figure 46.15 A view from inside the Large Electron–Positron (LEP) Collider tunnel, which is 27 km in circumference.

Figure 46.16 Computers at Fermilab create a pictorial representation such as this one of the paths of particles after a collision

shows that quarks participate in all the fundamental forces and that leptons participate in all except the strong force.

The Standard Model does not answer all questions. A major question still unanswered is why, of the two mediators of the electroweak interaction, the photon has no mass but the W and Z bosons do. Because of this mass difference, the electromagnetic and weak forces are quite distinct at low energies but become similar at very high energies, when the rest energy is negligible relative to the total energy. The behavior as one goes from high to low energies is called *symmetry breaking* because the forces are similar, or symmetric, at high energies but are very different at low energies. The nonzero rest energies of the W and Z bosons raise the question of the origin of particle masses. To resolve this problem, a hypothetical particle called the **Higgs boson,** which provides a mechanism for breaking the electroweak symmetry, has been proposed. The Standard Model modified to include the Higgs boson provides a logically consistent explanation of the massive nature of the W and Z bosons. Unfortunately, the Higgs boson has not yet been found, but physicists know that its rest energy should be less than 1 TeV. To determine whether the Higgs boson exists, two quarks each having at least 1 TeV of energy must collide. Calculations show that such a collision requires injecting 40 TeV of energy within the volume of a proton, however.

Because of the limited energy available in conventional accelerators using fixed targets, it is necessary to employ colliding-beam accelerators called **colliders.** The concept of colliders is straightforward. Particles that have equal masses and equal kinetic energies, traveling in opposite directions in an accelerator ring, collide head-on to produce the required reaction and form new particles. Because the total momentum of the interacting particles is zero, all their kinetic energy is available for the reaction. The Large Electron–Positron (LEP) Collider at CERN (Fig. 46.15) and the Stanford Linear Collider collide both electrons and positrons. The Super Proton Synchrotron at CERN accelerates protons and antiprotons to energies of 270 GeV. The world's highest-energy proton accelerator, the Tevatron at the Fermi National Laboratory in Illinois, produces protons at almost 1 000 GeV (1 TeV). The Relativistic Heavy Ion Collider at Brookhaven National Laboratory collides heavy ions to search for the quark–gluon plasma as discussed earlier. CERN expects a late 2007 completion date for the Large Hadron Collider, a proton–proton collider that will provide a center-of-mass energy of 14 TeV and enable exploration of Higgs-boson physics. The accelerator occupies the same 27-km circumference tunnel now housing the LEP Collider, and many countries will participate in the project.

In addition to increasing energies in modern accelerators, detection techniques have become increasingly sophisticated. We saw simple bubble-chamber photographs earlier in this chapter that required hours of analysis by hand. Figure 46.16 shows a modern detection display of particle tracks after a reaction; the tracks are analyzed rapidly by computer. The photograph at the beginning of this chapter shows a complex set of tracks from a collision of gold nuclei.

46.11 The Cosmic Connection

In this section, we describe one of the most fascinating theories in all science—the Big Bang theory of the creation of the Universe—and the experimental evidence that supports it. This theory of cosmology states that the Universe had a beginning and furthermore that the beginning was so cataclysmic that it is impossible to look back beyond it. According to this theory, the Universe erupted from an infinitely dense singularity between 10 and 20 billion years ago. The first few moments after the Big Bang saw such extremely high energy that it is believed that all four interactions of physics were unified and all matter was contained in a quark–gluon plasma.

The evolution of the four fundamental forces from the Big Bang to the present is shown in Figure 46.17. During the first 10^{-43} s (the ultrahot epoch, $T \sim 10^{32}$ K),

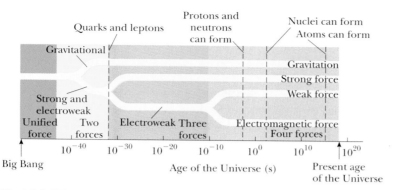

Figure 46.17   A brief history of the Universe from the Big Bang to the present. The four forces became distinguishable during the first nanosecond. Following that, all the quarks combined to form particles that interact via the nuclear force. The leptons, however, remained separate and to this day exist as individual, observable particles.

it is presumed the strong, electroweak, and gravitational forces were joined to form a completely unified force. In the first 10^{-35} s following the Big Bang (the hot epoch, $T \sim 10^{29}$ K), symmetry breaking occurred for gravity while the strong and electroweak forces remained unified. It was a period when particle energies were so great ($> 10^{16}$ GeV) that very massive particles as well as quarks, leptons, and their antiparticles existed. Then, after 10^{-35} s, the Universe rapidly expanded and cooled (the warm epoch, $T \sim 10^{29}$ to 10^{15} K) and the strong and electroweak forces parted company. As the Universe continued to cool, the electroweak force split into the weak force and the electromagnetic force approximately 10^{-10} s after the Big Bang.

After a few minutes, protons and neutrons condensed out of the plasma. For half an hour, the Universe underwent thermonuclear detonation, exploding as a hydrogen bomb and producing most of the helium nuclei that now exist. The Universe continued to expand, and its temperature dropped. Until about 700 000 years after the Big Bang, the Universe was dominated by radiation. Energetic radiation prevented matter from forming single hydrogen atoms because collisions would instantly ionize any atoms that happened to form. Photons experienced continuous Compton scattering from the vast numbers of free electrons, resulting in a Universe that was opaque to radiation. By the time the Universe was about 700 000 years old, it had expanded and cooled to approximately 3 000 K and protons could bind to electrons to form neutral hydrogen atoms. Because of the quantized energies of the atoms, far more wavelengths of radiation were not absorbed by atoms than were absorbed, and the Universe suddenly became transparent to photons. Radiation no longer dominated the Universe, and clumps of neutral matter steadily grew: first atoms, then molecules, gas clouds, stars, and finally galaxies.

Observation of Radiation from the Primordial Fireball

In 1965, Arno A. Penzias (b. 1933) and Robert W. Wilson (b. 1936) of Bell Laboratories were testing a sensitive microwave receiver and made an amazing discovery. A pesky signal producing a faint background hiss was interfering with their satellite communications experiments. The microwave horn that served as their receiving antenna is shown in Figure 46.18. Evicting a flock of pigeons from the 20-ft horn and cooling the microwave detector both failed to remove the signal.

The intensity of the detected signal remained unchanged as the antenna was pointed in different directions. That the radiation had equal strengths in all directions suggested that the entire Universe was the source of this radiation. Ultimately, it became clear that they were detecting microwave background radiation (at a wavelength of 7.35 cm), which represented the leftover "glow" from the Big Bang. Through a casual conversation, Penzias and Wilson discovered that a group

Figure 46.18   Robert W. Wilson (*left*) and Arno A. Penzias with the Bell Telephone Laboratories horn-reflector antenna.

Figure 46.19 Theoretical blackbody (brown curve) and measured radiation spectra (blue points) of the Big Bang. Most of the data were collected from the Cosmic Background Explorer, or COBE, satellite. The datum of Penzias and Wilson is indicated in red.

at Princeton University had predicted the residual radiation from the Big Bang and were planning an experiment to attempt to confirm the theory. The excitement in the scientific community was high when Penzias and Wilson announced that they had already observed an excess microwave background compatible with a 3-K blackbody source, which was consistent with the predicted temperature of the Universe at this time after the Big Bang.

Because Penzias and Wilson made their measurements at a single wavelength, they did not completely confirm the radiation as 3-K blackbody radiation. Subsequent experiments by other groups added intensity data at different wavelengths as shown in Figure 46.19. The results confirm that the radiation is that of a black body at 2.7 K. This figure is perhaps the most clear-cut evidence for the Big Bang theory. The 1978 Nobel Prize in Physics was awarded to Penzias and Wilson for this most important discovery.

The discovery of the cosmic background radiation brought with it a problem, however: the radiation was too uniform. Scientists believed that slight fluctuations in this background had to occur to act as nucleation sites for the formation of the galaxies and other objects we now see in the sky. In 1989, NASA launched a satellite called COBE (KOH-bee), for Cosmic Background Explorer, to study this radiation in greater detail. In 1992, George Smoot (b. 1945) at the Lawrence Berkeley Laboratory found, on the basis of the data collected, that the background was not perfectly uniform but instead contained irregularities that corresponded to temperature variations of 0.000 3 K. Smoot won the 2006 Nobel Prize in Physics for this work. The Wilkinson Microwave Anisotropy Probe, launched in June 2001, exhibits data that allows observation of temperature differences in the cosmos in the microkelvin range.

Other Evidence for an Expanding Universe

The Big Bang theory of cosmology predicts that the Universe is expanding. Most of the key discoveries supporting the theory of an expanding Universe were made in the 20th century. Vesto Melvin Slipher (1875–1969), an American astronomer, reported in 1912 that most nebulae are receding from the Earth at speeds up to several million miles per hour. Slipher was one of the first scientists to use Doppler shifts (see Section 17.4) in spectral lines to measure galaxy velocities.

In the late 1920s, Edwin P. Hubble (1889–1953) made the bold assertion that the whole Universe is expanding. From 1928 to 1936, until they reached the limits of the 100-inch telescope, Hubble and Milton Humason (1891–1972) worked at Mount Wilson in California to prove this assertion. The results of that work and of its continuation with the use of a 200-inch telescope in the 1940s showed that the speeds at which galaxies are receding from the Earth increase in direct proportion to their distance R from us (Fig. 46.20). This linear relationship, known as **Hubble's law,** may be written

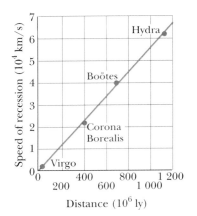

Figure 46.20 Hubble's law: a plot of speed of recession versus distance for four galaxies.

Hubble's law ▶

$$v = HR \qquad (46.4)$$

where H, called the **Hubble constant,** has the approximate value

$$H \approx 17 \times 10^{-3} \ \text{m/s} \cdot \text{ly}$$

EXAMPLE 46.5 | **Recession of a Quasar**

A quasar is an object that appears similar to a star and is very distant from the Earth. Its speed can be determined from Doppler-shift measurements in the light it emits. A certain quasar recedes from the Earth at a speed of $0.55c$. How far away is it?

SOLUTION

Conceptualize A common mental representation for the Hubble law is that of raisin bread cooking in an oven. Imagine yourself at the center of the loaf of bread. As the entire loaf of bread expands upon heating, raisins near you move slowly with respect to you. Raisins far away from you on the edge of the loaf move at a higher speed.

Categorize   We use a concept developed in this section, so we categorize this example as a substitution problem.

Find the distance through Hubble's law:

$$R = \frac{v}{H} = \frac{(0.55)(3.00 \times 10^8 \text{ m/s})}{17 \times 10^{-3} \text{ m/s} \cdot \text{ly}} = 9.7 \times 10^9 \text{ ly}$$

What If?   Suppose the quasar has moved at this speed ever since the Big Bang. With this assumption, estimate the age of the Universe.

Answer   Let's approximate the distance from Earth to the quasar as the distance the quasar has moved from the singularity since the Big Bang. We can then find the time interval from the particle under constant speed model: $\Delta t = d/v = R/v = 1/H \approx 18$ billion years, which is in approximate agreement with other calculations.

Will the Universe Expand Forever?

In the 1950s and 1960s, Allan R. Sandage (b. 1926) used the 200-inch telescope at Mount Palomar to measure the speeds of galaxies at distances of up to 6 billion light-years away from the Earth. These measurements showed that these very distant galaxies were moving approximately 10 000 km/s faster than Hubble's law predicted. According to this result, the Universe must have been expanding more rapidly 1 billion years ago, and consequently we conclude from these data that the expansion rate is slowing.[6] Today, astronomers and physicists are trying to determine the rate of expansion. If the average mass density of the Universe is less than some critical value, the galaxies will slow in their outward rush but still escape to infinity. If the average density exceeds the critical value, the expansion will eventually stop and contraction will begin, possibly leading to a superdense state followed by another expansion. In this scenario, we have an oscillating Universe.

EXAMPLE 46.6   **The Critical Density of the Universe**

(A) Starting from energy conservation, derive an expression for the critical mass density of the Universe ρ_c in terms of the Hubble constant H and the universal gravitational constant G.

SOLUTION

Conceptualize   Figure 46.21 shows a large section of the Universe, contained within a sphere of radius R. The total mass in this volume is M. A galaxy of mass $m \ll M$ that has a speed v at a distance R from the center of the sphere escapes to infinity (at which its speed approaches zero) if the sum of its kinetic energy and the gravitational potential energy of the system is zero.

Categorize   The Universe may be infinite in spatial extent, but Gauss's law for gravitation (Problem 61 in Chapter 24) implies that only the mass M inside the sphere contributes to the gravitational potential energy of the galaxy–sphere system. Therefore, we categorize this problem as one in which we apply Gauss's law for gravitation. We model the sphere in Figure 46.21 and the escaping galaxy as an isolated system.

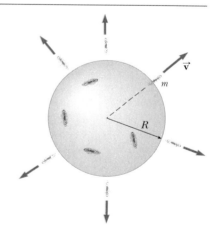

Figure 46.21   (Example 46.6) The galaxy marked with mass m is escaping from a large cluster of galaxies contained within a spherical volume of radius R. Only the mass within R slows the galaxy.

Analyze   Write an expression for the total mechanical energy of the system and set it equal to zero, representing the galaxy moving at the escape speed:

$$E_{\text{total}} = K + U = \tfrac{1}{2}mv^2 - \frac{GmM}{R} = 0$$

---

[6] The data at large distances have large observational uncertainties and may be systematically in error from effects such as abnormal brightness in the most distant visible clusters.

Substitute for the mass M contained within the sphere the product of the critical density and the volume of the sphere:

$$\tfrac{1}{2}mv^2 = \frac{Gm(\tfrac{4}{3}\pi R^3 \rho_c)}{R}$$

Solve for the critical density:

$$\rho_c = \frac{3v^2}{8\pi GR^2}$$

From Hubble's law, substitute for the ratio $v/R = H$:

(1) $$\rho_c = \frac{3}{8\pi G}\left(\frac{v}{R}\right)^2 = \frac{3H^2}{8\pi G}$$

(B) Estimate a numerical value for the critical density in grams per cubic centimeter.

SOLUTION

In Equation (1), substitute numerical values for H and G:

$$\rho_c = \frac{3H^2}{8\pi G} = \frac{3(17 \times 10^{-3}\ \text{m/s} \cdot \text{ly})^2}{8\pi(6.67 \times 10^{-11}\ \text{N} \cdot \text{m}^2/\text{kg}^2)} = 5.17 \times 10^5\ \text{kg/m} \cdot (\text{ly})^2$$

Reconcile the units by converting light-years to meters:

$$\rho_c = 5.7 \times 10^5\ \text{kg/m} \cdot (\text{ly})^2 \left(\frac{1\ \text{ly}}{9.46 \times 10^{15}\ \text{m}}\right)^2$$

$$= 6 \times 10^{-27}\ \text{kg/m}^3 = \boxed{6 \times 10^{-30}\ \text{g/cm}^3}$$

Finalize   Because the mass of a hydrogen atom is 1.67×10^{-24} g, this value of ρ_c corresponds to 3×10^{-6} hydrogen atoms per cubic centimeter or 3 atoms per cubic meter.

Missing Mass in the Universe?

The luminous matter in galaxies averages out to a Universe density of 5×10^{-33} g/cm^3. The radiation in the Universe has a mass equivalent of approximately 2% of the luminous matter. The total mass of all nonluminous matter (such as interstellar gas and black holes) may be estimated from the speeds of galaxies orbiting each other in a cluster. The higher the galaxy speeds, the more mass in the cluster. Measurements on the Coma cluster of galaxies indicate, surprisingly, that the amount of nonluminous matter is 20 to 30 times the amount of luminous matter present in stars and luminous gas clouds. Yet even this large, invisible component of *dark matter*, if extrapolated to the Universe as a whole, leaves the observed mass density a factor of 10 less than ρ_c calculated in Example 46.6. The deficit, called *missing mass*, has been the subject of intense theoretical and experimental work, with exotic particles such as axions, photinos, and superstring particles suggested as candidates for the missing mass. Some researchers have made the more mundane proposal that the missing mass is present in neutrinos. In fact, neutrinos are so abundant that a tiny neutrino rest energy on the order of only 20 eV would furnish the missing mass and "close" the Universe. Current experiments designed to measure the rest energy of the neutrino will have an effect on predictions for the future of the Universe.

Mysterious Energy in the Universe?

A surprising twist in the story of the Universe arose in 1998 with the observation of a class of supernovae that have a fixed absolute brightness. By combining the apparent brightness and the redshift of light from these explosions, their distance and speed of recession from the Earth can be determined. These observations led to the conclusion that the expansion of the Universe is not slowing down, but is accelerating! Observations by other groups also led to the same interpretation.

To explain this acceleration, physicists have proposed *dark energy*, which is energy possessed by the vacuum of space. In the early life of the Universe, gravity dominated over the dark energy. As the Universe expanded and the gravitational force between galaxies became smaller because of the great distances between them, the dark energy became more important. The dark energy results in an effective repulsive force that causes the expansion rate to increase.[7]

Although there is some degree of certainty about the beginning of the Universe, we are uncertain about how the story will end. Will the Universe keep on expanding forever, or will it someday collapse and then expand again, perhaps in an endless series of oscillations? Results and answers to these questions remain inconclusive, and the exciting controversy continues.

46.12   Problems and Perspectives

While particle physicists have been exploring the realm of the very small, cosmologists have been exploring cosmic history back to the first microsecond of the Big Bang. Observation of the events that occur when two particles collide in an accelerator is essential for reconstructing the early moments in cosmic history. For this reason, perhaps the key to understanding the early Universe is to first understand the world of elementary particles. Cosmologists and physicists now find that they have many common goals and are joining hands in an attempt to understand the physical world at its most fundamental level.

Our understanding of physics at short distances is far from complete. Particle physics is faced with many questions. Why does so little antimatter exist in the Universe? Is it possible to unify the strong and electroweak theories in a logical and consistent manner? Why do quarks and leptons form three similar but distinct families? Are muons the same as electrons apart from their difference in mass, or do they have other subtle differences that have not been detected? Why are some particles charged and others neutral? Why do quarks carry a fractional charge? What determines the masses of the elementary constituents of matter? Can isolated quarks exist?

An important and obvious question that remains is whether leptons and quarks have an underlying structure. If they do, we can envision an infinite number of deeper structure levels. If leptons and quarks are indeed the ultimate constituents of matter, however, scientists hope to construct a final theory of the structure of matter, just as Einstein dreamed of doing. This theory, whimsically called the Theory of Everything, is a combination of the Standard Model and a quantum theory of gravity.

String Theory: A New Perspective

Let's briefly discuss one current effort at answering some of these questions by proposing a new perspective on particles. While reading this book, you may recall starting off with the particle model and doing quite a bit of physics with it. In Part 2, we introduced the wave model, and there was more physics to be investigated via the properties of waves. We used a wave model for light in Part 5. Early in Part 6, however, we saw the need to return to the particle model for light. Furthermore, we found that material particles had wave-like characteristics. The quantum particle model discussed in Chapter 40 allowed us to build particles out of waves, suggesting that a wave is the fundamental entity. In this final chapter, however, we introduced elementary particles as the fundamental entities. It seems as if we cannot make up our mind! In this final section, we discuss a current research effort to build particles out of waves and vibrations.

---

[7] For an overview of dark energy, see S. Perlmutter, "Supernovae, Dark Energy, and the Accelerating Universe," *Physics Today* **56**(4): 53–60, April 2003.

Figure 46.22   (a) A piece of paper is cut into a rectangular shape. As a rectangle, the shape has two dimensions. (b) The paper is rolled up into a soda straw. From far away, it appears to be one-dimensional. The curled-up second dimension is not visible when viewed from a distance that is large compared with the diameter of the straw.

String theory is an effort to unify the four fundamental forces by modeling all particles as various quantized vibrational modes of a single entity, an incredibly small string. The typical length of such a string is on the order of 10^{-35} m, called the **Planck length.** We have seen quantized modes before in the frequencies of vibrating guitar strings in Chapter 18 and the quantized energy levels of atoms in Chapter 42. In string theory, each quantized mode of vibration of the string corresponds to a different elementary particle in the Standard Model.

One complicating factor in string theory is that it requires space–time to have ten dimensions. Despite the theoretical and conceptual difficulties in dealing with ten dimensions, string theory holds promise in incorporating gravity with the other forces. Four of the ten dimensions—three space dimensions and one time dimension—are visible to us. The other six are said to be *compactified;* that is, the six dimensions are curled up so tightly that they are not visible in the macroscopic world.

As an analogy, consider a soda straw. You can build a soda straw by cutting a rectangular piece of paper (Fig. 46.22a), which clearly has two dimensions, and rolling it into a small tube (Fig. 46.22b). From far away, the soda straw looks like a one-dimensional straight line. The second dimension has been curled up and is not visible. String theory claims that six space–time dimensions are curled up in an analogous way, with the curling being on the size of the Planck length and impossible to see from our viewpoint.

Another complicating factor with string theory is that it is difficult for string theorists to guide experimentalists as to what to look for in an experiment. The Planck length is so small that direct experimentation on strings is impossible. Until the theory has been further developed, string theorists are restricted to applying the theory to known results and testing for consistency.

One of the predictions of string theory, called **supersymmetry,** or SUSY, suggests that every elementary particle has a superpartner that has not yet been observed. It is believed that supersymmetry is a broken symmetry (like the broken electroweak symmetry at low energies) and the masses of the superpartners are above our current capabilities of detection by accelerators. Some theorists claim that the mass of superpartners is the missing mass discussed in Section 46.11. Keeping with the whimsical trend in naming particles and their properties, superpartners are given names such as the *squark* (the superpartner to a quark), the *selectron* (electron), and the *gluinos* (gluon).

Other theorists are working on **M-theory,** which is an eleven-dimensional theory based on membranes rather than strings. In a way reminiscent of the correspondence principle, M-theory is claimed to reduce to string theory if one compactifies from eleven dimensions to ten dimensions.

The questions listed at the beginning of this section go on and on. Because of the rapid advances and new discoveries in the field of particle physics, many of these questions may be resolved in the next decade and other new questions may emerge.

Summary

CONCEPTS AND PRINCIPLES

Before quark theory was developed, the four fundamental forces in nature were identified as nuclear, electromagnetic, weak, and gravitational. All the interactions in which these forces take part are mediated by **field particles.** The electromagnetic interaction is mediated by photons; the weak interaction is mediated by the $W^{\pm}$ and Z^0 bosons; the gravitational interaction is mediated by gravitons; and the nuclear interaction is mediated by gluons.

A charged particle and its **antiparticle** have the same mass but opposite charge, and other properties will have opposite values, such as lepton number and baryon number. It is possible to produce particle–antiparticle pairs in nuclear reactions if the available energy is greater than $2mc^2$, where m is the mass of the particle (or antiparticle).

Particles other than field particles are classified as hadrons or leptons. **Hadrons** interact via all four fundamental forces. They have size and structure and are not elementary particles. There are two types, **baryons** and **mesons.** Baryons, which generally are the most massive particles, have nonzero **baryon number** and a spin of $\frac{1}{2}$ or $\frac{3}{2}$. Mesons have baryon number zero and either zero or integral spin.

Leptons have no structure or size and are considered truly elementary. They interact only via the weak, gravitational, and electromagnetic forces. Six types of leptons exist: the electron e^-, the muon μ^-, and the tau τ^-, and their neutrinos ν_e, ν_μ, and ν_τ.

In all reactions and decays, quantities such as energy, linear momentum, angular momentum, electric charge, baryon number, and lepton number are strictly conserved. Certain particles have properties called **strangeness** and **charm.** These unusual properties are conserved in all decays and nuclear reactions except those that occur via the weak force.

Theorists in elementary particle physics have postulated that all hadrons are composed of smaller units known as **quarks,** and experimental evidence agrees with this model. Quarks have fractional electric charge and come in six **flavors:** up (u), down (d), strange (s), charmed (c), top (t), and bottom (b). Each baryon contains three quarks, and each meson contains one quark and one antiquark.

According to the theory of **quantum chromodynamics,** quarks have a property called **color;** the force between quarks is referred to as the **strong force** or the **color force.** The strong force is now considered to be a fundamental force. The nuclear force, which was originally considered to be fundamental, is now understood to be a secondary effect of the strong force due to gluon exchanges between hadrons.

The electromagnetic and weak forces are now considered to be manifestations of a single force called the **electroweak force.** The combination of quantum chromodynamics and the electroweak theory is called the **Standard Model.**

The background microwave radiation discovered by Penzias and Wilson strongly suggests that the Universe started with a Big Bang 10 to 20 billion years ago. The background radiation is equivalent to that of a black body at 3 K. Various astronomical measurements strongly suggest that the Universe is expanding. According to **Hubble's law,** distant galaxies are receding from the Earth at a speed $v = HR$, where H is the **Hubble constant,** $H \approx 17 \times 10^{-3}$ m/s · ly, and R is the distance from the Earth to the galaxy.

Questions

☐ denotes answer available in *Student Solutions Manual/Study Guide;* **O** denotes objective question

1. Name the four fundamental interactions and the field particle that mediates each.

2. **O** When an electron and a positron meet at low speed in empty space, they annihilate each other to produce two 0.511-MeV gamma rays. What law would be violated if they produced one gamma ray with an energy of 1.02 MeV? (a) conservation of energy (b) conservation of momentum (c) conservation of charge (d) conservation of baryon number (e) conservation of electron-lepton number (f) none of these answers

3. What are the differences between hadrons and leptons?

4. Describe the properties of baryons and mesons and the important differences between them.

5. **O** An isolated stationary muon decays into an electron, an electron antineutrino, and a muon neutrino. Is the total kinetic energy of these three particles (a) zero, (b) small, or (c) large compared to their rest energies, or (d) is any of these choices possible?

6. **O** The Ω^- particle is a baryon with spin $\frac{3}{2}$. The Ω^- particle has (a) three possible spin states in a magnetic field, (b) four possible spin states, (c) three times the charge of a spin-$\frac{1}{2}$ particle, (d) three times the mass of a spin-$\frac{1}{2}$ particle, or (e) none of these choices.

7. Kaons all decay into final states that contain no protons or neutrons. What is the baryon number for kaons?

8. The Ξ^0 particle decays by the weak interaction according to the decay mode $\Xi^0 \rightarrow \Lambda^0 + \pi^0$. Would you expect this decay to be fast or slow? Explain.

9. **O** What interactions affect protons in an atomic nucleus? Choose all correct answers. (a) the nuclear interaction (b) the weak interaction (c) the electromagnetic interaction (d) the gravitational interaction

10. Discuss the following conservation laws: energy, linear momentum, angular momentum, electric charge, baryon number, lepton number, and strangeness. Are all these laws based on fundamental properties of nature? Explain.

11. An antibaryon interacts with a meson. Can a baryon be produced in such an interaction? Explain.

12. Describe the essential features of the Standard Model of particle physics.

13. How many quarks are in each of the following: (a) a baryon, (b) an antibaryon, (c) a meson, (d) an antimeson? How do you explain that baryons have half-integral spins whereas mesons have spins of 0 or 1? *Note:* Quarks have spin $\frac{1}{2}$.

14. In the theory of quantum chromodynamics, quarks come in three colors. How would you justify the statement that "all baryons and mesons are colorless"?

15. The W and Z bosons were first produced at CERN in 1983 by causing a beam of protons and a beam of antiprotons to meet at high energy. Why was this discovery important?

16. **O** In one experiment, two balls of clay of the same mass travel with the same speed v toward each other. They collide head-on and come to rest. In a second experiment, two clay balls of the same mass are again used. One ball hangs at rest, suspended from the ceiling by a thread. The second ball is fired toward the first at speed v, to collide, stick to the first ball, and continue to move forward. Is the kinetic energy that is transformed into internal energy in the first experiment (a) one-fourth as much as in the second experiment, (b) one-half as much as in the second experiment, (c) the same as in the second experiment, (d) twice as much as in the second experiment, (e) four times as much as in the second experiment, or (f) none of these choices?

17. How did Edwin Hubble determine in 1928 that the Universe is expanding?

18. **O** Place the following events into the correct sequence from the earliest in the history of the Universe to the latest. (a) Neutral atoms form. (b) Protons and neutrons are no longer annihilated as fast as they form. (c) The Universe is a quark–gluon soup. (d) The Universe is like the core of a normal star today, forming helium by nuclear fusion. (e) The Universe is like the surface of a hot star today, consisting of a plasma of ionized atoms. (f) Polyatomic molecules form. (g) Solid materials form.

19. Neutral atoms did not exist until hundreds of thousands of years after the Big Bang. Why?

20. **O** Define the average density of the Solar System ρ_{SS} as the total mass of the Sun, planets, satellites, rings, asteroids, icy outliers, and comets, divided by the volume of a sphere around the Sun large enough to contain all these objects. The sphere extends about halfway to the nearest star, with a radius of approximately 2×10^{16} m, about two light-years. How does this average density of the Solar System compare with the critical density ρ_c required for the Universe to stop its Hubble's-law expansion? (a) ρ_{SS} is much greater than ρ_c. (b) ρ_{SS} is approximately or precisely equal to ρ_c. (c) ρ_{SS} is much less than ρ_c. (d) It is impossible to determine.

21. **Review question.** A girl and her grandmother grind corn while the woman tells the girl stories about what is most important. A boy keeps crows away from ripening corn while his grandfather sits in the shade and explains to him the Universe and his place in it. What the children do not understand this year they will better understand next year. Now you must take the part of the adults. State the most general, most fundamental, most universal truths you know. If you need to repeat someone else's ideas, get the best version of those ideas you can and state your source. If you do not understand something, make a plan to understand it better within the next year.

Problems

1, 2, 3 denotes straightforward, intermediate, challenging;   ☐ denotes full solution available in *Student Solutions Manual/Study Guide;*   ▲ denotes coached solution with hints available at **www.thomsonedu.com;**   denotes developing symbolic reasoning; ● denotes asking for qualitative reasoning;   ▪ denotes computer useful in solving problem

Section 46.1 The Fundamental Forces in Nature

Section 46.2 Positrons and Other Antiparticles

1. A photon produces a proton–antiproton pair according to the reaction $\gamma \rightarrow p + \bar{p}$. What is the minimum possible frequency of the photon? What is its wavelength?

2. At some time in your life, you may find yourself in a hospital to have a PET, or positron-emission tomography, scan. In the procedure, a radioactive element that undergoes e^+ decay is introduced into your body. The equipment detects the gamma rays that result from pair annihilation when the emitted positron encounters an electron in your body's tissue. Suppose you receive an injection of glucose containing on the order of 10^{10} atoms of ^{14}O, with half-life 70.6 s. Assume the oxygen remaining after 5 min is uniformly distributed through 2 L of blood. What is then the order of magnitude of the oxygen atoms' activity in 1 cm^3 of the blood?

3. *Your two cents' worth.* Model a penny as 3.10 g of copper. Consider an anti-penny minted from 3.10 g of copper anti-atoms, each with 29 positrons in orbit around a nucleus comprising 29 antiprotons and 34 or 36 antineutrons. (a) Find the energy released if the two coins collide. (b) Find the value of this energy at the unit price of $0.14/kWh, a representative retail rate for energy from the electric company.

4. Two photons are produced when a proton and antiproton annihilate each other. In the reference frame in which the center of mass of the proton–antiproton system is stationary, what are the minimum frequency and corresponding wavelength of each photon?

5. A photon with an energy $E_\gamma = 2.09$ GeV creates a proton–antiproton pair in which the proton has a kinetic energy of 95.0 MeV. What is the kinetic energy of the antiproton? *Note:* $m_p c^2 = 938.3$ MeV.

Section 46.3 Mesons and the Beginning of Particle Physics

6. Occasionally, high-energy muons collide with electrons and produce two neutrinos according to the reaction $\mu^+ + e^- \rightarrow 2\nu$. What kind of neutrinos are they?

7. ▲ One mediator of the weak interaction is the Z^0 boson, with mass 91 GeV/c^2. Use this information to find the order of magnitude of the range of the weak interaction.

8. ● (a) Prove that the exchange of a virtual particle of mass m can be associated with a force with a range given by

$$d = \frac{1\,240\ \text{eV} \cdot \text{nm}}{4\pi\,mc^2} = \frac{98.7\ \text{eV} \cdot \text{nm}}{mc^2}$$

(b) State the pattern of dependence of the range on the mass. (c) Of the interactions listed in Table 46.1, which are associated with field particles according to this rule and which are not? Explain your answer. (d) What is the range of the force that might be produced by the virtual exchange of a proton?

9. ▲ A neutral pion at rest decays into two photons according to $\pi^0 \rightarrow \gamma + \gamma$. Find the energy, momentum, and frequency of each photon.

10. When a high-energy proton or pion traveling near the speed of light collides with a nucleus, it travels an average distance of 3×10^{-15} m before interacting. From this information, find the order of magnitude of the time interval required for the strong interaction to occur.

11. ● A free neutron beta decays by creating a proton, an electron, and an antineutrino according to the reaction $n \rightarrow p + e^- + \bar{\nu}$. **What If?** Imagine that a free neutron were to decay by creating a proton and electron according to the reaction $n \rightarrow p + e^-$ and assume the neutron is initially at rest in the laboratory. (a) Determine the energy released in this reaction. (b) Energy and momentum are conserved in the reaction. Determine the speeds of the proton and electron after the reaction. (c) Is either of these particles moving at a relativistic speed? Explain.

Section 46.4 Classification of Particles

12. Identify the unknown particle on the left side of the reaction $? + p \rightarrow n + \mu^+$.

Section 46.5 Conservation Laws

13. Each of the following reactions is forbidden. Determine a conservation law that is violated for each reaction.
(a) $p + \bar{p} \rightarrow \mu^+ + e^-$
(b) $\pi^- + p \rightarrow p + \pi^+$
(c) $p + p \rightarrow p + \pi^+$
(d) $p + p \rightarrow p + p + n$
(e) $\gamma + p \rightarrow n + \pi^0$

14. ● (a) Show that baryon number and charge are conserved in the following reactions of a pion with a proton:

$$(1)\quad \pi^+ + p \;\rightarrow\; K^+ + \Sigma^+$$
$$(2)\quad \pi^+ + p \;\rightarrow\; \pi^+ + \Sigma^+$$

(b) The first reaction is observed, but the second never occurs. Explain.

15. ▲ The following reactions or decays involve one or more neutrinos. In each case, supply the missing neutrino (ν_e, ν_μ, or ν_τ) or antineutrino.
(a) $\pi^- \rightarrow \mu^- + ?$ (b) $K^+ \rightarrow \mu^+ + ?$
(c) $? + p \rightarrow n + e^+$ (d) $? + n \rightarrow p + e^-$
(e) $? + n \rightarrow p + \mu^-$ (f) $\mu^- \rightarrow e^- + ? + ?$

16. The first of the following two reactions can occur, but the second cannot. Explain.

$$K_S^0 \rightarrow \pi^+ + \pi^- \quad \text{(can occur)}$$

$$\Lambda^0 \rightarrow \pi^+ + \pi^- \quad \text{(cannot occur)}$$

17. ▲ Determine which of the following reactions can occur. For those that cannot occur, determine the conservation law (or laws) violated.
    (a) $p \rightarrow \pi^+ + \pi^0$
    (b) $p + p \rightarrow p + p + \pi^0$
    (c) $p + p \rightarrow p + \pi^+$
    (d) $\pi^+ \rightarrow \mu^+ + \nu_\mu$
    (e) $n \rightarrow p + e^- + \bar{\nu}_e$
    (f) $\pi^+ \rightarrow \mu^+ + n$

18. A K_S^0 particle at rest decays into a π^+ and a π^-. What is the speed of each of the pions? The mass of the K_S^0 is 497.7 MeV/c^2, and the mass of each π is 139.6 MeV/c^2.

19. (a) Show that the proton-decay reaction $p \rightarrow e^+ + \gamma$ cannot occur because it violates conservation of baryon number. (b) **What If?** Imagine that this reaction does occur, and the proton is initially at rest. Determine the energy and momentum of the positron and photon after the reaction. *Suggestion:* Recall that energy and momentum must be conserved in the reaction. (c) Determine the speed of the positron after the reaction.

20. Determine the type of neutrino or antineutrino involved in each of the following processes:
    (a) $\pi^+ \rightarrow \pi^0 + e^+ + ?$
    (b) $? + p \rightarrow \mu^- + p + \pi^+$
    (c) $\Lambda^0 \rightarrow p + \mu^- + ?$
    (d) $\tau^+ \rightarrow \mu^+ + ? + ?$

Section 46.6 Strange Particles and Strangeness

21. Determine whether or not strangeness is conserved in the following decays and reactions:
    (a) $\Lambda^0 \rightarrow p + \pi^-$
    (b) $\pi^- + p \rightarrow \Lambda^0 + K^0$
    (c) $\bar{p} + p \rightarrow \bar{\Lambda}^0 + \Lambda^0$
    (d) $\pi^- + p \rightarrow \pi^- + \Sigma^+$
    (e) $\Xi^- \rightarrow \Lambda^0 + \pi^-$
    (f) $\Xi^0 \rightarrow p + \pi^-$

22. ● The neutral meson ρ^0 decays by the strong interaction into two pions:

$$\rho^0 \rightarrow \pi^+ + \pi^- \quad (T_{1/2} \sim 10^{-23} \text{ s})$$

The neutral kaon also decays into two pions:

$$K_S^0 \rightarrow \pi^+ + \pi^- \quad (T_{1/2} \sim 10^{-10} \text{ s})$$

How do you explain the difference in half-lives?

23. For each of the following forbidden decays, determine which conservation law is violated:
    (a) $\mu^- \rightarrow e^- + \gamma$
    (b) $n \rightarrow p + e^- + \nu_e$
    (c) $\Lambda^0 \rightarrow p + \pi^0$
    (d) $p \rightarrow e^+ + \pi^0$
    (e) $\Xi^0 \rightarrow n + \pi^0$

24. Which of the following processes are allowed by the strong interaction, the electromagnetic interaction, the weak interaction, or no interaction at all?
    (a) $\pi^- + p \rightarrow 2\eta$
    (b) $K^- + n \rightarrow \Lambda^0 + \pi^-$
    (c) $K^- \rightarrow \pi^- + \pi^0$
    (d) $\Omega^- \rightarrow \Xi^- + \pi^0$
    (e) $\eta \rightarrow 2\gamma$

25. Identify the conserved quantities in the following processes:
    (a) $\Xi^- \rightarrow \Lambda^0 + \mu^- + \bar{\nu}_\mu$
    (b) $K_S^0 \rightarrow 2\pi^0$
    (c) $K^- + p \rightarrow \Sigma^0 + n$
    (d) $\Sigma^0 \rightarrow \Lambda^0 + \gamma$
    (e) $e^+ + e^- \rightarrow \mu^+ + \mu^-$
    (f) $\bar{p} + n \rightarrow \bar{\Lambda}^0 + \Sigma^-$

26. Fill in the missing particle. Assume (a) occurs via the strong interaction and (b) and (c) involve the weak interaction.
    (a) $K^+ + p \rightarrow ? + p$
    (b) $\Omega^- \rightarrow ? + \pi^-$
    (c) $K^+ \rightarrow ? + \mu^+ + \nu_\mu$

27. If a K_S^0 meson at rest decays in 0.900×10^{-10} s, how far does a K_S^0 meson travel if it is moving at $0.960c$?

28. The particle decay $\Sigma^+ \rightarrow \pi^+ + n$ is observed in a bubble chamber. Figure P46.28 represents the curved tracks of the particles Σ^+ and π^+, and the invisible track of the neutron, in the presence of a uniform magnetic field of 1.15 T directed out of the page. The measured radii of curvature are 1.99 m for the Σ^+ particle and 0.580 m for the π^+ particle. (a) Find the momenta of the Σ^+ and the π^+ particles in units of MeV/c. (b) The angle between the momenta of the Σ^+ and the π^+ particles at the moment of decay is 64.5°. Find the momentum of the neutron. (c) Calculate the total energy of the π^+ particle and of the neutron from their known masses ($m_\pi = 139.6$ MeV/c^2, $m_n = 939.6$ MeV/c^2) and the relativistic energy–momentum relation. What is the total energy of the Σ^+ particle? (d) Calculate the mass and speed of the Σ^+ particle.

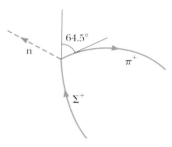

Figure P46.28

29. A particle of mass m_1 is fired at a stationary particle of mass m_2, and a reaction takes place in which new particles are created out of the incident kinetic energy. Taken together, the product particles have total mass m_3. The minimum kinetic energy the bombarding particle must have so as to induce the reaction is called the threshold energy. At this energy, the kinetic energy of the products is a minimum, so the fraction of the incident kinetic energy that is available to create new particles is a maximum. This condition is met when all the product particles have the same velocity and the particles have no kinetic energy of motion relative to one another. (a) By using conservation of relativistic energy and momentum, and the relativistic energy–momentum relation, show that the threshold energy is

$$K_{\text{min}} = \frac{[m_3{}^2 - (m_1 + m_2)^2]c^2}{2m_2}$$

Calculate the threshold energy for each of the following reactions: (b) $p + p \rightarrow p + p + p + \bar{p}$ (One of the initial protons is at rest. Antiprotons are produced.) (c) $\pi^- + p \rightarrow K^0 + \Lambda^0$ (The proton is at rest. Strange particles are produced.) (d) $p + p \rightarrow p + p + \pi^0$ (One of the initial protons is at rest. Pions are produced.) (e) $p + \bar{p} \rightarrow Z^0$ (One of the initial particles is at rest. Z^0 particles of mass 91.2 GeV/c^2 are produced.)

Section 46.7 Finding Patterns in the Particles

Section 46.8 Quarks

Section 46.9 Multicolored Quarks

Section 46.10 The Standard Model

Note: Problem 64 in Chapter 39 can be assigned with Section 46.10.

30. (a) Find the number of electrons and the number of each species of quarks in 1 L of water. (b) Make an order-of-magnitude estimate of the number of each kind of fundamental matter particle in your body. State your assumptions and the quantities you take as data.

31. The quark composition of the proton is uud and that of the neutron is udd. Show that in each case the charge, baryon number, and strangeness of the particle equal, respectively, the sums of these numbers for the quark constituents.

32. What If? Imagine that binding energies could be ignored. Find the masses of the u and d quarks from the masses of the proton and neutron.

33. The quark compositions of the K^0 and Λ^0 particles are $\bar{s}d$ and uds, respectively. Show that the charge, baryon number, and strangeness of these particles equal, respectively, the sums of these numbers for the quark constituents.

34. The reaction $\pi^- + p \rightarrow K^0 + \Lambda^0$ occurs with high probability, whereas the reaction $\pi^- + p \rightarrow K^0 + n$ never occurs. Analyze these reactions at the quark level. Show that the first reaction conserves the total number of each type of quark and the second does not.

35. Analyze each of the following reactions in terms of constituent quarks:
(a) $\pi^- + p \rightarrow K^0 + \Lambda^0$
(b) $\pi^+ + p \rightarrow K^+ + \Sigma^+$
(c) $K^- + p \rightarrow K^+ + K^0 + \Omega^-$
(d) $p + p \rightarrow K^0 + p + \pi^+ + ?$
In the last reaction, identify the mystery particle.

36. A Σ^0 particle traveling through matter strikes a proton; then a Σ^+ and a gamma ray, as well as a third particle, emerge. Use the quark model of each to determine the identity of the third particle.

37. Identify the particles corresponding to the quark combinations (a) suu, (b) $\bar{u}d$, (c) $\bar{s}d$, and (d) ssd.

38. What is the electrical charge of the baryons with the quark compositions (a) $\bar{u}\,\bar{u}d$ and (b) $\bar{u}d\,\bar{d}$? What are these baryons called?

Section 46.11 The Cosmic Connection

Note: Problem 17 in Chapter 39 can be assigned with this section.

39. Imagine that all distances expand at a rate described by the Hubble constant of 17.0×10^{-3} m/s · ly. (a) At what rate would the 1.85-m height of a basketball player be increasing? (b) At what rate would the distance between the Earth and the Moon be increasing? In fact, gravitation and other forces prevent the Hubble's-law expansion from taking place except in systems larger than clusters of galaxies.

40. Review problem. Refer to Section 39.4. Prove that the Doppler shift in wavelength of electromagnetic waves is described by

$$\lambda' = \lambda \sqrt{\frac{1 + v/c}{1 - v/c}}$$

where λ' is the wavelength measured by an observer moving at speed v away from a source radiating waves of wavelength λ.

41. ▲ A distant quasar is moving away from the Earth at such high speed that the blue 434-nm H_γ line of hydrogen is observed at 510 nm, in the green portion of the spectrum (Fig. P46.41). (a) How fast is the quasar receding? You may use the result of Problem 40. (b) Edwin Hubble discovered that all objects outside the local group of galaxies are moving away from us, with speeds proportional to their distances. Hubble's law is expressed as $v = HR$, where Hubble's constant has the approximate value $H = 17 \times 10^{-3}$ m/s·ly. Determine the distance from the Earth to this quasar.

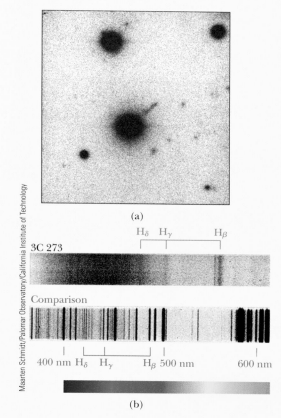

Figure P46.41   (a) Image of the quasar 3C273. (b) Spectrum of the quasar above a comparison spectrum emitted by stationary hydrogen and helium atoms. Both parts of the figure are printed as black-and-white photographic negatives to reveal detail.

42. The various spectral lines observed in the light from a distant quasar have longer wavelengths λ'_n than the wavelengths λ_n measured in light from a stationary source. Here n is an index taking different values for different spectral lines. The fractional change in wavelength toward

2 = intermediate; 3 = challenging;   ☐ = SSM/SG;   ▲ = ThomsonNOW;   = symbolic reasoning;   ● = qualitative reasoning the red is the same for all spectral lines. That is, the red-shift parameter Z defined by

$$Z = \frac{\lambda_n' - \lambda_n}{\lambda_n}$$

is common to all spectral lines for one object. In terms of Z, determine (a) the speed of recession of the quasar and (b) the distance from the Earth to this quasar. Use the result of Problem 40 and Hubble's law.

43. Using Hubble's law, find the wavelength of the 590-nm sodium line emitted from galaxies (a) 2.00×10^6 ly away from the Earth, (b) 2.00×10^8 ly away, and (c) 2.00×10^9 ly away. You may use the result of Problem 40.

44. ● The visible section of the Universe is a sphere centered on the bridge of your nose, with radius 13.7 billion light-years. (a) Explain why the visible Universe is getting larger, with its radius increasing by one light-year in every year. (b) Find the rate at which the volume of the visible section of the Universe is increasing.

45. ● Assume dark matter exists throughout space with a uniform density of 6.00×10^{-28} kg/m³. (a) Find the amount of such dark matter inside a sphere centered on the Sun, having the Earth's orbit as its equator. (b) Explain whether the gravitational field of this dark matter would have a measurable effect on the Earth's revolution.

46. **Review problem.** The cosmic background radiation is blackbody radiation from a source at a temperature of 2.73 K. (a) Use Wien's law to determine the wavelength at which this radiation has its maximum intensity. (b) In what part of the electromagnetic spectrum is the peak of the distribution?

47. **Review problem.** Use Stefan's law to find the intensity of the cosmic background radiation emitted by the fireball of the Big Bang at a temperature of 2.73 K.

48. It is mostly your roommate's fault. Nosy astronomers have discovered enough junk and clutter in your dorm room to constitute the missing mass required to close the Universe. After observing your floor, closet, bed, and computer files, they extrapolate to slobs in other galaxies and calculate the average density of the observable Universe as $1.20\rho_c$. How many times larger will the Universe become before it begins to collapse? That is, by what factor will the distance between remote galaxies increase in the future?

49. The early Universe was dense with gamma-ray photons of energy $\sim k_B T$ and at such a high temperature that protons and antiprotons were created by the process $\gamma \to p + \bar{p}$ as rapidly as they annihilated each other. As the Universe cooled in adiabatic expansion, its temperature fell below a certain value and proton pair production became rare. At that time, slightly more protons than antiprotons existed, and essentially all the protons in the Universe today date from that time. (a) Estimate the order of magnitude of the temperature of the Universe when protons condensed out. (b) Estimate the order of magnitude of the temperature of the Universe when electrons condensed out.

50. If the average density of the Universe is small compared with the critical density, the expansion of the Universe described by Hubble's law proceeds with speeds that are nearly constant over time. (a) Prove that in this case the age of the Universe is given by the inverse of Hubble's constant. (b) Calculate $1/H$ and express it in years.

51. Assume the average density of the Universe is equal to the critical density. (a) Prove that the age of the Universe is given by $2/3H$. (b) Calculate $2/3H$ and express it in years.

52. Hubble's law can be stated in vector form as $\vec{\mathbf{v}} = H\vec{\mathbf{R}}$. Outside the local group of galaxies, all objects are moving away from us with velocities proportional to their positions relative to us. In this form, it sounds as if our location in the Universe is specially privileged. Prove that Hubble's law is equally true for an observer elsewhere in the Universe. Proceed as follows. Assume we are at the origin of coordinates, one galaxy cluster is at location $\vec{\mathbf{R}}_1$ and has velocity $\vec{\mathbf{v}}_1 = H\vec{\mathbf{R}}_1$ relative to us, and another galaxy cluster has position vector $\vec{\mathbf{R}}_2$ and velocity $\vec{\mathbf{v}}_2 = H\vec{\mathbf{R}}_2$. Suppose the speeds are nonrelativistic. Consider the frame of reference of an observer in the first of these galaxy clusters. Show that our velocity relative to her, together with the position vector of our galaxy cluster from hers, satisfies Hubble's law. Show that the position and velocity of cluster 2 relative to cluster 1 satisfy Hubble's law.

Section 46.12 Problems and Perspectives

53. ● Classical general relativity views the structure of space–time as deterministic and well defined down to arbitrarily small distances. On the other hand, quantum general relativity forbids distances smaller than the Planck length given by $L = (\hbar G/c^3)^{1/2}$. (a) Calculate the value of the Planck length. The quantum limitation suggests that after the Big Bang, when all the presently observable section of the Universe was contained within a point-like singularity, nothing could be observed until that singularity grew larger than the Planck length. Because the size of the singularity grew at the speed of light, we can infer that no observations were possible during the time interval required for light to travel the Planck length. (b) Calculate this time interval, known as the Planck time T, and state how it compares with the ultrahot epoch mentioned in the text. (c) Does this reasoning suggest we may never know what happened between the time $t = 0$ and the time $t = T$? Explain.

Additional Problems

54. **Review problem.** Supernova Shelton 1987A, located approximately 170 000 ly from the Earth, is estimated to have emitted a burst of neutrinos carrying energy $\sim 10^{46}$ J (Fig. P46.54). Suppose the average neutrino energy was 6 MeV and your mother's body presented cross-sectional area 5 000 cm². To an order of magnitude, how many of these neutrinos passed through her?

55. The most recent naked-eye supernova was Supernova Shelton 1987A (Fig. P46.54). It was 170 000 ly away in the next galaxy to ours, the Large Magellanic Cloud. Approximately 3 h before its optical brightening was noticed, two neutrino detection experiments simultaneously registered the first neutrinos from an identified source other than the Sun. The Irvine–Michigan–Brookhaven experiment in a salt mine in Ohio registered 8 neutrinos over a 6-s period, and the Kamiokande II experiment in a zinc mine in Japan counted 11 neutrinos in 13 s. (Because the supernova is far south in the sky, these neutrinos entered

Figure P46.54   (Problems 54 and 55) The giant star Sanduleak
−69° 202 in the "before" picture became Supernova Shelton 1987A
in the "after" picture.

the detectors from below. They passed through the Earth before they were by chance absorbed by nuclei in the detectors.) The neutrino energies were between approximately 8 MeV and 40 MeV. If neutrinos have no mass, neutrinos of all energies should travel together at the speed of light, and the data are consistent with this possibility. The arrival times could vary simply because neutrinos were created at different moments as the core of the star collapsed into a neutron star. If neutrinos have nonzero mass, lower-energy neutrinos should move comparatively slowly. The data are consistent with a 10-MeV neutrino requiring at most approximately 10 s more than a photon would require to travel from the supernova to us. Find the upper limit that this observation sets on the mass of a neutrino. Other evidence sets an even tighter limit.

56. Name at least one conservation law that prevents each of the following reactions: (a) $\pi^- + p \rightarrow \Sigma^+ + \pi^0$ (b) $\mu^- \rightarrow \pi^- + \nu_e$ (c) $p \rightarrow \pi^+ + \pi^+ + \pi^-$

57. ▲ The energy flux carried by neutrinos from the Sun is estimated to be on the order of 0.4 W/m² at the Earth's surface. Estimate the fractional mass loss of the Sun over 10^9 yr due to the emission of neutrinos. The mass of the Sun is 2×10^{30} kg. The Earth–Sun distance is 1.5×10^{11} m.

58. Two protons approach each other head-on, each with 70.4 MeV of kinetic energy, and engage in a reaction in which a proton and positive pion emerge at rest. What third particle, obviously uncharged and therefore difficult to detect, must have been created?

59. A rocket engine for space travel using photon drive and matter–antimatter annihilation has been suggested. Suppose the fuel for a short-duration burn consists of N protons and N antiprotons, each with mass m. (a) Assume all the fuel is annihilated to produce photons. When the

photons are ejected from the rocket, what momentum can be imparted to it? (b) **What If?** If half of the protons and antiprotons annihilate each other and the energy released is used to eject the remaining particles, what momentum could be given to the rocket? (c) Which scheme results in the greater change in speed for the rocket?

60. A gamma-ray photon strikes a stationary electron. Determine the minimum gamma-ray energy to make the following reaction occur:

$$\gamma + e^- \rightarrow e^- + e^- + e^+$$

61. Determine the kinetic energies of the proton and pion resulting from the decay of a Λ^0 at rest:

$$\Lambda^0 \rightarrow p + \pi^-$$

62. Two protons approach each other with velocities of equal magnitude in opposite directions. What is the minimum kinetic energy of each proton if they are to produce a π^+ meson at rest in the following reaction?

$$p + p \rightarrow p + n + \pi^+$$

63. A Σ^0 particle at rest decays according to $\Sigma^0 \rightarrow \Lambda^0 + \gamma$. Find the gamma-ray energy.

64. An unstable particle, initially at rest, decays into a proton (rest energy 938.3 MeV) and a negative pion (rest energy 139.6 MeV). A uniform magnetic field of 0.250 T exists perpendicular to the velocities of the created particles. The radius of curvature of each track is found to be 1.33 m. What is the mass of the original unstable particle?

65. A π-meson at rest decays according to $\pi^- \rightarrow \mu^- + \bar{\nu}_\mu$. What is the energy carried off by the neutrino? Assume the neutrino has no mass and moves off with the speed of light. Take $m_\pi c^2 = 139.6$ MeV and $m_\mu c^2 = 105.7$ MeV.

66. **Review problem.** Use the Boltzmann distribution function e^{-E/k_BT} to calculate the temperature at which 1.00% of a population of photons has energy greater than 1.00 eV. The energy required to excite an atom is on the order of 1 eV. Therefore, as the temperature of the Universe fell below the value you calculate, neutral atoms could form from plasma and the Universe became transparent. The cosmic background radiation represents our vastly red-shifted view of the opaque fireball of the Big Bang as it was at this time and temperature. The fireball surrounds us; we are embers.

67. What processes are described by the Feynman diagrams in Figure P46.67? What is the exchanged particle in each process?

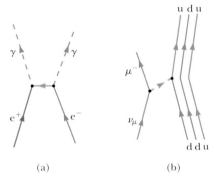

Figure P46.67

2 = intermediate;   3 = challenging;   □ = SSM/SG;   ▲ = ThomsonNOW;   ▬ = symbolic reasoning;   ● = qualitative reasoning

68. Identify the mediators for the two interactions described in the Feynman diagrams shown in Figure P46.68.

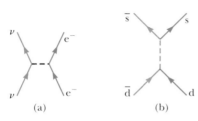

Figure P46.68

69. The cosmic rays of highest energy are mostly protons, accelerated by unknown sources. Their spectrum shows a cutoff at an energy on the order of 10^{20} eV. Above that energy, a proton interacts with a photon of cosmic microwave background radiation to produce mesons, for example according to $p + \gamma \rightarrow p + \pi^0$. Demonstrate this fact by taking the following steps. (a) Find the minimum photon energy required to produce this reaction in the reference frame where the total momentum of the photon–proton system is zero. The reaction was observed experimentally in the 1950s with photons of a few hundred MeV. (b) Use Wien's displacement law to find the wavelength of a photon at the peak of the blackbody spectrum of the primordial microwave background radiation, with a temperature of 2.73 K. (c) Find the energy of this photon. (d) Consider the reaction in part (a) in a moving reference frame so that the photon is the same as that in part (c). Calculate the energy of the proton in this frame, which represents the Earth reference frame.

Answers to Quick Quizzes

46.1 (a). The right-hand rule for the positive particle tells you that into the page is the direction that leads to a force directed toward the path's center of curvature.

46.2 **(i),** (c), (d). There is a baryon, the neutron, on the left of the reaction, but no baryon on the right. Therefore, baryon number is not conserved. The neutron has spin $\frac{1}{2}$. On the right side of the reaction, the pions each have integral spin and the combination of two muons must also have integral spin. Therefore, the total spin of the particles on the right-hand side must be integral and angular momentum is not conserved. **(ii),** (a). The sum of the proton and pion masses is larger than the mass of the neutron, so energy conservation is violated.

46.3 (b), (e), (f). The pion on the left has integral spin, whereas the three spin-$\frac{1}{2}$ leptons on the right must result in a total spin that is half-integral. Therefore, angular momentum is not conserved. Electron lepton number is zero on the left and -1 on the right. There are no muons on the left, but a muon and its neutrino on the right (both with $L_\mu = +1$). Therefore, muon lepton number is not conserved.

46.4 (b), (e). There is one spin-$\frac{1}{2}$ particle on the left and two on the right, so angular momentum is not conserved. There are no leptons on the left and one electron on the right, so electron lepton number is not conserved.

46.5 The diagram is

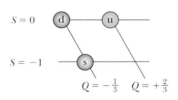

46.6 False. Because the charges on quarks are $+2e/3$ and $-e/3$, the maximum possible charge of a combination of a quark and an antiquark is $\pm e$.

TABLE A.1

Conversion Factors

Length

	m	cm	km	in.	ft	mi
1 meter	1	10^2	10^{-3}	39.37	3.281	6.214×10^{-4}
1 centimeter	10^{-2}	1	10^{-5}	0.393 7	3.281×10^{-2}	6.214×10^{-6}
1 kilometer	10^3	10^5	1	3.937×10^4	3.281×10^3	0.621 4
1 inch	2.540×10^{-2}	2.540	2.540×10^{-5}	1	8.333×10^{-2}	1.578×10^{-5}
1 foot	0.304 8	30.48	3.048×10^{-4}	12	1	1.894×10^{-4}
1 mile	1 609	1.609×10^5	1.609	6.336×10^4	5 280	1

Mass

	kg	g	slug	u
1 kilogram	1	10^3	6.852×10^{-2}	6.024×10^{26}
1 gram	10^{-3}	1	6.852×10^{-5}	6.024×10^{23}
1 slug	14.59	1.459×10^4	1	8.789×10^{27}
1 atomic mass unit	1.660×10^{-27}	1.660×10^{-24}	1.137×10^{-28}	1

Note: 1 metric ton = 1 000 kg.

Time

	s	min	h	day	yr
1 second	1	1.667×10^{-2}	2.778×10^{-4}	1.157×10^{-5}	3.169×10^{-8}
1 minute	60	1	1.667×10^{-2}	6.994×10^{-4}	1.901×10^{-6}
1 hour	3 600	60	1	4.167×10^{-2}	1.141×10^{-4}
1 day	8.640×10^4	1 440	24	1	2.738×10^{-5}
1 year	3.156×10^7	5.259×10^5	8.766×10^3	365.2	1

Speed

	m/s	cm/s	ft/s	mi/h
1 meter per second	1	10^2	3.281	2.237
1 centimeter per second	10^{-2}	1	3.281×10^{-2}	2.237×10^{-2}
1 foot per second	0.304 8	30.48	1	0.681 8
1 mile per hour	0.447 0	44.70	1.467	1

Note: 1 mi/min = 60 mi/h = 88 ft/s.

Force

	N	lb
1 newton	1	0.224 8
1 pound	4.448	1

(*Continued*)

TABLE A.1

Conversion Factors (*Continued*)

Energy, Energy Transfer

	J	ft · lb	eV
1 joule	1	0.737 6	6.242×10^{18}
1 foot-pound	1.356	1	8.464×10^{18}
1 electron volt	1.602×10^{-19}	1.182×10^{-19}	1
1 calorie	4.186	3.087	2.613×10^{19}
1 British thermal unit	1.055×10^3	7.779×10^2	6.585×10^{21}
1 kilowatt-hour	3.600×10^6	2.655×10^6	2.247×10^{25}

	cal	Btu	kWh
1 joule	0.238 9	9.481×10^{-4}	2.778×10^{-7}
1 foot-pound	0.323 9	1.285×10^{-3}	3.766×10^{-7}
1 electron volt	3.827×10^{-20}	1.519×10^{-22}	4.450×10^{-26}
1 calorie	1	3.968×10^{-3}	1.163×10^{-6}
1 British thermal unit	2.520×10^2	1	2.930×10^{-4}
1 kilowatt-hour	8.601×10^5	3.413×10^2	1

Pressure

	Pa	atm
1 pascal	1	9.869×10^{-6}
1 atmosphere	1.013×10^5	1
1 centimeter mercury[a]	1.333×10^3	1.316×10^{-2}
1 pound per square inch	6.895×10^3	6.805×10^{-2}
1 pound per square foot	47.88	4.725×10^{-4}

	cm Hg	lb/in.²	lb/ft²
1 pascal	7.501×10^{-4}	1.450×10^{-4}	2.089×10^{-2}
1 atmosphere	76	14.70	2.116×10^3
1 centimeter mercury[a]	1	0.194 3	27.85
1 pound per square inch	5.171	1	144
1 pound per square foot	3.591×10^{-2}	6.944×10^{-3}	1

[a]At 0°C and at a location where the free-fall acceleration has its "standard" value, 9.806 65 m/s².

TABLE A.2

Symbols, Dimensions, and Units of Physical Quantities

Quantity	Common Symbol	Unit[a]	Dimensions[b]	Unit in Terms of Base SI Units
Acceleration	$\vec{a}$	m/s²	L/T²	m/s²
Amount of substance	n	MOLE		mol
Angle	θ, ϕ	radian (rad)	1	
Angular acceleration	$\vec{\alpha}$	rad/s²	T⁻²	s⁻²
Angular frequency	ω	rad/s	T⁻¹	s⁻¹
Angular momentum	$\vec{L}$	kg · m²/s	ML²/T	kg · m²/s
Angular velocity	$\vec{\omega}$	rad/s	T⁻¹	s⁻¹
Area	A	m²	L²	m²
Atomic number	Z			
Capacitance	C	farad (F)	Q²T²/ML²	A² · s⁴/kg · m²
Charge	q, Q, e	coulomb (C)	Q	A · s

(*Continued*)

TABLE A.2

Symbols, Dimensions, and Units of Physical Quantities *(Continued)*

Charge density				
Line	λ	C/m	Q/L	A · s/m
Surface	σ	C/m^2	Q/L^2	A · s/m^2
Volume	ρ	C/m^3	Q/L^3	A · s/m^3
Conductivity	σ	$1/\Omega \cdot$m	Q^{2}T/ML^3	A^2 · s^3/kg · m^3
Current	I	AMPERE	Q/T	A
Current density	J	A/m^2	Q/TL^2	A/m^2
Density	ρ	kg/m^3	M/L^3	kg/m^3
Dielectric constant	κ			
Electric dipole moment	$\vec{\mathbf{p}}$	C · m	QL	A · s · m
Electric field	$\vec{\mathbf{E}}$	V/m	ML/QT^2	kg · m/A · s^3
Electric flux	Φ_E	V · m	ML^3/QT^2	kg · m^3/A · s^3
Electromotive force	$\mathcal{E}$	volt (V)	ML^2/QT^2	kg · m^2/A · s^3
Energy	E, U, K	joule (J)	ML^2/T^2	kg · m^2/s^2
Entropy	S	J/K	ML^2/T^{2}K	kg · m^2/s^2 · K
Force	$\vec{\mathbf{F}}$	newton (N)	ML/T^2	kg · m/s^2
Frequency	f	hertz (Hz)	T^{-1}	s^{-1}
Heat	Q	joule (J)	ML^2/T^2	kg · m^2/s^2
Inductance	L	henry (H)	ML^2/Q^2	kg · m^2/A^2 · s^2
Length	ℓ, L	METER	L	m
Displacement	$\Delta x, \Delta \vec{\mathbf{r}}$			
Distance	d, h			
Position	$x, y, z, \vec{\mathbf{r}}$			
Magnetic dipole moment	$\vec{\boldsymbol{\mu}}$	N · m/T	QL^2/T	A · m^2
Magnetic field	$\vec{\mathbf{B}}$	tesla (T) (= Wb/m^2)	M/QT	kg/A · s^2
Magnetic flux	Φ_B	weber (Wb)	ML^2/QT	kg · m^2/A · s^2
Mass	m, M	KILOGRAM	M	kg
Molar specific heat	C	J/mol · K		kg · m^2/s^2 · mol · K
Moment of inertia	I	kg · m^2	ML^2	kg · m^2
Momentum	$\vec{\mathbf{p}}$	kg · m/s	ML/T	kg · m/s
Period	T	s	T	s
Permeability of free space	μ_0	N/A^2 (= H/m)	ML/Q^2	kg · m/A^2 · s^2
Permittivity of free space	ϵ_0	C^2/N · m^2 (= F/m)	Q^{2}T^2/ML^3	A^2 · s^4/kg · m^3
Potential	V	volt (V) (= J/C)	ML^2/QT^2	kg · m^2/A · s^3
Power	$\mathcal{P}$	watt (W) (= J/s)	ML^2/T^3	kg · m^2/s^3
Pressure	P	pascal (Pa) (= N/m^2)	M/LT^2	kg/m · s^2
Resistance	R	ohm (Ω) (= V/A)	ML^2/Q^{2}T	kg · m^2/A^2 · s^3
Specific heat	c	J/kg · K	L^2/T^{2}K	m^2/s^2 · K
Speed	v	m/s	L/T	m/s
Temperature	T	KELVIN	K	K
Time	t	SECOND	T	s
Torque	$\vec{\boldsymbol{\tau}}$	N · m	ML^2/T^2	kg · m^2/s^2
Velocity	$\vec{\mathbf{v}}$	m/s	L/T	m/s
Volume	V	m^3	L^3	m^3
Wavelength	λ	m	L	m
Work	W	joule (J) (= N · m)	ML^2/T^2	kg · m^2/s^2

[a]The base SI units are given in uppercase letters.

[b]The symbols M, L, T, K, and Q denote mass, length, time, temperature, and charge, respectively.

This appendix in mathematics is intended as a brief review of operations and methods. Early in this course, you should be totally familiar with basic algebraic techniques, analytic geometry, and trigonometry. The sections on differential and integral calculus are more detailed and are intended for students who have difficulty applying calculus concepts to physical situations.

B.1　Scientific Notation

Many quantities used by scientists often have very large or very small values. The speed of light, for example, is about 300 000 000 m/s, and the ink required to make the dot over an i in this textbook has a mass of about 0.000 000 001 kg. Obviously, it is very cumbersome to read, write, and keep track of such numbers. We avoid this problem by using a method incorporating powers of the number 10:

$$10^0 = 1$$

$$10^1 = 10$$

$$10^2 = 10 \times 10 = 100$$

$$10^3 = 10 \times 10 \times 10 = 1\ 000$$

$$10^4 = 10 \times 10 \times 10 \times 10 = 10\ 000$$

$$10^5 = 10 \times 10 \times 10 \times 10 \times 10 = 100\ 000$$

and so on. The number of zeros corresponds to the power to which ten is raised, called the **exponent** of ten. For example, the speed of light, 300 000 000 m/s, can be expressed as 3.00×10^8 m/s.

In this method, some representative numbers smaller than unity are the following:

$$10^{-1} = \frac{1}{10} = 0.1$$

$$10^{-2} = \frac{1}{10 \times 10} = 0.01$$

$$10^{-3} = \frac{1}{10 \times 10 \times 10} = 0.001$$

$$10^{-4} = \frac{1}{10 \times 10 \times 10 \times 10} = 0.000\ 1$$

$$10^{-5} = \frac{1}{10 \times 10 \times 10 \times 10 \times 10} = 0.000\ 01$$

In these cases, the number of places the decimal point is to the left of the digit 1 equals the value of the (negative) exponent. Numbers expressed as some power of ten multiplied by another number between one and ten are said to be in **scientific notation.** For example, the scientific notation for 5 943 000 000 is 5.943×10^9 and that for 0.000 083 2 is 8.32×10^{-5}.

When numbers expressed in scientific notation are being multiplied, the following general rule is very useful:

$$10^n \times 10^m = 10^{n+m} \tag{B.1}$$

where n and m can be *any* numbers (not necessarily integers). For example, $10^2 \times 10^5 = 10^7$. The rule also applies if one of the exponents is negative: $10^3 \times 10^{-8} = 10^{-5}$.

When dividing numbers expressed in scientific notation, note that

$$\frac{10^n}{10^m} = 10^n \times 10^{-m} = 10^{n-m} \tag{B.2}$$

Exercises

With help from the preceding rules, verify the answers to the following equations:

1. $86\ 400 = 8.64 \times 10^4$
2. $9\ 816\ 762.5 = 9.816\ 762\ 5 \times 10^6$
3. $0.000\ 000\ 039\ 8 = 3.98 \times 10^{-8}$
4. $(4.0 \times 10^8)(9.0 \times 10^9) = 3.6 \times 10^{18}$
5. $(3.0 \times 10^7)(6.0 \times 10^{-12}) = 1.8 \times 10^{-4}$
6. $\dfrac{75 \times 10^{-11}}{5.0 \times 10^{-3}} = 1.5 \times 10^{-7}$
7. $\dfrac{(3 \times 10^6)(8 \times 10^{-2})}{(2 \times 10^{17})(6 \times 10^5)} = 2 \times 10^{-18}$

B.2   Algebra

Some Basic Rules

When algebraic operations are performed, the laws of arithmetic apply. Symbols such as x, y, and z are usually used to represent unspecified quantities, called the **unknowns.**

First, consider the equation

$$8x = 32$$

If we wish to solve for x, we can divide (or multiply) each side of the equation by the same factor without destroying the equality. In this case, if we divide both sides by 8, we have

$$\frac{8x}{8} = \frac{32}{8}$$

$$x = 4$$

Next consider the equation

$$x + 2 = 8$$

In this type of expression, we can add or subtract the same quantity from each side. If we subtract 2 from each side, we have

$$x + 2 - 2 = 8 - 2$$

$$x = 6$$

In general, if $x + a = b$, then $x = b - a$.

Now consider the equation

$$\frac{x}{5} = 9$$

If we multiply each side by 5, we are left with x on the left by itself and 45 on the right:

$$\left(\frac{x}{5}\right)(5) = 9 \times 5$$

$$x = 45$$

In all cases, *whatever operation is performed on the left side of the equality must also be performed on the right side.*

The following rules for multiplying, dividing, adding, and subtracting fractions should be recalled, where a, b, c, and d are four numbers:

	Rule	Example
Multiplying	$\left(\dfrac{a}{b}\right)\left(\dfrac{c}{d}\right) = \dfrac{ac}{bd}$	$\left(\dfrac{2}{3}\right)\left(\dfrac{4}{5}\right) = \dfrac{8}{15}$
Dividing	$\dfrac{(a/b)}{(c/d)} = \dfrac{ad}{bc}$	$\dfrac{2/3}{4/5} = \dfrac{(2)(5)}{(4)(3)} = \dfrac{10}{12}$
Adding	$\dfrac{a}{b} \pm \dfrac{c}{d} = \dfrac{ad \pm bc}{bd}$	$\dfrac{2}{3} - \dfrac{4}{5} = \dfrac{(2)(5) - (4)(3)}{(3)(5)} = -\dfrac{2}{15}$

Exercises

In the following exercises, solve for x.

Answers

1. $a = \dfrac{1}{1 + x}$ $x = \dfrac{1 - a}{a}$

2. $3x - 5 = 13$ $x = 6$

3. $ax - 5 = bx + 2$ $x = \dfrac{7}{a - b}$

4. $\dfrac{5}{2x + 6} = \dfrac{3}{4x + 8}$ $x = -\dfrac{11}{7}$

Powers

When powers of a given quantity x are multiplied, the following rule applies:

$$x^n x^m = x^{n+m} \tag{B.3}$$

For example, $x^2 x^4 = x^{2+4} = x^6$.

When dividing the powers of a given quantity, the rule is

$$\frac{x^n}{x^m} = x^{n-m} \tag{B.4}$$

For example, $x^8/x^2 = x^{8-2} = x^6$.

A power that is a fraction, such as $\frac{1}{3}$, corresponds to a root as follows:

$$x^{1/n} = \sqrt[n]{x} \tag{B.5}$$

For example, $4^{1/3} = \sqrt[3]{4} = 1.587\ 4$. (A scientific calculator is useful for such calculations.)

Finally, any quantity x^n raised to the mth power is

$$(x^n)^m = x^{nm} \tag{B.6}$$

Table B.1 summarizes the rules of exponents.

TABLE B.1

Rules of Exponents

$x^0 = 1$
$x^1 = x$
$x^n x^m = x^{n+m}$
$x^n/x^m = x^{n-m}$
$x^{1/n} = \sqrt[n]{x}$
$(x^n)^m = x^{nm}$

Exercises

Verify the following equations:

1. $3^2 \times 3^3 = 243$

2. $x^5 x^{-8} = x^{-3}$

3. $x^{10}/x^{-5} = x^{15}$

4. $5^{1/3} = 1.709\ 976$ (Use your calculator.)

5. $60^{1/4} = 2.783\ 158$ (Use your calculator.)

6. $(x^4)^3 = x^{12}$

Factoring

Some useful formulas for factoring an equation are the following:

$$ax + ay + az = a(x + y + z) \quad \text{common factor}$$

$$a^2 + 2ab + b^2 = (a + b)^2 \quad \text{perfect square}$$

$$a^2 - b^2 = (a + b)(a - b) \quad \text{differences of squares}$$

Quadratic Equations

The general form of a quadratic equation is

$$ax^2 + bx + c = 0 \tag{B.7}$$

where x is the unknown quantity and a, b, and c are numerical factors referred to as **coefficients** of the equation. This equation has two roots, given by

$$x = \frac{-b \pm \sqrt{b^2 - 4ac}}{2a} \tag{B.8}$$

If $b^2 \geq 4ac$, the roots are real.

EXAMPLE B.1

The equation $x^2 + 5x + 4 = 0$ has the following roots corresponding to the two signs of the square-root term:

$$x = \frac{-5 \pm \sqrt{5^2 - (4)(1)(4)}}{2(1)} = \frac{-5 \pm \sqrt{9}}{2} = \frac{-5 \pm 3}{2}$$

$$x_+ = \frac{-5 + 3}{2} = -1 \qquad x_- = \frac{-5 - 3}{2} = -4$$

where x_+ refers to the root corresponding to the positive sign and x_- refers to the root corresponding to the negative sign.

Exercises

Solve the following quadratic equations:

	Answers	
1. $x^2 + 2x - 3 = 0$	$x_+ = 1$	$x_- = -3$
2. $2x^2 - 5x + 2 = 0$	$x_+ = 2$	$x_- = \frac{1}{2}$
3. $2x^2 - 4x - 9 = 0$	$x_+ = 1 + \sqrt{22}/2$	$x_- = 1 - \sqrt{22}/2$

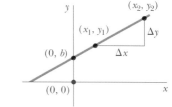

Figure B.1 A straight line graphed on an xy coordinate system. The slope of the line is the ratio of Δy to Δx.

Linear Equations

A linear equation has the general form

$$y = mx + b \tag{B.9}$$

where m and b are constants. This equation is referred to as linear because the graph of y versus x is a straight line as shown in Figure B.1. The constant b, called the **y-intercept,** represents the value of y at which the straight line intersects the y axis. The constant m is equal to the **slope** of the straight line. If any two points on the straight line are specified by the coordinates (x_1, y_1) and (x_2, y_2) as in Figure B.1, the slope of the straight line can be expressed as

$$\text{Slope} = \frac{y_2 - y_1}{x_2 - x_1} = \frac{\Delta y}{\Delta x} \tag{B.10}$$

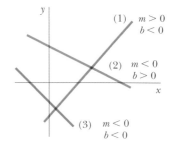

Figure B.2 The brown line has a positive slope and a negative y-intercept. The blue line has a negative slope and a positive y-intercept. The green line has a negative slope and a negative y-intercept.

Note that m and b can have either positive or negative values. If $m > 0$, the straight line has a *positive* slope as in Figure B.1. If $m < 0$, the straight line has a *negative* slope. In Figure B.1, both m and b are positive. Three other possible situations are shown in Figure B.2.

Exercises

1. Draw graphs of the following straight lines: (a) $y = 5x + 3$ (b) $y = -2x + 4$ (c) $y = -3x - 6$

2. Find the slopes of the straight lines described in Exercise 1.

Answers   (a) 5   (b) -2 (c) -3

3. Find the slopes of the straight lines that pass through the following sets of points:
   (a) $(0, -4)$ and $(4, 2)$ (b) $(0, 0)$ and $(2, -5)$ (c) $(-5, 2)$ and $(4, -2)$

Answers   (a) $\frac{3}{2}$ (b) $-\frac{5}{2}$ (c) $-\frac{4}{9}$

Solving Simultaneous Linear Equations

Consider the equation $3x + 5y = 15$, which has two unknowns, x and y. Such an equation does not have a unique solution. For example, $(x = 0, y = 3)$, $(x = 5, y = 0)$, and $(x = 2, y = \frac{9}{5})$ are all solutions to this equation.

If a problem has two unknowns, a unique solution is possible only if we have *two* equations. In general, if a problem has n unknowns, its solution requires n equations. To solve two simultaneous equations involving two unknowns, x and y, we solve one of the equations for x in terms of y and substitute this expression into the other equation.

EXAMPLE B.2

Solve the two simultaneous equations

$$(1) \qquad 5x + y = -8$$

$$(2) \qquad 2x - 2y = 4$$

Solution From Equation (2), $x = y + 2$. Substitution of this equation into Equation (1) gives

$$5(y + 2) + y = -8$$

$$6y = -18$$

$$y = \boxed{-3}$$

$$x = y + 2 = \boxed{-1}$$

Alternative Solution Multiply each term in Equation (1) by the factor 2 and add the result to Equation (2):

$$10x + 2y = -16$$

$$\underline{2x - 2y = 4}$$

$$12x \qquad = -12$$

$$x = \boxed{-1}$$

$$y = x - 2 = \boxed{-3}$$

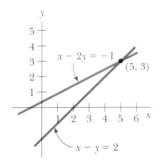

Figure B.3 A graphical solution for two linear equations.

Two linear equations containing two unknowns can also be solved by a graphical method. If the straight lines corresponding to the two equations are plotted in a conventional coordinate system, the intersection of the two lines represents the solution. For example, consider the two equations

$$x - y = 2$$

$$x - 2y = -1$$

These equations are plotted in Figure B.3. The intersection of the two lines has the coordinates $x = 5$ and $y = 3$, which represents the solution to the equations. You should check this solution by the analytical technique discussed earlier.

Exercises

Solve the following pairs of simultaneous equations involving two unknowns:

Answers

1. $x + y = 8$ $x = 5, y = 3$
    $x - y = 2$

2. $98 - T = 10a$ $T = 65, a = 3.27$
    $T - 49 = 5a$

3. $6x + 2y = 6$ $x = 2, y = -3$
    $8x - 4y = 28$

Logarithms

Suppose a quantity x is expressed as a power of some quantity a:

$$x = a^y \tag{B.11}$$

The number a is called the **base** number. The **logarithm** of x with respect to the base a is equal to the exponent to which the base must be raised to satisfy the expression $x = a^y$:

$$y = \log_a x \tag{B.12}$$

Conversely, the **antilogarithm** of y is the number x:

$$x = \text{antilog}_a y \tag{B.13}$$

In practice, the two bases most often used are base 10, called the *common* logarithm base, and base $e = 2.718\,282$, called Euler's constant or the *natural* logarithm base. When common logarithms are used,

$$y = \log_{10} x \quad (\text{or } x = 10^y) \tag{B.14}$$

When natural logarithms are used,

$$y = \ln x \quad (\text{or } x = e^y) \tag{B.15}$$

For example, $\log_{10} 52 = 1.716$, so $\text{antilog}_{10} 1.716 = 10^{1.716} = 52$. Likewise, $\ln 52 = 3.951$, so $\text{antiln } 3.951 = e^{3.951} = 52$.

In general, note you can convert between base 10 and base e with the equality

$$\ln x = (2.302\,585) \log_{10} x \tag{B.16}$$

Finally, some useful properties of logarithms are the following:

$$\left. \begin{array}{l} \log(ab) = \log a + \log b \\ \log(a/b) = \log a - \log b \\ \log(a^n) = n \log a \end{array} \right\} \text{ any base}$$

$$\ln e = 1$$
$$\ln e^a = a$$
$$\ln\left(\frac{1}{a}\right) = -\ln a$$

B.3   Geometry

The **distance** d between two points having coordinates (x_1, y_1) and (x_2, y_2) is

$$d = \sqrt{(x_2 - x_1)^2 + (y_2 - y_1)^2} \tag{B.17}$$

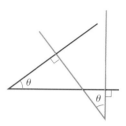

Figure B.4   The angles are equal because their sides are perpendicular.

Figure B.5   The angle θ in radians is the ratio of the arc length s to the radius r of the circle.

TABLE B.2

Useful Information for Geometry

Shape	Area or Volume	Shape	Area or Volume
Rectangle	Area $= \ell w$	Sphere	Surface area $= 4\pi r^2$ Volume $= \dfrac{4\pi r^3}{3}$
Circle	Area $= \pi r^2$ Circumference $= 2\pi r$	Cylinder	Lateral surface area $= 2\pi r \ell$ Volume $= \pi r^2 \ell$
Triangle	Area $= \frac{1}{2}bh$	Rectangular box	Surface area $=$ $2(\ell h + \ell w + hw)$ Volume $= \ell wh$

Two angles are equal if their sides are perpendicular, right side to right side and left side to left side. For example, the two angles marked θ in Figure B.4 are the same because of the perpendicularity of the sides of the angles. To distinguish the left and right sides of an angle, imagine standing at the angle's apex and facing into the angle.

Radian measure: The arc length s of a circular arc (Fig. B.5) is proportional to the radius r for a fixed value of θ (in radians):

$$s = r\theta$$
$$\theta = \frac{s}{r}$$

(B.18)

Figure B.6   A straight line with a slope of m and a y-intercept of b.

Table B.2 gives the **areas** and **volumes** for several geometric shapes used throughout this text.

The equation of a **straight line** (Fig. B.6) is

$$y = mx + b$$

(B.19)

where b is the y-intercept and m is the slope of the line.

The equation of a **circle** of radius R centered at the origin is

$$x^2 + y^2 = R^2$$

(B.20)

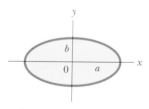

Figure B.7   An ellipse with semimajor axis a and semiminor axis b.

The equation of an **ellipse** having the origin at its center (Fig. B.7) is

$$\frac{x^2}{a^2} + \frac{y^2}{b^2} = 1$$

(B.21)

where a is the length of the semimajor axis (the longer one) and b is the length of the semiminor axis (the shorter one).

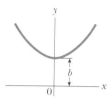

Figure B.8   A parabola with its vertex at $y = b$.

The equation of a **parabola** the vertex of which is at $y = b$ (Fig. B.8) is

$$y = ax^2 + b$$

(B.22)

The equation of a **rectangular hyperbola** (Fig. B.9) is

$$xy = \text{constant}$$

(B.23)

B.4   Trigonometry

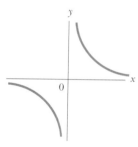

Figure B.9   A hyperbola.

That portion of mathematics based on the special properties of the right triangle is called trigonometry. By definition, a right triangle is a triangle containing a 90° angle. Consider the right triangle shown in Figure B.10, where side a is opposite the angle θ, side b is adjacent to the angle θ, and side c is the hypotenuse of the triangle. The three basic trigonometric functions defined by such a triangle are the sine (sin), cosine (cos), and tangent (tan). In terms of the angle θ, these functions are defined as follows:

$$\sin \theta = \frac{\text{side opposite } \theta}{\text{hypotenuse}} = \frac{a}{c} \qquad \text{(B.24)}$$

$$\cos \theta = \frac{\text{side adjacent to } \theta}{\text{hypotenuse}} = \frac{b}{c} \qquad \text{(B.25)}$$

$$\tan \theta = \frac{\text{side opposite } \theta}{\text{side adjacent to } \theta} = \frac{a}{b} \qquad \text{(B.26)}$$

a = opposite side
b = adjacent side
c = hypotenuse

Figure B.10 A right triangle, used to define the basic functions of trigonometry.

The Pythagorean theorem provides the following relationship among the sides of a right triangle:

$$c^2 = a^2 + b^2 \qquad \text{(B.27)}$$

From the preceding definitions and the Pythagorean theorem, it follows that

$$\sin^2 \theta + \cos^2 \theta = 1$$

$$\tan \theta = \frac{\sin \theta}{\cos \theta}$$

The cosecant, secant, and cotangent functions are defined by

$$\csc \theta = \frac{1}{\sin \theta} \quad \sec \theta = \frac{1}{\cos \theta} \quad \cot \theta = \frac{1}{\tan \theta}$$

The following relationships are derived directly from the right triangle shown in Figure B.10:

$$\sin \theta = \cos (90° - \theta)$$

$$\cos \theta = \sin (90° - \theta)$$

$$\cot \theta = \tan (90° - \theta)$$

Some properties of trigonometric functions are the following:

$$\sin (-\theta) = -\sin \theta$$

$$\cos (-\theta) = \cos \theta$$

$$\tan (-\theta) = -\tan \theta$$

The following relationships apply to *any* triangle as shown in Figure B.11:

$$\alpha + \beta + \gamma = 180°$$

Law of cosines $\begin{cases} a^2 = b^2 + c^2 - 2bc \cos \alpha \\ b^2 = a^2 + c^2 - 2ac \cos \beta \\ c^2 = a^2 + b^2 - 2ab \cos \gamma \end{cases}$

Law of sines $\quad \dfrac{a}{\sin \alpha} = \dfrac{b}{\sin \beta} = \dfrac{c}{\sin \gamma}$

Table B.3 (page A-12) lists a number of useful trigonometric identities.

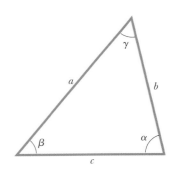

Figure B.11 An arbitrary, nonright triangle.

EXAMPLE **B.3**

Consider the right triangle in Figure B.12 in which $a = 2.00$, $b = 5.00$, and c is unknown. From the Pythagorean theorem, we have

$$c^2 = a^2 + b^2 = 2.00^2 + 5.00^2 = 4.00 + 25.0 = 29.0$$

$$c = \sqrt{29.0} = 5.39$$

Figure B.12 (Example B.3)

To find the angle θ, note that

$$\tan \theta = \frac{a}{b} = \frac{2.00}{5.00} = 0.400$$

Using a calculator, we find that

$$\theta = \tan^{-1}(0.400) = \boxed{21.8°}$$

where $\tan^{-1}(0.400)$ is the notation for "angle whose tangent is 0.400," sometimes written as arctan (0.400).

TABLE B.3
Some Trigonometric Identities

$\sin^2 \theta + \cos^2 \theta = 1$	$\csc^2 \theta = 1 + \cot^2 \theta$
$\sec^2 \theta = 1 + \tan^2 \theta$	$\sin^2 \dfrac{\theta}{2} = \frac{1}{2}(1 - \cos \theta)$
$\sin 2\theta = 2 \sin \theta \cos \theta$	$\cos^2 \dfrac{\theta}{2} = \frac{1}{2}(1 + \cos \theta)$
$\cos 2\theta = \cos^2 \theta - \sin^2 \theta$	$1 - \cos \theta = 2 \sin^2 \dfrac{\theta}{2}$
$\tan 2\theta = \dfrac{2 \tan \theta}{1 - \tan^2 \theta}$	$\tan \dfrac{\theta}{2} = \sqrt{\dfrac{1 - \cos \theta}{1 + \cos \theta}}$

$$\sin(A \pm B) = \sin A \cos B \pm \cos A \sin B$$
$$\cos(A \pm B) = \cos A \cos B \mp \sin A \sin B$$
$$\sin A \pm \sin B = 2 \sin \left[\tfrac{1}{2}(A \pm B)\right] \cos \left[\tfrac{1}{2}(A \mp B)\right]$$
$$\cos A + \cos B = 2 \cos \left[\tfrac{1}{2}(A + B)\right] \cos \left[\tfrac{1}{2}(A - B)\right]$$
$$\cos A - \cos B = 2 \sin \left[\tfrac{1}{2}(A + B)\right] \sin \left[\tfrac{1}{2}(B - A)\right]$$

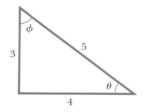

Figure B.13 (Exercise 1)

Exercises

1. In Figure B.13, identify (a) the side opposite θ (b) the side adjacent to ϕ and then find (c) $\cos \theta$, (d) $\sin \phi$, and (e) $\tan \phi$.

Answers   (a) 3   (b) 3   (c) $\frac{4}{5}$ (d) $\frac{4}{5}$ (e) $\frac{4}{3}$

2. In a certain right triangle, the two sides that are perpendicular to each other are 5.00 m and 7.00 m long. What is the length of the third side?

Answer   8.60 m

3. A right triangle has a hypotenuse of length 3.0 m, and one of its angles is 30°. (a) What is the length of the side opposite the 30° angle? (b) What is the side adjacent to the 30° angle?

Answers   (a) 1.5 m   (b) 2.6 m

B.5   Series Expansions

$$(a + b)^n = a^n + \frac{n}{1!} a^{n-1} b + \frac{n(n-1)}{2!} a^{n-2} b^2 + \cdots$$

$$(1 + x)^n = 1 + nx + \frac{n(n-1)}{2!} x^2 + \cdots$$

$$e^x = 1 + x + \frac{x^2}{2!} + \frac{x^3}{3!} + \cdots$$

$$\ln(1 \pm x) = \pm x - \tfrac{1}{2}x^2 \pm \tfrac{1}{3}x^3 - \cdots$$

$$\left. \begin{array}{l} \sin x = x - \dfrac{x^3}{3!} + \dfrac{x^5}{5!} - \cdots \\[2ex] \cos x = 1 - \dfrac{x^2}{2!} + \dfrac{x^4}{4!} - \cdots \\[2ex] \tan x = x + \dfrac{x^3}{3} + \dfrac{2x^5}{15} + \cdots \quad |x| < \dfrac{\pi}{2} \end{array} \right\} \quad x \text{ in radians}$$

For $x \ll 1$, the following approximations can be used:[1]

$$(1 + x)^n \approx 1 + nx \qquad \sin x \approx x$$

$$e^x \approx 1 + x \qquad \cos x \approx 1$$

$$\ln(1 \pm x) \approx \pm x \qquad \tan x \approx x$$

B.6   Differential Calculus

In various branches of science, it is sometimes necessary to use the basic tools of calculus, invented by Newton, to describe physical phenomena. The use of calculus is fundamental in the treatment of various problems in Newtonian mechanics, electricity, and magnetism. In this section, we simply state some basic properties and "rules of thumb" that should be a useful review to the student.

First, a **function** must be specified that relates one variable to another (e.g., a coordinate as a function of time). Suppose one of the variables is called y (the dependent variable), and the other x (the independent variable). We might have a function relationship such as

$$y(x) = ax^3 + bx^2 + cx + d$$

If a, b, c, and d are specified constants, y can be calculated for any value of x. We usually deal with continuous functions, that is, those for which y varies "smoothly" with x.

The **derivative** of y with respect to x is defined as the limit as Δx approaches zero of the slopes of chords drawn between two points on the y versus x curve. Mathematically, we write this definition as

$$\frac{dy}{dx} = \lim_{\Delta x \to 0} \frac{\Delta y}{\Delta x} = \lim_{\Delta x \to 0} \frac{y(x + \Delta x) - y(x)}{\Delta x} \tag{B.28}$$

where Δy and Δx are defined as $\Delta x = x_2 - x_1$ and $\Delta y = y_2 - y_1$ (Fig. B.14). Note that dy/dx does *not* mean dy divided by dx, but rather is simply a notation of the limiting process of the derivative as defined by Equation B.28.

A useful expression to remember when $y(x) = ax^n$, where a is a *constant* and n is *any* positive or negative number (integer or fraction), is

$$\frac{dy}{dx} = nax^{n-1} \tag{B.29}$$

If $y(x)$ is a polynomial or algebraic function of x, we apply Equation B.29 to *each* term in the polynomial and take $d[\text{constant}]/dx = 0$. In Examples B.4 through B.7, we evaluate the derivatives of several functions.

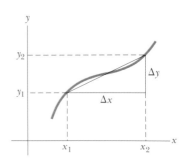

Figure B.14 The lengths Δx and Δy are used to define the derivative of this function at a point.

[1]The approximations for the functions $\sin x$, $\cos x$, and $\tan x$ are for $x \leq 0.1$ rad.

TABLE B.4

Derivative for Several Functions

$$\frac{d}{dx}(a) = 0$$

$$\frac{d}{dx}(ax^n) = nax^{n-1}$$

$$\frac{d}{dx}(e^{ax}) = ae^{ax}$$

$$\frac{d}{dx}(\sin ax) = a\cos ax$$

$$\frac{d}{dx}(\cos ax) = -a\sin ax$$

$$\frac{d}{dx}(\tan ax) = a\sec^2 ax$$

$$\frac{d}{dx}(\cot ax) = -a\csc^2 ax$$

$$\frac{d}{dx}(\sec x) = \tan x \sec x$$

$$\frac{d}{dx}(\csc x) = -\cot x \csc x$$

$$\frac{d}{dx}(\ln ax) = \frac{1}{x}$$

$$\frac{d}{dx}(\sin^{-1} ax) = \frac{a}{\sqrt{1 - a^2x^2}}$$

$$\frac{d}{dx}(\cos^{-1} ax) = \frac{-a}{\sqrt{1 - a^2x^2}}$$

$$\frac{d}{dx}(\tan^{-1} ax) = \frac{a}{1 + a^2x^2}$$

Note: The symbols a and n represent constants.

Special Properties of the Derivative

A. Derivative of the product of two functions   If a function $f(x)$ is given by the product of two functions—say, $g(x)$ and $h(x)$—the derivative of $f(x)$ is defined as

$$\frac{d}{dx}f(x) = \frac{d}{dx}[g(x)h(x)] = g\frac{dh}{dx} + h\frac{dg}{dx} \tag{B.30}$$

B. Derivative of the sum of two functions   If a function $f(x)$ is equal to the sum of two functions, the derivative of the sum is equal to the sum of the derivatives:

$$\frac{d}{dx}f(x) = \frac{d}{dx}[g(x) + h(x)] = \frac{dg}{dx} + \frac{dh}{dx} \tag{B.31}$$

C. Chain rule of differential calculus   If $y = f(x)$ and $x = g(z)$, then dy/dz can be written as the product of two derivatives:

$$\frac{dy}{dz} = \frac{dy}{dx}\frac{dx}{dz} \tag{B.32}$$

D. The second derivative   The second derivative of y with respect to x is defined as the derivative of the function dy/dx (the derivative of the derivative). It is usually written as

$$\frac{d^2y}{dx^2} = \frac{d}{dx}\left(\frac{dy}{dx}\right) \tag{B.33}$$

Some of the more commonly used derivatives of functions are listed in Table B.4.

EXAMPLE B.4

Suppose $y(x)$ (that is, y as a function of x) is given by

$$y(x) = ax^3 + bx + c$$

where a and b are constants. It follows that

$$y(x + \Delta x) = a(x + \Delta x)^3 + b(x + \Delta x) + c$$

$$= a(x^3 + 3x^2\,\Delta x + 3x\,\Delta x^2 + \Delta x^3) + b(x + \Delta x) + c$$

so

$$\Delta y = y(x + \Delta x) - y(x) = a(3x^2\,\Delta x + 3x\,\Delta x^2 + \Delta x^3) + b\,\Delta x$$

Substituting this into Equation B.28 gives

$$\frac{dy}{dx} = \lim_{\Delta x \to 0}\frac{\Delta y}{\Delta x} = \lim_{\Delta x \to 0}[3ax^2 + 3ax\,\Delta x + a\,\Delta x^2] + b$$

$$\frac{dy}{dx} = 3ax^2 + b$$

EXAMPLE B.5

Find the derivative of

$$y(x) = 8x^5 + 4x^3 + 2x + 7$$

Solution Applying Equation B.29 to each term independently and remembering that d/dx (constant) $= 0$, we have

$$\frac{dy}{dx} = 8(5)x^4 + 4(3)x^2 + 2(1)x^0 + 0$$

$$\frac{dy}{dx} = \boxed{40x^4 + 12x^2 + 2}$$

EXAMPLE B.6

Find the derivative of $y(x) = x^3/(x+1)^2$ with respect to x.

Solution We can rewrite this function as $y(x) = x^3(x+1)^{-2}$ and apply Equation B.30:

$$\frac{dy}{dx} = (x+1)^{-2}\frac{d}{dx}(x^3) + x^3\frac{d}{dx}(x+1)^{-2}$$

$$= (x+1)^{-2}3x^2 + x^3(-2)(x+1)^{-3}$$

$$\frac{dy}{dx} = \boxed{\frac{3x^2}{(x+1)^2} - \frac{2x^3}{(x+1)^3}}$$

EXAMPLE B.7

A useful formula that follows from Equation B.30 is the derivative of the quotient of two functions. Show that

$$\frac{d}{dx}\left[\frac{g(x)}{h(x)}\right] = \frac{h\dfrac{dg}{dx} - g\dfrac{dh}{dx}}{h^2}$$

Solution We can write the quotient as gh^{-1} and then apply Equations B.29 and B.30:

$$\frac{d}{dx}\left(\frac{g}{h}\right) = \frac{d}{dx}(gh^{-1}) = g\frac{d}{dx}(h^{-1}) + h^{-1}\frac{d}{dx}(g)$$

$$= -gh^{-2}\frac{dh}{dx} + h^{-1}\frac{dg}{dx}$$

$$= \frac{h\dfrac{dg}{dx} - g\dfrac{dh}{dx}}{h^2}$$

B.7   Integral Calculus

We think of integration as the inverse of differentiation. As an example, consider the expression

$$f(x) = \frac{dy}{dx} = 3ax^2 + b \tag{B.34}$$

which was the result of differentiating the function

$$y(x) = ax^3 + bx + c$$

in Example B.4. We can write Equation B.34 as $dy = f(x)\,dx = (3ax^2 + b)\,dx$ and obtain $y(x)$ by "summing" over all values of x. Mathematically, we write this inverse operation as

$$y(x) = \int f(x)\,dx$$

For the function $f(x)$ given by Equation B.34, we have

$$y(x) = \int (3ax^2 + b)\,dx = ax^3 + bx + c$$

where c is a constant of the integration. This type of integral is called an *indefinite integral* because its value depends on the choice of c.

A general **indefinite integral** $I(x)$ is defined as

$$I(x) = \int f(x)\,dx \tag{B.35}$$

where $f(x)$ is called the *integrand* and $f(x) = dI(x)/dx$.

For a *general continuous* function $f(x)$, the integral can be described as the area under the curve bounded by $f(x)$ and the x axis, between two specified values of x, say, x_1 and x_2, as in Figure B.15.

The area of the blue element in Figure B.15 is approximately $f(x_i)\,\Delta x_i$. If we sum all these area elements between x_1 and x_2 and take the limit of this sum as $\Delta x_i \rightarrow 0$, we obtain the *true* area under the curve bounded by $f(x)$ and the x axis, between the limits x_1 and x_2:

$$\text{Area} = \lim_{\Delta x_i \rightarrow 0} \sum_i f(x_i)\Delta x_i = \int_{x_1}^{x_2} f(x)\,dx \tag{B.36}$$

Integrals of the type defined by Equation B.36 are called **definite integrals.**

Figure B.15   The definite integral of a function is the area under the curve of the function between the limits x_1 and x_2.

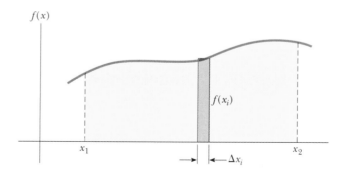

One common integral that arises in practical situations has the form

$$\int x^n \, dx = \frac{x^{n+1}}{n+1} + c \quad (n \neq -1) \tag{B.37}$$

This result is obvious, being that differentiation of the right-hand side with respect to x gives $f(x) = x^n$ directly. If the limits of the integration are known, this integral becomes a *definite integral* and is written

$$\int_{x_1}^{x_2} x^n \, dx = \frac{x^{n+1}}{n+1}\Bigg|_{x_1}^{x_2} = \frac{x_2^{n+1} - x_1^{n+1}}{n+1} \quad (n \neq -1) \tag{B.38}$$

EXAMPLES

1. $\displaystyle\int_0^a x^2 \, dx = \frac{x^3}{3}\Bigg]_0^a = \frac{a^3}{3}$

2. $\displaystyle\int_0^b x^{3/2} \, dx = \frac{x^{5/2}}{5/2}\Bigg]_0^b = \frac{2}{5}b^{5/2}$

3. $\displaystyle\int_3^5 x \, dx = \frac{x^2}{2}\Bigg]_3^5 = \frac{5^2 - 3^2}{2} = 8$

Partial Integration

Sometimes it is useful to apply the method of *partial integration* (also called "integrating by parts") to evaluate certain integrals. This method uses the property

$$\int u \, dv = uv - \int v \, du \tag{B.39}$$

where u and v are *carefully* chosen so as to reduce a complex integral to a simpler one. In many cases, several reductions have to be made. Consider the function

$$I(x) = \int x^2 e^x \, dx$$

which can be evaluated by integrating by parts twice. First, if we choose $u = x^2$, $v = e^x$, we obtain

$$\int x^2 e^x \, dx = \int x^2 \, d(e^x) = x^2 e^x - 2\int e^x x \, dx + c_1$$

Now, in the second term, choose $u = x$, $v = e^x$, which gives

$$\int x^2 e^x \, dx = x^2 e^x - 2x e^x + 2\int e^x \, dx + c_1$$

or

$$\int x^2 e^x \, dx = x^2 e^x - 2xe^x + 2e^x + c_2$$

TABLE B.5

Some Indefinite Integrals (An arbitrary constant should be added to each of these integrals.)

$$\int x^n \, dx = \frac{x^{n+1}}{n+1} \quad \text{(provided } n \neq 1\text{)}$$

$$\int \ln ax \, dx = (x \ln ax) - x$$

$$\int \frac{dx}{x} = \int x^{-1} \, dx = \ln x$$

$$\int xe^{ax} \, dx = \frac{e^{ax}}{a^2} \, (ax - 1)$$

$$\int \frac{dx}{a + bx} = \frac{1}{b} \ln (a + bx)$$

$$\int \frac{dx}{a + be^{cx}} = \frac{x}{a} - \frac{1}{ac} \ln (a + be^{cx})$$

$$\int \frac{x \, dx}{a + bx} = \frac{x}{b} - \frac{a}{b^2} \ln (a + bx)$$

$$\int \sin ax \, dx = -\frac{1}{a} \cos ax$$

$$\int \frac{dx}{x(x + a)} = -\frac{1}{a} \ln \frac{x + a}{x}$$

$$\int \cos ax \, dx = \frac{1}{a} \sin ax$$

$$\int \frac{dx}{(a + bx)^2} = -\frac{1}{b(a + bx)}$$

$$\int \tan ax \, dx = -\frac{1}{a} \ln (\cos ax) = \frac{1}{a} \ln (\sec ax)$$

$$\int \frac{dx}{a^2 + x^2} = \frac{1}{a} \tan^{-1} \frac{x}{a}$$

$$\int \cot ax \, dx = \frac{1}{a} \ln (\sin ax)$$

$$\int \frac{dx}{a^2 - x^2} = \frac{1}{2a} \ln \frac{a + x}{a - x} \quad (a^2 - x^2 > 0)$$

$$\int \sec ax \, dx = \frac{1}{a} \ln (\sec ax + \tan ax) = \frac{1}{a} \ln \left[\tan \left(\frac{ax}{2} + \frac{\pi}{4} \right) \right]$$

$$\int \frac{dx}{x^2 - a^2} = \frac{1}{2a} \ln \frac{x - a}{x + a} \quad (x^2 - a^2 > 0)$$

$$\int \csc ax \, dx = \frac{1}{a} \ln (\csc ax - \cot ax) = \frac{1}{a} \ln \left(\tan \frac{ax}{2} \right).$$

$$\int \frac{x \, dx}{a^2 \pm x^2} = \pm \frac{1}{2} \ln (a^2 \pm x^2)$$

$$\int \sin^2 ax \, dx = \frac{x}{2} + \frac{\sin 2ax}{4a}$$

$$\int \frac{dx}{\sqrt{a^2 - x^2}} = \sin^{-1} \frac{x}{a} = -\cos^{-1} \frac{x}{a} \quad (a^2 - x^2 > 0)$$

$$\int \cos^2 ax \, dx = \frac{x}{2} + \frac{\sin 2ax}{4a}$$

$$\int \frac{dx}{\sqrt{x^2 + a^2}} = \ln (x + \sqrt{x^2 \pm a^2})$$

$$\int \frac{dx}{\sin^2 ax} = -\frac{1}{a} \cot ax$$

$$\int \frac{x \, dx}{\sqrt{a^2 - x^2}} = -\sqrt{a^2 - x^2}$$

$$\int \frac{dx}{\cos^2 ax} = \frac{1}{a} \tan ax$$

$$\int \frac{x \, dx}{\sqrt{x^2 \pm a^2}} = \sqrt{x^2 \pm a^2}$$

$$\int \tan^2 ax \, dx = \frac{1}{a} (\tan ax) - x$$

$$\int \sqrt{a^2 - x^2} \, dx = \frac{1}{2} \left(x\sqrt{a^2 - x^2} + a^2 \sin^{-1} \frac{x}{a} \right)$$

$$\int \cot^2 ax \, dx = -\frac{1}{a} (\cot ax) - x$$

$$\int x\sqrt{a^2 - x^2} \, dx = -\frac{1}{3}(a^2 - x^2)^{3/2}$$

$$\int \sin^{-1} ax \, dx = x(\sin^{-1} ax) + \frac{\sqrt{1 - a^2 x^2}}{a}$$

$$\int \sqrt{x^2 \pm a^2} \, dx = \frac{1}{2} \left[x\sqrt{x^2 \pm a^2} \pm a^2 \ln (x + \sqrt{x^2 \pm a^2}) \right]$$

$$\int \cos^{-1} ax \, dx = x(\cos^{-1} ax) - \frac{\sqrt{1 - a^2 x^2}}{a}$$

$$\int x(\sqrt{x^2 \pm a^2}) \, dx = \frac{1}{3}(x^2 \pm a^2)^{3/2}$$

$$\int \frac{dx}{(x^2 + a^2)^{3/2}} = \frac{x}{a^2\sqrt{x^2 + a^2}}$$

$$\int e^{ax} \, dx = \frac{1}{a} e^{ax}$$

$$\int \frac{x \, dx}{(x^2 + a^2)^{3/2}} = -\frac{1}{\sqrt{x^2 + a^2}}$$

TABLE B.6

**Gauss's Probability Integral
and Other Definite Integrals**

$$\int_0^\infty x^n e^{-ax}\,dx = \frac{n!}{a^{n+1}}$$

$$I_0 = \int_0^\infty e^{-ax^2}\,dx = \frac{1}{2}\sqrt{\frac{\pi}{a}} \qquad \text{(Gauss's probability integral)}$$

$$I_1 = \int_0^\infty x e^{-ax^2}\,dx = \frac{1}{2a}$$

$$I_2 = \int_0^\infty x^2 e^{-ax^2}\,dx = -\frac{dI_0}{da} = \frac{1}{4}\sqrt{\frac{\pi}{a^3}}$$

$$I_3 = \int_0^\infty x^3 e^{-ax^2}\,dx = -\frac{dI_1}{da} = \frac{1}{2a^2}$$

$$I_4 = \int_0^\infty x^4 e^{-ax^2}\,dx = \frac{d^2 I_0}{da^2} = \frac{3}{8}\sqrt{\frac{\pi}{a^5}}$$

$$I_5 = \int_0^\infty x^5 e^{-ax^2}\,dx = \frac{d^2 I_1}{da^2} = \frac{1}{a^3}$$

$$\vdots$$

$$I_{2n} = (-1)^n \frac{d^n}{da^n} I_0$$

$$I_{2n+1} = (-1)^n \frac{d^n}{da^n} I_1$$

The Perfect Differential

Another useful method to remember is that of the *perfect differential*, in which we look for a change of variable such that the differential of the function is the differential of the independent variable appearing in the integrand. For example, consider the integral

$$I(x) = \int \cos^2 x \, \sin x \, dx$$

This integral becomes easy to evaluate if we rewrite the differential as $d(\cos x) = -\sin x \, dx$. The integral then becomes

$$\int \cos^2 x \, \sin x \, dx = -\int \cos^2 x \, d(\cos x)$$

If we now change variables, letting $y = \cos x$, we obtain

$$\int \cos^2 x \, \sin x \, dx = -\int y^2 \, dy = -\frac{y^3}{3} + c = -\frac{\cos^3 x}{3} + c$$

Table B.5 (page A-18) lists some useful indefinite integrals. Table B.6 gives Gauss's probability integral and other definite integrals. A more complete list can be found in various handbooks, such as *The Handbook of Chemistry and Physics* (Boca Raton, FL: CRC Press, published annually).

B.8   Propagation of Uncertainty

In laboratory experiments, a common activity is to take measurements that act as raw data. These measurements are of several types—length, time interval, temperature, voltage, and so on—and are taken by a variety of instruments. Regardless of the measurement and the quality of the instrumentation, **there is always uncertainty associated with a physical measurement.** This uncertainty is a combination of that associated with the instrument and that related to the system being measured. An example of the former is the inability to exactly determine the position of a length measurement between the lines on a meterstick. An example of uncertainty related to the system being measured is the variation of temperature within a sample of water so that a single temperature for the sample is difficult to determine.

Uncertainties can be expressed in two ways. **Absolute uncertainty** refers to an uncertainty expressed in the same units as the measurement. Therefore, the length of a computer disk label might be expressed as (5.5 ± 0.1) cm. The uncertainty of ± 0.1 cm by itself is not descriptive enough for some purposes, however. This uncertainty is large if the measurement is 1.0 cm, but it is small if the measurement is 100 m. To give a more descriptive account of the uncertainty, **fractional uncertainty** or **percent uncertainty** is used. In this type of description, the uncertainty is divided by the actual measurement. Therefore, the length of the computer disk label could be expressed as

$$\ell = 5.5 \text{ cm } \pm \frac{0.1 \text{ cm}}{5.5 \text{ cm}} = 5.5 \text{ cm } \pm 0.018 \quad \text{(fractional uncertainty)}$$

or as

$$\ell = 5.5 \text{ cm } \pm 1.8\% \quad \text{(percent uncertainty)}$$

When combining measurements in a calculation, the percent uncertainty in the final result is generally larger than the uncertainty in the individual measurements. This is called **propagation of uncertainty** and is one of the challenges of experimental physics.

Some simple rules can provide a reasonable estimate of the uncertainty in a calculated result:

Multiplication and division: When measurements with uncertainties are multiplied or divided, add the *percent uncertainties* to obtain the percent uncertainty in the result.

Example: The Area of a Rectangular Plate

$$A = \ell w = (5.5 \text{ cm } \pm 1.8\%) \times (6.4 \text{ cm } \pm 1.6\%) = 35 \text{ cm}^2 \pm 3.4\%$$

$$= (35 \pm 1) \text{ cm}^2$$

Addition and subtraction: When measurements with uncertainties are added or subtracted, add the *absolute uncertainties* to obtain the absolute uncertainty in the result.

Example: A Change in Temperature

$$\Delta T = T_2 - T_1 = (99.2 \pm 1.5)°\text{C} - (27.6 \pm 1.5)°\text{C} = (71.6 \pm 3.0)°\text{C}$$

$$= 71.6°\text{C} \pm 4.2\%$$

Powers: If a measurement is taken to a power, the percent uncertainty is multiplied by that power to obtain the percent uncertainty in the result.

Example: The Volume of a Sphere

$$V = \tfrac{4}{3}\pi r^3 = \tfrac{4}{3}\pi (6.20 \text{ cm } \pm 2.0\%)^3 = 998 \text{ cm}^3 \pm 6.0\%$$

$$= (998 \pm 60) \text{ cm}^3$$

For complicated calculations, many uncertainties are added together, which can cause the uncertainty in the final result to be undesirably large. Experiments should be designed such that calculations are as simple as possible.

Notice that uncertainties in a calculation always add. As a result, an experiment involving a subtraction should be avoided if possible, especially if the measurements being subtracted are close together. The result of such a calculation is a small difference in the measurements and uncertainties that add together. It is possible that the uncertainty in the result could be larger than the result itself!

Group I	Group II					Transition elements			

H 1
1.007 9
$1s$

	Symbol — **Ca** 20 — Atomic number
	Atomic mass[†] — 40.078
	$4s^2$ — Electron configuration

Li 3	**Be** 4
6.941	9.0122
$2s^1$	$2s^2$

Na 11	**Mg** 12
22.990	24.305
$3s^1$	$3s^2$

K 19	**Ca** 20	**Sc** 21	**Ti** 22	**V** 23	**Cr** 24	**Mn** 25	**Fe** 26	**Co** 27
39.098	40.078	44.956	47.867	50.942	51.996	54.938	55.845	58.933
$4s^1$	$4s^2$	$3d^14s^2$	$3d^24s^2$	$3d^34s^2$	$3d^54s^1$	$3d^54s^2$	$3d^64s^2$	$3d^74s^2$

Rb 37	**Sr** 38	**Y** 39	**Zr** 40	**Nb** 41	**Mo** 42	**Tc** 43	**Ru** 44	**Rh** 45
85.468	87.62	88.906	91.224	92.906	95.94	(98)	101.07	102.91
$5s^1$	$5s^2$	$4d^15s^2$	$4d^25s^2$	$4d^45s^1$	$4d^55s^1$	$4d^55s^2$	$4d^75s^1$	$4d^85s^1$

Cs 55	**Ba** 56	57–71*	**Hf** 72	**Ta** 73	**W** 74	**Re** 75	**Os** 76	**Ir** 77
132.91	137.33		178.49	180.95	183.84	186.21	190.23	192.2
$6s^1$	$6s^2$		$5d^26s^2$	$5d^36s^2$	$5d^46s^2$	$5d^56s^2$	$5d^66s^2$	$5d^76s^2$

Fr 87	**Ra** 88	89–103**	**Rf** 104	**Db** 105	**Sg** 106	**Bh** 107	**Hs** 108	**Mt** 109
(223)	(226)		(261)	(262)	(266)	(264)	(277)	(268)
$7s^1$	$7s^2$		$6d^27s^2$	$6d^37s^2$				

*Lanthanide series

La 57	**Ce** 58	**Pr** 59	**Nd** 60	**Pm** 61	**Sm** 62
138.91	140.12	140.91	144.24	(145)	150.36
$5d^16s^2$	$5d^14f^16s^2$	$4f^36s^2$	$4f^46s^2$	$4f^56s^2$	$4f^66s^2$

**Actinide series

Ac 89	**Th** 90	**Pa** 91	**U** 92	**Np** 93	**Pu** 94
(227)	232.04	231.04	238.03	(237)	(244)
$6d^17s^2$	$6d^27s^2$	$5f^26d^17s^2$	$5f^36d^17s^2$	$5f^46d^17s^2$	$5f^66d^07s^2$

Note: Atomic mass values given are averaged over isotopes in the percentages in which they exist in nature.
[†] For an unstable element, mass number of the most stable known isotope is given in parentheses.
[††] Elements 112 and 114 have not yet been named.
[†††] For a description of the atomic data, visit *physics.nist.gov/PhysRefData/Elements/per_text.html*

	Group III	Group IV	Group V	Group VI	Group VII	Group 0
					H 1 1.007 9 $1s^1$	**He** 2 4.002 6 $1s^2$
	B 5 10.811 $2p^1$	**C** 6 12.011 $2p^2$	**N** 7 14.007 $2p^3$	**O** 8 15.999 $2p^4$	**F** 9 18.998 $2p^5$	**Ne** 10 20.180 $2p^6$
	Al 13 26.982 $3p^1$	**Si** 14 28.086 $3p^2$	**P** 15 30.974 $3p^3$	**S** 16 32.066 $3p^4$	**Cl** 17 35.453 $3p^5$	**Ar** 18 39.948 $3p^6$

Ni 28 58.693 $3d^84s^2$	**Cu** 29 63.546 $3d^{10}4s^1$	**Zn** 30 65.41 $3d^{10}4s^2$	**Ga** 31 69.723 $4p^1$	**Ge** 32 72.64 $4p^2$	**As** 33 74.922 $4p^3$	**Se** 34 78.96 $4p^4$	**Br** 35 79.904 $4p^5$	**Kr** 36 83.80 $4p^6$
Pd 46 106.42 $4d^{10}$	**Ag** 47 107.87 $4d^{10}5s^1$	**Cd** 48 112.41 $4d^{10}5s^2$	**In** 49 114.82 $5p^1$	**Sn** 50 118.71 $5p^2$	**Sb** 51 121.76 $5p^3$	**Te** 52 127.60 $5p^4$	**I** 53 126.90 $5p^5$	**Xe** 54 131.29 $5p^6$
Pt 78 195.08 $5d^96s^1$	**Au** 79 196.97 $5d^{10}6s^1$	**Hg** 80 200.59 $5d^{10}6s^2$	**Tl** 81 204.38 $6p^1$	**Pb** 82 207.2 $6p^2$	**Bi** 83 208.98 $6p^3$	**Po** 84 (209) $6p^4$	**At** 85 (210) $6p^5$	**Rn** 86 (222) $6p^6$
Ds 110 (271)	**Rg** 111 (272)	**112**†† (285)		**114**†† (289)				

Eu 63 151.96 $4f^76s^2$	**Gd** 64 157.25 $4f^75d^16s^2$	**Tb** 65 158.93 $4f^85d^16s^2$	**Dy** 66 162.50 $4f^{10}6s^2$	**Ho** 67 164.93 $4f^{11}6s^2$	**Er** 68 167.26 $4f^{12}6s^2$	**Tm** 69 168.93 $4f^{13}6s^2$	**Yb** 70 173.04 $4f^{14}6s^2$	**Lu** 71 174.97 $4f^{14}5d^16s^2$
Am 95 (243) $5f^77s^2$	**Cm** 96 (247) $5f^76d^17s^2$	**Bk** 97 (247) $5f^86d^17s^2$	**Cf** 98 (251) $5f^{10}7s^2$	**Es** 99 (252) $5f^{11}7s^2$	**Fm** 100 (257) $5f^{12}7s^2$	**Md** 101 (258) $5f^{13}7s^2$	**No** 102 (259) $5f^{14}7s^2$	**Lr** 103 (262) $6d^15f^{14}7s^2$

TABLE D.1

SI Units

Base Quantity	SI Base Unit	
	Name	Symbol
Length	meter	m
Mass	kilogram	kg
Time	second	s
Electric current	ampere	A
Temperature	kelvin	K
Amount of substance	mole	mol
Luminous intensity	candela	cd

TABLE D.2

Some Derived SI Units

Quantity	Name	Symbol	Expression in Terms of Base Units	Expression in Terms of Other SI Units
Plane angle	radian	rad	m/m	
Frequency	hertz	Hz	s^{-1}	
Force	newton	N	$kg \cdot m/s^2$	J/m
Pressure	pascal	Pa	$kg/m \cdot s^2$	N/m^2
Energy	joule	J	$kg \cdot m^2/s^2$	$N \cdot m$
Power	watt	W	$kg \cdot m^2/s^3$	J/s
Electric charge	coulomb	C	$A \cdot s$	
Electric potential	volt	V	$kg \cdot m^2/A \cdot s^3$	W/A
Capacitance	farad	F	$A^2 \cdot s^4/kg \cdot m^2$	C/V
Electric resistance	ohm	Ω	$kg \cdot m^2/A^2 \cdot s^3$	V/A
Magnetic flux	weber	Wb	$kg \cdot m^2/A \cdot s^2$	$V \cdot s$
Magnetic field	tesla	T	$kg/A \cdot s^2$	
Inductance	henry	H	$kg \cdot m^2/A^2 \cdot s^2$	$T \cdot m^2/A$

CHAPTER 23

1. (a) $+160$ zC, 1.01 u   (b) $+160$ zC, 23.0 u   (c) -160 zC, 35.5 u   (d) $+320$ zC, 40.1 u   (e) -480 zC, 14.0 u   (f) $+640$ zC, 14.0 u   (g) $+1.12$ aC, 14.0 u   (h) -160 zC, 18.0 u

3. The force is $\sim 10^{26}$ N.

5. (a) 1.59 nN away from the other   (b) 1.24×10^{36} times larger   (c) 8.61×10^{-11} C/kg

7. 0.872 N at $330°$

9. (a) 2.16×10^{-5} N toward the other   (b) 8.99×10^{-7} N away from the other

11. (a) 82.2 nN toward the other particle   (b) 2.19 Mm/s

13. (a) 55.8 pN/C down   (b) 102 nN/C up

15. The field at the origin can be to the right if the unknown charge is $-9Q$, or the field can be to the left if and only if the unknown charge is $+27Q$.

17. (a) $5.91k_eq/a^2$ at $58.8°$ (b) $5.91k_eq^2/a^2$ at $58.8°$

19. (a) $k_eQx\hat{\mathbf{i}}/(R^2+x^2)^{3/2}$ (b) As long as the charge is symmetrically placed, the number of charges does not matter. A continuous ring corresponds to n becoming larger without limit.

21. 1.59×10^6 N/C toward the rod

23. (a) $6.64\hat{\mathbf{i}}$ MN/C   (b) $24.1\hat{\mathbf{i}}$ MN/C   (c) $6.40\hat{\mathbf{i}}$ MN/C   (d) $0.664\hat{\mathbf{i}}$ MN/C, taking the axis of the ring as the x axis

25. (a) 93.6 MN/C; the near-field approximation is 104 MN/C, about 11% high.   (b) 0.516 MN/C; the charged-particle approximation is 0.519 MN/C, about 0.6% high.

27. $-21.6\hat{\mathbf{i}}$ MN/C

31. (a) 86.4 pC for each   (b) 324 pC, 459 pC, 459 pC, 432 pC   (c) 57.6 pC, 106 pC, 154 pC, 96.0 pC

33.

35. (a)

The field is zero at the center of the triangle.   (b) $1.73\,k_eq\hat{\mathbf{j}}/a^2$

37. (a) 61.3 Gm/s²   (b) 19.5 μs   (c) 11.7 m   (d) 1.20 fJ

39. K/ed in the direction of motion

41. (a) 111 ns   (b) 5.68 mm   (c) $(450\hat{\mathbf{i}} + 102\hat{\mathbf{j}})$ km/s

43. (a) 21.8 μm   (b) 2.43 cm

45. (a) 10.9 nC   (b) 5.44 mN

47. 40.9 N at $263°$

49. $Q = 2L\sqrt{\dfrac{k(L-L_i)}{k_e}}$

53. $-707\hat{\mathbf{j}}$ mN

55. (a) $\theta_1 = \theta_2$

57. (a) 0.307 s   (b) Yes. Ignoring gravity makes a difference of 2.28%.

59. (a) $\vec{\mathbf{F}} = 1.90(k_eq^2/s^2)(\hat{\mathbf{i}} + \hat{\mathbf{j}} + \hat{\mathbf{k}})$ (b) $\vec{\mathbf{F}} = 3.29(k_eq^2/s^2)$ in the direction away from the diagonally opposite vertex

65. $\dfrac{k_e\lambda_0}{2x_0}(-\hat{\mathbf{i}})$

CHAPTER 24

1. 4.14 MN/C

3. (a) aA (b) bA (c) 0

5. 1.87 kN·m²/C

7. (a) -6.89 MN·m²/C   (b) The number of lines entering exceeds the number leaving by 2.91 times or more.

9. $-Q/\epsilon_0$ for S_1; 0 for S_2; $-2Q/\epsilon_0$ for S_3; 0 for S_4

11. (a) $+Q/2\epsilon_0$ (b) $-Q/2\epsilon_0$

13. -18.8 kN·m²/C

15. 0 if $R \leq d$; $(2\lambda/\epsilon_0)\sqrt{R^2-d^2}$ if $R > d$

17. (a) 3.20 MN·m²/C   (b) 19.2 MN·m²/C   (c) The answer to part (a) could change, but the answer to part (b) would stay the same.

19. 2.33×10^{21} N/C

21. 508 kN/C up

23. -2.48 μC/m²

25. 5.94×10^5 m/s

27. $\vec{\mathbf{E}} = \rho r/2\epsilon_0$ away from the axis

29. (a) 0   (b) 7.19 MN/C away from the center

31. (a) 51.4 kN/C outward   (b) 646 N·m²/C

33. (a) 0   (b) 5 400 N/C outward   (c) 540 N/C outward

35. (a) $+708$ nC/m² and -708 nC/m²   (b) $+177$ nC and -177 nC

37. 2.00 N

39. (a) $-\lambda, +3\lambda$ (b) $3\lambda/2\pi\epsilon_0 r$ radially outward

41. (a) 80.0 nC/m² on each face   (b) $9.04\hat{\mathbf{k}}$ kN/C   (c) $-9.04\hat{\mathbf{k}}$ kN/C

43. (b) $Q/2\epsilon_0$ (c) Q/ϵ_0

45. (a) The charge on the exterior surface is -55.7 nC distributed uniformly.   (b) The charge on the interior surface is $+55.7$ nC. It might have any distribution.   (c) The charge within the shell is -55.7 nC. It might have any distribution.

47. (a) $\rho r/3\epsilon_0$, $Q/4\pi\epsilon_0 r^2$, 0, $Q/4\pi\epsilon_0 r^2$, all radially outward   (b) $-Q/4\pi b^2$ and $+Q/4\pi c^2$

49. $\theta = \tan^{-1}[qQ/(2\pi\epsilon_0 dmv^2)]$

51. (a) σ/ϵ_0 away from both plates   (b) 0   (c) σ/ϵ_0 away from both plates

53. $\sigma/2\epsilon_0$ radially outward

57. $\vec{\mathbf{E}} = a/2\epsilon_0$ radially outward

61. (b) $\vec{\mathbf{g}} = GM_E r/R_E^3$ radially inward

63. (a) -4.00 nC   (b) $+9.56$ nC   (c) $+4.00$ nC and $+5.56$ nC

65. (a) If the volume charge density is nonzero, the field cannot be uniform in magnitude.   (b) The field must be uniform in magnitude along any line in the direction of the field. The field magnitude can vary between points in a plane perpendicular to the field lines.

CHAPTER 25

1. (a) 152 km/s   (b) 6.49 Mm/s

3. 1.67 MN/C

5. 38.9 V; the origin

7. (a) $2QE/k$ (b) QE/k (c) $2\pi\sqrt{m/k}$ (d) $2(QE - \mu_k mg)/k$

9. (a) 0.400 m/s   (b) It is the same. Each bit of the rod feels a force of the same size as before.

11. (a) 1.44×10^{-7} V   (b) -7.19×10^{-8} V   (c) -1.44×10^{-7} V, $+7.19 \times 10^{-8}$ V

13. (a) 6.00 m   (b) -2.00 μC

15. -11.0 MV

17. 8.95 J

21. (a) no point at a finite distance from the particles
(b) $2k_eq/a$

23. (a) 10.8 m/s and 1.55 m/s (b) Greater. The conducting spheres will polarize each other, with most of the positive charge of one and of the negative charge of the other on their inside faces. Immediately before they collide, their centers of charge will be closer than their geometric centers, so they will have less electric potential energy and more kinetic energy.

25. $5k_eq^2/9d$

27. $\left[(1 + \sqrt{\tfrac{1}{8}})\dfrac{k_eq^2}{mL}\right]^{1/2}$

29. (a) 10.0 V, -11.0 V, -32.0 V (b) 7.00 N/C in the $+x$ direction

31. $\vec{E} = (-5 + 6xy)\hat{\imath} + (3x^2 - 2z^2)\hat{\jmath} - 4yz\hat{k}$; 7.07 N/C

33. $E_y = \dfrac{k_eQ}{y\sqrt{\ell^2 + y^2}}$

35. (a) C/m^2 (b) $k_e\alpha[L - d\ln(1 + L/d)]$

37. -1.51 MV

39. (a) 0, 1.67 MV (b) 5.84 MN/C away, 1.17 MV (c) 11.9 MN/C away, 1.67 MV

41. (a) 248 nC/m^2 (b) 496 nC/m^2

43. (a) 450 kV (b) 7.51 μC

45. (a) 1.42 mm (b) 9.20 kV/m

47. 253 MeV

49. (a) -27.2 eV (b) -6.80 eV (c) 0

51. (a) Yes. The inverse proportionality of potential to radius is sufficient to show that $200R = 150(R + 10$ cm$)$, so $R = 30.0$ cm. Then $Q = 6.67$ nC. (b) Almost but not quite. Two possibilities exist: $R = 29.1$ cm with $Q = 6.79$ nC and $R = 3.44$ cm with $Q = 804$ pC.

53. 4.00 nC at $(-1.00$ m, 0$)$ and -5.01 nC at $(0, 2.00$ m$)$

55. $k_eQ^2/2R$

57. $V_2 - V_1 = (-\lambda/2\pi\epsilon_0)\ln(r_2/r_1)$

61. (b) $E_r = 2k_e\,p\cos\theta/r^3$; $E_\theta = k_e\,p\sin\theta/r^3$; yes; no
(c) $V = k_e py(x^2 + y^2)^{-3/2}$;
$\vec{E} = 3k_e pxy(x^2 + y^2)^{-5/2}\hat{\imath} + k_e p(2y^2 - x^2)(x^2 + y^2)^{-5/2}\hat{\jmath}$

63. $V = \pi k_e C\left[R\sqrt{x^2 + R^2} + x^2\ln\left(\dfrac{x}{R + \sqrt{x^2 + R^2}}\right)\right]$

65. (a) 488 V (b) 78.1 aJ (c) 306 km/s (d) 390 Gm/s^2 toward the negative plate (e) 651 aN toward the negative plate (f) 4.07 kN/C

67. Outside the sphere, $E_x = 3E_0a^3xz(x^2 + y^2 + z^2)^{-5/2}$, $E_y = 3E_0a^3yz(x^2 + y^2 + z^2)^{-5/2}$, and $E_z = E_0 + E_0a^3(2z^2 - x^2 - y^2)(x^2 + y^2 + z^2)^{-5/2}$. Inside the sphere, $E_x = E_y = E_z = 0$.

CHAPTER 26

1. (a) 48.0 μC (b) 6.00 μC

3. (a) 1.33 μC/m^2 (b) 13.3 pF

5. (a) 11.1 kV/m toward the negative plate (b) 98.3 nC/m^2 (c) 3.74 pF (d) 74.7 pC

7. 4.42 μm

9. (a) 2.68 nF (b) 3.02 kV

11. (a) 15.6 pF (b) 256 kV

13. (a) 3.53 μF (b) 6.35 V and 2.65 V (c) 31.8 μC on each

15. 6.00 pF and 3.00 pF

17. (a) 5.96 μF (b) 89.5 μC on 20 μF, 63.2 μC on 6 μF, 26.3 μC on 15 μF and on 3 μF

19. 120 μC; 80.0 μC and 40.0 μC

21. ten

23. 6.04 μF

25. 12.9 μF

27. (a) 216 μJ (b) 54.0 μJ

31. (a) 1.50 μC (b) 1.83 kV

35. 9.79 kg

37. (a) 81.3 pF (b) 2.40 kV

39. 1.04 m

41. 22.5 V

43. (b) -8.78×10^6 N/C·m; $-5.53 \times 10^{-2}\hat{\imath}$ N

45. 19.0 kV

47. (a) 11.2 pF (b) 134 pC (c) 16.7 pF (d) 66.9 pC

49. (a) 40.0 μJ (b) 500 V

51. 0.188 m^2

55. Gasoline has 194 times the specific energy content of the battery and 727 000 times that of the capacitor.

57. (a) $Q_0^2 d(\ell - x)/(2\ell^3\epsilon_0)$ (b) $Q_0^2 d/(2\ell^3\epsilon_0)$ to the right (c) $Q_0^2/(2\ell^4\epsilon_0)$ (d) $Q_0^2/(2\ell^4\epsilon_0)$; they are precisely the same.

59. 4.29 μF

61. (a) The additional energy comes from work done by the electric field in the wires as it forces more charge onto the already-charged plates. (b) The charge increases according to $Q/Q_0 = \kappa$.

63. 750 μC on C_1 and 250 μC on C_2

65. $\tfrac{4}{3}C$

CHAPTER 27

1. 7.50×10^{15} electrons

3. (a) $0.632I_0\tau$ (b) $0.999\,95I_0\tau$ (c) $I_0\tau$

5. (a) 17.0 A (b) 85.0 kA/m^2

7. (a) 2.55 A/m^2 (b) 5.31×10^{10} m^{-3} (c) 1.20×10^{10} s

9. (a) 221 nm (b) No. The deuterons are so far apart that one does not produce a significant potential at the location of the next.

11. 6.43 A

13. (a) 1.82 m (b) 280 μm

15. $6.00 \times 10^{-15}/\Omega \cdot$m

17. 0.180 V/m

19. (a) 31.5 n$\Omega \cdot$m (b) 6.35 MA/m^2 (c) 49.9 mA (d) 659 μm/s (e) 0.400 V

21. 0.125

23. 5.00 A, 24.0 Ω

25. 5.49 Ω

27. 36.1%

29. (a) 3.17 m (b) 340 W

31. (a) 0.660 kWh (b) $0.039 6

33. $0.232

35. $0.269/day

37. (a) 184 W (b) 461°C

39. ~ $1

41. Any diameter d and length ℓ related by $d^2 =$ $(4.77 \times 10^{-8} \text{ m})\ell$, such as length 0.900 m and diameter 0.207 mm. Yes.

45. Experimental resistivity = 1.47 $\mu\Omega \cdot$ m $\pm 4\%$, in agreement with 1.50 $\mu\Omega \cdot$ m

47. (a) 8.00 V/m in the x direction   (b) 0.637 Ω (c) 6.28 A   (d) 200 MA/m^2 in the x direction

49. (a) 667 A   (b) 50.0 km

51.

Material	$\alpha' = \alpha/(1 - 20\alpha)$
Silver	$4.1 \times 10^{-3}/°\text{C}$
Copper	$4.2 \times 10^{-3}/°\text{C}$
Gold	$3.6 \times 10^{-3}/°\text{C}$
Aluminum	$4.2 \times 10^{-3}/°\text{C}$
Tungsten	$4.9 \times 10^{-3}/°\text{C}$
Iron	$5.6 \times 10^{-3}/°\text{C}$
Platinum	$4.25 \times 10^{-3}/°\text{C}$
Lead	$4.2 \times 10^{-3}/°\text{C}$
Nichrome	$0.4 \times 10^{-3}/°\text{C}$
Carbon	$-0.5 \times 10^{-3}/°\text{C}$
Germanium	$-24 \times 10^{-3}/°\text{C}$
Silicon	$-30 \times 10^{-3}/°\text{C}$

53. It is exact. The resistance can be written $R = \rho L^2/V$ and the stretched length as $L = L_i(1 + \delta)$. Then the result follows directly.

55. (b) Charge is conducted by current in the direction of decreasing potential. Energy is conducted by heat in the direction of decreasing temperature.

59. Coat the surfaces of entry and exit with a material of much higher conductivity than the bulk material of the object. The electric potential will be essentially uniform over each of these electrodes. Current will be distributed over the whole area where each electrode is in contact with the resistive object.

61. (a) $\dfrac{\epsilon_0 \ell}{2d}(\ell + 2x + \kappa\ell - 2\kappa x)$

(b) $\dfrac{\epsilon_0 \ell v \Delta V(\kappa - 1)}{d}$ clockwise

63. 2.71 MΩ

65. 2 020°C

19. (a) -222 J and 1.88 kJ   (b) 687 J, 128 J, 25.6 J, 616 J, 205 J   (c) 1.66 kJ of chemical energy is transformed into internal energy.

21. 0.395 A and 1.50 V

23. 1.00 A up in 200 Ω, 4.00 A up in 70 Ω, 3.00 A up in 80 Ω, 8.00 A down in 20 Ω, 200 V

25. (a) 909 mA   (b) -1.82 V $= V_b - V_a$

27. (a) 5.00 s   (b) 150 μC   (c) 4.06 μA

29. (a) -61.6 mA   (b) 0.235 μC   (c) 1.96 A

31. (a) 6.00 V   (b) 8.29 μs

33. 0.302 Ω

35. 16.6 kΩ

37. (a) 12.5 A, 6.25 A, 8.33 A   (b) No. Together they would require 27.1 A.

39. (a) 1.02 A down   (b) 0.364 A down   (c) 1.38 A up   (d) 0   (e) 66.0 μC

41. 2.22 h

43. a is 4.00 V higher

45. 87.3 %

47. 6.00 Ω, 3.00 Ω

49. (a) $I_1 = \dfrac{IR_2}{R_1 + R_2}$, $I_2 = \dfrac{IR_1}{R_1 + R_2}$

51. (a) $R \leq 1\,050 \ \Omega$ (b) $R \geq 10.0 \ \Omega$

53. (a) 9.93 μC   (b) 33.7 nA   (c) 334 nW   (d) 337 nW

55. (a) 40.0 W   (b) 80.0 V, 40.0 V, 40.0 V

57. (a) 9.30 V, 2.51 Ω (b) 186 V and 3.70 A   (c) 1.09 A   (d) 143 W   (e) 0.162 Ω (f) 3.00 mW   (g) 2.21 W   (h) The power output of the emf depends on the resistance connected to it. A question about "the rest of the power" is not meaningful when it compares circuits with different currents. The net emf produces more current in the circuit where the copper wire is used. The net emf delivers more power when the copper wire is used, 687 W rather than 203 W without the wire. Nearly all this power results in extra internal energy in the internal resistance of the batteries, which rapidly rise to a high temperature. The circuit with the copper wire is unsafe because the batteries overheat. The circuit without the copper wire is unsafe because it delivers an electric shock to the experimenter.

61. (a) 0 in 3 kΩ and 333 μA in 12 kΩ and 15 kΩ (b) 50.0 μC   (c) $(278 \ \mu\text{A})e^{-t/180 \text{ ms}}$ (d) 290 ms

63. (a) $R_x = R_2 - R_1/4$ (b) $R_x = 2.75 \ \Omega$. The station is inadequately grounded.

65. (a) $2\Delta t/3$ (b) $3\Delta t$

CHAPTER 28

1. (a) 6.73 Ω (b) 1.97 Ω

3. (a) 12.4 V   (b) 9.65 V

5. (a) 17.1 Ω (b) 1.99 A for 4 Ω and 9 Ω, 1.17 A for 7 Ω, 0.818 A for 10 Ω

7. (a) 227 mA   (b) 5.68 V

9. (a) 75.0 V   (b) 25.0 W, 6.25 W, and 6.25 W; 37.5 W

11. $R_1 = 1.00$ kΩ, $R_2 = 2.00$ kΩ, $R_3 = 3.00$ kΩ

13. It decreases. Closing the switch opens a new path with resistance of only 20 Ω. $R = 14.0 \ \Omega$

15. 14.2 W to 2 Ω, 28.4 W to 4 Ω, 1.33 W to 3 Ω, 4.00 W to 1 Ω

17. 846 mA down in the 8-Ω resistor; 462 mA down in the middle branch; 1.31 A up in the right-hand branch

CHAPTER 29

1. (a) up   (b) toward you, out of the plane of the paper   (c) no deflection   (d) into the plane of the paper

3. $(-20.9\hat{\mathbf{j}})$ mT

5. 48.9° or 131°

7. 2.34 aN

9. (a) 49.6 aN south   (b) 1.29 km

11. $r_\alpha = r_d = \sqrt{2}r_p$

13. (a) 5.00 cm   (b) 8.78×10^6 m/s

15. 7.88 pT

17. 244 kV/m

19. 0.278 m

21. (a) 4.31×10^7 rad/s (b) 51.7 Mm/s

23. 70.1 mT

25. 0.245 T east

27. (a) 4.73 N (b) 5.46 N (c) 4.73 N

29. 1.07 m/s

31. $2\pi rIB \sin\theta$ up

33. 2.98 μN west

35. 9.98 N·m clockwise as seen looking down from above

37. (a) Minimum: pointing north at 48.0° below the horizontal; maximum: pointing south at 48.0° above the horizontal. (b) 1.07 μJ

39. The magnetic moment cannot go to infinity. Its maximum value is 5.37 mA·m² for a single-turn circle. Smaller by 21% and by 40% are the magnetic moments for the single-turn square and triangle. Circular coils with several turns have magnetic moments inversely proportional to the number of turns, approaching zero as the number of turns goes to infinity.

41. 43.1 μT

43. (a) The electric current experiences a magnetic force. (c) no, no, no

45. 12.5 km. It will not hit the Earth, but it will perform a hairpin turn and go back parallel to its original direction.

47. (a) -8.00×10^{-21} kg·m/s (b) 8.90°

49. (a) $(3.52\hat{\mathbf{i}} - 1.60\hat{\mathbf{j}})$ aN (b) 24.4°

51. 128 mT north at an angle of 78.7° below the horizontal

53. 0.588 T

55. 0.713 A counterclockwise as seen from above

57. 2.75 Mrad/s

59. 3.70×10^{-24} N·m

61. (a) 1.33 m/s (b) Positive ions moving toward you in magnetic field to the right feel upward magnetic force and migrate upward in the blood vessel. Negative ions moving toward you feel downward magnetic force and accumulate at the bottom of this section of vessel. Therefore, both species can participate in the generation of the emf.

63. (a) $v = qBh/m$. If its speed is slightly less than the critical value, the particle moves in a semicircle of radius h and leaves the field with velocity $-v\hat{\mathbf{j}}$. If its speed is incrementally greater, the particle moves in a quarter circle of the same radius and moves along the boundary outside the field with velocity $v\hat{\mathbf{i}}$. (b) The particle moves in a smaller semicircle of radius mv/qB, attaining final velocity $-v\hat{\mathbf{j}}$. (c) The particle moves in a circular arc of radius $r = mv/qB$, leaving the field with velocity $v\sin\theta\hat{\mathbf{i}} + v\cos\theta\hat{\mathbf{j}}$, where $\theta = \sin^{-1}(h/r)$.

65. (a) For small angular displacements, the torque on the dipole is equal to a negative constant times the displacement.

(b) $f = \dfrac{1}{2\pi}\sqrt{\dfrac{\mu B}{I}}$

(c) The equilibrium orientation of the needle shows the direction of the field. In a stronger field, the frequency is higher. The frequency is easy to measure precisely over a wide range of values. 2.04 mT.

CHAPTER 30

1. 12.5 T

3. (a) 28.3 μT into the paper (b) 24.7 μT into the paper

5. $\dfrac{\mu_0 I}{4\pi x}$ into the paper

7. (a) $2I_1$ out of the page (b) $6I_1$ into the page

9. (a) along the line ($y = -0.420$ m, $z = 0$) (b) $(-34.7\hat{\mathbf{j}})$ mN (c) $(17.3\hat{\mathbf{j}})$ kN/C

11. at A, 53.3 μT toward the bottom of the page; at B, 20.0 μT toward the bottom of the page; at C, zero.

13. (a) $4.5\dfrac{\mu_0 I}{\pi L}$

(b) Stronger. Each of the two sides meeting at the nearby vertex contributes more than twice as much to the net field at the new point.

15. $(-13.0\hat{\mathbf{j}})$ μT

17. $(-27.0\hat{\mathbf{i}})$ μN

19. parallel to the wires and 0.167 m below the upper wire

21. (a) opposite directions (b) 67.8 A (c) Smaller. A smaller gravitational force would be pulling down on the wires, therefore tending to reduce the angle.

23. 20.0 μT toward the bottom of the page

25. at a, 200 μT toward the top of the page; at b, 133 μT toward the bottom of the page

27. (a) 6.34 mN/m inward (b) Greater. The magnetic field increases toward the outside of the bundle, where more net current lies inside a particular radius. The larger field exerts a stronger force on the strand we choose to monitor.

29. (a) 0

(b) $\dfrac{\mu_0 I}{2\pi R}$ tangent to the wall in a counterclockwise sense

(c) $\dfrac{\mu_0 I^2}{(2\pi R)^2}$ inward

31. (a) $\mu_0 b r_1^2/3$ (b) $\mu_0 b R^3/3r_2$

35. 31.8 mA

37. 226 μN away from the center of the loop, 0

39. (a) 3.13 mWb (b) 0

41. (a) 7.40 μWb (b) 2.27 μWb

43. 2.02

45. (a) 8.63×10^{45} electrons (b) 4.01×10^{20} kg

47. $\dfrac{\mu_0 I}{2\pi w} \ln\left(1 + \dfrac{w}{b}\right)\hat{\mathbf{k}}$

49. $(-12.0\hat{\mathbf{k}})$ mN

51. 143 pT

57. (a) 2.46 N upward (b) The magnetic field at the center of the loop or on its axis is much weaker than the magnetic field immediately outside the wire. The wire has negligible curvature on the scale of 1 mm, so we model the lower loop as a long, straight wire to find the field it creates at the location of the upper wire. (c) 107 m/s² upward

59. (a) 274 μT (b) $(-274\hat{\mathbf{j}})$ μT (c) $(1.15\hat{\mathbf{i}})$ mN (d) $(0.384\hat{\mathbf{i}})$ m/s² (e) acceleration is constant (f) $(0.999\hat{\mathbf{i}})$ m/s

61. $\dfrac{\mu_0 I_1 I_2 L}{\pi R}$ to the right

65. $\frac{1}{3}\rho\mu_0\omega R^2$

67. (a) $\dfrac{\mu_0 I(2r^2 - a^2)}{\pi r(4r^2 - a^2)}$ to the left

(b) $\dfrac{\mu_0 I(2r^2 + a^2)}{\pi r(4r^2 + a^2)}$ toward the top of the page

CHAPTER 31

1. (a) 101 μV tending to produce clockwise current as seen from above (b) It is twice as large in magnitude and in the opposite sense.

3. 9.82 mV

5. (b) 3.79 mV   (c) 28.0 mV

7. 160 A

9. (a) 1.60 A counterclockwise   (b) 20.1 μT   (c) left

11. $-(14.2\text{ mV})\cos(120t)$

13. 283 μA upward

15. $(68.2\text{ mV})e^{-1.6t}$, tending to produce counterclockwise current

17. 272 m

19. 13.3 mA counterclockwise in the lower loop and clockwise in the upper loop

21. (a) 1.18 mV. The wingtip on the pilot's left is positive. (b) no change   (c) No. If we try to connect the wings into a circuit with the lightbulb, we run an extra insulated wire along the wing. In a uniform field, the total emf generated in the one-turn coil is zero.

23. (a) 3.00 N to the right   (b) 6.00 W

25. 24.1 V with the outer contact positive

27. 2.83 mV

29. (a) $F = N^2 B^2 w^2 v/R$ to the left   (b) 0
(c) $F = N^2 B^2 w^2 v/R$ to the left

31. 145 μA upward in the picture

33. 1.80 mN/C upward and to the left, perpendicular to $\vec{r}_1$

35. (a) 7.54 kV   (b) The plane of the loop is parallel to $\vec{\mathbf{B}}$.

37. $(28.6\text{ mV})\sin(4\pi t)$

39. (a) 110 V   (b) 8.53 W   (c) 1.22 kW

41. Both are correct. The current in the magnet creates an upward magnetic field 🖑, so the N and S poles on the solenoid core are shown correctly. On the rail in front of the brake, the upward magnetic flux increases as the coil approaches, so a current is induced here to create downward magnetic field 🖐. This current is clockwise, so the S pole on the rail is shown correctly. On the rail behind the brake, the upward magnetic flux is decreasing. The induced current in the rail will produce upward magnetic field by being counterclockwise 🖑 as the picture correctly shows.

43. (b) Larger R makes current smaller, so the loop must travel faster to maintain equality of magnetic force and weight. (c) The magnetic force is proportional to the product of field and current, while the current is itself proportional to field. If B becomes two times smaller, the speed must become four times larger to compensate.

45. $-(7.22\text{ mV})\cos(2\pi\,523t/\text{s})$

47.

(a) Doubling N doubles amplitude. (b) Doubling ω doubles the amplitude and halves the period. (c) Doubling ω and halving N leaves the amplitude the same and cuts the period in half.

49. (a) 3.50 A up in 2 Ω, and 1.40 A up in 5 Ω (b) 34.3 W
(c) 4.29 N

51. ~10^{-4} V, by reversing a 20-turn coil of diameter 3 cm in 0.1 s in a field of 10^{-3} T

53. 1.20 μC

55. (a) 0.900 A from b toward a (b) 0.108 N   (c) b (d) No. Instead of decreasing downward magnetic flux to induce clockwise current, the new loop will see increasing downward flux to cause counterclockwise current, but the current in the resistor is still from b to a.

57. (a) $C\pi a^2 K$ (b) the upper plate   (c) The changing magnetic field within the loop induces an electric field around the circumference, which pushes on charged particles in the wire.

59. (a) 36.0 V   (b) 600 mWb/s   (c) 35.9 V   (d) 4.32 N $\cdot$ m

63. 6.00 A

67. $(-87.1\text{ mV})\cos(200\pi t + \phi)$

CHAPTER 32

1. 100 V

3. $-(18.8\text{ V})\cos(377t)$

5. -0.421 A/s

7. (a) 188 μT   (b) 33.3 nT $\cdot$ m^2 (c) 0.375 mH   (d) B and Φ_B are proportional to current; L is independent of current

9. $\mathcal{E}_0/k^2 L$

11. (a) 0.139 s   (b) 0.461 s

13. (a) 2.00 ms   (b) 0.176 A   (c) 1.50 A   (d) 3.22 ms

15. (a) 0.800   (b) 0

17. (a) 6.67 A/s   (b) 0.332 A/s

19. (a) 1.00 kΩ (b) 3.00 ms

21. (a) 5.66 ms   (b) 1.22 A   (c) 58.1 ms

23. 2.44 μJ

25. 44.2 nJ/m^3 for the $\vec{\mathbf{E}}$ field and 995 μJ/m^3 for the $\vec{\mathbf{B}}$ field

27. (a) 66.0 W   (b) 45.0 W   (c) 21.0 W   (d) At all instants after the connection is made, the battery power is equal to the sum of the power delivered to the resistor and the power delivered to the magnetic field. Immediately after $t = 0$, the resistor power is nearly zero and nearly all the battery power is going into the magnetic field. Long after the connection is made, the magnetic field is absorbing no more power and the battery power is going into the resistor.

29. $\dfrac{2\pi B_0^2 R^3}{\mu_0} = 2.70 \times 10^{18}\,\text{J}$

31. 1.00 V

33. (a) 18.0 mH   (b) 34.3 mH   (c) −9.00 mV

35. $M = \dfrac{N_1 N_2 \pi \mu_0 R_1^2 R_2^2}{2(x^2 + R_1^2)^{3/2}}$

37. 400 mA

39. 281 mH

41. 608 pF

43. (a) 6.03 J   (b) 0.529 J   (c) 6.56 J

45. (a) 4.47 krad/s   (b) 4.36 krad/s   (c) 2.53%

47. (a) $0.693(2L/R)$ (b) $0.347(2L/R)$

49. (a) −20.0 mV   (b) $-(10.0\,\text{MV/s}^2)t^2$ (c) 63.2 μs

51. $(Q/2N)(3L/C)^{1/2}$

53. (a) Immediately after the circuit is connected, the potential difference across the resistor is zero and the emf across the coil is 24.0 V. (b) After several seconds, the potential difference across the resistor is 24.0 V and that across the coil is 0. (c) The two voltages are equal to each other, both being 12.0 V, only once, at 0.578 ms after the circuit is connected. (d) As the current decays, the potential difference across the resistor is always equal to the emf across the coil.

55.

57. (b) 91.2 μH   (c) 90.9 μH is only 0.3% smaller

61. (a) 72.0 V; b
   (b)

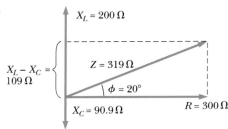

   (c) 75.2 μs

63. 300 Ω

65. (a) 62.5 GJ   (b) 2 000 N

67. (a) 2.93 mT up   (b) 3.42 Pa   (c) The supercurrents must be clockwise to produce a downward magnetic field that cancels the upward field of the current in the windings.

(d) The field of the windings is upward and radially outward around the top of the solenoid. It exerts a force radially inward and upward on each bit of the clockwise supercurrent. The total force on the supercurrents in the bar is upward. (e) 1.30 mN

CHAPTER 33

1. $\Delta v(t) = (283\,\text{V})\sin(628t)$

3. 2.95 A, 70.7 V

5. 14.6 Hz

7. (a) 42.4 mH   (b) 942 rad/s

9. 5.60 A

11. 0.450 Wb

13. (a) 141 mA   (b) 235 mA

15. 100 mA

17. (a) 194 V   (b) current leads by 49.9°

19. (a) 78.5 Ω (b) 1.59 kΩ (c) 1.52 kΩ (d) 138 mA
   (e) −84.3°

21. (a) 17.4°   (b) The voltage leads the current.

23. 1.88 V

25.

The figure shows a phasor diagram with $X_L = 200\,\Omega$, $X_C = 90.9\,\Omega$, $X_L - X_C = 109\,\Omega$, $R = 300\,\Omega$, $Z = 319\,\Omega$, and $\phi = 20°$.

27. 8.00 W

29. (a) 16.0 Ω (b) −12.0 Ω

31. (a) 39.5 V·m/ΔV (b) The diameter is inversely proportional to the potential difference. (c) 26.3 mm   (d) 13.2 kV

33. $11(\Delta V_{\text{rms}})^2/14R$

35. 1.82 pF

37. 242 mJ

39. 0.591 and 0.987; the circuit in Problem 21

41. 687 V

43. (a) 29.0 kW   (b) 5.80×10^{-3} (c) If the generator were limited to 4 500 V, no more than 17.5 kW could be delivered to the load, never 5 000 kW.

45. (b) 0; 1   (c) $f_h = (10.88RC)^{-1}$

47. (a) 613 μF   (b) 0.756

49. (a) 580 μH and 54.6 μF   (b) 1   (c) 894 Hz   (d) Δv_{out} leads Δv_{in} by 60.0° at 200 Hz. Δv_{out} and Δv_{in} are in phase at 894 Hz. Δv_{out} lags Δv_{in} by 60.0° at 4 000 Hz. (e) 1.56 W, 6.25 W, 1.56 W   (f) 0.408

51. (a) X_C could be 53.8 Ω or it could be 1.35 kΩ. (b) X_C must be 53.8 Ω. (c) X_C must be 1.43 kΩ.

53. 56.7 W

55. Tension T and separation d must be related by $T = (274\,\text{N/m}^2)d^2$. One possibility is $T = 10.9$ N and $d = 0.200$ m.

57. (a) 225 mA   (b) 450 mA

59. (a) 1.25 A   (b) The current lags the voltage by 46.7°.

61. (a) 200 mA; voltage leads by 36.8°   (b) 40.0 V; $\phi = 0°$
   (c) 20.0 V; $\phi = -90.0°$ (d) 50.0 V; $\phi = +90.0°$

63. (b) 31.6

67. (a)

f (Hz)	X_L (Ω)	X_C (Ω)	Z (Ω)
300	283	12 600	12 300
600	565	6 280	5 720
800	754	4 710	3 960
1 000	942	3 770	2 830
1 500	1 410	2 510	1 100
2 000	1 880	1 880	40.0
3 000	2 830	1 260	1 570
4 000	3 770	942	2 830
6 000	5 650	628	5 020
10 000	9 420	377	9 040

(b) Impedance (kΩ)

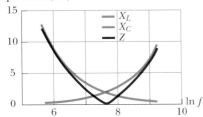

69. (a) and (b) 19.7 cm at 35.0°. The answers are identical.
(c) 9.36 cm at 169°

CHAPTER 34

1. (a) 11.3 GV·m/s   (b) 0.100 A

3. 1.85 aT up

5. $(-2.87\hat{\mathbf{j}} + 5.75\hat{\mathbf{k}})$ Gm/s²

7. (a) the year 2.69×10^3 (b) 499 s   (c) 2.56 s
(d) 0.133 s   (e) 33.3 μs

9. (a) 6.00 MHz   (b) $(-73.3\hat{\mathbf{k}})$ nT
(c) $\vec{\mathbf{B}} = [(-73.3\hat{\mathbf{k}})\, \text{nT}] \cos (0.126x - 3.77 \times 10^7 t)$

11. (a) 0.333 μT   (b) 0.628 μm   (c) 477 THz

13. 75.0 MHz

15. 3.33 μJ/m³

17. 307 μW/m²

19. 3.33×10^3 m²

21. (a) 332 kW/m² radially inward   (b) 1.88 kV/m and 222 μT

23. (a) $\vec{\mathbf{E}} \cdot \vec{\mathbf{B}} = 0$ (b) $(11.5\hat{\mathbf{i}} - 28.6\hat{\mathbf{j}})$ W/m²

25. (a) 2.33 mT   (b) 650 MW/m²   (c) 510 W

27. (a) 88.8 nW/m²   (b) 11.3 MW

29. 83.3 nPa

31. (a) 1.90 kN/C   (b) 50.0 pJ   (c) 1.67×10^{-19} kg·m/s

33. (a) 590 W/m²   (b) 2.10×10^{16} W   (c) 70.1 MN
(d) The gravitational force is ∼ 10^{13} times stronger and in the opposite direction. (e) On the Earth, the Sun's gravitational force is also ∼ 10^{13} times stronger than the light-pressure force and in the opposite direction.

35. (a) 134 m   (b) 46.9 m

37. (a) away along the perpendicular bisector of the line segment joining the antennas   (b) along the extensions of the line segment joining the antennas

39. (a) $\vec{\mathbf{E}} = \frac{1}{2}\mu_0 c J_{\text{max}} [\cos (kx - \omega t)]\hat{\mathbf{j}}$
(b) $\vec{\mathbf{S}} = \frac{1}{4}\mu_0 c J_{\text{max}}^2 [\cos^2 (kx - \omega t)]\hat{\mathbf{i}}$
(c) $I = \dfrac{\mu_0 c J_{\text{max}}^2}{8}$ (d) 3.48 A/m

41. (a) 6.00 pm   (b) 7.50 cm

43. (a) 4.17 m to 4.55 m   (b) 3.41 m to 3.66 m
(c) 1.61 m to 1.67 m

45. 1.00 Mm = 621 mi; not very practical

47. (a) 3.85×10^{26} W   (b) 1.02 kV/m and 3.39 μT

49. (a)

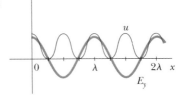

(b), (c) $u_E = u_B = \frac{1}{2}\epsilon_0 E_{\text{max}}^2 \cos^2 (kx)$
(d) $u = \epsilon_0 E_{\text{max}}^2 \cos^2 (kx)$ (e) $E_\lambda = \frac{1}{2}A\lambda\epsilon_0 E_{\text{max}}^2$
(f) $I = \frac{1}{2}c\epsilon_0 E_{\text{max}}^2 = \frac{1}{2}\sqrt{\dfrac{\epsilon_0}{\mu_0}} E_{\text{max}}^2$. This result agrees with
$I = \dfrac{E_{\text{max}}^2}{2\mu_0 c}$ in Equation 34.24.

51. (a) 6.67×10^{-16} T   (b) 5.31×10^{-17} W/m²
(c) 1.67×10^{-14} W   (d) 5.56×10^{-23} N

53. 95.1 mV/m

55. (a) $B_{\text{max}} = 583$ nT, $k = 419$ rad/m, $\omega = 126$ Grad/s;
$\vec{\mathbf{B}}$ vibrates in xz plane   (b) $\vec{\mathbf{S}}_{\text{avg}} = (40.6\hat{\mathbf{i}})$ W/m²
(c) 271 nPa   (d) $(406\hat{\mathbf{i}})$ nm/s²

57. (a) 22.6 h   (b) 30.6 s

59. (a) 8.32×10^7 W/m²   (b) 1.05 kW

61. (b) 17.6 Tm/s², 1.75×10^{-27} W   (c) 1.80×10^{-24} W

63. (a) $2\pi^2 r^2 f B_{\text{max}} \cos \theta$, where θ is the angle between the magnetic field and the normal to the loop   (b) The loop should be in the vertical plane containing the line of sight to the transmitter.

65. (a) 388 K   (b) 363 K

CHAPTER 35

1. 300 Mm/s. The sizes of the objects need to be accounted for; otherwise, the answer would be too large by 2%.

3. 114 rad/s

5. (a) 1.94 m   (b) 50.0° above the horizontal

7. 23.3°

9. 25.5°, 442 nm

11. 19.5° above the horizon

13. 22.5°

15. (a) 181 Mm/s   (b) 225 Mm/s   (c) 136 Mm/s

17. 30.0° and 19.5° at entry; 19.5° and 30.0° at exit

19. 3.88 mm

21. 30.4° and 22.3°

23. (a) yes, if the angle of incidence is 58.9° (b) No. Both the reduction in speed and the bending toward the normal reduce the component of velocity parallel to the interface. This component cannot remain constant unless the angle of incidence is 0°.

25. 86.8°

27. 27.9°

29. (b) 37.2° (c) 37.3° (d) 37.3°

31. 4.61°

33. (a) 24.4° (b) 37.0° (c) 49.8°

35. 1.000 08

37. (a) $dn/(n-1)$ (b) yes; yes; yes (c) 350 μm

39. Skylight incident from above travels down the plastic. If the index of refraction of the plastic is greater than 1.41, the rays close in direction to the vertical are totally reflected from the side walls of the slab and from both facets at the bottom of the plastic, where it is not immersed in gasoline. This light returns up inside the plastic and makes it look bright. Unless the index of refraction of the plastic is unrealistically large (greater than about 2.1), total internal reflection is frustrated where the plastic is immersed in gasoline. There the downward-propagating light passes from the plastic out into the gasoline. Little light is reflected up, and the gauge looks dark.

41. Scattered light leaving the photograph in all forward horizontal directions in air is gathered by refraction into a fan in the water of half-angle 48.6°. At larger angles, you see things on the other side of the globe, reflected by total internal reflection at the back surface of the cylinder.

43. 77.5°

45. 2.27 m

47. 62.2%

49. 82 reflections

51. 27.5°

53. (a) Total internal reflection occurs for all values of θ, or the maximum angle is 90°. (b) 30.3° (c) Total internal reflection never occurs as the light moves from lower-index polystyrene into higher-index carbon disulfide.

55. 2.36 cm

57. $\theta = \sin^{-1}\left[\dfrac{L}{R^2}(\sqrt{n^2R^2 - L^2} - \sqrt{R^2 - L^2})\right]$

61. (a) nR_1 (b) R_2

63. (a) 1.20 (b) 3.40 ns

CHAPTER 36

1. ~ 10^{-9} s younger

3. 35.0 in.

5. (a) $-(p_1 + h)$ (b) virtual (c) upright (d) 1.00 (e) no

7. (a) −12.0 cm; 0.400 (b) −15.0 cm; 0.250 (c) upright

9. (a) $q = 45.0$ cm; $M = -0.500$
(b) $q = -60.0$ cm; $M = 3.00$
(c) Image (a) is real, inverted, and diminished. Image (b) is virtual, upright, and enlarged.

11. (a) 2.22 cm (b) 10.0

13. (a) 160 mm (b) $R = -267$ mm

15. (a) convex (b) at the 30.0 cm mark (c) −20.0 cm

17. (a) a concave mirror with radius of curvature 2.08 m (b) 1.25 m from the object

19. (a) The image starts 60.0 cm above the mirror and moves up faster and faster, running out to an infinite distance above the mirror. At that moment, the image rays are parallel and the image is equally well described as infinitely far below the mirror. From there, the image moves up, slowing down as it moves, to reach the mirror vertex. (b) at 0.639 s and 0.782 s

21. 38.2 cm below the top surface of the ice

23. 8.57 cm

25. (a) inside the tank, 24.9 cm behind the front wall; virtual, right side up, enlarged (b) inside the tank, 93.9 cm behind the front wall; virtual, right side up, enlarged (c) 1.10 and 1.39 (d) 9.92 cm and 12.5 cm (e) The plastic has uniform thickness, so the surfaces of entry and exit for any particular ray are very nearly parallel. The ray is slightly displaced, but it would not be changed in direction by going through the plastic wall with air on both sides. Only the difference between the air and water is responsible for the refraction of the light.

27. (a) 16.4 cm (b) 16.4 cm

29. (a) $q = 40.0$ cm, real and inverted, actual size $M = -1.00$ (b) $q = \infty$, $M = \infty$, no image is formed (c) $q = -20.0$ cm, upright, virtual, enlarged $M = +2.00$

31. 2.84 cm

33. (a) −12.3 cm, to the left of the lens (b) 0.615
(c)

35. (a) 5.36 cm (b) −18.8 cm (c) virtual, right side up, enlarged (d) A magnifying glass with focal length 7.50 cm is used to form an image of a stamp, enlarged 3.50 times. Find the object distance. Locate and describe the image.

37. (a) $p = \dfrac{d}{2} \pm \sqrt{\dfrac{d^2}{4} - fd}$

(b) Both images are real and inverted. One is enlarged, the other diminished.

39. 2.18 mm away from the film plane

41. 21.3 cm

43. −4.00 diopters, a diverging lens

45. (a) at 4.17 cm   (b) 6.00

47. (a) −800   (b) image is inverted

49. 3.38 min

51. −40.0 cm

53. −25.0 cm

55. $x' = (1\ 024\ \text{cm} - 58x)\ \text{cm}/(6\ \text{cm} - x)$. The image starts at the position $x_i' = 171$ cm and moves in the positive x direction, faster and faster, until it is out at infinity when the object is at the position $x = 6$ cm. At this instant, the rays from the top of the object are parallel as they leave the lens. Their intersection point can be described as at $x' = \infty$ to the right or equally well as at $x' = -\infty$ on the left. From $x' = -\infty$, the image continues moving to the right, now slowing down. It reaches, for example, −280 cm when the object is at 8 cm and −55 cm when the object is finally at 12 cm. The image has traveled always to the right, to infinity and beyond.

57. Align the lenses on the same axis and 9.00 cm apart. Let the light pass first through the diverging lens and then through the converging lens. The diameter increases by a factor of 1.75.

59. 0.107 m to the right of the vertex of the hemispherical face

61. 8.00 cm. Ray diagram:

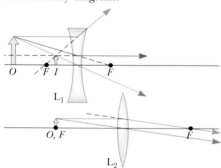

63. 1.50 m in front of the mirror; 1.40 cm (inverted)

65. (a) 30.0 cm and 120 cm   (b) 24.0 cm   (c) real, inverted, diminished with $M = -0.250$

67. (a) 20.0 cm to the right of the second lens, −6.00
(b) inverted   (c) 6.67 cm to the right of the second lens, −2.00, inverted

CHAPTER 37

1. 1.58 cm

3. (a) 55.7 m   (b) 124 m

5. 1.54 mm

7. (a) 2.62 mm   (b) 2.62 mm

9. 36.2 cm

11. (a) 10.0 m (b) 516 m (c) Only the runway centerline is a maximum for the interference patterns for both frequencies. If the frequencies were related by a ratio of small integers k/ℓ, the plane could by mistake fly along the kth side maximum of one signal, where it coincides with the ℓth side maximum of the other. The plane cannot make a sharp turn at the end of the runway, so it would then be headed in the wrong direction to land.

13. (a) 13.2 rad (b) 6.28 rad (c) 0.012 7 degree
(d) 0.059 7 degree

15. (a) 1.93 μm (b) 3.00λ (c) It corresponds to a maximum. The path difference is an integer multiple of the wavelength.

17. 48.0 μm

19. $E_1 + E_2 = 10.0 \sin (100\pi t + 0.927)$

21. (a) 7.95 rad (b) 0.453

23. (a) green (b) violet

25. 512 nm

27. 96.2 nm

29. (a) 238 nm (b) The wavelength increases because of thermal expansion of the filter material. (c) 328 nm

31. 4.35 μm

33. 1.20 mm

35. 39.6 μm

37. 1.62 cm

39. 1.25 m

41. (a) $\sim 10^{-3}$ degree   (b) $\sim 10^{11}$ Hz, microwave

43. 2.52 cm

45. $20.0 \times 10^{-6}\ °\text{C}^{-1}$

47. 3.58°

49. 1.62 km

51. 421 nm

55. (b) 266 nm

57. $y' = (n - 1)tL/d$

59. (a) 70.6 m   (b) 136 m

61. (a) 14.7 μm   (b) 1.53 cm   (c) −16.0 m

63. (a) 4.86 cm from the top   (b) 78.9 nm and 128 nm
(c) 2.63×10^{-6} rad

65. 0.505 mm

67. 0.498 mm

CHAPTER 38

1. 4.22 mm

3. 0.230 mm

5. three maxima, at 0° and near 46° on both sides

7. 0.016 2

9.

11. 1.00 mrad

13. 3.09 m

15. 13.1 m

17. Neither. It can resolve no objects closer than several centimeters apart.

19. 7.35°

21. 5.91° in first order, 13.2° in second order, 26.5° in third order

23. (a) 478.7 nm, 647.6 nm, and 696.6 nm (b) 20.51°, 28.30°, and 30.66°

25. three, at 0° and at 45.2° to the right and left.

27. (a) five orders (b) ten orders in the short-wavelength region

29. 2

31. 14.4°

33. The crystal cannot produce diffracted beams of visible light. Bragg's law cannot be satisfied for a wavelength much larger than the distance between atomic planes in the crystal.

35. (a) 54.7° (b) 63.4° (c) 71.6°

37. 60.5°

39. (b) For light confined to a plane, yes.

$$\left| \tan^{-1}\left(\frac{n_3}{n_2}\right) - \tan^{-1}\left(\frac{n_1}{n_2}\right) \right|$$

41. (a) 0.875 (b) 0.789 (c) 0.670 (d) You can get more and more of the incident light through the stack of ideal filters, approaching 50%, by reducing the angle between the axes of each one and the next.

43. (a) 6 (b) 7.50°

45. (a) 0.045 0 (b) 0.016 2

47. 632.8 nm

49. (a) 25.6° (b) 19.0°

51. 545 nm

53. (a) 3.53×10^3 grooves/cm (c) 11 maxima

55. 4.58 μm < d < 5.23 μm

57. 15.4

59. (a) 41.8° (b) 0.593 (c) 0.262 m

61. (b) 3.77 nm/cm

63. (b) 15.3 μm

65. $\phi = 1.391\ 557\ 4$ after 17 steps or fewer

67. $a = 99.5$ μm ± 1%

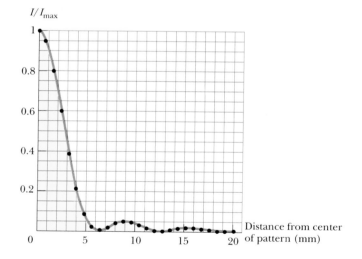

CHAPTER 39

5. 0.866c

7. 1.54 ns

9. 0.800c

11. (a) 39.2 μs (b) It was accurate to one digit. Cooper aged 1.78 μs less in each single orbit.

13. (a) 20.0 m (b) 19.0 m (c) 0.312c

15. 11.3 kHz

17. (b) 0.050 4c

19. (a) 0.943c (b) 2.55 km

21. B occurred 4.44×10^{-7} s before A

23. 0.960c

25. (a) 2.73×10^{-24} kg · m/s (b) 1.58×10^{-22} kg · m/s (c) 5.64×10^{-22} kg · m/s

27. 4.50×10^{-14}

29. 0.285c

31. (a) 5.37×10^{-11} J (b) 1.33×10^{-9} J

33. 1.63×10^3 MeV/c

35. (a) 938 MeV (b) 3.00 GeV (c) 2.07 GeV

39. (a) 0.979c (b) 0.065 2c (c) 0.914c = 274 Mm/s (d) 0.999 999 97c; 0.948c; 0.052 3c = 15.7 Mm/s

41. 4.08 MeV and 29.6 MeV

43. smaller by 3.18×10^{-12} kg, which is too small a fraction of 9 g to be measured

45. 4.28×10^9 kg/s

47. (a) 26.6 Mm (b) 3.87 km/s (c) -8.34×10^{-11} (d) 5.29×10^{-10} (e) $+4.46 \times 10^{-10}$

49. (a) a few hundred seconds (b) $\sim 10^8$ km

51. (a) $u = c\left(\dfrac{2H + H^2}{1 + 2H + H^2}\right)^{1/2}$, where $H = K/mc^2$

(b) u goes to 0 as K goes to 0. (c) u approaches c as K increases without limit.

(d) $a = \dfrac{\mathcal{P}}{mcH^{1/2}(2 + H)^{1/2}(1 + H)^2}$

(e) $a = \dfrac{\mathcal{P}}{mc\,(2H)^{1/2}} = \dfrac{\mathcal{P}}{(2mK)^{1/2}}$, in agreement with the non-relativistic case.

(f) a approaches $\mathcal{P}/mcH^3 = \mathcal{P}m^2c^5/K^3$ (g) As energy is steadily imparted to the particle, the particle's acceleration decreases. It decreases steeply, proportionally to $1/K^3$ at high energy. In this way, the particle's speed cannot reach or surpass a certain upper limit, which is the speed of light in vacuum.

53. 0.712%

55. (a) 76.0 min (b) 52.1 min

57. (a) 0.946c (b) 0.160 ly (c) 0.114 yr (d) 7.50×10^{22} J

59. yes, with 18.8 m to spare

61. (b) For u small compared with c, the relativistic expression agrees with the classical expression. As u approaches c, the acceleration approaches zero, so the object can never reach or surpass the speed of light.
(c) Perform the operation $\int(1 - u^2/c^2)^{-3/2}du = (qE/m)\int dt$ to obtain $u = qEct(m^2c^2 + q^2E^2t^2)^{-1/2}$ and then $\int dx = \int qEct(m^2c^2 + q^2E^2t^2)^{-1/2}dt$ to obtain $x = (c/qE)\left[(m^2c^2 + q^2E^2t^2)^{1/2} - mc\right]$

63. (a) $M = \dfrac{2m\sqrt{4 - u^2/c^2}}{3\sqrt{1 - u^2/c^2}}$

(b) M = 4m/3. The result agrees with the arithmetic sum of the masses of the two colliding particles.

65. (a) The refugees conclude that Tau Ceti exploded 16.0 yr before the Sun. (b) A stationary observer at the midpoint concludes that Tau Ceti and the Sun exploded simultaneously.

67. 1.82×10^{-3} eV

CHAPTER 40

1. 5.18×10^3 K

3. approximately 5 200 K. A firefly cannot be at this temperature, so its light cannot be blackbody radiation.

5. $1.30 \times 10^{15}/$s

7. (a) 5.78×10^3 K   (b) 501 nm

9. (a) 2.57 eV   (b) 12.8 μeV   (c) 191 neV   (d) 484 nm (visible), 9.68 cm, and 6.52 m (radio waves)

11. 2.27×10^{30} photons/s

15. (a) 1.90 eV   (b) 0.216 V

17. (a) We find the energy of a photon with wavelength 400 nm and check whether it exceeds the work function. Of these metals, only lithium shows the photoelectric effect. (b) 0.808 eV

19. 8.41 pC

21. 1.78 eV, 9.47×10^{-28} kg·m/s

23. 70.0°

25. (a) 43.0°   (b) 602 keV, 3.21×10^{-22} kg·m/s   (c) 278 keV, 3.21×10^{-22} kg·m/s

27. (a) 2.88 pm   (b) 101°

29. It is sufficient because Compton's equation and the conservation of vector momentum give three independent equations in the unknowns λ', λ_0, and u. Wavelength is 3.82 pm

31. (a) 0.667   (b) 0.001 09

33. (a) 14.0 kV/m, 46.8 μT   (b) 4.19 nN   (c) 10.2 g

35. (a) 0.174 nm   (b) 5.37 pm or 5.49 pm ignoring relativistic correction

37. (a) ~ 100 MeV or more   (b) ~ −1 MeV. No. With kinetic energy much larger than the magnitude of its negative electric potential energy, the electron would immediately escape.

39. (b) No. $\lambda^{-2} + \lambda_C^{-2}$ cannot be equal to λ^{-2}

41. (a) 14.9 keV or, ignoring relativistic correction, 15.1 keV   (b) 124 keV

43. (a) 3.91×10^4 (b) 20.0 GeV/c = 1.07×10^{-17} kg·m/s   (c) 6.22×10^{-17} m, small compared with the size of the nucleus. The scattering of the electrons can give information about the particles forming the nucleus.

47. (a) 993 nm   (b) 4.96 mm   (c) If its detection forms part of an interference pattern, the neutron must have passed through both slits. If we test to see which slit a particular neutron passes through, the neutron will not form part of the interference pattern.

49. Within 1.16 mm for the electron, 5.28×10^{-32} m for the bullet

51. 3.79×10^{28} m, 190 times the diameter of the observable Universe

53. (b) 519 am

55. (a)

Maximum photoelectron energy increasing linearly with photon frequency

(b) 6.4×10^{-34} J·s $\pm$ 8%   (c) 1.4 eV

57. $\dfrac{hc}{\lambda} - \dfrac{e^2 B^2 R^2}{2m_e}$

63. 0.143 nm. The wavelength is comparable to the distance between atoms in a crystal, so diffraction can be observed.

67. 2.81×10^{-8}

CHAPTER 41

1. (a) 126 pm   (b) 5.27×10^{-24} kg·m/s   (c) 95.5 eV

3. (a) 0.434 nm   (b) 6.00 eV

5. (a)

(b) 2.20 nm, 2.75 nm, 4.12 nm, 4.71 nm, 6.60 nm, 11.0 nm

7. 0.793 nm

9. 6.16 MeV, 202 fm, a gamma ray

11. 0.513 MeV, 2.05 MeV, 4.62 MeV. They do; the MeV is the natural unit for energy radiated by an atomic nucleus.

13. (a) $\Delta p \approx \hbar/2L$ (b) $E \approx \hbar^2/8mL^2$. This estimate is too low by $4\pi^2 \approx 40$ times. It correctly displays the pattern of dependence of the energy on the mass and on the length of the well.

17. At $L/4$ and at $3L/4$. We look for $\sin(2\pi x/L)$ taking on its extreme values 1 and −1 so that the squared wave function is as large as it can be.

19. (a) 0.196   (b) 0.609   (c) The classical probability, $\frac{1}{3}$, is very different.

23. (a) $E = \hbar^2/mL^2$

(b) $\displaystyle\int_{-L}^{L} A^2(1 - x^2/L^2)^2 dx = 1$ gives $A = \left(\dfrac{15}{16L}\right)^{1/2}$.

(c) $\frac{47}{81} = 0.580$

27. (a)

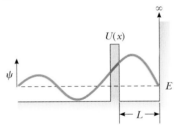

(b) $2L$

29. By 0.959 nm, to 1.91 nm

31. (a) 0.010 3   (b) 0.990

33. 1.35

37. 600 nm

39. (b) The acceleration is equal to a negative constant times the excursion from equilibrium. The frequency is $\dfrac{1}{2\pi}\sqrt{\dfrac{k}{\mu}}$.

41. $\sim 10^{-10^{30}}$

43. (a)

(b)

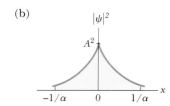

(c) The wave function is continuous. It shows localization by approaching zero as $x \to \pm\infty$. It is everywhere finite and can be normalized. (d) $A = \sqrt{\alpha}$ (e) 0.632

45. (b) For the transition from level 2 to level 1, $\lambda = 1.38$ μm infrared. For level 4 to level 1, $\lambda = 276$ nm ultraviolet. For level 3 to level 2, $\lambda = 827$ nm infrared. For level 4 to level 2, $\lambda = 344$ nm near ultraviolet. For level 4 to level 3, $\lambda = 590$ nm yellow-orange visible.

47. 0.029 4

49. (a) 434 THz   (b) 691 nm   (c) 165 peV or more

53. (a)

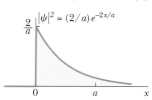

(b) 0   (c) 0.865

55. (a) $-7k_e e^2/3d$ (b) $h^2/36m_e d^2$ (c) 49.9 pm   (d) The lithium interatomic spacing of 280 pm is 5.62 times larger. Therefore, it is of the same order of magnitude as the interatomic spacing $2d$ here.

57. (a) $A = \left(\dfrac{2}{17L}\right)^{1/2}$ (b) $|A|^2 + |B|^2 = 1/a$

59. 2.25

CHAPTER 42

1. (a) 5   (b) no; no

3. (b) 0.846 ns

5. (a) 1.89 eV, 656 nm   (b) 3.40 eV, 365 nm

7. (a) 0.212 nm   (b) 9.95×10^{-25} kg·m/s
(c) 2.11×10^{-34} kg·m^2/s   (d) 3.40 eV   (e) -6.80 eV
(f) -3.40 eV

9. (a) $E_n = -54.4$ eV/n^2 for $n = 1, 2, 3, \ldots$

n		E (eV)
∞		0
4		-3.40
3		-6.05
2		-13.6
1		-54.4

(b) 54.4 eV

11. $r_n = (0.106 \text{ nm})n^2$, $E_n = -6.80$ eV/n^2, for $n = 1, 2, 3, \ldots$

13. (a) 1.31 μm   (b) 164 nm

15. (a) $\Delta p \geq \hbar/2r$ (b) Choosing $p \approx \hbar/r$, we find that $E = K + U = \hbar^2/2m_e r^2 - k_e e^2/r$. (c) $r = \hbar^2/m_e k_e e^2 = a_0$ and $E = -13.6$ eV, in agreement with the Bohr theory

17. (b) 0.497

19. It does, with $E = -k_e e^2/2a_0$.

21. (a)

n	ℓ	m_ℓ	m_s
3	2	2	$\frac{1}{2}$
3	2	2	$-\frac{1}{2}$
3	2	1	$\frac{1}{2}$
3	2	1	$-\frac{1}{2}$
3	2	0	$\frac{1}{2}$
3	2	0	$-\frac{1}{2}$
3	2	-1	$\frac{1}{2}$
3	2	-1	$-\frac{1}{2}$
3	2	-2	$\frac{1}{2}$
3	2	-2	$-\frac{1}{2}$

(b)

n	ℓ	m_ℓ	m_s
3	1	1	$\frac{1}{2}$
3	1	1	$-\frac{1}{2}$
3	1	0	$\frac{1}{2}$
3	1	0	$-\frac{1}{2}$
3	1	-1	$\frac{1}{2}$
3	1	-1	$-\frac{1}{2}$

23. (a) Find the orbital quantum number of an electron in a state in which it has orbital angular momentum equal to 4.714×10^{-34} J·s. (b) orbital quantum number = 4

25. (a) 2   (b) 8   (c) 18   (d) 32   (e) 50

27. (a) 3.99×10^{17} kg/m^3 (b) 81.7 am   (c) 1.77 Tm/s
(d) It is $5.91 \times 10^3 c$, which is huge compared with the speed of light and impossible.

29. $n = 3$; $\ell = 2$; $m_\ell = -2, -1, 0, 1,$ or 2; $s = 1$; $m_s = -1, 0,$ or 1, for a total of 15 states

31. The $4s$ subshell is filled first. We would expect $[Ar]3d^44s^2$ to have lower energy, but $[Ar]3d^54s^1$ has more unpaired spins and lower energy according to Hund's rule. It is the ground-state configuration of chromium.

33. aluminum

35. (a) $1s, 2s, 2p, 3s, 3p, 4s, 3d, 4p, 5s, 4d, 5p, 6s, 4f, 5d, 6p, 7s$ (b) Element 15 should have valence $+5$ or -3, and it does. Element 47 should have valence -1, but it has valence $+1$. Element 86 should be inert, and it is.

37. 18.4 T

39. 1.4 and 1.0. When the outermost electron in sodium is promoted from the $3s$ state into a $3p$ state, its wave function still overlaps somewhat with the ten electrons below it. It therefore sees the $+11e$ nuclear charge not fully screened and on average moves in an electric field like that created by a particle with charge $+11e - 9.6e = 1.4e$. When this valence electron is lifted farther to a $3p$ state, it is essentially entirely outside the cloud of ten electrons below it and moves in the field of a net charge $+11e - 10e = 1e$.

41. 0.072 5 nm

43. iron

45. 28.2 THz, 10.6 μm, infrared

47. 3.49×10^{16} photons

49. (a) 217 nm (b) 93.1 nm

51. (a) 609 μeV (b) 6.9 μeV (c) 147 GHz, 2.04 mm

53. The classical frequency is $4\pi^2 m_e k_e^2 e^4 / h^3 n^3$.

55. (a) -8.16 eV, -2.04 eV, -0.902 eV, -0.508 eV, -0.325 eV (b) 1 090 nm, 811 nm, 724 nm, 609 nm (d) The spectrum could be that of hydrogen, Doppler-shifted by motion away from us at speed $0.471c$.

57. (a) 1.57×10^{14} m$^{-3/2}$ (b) 2.47×10^{28} m^{-3} (c) 8.69×10^8 m^{-1}

59. (a) 4.20 mm (b) 1.05×10^{19} photons (c) 8.82×10^{16}/mm^3

61. $3h^2/4mL^2$

65. 5.39 keV

67. 0.125

69. 9.79 GHz

CHAPTER 43

1. (a) 921 pN toward the other ion (b) -2.88 eV

3. (a) $(2A/B)^{1/6}$ (b) $B^2/4A$ (c) 74.2 pm, 4.46 eV

5. ~ 10 K

7. (a) 40.0 μeV, 9.66 GHz (b) If r is 10% too small, then f is 20% too large.

9. 5.63 Trad/s

13. 2.72×10^{-47} kg $\cdot$ m^2

15. 0.358 eV

17. $(18.4 \ \mu\text{eV})J(J + 1)$, where $J = 0, 1, 2, 3, \ldots$

19. 2.9×10^{-47} kg $\cdot$ m^2

21. only 64.1 THz

23. -7.84 eV

27. An average atom contributes 0.981 electron to the conduction band.

29. (a) 1.57 Mm/s (b) It is larger by ten orders of magnitude.

31. (a) 4.23 eV (b) 3.27×10^4 K

33. 5.28 eV

37. (a) 1.10 (b) 1.47×10^{-25}. It is vastly smaller. Very few states well above the Fermi energy are occupied at room temperature.

39. (a) 275 THz (b) 1.09 μm

41. The gap should be less than or equal to 1.24 eV. Silicon's energy gap of 1.14 eV means that it can absorb the energy of nearly all the photons in sunlight and is an appropriate material for a solar energy collector.

43. 226 nm

45. (a) 59.5 mV (b) -59.5 mV

47. 4.19 mA

49. 203 A to produce a magnetic field in the direction of the original field

51. (a)

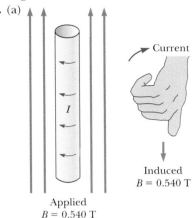

Applied $B = 0.540$ T

Current

Induced $B = 0.540$ T

(b) 10.7 kA

53. (a) 6.15×10^{13} Hz (b) 1.59×10^{-46} kg $\cdot$ m^2 (c) 4.79 μm or 4.96 μm

55. 7

59. (a) 0.350 nm (b) -7.02 eV (c) $-1.20\hat{\mathbf{i}}$ nN

61. (a) r_0 (b) B (c) $(a/\pi)[B/2\mu]^{1/2}$ (d) $B - (ha/\pi)[B/8\mu]^{1/2}$

CHAPTER 44

1. $\sim 10^{28}$; $\sim 10^{28}$; $\sim 10^{28}$

3. (a) 2.81×10^{-5} (b) 1.38×10^{-14}

5. (a) 455 fm (b) 6.04 Mm/s

7. 16.0 km

9. (a) 1.11 MeV/nucleon (b) 7.07 MeV/nucleon (c) 8.79 MeV/nucleon (d) 7.57 MeV/nucleon

11. (a) $^{139}_{55}$Cs (b) $^{139}_{57}$La (c) $^{139}_{55}$Cs

13. greater for $^{15}_{7}$N by 3.54 MeV

15. 7.93 MeV

17. 200 MeV

19. 1.16 ks

21. (a) 1.55×10^{-5}/s, 12.4 h (b) 2.39×10^{13} atoms (c) 1.88 mCi

23. 9.47×10^9 nuclei

25. 1.41

27. 2.66 d

29. 4.27 MeV

31. (a) $N_d = 10^6(1 - e^{-0.069 \, 3t})$, where t is in hours   (b) The number of daughter nuclei starts from zero at $t = 0$. It first increases most rapidly, at $6.93 \times 10^4/\text{h}$, and then more and more slowly. Its rate of change approaches zero while the number asymptotically approaches 1.00×10^6 as t increases without limit.

33. (a) $e^- + p \to n + \nu$ (b) $^{15}_{8}\text{O}$ atom $\to$ $^{15}_{7}\text{N}$ atom $+ \nu$ (c) 2.75 MeV

35.

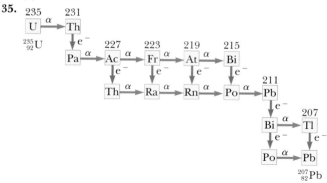

37. (a) 148 Bq/m^3 (b) $7.05 \times 10^7 \text{ atoms/m}^3$ (c) 2.17×10^{-17}

39. (a) $^{21}_{10}\text{Ne}$ (b) $^{144}_{54}\text{Xe}$ (c) $^{0}_{1}e^+ + \nu$

41. $^{197}_{79}\text{Au} + ^{1}_{0}\text{n} \to ^{198}_{80}\text{Hg} + ^{0}_{-1}e^- + \bar{\nu}$ (b) 7.89 MeV

43. 8.005 3 u; 10.013 5 u

45. (a) 29.2 MHz   (b) 42.6 MHz   (c) 2.13 kHz

47. (a) 5.70 MeV   (b) 3.27 MeV, exothermic

49. (b) 1.53 MeV

51. Find the reaction energy (Q value) of the reaction $^{10}_{5}\text{B} + ^{4}_{2}\text{He} \to ^{13}_{6}\text{C} + ^{1}_{1}\text{H}$. Solving, $Q = 4.06$ MeV is the energy released by the reaction as it is converted from rest energy into other forms.

53. (a) The process cannot occur because energy input would be required.   (b) can occur   (c) K_e between 0 and 156 keV

55. (a) conservation of energy   (b) Electric potential energy of the parent nucleus can supply the required energy.   (c) 1.20 MeV

57. (a) 61.8 Bq/L   (b) 40.3 d

59. 5.94 Gyr

61. 2.20 μeV

63. 0.400%

65. (a) 3.91 Gyr   (b) It could be no older. We must make some assumption about the original quantity of radioactive material. In part (a), we assumed the rock originally contained no strontium.

CHAPTER 45

1. 0.387 g

3. $^{144}_{54}\text{Xe}$, $^{143}_{54}\text{Xe}$, and $^{142}_{54}\text{Xe}$

5. $^{1}_{0}\text{n} + ^{232}\text{Th} \to ^{233}\text{Th} \to ^{233}\text{Pa} + e^- + \bar{\nu}$, $^{233}\text{Pa} \to ^{233}\text{U} + e^- + \bar{\nu}$

7. 5.80 Mm

9. approximately 3 000 yr

11. 2.68×10^5

13. (a) $4.55 \times 10^{-24} \text{ kg} \cdot \text{m/s}$ (b) 0.146 nm. This size is the same order of magnitude as an atom's outer electron cloud and is vastly larger than a nucleus.

15. (a) 31.9 g/h   (b) 122 g/h

17. (a) 3.24 fm   (b) 444 keV   (c) $\frac{2}{5}v_i$ (d) 740 keV   (e) possibly by tunneling

19. (a) 2.22 Mm/s   (b) $\sim 10^{-7}$ s

21. (a) 10^{14} cm^{-3} (b) $1.24 \times 10^5 \text{ J/m}^3$ (c) 1.77 T

23. 1.66×10^3 yr

25. (a) 2.5 mrem per x-ray image   (b) The technician's occupational exposure is high compared to background radiation; it is 38 times 0.13 rem/yr.

27. 2.09×10^6 s

29. 1.14 rad

31. 3.96×10^{-4} J/kg

33. (a) 3.12×10^7 (b) 3.12×10^{10} electrons

35. 4.45×10^{-8} kg/h

37. (b) R/λ

39. (a) 8×10^4 eV   (b) 4.62 MeV and 13.9 MeV   (c) 1.03×10^7 kWh

41. (a) 177 MeV   (b) $K_{\text{Br}} = 112$ MeV, $K_{\text{La}} = 65.4$ MeV   (c) $v_{\text{Br}} = 15.8 \text{ Mm/s}$, $v_{\text{La}} = 9.20 \text{ Mm/s}$

43. 482 Ci, less than the fission inventory by on the order of a hundred million times

45. 2.56×10^4 kg

47. (a) 333 μCi, 15.5 μCi, 312 μCi   (b) 50.4%, 2.35%, 47.3%   (c) It is potentially dangerous, notably if the material is inhaled as a powder. With precautions to minimize human contact, microcurie sources can be routinely used in laboratories.

49. (a) 2.65 GJ   (b) The fusion energy is 78.0 times larger.

51. (a) $4.91 \times 10^8 \text{ kg/h} = 4.91 \times 10^5 \text{ m}^3/\text{h}$ (b) 0.141 kg/h

53. (a) 15.5 cm   (b) 51.7 MeV   (c) The number of decays per second is the decay rate R, and the energy released in each decay is Q. Then the energy released per unit time interval is $\mathcal{P} = QR$.   (d) 227 kJ/yr   (e) 3.18 J/yr

55. 14.0 MeV or, ignoring relativistic correction, 14.1 MeV

57. (a) 2.24×10^7 kWh   (b) 17.6 MeV   (c) 2.34×10^8 kWh   (d) 9.36 kWh   (e) Coal is cheap at this moment in human history. We hope that safety and waste disposal problems can be solved so that nuclear energy can be affordable before scarcity drives up the price of fossil fuels.

59. (b) 26.7 MeV

61. (a) 5×10^7 K   (b) 1.94 MeV, 1.20 MeV, 1.02 MeV, 7.55 MeV, 7.30 MeV, 1.73 MeV, 1.02 MeV, 4.97 MeV, 26.7 MeV   (c) Most of the neutrinos leave the star directly after their creation, without interacting with any other particles.

63. (b) 1.00 μs   (c) 2.83 km/s   (d) 14.6 μs   (e) yes; 107 kilotons of TNT

CHAPTER 46

1. 453 ZHz; 662 am

3. (a) 558 TJ   (b) $\$2.17 \times 10^7$

5. 118 MeV

7. $\sim 10^{-18}$ m

9. 67.5 MeV, 67.5 MeV/c, 16.3 ZHz

11. (a) 0.782 MeV   (b) $v_e = 0.919c$, $v_p = 380$ km/s   (c) The electron is moving faster than one-tenth the speed of light, and the proton is not. Relativistic kinetic energy and momentum equations need be used only for the electron.

13. (a) muon lepton number L_μ and electron lepton number L_e (b) charge   (c) baryon number   (d) baryon number   (e) charge

15. (a) $\overline{\nu}_\mu$ (b) ν_μ (c) $\overline{\nu}_e$ (d) ν_e (e) ν_μ (f) $\overline{\nu}_e + \nu_\mu$

17. Reactions (a), (c), and (f) violate baryon number conservation. Reactions (b), (d), and (e) can occur. Reaction (f) violates muon lepton number conservation.

19. (b) $E_e = E_\gamma = 469$ MeV, $p_e = p_\gamma = 469$ MeV/c (c) 0.999 999 4c

21. Reactions (b) and (c) conserve strangeness. Reactions (a), (d), (e), and (f) violate strangeness conservation.

23. (a) electron and muon lepton numbers   (b) electron lepton number   (c) charge and strangeness   (d) baryon number   (e) strangeness

25. (a) baryon number, charge, L_e and L_τ (b) baryon number, charge, L_e, L_μ, and L_τ (c) strangeness, charge, L_e, L_μ, and L_τ (d) baryon number, strangeness, charge, L_e, L_μ, and L_τ (e) baryon number, strangeness, charge, L_e, L_μ, and L_τ (f) baryon number, strangeness, charge, L_e, L_μ, and L_τ

27. 9.26 cm

29. (b) 5.63 GeV   (c) 768 MeV   (d) 280 MeV   (e) 4.43 TeV

35. (a) The reaction $\overline{u}d + uud \rightarrow \overline{s}d + uds$ has a total of 1 u, 2 d, and 0 s quarks originally and finally. (b) The reaction $\overline{d}u + uud \rightarrow \overline{s}u + uus$ has a net of 3 u, 0 d, and 0 s before and after. (c) The reaction $\overline{u}s + uud \rightarrow$ $\overline{s}u + \overline{s}d + sss$ shows conservation at 1 u, 1 d, and 1 s quark. (d) The process $uud + uud \rightarrow \overline{s}d + uud + \overline{d}u + uds$ nets 4 u, 2 d, and 0 s quarks initially and finally; the mystery particle is a Λ^0 or a Σ^0.

37. (a) Σ^+ (b) π^- (c) K^0 (d) Ξ^-

39. (a) 3.32 am/s   (b) 690 pm/s

41. (a) $0.160c$ (b) 2.82×10^9 ly

43. (a) 590.07 nm   (b) 597 nm   (c) 661 nm

45. (a) 8.41×10^6 kg   (b) No. It is only the fraction 4.23×10^{-24} of the mass of the Sun.

47. 3.15 μW/m^2

49. (a) $\sim 10^{13}$ K   (b) $\sim 10^{10}$ K

51. (b) 11.8 Gyr

53. (a) 1.61×10^{-35} m   (b) 5.38×10^{-44} s, of the same order of magnitude as the ultrahot epoch   (c) Yes. The opaque fireball of the Big Bang, measured as the cosmic microwave background radiation, prevents us from receiving visible light from things before the Universe was a few hundred thousand years old. Walls of more profound fire hide all information from still earlier times.

55. 19 eV/c^2

57. one part in 50 000 000

59. (a) $2Nmc$ (b) $3^{1/2}Nmc$ (c) method (a)

61. 5.35 MeV and 32.3 MeV

63. 74.4 MeV

65. 29.8 MeV

67. Diagram (a): electron–positron annihilation; e$^-$. Diagram (b): a neutrino collides with a neutron, producing a proton and a muon; W$^+$

69. (a) 127 MeV   (b) 1.06 mm   (c) 1.17 meV   (d) 5.81×10^{19} eV

Locator note: **boldface** indicates a definition; *italics* indicates a figure; *t* indicates a table

Aberrations, in lenses, *1030*, **1030**–1031, 1039–1040

Absolute uncertainty, **A-20**

Absorption, stimulated, *1244*, **1244**

Absorption spectroscopy, *1216*, **1216**–1217

AC. *See* Alternating current

Acceleration (*a*), in relativistic conditions, 1135

Accelerator, at Fermilab, *1111*

Acceptor atoms, **1279**

Accommodation, **1033**, 1034–1035

Activity, of radioactive substance, **1305**, 1306–1307
  units of, 1306

Addition
  and differentiation, A-14
  and uncertainty, A-20

Air
  dielectric constant and dielectric strength of, 736*t*
  index of refraction, 987*t*

Algebra, A-5–A-9

Alkali metals, 1240

Allowed transitions, 1241–1242

Alloys, metal, 1270

Alpha (*α*) decay, 1304, *1304*
  decay pathways, 1317*t*
  as example of tunneling, 1202, 1312
  and radiation damage, 1343, 1344*t*

Alternating current (AC)
  AC-DC converters, *941*, 942–943
  advantages of, 942
  voltage of, 926

Alternating current (AC) circuits
  average power in, 935–937, **936**, 938, *938*
  capacitors in, *929*, 929–931, *930*, 936
  household wiring, 796
    electrical safety, 796–798
  inductors in, *927*, 927–929, *928*, 936
  RC circuits, as filters, 943, *943*
  resistors in, *924*, 924–927, 936

Alternating current (AC) generators, *880*, 880–882

Alternating current (AC) sources, 923–924, *924*
  symbol for, 924

Alternating current (AC) transformers, 939–942, *940*, **940**, *941*

Aluminum (Al), isotopes of, 1310*t*

Alzheimer's disease, 1360, *1361*

Ammeter, *794*, **794**–795

Amorphous solids, 1096

Ampere (A), 753, **843**

Ampère, Andre-Marie, *845*

Ampère-Maxwell law, 953–**954**, 955

Ampère's law, 844–848, **845**
  general form of (Ampère-Maxwell law), 953–**954**, 955

Amperian loop, 845

Analysis models
  quantum particle under boundary conditions, 1196, 1208
  waves in interference, **1055**
  wave under reflection, **982**
  wave under refraction, **988**

Analyzer, *1093*, 1093–1094

Anderson, Carl, 1359, 1361

Angle(s), critical, **993**–995, *994*

Angle of deviation, *989*, 989–990

Angle of incidence, **982**–983

Angle of reflection, **982**–983

Angle of refraction, *985*, **985**–986, 988

Angular frequency (*ω*)
  of AC voltage, 924
  of electromagnetic waves, 959
  of oscillation in *LC* circuit, 909

Angular magnification, **1035**–1036, 1037, 1038–1039

Angular momentum ($\vec{\mathbf{L}}$)
  orbital
    and magnetic moment, 852
    quantization of, 852–853, 1231–1233
  of photon, 1241

Angular speed, of charge in magnetic field, 813–814, 815–816

Angular wave number. *See* Wave number

Antenna
  dipole, 966
  electromagnetic wave production by, 965–966, *966*
  half-wave, 965–966, *966*

Antibaryons, 1365, 1373

Anti-bottom quark, 1373*t*, 1374*t*

Anti-charmed quark, 1373*t*, 1374*t*

Antiderivative. *See* Integration

Anti-down quark, 1373*t*, 1374*t*

Antileptons, 1367

Antilogarithm, **A-9**

Antimuons, 1368

Antineutrinos ($\bar{\nu}$), *1314*, **1314**

Antineutrons, 1360

Antiparticles, *1359*, 1359–1360. *See also specific particles*
  of leptons, 1365
  notation, 1362

Antiprotons, 1360

Antiquarks, 1372, 1373*t*, 1374*t*

Anti-strange quark, 1373*t*, 1374*t*

Anti-top quark, 1373*t*

Anti-up quark, 1373*t*, 1374*t*

Apex angle (Φ), *989*, 989–990

Apollo 11, 984

Aqueous humor, 1033, *1033*

Arago, Dominique, 1078

Area, of geometric shapes, A-10*t*

Art forgery detection, 1348

Astigmatism, **1035**

Astronomy and astrophysics. *See also* Moon; Stars; Sun; Telescopes
  Doppler effect, 1129–1130
  Einstein's cross, *1142*
  space-time distortion by gravity, 1140–1142

Atom(s)
  classical model of, *852*, 852–853
  etymology of, 1357
  history of concept, 1357
  ionization of, **1221**
  magnetic dipole moment of, *852*, 852–853, 853*t*
  models
    Bohr (semiclassical), *1219*, 1219–1224, *1221*, 1230
    early, *1218*, 1218–1219, *1219*
    planetary, 1219, *1219*
    quantum, 1224–1230
  quantization of energy in. *See* Energy levels, quantization of
  shells, **1226**, 1227*t*
    filling of, 1237–1238, *1238*
  spin angular momentum in, 853
  subshells, **1226**, 1227*t*
    filling of, 1237–1238, *1238*

Atomic cohesive energy
  of covalent solids, 1270, *1270*
  of ionic solids, **1269**

Atomic mass, A-22*t*–A-23*t*
  vs. mass number, 1294

Atomic mass unit (u), **1294**–1295
  rest-energy equivalent, 1295

Atomic number (*Z*), **1222**, **1294**, A-22*t*–A-23*t*
  vs. ionization energy, 1240, *1241*, 1242–1244, *1243*
  vs. neutron number, for stable nuclei, *1298*

Atomic spectra, of gases, *1216*, 1216–1218, *1217*, 1241–1244

Atomic spectroscopy, 1089

Audio systems
  impedance matching in, 940
  speaker crossover networks, 943

Audio tape, 854

Aurora Australis, 816

Aurora Borealis, 816

Automobiles
  hybrid drive system, 883, *883*
  rearview mirrors
    convex, 1017, *1017*
    day/night setting on, 1010, *1010*
  taillight reflectors, 984, *984*

Average current, **753**

Average energy density, of electromagnetic waves, 962

Average power ($\mathcal{P}_{avg}$), in AC circuits, 935–937, **936**, 938, *938*

Back emf, 882–884, 898
Background radiation, 1344
Bainbridge mass spectrometer, 817
Balmer, Johann Jacob, 1217, 1222
Balmer series, *1217,* **1217,** *1221*
Band, **1275**
Band theory of solids, 1274–1276, *1275, 1276*
  and electrical conduction, *1276,* 1276–1279, *1277, 1278*
Bardeen, John, 1282, 1284
Barrier height, **1200**
Barrier penetration. *See* Tunneling
Barriers, tunneling through, *1200,* **1200**–1202
Baryon(s)
  antiparticles, 1365
  composition, *1372,* 1373, 1374*t,* 1376, *1376*
  properties, **1364**–1365, 1364*t*
    patterns in, 1370–1371, *1371*
Baryon number, **1365**
  law of conservation of, **1365**–1366
Base of logarithms, **A-9**
Battery
  circuit symbol for, *727*
  emf of, 775–778, **776**
  function of, 753
  internal resistance, 776, **776,** 777
  terminal voltage, 776, *776,* 777
BCS theory, 1284, 1285
Beam splitters, 1064, *1064*
Beauty quark, 1374
Becquerel, Antoine-Henri, 1293, 1304, *1304*
Bednorz, J. Georg, 1284–1285
Beryllium (Be)
  electronic configuration, 1238, *1239*
  isotopes, 1310*t*
Beta (β) decay, 1304, *1313,* 1313–1315, *1314*
  and carbon dating, 1315–1316
  and cellular damage, 1344*t*
  decay pathways, 1317*t*
  and neutron activation analysis, 1348
Betelgeuse (star), color of, 1155, *1155*
Biconcave lens, *1023*
Biconvex lens, *1023*
Big Bang theory, 1378–1379
  cosmic background radiation from, 1379–1380, *1380*
  and expansion of universe, 1380–1383, *1381*
Binding energy, nuclear, *1299,* **1299**–1300
  in liquid-drop model, 1300–1302, *1302*
Binomial expansions, A-12–A-13
Biophysics, radiation damage, 1342–1344, 1344*t*
Biot, Jean-Baptiste, 837–838
Biot-Savart law, 837–842, **838**

Birefringent materials, *1096,* **1096**–1097, 1096*t*
Black body, *1154,* **1154**
Blackbody radiation, **1154**
  classical prediction *vs.* experimental results, 1154–1156, *1155, 1156*
  quantum-mechanical approach, 1156–1158
Black holes, 1142
Bohr, Niels, 1153, 1219–1224, *1220, 1234*
Bohr magneton (μ_B), **853**
Bohr radius, 1221
Boltzmann distribution law, 1157
Bonds. *See* Molecular bonds
Born, Max, 1188
Boron (B)
  electronic configuration, 1238, *1239*
  isotopes, 1310*t*
Bosons, 1284
Bottom quark (b), 1373*t,* **1374,** 1374*t*
Boundary conditions
  classical particles under, 1191
  quantum particles under, *1191,* 1191–1196, *1192, 1193*
    analogy to standing waves, 1195–1196
    free electron theory of metals, 1270–1274
    Schrödinger equation and, 1197–1198, 1199
    well of finite height, *1198,* 1198–1200, *1199*
    well of infinite height, *1191,* 1191–1196, *1192, 1193*
Brackett series, 1217–1218
Bragg, W. L., 1092
Bragg's law, **1092**
Brattain, Walter, 1282
Bremsstrahlung, 1242, *1242*
Brewster, David, 1095
Brewster's angle, **1095**
Brewster's law, **1095**
Bubble chambers, *1345,* **1345**
Buckminsterfullerene (buckyballs), 1269–1270, *1270*
Buckyballs. *See* Buckminsterfullerene
Buckytubes, 1270

Calcium (Ca), isotopes, 1310*t*
Calculus, A-13–A-19
Camera(s), *1031,* **1031**–1032
  digital, *1031,* **1031**–1032, 1164
  lens coatings, *1063,* 1064
  light meter in, 1163
  spherical aberration in, 1030
Cancer, radiation and, 1343
Capacitance, 722–**723**
  calculation of, 724–727
  of capacitor with dielectric, *735,* 735–736

Capacitance (*Continued*)
  mechanical analogy to, 907–909, 912*t*
  units of, 723
Capacitive reactance, 930, 931, 934
Capacitor(s), **722,** *722, 723. See also RC* circuits; *RLC* circuits
  in AC circuits, *929,* 929–931, *930,* 936
  applications of, 722, 725, 734–735, 936, 939, 943
  breakdown voltage of, 736
  capacitance of. *See* Capacitance
  charge on, 722–723
  charging of, *789,* 789–790, *790,* 791–792
  circuit symbol for, **727**
  cylindrical, capacitance of, *725,* 725–726
  with dielectrics, *735,* 735–738, *737,* 740–743
  discharging of, 790–794
  displacement current in, **953**–955
  electrolytic, 737, *737*
  energy stored in, 731–735, *732,* 903
  equivalent, 728, 729–731
  labels on, 723
  parallel combination of, **727**–728, *728*
  parallel-plate, 722–723, *723*
    capacitance of, 724–725
  rated voltage of, 736
  series combination of, *729,* 729–731
  spherical, capacitance of, *726,* 726–727
  types of, 737, *737*
  variable, 737, *737*
  working voltage of, 736
Carbon (C). *See also* Diamonds
  C_{60} (buckminsterfullerene), 1269–1270, *1270*
  covalent bonding of, *1269,* 1269–1270
  electronic configuration, 1238, *1239*
  isotopes, 1294, 1310*t*
    and carbon dating, 1315–1316
    decay of, 1306–1307, *1314*
  mass of nucleus, 1295*t*
Carbon dating, 1315–1316
Carbon dioxide, index of refraction, 987*t*
Carlson, Chester, 712
CCD. *See* Charge-coupled device
CDs. *See* Compact discs
Central maximum, **1078**
CERN. *See* European Laboratory for Particle Physics
Chadwick, James, 1296
Chain reaction, nuclear, *1332*
  critical, subcritical, and supercritical, **1333**
  self-sustained, 1332–1333, *1333*
Challenger shuttle disaster, *1362*
Chamberlain, Owen, 1360
Characteristic x-rays, 1242, *1242*

Charcoal briquettes, color of glow from, 1154, *1154*
Charge (*q*)
  conservation of, 643, *643*, 785
  mechanical analogy to, 912*t*
  positive and negative, **643**
  properties of, 642–644
  quantization of, 643, 710
  source, 651, *651*
  test, 651, *651*
  units of, 646
Charge carrier, 753
  drift speed of, 754–755
Charge-coupled device (CCD), 1031, *1031*, 1164
Charge density
  linear, **655**
  surface, **655**
  volume, **655**
Charging
  of capacitors, *789*, 789–790, *790*, 791–792
  by conduction, 645
  of conductors, 644–645
  by induction, *644*, **644**–645
  of insulators, 645, *645*
Charm (*C*), 1373
Charmed quark (c), 1373–1374, 1373*t*, 1374*t*
Chernobyl nuclear power plant, 1334–1335
Chip technology, advances in, 1283, *1283*
Christmas tree light strings, 781–782, *782*
Chromatic aberration, *1030*, 1030–**1031**
Chu, Steven, 1247
Circles, **A-10**
Circuit(s). *See* Alternating current (AC) circuits; Direct current (DC) circuits
Circuit diagram, **727**
Circuit elements. *See* Capacitor(s); Diode(s); Inductor(s); Resistor(s); Transistors
Circuit symbols, **727**
  AC source, 924
  battery, *727*
  capacitor, *727*
  diode, 942
  inductor, 900
  resistor, 763
  switch, *727*, 900, *900*
Circular aperture, resolution through, *1084*, 1084–1086
Cladding, 995, *995*
"Climate Change 2007: The Physical Science Basis" (UN Intergovernmental Panel on Climate Change), 1267
Clocks. *See* Time

Cloud chambers, **1345**
Coaxial cable
  capacitance, 726
  inductance of, 905, *905*
  radial resistance, 759–760
Cobalt (Co), isotopes, 1311*t*
  in radiation therapy, 1348, *1348*
COBE satellite, 1380
Coherent light sources, **1052**
Cohesive energy, atomic
  of covalent solids, 1270, *1270*
  of ionic solids, **1269**
Coil
  Faraday's law for, 869
  inductance of, 898–900
  magnetic dipole moment of, 823–825
Collective model of nucleus, 1303
Colliders, 1357, *1357*, *1378*, **1378**
Color charge, **1375**–1377, *1376*
Color force, **1376**
Color television, 1034
Common logarithms, **A-19**
Commutator, 882, *882*
Compact discs (CDs), as diffraction grating, 1088, *1088*
Compactified dimensions, 1384
Compass
  Earth's magnetic field and, 855–856, *856*
  history of, 808
Composition resistor, 757
Compton, Arthur Holly, 1165, *1165*
Compton effect, *1165*, **1165**–1167, *1166*
Compton shift equation, **1166**–1167
Compton wavelength, **1166**
Computer disks, magnetic, 854
Computer keyboard buttons, 725
Concave mirrors, **1010**–1013, *1011*, *1012*, *1013*, 1015–1017
  ray diagrams for, 1014, *1014*
  in telescopes, *1039*, 1039–1040
Condon, E. U., 1312
Conduction. *See* Electrical conduction
Conduction band, **1276**–1279, *1277*, *1278*, *1279*
Conduction electrons, 761
Conductivity (σ), **756**, 761
Conductor(s), **644**
  charged, electric potential due to, *707*, 707–709
  charging of, 644–645
  current-carrying
    in magnetic field, *819*, 819–821, *820*
    Hall effect, 825–827, *826*
    magnetic field created by, 837–842
    (*See also* Ampère's law)
    magnetic force on, *819*, 819–821, *820*, *821*
  electric field of
    in cavities, 708–709
    at sharp points, 708, 709

Conductor(s) (*Continued*)
  in electrostatic equilibrium
    properties of, 682, 682–685, *683*
    surface electric potential of, 707
  parallel, magnetic force between, 842–844, *843*
Cones, in eye, 1033–1034, *1034*
Confinement time (τ), **1338**, 1339
Conservation of baryon number, **1365**–1366
Conservation of charm, 1373
Conservation of electron lepton number, 1367–1368
Conservation of energy
  in relativistic situations, 1139
  uncertainty principle and, 1362
Conservation of lepton numbers, 1367–1368
Conservation of muon lepton number, 1368
Conservation of strangeness, **1369**–1370
Conservation of tau lepton number, 1368
Conservative field, 697
Constructive interference, of light waves, **1054**, 1057, 1058, 1060, 1082
Control rods, of nuclear reactor, *1334*, **1334**
Conversion of units, A-1*t*–A-2*t*
Convex-concave lens, *1023*
Convex mirror(s), *1013*, **1013**, 1017, *1017*
  ray diagrams for, *1014*, 1015
Cooper, L. N., 1284
Cooper pair, 1284
Coordinate systems
  space-time coordinates, 1130–1131
  spherical polar coordinates, 1225, *1225*
Copper (Cu)
  Fermi energy, 1272*t*
  isotopes, 1311*t*
Cornea, 1033, *1033*
Correspondence principle, **1223**
Cosine function, A-11
Cosmic background radiation, 1379–1380, *1380*
Cosmic rays, 816
Coulomb (C), **646**, *843*
Coulomb, Charles, 645–646, *646*
Coulomb constant (*k*ₑ), **646**
Coulomb force. *See* Electric force
Coulomb repulsion effect, in liquid-drop model of nucleus, 1301
Coulomb's law, 645–651, **646**
  vector form of, 647–648
Covalent bonding, *1259*, **1259**–1260
Crab Nebula, 952
Critical angle, **993**–995, *994*
Critical ignition temperature (*T*ᵢgₙᵢₜ), 1338–1340

Critically damped oscillation, in *RLC* circuit, 913
Critical reaction, **1333**
Critical temperature, **762**, 1283
Crossover networks, 943
Crystal(s)
  birefringent, *1096*, 1096–1097, *1096t*
  energy gap in, *1277t*
  ionic bonding in, 1268, *1268*
  X-ray diffraction by, *1091*, 1091–1092, *1092*
Crystalline lens, 1033, *1033*
Crystalline solids, 1096
Cubic zirconia, 994
Curie (Ci), 1306
Curie, Marie, 1304, *1304*
Curie, Pierre, 1304, *1304*
Curie temperature, **854**, 854t
Current (*I*), 752–755, **753**, *753. See also* Electrical conduction
  average (*I*avg), **753**
  in capacitor in AC circuit, 929–931, *930*
  direction of, 753
  displacement, **953**–955
  induced, 867–868, *868*, 897–**898** (*See also* Induction)
  in inductor in AC circuit, 927–929, *928*
  instantaneous, **753**
  mechanical analogy to, *912t*
  microscopic model of, 753–755, *754*
  misconceptions about, 763
  in resistor in AC circuit, *924*, 924–927, *926*
  in series *RLC* circuit, *932*, 932–933, 937–938, *938*
Current balance, 843
Current density (*J*), **756**, 761
Current loop
  magnetic dipole moment of, *823*, **823**, 852
  in magnetic field
    magnetic force on, 821, *821*
    torque on, 821–825, *822*
Cutoff frequency, **1161**, 1162–1163, *1163t*
Cutoff wavelength, 1163
Cyclotron, 817–819, *818*
Cyclotron frequency, 814
Cylindrical capacitors, capacitance of, *725*, 725–726

Damped oscillation, in *RLC* circuit, 913, *913*
Dark energy, in universe, 1382–1383
Dark matter, 1382
D'Arsonval galvanometer, 794
Daughter nucleus, **1308**
Davisson, C. J., 1169
Davisson-Germer experiment, 1169
DC. *See* Direct current

Dead Sea Scrolls, dating of, 1315
De Broglie, Louis, 1153, *1168*, 1168–1169
De Broglie wavelength, 1168–1170
Debye, Peter, 1165
Decay, spontaneous, **1308**
Decay constant, **1305**
Decay rate, **1305**
Dees, of cyclotron, 818, *818*
Defibrillator, 734, *734*
Delta (Δ), *1364t*, *1374t*
Density-of-states function, **1272**
Depletion region, of junction diode, 1280, *1280*
Depth of field, **1032**
Derivatives, **A-13**, A-14t
  second, of position curve, A-14
Derived quantities, A-24
Destructive interference, of light waves, **1054**, 1057, 1061, 1079–1080, 1081, 1082
Deuterium, fusion and, 1336–1339, *1338*, 1341, *1341*, *1342*
Diamagnetism, 854–855, *855*
  superconductors and, 1284
Diamonds
  sparkling of, 994
  structure and bonding of, *1269*, 1269–1270
Dielectric constant (κ), **735**, 736, *736t*
Dielectrics, **735**
  atomic description of, 740–741, *741*
  capacitors with, *735*, 735–738, *737*, 740–743
  polarization of, 740–741, *741*
Dielectric strength, 736, *736t*
Differentials, perfect, A-19
Differentiation, A-13–A-15
Diffraction, **1053**, 1077–1078
  of electrons, 1169
  of X-rays by crystals, *1091*, 1091–1092, *1092*
Diffraction grating(s), **1086**–1089, *1087*
  applications, *1089*, 1089–1091, *1090*, *1091*
  light intensity distribution, 1087, *1087*
  position of fringes, 1087, *1087*, 1088–1089
Diffraction grating spectrometer, 1089, *1089*
Diffraction patterns, **1078**
  of diffraction grating. *See* Diffraction grating(s)
  as interference, 1078, 1079
  of light passing circular object, 1078, *1078*
  of light passing edge of object, 1078, *1078*
  multiple slits, 1058, *1059*
  single slits, *1078*, 1078–1083, *1079*, *1081*, *1082*

Diffraction patterns (*Continued*)
  light intensity distribution, 1056–1058, *1057*, *1058*, 1081, *1081*
  position of fringes, 1079–1081, *1082*, 1082–1083
Diffuse reflection, *982*, **982**
Digital cameras, *1031*, **1031**–1032, 1164
Digital micromirror device, *984*, 984–985
Digital movie projectors, 984–985
Dimension, A-2–A-3
Diode(s), **942**
  applications, 942, *942*
  circuit symbol for, 942
  junction, 758, *758*, *1280*, **1280**
  light-absorbing, 1281, *1281*
  light-emitting (LEDs), 1281, *1281*
Diopters, **1035**
Dipole antenna, 966
Dipole-dipole force, 1260
Dipole-induced dipole force, 1260
Dirac, Paul A. M., 1234, *1358*, 1358–1359, *1359*
Dirac sea, 1359, *1359*
Direct current (DC), **775**
  AC-DC converters, *941*, 942–943
  disadvantages of, 942
Direct current (DC) circuits
  emf in, 775–778, **776**
  Kirchhoff's rules, **785**–788
    problem-solving strategy for, 786–787
  RC circuits, **788**–794
  resistors in parallel, *780*, **780**–782, 782–785
  resistors in series, **778**–780, *779*, 781–782, 783–784
  RLC circuits. *See RLC* circuits
  RL circuits, *900*, **900**–905, *901*, *902*
  time response of, 901
Direct current (DC) generators, 882, *882*
Direction of polarization, **1093**
Direction of propagation, 957
Disintegration energy (*Q*), **1308**–1312, 1313, *1313*, 1315
  in fission, 1331
Dispersion, of light waves, *992*, 992–993, *993*
Dispersion force, 1260
Displacement current (*I*d), **953**–955
Dissociation energy, **1259**
Distance (*d*), **A-9**
Division, and uncertainty, A-20
DNA molecules, hydrogen bonding in, *1260*, 1260–1261
Domains, in ferromagnetic material, *854*, **854**
Domain walls, **854**
Donor atom, **1278**
Doped semiconductors, **1278**–1279, *1279*

Doping, **1278**
Doppler effect, relativistic, 1129–1130
Double-refracting materials. *See*
  Birefringent materials
Double-slit interference patterns, *1052,*
  1052–1053
  in electron beams, *1174,* 1174–1175,
  1175
  light intensity distribution, 1056–1058,
  1057, 1058, 1081–1083, *1082*
  position of fringes, *1054,* 1054–1056,
  1082, 1082–1083
  Young's double-slit experiment, *1052,*
  1052–1053
Doublet, 1234
Down quark (d), **1372,** 1373*t,* 1374*t*
Drift chamber, **1346**–1347
Drift speed, 754–755
Drift velocity, 761
Drude, Paul, 760
Dynode, of photomultiplier tube, *1163,*
  1163–1164

e (Euler's number), A-9
e (fundamental charge), 643
ϵ_0 (permittivity of free space), **644**
Earth, magnetic field of, 855–856, *856*
Ear thermometers, 1158, *1158*
Eddy currents, induced, *884,* **884**–885,
  885
Eightfold way, *1371,* **1371**
Einstein, Albert, *1112, 1118*
  and general theory of relativity, *1118,*
  1140–1142
  on photoelectric effect, 979,
  1161–1163
  and quantum mechanics, *1118,* 1153,
  1158
  and special theory of relativity, 955,
  1111, 1112–1113, 1116, 1118,
  1118, 1119–1120, 1130
Einstein's cross, *1142*
Elastic scattering, 1318–1319
Electrical conduction. *See also*
  semiconductors
  band theory of solids and, *1276,*
  1276–1279, *1277, 1278*
  charging objects by, 645
  metals, *1276,* 1276–1277
  models of
    classical, 760–762
    quantum, 1270–1274
  superconductors, **762**–763, *763,* 763*t*
Electrically charged, **642**
Electrical meters, 794–795
Electrical power, 763–766
  average, in AC circuits, 935–937, **936,**
  938, *938*
  delivered by electromagnetic waves,
  962
  to device, 764

Electrical power (*Continued*)
  to resistor, 764–765, 793–794, 925–926
  transmission of, *752,* 765, 939, *941,*
  941–942
Electrical safety, in household wiring,
  796–798
Electrical systems, analogies with
  mechanical systems, 907–909,
  912–913, 912*t*
Electric charge. *See* Charge
Electric current. *See* Current
Electric dipole, **653**
  in electric field, *738,* 738–740
    potential energy of, 738–739
  electric field lines of, *660,* 661
  electric field of, **654**
  electric potential around, 698, *698,*
  702–704
  equipotential surfaces for, 701, *701*
Electric dipole moment, *738,* 738–739
Electric field(s) ($\vec{E}$), **651**–654
  of conductor
    in cavities, 708–709
    at sharp points, 708, 709
  of continuous charge distribution,
  654–659
  direction of, 652, *652*
  of electric dipole, **654**
  electric dipole in, *738,* 738–740
    potential energy of, 738–739
  energy density of, **733,** 961–962
  induced by magnetic flux change, *878,*
  878–880
  Lorentz force and, 816
  vs. magnetic field, 838
  motion of charged particle in,
  (uniform field), 661–664
  problem-solving strategy for, 656
  as rate of change of electrical potential,
  694
  of sinusoidal electromagnetic wave,
  957, 959–960
  superposition principle for, 652–654
  units of, 694
  value of, from electric potential,
  701–703
Electric field lines, *659,* **659**–661, *660,*
  661
  of electric dipole, *660,* 661
  and electric flux, *673,* 673–674, *674*
  electric potential and, 694, *695*
  vs. magnetic field lines, 840, 850, *850*
  rules for drawing, 660
Electric flux (Φ_E), *673,* **673**–676, *674*
  net, **674**–675
    through closed surface, 674–675, *675*
    (*See also* Gauss's law)
  units of, 674
Electric force (Coulomb force),
  646–651
  conservative, 646
  and fusion, 1337, *1337*

vs. magnetic force, 811
  and nuclear stability, 1297–1299, *1298,*
  1303
Electric guitar, 869–870, *870*
Electricity, etymology of, 641
Electric potential (*V*), 692–694, **693**
  applications of, 710–712
  in capacitors, 731–735, *732*
  in DC circuit, 776
  due to charged conductor, *707,*
  707–709
  due to continuous charge distribution,
  703–707
  due to electric dipole, 698, *698,*
  702–704
  due to point charges, *697,* 697–700,
  698
  problem-solving strategy for, 703–704
  in uniform electric field, 694–697
  units of, 694
  value of electric field from, 701–703
Electric shock, 797
Electrolyte, 737
Electrolytic capacitor, 737, *737*
Electromagnetic field, force on particle
  in, 956
Electromagnetic plane waves, *957,*
  957–960
  wave intensity of, 961–963
Electromagnetic waves
  electromagnetic spectrum, *952,*
  966–968, *967*
  energy carried by, 961–963
  energy density of, 962
  light as, 956–957, 959, 979, 1116
  medium for, 952
  momentum transferred by, 963
  plane, *957,* **957**–960
    wave intensity of, 961–963
  production by antenna, 965–966, *966*
  properties of, 956, 959–960
  radiation pressure of, 963–965
  sinusoidal, *957,* 959–960
  speed of, 956–957, 959
  wave intensity of, 961–963
Electromagnetism
  definition of, 641
  electroweak theory and, 1377
  evolution of, at origin of universe,
  1379, *1379*
  field particles for, 1358, 1359*t,* 1362,
  1362
  as fundamental force, 642, 1358
  in Standard Model, *1377,* 1377–1378
Electromotive force. *See* emf
Electron(s). *See also* Free-electron theory
  of metals
  antiparticles, 1359
  charge of, 643, 646, 646*t,* 1294
  conduction, 761
  double-slit interference patterns, *1174,*
  1174–1175, *1175*

Electron(s) (*Continued*)
e/*m*$_e$ ratio for, 817
as lepton, 1365
linear momentum of, 1135
mass of, 1294–1295, 1295*t*
nuclear force and, 1298
orbital magnetic moment of, *852,* 852–853
properties, 1364*t*
rest energy of, 1137
spin angular momentum of ($\vec{\mathbf{S}}$), **1235**–1236, *1236*
spin magnetic moment of ($\vec{\boldsymbol{\mu}}_{\text{spin}}$), 1236
spin of, *853,* **853,** **1233**–1235, *1234*
   and spin magnetic quantum number, 1234–1237
wavelength of, 1169
Electron affinity, **1258**
Electron beam, bending of, *815,* 815–816
Electron bombardment CCD camera, 1164
Electron capture, **1314**–1315, 1317*t*
Electron cloud, 1228, *1228*
Electron configurations, A-22*t*–A-23*t*
Electron gas, 761
Electron-hole pairs, **1278**
Electronic devices. *See also* Generators, electric; Motor(s), electric
   AC-DC converters, *941,* 942–943
   chip technology, advances in, 1283, *1283*
   color televisions, 1034
   defibrillators, 734, *734*
   digital cameras, *1031,* **1031**–1032
   electric guitars, 869–870, *870*
   laser pointers, 964–965
   laser printers, 712
   light meter, in camera, 1163
   metal detectors, 897
   photocopy machines, 712, *712*
   smoke detectors, *1312,* 1312–1313
   solar cells, 1063–1064
   stud finders, *737*
   toothbrushes, *906,* **906**–907
Electron lepton number, conservation of, 1367–1368
Electron microscopes, *1170,* **1170**–1171, *1171*
Electron-neutrino (ν_e), 1364*t,* 1365, 1367
Electron spin resonance, 1320
Electron volt (eV), 694
Electrostatic equilibrium, conductors in
   properties of, *682,* **682**–685, *683*
   surface electric potential of, 707
Electrostatic precipitators, *711,* 711–712
Electrostatics, applications of, 710–712
Electroweak force, 1377, 1379, *1379*
Electroweak theory, **1377**

Elements. *See also* Periodic table
   ionization energy *vs.* atomic number, 1240, *1241,* 1242–1244, *1243*
   origin of, 1300
Ellipse, **A-10,** *A-10*
emf ($\mathcal{E}$), 775–778, **776**
   back, 882–884, 898
   induced, 867–868, *868,* 897–**898** (*See also* Induction)
   motional, *871,* **871**–876, *872*
   self-induced, **898**
Emissions
   line width of, 1177
   spontaneous, *1244,* **1244**
   stimulated, *1244,* **1244**–1246
Emission spectroscopy, *1216,* **1216**–1218, *1217,* 1241–1244
Endothermic reactions, **1318**
Energetically favorable bonds, **1259**
Energy (*E*). *See also* Conservation of energy; Kinetic energy; Potential energy
   atomic cohesive
      of covalent solids, 1270, *1270*
      of ionic solids, **1269**
   in capacitor, 731–735, *732,* 903
   carried by electromagnetic waves, 961–963
   electrical, mechanical analogies to, 907–909, 912–913, 912*t*
   ground-state, **1193**
   in inductor, 903–905
   mass as form of, 1137, 1139
   quantization of. *See* Quantum mechanics
   of quantum particle in a box, *1193,* 1193–1194
   relativistic, 1135–1140
   and resistor
      delivered, 764–765, 793–794, 925–926
      stored, 903
   units of, A-2*t*
Energy density
   of electric field, **733,** 961–962
   of electromagnetic wave, 962
   of magnetic field, 904, 961–962
Energy gap, *1277,* **1277**–1279, 1277*t,* *1278, 1279*
Energy-level diagram, **1156,** *1157,* 1193, *1193*
   molecular, 1242, *1242*
Energy levels
   quantization of, **1156,** *1157,* 1193, *1193* (*See also* Transitions)
   in Bohr model, 1219–1224, *1221,* 1230
   in quantum model, 1225–1227
   of simple harmonic oscillator, 1205–1206, *1206*
Engines. *See* Motor(s)

Equation(s)
   coefficients of, A-7
   Compton shift equation, **1166**–1167
   Galilean transformation equations, **1114**–1115
   Lorentz space-time transformation equations, **1130**–1131
   Lorentz velocity transformation equations, 1131–1134, **1132**
   Maxwell's equations, **955**–956, 1116
   quadratic, A-7
   Schrödinger equation, 1188, 1196–1198, **1197,** 1199
   and quantum model of hydrogen atom, 1224–1227
Equilibrium, electrostatic. *See* Electrostatic equilibrium
Equipotential surfaces, 695, 701, *701*
Equivalence, principle of, **1141**
Equivalent resistance, **779,** 780–781
Eta (η), properties, 1364*t*
Ether, 1115
Ether wind theory, *1115,* 1115–1116, *1116*
Euler's number (*e*), A-9
European Laboratory for Particle Physics (CERN), 1123, 1363, 1375, 1378
Event, coordinates of, **1114**
Exchange particles. *See* Field particles
Excited states, *1193,* **1193**
Exclusion principle, **1237**–1238
   bosons and, 1284
   quarks and, 1375
Exothermic reactions, **1318**
Expectation value, **1189,** 1190–1191, 1195
Exponents, **A-4,** A-6*t*
   and uncertainty, A-20
Exposure times, of camera, 1031, 1032
Extraordinary (E) rays, *1096,* **1096**
Eye(s)
   anatomy of, *1033,* 1033–1034
   conditions of, *1034,* 1034–1035, *1035*
   vision in, 1033–1034, *1034*
Eyepiece, 1037, *1037,* 1038, *1038*

Factoring, A-7
Farad (F), **723**
Faraday, Michael, 641, 651, 659, 809, 867, *867,* 868, *868*
Faraday's law of induction, 867–871, **869,** 955
   applications, 869–871
   general form, 879
Far point, of eye, **1033**
Farsightedness (hyperopia), *1034,* **1034**
Femtometer (Fermi; fm), 1296
Fermi (femtometer; fm), 1296

Fermi, Enrico, 1313, 1330, 1332–1333, *1333*

Fermi-Dirac distribution function ($f(E)$), *1271*, **1271**–1272

Fermi energy, *1271*, **1271**–1272, 1272*t*
  insulators, 1277, *1277*
  metals, *1271*, **1271**–1272, 1272*t*, 1276, *1276*

Fermilab, *1111*

Fermions, 1372, 1375

Ferromagnetism, **853**–854, *854*, 854*t*

Feynman, Richard P., 1362, *1362*

Feynman diagram, *1362*, **1362**, *1363*

Fiber optics, *995*, 995–996

Field-effect transistor, **1282**

Field forces, 653. *See also* Electric field; Gravitational field; Magnetic field

Field particles (exchange particles; gauge bosons), **1358**, 1359*t*, 1361–1362, *1362*, *1363*, 1377

Films, thin, interference in, *1060*, 1060–1064, *1061*, *1062*
  problem-solving strategy, 1062

Filter circuits, 943, *943*

First-order maximum, 1054

Fish
  apparent depth of, *1020*, 1020–1021
  view from underwater, 994–995, *995*

Fission, **1330**–1332, *1331*
  energy released in, 1331–1332
  nuclear binding energy and, 1299–1300

Fission fragments, **1331**

Fission reactors, 1139, *1329*, 1332–1335
  advantages of, 1335
  control of, 1334, *1334*
  core design, 1334, *1334*
  history of, 1332–1333, *1333*
  safety and waste disposal, 1334–1335

Fizeau, Armand H. L., 980, *980*

Flavors, of quarks, **1372**

Fluorine (F)
  electronic configuration, *1239*
  isotopes, 1310*t*

Flux
  electric, *673*, **673**–676, *674*
    net, **674–675**
    through closed surface, 674–675, *675* (*See also* Gauss's law)
    units of, 674
  magnetic, **850**
    electric field induced by change in, *878*, 878–880
    Gauss's law for, 850–852

f-number
  of camera lens, **1031**–1032
  of human eye, 1033

Focal length (f)
  of combination of thin lenses, 1029
  of concave mirror, *1012*, **1012**

Focal length (f) (*Continued*)
  sign conventions, 1013*t*
  of thin lens, **1022**

Focal point(s)
  of concave mirror, *1012*, **1012**
  of thin lens, 1023, *1023*

Food
  analysis of heavy metal contamination in, 1217
  irradiating to preserve, 1348–1349, *1349*

Forbidden transitions, 1241

Force(s) ($\vec{F}$), *See also* Electric force; Electroweak force; Fundamental forces; Gravitational force; Magnetic force; Nuclear force; Strong force; Weak force
  on charged object, 652
  on charged particle, 652
  color force, **1376**
  dipole-dipole force, 1260
  dipole-induced dipole force, 1260
  dispersion force, 1260
  field, 651
  on particle in electromagnetic field, 956
  relativistic, 1135
  units of, A-1*t*

Fourier transform, 1065

Fourier transform infrared (FTIR) spectroscopy, 1064–1065

Fractional uncertainty, A-20

Franklin, Benjamin, 643

Fraunhofer diffraction pattern, **1078**–1083, *1079*, *1081*

Free-electron theory of metals
  classical, 760–762
  quantum, 1270–1274

Frequency (f)
  of electromagnetic waves, 959
  of light, and particle model, 1168

Fresnel, Augustin, 1078

Fresnel lens, 1024–1025, *1025*

Fringes, **1052**, *1053*

Frisch, Otto, 1330

FTIR. *See* Fourier transform infrared (FTIR) spectroscopy

Fuel elements, of reactor, 1333, 1334, *1334*

Fuller, R. Buckminster, 1269

Functions, **A-13**

Fundamental forces, 1358
  electromagnetism as, 642
  evolution of, at origin of universe, 1378–1379, *1379*
  field particles for, **1358**, 1359*t*, 1361–1362, *1362*, *1363*, 1377
  string theory and, 1384
  strong force as, 1363–1364, 1374–1375

Fusion, nuclear, **1335**–1336
  critical ignition temperature (T_{ignit}), 1338–1340

Fusion, nuclear (*Continued*)
  energy released in, 1335–1336, 1336, 1338, 1339
  tunneling and, 1203

Fusion reactors, 1139, 1336–1339
  advantages and disadvantages of, 1342
  design of, 1341–1342
  inertial confinement of plasma, *1341*, **1341**
  magnetic confinement of plasma, *1339*, **1339**–1341

Gabor, Dennis, 1089–1090

Galilean relativity, *1113*, **1113**–1116, *1114*
  limitations of, 1115–1116

Galilean transformation equations, **1114**–1115

Galilei, Galileo, 979

Galvanometer, **794**, 795, *795*

γ (gamma), 1121, 1121*t*

Gamma (γ) decay, 1304, *1304*, 1317, *1317*
  decay pathways, 1317*t*
  and food preservation, 1348
  and radiation damage, 1343

Gamma rays, *967*, **968**, 1293

Gamow, George, 1312

Gauge bosons. *See* Field particles

Gauss (G), **812**

Gauss, Karl Friedrich, *676*

Gaussian surface
  definition of, 676
  flux through. *See* Gauss's law
  as imaginary surface, 678

Gauss's law, 676–678, *677*, **677**, 955
  applications of, 678–682
  in magnetism, 850–**852**, 955
  problem-solving strategy for, 678

Gauss's probability integral, **A-19**t

Geiger, Hans, 1218, 1293

Geiger counter, *1346*, **1346**

Gell-Mann, Murray, 1371, *1371*, 1372, 1373

General theory of relativity. *See* Relativity, general

General wave equation, 959

Generators, electric
  AC, *880*, 880–882
  DC, 882, *882*

Genetic radiation damage, 1343

GEO, 1066

Geometric optics, *1051*
  ray approximation in, 981, *981*

Geometry, A-9–A-10

Gerlach, Walter, 1234, *1235*

Germer, L. H., 1169

GFI. *See* Ground fault interrupter

Gilbert, William, 809

Ginza district, Tokyo, 1215

Glashow, Sheldon, 1377
Glass
   birefringent qualities, 1097
   dielectric constant and dielectric
      strength of, 736*t*
   index of refraction, 987*t*
Global warming, greenhouse effect and,
   1267
Gluinos, 1384
Gluons, 1358, 1359*t*, **1376**
GLV. *See* Grating light valve
Goeppert-Mayer, Maria, 1303, *1303*
Gold (Au)
   Fermi energy, 1272*t*, 1273
   isotopes, 1311*t*
Goudsmit, Samuel, 1234, 1235
Graphical representation, A-8, *A-8*
Graphite
   microscopic view of surface, *1203*
   structure of, *1269*
Grating light valve (GLV), 1089, *1089*
Gratings, diffraction. *See* Diffraction
   grating(s)
Gravitational field ($\vec{g}$), deflection of
   light by, *1118*, 1142, *1142*
Gravitational force ($\vec{F}_g$)
   evolution of, at origin of universe,
      1379, *1379*
   field particles for, 1358, 1359*t*, 1363
   as fundamental force, 1358
   in Standard Model, 1377, *1377*
Gravitational lens, *1142*
Gravitational waves, 1065–1066
Gravitons, 1358, 1359*t*, 1363
Gravity, and general relativity,
   1140–1142
Gray (Gy), 1344, 1344*t*
Greenhouse effect, and global warming,
   1267
Ground, **644**
   neutral wire, in household wiring, 796,
      796, 797, 797–798
   three-pronged electrical cords, *797,*
      797–798
Ground fault interrupter (GFI), 798,
   869
Ground state, *1193,* **1193**
Ground-state energy, **1193**
Group speed, of wave packet, 1173–1174
Guitar, electric, 869–870, *870*
Gurney, R. W., 1312

Hadrons, 1363, **1364**–1365, 1376
   properties, 1364*t*
Hafele, J. C., 1122
Hahn, Otto, 1330
Half-life ($T_{1/2}$), **1305**–1306, 1312
   of selected isotopes, 1310*t*–1311*t*
Half-power points, **939**
Half-wave antenna, electromagnetic
   wave production by, 965–966, *966*

Half-wave rectifier, 942, *942*
Hall, Edwin, 825
Hall coefficient, **827**
Hall effect, 825–827, *826*
Hall field, 826
Hall voltage, *826,* **826**–827
Halogens, 1239–1240
HCl molecule, absorption spectrum,
   1242–1243, *1243*
Heat (Q), units of, A-2*t*
Heat lamps, 967
Heat rays, 967
Heat sinks, 764
Heavy metal contamination of food,
   analysis of, 1217
Heisenberg, Werner, 1153, 1175, *1176,*
   1186
Heisenberg uncertainty principle,
   1175–**1177**
   and conservation of energy, 1362
   orbital magnetic quantum number
      and, **1232**
Helium (He)
   discovery of, 1217
   electronic configuration, 1238
   isotopes, 1310*t*
   mass of nucleus, 1295*t*
Helium-neon gas laser, 1246, *1246*
Henry (H), *898,* 899
Henry, Joseph, 643, 809, 867, *898*
Hertz, Heinrich, 952, 956, *956,* 979
Higgs boson, **1378**
Hole, in valence band, 1277–1278, *1278,*
   1279, *1279*
Holography, **1089**–1091, *1090, 1091*
Household wiring, 796
   electrical safety, 796–798
Hubble, Edwin, 1130, 1380
Hubble constant, **1380**
Hubble's law, *1380,* **1380**
Hubble Space Telescope, 1077, 1086,
   1086
Humason, Milton, 1380
Hummingbird, feather color on, *1051*
Hund's rule, 1238
Huygens, Christian, 978, *978*
Huygens's principle, **990**–992, *991*
Hybrid drive system, 883, *883*
Hydrogen (H)
   absorption spectrum, *1216*
   emission spectrum, *1216,* 1217,
      1217–1218
   frequency of photon emitted from,
      1222, 1223, 1241
   isotopes, 1294, 1310*t*
   space quantization for, 1233
Hydrogen atom
   allowed transitions, 1241, *1241*
   Bohr (semiclassical) model of, *1219,*
      1219–1224, *1221, 1230*
   electric and gravitational force within,
      647

Hydrogen atom (*Continued*)
   importance of understanding,
      1215–1216
   mass of, 1295*t*
   quantum model of, 1224–1227
   quantum numbers, 1225–1226, 1226*t*
      for *n*=2 state, 1236, 1237*t*
   physical interpretation of, 1230–1237
   radial probability density function,
      1228, 1228–1230, *1230*
Hydrogen bonding, *1260,* **1260**–1261
Hydrogen gas, molar specific heat of,
   1206–1207
Hydrogen molecules, exclusion
   principle and, 1259–1260
Hyperbolas, A-10, *A-10*
Hyperopia (farsightedness), *1034,* **1034**

I^2R loss, 765
Ice, index of refraction, 987*t*
Ice Man (Bronze Age remains), *1293,*
   1315–1316
Ideal solenoid, 848–849, *849*
Image(s), **1008**
   real, **1009**
   virtual, **1009**
Image distance, *1008,* **1008**
Image formation. *See also* Lens(es);
      Resolution
   in concave mirrors, **1010**–1013, *1011,
      1012, 1013,* 1015–1017
   ray diagrams, 1014, *1014*
   in convex mirrors, *1013,* **1013,** 1017,
      1017
   ray diagrams, *1014,* 1015
   in flat mirrors, *1008,* 1008–1010, *1009,
      1010*
   lateral magnification, **1009,** 1037
   ray diagram analysis of
      for mirrors, 1013–1015, *1014*
      for thin lenses, *1024,* 1024–1028, 1028
   by refraction, *1017,* 1017–1021, *1018,
      1019t*
   sign conventions
      reflection, 1013, *1013,* 1013*t*
      refraction, 1018, 1019*t*
      for thin lens, 1023, *1023,* 1023*t*, 1024
   in telescopes, *1039,* 1039–1040
Impedance, **933,** 934
Impedance matching, 940
Incoherent light sources, **1052**
Indefinite integrals, **A-16,** A-18*t*
Independent-particle model of nucleus.
      See Shell model of nucleus
Index of refraction (*n*), **986**–990, *987,*
   987*t*
   birefringent materials, **1096,** 1096*t*
   lens-makers' equation and, 1022–1023
   and polarizing angle (Brewster's law),
      1095
   and wavelength, 987, *987,* 992, *992*

Induced polarization, 740, 741, *741*
Inductance (*L*), **898**–900
  mechanical analogy to, 907–909,
    912–913, 912*t*
  mutual, *906*, **906**–907
  units of, 899
Induction, 867–868, *868*
  charging objects by, *644*, **644**–645
  eddy currents, *884*, **884**–885, *885*
  electric field created by, *878*, 878–880
  Faraday's law of, 867–871, **869**, 955
    applications, 869–871
    general form, 879
  in generators and motors, 880–884
  Lenz's law, *876*, **876**–878
  motional emf, *871*, **871**–876, *872*
  mutual, *906*, 906–907
  self-induction, **898**
Inductive reactance, **928**–929, 934
Inductor(s), *900*, **900**. *See also RLC*
    circuits; RL circuits
  in AC circuits, *927*, 927–929, *928*, 936
  circuit symbol for, 900
  energy stored in, 903–905
Inelastic scattering, 1319
Inertial confinement, of plasma, *1341*,
    1341
Infrared waves, *967*, **967**
In phase, **925**
In resonance, **937**
Instantaneous current (*I*), **753**
Institute for Advanced Studies,
    Copenhagen, *1220*
Instrumentation. *See also* Microscopes;
    Telescopes; Thermometers
  accelerator (Fermilab), *1111*
  ammeters, *794*, **794**–795
  beam splitters, 1064, *1064*
  colliders, 1357, *1357*, *1378*, **1378**
  compasses, 808, 855–856, *856*
  cyclotrons, 817–819, *818*
  electrostatic precipitators, *711*,
    711–712
  galvanometers, **794**, 795, *795*
  interferometers, *1064*, **1064**–1066,
    1065, *1116*, 1116–1117
  magnetic bottles, 816, *816*
  neutrino detectors, 1366, *1366*
  synchrotrons, 818
  telescopes, 1038–1040, *1038*, *1039*
  torsion balance, 645, *645*
  Van de Graaff generators, 710–711, *711*
  voltmeters, *795*, **795**
Insulators, **644**
  band theory and, 1276–1277, *1277*
  charging of, 645, *645*
Integrated circuits, *1257*, 1282–**1283**,
    1283
Integration, A-16–A-19*t*
  line integral, 693
  path integral, 693
  surface integral, 674

Intel, and chip technology, 1283, *1283*
Intensity (*I*)
  of diffraction grating interference
    pattern, 1087, *1087*
  in double-slit interference pattern,
    1056–1058, *1057*, *1058*,
    1081–1083, *1082*
  of electromagnetic waves, 961–963
  of polarized beam (Malus's law), **1094**
  in single slit interference pattern,
    1081, *1081*
Interference
  constructive, of light waves, **1054**, 1057,
    1058, 1060, 1082
  destructive, of light waves, **1054**, 1057,
    1061, 1079–1080, 1081, 1082
  of electrons, *1174*, 1174–1175, *1175*
  of light waves, 1051–1066 (*See also*
    Diffraction patterns)
    conditions for, 1051–1053, *1054*,
      1057, 1058, 1060–1061, 1081,
      1082
    constructive, **1054**, 1057, 1058, 1060,
      1082
    destructive, **1054**, 1057, 1061,
      1079–1080, 1081, 1082
    diffraction pattern as, 1078, 1079
    double-slit experiment (Young), *1052*,
      1052–1053
    double-slit interference patterns,
      1052–1058, *1053*, *1054*, *1057*,
      1058
    multiple slits, 1058, *1059*
    Newton's rings, *1061*, **1061**–1064
    in thin films, *1060*, 1060–1064, *1061*,
      1062
      problem-solving strategy, 1062
  sound waves, 1052
  water waves, *1053*
Interferogram, 1065
Interferometer, *1064*, **1064**–1066, *1065*,
    1116, 1116–1117
Intergovernmental Panel on Climate
    Change (UN), 1267
Internal resistance, *776*, **776**, 777
Intrinsic semiconductor, **1278**
Invariant mass, **1137**
Io, moon of Jupiter, 979–980
Iodine (I), isotopes
  activity of, 1307
  in radiation therapy, 1348
  in radioactive tracing, 1347
Ion(s), heavy, radiation damage from,
    1344*t*
Ion chambers, *1345*, **1345**–1346
Ion density (*n*), **1338**
Ionic bonding, **1258**–1259, *1259*
  in solids, *1268*, 1268–1269
Ionic cohesive energy, of solids, **1269**
Ionization energy, **1221**
  vs. atomic number, 1240, *1241*,
    1242–1244, *1243*

Ionization of atom, **1221**
Iridescence, *1051*
Iris, of eye, 1033, *1033*
Iron (Fe), isotopes, 1310*t*
Isotones, 1302
Isotopes, **1294**
  chemical and nuclear information,
    1310*t*–1311*t*
ITER fusion reactor, 1340

Jensen, Hans, 1303, *1303*
Joule heating, 765
JT-60 U tokamak (Japan), 1340–1341
Junction, **780**
Junction diodes, 757, *758*, *1280*, **1280**
Junction rule, *785*, **785**–788
Junction transistor, **1282**
Jupiter, magnetic field of, 856
J/Ψ particle, 1373–1374

Kamerlingh-Onnes, Heike, 763
Kaons (K), 1364*t*, 1369, *1369*
K capture, 1314
Keating, R. E., 1122
Keck Observatory, 1040, 1085–1086
Keyboard buttons, 725
Kilby, Jack, 1282–1283, *1283*
Kinetic energy (*K*)
  of electron, in photoelectric effect,
    1160–1162, *1163*
  relativistic, 1135–1136, *1136*, 1138
  and work. *See* Work-kinetic energy
    theorem
Kirchhoff, Gustav, *786*
Kirchhoff's rules, **785**–788
  problem-solving strategy for, 786–787

Laguerre polynomials, 1274
Lambda (Λ⁰), 1364*t*, 1369, *1369*, 1374*t*
Land, E. H., 1093
Large Electron-Positron (LEP) Collider,
    1378, *1378*
Large Hadron Collider, 1378
Laser cooling, 1248
Laser fusion, 1341, *1341*
Laser interferometer gravitational-wave
    observatory (LIGO), *1065*,
    1065–1066
Laser pointers, radiation pressure from,
    964–965
Laser printers, 712
Lasers
  applications, 1246–1248, *1247*, *1248*
  generation of light by, **1245**–1246,
    1246
  trapping of atoms with, *1247*,
    1247–1248
Lateral magnification, **1009**, 1037
Laue, Max von, 1091

Laue pattern, *1091,* 1092, *1092*
Law of conservation of baryon number, **1365**–1366
Law of conservation of electron lepton number, 1367–1368
Law of conservation of muon lepton number, 1368
Law of conservation of strangeness, **1369**–1370
Law of conservation of tau lepton number, 1368
Law of cosines, A-11
Law of reflection, **982**–983, 991, *991*
  applications, *984,* 984–985
Law of sines, A-11
Lawrence, E. O., *818*
Lawson, J. D., 1338
Lawson number ($n\tau$), **1338**–1339, *1339*
Lawson's criterion, *1339,* **1339**
LC circuit
  energy stored in, 908–909
  oscillations in, *907,* **907**–911, *908, 910*
Lead, isotopes, 1311*t*
Lederman, Leon, 1374
LEDs. *See* Light-emitting diodes
Length
  contraction of, in special relativity, **1125**–1126, *1126, 1128,* 1128–1129
  proper, **1125**–1126
  units of, A-1*t*
Lens(es)
  aberrations in, *1030,* **1030**–1031, 1039–1040
  in cameras, *1031,* **1031**–1032
  converging
    in cameras, 1031, *1031*
    focal points, 1023, *1023*
    image formation in, 1024, *1024,* 1025–1026
    magnification by, 1035
    for vision correction, *1034,* **1034,** 1035
  diverging
    chromatic aberration and, 1031
    focal points, 1023, *1023*
    image formation in, 1024, *1024,* 1026–1028, *1027*
    for vision correction, **1034,** *1035*
  of eye, *1033,* 1033–1035, *1034, 1035*
  Fresnel, 1024–1025, *1025*
  gravitational, *1142*
  lens-makers' equation, **1022**–1023
  power (*P*) of, **1035**
  shapes, 1023, *1023*
  thick, 1021–1022, *1022*
  thin, 1021–1030, *1022*
    combination of, 1028–1030, *1029*
    magnification by, 1023, *1024,* 1028
    ray diagrams for, *1024,* 1024–1028
    sign conventions for, 1023, *1023,* 1023*t,* 1024
    thin lens equation, **1023**

Lens-makers' equation, **1022**–1023
Lenz's law, *876,* **876,** 876–878
Lepton(s), 1363, **1365,** 1375*t,* 1377
  properties, 1364*t*
  in Standard Model, *1377,* 1377–1378
Lepton numbers, conservation laws, 1367–1368
Levitation, magnetic, superconductors and, 1284
Lifetime, of excited state, **1177**
Light
  coherent sources of, **1052**
  deflection by gravitational field, *1118,* 1142, *1142*
  early models of, 978–979
  incoherent sources of, **1052**
  and life on earth, 977
  monochromatic sources of, **1052**
  nature of, 978–979
  quantization of, 979
  speed of (*c*)
    measurement of, 979–980
    Michelson-Morley experiment, *1116,* 1116–1119
    relativity and, 1115–1119
  wave model of, 956–957, 959, 979, 1116
Light-absorbing diodes, 1281, *1281*
Lightbulbs
  in Christmas tree light strings, 781–782, *782*
  failure of, 779
  filament color, *1153*
  three-way, 781, *781*
Light-emitting diodes (LEDs), 1281, *1281*
Lightning, *641, 651, 692*
Light meter, in camera, 1163
Light waves
  diffraction of. *See* Diffraction
  dispersion of, *992,* **992**–993, *993*
  as electromagnetic radiation, 956–957, 959, 979, 1116
  interference of. *See* Interference
  polarization of. *See* Polarization
  reflection of. *See* Reflection
  refraction of. *See* Refraction
  unpolarized, *1093,* **1093**
  visible light spectrum, *967,* **967**–968, *992,* **992**–993, *993*
  wavelength of, measuring, 1055, 1062, 1064, 1087, 1089
LIGO. *See* Laser interferometer gravitational-wave observatory
Linear charge density (λ), **655**
Linear equations, A-7–A-9
Linearly polarized waves, *957,* **957**
Linear momentum ($\vec{p}$), relativistic, **1134**–1135, 1137, 1138–1139
Line integrals, 693
Line of stability, 1298, *1298*
Lines, **A-10,** *A-10*

Line spectra
  atomic, *1216,* **1216**–1218, *1217,* 1241–1244
    splitting of, in magnetic field, 1222, 1231–1233, *1232*
  molecular, *1262, 1266,* 1266–1268, *1267*
  X-ray, *1242,* 1242–1244
Liquid-drop model of nucleus, **1300**–1302
Lithium (Li)
  electronic configuration, 1238, *1239*
  Fermi energy, 1272*t*
  fusion reactors and, 1341–1342, *1342*
  isotopes, 1310*t*
Live wire, 796, *796*
Livingston, M. S., *818*
Lloyd's mirror, 1059, *1059*
Load, 776
Load resistance, **776,** 778
Logarithms, **A-9**
Loop rule, **785**–788
Lorentz, Hendrik A., 1130
Lorentz force, 816
Lorentz force law, **956**
Lorentz space-time transformation equations, **1130**–1131
Lorentz velocity transformation equations, 1131–1134, **1132**
Lyman series, 1217–1218, *1221*

Madelung constant (α), **1268**
Magic numbers, **1302**
Magnesium (Mg), isotopes, 1310*t*
Magnet(s)
  polarity of, 808–809
  superconducting, 763
Magnetic bottle, 816, *816*
Magnetic confinement, of plasma, *1339,* **1339**–1341
Magnetic declination, 856
Magnetic dipole, in magnetic field, potential energy of, 823
Magnetic dipole moment ($\vec{\mu}$)
  of atoms, *852,* 852–853, 853*t*
  of coil, 823–825
  of current loop, *823,* **823,** 852
  nuclear, 1319–1320
Magnetic energy density, 904
Magnetic field(s) ($\vec{B}$). *See also* Magnetic force
  Ampère's law, 844–848, **845**
    general form of (Ampère-Maxwell law), 953–**954,** 955
  characteristics of, 809–813
  charge moving in
    applications, 816–819
    magnetic force on, *810,* 810–813, **811**
    nonuniform field, 816, *816*
    uniform field, *813,* 813–816, *814*

Magnetic field(s) ($\vec{B}$) (*Continued*)
conductor, current-carrying, in, *819*, 819–821, *820*
  Hall effect, 825–827, *826*
from conductor, current-carrying, 837–842
  Ampère's law, 844–848, **845**
    general form of (Ampère-Maxwell law), 953–**954,** 955
current loop in
  magnetic force on, 821, *821*
  torque on, 821–825, *822*
direction of, 809, 813, *813* (*See also* Biot-Savart law)
of Earth, 855–856, *856*
vs. electric field, 838
energy density of, 904, 961–962
Gauss's law for, 850–**852**
of inductor, energy stored in, 903–905
Lorentz force and, 816
magnitude of, 811–812, 812*t*, 815 (*See also* Biot-Savart law)
operational definition of, **811**
of sinusoidal electromagnetic wave, *957*, 959–960
of solenoid, *848*, 848–849
sources of, 954
of toroid, *847*, 847–848
units of, 812
Magnetic field lines, 809, *809*, *810*, 813, *813*
vs. electric field lines, 840, 850, *850*
Magnetic fingerprinting, *808*
Magnetic flux (Φ_B), **850**
electric field induced by change in, *878*, 878–880
Gauss's law for, 850–852
Magnetic force
on current-carrying conductor, *819*, 819–821, *820*, *821*
on current loop, 821, *821*
vs. electric force, 811
on moving charge, *810*, 810–813, **811**
  right hand rule for, 811, *811*
between parallel conductors, 842–844, *843*
Magnetic poles, 808–809
Magnetic resonance imaging. *See* MRI
Magnetism
etymology of, 641
historical awareness of, 641, 808–809
in matter, 852–855
Magnetite, 641, 808
Magnification
angular, **1035**–1036, 1037, 1038–1039
by combinations of thin lenses, 1028
lateral, **1009**, 1037
magnifying glass (simple magnifier), *1035*, 1035–1037, *1036*
microscope, *1037*, **1037**–1038 (*See also* Microscopes)
sign conventions, 1013*t*

Magnification (*Continued*)
telescope, *1038*, **1038**–1040, *1039*
by thin lenses, 1023, *1024*
Malus's law, **1094**
Manganese (Mn), isotopes, 1310*t*
Maricourt, Pierre de, 808
Mariner 10 spacecraft, 964
Marsden, Ernest, 1218, 1293
Mass (*m*)
as form of energy, 1137, 1139
invariant, **1137**
in radioactive decay, 1139–1140
relativistic, 1134
relativity and, 1134, 1137, 1139–1142
units of, A-1*t*
Mass defect, **1330**
Mass number (*A*), **1294**
Mass spectrometer, *817*, **817**, *818*
Materials science. *See also* Metal(s); Solid(s)
birefringent materials, *1096*, **1096**–1097, 1096*t*
neutron activation analysis, **1348**
nonohmic materials, 756
  current-potential difference curve, 758, *758*
ohmic materials, 756
  current-potential difference curve, 758, *758*
  resistivity, **757**
optical activity, 1098
Mathematics, A-4–A-21
Maximum
central, **1078**
secondary, **1078**
side, **1078**
Maxwell, James Clerk, 641, 953, *953*, 956, 979, 1115
Maxwell's equations, **955**–956, 1116
Measurement. *See also* Instrumentation; SI (*Système International*) units
uncertainty in, A-20–A-21
Mechanical devices
computer keyboard buttons, 725
photocopy machines, 712, *712*
Mechanical engineering. *See* Automobiles; Railroad; Spacecraft
Mechanical systems, analogies with electrical systems, 907–909, 912–913, 912*t*
Mechanics
classical, limitations of, 1112
quantum. *See* Quantum mechanics
Medicine
MRI (magnetic resonance imaging), *1320*, **1320**–1321
positron-emission tomography (PET), *1360*, *1361*
radiation therapy, 1348, *1348*
radioactive tracers, 1347, *1347*
X-rays, and cellular damage, 1320, 1343, 1344*t*

Meissner effect, 855, *855*, 1283–1284, *1284*
Meitner, Lise, 1330
Mendeleev, Dmitri, 1239
Mercury (Hg)
emission spectrum, *1216*
isotopes, 1311*t*
Mesons
composition, *1372*, 1373–1374, 1374*t*, 1375–1376, *1376*
discovery of, **1361**
properties, **1364**, 1364*t*
patterns in, 1371, *1371*
Metal(s)
electrical conduction in
  band theory and, *1276*, 1276–1277
  classical view, 760–762
  quantum view, 1270–1274
Fermi energy, *1271*, **1271**–1272, 1272*t*, 1276, *1276*
Hall-effect measurements in, 827
metallic bonds, 1270, *1270*
resistivity, 758*t*
superconductors, 763, 763*t*
temperature coefficients of resistivity, 758*t*, 762, *762*
work function of, 1162, 1162*t*, 1163
Metal detectors, 897
Michelson, Albert A., 1064, 1116
Michelson interferometer, *1064*, 1064–1066, *1065*, *1116*, 1116–1117
Michelson-Morley experiment, *1116*, 1116–1119
Microchips, *1257*
Microscopes, **1037**–1038
compound, *1037*
electron, *1170*, **1170**–1171, *1171*
scanning tunneling (STM), *1186*, 1203, *1203*
Microwaves, *967*, **967**
Millikan, Robert, 643, 709
Millikan oil-drop experiment, *709*, 709–710
Mirror(s). *See also* Reflection
concave, **1010**–1013, *1011*, *1012*, *1013*, 1015–1017
  ray diagrams for, 1014, *1014*
  in telescopes, *1039*, 1039–1040
convex, *1013*, **1013**, 1017, *1017*
  ray diagrams for, *1014*, 1015
diverging, *1013*, **1013**
flat, *1008*, 1008–1010, *1009*, *1010*
lateral magnification, **1009**
mirror equation, **1012**
multiple images in, 1010, *1010*
ray diagrams for, 1013–1015, *1014*
reversal of image in, 1009
sign conventions for, 1013, *1013*, 1013*t*
spherical aberration in, 1030
Missing mass, in Universe, 1382

Models. *See also* Analysis models; Atom(s), models; Nucleus, atomic, models; Particle model; Standard Model; Wave model
of electrical conduction
classical, 760–762
quantum, 1270–1274
of light
early models, 978–979
wave model, 956–957, 959, 979, 1116
Moderators, **1330**
Molar specific heat, of hydrogen gas, 1206–1207
Molecular bonds, 1258–1261
covalent, *1259*, **1259**–1260, *1269*, 1269–1270
hydrogen, *1260*, **1260**–1261
ionic, **1258**–1259, *1259*, *1268*, 1268–1269
metallic, 1270, *1270*
in solids, 1268–1270
Van der Waals, 1260
Molecular spectra, *1262*, *1266*, 1266–1268, *1267*
Molecule(s)
nonpolar, 739
polar, 739, *739*
induced polarization, 740, 741, *741*
rotational motion of, 1261–1263, *1262*
symmetric, 740, *740*
vibrational motion of, *1264*, 1264–1266, *1265*
Moment of inertia (*I*), for diatomic molecule, 1261
Momentum ($\vec{p}$). *See also* Angular momentum; Linear momentum
of electromagnetic waves, 963
orbital angular
and magnetic moment, 852
quantization of, 852–853, 1231–1233
of quantum particle in a box, 1193
Monochromatic light sources, **1052**
Moon, measurement of distance to, 984
Morley, Edward W., 1116
Moseley, Henry G. J., 1243
MOSFET (metal-oxide-semiconductor field-effect transistor), *1282*, **1282**
Motion. *See* Simple harmonic motion; Wave(s)
Motional emf, *871*, **871**–876, *872*
Motor(s)
electric, 882–884
eddy currents in, 885
torque generation in, 823
Movie projectors, digital, 984–985
MRI (magnetic resonance imaging), *1320*, **1320**–1321
M-theory, **1384**
Müller, K. Alex, 1284–1285
Multiplication
and differentiation, A-14
and uncertainty, A-20

Muon lepton number, conservation of, 1368
Muon-neutrino (ν_μ), 1364*t*, 1365, 1368
Muons (μ^-), **1361**–1362
and length contraction, 1126
as lepton, 1365
properties, 1364*t*
and time dilation, *1122*, 1122–1123
Mutual inductance, *906*, **906**–907
Myopia (nearsightedness), **1034**, *1035*

Nanotechnology, *1186.* **1200,** 1279
Napoleon, cause of death, 1348
National Ignition Facility (Livermore National Laboratory), 1341
National Institute of Standards and Technology (NIST), 843
National Spherical Torus Experiment (NSTX), *1339*, 1340
Natural frequency (ω_0), of *LC* circuit, 909, 937
Natural logarithms, A-9
Natural radioactivity, 1317–1318, 1317*t*, *1318*
Near point, **1033**
Nearsightedness (myopia), **1034**, *1035*
Ne'eman, Yuval, 1371
Negative electric charge, **643**
Neon (Ne)
electronic configuration, *1239*
emission spectrum, *1216*
isotopes, 1310*t*
Neon signs, 1215, 1217
Neptunium (Np), radioactive series, 1317–1318, 1317*t*
Neutral wire, 796, *796*
Neutrinos (ν), **1313**–1315, *1314*, 1364*t*, 1365, 1382
Neutron(s)
charge of, 643, 646*t*, 1294
composition of, 1320, 1374*t*
decay of, 1367–1368
discovery of, 1296
interactions with nuclei, 1329–1330 (*See also* Fission)
magnetic dipole moment of, 853
mass of, 1294–1295, 1295*t*
properties, 1364*t*
thermal, **1330**
Neutron activation analysis, **1348**
Neutron capture, **1330**
Neutron number (*N*), **1294**
vs. atomic number, for stable nuclei, *1298*
Neutron radiation
damage from, 1343, 1344*t*
and neutron activation analysis, 1348
Newton, Isaac, 978, 992, 1140, A-13
Newtonian mechanics. *See* Mechanics
Newton's rings, **1061**–1064, *1062*
Nickel (Ni), isotopes, 1311*t*

NIST. *See* National Institute of Standards and Technology
Nitrogen (N)
electronic configuration, *1239*
isotopes, 1310*t*
decay of, *1314*
NMR (nuclear magnetic resonance), *1320*, **1320**
Nobel Prize, 710, 1089–1090, *1156*, 1161, *1165*, *1168*, *1218*, *1220*, 1248, 1282, 1284, 1285, 1296, 1303, *1303*, 1304, *1304*, 1313, *1333*, *1358*, 1359, 1361, *1361*, *1362*, 1363, *1371*, 1374, 1377, 1380
Noble gases, 1239, 1240–1241
Van der Waals bonding in, 1260
Nonohmic materials, 756
current-potential difference curve, 758, *758*
Nonpolar molecules, 739
Normalization condition, **1207**
Normalized wave function, **1189**
for particle in a box, 1192
Northern Lights, 816
North pole
of Earth, 855–856, *856*
of magnet, 808–809
Notation
for antiparticles, 1362
for atomic nuclei, 1294
for nuclear reactions, 1318
for quantities, A-2*t*–A-3*t*
Noyce, Robert, 1282
n-type semiconductors, **1278**, 1280
Nuclear binding energy, *1299*, **1299**–1300
in liquid-drop model, 1300–1302, *1302*
Nuclear bomb, development of, *1220*, *1362*
Nuclear force, **1297**–1299, *1298*, 1303, 1358
discovery of, 1361
field particles for, 1358, 1359*t*, 1361–1363, *1363*
source of, *1376*, 1376–1377
vs. strong force, 1363–1364, 1374–1375
Nuclear magnetic resonance. *See* NMR
Nuclear magneton (μ_n), 1319–1320
Nuclear reactions, **1318**–1319
Nuclear reactors
fission, 1139, *1329*, 1332–1335
advantages of, 1335
control of, 1334, *1334*
core design, 1334, *1334*
history of, 1332–1333, *1333*
safety and waste disposal, 1334–1335
fusion, 1139, 1336–1339
advantages and disadvantages of, 1342
design of, 1341–1342
inertial confinement of plasma, *1341*, **1341**

Nuclear reactors (*Continued*)
  magnetic confinement of plasma,
    1339, **1339**–1341
  radiation damage in, 1342
Nuclear spin quantum number (*I*), **1319**
Nucleons, **1294**. *See also* Neutron(s);
    Proton(s)
  charge and mass, 1294–1295
  quantization of energy states, 1303
Nucleus, atomic
  charge and mass, 1294–1295
  density, 1296
  energy states in external magnetic
    field, 1320, *1320*
  magnetic dipole moment of, 853
  models
    collective model, 1303
    history of, 1219
    independent-particle model. *See* shell
      model
    liquid-drop model, **1300**–1302
    shell model, 1302–**1303**
  properties of, 1294–1299
  radius of, 1296, 1302
  size and structure, *1296*, 1296–1297
  stability of, 1297–1299, *1298* (*See also*
    Radioactive decay)
  symbols for, 1294

Object distance, *1008*, **1008**
Objective, 1037, *1037*, 1038, *1038*
Oersted, Hans Christian, 641, 809, *809*
Ohm (Ω), 757
Ohm, Georg Simon, 756, *756*
Ohmic materials, 756
  current-potential difference curve, 758,
    758
  resistivity, **757**
Ohm's law, **756**
Omega (Ω), 1364*t*, 1374*t*, 1375
Omega laser fusion lab, 1341
Omega minus (Ω⁻), 1371–1372
Open-circuit voltage, **776**
Optical activity, 1098
Optical fibers, *995*, **995**–996
Optical molasses, 1247
Optical stress analysis, 1097, *1097*
Optical tweezers, 1248
Optic axis, 1096
Optics. *See also* Image(s); Image
    formation; Lens(es); Light; Light
    waves
  geometric, *1051*
    ray approximation in, 981, *981*
  physical, *1051*
  wave, *1051*
Orbital, **1237**
  exclusion principle and, **1237**
Orbital angular momentum
  and magnetic moment, 852
  quantization of, 852–853, 1231–1233

Orbital magnetic moment, 852
Orbital magnetic quantum number
    (*m_ℓ*), **1226**
  allowed values, 1226*t*, 1231–1232,
    1232
  physical interpretation of, 1231–1233
Orbital quantum number (ℓ), **1226**
  allowed values, 1226*t*, 1231
  physical interpretation of, 1230–1231
Order number, **1054**
Ordinary (O) ray, *1096*, **1096**
Orion (constellation), color of stars in,
    1155, *1155*
Oscillation. *See also* Simple harmonic
    motion
  damped, in *RLC* circuit, 913, *913*
  eddy current damping and, 884, *884*,
    885, *885*
  in *LC* circuit, *907*, **907**–911, *908*, *910*,
    937
  overdamped, in *RLC* circuit, 913
  in series *RLC* circuit, *911*, **911**–913, *913*
Overdamped oscillation, in *RLC* circuit,
    913
Oxygen (O)
  electronic configuration, *1239*
  isotopes, 1310*t*

Pair annihilation, 1360
Pair production, *1359*, *1360*
Paper, dielectric constant and dielectric
    strength of, 736*t*
Parabolas, **A-10**, *A-10*
Parallel combination
  of capacitors, **727**–728, *728*
  of resistors, *780*, **780**–782, *782*–785
Parallel-plate capacitors, 722–723, *723*
  capacitance of, 724–725
Paramagnetism, 854, *855*
Paraxial rays, 1011
Parent nucleus, **1308**
Partial integration, **A-17**
Particle(s). *See also* Quantum particles
  classification of, 1363–1365
  constructing from wave, *1171*,
    1171–1173, *1172*, *1173*, 1383–1384
  detection of, 1344–1347, *1345*, *1346*,
    1347, 1378, *1378*
  elementary
    classifications of, 1377
    search for, 1374
  properties, 1364*t*
  rest energy of, **1137**
  search for patterns in, 1370–1372
  total energy of, **1137**
  wave properties of, 1168–1171
Particle-in-a-box problem
  classical particles, 1191
  quantum particles, *1191*, 1191–1196,
    1192, *1193*
    analogy to standing waves, 1195–1196

Particle-in-a-box problem (*Continued*)
  free electron theory of metals,
    1270–1274
  Schrödinger equation and, 1197–1198
Particle model
  of light, *vs.* wave model, 1167–1168
  and principle of complementarity,
    1169
  wave properties of particles and,
    1168–1171
Particle physics, questions remaining in,
    1383
Paschen series, 1217–1218, *1221*
Path difference, **1054**
Path integral, 693
Pauli, Wolfgang, 1234, *1234*, 1313
Pentium chips, technological
    improvement in, 1283, *1283*
Penzias, Arno A., *1379*, 1379–1380
Percent uncertainty, **A-20**
Period (*T*), of electromagnetic wave,
    959
Periodic motion. *See* Oscillation; Simple
    harmonic motion; Waves
Periodic table, **1239**–1241, *1240*, 1243,
    1370, A-22*t*–A-23*t*
Permeability of free space (*μ₀*), **838**
Permittivity of free space (*ε₀*), **646**
PET. *See* Positron-emission tomography
Phase angle, **932**, 933, 934–935
Phase change, in reflection, *1059*,
    1059–1060, *1060*
Phase difference, in double-slit
    experiment, 1057
Phase speed, of wave packet, 1172
Phasor, **925**
Phasor diagram, 925
Phipps, T. E., 1234–1235
Phipps-Taylor experiment, 1234–1235,
    1236
Phosphorus (P), isotopes, 1310*t*
Photoconductors, 712
Photocopy machines, 712, *712*
Photoelectric effect, 979, *1160*,
    1160–1165, *1163*
  applications of, *1163*, 1163–1164
  classical prediction *vs.* experimental
    results, 1161
  equation for, 1162
  quantum-mechanical approach,
    1162–1164
Photoelectric photometry, 1164
Photoelectrons, **1160**
Photographic emulsion, radiation
    detection with, **1345**
Photomultiplier tubes, *1163*, 1163–1164
Photon(s), 979, **1161**
  angular momentum of, 1241
  antiparticle, lack of, 1360
  energy of, 979
  as field particle, 1358, 1359*t*, 1362,
    1362

Photon(s) (*Continued*)
  history of concept, 1158, 1161
  and photoelectric effect, 1162–1164
  total energy of, **1137**
  virtual, 1362
  wave model of light and, 1167–1168
Phototubes, 1163
Photovoltaic solar cells. *See* Solar cells
Physical optics, *1051*
Physics. *See also* Astronomy and
    astrophysics
  particle, questions remaining in, 1383
Pickup coil, 869–870, *870*
Pi mesons (π). *See* Pions
Pions (π), **1361**–1363, *1363*
  neutral, lack of antiparticle, 1360
  pion-exchange model, 1362–1363,
    1363, 1376, 1376–1377
  properties, 1364, 1364*t*
Planck, Max, 979, 1111, 1153, *1156,*
    1156–1158
Planck length, **1384**
Planck's constant, 979, **1156,** 1157, 1163
Plane of polarization, **1093**
Plane-polarized light, **1093.** *See also*
    Polarization
Planet(s). *See specific planets*
Plane waves, electromagnetic, *957,*
    957–960
  wave intensity of, 961–963
Plano-concave lens, 1023, *1023*
Plano-convex lens, 1023, *1023,* 1061,
    1061
Plasma, 816, 1337
  inertial confinement of, *1341,* **1341**
  magnetic confinement of, *1339,*
    1339–1341
  quark-gluon, **1375,** 1378
Plasma balls, *673*
Plastic, birefringent qualities, 1097,
    1097
Plates, of capacitor, 722
Pluto, telescope imaging of, 1086, *1086*
Plutonium (Pu), isotopes, 1311*t*
p-n junction, *1280,* **1280,** 1282
Point charge, **646**
Poisson, Simeon, 1078
Polarization
  of dielectrics, 740–741, *741*
  of light waves, 1093–1098
    direction of polarization, **1093**
    by double refraction, *1096,*
      1096–1097
    Malus's law, **1094**
    plane of polarization, **1093**
    polarizing angle, 1095, *1095*
    by reflection, 1094–1096, *1095, 1096*
    by scattering, *1097,* **1097**–1098
    by selective absorption, 1093–1094,
      1094
    sunlight, *1097,* **1097**–1098
  and optical activity, 1098

Polarizer, *1093,* 1093–1094
Polarizing angle, 1095, *1095*
Polar molecules, 739, *739*
  induced polarization, 740, 741, *741*
Polaroid, 1093
Polonium (Po)
  discovery of, 1304
  isotopes, 1311*t*
    decay of, 1139–1140
Population inversion, **1245**–1246
Positive electric charge, **643**
Positron-emission tomography (PET),
    1360, *1361*
Positrons (e⁺), 1304, *1304, 1359,*
    1359–1360, *1360*
Potassium (K)
  Fermi energy, 1272*t*
  isotopes, 1310*t*
Potential. *See* Electric potential
Potential difference (ΔV), **693**
  applications of, 710–712
  mechanical analogy to, 912*t*
  in uniform electric field, 694–697
  units of, 694
  value of electric field from, 701–703
Potential energy. *See also* Electric
    potential
  of crystal, 1268
  of electric dipole in electric field,
    738–739
  in electric fields, 693
  of magnetic dipole in magnetic field,
    823
  of pair of point charges, 698–700
Potential energy function (*U*)
  for diatomic molecule, 1264, *1264*
  for system of two atoms, 1258, *1258*
Power (𝒫)
  electrical. *See* Electrical power
  of lens, **1035**
Power factor, **936**
Power lines, power transmission
    through, 752, 765, 939, *941,*
    941–942
Power plants, commercial, *867,* 881,
    924. *See also* Nuclear reactors
Poynting vector, **961**
Presbyopia, **1034**–1035
Pressure (*P*), units of, A-2*t*
Pressurized-water reactor, 1333–1334,
    1334
Primary winding, 940, *940*
Princeton Plasma Physics Laboratory,
    1339, 1340
Principal axis, of mirror, **1010,** *1011*
Principal quantum number (*n*),
    1225–1226
  allowed values, 1226*t*
Principle of complementarity, **1169**
Principle of equivalence, **1141**
Principle of Galilean relativity, *1113,*
    1113–1116, *1114*

Principle of relativity, **1118**–1119
Prism, refraction in, 989, *989*
Probability
  and Gauss's probability integral, A-19*t*
  probabilistic interpretation of
    quantum mechanics, 1186–1191
Probability amplitude. *See* Wave function
    (Ψ)
Probability density, **1188**
Problem-solving strategies. *See also*
    Analysis models
  for electric field calculation, 656
  for electric potential, 703–704
  for Gauss's law problems, 678
  for interference in thin films, 1062
  for Kirchhoff's rules, 786–787
Projectors, digital, 984–985
Propagation, of electromagnetic waves,
    957
Propagation of uncertainty, A-20–A-21,
    A-21
Proper length, **1125**–1126
Proper time interval, **1122**
Proportional counters, 1346
Proton(s)
  charge of, 643, 646, 646*t,* 1294
  composition of, 1374*t*
  decay, detection of, *1366,* 1366–1367
  magnetic dipole moment of, 853
  mass of, 1294–1295, 1295*t*
  properties, 1364*t*
  rest energy of, 1138
  stability of, 1365, 1366–1367
Proton-proton cycle, **1335**–1336
p-type semiconductors, **1279,** 1280
Pupil of eye, 1033, *1033*
Pythagorean theorem, A-11

QCD. *See* Quantum chromodynamics
Quality factor, **938**–939
Quantization
  of atomic orbital angular momentum,
    852–853, 1231–1233
  of electric charge, 643, 710
  of energy levels, **1156,** *1157,* 1193,
    1193 (*See also* Transitions)
  in Bohr model, 1219–1224, *1221,*
    1230
  in quantum model, 1225–1227
  of energy of particle in a box, *1193,*
    1193–1194
  of light, 979
  of molecular rotational motion,
    1261–1263, *1262*
  of molecular vibrational motion, *1264,*
    1264–1266, *1265*
  of nucleon energy states, 1303
  space, **1231**–1232, *1232,* 1233–1235,
    1235
Quantum chromodynamics (QCD),
    1376

Quantum corral, *1186,* 1200
Quantum dot, **1200**
Quantum electrodynamics theory, *1362*
Quantum mechanics. *See also*
　　Quantization; Quantum particles
　on blackbody radiation, 1156–1158
　Compton effect, *1165,* **1165**–1167,
　　1166
　correspondence principle, **1223**
　free-electron theory of metals,
　　1270–1274
　history of, 1153, *1156,* 1156–1158
　impact of, 1111
　model of atom, 1224–1230
　orbital angular momentum in,
　　852–853, 1231–1233
　particle and wave models of light,
　　1167–1168
　photoelectric effect, 1162–1164
　probabilistic interpretation of,
　　1186–1191
　on simple harmonic motion,
　　1159–1160, 1205–1207, *1206*
　spin angular momentum in, 853
　strangeness of, 1153
　wave properties of particles, 1168–1171
Quantum number(s), **1156,** 1159
　bottomness, 1374
　charm, 1373
　exclusion principle and, 1237
　of hydrogen atom, 1225–1226, 1226*t*
　　for *n* = 2 state, 1236, 1237*t*
　　physical interpretation of, 1230–1237
　nuclear spin (*I*), **1319**
　physical interpretation of, 1230–1237
　topness, 1374
Quantum number space, 1273, *1273*
Quantum particles, **1171**–1173
　under boundary conditions, *1191,*
　　1191–1196, *1192, 1193*
　analogy to standing waves, 1195–1196
　free electron theory of metals,
　　1270–1274
　Schrödinger equation and,
　　1197–1198, 1199
　well of finite height, *1198,* 1198–1200,
　　1199
　well of infinite height, *1191,*
　　1191–1196, *1192, 1193*
　electron double-slit experiment, *1174,*
　　1174–1175, *1175*
　expectation value of, **1189,** 1190–1191,
　　1195
　Heisenberg uncertainty principle and,
　　1175–**1177**
　probability density of, 1188
　quantization of energy, *1193,*
　　1193–1194
　tunneling by, *1200,* **1200**–1202
　applications, 1202–1205
　wave equation for (Schrödinger
　　equation), 1188, 1196–1198, **1197**

Quantum particles (*Continued*)
　wave function of. *See* Wave function
　　(Ψ)
　wave properties of, 1168–1171
Quantum particle under boundary
　　conditions model, 1196, 1208
Quantum states, **1156,** *1157. See also*
　　Quantization, of energy levels;
　　Quantum number(s)
　allowed, 1237–1238, *1238,* 1241–1242
Quantum statistics, 1271
Quark-gluon plasma, **1375,** 1378
Quarks, *1372,* **1372**–1375
　in baryons, 1374*t*
　color charge, **1375**–1377, *1376*
　flavors, **1372**
　interaction of (quantum
　　chromodynamics), 1376
　in mesons, 1374*t*
　original model, 1372–1373
　properties, 1373*t*
　rest energy and charge, 1375*t*
　in Standard Model, *1377,* 1377–1378
Q value. *See* Disintegration energy;
　　Reaction energy

Rad (radiation absorbed dose),
　　1343–1344
Radial probability density function,
　　1228, 1228–1230, *1230*
Radian (rad), **A-10**
Radiation, particle. *See also* Radioactive
　　decay; Radioactivity
　background, 1344
　damage from, 1342–1344
　detection of, 1344–1347, *1345, 1346,*
　　1347, 1378, *1378*
　discovery of, 1293
　dose limit recommendations, 1344
　fatal doses, 1344
　units for, 1343–1344, 1344*t*
　uses of, *1347,* 1347–1349, *1348, 1349*
Radiation pressure, of electromagnetic
　　waves, 963–965
Radiation therapy, 1348, *1348*
Radio, receiving circuit in, 939
Radioactive decay
　alpha (α) decay, 1304, *1304, 1308,*
　　1308–1313
　decay pathways, 1317*t*
　as example of tunneling, 1202, 1312
　and radiation damage, 1343, 1344*t*
　beta (β) decay, 1304, *1313,* 1313–1315,
　　1314
　and carbon dating, 1315–1316
　and cellular damage, 1344*t*
　decay pathways, 1317*t*
　and neutron activation analysis, 1348
　gamma (γ) decay, 1304, *1304,* 1317,
　　1317
　decay pathways, 1317*t*

Radioactive decay (*Continued*)
　and food preservation, 1348
　and radiation damage, 1343
　mass change in, 1139–1140
　radioactive series, 1317–1318, 1317*t*
　rate of, 1304–1305, *1305*
　types of, 1304
Radioactive tracers, 1347, *1347*
Radioactivity, **1304**–1307. *See also*
　　Radiation, particle; Radioactive
　　decay
　artificial, **1317,** 1317*t*
　natural, **1317**–1318, 1317*t, 1318*
Radio waves, *967,* **967**
Radium (Ra)
　decay of, 1308, *1308,* 1309, 1312, 1318,
　　1318t
　discovery of, 1304
　isotopes, 1311*t*
Radon (Rn), isotopes, 1311*t*
Rail guns, 837
Railroad electromagnetic braking
　　systems, 884–885
Rainbow(s), *978, 993, 993, 1055*
Rainbow hologram, 1089–1090
Ray(s), **957**
　extraordinary (E), *1096,* **1096**
　ordinary (O), *1096,* **1096**
Ray approximation, 981, *981*
Ray diagrams
　for mirrors, 1013–1015, *1014*
　for thin lenses, *1024,* 1024–1028, 1028
Rayleigh-Jeans law, 1155–1156,
　　1156–1157, *1157*
Rayleigh's criterion, **1083**–1084
RBE (relative biological effectiveness),
　　1343, 1344*t*
RC circuits
　alternating current, 943, *943*
　direct current, **788**–794
RC high pass filter, 943, *943*
RC low pass filter, 943, *943*
Reaction energy (*Q*), 1318
Real image, **1009**
Rectification, **942**–943
Rectifier(s), *942,* **942**–943
Red shift
　of astronomical objects, 1129–1130
　of light in gravitational field, 1141
Reduced mass, of diatomic molecule,
　　1261
Reflecting telescope, **1038,** *1039,*
　　1039–1040
Reflection, 981–985, *982. See also*
　　Mirror(s)
　change of phase in, *1059,* 1059–1060,
　　1060
　diffuse, *982,* **982**
　polarization of light by, 1094–1096,
　　1095, 1096
　and radiation pressure, 964
　retroreflection, 984, *984*

Reflection (*Continued*)
  sign conventions for, 1013, *1013*, 1013*t*
  specular, *982*, **982**
  total internal, **993**–995, *994*
    applications, *995*, 995–996
Reflection coefficient (*R*), **1200**–1201
Refracting telescope, *1038*, **1038**–1039,
    1040
Refraction, *985*, 985–990, *986*. *See also*
    Index of refraction; Lens(es)
  by flat surface, 1019, *1019*
  image formation by, *1017*, 1017–1021,
    1018, 1019*t*
  polarization by double, *1096*,
    1096–1097
  Snell's law of, **988**, 989, *991*, 991–992
Reines, Frederick, 1313
Relativistic Doppler effect, 1129–1130
Relativistic Heavy Ion Collider (RHIC),
    1357, 1375, 1378
Relativistic mass, 1134
Relativity, Galilean, *1113*, **1113**–1116,
    1114
  limitations of, 1115–1116
Relativity, general, 1140–1142, *1141*,
    1141
  on gravitational waves, 1065
  history of theory, *1118*, 1140
Relativity, special
  applications, 1113
  conservation of energy and, 1139
  energy-momentum relationship, 1137
  history of theory, 1111, 1112–1113,
    1116, 1118, *1118*
  length contraction, **1125**–1126, *1126*,
    1128, 1128–1129
  Lorentz space-time transformation
    equations, **1130**–1131
  Lorentz velocity transformation
    equations, 1131–1134, **1132**
  mass and, 1134, 1137, 1139–1140
  Michelson-Morley experiment, *1116*,
    1116–1119
  observer agreements and
    disagreements, 1132
  principle of relativity, **1118**–1119
  relativistic Doppler effect, 1129–1130
  relativistic energy, 1135–1140
  relativistic force, 1135
  relativistic kinetic energy, 1135–1136,
    1136, 1138
  relativistic linear momentum,
    1134–1135, 1137, 1138–1139
  space-time graphs, *1126*, 1126–1127,
    1128, 1128–1129
  and speed of light, 1115–1119
  and time
    dilation of, *1120*, 1120–1124, **1121**,
      1127, 1131
    relativity of, *1119*, 1119–1120
  twin paradox, *1124*, 1124–1125, 1126,
    1126

Rem (radiation equivalent in man),
    1343–1344
Reproduction constant (*K*), 1333, 1334
Resistance (*R*), 756–760, **757**
  equivalent, **779**, **780**–781
  internal, *776*, **776**, 777
  mechanical analogy to, 912–913, 912*t*
  temperature and, *762*, 762–763
Resistivity (*ρ*), **757**, 758*t*, 762
Resistor(s), *757*, **757**. *See also RC* circuits;
    RLC circuits; *RL* circuits
  in AC circuit, *924*, 924–927, 936
  circuit symbol for, 763
  color coding of, 757, *758*, 758*t*
  composition, 757
  energy delivered to, 764–765, 793–794,
    925–926
  energy stored in, 903
  in parallel combination, *780*, **780**–782,
    782–785
  power delivered to, 764–765, 793–794,
    925–926, 936
  in series combination, **778**–780, *779*,
    781–782, 783–784
  wire-wound, 757
Resolution
  circular aperture, *1084*, 1084–1086
  single-slit aperture, *1083*, 1083–1084
Resonance, in series *RLC* circuits,
    937–939, *938*
Resonance frequency (*ω*₀), of series
    RLC circuit, **937**–939, *938*
Resonant tunneling devices, **1203**–1204,
    1204
Resonant tunneling transistors, *1204*,
    1204–1205
Rest energy, **1137**–1138
Rest-energy equivalent, of atomic mass
    unit, 1295
Retina, 1033, *1033*
Retroreflection, 984, *984*
RHIC (Relativistic Heavy Ion Collider),
    1357
Richter, Burton, 1374
Rigel (star), color of, 1155, *1155*
Right hand rule
  for Ampère's law, 845
  for direction of magnetic field, 840,
    840
  for force on charge in magnetic field,
    811, *811*
  for torque on current loop in magnetic
    field, 823, *823*
Ripple, 943
RLC circuits, series
  alternating current, *932*, 932–935, *933*
  average power in, 935–937, 938, *938*
  resonance in, 937–939, *938*
  oscillations in, *911*, **911**–913, *913*
RL circuits, direct current, *900*,
    900–905, *901*, *902*
rms current, 925–927, **926**

rms voltage, 926
Rods, 1033
Roemer, Ole, *979*, 979–980
Roentgen (R), **1343**
Roentgen, Wilhelm, 1091
Root-mean-square. *See* rms
Rotational quantum number, **1262**
Rubber, dielectric constant and
    dielectric strength of, 736*t*
Rubbia, Carlo, 1363
Rutherford, Ernest, 1218–1219, 1293,
    1296, 1318
Rydberg, Johannes, 1217, 1222
Rydberg atoms, 1223–1224
Rydberg constant, **1217**

Safety, electrical, household wiring and,
    796–798
Salam, Abdus, 1377
Sandage, Allan R., 1381
San Onofre Nuclear Generating Station,
    1329
Savart, Félix, 837–838
s bands, 1274–1276, *1275*, *1276*
Scanning tunneling microscope (STM),
    1186, 1203, *1203*
Scattering event, **1318**–1319
  polarization by, *1097*, **1097**–1098
Schrieffer, J. R., 1284
Schrödinger, Erwin, 1153, 1186, 1188,
    1197, 1202
Schrödinger equation, 1188, 1196–1198,
    1197, 1199
  and quantum model of hydrogen
    atom, 1224–1227
Schwinger, Julian, *1362*
Scientific notation, **A-4**, A-4–A-5
Scintillation counter, **1346**
Secondary maxima, **1078**
Secondary winding, 940, *940*
Seeds (radiation therapy devices), 1348
Segré, Emilio, 1360
Selection rules, for allowed transitions,
    1241
Selectrons, 1384
Self-induced emf, **898**
Self-induction, **898**
Self-sustained chain reaction,
    1332–1333, *1333*
Semiconductor devices, *1279*,
    1279–1283
Semiconductor-diode detector, **1346**
Semiconductor lasers, 1281
Semiconductors, **644**, 762, *1277*,
    1277–1278, *1278*
  doped, **1278**–1279, *1279*
  intrinsic, **1278**
  n-type, **1278**, 1280
  p-type, **1279**, 1280
Semiempirical binding-energy formula,
    1301–1302

Series combination
  of capacitors, *729*, 729–731
  of resistors, **778**–780, *779*, 781–782,
    783–784
Series limit, **1217**
Series *RLC* circuits. *See RLC* circuits,
    series
Shell model of nucleus, 1302–**1303**
Shells, atomic, **1226**, 1227*t*
  filling of, 1237–1238, *1238*
Shockley, William, 1282
Short-circuit condition, 796–797
Side maxima, **1078**
Sievert (Sv), 1344, 1344*t*
Sigma (Σ) [particle], 1364*t*, 1369, 1374*t*
Silicone (Si)
  crystals, *1269*
  isotopes, 1310*t*
Silver (Ag), Fermi energy, 1272*t*
Simple harmonic motion, quantum-
    mechanical viewpoint on,
    1159–1160, 1205–1207, *1206*
Simultaneity, and theory of relativity,
    1119, 1119–1120, *1128*, 1128–1129,
    1131
Single-slit aperture, resolution through,
    1083, 1083–1084
Single-slit diffraction patterns, *1078*,
    1078–1083, *1079, 1081, 1082*
  light intensity distribution, 1056–1058,
    1057, 1058, 1081, *1081*
  position of fringes, 1079–1081, *1082*,
    1082–1083
Sinusoidal electromagnetic waves, *957*,
    959–960
SI (*Système International*) units, A-1*t*–A-3*t*,
    A-24*t*
  of activity, 1306
  of capacitance, 723
  of charge, 843
  of current, 753
  of current density, 756
  of electric charge, 646
  of electric field, 694
  of electric flux, 674
  of electric potential, 694
  of inductance, 899
  of magnetic dipole moment, 823
  of magnetic field, 812
  of potential difference, 694
  of power, 765
  of Poynting vector, 961
  for radiation, 1344, 1344*t*
  of resistance, 757
Sky, color of, 977, 1097–1098, *1098*
Slipher, Vesto Melvin, 1380
Slit(s), diffraction and interference
    from. *See* Diffraction; Double-slit
    interference patterns; Interference;
    Single-slit diffraction patterns
Slope, of graphs, A-7, **A-7**
Smoke detectors, *1312*, 1312–1313

Smoot, George, 1380
Snell, Willebrord, 988
Snell's law of refraction, **988,** 989, *991*,
    991–992
Soap, surfactants in, 739–740
Sodium (Na)
  Fermi energy, 1272*t*
  isotopes, 1310*t*
  photoelectric effect for, 1164–1165
  s bands, 1274–1276, *1275, 1276*
Sodium chloride (NaCl)
  crystals, 1092, *1092*, 1268, 1268–1269
  ionic bonding in, 1258, *1259*
Solar cells
  coating on, 1063–1064
  photon absorption in, 1281
Solar energy, 962
Solar sailing, 964
Solar system, dust particles in, 964
Solenoid
  electric field induced in, *879*, 879–880
  ideal, 848–849, *849*
  inductance of, 899–900
  magnetic field of, *848*, 848–849
Solenoidal Tracker at RHIC (STAR)
    detector, 1347, *1347, 1357*
Solid(s)
  amorphous, 1096
  band theory of, 1274–1276, *1275, 1276*
    and electrical conduction, *1276*,
      1276–1279, *1277, 1278*
  bonding in, 1268–1270
    covalent, *1269*, 1269–1270
    ionic, *1268*, 1268–1269
    metallic solids, 1270, *1270*
  crystalline, 1096 (*See also* Crystal(s))
Solid solutions, metal, 1270
Somatic radiation damage, 1343
Source charge, 651, *651*
South pole
  of Earth, 855–856, *856*
  of magnet, 808–809
Spacecraft
  Hubble Space Telescope, 1077, 1086,
    1086
  Mariner 10, 964
Space quantization, **1231**–1232, *1232*,
    1233–1235, *1235*
Space-time
  distortion by gravity, 1140–1142
  string theory and, 1384
Space-time graphs, *1126*, 1126–1127,
    1128, 1128–1129
Space-time transformation equations
  Galilean, **1114**–1115
  Lorentz, **1130**–1131
Spark chamber, **1346**
Speakers, crossover networks in, 943
Special theory of relativity. *See also*
    Relativity, special
  limitations of, 1153
  Maxwell's equations and, 955

Spectral lines. *See* Line spectra
Spectroscopy
  atomic, 1089
  Fourier transform infrared (FTIR),
    1064–1065
Spectrum. *See also* Line spectra
  electromagnetic, 952, **966**–968, *967*
  visible light, *967*, 967–968, *992*,
    992–993, *993*
Specular reflection, *982*, **982**
Speed (*v*)
  of light (*c*)
    measurement of, 979–980
    Michelson-Morley experiment, *1116*,
      1116–1119
    relativity and, 1115–1119
    units of, A-1*t*
Spherical aberration, 1011, *1011, 1030*,
    1030
Spherical capacitors, capacitance of,
    726, 726–727
Spherical polar coordinates, 1225,
    1225
Spherical waves, **957**
Spin, of atomic particles, *853*, **853**
Spin angular momentum ($\vec{\mathbf{S}}$),
    1235–1236, *1236*
  of nucleus, *1319*, 1319–1320
Spin down, 1234
Spin magnetic moment ($\vec{\boldsymbol{\mu}}_{spin}$), of
    electron, 1236
Spin magnetic quantum number (m_s),
    1234, 1235
  physical interpretation of, 1233–1237
Spin-orbit effects
  atomic, 1234
  nuclear, 1303
Spin up, 1234
Spontaneous decay, **1308**
Spontaneous emission, *1244*, **1244**
Square barriers, *1200*, **1200**
Square well, **1197**
  of finite height, particle in, *1198*,
    1198–1200, *1199*
  of infinite height, particle in. *See*
    Particle-in-a-box problem
Squarks, 1384
Standard Model, *1377*, **1377**–1378
Stanford Linear Collider, 1378
STAR (Solenoidal Tracker at RHIC)
    detector, 1347, *1347, 1357*
Stars, fusion in, 1335–1336
Stationary states, **1219**
Stefan's law, 1154, 1155
Step-down transformers, 940, *941*
Step-up transformers, 940
Stern, Otto, 1234, *1235*
Stern-Gerlach experiment, 1234, *1235*,
    1236
Stimulated absorptions, *1244*, **1244**
Stimulated emission, *1244*, **1244**–1246
Stopping potential, *1160*, **1160**

Strangeness, **1369**
  conservation of, **1369**–1370
Strange particles, *1369*, 1369–1370
Strange quark (s), **1372**, 1373*t*, 1374*t*
Strassman, Fritz, 1330
Stress analysis, optical, 1097, *1097*
String theory, 1383–**1384**
Strong force
  evolution of, at origin of universe, 1379, *1379*
  field particles for, 1376
  as fundamental force, 1363–1364, 1374–1375
  in Standard Model, 1377, *1377*
Stud finders, *737*
Subcritical reaction, 1333
Subshells, atomic, **1226**, 1227*t*
  filling of, 1237–1238, *1238*
Subtraction, and uncertainty, A-20
Sulfur (S), isotopes, 1310*t*
Sun
  elements in atmosphere of, 1217
  fusion in, 1335–1336
Sunglasses
  polarized, 1096
  UV protection and, *968*
Sunlight
  energy delivered to Earth by, 962
  polarization of, *1097*, **1097**–1098
Superconductors, **762**–763, *763*, 763*t*, **1283**–1285, *1284*
  high-temperature, 1284–1285
  Meissner effect in, 855, *855*, 1283–1284, *1284*
Supercritical reaction, 1333
Super Kamiokande neutrino detection facility, 1366, *1366*
Supernumerary bows, *1055*
Superposition principle, for electric fields, 652–654
Super Proton Synchrotron, 1378
Supersymmetry (SUSY), 1384
Surface charge density (σ), **655**
  of nonspherical conductor, 707–708
  of spherical conductor, 707
Surface effect, in liquid-drop model of nucleus, 1300
Surface integral, 674
Surfactants, 739–740
SUSY. *See* Supersymmetry
Switch, circuit symbol for, *727*, 900, *900*
Symbols. *See* Notation
Symmetric molecules, 740, *740*
Symmetry breaking, 1378
Symmetry effect, in liquid-drop model of nucleus, 1301
Synchrotrons, 818

TAMA, 1066
Tau (τ⁻), 1364*t*, 1365, 1374

Tau lepton number, conservation of, 1368
Tau-neutrino (ν_τ), 1364t, 1365, 1368
Taylor, J. B., 1234–1235
Telescopes
  atmospheric blurring, 1086, *1086*
  Hubble Space Telescope, 1077, 1086, *1086*
  Keck Observatory, 1040, 1085–1086
  magnification in, *1038*, **1038**–1040, *1039*
  radio, 1086
  resolution of, 1085–1086
  Yerkes Observatory, 1040
Television, color, 1034
Television remote control, infrared LED in, 1281
Temperature (*T*)
  critical, **762**
  and resistance, *762*, 762–763
Temperature coefficient of resistivity, 758*t*, **762**
Terminal voltage, 776, *776*, 777
Tesla (T), **812**
Tesla, Nikola, *940*
Test charge, 651, *651*
Tevatron, 1378
TFTR. *See* Tokamak Fusion Test Reactor
Theory of Everything, 1383
Thermalization, of neutrons, 1330
Thermal neutrons, **1330**
Thermal radiation. *See* Blackbody radiation
Thermometers, ear, 1158, *1158*
Thermonuclear fusion reactions, **1336**
Thin films. *See* Films
Thin lens. *See* Lens(es)
Thin lens equation, **1023**
Thomson, G. P., 1169
Thomson, Joseph John, 817, *818*, 1218, *1218*
Thorium (Th), radioactive series, 1317–1318, 1317*t*, *1318*
Three Mile Island nuclear power plant, 1334
Three-pronged electrical cords, *797*, 797–798
Three-way lightbulbs, 781, *781*
Threshold energy, **1318**
Time (*t*)
  dilation of, in special relativity, *1120*, 1120–1124, **1121**, 1127, 1131
  gravity and, in general relativity, 1141–1142
  probability, and indeterminacy of future, 1202
  proper time interval, **1122**
  relativity of, *1119*, 1119–1120
Time constant (τ)
  of *RC* circuit, 790
  of *RL* circuit, 901, 902–903

Time-independent Schrödinger equation. *See* Schrödinger equation
Time response, of circuits, 901
Ting, Samuel, 1374
Tokamak, *1339*, **1339**–1340
Tokamak Fusion Test Reactor (TFTR), *1339*, 1340
Tomonaga, Sin Itiro, *1362*
Toothbrush, electric, induced current charge system in, *906*, **906**–907
Top quark (*t*), 1373*t*, **1374**
Toroid, magnetic field of, *847*, 847–848
Torque (τ̄)
  on current loop in magnetic field, 821–825, *822*
  on electric dipole in electric field, 738, *738*
Torsion balance, 645, *645*
Total energy, **1137**, 1139
Total instantaneous energy density, of electromagnetic waves, 962
Total internal reflection, **993**–995, *994*
  applications, *995*, 995–996
Tracers, radioactive, 1347, *1347*
Track detectors, **1346**
Transformation equations
  space-time
    Galilean, **1114**–1115
    Lorentz, **1130**–1131
  velocity
    Galilean, 1114–**1115**
    Lorentz, 1131–1134, **1132**
Transformer(s), 765, *923*
  AC, 939–942, *940*, **940**, *941*
  eddy currents in, 885
Transistors, 1282, *1282*
Transitions
  allowed, 1241–1242
  of molecules, between rotational energy levels, 1262
  spontaneous emissions, *1244*, **1244**
  stimulated absorption, *1244*, **1244**
  stimulated emissions, *1244*, **1244**–1246
Transmission axis, 1093
Transmission coefficient (*T*), **1200**–1202
Transmission electron microscope, *1170*, 1170–1171
Transmission grating, 1086–1087
Transportation. *See* Automobiles; Railroad; Spacecraft
Transuranic elements, 1317–1318, *1333*
Trigonometric functions, A-10–A-12
  identities for, A-12*t*
Tritium, fusion and, 1337, 1338, *1338*, 1341, *1341, 1342*
Truth quark, 1374
Tunneling, *1200*, **1200**–1202
  applications, 1202–1205
Twin paradox, *1124*, 1124–1125, 1126, *1126*

Uhlenbeck, George, 1234, 1235
Ultraviolet catastrophe, 1156
Ultraviolet waves, *967*, **968**, *968*
Uncertainty, **A-20**, A-20–A-21
Uncertainty principle. *See* Heisenberg
     uncertainty principle
Units. *See also* SI (*Système International*)
     units
  conversion of, A-1*t*–A-2*t*
Universe
  critical density of, 1381–1382
  dark energy in, 1382–1383
  expansion of, 1380–1382, *1381*,
     1382–1383
  missing mass in, 1382
  origin of. *See* Big Bang theory
Unknowns, **A-5**
Unpolarized light beams, *1093*, **1093**
Up quark (u), **1372**, 1373*t*, 1374*t*
Uranium (U)
  decay of, 1308, 1312
  enrichment of, 1333
  fission of, *1331*, 1331–1332
  in fission reactors, 1139, 1333
  isotopes, 1311*t*
  radioactive series, 1317–1318, 1317*t*

Vacuum, dielectric constant and
     dielectric strength of, 736*t*
Vacuum tubes, 1279
Valence band, **1276**–1279, *1277*, *1278*,
     1279
Van Allen radiation belt, 816, *816*
Van de Graff, Robert J., 710
Van de Graff generators, 710–711,
     711
Van der Meer, Simon, 1363
Van der Waals bonding, 1260
Van der Waals forces, **1260**
Variable capacitors, 737, *737*
Vector model, *1232*, **1232**
Velocity selector, 817, *817*
Velocity transformation equations
  Galilean, 1114–**1115**
  Lorentz, 1131–1134, **1132**
Venus, magnetic field of, 856
Vibrational quantum number, 1264
Videotape, 854
VIRGO, 1066
Virtual image, **1009**
Virtual object, **1028**
Virtual photons, 1362
Visible light spectrum, *967*, **967**–968,
     992, **992**–993, *993*
Volt (V), **694**
Voltage (Δ*V*), **693**
  across capacitor in AC circuit, *930*,
     930–931
  across inductor in AC circuit, 927–928,
     928

Voltage (Δ*V*) (*Continued*)
  across resistor in AC circuit, *924*,
     924–925
  of alternating current, 926
  open-circuit, **776**
  in *RLC* series circuit, *932*, 932–933
Voltage amplitude, of AC source, **923**
Voltmeter, *795*, **795**
Volume (*V*), of geometric shapes,
     A-10*t*
Volume charge density (ρ), **655**
Volume effect, in liquid-drop model of
     nucleus, 1300

Water
  dielectric constant and dielectric
     strength of, 736*t*
  index of refraction, 987*t*, 1019
  view from underneath, 994–995,
     995
  waves in, *1053*
Water molecule
  hydrogen bonding of, 1260
  polarization in, 739, *739*, 740
Watt (W), 765
Wave(s). *See also* Electromagnetic waves;
     Light waves
  constructing particles from, *1171*,
     1171–1173, *1172*, *1173*,
     1383–1384
  linearly polarized, *957*, **957**
  spherical, **957**
  water, *1053*
Wave equation, general, 959
Wave front, **957**
Wave function (Ψ), **1187**–1188
  composite, band theory and,
     1274–1276, *1275*, *1276*
  of covalent bond, 1259, *1259*
  and expectation value, 1189,
     1190–1191, 1195
  for free particle, 1188
  for hydrogen, 1227–1230
     ground state, 1227
     2*s* state, 1230
  normalized, **1189**
  one-dimensional, 1188–1191
  for particle in finite well, 1191–1193
  for particle in infinite well, 1198–1199,
     1199
  of simple harmonic oscillator, 1205
Wave intensity, of electromagnetic
     waves, 961–963
Wavelength (λ)
  of blackbody radiation, 1154–1155,
     1155
  cutoff, 1163
  of electromagnetic waves, 959
  index of refraction and, 987, *987*, 992,
     992

Wavelength (λ) (*Continued*)
  of light
     measuring, 1055, 1062, 1064, 1087,
        1089
     and particle model, 1168
  of quantum particle in a box, 1192
Wavelets, **990**
Wave model
  of light, *vs.* particle model,
     1167–1168
  of particles, 1168–1171 (*See also*
     Quantum particles)
  and principle of complementarity,
     1169
Wave number (*k*)
  of electromagnetic waves, 959
  of particle wave function, 1188
Wave optics, *1051*
Wave packet, 1172–1173
  group speed of, 1173–1174
  phase speed of, 1172
Waves in interference model, **1055**
Wave under reflection model, **982**
Wave under refraction model, **988**
W bosons, 1358, 1359*t*, 1363, 1377,
     1378
Weak charge, 1377
Weak force
  electroweak theory and, 1377
  evolution of, at origin of universe,
     1379, *1379*
  field particles for, 1358, 1359*t*, 1363
  as fundamental force, 1358
  in Standard Model, *1377*, 1377–1378
Weber (Wb), 850
Weinberg, Steven, 1377
Wells, **1197**
  nanotechnology and, 1200
  square, **1197**
     of finite height, particle in, *1198*,
        1198–1200, *1199*
     of infinite height, particle in. *See*
        Particle-in-a-box problem
White light, dispersion and, *992*,
     992–993, *993*
Wien's displacement law, **1154**–1155,
     1158
Wilkinson Microwave Anisotropy probe,
     1380
Wilson, Charles, *1165*
Wilson, Robert W., *1379*, 1379–1380
Wire-wound resistors, 757
Work (*W*)
  in electric field, 693–694
  and kinetic energy. *See* Work-kinetic
     energy theorem
Work function, of metal, 1162, 1162*t*,
     1163
Work-kinetic energy theorem, relativistic
     form of, 1136
World line, *1126*, **1126**

Xerography, 712, *712*
Xi (Ξ), 1364*t*, 1374*t*
X-rays, *967*, **968**
    and cellular damage, 1320, 1343,
        1344*t*
    characteristic, 1242, *1242*
    diffraction by crystals, *1091*,
        1091–1092, *1092*
    and food preservation, 1348

line spectra, *1242*, 1242–1244
    in radiation therapy, 1348
X-ray spectra, *1242*, 1242–1244

Yerkes Observatory, 1040
y-intercepts, **A-7**, *A-7*
Young, Thomas, 979, 1052, *1053*,
    1055

Yukawa, Hideki, 1361, *1361*, 1362, 1376,
    1376

Z bosons, 1358, 1359*t*, 1363, 1377, 1378
Zeeman effect, 1232–1233, *1233*
Zeroth-order maximum, 1054
Zinc (Zn), isotopes, 1311*t*
Zweig, George, 1372, 1373

Standard Abbreviations and Symbols for Units

Symbol	Unit	Symbol	Unit
A	ampere	K	kelvin
u	atomic mass unit	kg	kilogram
atm	atmosphere	kmol	kilomole
Btu	British thermal unit	L	liter
C	coulomb	lb	pound
°C	degree Celsius	ly	lightyear
cal	calorie	m	meter
d	day	min	minute
eV	electron volt	mol	mole
°F	degree Fahrenheit	N	newton
F	farad	Pa	pascal
ft	foot	rad	radian
G	gauss	rev	revolution
g	gram	s	second
H	henry	T	tesla
h	hour	V	volt
hp	horsepower	W	watt
Hz	hertz	Wb	weber
in.	inch	yr	year
J	joule	Ω	ohm

Mathematical Symbols Used in the Text and Their Meaning

Symbol	Meaning
$=$	is equal to
$\equiv$	is defined as
$\neq$	is not equal to
$\propto$	is proportional to
$\sim$	is on the order of
$>$	is greater than
$<$	is less than
$\gg (\ll)$	is much greater (less) than
$\approx$	is approximately equal to
Δx	the change in x
$\sum_{i=1}^{N} x_i$	the sum of all quantities x_i from $i = 1$ to $i = N$
$|x|$	the magnitude of x (always a nonnegative quantity)
$\Delta x \rightarrow 0$	Δx approaches zero
$\dfrac{dx}{dt}$	the derivative of x with respect to t
$\dfrac{\partial x}{\partial t}$	the partial derivative of x with respect to t
$\displaystyle\int$	integral